LATTICE GAUGE THEORIES

AND

MONTE CARLO SIMULATIONS

Lattice Gauge Theories and Monte Carlo Simulations

Claudio Rebbi
Brookhaven National Laboratory

World Scientific

World Scientific Publishing Co Pte Ltd
P O Box 128
Farrer Road
Singapore 9128

The author and publisher are indebted to the original authors and publishers of the various journals for their assistance and permission to reproduce the selected papers found in this volume.

ISBN 9971-950-70-7
9971-950-71-5 pbk

Printed in Singapore by Richard-Clay (S. E. Asia) Pte. Ltd.

PREFACE

Quantized gauge theories play a central role in the description of particle interactions at a fundamental level. As for all quantum fields, quantized gauge systems must be regularized for a proper mathematical definition. Conventional schemes of regularization are based on a perturbative expansion: they can be used to derive a variety of theoretical predictions, but are not suitable for a study of those phenomena which are governed by a strong coupling constant or are of nonperturbative nature. The formulation of gauge theories on a space-time lattice was proposed by Wilson in 1974 to overcome the limitations of perturbative regularization. The lattice regularized quantum gauge theory is a very elegant mathematical construct, which does not rely on any weak coupling for its definition. In particular, strong coupling techniques can be applied and in the strong coupling regime sought for phenomena such as quark confinement are seen to occur. The regularization given by the lattice must, however, eventually be removed by letting the lattice spacing go to zero. In this process the coupling constant gets renormalized and results originally obtained for strong coupling may have to be extrapolated to weak coupling. While on the basis of strong coupling expansions alone this extrapolation may be problematic, information on the outcome of the renormalization procedure can be obtained exploiting another important property of lattice gauge models. These systems, which have many formal analogies with statistical systems used in the description of thermodynamical behavior, like their statistical counterparts can be investigated by numerical methods known as Monte Carlo simulations. Use of the lattice regularization in conjunction with Monte Carlo simulations has produced invaluable results during the last few years, especially for the gauge theory of strong interactions known as Quantum Chromodynamics.

The purpose of this book is to illustrate the lattice formulation of a gauge theory, the techniques which can be used to derive its predictions, with particular emphasis on Monte Carlo simulations, and the major results obtained. A collection of papers, selected so as to provide a global view of what has been accomplished in the field, is presented to the reader. The selection is however not exhaustive of all important contributions: there are many more than could have fit into a volume of reasonable size. A few introductory chapters precede the reprints. Chapter 2, on the formulation of a quantum gauge theory, and Chapter 3, on the Monte Carlo method, have been written to provide the reader who is not familiar with the subject with basic notions about the topics considered in this book. Chapter 4 ought to serve as a guide through the collected papers. Also, additional references can be found in Chapter 4.

I wish to express my gratitude to Dr. Phua and WSPC for inviting me to edit this book, to my colleague Mike Creutz for assistance in the selection of the reprints, to Mrs. Isabell Harrity for her prompt and accurate typing of the manuscript and to the authors and publishers of the journals for their permission to reproduce the papers.

CONTENTS

Introductory Chapters

MONTE CARLO COMPUTATIONS IN LATTICE GAUGE THEORIES

1. INTRODUCTION

Gauge theories are vital for our understanding of particle systems. Almost all models currently employed to describe particle interactions at a fundamental level make use of the notion of a gauge field. The corresponding quantum systems must be regularized to be given a well defined mathematical meaning and techniques, which are both elegant and powerful, have been developed for the purpose. However, most of these techniques are based on the existence of one or more weak coupling parameters in which the theory can be expanded perturbatively. As such they are not suited for the analysis of phenomena governed by an intrinsically large coupling constant, or, an even worse case, where the behavior at the origin, in the space of complex coupling parameters, is non-analytic. To overcome these difficulties a different method of regularization has been advocated by Wilson [R1]*. It consists of formulating the gauge theory on a discrete lattice of points in Euclidean space-time. The ultraviolet divergences are thus removed and, if the space-time volume of the whole system is made finite to proceed only later to an infinite volume limit, all quantum averages are given by mathematically well-defined expressions, irrespective of the value of the coupling constant. Thus, the lattice regularized theory lends itself to strong coupling expansions, quite analogous to the high temperature expansions for thermodynamical systems, opening a new domain of analytical investigations.

The lattice formulation, however, does not solve all problems of strong coupling. The regularization must eventually be removed

*For notation regarding references see the end of the section.

by letting the spacing between contiguous lattice sites gradually decrease to zero and in the process the coupling parameters must be renormalized: a well-defined continuum limit can be obtained only if these parameters approach a scaling fixed point. The problem remains therefore of demonstrating the existence of such fixed points and of extrapolating there whatever results may have been established within the domain of strong couplings. While expansion techniques may not be powerful enough to achieve these goals, another very useful feature of the lattice formulation succours the theorists. The lattice regularized gauge systems, like their thermodynamical counterparts, allow numerical computations by the technique of importance sampling known as Monte Carlo simulation [A1–4]. The numerical analysis reaches a little further from where the insight obtained by strong coupling expansions seems to fail and bridges the gap between the strong coupling domain and the domain of scaling toward the continuum limit.

Monte Carlo simulations have been extensively applied to the study of lattice gauge theories during the last few years and several important results have been achieved. Some have demonstrated, albeit numerically, the validity of long-standing theoretical conjectures, some have allowed the actual determination of interesting physical observables. The whole lattice approach being motivated by the instance of getting new clues to strong coupling phenomena, it is not surprising that the majority, as well as the most relevant, of Monte Carlo results have been obtained for the gauge theory of strong interactions known as Quantum Chromodynamics (QCD).

The purpose of this book is to gather together a collection of reprints providing a path through the formulation of a gauge theory on a lattice all the way to some of the most recent Monte Carlo results for QCD. Once the basic ideas underlying the lattice regularization and the Monte Carlo numerical technique are assimilated, it is really straightforward to proceed through the literature in this field. However, the fundamental notions are not found all together, other than in review articles or lecture notes. Hence, to facilitate the task of the reader not already familiar with the subject, two introductory chapters, one on the definition of a lattice gauge system, the other on the Monte Carlo method, precede the reprints. A detailed account of the results achieved would be too cumbersome as an introductory chapter, and these are best learnt from the original articles. Thus, only a rather concise guide through the literature concludes, in Chapter 4, the introductory notes.

Finally, a word about references. Those to be found reprinted in this volume are indicated by an R followed by the serial number. Additional references are indicated by A with a serial number and listed at the end of the introductory chapters. Given the very large number of articles published on the topic of lattice gauge theories and, more specifically, on Monte Carlo simulations of such systems, any collection of reprints which can be put together in a volume of manageable size is bound to exclude a large number of valuable works. No judgement of quality is implied in the selection made in this book, which has rather been based on the criterion of providing a continuous and possibly smooth pathway through the subject of lattice gauge theories and Monte Carlo simulations. To a lesser extent these remarks apply also to the list of references appended at the end of the introductory chapters, which in no way claims to be complete, or exhaustive of the relevant contributions to the field.

2. FORMULATION OF A QUANTIZED LATTICE GAUGE SYSTEM

There are several ways to define a quantum field theory, but the method which is of most interest for the topic considered in this book proceeds through the following formal steps:

i) Time is rotated to the imaginary axis, to achieve a positive definite Euclidean metric in space-time.

ii) A suitable action functional $S(\phi, g, m)$ of the field configuration is defined. With very symbolic notation ϕ denotes here the collection of all fields, g stands for all coupling parameters, m for all masses and other possible parameters of the theory.

iii) The quantum expectation value of any observable $\mathcal{O}(\phi)$ is defined by averaging the value it takes on a given field configuration over all field configurations, with measure proportional to $\exp\{-S\}$ (we adopt units where $\hbar = c = 1$):

$$\langle \mathcal{O} \rangle = Z^{-1} \int D\phi \, \mathcal{O}(\phi) \, e^{-S(\phi, g, m)} , \tag{2.1}$$

$$Z = \int D\phi \, e^{-S(\phi, g, m)} . \tag{2.2}$$

The quantity Z in Eq. (2.2) is the vacuum to vacuum permanence amplitude, also referred to as partition function because of the analogy with the statistical formulation of a thermodynamical system.

iv) The quantum averages of Eq. (2.1) are continued back, if necessary, to Minkowskian space-time.

The formal expressions in the r.h.s. of Eqs. (2.1), (2.2) must be given well-defined mathematical meaning, and this is achieved by the lattice regularization. The continuum of space-time points is replaced by the vertices of a lattice, which is almost always taken to be hypercubical [see however R8, 9; A5]. (We shall denote by indices $i, j \ldots$ the generic points and by a the spacing of the hypercubical lattice). The functional integrals in Eqs. (2.1), (2.2) can then be converted into ordinary multiple integrals, by limiting the overall size of the lattice to a finite volume V and by defining the fields only at the lattice sites, $\phi(x) \to \phi_i$, replacing at the same time the partial derivatives occurring in S with finite difference operators. Eventually the limits $V \to \infty$, $a \to 0$ must be taken. However, this straightforward method of regularization, which is indeed followed for the matter fields present in the system, does not lead to the most elegant or convenient definition of a lattice gauge theory. To see how a gauge quantum field is best put on a lattice, it is necessary to reconsider briefly the geometrical meaning of the gauge potentials $A_\mu{}^\alpha(x)$ and of the field strength $F_{\mu\nu}{}^\alpha(x)$.

The gauge potentials specify "transport operators", which can be used to compare orientations in some intrinsic space of matter fields at neighboring points x^μ and $x^\mu + \mathrm{d}x^\mu$. Let us assume that the field $\phi(x)$ belongs to a definite representation of the gauge group $\mathscr{G}$ and that τ_α are the infinitesimal generators in the same representation. A scalar product $\phi^\dagger(x + \mathrm{d}x)\phi(x)$ would not be invariant under local gauge transformations

$$\phi(x) \to G(x)\phi(x)\,, \tag{2.3}$$

where $G(x)$ is a position dependent, finite element of $\mathscr{G}$. To construct an invariant scalar product, the field must first be "transported" from x^μ to $x^\mu + \mathrm{d}x^\mu$, and this done multiplying it by the transport operator

$$\mathrm{e}^{igA_\mu{}^\alpha(x)\tau_\alpha \mathrm{d}x^\mu}\,, \tag{2.4}$$

where g is the gauge coupling constant. The quantity

$$\phi^\dagger(x + \mathrm{d}x)\,\mathrm{e}^{igA_\mu{}^\alpha \tau_\alpha \mathrm{d}x^\mu}\,\phi(x)\,, \tag{2.5}$$

given suitable transformation properties of $A_\mu{}^\alpha(x)$ itself, is invariant under the local gauge transformation (2.3).

All this suggests that on the lattice the gauge dynamical variables be associated with the oriented links between neighboring vertices i and j rather than with the vertices themselves: on the lattice, indeed, the displacements between neighboring sites play the role of the infinitesimal displacements $x^\mu \to x^\mu + dx^\mu$ of the continuum system. Thus, in the lattice regularization we take as basic dynamical variables finite group elements $U_{ji} \in \mathcal{G}$ defined over the oriented links of the lattice. U_{ji} is a finite element because the displacement is now over a finite length a. Notice that this opens the interesting possibility of considering discrete gauge groups as well. In a gauge transformation the matter fields, still defined at the lattice sites, change as follows

$$\phi_i \to G_i \phi_i \quad (G_i \in \mathcal{G}), \tag{2.6}$$

whereas the gauge dynamical variables, which must satisfy

$$U_{ij} = U_{ji}^{-1}, \tag{2.7}$$

transform as follows

$$U_{ji} \to G_j U_{ji} G_i^{-1}. \tag{2.8}$$

A scalar product such as

$$\phi_j^\dagger U_{ji} \phi_i \tag{2.9}$$

is then gauge invariant.

In the continuum theory the transport along a path γ is defined by a path-ordered exponentiated line integral of $A_\mu{}^\alpha(x)$:

$$U_\gamma = \mathrm{P}\, e^{ig\int_\gamma A_\mu{}^\alpha \tau_\alpha dx^\mu}. \tag{2.10}$$

If, on the lattice, the path goes through the sequence of neighboring sites $i_1, i_2 \ldots i_N$, the corresponding transport operator is

$$U_\gamma = U_{i_N i_{N-1}} \ldots U_{i_3 i_2} U_{i_2 i_1}, \tag{2.11}$$

and in a gauge transformation

$$U_\gamma \to G_{i_N} U_\gamma G_{i_1}^{-1}. \tag{2.12}$$

Of particular importance are the transport operators associated with closed loops λ, starting from a point i and returning to the same point. From Eq. (2.12) we see that in a gauge transformation U_λ undergoes a similarity transformation

$$U_\lambda \to G_i U_\lambda G_i^{-1} . \tag{2.13}$$

It follows that the trace of U_λ (or, for that matter, any class function of U_λ)

$$W_\lambda = \mathrm{Tr}\, U_\lambda \tag{2.14}$$

is gauge invariant. The quantities W_λ are often called Wilson's loop factors.

In the continuum theory the all important $F_{\mu\nu}{}^\alpha$ are also associated with a sort of closed path: namely, the one defined by the infinitesimal parallelogram of sides dx^μ and dx^ν. Transported along such a contour a field undergoes a transformation given by

$$e^{igF_{\mu\nu}{}^\alpha \tau_\alpha dx^\mu dx^\nu} \quad \text{(no sum over } \mu, \nu). \tag{2.15}$$

The lattice equivalent of the elementary closed loops defined by dx^μ and dx^ν are the perimeters of the squares (commonly called plaquettes) having vertices in 4 (sequentially) neighboring sites i_1, i_2, i_3, i_4. We shall denote by

$$U_P = U_{i_1 i_4} U_{i_4 i_3} U_{i_3 i_2} U_{i_2 i_1} \tag{2.16}$$

the corresponding transport operators and, so as $F_{\mu\nu}{}^\alpha$ is used in the continuum theory to define the action, likewise we shall proceed from U_P for the construction of the lattice gauge action.

For clarity's sake, it pays at this point to be specific and to consider a definite group, which we take to be SU(2). We parametrize

$$U_{ji} = \mathrm{I} \cos\theta_{ji} + \sigma \cdot \mathbf{n}_{ji} \sin\theta_{ji} ,$$
$$U_P = \mathrm{I} \cos\theta_P + \sigma \cdot \mathbf{n}_P \sin\theta_P , \tag{2.17}$$

where $\mathbf{n}_{ji}$ and $\mathbf{n}_P$ are unit vectors, I is the identity matrix and σ are the Pauli matrices, related to the infinitesimal generators by

$$\tau_\alpha = \frac{\sigma_\alpha}{2} \quad . \tag{2.18}$$

Replacing dx^μ and dx^ν in expressions (2.4) and (2.15) with the lattice spacing a and taking the formal limit $a \to 0$, it is easy to verify that

$$\theta_{ji} = \frac{1}{2} ga |A_\mu{}^\alpha| \quad ,$$
$$n_{ji}{}^\alpha = A_\mu{}^\alpha / |A_\mu{}^\alpha| \tag{2.19}$$

and

$$\theta_P = \frac{1}{2} ga^2 |F_{\mu\nu}{}^\alpha| ,$$
$$n_P^\alpha = F_{\mu\nu}{}^\alpha / |F_{\mu\nu}{}^\alpha| . \tag{2.20}$$

This suggest that the action density of the continuum theory, namely

$$\mathscr{L} d^4x = \frac{1}{4} F_{\mu\nu}{}^\alpha F_\alpha{}^{\mu\nu} d^4x \ , \tag{2.21}$$

be replaced by some suitable gauge invariant function of the plaquette transport operator U_P, behaving for small θ_P in a way proportional to $\theta_P{}^2$. The simplest such function is $1 - \cos\theta_P = 1 - 1/2 \operatorname{Tr} U_P$. We are thus led to a plaquette action

$$S_P = \frac{4}{g^2}\left(1 - \frac{1}{2} \operatorname{Tr} U_P\right) \tag{2.22}$$

and to a total action just given by the sum over all plaquettes of S_P:

$$S = \sum_P S_P = \frac{4}{g^2} \sum_P \left(1 - \frac{1}{2} \operatorname{Tr} U_P\right) . \tag{2.23}$$

The numerical factor $4/g^2$ is inserted in such way that in the formal continuum limit S reduces to $\int d^4x (1/4) F_{\mu\nu}{}^\alpha F_\alpha{}^{\mu\nu}$. The action given by Eq. (2.23) is commonly referred to as Wilson's lattice gauge action.

Several remarks are in order here. First of all we notice that, in analogy to the continuum case, S_P takes its lowest value, namely 0, when the transport around the plaquette is trivial, reducing to the identity and grows larger as U_P deviates from I. The coupling constant weights the role of the plaquette excitations (i.e. $U_P \neq I$)

in the formation of the action: small g suppresses the effect of small excitations; the opposite happens for g large. To put into evidence the role of g it is customary to define a coupling independent "internal energy"* of the plaquette

$$E_P = 1 - \frac{1}{2}\,\mathrm{Tr}\,U_P\,. \tag{2.24}$$

This generalizes to

$$E_P = 1 - \frac{1}{2N}\,\mathrm{Tr}(U_P + U_P{}^{\dagger}) \tag{2.25}$$

for the SU(N) gauge groups and to

$$E_P = 1 - \mathrm{Re}\,U_P \tag{2.26}$$

for the U(1) groups, whose elements can be parametrized as complex numbers of unit module ($U_P = e^{i\theta_P}$). The plaquette action is then obtained multiplying E_P by a coupling parameter β (the notation derives from the analogy with thermodynamical systems, where the corresponding coupling parameter is indeed $\beta = 1/kT$):

$$S = \sum_P S_P = \sum_P E_P\,. \tag{2.27}$$

The relation between β and g, which is determined by various normalization conventions of the continuum theories, is

$$\beta = \frac{1}{g^2} \quad \text{for U(1)}\,, \tag{2.28}$$

$$\beta = \frac{2N}{g^2} \quad \text{for SU}(N)\,. \tag{2.29}$$

As a final remark, we notice that the requirement that the lattice theory should formally reduce to the continuum theory as $a \to 0$ poses constraints only on the behavior of E_P for U_P in the neighborhood of I. Thus several alternative forms for the action can be and have been considered in the literature [see R42-46].

---

*This "internal energy" is actually never a true physical energy, but rather is a rescaled action. The analogy with thermodynamical systems, to be illustrated later, motivates the name.

The inclusion into the action of terms involving the matter fields is straightforward. The total lattice S will consist of a sum

$$S = S_G(U, g) + S_M(\phi, U, g', m) \ , \tag{2.30}$$

where S_G is the gauge field action described above and S_M is the action for the matter fields. This will in general be constructed from the continuum action replacing the covariant partial derivatives with covariant finite differences, i.e., differences where fields at different sites are not subtracted directly but only after they have been transported to a common location (normally the site of either field variable) by a suitable gauge variable U_{ji}. Thus, S_M will typically contain couplings between neighboring matter field variables which are mediated by the gauge dynamical variables (as in the expression (2.9)). Contrary to the case of the continuum theory, the gauge coupling will not appear in S_M, because it has been in a sense re-absorbed in the definition of the lattice gauge variables. Other coupling constants, specifying the strength of the various self-couplings of the matter fields, will generally appear in S_M (they are denoted by g'), together with other possible parameters, such as masses of constituent fields. Lattice gauge theories with bosonic matter fields are considered in [R47 and R48].

A problem encountered with fermionic systems is that a straightforward transcription onto the lattice of the continuum action generates an unwanted multiplication of states, in the sense that, as one proceeds to the continuum limit $a \to 0$, one finds that beyond the original continuum system, additional low frequency limits can be extracted from the lattice gauge theory, with all the properties of continuum fields. It would take too long here to go deeper into the origin of such a problem, which the reader will find discussed in [R49 and R50], or to illustrate possible ways to go around the difficulty, which however the reader can find in all the reprints dealing with fermions.

The lattice regularized quantum expectation value of an observable $\mathcal{O}(U_{ij})$ is given by

$$\langle \mathcal{O} \rangle = Z^{-1} \int \prod_{\{ij\}} dU_{ji} \, \mathcal{O}(U_{ji}) \, e^{-S(U_{ji}, g)} \ , \tag{2.31}$$

$$Z = \int \prod_{\{ij\}} \mathrm{d}U_{ji}\, \mathrm{e}^{-S(U_{ji},\, g)}\,, \tag{2.32}$$

where the integrals are invariant integrals over the group manifolds, or simply sums over all group elements if the gauge group is discrete. In Eqs. (2.31) and (2.32) we have assumed, as we shall do in the rest of this section, that the only dynamical variables are those of the gauge field. The extension to the case where matter fields are also present is however straightforward.

The integrals in the r.h.s of Eqs. (2.31) and (2.32) are ordinary multiple integrals (remember that the total volume V of the lattice is at first taken to be finite) and so the formula for $\langle \mathcal{O} \rangle$ is mathematically well defined. Notice also that the integrations are done over the whole group manifold and configurations which are gauge equivalent (i.e., related by a gauge transformation as in Eq. (2.8)) are all included in the sum. Contrary to the continuum case, this integration over pure gauge degrees of freedom does not make the quantum averages ill defined: thus in the lattice regularization there is no need to fix gauge and an explicitly gauge invariant formulation of the quantum averages is achieved.

Eqs. (2.31) and (2.32) bear a striking resemblance to similar expressions, which are encountered in the definition of thermodynamical averages. The analogy is made even clearer if the definitions of Eqs. (2.24–2.29) are used to rewrite the measure factor $\exp\{-S\}$ as $\exp\{-\beta \Sigma E_P\}$. This is formally the same measure as in a thermodynamical average, if β is identified with $1/kT$ and ΣE_P with the internal energy of the system. But one should remember that in the lattice quantum field theory β is related to a coupling constant, rather than to a physical temperature, and, as already remarked, the "internal energy" is a rescaled action rather than a true energy. For the modifications which must be made to include genuine temperature effects see, for instance, [R27–29].

The analogy with the statistical treatment of thermodynamics reveals in particular that the regimes of weak coupling and low temperature correspond, as do the regimes of strong coupling and high temperature. Thus the expansion techniques for either high values of β (low temperature $\leftrightarrow$ weak coupling) or small values of β (high temperature $\leftrightarrow$ strong coupling), which have produced several important results for thermodynamical systems, can be carried over to quantum lattice gauge theories.

The weak coupling expansion of a lattice gauge system actually consists of nothing more (nor less) than the standard perturbative expansion of a quantum field theory, made however more complicated by the loss of Lorentz covariance and by the additional interactions introduced by the discreteness of the lattice. In general, problems which admit perturbative solutions are better treated with more conventional, continuum-based schemes of regularization. Weak coupling expansions of the lattice theory become necessary, however, when one wants to compare results obtained by this method of regularization with results obtained by perturbative methods in the continuum, since the weak coupling domain is the only one where the two schemes overlap. For analyses of this type see [R19 and R20].

The possibility of performing strong coupling expansions constitutes instead a novel and powerful tool provided by the lattice regularization. Strong coupling expansions are performed by expanding the measure $\exp\{-\beta\Sigma_P E_P\}$ in powers of β and integrating the terms of the series, which are generally now polynomials in the dynamical variables U_{ji}. Frequently resummations are performed and suitable functions of β are used as substitute expansion parameters in order to simplify the series or to achieve better properties of convergence. Strong coupling expansion techniques are explained and used in [R1, 5, 21, 25, 26, 51, 52].

Strong coupling expansions, however, do not necessarily provide a solution to the problem of describing strong interactions and consequent phenomena in the actual physical world. The reason is that the parameter which is used for the expansion, $\beta = \text{const}/g^2$, does not represent the physical strength of the interaction, but rather its magnitude within the regularized lattice theory. In other words, g is an unrenormalized coupling constant. Contact with the physical world, via a suitable continuum limit, must still be made, and, as the regularized theory is renormalized towards its continuum limit, the value of the lattice coupling constant g will be modified. To obtain physical results from the strong coupling series, then, extrapolations in β have to be made.

Let us discuss here the very important notion of renormalization. Suppose that physical quantities q_1, q_2 . . . , bearing dimensions $-d_1$, $-d_2$ . . . in units of the lattice spacing, have been calculated in the lattice regularized version of the theory. (Adopting, as we do, units where $\hbar = c = 1$, q_1, q_2 have dimension of $[\text{mass}]^{d_1}$, $[\text{mass}]^{d_2}$). The result of the calculation will take the form

$$q_1 = a^{-d_1} f_1(g) \ ,$$

$$q_2 = a^{-d_2} f_2(g) \ , \ \ldots \tag{2.33}$$

where the dependence on the lattice spacing is trivial, and the whole content of the theory is expressed by the mathematical functions f_1, f_2 . . . of the dimensionless coupling constant g. If some of our observables, say $q_{i_1} \equiv \ell_1$, $q_{i_2} \equiv \ell_2$ . . . , are for instance correlation lengths, they will be given by

$$\ell_1 = a f_{i_1}(g) \ ,$$

$$\ell_2 = a f_{i_2}(g) \ , \ \ldots, \tag{2.34}$$

$f_{i_1}(g)$, $f_{i_2}(g)$ etc. measuring the correlation lengths in number of lattice spacings.

It is now obvious that merely letting $a \to 0$ will not produce a meaningful continuum limit. g must be changed at the same time as a is set to zero, (renormalization) in such a way as to make the observables approach well-defined finite limits, if one wants to recover a continuum field theory at the end of the process. That this is possible, and therefore that the lattice theory admits a continuum limit, is a priori not certain. Indeed, the existence of a continuum limit puts very stringent requirements on the theory. We see it first examining any correlation length, e.g. ℓ_1. To make ℓ_1 approach a finite limit by suitably changing g

as $a \to 0$, there must be a critical value g_{cr} of g such that $f_{i_1}(g) \to \infty$ as $g \to g_{cr}$. Values of the coupling constant which make the correlation go to infinity on the lattice define critical points of the system. Moreover, as $g \to g_{cr}$, all the functions $f_i(g)$ must tend either to infinity or to zero (according to whether d_i is negative or positive) and in such a way that the same rate of approach of g to g_{cr}, as $a \to 0$, which makes ℓ_1 tend to a constant value, also makes *all* the observables q_i tend to constant values. We say then that the critical point defined by $g = g_{cr}$ is a scaling critical point.

From the above discussion it is clear that the passage to the continuum limit requires a definite functional relationship between a and g. This could be obtained by demanding that any of the observables, e.g. q_1, remains strictly constant throughout the process of renormalization. Then, from Eq. (2.33) one would infer

$$a \equiv a(g) = [q_1]^{1/d} [f_1(g)]^{1/d} \quad . \tag{2.35}$$

It is useful however to allow more freedom in the rule for relating a and g which might be derived, for instance, from theoretical considerations. One therefore expresses the functional relationship between a and g in the general form

$$a = \frac{1}{\Lambda} f(g) , \tag{2.36}$$

where f is a definite function embodying the correct scaling behavior of a as $g \to g_{cr}$ and Λ is a scale parameter. Λ does not have a direct physical significance and in general it will depend on the scheme of renormalization (for example, on the specific choice of the lattice gauge action, within the class of actions leading to the same continuum theory).

The scaling properties of the critical point now manifest themselves in the fact that all the functions $f_i(g)$ must behave as

$$f_i(g) \sim c_i [f(g)]^{d_i} \tag{2.37}$$

for $g \to g_{cr}$, c_i being definite constants. Substituting Eqs. (2.36) and (2.37) into Eq. (2.33), we find that the continuum values of the observables q_i are given by

$$g_i = c_i \Lambda^{d_i} \quad , \tag{2.38}$$

i.e., all physical observables are expressed in terms of the scale parameter Λ. Any observable might then be used to establish the scale of masses and, eliminating Λ, all other observables expressed in terms of that one.

If $g_{cr} \neq 0$, non-perturbative methods will be required to determine the functional relationship of Eq. (2.36). However, for non-Abelian quantum gauge theories it has been shown [see R17, 18] that $g_{cr} = 0$ has the properties of a scaling critical point. The functional form of Eq. (2.36) can then be determined by perturbative arguments and this leads to [R17, 18; A6, 7]

$$f(g) = (g^2 \gamma_0)^{\gamma_1/2\gamma_0} \, e^{-1/2\gamma_0 g^2} \, (1 + O(g^2)) \quad ,$$

$$\gamma_0 = \frac{1}{16\pi^2}\left[\frac{11N}{3}\right] , \quad \gamma_1 = \left[\frac{1}{16\pi^2}\right]^2 \left[\frac{34N^2}{3}\right] \tag{2.39}$$

for a pure SU(N) theory. In order to relate results obtained in the lattice formulation to results which can be obtained by perturbative methods (for instance, behaviors of inclusive cross sections in deep inelastic scattering or high energy e^+e^- collisions; electroproduction in high energy pp and $p\bar{p}$ collisions etc.) the lattice scale factor Λ must be compared with scale factors defined in the conventional scheme of perturbative renormalization. This computation can be performed exactly (the only approximation being the need for performing a few integrals numerically) and it can be found in [R19, 20].

From this whole discussion it emerges that, in the most relevant case of Quantum Chromodynamics, the strong coupling results obtained for small values of β must be extrapolated, if not all the way to $\beta_{cr} = \infty$ at least to a value of β large enough that scaling toward the continuum limit is seen to take place. Also, the extrapolation should provide a self-consistency check for the theory, giving indications that no intervening critical points make the properties of the system totally different in the strong coupling domain and in the scaling domain: thus features such as confinement of quarks, which can easily be demonstrated in the strong coupling domain, would survive the passage to the continuum limit. From this point of view strong coupling expansions have not proven adequate. Although not inconsistent with scaling at large values of β, the series expansions, carried as far as it has been up to now possible, begin to produce uncertain results precisely where the scaling behavior (as seen in numerical computations) appear to set in.

The lattice formulation of a gauge theory, however, offers another valuable advantage. As in the statistical formulation of thermodynamical systems, the averages which define the expectation values of physical observables can be computed approximately by numerical methods, based on importance samplings. These numerical methods go under the name of Monte Carlo simulations. Monte Carlo simulations have been very profitably applied to latttice gauge theories during the last few years and, indeed, the major goal of this book is to provide the reader with a compact view of what has been achieved.

Before concluding this chapter I ought to mention that although the main emphasis of this book is on so-called Euclidean formulation of a lattice gauge theory, where both space and time are made discrete, it is also possible to regularize the theory by discretizing only space, maintaining a continuous evolution in time (no rotation $t \rightarrow it$ is performed either). A Hamiltonian operator generates the time evolution, hence the name Hamiltonian formulation of the lattice gauge system. Many interesting results have been obtained in this way and although reasons of space preclude the inclusion of the corresponding articles in this book, the fundamentals of the method and a very interesting application can be found in [R6 and R21].

3. THE MONTE CARLO METHOD

The integrals appearing in the definition of the regularized quantum expectation values (Eqs. (2.31) and (2.32)) are ordinary multiple integrals. If the integrals are approximated by sums over a sufficiently dense set of points or if the gauge group is discrete, in which case the integration symbol really stands for a sum over all group elements, the r.h.s. of Eqs. (2.31) and (2.32) reduce to sums over a finite number of terms. We shall say that each of these terms specifies a configuration C of the system. The fact that quantum averages are exactly or approximately expressed by sums over finite sets of configurations leads one to think that they ought to be calculable by numerical methods. This is indeed the case, but not through a straightforward summation procedure: even for the simplest gauge model, the one defined over the group Z_2 with elements equal to +1 and −1, a four-dimensional lattice of any meaningful extent implies such a large number of configurations that a direct numerical sum is beyond the possibilities of the most powerful computer (the total number of configurations is $2^{1024} \approx 10^{310}$ for a modest 4^4 lattice). Instead, one may exploit the fact that, as in the case of thermodynamical systems, only a comparatively small subset of configurations effectively contributes to the quantum averages. If, by means of a suitable stochastic procedure, a large sample of configurations is selected among the relevant ones with a probability distribution proportional to the measure factor $\exp\{-S\}$, the exact quantum averages may be approximated by averages taken over this large sample of configurations.

The numerical method outlined above goes under the name of Monte Carlo (MC) simulation. In more detail, the working of the algorithm is as follows [A1–4]. An initial configuration $C^{(1)}$ is stored into the memory of the computer (i.e., the values of all gauge dynamical variables U_{ji} and of bosonic matter fields ϕ_i, if present, are recorded in memory; fermionic matter fields are represented by anticommuting variables, i.e., elements of a Grassmann algebra, and MC simulations involve then subtleties which will be commented upon in the next chapter). From $C^{(1)}$ the computer generates a new configuration $C^{(2)}$, which replaces $C^{(1)}$ in the memory, by a stochastic procedure: $C^{(2)}$ is not obtained deterministically from $C^{(1)}$, but according to an algorithm which involves the extraction of random numbers. Thus, only a transition probability $p(C \to C')$ is defined for the passage between one configuration and the next. (As a matter of fact, computers generate only pseudorandom numbers, so the whole procedure is deterministic and reproducible, but it approximates very well a true stochastic algorithm). From $C^{(2)}$ the computer generates a new configuration $C^{(3)}$, and so on, producing eventually a very large number of configurations. The stochastic process is designed in such a way that the probability of encountering any definite configuration C at the kth step in the sequence converges, as k grows larger, to a distribution proportional to the correct measure factor $\exp\{-S(C)\}$, ($S(C)$ denoting the action of the configuration C). Assuming that after n_0 steps the probability distribution has come close enough to the correct limiting one, we approximate the exact quantum expectation values defined in Eqs. (2.31, 2.32) by

$$\langle \mathcal{O} \rangle = \frac{1}{n} \sum_{k=n_0+1}^{n} \mathcal{O}(C^{(k)}) \,, \tag{3.1}$$

where $\mathcal{O}(C)$ stands for the value taken by the observable $\mathcal{O}$ in the configuration C.

The Boltzmann distribution $p(C) \propto \exp\{-S(C)\}$ must be an eigenvector of the stochastic matrix $p(C \to C')$. This is guaranteed if p obeys detailed balance condition

$$\frac{p(C \to C')}{p(C' \to C)} = \frac{e^{-S(C')}}{e^{-S(C)}} \,. \tag{3.2}$$

Indeed, the general property of stochastic transition matrices

$$\sum_{C'} \mathrm{p}(C \rightarrow C') = 1 \tag{3.3}$$

and Eq. (3.2) imply

$$\sum_{C} \mathrm{e}^{-S(C)}\, \mathrm{p}(C \rightarrow C') = \sum_{C} \mathrm{e}^{-S(C')}\, \mathrm{p}(C' \rightarrow C) = \mathrm{e}^{-S(C')} \ .$$

The detailed balance condition is not a necessary requirement for convergence to the Boltzmann distribution, but it is satisfied in the most commonly implemented algorithms. Also, it does not fix the algorithm completely, and variations are possible. The procedure which is probably used most frequently is essentially the one introduced in the original paper of Metropolis, Rosenbluth, Rosenbluth, Teller and Teller [A1] and goes under the name of Metropolis procedure. It works as follows.

The transition matrix $\mathrm{p}(C \rightarrow C')$ is determined in two steps. First a new candidate configuration C' is selected starting from C according to some probability distribution $\mathrm{p}_0(C \rightarrow C')$, satisfying the equality

$$\mathrm{p}_0(C \rightarrow C') \rightarrow \mathrm{p}_0(C' \rightarrow C) \ . \tag{3.4}$$

The variation in action

$$\Delta S = S(C') - S(C) \tag{3.5}$$

that would be induced by the change is calculated. A pseudorandom number is then selected with uniform probability distribution between 0 and 1 and:

if $r < \mathrm{e}^{-\Delta S}$ the change is accepted – the new configuration in the sequence is C';

if $r > \mathrm{e}^{-\Delta S}$ the change is rejected and the new configuration is again C.

Clearly, if the passage from C to C' lowers the action, $\mathrm{e}^{-\Delta S}$ is larger than one and the change is always accepted. Vice versa, if the action is increased by the proposed change of configuration, this is accepted only with conditional probability $\mathrm{e}^{-\Delta S}$. It is this occasional acceptance of changes which increase the action which simulates the effects of quantum fluctuations. It is straightforward to verify that the matrix $\mathrm{p}(C \rightarrow C')$ defined by the above procedure satisfies the condition of detailed balance: indeed, assuming for instance $S(C') < S(C)$, we have $p(C \to C') = p_0(C \to C')e^{-\Delta S}$ and $p(C' \to C) = p_0(C' \to C)$. Hence

$$\frac{p(C \to C')}{p(C' \to C)} = \frac{p_0(C \to C')}{p_0(C' \to C)} e^{-\Delta S}$$

$$= e^{-\Delta S} = \frac{e^{-S(C')}}{e^{-S(C)}} \quad \text{(by Eq. 3.3)}. \tag{3.6}$$

In practice the passage from a configuration to the next is obtained by changing, or "upgrading", just one of the dynamical variables (or at most a limited set of neighboring variables). In this way the variation ΔS is kept small (with respect to the order of magnitude of S itself) and one avoids drastic changes of configuration which, implying a very large positive ΔS, would almost invariably lead to a rejection of the move. Also, the locality properties of S (it is a sum of several terms, like S_P, where neighboring variables only are coupled) make the computation of ΔS induced by the variation of a single variable arithmetically rather simple, which is crucial when the computer has to repeat such a calculation an enormous number of times.

Thus, in the Metropolis procedure, the transitions $p_0(C \to C')$ and $p(C \to C')$ can be further qualified, for a gauge system, as $p_0(U_{ji} \to U'_{ji})$ and $p(U_{ji} \to U'_{ji})$, all other U_{ji} being kept fixed. If the variable U_{ji} to be upgraded is selected at random, the probability that a definite U_{ji} should be proposed for the upgrading also enters in the determination of the transition probability, and one can correctly speak of a definite $p(C \to C')$ associated with each step. Very frequently, however, the variables to be upgraded are not randomly selected, but, rather, one proceeds according to some definite pattern throughout the lattice. In this case $U_{j_1 i_1}$ is upgraded first, then $U_{j_2 i_2}$ . . . and, correspondingly, one ought to talk of several, distinct matrices $p_{j_1 i_1}(C \to C')$, $p_{j_2 i_2}(C \to C')$ . . . . Every individual transition matrix satisfies however the principle of detailed balance and the Boltzmann distribution $\exp\{-S(C)\}$ remains an eigenvector

Recapitulating, in most of the applications the variables U_{ji} are upgraded one by one and the passage between two consecutive configurations $C^{(k)}$ and $C^{(k+1)}$ in the MC sequence, i.e. one MC step, involves at most the replacement of one U_{ji} in memory with its upgraded value. When one step per dynamical variable has been performed, one says that a MC iteration (or one sweep of the lattice) has been completed.

A variation of the procedure just described consists in performing several upgrading steps on a definite variable before proceeding to the next. This is particularly convenient when the rate of rejection is high. A large fraction of the arithemetics leading to ΔS may not have to be repeated when the same variable is subjected to several attempted upgrades. Thus, performing multiple upgrades of the same variable may achieve a faster rate of convergence to statistical equilibrium. This slight variation of the original method is sometimes referred to as improved Metropolis algorithm. For an explicit program implementing this algorithm with a finite gauge group see [R16].

In the limit where the number of upgrades of an individual variable U_{ji} becomes infinite, eventually the new value U'_{ji} is determine by a distribution proportional to exp $-S(U_{ji};\, U_{j'i'},\, U_{j''i''} \ldots)$; where all the variables $U_{j'i'}$, $U_{j''i''} \ldots$, but for U_{ji} are kept fixed. The MC procedure where the new value for a definite variable is selected according to this probability distribution [A8; R22] has been called "the heat bath algorithm". In general selecting a new value U'_{ji} on the basis of the heat bath probability distribution is computationally demanding and the improved Metropolis procedure leads to a more efficient algorithm. There are notable cases, however, where the structure of the gauge group manifold and the form of the action allow a fast and elegant implementation of the heat bath procedure. For detailed examples see [R22, 24; A9].

Beyond the general feature of the MC method which have been described above, there is little that can be said with equal generality about the applications of the technique to specific computations. In particular, one shall have to make a choice on the size of the lattice (dictated by the conflicting requirements of dealing with a sufficiently large volume and yet of containing the computation within a reasonable time), on the boundary conditions to impose (which are mostly taken to be periodic, in order to reduce boundary effects), on the initial configuration and on the total number of MC iterations to be performed. A discussion on some of the available choices can be found in [R11]. The reader will find there how MC simulations where the value of the coupling parameter β is changed slightly from one to the next (simulations of "thermal cycles") can provide information on the presence of critical points through the appearance of hysteresis loops; and also how simulations starting from different choices of the initial configuration (cold, with all U_{ji} equal to the identity – hot, with the U_{ji} selected at random – mixed phase configurations, with part of the lattice cold, part hot) can be used to refine that information. Doing a Monte Carlo simulation is, however, very much like performing an experiment and the detail of the computation will eventually be determined by the goals, resources and ingenuity of the performer. The variety of possible simulations is enormous; rather than trying to categorize here on what can and what cannot be done, it is probably better to let the reader ascertain for himself the possibilities of the technique by progressing through the reprinted literature.

4. A GUIDE THROUGH THE LITERATURE

Wilson's article "Confinement of quarks", published in 1974, laid the foundation for the applications of Euclidean lattice gauge theories to particle systems and constitutes a landmark in the development of the theory of strong interactions. Appropriately, it opens the collection of reprints [R1]. The reader will find there the definition of an Euclidean lattice gauge theory, in the form which has been outlined in Chapter 2 and which has basically been adopted in almost all later investigations. The confining properties of the lattice gauge theory in its strong coupling limit are also illustrated in [R1]. In a sense, Wilson's article contains all the ingredients for a successful theory of strong interactions [see also A10]: what is missing is a demonstration that the cut-off introduced by the lattice can be removed and that the theory would then exhibit the expected short distance behavior. Such a demonstration was to be achieved, by numerical methods, a few years later.

Wilson's article, however, was not the first to present the definition of an Euclidean lattice gauge theory. A generalization of the Ising model with local gauge invariance had been described by Wegner a few years earlier. Wegner's article is the second of the reprints [R2].

The next three articles [R3–5], by Balian, Drouffe and Itzykson, contain a detailed analysis of the general properties of lattice gauge theories, including strong coupling expansions. The topic of strong coupling expansions will not be considered much further in this book, mainly for reasons of space. The literature on applications of strong coupling techniques to lattice gauge theory is however very rich. For a few examples see [R21, 25, 26, 51, 52; A11–15].

As I have mentioned in Chapter 2, the formulation of a gauge theory on a space-time lattice is not the only possible way to introduce a cut-off through a discretization of space. Another interesting possibility consists in leaving time as a continuous variable and to introduce a lattice in three-dimensional space only. One thus arrives at the so-called Hamiltonian formulation of a lattice gauge theory, which is illustrated in [R6]. The relation between the Euclidean and the Hamiltonian formulations is further elucidated in [R7]. The Hamiltonian formulation, less suitable than the Euclidean one for the kind of numerical computations considered in this book, will not be followed up in the rest of the volume, but one application may be found in [R21].

The series of papers on general aspects of lattice gauge theories is closed with two articles by Christ, Friedberg and Lee, [R8, 9], where it is shown how a gauge theory can be defined on an arbitrary, not necessarily regular, lattice. Indeed, allowing the vertices of the lattice to occupy random positions, selected according to a translationally and rotationally invariant distribution, one can preserve the symmetry properties of continuous space-time even in the lattice formulation. While in a numerical computation the advantage of preserving such symmetries may be offset by the complexity of dealing with an irregular lattice, the approach of [R8–9] is extremely elegant and conceptually very interesting.

With the next reprint [R10] we enter into the domain of Monte Carlo simulations. The simplest lattice gauge theories are those based on the Abelian U(1) gauge group and on its finite subgroups Z_N (Z_N is the subgroup of U(1) consisting of the elements $\exp\{2\pi n/N\}$ or, equivalently, the group of addition of integers mod N). The gauge invariant generalization of the Ising model (see [R2, 4]), is, in turn, the most elementary among Z_N – lattice gauge theories. Thus the gauge invariant Ising system constituted a good model to test the applicability of Monte Carlo methods to lattice gauge theories. Such numerical experiments were carried by Creutz, Jacobs and the writer in 1979 [10]. Part of the motivation was just to verify how effectively Monte Carlo computations could be done for a field theoretical system, where the requirement of dealing with a four-dimensional medium implies an extremely large number of dynamical variables. However, something was to be learned also from the analysis of a system as simple as the Z_2 – gauge model. Indeed from the analogy with the ordinary two-dimensional Ising model, based on the Migdal-Kadanoff recursion relations [A16–17], the theoretical prejudices were that the Z_2 – lattice gauge system should exhibit a second-order phase transition at its self-duality point (for duality transformations see [R4, A18]). The MC simulation gave instead very strong evidence for a presence of a first-order phase transition, which constituted the major physical result obtained in [R10].

About the other Abelian models, studied numerically in R[11], from the point of view of a continuum theory the interest of these systems is marginal. In continuous space-time the U(1) – gauge theory (without matter fields) is the theory of free photons. However, the Abelian models are non-trivial on the lattice, where the formulation based on finite group elements as dynamical variables induces self-couplings of all orders. In particular, in the strong coupling regime, all these models exhibit confining properties. Consistency of the lattice formulation demands then the existence, in the U(1) – theory, of at least one critical point, separating a strong coupling phase from a phase dominated by spin-wave excitations: if such two-phase structure exists, the continuum limit may be recovered on the spin-wave phase (the spin-wave excitations becoming the quantized photons), with no contradiction arising from the existence of a strong coupling, confining phase. Monte Carlo simulations [R11] showed that the Abelian models conform to the expectation outlined above. Indeed, the MC analysis of the Z_N – systems revealed a rather rich and interesting phase structure: N large enough, two critical points separate a strong-coupling phase from an intermediate phase dominated by spin-wave excitations and, finally, from a weak coupling phase. In the limit $N \to \infty$, in which Z_N tends to U(1), the second critical point moves to zero coupling and a structure of two phases only (strong coupling phase and spin-wave phase) survives [R11]. It is remarkable that the three-phase structure of the Z_N – models was discovered numerically while, independently, theoretical arguments suggesting precisely the same phase structure were being proposed [A19–21]. The reprints [R12] and [R13] present additional MC studies of the Abelian models. An accurate determination of the specific heat (through the measurement of fluctuations) and finite scaling arguments are used in [R12] to locate precisely the critical point in the U(1) – system and to investigate its properties. The analysis in [R13] focuses instead on the occurrence and relevance of topological excitations. The existence of a phase transition in the U(1) – model has eventually been rigorously demonstrated in [A22].

Shortly after the MC analysis of the Abelian systems [R10–11] was completed, the phase structure of the SU(2) – gauge system was studied by Creutz [R14]. His MC simulations produced evidence for the absence of any phase transition in the four-dimensional SU(2) – gauge theory, a very important result, because it implied that the strong coupling features of the non-Abelian model (in particular, confinement of quarks) were likely to persist all the way to the limit of vanishing bare coupling, where the continuum theory might be recovered. [R15] presents a MC study of gauge systems based on finite non-Abelian subgroups of SU(2). A similar investigation can be found in [A23]; [R16] illustrates in detail a computer code which can be used to simulate systems with finite gauge groups. Apart from the general interest in the properties of the systems considered in [R15, 16; A23] as statistical systems, the study of models with finite subgroups of SU(2) as gauge is of practical importance because it is found that the models defined with the largest subgroups give results for the quantum averages which are essentially indistinguishable from the results obtained with the SU(2) – system directly, throughout a range of values of the coupling constant extending well into the scaling domain. On the other hand the replacement of SU(2) with a finite subgroup in the MC algorithm implies substantial savings in computer time; for these reasons several of the numerical studies of the SU(2) system have been done approximating SU(2) with its largest non-Abelian subgroup, i.e. the 120-element group

$\tilde{I}$, the covering of the group rotational symmetries of the icosahedron. Unfortunately, an analogous approximation is not available for the SU(3) – lattice gauge model [A24].

The absence of a phase transition, although welcome, is not sufficient to demonstrate that the SU(2) lattice gauge theory can be used to define a consistent continuum limit. For this, as discussed in Chapter 2, one needs to show that as the lattice coupling constant g is sent to zero all physical quantities scale in the appropriate way. As for the "appropriate" scaling behavior, if the scaling critical point is indeed at $g = 0$, this can be determined by perturbative arguments. The discovery, made in 1973 by Gross and Wilczek and by Politzer, that $g = 0$ is a possible scaling critical point of non-Abelian quantized gauge theories and the determination of the expected scaling behavior have been steps of fundamental importance for the development of the theory of strong interactions. The original papers are reprinted in [R17] and [R18] (see also [A24]).

The functional relation between lattice spacing a and coupling constant g (see Eqs. 2.36 and 2.39) involve a dimensional scale parameter Λ. All physical quantities can, in principle, be expressed in terms of Λ or, vice versa, the value of Λ can be determined from the calculation of any physical observable. Λ itself, however, has no direct physical significance and its value, in terms of some definite observable, will change if the scheme of renormalization is modified. It is then useful to know how the scale parameters corresponding to different procedures of renormalization are related. In particular, it is quite important to determine the relation between the lattice scale parameter and other scale parameters, such as Λ_{MS} or Λ_{MOM}, used in perturbative regularizations, because this will allow a comparison between predictions from the lattice gauge theory and calculations in perturbative Quantum Chromodynamics. The computation of ratios between lattice and perturbative scale parameters can be found in [R19] and [R20]. These two reprints are followed by several papers where the scaling properties of various non-Abelian lattice gauge models are verified and used to calculate physical observables.

A quantity of primary importance in the theory of strong interactions is the so-called string tension, i.e., the asymptotic value of the force between quark and antiquark at large separation. The confining properties of the theory manifest themselves in non-vanishing tension. In lattice gauge theory the value of the string tension can be determined from the expectation value $\langle W_\lambda \rangle$ of Wilson loop factors (see Chapter 2, Eq. (2.14)). $\langle W_\lambda \rangle$ can be considered as the ratio between the vacuum to vacuum amplitudes in the presence and respectively absence of a source circulating along the loop λ (see Eq. (2.10)). If λ is taken to be a rectangular path extending for m sites along a space direction and n sites along the time axis, for $n \gg m$ the above ratio of amplitudes can be expressed in terms of the potential energy $V(r)$, of a pair of sources separated by a distance $r = ma$ (a denotes the lattice spacing):

$$\langle W_\lambda \rangle = \exp\{-na\, V(ma)\} . \tag{4.1}$$

It is then clear that a non-vanishing string tension, of magnitude σ, implies a fall-off of $\langle W_\lambda \rangle$ proportional to the area of mna^2 enclosed by the path when m is also made large (because $V(r) \sim \sigma r$ as $r \to \infty$). This result extends to contours of general shape: if σ is non-vanishing $\langle W_\lambda \rangle$ falls off as

$$\langle W_\lambda \rangle \sim \exp\{-\sigma A\} , \tag{4.2}$$

where the A is the area enclosed by the contour, as the loop is increased in size. Thus σ can be determined from the rate of fall-off of $\langle W_\lambda \rangle$.

A lattice calculation will not produce σ directly, but rather a dimensionless function K of the lattice coupling constant related to σ by

$$K(g) = \sigma a^2 . \tag{4.3}$$

For instance, if the loop λ is a rectangular loop of sides ma and na, the fall-off of Eq. (4.2) would manifest itself in a behavior

$$\langle W_\lambda \rangle \underset{m,n \to \infty}{\sim} \exp\{-K(g)mn\} .$$

Eq. (4.3) then follows immediately from the fact that the physical value of the area enclosed by λ is $A = mna^2$. According to the general discussion of renormalization presented in Chapter 2, $K(g)$ should scale as

$$K(g) \sim c[f(g)]^2 \ , \tag{4.5}$$

with c a constant coefficient and $f(g)$ as in Eq. (2.39), if $g \to 0$ defines a consistent continuum limit. If Eq. (4.5) is found to be valid, then the physical value of the string tension is given by

$$\sigma = c\Lambda^2 \ . \tag{4.6}$$

The investigations presented in [R21] through [R26] are all concerned with verifying the scaling behavior of $K(g)$ as $g \to 0$ and with using such a scaling behavior to find the relation between string tension and lattice scale parameter. What emerges from [R21–26] is that the strong coupling results appear indeed to extrapolate into the appropriate scaling behavior, both in the SU(2) and the SU(3) models, as $g \to 0$, and therefore that a continuum limit with confining properties does exist. The studies in [R21] and [R25, 26] are based on strong coupling expansions and rely on some method of extrapolation to probe the behavior of the theory as g becomes small. The results presented in [R22–24] are derived from Monte Carlo simulations and are independent of any extrapolation technique: from this point of view the evidence for scaling derived from Monte Carlo simulations appears particularly impressive. Limitations apply, however, to the Monte Carlo results as well: because of constraints on the maximum lattice size and because of the statistical nature of the approximation, which prevents the calculation of very small effects, the force between sources can be evaluated only at separations of a few lattice sites. As g becomes smaller, the physical distance between sources at fixed lattice separation also decreases ($r = ma(g)$ goes to zero as $a(g)$ for $g \to 0$, at m fixed) and eventually the MC simulation probes the short distance behavior of the force rather than its asymptotic value. Thus, only a window of values of g for which scaling can be numerically verified is left. But this window, straddling the strong coupling and the weak coupling regimes, is large enough that when Creutz presented the first MC data on the string tensions [R22], his results were recognized as providing fundamental evidence for the validity of the non-Abelian theory of strong interactions. Creutz's results also opened the way for the numerical determination of a variety of physical quantities.

The reprints [R27] through [R37] all present evaluations of observables of the theory of strong interactions by means of numerical simulations. The computations are done either in the SU(3) theory or in the simplified SU(2) model. The quantities which are being calculated include the temperature at which quarks become deconfined, the potential between static quarks at intermediate and short distances, and the mass-gap of the theory, i.e., the mass of the lowest among the quantized excitations of the pure gauge medium (such quantum states are often called "glueballs"). It would take too long here to discuss how the values of the physical quantities are derived from the various averages calculated by MC simulations, and the reader is referred to the reprints for that purpose. Here I only wish to comment that the results on temperature effects [R27–29] nicely confirmed earlier work [A26–27] on the existence of a deconfining transition and allowed one to determine the deconfining temperature; that the properties of the potential between static sources [R30–32] agree with the expectations from asymptotic freedom and with rotational symmetry in the continuum limit; that the limitations deriving from the numerical nature of MC simulations are felt rather severely in the calculation of the mass-gap [R33–37] and that, although estimates of the masses of excited glueballs have been attempted, the results are probably not conclusive. Also, the number of published works using MC determinations to determine pure gauge observables is very high, and the selection presented in this volume is necessarily quite restricted. For a (non-exhaustive) list of additional references see [A28–39].

Among the various observables of a gauge theory some are singled out by their topological character. These observables assume non-vanishing values in correspondence to field configurations which cannot be continuously deformed into the configuration of least action. There are topological observables that admit a rigorous definition in lattice system (vortices, monopoles); others, such as instantons, which depend on the continuity properties of space-time and can be defined only approximately on the lattice. All have been associated with a variety of physically relevant effects. A MC study of topological observables in Abelian models has already been presented in [R13]. The reprints [R38–41] investigate topological excitations occurring in non-Abelian systems. [R38] gives a general discussion of vortex-like structures [see also A40]. Further considerations on monopoles and vortices in non-Abelian models, together with the results of MC studies of such quantities, are presented in [R39]. For additional investigations on this subject see [A41–43]. The papers [R40, 41] analyze instead, by MC simulations, topological observables which are associated with instanton excitations of the continuum theory. The numerical analysis is made difficult by the need of isolating a rather small non-perturbative effect, attributed to instantons, from a larger perturbative background, which is present because, on the lattice, the topological character of the observables is lost. It would be very useful to be able to measure, by MC simulations, quantities which can be associated to instanton contributions in the continuum limit, but which retain a non-perturbative character also in the regularized lattice theory. A quantity of this kind has been defined in [A44], but it has not been used as yet in actual MC simulations because of its mathematical complexity.

A lot of freedom exists in the definition of the action for a lattice gauge system. Wilson's form, given in Eq. (2.23), represents only one specific choice among an infinite variety, although it is the one that has been most frequently used in application. Also, it is expected that a wide class of actions may be used to define the same continuum limit. Changing the action of the lattice system corresponds to modifying the scheme of regularization. The continuum limit should not be affected, although different values for the lattice scale parameter Λ are expected. Several papers have investigated the universality properties (i.e., independence from the choice of a specific action) of the continuum limit of non-Abelian models, in particular of the SU(2) – lattice gauge theory. [R42] illustrates a lattice action, the so-called heat-kernel action, which is singled out by special geometrical significance [see also A45]. Another form of the action with very simple geometrical interpretation although not everywhere regular, has been proposed by Manton [A46]. SU(2) systems defined with the heat-kernel and Manton's action have been numerically studied in [R43] and found to display a scaling behavior for the string tension quite analogous to the one encountered with Wilson's action. Similar results for the mass-gap have been found in [A47]. The ratios of the scale parameters have been evaluated. The same ratios can be computed analytically, by perturbative methods. Numerical and analytical results have been compared in [A48] and found to be in reasonable agreement, with discrepancies which, although non-negligible, can be justified by the neglect of perturbative corrections of higher order.

One of the simplest ways to test universality is to add to Wilson's action a term of the same form (see Eq. (2.23)), but with the trace acting on matrices in the adjoint rather than in the fundamental representation. One thus defines a two-parameter class of actions, the parameters being the coefficients β_F and β_A of the terms in the fundamental and adjoint representations. The $\beta_F = 0$ axis defines an O(3) – gauge theory. Although the O(3) and SU(2) – models (in the absence of matter fields in the fundamental representation) are expected to define the same continuum limit, it was a rather startling numerical discovery that in the O(3) – system the strong coupling and weak coupling regimes are separated by a phase transition [R44, 45]. In [R46] MC simulations were used to study the properties of the SU(2) – system in an extended region of the β_F, β_A plane and an interesting phase structure was found. In particular, it emerged that the phase transition of the O(3) – model extends quite a way in the β_F, β_A plane, and that the line of critical points comes very close to the $\beta_A = 0$ axis which defines the system originally considered by Creutz [R14]. This rich phase structure does not invalidate the conclusions on the confining properties of the theory in the continuum limit, because of the existence of paths from the strong coupling to the weak coupling domain which go around the singularities. Nevertheless, one wonders what implications would have been drawn from the first numerical simulations if the line of phase transitions had crossed the $\beta_A = 0$ axis and the MC analysis had revealed a clear two-phase structure. The phase structure of SU(N) – models has also been studied (see for instance [A49, 50] and it has been found that, for $N < 4$, the line of phase transitions does indeed cross the $\beta_A = 0$ axis. The universality properties of the SU(2) – models in the β_F, β_A plane have been studied in detail [A51 – 53]: the conclusions are that some discrepancy exists between numerically determined values for the scale parameters and values one would infer from a direct perturbative calculation. The discrepancies may be however attributed to shortcomings of the perturbative analysis and better agreement can be obtained if non-perturbative effects are incorporated by a self-consistent technique.

The papers [R47] and R[48] investigate by MC simulation the properties of systems containing bosonic matter fields coupled to gauge fields. These kinds of studies [see also A54–56] have revealed interesting phase structure in the plane defined by the gauge and matter field coupling constants, generally in agreement with theoretical expectations. The interest in mixed gauge bosonic matter field systems is however marginal for the theory of strong interactions, where the relevant matter fields (the fields of the quarks) are fermionic in nature.

All of the remaining reprints in the collection deal with lattice fermionic systems. In the studies of these systems one encountered problems at two separate levels. Difficulties arise in the very definition of the lattice regularization. Additional problems are met when one tries to apply MC simulations. About the first more conceptual point, one finds that a straightforward generalization of the continuum Dirac Lagrangian to the lattice produces an unwanted degeneracy of fermions when one tries to recover the continuum limit. The origin of the problem may be seen in the fact that in the fermionic case, the equation of propagation involves first order derivatives, rather than derivatives of second order as in the case with bosons. On the lattice these derivatives are naturally transcribed into central differences: $\partial_x \psi(x) \rightarrow (\psi(x + a) - \psi(x - a))/2a$, but this procedure effectively makes the size of the unit cell equal to $2a$, and its volume equal to $16a^4$. The unit cell then contains 16 times more degrees of freedom than one would expect and these reappear as degenerate modes in the continuum limit. What happens is that the lattice Dirac operator admits eigenvalues close to the fermions' mass not only in correspondence to eigenmodes where the variation of ψ from point to point is smooth, but also in correspondence to eigenmodes where such a smooth variation is modulated by factors of the type $\exp(2\pi i x_1/a)$, $\exp(2\pi i x_2/a)$, . . . which change from site to site. In other words, a continuum limit can be recovered not only in correspondence to the neighborhood of the origin in momentum space, but also at various corners of the Brillouin zone.

The argument given above is not inescapable: one might think that a more sophisticated transcription of the derivatives on the lattice might avoid the multiplication of modes. This is true only to a certain extent. Considerations based on the anomaly of the axial current [R49] or on the topology of the Brillouin zone [R50] show that one cannot formulate a lattice theory with fermions of a single chirality; moreover, if for massless fermions a continuous chiral symmetry is to be preserved, this symmetry cannot be the singlet U(1) chiral symmetry, and must be realized only as a symmetry in which different modes rotate in opposite directions so as to cancel the anomaly. In particular, a theory describing a single massless fermionic species on the lattice must break chiral invariance.

Thus, there are formulations which reduce the unwanted degeneracy mode, but one cannot have a lattice theory of fermions with all the formal properties of the continuum theory. In MC simulations the two formulations which have been used most often are one introduced by Wilson [A10] and one which constitutes a generalization to Euclidean space-time of a formalism introduced by Susskind and collaborators in the Hamiltonian context [A57, 58]. The reader will find these two formulations detailed in almost all the reprints dealing with fermionic systems. Further elaborations on the problem of putting fermions on the lattice may be found in [A59–66].

The reprints [R51] and]R52] present interesting results on lattice systems with gauge fields and fermionic fields, derived by strong coupling expansions. All of the remaining reprints deal with MC simulations for systems with fermions. Here one encounters another serious difficulty of a more technical nature. The variables representing fermionic fields in the functional integral over the configurations are not ordinary numbers but elements of a Grassmann algebra (anticommuting numbers). These anticommuting numbers really serve to express sums over fermionic occupation numbers which can be 0 or 1. Thus in principle it ought to be possible to formulate the sum over configurations in terms of binary variables and to proceed to a MC simulation. But, because of the anticommuting nature of the fermionic variables, the measure then becomes not positive definite and the standard Monte Carlo simulations become unfeasible. (Only for 2-dimensional systems it has been possible to formulate the sum over fermionic degrees of freedom in such a way as to maintain the positivity of the measure [see A67]). To perform MC simulations use has been made of the fact that for most systems of interest the fermionic fields appear bilinearly in the action and the integration over these variables can be explicitly carried out [A68]. This leaves an integration over gauge variables only, but with a new measure given by the product of the pure gauge measure, $\exp\{-S_G\}$, times the determinant of the lattice Dirac operator (fermionic determinant). This measure, contrary to the measures encountered with pure gauge systems or with gauge systems coupled to bosonic matter fields, is non-local in the U_{ji} variables: it then becomes very difficult to evaluate the change of measure induced by the variation of a definite U_{ji}, a crucial step for MC simulations. Recent research efforts have gone into finding efficient ways to calculate the variation of the non-local measure factor. The reprints [R53–56] illustrate three techniques (and one application to the Schwinger model) which achieve a rather rapid calculation of the change of measure i at the cost of some degree of approximation (another method, less approximate but more time consuming, has been presented in [A69, 70]).

The simplest possibility, and at the same time the most drastic approximation, consists of neglecting the effects of the fermionic determinant altogether, using the pure gauge measure to average the various observables, *including* those involving fermionic fields. Such an approximation is not necessarily as violent as it may a priori seem: it maintains the interaction between the gauge field and the propagating fermions and the self interaction of the gauge field. What is neglected is the contribution from internal closed fermionic loops. On the basis of various arguments it has been suggested that this loopless approximation should produce reasonable results for many states in the hadronic spectrum (exceptions are splittings such as those between η and π, or χ and η, which are particularly sensitive to inner fermionic loops through the axial anomaly).

The loopless approximation (called variously quenched approximation or valence approximation), in conjunction with MC simulations, has been applied by several authors to the study of the spectrum of QCD. The earliest and some of the most relevant results can be found in the final reprints of the collection [R57–62]. Results on magnetic moments [A71, 72] and on the restoration of chiral symmetry with temperature [A73] have also been obtained. This is a field where interesting investigations continue to appear, not without some controversy on the validity of the results, which have generally been remarkably good, especially if one takes into account all limitations in the calculation. A MC simulation of QCD including the effect of the fermionic determinant has been presented in [A74].

The MC studies of systems with fermions conclude the series of reprints published in this book. Reasons of space only have not allowed me to include any representative of two sets of investigations, which have also recently produced quite interesting results for gauge theories. These are the studies based on mean field techniques and on the reduction of the whole lattice to a one-site, four-plaquette lattice, which becomes possible in the large N limit. The interested reader can find these two topics investigated, for instance, in [A75–78] and [A79–84].

I hope that the readers of this volume will be convinced that the lattice formulation of gauge field theories has allowed enormous progress to be made toward an understanding of the properties of the corresponding continuum systems. In particular, the lattice regularization has been quite fruitful for the study of QCD, the gauge theory of strong interactions. Many of the results have been achieved by numerical methods, through Monte Carlo simulations. It is remarkable that these computations could produce sensible results in spite of the limitations which the 4-dimensional nature of space-time imposes on the size of the lattice. What happens is that the transition from the strong coupling regime to the scaling regime, where one can see features of the continuum limit, is abrupt and occurs at physical values of the lattice spacing large enough so that the small lattice still provides a representation of a sufficiently large volume of space-time. And here nature seems to have been helpful, because the abruptness of the transition on the lattice most likely reflects the abruptness in the transition between quarks bound inside hadrons and quarks moving asymptotically free, which is experimentally observed in the real world. Still, a lot remains to made. Some of the results which have been obtained are marginal; all need to be based on a firmer analytical foundation. Advances, by orders of magnitude, in the calculational power of computers will be helpful; progress in theoretical understanding will be needed; an intelligent combination of improved computational and analytical capabilities will produce the best future results.

ADDITIONAL REFERENCES

A1 N. Metropolis, A. W. Rosenbluth, M. N. Rosenbluth, A. H. Teller and E. Teller, *J. Chem. Phys.* **21** (1953) 1087.

A2 J. M. Hammersley and D. C. Handscomb, *Monte Carlo Methods* (Methuen, London, 1964).

A3 K. Binder, in *Phase Transitions and Critical Phenomena*, C. Domb and M. S. Green, eds., Vol. **5**B (Academic Press, New York, 1976).

A4 K. Binder (ed.), *Monte Carlo Methods* (Springer, Berlin-Heidelberg-New York, 1979).

A5 W. Celmaster, *Phys. Rev.* **D26** (1982) 2955.

A6 W. E. Caswell, *Phys. Rev. Lett.* **33** (1974) 244.

A7 D. R. T. Jones, *Nucl. Phys.* **B75** (1974) 531.

A8 C.-P. Yang, *Proc. of Symposia in Applied Mathematics*, Vol. **XV**, p. 351, Amer. Math. Soc., Providence RI, 1963.

A9 N. Cabibbo and E. Marinari, *Phys. Lett.* **119**B (1982) 387.

A10 K. Wilson, in *New Phenomena in Subnuclear Physics*, A. Zichichi, ed. (Plenum, New York, 1977).

A11 G. Münster, *Nucl. Phys.* **B140**[FS3] (1981) 439; **B200**, 536(E); **B205**, 648(E).

A12 J. M. Drouffe, *Nucl. Phys.* **B170** (1980) 91.

A13 G. Münster, *Nucl. Phys.* **B180** (1981) 23.

A14 H. Kawai and R. Nakayama, *Phys. Lett.* **113**B (1982) 329.

A15 J. Smit, *Nucl. Phys.* **B206** (1982) 309.

A16 A. A. Migdal, *Zh Eksp. Teor. Fiz.* **69** (1975), 457, 810 [*Sov. Phys. JETP* **42** (1975) 413, 743].

A17 L. P. Kadanoff, *Rev. Mod. Phys.* **49** (1977) 267.

A18 R. Savit, *Rev. Mod. Phys.* **52** (1980) 453.

A19 S. Elitzur, R. Pearson and J. Shigemitsu, *Phys. Rev.* **D19** (1979) 3698;

A20 D. Horn, M. Weinstein and S. Yankieldwics, *Phys. Rev.* **D19** (1979) 3715;

A21 A. Ukawa, P. Windey and A. Guth, *Phys. Rev.* **D21** (1980) 1013.

A22 A. H. Guth, *Phys. Rev.* **D21** (1980) 2291.

A23 D. Petcher and D. Weingarten, *Phys. Rev.* **D22** (1980) 2465.

A24 G. Bhanot and C. Rebbi, *Phys. Rev.* **D24** (1981) 3319.
A25 D. Gross and F. Wilczek, *Phys. Rev.* D8 (1976) 3633.
A26 A. M. Polyakov, *Phys. Lett.* 72B (1978) 477.
A27 L. Susskind, *Phys. Rev.* **D20** (1979) 2610.
A28 K. Kajantie, C. Montonen and E. Pietarinen, *Zeit. Phys.* C9 (1981) 253.
A29 J. Engels, F. Karsch, H. Satz and I. Montvay, *Phys. Lett.* **101B** (1981) 89; **102B** (1981) 332.
A30 L. McLerran and B. Svetisky, *Phys. Rev.* **D24** (1981) 450.
A31 R. Brower, M. Creutz and M. Nauenberg, *Nucl. Phys.* **B210** (1982) 133.
A32 B. Berg and A. Billoire, *Phys. Lett.* **114B** (1982) 324.
A33 K. Ishikawa, M. Teper and G. Schierholz, *Phys. Lett.* **110B** (1982) 399.
A34 A. DiGiacomo and G. C. Rossi, *Phys. Lett.* **100B** (1981) 418.
A35 A. DiGiacomo and G. Paffuti, *Phys. Lett.* **108B** (1982) 327.
A36 J. Kripfganz, *Phys. Lett.* **101B** (1981) 169.
A37 T. Banks, R. Horsley, H. R. Rubinstein and U. Wolff, *Nucl. Phys.* **B190** (1981) 692.
A38 R. Kirschner, J. Kripfganz, J. Ranft and A. Schiller, *Nucl. Phys.* **B210** (1982) 567.
A39 E. Kovacs, *Phys. Rev.* **D25** (1982) 3312.
A40 G. Mack and V. B. Petkova, *Ann. Phys.* **123** (1979) 442.
A41 R. C. Brower, D. A. Kessler and H. Levine, *Phys. Rev. Lett.* **47** (1981) 621.
A42 G. Mack ad E. Pietarinen, *Phys. Lett.* **94B** (1980) 397.
A43 J. Groeneveld, J. Jurkiewicz and C. P. Korthals-Altes, *Physica Scripta* **23** (1981) 1022.
A44 M. Lüscher, *Comm. Math. Phys.* **85** (1982) 39.
A45 J. M. Drouffe, *Phys. Rev.* **D18**, (1978) 1174.
A46 N. S. Manton, *Phys. Lett.* **96B** (1980) 328.
A47 K. H. Mütter, K. Schilling, *Phys. Lett.* **121B** (1983) 267.
A48 C. B. Lang, C. Rebbi, P. Salomonson, and B. S. Skagerstam, *Phys. Rev.* **D26** (1982) 2028.
A49 M. Creutz, *Phys. Rev. Lett.* **46** (1981) 1441.
A50 M. Creutz and K. J. M. Moriarty, *Nucl. Phys.* **B210** (1982) 50.

A51 G. Bhanot and R. Dashen, *Phys. Lett.* **113**B (1982) 299.

A52 A. Gonzales-Arroyo, C. P. Korthals-Altes, J. Peiro, and M. Perrottet, *Phys. Lett.* **116**B (1982) 414.

A53 B. Grossmann and S. Samuel, *Phys. Lett.* **120**B (1983) 383.

A54 K. C. Bowler, G. X. Pawley, B. J. Pendleton, D. J. Wallace and G. W. Thomas, *Phys. Lett.* **104**B (1981) 491.

A55 D. Callaway and L. Carson, *Phys. Rev.* D**25** (1982) 531.

A56 R. Brower, D. Kessler, T. Schalk, H. Levine and M. Nauenberg, *Phys. Rev.* D**25** (1982) 3319.

A57 L. Susskind, *Phys. Rev.* D**16** (1977) 3031.

A58 T. Banks, S. Raby, L. Susskind, J. Kogut, D. R. T. Jones, P. Scharbach and D. Sinclair, *Phys. Rev.* D**15** (1977) 1111.

A59 and J. B. Healy, *Phys. Rev.* D**16** (177) 387;

A60 H. Nielsen and M. Ninomiya, *Nucl. Phys.* B**195** (1981) 20; B**193** (1981) 173;

A61 W. Kerler, *Phys. Rev.* D**23** (1981) 2384;

A62 J. Rabin, *Nucl. Phys.* B**201** (1982) 315;

A63 T. Banks, Y. Dothan and O. Horn, *Phys. Lett.* **117**B (1982) 413.

A64 P. Becher and H. Joos, DESY preprint 1982.

A65 H. Kluberg-Stern, A. Morel, O. Napoly and B. Peterson, *Proceedings of the 21st Int. Conf. on High Energy Physics,* Paris 1982.

A66 F. Gliozzi, *Nucl. Phys.* B**204** (1982) 419.

A67 R. Blankenbeckler, J. Hirsch, D. Scalapino and R. Sugar, *Phys. Rev. Lett.* **47** (1982) 1628.

A68 P. Matthews and A. Salam, *Nuovo Cim.* **12** (1954) 563; *Nuovo Cimento* **2** (1955) 120.

A69 D. Weingarten and D. Petcher, *Phys. Lett.* **99**B (1981) 333.

A70 H. Hamber, *Phys. Rev.* D**24** (1981) 951.

A71 G. Martinelli, G. Parisi, R. Petronzio and F. Rapuano, *Phys. Lett.* **116**B (1982) 434.

A72 C. Bernard, T. Draper, K. Olynyk and M. Rushton, *Phys. Rev. Lett.* **49** (1982) 1076.

A73 J. Kogut, M. Stone, H. Wyld, J. Shigemitsu, S. Shenker, and D. Sinclair, *Phys. Rev. Lett.* **48** (1982) 1140.

A74 H. Hamber, E. Marinari, G. Parisi and C. Rebbi, *Phys. Lett.* B, to the published.

A75 J. Greensite and B. Lautrup, *Phys. Lett.* **104**B (1981) 41.

A76 J. M. Drouffe, *Phys. Lett.* **105B** (1981) 46.
A77 H. Flyvbjerg, B. Lautrup and J. B. Zuber, *Phys. Lett.* **110B** (1982) 279.
A78 V. Alessandrini, *Phys. Lett.* **117B** (1982) 423.
A79 T. Eguchi and H. Kawai, *Phys. Rev. Lett.* **48** (1982) 1063.
A80 G. Bhanot, U. Heller and H. Neuberger, *Phys. Lett.* **113B** (1982) 47 and **115B** (1982) 237.
A81 G. Parisi, *Phys. Lett.* **112B** (1982) 463.
A82 D. J. Gross and Y. Kitazawa, *Nucl. Phys.* **B206** (1982) 440.
A83 M. Okawa, *Phys. Rev. Lett.* **49** (1982) 353, 705.
A84 A. Gonzales-Arroyo and M. Okawa, *Phys. Lett.* **120B** (1983) 174.

REPRINTED PAPERS

PHYSICAL REVIEW D VOLUME 10, NUMBER 8 15 OCTOBER 1974

Confinement of quarks*

Kenneth G. Wilson
Laboratory of Nuclear Studies, Cornell University, Ithaca, New York 14850
(Received 12 June 1974)

A mechanism for total confinement of quarks, similar to that of Schwinger, is defined which requires the existence of Abelian or non-Abelian gauge fields. It is shown how to quantize a gauge field theory on a discrete lattice in Euclidean space-time, preserving exact gauge invariance and treating the gauge fields as angular variables (which makes a gauge-fixing term unnecessary). The lattice gauge theory has a computable strong-coupling limit: in this limit the binding mechanism applies and there are no free quarks. There is unfortunately no Lorentz (or Euclidean) invariance in the strong-coupling limit. The strong-coupling expansion involves sums over all quark paths and sums over all surfaces (on the lattice) joining quark paths. This structure is reminiscent of relativistic string models of hadrons.

I. INTRODUCTION

The success of the quark-constituent picture both for resonances and for deep-inelastic electron and neutrino processes makes it difficult to believe quarks do not exist. The problem is that quarks have not been seen. This suggests that quarks, for some reason, cannot appear as separate particles in a final state. A number of speculations have been offered as to how this might happen.[1]

Independently of the quark problem, Schwinger observed many years ago[2] that the vector mesons of a gauge theory can have a nonzero mass if vacuum polarization totally screens the charges in a gauge theory. Schwinger illustrated this result with the exact solution of quantum electrodynamics in one space and one time dimension, where the photon acquires a mass $\sim e^2$ for any nonzero charge e [e has dimensions of (mass)$^{1/2}$ in this theory]. Schwinger suggested that the same effect could occur in four dimensions for sufficiently large couplings.

Further study of the Schwinger model by Lowenstein and Swieca[3] and Casher, Kogut, and Susskind[4] has shown that the asymptotic states of the model contain only massive photons, not electrons. Nevertheless, as Casher *et al.* have shown in detail, the electrons are present in deep-inelastic processes and behave like free pointlike particles over short times and short distances. The polarization effects which prevent the appearance of electrons in the final state take place on a longer time scale (longer than $1/m_\gamma$, where m_γ is the photon mass).

A new mechanism which keeps quarks bound will be proposed in this paper. The mechanism applies to gauge theories only. The mechanism will be illustrated using the strong-coupling limit of a gauge theory in four-dimensional space-time. However, the model discussed here has a built-in ultraviolet cutoff, and in the strong-coupling limit all particle masses (including the gauge field masses) are much larger than the cutoff; in consequence the theory is far from covariant.

The confinement mechanism proposed here is soft (long-time scale). However, in the model discussed here the cutoff spoils the possibility of free pointlike behavior for the quarks.

The model discussed in this paper is a gauge theory set up on a four-dimensional Euclidean lattice. The inverse of the lattice spacing a serves as an ultraviolet cutoff. The use of a Euclidean space (i.e., imaginary instead of real times) instead of a Lorentz space is not a serious restriction; the energy eigenstates (including scattering states) of the lattice theory can be determined from the "transfer-matrix" formalism as has been discussed by suri[5] and reviewed by Wilson and Kogut.[6] A brief discussion of the transfer-matrix method is given in Sec. III.

In Schwinger's speculations about four dimensions, the photon mass would be zero for any charge e less than a critical coupling e_c; for $e > e_c$ the photon mass would be nonzero and vary with e. Figure 1 shows how a plot of m_γ vs e might look. The point e_c is a point of nonanalyticity. Similar nonanalytic points, called critical points, occur in solid-state physics at certain types of phase transitions. Consider, for instance, a ferromagnet in the absence of an external field. For any temperature above the Curie temperature T_C, the spontaneous magnetization M is 0. Below T_C, M is nonzero and a function of temperature. At T_C there may be either a first-order phase transition (in which case M is discontinuous at T_C) or a second-order phase transition (critical point) for which $M \to 0$ as $T \to T_C$ from either side.

By analogy with the solid-state situation one can think of the transition from zero to nonzero photon mass as a change of phase: this analogy is best understood by imagining the particles of quantum electrodynamics to be the excitations of a medium (the ether). In this case it is the ether which undergoes a change of phase at e_c. There is again a question whether this change of phase is first-order (cf. Fig. 2) or second-order (Fig. 1). (Coleman and Weinberg[7] have found a nontrivial example of a first-order transition in another context.)

The model discussed in this paper is a single Abelian gauge field coupled (with strength g) to massive quarks. In weak coupling the gauge field behaves like a normal free zero-mass field (despite modifications introduced in the lattice quantization) and the quarks are unbound. In strong coupling the gauge field is massive and the quarks are bound, showing the existence of the second phase. Thus there should be a phase transition at some intermediate value of g. Nothing is known about this transition at the present time.

The quantization procedure and strong-coupling approximation described in this paper can be applied to non-Abelian gauge theories also. This will be explained briefly in Sec. III.

An extraordinary feature of the strong-coupling expansion of the lattice theory (see Sec. IV) is that it has the same general structure as the relativistic string models of hadrons.[8] The vacuum expectation values of the gauge theory involve (in the strong-coupling expansion) sums over all quark paths and sums over all surfaces connecting these paths; the surfaces are generated by the gauge field treated in strong coupling. The paths and surfaces are defined on a discrete lattice. There are geometrical difficulties in relating the surfaces on the lattice to the continuum surfaces of the string models; it is not known at present whether these difficulties can be overcome.

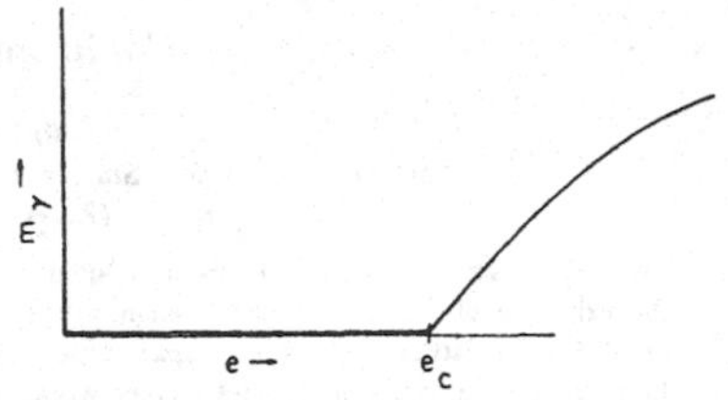

FIG. 1. Speculative plot of photon mass vs renormalized charge e, in unknown units. The transition at e_c is second-order (see text).

In Sec. II the nature of the quark binding mechanism will be discussed, qualitatively. In Sec. III the gauge theory will be formulated on a discrete lattice, both classically and quantum mechanically. In Sec. IV the strong-coupling expansion for the lattice gauge theory is explained. In Sec. V a cursory discussion of weak coupling is given. In Sec. VI there is a brief discussion of the problem of Lorentz invariance and the relation to string models.

II. QUARK BINDING MECHANISM

The binding mechanism will be explained in this section using the Feynman path-integral picture. The path-integral framework will be used in an intuitive rather than a formal way. Consider the current-current propagator

$$D_{\mu\nu}(x) = \langle \Omega | T J_\mu(x) J_\nu(0) | \Omega \rangle \, , \qquad (2.1)$$

whose Fourier transform determines the e^+-e^- annihilation cross section into hadrons. Assume that the currents $J_\mu(x)$ are built from quark fields as in the quark-parton model. Assume that the quarks interact through a single gauge field. (The restriction to one field is only for simplicity.) In the Feynman path-integral picture the propagator

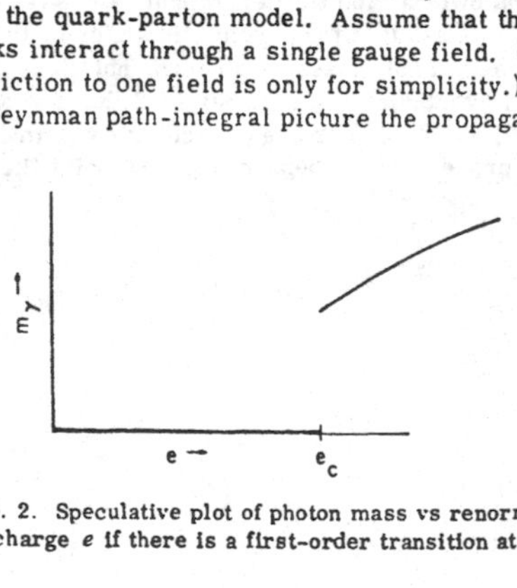

FIG. 2. Speculative plot of photon mass vs renormalized charge e if there is a first-order transition at e_c.

$D_{\mu\nu}(x)$ is given by a weighted average over all possible classical quark paths and all possible classical values of the gauge field. The currents $J_\mu(x)$ and $J_\nu(0)$ are thought of as producing a quark pair at the origin which later annihilate at the point x: One has to sum over all paths joining the points 0 and x for each of the pair of quarks.

An example of paths for the quark and antiquark are shown in Fig. 3. The vacuum can also emit and absorb quark pairs; this leads to further closed loops as illustrated in Fig. 4. All possible loops must be summed over too. It is also possible to have independent loops for the points 0 and x (Fig. 5), but this possibility will not be important here.

The weight associated with a given quark path or set of paths includes a factor of $\exp[ig\oint A_\mu(x) \times ds^\mu]$, where $A_\mu(x)$ is the gauge field. Here $\oint \cdots ds^\mu$ is a line integral or a sum over line integrals for each of the quark-antiquark loops, including the loop joining the points 0 and x. The constant g is the coupling constant of the gauge theory. There are further weight factors independent of A. Finally, independently of the quark paths there is another weight factor, namely, the exponential of the free action for the gauge field. The combined weight factor is then averaged over all quark paths and all gauge fields $A_\mu(x)$ to give the current-current propagator.

In order that quarks exist as separate final-state particles it must be possible to have quark-antiquark loops with well-separated quark and antiquark lines, at least when x and 0 are far apart. This is illustrated in Fig. 6(a). If the quark and antiquark paths are unlikely to separate beyond a fixed size, say 10^{-13} cm [see Fig. 6(b)], then clearly no detector will see a quark or antiquark in isolation.

It is assumed in this discussion that vacuum loops are not important. If vacuum loops are important enough then space will be filled with a high *density* of vacuum-produced quark-antiquark pairs. In particular, there will be many quark-antiquark pairs inside a detector of macroscopic size. The question then is whether there can be an excess of quarks over antiquarks, or vice versa, in a region of macroscopic size. This is

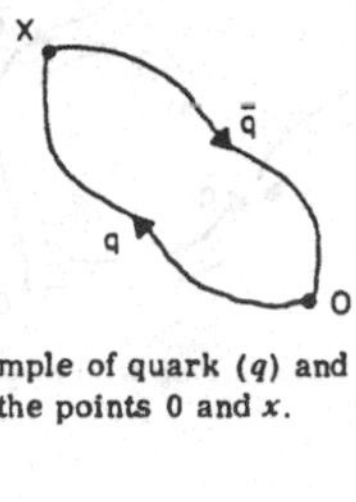

FIG. 3. An example of quark (q) and antiquark ($\bar{q}$) paths connecting the points 0 and x.

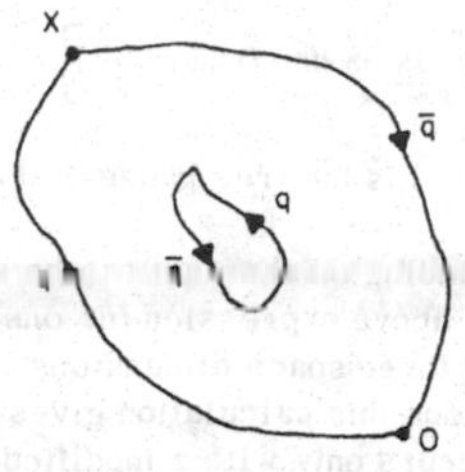

FIG. 4. Example of current loop (as in Fig. 3) with an extra vacuum loop.

a more difficult question to answer and will not be discussed in this paper.

Note that x must be large: If x is small there is little likelihood of finding a large size loop. This may seem a bit peculiar: One expects quarks to appear in the final state of e^+-e^- annihilation only at large virtual-photon momentum q if they appear at all, and large q means small x, not large x. The answer to this paradox has two parts. First, the important paths in the Feynman path integral bear no detailed relation to possible physical final states (the paths are paths of bare particles, not physical particles). Secondly, large x does not necessarily mean small q. In fact the study of whether well-separated quark-antiquark paths exist for large x is really a search for a quark-antiquark threshold in e^+-e^- annihilation, which would contribute a term $\sim\exp(2mi\sqrt{x^2})$ to the current-current propagator for *large* x, where m is the quark mass. (Here ~ means up to a power of x^2.) Such a term corresponds in momentum space to the singularity at the threshold $q^2 = (2m)^2$.

Suppose the gauge-field averaging is performed before the quark-paths averaging. Then one determine the average over all gauge fields of the weight factor $\exp[i\oint gA_\mu(y)ds^\mu]$ weighted further with the exponential of the free gauge-field action. For an Abelian gauge theory this average can be computed explicitly: It is

FIG. 5. Example of separate quark loops for the points 0 and x. (Integration over the gauge field produces gauge propagators which connect these loops.)

$$\exp\left[-g^2\oint ds^\mu \oint ds'^\nu D_{\mu\nu}(y-y')\right],$$

where $D_{\mu\nu}(y-y')$ is the free gauge-field propagator.

The quark binding mechanism can be seen by comparing the above expression for one space dimension and three space dimensions. In three space dimensions this calculation gives no binding (the binding occurs only with a modified gauge-field action: see Sec. IV), while there is binding in one space dimension. In three space dimensions $D_{\mu\nu}(y-y')$ behaves as $(y-y')^{-2}$. In consequence large values of $(y'-y)$ are negligible in the double line integral. Hence, the double integral is proportional to P, where P is the length of the loop. Unfortunately the integral is divergent at $y'=y$; a cutoff is needed for the integral to make sense. Since a cutoff will be introduced anyway in this paper, this is not a major concern. For simple loops, the perimeter P is roughly of order $(x^2)^{1/2}$ (ignoring the case that x is close to the light cone). Thus one has an exponential of the type one expects when free quarks are present.

In one space dimension, $D_{\mu\nu}(y-y')$ behaves as $\ln[(y-y')^2]$ for $y'-y$ large, and y' and y can freely range separately over the loop. In this case the double integral is proportional to P^2. Now the gauge-field average behaves as e^{-icP^2}, where c is a constant. In this case the contribution of large loops is heavily suppressed and there are no free quarks. [The case of nearby quark-antiquark pairs as in Fig. 6(b) is special—in this case large $y-y'$ is unimportant due to cancellation between the quark path and the nearby and oppositely directed antiquark path. In this case the double integral behaves as P, not P^2, but in this case there are no isolated quarks.]

In the strongly coupled lattice gauge theory described in later sections, the gauge-field average of $\exp[ig\oint A_\mu(x)ds^\mu]$ behaves as $\exp(ic'A)$, where A is the *enclosed area* of the loop. This heavily suppresses large loops, such as in Fig. 6(a), where A is of order P^2. One can think of one factor P as being roughly $(x^2)^{1/2}$, the other P as being analogous to a mass multiplying $(x^2)^{1/2}$. Since $P\to\infty$ as $x\to\infty$, the quark-antiquark threshold is at infinite mass.

In all these calculations one can have a large loop if there is a nearby vacuum loop (Fig. 7). In this case one always gets $e^{ic''P}$ behavior. For example, in the strong-coupling case the relevant enclosed area is the area between the two loops which is proportional to the perimeter P provided the separation of the two loops is fixed independently of P. This is in accord with Schwinger's picture. While an isolated well-separated loop

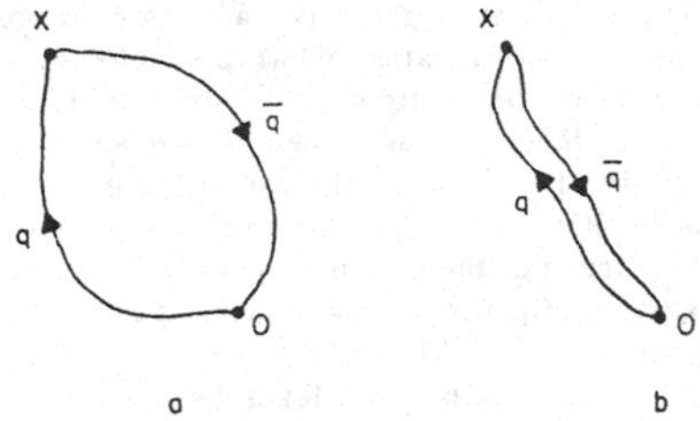

FIG. 6. (a) Loop with well-separated quark and antiquark. (b) Loop with small separation between quark and antiquark.

may be suppressed (due to P^2 or A dependence in the exponential) a loop closely shielded by a vacuum-polarization loop is always unsuppressed.

The binding mechanism proposed here is soft: The exponential damping is associated with large size loops having large areas. The behavior of small loops is irrelevant to the binding mechanism. Also for small loops both their area and perimeter are small and neither is of great importance in an exponential.

The mechanism discussed here works equally well for Dirac quarks or scalar quarks. This is in contrast to the Higgs mechanism which can wipe out the charge of scalar particles only.

III. LATTICE QUANTIZATION OF GAUGE FIELDS

A. Classical action on a lattice

In this section the gauge-field action (space-time integral of the Lagrangian) will be defined on a discrete lattice with spacing a in Euclidean space-time. The simplest way to proceed is to consider a continuum action, substitute finite-difference approximations for derivatives, and replace the space-time integral by a sum over the lattice sites. However, the result of this is an action which is not gauge-invariant for nonzero a. Because of the vagaries of renormalization this is likely to mean that the quantized theory still lacks

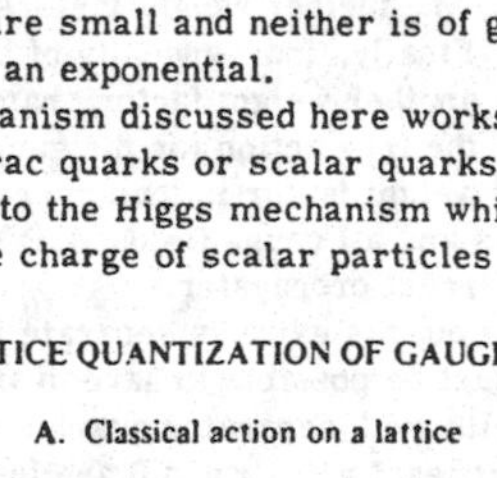

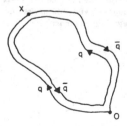

FIG. 7. Quark-antiquark loop with nearby vacuum loop.

gauge invariance in the limit $a\to 0$ (if such a limit exists). The alternative is to formulate gauge invariance for a lattice theory, and tinker with the action so that it is gauge invariant for any a. This alternative will be pursued here.

For convenience a charged Dirac field ψ coupled to a single gauge field A_μ will be discussed in detail. Generalizations to non-Abelian gauge groups will be noted later.

On the lattice the fields are ψ_n, $\bar\psi_n$, and $A_{n\mu}$, where n is a four-vector with integer components referring to points on a simple hypercubic lattice. A simple action on the lattice for the Dirac field (ignoring the gauge field for now) has the form

$$A_\psi=-a^4\sum_n m_0\bar\psi_n\psi_n+\tfrac12 a\sum_n\sum_\mu\bar\psi_n\gamma_\mu(\psi_{n+\hat\mu}-\psi_{n-\hat\mu}),\qquad(3.1)$$

where m_0 is the bare mass; $\hat\mu$ is a unit vector along the axis μ; $a^4\sum_n$ replaces the space-time integration of the continuum theory, and $(\psi_{n+\hat\mu}-\psi_{n-\hat\mu})/2a$ replaces $\nabla\psi$. There is no over-all factor of i due to the Euclidean metric. A gauge transformation on the lattice can be defined as follows:

$$\psi_n\to e^{iy_n g}\psi_n\,,\qquad(3.2)$$

$$\bar\psi_n\to e^{-iy_n g}\bar\psi_n\,,\qquad(3.3)$$

$$A_{n\mu}\to A_{n\mu}-(y_{n+\hat\mu}-y_n)/a\,,\qquad(3.4)$$

where g is the coupling constant and y_n is arbitrary. The terms $\bar\psi_n\psi_{n+\hat\mu}$ and $\bar\psi_n\psi_{n-\hat\mu}$ are not invariant to this transformation; the corresponding gauge-invariant expressions are $\bar\psi_n\psi_{n+\hat\mu}\times\exp(iagA_{n\mu})$ and $\bar\psi_n\psi_{n-\hat\mu}\exp(-iagA_{n-\hat\mu,\mu})$. Thus, a gauge-invariant form for A_ψ is

$$A_\psi=\tfrac12 a^3\sum_n\sum_\mu(\bar\psi_n\gamma_\mu\psi_{n+\hat\mu}e^{iagA_{n\mu}}-\bar\psi_{n+\hat\mu}\gamma_\mu\psi_n e^{-iagA_{n\mu}})-a^4\sum_n m_0\bar\psi_n\psi_n\,.\qquad(3.5)$$

It is convenient to define

$$B_{n\mu}=agA_{n\mu}\,.\qquad(3.6)$$

In the action A_ψ, the field $B_{n\mu}$ acts like an angular variable: A_ψ is periodic in $B_{n\mu}$ with period 2π. The free gauge-field action will be defined to preserve this property. This does not mean that $A_{n\mu}$ is an angular variable in the continuum limit. Owing to the relation (3.6), $A_{n\mu}$ becomes an infinitesimal angular variable $A_\mu(x)$ for $a\to 0$; such a variable has the range $-\infty<A_\mu(x)<\infty$ without any periodicity.

A gauge-invariant lattice approximation for $\nabla_\mu A_\nu-\nabla_\nu A_\mu$ is

$$F_{n\mu\nu}=(A_{n+\hat\mu,\nu}-A_{n\nu}-A_{n+\hat\nu,\mu}+A_{n\mu})/a\,.\qquad(3.7)$$

It is convenient to define a rescaled form of $F_{n\mu\nu}$:

$$f_{n\mu\nu}=a^2gF_{n\mu\nu}=B_{n\mu}+B_{n+\hat\mu,\nu}-B_{n+\hat\nu,\mu}-B_{n\nu}\,.\qquad(3.8)$$

A simple lattice action for the gauge field which preserves periodicity is

$$A_B=\frac{1}{2g^2}\sum_{n\mu\nu}e^{if_{n\mu\nu}}.\qquad(3.9)$$

In the continuum limit, $f_{n\mu\nu}\to 0$ due to the factor a^2 in the definition of $f_{n\mu\nu}$. Thus for small a, one can write

$$A_B\simeq\frac{1}{2g^2}\sum_{n\mu\nu}(1+if_{n\mu\nu}-\tfrac12 f_{n\mu\nu}{}^2-\cdots)\,.\qquad(3.10)$$

The constant term is irrelevant. The linear term in $f_{n\mu\nu}$ is 0 because $f_{n\mu\nu}$ is odd in the indices μ and ν. The quadratic term gives

$$A_B\simeq-\tfrac14 a^4\sum_{n\mu\nu}F_{n\mu\nu}{}^2\,,\qquad(3.11)$$

which is the conventional gauge-field action in a lattice approximation. The terms involving $f_{n\mu\nu}{}^3$, $f_{n\mu\nu}{}^4$, etc. all vanish for $a\to 0$ even after removing a factor a^4 to convert $\sum_n$ into an integral.

The full action may now be written

$$A=-c\sum_n\bar\psi_n\psi_n+K\sum_n\sum_\mu(\bar\psi_n\gamma_\mu\psi_{n+\hat\mu}e^{iB_{n\mu}}-\bar\psi_{n+\hat\mu}\gamma_\mu\psi_n e^{-iB_{n\mu}})+\frac{1}{2g^2}\sum_n\sum_{\mu\nu}e^{if_{\mu\nu}}\,,\qquad(3.12)$$

with $c=m_0a^4$, $K=a^3/2$. This action reduces to the usual continuum action for $a\to 0$; for finite a it is gauge invariant and periodic in the gauge field. Note, however, that the continuum limit is a *classical* limit in which the lattice variables ψ_n, $\bar\psi_n$, and $A_{n\mu}$ approach continuum functions $\psi(x)$, $\bar\psi(x)$, and $A_\mu(x)$ with $x=na$. The continuum limit of the quantized theory is much harder to discuss owing to renormalization problems.

B. Quantization

The problem of principal interest here is the quantization of the gauge field. Therefore, the gauge field will be quantized by itself to start with. Later the quantization of ψ will be discussed. At the end of this section the generalizations to non-Abelian gauge theories will also be described.

The quantization of the lattice gauge theory will

be carried out in two steps. The first step will be to define a lattice version of Euclidean vacuum expectation values, starting from a lattice version of the Feynman path integral. The second step will be to define a quantum theory on the lattice, which will allow the introduction of a real time variable and the definition of particle states and scattering amplitudes. In both cases the lattice provides an ultraviolet cutoff and there is no Lorentz invariance. Lorentz invariance can only be achieved in the limit a (lattice spacing) $\to 0$, if at all, and in practice this is a difficult limit to evaluate.

As discussed in Sec. I, one would like to calculate the gauge-field average of $\exp[ig\oint A_\mu(x)ds^\mu]$ weighted with the gauge-field action. On a lattice the line integral becomes a sum over a closed path P on the lattice (see, e.g., Fig. 8). The sum has the form: $i\sum_P(\pm)B_{n\mu}$, where a particular $B_{n\mu}$ is present in the sum if the path connects the sites n and $n+\hat{\mu}$ ($-B_{n\mu}$ appears if the path goes from $n+\hat{\mu}$ to n).

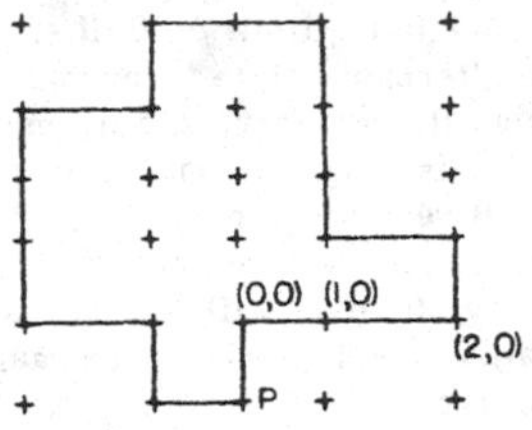

FIG. 8. Example of a lattice path P.

On the lattice, an average over all gauge fields involves integrating over all values of the $B_{n\mu}$ for all n and μ. Normally one would have integrals over an infinite range: $-\infty < B_{n\mu} < \infty$, but because of the periodicity in $B_{n\mu}$ there is no point to integrating over more than a single period. Thus the lattice version of the gauge-field average is

$$I(P)=Z^{-1}\left(\prod_m\prod_\nu\int_{-\pi}^{\pi} dB_{m\nu}\right)\times\exp\left(i\sum_P(\pm)B_{n\mu}+\frac{1}{2g^2}\sum_{n\mu\nu}e^{if_{n\mu\nu}}\right), \tag{3.13}$$

with

$$Z=\left(\prod_m\prod_\nu\int_{-\pi}^{\pi} dB_{m\nu}\right)\exp\left(\frac{1}{2g^2}\sum_{n\mu\nu}e^{if_{n\mu\nu}}\right). \tag{3.14}$$

Note that no gauge-fixing term has been added to the action. The finite range of $B_{n\mu}$ makes a gauge-fixing term unnecessary. In continuum gauge theories where $A_\mu(x)$ has an infinite range $[-\infty < A_\mu(x) < \infty]$ a gauge-fixing term is essential to have a convergent functional integral. The reason for this is that the volume in path-integral space generated by all possible gauge transformations is infinite; the gauge-fixing term provides a convergence factor in this volume.[10] In the lattice theory the total volume of integration is finite if the lattice itself is of finite extent; no convergence factor is required. For a lattice of infinite extent there are divergences due to the infinite number of integrations (in other words, the infinite lattice volume), but these divergences are normally removed by the division by Z (this division is equivalent to removing all vacuum loops in perturbation theory).

One can define more conventional vacuum expectation values in a similar manner. For instance, one can define a propagator. In the absence of a gauge-fixing term it is awkward to define a propagator for the gauge field $B_{n\mu}$ itself; instead one can define a gauge-invariant propagator as

$$D_{n\mu\pi,\sigma\tau}=Z^{-1}\left(\prod_m\prod_\nu\int_{-\pi}^{\pi} dB_{m\nu}\right)\times\exp\left(if_{n\mu\pi}-if_{0\sigma\tau}+\frac{1}{2g^2}\sum_{m\nu\rho}e^{if_{m\nu\rho}}\right). \tag{3.15}$$

This is a propagator for the operator $e^{if_{n\mu\nu}}$; it is defined only for the lattice points n of a Euclidean space-time lattice. If the lattice spacing is a, this means the propagator is defined only for imaginary times of the form in_0a, where n_0 is an integer.

A theory defined only for discrete imaginary values of the time leaves much to be desired. Fortunately, one can generalize the theory to define a Hamiltonian for a quantized theory. The particle eigenstates and scattering amplitudes of the theory can then be obtained, in principle, by diagonalizing the Hamiltonian. The Hamiltonian will be defined using the transfer-matrix formalism. Only a brief discussion of the transfer-matrix approach will be given here. For a review of the ideas see Wilson and Kogut.[6] A detailed discussion including approximate calculation of single-particle energies and scattering amplitudes in a simple scalar field theory is given by suri.[5]

Consider the expression for Z. Introduce finite bounds on the lattice coordinate n_0, say

$$-N \leq n_0 \leq N. \tag{3.16}$$

Introduce periodic boundary conditions (see below). Then one can write Z as the trace of a matrix V, more precisely

$$Z = \mathrm{Tr}\, V^{2N+1} . \tag{3.17}$$

This formula is made possible by the fact that each term in the action A involves no more than two adjacent values of n_0.

To set up the matrix V, one must first understand the space on which it acts. The space used here is the space of all functions $\psi(B_{\vec{n}i})$ (periodic in each $B_{\vec{n}i}$ with period 2π), where the index i runs from 1 to 3 only, and the lattice variable $\vec{n}$ has only three components (n_1, n_2, n_3). The matrix V will be defined as a function of two sets of arguments, say $B_{\vec{n}i}$ and $B'_{\vec{n}i}$, these two sets of arguments referring to the space-time fields B_{ni} for two adjacent values of n_0. Matrix multiplication of two V's involves integrations over a set of variables $\{B_{\vec{n}i}\}$. Define V as

$$V = \prod_{\vec{m}} \int_{-\pi}^{\pi} dB_{\vec{m}0} e^{U} . \tag{3.18}$$

The quantity U as written out below looks complicated, but all it is is that part of the action A referring to a given nearest-neighbor pair of values for n_0; terms in A referring to a single value of n_0 are divided equally between the matrices connecting n_0 to n_0+1 and n_0 to n_0-1. The result is

$$U = \frac{1}{4g^2} \sum_{\vec{n}} \sum_{i} \sum_{j} (e^{if'_{\vec{n}ij}} + e^{if_{\vec{n}ij}}) + \frac{1}{2g^2} \sum_{\vec{n}} \sum_{i} (e^{if_{\vec{n}i0}} + e^{if_{\vec{n}0i}}) , \tag{3.19}$$

where

$$f_{\vec{n}ij} = B_{\vec{n}i} + B_{\vec{n}+\hat{i},j} - B_{\vec{n}+\hat{j},i} - B_{\vec{n}j} \tag{3.20}$$

and

$$f_{\vec{n}i0} = B_{\vec{n}i} + B_{\vec{n}+\hat{i},0} - B'_{\vec{n}i} - B_{\vec{n}0} . \tag{3.21}$$

With the definition of V given here, the trace $\mathrm{Tr}\, V^{2N+1}$ is easily seen to reproduce all the integrations involved in the equation for Z, and the sum of the $2N+1$ exponents U reproduces the action A except for some additional terms coupling the boundary $n_0 = N$ to the boundary $n_0 = -N$ to achieve a periodic structure.

Note that the 0 components of $B_{n\mu}$ have received special treatment. This is because there are no terms in A involving B_{n0} for more than one value of n_0; this makes it possible to include integrations over B_{n0} in the definition of V rather than in the definition of matrix multiplication.

The matrix V is used to define the quantized theory. Briefly, this is accomplished as follows. V is a Hermitian matrix, i.e.,

$$V(B, B')^* = V(B', B) . \tag{3.22}$$

[This result can be verified by close examination of Eqs. (3.18)–(3.21). One must remember that the variables $B_{\vec{n}0}$ are integrated out in the definition of V. This means that in forming the complex conjugate of V one can also make the change of variable $B_{\vec{n}0} \to -B_{\vec{n}0}$.] Hence V has a complete orthogonal set of eigenstates Ψ and eigenvalues λ. The Hamiltonian H is now *defined* as follows: The eigenstates of H are the eigenstates of V; the eigenvalues of H are given by

$$E = -a^{-1} \ln\lambda , \tag{3.23}$$

where λ is the corresponding eigenvalue of V. The reason for the factor a^{-1} will be evident shortly. The reason for using the logarithm is so that the energies of multiparticle scattering states will be the sum of single-particle energies (see suri[5]).

A problem arises with this definition if V has any negative eigenvalues λ. If this were to happen, H would have complex eigenvalues. This did not happen in the case studied by suri[5]; whether it happens here the author does not know; this question must be studied further. Even if V has negative eigenvalues, they may be irrelevant in the limit $a \to 0$ if such a limit exists.

The definition (3.23) means that

$$V = e^{-aH} . \tag{3.24}$$

This means V is the operator which propagates a state through an imaginary time ia. It is a consequence of this that the propagator $D_{n\mu\pi,\sigma\tau}$ is a vacuum expectation value for imaginary time in_0a in the theory with Hamiltonian H. For proof of this statement see Refs. 5 and 6.

The lattice quantization procedure can be extended to non-Abelian gauge theories. This is done as follows. In place of the single variable $B_{n\mu}$, one has a set of variables $B^{\alpha}_{n\mu}$ where α is an internal index. For each n and μ, $B^{\alpha}_{n\mu}$ is to parametrize an element $b_{n\mu}$ of the gauge group. In place of $\exp(iB_{n\mu})$ one substitutes $U(b_{n\mu})$, where U is the unitary matrix representing $b_{n\mu}$ in the quark representation. The product $\bar{\psi}_n\gamma_\mu\psi_{n+\hat{\mu}}\exp(iB_{n\mu})$ is replaced by $\bar{\psi}_n\gamma_\mu U(b_{n\mu})\psi_{n+\hat{\mu}}$. A gauge transformation is defined by a set of group transformations y_n; under these transformations

$$\psi_n \to U(y_n)\psi_n , \tag{3.25}$$

$$\bar{\psi}_n \to \bar{\psi}_n U^{\dagger}(y_n) , \tag{3.26}$$

$$b_{n\mu} \to y_n b_{n\mu} y_{n+\hat{\mu}}^{-1} , \tag{3.27}$$

$$U(b_{n\mu}) \to U(y_n) U(b_{n\mu}) U^{\dagger}(y_{n+\hat{\mu}}) , \tag{3.28}$$

where $y_n b_{n\mu} y_{n+\hat{\mu}}^{-1}$ is computed according to the multiplication law of the group. A simple gauge-invariant action for the gauge field is

$$A_B = \frac{1}{2g^2} \sum_{n\mu\nu} \mathrm{Tr}\, U_A(b_{n\mu} b_{n+\hat{\mu},\nu} b_{n+\hat{\nu},\mu}^{-1} b_{n\nu}^{-1}), \tag{3.29}$$

where the unitary transformation U_A is taken from the adjoint representation of the group. (Any representation will do as well.) In the classical continuum limit, $b_{n\mu}$ becomes an infinitesimal group transformation $[B^{\alpha}_{n\mu} = gaA^{\alpha}_{\mu}(na)$ with A^{α}_{μ} fixed as $a\to 0]$; in this limit one can show, with some effort, that A_B reduces to the standard continuum Yang-Mills action. When the non-Abelian theory is quantized, the integrations over $b_{m\nu}$ are group integrations over all elements of the group. These are compact integrations (for compact groups) and no gauge-fixing term is required.

Since quark binding can be illustrated using Abelian theory, the non-Abelian theory will not be studied further.

Finally, the quantization of the Dirac field will be discussed. It is convenient initially to quantize the Dirac field in an analogous manner to the gauge field. The only problem that occurs is to define "integration" for a Fermi field. This can be done.[11]

The property of the integral that is crucial for quantization is translational invariance in the integration variable. For example, when quantizing a scalar field ϕ on a lattice the field averaging involves the integral $\int_{-\infty}^{\infty} d\phi_n$ which has the translational invariance

$$\int_{-\infty}^{\infty} d\phi_n f(\phi_n + J_n) = \int_{-\infty}^{\infty} f(\phi_n) d\phi_n \tag{3.30}$$

for any integrable function f and any constant J_n. It is this translational invariance that makes the Feynman path integral provide a realization of Schwinger's action principle (see, e.g., Ref. 12 for further discussion). Analogously one needs to define an integration over Fermi fields with the same translational invariance. Stated abstractly, one wants to define a bracket operation $\langle\cdots\rangle$ defined on functions of purely anticommuting Fermi fields ψ_n with the property

$$\langle f(\psi_n + \eta_n, \bar{\psi}_n + \bar{\eta}_n)\rangle = \langle f(\psi_n, \bar{\psi}_n)\rangle, \tag{3.31}$$

where η_n and $\bar{\eta}_n$ are anticommuting c-numbers (these have been introduced by Schwinger). The bracket operation should produce a number for every function f; it should also be a linear operation. Thus for a finite lattice it is sufficient to specify the bracket $\langle\cdots\rangle$ for all monomials in the ψ_n and $\bar{\psi}_n$. Because of the anticommutation rules, ψ_n^2 and $\bar{\psi}_n^2$ vanish (more correctly, $\psi_{n\alpha}^2$ and $\bar{\psi}_{n\alpha}^2$ vanish where $\psi_{n\alpha}$ and $\bar{\psi}_{n\alpha}$ are any component of ψ_n and $\bar{\psi}_n$), therefore, there are only a finite number of possible monomials. It is now easy to see that the bracket $\langle\cdots\rangle$ must vanish for all products of ψ's and $\bar{\psi}$'s, except the product containing all possible ψ's and $\bar{\psi}$'s. For example, suppose there are two lattice sites 0 and 1, and ψ_n and $\bar{\psi}_n$ are single component fields. Then the brackets must be

$$\begin{aligned}
&\langle 1\rangle = 0,\\
&\langle\psi_0\rangle = \langle\psi_1\rangle = \langle\bar{\psi}_0\rangle = \langle\bar{\psi}_1\rangle = 0,\\
&\langle\psi_0\psi_1\rangle = \langle\psi_0\bar{\psi}_1\rangle = \cdots = 0,\\
&\langle\bar{\psi}_0\psi_0\psi_1\rangle = \cdots = 0,\\
&\langle\bar{\psi}_0\psi_0\bar{\psi}_1\psi_1\rangle = 1;
\end{aligned} \tag{3.32}$$

where 1 is a constant which was chosen arbitrarily. This definition of the bracket operation satisfies translational invariance: for example,

$$\langle(\bar{\psi}_0 + \bar{\eta}_0)(\psi_0 + \eta_0)(\bar{\psi}_1 + \bar{\eta}_1)(\psi_1 + \eta_1)\rangle = \langle\bar{\psi}_0\psi_0\bar{\psi}_1\psi_1\rangle = 1$$

because the terms multiplying the η's are all 0. Note also that the anticommutation rules mean that, for example,

$$\langle\psi_0\bar{\psi}_0\bar{\psi}_1\psi_1\rangle = -1. \tag{3.33}$$

(In analogy to the scalar case, one requires $\psi_0\bar{\psi}_0 = -\bar{\psi}_0\psi_0$, not $\psi_0\bar{\psi}_0 = 1 - \bar{\psi}_0\psi_0$.)

One can now define the Feynman path integral on a lattice for the complete gauge theory including the Dirac fields. For example, the current-current propagator on the lattice is

$$D_{n\mu\nu} = Z_{\mathrm{tot}}^{-1}\left(\prod_m \prod_\nu \int_{-\pi}^{\pi} dB_{m\nu}\right)\langle\bar{\psi}_n\gamma_\mu\psi_n\bar{\psi}_0\gamma_\nu\psi_0 e^A\rangle, \tag{3.34}$$

where A is the full action of Eq. (3.12) and

$$Z_{\mathrm{tot}} = \left(\prod_m \prod_\nu \int_{-\pi}^{\pi} dB_{m\nu}\right)\langle e^A\rangle. \tag{3.35}$$

This formulation of the path integral is different from the formulation discussed in Sec. II. However, one can easily derive a lattice form of the path integrals of Sec. II from the present expression. The procedure is to expand Eq. (3.34) in powers of K, where K is the coefficient of the nearest-neighbor coupling terms $\bar{\psi}_n\gamma_\mu\psi_{n+\hat{\mu}}e^{iB_{n\mu}}$ etc. This nearest-neighbor coupling term can be represented diagrammatically by a line from the site n to the site $n+\hat{\mu}$. The expansion is best described by studying an example of a term from the expansion of the numerator of Eq. (3.34), which will now be discussed. An example of a term in the expansion is represented diagrammatically in Fig. 9. The expression for this term is

$$\left(\prod_m \prod_\nu \int_{-\pi}^{\pi} dB_{m\nu}\right) K^4 \langle \bar\psi_{11}\gamma_\mu \psi_{11} \bar\psi_{00}\gamma_\nu \psi_{00} \bar\psi_{00}\gamma_0 \psi_{10} e^{iB_{00,0}} \bar\psi_{10}\gamma_1 \psi_{11} e^{iB_{10,1}} \bar\psi_{11}\gamma_0 \psi_{01} e^{-iB_{01,0}} \bar\psi_{01}\gamma_1 \psi_{00} e^{-iB_{00,1}} e^{A_0}\rangle = D \tag{3.36}$$

(D for diagram), where the four lattice sites involved are $(n_0, n_1) = (0, 0)$, $(1, 0)$, $(0, 1)$, and $(1, 1)$; the values of n_2 and n_3 are constant and have been suppressed in the notation of Eq. (3.36). The action A_0 omits the K term, and is

$$A_0 = -c \sum_n \bar\psi_n \psi_n + \frac{1}{2g^2} \sum_n \sum_{\mu\nu} e^{if_{n\mu\nu}} . \tag{3.37}$$

The calculation of D has two parts, one being the integration over all gauge fields $B_{m\mu}$, the other being the calculation of the $\psi, \bar\psi$ bracket. These are independent calculations, i.e., D factors into $D_\psi D_B$. The quantity D_B is

$$D_B = \left(\prod_m \prod_\nu \int_{-\pi}^{\pi} dB_{m\nu}\right) \exp\left[i(B_{00,0} + B_{10,1} - B_{01,0} - B_{00,1}) + \frac{1}{2g^2} \sum_{n\mu\nu} e^{if_{n\mu\nu}}\right] . \tag{3.38}$$

This is an example of a guage-field average of the exponential of a line integral over a closed loop, the loop being the loop of Fig. 9. The ψ bracket calculation can be factorized further into separate bracket calculations for each lattice site, since A_0 contains no terms involving ψ or $\bar\psi$ and coupling different lattice sites. Consider only the four lattice sites on the loop, for simplicity. By moving the ψ's around some (using the anticommuting rule) the bracket becomes

$$\langle \bar\psi_{00}\gamma_0 \psi_{10} e^{-c\bar\psi_{10}\psi_{10}} \bar\psi_{10}\gamma_1 \psi_{11} e^{-c\bar\psi_{11}\psi_{11}} \bar\psi_{11}\gamma_\mu \psi_{11} \bar\psi_{11}\gamma_0 \psi_{01} e^{-c\bar\psi_{01}\psi_{01}} \bar\psi_{01}\gamma_1 \psi_{00} e^{-c\bar\psi_{00}\psi_{00}} \bar\psi_{00}\gamma_\nu \psi_{00}\rangle .$$

To make a product of all possible ψ's and $\bar\psi$'s means one must have products of all possible ψ_{00}'s and $\bar\psi_{00}$'s, all possible ψ_{01}'s and $\bar\psi_{01}$'s, etc. In summary the complete bracket may be written as a product of four separate brackets. Define

$$D_\psi^1 = \langle \psi_{10} e^{-c\bar\psi_{10}\psi_{10}} \bar\psi_{10}\rangle , \tag{3.39}$$

$$D_{\psi\mu} = \langle \psi_{11} e^{-c\bar\psi_{11}\psi_{11}} \bar\psi_{11}\gamma_\mu \psi_{11} \bar\psi_{11}\rangle . \tag{3.40}$$

Both D_ψ^1 and $D_{\psi\mu}$ are matrices in spin space due to the spinor indices implied for ψ_{10}, $\bar\psi_{10}$, ψ_{11}, and $\bar\psi_{11}$. The full bracket is simply

$$D_\psi = -K^4 \,\mathrm{Tr}(D_\psi^1 \gamma_1 D_{\psi\mu} \gamma_0 D_\psi^1 \gamma_1 D_{\psi\nu} \gamma_0) . \tag{3.41}$$

The matrices D_ψ^1 and $D_{\psi\mu}$ are easily determined. For example, D_ψ^1 explicitly is

$$D_{\psi\alpha\beta}^1 = \left\langle \psi_{10\alpha} \exp\left(-c \sum_\gamma \bar\psi_{10\gamma}\psi_{10\gamma}\right) \bar\psi_{10\beta}\right\rangle . \tag{3.42}$$

The exponential can be expanded in powers of c; assuming the spinors have four components only the c^3 term can produce a product of all four ψ_{10}'s times all four $\bar\psi_{10}$'s; the result is

$$D_{\psi\alpha\beta}^1 = -\delta_{\alpha\beta}(-c)^3 \tag{3.43}$$

(the minus sign comes from the convention that the bracket is positive when $\bar\psi_{10\beta}$ appears to the left of $\psi_{10\beta}$). A similar calculation gives

$$D_{\psi\mu} = c^2 \gamma_\mu . \tag{3.44}$$

The results of this example are easily generalized. A term of general order K^l is nonzero only if the nearest-neighbor couplings combine to form closed loops (the lattice site at the endpoint of an open line would have an extra ψ_n or $\bar\psi_n$ so the bracket at n would give 0). The bracket calculation for a closed loop gives a trace involving K times a γ matrix for each line in the loop and D's for each lattice site in the loop (except the points n and 0 where there are currents). The average over gauge fields involves an exponential of a sum of $B_{n\mu}$'s around each loop. There can be any number of loops.

IV. STRONG-COUPLING APPROXIMATION

The gauge field average $I(P)$ which determines whether quarks are bound was defined on a lattice in Sec. III (Eq. 3.13). There are two limits in which this average can be calculated. The most interesting limit is the strong-coupling limit $g \to \infty$. This is the limit which exhibits quark binding. A strong-coupling expansion will be derived in this section.

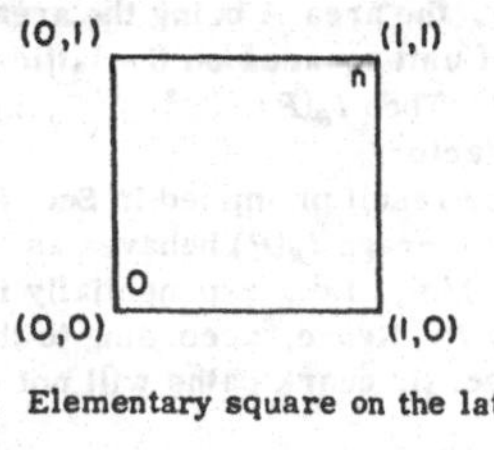

FIG. 9. Elementary square on the lattice.

The strong-coupling expansion will be the basis for a reformulation of the gauge-field theory as a string model. This will also be explained in this section.

Consider specifically the numerator of Eq. (3.13), to be denoted $I_N(P)$:

$$I_N(P)=\left(\prod_m\prod_\nu\int_{-\pi}^{\pi}dB_{m\nu}\right)\times\exp\left[i\sum_P(\pm)B_{n\mu}+\frac{1}{2g^2}\sum_{n\mu\nu}e^{if_{n\mu\nu}}\right]. \tag{4.1}$$

Expanding in powers of $1/g^2$, the zeroth-order term is

$$I_N^{(0)}(P)=\left(\prod_m\prod_\nu\int_{-\pi}^{\pi}dB_{m\nu}\right)\exp\left[i\sum_P(\pm)B_{n\mu}\right]. \tag{4.2}$$

This term vanishes, since for any $B_{n\mu}$ which appears in $\sum_P$, there is an integral $\int_{-\pi}^{\pi}dB_{n\mu}\times\exp(\pm iB_{n\mu})$ which is zero. Thus one must seek higher-order terms in g^{-2} which cancel the $B_{n\mu}$ in the line integral. The first-order term is

$$I_N^{(1)}(P)=\frac{1}{2g^2}\left(\prod_m\prod_\nu\int_{-\pi}^{\pi}dB_{m\nu}\right)\sum_{l\tau\sigma}\exp\left[i\sum_P(\pm)B_{n\mu}+if_{l\tau\sigma}\right]. \tag{4.3}$$

The quantity $f_{l\tau\sigma}$ is itself a line integral of the gauge field; it is the line integral around a square originating at the lattice site l of size a (unit square). The integral for $I_N^{(1)}(P)$ will vanish unless it is possible to find a unit square such that $f_{l\tau\sigma}$ cancels completely the line integral $\sum_P(\pm)B_{n\mu}$. This is possible only if the path P is itself a unit square. Otherwise the first-order term vanishes and one must study the terms of order g^{-4} or higher.

The term of order g^{-2k} has the form

$$I_N^{(k)}(P)=\frac{1}{k!}\left(\frac{1}{2g^2}\right)^k\left(\prod_m\prod_\nu\int_{-\pi}^{\pi}dB_{m\nu}\right)\sum_{l_1\tau_1\sigma_1}\cdots\sum_{l_k\tau_k\sigma_k}\exp\left[i\sum_P(\pm)B_{n\mu}+if_{l_1\tau_1\sigma_1}+\cdots+if_{l_k\tau_k\sigma_k}\right]. \tag{4.4}$$

The only nonzero terms in this sum are those for which

$$\sum_P(\pm)B_{n\mu}+f_{l_1\tau_1\sigma_1}+\cdots+f_{l_k\tau_k\sigma_k}=0. \tag{4.5}$$

[See Eq. (3.8) for the definition of $f_{l\tau\sigma}$ in terms of $B_{n\mu}$.] This equation can be understood geometrically. Each $f_{l\tau\sigma}$ corresponds to a square of size a on the lattice. For this sum to vanish the set of squares defined by $f_{l_1\tau_1\sigma_1}\cdots f_{l_k\tau_k\sigma_k}$ must combine to make a surface with boundary P. (To be precise, each $f_{l\tau\sigma}$ corresponds to a line integral around a square, and when these squares are joined to make a surface the line integrals must cancel along all internal lines of the surface. The line integrals along the boundary P of the surface must run in the opposite direction to the original path P.) See Fig. 10.

For a given path P the lowest nonzero order in $I_N(P)$ is determined by the minimal area A enclosed by P, the area A being the area of any surface built of unit squares on the lattice with boundary P. Then $I_N(P)\sim(g^2)^{-A/a^2}$, apart from a numerical factor.

This is the result promised in Sec. II: The gauge-field average $I_N(P)$ behaves as $\exp[-A(\ln g^2)/a^2]$, i.e., exponentially in the area enclosed by P. Hence, according to the arguments of Sec. II, quark paths will not separate macroscopically, and there will be no quarks among the final-state particles.

Consider higher-order terms in the expansion of $I_N(P)$ for given P. There are many such terms because there are many surfaces with boundary P. In particular, there are many ways to combine subsets of f's to add to zero so such subsets can be added to any minimal sum of f's which forms a surface with boundary P. The simplest example of a set of f's which add to zero are the set of f's corresponding to the six faces of a unit cube. Written out, this gives

$$0=f_{n\mu\nu}-f_{n+\hat{\tau},\mu\nu}+f_{n\nu\tau}-f_{n+\hat{\mu},\nu\tau}+f_{n\tau\mu}-f_{n+\hat{\nu},\tau\mu}, \tag{4.6}$$

which is easily checked using Eq. (3.8). [Equation

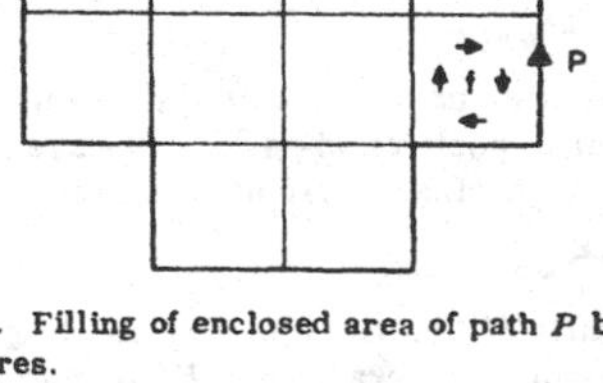

FIG. 10. Filling of enclosed area of path P by elementary squares.

(4.6) is the lattice analog of the equation $\epsilon^{\mu\nu\tau\sigma}\nabla_\tau F_{\mu\nu}(x)=0$.]

Let $A(P)$ be the minimal area as defined above enclosed by P. Since one can place a unit cube anywhere on this minimal area, it means that there are roughly $A(P)/a^2$ more terms of order $g^{-12}(g^2)^{-A(P)/a^2}$ in the expansion of $I_N(P)$ than there are terms of order $(g^2)^{-A(P)/a^2}$. [One can place unit cubes anywhere in space, not just on the minimal surface; but when one divides $I_N(P)$ by Z all disconnected terms cancel, as usual.] This suggests that the $1/g^2$ expansion is not very useful in the limit $A(P)\to\infty$, which is the limit of interest for quark binding. However, experience with related problems suggests that $I_N(P)$ is not the appropriate quantity to expand; instead one should try writing

$$I(P)=Z^{-1}I_N(P)$$

$$=(g^2)^{-A(P)}e^{-c(P,g^2)}\,, \tag{4.7}$$

and expand $c(P,g^2)$ in powers of g^{-2} instead. One would expect $c(P,g^2)$ to be dominated by a term proportional to $A(P)$, say

$$c(P,g^2)=A(P)f(g^{-2})+O(P) \tag{4.8}$$

(where P is the length of the path P). The crucial question is the nature of the series for $f(g^{-2})$. Past experience with similar types of expansions (namely, the high-temperature expansions of statistical mechanics: see, e.g., Ref. 13) suggests that $f(g^{-2})$ will have a convergent expansion at least for g^{-2} less than a critical value g_c^{-2}. However, no calculations have been done in the gauge-field theory for $f(g^{-2})$ as yet.

Consider the complete expansion of $I(P)$. Each nonzero term in the expansion corresponds to a surface with perimeter P. The complete expansion corresponds to a sum over all possible surfaces with given perimeter P. "All possible" surfaces include surfaces which intersect themselves (to take into account terms where a given $f_{n\mu\nu}$ appears several times in the sum $f_{l_1\tau_1\sigma_1}+\cdots+f_{l_k\tau_k\sigma_k}$). There is a weight factor for each surface, aside from the power of g^{-2} determined by the area of the surface. For a simple surface, the weight factor is 1; the weight is more complicated for self-intersecting surfaces.

Thus, the strong-coupling expansion for the current-current propagator has the same general structure as in string models of hadrons. One is actually dealing here with a double expansion. An expansion in the coefficient K (appearing in the Dirac field action) was needed to define quark loops on the lattice; the sum of the K expansion is a sum over all possible quark loops. The g^{-2} expansion is needed to define surfaces filling in the quark loops. The sum of the g^{-2} expansion is a sum over all such surfaces. This is precisely the structure appearing in string models: combined sums over quark loops and interpolating surfaces. However, the loops and surfaces of the gauge-field theory are defined on a lattice whereas the loops and surfaces of the string models are defined on a continuum. It may not be easy to derive quantitative relations between the two types of surfaces.

V. WEAK-COUPLING APPROXIMATION

The weak-coupling approximation will be discussed briefly, leaving many questions open. Only the pure gauge field will be discussed. Consider again the expression

$$I_N(P)=\left(\prod_m\prod_\nu\int_{-\pi}^{\pi}dB_{m\nu}\right)\times\exp\left[i\sum_P(\pm)B_{n\mu}+\frac{1}{2g^2}\sum_{n\mu\nu}e^{if_{n\mu\nu}}\right]. \tag{5.1}$$

Suppose the integration variables were $f_{n\mu\nu}$ rather than $B_{m\nu}$. For small g, only small values of $f_{n\mu\nu}$ would be important in the integral, in order that $\mathrm{Re}\,e^{if_{n\mu\nu}}$ be near its maximum value 1. One would then expand:

$$\frac{1}{2g^2}\sum_{n\mu\nu}e^{if_{n\mu\nu}}\simeq\frac{1}{2g^2}\sum_{n\mu\nu}(1-\tfrac{1}{2}f_{n\mu\nu}{}^2)\,. \tag{5.2}$$

With this approximation one could extend the limits of integration on $f_{n\mu\nu}$ from $\pm\pi$ to $\pm\infty$, with negligible error; one would then have a set of Gaussian integrals to evaluate.

In practice the integration variables are the $B_{m\nu}$, not the $f_{n\mu\nu}$. However, one can make a change of variable from the $B_{m\nu}$ to the $f_{n\mu\nu}$. It is not possible to eliminate all the $B_{m\nu}$ by this transformation, and not all the variables $f_{n\mu\nu}$ are independent. Nevertheless, the transformation is sufficient to make $I_N(P)$ calculable for small g^2.

To make the change of variables precise, consider a system of finite size ($1\le n_i\le N$) with periodic boundary conditions. Then one can change variables from the $B_{m\nu}$ to a subset of the $f_{n\mu\nu}$ plus some gauge transformation variables ϕ_n, plus four extra variables ζ_μ, as follows:

(i) For $n_0\ne N$, n_1, n_2, n_3 arbitrary, $B_{n\mu}$ ($\mu=1,2,3$) is replaced by $f_{n\mu0}$. For B_{n0}, one writes

$$B_{n0}=\phi_{n+\hat{0}}-\phi_n \tag{5.3}$$

and replaces B_{n0} by ϕ_n. This is the essence of the transformation from $B_{n\mu}$ to $f_{n\mu0}$ for $\mu\ne0$ and from B_{n0} to ϕ_n. To complete the transformation one must discuss the surface $n_0=N$.

(ii) For $n_0 = N$, $n_1 \neq N$, and n_2 and n_3 arbitrary, $B_{n\mu}$ $(\mu \neq 1)$ is replaced by $f_{n\mu 1}$. B_{n1} is replaced by ϕ_n, with

$$B_{n1} = \phi_{n+\hat{1}} - \phi_n \ . \tag{5.4}$$

(iii) For $n_0 = n_1 = N$, $n_2 \neq N$, and n_3 arbitrary, $B_{n\mu}$ $(\mu \neq 2)$ is replaced by $f_{n\mu 2}$. B_{n2} is replaced by ϕ_n, with $B_{n2} = \phi_{n+\hat{2}} - \phi_n$.

(iv) For $n_0 = n_1 = n_2 = N$, and $n_3 \neq N$, $B_{n\mu}$ $(\mu \neq 3)$ is replaced by $f_{n\mu 3}$, B_{n3} by ϕ_n, where $B_{n3} = \phi_{n+\hat{3}} - \phi_n$.

(v) For $n_0 = n_1 = n_2 = n_3 = N$ one writes $B_{n\mu} = \zeta_\mu$ with ζ_μ being the new variables. One also sets $\phi_n = 0$ for this value of n.

The variables $f_{n\mu\nu}$ which are not integration variables (for example, $f_{n\mu\nu}$ with $\mu \neq 0$ and $\nu \neq 0$, for $n_0 \neq N$) can be expressed in terms of the independent $f_{n\mu\nu}$ variables using Eq. (4.6); neither the ϕ_n nor ζ_μ appear in these expressions.

When an arbitrary $B_{m\nu}$ is expressed in terms of the new independent variables, one finds (ζ_ν is present only if $m_\nu = N$):

$$B_{m\nu} = \zeta_\nu + \phi_{m+\hat{\nu}} - \phi_m + (\text{linear combination of } f_{n\tau\sigma}) \ , \tag{5.5}$$

i.e., the ϕ variables define a gauge transformation and ζ_ν represents a translation of some of the B's. It is easily verified that the integrand of $I_N(P)$ involves only the f's: It is independent of both the ϕ's and ζ_ν (the latter does not appear because any closed path P has as many $-B_{n\nu}$ terms as $+B_{n\nu}$ terms on the sublattice $n_\nu = N$). Hence the ϕ_n and ζ_ν integrations can be computed trivially. The f integrations are nontrivial because of the constraint (4.6).

What one wants to accomplish is to reduce the lattice theory for small g to something like a conventional free gauge-field theory. This means restoring the $B_{m\nu}$ as the integration variables, but with infinite limits of integration, and with a gauge-fixing term included. Suppose, for example, one starts with

$$I_N'(P) = \left(\prod_m \prod_\nu \int_{-\infty}^{\infty} dB_{m\nu}\right) \exp\left(i \sum_P (\pm) B_{n\mu} - \frac{\alpha}{2} \sum_n \left[\sum_\mu (B_{n\mu} - B_{n-\hat{\mu},\mu})\right]^2 - \frac{1}{4g^2} \sum_{n\mu\nu} f_{n\mu\nu}{}^2\right) \ , \tag{5.6}$$

where the α term is a lattice version of a $(\nabla_\mu A_\mu)^2$ gauge-fixing term. This integral can be computed by explicit Gaussian integration methods rather more easily than the $f_{n\mu\nu}$ integrations for $I_N(P)$. In addition, $I_N'(P)$ can also be reduced to an integral over a subset of the $f_{n\mu\nu}$, using the same change of variables as for $I_N(P)$. The result is different in this case due to the α term which couples the ϕ's to the f's; also there is no convergence factor for the ζ integration. To make $I_N'(P)$ well defined and equal to $I_N(P)$, one must (a) put in a convergence factor for the ζ integral, i.e., a term $-\frac{1}{2}\beta(\sum_n B_{n\mu})^2$ and (b) add a quadratic form in the f's to compensate for the result of the ϕ integration of the gauge-fixing term. The author has not carried through this calculation; but since the net result is still that $I_N(P) = I_N'(P)$ is a Gaussian integration in the B's, the result will presumably be similar to the conventional free-field calculation reported in Sec. II.

VI. PHASE TRANSITIONS

In the strong-coupling limit ($g \to \infty$, $K \to 0$) the gauge theory is far from being Lorentz-invariant. More precisely, since the action was defined on a Euclidean metric, it is Euclidean invariance that is missing. In the strong-coupling limit, vacuum expectation values decrease rapidly at separations of only a few lattice sites (there is a factor g^{-2} or K or both for each unit lattice spacing of separation). This corresponds to the existence of masses much larger than the cutoff. [The usual rule is that if a propagator falls as $e^{-x/\xi}$ for x large then the lowest mass intermediate state contributing to the propagator has mass $1/\xi$. If the propagator behaves as g^{-2n} for distances $x = na$, then the corresponding mass is $2(\ln g)/a$. This is larger than the cutoff momentum π/a if g is large.]

Thus, one is interested in practice in values of g and K such that the correlation length ξ is much larger than the lattice spacing a, in order that the corresponding mass is much less than the cutoff. One knows from statistical mechanics that large correlation lengths are associated with second-order phase transitions (critical points). Thus one seeks special values g_c and K_c for g and K at which there is a phase transition.[14]

It has already been argued that there are two distinct phases for the gauge field, a strong-coupling phase for large g which binds quarks, and a weak-coupling phase for small g which does not bind quarks. The arguments given neglected quark vacuum loops, which is reasonable if K is small. There should be a transition between these two phases which would occur at a critical value g_c for any g and any small value of K. This is one possible phase transition; it is this transition which was discussed in Sec. II. But, as will be argued below, this is probably a first-order transition rather than second order.

Suppose one wishes to construct a model of strong interactions using the lattice theory of this paper with the gauge group separate from ordinary SU(3)×SU(3) symmetry. Then the gauge fields would all be SU(3)×SU(3) singlets, while the quark fields would carry SU(3)×SU(3) quantum numbers as well as gauge-group quantum numbers. A little thought shows that in the strong-coupling limit ($g\to\infty$, $K\to 0$), SU(3)× SU(3) is an exact symmetry rather than a spontaneously broken symmetry. Varying g does not change this situation; so one must hope that by increasing K one can change the exact SU(3)×SU(3) into broken SU(3) ×SU(3). If this does not work one is free to introduce additional terms into the quark field action in hopes of forcing a spontaneous breaking of SU(3)×SU(3). Suppose, for simplicity, that SU(3) ×SU(3) can be broken by increasing K. Then there will be a phase transition at a critical value K_c for K where one changes from exact SU(3)×SU(3) to spontaneously broken SU(3)×SU(3). If this transition is a second-order transition then there will be a large correlation length for K near K_c; in this case the theory might be a realistic model of broken SU(3)×SU(3) for K slightly greater than K_c (with g large enough to maintain quark binding).

In summary, the transition of real interest is a transition in K (or some other parameter introduced into the quark action) rather than g.

Apart from special limits ($g\to\infty$ and $K\to 0$, or g small) it is very difficult to solve the lattice theory. It is especially difficult to solve the lattice theory near a critical point with a large correlation length. Various methods have been developed by statistical mechanicians to deal with this problem. In the remainder of this section these methods will be discussed briefly. There are essentially three approaches to consider: (1) mean-field techniques, (2) series expansions, and (3) the renormalization-group approach.

Mean-field techniques[15] are the simplest and crudest methods for studying a critical point; invariably they are the methods one uses first in studying a new situation. They are used to determine if there is a phase transition, whether it is first or second order, and to give rough estimates of the behavior near the transition. None of the results of a mean-field calculation are entirely trustworthy. Examples of mean-field calculations will be given later.

An example of a series expansion would be the expansion of the current-current propagator for small momentum (momentum $\ll 1/a$) in powers of g^{-2} and K, to high order in g^{-2} and K. One then uses Padé-approximant techniques to look for singularities in either g or K that would be associated with a mass approaching 0. In simple statistical-mechanical problems one can generate 12 terms or so in analogous expansions. The expansion for the lattice theory of this paper is more complicated, but one could hope to generate maybe 6 or 7 orders with some practice. Series expansions require considerably more effort than mean-field calculations; they apply mainly to propagators, being very awkward to perform on three- and four-point functions, and one must have a clear idea of what one is trying to learn before attempting such calculations. See Ref. 16 for one of the best series-expansion formalisms; see Ref. 13 for a general review.

The renormalization-group approach is potentially the most powerful and accurate method for studying lattice theories near a critical point, but at present the renormalization-group techniques are too limited in scope to be applicable to the present problem. See Refs. 6 and 17.

Return to mean-field ideas.[15] The prototype mean-field calculation is a calculation of the magnetization as a function of the external field for an Ising ferromagnet. Let s_n be the spin at site n with values ±1 only; let the interaction be

$$\frac{-H}{kT}=K\sum_n\sum_\mu s_n s_{n+\hat{\mu}}+h\sum_n s_n\,, \tag{6.1}$$

where K is related to the spin-spin coupling and h is proportional to the external field. Then

$$M=Z^{-1}\left\langle s_0\exp\left(K\sum_n\sum_\mu s_n s_{n+\hat{\mu}}+h\sum_n s_n\right)\right\rangle, \tag{6.2}$$

where $\langle\cdots\rangle$ means a sum over all configurations of all spins, and

$$Z=\left\langle\exp\left(K\sum_n\sum_\mu s_n s_{n+\hat{\mu}}+h\sum_n s_n\right)\right\rangle\,. \tag{6.3}$$

In the mean-field approximation, one assumes that the spins $s_{\hat{\mu}}$ coupled to s_0 can be replaced by their average value M. As a result, the formula for M simplifies to a sum over s_0 only, namely

$$M=Z_0^{-1}\sum_{s_0=\pm1}s_0e^{(2dKM+h)s_0}\,, \tag{6.4}$$

with d being the dimensionality (3 usually) and

$$Z_0=\sum_{s_0=\pm1}e^{(2dKM+h)s_0}\,. \tag{6.5}$$

The result is

$$M=\tanh(2dKM+h)\,. \tag{6.6}$$

If $2dK<1$ this equation has a unique solution for M as a function of h; in particular, $M=0$ for $h=0$. For $2dK>1$ the solution is multiple-valued; stability considerations show that one must choose a solution with $M\neq 0$ when $h=0$.

In this approximation one has actually replaced $\sum_\mu s_\mu^2$ by $2dM$, which is a good approximation if d is large. This is generally true of mean-field theories.

An analogous mean-field calculation can be performed for the lattice gauge theory. In this case a simple external field term has the form

$$h\sum_{n\mu}(e^{iB_{n\mu}}+e^{-iB_{n\mu}})$$

(this is to be added to the gauge-field action), and M can be defined to be the expectation value of $e^{iB_{0\mu}}$. The question is whether M is zero in the limit $h\to 0$. In the gauge-field theory $e^{iB_{0\mu}}$ couples to a product of three other exponentials; as a mean-field approximation one replaces this product by M^3. The result of this is that

$$M=Z_0^{-1}\int_{-\pi}^{\pi}dB_{0\mu}e^{iB_{0\mu}}\times\exp\left[\left(\frac{2(d-1)}{2g^2}M^3+h\right)(e^{iB_{0\mu}}+e^{-iB_{0\mu}})\right], \tag{6.7}$$

where d is the space-time dimensionality, and

$$Z_0=\int_{-\pi}^{\pi}\exp\left[\left(\frac{2(d-1)}{g^2}M^3+h\right)(e^{iB_{0\mu}}+e^{-iB_{0\mu}})\right]. \tag{6.8}$$

The result of this is that

$$M=f\left(\frac{(d-1)}{g^2}M^3+h\right), \tag{6.9}$$

where f is a ratio of Bessel's functions. If g is large the solution to this equation is unique and $M=0$ for $h=0$. If g is small then there are solutions with $M\neq 0$ for $h=0$, and stability considerations show again that the $M\neq 0$ solutions are preferred.

In the magnetic case, one finds that the spontaneous magnetization M goes to zero for $2dk-1$ [from Eq. (6.6)]. However, the gauge-field case never has a solution for $h=0$ with M small but nonzero. Thus there is a first-order transition at the value of g for which M changes from zero to being nonzero.

A nonzero value of M in the limit $h\to 0$ means one has spontaneous breaking of the gauge-field symmetry. So for small g the theory shows spontaneous breaking.

A much more thorough discussion of the mean-field approximation has been given by Balian, Drouffe, and Itzykson.[18] A Hamiltonian formulation of the lattice gauge theory has been given by Kogut and Susskind.[19] A clear review of quark confinement in the lattice theory is given in Ref. 20. Another formulation of the connection between strongly coupled gauge theories and string models is given in Ref. 21.

ACKNOWLEDGMENTS

The author has benefited from conversations with many people in developing and especially in understanding the lattice gauge theory. Persons I am indebted to include J. Bjorken, R. P. Feynman, M. E. Fisher, D. Gross, J. Kogut, L. Susskind, T.-M. Yan, and members of a seminar at Orsay.

*Work supported in part by the National Science Foundation.

[1]See, e.g., J. M. Cornwall and R. E. Norton, Phys. Rev. D 8, 3338 (1973); R. Jackiw and K. Johnson, *ibid.* 8, 2386 (1973); J. Kogut and L. Susskind, *ibid.* 9, 3501 (1974); D. Amati and M. Testa, Phys. Lett. 48B, 227 (1974); G. 't Hooft, Nucl. Phys. B (to be published); P. Olesen, Phys. Lett. 50B, 255 (1974); A. Chodos, R. L. Jaffe, K. Johnson, C. B. Thorn, and V. F. Weisskopf, Phys. Rev. D 9, 3471 (1974).

[2]J. Schwinger, Phys. Rev. 125, 397 (1962); 128, 2425 (1962).

[3]J. H. Lowenstein and J. A. Swieca, Ann. Phys. (N.Y.) 68, 172 (1971).

[4]A. Casher, J. Kogut, and L. Susskind, Phys. Rev. Lett. 31, 792 (1973); Phys. Rev. D 10, 732 (1974).

[5]A. suri, Ph.D. thesis, Cornell University, 1969 (unpublished).

[6]K. G. Wilson and J. Kogut, Phys. Rep. 12C, 75 (1974). Sec. X.

[7]S. Coleman and E. Weinberg, Phys. Rev. D 7, 1888 (1973).

[8]Y. Nambu, in *Symmetries and Quark Models*, proceedings of the International Conference on Symmetries and Quark Models, Wayne State Univ., 1969, edited by Ramesh Chand (Gordon and Breach, New York, 1970); L. Susskind, Nuovo Cimento 69A, 457 (1970); G. Konisi, Prog. Theor. Phys. 48, 2008 (1972); P. Goddard, J. Goldstone, C. Rebbi, and C. B. Thorn, Nucl. Phys. B56, 109 (1972); J.-L. Gervais and B. Sakita, Phys. Rev. Lett. 30, 716 (1973); S. Mandelstam, Nucl. Phys. B64, 205 (1973).

[9]R. P. Feynman, Phys. Rev. 80, 440 (1950).

[10]See, e.g., E. S. Abers and B. W. Lee, Phys. Rep. 9C, 1 (1973).

[11]See, e.g., F. A. Berezin, *The Method of Second Quantization* (Academic, New York, 1966), pp. 52ff.

[12]K. G. Wilson, Phys. Rev. D 6, 419 (1972).

[13]M. E. Fisher, Rep. Prog. Phys. 30, 615 (1967).

[14]For the relation of relativistic field theory to critical

points see Ref. 6, Sec. X, and references cited therein.
[15]For a review see R. Brout, *Phase Transitions* (Benjamin, New York, 1965), p. 8.
[16]D. Jasnow and M. Wortis, Phys Rev. 176, 739 (1968).
[17]K. G. Wilson, Cargèse (1973) Lecture Notes, in preparation.
[18]R. Balian, J. M. Drouffe, and C. Itzykson, Phys. Rev. D (to be published).
[19]J. Kogut and L. Susskind, Phys. Rev. D (to be published).
[20]K. Wilson, Cornell Report No. CLNS-271 (to be published in the proceedings of the conference on Yang-Mills Fields, Marseille, 1974).
[21]H. B. Nielsen and P. Olesen, Nucl. Phys. B61, 45 (1973); L. J. Tassie, Phys. Lett. 46B, 397 (1973).

Reprinted from:

JOURNAL OF MATHEMATICAL PHYSICS VOLUME 12, NUMBER 10 OCTOBER 1971

Duality in Generalized Ising Models and Phase Transitions without Local Order Parameters*

Franz J. Wegner†

Department of Physics, Brown University, Providence, Rhode Island 02912

(Received 29 March 1971)

It is shown that any Ising model with positive coupling constants is related to another Ising model by a duality transformation. We define a class of Ising models M_{dn} on d-dimensional lattices characterized by a number $n = 1, 2, \ldots, d$ ($n = 1$ corresponds to the Ising model with two-spin interaction). These models are related by two duality transformations. The models with $1 < n < d$ exhibit a phase transition without local order parameter. A nonanalyticity in the specific heat and a different qualitative behavior of certain spin correlation functions in the low and the high temperature phases indicate the existence of a phase transition. The Hamiltonian of the simple cubic dual model contains products of four Ising spin operators. Applying a star square transformation, one obtains an Ising model with competing interactions exhibiting a singularity in the specific heat but no long-range order of the spins in the low temperature phase.

1. INTRODUCTION

This paper deals with a general concept of duality and with phase transitions without a local order parameter.

Duality[1-5] is an inherent symmetry of the two-dimensional Ising model without crossing interaction bonds. This symmetry relates the partition function and the correlation functions[6-8] of a two-dimensional Ising model at temperature T to the partition function and the correlation functions of its dual Ising model at temperature T^*, where T^* is a decreasing function of T. In this paper the duality transformation is generalized to arbitrary Ising models with positive interaction constants (Sec. 2). This concept of duality is applied to a class of Ising models M_{dn} on d-dimensional lattices (Sec. 3). To obtain the Hamiltonian of the model M_{dn}, one takes the product of all spins located at the two ends of lines ($n = 1$), at the perimeter of surfaces ($n = 2$), and so on. Therefore, $n = 1$ describes the usual Ising model with two-spin interactions. The systems M_{dn} and $M_{d\,d-n}$ on dual lattices without external magnetic field are connected by a duality relation (Sec. 3A). For even dimensions $d = 2n$, one obtains self-dual models (models which are identical with their dual models). If there is only one singularity in the partition function of a self-dual model, then it must occur at $T = T^*$. Self-duality implies a symmetric singularity of the specific heat around the critical temperature (Sec. 3C). If an external magnetic field is present, the systems M_{dn} and $M_{d\,d-n+1}$ on dual lattices are connected by duality relations (Sec. 3A, 3C).

Most known phase transitions can be described by a local order parameter.[9-14] The models M_{dn} with $1 < n < d$ exhibit a phase transition without a local order parameter (Sec. 3B). The existence of a phase transition is indicated by a singularity in the specific heat (at least for $n = d - 1$) and by a qualitatively different asymptotic behavior of certain correlation functions at high and at low temperatures (Sec. 3B). For $n > 1$ the Hamiltonian consists of products of more than two spins. Applying the decoration,[15,16] the star triangle[3-5,17] and/or the star square[18] transformations, one reduces these models to Ising models with two-spin interactions (Sec. 2D). Thus the simple cubic dual model can be transformed to an Ising model with competing two-spin interactions. This model exhibits a singularity in the specific heat, but below the critical temperature there is no long range ordering of the spins (Sec. 4).

2. THE DUALITY TRANSFORMATION

The duality transformation for general Ising models is derived in this section. First (Sec. 2A) the Ising models with general interactions are defined, and some properties, like the degeneracy of the ground state and the spin correlation functions which vanish for all temperatures, are discussed. In Sec. 2B the duality relation for the partition function is stated and proved. The dislocation correlation functions are expressed both in terms of spin correlation functions of the original model and of the dual model in Sec. 2C. We show that a dual model exists for any Ising model (with positive interactions) and that this model can be reduced to an Ising model with only two-spin interactions and an external magnetic field (Sec. 2D).

A. The Model

The most general interaction of a system of N_s Ising spins $S(r) = \pm 1$, located at sites r of a lattice, is

$$H = -\textstyle\sum_b I(b) R(b), \qquad (2.1)$$

in which $I(b)$ is the coupling constant of the interaction bond labeled by the index b and

$$R(b) = \textstyle\prod_r S(r)^{\theta(r,b)}, \qquad \theta(r,b) \in \{0,1\}. \qquad (2.2)$$

We express all quantities which may assume two values by the two elements of the set $\{0,1\}$,

$$S(r) = (-1)^{\sigma(r)}, \qquad \sigma(r) \in \{0,1\}, \qquad (2.3)$$

$$R(b) = (-1)^{\rho(b)}, \qquad \rho(b) \in \{0,1\}. \qquad (2.4)$$

We define the field operations of addition (modulo 2)

$$0 \oplus 0 = 1 \oplus 1 = 0, \quad 0 \oplus 1 = 1 \oplus 0 = 1 \qquad (2.5)$$

and multiplication (modulo 2)

$$0\cdot 0 = 0\cdot 1 = 1\cdot 0 = 0, \quad 1\cdot 1 = 1 \qquad (2.6)$$

for the set $\{0,1\}$.

Then Eq. (2.2) can be written

$$\rho(b) = \oplus_r \theta(r,b)\,\sigma(r). \tag{2.7}$$

The operation symbol $\oplus$ with an index denotes summation (2.5) over this index. Let N_θ be the rank of the matrix $\theta(r,b)$. Then there are 2^{N_θ} different configurations[19] $\{\rho(b)\}$. We now restrict ourselves to systems with positive interaction constants, $I(b) > 0$. The ground states of the system (2.1) are defined by $R(b) = 1$ for all b. Therefore, the ground states are determined by the solutions $\{\sigma_0(r)\}$ of the homogeneous equations

$$\oplus_r \theta(r,b)\,\sigma_0(r) = 0 \quad \text{for all } b. \tag{2.8}$$

This system of equations has 2^{N_g} solutions with

$$N_g = N_s - N_\theta. \tag{2.9}$$

Therefore, the ground state is 2^{N_g}-fold degenerate. We associate the unitary operators

$$U\{\sigma_0\} = \Pi_r S_x(r)^{\sigma_0(r)} \tag{2.10}$$

with all ground states $\{\sigma_0(r)\}$. The operator $S_x(r)$ flips the spin at site r,

$$S_x^2(r) = 1, \quad S_x(r)S(r)S_x(r) = -S(r). \tag{2.11}$$

The operators U commute with all operators R:

$$U\{\sigma_0\}\,R(b)\,U\{\sigma_0\}^{-1} = R(b)(-1)^{\oplus_r\theta(r,b)\sigma_0(r)} = R(b). \tag{2.12}$$

Therefore, all the operators U commute with the Hamiltonian:

$$U\{\sigma_0\}\,HU\{\sigma_0\}^{-1} = H. \tag{2.13}$$

A product of spins $\Pi_r S(r)^{\psi(r)}$, $\psi(r) \in \{0,1\}$, is transformed by $U\{\sigma_0\}$ into

$$U\{\sigma_0\}\,\Pi_r S(r)^{\psi(r)} U\{\sigma_0\}^{-1} = \Pi_r S(r)^{\psi(r)}(-1)^{\oplus_r\psi(r)\sigma_0(r)} \tag{2.14}$$

This product of spins commutes with all operators U if and only if

$$\oplus_r \sigma_0(r)\psi(r) = 0 \tag{2.15}$$

for all configurations $\{\sigma_0(r)\}$. There are $2^{N_s-N_g} = 2^{N_\theta}$ solutions $\{\psi(r)\}$, since the configurations $\{\sigma_0(r)\}$ form an N_g-dimensional linear manifold.

The product of operators $\Pi_b R(b)^{\phi(b)}$, $\phi(b) \in \{0,1\}$, can be expressed as a product of spin operators,

$$\Pi_b R(b)^{\phi(b)} = \Pi_r S(r)^{\psi(r)} \tag{2.16}$$

with

$$\psi(r) = \oplus_b \theta(r,b)\,\phi(b). \tag{2.17}$$

Since the rank of the matrix $\theta(r,b)$ is N_θ, the products of $R(b)$ in Eq. (2.16) represent 2^{N_θ} different products of spin operators characterized by the sets $\{\psi(r)\}$ of Eq. (2.17). The products of Eq. (2.16) commute with all operators U. Since there are only 2^{N_θ} different spin products which commute with all U, it follows that a product of spin operators commutes with all operators U if and only if it is a product of operators R. A product of spin operators which does not commute with all operators $U\{\sigma_0\}$ vanishes, since from

$$U\Pi_r S(r)^{\psi(r)} U^{-1} = -\Pi_r S(r)^{\psi(r)} \tag{2.18}$$

and from Eq. (2.13) it follows that

$$\langle\Pi_r S(r)^{\psi(r)}\rangle = \langle\Pi_r S(r)^{\psi(r)} U^{-1}U\rangle = \langle U\Pi_r S(r)^{\psi(r)} U^{-1}\rangle = -\langle\Pi_r S(r)^{\psi(r)}\rangle = 0. \tag{2.19}$$

Therefore, the expectation value of a product of spin operators vanishes if this product cannot be represented by a product of operators R.

It follows from Eqs. (2.16) and (2.17) that those products of operators R which are unity for each spin configuration $\{S(r)\}$ are determined by the $2^{N_b-N_\theta}$ solutions $\{\phi_0(b)\}$ of the system of homogeneous equations

$$\oplus_b \theta(r,b)\,\phi_0(b) = 0 \quad \text{for all } r. \tag{2.20}$$

Hereafter we will call any product of operators which is unity for each spin configuration the unit element.

B. The Duality Relation for the Partition Functions

We call two Ising models which are characterized by matrices $\theta(r,b)$, $\theta^*(r^*,b)$ and coupling parameters $K(b) = \beta I(b)$, $K^*(b) = \beta^* I^*(b)$ (β and β^* are the inverse temperatures of these systems) dual to each other if they fulfill these three conditions:

(a) the closure condition

$$\oplus_b \theta(r,b)\,\theta^*(r^*,b) = 0 \tag{2.21}$$

for all pairs of r, r^*,

(b) the completeness relation

$$N_\theta + N_\theta^* = N_b, \tag{2.22}$$

in which N_θ and N_θ^* are the ranks of the matrices θ and θ^* and N_b is the number of bonds b, and

(c)

$$\tanh K(b) = e^{-2K^*(b)} \tag{2.23}$$

for all bonds b.

The symmetric partition functions $Y\{K\}$ and $Y^*\{K^*\}$,

$$Y\{K\} = Z\{K\}\,2^{-(N_s+N_g)/2}\,\Pi_b[\cosh 2K(b)]^{-1/2}, \tag{2.24}$$

$$Z\{K\} = \sum_{\{S(r)\}} e^{-\beta H\{S\}}, \tag{2.25}$$

(and similarly for $Y^*\{K^*\}$) of two dual Ising models obey

$$Y\{K\} = Y^*\{K^*\}. \tag{2.26}$$

For the particular case of a planar Ising model without crossing bonds, this relation was proved by Wannier.[3] We prove now Eq. (2.26) for the general case, comparing the high-temperature expansion for Z with the low-temperature expansion for Z^*. From

$$Z\{K\} = \sum_{\{S(r)\}} e^{-\beta H\{S\}} = \sum_{\{S(r)\}} \Pi_b e^{K(b)R(b)} \tag{2.27}$$

one obtains

$$Z\{K\} = \Pi_b \cosh K(b) \sum_{\{\varphi(b)\}} \Pi_b \tanh K(b)^{\varphi(b)} \times \sum_{\{S(r)\}} \Pi_b R(b)^{\varphi(b)}, \tag{2.28}$$

since

$$e^{K(b)R(b)} = \cosh K(b)[1 + R(b)\tanh K(b)] = \cosh K(b) \sum_{\{\varphi(b)\}} \tanh K(b)^{\varphi(b)} R(b)^{\varphi(b)} \tag{2.29}$$

follows from $R(b) = \pm 1$.

If the product of the operators R in Eq. (2.28) is the unit element, then the sum over all spin configurations yields 2^{N_s}; otherwise the sum vanishes. The product of the operators R is the unit element for all sets $\{\phi_0(b)\}$ of Eq. (2.20) and only these sets. Therefore, it follows that

$$Z\{K\} = 2^{N_s} \Pi_b \cosh K(b) \sum_{\{\varphi_0(b)\}} \Pi_b \tanh K(b)^{\varphi_0(b)}. \tag{2.30}$$

The partition function $Z^*\{K^*\}$ can be written

$$Z^*\{K^*\} = \sum_{\{S(r^*)\}} e^{-\beta H^*\{S\}} = \sum_{\{S(r^*)\}} \Pi_b e^{K^*(b)R^*(b)} = \Pi_b e^{K^*(b)} \sum_{\{S(r^*)\}} \Pi_b e^{-2K^*(b)\rho^*(b)}, \tag{2.31}$$

since $R^*(b) = 1 - 2\rho^*(b)$. From the closure condition (2.21), one obtains

$$\oplus_b \theta(r,b)\,\rho^*(b) = \oplus_b \oplus_{r^*} \theta(r,b)\theta^*(r^*,b)\sigma(r^*) = 0. \tag{2.32}$$

Therefore, each set $\{\rho^*(b)\}$ obeys Eq. (2.20) with $\rho^*(b) = \phi_0(b)$. It follows that

$$Z^*\{K^*\} = \Pi_b e^{K^*(b)} \sum_{\{\varphi_0(b)\}} N\{\phi_0\}\ \Pi_b e^{-2K^*(b)\varphi_0(b)}. \tag{2.33}$$

Here $N\{\phi_0\}$ denotes the number of configurations $\{S(r^*)\}$ which obey

$$\phi_0(b) = \oplus_{r^*} \theta(r^*,b)\sigma(r^*) \quad \text{for all } b. \tag{2.34}$$

If for a given set $\{\phi_0(b)\}$ Eq. (2.34) has no solutions, then $N\{\phi_0\} = 0$; otherwise $N\{\phi_0\} = 2^{N_s*-N_0*}$. In particular, for $\beta^* = 0$ it follows that

$$Z^* = 2^{N_s^*} = \sum_{\{\varphi_0(b)\}} N\{\phi_0\}. \tag{2.35}$$

There are $2^{N_b-N_\theta} = 2^{N_\theta^*}$ sets $\{\phi_0(b)\}$ [we used the completeness relation (2.22)]. Therefore all N obey $N\{\phi_0\} = 2^{N_s*-N_\theta*}$. From Eq. (2.33) one obtains

$$Z^*\{K^*\} = 2^{N_g^*} \Pi_b e^{K^*(b)} \sum_{\{\varphi_0(b)\}} \Pi_b e^{-2K^*(b)\varphi_0(b)}. \tag{2.36}$$

From Eqs. (2.30) and (2.36) one obtains $Y\{K\} = Y^*\{K^*\}$, Eq. (2.26), using Eq. (2.23).

If the completeness relation (2.22) is not fulfilled, but

$$N_b - N_\theta - N_\theta^* = N_m > 0, \tag{2.37}$$

and if all $K(b)$ and $K^*(b)$ are positive, then it follows from Eq. (2.33) that

$$Z^*\{K^*\} \le 2^{N_g^*} \Pi_b e^{K^*(b)} \sum_{\{\varphi_0(b)\}} \Pi_b e^{-2K^*(b)\varphi_0(b)}. \tag{2.38}$$

Using the analogous inequality for $Z\{K\}$, one obtains the inequality

$$2^{-N_m/2}\, Y\{K\} \le Y^*\{K^*\} \le 2^{N_m/2}\, Y\{K\}. \tag{2.39}$$

C. Dislocations

We now consider systems with magnetic dislocations. Let the operator $M(b)$ change the sign of the interaction constant $I(b)$ in the Hamiltonian. Then one obtains

$$\langle \Pi_b M(b)^{\varphi^*(b)} \rangle \{K\} = \langle \Pi_b e^{-2\varphi^*(b)K(b)R(b)} \rangle = \langle \Pi_b [\cosh 2K(b) - R(b)\sinh 2K(b)]^{\varphi^*(b)} \rangle \tag{2.40}$$

and

$$\langle \Pi_b M(b)^{\varphi^*(b)} \rangle \{K\} = Z\{(-1)^{\varphi^*} K\}/Z\{K\} = Y\{(-1)^{\varphi^*} K\}/Y\{K\} \tag{2.41}$$

with $\phi^*(b) \in \{0,1\}$. From $\tanh K = e^{-2K^*}$, Eq. (2.23), it follows that

$$\tanh (-1)^{\varphi^*} K = e^{-2K^* - i\pi\varphi^*}. \tag{2.42}$$

Substituting Eq. (2.26) into (2.41) and using (2.42), one obtains

$$\langle \Pi_b M(b)^{\varphi^*(b)} \rangle \{K\} = Y^*\{K^* + \tfrac{1}{2} i\pi\phi^*\}/Y^*\{K^*\} = i^{-\Sigma_b \varphi^*(b)} \langle \Pi_b e^{i\pi\varphi^*(b)R^*(b)/2} \rangle \{K^*\} = \langle \Pi_b R^*(b)^{\varphi^*(b)} \rangle \{K^*\}. \tag{2.43}$$

Therefore, the expectation value of a product of dislocation operators equals the expectation value of the corresponding product of operators R^* in the dual lattice. Since R^* is a product of spin operators $S(r^*)$, one may introduce corresponding operators $\mu(r^*)$ in the original model and represent $M(b)$ by

$$M(b) = \Pi_{r^*} \mu(r^*)^{\theta^*(r^*,b)}, \qquad \mu^2(r^*) = 1. \tag{2.44}$$

Then one obtains

$$\langle \Pi_{r^*}\mu(r^*)^{\nu(r^*)}\rangle \{K\} = \langle \Pi_{r^*} S(r^*))^{\psi(r^*)}\rangle \{K^*\}. \tag{2.45}$$

For the particular case of the two-dimensional Ising model without crossing interactions this was derived by Kadanoff and Ceva.[7]

D. Construction of a Dual Ising Model: Reduction to Two-Spin Interactions

A dual Ising model exists for any given Ising model (with positive interactions) of Eq. (2.1). To obtain this dual model, one has to find a complete set of solutions $\{\phi_0(b)\}$ of Eq. (2.20). This set is complete if each solution $\{\phi_0(b)\}$ of Eq. (2.20) is a linear combination of the solutions of that set. Associate with each solution of the set a point $r^*\{\phi_0\}$. Then the lattice which is defined by the matrix

$$\theta^*(r^*\{\phi_0\}, b) = \phi_0\{b\} \tag{2.46}$$

is dual to the original lattice.

The Hamiltonian of the dual lattice may contain products of a large number of spins $S(r^*)$. We list three transformations[20] which reduce these systems with many-spin interactions to Ising models with two-spin interactions and possibly a magnetic field.

The Decoration Transformation[15,16]

The interaction $-lR_1R_2$, in which R_1 and R_2 are products of spins, can be reduced to an interaction $-I_1R_1S - I_2R_2S$, in which S is a new spin or a product of new spins

$$e^{KR_1R_2} = \tfrac{1}{2} f \textstyle\sum_S e^{K_1R_1S + K_2R_2S} \tag{2.47}$$

with

$$f^2 = [\cosh(K_1 + K_2)\cosh(K_1 - K_2)]^{-1}, \tag{2.48}$$

$$\tanh K = \tanh K_1 \tanh K_2. \tag{2.49}$$

This transformation reduces products of more than three spins in the Hamiltonian to products of three spins.

A Generalized Triangle Transformation[3-5,17]

An interaction $-IS_1S_2S_3$ can be reduced to two-spin interactions and an interaction with a magnetic field by the transformation

$$\exp(KS_1S_2S_3) = \tfrac{1}{2} f \textstyle\sum_S \exp[K_0S + (K_1S + K_2) \times (S_1 + S_2 + S_3) + K_3(S_1S_2 + S_1S_3 + S_2S_3)] \tag{2.50}$$

with

$$f^8 = f_0^{-1} f_1^{-3} f_2^{-3} f_3^{-1}, \quad e^{8K} = f_0^{-1} f_1^3 f_2^{-3} f_3, \tag{2.51}$$

$$e^{8K_2} = f_0 f_1 f_2^{-1} f_3^{-1}, \quad e^{8K_3} = f_0^{-1} f_1 f_2 f_3^{-1},$$
$$f_n = \cosh[K_0 + (2n-3)K_1]. \tag{2.52}$$

For the particular choice $K_1 = -K_0$ the equations simplify to

$$f^8 = [\cosh^4(2K_0)\cosh(4K_0)]^{-1},$$
$$e^{8K} = \cosh^4(2K_0)/\cosh(4K_0),$$
$$e^{8K_2} = e^{-8K_3} = \cosh(4K_0). \tag{2.53}$$

A Star Square Transformation[18]

If the Hamiltonian is invariant under flipping of all spins, then one may prefer to conserve this invariance. Products of more than four spins in the interaction can be reduced to four-spin interactions by the decoration transformation (2.47). The four-spin interactions are reduced to two-spin interactions by a star square transformation

$$\exp(KS_1S_2S_3S_4) = \tfrac{1}{2} f \textstyle\sum_S \exp[K_0S(S_1 + S_2 + S_3 + S_4) + K_1(S_1S_2 + S_1S_3 + S_1S_4 + S_2S_3 + S_2S_4 + S_3S_4)] \tag{2.54}$$

with

$$e^{8K} = \cosh(4K_0)/\cosh^4(2K_0), \tag{2.55a}$$

$$e^{-8K_1} = \cosh(4K_0), \tag{2.55b}$$

$$f^8 = 1/[\cosh(4K_0)\cosh^4(2K_0)]. \tag{2.55c}$$

For real K_0 the right-hand side of Eq. (2.55a) is less than or equal to 1. Therefore, K must be negative or zero. To obtain negative K's, one may apply the decoration transformation with negative K_1 and K_2, Eq. (2.49).

Therefore, we have shown that there exists a dual Ising model (2.46) to any Ising model and that this can be reduced to an Ising model with only two-spin interactions and possibly an interaction with a magnetic field.

3. THE MODELS M_{dn} AND THEIR PROPERTIES

In this section we consider the models M_{dn}. In Sec. 3A we define the models and derive the duality relations which relate the systems M_{dn} and $M_{d\,d-n+1}$ in an external magnetic field and the duality relation between the systems M_{dn} and $M_{d\,d-n}$ without an external magnetic field. The behavior of the spin correlation functions at high and low temperatures is discussed in Sec. 3B. We prove that there is no local order parameter in the systems with $n > 1$. In Sec. 3C we discuss the thermodynamic properties of the systems.

A. The Models, Duality

We consider a d-dimensional hypervolume divided into C_d hypercells $B^{(d)}$. These are bounded by $(d-1)$-dimensional hypercells $B^{(d-1)}$ (total number C_{d-1}), these again by $(d-2)$-dimensional hypercells

$B^{(d-2)}$ (total number C_{d-2}), and so on, until we arrive at 0-dimensional hypercells which are simply the C_0 corners $B^{(0)}$ of the d-dimensional hypercells. For this original lattice L, one may construct a dual lattice L^* by placing one dual corner $B^{(0)*}$ in each original hypercell $B^{(d)}$, then connecting the dual corners by dual edges $B^{(1)*}$, each of which intersects one hypercell $B^{(d-1)}$, then connecting these dual edges by dual faces $B^{(2)*}$, each of which intersects one hypercell $B^{(d-2)}$, and so on, until we obtain the d-dimensional dual hypercells $B^{(d)*}$, each of which contains one original corner $B^{(0)}$. Denoting the number of the m-dimensional hypercells by C^*_m, we obtain

$$C^*_m = C_{d-m}, \qquad (3.1)$$

since by construction there is a one-to-one correspondence of the m-dimensional dual hypercells to the $(d-m)$-dimensional original hypercells. Let us denote the intersection point of $B^{(m)}$ and its dual hypercell $B^{(d-m)*}$ by $r^{(m)} = r^{(d-m)*}$. Then a hypercell $B(r^{(m)})$ and a dual hypercell $B^*(r^{(m)})$ is associated with each point $r^{(m)}$.

Let us consider some examples. A *linear chain* ($d = 1$, Fig. 1) of points $r^{(0)} = i$ (black circles) (we denote integers by i, j, k) divides the line into one-dimensional cells (segments). The dual lattice consists of the segments between the points $r^{(1)} = i + \frac{1}{2}$ (open circles). The *square lattice* ($d = 2$, Fig. 2) consists of the squares bounded by the continuous lines; its dual lattice consists of the squares bounded by the broken lines. The corners $r^{(0)} = (i, j)$ of the original lattice are denoted by black circles, the corners $r^{(2)} = (i + \frac{1}{2}, j + \frac{1}{2})$ of the dual lattice are denoted by open circles and the edges of the original lattice intersect the edges of the dual lattice at the points $r^{(1)}$ (triangles). In Fig. 3 a cube of the original *cubic lattice* ($d = 3$) and a cube of the dual lattice are drawn. The corners $r^{(0)} = (i, j, k)$ of the original lattice are denoted by black circles and the corners $r^{(3)} = (i + \frac{1}{2}, j + \frac{1}{2}, k + \frac{1}{2})$ by open circles. The edges (continuous lines) of the original lattice and the faces of the dual lattice intersect at points $r^{(1)}$ (open squares), whereas the faces of the original lattice and the edges (broken lines) of the dual lattice intersect at points $r^{(2)}$ (black squares). We have considered only self-dual lattices, that is, lattices which are topologically equivalent to their dual lattice. Not all lattices are self-dual.

Now let us return to the general case and introduce the functions Θ and Θ^*. Let $\Theta(r^{(m-1)}, r^{(m)}) = 1$, if $r^{(m-1)}$ lies on the boundary of $B(r^{(m)})$; otherwise, $\Theta(r^{(m-1)}, r^{(m)}) = 0$. Let $\Theta^*(r^{(m)}, r^{(m-1)}) = 1$ if $r^{(m)}$ lies on the boundary of $B^*(r^{(m-1)})$; otherwise

$$\Theta^*(r^{(m)}, r^{(m-1)}) = \Theta(r^{(m-1)}, r^{(m)}), \qquad (3.2)$$

that is, if $r^{(m-1)}$ lies on the boundary of $B(r^{(m)})$, then $r^{(m)}$ lies on the boundary of $B^*(r^{(m-1)})$.

Since the m-dimensional boundaries of $B^{(m+1)}$ form a *closed* m-dimensional hypersurface, two m-dimensional boundaries $B^{(m)}$ of $B^{(m+1)}$ meet in each $(m-1)$-dimensional hypercell at the boundary of $B^{(m+1)}$. Therefore, one obtains

$$\sum_{r^{(m)}} \Theta(r^{(m-1)}, r^{(m)})\Theta(r^{(m)}, r^{(m-1)}) = 0. \qquad (3.3)$$

The Ising model M_{dn} on the lattice L with n dimensional bonds consists of Ising spins $S(r) = \pm 1$ at all sites $r = r^{(n-1)}$ interacting via

$$-\beta H_{dn} = K\sum_{r^{(n)}} \Pi_r S(r)^{\theta(r, r^{(n)})} + h\sum_r S(r). \qquad (3.4)$$

The product in the first term of the Hamiltonian runs over all spins $S(r)$ lying on the boundary of the n-dimensional hypercell $B(r^{(n)})$. For $n = 1$, the model (3.4) describes the Ising model with two-spin interactions between spins lying at the two ends of an edge and an external magnetic field

$$B = k_B \, \text{Th}/\mu_B. \qquad (3.5)$$

FIG. 1. The linear chain.

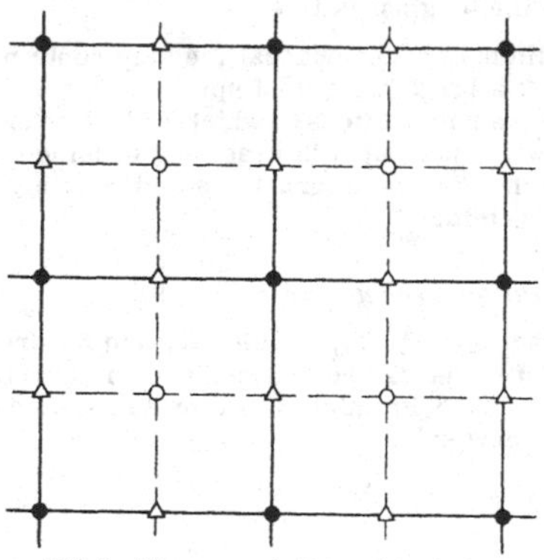

FIG. 2. The square lattice and its dual.

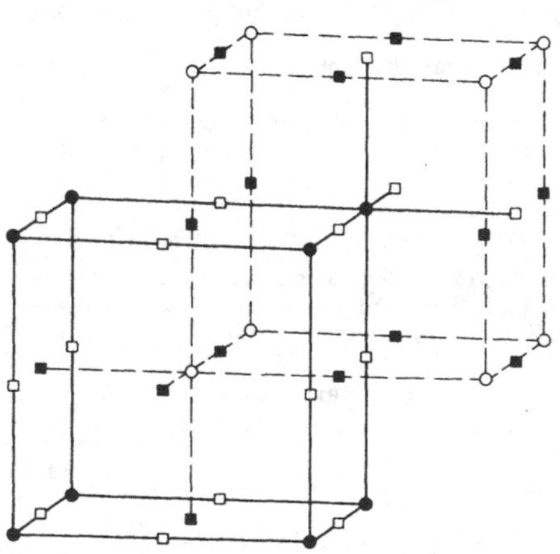

FIG. 3. The simple cubic lattice and its dual.

For the lattices considered above we obtain the Hamiltonians

$$-\beta H_{11} = K\sum_i S(i)S(i+1) + h\sum_i S(i), \tag{3.6}$$

$$-\beta H_{21} = K\sum_{ij} S(i,j)[S(i+1,j) + S(i,j+1)] + h\sum_{ij} S(i,j), \tag{3.7}$$

$$-\beta H_{31} = K\sum_{ijk} S(i,j,k)[S(i+1,j,k) + S(i,j+1,k) + S(i,j,k+1)] + h\sum_{ijk} S(i,j,k). \tag{3.8}$$

For $n = 2$ the Hamiltonian contains the products of spins lying at the boundary of the faces $B^{(2)}$,

$$\begin{aligned}-\beta H_{22} = {} & K\sum_{ij} S(i,j+\tfrac{1}{2})S(i+1,j+\tfrac{1}{2}) \times S(i+\tfrac{1}{2},j)S(i+\tfrac{1}{2},j+1) \\ & + h\sum_{ij}[S(i,j+\tfrac{1}{2}) + S(i+\tfrac{1}{2},j)],\end{aligned} \tag{3.9}$$

$$\begin{aligned}-\beta H_{32} = {} & K\sum_{ijk}[S(i,j+\tfrac{1}{2},k)S(i+1,j+\tfrac{1}{2},k) \\ & \times S(i+\tfrac{1}{2},j,k)S(i+\tfrac{1}{2},j+1,k) \\ & + S(i,j,k+\tfrac{1}{2})S(i+1,j,k+\tfrac{1}{2})S(i+\tfrac{1}{2},j,k) \\ & \times S(i+\tfrac{1}{2},j,k+1) + S(i,j,k+\tfrac{1}{2}) \\ & \times S(i,j+1,k+\tfrac{1}{2})S(i,j+\tfrac{1}{2},k)S(i,j+\tfrac{1}{2},k+1)] \\ & + h\sum_{ijk}[S(i,j,k+\tfrac{1}{2}) + S(i,j+\tfrac{1}{2},k) \\ & + S(i+\tfrac{1}{2},j,k)].\end{aligned} \tag{3.10}$$

For $n = 3$ the Hamiltonian contains the products of spins lying at the boundary of the volumes $B^{(3)}$:

$$\begin{aligned}-\beta H_{33} = {} & K\sum_{ijk} S(i,j+\tfrac{1}{2},k+\tfrac{1}{2})S(i+1,j+\tfrac{1}{2},k+\tfrac{1}{2}) \\ & \times S(i+\tfrac{1}{2},j,k+\tfrac{1}{2})S(i+\tfrac{1}{2},j+1,k+\tfrac{1}{2}) \\ & \times S(i+\tfrac{1}{2},j+\tfrac{1}{2},k)S(i+\tfrac{1}{2},j+\tfrac{1}{2},k+1) \\ & + h\sum_{ijk}[S(i,j+\tfrac{1}{2},k+\tfrac{1}{2}) \\ & + S(i+\tfrac{1}{2},j,k+\tfrac{1}{2}) + S(i+\tfrac{1}{2},j+\tfrac{1}{2},k)].\end{aligned} \tag{3.11}$$

In general, the model M_{dn} on a hypercubic lattice consists of $N_s = \binom{d}{n-1}N$ Ising spins located at the centers of the $(n-1)$-dimensional hypercubes. (N is the number of the d-dimensional hypercubes). The Hamiltonian consists of the sum of the products of the $2n$ spins at the $(n-1)$-dimensional hypersurfaces of the $N_b = \binom{d}{n}N$ hypercubes $B^{(n)}$. Let us denote a subset of n unit vectors e_i along the main axes by E_n; then the model M_{dn} for the d-dimensional hypercubic lattice is defined by

$$-\beta H_{dn} = K\sum_{r^{(0)},E_n} R(r^{(0)},E_n) + h\sum_{r^{(0)},E_{n-1}} S(r^{(0)} + v(E_{n-1})) \tag{3.12}$$

with

$$v(E_n) = \tfrac{1}{2}\sum_{e\in E_n} e, \tag{3.13}$$

$$R(r^{(0)},E_n) = \prod_{e\in E_n} S(r^{(0)} + v(E_n) - \tfrac{1}{2}e) \times S(r^{(0)} + v(E_n) + \tfrac{1}{2}e). \tag{3.14}$$

Similarly one defines the Ising model M^*_{dn} on the dual lattice L^*. The Ising spins $S(r^*) = \pm 1$ are located at the sites $r^* = r^{(n-1)*} = r^{(d-n+1)}$ and interact via

$$-\beta^* H^*_{dn} = K^*\sum_{r^{(n)*}}\prod_{r^*} S(r^*)^{\upsilon^*(r^*,r^{(n)*})} + h^*\sum_{r^*} S(r^*). \tag{3.15}$$

We now show that the models M_{dn} and $M^*_{d,d-n+1}$ are related by the duality relation

$$Y_{dn}(K,h) = Y^*_{d,d-n+1}(K^*,h^*) \tag{3.16}$$

with

$$\tanh K = e^{-2h^*}, \quad \tanh h = e^{-2K^*}. \tag{3.17}$$

If we label the interaction of the spin $S(r^{(n-1)})$ with the external magnetic field by $b(r^{(n-1)})$ and the interaction of the spins on the boundary of $B(r^{(n)})$ by $b(r^{(n)})$, then we have

$$\theta(r^{(n-1)},b(r^{(n)})) = \Theta(r^{(n-1)},r^{(n)}), \tag{3.18a}$$

$$\theta(r^{(n-1)},b(r^{(n-1)\prime})) = \delta_{r^{(n-1)},r^{(n-1)\prime}}, \tag{3.18b}$$

$$\theta^*(r^{(n)},b(r^{(n)\prime})) = \delta_{r^{(n)},r^{(n)\prime}}, \tag{3.19a}$$

$$\theta^*(r^{(n)},b(r^{(n-1)})) = \Theta^*(r^{(n)},r^{(n-1)}). \tag{3.19b}$$

Substituting Eqs. (3.18) and (3.19) into Eq. (2.21) and using Eq. (3.2), we find that the closure condition is fulfilled. From Eqs. (3.18b) and (3.19a) it follows that $N_\theta = N_s$ and $N^*_\theta = N^*_s$. Since $N_b = N_s + N^*_s$, the completeness relation is fulfilled.

We now compare the models M_{dn} and $M^*_{d\,d-n}$ without external magnetic fields. Then the bonds are connected with the sites $r^{(n)}$ by Eq. (3.18a) and

$$\theta^*(r^{(n+1)},b(r^{(n)})) = \Theta^*(r^{(n+1)},r^{(n)}) = \Theta(r^{(n)},r^{(n+1)}). \tag{3.20}$$

From Eq. (3.3) it follows that the closure condition is fulfilled. We now discuss the completeness relation (2.22). In the Appendix we derive relations between the N's, C's, and the topology of the lattice. Here we summarize the results: The exponents N_g and N^*_g of the orders of the degeneracy 2^{N_g} and $2^{N_g^*}$ of the models (3.4) and (3.15) are

$$N_g = t_g + \sum_{m=0}^{n-2}(-1)^{n-m}C_m, \tag{3.21}$$

$$N^*_g = t^*_g + \sum_{m=n+2}^{d}(-1)^{m-n}C_m, \tag{3.22}$$

in which t_g and t^*_g depend only on the boundary conditions and on n. From a generalization of Euler's theorem[21]

$$\sum_{m=0}^{d}(-1)^m C_m = t, \tag{3.23}$$

in which l depends only on the topology (boundary conditions), from

$$N_s = C_{n-1}, \quad N_b = C_n, \quad N_s^* = C_{n+1} \tag{3.24}$$

and from

$$N_m = N_g - N_s + N_b - N_s^* + N_g^*, \tag{3.25}$$

which is derived from Eqs. (2.9) and (2.37), it follows that

$$N_m = l_g + l_g^* + (-)^{d-n}\, l. \tag{3.26}$$

Therefore, N_m depends only on the topology of the system and on n. For a d-dimensional hypersurface wrapped on a $(d+1)$-dimensional hypersphere, one obtains $N_m = 0$ for $1 \le n \le d-1$. Therefore, the duality relation

$$Y_{dn}(K, 0) = Y^*_{d\,d-n}(K^*, 0), \tag{3.27}$$

with

$$\tanh K = e^{-2K^*}, \tag{3.28}$$

holds for this boundary condition. For the two-dimensional Ising model ($d = 2, n = 1$) this was shown in Ref. 3. For systems with periodic boundary conditions one obtains $N_m = \binom{d}{n}$. In the thermodynamic limit the factors $2^{N_m/2}$ in Eq. (2.39) can be neglected, and, using Eqs. (2.24), (2.25), and

$$-\beta F(K) = \ln Z(K), \tag{3.29}$$

we obtain for the free energy

$$\beta^* F^*_{d\,d-n}(K^*) = \beta F_{dn}(K) - \tfrac{1}{2}(N_g^* + N_s^* - N_g - N_s) \times \ln 2 + \tfrac{1}{2} N_b \ln \sinh 2K. \tag{3.30}$$

B. Correlation Functions

In this section we discuss the behavior of the spin correlation functions of the systems M_{dn} without an external magnetic field. We showed in Sec. 2 that an operator

$$U\{\sigma_0\} = \Pi_r S_x(r)^{\sigma_0(r)} \tag{2.10}$$

commutes with the Hamiltonian H and all operators R if

$$\oplus_r\, \theta(r, b)\, \sigma_0(r) = 0 \quad \text{for all } b. \tag{2.8}$$

The only solution for $n = 1$ besides the trivial solution $\sigma_0(r) = 0$ is

$$\sigma_0(r) = 1. \tag{3.31}$$

For $n > 1$ we obtain solutions

$$\sigma_0(r) = \Theta(r^{(n-2)}, r), \tag{3.32}$$

which can be verified using Eqs. (3.3) and (3.18a). Therefore, each operator R is invariant under flipping of all spins lying on the $(n-1)$-dimensional hypercells $B(r)$, which meet in the hypercell $B(r^{(n-2)})$. This leads to the high degeneracy 2^{N_g}, where N_g is given by Eq. (3.21). Since, for $r \neq r'$, there exists a neighbor $r^{(n-2)}$ of r with $\Theta(r^{(n-2)}, r) = 1$ and $\Theta(r^{(n-2)}, r') = 0$, we obtain from Eq. (2.19)

$$\langle S(r) S(r') \rangle = \delta_{rr'}. \tag{3.33}$$

Therefore, there is no long-range spin autocorrelation at any temperature. The only products of spins whose expectation values do not vanish can be represented by a product of operators R. These products are the products of all spins lying on the $(n-1)$-dimensional boundary of an n-dimensional hypervolume which consists of n-dimensional hypercells.

We now consider the long-range behavior of $\langle \Pi_r S(r) \rangle$ of the model M_{dn}, Eq. (3.12), where the spins of the product lie on the boundary of an n-dimensional hypercube. From the high temperature expansion it follows that

$$\langle \Pi_r S(r) \rangle = [\tanh K + 2(d-n)(\tanh K)^{1+2n} + \cdots]^v \quad \text{for } n > 1. \tag{3.34}$$

and

$$\langle \Pi_r S(r) \rangle = \tfrac{1}{2}\{\tanh K + [2(d-1)]^{1/2} \times (\tanh K)^2 + \cdots\}^v + \tfrac{1}{2}\{\tanh K - [2(d-1)]^{1/2} (\tanh K)^2 + \cdots\}^v \quad \text{for } n = 1, \tag{3.35}$$

where v is the volume of the hypercube (for $n = 1$, v is the distance between the two spins; for $n = 2$, v is the area of the square spanned by the spins). From the low temperature expansion one obtains

$$\langle \Pi_r S(r) \rangle = (1 - e^{-4(d-n+1)K} + \cdots)^f \quad \text{for } n < d, \tag{3.36}$$

$$\langle \Pi_r S(r) \rangle = (1 - 2e^{-2K} + \cdots)^v \quad \text{for } n = d, \tag{3.37}$$

where f is the hyperarea of the hypercube (for $n = 1$, f is the number of the ends of the line, that is, $f = 2$; for $n = 2$, f is the perimeter of the square). Therefore, we deduce that the behavior of these correlation functions in the limit of large hypercubes is different in the low and the high temperature phases, and we expect

$$\langle \Pi_r S(r) \rangle \propto \begin{cases} \exp[-v/v_0(T)] & \text{for } T > T_c, n < d, \quad (3.38) \\ \exp[-f/f_0(T)] & \text{for } T < T_c, n < d. \quad (3.39) \end{cases}$$

We attribute the qualitatively different assymptotic behavior in both temperature regions to different states of the system above and below a critical temperature T_c.

For $n = d - 1$ the different behavior in both temperature regions becomes more evident if one makes use of the duality relation for dislocations, Eq. (2.43). One obtains

$$\langle \Pi_r S(r) \rangle \{K\} = \langle \Pi_b R(b) \rangle \{K\} = \langle \Pi_b M^*(b) \rangle \{K\}, \tag{3.40}$$

where the product runs over all b's in the $(d-1)$-dimensional hypercube. The expectation value on the right-hand side of Eq. (3.40) is to be taken in the model M^*_{d1}. The logarithm of this expectation value is proportional to the change in free energy due to the dislocations. This free energy is proportional to the $(d-2)$-dimensional hyperarea of the boundary in the disordered state ($T^* > T^*_c$, that is, for $T < T_c$), and it is proportional to the $(d-1)$-dimensional hypervolume in the ordered state of the dual system ($T^* < T^*_c$, that is, for $T > T_c$). This is in agreement with Eqs. (3.38) and (3.39).

We now compare the systems M_{dn} and $M_{d+1,n}$ on a hypercubic lattice. From the theorem of Griffiths generalized by Kelly and Sherman[22] it follows that any expectation value $\langle \Pi_r S(r) \rangle$ in the system M_{dn} is less or equal to the expectation value in the system $M_{d+1,n}$,

$$\langle \Pi_r S(r) \rangle_d \le \langle \Pi_r S(r) \rangle_{d+1}, \tag{3.41}$$

since the $(d+1)$-dimensional system consists of layers of the system M_{dn} plus an additional interaction between the layers. Therefore, if this expectation value shows the long-range behavior, Eq. (3.39), for M_{dn}, then this long-range behavior is also apparent in $M_{d+1,n}$, and we obtain $T_{c,dn} \le T_{c,d+1,n}$, that is,

$$K_{c,d+1,n} \le K_{c,dn}. \tag{3.42}$$

The systems M_{dn} with $n > 1$ exhibit an unusually high ground-state entropy $S_0 \propto N$. Taking

$$S(r^{(0)} + v(E_{n-1})) = 1, \quad \text{if } e_d \in E_{n-1}, \tag{3.43}$$

in the hypercubic models (3.12), we may eliminate all spins with half-valued r_d-component. These systems (we denote them by M'_{dn}) consist of $N_s = N\binom{d-1}{n-1}$ spins and have a much smaller degeneracy,

$$N_g = \frac{N}{N_d}\binom{d-2}{n-2} + \binom{d-2}{n-1},$$

N_d denoting the length of the periodicity in the r_d direction.

For $n = d$ the system disintegrates into linear chains. For $n = 1$ the system is unchanged. For $d = 3, n = 2$ one obtains the Hamiltonian

$$\begin{aligned} -\beta H'_{32} = K \textstyle\sum_{ijk} [& S(i, j+\tfrac12, k) S(i+1, j+\tfrac12, k) \\ & \times S(i+\tfrac12, j, k) S(i+\tfrac12, j+1, k) \\ & + S(i+\tfrac12, j, k) S(i+\tfrac12, j, k+1) \\ & + S(i, j+\tfrac12, k) S(i, j+\tfrac12, k+1)]. \end{aligned} \tag{3.44}$$

These systems obey the closure condition (2.21) if one chooses the model $M^*_{d\,d-n}$ on the hypercubic lattice as the dual model. One obtains

$$N_m = \frac{N}{N_d}\binom{d-2}{n-2} + \binom{d-1}{n} + \binom{d-2}{n-1}. \tag{3.45}$$

Therefore, in the thermodynamic limit $N \to \infty$, $N_d \to \infty$, the duality relation (3.30) holds, and the free energies of M_{dn} and M'_{dn} show the same nonanalyticities. In the systems M'_{dn} the spins separated by a vector pointing in the e_d direction are correlated. At high temperatures one obtains

$$\begin{aligned} \langle S(r) S(r + r_d e_d) \rangle = & [\tanh K + 2(d-n) \\ & \times (\tanh K)^{1+2n} + \cdots]^{|r_d|} \quad \text{for } n > 1, \end{aligned} \tag{3.46}$$

and for low temperatures it follows that

$$\begin{aligned} \langle S(r) S(r + r_d e_d) \rangle = & (1 - e^{-4(d-n+1)K} \\ & + \cdots)^{2+2(n-1)|r_d|} \quad \text{for } n < d. \end{aligned} \tag{3.47}$$

Therefore, we expect an exponential decay of the correlation function for large r_d at all temperatures if $n > 1$. Here again we find no long range order.

Absence of a Local Order Parameter

A second-order phase transition with local order parameter is characterized as follows: Let us add local operators $\psi_i(r)$ to the Hamiltonian H_0,

$$-\beta H = K H_0 + \textstyle\sum_{ir} h_i \psi_i(r).$$

Then there is a discontinuity of the first-order derivatives with respect to h of the free energy $F(K, \{h\})$ along a ν'-dimensional hypervolume known as a first-order transition line in the ν-dimensional (K, h_i) space. This hypervolume is bounded by a $(\nu' - 1)$-dimensional λ-hypersurface commonly known as λ-point or λ-line, where the second-order phase transition takes place. Any local operator $\psi(r) = \sum h_i \psi_i(r)$ with a discontinuity of $\sum_r \langle \psi(r) \rangle = -\beta \Sigma_i h_i \partial F / \partial h_i$ along the first-order transition can be considered as an order parameter. In the homogeneous phase the limit

$$\lim_{r \to \infty} [\langle \psi(0) \psi(r) \rangle - \langle \psi(0) \rangle \langle \psi(r) \rangle] \tag{3.48}$$

vanishes. If the expectation values in expression (3.48) are averaged over *all* states along a first-order transition, then the limit (3.48) does not vanish. In the Ising model ($n = 1$) with ferromagnetic interactions $\psi(r) = S(r)$ is such a local operator. For $T = 0$ we have $\langle S(0) S(r) \rangle = 1$, whereas $\langle S(r) \rangle = 0$. In the models M_{dn} with $n > 1$ there is no first-order transition for $T < T_{c,dn}$, $h_i = 0$ associated with a local order parameter $\langle\psi(r)\rangle$ if we confine ourselves to operators which are polynomials of spin operators located in a finite region about r. We can see this as follows: Any product of spins $S(r)$ which do not lie on a closed $(n-1)$-dimensional hypersurface of hypercells $B^{(n-1)}$ gives vanishing contributions for a sufficiently large distance r. Therefore, we may confine ourselves to expressions for ψ, which are polynomials of $R(b)$,

$$\psi(r) = P(r; R(b)). \tag{3.49}$$

Applying Eqs. (2.40) and (2.43), one obtains

$$[\langle\psi(0)\psi(r)\rangle - \langle\psi(0)\rangle\langle\psi(r)\rangle](K) = [\langle\psi^*(0)\psi^*(r)\rangle - \langle\psi^*(0)\rangle\langle\psi^*(r)\rangle](K^*) \tag{3.50}$$

with

$$\psi^*(r) = P(r; \cosh 2K^* - R^*(b)\sinh 2K^*). \tag{3.51}$$

Therefore, the correlation of the ψ's in the model M_{dn} below T_c is related to the correlation of the ψ^*'s in the dual model $M^*_{d\,d-n}$ above T_c^*. According to the cluster property of the Ising model, proved rigorously by Ruelle[23] for Ising models with $n = 1$, the right-hand side of Eq. (3.49) vanishes for $r \to \infty$. Therefore, there is no first-order transition characterized by a local order parameter in the models M_{dn} with $n > 1$ along the K axis.[24]

C. Thermodynamic Properties

In this section we consider the thermodynamic properties of the systems M_{dn}.

The Linear Chain M_{11}

The partition function of the linear chain (3.6) of N Ising spins with nearest neighbor interaction and the periodic boundary condition $S(N+1) = S(1)$ can be calculated[5] explicitly:

$$Z(K,h) = \mathrm{tr}\begin{pmatrix} e^{K+h} & e^{-K} \\ e^{-K} & e^{K-h}\end{pmatrix}^N. \tag{3.52}$$

With Eq. (2.24),

$$Y(K,h) = f^N Z(K,h), \qquad f = (2\cosh 2K \cosh 2h)^{-1/2}, \tag{3.53}$$

one obtains

$$Y(K,h) = \mathrm{tr}\begin{pmatrix} fe^{K+h} & fe^{-K} \\ fe^{-K} & fe^{K-h}\end{pmatrix} = F_N(fe^K\cosh h, f^2(e^{2K} - e^{-2K})). \tag{3.54}$$

The first argument of F_N is half the trace of the 2×2 matrix; the second argument is its determinant. It follows that

$$F_N(t,d) = (t + \sqrt{t^2 - d})^N + (t - \sqrt{t^2-d})^N, \tag{3.55}$$

with

$$t = [2(1+e^{-4K})(1+\tanh^2 h)]^{-1/2}, \tag{3.56}$$

$$d = \tfrac{1}{2}(1-e^{-4K})(1-\tanh^2 h)(1+e^{-4K})^{-1} \times (1+\tanh^2 h)^{-1}. \tag{3.57}$$

Since the linear chain is a self-dual lattice, one obtains from Eq. (3.16) that

$$Y_{11}(K,h) = Y_{11}(K^*,h^*) \tag{3.58}$$

with Eq. (3.17), which is fulfilled since t and d, Eqs. (3.56) and (3.57), are invariant under this transformation.

The Models M_{dd}

The partition functions $Z_{dd}(K,0)$ of the models M_{dd} without external magnetic field can be calculated from the duality relation (3.16), (3.17):

$$Y_{dd}(K,0) = Y^*_{d1}(\infty, h^*), \quad \tanh h^* = e^{-2K}. \tag{3.59}$$

Since in the model M^*_{d1} all spins are coupled by a two-spin interaction with infinite K^*, only the two configurations $\{S(r^*) = 1\}$ and $\{S(r^*) = -1\}$ contribute

$$Z^*_{d1}(K^*,h^*) \sim \exp(N_b K^* + N^*_s h^*) + \exp(N_b K^* - N^*_s h^*). \tag{3.60}$$

It follows that

$$Z_{dd}(K,0) = 2^{N_s}[(\cosh K)^{N_b} + (\sinh K)^{N_b}]. \tag{3.61}$$

The partition functions of the models M_{dd} are analytic in K for all finite K and $h = 0$. Since the Ising model $M^*_{d1}, d > 1$, shows a phase transition for $h^* = 0$ at $K^* = K^*_{c,d1}$, a nonanalyticity is apparent in the partition function $Y_{dd}(K,h)$ for $K \to \infty$ at $h = -\frac{1}{2}\ln[\tanh(K^*_{c,d1})]$.

The Models M_{dn} with $n < d$ without External Magnetic Field

The nonanalyticity which is apparent in the free energy F_{d1} at the critical $K_{c,d1} = \beta_c J$ (β_c is the inverse critical temperature) also occurs in the free energy $F^*_{d\,d-1}$ [Eq. (3.30)]. Since the correlation functions, Eqs. (3.38) and (3.39), show a qualitatively different assymptotic behavior at low and high temperatures, we expect a phase transition for all infinite systems M_{dn} with $1 \le n < d$ at some $K = K_{c,dn}$ accompanied by a nonanalyticity of the free energy. The critical K's of the model and its dual model are related by Eq. (3.28), which can be cast in the symmetric form

$$\sinh 2K_{c,dn} \sinh 2K^*_{c,d\,d-n} = 1. \tag{3.62}$$

In particular, for self-dual lattices like the hypercubic lattice, one obtains

$$\sinh 2K_{c,dn} \sinh 2K^*_{c,d\,d-n} = 1. \tag{3.63}$$

TABLE I. Critical parameters of some three-dimensional Ising models and their dual models.

original lattice	diamond	simple cubic	body-centered cubic	face-centered cubic
K_c	0.3698	0.2217	0.1575	0.1021
K_c^*	0.5195	0.7613	0.9284	1.1426
E_c/E_0	0.432	0.3284	0.270	0.245
E_c^*/E_0^*	0.937	0.9495	0.964	0.971
S_c/S_∞	0.737	0.808	0.845	0.853
$(S_c^* - S_0^*)/(S_\infty^* - S_0^*)$	0.100	0.092	0.072	0.063

For a self-dual model ($d = 2$, self-dual lattice) it follows that

$$K_{c,d,d/2} = \tfrac{1}{2}\ln(\sqrt{2} + 1). \tag{3.64}$$

We derived the inequality $K_{c,d+1,n} \leq K_{c,dn}$ for hypercubic lattices, Eq. (3.42). From Eq. (3.63) one obtains

$$K_{c,d,n-1} \leq K_{c,d+1,n}, \tag{3.65}$$

and from Eq. (3.42) and (3.65) it follows that

$$K_{c,d,n-1} \leq K_{c,d,n}. \tag{3.66}$$

The critical temperature of the hypercubic systems is a decreasing function of n.

Since the duality relation (3.30) relates the free energy F_{dn} at high temperatures to the free energy $F^*_{d\,d-n}$ of its dual model at low temperatures, we deduce that the critical exponent $\alpha_{d\,d-n}$ of the specific heat of the model $M^*_{d\,d-n}$ above T_c^* is

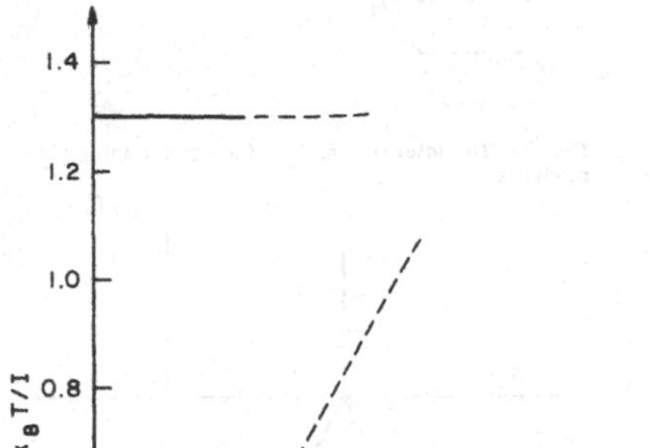

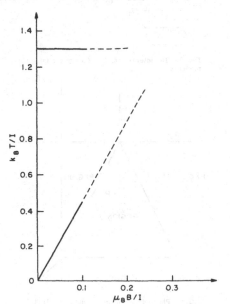

FIG. 4. Phase diagram of the system (3.10).

given by the critical exponent α'_{dn} of the model M_{dn} below T_c and vice versa:

$$\alpha_{d\,d-n} = \alpha'_{dn}. \tag{3.67}$$

Therefore, any asymetry in the specific heat of the model M_{dn} near T_c is also apparent in the specific heat of the dual model, but the high temperature and the low temperature regions are interchanged. Self-dual systems exhibit a symmetric singularity of the specific heat around the critical temperature.

From the thermodynamic relation

$$E = \frac{\partial(\beta F)}{\partial \beta} = \frac{\partial(KF)}{\partial K}, \tag{3.68}$$

it follows that

$$E^*(K^*)/E_0^* = \cosh(2K) - \sinh(2K)E(K)/E_0, \tag{3.69}$$

in which E_0 denotes the ground state energy $E_0 = -IN_b$. Therefore, using Eqs. (3.30), (3.68), and

$$F = E - TS, \tag{3.70}$$

one is able to calculate the energy E^* and the entropy S^* of the dual model from E and S. From the critical parameters[13] of the Ising model on the diamond, the simple cubic, the body-centered cubic, and the face-centered cubic lattice, we have calculated the critical parameters of their dual models. The results are listed in Table I. The binding energies of the dual models at critical temperature are unusually large [for example, 95% of the ground state energy for the model (3.10)]. This is in agreement with the unusually low critical entropy. For the model (3.10) we obtain $S_c^*/k_B N = 0.82$ which is to be compared with the zero temperature entropy $S_0^*/k_B N = \ln 2 = 0.69$ and the entropy at infinite temperature $S_\infty^*/k_B N = 3\ln 2 = 2.08$.

The Systems M_{dn} with $1 < n < d$ in an External Magnetic Field

Near the critical temperature the Ising models M_{d1} are very sensitive to an external magnetic field, since the spins exhibit a long range correlation. This does not apply to systems M_{dn} with $n > 1$. Therefore, a phase transition line $K = K_c(h)$

is expected. This function $K_c(h)$ can be calculated for small h if one assumes that the nonanalytic part of the free energy depends on $K - K_c(h)$ only[25,26]:

$$F_{\text{sing}}(K,h) = F_{\text{sing}}(K - K_c(h)). \tag{3.71}$$

From $KF = -\bar{I} \ln Z$, one obtains Eq. (3.68). The mth derivative of KF with respect to h can be expressed as spin correlation functions involving up to m spins. Only products of spins which can be expressed as products of $R(b)$ yield nonvanishing expectation values. Since $R(b)$ is a product of $2n$ spins for the hypercubic systems, one obtains for $m < 2n$ only constant contributions of the type $\langle S^2(r) S^2(r') \cdots \rangle$. For $m = 2n$ expectation values $\langle R(b) \rangle$ also occur yielding $(2n)!\,E$:

$$\left.\frac{\partial^m (KF)}{\partial h^m}\right|_{h\,0} = \begin{cases} \text{const} & \text{for } m < 2n \\ \text{const} + (2n)!\,E & \text{for } m = 2n. \end{cases} \tag{3.72}$$

From Eq. (3.71) and (3.72) it follows for the hypercubic systems that

$$K_{c,dn}(h) = K_{c,dn}(0) \quad - h^{2n} + \cdots . \tag{3.73}$$

From the duality relation, Eqs. (3.16), (3.17), one obtains for large K the phase transition line

$$h_{c,dn}(K) = K_{c,d\,d-n}(0) - \sinh 2K_{c,d\,d-n}(0) e^{-4nK} + \cdots . \tag{3.74}$$

The reduced critical temperature $K_c^{-1} = k_B T_c / I$ is plotted as a function of the reduced magnetic field $h/K_c = \mu_B / I$ in Fig. 4 for the cubic model M_{32}, Eq. (3.10).

In an external magnetic field the systems M_{dn} with $d = 2n - 1$ on self-dual lattices are self-dual [Eq. (3.16)].

4. PHASE TRANSITION IN AN ISING MODEL WITH COMPETING INTERACTIONS

In this section we describe an Ising model with competing two-spin interactions. For special values of temperature and interaction parameters this model is related to the model (3.10) by the star square transformation (2.54). This system shows a singularity in the specific heat, but it shows no long range order below the critical temperature.

As in model (3.10) spins are located at the centers $r^{(1)}$ of the edges of the cubes (open squares in Fig. 3). Moreover, spins are located at the centers $r^{(2)}$ of the faces of these cubes (black squares in Fig. 3). We assume an interaction strength I_1 for nearest neighbor pairs $S(r^{(1)})$ and $S(r^{(2)})$, an interaction strength I_2 for next nearest neighbor pairs of spins $S(r^{(1)})$ and $S(r^{(1)\prime})$, and an interaction strength I_3 for pairs of spins $S(r^{(1)})$ lying opposite a spin $S(r^{(2)})$ (Fig. 5). Denoting the central spin $S(r^{(2)})$ of a face by S_5 and its four nearest neighbor spins by S_1, S_2, S_3, S_4, then the Hamiltonian H' of our model is the sum over all faces

$$\begin{aligned} H' = -\sum [& I_1 S_5 (S_1 + S_2 + S_3 + S_4) \\ & + I_2 (S_1 S_2 + S_2 S_3 + S_3 S_4 + S_4 S_1) \\ & + I_3 (S_1 S_3 + S_2 S_4)]. \end{aligned} \tag{4.1}$$

We discuss the ground state of this model. The system is invariant under simultaneous reversal of I_1 and S_5. We assume I_1 to be positive. The ground state depends on the ratios I_2/I_1 and I_3/I_1. In Fig. 6 we plot the phase diagram at zero temperature. In region 0 the system is ferromagnetic, that is, all spins $S(r^{(1)})$ point in the same direction. In region 1 one of the four spins $S(r^{(1)})$ of a face points in one direction, all three other spins of the face point in the opposite direction. In region 2 a

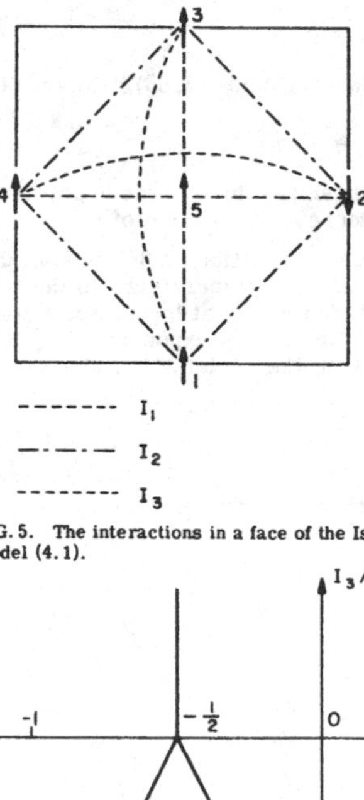

FIG. 5. The interactions in a face of the Ising model (4.1).

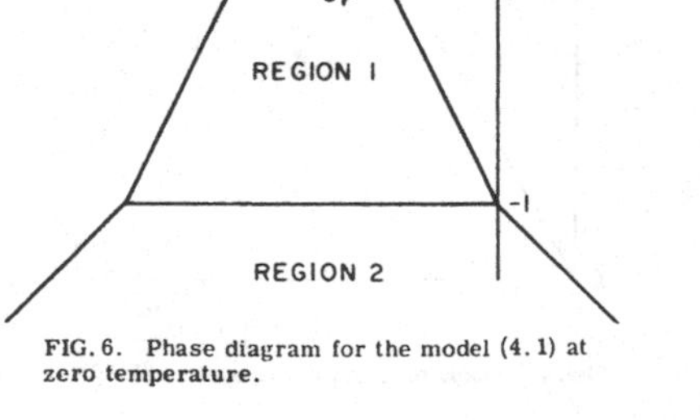

FIG. 6. Phase diagram for the model (4.1) at zero temperature.

pair of neighbored spins $S(r^{(1)})$ points in one direction, the other pair of spins $S(r^{(1)})$ of this face points in the other direction. In region 2′ two opposite spins $S(r^{(1)})$ at a face should point in one direction, the other pair in the other direction, but such an ordering is not possible in three dimensions. Therefore, in this region the ground state cannot be determined by looking merely for the ground state of one face.

Here we are interested in region 1. The ground states of system (4.1) and (3.10) for negative I are the same. The spins S_5 are determined by the surrounding spins S_1, S_2, S_3, S_4. Since the partition function of system (3.10) is invariant under change of sign of I, we obtain the ground state entropy of the system (4.1) in region 1,

$$S_0 = Nk_B \ln 2. \tag{4.2}$$

From the star square transformation (2.54) we find that the partition function Z' of system (4.1) and the partition function $\hat{Z}$ of the Hamiltonian

$$\hat{H} = -\sum [\hat{I}_1 S_1S_2S_3S_4 + \hat{I}_2(S_1S_2 + S_2S_3 + S_3S_4 + S_4S_1) + \hat{I}_3(S_1S_3 + S_2S_4)] \tag{4.3}$$

are related by

$$Z'(K_1, K_2, K_3) = (2e^{2\hat{K}-\hat{K}_1})^{3N}\hat{Z}(\hat{K}_1, \hat{K}_2, \hat{K}_3), \tag{4.4}$$

where

$$\hat{K}_2 = K_2 + \hat{K}, \quad \hat{K}_3 = K_3 + \hat{K}, \tag{4.5}$$

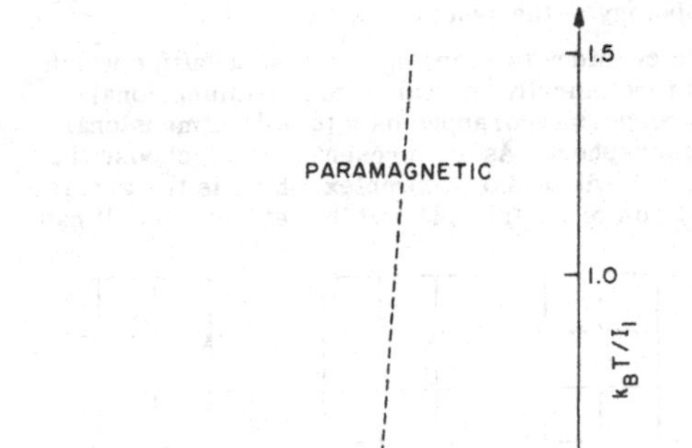

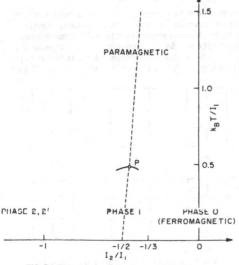

FIG. 7. Phase diagram for the model (4.1) for $I_2 = I_3$. Along the broken line the free energy can be calculated from that of the simple cubic Ising model. The heavy line denotes the phase transition line from Eq. (4.14).

$$\cosh 4K_1 = e^{8\hat{K}}, \cosh 2K_1 = e^{2\hat{K}-2\hat{K}_1}. \tag{4.6}$$

Along the line $\hat{K}_2 = \hat{K}_3 = 0$ (broken line in Fig. 7) the partition function can be expressed in terms of that of the simple cubic Ising model. In particular from the critical singularity of the partition function of the simple cubic Ising model[13] at $I/k_BT_c = 0.2217$ we obtain a singularity of the partition function Z' at $K_{1c} = 2.039, K_{2c} = K_{3c} = -0.9344$, that is, for $I_2/I_1 = I_3/I_1 = -0.4582$, $k_BT_c/I_1 = 0.4904$ (point P of Figs. 6 and 7).

Now let us expand $\hat{Z}(\hat{K}_1, \hat{K}_2, \hat{K}_3)$ into powers of $\hat{K}_2$ and $\hat{K}_3$:

$$\ln \hat{Z}(\hat{K}_1, \hat{K}_2, \hat{K}_3) = \ln Z_{32}(\hat{K}_1) + \sum_{ij} a_{ij}\hat{K}_2^i\hat{K}_3^j. \tag{4.7}$$

The coefficients a_{ij} can be expressed in terms of the spin correlation functions of system (3.10):

$$a_{10} = a_{01} = 0, \tag{4.8}$$

$$a_{20} = \tfrac{1}{2}\partial^2 \ln \hat{Z}/\partial\hat{K}_2^2$$
$$= \tfrac{1}{2}\sum\langle(S_1S_2 + S_2S_3 + S_3S_4 + S_4S_1)^2\rangle$$
$$= 2\sum(1 + \langle S_1S_2S_3S_4\rangle) = 6N - 2E/I,$$

$$a_{02} = 3N - E/I, \quad a_{11} = 0. \tag{4.9}$$

In general a $2n$th or $(2n+1)$th derivative of $\ln Z$ can be expressed in terms of cumulants of at most n operators $R(b)$ or products of $R(b)$'s. We expect that such a cumulant shows a singularity at the critical temperature of the form $\epsilon^{2-\alpha-n}$ [with $\epsilon = (T-T_c)/T_c$], since such cumulants occur in the nth derivative of the free energy with respect to the interaction constants in a system with Hamiltonian $-\hat{I}\sum R(b) - \hat{I}_2\sum RR - \cdots$. The nth derivative with respect to $\hat{I}$ is proportional to the nth temperature derivative of the free energy and is thus proportional to $\epsilon^{2-\alpha-n}$. We assume that the cumulants of products of R's show no stronger singular behavior. Since $\hat{K}_2$ and $\hat{K}_3$ are regular functions of T for fixed $I_2/I_1, I_3/I_1$, we obtain for $I_2/I_1 = I_3/I_1 = -0.4582$

$$\ln Z' = \ln Z_{32}(\hat{K}_1) + \text{regular terms} + O(\epsilon^{3-\alpha}). \tag{4.10}$$

Therefore, we obtain for the singular part of the specific heat $c'_{\text{sing}}(T)$,

$$c'_{\text{sing}}(T_c(1+\epsilon))$$
$$= q^{-2}c_{\text{sing},31}(T_{c,31}(1-q\epsilon)) \cdot[1 + O(\epsilon)]$$
$$= 2.088\, c_{\text{sing},31}(T_{c,31}(1 - 0.6924\epsilon)) \cdot[1 + O(\epsilon)], \quad q = (\partial\ln K_1/\partial\ln K)_{Tc}. \tag{4.11}$$

If we assume that the singular part of the free energy depends only on $T - T_c(I_1, I_2, I_3)$ [compare

Eq. (3.71)], then from Eqs. (4.8) and (4.9) and from

$$\frac{\partial \ln \hat{Z}}{\partial \hat{K}_1} = -\frac{E}{I} \tag{4.12}$$

we obtain

$$\hat{F}_{\text{sing}}(\hat{K}_1, \hat{K}_2, \hat{K}_3) = F_{\text{sing},32}(\hat{K}_1 + 2\hat{K}_2^2 + \hat{K}_3^2 + \cdots). \tag{4.13}$$

Therefore, we may expand the critical temperature in powers of $I_2/I_1 + 0.4582$ and $I_3/I_1 + 0.4582$ (Fig. 7):

$$k_B T_c(I_1, I_2, I_3)/I_1 = 0.4904 - 4.00(I_2/I_1 + 0.4582)^2 - 2.00(I_3/I_1 + 0.4582)^2 - \cdots. \tag{4.14}$$

Since the ground state of this system is the same as for the model (3.10), the two-spin correlations at $T = 0$ vanish and no long range order is expected below T_c.

5. CONCLUSION

In 1966 Mermin and Wagner[27] proved that there is no spontaneous magnetization in the two-dimensional isotropic Heisenberg model. On the other hand, there is evidence from high temperature expansions of the magnetic susceptibility[28] that this system undergoes a phase transition. This raises the question of whether or not it is possible to have a phase transition without a local order parameter. In this paper we have exhibited systems which undergo phase transitions but which do not have a local order parameter. The specific systems were certain classes of Ising models. It would be of some interest to generalize this concept to other types of systems.

ACKNOWLEDGMENTS

It is a pleasure to thank Professor L. P. Kadanoff and Professor H. Wagner for helpful discussions.

APPENDIX

In this appendix we derive the Eqs. (3.21)–(3.23). Any lattice can be created from another lattice with the same boundary conditions by applying one of the following steps (see Fig. 8) as many times as needed.

Step 1: Divide an m-dimensional hypercell into two parts by creating an $(m + 1)$-dimensional hypercell.

Step 2: Collapse an $(m + 1)$-dimensional hypercell by merging two m-dimensional hypercells with the same boundary together into one.

By applying any of these steps, the left-hand side of

$$\sum_{m=0}^{d} (-1)^{d-m} C_m = t \tag{3.23}$$

remains unchanged. Therefore, t depends only on the boundary conditions. This is a generalization of Euler's theorem.

Next we consider the change of N_g resulting from the application of Step 1. If $m > n$, then the Hamiltonian does not change. For $m = n$ one interaction is effectively duplicated, since one boundary $B^{(n)}$ is duplicated. For $m = n - 1$ one spin is replaced by two spins, but for the ground state both must be equal. For $m = n - 2$ there is also one additional spin. Taking this spin aligned upwards, one obtains a one-to-one correspondence with the ground state of the original system. But changing the signs of all spins lying on bonds adjacent to one $(n - 1)$-dimensional hypercell at the boundary of the new bond, we obtain another ground state. Therefore, the new system has twice the degeneracy of the original system. For $m < n - 2$ the Hamiltonian does not change. Therefore, we obtain (Step 2 is just the inverse of Step 1)

$$N_g = t_g + \sum_{m=0}^{n-2} (-1)^{n-m} C_m. \tag{3.21}$$

Since this expression changes only by + 1 after application of Step 1 and by − 1 after application of Step 2 for $m = n - 2$, t_g depends only on the boundary conditions and on n. Similarly, we obtain

$$N_g^* = t_g^* + \sum_{m=n+2}^{d} (-1)^{m-n} C_m. \tag{3.22}$$

Therefore, N_m, Eq. (3.16), depends only on the topology of the system and on n.

We consider two topologies: first, a lattice which is topologically equivalent to a d-dimensional hypersurface wrapped on a $(d + 1)$-dimensional hypersphere. As a representative we choose the $(d + 1)$-dimensional simplex, which is the generalization of the triangle and the tetrahedron. It has

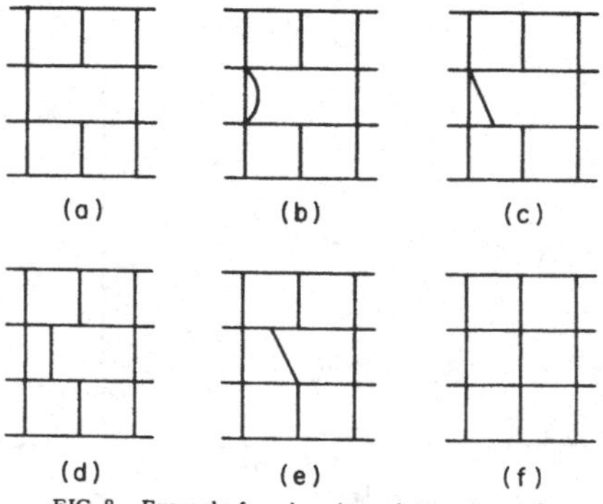

FIG. 8. Example for changing a lattice by applying the Steps 1 and 2. From the lattice (a) the lattice (f) is created by applying once the Step 1 with $m = 1$ (b), twice Step 1 with $m = 0$ (c), (d), twice the Step 2 with $m = 0$ (e), (f). The number $C_2 - C_1 + C_0$ remains unchanged.

$C_0 = d + 2$ corners. Any two corners are connected by an edge. Any three edges span a face and so on. It follows that $C_m = \binom{d+2}{m+1}$. Using Eq. (3.13), one obtains

$$t = 1 + (-1)^d. \tag{A1}$$

We may number the spins of the model M_{un} on this lattice by n indices $1 \le i_1 < i_2 \cdots < i_n \le d + 2$. For the ground state all spins with $i_n = d + 2$ can be chosen arbitrarily. Then all other spins are given by $S(i_1 \cdots i_n) = S(i_2 \cdots i_n, d + 2) \cdot S(i_1 i_3 \cdots i_n, d + 2) \cdots S(i_1 \cdots i_{n-1}, d + 2)$. Therefore it follows that $N_g = \binom{d+1}{n-1}$. From the Eqs. (3.21), (3.22), (3.26), and (A1) one obtains

$$t_g = (-1)^{n+1}, \quad t_g^* = (-1)^{d+n+1} \quad \text{for } n \le d - 1, \tag{A2}$$

$$N_m = 0 \quad \text{for } n \le d - 1. \tag{A3}$$

Since $N_m = 0$, the duality relation (3.17) holds for all lattices wrapped on a $(d + 1)$-dimensional hypersphere.

Secondly, we consider a lattice with periodic boundary conditions. As a representative we choose the d-dimensional hypercube. Then it follows that $C_m = \binom{d}{m}$, and one obtains from Eq. (3.23)

$$t = 0. \tag{A4}$$

Because of the periodic boundary conditions, all spins occur twice in the products R. Therefore, all spins can be chosen arbitrarily, $N_g = \binom{d}{n-1}$. Then one obtains from the Eqs. (3.21), (3.22), (3.26), and (A1)

$$t_g = \binom{d-1}{n-1}, \quad t_g^* = \binom{d-1}{n}, \quad N_m = \binom{d}{n}. \tag{A5}$$

* Work supported in part by the National Science Foundation.

† On leave from the Kernforschungsanlage Jülich, Germany, where part of this work was done.

[1] H. A. Kramers and G. H. Wannier, Phys. Rev. **60**, 252 (1941).

[2] L. Onsager, Phys. Rev. 65, 117 (1944).

[3] G. H. Wannier, Rev. Mod. Phys. **17**, 50 (1945).

[4] Compare also the review articles G. F. Newell and E. W. Montroll, Rev. Mod. Phys. **25**, 353 (1953) and Ref. 5.

[5] C. Domb, Advan. Phys. **9**, 149 (1960).

[6] L. P. Kadanoff, Phys. Rev. Letters **23**, 1430 (1969).

[7] L. P. Kadanoff and H. Ceva, Phys. Rev. B 3, 3918 (1971).

[8] Compare also M. E. Fisher and A. E. Ferdinand, Phys. Rev. Letters **19**, 169 (1967).

[9] R. Brout, *Phase Transitions* (Benjamin, New York, 1965).

[10] P. W. Anderson, Rev. Mod. Phys. **38**, 298 (1966).

[11] L. D. Landau and E. M. Lifshitz, *Statistical Physics* (Pergamon, London, 1958), Chap. 14.

[12] L. P. Kadanoff *et al.*, Rev. Mod. Phys. **39**, 395 (1967).

[13] M. E. Fisher, Rept. Progr. Phys. **30**, 615 (1967).

[14] P. Heller, Rept. Progr. Phys. **30**, 731 (1967).

[15] I. Syozi, Progr. Theoret. Phys. (Kyoto) **6**, 306 (1951).

[16] S. Naya, Progr. Theoret. Phys. (Kyoto) **11**, 53 (1954).

[17] L. Onsager, Ref. 2.

[18] A. Pais, Proc. Natl. Acad. Sci. (U.S.) **49**, 34 (1963).

[19] Since the algebra on the set $\{0, 1\}$ with the operations (2.5) and (2.6) is a field, all theorems for vectors and matrices over a field apply. Compare, e.g., G. Birkhoff and S. MacLane, *A Survey of Modern Algebra* (MacMillan, New York, 1959), Chap. VII, Secs. 3–8 and 12, Chap. VIII, Secs. 1–9. Note that n linear independent vectors span a space of 2^n vectors.

[20] M. E. Fisher, Phys. Rev. **113**, 969 (1959).

[21] D. M. Y. Sommerville, *An Introduction to the Geometry of N Dimensions* (Dover, New York, 1958), Chap. IX.

[22] D. G. Kelly and S. Sherman, J. Math. Phys. **9**, 466 (1968).

[23] D. Ruelle, Rev. Mod. Phys. **36**, 580 (1964), proves for Ising models with two-spin interactions that any cumulant $\chi(r_1, r_2, \cdots)$ vanishes in the limit $|r_1 - r_2| \to \infty$ for $0 \le \beta \le \beta_0$, where β_0 depends on the system. (This proof can also be applied to lattices). Since the expectation value (3.50) can be expressed by a sum of products of cumulants and any product contains at least one cumulant $\chi(r_1^0, r_2^0 + r, \cdots)$, it can be proven exactly that the right-hand side of Eq. (3.50) vanishes for $n = d - 1$ for $0 \le \beta \le \beta_0$.

[24] Assuming that a product of operators R can be expressed by the sum of a constant, an energy density, and operators with a less critical behavior in the sense of the operator algebra due to Kadanoff (Refs. 6 and 7), we find from the assumption $F_{\rm sing}(K, \{h\}) = F_{\rm sing}(K - K_c\{h\})$ (Refs. 25 and 26) finite derivatives $\partial K_c\{h\}/\partial h_i$. This gives some evidence that a $(\nu - 1)$-dimensional λ-hypersurface $K = K_c\{h\}$ exists which separates the $(K, \{h\})$ space into two phases (in some region around $h_i = 0$) and that this λ-hypersurface is not the boundary of a first-order phase transition.

[25] M. E. Fisher, Phil. Mag. **7**, 1731 (1962).

[26] E. Riedel and F. Wegner, Z. Physik 225, 195 (1969).

[27] N. D. Mermin and H. Wagner, Phys. Rev. Letters 17, 1133 (1966).

[28] H. E. Stanley and T. A. Kaplan, Phys. Rev. Letters 17, 913 (1966).

PHYSICAL REVIEW D VOLUME 10, NUMBER 10 15 NOVEMBER 1974

Gauge fields on a lattice. I. General outlook

R. Balian, J. M. Drouffe, and C. Itzykson
Service de Physique Théorique, Centre d'Études Nucléaires de Saclay, B.P. no. 2, 91190 Gif-sur-Yvette, France
(Received 3 June 1974)

We present Wilson's model of gauge-field theory on a lattice, including a coupling to a matter field. The algebraic structure is surveyed for both commutative and noncommutative groups. Various regimes are suggested by mean-field theory according to the relative values of coupling constants. In particular the gauge field undergoes a first-order transition while the matter-field transition is of second order.

I. INTRODUCTION

The reasons for theorists' interest in gauge fields need not be explained here in detail. Apart from an aesthetical appeal, they seem to provide the most promising models for elementary-particle interactions. This seems to be the case not only in the realm of weak and electromagnetic forces, but also in the domain of strong interactions. Their dynamics in this regime is, however, poorly understood. Hence any indication of the strong-coupling region is *a priori* interesting.

Recently Wilson has introduced a model and various techniques that are novel in this game.[1] The purpose of this work is to comment on Wilson's model and present some numerical results.

It must be made clear that the model itself is to a large extent unrealistic. A noncovariant ultraviolet cutoff procedure is introduced which breaks Lorentz invariance (or rather Euclidean invariance after rotation to an imaginary time). Nevertheless, since it allows an investigation in an unfamiliar regime, it is interesting.

Let us first briefly present the main line of reasoning. Assume that a field theory is studied in the Euclidean region and replace continuous space-time points by a discrete lattice. According to standard practice the full contents of the theory can be thought in terms of a Feynman path integral of e^S, where S is the action, as a functional of classical field variables defined now on the discrete lattice. This is to be integrated over with an appropriate measure on the field variables. The analogy with statistical mechanics is evident. A set of values for the classical field corresponds to a configuration, and S appropriately scaled corresponds to "energy divided by temperature" for this configuration. The sum over configurations provides the analog of the partition function, the logarithm of which is essentially the free energy. Note that the E/T of the statistical interpretation is *not* the energy of the field theory. The latter can be recovered if one wishes through a study of the transfer matrix; we shall not enter into these matters here.

Discretization introduces a fundamental length in the problem, the inverse of which is a natural ultraviolet cutoff. The price paid is that Euclidean invariance is lost, and at best can be recovered when the spacing becomes immaterial. We shall see, however, in which precise sense this statement has to be made; a naive idea would be that no matter what the spacing a is, if we look at "soft phenomena," i.e., for distances $r \gg a$, the breaking of invariance ought not to be too disastrous. This, however, might be superficial, since, as we shall see by investigating more closely the types of distortions introduced by discretization, there exists some kind of built-in breaking of invariance which is hard to control even on a large scale.

If no external source is present the path integral describes the vacuum-to-vacuum amplitude. Hence the problem is: What is the stable vacuum; what are the elementary excitations? Of course, in order for us to be able to answer these questions, external sources have to be introduced at some later stage.

It is clear that with an ultraviolet cutoff present, the usual problems of renormalization of field theory disappear at first. This is, however, only an illusion if we look for small-distance phenomena where one would have to control the behavior as the spacing goes to zero. This is not, however, the purpose of the present approach at this stage; hence this aspect of the question will not be in the forefront. On the other hand, it is well known that gauge fields introduce long-range strong forces. In other words, the infrared problem is indeed catastrophic. This is what we really wish to investigate.

The first technical problem solved by Wilson was to present a gauge theory directly on the lattice that is not a straightforward and unimaginative replacement of a continuous space-time set of functions by their values at discrete points. The clue to this is in the original formulation of gauge invariance. By following essentially the steps that lead from a global invariance under an internal symmetry group to its local realization, one obtains Wilson's correct answer. What this amounts to is replacing a field with values in a *Lie algebra* (the continuous case) by a field variable taking its value in a *Lie group*. Even though this appears at first as a new technical difficulty, it has an unexpected bonus which is easily understood. Indeed the action is *gauge invariant*. This freedom of gauge is ordinarily a headache for quantization, or what is equivalent to formulating a Feynman path integral. For, notwithstanding the fact that the gauge functions vary from point to point, at each point they take values over an infinite interval, with a result that they yield infinite factors right at the beginning of the theory. The remedy to that is usually to *break gauge invariance* at the level of the Lagrangian and go through a complex procedure involving Ward identities to show that the final theory has the desirable covariance properties under gauge transformations. Here, however, the gauge degree of freedom takes its values in a Lie group which, for all practical purposes, is a *compact* group. If, furthermore, a spatial cutoff is introduced by considering a large but finite lattice, the gauge degrees of freedom will introduce no infinity. Consequently *no explicit breaking* of this invariance is needed.

This very interesting consequence of discretization suggests that new phenomena take place. To illustrate this possibility let us turn to a much simpler example that will, in fact, be discussed in detail below. Let us assume that instead of dealing with a local group, our dynamical system exhibits an internal global symmetry of the usual type at the level of the evolution equations. To be specific, let us think of a very simplified σ model. It has two degrees of freedom, the pion and the σ meson, with an O(2) invariance. We can think of the field as a two-component vector in some "isospin space." Let us further discretize the model and impose the restrictions that at each point this vector is of fixed length (to mimic the interaction) and that at neighboring sites the vectors interact by pairs through a term proportional to the scalar product (to mimic the kinetic term in the action). This model then has a global invariance. The scalar product is proportional to the square of the length of our vectors and can be thought of as the ratio between kinetic terms and interaction terms, hence inversely proportional to a fictitious coupling constant g. On the other hand, in the "statistical" interpretation this same strength can be described as inversely proportional to temperature T. The conclusion is that $T \propto g$. Now in the statistical view we can think of our vectors as giving the direction of two-dimensional magnets. Depending on a sign the whole system is of the type ferro- or antiferromagnetic. It turns out here that the correct sign is of the ferromagnetic type, whereupon at low temperature ($g \to 0$) "spins" tend to align. Statistical mechanics even suggests that a transition occurs. At low T (low coupling) an ordered phase is created. This means that the symmetry is spontaneously broken. A Goldstone long-range excitation is present (the massless pion) and a shorter-range branch is also present (the massive σ). As $T \to 0$ ($g \to 0$) the shorter-range excitation essentially disappears (the σ mass goes to infinity) and what are left are noninteracting transverse spin waves: the free massless pion field. Above the critical temperature T_c, however, disorder appears; the symmetry is really implemented in the ordinary way and we have a degenerate massive doublet of excitations. The region of the second-order phase transition is the most interesting from the statistical-mechanics point of view as well as for the particle interpretation. In this region we study the transition between the two possible types of symmetry. The specific dynamics as well as the precise value of the ultraviolet cutoff is indeed inessential since the "soft" scale is set by the correlation length, i.e., the inverse of the very small mass of the σ. In this region we might hope, as far as infrared behavior is concerned, that Euclidean invariance is restored.

The above description justifies the fact that we have devoted Sec. II of this work to this model. We call it the *scalar model*. The reason for this denomination is that in the limit of zero spacing it degenerates into the field theory for a massless scalar field corresponding to the spin waves of statistical mechanics. In this limit all quantities vary continuously from point to point and hence correspond to an almost perfect ordering. The model is then naturally associated with low temperatures or small coupling constants.

We shall then define a *gauge-field model* on the lattice to implement a *local* symmetry.

In the light of the previous discussion, it seems natural to expect that a transition takes place. At low temperature, gauge invariance would be spontaneously broken, corresponding to the usual "free field" case. For a large coupling constant, however, a disordered phase would appear, with local gauge invariance strictly enforced, and without any Goldstone "photon."

It is clear that any quantization of the photon field implies a choice of gauge even though physical quantities ought to be independent of this choice. However, the disordered phase where local invariance is strictly enforced is an unusual situation where quantities such as $\langle A_\mu(x)A_\nu(y)\rangle$ are obviously zero. Furthermore, there is some subtlety involved in choosing an order parameter. We find indication for a *first-order* transition in this system, in contradistinction with the previous case. Furthermore, the high-temperature phase exhibits interesting phenomena which will be described below. These support Wilson's idea that such a phase might provide a mechanism for charged particles binding, which would prevent them from escaping from each other.

We shall first present the various models, introducing notations and describing their main features. In the next parts the scalar, Abelian, and non-Abelian fields are studied in some detail. We shall also study the interesting *coupled system* to be introduced in the text. In each case we try to convince ourselves (and hopefully the reader) that a transition really takes place.

In particular we can be guided by mean field theory which is physically motivated as the dimension d of the lattice gets large. In all cases we find that the transition temperature, or coupling constant, grows linearly with d. Apart from trivial exact solutions in very low dimension (where no transition really takes place but ordering sets in as we approach $T=0$), one has no recourse but to turn to numerical calculations. These calculations involve an adaptation of the techniques familiar to devotees of the Ising model. Since this approach is by itself interesting, we shall postpone its discussion to a forthcoming paper.

It is a pleasant feature of the model to see how naturally non-Abelian gauge fields are incorporated in the formalism at a very minimal cost. Furthermore, insofar as the high-temperature expansion is concerned, they only add a touch of group theory to the bulk of the preceding numerical work.

Let us finally stress that it would certainly be very interesting to be able to formulate the analog of the present strong-coupling expansion in a more realistic field-theoretic case.

II. THE MODELS

A. Transition amplitudes as path integrals

The dynamics of a field theory is conveniently described by the vacuum-to-vacuum transition amplitude in the presence of suitably chosen external sources coupled to the system. These we denote collectively by J, and the above transition amplitude contains all information on the system. Let us call it $\langle 0|0\rangle_J$. Stability of the vacuum is included in the statement that for $J=0$, $\langle 0|0\rangle=1$, and that to first order $(\delta/\delta J)\langle 0|0\rangle_{J=0}=0$. If calculations indicate that this is not the case, this means most likely that the state called vacuum has not been correctly identified and the theory has to be modified accordingly. Even though infinities plague field theory, intuitive approaches can be used as formal tools at unsophisticated levels, to be made more precise at a later stage of calculation. Such a tool is provided by the Feynman path-integral formulation. Since by now the subject has become more familiar, we shall be using it without the usual apology.

In essence what it involves is the following. There exists an underlying classical field theory in terms of classical fields collectively denoted by $\phi(x)$ with an action integral $S(\phi)$. Classical equations of motion would emerge by requiring $S(\phi)$ to be stationary with respect to variations of ϕ: $\delta S(\phi)=0$. In quantum theory we evaluate transition amplitudes. These amplitudes involve summing over all possible "paths" the elementary contributions of the form

$$e^{iS(\phi)}.$$

A point of physics seldom stressed is that the only trace of the asymptotic states between which the amplitude is evaluated is essentially in the boundary conditions for large times on the fields ϕ. In particular for the vacuum-to-vacuum amplitude it is generally assumed that the field ϕ vanishes. This assumption might reveal itself faulty. Thus we write

$$\langle 0|0\rangle_J = A\int \mathcal{D}\phi\, e^{iS(\phi,J)}. \tag{2.1}$$

The normalization factor A is chosen by requiring that for $J=0$, $\langle 0|0\rangle=1$. To be specific let us think in terms of a self-coupled scalar field ϕ where

$$S(\phi,J)=\int dx_0 d^3x\{\tfrac{1}{2}[(\partial_0\phi)^2-(\vec{\nabla}\phi)^2]-V(\phi,J)\}, \tag{2.2}$$

with V a polynomial in $\phi(x)$ and $J(x)$. We have distinguished spatial and time arguments in order to be able to jump immediately to a Euclidean space by assuming that an analytic continuation to pure imaginary times ($x_0\to -ix_0$) can be performed. The underlying assumption is rather the reverse: that the Euclidean quantities stand a better chance to be well defined, in such a way that at a final stage an analytic continuation back to physical Minkowski space will be performed for meaningful quantities such as Green functions. Thus, keeping the same notations, we write

$$Z(J) = \int \mathfrak{D}\phi e^{S(\phi, J)},$$
$$S(\phi, J) = -\int d^4x[\tfrac{1}{2}(\partial\phi)^2 + V(\phi, J)] , \qquad (2.3)$$

with $(\partial\phi)^2$ standing now for the Euclidean square of the gradient of ϕ:

$$(\partial\phi)^2 = (\partial_0\phi)^2 + (\partial_1\phi)^2 + (\partial_2\phi)^2 + (\partial_3\phi)^2 .$$

This expression has been written for simplicity as if we were dealing with a single scalar field $\phi(x)$. It is clear that (2.3) is not very well defined at this stage until we make precise (i) the test function space of classical fields and (ii) the integration procedure.

One way to proceed is to replace the continuum space-time by a discrete set of points of a lattice. We assume the latter to be hypercubical. That is, we restrict x to

$$x = \sum_0^3 x_i n_i a , \qquad (2.4)$$

with x_i an integer, $n_i \cdot n_j = \delta_{ij}$. The quantity a is the lattice spacing.

Any expression for S that tends in the limit $a \to 0$ to the one given in (2.3) is *a priori* a good candidate. As far as the potential term V is concerned, there is *a priori* no freedom. Furthermore, we shall for the moment suppress the external source J, and approximate V in such a way that it limits at every point x the range of values for $\phi(x)$ in a way to be made precise later on. We now concentrate on the kinetic term $[\partial\phi(x)]^2$ alone.

Our choice, which admittedly appears rather artificial at this stage, is the following. We set

$$S(\phi) = \sum_{(xx')} \frac{d^{d-4}}{g} \cos a[\phi(x) - \phi(x')] , \qquad (2.5)$$

with $\sum_{(xx')}$ meaning summation over all *pairs of neighboring lattice points*. We insist that the dimension of ϕ be the traditional one of inverse length, hence the occurrence of $a\phi(x)$ as a dimensionless quantity. We have furthermore introduced a new parameter d as the dimension of space-time. Up to now $d = 4$, but we shall allow d to be an arbitrary integer, and for some analytical calculations to be performed later d will take arbitrary real values. Finally, we have added a factor $1/g$ $(g > 0)$ as the ratio of the kinetic to the potential term in the action, the latter restricting the values of $\phi(x)$ to a range $-\pi/a$ to $+\pi/a$.

It is readily seen what the limit $a \to 0$ means. We can arrange the sum $\sum_{(x,x')}$ in the following way. First keeping x fixed we let x' run over $x' = x + an_i$, n_i being any one of the (positive) unit coordinate vectors. Then we sum over x. If a is small and if $\phi(x)$ varies smoothly from point to point we have approximately

$$\phi(x') \simeq \phi(x) + a\frac{\partial}{\partial x_i}\phi(x) ,$$
$$S(\phi) \simeq \frac{1}{g}\sum_x \{da^{d-4} - \tfrac{1}{2}a^d[\partial\phi(x)]^2\}$$
$$= \text{const} - \frac{1}{2g}\int d^dx[\partial\phi(x)]^2 .$$

Apart from an (infinite) irrelevant constant, we have thus as a limit the kinetic term of our previous action. However, for a small but finite a the restriction to $|a\phi| < \pi$ together with the higher-order terms in the expansion of the cosine lead to nontrivial interactions (essentially nonrenormalizable, since they involve higher and higher derivatives in the continuous limit where they are damped by powers of a^{2n}). Thus the length a plays a double role; it is an ultraviolet cutoff and scales the various interaction terms. Another way of identifying g as a coupling constant is to rescale ϕ into $\sqrt{g}\tilde{\phi}$ and replace the interaction $\theta(\pi^2 - a^2 g\tilde{\phi}^2)$ by a quantity proportional to $\exp[-\text{const} \times (-\tilde{\phi}^2 + ga^2\tilde{\phi}^4)]$.

B. The scalar model and global invariance

Our first discrete model is thus described by the action (2.5). Since we restrict the range of $|a\phi|$ to an interval of 2π and since the cosine is a periodic function, we can rescale ϕ into ϕ/a and write

$$Z = \int \prod_x \frac{d\phi(x)}{2\pi} \exp\Big(\beta \sum_{(x,x')} \cos[\phi(x) - \phi(x')]\Big), \qquad (2.6)$$

where β stands for $1/ga^{4-d}$. Each integral over ϕ runs in an interval of 2π and we have normalized Z in such a way that $Z = 1$ for $\beta \to 0$ (i.e., for $g \propto T \to \infty$). Let us recall that x stands for an arbitrary point on the lattice, i.e., an ordered set of d integers.

Finally, in order to give a meaning to the infinite integral (2.6), we introduce a spatial cutoff by restricting the lattice to a torus (imposing periodic boundary conditions) as follows: The variables x_i take integer values from 0 to $L-1$ with $x_{i+L} \equiv x_i$. In this way we have $N \equiv L^d$ sites. We shall measure every quantity per site (i.e., per unit volume up to a proportionality constant a^d) and then let N grow infinitely large. In particular F will stand for

$$F = \lim_{N \to \infty} \frac{1}{N} \ln Z . \qquad (2.7)$$

It is related to the generating functional of connected vacuum-to-vacuum diagrams of field theory. We shall call it, by analogy with statistical mechanics, the free energy, even though it differs by a factor $-\beta$ from the traditional quantity.

A suggestive way to rewrite (2.6) is to introduce *a complex number of unit modulus* $k(x) \equiv e^{i\phi(x)}$ or *a unit two-dimensional vector* $\vec{k}(x)$ at each site. Thus we can write

$$\beta \sum_{(x,x')} \cos[\phi(x)-\phi(x')] \equiv \beta \mathrm{Re} \sum_{(x,x')} k^*(x)k(x') \equiv \beta \sum_{(x,x')} \vec{k}(x)\cdot\vec{k}(x') . \tag{2.8}$$

With this notation we observe that our model is identical with a classical planar Heisenberg model on a d-dimensional lattice. It has "spins" of unit length at each site (i.e., restricted to vary on a circle) and a "statistical mechanical energy" involving nearest neighbors

$$-\sum_{(x,x')} \vec{k}(x)\cdot\vec{k}(x'),$$

i.e., of ferromagnetic character, which mimics an exchange energy favoring alignment of the spins.

The N-integral (2.6) is obviously invariant under a global rotation of all spins

$$\vec{k}(x) \to R\vec{k}(x) \text{ or } k(x) \to e^{i\phi}k(x) , \tag{2.9}$$

where R is a rotation independent of x. In complex notation, it is written $e^{i\phi}$ with constant ϕ. It is known (folklore) that for $d>2$ such a system exhibits a transition.[2] For small β there exists a disordered phase exhibiting this O(2) symmetry. Within the context of field theory, this means that the vacuum is O(2) invariant. The wave excitations have a finite mass (a finite range) and possess the same symmetry. These will be identified with an ("isotopic") doublet of particles (the σ and π particles). Above some critical value β_c a different situation prevails. In this case the vacuum is no longer O(2)-invariant nor are the excitations. According to Goldstone's theorem[3] a long-range (zero-mass) excitation (the π) appears together with a finite-range one (the σ). We shall see that the range of the latter goes to zero (the mass goes to infinity) as $\beta \to \infty$ $(g \to 0)$ and we are then left with a scalar massless field which can be identified with the situation we started with in the preceding paragraph.

Mutatis mutandis we could generalize the model to an invariance group SO(n) by replacing the two-dimensional "spin" $\vec{k}$ by an n-dimensional vector of unit length. If we continue to call this vector $\vec{k}(x)$ the action will still be written

$$S=\beta \sum_{(x;x')} \vec{k}(x)\cdot\vec{k}(x') , \tag{2.10}$$

and the only change to make in the definition of Z is to replace the integration volume

$$\frac{d\phi(x)}{2\pi}$$

by the SO(n)-invariant normalized volume element $d^{n-1}\hat{k}(x)$ on the unit sphere in n-dimensional space, the normalization being such that

$$\int d^{n-1}\hat{k}=1 .$$

The basic phase-transition phenomena we just described will be identical: below β_c, SO(n)-invariant vacuum and n-plet of degenerate massive excitations; above β_c, $(n-1)$-plet of massless "transverse" π's and a massive longitudinal σ.

Of course, the above statements will be made more precise as we proceed to explicit computations.

C. Local invariance, gauge field, and minimal coupling

It is clear that if we allow a rotation R [be it in SO(2) in the simplest case or in SO(n) in the generalized one] to depend on x, the action (2.10) will not be invariant. This is due to the fact that we couple nearest neighbors in (2.10). Thus in the transformation

$$\vec{k}(x) \to R(x)\vec{k}(x) \tag{2.11}$$

the coupling term $\vec{k}(x)\cdot\vec{k}(x')$ becomes

$$\vec{k}(x)\cdot\vec{k}(x') \to k(x)^T R^{-1}(x)R(x')k(x') ,$$

with $R^{-1} \equiv R^T$ for an orthogonal group.

In order to implement a local invariance one introduces, by analogy with the familiar continuous case, a gauge field. As the scalar field was a map, "point x on the lattice $\to$ point on the unit sphere in n-dimensional space," the gauge field is a map, "ordered link (x, x') on the lattice $\to A(x,x')$ element of the group SO(n)." To make precise what we mean by ordered link we introduce a *semi-order* on the lattice as follows. First we choose a positive sign on each axis of coordinate, then we say that

$$x' \geq x \text{ if for each } i \ (0 \leq i \leq d-1), \quad x_i' \geq x_i .$$

Each pair of neighbors on the lattice then defines an ordered link (x, x') by requiring that $x' > x$ (this is obviously possible since x and x' differ only in one coordinate).

To the link in the reverse order (x', x) the associated rotation is then taken to be the inverse one. In other words,

$$A(x',x) \equiv A^{-1}(x,x') . \tag{2.12}$$

Under a global rotation R we assume the transformation law to be

$$A(x, x') \to RA(x, x')R^{-1},$$

while under a *local* gauge transformation $R(x)$ we require that

$$A(x, x') \to R(x)A(x, x')R^{-1}(x') . \tag{2.13}$$

We notice immediately that (2.13) is compatible with (2.12) since

$$\begin{aligned} A(x', x) &\equiv A^{-1}(x, x') \\ &\to [R(x)A(x, x')R^{-1}(x')]^{-1} \\ &= R(x')A^{-1}(x, x')R^{-1}(x) \\ &= R(x')A(x', x)R^{-1}(x) . \end{aligned}$$

Having defined a gauge field $A(x, x')$, the coupling between nearest neighbors is changed in a *minimal* way through

$$\vec{k}(x) \cdot \vec{k}(x') \to k^T(x)A(x, x')k(x') . \tag{2.14}$$

The notation $k^T(x)$ is to indicate a row vector. If we combine (2.11) and (2.13) we see that this new coupling is obviously invariant under local transformations.

We pause to stress the analogy of the preceding procedure with conventional minimal coupling. The line of thought is entirely parallel and the results similar. In both cases the introduction of a gauge field is necessitated by a slight nonlocality of the "Lagrangian" generally in the kinetic term. In fact, we even recover the usual case by going to the continuous limit. Recall that the lattice spacing a had been eliminated by scaling. Then with $x' > x$ we have $x' = x + an_\mu$ for some direction n_μ kept fixed:

$$k(x') = k(x) + a\partial_\mu k(x) + \tfrac{1}{2}a^2\partial_\mu{}^2 k(x) + \cdots ,$$

while we may assume $A(x, x')$ sufficiently close to unity to write it as

$$A(x, x') = I + iea\mathcal{A}_\mu(x) - \tfrac{1}{2}e^2a^2\mathcal{A}_\mu(x)^2 + \cdots , \tag{2.15}$$

where $\mathcal{A}_\mu(x)$ is a convenient way to write $\mathcal{A}(x, x')$ and *belongs to the Lie algebra of* SO(n). To be more precise, it is a representative of this element in the representation which acts on the vectors $\vec{k}$. If we take a basis $\lambda^\alpha{}_{\beta\gamma}$ of this Lie algebra, we can write in detail

$$\begin{aligned} A(x, x')_{\beta\gamma} = {} & \delta_{\beta\gamma} + iea\mathcal{A}_\mu(x)_\alpha\lambda^\alpha_{\beta\gamma} \\ & - \tfrac{1}{2}e^2a^2\mathcal{A}_\mu(x)_\alpha\mathcal{A}_\mu(x)_{\alpha'}\lambda^\alpha_{\beta\delta}\lambda^{\alpha'}_{\delta\gamma} + \cdots . \end{aligned}$$

For each index α, $\mathcal{A}_\mu(x)_\alpha$ is thus a vector field. The "charge" e has been introduced according to usual practice. We rewrite formula (2.14) to second order in the lattice spacing. Remembering that $\vec{k}(x)^2 = 1$, it follows that $\vec{k}(x)\partial_\mu\vec{k}(x) = 0$ and $\vec{k}(x) \cdot \mathcal{A}\vec{k}(x) = 0$ due to the antisymmetry of $\mathcal{A}$ [since it belongs to the Lie algebra of SO(n) which preserves the norm $\vec{k}^2$]. Then

$$k^T(x)[I + iea\mathcal{A}_\mu(x) - \tfrac{1}{2}e^2a^2\mathcal{A}_\mu(x)^2 + \cdots][k(x) + a\partial_\mu k(x) + \tfrac{1}{2}a^2\partial_\mu{}^2 k(x) + \cdots]$$
$$= 1 + \tfrac{1}{2}a^2 k(x)^T\{\partial_\mu{}^2 + 2ie\mathcal{A}_\mu(x)\partial_\mu + [ie\mathcal{A}_\mu(x)]^2\}k(x) .$$

Now

$$k^T(x)\partial_\mu{}^2 k(x) = -\partial_\mu k(x)^T\partial_\mu k(x) ,$$

while due to the antisymmetry of $\mathcal{A}$

$$k^T(x)2ie\mathcal{A}_\mu(x)\partial_\mu k(x) = k(x)^T ie\mathcal{A}(x)\partial_\mu k(x) - ie[\partial_\mu k(x)]^T\mathcal{A}(x)k(x) .$$

If we introduce the covariant derivative D_μ as

$$(D_\mu k)(x) = (\partial_\mu - ie\mathcal{A}_\mu)k(x)$$

acting on a column vector we shall have for a row vector

$$(D_\mu k)(x)^T \equiv \partial_\mu k(x)^T + iek(x)^T\mathcal{A}(x) .$$

Finally then

$$k^T(x)A(x, x')k(x) = 1 - \tfrac{1}{2}a^2 D_\mu k(x)^T D_\mu k(x) + \cdots . \tag{2.16}$$

Again apart from an inessential constant factor, if we sum over all pairs (x, x'), i.e., over all directions μ and over all x's, we recover the modified kinetic term in the presence of a gauge field $\mathcal{A}_\mu(x)_\alpha$ in the continuous model.

If $n = 2$ we have an *Abelian* gauge field and a situation very similar to electromagnetism. We leave it to the reader to rewrite (2.16) in this case with $k_1(x) = \cos\phi(x)$ and $k_2(x) = \sin\phi(x)$; the index α takes only one value as there is one generator only and $(\lambda_{\beta\gamma}) = \begin{pmatrix} 0 & i \\ -i & 0 \end{pmatrix}$ with β and γ taking two values only.

If $n > 2$ we have a typical *non-Abelian* theory. Later on, to present specific calculations, we shall choose, for instance, $n = 3$, with SO(3) $= \mathrm{SU}(2)/Z_2$, or $n \to \infty$.

The above construction for orthogonal groups can in fact be generalized to other representations or to other types of compact groups (in particular to unitary groups) without any difficulty.

We now return to our discrete lattice. Having exhibited the minimal coupling to the gauge field in order to build up in the theory a local invariance we stress now that formulas (2.11) and (2.13) define the gauge transformations. In particular (2.13) is the integrated form of the gauge transformation on the gauge field $A(x, x')$. The gauge group is thus the Nth tensor product of SO(n) groups, i.e., as long as the spatial cutoff N is kept finite, a compact group

$$\overset{N}{\otimes}\mathrm{SO}(n) .$$

Beyond minimal coupling it would seem natural to introduce in the action S an extra term designed to produce some dynamics for the gauge degrees of freedom. It is then also obvious that we want to preserve the local invariance. By inspection of (2.13) it is seen that a product

$$A(x_1x_2)A(x_2x_3)\cdots A(x_kx_1)$$

is gauge invariant. In this product the set $x_1x_2x_3\cdots x_kx_1$ constitutes a "closed curve" on the lattice, that is, a set of nearest neighbors (x_1x_2) $(x_2x_3)\cdots$ with the last point identical with the first. This is still not a number but an element of the group SO(n), or if we prefer, a matrix representative. In order to define an additional term in the action we proceed as follows. First we choose *a real irreducible character* of the group, χ. This means that we pick an irreducible representation of SO(n) in our case, such that the trace of the representatives is real. We denote this trace by χ. Then, instead of an arbitrary closed curve on the lattice, we choose the simplest one that we call (from the French) a *plaquette*. This is a set of four nearest neighbors (x_1x_2), (x_2x_3), (x_3x_4), (x_4x_1). Such a plaquette can be identified with a two-dimensional face of a hypercube on our lattice. It will be given, for instance, by

$$x_1 = x ,$$

$$x_2 = x + n_\mu ,$$

$$x_3 = x + n_\mu + n_\nu ,$$

$$x_4 = x + n_\nu .$$

The added contribution to the action due to the gauge field will be taken proportional to

$$\sum_{\text{plaquettes}} \chi(A(x_1x_2)A(x_2x_3)A(x_3x_4)A(x_4x_1)) ,$$

where the sum runs over all distinct plaquettes of the lattice. By definition $\chi = \chi^*$. All irreducible representations of the compact group SO(n) can be taken to be equivalent to unitary ones, while the Hermitian conjugate of a unitary matrix is its inverse. The representative of an inverse is the inverse of a representative. Consequently

$$\chi(A(x_1, x_2)A(x_2, x_3)A(x_3, x_4)A(x_4, x_1))$$

$$= \chi(A^{-1}(x_4, x_1)A^{-1}(x_3, x_4)A^{-1}(x_2, x_3)A^{-1}(x_1, x_2))$$

$$= \chi(A(x_1, x_4)A(x_4, x_3)A(x_3, x_2)A(x_2, x_1)) ,$$

where we have used (2.12). This property means that the order in which we orient the closed curve $(x_1x_2x_3x_4x_1)$ is irrelevant.

Finally, to sum over all configurations we shall use the invariant measure on the group SO(n) which we denote by dA. We assume that it is normalized to unity:

$$\int dA = 1 .$$

To make notations shorter we use the symbols s for site, l for link, and p for plaquette. To each site is assigned an "isotopic" vector $\vec{k}$, to each ordered link a gauge field A and an interaction kAk, and to each plaquette a term $\chi(AAAA)$. Since $dA = dA^{-1}$ it is immaterial in fact how we orient the links, as long as we are consistent over all of the lattice.

The Abelian group SO(2) is a slight exception to the above formalism since all its irreducible representations are one-dimensional and complex. One has

$$A \equiv e^{i\alpha},$$

$$\chi(A) = e^{im\alpha}, \quad m \text{ integer} .$$

In this case it is understood that we take for simplicity $m = 1$ and replace χ by Reχ, i.e., $\chi(A) = \cos\alpha$.

We call β_l the coefficient in front of the interaction term kAk, and β_p the one of $\chi(AAAA)$ in the full action S, which we now write:

$$S = \beta_l \sum_l k^T(x)A(x, x')k(x') + \beta_p \sum_p \chi(A(x_1, x_2)A(x_2, x_3)A(x_3, x_4)A(x_4, x_1)) , \tag{2.17}$$

$$Z = e^{NF} = \int \prod_s dk_x \prod_l dA(x, x')e^S . \tag{2.18}$$

As previously, we have assumed a spatial cutoff $N \to \infty$ and noticed that there are N sites and Nd links. The notation $Z = e^{NF}$ is a shorthand for

$$F = \lim_{N\to\infty} \frac{1}{N} \ln Z .$$

In the Abelian case of SO(2), writing $k = (\cos\phi, \sin\phi)$, $A = e^{i\alpha}$, the above formulas become

$$S_{Ab}=\beta_l\sum_l \cos[\phi(x)-\phi(x')+\mathcal{a}(x,x')]+\beta_p\sum_p\cos[\mathcal{a}(x_1,x_2)+\mathcal{a}(x_2,x_3)+\mathcal{a}(x_3,x_4)+\mathcal{a}(x_4,x_1)]\ , \tag{2.17$_{Ab}$}$$

$$Z_{Ab}=e^{NF_{Ab}}=\int^{N+dN}\prod_s\frac{d\phi(x)}{2\pi}\prod_l\frac{d\mathcal{a}(x,x')}{2\pi}e^{S_{Ab}}. \tag{2.18$_{Ab}$}$$

All integrals run over angles in an interval of 2π.

The case presented in Sec. II B corresponds to the freezing of all gauge degrees of freedom at the unit value $A=1$. Another interesting limit is obtained by taking $\beta_l=0$, in which case we have the pure Yang-Mills dynamics of the gauge field.

As for the interaction term, it is nice to observe that in the continuous limit one again recovers known formulas for the pure Yang-Mills interaction $\chi(AAAA)$. Again we restore the lattice spacing a. In the continuous limit, we write

$$A(x,x+an_\mu)=e^{iae\mathcal{a}_\mu(x)}\ .$$

Then, the contribution of the plaquette $(\mu\nu)$ issued from the corner x is of the form

$$\chi(AAAA)=\chi(e^{ia^2e\mathfrak{F}_{\mu\nu}(x)})\ , \tag{2.19}$$

where $\mathfrak{F}_{\mu\nu}$ belongs to the Lie algebra of the group.

In order to evaluate $\mathfrak{F}_{\mu\nu}$ to lowest order in a, we expand $\mathcal{a}_\mu(x+an_\nu)$ and $\mathcal{a}_\nu(x+an_\mu)$ in (19), and apply the Baker-Hausdorff formula:

$$e^Xe^Y=\exp(X+Y+\tfrac{1}{2}[X,Y]+\cdots)\ . \tag{2.20}$$

This yields, dropping the variable x,

$$\begin{aligned}AAAA&=e^{iae\mathcal{a}_\mu}e^{iae(\mathcal{a}_\nu+a\partial_\mu\mathcal{a}_\nu)}e^{-iae(\mathcal{a}_\mu+a\partial_\nu\mathcal{a}_\mu)}e^{-iae\mathcal{a}_\nu}\\&\simeq\exp\{iae(\mathcal{a}_\mu+\mathcal{a}_\nu+a\partial_\mu\mathcal{a}_\nu)-\tfrac{1}{2}a^2e^2[\mathcal{a}_\mu,\mathcal{a}_\nu]\}\\&\quad\times\exp\{-iae(\mathcal{a}_\mu+\mathcal{a}_\nu+a\partial_\nu\mathcal{a}_\mu)-\tfrac{1}{2}a^2e^2[\mathcal{a}_\mu,\mathcal{a}_\nu]\}\\&\simeq\exp\{ia^2e(\partial_\mu\mathcal{a}_\nu-\partial_\nu\mathcal{a}_\mu)-a^2e^2[\mathcal{a}_\mu,\mathcal{a}_\nu]\}\ .\end{aligned}$$

Hence

$$\mathfrak{F}_{\mu\nu}=\partial_\mu\mathcal{a}_\nu-\partial_\nu\mathcal{a}_\mu+ie[\mathcal{a}_\mu,\mathcal{a}_\nu] \tag{2.21}$$

is recovered as the generalization of the electromagnetic field.

Introducing the complete basis $\mathcal{L}_\alpha$ of generators of the group, and their representatives L_α in the chosen irreducible representation, we write

$$\mathfrak{F}_{\mu\nu}=\mathfrak{F}^\alpha_{\mu\nu}\mathcal{L}_\alpha\ ,$$

and (2.19) becomes

$$\begin{aligned}\chi(AAAA)&\simeq \mathrm{Tr}(1+ia^2e\mathfrak{F}^\alpha_{\mu\nu}L_\alpha-\tfrac{1}{2}a^4e^2\mathfrak{F}^\alpha_{\mu\nu}\mathfrak{F}^\beta_{\mu\nu}L_\alpha L_\beta)\\&=\text{dimension of the representation}\\&\quad-\tfrac{1}{2}a^4e^2\mathfrak{F}^\alpha_{\mu\nu}\mathfrak{F}^\beta_{\mu\nu}\mathrm{Tr}\,L_\alpha L_\beta\ .\end{aligned}$$

The term linear in $\mathfrak{F}$ has been dropped because L_α is traceless. The calculation is up to now entirely general. If the group is semisimple [as SO(n) or SU(n) is], we have by a proper choice of basis

$$\mathrm{Tr}\,L_\alpha L_\beta=\frac{\delta_{\alpha\beta}(\sum_\alpha \mathrm{Tr}\,L_\alpha^{\,2})}{\sum_\alpha 1}\ ,$$

so that for any semisimple group and real irreducible representation the term $\chi(AAAA)$ in the action reduces in the continuous limit to

$$\chi(AAAA)=C_1-\tfrac{1}{2}a^4e^2C_2\sum_\alpha(\mathfrak{F}^\alpha_{\mu\nu})^2 \tag{2.22}$$

For instance, for SU(2), we have $A=u_0+i\vec{u}\cdot\vec{\sigma}$ with $u_0^{\,2}+\vec{u}^{\,2}=1$ and $\vec{\sigma}$ the Pauli matrices. Hence u runs on the unit sphere in the 4-dimensional space, and $dA=(1/2\pi^2)d^4u\,2\delta(u^2-1)$. If we denote by ψ the angle of the associated rotation of SO(3) $=\mathrm{SU}(2)/Z_2$, we have $u_0=\cos\frac{1}{2}\psi$ and $|\vec{u}|=\sin\frac{1}{2}\psi$. Then we take for χ the trace in the representation of spin j, which yields

$$\chi(A)=\frac{\sin(2j+1)\frac{1}{2}\psi}{\sin(\frac{1}{2}\psi)}\ ,$$

and finally for a plaquette in the continuous limit

$$\chi(AAAA)=(2j+1)-\tfrac{1}{2}a^4e^2\tfrac{1}{3}j(j+1)\sum_{\alpha=1}^{3}(\mathfrak{F}^\alpha_{\mu\nu})^2\ . \tag{2.23}$$

In any case, apart from an inessential constant, the sum over all the plaquettes reproduces up to a choice of scale (which as we see implies $\beta_p\propto 1/e^2$) the conventional action for the Yang-Mills field

$$-\tfrac{1}{4}\int dx\,\mathfrak{F}^2\ .$$

This applies obviously also to the commutative case.

Thus the action (2.17) is a good candidate on our discrete lattice for a matter field (k) coupled to a gauge field (A). The usual Yang-Mills construction might seem even simpler in that case.

While the discretization in the matter-field case yields a model well known in statistical mechanics, it does not seem to be so in the gauge-field case to our knowledge. Interactions involve four links at once, and on the other hand there is a very large built-in local invariance. As we proceed further we shall see that the kind of "stuff" that this might represent is in a sense more like a *liquid* in its disordered phase. Again a phase transition occurs for some β_p^c. Beyond this value an ordered phase corresponding to the "breaking" of gauge invariance comes into play. A long-range excitation is present corresponding to the usual massless Yang-Mills boson. It might perhaps, at first, seem surprising that, at least for $\beta_p \to \infty$ corresponding to the continuous limit discussed above, gauge invariance is spontaneously broken. On second thought, however, one realizes that any known quantization procedure in the usual theory breaks gauge invariance, which is restored by asking "gauge-invariant questions." Thus, while the discrete case does not involve any breaking at the level of the action, this occurs in a natural way in its ordered (large-β_p) phase.

Correspondingly, for $\beta_p < \beta_p^c$ only gauge-invariant quantities have nonvanishing expectation values. Typical of these are expressions of the form (we assume for the moment that $\beta_l = 0$)

$$\langle \chi(A(x_1, x_2)A(x_2, x_3)\cdots A(x_k, x_1))\rangle ,$$

where the points $(x_1, x_2)\cdots(x_k, x_1)$ form a closed curve on the lattice. In the commutative case and in the continuous limit this expression is essentially

$$\left\langle \exp\left[ie \oint_C dx_\mu \mathfrak{A}_\mu(x)\right]\right\rangle .$$

(In the noncommutative case a T-ordering symbol along the curve should appear in front of the exponential.) This average might be taken as a representative of the effect of a closed loop for a charged particle interacting with the gauge field. Now for $\beta_p \gg \beta_p^c$ (or $e \to 0$) we expect lowest order in perturbation theory to be a reliable guide, in which case one would have

$$\left\langle \exp\left[ie \oint_C dx_\mu \mathfrak{A}_\mu(x)\right]\right\rangle \sim \exp\left[-e^2 \oint_C \oint_C dx_{1\mu}\Delta_{\mu\nu}(x_1 - x_2)dx_{2\nu}\right] ,$$

with $\Delta_{\mu\nu}$ the free massless propagator (in Euclidean space) which behaves like $\delta_{\mu\nu}/|x_1 - x_2|^{d-2}$. Thus the above exponential decreases most likely as exp(−const × length of C), up to logarithms. A proper evaluation requires the ultraviolet cutoff.

We shall see, however, that for e large enough or $\beta_p < \beta_p^c$ the corresponding expectation value behaves rather like

$$\exp(-\text{const} \times \text{minimal area enclosed by } C) .$$

This is an indication that in the strong-coupling limit a pair of oppositely charged particles have a very hard time to separate themselves in the presence of the gauge field. The long-range part of the forces seem to have built up a strong attractive barrier. This will be elaborated later.

Let us add a remark on the structure of "interactions" provided by the Yang-Mills Lagrangian in its discrete version. For that purpose consider the commutative case of SO(2) where in an appropriate scale the relevant term is for each plaquette

$$\mathfrak{L}_p = \cos(\mathfrak{A}_{12} + \mathfrak{A}_{23} + \mathfrak{A}_{34} + \mathfrak{A}_{41}) .$$

Let us identify $e^{i\mathfrak{A}}$ with a unit two-dimensional vector that we denote by $\vec{A}$. We set

$$e^{i\mathfrak{A}_{12}} \to \vec{A}^1 ,$$
$$e^{i\mathfrak{A}_{23}} \to \vec{A}^3 ,$$
$$e^{-i\mathfrak{A}_{34}} = e^{i\mathfrak{A}_{43}} \to \vec{A}^2 ,$$
$$e^{-i\mathfrak{A}_{41}} = e^{i\mathfrak{A}_{14}} \to \vec{A}^4 ,$$

and find

$$\begin{aligned}\mathfrak{L}_p &= \mathrm{Re}[(A_1^1 + iA_2^1)(A_1^3 + iA_2^3)(A_1^2 - iA_2^2)(A_1^4 - iA_2^4)]\\ &= \mathrm{Re}\{[A_1^1A_1^2 + A_2^1A_2^2 + i(A_2^1A_1^2 - A_1^1A_2^2)]\\ &\quad\times[A_1^3A_1^4 + A_2^3A_2^4 + i(A_2^3A_1^4 - A_1^3A_2^4)]\}\\ &= (\vec{A}^1\cdot\vec{A}^2)(\vec{A}^3\cdot\vec{A}^4) - (\vec{A}^2\times\vec{A}^1)\cdot(\vec{A}^4\times\vec{A}^3)\\ &= (\vec{A}^1\cdot\vec{A}^2)(\vec{A}^3\cdot\vec{A}^4) - (\vec{A}^1\cdot\vec{A}^3)(\vec{A}^2\cdot\vec{A}^4)\\ &\quad + (\vec{A}^1\cdot\vec{A}^4)(\vec{A}^2\cdot\vec{A}^3) .\end{aligned} \qquad (2.24)$$

In other words, $\mathfrak{L}_p$ is a Pfaffian,

$$\mathfrak{L}_p = \mathrm{Pf}\begin{Bmatrix} (\vec{A}^1\cdot\vec{A}^2) & (\vec{A}^1\cdot\vec{A}^3) & (\vec{A}^1\cdot\vec{A}^4)\\ & (\vec{A}^2\cdot\vec{A}^3) & (\vec{A}^2\cdot\vec{A}^4)\\ & & (\vec{A}^3\cdot\vec{A}^4)\end{Bmatrix} , \qquad (2.24')$$

or the root of the antisymmetric determinant obtained by completing the Pfaffian. This alternative expression (useful in the sequel) shows even in the Abelian case the complexity of the "interaction." If to each ordered link we associate a "spin" $\vec{A}$ we see how these spins combine four by four. The disadvantage of this notation is that one loses track of the local SO(2) invariance.

Finally, we have only considered up to now continuous gauge groups. However, since we are directly working on the group and not on its Lie algebra we can extend these models to include discrete groups. In particular we can look at the case when this group is

$$\bigotimes^N Z_2$$

(Z_2 being the group with two elements ±1). It is the natural extension to $n=1$ of the previous models described by Eqs. (2.17) and (2.18). More precisely, $k(x)$ takes the values ±1 as well as $A(x, y)$, while $\chi(AAAA) \equiv AAAA$. As a result, when one freezes $A(x, y)$ to the value unity, one recovers the ordinary Ising model. Thus one obtains a gauge-invariant generalization of the Ising model with a discrete gauge group with no continuous version. It turns out that precise and nontrivial results can be obtained in this case, which we shall present in another paper of this series.

III. EXACT SOLUTIONS IN LOW DIMENSION

It is well known from the Ising case that some statistical models are soluble in small dimension. The analogy here is for the scalar or σ model for $d=1$. No explicit solution corresponding to the Onsager one is available for $d=2$. This is unfortunate, since for the gauge field the model only makes sense for $d \geq 2$. For the latter, the solution can be found for $d=2$. We shall thus present these various solutions but will be unable here to treat an example of the coupled case (2.17), (2.18).

Apart from being useful in that one gets some familiarity with the manipulation of the expressions, these soluble cases are expected to behave near $\beta \to \infty$ as will the more "realistic" ones in higher dimension near their critical point. Hence for these models $\beta=\infty$ or $T=0$ (or zero coupling) can be identified in a certain sense with a critical point.

At the other extreme when d gets very large we shall see that we can essentially treat the coupled model exactly. This will be postponed until the next section. Numerical calculations at intermediate d will be presented in another article.

A. Scalar model for $d=1$

For $d=1$, the "free energy" is defined by

$$Z=e^{NF}=\int \prod_0^{N-1} \frac{d\phi_i}{2\pi} \exp\left[\beta \sum_0^{N-1} \cos(\phi_{i+1}-\phi_i)\right]. \tag{3.1}$$

More generally, for SO(n), we have to integrate the vector k_i associated to each site over a unit sphere:

$$Z=e^{NF}=\int \prod_0^{N-1} d^{n-1}k_i \exp\left(\beta \sum_0^{N-1} k_i \cdot k_{i+1}\right). \tag{3.2}$$

This model is well known in the context of statistical mechanics.[2] The "free energy" is given by

$$F=\ln[\Gamma(\tfrac{1}{2}n)(\beta)^{1-n/2} I_{n/2-1}(\beta)] , \tag{3.3}$$

where $I_\nu(x)$ is the modified Bessel function. In particular, for $n=2$, we have

$$F=u(\beta)=\ln I_0(\beta) , \tag{3.4}$$

a function which will play a crucial role in the remaining part of this work. The main properties of $u(x)$ and its first two derivatives are displayed in Table I. No transition, of course, occurs in this system, the nearest singularity being the complex zero of I_0, $\beta=\pm i\,2.405$.

More interesting is the behavior of the correlation function[2]

TABLE I. The function $u(x)=\ln I_0(x)$ and its first two derivatives.

Function		x small	x large
$u(x)$	even	$(\frac{1}{2}x)^2-\frac{1}{4}(\frac{1}{2}x)^4+\cdots$	$x-\frac{1}{2}\ln(2\pi x)+\frac{1}{8x}-\frac{4}{(8x)^2}+\cdots$
$u'(x)$	odd	$(\frac{1}{2}x)-\frac{1}{2}(\frac{1}{2}x)^3+\cdots$	$1-\frac{1}{2x}-\frac{1}{8x^2}-\frac{1}{8x^3}+\cdots$
$u''(x)$	even	$\frac{1}{2}-\frac{3}{4}(\frac{1}{2}x)^2+\cdots$	$\frac{1}{2x^2}+\frac{1}{4x^3}+\frac{3}{8x^4}+\cdots$

$$G(r,\beta)=\langle k_i \cdot k_{i+r}\rangle=\left(\frac{dF}{d\beta}\right)^r , \tag{3.5}$$

which behaves for β large as

$$G(r,\beta)\sim \exp\left(-r\frac{n-1}{2\beta}\right) . \tag{3.6}$$

At low temperature, the correlation range $2\beta/(n-1)$ becomes very large. In other words, the mass $\mu=(n-1)/2\beta$ goes to zero as if the system were more and more ordered near $\beta=\infty$. An ordered regime with zero mass will set in below some critical temperature for $d>2$, as discussed later.

The generalization to $n>2$ acquires an additional interest because an exact solution exists in another limit. The latter is the Stanley[4] limit obtained by keeping d fixed and letting $n\to\infty$. One obtains then a nontrivial behavior with a transition.

B. Abelian gauge field for $d=2$

We turn now to a pure Abelian gauge field. That is, we set $\beta_l=0$ and drop the index p on β_p. Such a model only makes sense for $d \geq 2$, hence we expect that it is simple for $d=2$. This is indeed the case. Recall the formulas $(2.17)_{\rm Ab}$ and $(2.18)_{\rm Ab}$:

$$Z=e^{NF}=\int \prod_l \frac{d\mathcal{a}_{ij}}{2\pi} e^S ,$$
$$S=\beta \sum_p \cos(\mathcal{a}_{12}+\mathcal{a}_{23}+\mathcal{a}_{34}+\mathcal{a}_{41}) . \tag{3.7}$$

To be specific assume the lattice to be a square of $N=L^2$ lattice sites. Note that while there are $2N$ links there are N plaquettes. If we *do not* assume periodic boundary conditions it is readily seen that the plaquette variables $\mathcal{a}_{12}+\mathcal{a}_{23}+\mathcal{a}_{34}+\mathcal{a}_{41}$ are independent. Consequently, taking them among the integration variables, we obtain

$$e^{NF}=I_0(\beta)^N .$$

Hence

$$F=u(\beta) \tag{3.8}$$

exactly as before for the scalar model and $n=2$, $d=1$.

As we discussed in Sec. II the interesting correlations are expressed not in terms of Green's functions but in terms of averages of gauge-invariant quantities of the type

$$\mathcal{C} = \left\langle \exp\left(i \sum_C \mathcal{Q}_{ij}\right)\right\rangle$$

$$= Z^{-1} \int^{2N} \prod_l \frac{d\mathcal{Q}_{ij}}{2\pi} \exp\left(i \sum_C \mathcal{Q}_{ij} + S\right) , \quad (3.9)$$

where $\sum_C \mathcal{Q}_{ij}$ means a sum along a simple closed curve on the lattice. We assume the curve not to be self-intersecting, that is, to separate an internal region from an external one. Now due to the fact that $\exp(i\mathcal{Q}_{ij}) = \exp(-i\mathcal{Q}_{ji})$ one sees (as in the traditional proof of Cauchy's theorem) that

$$\exp\left(i \sum_C \mathcal{Q}_{ij}\right) = \exp\left(i \sum_p \mathcal{Q}_p\right) ,$$

where $\mathcal{Q}_p$ is a shorthand notation for the sum of four $\mathcal{Q}$'s pertaining to a plaquette and the sum runs over all internal plaquettes enclosed by C. This manipulation has been possible (i) because of the topology of the plane (a simple nonintersecting curve defines an interior and an exterior), and (ii) because of the Abelian character of the field. The number of plaquettes enclosed by C is nothing but the *area* $s(C)$ bounded by this curve. Consequently

$$\mathcal{C} = \left[\frac{\int_0^{2\pi} \frac{d\mathcal{Q}}{2\pi} \exp(\beta\cos\mathcal{Q} + i\mathcal{Q})}{\int_0^{2\pi} \frac{d\mathcal{Q}}{2\pi} \exp(\beta\cos\mathcal{Q})}\right]^{s(C)}$$

$$= \exp\left(-s(C) \ln \frac{1}{u'(\beta)}\right) . \quad (3.10)$$

We observe a similitude with the behavior of the correlation function in the scalar case for $d = 1$. The average value $\mathcal{C}$ decreases exponentially with the area enclosed by the curve C. This is what we expected in general in the disordered phase.

Note that the coefficient of this decrease behaves exactly as we discussed in the preceding paragraph. Namely, as $\beta \to \infty$ it goes to zero. This type of "binding" becomes less and less effective as we approach the "pseudocritical point" $\beta = \infty$.

C. Non-Abelian gauge field for $d = 2$

Let us see how far one can go in the non-Abelian gauge field case for the simple topology of the plane. The function to be computed first is

$$Z = e^{NF} = \int \prod_l dA \exp\left[\beta \sum_p \chi(A_{12}A_{23}A_{34}A_{41})\right] , \quad (3.11)$$

where we recall that dA stands for the normalized invariant measure on the group with

$$dA = dAB = dBA = dA^{-1}$$

for a fixed element B of the compact group. Again this allows us to consider on a square of $N = L^2$ lattice points the plaquette variables $A_p = A_{12}A_{23}A_{34}A_{41}$ as independent ones. To see this one might start from the edges of the square and integrate successively on the free variables pertaining to the boundary links. This leads to

$$e^F = \int dA\, e^{\beta\chi(A)} . \quad (3.12)$$

Qualitatively, F will have the same properties as in the Abelian case. To illustrate this point consider the case of the group SO(3) and the jth character (cf. Sec. II C). Let θ be the polar angle on the unit sphere in four-dimensional space; then

$$e^F = \frac{2}{\pi} \int_0^\pi d\theta \sin^2\theta \exp\left(\beta \frac{\sin(2j+1)\theta}{\sin\theta}\right) . \quad (3.13)$$

For j arbitrary this is still quite complicated, while for $j = \frac{1}{2}$, corresponding to the simplest character, we have

$$e^F = \frac{2}{\pi} \int_0^\pi d\theta \sin^2\theta \exp(2\beta\cos\theta) ,$$

$$= \frac{1}{\pi} \int_0^\pi d\theta (1 - \cos 2\theta) \exp(2\beta\cos\theta)$$

$$= I_0(2\beta) - I_2(2\beta) = \frac{I_1(2\beta)}{\beta} .$$

Finally then

$$F = \ln \frac{I_1(2\beta)}{\beta} \quad [\mathrm{SO}(3),\ j = \tfrac{1}{2}] . \quad (3.14)$$

A general expression can in fact be obtained for any SO(n) and any character χ. The handling of the average

$$\mathcal{C} = \left\langle \chi\left(\prod_C A_{ij}\right)\right\rangle , \quad (3.15)$$

where $\Pi_C A_{ij}$ denotes an ordered product along a simple closed curve C, is not as simple due to the noncommutativity of the group. The "Cauchy trick" does not seem to apply simply for this reason. Nevertheless, it is also possible to prove that $\ln\mathcal{C}$ is proportional to the area enclosed by C.

IV. MEAN-FIELD APPROXIMATION

At the extreme opposite of the low-dimensionality case discussed in the previous section, one expects another type of simplification when $d \to \infty$. For $d \to \infty$ the number of neighbors "interacting" with a spin of our scalar model, say, grows like

d. Hence it might be expected that their over-all effect is equivalent to a mean field to be determined consistently. This is the mean-field approximation. We shall show, using Peierls's inequality,[5] that one obtains in this way a lower bound on F. This bound is assumed exact as $d \to \infty$. We do not know a rigorous proof of this fact. However, a power-series expansion in $1/d$ is readily obtained, as we shall see in a forthcoming paper.

While the mean-field approximation is a standard device in statistical mechanics, we do not expect every reader to be familiar with it. Furthermore, some caution has to be exercised owing to gauge invariance. Consequently we proceed by steps. First we present the mean field in detail in the scalar model. We then extend it, without special care, first to the pure Abelian gauge field, then to the coupled system. (We ignore in this section non-Abelian gauge fields.) Finally we refine the analysis to justify this rather blunt procedure in order to meet possible objections on the role of gauge invariance.

Our most interesting result is the phase diagram (Fig. 1) obtained for the coupled system. Perturbation theory is possible around the mean-field approximation. As a consequence, the over-all picture obtained here is likely to be close to the exact solution except for fine details, especially near critical curves.

A. Scalar model

We want to compute

$$Z = e^{NF} = \int \prod_0^{N-1} d\vec{k}\, \exp\left(\beta \sum_{(ij)} \vec{k}_i \cdot \vec{k}_j\right) \tag{4.1}$$

for d very large. A given $\vec{k}_i$ interacts with $\sum_{j(i)} \vec{k}_j$, where $j(i)$ denotes the $2d$ neighbors of the site i.

The average

$$\frac{1}{2d} \sum_{j(i)} \vec{k}_j$$

can be expected to behave like some mean field when $d \to \infty$. To present the matter on a firmer basis, one proceeds as follows. If we give ourselves the N values of the two-dimensional vectors $\vec{k}_i$ we call it a configuration $\{k_i\}$. The configuration space is thus

$$\overset{N}{\otimes} S_1 ,$$

the Cartesian product of N unit circles. This is

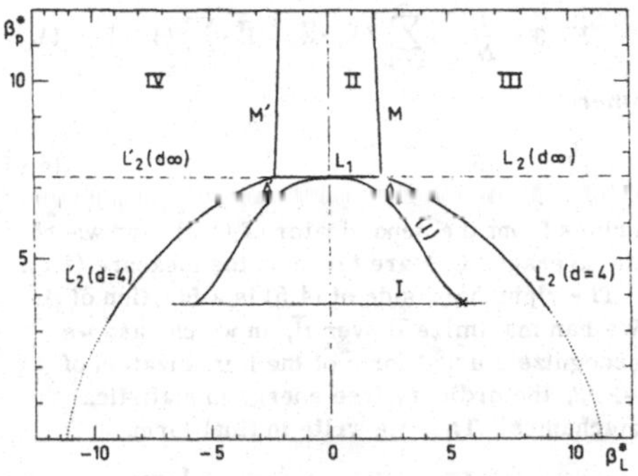

FIG. 1. The phase diagram for the coupled system. The phases are (I) disordered $H = K = 0$; (II) Yang-Mills order $H = 0$, $K \neq 0$; (III) "ferromagnetic" order $H \neq 0$, $K \neq 0$; (IV) "antiferromagnetic" order $H \neq 0$, $K \neq 0$. The curves (L_2) and (L_2') (drawn for $d = 4$) are speculative as one moves away from the triple points A, A'. The curve (l) arises from the poor mean-field approximation to the "B model" of Sec. IV D.

a compact space (identical with an N-torus) on which we can define normalized measures

$$d\mu\{k_i\}, \quad \int_{\overset{N}{\otimes} S_1} d\mu\{k_i\} = 1 .$$

Rather than the actual measure

$$\frac{1}{Z} \prod_i d\hat{k}_i e^S ,$$

the above discussion suggests to introduce the easily tractable one

$$d\mu\{k_i\} = \frac{\prod_0^{N-1} d\hat{k}_i \exp(\vec{H} \cdot \vec{k}_i)}{\int \prod_0^{N-1} d\hat{k}_i \exp(\vec{H} \cdot \vec{k}_i)} , \tag{4.2}$$

where $\vec{H}$ represents a mean field (up to a scaling factor) which remains to be determined. We shall assume this field to be uniform for the moment.

Now we have on the measure μ Peierls's inequality (due to the convexity of the exponential function),

$$\langle e^A \rangle \geq e^{\langle A \rangle} , \tag{4.3}$$

so that we may write

$$\int d\mu\, \exp\left(\beta \sum_{(ij)} \vec{k}_i \cdot \vec{k}_j - \sum_i \vec{H} \cdot \vec{k}_i\right) \geq \exp\left[\int d\mu \left(\beta \sum_{(ij)} \vec{k}_i \cdot \vec{k}_j - \sum_i \vec{H} \cdot \vec{k}_i\right)\right] . \tag{4.4}$$

Inserting (4.1) and (4.2) in (4.4), we finally obtain

$$F \geq \mathfrak{F} + \frac{1}{N}\left(\beta \sum_{(ij)} \langle \vec{k}_i \cdot \vec{k}_j \rangle - \vec{H} \cdot \sum_i \langle \vec{k}_i \rangle\right), \quad (4.5)$$

where

$$e^{\mathfrak{F}} = \int d\hat{k}\, e^{\vec{H} \cdot \vec{k}} \quad (4.6)$$

arises from the denominator of (4.2), and where the averages $\langle\ \rangle$ are taken on the measure (4.2).

The right-hand side of (4.5) is a function of $\vec{H}$. We can maximize it over $\vec{H}$, in which case we recognize a usual form of the minimization of $-F/\beta$, the ordinary free energy in statistical mechanics. Thus we write in final form

$$F \geq \operatorname*{Sup}_{\vec{H}} \left[\beta \frac{1}{N} \sum_{(ij)} \langle \vec{k}_i \cdot \vec{k}_j \rangle + \left(\mathfrak{F}(\vec{H}) - \frac{1}{N} \sum_i \langle \vec{H} \cdot \vec{k}_i \rangle\right)\right]. \quad (4.7)$$

Up to a constant the first term is proportional to energy, the second to entropy. Both terms are readily evaluated.

We have

$$\mathfrak{F}(\vec{H}) = \ln\left(\int \frac{d\phi}{2\pi} e^{|\vec{H}|\cos\phi}\right) = u(H), \quad H \equiv |\vec{H}|$$

$$\langle \vec{k}_i \rangle = \frac{\vec{H}}{H} u'(H) \equiv \hat{H} u'(H), \quad (4.8)$$

$$\langle \vec{k}_i \cdot \vec{k}_j \rangle = \langle \vec{k}_i \rangle \cdot \langle \vec{k}_j \rangle = u'(H)^2 .$$

Finally, using the fact that we have N sites on the lattice and Nd links or pairs of interacting neighbors,

$$F \geq \operatorname*{Sup}_{\vec{H}} \{\beta d u'(H)^2 + [u(H) - H u'(H)]\} . \quad (4.9)$$

The idea is now that for $d \to \infty$, the right-hand side is in fact the value of F; thus in this limit we replace the inequality by an equality sign. The natural scale of inverse "temperature" or inverse coupling constant thus appears to be βd.

For small H the right-hand side of (4.9) exhibits the competition of two terms in H^2, the first with a positive coefficient and the second with a negative one, while for H large the second dominates and gets large and negative like $-\frac{1}{2}\ln H$ (see Table I). Consequently, for small β the maximum is reached for $H=0$ (in which case it is zero), while for β large it is obtained for some finite value H_{eff} at which point F is positive. The dividing line is obtained by requiring the coefficient of H^2 for small H to vanish. That is,

$$(\tfrac{1}{2}H)^2(\beta_c d - 1) = 0 .$$

Thus the critical value of β is

$$\beta_c d = 1 . \quad (4.10)$$

For $\beta d < 1$,

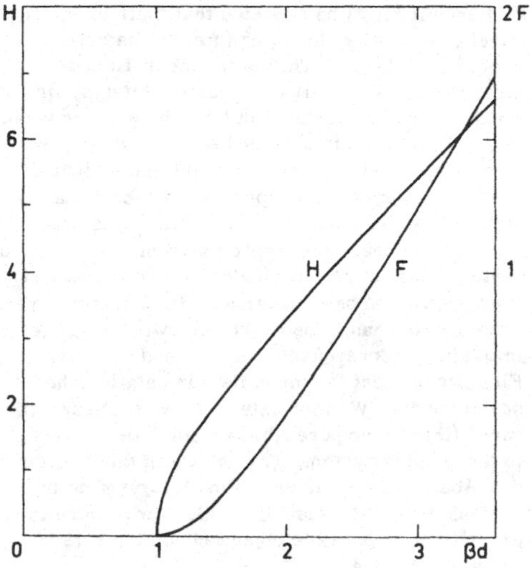

FIG. 2. The functions F and H_{eff} of the scalar model in the mean-field theory ($n = 2$).

$$H_{\text{eff}} = 0 \text{ and } F = 0 ;$$

for $\beta_c d = 1 + \epsilon$,

$$H_{\text{eff}} = 2(2\epsilon)^{1/2} + \cdots \quad \text{and } F = \epsilon^2 + \cdots ;$$

finally, for $\beta_c d > 1$, H_{eff} is given by

$$2\beta d u'(H_{\text{eff}}) = H_{\text{eff}}, \quad (4.11)$$

and F is positive:

$$F = u(H_{\text{eff}}) - \tfrac{1}{2} H_{\text{eff}}\, u'(H_{\text{eff}}) . \quad (4.12)$$

This behavior is summarized in Fig. 2. Since beyond the critical points H_{eff} starts from a zero value and since F and $\partial F/\partial\beta$ are continuous, we find a typical *second-order* transition.

Within the mean-field approximation it is also possible to compute the inverse correlation length μ. Consider the response of the system to the coupling to an external field Q_i. We write

$$Z(\beta, Q_i) = \int \prod_0^{N-1} d\hat{k}_i \exp\left(\beta \sum_{(ij)} \vec{k}_i \cdot \vec{k}_j + \sum_i \vec{Q}_i \cdot \vec{k}_i\right) . \quad (4.13)$$

The connected two-point function is

$$G_{\mu\nu}(i,j;\beta) = \frac{\partial^2}{\partial Q_i^\mu \partial Q_j^\nu} \ln Z(\beta, Q_i)\bigg|_{Q_i=0} . \quad (4.14)$$

Within the mean-field approximation we take a varying mean field $\vec{H}_i$ and find ($\hat{H}_i = \vec{H}_i / H_i$)

$$\ln Z(\beta, Q_i) = \operatorname*{Sup}_{\vec{H}_i}\left\{\sum_{ij} \beta \hat{H}_i \cdot \hat{H}_j u'(H_i) u'(H_j) + \sum_i [u(H_i) + (\vec{Q}_i - \vec{H}_i)\cdot \hat{H}_i u'(H_i)]\right\}. \tag{4.15}$$

Q_i is infinitesimal since we are only interested in the second derivative in Q at $Q=0$. The condition for a maximum in H is obtained by setting the gradient in H of the right-hand side equal to zero. This is

$$\beta\left[\sum_{j(i)} (I - \hat{H}_i \otimes \hat{H}_i)\frac{\hat{H}_j}{H_i} u'(H_i)u'(H_j) + (\hat{H}_j \cdot \hat{H}_i)\hat{H}_i u''(H_i)u'(H_j)\right] + u'(H_i)(I - \hat{H}_i \otimes \hat{H}_i)\frac{\vec{Q}_i}{H_i} + (\vec{Q}_i - \vec{H}_i)\cdot \hat{H}_i \hat{H}_i u''(H_i) = 0 . \tag{4.16}$$

Not only is $\vec{Q}_i$ very small, but since we are interested in long-wavelength excitations, we can assume its direction to vary very slowly around a mean direction. For $\beta d > 1$ its main effect will be to drag the mean field along this direction. We can thus speak of longitudinal and transverse excitations (with respect to this direction in "isospin space"). As we can verify using (4.16) and in accordance with Goldstone's theorem the transverse excitation (the π) is of infinite range, $\mu_\pi = 0$ ($\beta d > 1$). The longitudinal excitations are obtained as follows. Take the scalar product of (4.16) with $\hat{H}_i$:

$$\sum_{j(i)} \beta(\hat{H}_j \cdot \hat{H}_i)^2 u''(H_i)u'(H_j) + u''(H_i)\hat{H}_i \cdot (\vec{Q}_i - \vec{H}_i) = 0 . \tag{4.17}$$

Since $\vec{Q}_i$ is small we can write for the longitudinal part

$$\vec{H}_i = \vec{H} + \sum_s \chi_{is}\vec{Q}_s ,$$

with χ_{is} essentially identical with the longitudinal part of the Green's function (4.14). Expanding now (4.17) to first order in Q we find (with $|\vec{H}| = H_{\text{eff}}$ and suppressing the index "eff")

$$\beta\sum_{j(i)}\left[u'''(H)u'(H)\sum_s \chi_{is}\hat{H}\cdot\vec{Q}_s + u''(H)^2\sum_s \chi_{js}\hat{H}\cdot\vec{Q}_s\right] + u''(H)\hat{H}\cdot\vec{Q}_i + [Hu'''(H) - u''(H)]\sum_s \chi_{is}\hat{H}\cdot\vec{Q}_s = 0 . \tag{4.18}$$

The elementary solution of this equation fulfills then

$$2\beta d\left\{[u'''(H)u'(H) + u''(H)^2]\chi_{is} + u''(H)^2\frac{1}{2d}\sum_{j(i)}(\chi_{js} - \chi_{is})\right\} - [u''(H) - Hu'''(H)]\chi_{is} + u''(H)\delta_{is} = 0 .$$

This is a second-order difference equation on the lattice with $\sum_{j(i)}(\chi_{js} - \chi_{is})$ playing the role of the Laplacian. For i very far from s it can be approximated by an ordinary partial differential equation.

Recalling (4.11) and expressing everything as functions of H instead of β we find with Δ_d the Laplacian in d dimensions

$$\left\{\Delta_d + \frac{2d}{u''(H)^2}\left[u'''(H)u'(H) + u''(H)^2 - \frac{u'(H)}{H}[u''(H) - Hu'''(H)]\right]\right\}\chi(x, x') = -\frac{2d}{u''(H)}\frac{u'(H)}{H}\delta(x - x') . \tag{4.19}$$

This is a typical free-particle wave equation. If we call

$$\mu_\sigma^2 = -\frac{2d}{u''(H)^2}\left[u'''(H)u'(H) + u''(H)^2 - \frac{u'(H)}{H}[u''(H) - Hu'''(H)]\right], \quad \beta d > 1 \tag{4.20}$$

μ_σ^2 is an effective-mass square for longitudinal excitations. Asymptotically for large $|x - x'|$ the behavior of χ is

$$\chi(x, x') \sim \text{const} \times \frac{\exp(-\mu_\sigma|x - x'|)}{|x - x'|^{d-2}} . \tag{4.21}$$

As $H \to 0$ or $\beta d \to 1$ we find from (4.20)

$$\mu_\sigma^2 \sim 28d(\beta d - 1), \quad \beta d > 1 . \tag{4.22}$$

For $\beta d < 1$ the Goldstone solution disappears, H_i is of order Q_i. We again expand (4.16) to first order in Q with

$$\vec{H}_i = \sum_s \chi_{is}\vec{Q}_s ,$$

and we find

$$\tfrac{1}{2}\beta\sum_{j(i)}\chi_{js}\vec{Q}_s - \sum_s \chi_{is}\vec{Q}_s + \vec{Q}_i = 0 . \tag{4.23}$$

Thus χ satisfies

$$\beta d\left[\frac{1}{2d}\sum_{j(i)}(\chi_{js} - \chi_{is})\right] - (1 - \beta d)\chi_{is} = -\delta_{is} . \tag{4.24}$$

The same comments as above apply to

$$\mu_{\sigma,\pi}{}^2 = \frac{1-\beta d}{\beta d}, \quad \beta d<1 . \tag{4.25}$$

The two masses are now degenerate. We observe in the vicinity of $\beta d=1$ a change of slope by comparing (4.22) and (4.25). These results, however, are rough, and in the vicinity of $d=4$ one knows from Wilson's theory of critical exponents[6] that modifications appear.

Finally, by studying the expression (4.20), one can show that for $\beta d\to\infty$ (limit of zero coupling) $\mu_\sigma{}^2\to\infty$ (in fact, $\mu_\sigma{}^2$ grows like β), as we stated earlier.

In Fig. 3 we have given the global picture for the masses as functions of β.

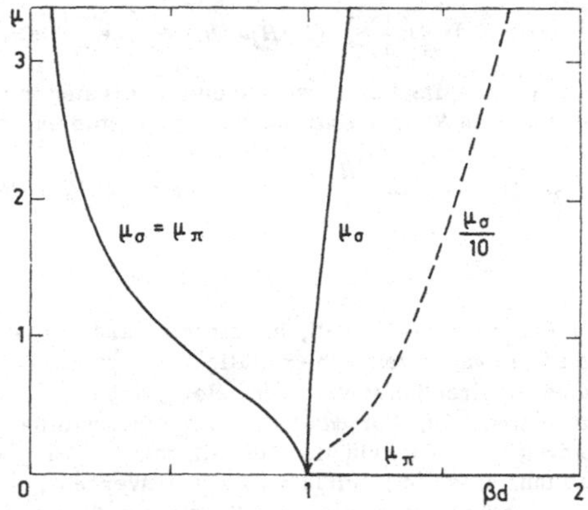

FIG. 3. Behavior of μ_σ and μ_π as functions of βd. Owing to the sharp rise of μ_σ beyond the critical point, we have also plotted on the same scale $\frac{1}{10}\mu_\sigma$.

B. Pure gauge field

We now undertake a parallel discussion for the Abelian gauge-field models $(2.17)_{\rm Ab}$ and $(2.18)_{\rm Ab}$. It helps to use the "spins" introduced at the end of Sec. II C with the expression (2.24) for a plaquette contribution.

A nonvanishing mean field will correspond to a nonvanishing average value $\langle A_{ij}\rangle$. This in turn implies breaking of gauge invariance. If we have a uniform mean field, $\langle A_{ij}\rangle$ will be independent of the link (ij). Of course, this is not the most general way in which gauge invariance can be broken. However, it is good enough for our purpose, and we shall not yet elaborate this subtle point further.

We denote by $\vec{K}$ the mean field associated with the spins $\vec{A}_{ij}$. This is again a two-dimensional vector. If we recall that there are $Nd(d-1)/2$ plaquettes on the lattice while there are Nd links, we readily find

$$F \geq \operatorname*{Sup}_{\vec{K}} \{\tfrac{1}{2}\beta d(d-1)[u'(K)]^4 + d[u(K)-Ku'(K)]\} . \tag{4.26}$$

As in the previous paragraph we take the right-hand side of (4.26) as representative of F for $d\to\infty$. We note, however, that the situation is different from the scalar case due to the occurrence of u'^4 instead of u'^2 in the "energy" term. For small as well as large K values, the entropy term $u(K)-Ku'(K)$ always dominates irrespective of the value of β. The expression $F(K,\beta)$ to be maximized always has a local maximum at $K=0$ and possibly a second maximum at some other finite value of K. For various values of β the expression $F(K,\beta)$ is sketched in Fig. 4 as function of K. For some critical value β_c the relevant maximum at $K=0$, where $F=0$, jumps to $F=0$ at some critical finite field K_c. As β gets larger than β_c the maximum is obtained for some $K>K_c$ with $F>0$. This sudden jump in K will correspond to a discontinuity in $dF/d\beta$ at β_c and hence is characteristic of a *first-order transition*.

First-order transitions are in fact common in thermodynamical systems even though mostly second-order ones have been scrutinized recently[6] due to interesting fluctuation phenomena. However, it is to some point slightly surprising to find this behavior for the gauge field. We believe this is due to the very different invariance that is broken when "order" sets in. As we said earlier, and as will appear more clearly in the next paragraph, the kind of transition that we observe can be roughly compared to a liquid-solid one.

Next we evaluate K_c and β_c. They are obtained from the pair of equations expressing that $F=0$ and $\partial F/\partial K=0$ excluding the solution $K=0$. Thus

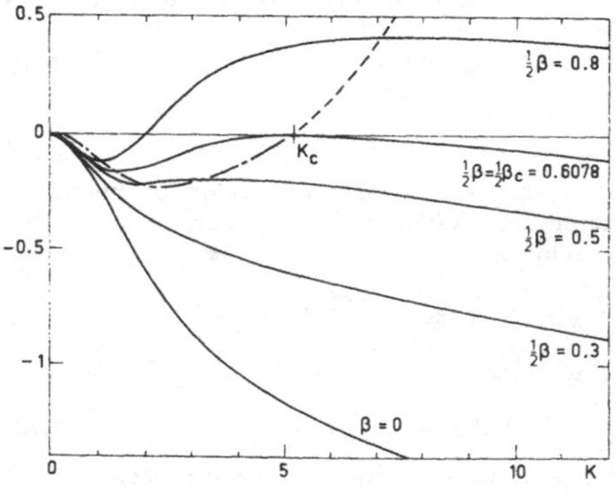

FIG. 4. The function $F(K,\beta)$ for the Abelian Yang-Mills field as a function of the mean field K.

from (4.26)

$$\tfrac{1}{2}\beta_c(d-1)[u'(K_c)]^4 - [u(K_c) - K_c u'(K_c)] = 0\,,$$
$$2\beta_c(d-1)[u'(K_c)]^3 - K_c = 0\,. \tag{4.27}$$

These equations can be rewritten

$$\beta_c(d-1) = \frac{K_c}{2[u'(K_c)]^3}\,,$$
$$4u(K_c) - 3K_c u'(K_c) = 0\,. \tag{4.28}$$

The second equation yields K_c; the first one then gives β_c. They can be solved numerically. The curve $y(x) = u(x) - \frac{3}{4}xu'(x)$ is drawn in Fig. 5. We find

$$K_c = 5.32\,,$$
$$\tfrac{1}{2}\beta_c(d-1) = 1.82\,. \tag{4.29}$$

C. Coupled system

It is interesting to treat now the complete system of 2-dimensional unit spins $\vec{k}_i$ coupled to an Abelian gauge field. The complete action is a sum

$$S_1 + S_2\,,$$
$$S_1 = \beta_p \sum_p \cos(\mathcal{A}_{12} + \mathcal{A}_{23} + \mathcal{A}_{34} + \mathcal{A}_{41})\,, \tag{4.30}$$
$$S_2 = \beta_l \sum_l k_i^T A_{ij} k_j\,.$$

The partial results of the previous two paragraphs indicate a competition between a second-order transition in the (σ, π) system and a first-order one of the gauge field.

It is indeed clear from the outset that one has the following limiting behaviors. If $\beta_l = 0$ we recover clearly the pure gauge field with its first-order transition. Next if $\beta_p \to \infty$ then the gauge

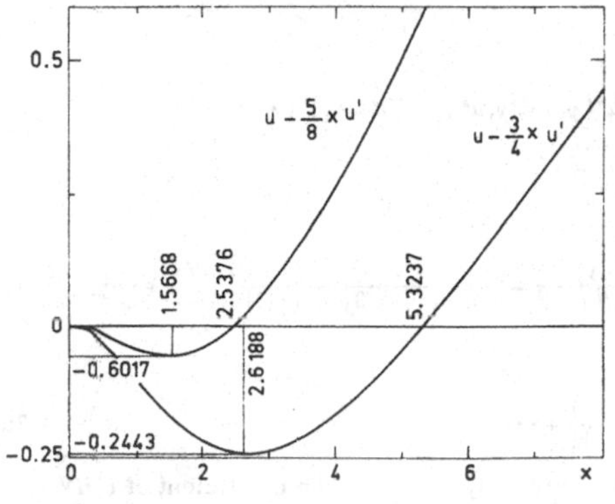

FIG. 5. The curves $u(x) - \frac{3}{4}xu'(x)$ and $u(x) - \frac{5}{8}xu'(x)$.

field A_{ij} is quenched to a pure gauge $A_{ij} = O_i O_j^{-1}$. This results in the Abelian as well as non-Abelian case (where it is in fact slightly less trivial) from the conditions $\chi(AAAA)$ = maximum value on any plaquette. Thus redefining $k_i \to O_i k_i$ one now recovers the scalar model with its second-order transition. Finally, if $\beta_p = 0$ one can in fact compute F exactly, which turns out to be analytic in β_l on this line. Indeed each k_i can be written $O_i q$ with q a fixed unit vector and O_i belongs to SO(n). We change variables from A_{ij} to $A_{ij} = O_i B_{ij} O_j^{-1}$; thus the integral over k_i is trivial, and now so is the one on the independent B_{ij}'s, with the result

$$F = d \ln\left(\int dB \exp(\beta_l q^T B q)\right)\,.$$

We assume for the moment that β_p and $\beta_l \geq 0$, and call $\vec{H}$ and $\vec{K}$ respectively the mean fields associated to the $\vec{k}$ and A degrees of freedom. By using the method of the previous two paragraphs we obtain inequalities analogous to (4.7) and (4.26) and take the right-hand side as the value of F for $d \to \infty$; thus

$$F = \sup_{H,K}\{\beta_p \tfrac{1}{2}d(d-1)[u'(K)]^4 + d[u(K) - Ku'(K)] + \beta_l d[u'(H)]^2 u'(K) + u(H) - Hu'(H)\}\,. \tag{4.31}$$

Owing to gauge invariance no reference to the relative direction of $\vec{H}$ and $\vec{K}$ remains; only their lengths appear. The expression to be maximized, which we shall call $F(H, K)$, is the sum of two terms pertaining to the gauge field and the $\vec{k}$ field, respectively. These resemble very much the terms we already studied. The only difference and the only place where the coupling occurs is in the replacement of $\beta_l d[u'(H)]^2$ by $\beta_l d u'(K)[u'(H)]^2$. Thus $\beta_l u'(K)$ [recall from Table I that $0 \leq u'(K) \leq 1$] plays the role of an effective coupling.

We have to maximize $F(H, K)$ in the quadrant $H \geq 0$, $K \geq 0$. Now the innocent looking replacement $\beta_l \to \beta_l u'(K)$ has the consequence that for H and K small enough the negative quadratic form

$$d[u(K) - Ku'(K)] + u(H) - Hu'(H) \simeq [d(\tfrac{1}{2}K)^2 + (\tfrac{1}{2}H)^2]$$

always dominates no matter what β_l and β_p are. Hence the origin (where $H = K = F = 0$) is always a local maximum and in fact *the* maximum for β_p, β_l small enough.

Consider the surface $(F(H, K), H, K)$ in a three-dimensional space. It is symmetric with respect to the plane $H = 0$. Its interaction with this plane looks like the curves sketched in Fig. 4. Since $F(H, K)$ is an even function of H, the extremum which appears as β_p grows is either a local maximum of the surface or a saddle point. Clearly, when β_l is small enough this is necessarily a local maximum. Then it will come in competition with the maximum at $H=K=0$ when F will cross the value 0, i.e., for $\frac{1}{2}(d-1)\beta_p = 1.82$ as obtained in equations (4.28) and (4.29). Thus we find in the β_l, β_p plane a first-order transition along a segment of line (a segment only since β_p varies from zero to some fixed value as we shall shortly see). We shall find it more convenient to use the variables $2\beta_p(d-1)$ and $2\beta_l d$, and will compute below the limiting value of β_l along this segment. Thus we have a transition line (L_1) along

$$\beta_p^* = 2\beta_p(d-1) = 7.29\ ,$$
$$0 \leqslant \beta_l^* = 2\beta_l d \leqslant 2.22\ ,$$
$$(L_1) - \text{first-order transition.} \tag{4.32}$$

The limiting value for β_l is obtained through the following consideration. The extremum found above in the plane $H=0$ at $K=K_c=5.32$ competing with the one at the origin is a true maximum as long as β_l is small enough. We can study its nature by examining the curvature in the H direction at this point. We observe that it is negative for small β_l and increases as β_l increases. It vanishes for the value $2\beta_l d = 2.22$ as was stated in (4.32).

Beyond this value the above extremum becomes a saddle point. Thus two new maxima at two opposite and nonvanishing values of H appear. Again by adjusting β_l and β_p we can bring these maxima at $F=0$. This defines a new first-order transition line (L_2) obtained by requiring that

$$F(H,K) = 0\ ,$$
$$\frac{\partial F}{\partial H}(H,K) = 0, \qquad H, K \neq 0 \tag{4.33}$$
$$\frac{\partial F}{\partial K}(H,K) = 0\ .$$

These equations can be written as

$$\tfrac{1}{2}\beta_p d(d-1)[u'(K)]^4 + \beta_l d u'(K)[u'(H)]^2 + d[u(K) - Ku'(K)] + [u(H) - Hu'(H)] = 0\ , \tag{4.34a}$$

$$2\beta_p d(d-1)[u'(K)]^3 + \beta_l d[u'(H)]^2 - dK = 0, \tag{4.34b}$$

$$2\beta_l d u'(H) u'(K) - H = 0\ . \tag{4.34c}$$

We can simplify them slightly by replacing (4.34a) by a combination (4.34a) $-\frac{1}{4}u'(K)$ (4.34b) $-\frac{3}{8}u'(H)$ $\times$(4.34c); thus we find

$$d[u(K) - \tfrac{3}{4}Ku'(K)] + [u(H) - \tfrac{5}{8}Hu'(H)] = 0\ , \tag{4.35a}$$

$$2\beta_p d(d-1) = \frac{dKu'(K) - \frac{1}{2}Hu'(H)}{[u'(K)]^4}\ , \tag{4.35b}$$

$$2\beta_l d = \frac{H}{u'(H)u'(K)} \tag{4.35c}$$

$$(L_2) - \text{first-order transition}\ .$$

Equation (4.35a) defines a curve in the (H,K) plane. Given this curve, (4.35b) and (4.35c) represent parametrically (L_2) in the β_l, β_p plane.

We have already plotted $y(x) = u(x) - \frac{3}{4}xu'(x)$ in Fig. 5. The quantity $y(x) = u(x) - \frac{5}{8}xu'(x)$ has a very similar shape and is reproduced on the same figure. From these two figures, we can sketch the plot of the critical curve in (H,K) space corresponding to (L_2). This is drawn in Fig. 6. The point A, $H=0$, $K=y_1=5.32$ is d-independent and corresponds to the extremity of (L_1) and the starting point of (L_2). Let us study (L_2) in the vicinity of this point. The two small quantities are H and $(K-y_1)$. We expand equations (4.35) around A in these two infinitesimals:

$$d[u(K) - \tfrac{3}{4}Ku'(K)] = \tfrac{1}{4}d[u'(y_1) - 3y_1u''(y_1)](K-y_1) - \tfrac{1}{8}d[2u''(y_1) + 3y_1u'''(y_1)](K-y_1)^2 + \cdots\ ,$$

$$u(H) - \tfrac{5}{8}Hu'(H) = -\tfrac{1}{4}(\tfrac{1}{2}H)^2 + \tfrac{3}{8}(\tfrac{1}{2}H)^4 + \cdots\ .$$

Thus (4.35a) yields

$$(K-y_1) = \frac{1}{d[u'(y_1) - 3y_1u''(y_1)]}\left(\frac{H}{2}\right)^2 + \left\{\frac{2u''(y_1) + 3y_1u'''(y_1)}{2d^2[u'(y_1) - 3y_1u''(y_1)]^3} - \frac{3}{2d[u'(y_1) - 3y_1u''(y_1)]}\right\}\left(\frac{H}{2}\right)^4 + \cdots\ .$$

From this and (4.35c):

$$2\beta_l d = \frac{2}{u'(y_1)} + \left(\frac{H}{2}\right)^2 \frac{1}{u'(y_1)}\left\{1 - \frac{2}{du'(y_1)[u'(y_1) - 3y_1u''(y_1)]}\right\} + \cdots\ . \tag{4.36}$$

The quantity $2/u'(y_1) = 2.22$, is just the value of $2\beta_l d$ at the extremity of (L_1). The coefficient of $(\frac{1}{2}H)^2$ is positive for d large enough. Similarly from (4.35b),

$$2\beta_p(d-1)=\frac{y_1}{[u'(y_1)]^3}-\left(\frac{H}{2}\right)^4\frac{1}{d[u'(y_1)]^4}\left[1+O\left(\frac{1}{d}\right)\right]+\cdots. \tag{4.37}$$

We see similarly that $y_1/[u'(y_1)]^3=7.29$ corresponds to the value of $2\beta_p(d-1)$ at the extremity of (L_1); then the coefficient of $(\frac{1}{2}H)^4$ is negative if $d\to\infty$. Comparing (4.36) and (4.37) we see that (L_2) has a parabolic shape towards negative β_p and positive β_l in the vicinity of A. Thus (L_1) and (L_2) join together smoothly [the tangents are identical at (A)] and constitute a unique first-order transition curve (L). We shall return later to this very smooth behavior. Note also that the curve (L_2) flattens as d grows (the curvature goes to zero).

In the vicinity of B in Fig. 6, $H\to y_2=2.54$, $K\to 0$,

$$\begin{aligned}\beta_l^*&=2\beta_l d\sim\frac{2}{K}\frac{y_2}{u'(y_2)}\to\infty\,,\\ \beta_p^*&=2\beta_p(d-1)\sim\frac{2}{(\frac{1}{2}K)^2}-\frac{1}{2d}\frac{y_2u'(y_2)}{(\frac{1}{2}K)^4}\to-\infty\,.\end{aligned} \tag{4.38}$$

This has a quartic shape in the β_l, β_p plane extending in the region $\beta_p\to-\infty$. We remark a very nonuniform behavior as d grows. The leading negative term in β_p is proportional to $1/d$.

The shape of this curve (L_2) is quite nonphysical in the region $\beta_l^*\to\infty$, since in the mean time, it crosses the axis $\beta_p^*=0$ where as we know no transition occurs. Furthermore, if we strictly adhere to the limit $d\to\infty$, then (L_2) degenerates into $\beta_p^*=7.29$ given in (4.32). A correct treatment will be to include in a systematic fashion *all* $1/d$ corrections, which presumably alter the shape of

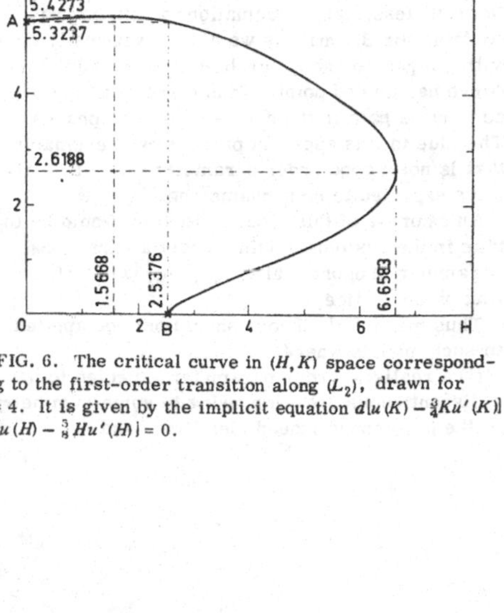

FIG. 6. The critical curve in (H,K) space corresponding to the first-order transition along (L_2), drawn for $d=4$. It is given by the implicit equation $d[u(K)-\frac{3}{8}Ku'(K)]+[u(H)-\frac{3}{8}Hu'(H)]=0$.

(L_2), bending it downwards, but not enough to cross the axis $\beta_p^*=0$. This is why in Fig. 1, where the phase diagram has been drawn in the mean-field approximation for $d=4$, the curve (L_2) appears as dotted in the large β_l^* direction.

This is, however, not the end of the story. Returning to point A we have followed two new bumps emerging in the surface $F(H,K)$ and reaching the plane $F=0$. But at A the local maximum was becoming a saddle point. Thus instead of comparing the new bumps $H\neq 0$, $K\neq 0$, with the maximum at $H=K=0$, we could have followed in the plane $H=0$ the transformation of the maximum into a saddle point at which point the maximum with $F>0$ divides itself into two new other ones. This clearly defines a new curve (M) corresponding to a *second order* transition. The equations governing this mechanism are

$$\begin{aligned}&\frac{\partial F}{\partial K}(H=0,K)=0\,,\\ &\frac{\partial^2 F}{\partial H^2}(H=0,K)=0\,.\end{aligned} \tag{4.39}$$

Explicitly we find from (4.31) a parametric expression for (M) in terms of K:

$$\begin{aligned}\beta_p^*&=2\beta_p(d-1)=\frac{K}{u'(K)^3}\,,\\ \beta_l^*&=2\beta_l d=\frac{2}{u'(K)}\,,\end{aligned}$$

$$(M)\to\text{second-order transition}\,. \tag{4.40}$$

The reader might notice that Eqs. (4.40) are obtainable from (4.35b) and (4.35c) by taking the limit $H=0$. The curve (M) starts from A for the value K_c and goes to infinity in the positive β_p direction with an asymptote at $\beta_l d=1$. This value corresponds to the transition in the $\vec{k}$ system alone, because the gauge field is frozen at its unit value (in the group) when $\beta_p\to\infty$ (or $T_p=0$). For amusement we can easily complete the picture in the $\beta_l<0$ half plane by symmetry. The difference is that the region bounded by (L_2') and (M') in which $H,K\neq 0$ corresponds rather to an antiferromagnetic ordering where at successive sites the effective field $\vec{H}_i$ jumps from a value $\vec{H}$ to $-\vec{H}$.

A careful study, the details of which need not be presented here, shows that we have exhausted all possible transitions.

On the whole, the phase diagram of the coupled Abelian system in the mean-field approximation is represented on Fig. 1. The phase I is fully disordered $(H=K=0)$; both the gauge field and the particle field have vanishing values. The first-order transition line (L) separates this phase from the other phases II, III, IV, in which the gauge invariance is spontaneously broken ($K \neq 0$ and $\langle A \rangle \neq 0$). In such phases, we have assumed $\vec{K}$ to be a constant on all links of the lattice, and found an overall rotational degeneracy in the solution; however, the actual degeneracy is much larger owing to gauge invariance, since independent rotations may be performed on each site of the lattice. In the phase II, rotational invariance of the field k remains unbroken ($H=0$, $\langle k \rangle = 0$). The second-order transition line (M) separates this phase from the fully ordered phase III in which all invariances are spontaneously broken, giving rise to a massless π and a massive σ. This phase III may be considered as "ferromagnetic," since if the self-consistent field $\vec{K}$ for the gauge field is taken as uniform, the order parameter $\langle \vec{k}_i \rangle$ is constant over the lattice. (Similarly, the symmetrical phase IV is "antiferromagnetic" with the same uniform choice for $\vec{K}$.) Clearly, the coupling $\beta_l \sum_l k_i A_{ij} k_j$ between neighboring fields k_i, k_j may become effective only when $\langle A_{ij} \rangle$ is nonzero. Thus the occurrence of a transition for the gauge field, leading (for β_p large enough) to a nonzero value for $\langle A_{ij} \rangle$, is a prerequisite to the ordering of the particle field k. This ordering takes place [along (M) or along (L_2)] when $\langle A_{ij} \rangle$ already has a finite value.

This phase diagram presents a great analogy with the (p, T) phase diagram of a material which may become magnetic, with β_p playing the role of pressure and β_l the role of inverse temperature. The disordered phase I is the equivalent of a liquid. When pressure (β_p) is increased, it crystallizes (phases II, III, IV) through a first-order transition, $\langle A_{ij} \rangle$ playing the role of the lattice order parameter in the solid. If the k_i are interpreted as the atomic spins, the term $\sum_l k_i A_{ij} k_j$ has the same features as an exchange interaction, which becomes effective only in the crystalline phase. We thus have a second-order magnetic transition (M) or (M') between the nonmagnetic crystal II and the ferro- (or antiferro-) magnetic crystal III (or IV) at low temperature ($1/|\beta_l|$ small). The curve (M) is very steep, because once it is settled, the crystalline order is not very sensitive to pressure, and the exchange interaction between spins does not vary much with pressure. In the field-theoretical interpretation, the situation is the same, since along (M) the value of $\langle A \rangle = u'(K)$ varies only from 0.90 to its maximum value 1.

The triple point A is the most interesting feature of the phase diagram. The fact that the slope of (L) remains continuous across A may easily be understood by standard thermodynamic arguments, since (M) is a second-order transition line. It reflects the fact that second-order effects are weak compared to first-order ones. Although the phase diagram has been established in the mean-field approximation, expected to be exact for infinite dimensionality, experience in statistical mechanics suggests that the qualitative features will remain unchanged by a more refined treatment.

D. Validity of mean-field theory

In the gauge-invariant theories one may at first question the validity of the qualitative results obtained through mean-field theory. The latter needs in fact to be made more precise. Indeed the mean field is a conjugate variable to some order parameter. *Order* can be defined as the prevalent situation as "temperature" goes to zero, i.e., here when β_p and $\beta_l \to \infty$. From the structure of the action S it is seen that for arbitrary O_i this amounts to $k_i = O_i q$ (fixed q), $A_{ij} = O_i O_j^{-1}$. Our previous treatment is seen to imply a choice among the highly degenerate "vacuum states." One could argue that it would be more reasonable to break gauge invariance first by integration over some subset of variables.

For instance, returning to the general expression for the action we might perform a change of variables as follows. For fixed $k_i = O_i q$ we set $A_{ij} = O_i B_{ij} O_j^{-1}$. It is then possible to take as new variables B's and O's and integrate over O's. The result is a B model

$$Z = \int \prod_{(ij)} dB_{ij}\, e^S ,$$
$$S = \beta_l \sum_l q^T B_{ij} q + \beta_p \sum_p \chi(B_{12}B_{23}B_{34}B_{41}) , \tag{4.41}$$

for which apparently no reference remains to the particle field.

Application of mean-field theory to (4.41) would yield the first-order transition line (l) of Fig. 1. Nevertheless, large fluctuations would be unavoidable for β_l small as we did not yet really cope with the genuine gauge problem. Moreover, this curve has an end point. Thus there is a seemingly continuous path in the diagram between phases. The clue to this apparent paradox is presumably that is not a good order parameter since it is always expected to be nonvanishing.

An example of this type of situation would be to take in the customary Ising model a variational parameter proportional to $S_i S_j = b_{ij}$ [with (ij) a link on the lattice].

Thus mean-field theory should only be applied in such models where

(i) a realistic order parameter expected to have a discontinuous behavior owing to some symmetry of the problem is indeed identified, and

(ii) the vacuum is not too degenerate in order to avoid wild fluctuations which would spoil any attempt to improve the approximation by a perturbation expansion.

In order to meet these qualitative prerequisites we can further proceed starting from Eq. (4.41) by defining a "Coulomb gauge" in the following way. We choose some direction, denoted the time axis, to play a particular role. To each lattice point we let correspond the timelike link that starts from the point. For the sake of clarity let T_i be the corresponding B_{ij} variables while $\tilde{S}_{ij}$ denote the other spacelike B_{ij} variables. Let further D stand for the unit time displacement operator on the lattice. It is easily seen that T_i can be written as $O_{Di}O_i^{-1}$. One can then change variables from T_i and $\tilde{S}_{ij}$, to $k_i = O_i q$ and $S_{ij} = O_i \tilde{S}_{ij} O_j^{-1}$. By integrating over the little groups of q one recovers "matter variables" k_i and a subset S_{ij} of the previous gauge field variables A_{ij} as follows:

$$Z = \int \prod_i d\hat{k}_i \int \prod_{(ij)} dS_{ij}\, e^S ,$$

$$S = \beta_l \Big(\sum_i k_i^T k_{Di} + \sum_{(ij)} k_i^T S_{ij} k_j \Big) + \beta_p \Big(\sum_{(ij)} \chi(S_{ij} S_{Dj,Di}) + \sum_p \chi(S_{12}S_{23}S_{34}S_{41}) \Big) . \tag{4.42}$$

All summations are carried over spacelike links and spacelike plaquettes. Clearly one could have performed the steps leading to (4.42) directly. The remaining gauge arbitrariness is now only a "surface" effect with a group

$$\bigotimes^{N^{(d-1)/d}} \mathrm{SO}(n)$$

and of course the infinite-volume limit ($N \to \infty$) is to be taken first. As a result this arbitrariness is presumably irrelevant

We see that (4.42) amounts to restricting all timelike links to the unit value. Application of mean-field theory now yields for the Abelian case $n = 2$

$$F = \operatorname*{Sup}_{(H,K)} F(H,K),$$

$$\begin{aligned} F(H,K) = {} & \tfrac{1}{2}\beta_p(d-1)(d-2)[u'(K)]^4 + \beta_p(d-1)[u'(K)]^2 \\ & + (d-1)[u(K) - Ku'(K)] \\ & + \beta_l(d-1)[u'(H)]^2 u'(K) \\ & + \beta_l[u'(H)]^2 + u(H) - Hu'(H) , \end{aligned} \tag{4.43}$$

to be compared with (4.31). There are clearly differences, which, however, are washed out by taking the variables β_p^* and β_l^* as previously and letting $d \to \infty$. One then recovers exactly the results of the previous section in this limit and qualitatively the same phase diagram for finite and large enough d. One may remark the presence of a new quadratic term in $[u'(K)]^2$. Thus there is a possibility for a second-order transition in the pure gauge field for low dimension. However, this term is doubtful even as a zeroth-order approximation. Hence one feels more confident about the mean-field approximation and it is suggested to collect systematically all $1/d$ corrections.

V. CONCLUSION

In this first paper we have only used rather unsophisticated mathematical means, and did not touch upon several problems pertaining to the noncommutative case. However, we have disclosed a very rich variety of interesting phenomena in this Wilson model. Apart from its speculative application to new binding modes in the domain of particle theory, it would certainly be amusing to find some physical system to which the thermodynamical version would apply. In the next papers we shall present perturbative and diagrammatic expansions and further develop the study of this model.

ACKNOWLEDGMENTS

It is a pleasure to acknowledge fruitful discussions with E. Brezin and J. Zinn-Justin, who introduced us to the problem.

[1]This model was presented by K. Wilson in a seminar at Orsay during the summer of 1973. E. Brezin communicated to us later some fragments of a manuscript by Wilson, which has now been published [Phys. Rev. D 10, 2445 (1974)].

[2]A useful review of phase-transition phenomena including a large wealth of references is in H. E. Stanley, *Introduction to Phase Transitions and Critical Phenomena* (Clarendon, Oxford, 1971).

[3]J. Goldstone, Nuovo Cimento 19, 154 (1961).

[4]H. E. Stanley, Phys. Rev. 176, 718 (1968).

[5]R. E. Peierls, Phys. Rev. 54, 918 (1938).

[6]K. G. Wilson and J. Kogut, Phys. Rep. 12C, 75 (1974).

PHYSICAL REVIEW D VOLUME 11, NUMBER 8 15 APRIL 1975

Gauge fields on a lattice. II. Gauge-invariant Ising model

R. Balian, J. M. Drouffe, and C. Itzykson
Service de Physique Théorique, Centre d'Etudes Nucléaires de Saclay, BP No. 2, 91190 Gif-sur-Yvette, France
(Received 29 August 1974)

We study the case of a discrete local gauge group Z_2 in order to discuss the existence of a transition in dimension $d \geq 3$. We compute the critical constant for $d = 3$ and 4 and show that in three dimensions the transition is a second-order one.

I. INTRODUCTION

We have described in a previous paper[1] [hereafter referred to as (I)] a gauge theory on a lattice according to Wilson's ideas.[2] It was suggested that the system undergoes phase transitions. For a small enough coupling constant the gauge field behaves qualitatively as ordinary perturbation theory in the continuous limit would indicate, while beyond some critical coupling a new phase sets in. The long-range forces become so strongly attractive that they provide a binding mechanism for charged particles.

We want to look here more closely at the nature of the transition. To simplify as much as possible we make use of a possibility afforded by discretization, namely the introduction of a finite local gauge group instead of the usual continuous Lie local gauge group. The simplest one is Z_2, the (multiplicative) group with two elements $\{1, -1\}$. In (I) the local group was $O(n)$, $n \geq 2$. The present case amounts to setting $n = 1$. A drawback of this choice might be the absence of Goldstone bosons in the ordered phase. We notice, however, with reference to (I), that as the dimension d of the lattice gets large, mean-field theory hardly distinguishes $n = 1$ from $n \geq 2$. In particular, it predicts in all cases a first-order transition. Consequently, as an instrument to investigate the validity of mean-field theory, the present simplification is not too drastic.

With Z_2 as gauge group, we can use several devices introduced in the context of Ising models. The first of these is specific to a system with configuration variables taking values ± 1. It is a duality transformation to be described in Sec. II. Combining this duality with some mathematical results (collected in an appendix) enables one to reduce the problem in three dimensions to a standard Ising model of which much is known. It also allows one to locate exactly the critical constant in four dimensions (Sec. III). We shall also use the Griffiths-Kelly-Sherman inequalities,[3] which express the fact that strengthening "ferromagnetic" couplings can only strengthen correlations. This is explained in Sec. IV where a discussion of a global-order parameter is given.

Once some exact results are known, approximation methods, such as perturbation theory around mean-field theory in the Coulomb gauge, can be checked. If they are found reliable, one can then proceed to use them in more general cases. This is another motivation for the present work.

We recall the model. Let there be given a hypercubical lattice in d dimensions with unit spacing. We introduce a spatial cutoff by retaining N sites (eventually we let $N \to \infty$). To each link (ij) of neighboring sites we assign a variable $A_{ij} = A_{ji}$ taking the values ± 1. A set of four neighboring links is a plaquette $p = (ijkl)$. We compute a partition function

$$Z \equiv 2^{-Nd} \sum_{\{A_{ij} = \pm 1\}} \exp\left(\beta_l \sum_l A_{ij} + \beta_p \sum_p A_{ij} A_{jk} A_{kl} A_{li} \right), \tag{1.1}$$

and define a free energy as

$$F \equiv \lim_{N \to \infty} \frac{1}{N} \ln Z . \tag{1.2}$$

We wish to study the occurrence and properties of phase transitions as β_l and β_p vary. Had we insisted on presenting a gauge-invariant version of the Ising model, we would have introduced extra variables $k_i = \pm 1$ at each site and replaced the β_l term by $\beta_l \sum_l k_i A_{ij} k_j$. However, the gauge transformation $A_{ij} \to k_i A_{ij} k_j$ eliminates the k's while leaving the plaquette coupling invariant, and reduces the problem to the study of (1.1). We shall mostly be interested in the case $\beta_l = 0$, which we call the pure-gauge-field case.

II. DUALITY

Application of duality transformations to Ising-type models is well known. It has recently been further developed by Wegner.[4] Owing to this circumstance, although our presentation is slightly different, we shall be rather brief.

Geometrical duality transforms q-dimensional manifolds into $(d-q)$-dimensional ones. Let us present it for $d = 3$. We introduce a dual lattice obtained from the original one through a translation by an amount $(\frac{1}{2}, \frac{1}{2}, \frac{1}{2})$. There is a natural correspondence between a site, a link, a plaquette, or a cube on the original lattice, and a cube, a plaquette, a link, or a site on the dual one. For instance, to a link is associated the dual plaquette that it intersects. This construction is easily generalized to any dimension and is used as follows.

(*i*) $d=2$. We write (1.1) as

$$Z=2^{-2N}(\cosh\beta_l)^N(\cosh\beta_p)^{2N}\sum_{\{A_{ij}=\pm1\}}\prod_l(1+\tanh\beta_l A_{ij})\prod_p(1+\tanh\beta_p A_{ij}A_{jk}A_{kl}A_{li}), \tag{2.1}$$

expand the products, and sum over $A_{ij}=\pm1$. Nonvanishing terms are in one-to-one correspondence with configurations of P distinct plaquettes selected on the lattice. The boundary of each configuration is defined as the set of L links which belong to one and only one of the selected plaquettes. A configuration contributes a term $(\tanh\beta_l)^L(\tanh\beta_p)^P$. If a plaquette is selected, let us set $s_i=-1$ at the corresponding site i of the dual lattice and $s_i=+1$ otherwise. Obviously $P=\sum_i \frac{1}{2}(1-s_i)$ and $L=\sum_l \frac{1}{2}(1-s_i s_j)$, the summations running over sites and links of the isomorphic dual lattice. Consequently, we have

$$Z=(\tfrac{1}{4}\cosh\beta_p\cosh^2\beta_l)^N\sum_{\{s_i=\pm1\}}\exp\left[\sum_l \tfrac{1}{2}\ln\tanh\beta_l(1-s_is_j)+\sum_i \tfrac{1}{2}\ln\tanh\beta_p(1-s_i)\right]. \tag{2.2}$$

Up to a factor we recognize the partition function Z_I for an Ising model in an external field H. Letting

$$Z_I\equiv 2^{-N}\sum_{\{s_i=\pm1\}}\exp\left(\sum_l \beta_* s_i s_j+\sum_i H s_i\right),$$
$$F_I\equiv\lim_{N\to\infty}\frac{1}{N}\ln Z_I, \tag{2.3}$$

we get the equality

$$F(\beta_l,\beta_p)=\tfrac{1}{2}\ln(\sinh^2 2\beta_p\,\sinh 2\beta_p)+F_I(\beta_*,H),$$
$$\beta_*=-\tfrac{1}{2}\ln\tanh\beta_l,\quad H=-\tfrac{1}{2}\ln\tanh\beta_p. \tag{2.4}$$

As long as $H\neq0$, the system exhibits no transition, while for $H=0$ there exists a critical value β_c of β_* $(\sinh 2\beta_c=1)$ separating two phases: a disordered one for $\beta_*<\beta_c$ and an ordered one for $\beta_*>\beta_c$.

The condition $H\neq0$ means β_p is finite. As $\beta_p\to\infty$ we find a transition for $\tanh\beta_l=e^{-2\beta_c}$, i.e., $\beta_l=\beta_c$. This is in agreement with the discussion given in (I): No transition occurs for finite β_l, β_p. We knew already that for $\beta_l=0$ the gauge-field model is trivial for $d=2$, with $F(0,\beta_p)=\ln\cosh\beta_p$.

(*ii*) $d=3$. Repeating the previous argument we find that the coupled model is self-dual with

$$F(\beta_l,\beta_p)=\tfrac{3}{4}\ln(\sinh 2\beta_l\,\sinh 2\beta_p)$$
$$+F(-\tfrac{1}{2}\ln\tanh\beta_p,-\tfrac{1}{2}\ln\tanh\beta_l). \tag{2.5}$$

Note the interchange of indices (l,p) between the two sides of this equality. We exhibit the symmetry of this self-duality by introducing the bounded variables

$$\xi_l=\ln(1+e^{-2\beta_l}),$$
$$\xi_p=\ln(1+e^{-2\beta_p}),$$
$$0\leq\xi_l,\ \xi_p\leq\ln 2,$$

and the function

$$f(\xi_l,\xi_p)\equiv F(\beta_l,\beta_p)-\tfrac{3}{2}\ln(1+e^{2\beta_l})(1+e^{2\beta_p}). \tag{2.6}$$

Then this function f is symmetric with respect to the line $\xi_l+\xi_p=\ln 2$:

$$f(\xi_l,\xi_p)=f(\ln2-\xi_p,\ln2-\xi_l). \tag{2.7}$$

Let C be a simple closed curve on the lattice (for d arbitrary). We defined in (I) the average $\mathfrak{C}$ of the product of the link variables A_{ij} along C:

$$\mathfrak{C}\equiv\left\langle\prod_C A_{ij}\right\rangle=Z^{-1}2^{-Nd}\sum_{\{A_{ij}=\pm1\}}\exp\left(\beta_l\sum_l A_{ij}+\beta_p\sum_p A_{ij}A_{jk}A_{kl}A_{li}\right)\prod_C A_{ij}. \tag{2.8}$$

This is also equal to the average of the product of all plaquette variables $A_{ij}A_{jk}A_{kl}A_{li}$ for a set S of plaquettes bounded by C. For $d=3$, we find by applying the duality transformation

$$\mathfrak{C}=\left\langle\exp\left(-2\beta_*\sum_{S_*}s_{ij}\right)\right\rangle_*. \tag{2.9}$$

The $\langle\ \rangle_*$ average is computed with the dual coupling constants: $\beta_p\to-\frac{1}{2}\ln\tanh\beta_p$ for plaquettes and $\beta_l\to\beta_*=-\frac{1}{2}\ln\tanh\beta_p$ for links of the dual lattice (to which are associated variables $s_{ij}=\pm1$). Formula (2.9) allows an interpretation of $\mathfrak{C}$ as

$$\mathfrak{C}=\exp[-(\mathfrak{F}-\mathfrak{F}')], \tag{2.10}$$

where $\mathcal{F}$ is the free energy of the dual model, and $\mathcal{F}'$ the similar quantity obtained by reversing the sign of the coupling constants on the links of the dual model belonging to S_*, that is, all the links intersecting the surface S.

(*iii*) $d=4$. In this case a cube is dual to a link, a plaquette to a plaquette. We can define a free energy $F_*(\beta_{*l},\beta_{*p})$ for a dual model by

$$\exp(NF_*)=2^{-6N}\sum_{\{s_{ijkl}=\pm 1\}}\exp\left(\beta_{*l}\sum_p s_{ijkl}+\beta_{*p}\sum_c\prod_{\alpha=1}^{6}s_{i_\alpha j_\alpha k_\alpha l_\alpha}\right). \tag{2.11}$$

Each plaquette $(ijkl)$ of the dual lattice carries a variable $s_{ijkl}=\pm 1$. These are combined six by six along the faces of three-dimensional cubes to give the interaction $\prod_{\alpha=1}^{6}s_{i_\alpha j_\alpha k_\alpha l_\alpha}$. We defined $\beta_{*l}=-\frac{1}{2}\ln\tanh\beta_p$, $\beta_{*p}=-\frac{1}{2}\ln\tanh\beta_l$. Between the original model and the dual one, duality yields the relation

$$F(\beta_l,\beta_p)=2\ln\tfrac{1}{2}\sinh 2\beta_l+3\ln 2\sinh 2\beta_p+F_*(\beta_{*l},\beta_{*p}). \tag{2.12}$$

III. CRITICAL COUPLINGS IN THREE AND FOUR DIMENSIONS

Duality has given a complete solution in two dimensions. In particular, we have recovered the fact that the pure gauge system undergoes no transition. The results of the previous section will enable us to show that a transition occurs in three and more dimensions. In the pure gauge system, the critical values of β_p are given by

$$d=3,\quad \beta_c=0.7613,$$
$$d=4,\quad \beta_c=0.4407. \tag{3.1}$$

Let us see how these values are obtained.

In three dimensions the coupled model is self-dual. If one sets $\beta_l=0$ in (2.5), infinities occur on the right side while the left side is obviously finite. The required cancellations are exhibited on the form (2.7) which reduces to

$$f(\ln 2,\xi_p)=f(\ln 2-\xi_p,0). \tag{3.2}$$

Thus, the study of the pure gauge model is equivalent to the study of the coupled system in the limit $\beta_{*p}\to\infty$ ($\xi_{*p}=0$). We then expect the gauge field to reduce to a pure gauge, as discussed in (I). This is indeed true, since

$$\tilde{Z}\equiv Z(\beta_{*l},\beta_{*p})(\cosh\beta_{*p})^{-3N}\Big|_{\tanh\beta_{*p}=1}=(\cosh\beta_{*l})^{3N}\sum_{\substack{\{s_{ij}=\pm 1\}\\ \{s_{ij}s_{jk}s_{kl}s_{li}=1\}}}\prod_l(1+\tanh\beta_{*l}s_{ij}). \tag{3.3}$$

The last sum is taken under the constraints that the product $s_{ij}s_{jk}s_{kl}s_{li}$ for every plaquette is equal to 1. A theorem (see Appendix) states that the general solution of these constraints is $s_{ij}=s_is_j$, with s_i defined up to an overall sign. Consequently, we obtain

$$\tilde{Z}=\tfrac{1}{2}\sum_{\{s_i=\pm 1\}}\exp\left(\beta_{*l}\sum_l s_is_j\right), \tag{3.4}$$

which means that for $d=3$ the pure-gauge-field model is related to the Ising model:

$$F(\beta_l=0,\beta_p)=-\tfrac{1}{2}\ln 2+\tfrac{3}{2}\ln\sinh 2\beta_p+F_I(-\tfrac{1}{2}\ln\tanh\beta_p), \tag{3.5}$$

where F_I is the free energy of the three-dimensional Ising model. The Ising model is known to have a unique second-order transition in three dimensions. The techniques for the proof are based on the arguments of Griffiths and Peierls[3] to be discussed in the next section. The value β_I at which F_I is singular is obtained numerically from the high-temperature expansion[5] and is given by

$$\frac{1}{2\beta_I}=2.255\,16. \tag{3.6}$$

From the relation $-\frac{1}{2}\ln\tanh\beta_c=\beta_I$, we find the value given in (3.1).

We turn to the case $d=4$. We use Eqs. (2.12) and (2.11) and set $\beta_l=0$ or $\tanh\beta_{*p}=1$. This amounts to sum over plaquette variables s_{ijkl} constrained by $\prod_{\alpha=1}^{6}s_{i_\alpha j_\alpha k_\alpha l_\alpha}=1$, the product running over the faces of every cube. The same theorem quoted in the Appendix states that, up to a gauge transformation, one has then $s_{ijkl}=s_{ij}s_{jk}s_{kl}s_{li}$. Taking into account this gauge arbitrariness, we sum freely over the variables s_{ij} and divide by the "volume" of the gauge group. By inspection of (2.11) it is seen that, as we let β_l approach zero or β_{*p} approach infinity, we can extract an infinite term from F_* (proportional to β_{*p}). Using then the above expression for s_{ijkl}, we recover the original pure-gauge-field model. The final result is expressed as a self-duality formula:

$$F(0,\beta_p)=3\ln\sinh 2\beta_p+F(0,-\tfrac{1}{2}\ln\tanh\beta_p). \tag{3.7}$$

This is a remarkable result,[4] analogous to the

Kramers-Wannier duality for the Ising model in two dimensions. If we assume that the gauge field undergoes a unique transition for $d=4$ as it did for $d=3$, then the critical constant follows from (3.7):

$$\beta_c = -\tfrac{1}{2}\ln\tanh\beta_c \quad \text{or} \quad \sinh 2\beta_c = 1\,. \tag{3.8}$$

This yields the value quoted in (3.1).

The trend exhibited by (3.1) (to which we can add $\beta_c=\infty$ for $d=2$) is a clear decrease of the critical value as d increases. In fact, we expect a behavior in $1/d$ for d large.

IV. GLOBAL ORDER PARAMETER

From now on we set $\beta_l=0$. In order to analyze further the nature of the transition, it is interesting to find a quantity with a qualitatively discontinuous behavior. Due to gauge invariance the choice of such an order parameter is not straightforward. For instance, if we look at the Green's function

$$\langle \{A_{ij}A_{jk}A_{kl}A_{li}\}\{A_{mn}A_{np}A_{pq}A_{qm}\}\rangle_{\text{connected}}$$

pertaining to two plaquettes far apart, we expect it to decrease exponentially with the distance both for β_p small and β_p large. However, Wilson[2] has suggested that the average $\mathfrak{C}$ along the closed curve C defined in (2.8) may be used to define ordering. We recall from (I) that for $d=2$, $-\ln\mathfrak{C}$ is proportional to $|S|$, the area of the set S of plaquettes enclosed by C. For β_p large enough, it was also made plausible that if $d\geq 3$, $-\ln\mathfrak{C}$ increases like the length $|C|$ of C. We prove the following result for the present model.

Theorem: Let $d\geq 3$ and $|S|$ denote the minimal area enclosed by C; if β_p is small enough, there exist two positive constants a_1 and a_2 such that

$$a_2 \leq -\frac{\ln\mathfrak{C}}{|S|} \leq a_1\,. \tag{4.1}$$

Set $\beta_l=0$ and expand both numerator and denominator of (2.8) in powers of $t=\tanh\beta_p$. Let a diagram D be a set of plaquettes chosen on the lattice (which we take at first as finite). The boundary of D, noted ∂D, is the set of links which belong to an odd number of plaquettes of D. If ∂D is empty, we say that D is closed. Denoting by $|D|$ the number of plaquettes of D, we have

$$\mathfrak{C}=\frac{\mathcal{P}}{\mathfrak{Q}}\,,\quad \mathcal{P}=\sum_{D:\,\partial D=C} t^{|D|}\,,$$
$$\mathfrak{Q}=\sum_{D:\,\partial D=\emptyset} t^{|D|}\,. \tag{4.2}$$

The sum in the denominator $\mathfrak{Q}$ includes the empty diagram $D=\emptyset$ which gives a contribution equal to 1. Among the diagrams involved in the numerator $\mathcal{P}$ of (4.2), let us distinguish the family of *irreducible* diagrams $\bar{D}$, defined as follows. A diagram with boundary C is called irreducible if it does not contain any closed subset of plaquettes. When a diagram with boundary C is reducible, it may be decomposed at least in one fashion into an irreducible part and a closed part, having no plaquettes in common. This is easily seen by repeatedly stripping the diagram from closed parts.

The first part of the proof consists in showing that $\mathfrak{C}$ is smaller than the contribution $\bar{\mathcal{P}}$ of irreducible diagrams:

$$\mathfrak{C}<\bar{\mathcal{P}}=\sum_{\substack{\bar{D}:\,\partial\bar{D}=C\\ \bar{D}\ \text{irreducible}}} t^{|\bar{D}|}\,. \tag{4.3}$$

Consider the product $\bar{\mathcal{P}}\mathfrak{Q}$. It contains the contribution to $\mathcal{P}$ of irreducible diagrams, coming from the term 1 in $\mathfrak{Q}$. It also contains the contribution to $\mathcal{P}$ of each reducible diagram counted as many times as this diagram may be decomposed into an irreducible part (contribution to $\bar{\mathcal{P}}$) and a closed part (contributing to $\mathfrak{Q}$), i.e., at least once. Finally, the product $\bar{\mathcal{P}}\mathfrak{Q}$ generates additional terms which do not appear in $\mathcal{P}$, and for which some plaquettes are repeated twice. All terms are positive, since we have "attractive" interactions. Therefore $\bar{\mathcal{P}}\mathfrak{Q}$ is larger than $\mathcal{P}$ and (4.3) holds.

An analogous result had been established[6] in the context of the Ising model, for which the two-point correlation function was shown to be bounded by the sum of all self-avoiding walks. The present diagrams appear as two-dimensional extensions of the Ising ones: Plaquettes replace links, and the boundary contour C replaces the two end points. Irreducible diagrams, which are the self-avoiding walks in the Ising case, have here a more complicated topology, since the plaquettes of the two dimensional irreducible manifold are not naturally ordered as are the links of a walk.

In the second part of the proof, we provide an upper bound for the number $\bar{n}_k$ of irreducible diagrams made of exactly k plaquettes and bordered by C. Since k is at least equal to the minimal area $|S|$ enclosed by C, we have

$$\bar{\mathcal{P}}=\sum_{k\geq|S|}\bar{n}_k t^k\,. \tag{4.4}$$

Let us define an iterative process designed to generate at least all irreducible diagrams. For this purpose, we number once for all the links of the lattice. The construction starts from the contour C. We pick along C the link of lowest rank, and select a plaquette p_1 adjacent to this link: There are $2d-2$ such possible choices. We now define a new contour C_1 along which we shall add the second plaquette:

$$C_1 = C\Delta\partial p_1\,.$$

This is the symmetric difference between the contour C and the boundary of p_1, i.e., the set of links which belong either to C or to ∂p_1 but not to both. The plaquette p_2 is chosen among those which border the link of lowest rank of C_1, and so on. At each step of this iterative process, we perform the following operations:

(i) Identify along C_{q-1} the link of lowest rank,
(ii) select a plaquette p_q adjacent to this link, and
(iii) introduce a new contour $C_q = C_{q-1}\Delta\partial p_q$.

This construction stops at some finite stage if the resulting contour is empty. We thus obtain a finite ordered set of plaquettes having C as boundary. We denote it by $\hat{D}$. Some of these sets $\hat{D}$ are genuine diagrams contributing to $\mathcal{P}$, but it may happen that such a set contains some plaquette more than once and thus cannot occur as a diagram D. If the ordering is ignored, a given set of plaquettes may, of course, be obtained several times. Let $\hat{n}_k$ be the number of sets $\hat{D}$ with k plaquettes (distinct or not). At each step there are $2d-2$ possibilities for adding a plaquette. Thus, we find at most $(2d-2)^k$ ordered sets at stage k (the construction might indeed have stopped before). Among them, those for which $C_k = \emptyset$ are obviously only a subclass, and therefore

$$\hat{n}_k < (2d-2)^k .$$

We can improve our bound by modifying the definition of the sets $\hat{D}$ (and correspondingly of their number $\hat{n}_k$); we exclude the sets with overlapping plaquettes, by changing the rule (ii). If the link of lowest rank on C_{q-1} belongs to C, there are still at most $2d-2$ possible choices for p_q. If, however, it does not, we have already selected an odd number of plaquettes adjacent to it, and hence the number of choices is then at most $2d-3$. Since at least $k-|C|$ steps involve such a link, we obtain now the bound

$$\hat{n}_k < (2d-2)^{|C|}(2d-3)^{k-|C|}. \tag{4.5}$$

It remains to show that any irreducible diagram $\tilde{D}$ is obtained (at least once) as an ordered set $\hat{D}$. This will imply $\tilde{n}_k < \hat{n}_k$. For this purpose, given an irreducible diagram $\tilde{D}$, let us order its k plaquettes by using the above procedure. Instead of performing our choice of plaquettes over all those of the lattice, we restrict this choice to those of $\tilde{D}$, keeping otherwise the same rules. At stage q, we note that C_{q-1} is the boundary of the remaining plaquettes of $\tilde{D}$. Hence, each link belonging to C_{q-1} borders an odd number of remaining plaquettes, and thus step (ii) is always possible, unless C_{q-1} is empty. This, however, cannot happen for $q \le k$, k being the number of plaquettes of $\tilde{D}$; otherwise the remaining plaquettes would form a closed subset, and $\tilde{D}$ would be reducible. Thus, indeed, we have $\tilde{n}_k < \hat{n}_k$.

Taking into account this result with (4.3), (4.4), and (4.5), we obtain

$$\mathcal{C} < \sum_{k \ge |S|} (2d-2)^{|C|}(2d-3)^{k-|C|} t^k$$
$$= \left(\frac{2d-2}{2d-3}\right)^{|C|} \frac{[(2d-3)t]^{|S|}}{1-(2d-3)t}, \tag{4.6}$$

provided that

$$t < \frac{1}{2d-3}. \tag{4.7}$$

If the curve C gets very large in such a way that $|C|/|S| \to 0$, we have established the left inequality (4.1).

In order to get the other inequality (4.1), we need the Griffiths-Kelly-Sherman (GKS) result[3] which we recall for completeness. Consider N sites with variables $\sigma_i = \pm 1$ attached to each one. Let $\Lambda = \{R, S, \dots\}$ be the family of all subsets of sites and write $\sigma_R \equiv \prod_{i\in R}\sigma_i$. Define

$$\langle \sigma_R \rangle = \left[\sum_{\{\sigma_i=\pm 1\}} \sigma_R \exp\left(\sum_{S\in\Lambda} J_S\sigma_S\right)\right]$$
$$\times \left[\sum_{\{\sigma_i=\pm 1\}} \exp\left(\sum_{S\in\Lambda} J_S\sigma_S\right)\right]^{-1}. \tag{4.8}$$

Then if $J_R \ge 0$ for all R, GKS state that

$$\langle \sigma_R \rangle \ge 0,$$
$$\langle \sigma_R\sigma_S \rangle \ge \langle \sigma_R \rangle \langle \sigma_S \rangle. \tag{4.9}$$

To apply this result to $\mathcal{C}$ we write it as

$$\mathcal{C} = \left\langle \prod_S A_{ij}A_{jk}A_{kl}A_{li} \right\rangle, \tag{4.10}$$

where the product runs over a minimal set S of plaquettes with $\partial S = C$ and area of $S = |S|$. Applying inequality (4.9) we find

$$\mathcal{C} \ge \langle A_{ij}A_{jk}A_{kl}A_{li}\rangle^{|S|}, \tag{4.11}$$

in fact, no matter what the dimension or the value of β is. This establishes the right inequality in (4.1) and the theorem is proved.

It is likely that (4.1) can be strengthened to yield

$$-\frac{\ln\mathcal{C}}{|S|} \to \text{constant}$$

for β_p small enough and C going regularly to infinity. From (4.7) we derive that, if β_c is the largest value for which (4.1) holds, then

$$\tanh\beta_c \ge \frac{1}{2d-3}. \tag{4.12}$$

Thus for $d=2$ there is no transition as expected, while (4.12) is easily verified for $d=3$ and 4 using the critical constants given in (3.1). For d large we expected $\beta_c \sim 1/d$, which is again in agreement with (4.12).

It would be nice to complete this theorem by proving that for β_p large enough and $d \geq 3$, $-\ln \mathcal{C}$ behaves like $|C|$. In fact, it is sufficient to establish it for $d=3$ (where a transition is known to occur). An argument based on inequality (4.9) will then show that it also applies to $d>3$. Using duality, this amounts to studying a small β_* property for a corresponding three-dimensional Ising problem. Perturbation expansion of $\mathfrak{F}-\mathfrak{F}'$ in Eq. (2.10) in powers of β_* shows that, to a finite order, $-\ln\mathcal{C}$ is indeed proportional to $|C|$. Although strong indications exist that the result holds beyond perturbation theory, we have not been able to find a completely satisfactory proof. This is unfortunate, for it would have demonstrated the existence of a transition in any dimension $d \geq 3$, with $-(\ln\mathcal{C})/|S|$ as an order parameter. The order of the transition remains questionable: We have shown above that a second-order transition exists for $d=3$, but mean-field predictions seem to indicate first-order transitions for $d \geq 4$.

APPENDIX

We sketch some results of cohomology on a lattice analogous to similar properties of differential forms in the continuous case.

On a hypercubical lattice in d dimensions we define $d+1$ sets $\mathcal{L}_p$ of functions with values in the group $Z_2 \equiv \{1, -1\}$. The set $\mathcal{L}_0$ contains functions defined at each site, the set $\mathcal{L}_1$ functions on links, $\mathcal{L}_2$ on plaquettes, $\mathcal{L}_3$ on cubes and so on. Sites, links, plaquettes, cubes, ... are simplexes of dimension 0, 1, 2, 3, ... In each $\mathcal{L}_p$ a privileged element e assigns the value $+1$ to all p-dimensional simplexes. A product of two elements in $\mathcal{L}_p$ is the function that assigns to each simplex the product of the values of the given elements: e is a unity for this product. Each element is idempotent and $\mathcal{L}_p$ is a group. Let ∂ be a map from $\mathcal{L}_p$ to $\mathcal{L}_{p+1}$ $(0 \leq p \leq d-1)$ defined as follows. For each $\varphi \in \mathcal{L}_p$, $\partial\varphi$ assumes on a $(p+1)$-dimensional simplex a value equal to the products of the values of φ on the p-dimensional simplexes of its boundary. For instance, if $\varphi \in \mathcal{L}_1$, $\partial\varphi_{ijkl} = \varphi_{ij}\varphi_{jk}\varphi_{kl}\varphi_{li}$. It is clear that $\partial(\varphi\psi) = \partial\varphi\partial\psi$, $\partial e = e$, and $\partial(\partial\varphi) = e$. It is convenient to define $\mathcal{L}_{-1} = Z_2$ and extend the definition of ∂ for an element $\pm$ of $\mathcal{L}_{-1}$ as the constant function which assigns this value at each site.

Theorem: if $\varphi \in \mathcal{L}_p$ and $\partial\varphi = e$, then $\varphi = \partial\psi$, for some $\psi \in \mathcal{L}_{p-1}$. Of course, ψ is arbitrary to the extent that it can be multiplied by an element of the form $\partial\psi'$, $\psi' \in \mathcal{L}_{p-2}$ (a generalized gauge transformation).

The most elementary case is with $\varphi \in \mathcal{L}_0$: $\partial\varphi = e$ means $\varphi = \text{const}$, which by its very definition means $\varphi = \partial\psi$, $\psi \in \mathcal{L}_{-1}$.

A general proof of the theorem is not very instructive. Let us rather discuss as an example the case $d=3$, $p=2$. On an ordinary cubic lattice we have a function φ with value ± 1 for each plaquette, such that the product of its values on the faces of each cube is $+1$. By duality, this provides us with a function $\bar{\varphi}$ defined on the links of a dual lattice, such that the product of the six values corresponding to the six links incident on a site is $+1$. Let us mark all the links where $\bar{\varphi}$ assumes the value -1. Because of the aforementioned condition this set can be decomposed (perhaps not uniquely) into elementary closed circuits. Each of those can be considered as the boundary of a surface made of plaquettes. This choice involves, of course, a large arbitrariness. We define a function $\bar{\psi}$ equal to -1 on these plaquettes and $+1$ otherwise. Consider the product of the values of $\bar{\psi}$ on the four plaquettes having a fixed link in common. An even number of these plaquettes carries a value $\bar{\psi} = -1$ if this link does not belong to the closed circuits, and an odd one if it does. Thus the product in question is precisely equal to the value of $\bar{\varphi}$ on this link. If we now return to the original lattice, there corresponds to $\bar{\psi}$ a function $\psi \subset \mathcal{L}_1$ and the property just proved means $\partial\psi = \varphi$.

---

[1]R. Balian, J. M. Drouffe, and C. Itzykson, Phys. Rev. D $\underline{10}$, 3376 (1974). This paper, to which we refer as (I), contains the notations and motivations for the present one.

[2]K. G. Wilson, Phys. Rev. D $\underline{10}$, 2445 (1974).

[3]R. B. Griffiths, J. Math. Phys. $\underline{8}$, 478 (1967); D. G. Kelly and S. Sherman, J. Math. Phys. $\underline{9}$, 466 (1968); R. E. Peierls, Phys. Rev. $\underline{54}$, 918 (1938). For an illuminating review of the mathematical aspects of the Ising models see Ginibre, in Cargèse Lectures in Physics, 1973 (Gordon and Breach, New York, to be published).

[4]F. J. Wegner, J. Math. Phys. $\underline{12}$, 2259 (1971).

[5]M. Fisher and D. S. Gaunt, Phys. Rev. $\underline{133}$, 225 (1964).

[6]M. Fisher, Phys. Rev. $\underline{162}$, 480 (1967).

PHYSICAL REVIEW D VOLUME 11, NUMBER 8 15 APRIL 1975

Gauge fields on a lattice. III. Strong-coupling expansions and transition points

R. Balian, J. M. Drouffe, and C. Itzykson
Service de Physique Théorique, Centre d'Etudes Nucléaires de Saclay, BP No. 2-91190 Gif-sur-Yvette, France
(Received 6 November 1974)

We discuss the principles of the high-temperature expansion leading to a variation-perturbation method. For pure gauge fields, diagrams are two-dimensional manifolds. As an application, we compute the critical coupling constants for discrete, Abelian, and SU(2) gauge groups and compare them with some earlier results.

I. INTRODUCTION

In previous papers,[1] we described a gauge theory on a lattice following Wilson's idea.[2] The motivations and the notations are discussed at length in papers I and II. This paper presents some numerical calculations in the disordered, fully symmetric, high-temperature phase of the gauge field system (i.e., in the strong coupling limit).

In Sec. II, we recall the formalism of high-temperature expansions and the analysis of diagrams in terms of their strongly irreducible parts. This leads to a variational method, associated with a perturbative expansion of the generalized free energy to be varied.[3] To lowest order this procedure yields results equivalent to the mean-field approximation. The outline is general and we apply it to the two models discussed in Ref. 1: the scalar model (Sec. III) and the gauge model (Sec. IV).

II. FORMALISM OF HIGH-TEMPERATURE EXPANSIONS

A. The partition function in terms of diagrams

Consider the generating functional (or partition function)

$$Z(\{h_i\})=\int\prod_i D\phi_i \exp\Big(\sum h_i\phi_i+\beta S\Big). \qquad (2.1)$$

Here ϕ_i are (possibly multicomponent) fields assigned to the site i of a discrete lattice in dimension d. The values assumed by ϕ_i can be continuous or discrete. In the latter case, integration is replaced by summation. Whenever the range of ϕ is compact, we shall normalize the measure to unity. The action βS, assumed to be translationally invariant, is written

$$\beta S=\beta\sum_k \frac{1}{k!}\sum_{i_1,\ldots,i_k} V_{(k)}^{i_1\cdots i_k}\phi_{i_1}\cdots\phi_{i_k}, \qquad (2.2)$$

and the "potential" $V_{(k)}$ is totally symmetric in its indices.

At infinite temperature ($\beta=0$), βS vanishes and fields at different sites are independent. Thus Z factorizes into $Z_0=\prod_i z(h_i)$, with

$$z(h)\equiv \exp u(h)=\int D\phi \exp(h\varphi). \qquad (2.3)$$

The unperturbed average $\langle X\rangle_0$ of a functional $X(\varphi_i)$ is defined as

$$\langle X\rangle_0=Z_0^{-1}\int\prod_i D\varphi_i \exp\Big(\sum_i h_i\varphi_i\Big)X(\varphi_i). \qquad (2.4)$$

For small β, we expand the exponential $e^{\beta S}$ in powers of β; thus we are led to the unsophisticated high-temperature series

$$Z(\{h_i\})=Z_0(\{h_i\})\sum_{n=0}^{\infty}\frac{\beta^n}{n!}\langle S^n\rangle_0. \qquad (2.5)$$

It is convenient[3,4] to interpret the terms in the series (2.5) as corresponding to *graphs*. Each graph consists of a finite subset of distinguished sites on the lattice and a set of vertices representing interactions. Lines are drawn joining sites to vertices. Each site is linked to at least one vertex. A contribution to Z is assigned to each graph in the following way:

(i) A factor $\beta V_{(k)}^{i_1\cdots i_k}$ corresponds to a vertex linking the k sites $i_1,\ldots,i_k$.

(ii) Each site i linked to n_i vertices yields a factor

$$\langle\phi_i^{n_i}\rangle_0=z(h_i)^{-1}\frac{d^{n_i}}{dh_i^{n_i}}z(h_i). \qquad (2.6)$$

This formula is generalized in an obvious way if ϕ_i is a multicomponent field.

(iii) If a graph remains invariant by an interchange of some of its *vertices*, its contribution has to be divided by the order of its symmetry group.

The sum over all possible distinct graphs (including the empty one, which contributes a term equal to 1) reproduces the expansion (2.5) for $Z(\{h_i\})/Z_0(\{h_i\})$.

If we set all h_i equal to h, we may group the contributions of a family of graphs differing only by the locations of the sites, but yielding identical contributions due to the properties of the potentials $V_{(k)}$. We represent such a family of graphs by a nonlabeled *diagram*. A diagram can thus be drawn independently of the structure of the lattice (for instance, of its dimensionality). Besides the previous rules, we then assign to a given diagram an extra factor, equal to the number of distinct corresponding graphs. Such a factor can be thought of as a number of configurations (n.c.), i.e., the number of ways a diagram can be mapped on the lattice. It may be evaluated by relabeling its sites under the condition that two sites cannot receive the same label.

In this type of calculations one is faced with two problems:

(i) Enumerate all diagrams of a given order in β, and

(ii) find the corresponding n.c. for each diagram.

A discussion of these points is given by Domb[4] for the case of Ising-type interactions ($V_{(k)}=0$ if $k>2$). In Sec. IV we shall present the more intricate case of the gauge model.

Because of translational invariance (with periodic boundary conditions), the n.c. is a polynomial in N, the number of sites, which vanishes for $N=0$. The degree of this polynomial is equal to the number of connected parts of the diagram. In calculating $F=\lim_{N\to\infty}(1/N)\ln Z$, we make use of the existence of the infinite-volume limit. As a result, the high-temperature series for $F-u(h)$ is defined in terms of the same diagrams and same rules as before, except for the replacement of the n.c. by the coefficient of order 1 in its power expansion in N. We call this coefficient the reduced number of configurations (r.n.c.).

With the above definition of diagrams, connected as well as disconnected ones contribute to F. In the subsequent sections, we shall also use more elaborate expansions in terms of suitably modified connected diagrams, which we now recall.

B. Connected, irreducible graphs

The unperturbed averages $\langle\phi^n\rangle_0=e^{-u}(d^n/dh^n)e^u$ can be expressed in terms of cumulants defined as

$$\langle\phi^n\rangle_c=\frac{d^n u}{dh^n}. \qquad (2.7)$$

The relation is

$$\langle\phi^n\rangle_0=\sum\langle\phi^{n_1}\rangle_c\cdots\langle\phi^{n_j}\rangle_c, \qquad (2.8)$$

where the summation extends over all partitions of n distinct objects. Substituting (2.8) in the expansion of Z/Z_0 defines new graphs, with the property that the lines arriving at each site are tied together in all possible ways, each one corresponding to one term in (2.8). This is to be contrasted with the previous rules, where a single factor $\langle\phi^n\rangle_0$ was assigned to a given site linked to n vertices. With these new rules, when summing over all diagrams, we are allowed to let the sites overlap. Consequently, summations over all connected parts are independent, and $\ln(Z/Z_0)$ is expressed entirely in terms of connected graphs. If we set all h_i equal to h, we obtain

$$F=u(h)+\sum(\text{connected diagrams}). \qquad (2.9)$$

To be precise, let us repeat how one computes the contributions of the diagrams entering (2.9). Such a diagram consists of points, vertices, and lines. Each point is joined to at least one vertex through a line. The diagram is to be connected. A vertex of k lines yields a factor $\beta V_{(k)}$, a point linked to n vertices gives a factor $\langle\phi^n\rangle_c$. This contribution is completed by including two weights:

(i) a symmetry factor $p(D)^{-1}$, where $p(D)$ is the order of the symmetry group of the diagrams by interchange of its vertices, and

(ii) a r.n.c., computed by dividing by N the number of distinct mappings of the diagram points onto the lattice sites (note that for connected diagrams the n.c. is simply equal to N times the r.n.c.).

The restriction that two points cannot be mapped on the same site is removed. Two maps differing only by a relabeling of the points of the diagram are not distinguished. Examples illustrating these rules will be worked out shortly.

A connected graph is *reducible* if it is cut out into k parts by removing one of its k-vertices. One allows among the possible parts a single site without any vertex. We might of course have enlarged the definition of the graphs to include such a case, to which corresponds the term $u(h_i)$ in the expression (2.9). A general connected graph is thus a tree of lines joining vertices to irreducible "bubbles" (see Fig. 1).

Call $B(\{h_i\})$ the sum over connected irreducible

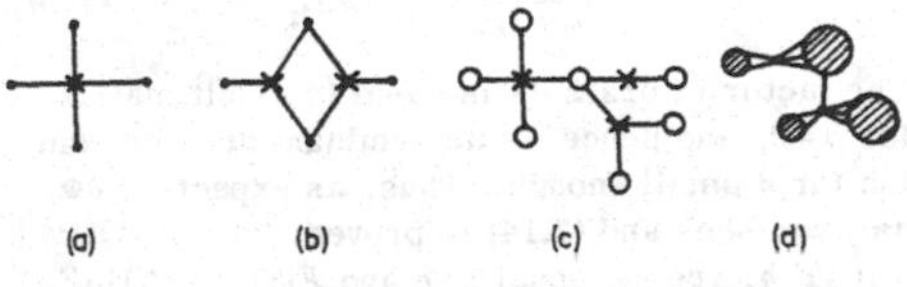

FIG. 1. Examples of graphs: sites are represented as points, vertices by crosses, irreducible parts by circles, strongly irreducible parts by hatched circles. (a) Reducible graph. (b) Irreducible graph. (c) General decomposition of a graph into a tree of irreducible parts. (d) Same decomposition as before but in strongly irreducible parts.

graphs. For a general connected graph, adding a k-vertex increases the number n_k of k-vertices by one unit and the number n_B of bubbles by $k-1$ units. We then get the topological identity

$$1 = n_B - \sum_k (k-1) n_k . \tag{2.10}$$

Set

$$\langle \phi_i \rangle \equiv m_i = \frac{\partial}{\partial h_i} \ln Z \tag{2.11}$$

and average the relation (2.10) by multiplying both sides by the contribution of each graph. The following equality is obtained:

$$\ln Z(\{h_i\}) = B\left(\left\{ h_i + \beta \sum_k \frac{1}{(k-1)!} \sum_{j_2 \cdots j_k} V_{(k)}^{i j_2 \cdots j_k} m_{j_2} \cdots m_{j_k} \right\}\right) - \beta \sum_k \frac{k-1}{k!} \sum_{i_1 \cdots i_k} V_{(k)}^{i_1 \cdots i_k} m_{i_1} \cdots m_{i_k} . \tag{2.12}$$

The combination of (2.11) and (2.12) yields a variational principle for the calculation of the total free energy. Namely, consider the right-hand side of (2.12) as a function Φ of the two independent sets of variables h_i and m_i; the h_i's being kept fixed, $\ln Z$ is the stationary value of Φ when varying the m_i's:

$$\ln Z(\{h_i\}) = \Phi(\{h_i\}, \{m_i\}) , \tag{2.13}$$

$$\frac{\partial \Phi}{\partial m_i} = 0 . \tag{2.14}$$

The proof may rely on the topological relation (2.10).[5] We may also check (2.14) directly. From the definition of Φ, we have

$$\frac{\partial \Phi}{\partial m_l} = \sum_j \frac{\partial B}{\partial h_j}\left(\left\{ h_i + \beta \sum_k \frac{1}{(k-1)!} \sum_{j_2 \cdots j_k} V_{(k)}^{i j_2 \cdots j_k} m_{j_2} \cdots m_{j_k} \right\}\right) \beta \sum_k \frac{1}{(k-2)!} \sum_{j_3 \cdots j_k} V_{(k)}^{j l j_3 \cdots j_k} m_{j_3} \cdots m_{j_k}$$
$$- \beta \sum_k \frac{1}{(k-2)!} \sum_{j_2 \cdots j_k} V_{(k)}^{l j_2 \cdots j_k} m_{j_2} \cdots m_{j_k} . \tag{2.15}$$

On the other hand, making use of (2.12) where m_i takes its actual value (2.11), we obtain

$$m_j = \frac{\partial B}{\partial h_j} + \sum_l \frac{\partial \Phi}{\partial m_l} \frac{\partial m_l}{\partial h_j} . \tag{2.16}$$

We now eliminate the explicit B term of (2.15) by use of (2.16), which yields

$$0 = \sum_l \left[\delta_{jl} + \beta \sum_k \frac{1}{(k-2)!} \times \sum_{j_2 \cdots j_k} V_{(k)}^{j j_2 \cdots j_k} \frac{\partial m_l}{\partial h_{j_2}} m_{j_3} \cdots m_{j_k} \right] \frac{\partial \Phi}{\partial m_l} .$$

The factor in brackets is equal to a unit matrix for $\beta = 0$, and hence its determinant does not vanish for β small enough. Thus, as expected, $\partial\Phi/\partial m_l$ vanishes and (2.14) is proved.

If all h_i are set equal to h and $F(h) = (1/N)\ln Z$, then all m_i are equal to m, $B = Nb(h)$, $\Phi = N\phi(h, m)$, and $m = dF/dh$. All previous formulas apply after omitting a factor N. In later applications, however, N will stand for the number of lattice nodes, whereas the field variables will be defined on the links, the number of which is Nd. The relation between m and dF/dh will therefore be modified into $md = dF/dh$.

Although the lowest-order approximation to B in (2.13) yields, as we shall see below, the mean-field results at the stationary point, it should be noticed that the stationary value of Φ is not necessarily a maximum in m.

If one sets the external field h equal to zero, the extremum lies at the point $m = 0$ for $\beta = 0$. If there is a second-order transition at β_c, m smoothly departs from $m = 0$ at that point. The critical value β_c is then obtained by requiring that

$$\frac{\partial^2}{\partial m^2} \phi(0, 0) = 0 . \tag{2.17}$$

If, on the other hand, the transition is a first-order one with a jump from 0 to m_c at β_c, these two quantities follow from the system of equations:

$$\phi(0, 0) = \phi(0, m_c) ,$$
$$\frac{\partial \phi}{\partial m}(0, m_c) = 0 . \tag{2.18}$$

Of course the whole method allows one to compute in a systematic way various other quantities of interest. In the sequel we shall mainly be interested in the calculation of the critical value of β.

C. Strong irreducibility

It is possible to go further in the classification of graphs by strengthening the concept of irreducibility.[3] Up to now, irreducibility was defined in terms of vertices. The new type, called strong irreducibility, is defined with respect to sites. A connected graph is *strongly irreducible* if omission of any one of its sites (and of the corresponding lines) leaves it connected. A strongly irreducible graph with more than one vertex is *a fortiori* irreducible. A general connected graph is thus a tree of strongly irreducible parts [Fig. 1(d)].

Considering this decomposition of a graph into $n_{\mathfrak{B}}$ strongly irreducible parts, connected by n_L sites, let n_M be equal to the sum of these connecting sites, each one counted as many times as the number of strongly irreducible parts to which it is linked. The following topological identity holds[3,5]:

$$1 = n_{\mathfrak{B}} + n_L - n_M . \quad (2.19)$$

Let $\mathfrak{B}(\{M_i^{(n)}, h_i\})$ be the sum of contributions over strongly irreducible graphs computed according to the preceding rules, except for the replacement of $\langle \phi_i^n \rangle_c$ by the variables $M_i^{(n)}$ defined consistently through

$$M_i^{(n)} = \left[\exp\left(\sum_{k=1}^{\infty} G_i^{(k)} \frac{\partial^k}{\partial h_i^k}\right)\right] \langle \phi_i^n \rangle_c ,$$

$$G_i^{(k)} = \frac{\partial \mathfrak{B}}{\partial M_i^{(k)}} (\{M_j^{(n)}, h_j\}) \quad (2.20)$$

(for the first of these equations, no summation over i is implied). Let $\mathfrak{L}_i$ be the extension of formula (2.20) to $n = 0$, in the form

$$\mathfrak{L}_i = \left[\exp\left(\sum_{k=1}^{\infty} G_i^{(k)} \frac{\partial^k}{\partial h_i^k}\right)\right] u(h_i) . \quad (2.21)$$

If one returns to (2.19) and averages both sides using the contribution of all connected graphs, one finds

$$\ln Z(\{h_i\}) = \mathfrak{B}(\{M_i^{(n)}, h_i\}) + \sum_i \mathfrak{L}_i - \sum_i \sum_{k=1}^{\infty} M_i^{(k)} G_i^{(k)} . \quad (2.22)$$

For a detailed justification of this result, see Refs. 3 and 5. As before, the solution of the self-consistent equations (2.20) can be cast into a variational principle. Namely, the right-hand side of (2.22) is considered as a function of independent h's, M's, and G's. Keeping h's fixed and varying M's and G's yields at the stationary point the value of $\ln Z$. All the discussion at the end of Sec. II B may be carried over to the present extension.

III. SCALAR MODEL

In this section, we apply the previous machinery to the scalar model with global symmetry O(n) described in paper I. For $n = 1$, this is nothing but the Ising model and for $n \geq 2$ it is the classical Heisenberg model. We shall also recover the Stanley model in the $n \to \infty$ limit using the technique described in Sec. II C.

A. Critical coupling constant as a power series in $1/d$

Fields are n-dimensional unit vectors $\vec{k}_i$ located at the nodes of a d-dimensional hypercubical lattice and interactions are between nearest neighbors. Thus, we have

$$S = \sum_{(i,j)} \vec{k}_i \cdot \vec{k}_j . \quad (3.1)$$

In the graphs, vertices join two neighboring sites and can just be represented by this bond. The function $z(\vec{h})$ in (2.3) is now ($h = |\vec{h}|$)

$$z(h) \equiv e^{u(h)} = \int d^{n-1}k \exp(\vec{k} \cdot \vec{h})$$

$$= a_n h^{1-n/2} I_{n/2-1}(h) , \quad (3.2)$$

where $I_\nu(x)$ is the modified Bessel function, and where a_n is adjusted in such a way that $u(0) = 0$.

To find β_c, we use the method of Sec. II B, where $b(h)$ is expanded in powers of the number of bonds. It is convenient to replace the variable m to be determined variationally by

$$H \equiv h + 2\beta d m .$$

In the absence of external field, Eqs. (2.13) and (2.14) then reduce to

$$F = b(H) - \frac{H^2}{4\beta d},$$

$$\frac{db}{dH} = \frac{H}{2\beta d} . \quad (3.3)$$

To lowest order, we have $b_0(h) = u(h)$, and (3.3) reduces to the mean-field result (see, for instance, paper I), with a second-order transition. The critical β_c, obtained by requiring the vanishing of the H^2 term in F, is found to be equal to $\beta_c = n/2d$ to this order. We can now consider in a systematic fashion the expansion of $b(h)$ in powers of β. In Table I, we have collected the contributions to the coefficient of $\frac{1}{2}H^2$ up to sixth order. A diagram of order p yields a contribution to b which is a polynomial of degree $[p/2]$ in d,

TABLE I. Coefficients of $\frac{1}{2}H^2$ in the contributions to $b(H)$, up to sixth order in β, for the scalar model.

	$\frac{1}{n}$		$\frac{36\beta^5}{n^6(n+2)}\,d(2d-1)$
	$-\frac{2\beta^2 d}{n^3}$		$-\frac{8\beta^6 d^3}{n^7(n+2)}\,(7n+16)$
	$\frac{4\beta^3 d}{n^4(n+2)}$		$-\frac{40\beta^6 d^3}{n^7}$
	$\frac{12\beta^4 d^2}{n^5}$	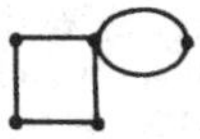	$\frac{96\beta^6}{n^7}\,d^2(2d-1)$
	$-\frac{6\beta^4}{n^5}\,d(2d-1)$		$\frac{4\beta^6 d}{n^7(n+2)}\,(4d^2+8d-7)$
	$-\frac{8\beta^4 d}{n^5(n+2)}$		$-\frac{20\beta^6}{n^7}\,d(6d^2-9d+4)$
	$-\frac{64\beta^5 d^2}{n^6(n+2)}$		

given by the n.c. As β_c is of order $1/d$ from the zeroth-order result, the correction to b and therefore to $\beta_c d$ is of order $d^{-p-[p/2]}$. Consequently the series for β_c appears as a $1/d$ expansion. Thus our calculation yields

$$(\beta_c^*)^{-1}=\left(\frac{2\beta_c d}{n}\right)^{-1}$$

$$=1-\frac{1}{2d}-\frac{2-2/(n+2)}{(2d)^2}-\frac{7-8/(n+2)}{(2d)^3}$$

$$+O\left(\frac{1}{d^4}\right). \qquad (3.4)$$

We recover the results of Ref. 6 for the case $n=1$. We shall also check this formula in the limit $n\to\infty$ at the end of this section. Some numerical values are listed in Table II for β_c^* to sixth order in the expansion of $b(H)$.

TABLE II. Values for $\beta_c^*=2\beta_c d/n$ in the scalar model obtained through the $1/d$ expansion. The numbers in parentheses are obtained using the alternative method of Sec. III B. Exact results are listed in brackets, either for $n=1$, $d=2$, or for the limit $n\to\infty$ (Sec. III C).

n \ d	2	3	4	5	∞
1	1.6696 [1.763]	1.2883	1.1824	1.1334	1
2		1.3012 (1.3170)	1.1879 (1.1924)	1.1364 (1.1357)	1
3		1.3091	1.1913	1.1382	1
∞	[∞]	1.3416 [1.4808]	1.2047 [1.2385]	1.1455 [1.1563]	1 [1]

B. Free energy and two-point function ($n=2$)

Picking the special case $n=2$ as a typical example, we now perform various high-temperature expansions. They will provide a cross-check for the value of β_c in reasonable agreement with the previous one. Up to twelfth order, the high-temperature expansion of the free energy (2.9) in zero external field yields

$$F(\beta)=d\beta^2+(d^2-\tfrac{5}{4}d)\beta^4+(\tfrac{16}{3}d^3-16d^2+\tfrac{97}{9}d)\beta^6+(54d^4-284d^3+\tfrac{22747}{48}d^2-\tfrac{15613}{64}d)\beta^8$$
$$+(\tfrac{3966}{5}d^5-6328d^4+\tfrac{55537}{3}d^3-\tfrac{92585}{4}d^2+\tfrac{6101009}{600}d)\beta^{10}$$
$$+(\tfrac{45280}{3}d^6-167836d^5+\tfrac{6677002}{9}d^4-\tfrac{14484205}{9}d^3+\tfrac{5457782167}{3240}d^2-\tfrac{17218423249}{25920}d)\beta^{12}\sim O(\beta^{14})\,. \tag{3.5}$$

The curve $F=F(\beta)$ is drawn in Fig. 2 for $d=4$. The zero-momentum propagator $\chi(\beta)$ (the susceptibility), computed up to seventh order, reads

$$\chi(\beta)=1+2\beta d+(4d^2-2d)\beta^2+(8d^3-8d^2+d)\beta^3+(16d^4-24d^3+4d^2+4d)\beta^4+(32d^5-64d^4+20d^3+18d^2-\tfrac{19}{3}d)\beta^5$$
$$+(64d^6-160d^5+80d^4+24d^3+\tfrac{146}{3}d^2-\tfrac{341}{6}d)\beta^6+(128d^7-384d^6+272d^5+8d^4+246d^3-\tfrac{2423}{6}d^2+\tfrac{1071}{8}d)\beta^7+O(\beta^8)\,. \tag{3.6}$$

The corresponding curve is also drawn in Fig. 2 for $d=4$. At $\beta=\beta_c$ a singularity occurs. Its position may be determined by computing either the limiting ratio of two consecutive terms of the series or the smallest positive zero of a Padé-approximant denominator. Both methods give the numbers quoted in parentheses in Table II.

For large distances, the propagator behaves as $e^{-\mu r}$ up to a power of r. The first few terms of the expansion of μ (in units of the inverse lattice spacing) are

$$\mu(S_2)^{1/2}=\ln\beta+(S_{-1}-2d+\tfrac{1}{2})\beta^2$$
$$+[5S_{-1}^{\;2}-S_{-1}S_{-2}-\tfrac{1}{2}S_{-3}+(\tfrac{5}{2}-12d)S_{-1}+(2d-\tfrac{1}{2})S_{-2}-\tfrac{1}{2}S_2-10d^2+12d-\tfrac{67}{24}]\beta^4+O(\beta^6)\,, \tag{3.7}$$

with

$$\alpha_i=|r_i|\Big/\sum_{j=1}^{d}|r_j|$$

and

$$S_n=\sum_{i=1}^{d}\alpha_i^{\,n}\,.$$

Away from the critical point, no Euclidean invariance appears, even in the large-distance limit. It is, nevertheless, generally believed that this invariance shows up near the critical point when the correlation distance becomes infinite.

C. The Stanley model ($n\to\infty$ limit)

It is well known that, in the $n\to\infty$ limit, the scalar model is soluble.[7] We briefly outline the use of the present techniques to find this result. For n large, we need an estimate of $z(h)=e^{u(h)}$ for h of order n in formula (3.2). This is given by

$$u(h)\simeq\frac{1}{2}n\left\{\left(1+\frac{4h^2}{n^2}\right)^{1/2}-1-\ln\left[\frac{1}{2}+\frac{1}{2}\left(1+\frac{4h^2}{n^2}\right)^{1/2}\right]\right\}, \tag{3.8}$$

where h^2 stands for $\sum_{\alpha=1}^{n}h_\alpha^{\,2}$. Returning to (2.22) and (2.21), it is easily shown that $G^{(k)}$, $k\geq 3$, can

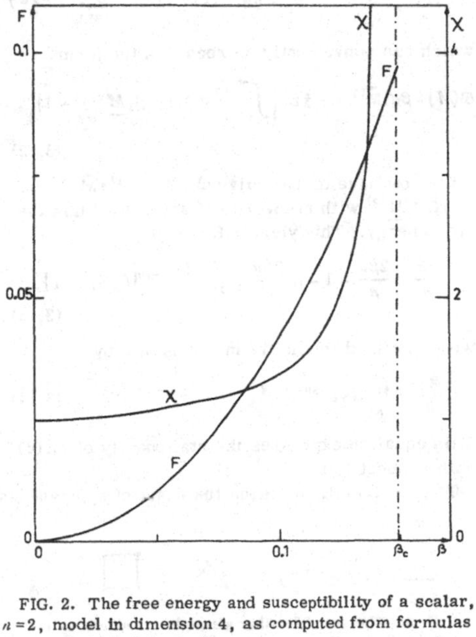

FIG. 2. The free energy and susceptibility of a scalar, $n=2$, model in dimension 4, as computed from formulas (3.5) and (3.6) of the text.

be dropped for $n\to\infty$ so that

$$\mathcal{L}(\vec{G}^{(1)},\underline{G}^{(2)},\vec{h})=\left[\exp\left(\frac{\partial}{\partial h_a}\underline{G}^{(2)}{}_{ab}\frac{\partial}{\partial h_b}\right)\right]\times u(|\vec{h}+\vec{G}^{(1)}|)\,. \tag{3.9}$$

The notation $\underline{G}^{(2)}$ is to recall that we are dealing with a $n\times n$ matrix. To leading order in n, (3.9) yields the approximation

$$\mathcal{L}\simeq u((|\vec{h}+\vec{G}^{(1)}|^2+2\,\mathrm{tr}\underline{G}^{(2)})^{1/2})\,. \tag{3.10}$$

Finally the relevant set of diagrams for the computation of $\mathcal{B}$ turns out to be those with no more than one loop (Fig. 3). Thus

$$\mathcal{B}(M)=\beta d(\vec{M}^{(1)})^2+\sum_{p=2}^{\infty}\frac{1}{2p}\mathrm{tr}[(\underline{M}^{(2)})^p]\frac{1}{N}\mathrm{Tr}[(\beta V)^p]\,. \tag{3.11}$$

Here $\underline{M}^{(2)}$ is a $n\times n$ matrix, while V is an $N\times N$ matrix with elements V_{ij} equal to one or zero according to whether (ij) are nearest neighbors or not. Using a Fourier series to compute the trace over $(\beta V)^p$ and noting that this quantity vanishes for $p=1$, we arrive at

$$\mathcal{B}(M)=\beta d(\vec{M}^{(1)})^2-\tfrac{1}{2}\mathrm{tr}\int_0^{2\pi}\frac{dq_1}{2\pi}\cdots\times\int_0^{2\pi}\frac{dq_d}{2\pi}\ln\left(1-2\beta\underline{M}^{(2)}\sum_{\alpha=1}^{d}\cos q_\alpha\right),$$

which can conveniently be recast in the form

$$\mathcal{B}(M)=\beta d(\vec{M}^{(1)})^2+\tfrac{1}{2}\mathrm{tr}\left\{\int_0^\infty\frac{ds}{s}e^{-s}[I_0(2\beta_s\underline{M}^{(2)})^d-1]\right\}. \tag{3.12}$$

We now have to maximize $\mathcal{B}+\mathcal{L}-\vec{G}^{(1)}\cdot\vec{M}^{(1)}-\mathrm{tr}\underline{G}^{(2)}\underline{M}^{(2)}$ with respect to M and G to obtain the free energy. This yields, for $\beta<\beta_c$,

$$\frac{2F}{n}=\frac{2\beta\nu}{n}-1-\ln\frac{2\beta\nu}{n}+\int_0^\infty\frac{ds}{s}e^{-s\nu}[I_0(s)^d-1]\,, \tag{3.13}$$

with ν defined implicitly in terms of β by

$$\frac{2\beta}{n}=\int_0^\infty ds\,e^{-s\nu}I_0(s)^d\,. \tag{3.14}$$

This equation expresses the stationarity of (3.13) with respect to ν.

The critical β_c, marking the onset of spontaneous magnetization, is obtained either from the variational equations giving $\vec{M}^{(1)}$ or by setting ν to its lowest possible value equal to d in (3.14). This gives

$$\beta_c^*=\frac{2\beta_c d}{n}=d\int_0^\infty ds[e^{-s}I_0(s)]^d\,. \tag{3.15}$$

Figure 4 shows the behavior of β_c^*, which blows up for $d=2$. An expansion in powers of $1/d$ is obtained by using the Taylor series of the Bessel function, and this agrees with the series (3.4), where one omits the terms in $1/(n+2)$. The values of β_c^* for $d=3,4,5$ are indicated between brackets in Table II.

IV. YANG-MILLS MODEL

We shall first discuss some technical points for obtaining the high-temperature expansion for the pure Yang-Mills field in the simple case of a discrete Z_2 gauge group. We further extend the analysis to a continuous group, Abelian [U(1)] or non-Abelian [SU(2)]. Finally, we present some numerical results.

A. Diagrammatic rules

We consider the action as a sum over all plaquettes

$$S=\sum_p A_{12}A_{23}A_{34}A_{41}\,, \tag{4.1}$$

where the fields $A_{ij}=\pm1$ and sources h are attached to each link of the hypercubical lattice. The unperturbed partition function relative to a link is then

$$z(h)\equiv e^{u(h)}=\tfrac{1}{2}\sum_{A=\pm1}e^{Ah}=\cosh h\,. \tag{4.2}$$

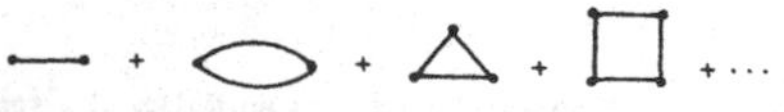

FIG. 3. The set of leading strongly irreducible diagrams for the calculation of the $n\to\infty$ limit.

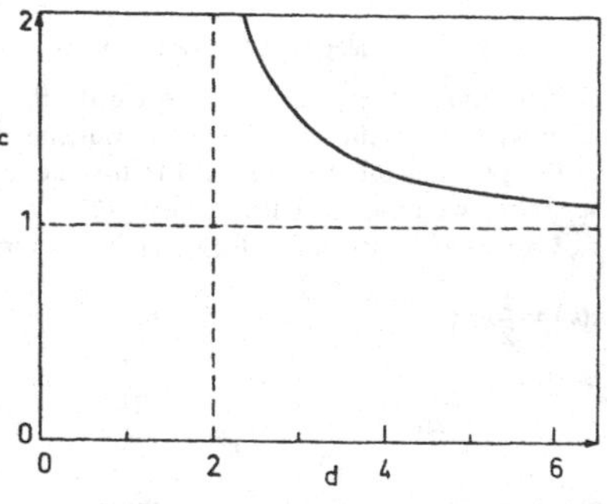

FIG. 4. The critical coupling β_c as a function of d in the $n\to\infty$ limit.

The expansion of Z in powers of β is performed as in Sec. II A. Graphs are drawn as *two-dimensional surfaces built of plaquettes* which now represent the four-vertices. When the graph is considered independently from the lattice by preserving the relation between links and plaquettes, it is called a diagram, as previously. The problem now is to analyze these diagrams and to compute the associate r.n.c.

We shall disregard the simplification introduced in the present case by the remark that a series in powers of $\tanh\beta$ obeys simpler rules. Hence we study the plain β expansion (2.5) of Z. To do this, we first introduce *skeleton diagrams*, which never contain the same plaquette of the lattice twice. Given a skeleton, one can reconstruct all associated diagrams by dressing each of its plaquettes once, twice, three times, ...

In Table III, we present all the connected skeletons up to four plaquettes. Computations in zero

TABLE III. The number of configurations for all connected skeletons up to fourth order.

1		$\frac{1}{2}d(d-1)$
2		$d(d-1)(2d-3)$
3	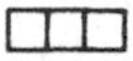	$d(d-1)(2d-3)^2$
3	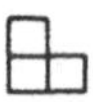	$2d(d-1)(4d^2-14d+13)$
3		$\frac{2}{3}d(d-1)(d-2)(2d-3)$
3		$\frac{4}{3}d(d-1)(d-2)$
4		$\frac{1}{2}d(d-1)(16d^3-72d^2+107d-52)$
4	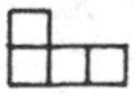	$4d(d-1)(2d-3)(4d^2-14d+13)$
4	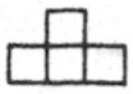	$2d(d-1)(8d^3-44d^2+84d-55)$
4	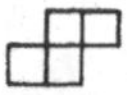	$2d(d-1)(8d^3-44d^2+84d-55)$
4		$8d(d-1)(d-2)(2d^2-8d+9)$
4		$\frac{1}{6}d(d-1)(d-2)(2d-3)(2d-5)$
4	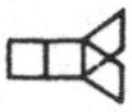	$2d(d-1)(d-2)(2d-3)^2$
4		$4d(d-1)(d-2)(4d^2-16d+17)$
4		$\frac{1}{2}d(d-1)(d-2)$
4		$16d(d-1)(d-2)^2$
4		$8d(d-1)(d-2)^2$
4	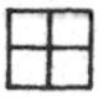	$\frac{1}{2}d(d-1)(4d^2-16d+17)$
4		$2d(d-1)(d-2)$

external field require the consideration of closed skeletons (because each link must occur an even number of times). Connected ones are classified up to sixteen plaquettes in Table IV.

On these same two tables appear the r.n.c. The latter is identical for a skeleton and for its dressed counterparts. We were unable to find a general rule to write the r.n.c. for the connected, strongly irreducible skeletons, and had to calculate them case by case. However, once these were obtained, the r.n.c. of an arbitrary diagram can readily be computed. Let us illustrate the reasoning on some

TABLE IV. The number of configurations for strongly irreducible skeletons contributing to zero-external-field expansions up to sixteenth order.

	(1)	$\frac{1}{2}d(d-1)$
	(6)	$\frac{1}{6}d(d-1)(d-2)$
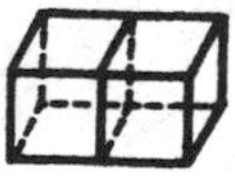	(10)	$\frac{1}{6}d(d-1)(d-2)(4d-9)$
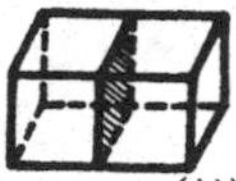	(11)	$\frac{1}{6}d(d-1)(d-2)(4d-9)$
	(12)	$\frac{4}{3}d(d-1)(d-2)(d-3)$
	(12)	$\frac{1}{3}d(d-1)(d-2)(d-3)$
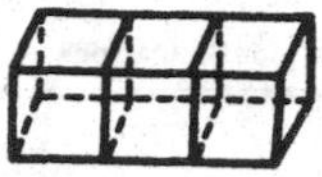	(14)	$\frac{1}{2}d(d-1)(d-2)(4d^2-20d+25)$
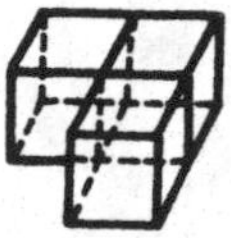	(14)	$2d(d-1)(d-2)(4d^2-20d+25)$
	(14)	$d(d-1)(d-2)(d-3)$
	(14)	$4d(d-1)(d-2)(d-3)$
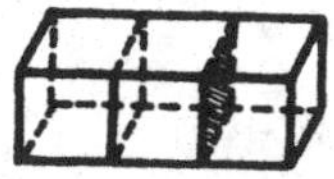	(15)	$d(d-1)(d-2)(4d^2-20d+25)$
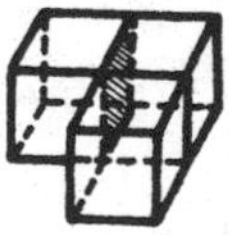	(15)	$4d(d-1)(d-2)(4d^2-22d+31)$

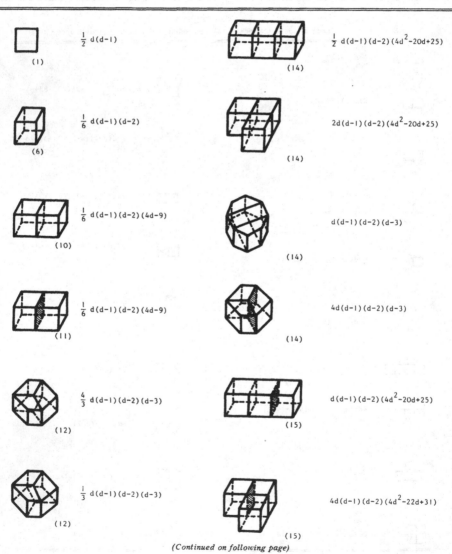

(Continued on following page)

Table IV (*continued*)

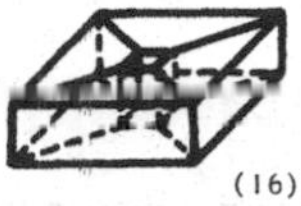 (16)	$8d(d-1)(d-2)(d-3)$	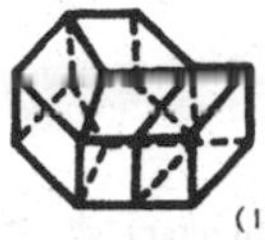 (16)	$16d(d-1)(d-2)(d-3)^2$
 (16)	$16d(d-1)(d-2)(d-3)^2$	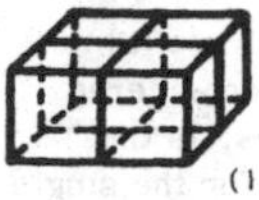 (16)	$\frac{1}{2}d(d-1)(d-2)(4d^2-24d+37)$
 (16)	$8d(d-1)(d-2)(d-3)^2$		

simple examples.

(a) Single plaquette. There are $N[d(d-1)/2]$ plaquettes on the lattice.

(b) Two plaquettes sharing one link. Choosing the link leads to Nd possibilities. The first plaquette can then be set in $2(d-1)$ directions, the second one in $2(d-1)-1$. Dividing by a factor 2, because of the indiscernability between the two plaquettes, finally leads to $Nd(d-1)(2d-3)$.

(c) Three plaquettes at the corner of a cube. There are $N[d(d-1)(d-2)/3!]$ cubes on the lattice and 8 corners on each cube. Hence the r.n.c. is $\frac{4}{3}d(d-1)(d-2)$.

It is easy to compute the r.n.c. of a disconnected skeleton S. We start from the set theoretic formula[4]

$$\{S_1\}\cdot\{S_2\}=\sum_{S=S_1\cup S_2} n_s\{S\}, \tag{4.3}$$

where $\{S\}$ denotes the n.c. of a diagram S, and where the summation runs over all skeletons which can be decomposed as $S_1\cup S_2$ in n_S different ways. The r.n.c. $[S]$, i.e., the coefficient of N in $\{S\}$, then satisfies

$$\sum_{S=S_1\cup S_2} n_s[S]=0\,. \tag{4.4}$$

Equation (4.4) allows a recursive calculation of $[S]$ if we isolate on its left-hand side the term corresponding to the maximal number of disconnected parts.

This method can be refined to compute the r.n.c. for connected, but not strongly irreducible skeletons. Let us distinguish a link of a given connected skeleton S, and denote by k the symmetry number pertaining to this link (meaning that there are k links on the skeleton playing the same topological role). Every link of the lattice will be occupied by the distinguished one of the skeleton an equal number of times (up to a choice among the k identical ones). This number of mappings is therefore, again up to the symmetry of order k,

$$(Nd)^{-1}k\{S\} = \frac{k[S]}{d}.$$

If S can be decomposed into $S_1 \cup S_2$ (both sharing the distinguished link) in n_s ways, formula (4.3) becomes

$$\frac{k_1[S_1]}{d}\frac{k_2[S_2]}{d} = \sum_{S=S_1\cup S_2} n_s \frac{k[S]}{d}. \tag{4.5}$$

Let us consider, for instance, the example (b) above. The distinguished link for the considered skeleton, made of two adjacent squares, is the central one ($k = 1$), while it is any one for the single square ($k = 4$). Equation (4.5) gives the diagrammatic equation shown in Fig. 5, where the factor $n_s = 2$ in front of the first term of the right-hand side represents the two ways in which the skeleton can be obtained as a union of two distinct plaquettes. We thus recover the r.n.c. $d(d-1)(2d-3)$ of the reducible diagram of example (b).

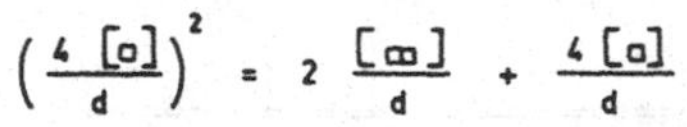

FIG. 5. The diagrammatic equivalent of Eq. (4.5) for two plaquettes sharing one link.

B. Continuous gauge groups

We discuss in turn the extra factors contributed by the gauge groups U(1) and SU(2) (see paper I).

1. *Abelian group* U(1)

The action is now

$$S = \sum_p \cos(\psi_{12} + \psi_{23} + \psi_{34} + \psi_{41})$$
$$= \sum_p{}' \zeta_{12}\zeta_{23}\zeta_{34}\zeta_{41}, \tag{4.6}$$

where $\sum_p'$ means a sum over *oriented* plaquettes, $\zeta_{ij} = \zeta^*_{ji} = e^{i\psi_{ij}}$. The sum over configurations involves integration over every angle ψ in an interval of 2π with a measure $(2\pi)^{-1}d\psi$. The source term is written

$$\sum_l{}' h_{ij}\zeta_{ji},$$

where again the sum runs over oriented links and $h_{ij} = h^*_{ji}$.

In the diagrammatic expansion, a vertex is now associated with an oriented plaquette, field variables being associated with the oriented links. The contribution of any link described n_1 times in the direction $i \to j$ and n_2 times in the reverse direction $j \to i$ is

$$\frac{\partial^{n_1+n_2}}{\partial h^{*n_1}_{ij}\partial h^{n_2}_{ij}} z(h_{ij}), \tag{4.7}$$

where

$$z(h) = \int \frac{d\zeta}{2i\pi\zeta}\exp(h^*\zeta + h\zeta^*)$$
$$= I_0(2|h|). \tag{4.8}$$

The general procedure of Sec. II then applies in a straightforward way, diagrams being now built with oriented plaquettes.

2. *Non-Abelian group* SU(2)

On the example of SU(2) as a gauge group, we illustrate the new technical points arising from the noncommutativity of the gauge field. The action is

$$S = \sum_p \chi(A_{12}A_{23}A_{34}A_{41}). \tag{4.9}$$

The matrices $A_{ij} = A^{-1}{}_{ji}$ belong to SU(2) and can be parametrized with an angle ϕ ($0 \le \phi \le 2\pi$) and a unit three-dimensional vector $\vec{n}$ as

$$A = \cos\phi + i\sin\phi\,\vec{n}\cdot\vec{\sigma}, \tag{4.10}$$

where $\vec{\sigma}$ stands for the Pauli matrices. The normalized measure on SU(2) is

$$dA = \frac{\sin^2\phi\, d\phi\, d^2n}{2\pi^2}.$$

We choose for χ the character associated with the spin-$\frac{1}{2}$ representation:

$$\chi(A) = \mathrm{tr}A = 2\cos\phi. \tag{4.11}$$

If we were to display the matrix elements of A and expand $\chi(AAAA)$ in terms of those, we could blindly use the formalism for multicomponent fields described previously. It is, however, wiser to try to take as much advantage as possible of the group-theoretic framework. For instance, in the case of a zero external field, we found it convenient to compute the contribution of a given diagram by using recursive formulas such as

$[A_1, A_2 \in \mathrm{SU}(2)]$

$$\rho_{pq}(A_1,A_2) \equiv \int dA[\chi(A_1A)]^p[\chi(A^+A_2)]^q$$

$$= \sum C_r^{pq}[\chi(A_1A_2)]^r \,. \quad (4.12)$$

A generating function for ρ_{pq} is

$$R(A_1,A_2) = \sum_{p,q} \rho_{pq}(A_1,A_2)\frac{u^p}{p!}\frac{v^q}{q!}$$

$$= \int dA \exp[u\,\mathrm{tr}(A_1A) + v\,\mathrm{tr}(A^+A_2)]$$

$$= \frac{I_1(2x)}{x}, \quad x = [u^2+v^2+uv\,\mathrm{tr}(A_1A_2)]^{1/2} \quad (4.13)$$

from which we derive the coefficients C_r^{pq}. The result is

$$C_r^{pq} = \frac{p!q!}{[1+\frac{1}{2}(p+q)]![\frac{1}{2}(p-r)]![\frac{1}{2}(q-r)]!r!} \quad (4.14)$$

or $C_r^{pq}=0$ if any of the arguments of the factorials is not a non-negative integer. (This implies in particular that $p=q=r \bmod 2$.)

Let us illustrate the use of formulas (4.12) and (4.14) by exhibiting the contribution to the free energy of a diagram consisting of a cube. It reads

$$\beta^6\tfrac{1}{6}[d(d-1)(d-2)]\int \prod_1^{12} dA \prod_1^6 \chi(AAAA)\,.$$

We have used a short-hand notation to indicate the integration over the twelve A_{ij} attached to the links and the six factors contributed by the plaquettes. This integral is calculated by use of the previous formula and yields

$$\beta^6\tfrac{1}{6}[d(d-1)(d-2)]\sum_{pqrst} C_p^{11}C_q^{p1}C_r^{q1}C_s^{r1}C_t^{s1}(\mathrm{tr}\underline{1})^t$$

$$= \beta^6\tfrac{1}{6}[d(d-1)(d-2)]\tfrac{1}{16}\,.$$

To give a slightly more tricky case, let us look at the eighth-order contribution arising from a cube dressed with two extra plaquettes on one of its six faces (observe that there is no such term of order seven, since each link variable must appear an even number of times, and this also explains why two extra plaquettes have to be added on the same face):

$$6\beta^8\frac{d(d-1)(d-2)}{6}\int\prod_1^{12} dA\,\frac{[\chi(AAAA)]^3}{3!}$$

$$\times\prod_1^5 \chi(AAAA)\,.$$

The extra factor 6 arises from the choice of the distinguished plaquette repeated three times, and the denominator 3! is the symmetry number associated with permutations of these three plaquettes. The calculation of the integral proceeds as before and gives

$$\sum_{pqrst} C_p^{31}C_q^{p1}C_r^{q1}C_s^{r1}C_t^{s1}(\mathrm{tr}\underline{1})^t = \tfrac{1}{8}\,.$$

The generalization to a cube dressed in an arbitrary way and to various other diagrams is obvious.

In order to be able to perform computations in a nonzero external field as was indicated in Sec. II, for instance to obtain the critical value of the coupling constant, it is necessary to introduce in the action a source term of the form

$$\sum_l \chi(A_{ij}h_{ji})\,.$$

A source should be an element of a real vector space; on the other hand, it is obvious from the above form of the source term that h_{ij} should be (like the average value of A_{ij}) a linear combination of elements of SU(2). Thus, each h_{ij} is a 2×2 matrix of the form

$$h = u_0 + i\vec{u}\cdot\vec{\sigma}\,, \quad u_0, \vec{u} \text{ real}\,. \quad (4.15)$$

Moreover, like the field A itself, the source satisfies $h_{ij}=h_{ji}^{+}$. If A_{ij} and h_{ij} are parametrized according to (4.10) and (4.15), the source term is

$$\chi(A_{ij}h_{ji}) = 2(u_0\cos\phi + \sin\phi\,\vec{n}\cdot\vec{u})\,,$$

i.e., twice the associated scalar product. In particular, from (4.13), we get

$$z(h) = \int dA \exp[\chi(Ah^+)]$$

$$= \frac{I_1(2\eta)}{\eta}, \quad (4.16)$$

where

$$\eta^2 = u_0^2 + \vec{u}^2 = \det h\,.$$

C. Numerical results

A straightforward application of the high-temperature expansion provides us with a formula for the *free energy* without external field. It turned out that the most economical method was also the least sophisticated one of Sec. II A. We present the results up to sixteenth order for the three groups discussed above.

For Z_2,

$$F=d(d-1)\left[\tfrac{1}{4}\beta^2-\tfrac{1}{24}\beta^4+\left(\tfrac{1}{6}d-\tfrac{29}{90}\right)\beta^6+\left(-\tfrac{1}{3}d+\tfrac{3343}{5040}\right)\beta^8+\left(d^2-\tfrac{184}{45}d+\tfrac{118\,471}{28\,350}\right)\beta^{10}+\left(-\tfrac{8}{3}d^2+\tfrac{121\,153}{11\,340}d-\tfrac{20\,022\,781}{1871\,100}\right)\beta^{12}$$
$$+\left(10d^3-\tfrac{208}{3}d^2+\tfrac{935\,561}{5670}d-\tfrac{5\,647\,451\,354}{42\,567\,525}\right)\beta^{14}+\left(-\tfrac{74}{3}d^3+\tfrac{120\,761}{840}d^2-\tfrac{345\,869\,921}{1\,247\,400}d+\tfrac{3\,612\,986\,481\,191}{20\,432\,412\,000}\right)\beta^{16}+O(\beta^{18})\right]. \tag{4.17}$$

We might also have written in this case an expansion in powers of $\tanh\beta$ involving fewer terms. Namely, the expression

$$Z=2^{-Nd}\sum_{\{A_{ij}=\pm1\}}\exp\left(\beta\sum_{p}AAAA\right)$$
$$=2^{-Nd}(\cosh\beta)^{Nd(d-1)/2}\sum_{\{A_{ij}\}}\left[\prod_{p}(1+AAAA\tanh\beta)\right]$$

is expanded as a sum over closed skeleton diagrams only, each plaquette carrying now a factor $\tanh\beta$. This remark provides a check on (4.17).

For the continuous gauge groups considered above, we find for U(1),

$$F=d(d-1)\left[\tfrac{1}{2}(\tfrac{1}{2}\beta)^2-\tfrac{1}{8}(\tfrac{1}{2}\beta)^4+\left(\tfrac{1}{3}d-\tfrac{11}{18}\right)(\tfrac{1}{2}\beta)^6+\left(-d+\tfrac{757}{384}\right)(\tfrac{1}{2}\beta)^8\right.$$
$$+\left(2d^2-\tfrac{85}{12}d+\tfrac{2473}{400}\right)(\tfrac{1}{2}\beta)^{10}+\left(-\tfrac{26}{3}d^2+\tfrac{6569}{192}d-\tfrac{1\,750\,513}{51\,840}\right)(\tfrac{1}{2}\beta)^{12}$$
$$\left.+\left(20d^3-\tfrac{749}{6}d^2+\tfrac{197\,803}{720}d-\tfrac{44\,476\,939}{211\,680}\right)(\tfrac{1}{2}\beta)^{14}+\left(-96d^3+\tfrac{4777}{8}d^2-\tfrac{43\,844\,513}{34\,560}d+\tfrac{9\,463\,083\,949}{10\,321\,920}\right)(\tfrac{1}{2}\beta)^{16}+O(\beta^{18})\right], \tag{4.18}$$

and for SU(2),

$$F=d(d-1)\left[\tfrac{1}{4}\beta^2-\tfrac{1}{48}\beta^4+\left(\tfrac{1}{96}d-\tfrac{7}{576}\right)\beta^6+\left(-\tfrac{1}{96}d+\tfrac{7}{384}\right)\beta^8+\left(\tfrac{1}{256}d^2-\tfrac{49}{4608}d+\tfrac{10\,481}{1\,612\,800}\right)\beta^{10}\right.$$
$$+\left(-\tfrac{3}{512}d^2+\tfrac{85\,189}{4\,976\,640}d-\tfrac{965\,807}{87\,091\,200}\right)\beta^{12}+\left(\tfrac{5}{2048}d^3-\tfrac{155}{12\,288}d^2+\tfrac{3229}{122\,880}d-\tfrac{288\,747\,853}{13\,412\,044\,800}\right)\beta^{14}$$
$$\left.+\left(-\tfrac{107}{24\,576}d^3+\tfrac{25\,657}{1\,179\,648}d^2-\tfrac{15\,024\,019}{371\,589\,120}d+\tfrac{1\,199\,262\,152\,197}{41\,845\,579\,776\,000}\right)\beta^{16}+O(\beta^{18})\right]. \tag{4.19}$$

The corresponding curves are drawn in dimension four in Fig. 6.

Let us now investigate the *critical couplings*. Variational expansions generalizing (3.3) may be written for each group, and will yield to lowest order in β the result of mean field theory, namely, the existence of a first-order transition for $\beta_c = O(1/d)$. However, higher-order terms do not provide a natural expansion of β_c in powers of $1/d$, as was the case for the scalar model. Indeed, the maximum dimensionality occuring in pth-order terms of the expansion (2.12) is now p. Since β_c is of order $1/d$, this yields a contribution to Φ, and hence to $\beta_c d^2$, of order 1. Successive terms of the perturbation expansion therefore all contribute to the corrections of order d^{-2} to the critical coupling as given by mean field theory.

Furthermore, due to gauge invariance, setting all m_i to an equal value m is, in fact, dangerous. This was discussed at length in paper I, where we proposed to integrate first over a subset of field variables in order to break formal gauge invariance. A possible strategy was to eliminate all "vertical" variables, resulting in an action

$$S=\sum_{p_\perp}\chi(A_{12}A_{23}A_{34}A_{41})+\sum_{p_\parallel}\chi(A_{12}A_{34}). \tag{4.20}$$

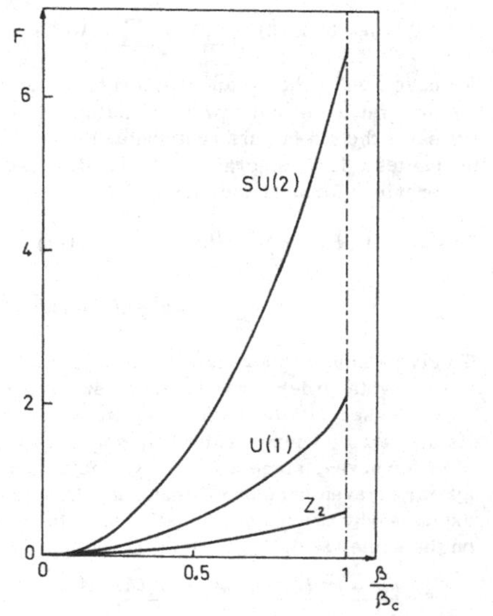

FIG. 6. The free energy of the gauge models for the three groups Z_2, U(1), and SU(2), plotted against β/β_c for $d=4$. We have used for β_c the best values of Table V.

Here $p_\perp$ denotes the plaquettes perpendicular to the "vertical," timelike direction, $p_\parallel$ denotes the parallel ones, and the variables A_{ij} only refer to perpendicular, spacelike links. This was called the action in the *Coulomb gauge*. Of course some invariance still remains, but it is now a surface effect rather than a volume effect. With this choice, we may set all remaining spacelike m_i variables equal to m in the variational procedure.

We have now all the necessary tools to perform the variational calculation of β_c along the lines of Sec. II B. We have pushed the evaluation of the B function occurring in (2.12) up to third order in β in the case of the Z_2 gauge group, and to second order in the cases of U(1) and SU(2). This was only designed to demonstrate the feasibility of these calculations, which could of course be pursued. The analytical expressions soon become rather cumbersome. As an example, let us display the simplest of them, relative to the Z_2 group. We express the variational function $\Phi(h,m)$ and the irreducible kernel $B(h)$ occurring in (2.13) and (2.12):

$$\begin{aligned}\varphi(h,m) &= \frac{1}{N(d-1)}\Phi(h,m)\\ &= b(h+2\beta(d-2)m^3+2\beta m)\\ &\quad -\tfrac{3}{2}(d-2)\beta m^4-\beta m^2\,,\end{aligned} \tag{4.21}$$

with $b(h)$ given, in terms of $u=u(h)=\ln\cosh h$ and its derivatives $u_i=d^iu(h)/dh^i$, by

$$\begin{aligned}b(h) &= \frac{1}{N(d-1)}B(h)\\ &= u+\tfrac{1}{2}\beta^2[\tfrac{1}{2}(d-2)(u_2{}^4+4u_2{}^3u_1{}^2+6u_2{}^2u_1{}^4)+u_2{}^2]+\beta^3[\tfrac{1}{2}(d-2)(\tfrac{1}{6}u_3{}^4+2u_1u_2u_3{}^3+\tfrac{2}{3}u_1{}^3u_3{}^3+9u_1{}^2u_2{}^2u_3{}^2+6u_1{}^4u_2u_3{}^2\\ &\qquad +u_1{}^6u_3{}^2+12u_1{}^5u_2{}^2u_3+16u_1{}^3u_2{}^3u_3+9u_1{}^4u_2{}^4+4u_1{}^6u_2{}^3)\\ &\qquad +\tfrac{1}{6}u_3{}^2+\tfrac{4}{3}(d-2)(d-3)u_1{}^6u_2{}^3]\\ &\quad +O(\beta^4)\,.\end{aligned} \tag{4.22}$$

We recall from paper I that, to lowest order, we find a first-order transition in F when varying m. Thus we use Eqs. (2.18), setting $h=0$ in (4.21) and in its analogs for the other groups, and looking for the stationary value of the approximate φ, up to various orders of the expansion (4.22). The results are displayed in Tables V and VI.

For the gauge group Z_2, the values obtained for $\beta_c^*=2\beta_c d$ in dimensions $d=3$ and $d=4$ may be compared with the exact results predicted in paper II. The agreement seems to be excellent in dimension 4, and seems to indicate that the transition is indeed a first-order one. However, for $d=3$, where one knows that the Yang-Mills field has a second-order transition (while the present calculation is performed in the framework of a first-order one), some discrepancy occurs. It was indeed expected that $d=3$ would appear as a limiting dimensionality. This has also been checked by treating with the present formalism the case $d=2$, known to be equivalent to a one-dimensional Ising model. Whereas the zeroth order (mean field) predicts a spurious (second-order) transition, the second-order terms are sufficient to rule out this transition. The perturbation variation treatment we are using thus does not seem to suffer from the

TABLE V. Critical couplings $\beta_c^*=2\beta_c d$ for gauge models.

Group	Approx.	$d=3$	$d=4$	$d=5$	$d=6$	$d=\infty$
Z_2	0	2.6028	2.6840	2.7104	2.7229	
	2	4.2970	3.5272	3.2905	3.1677	2.7552
	3	4.2941	3.5257	3.2897	3.1672	
	exact	4.5678	3.5254	?	?	
U(1)	0	5.6563	6.2133	6.4793	6.6384	
	2		11.3761	9.0530	8.4929	7.2934
	Padé	10.0934	8.9781	8.4968	8.2298	
SU(2)	0	5.8708	6.7270	7.1520	7.4096	
	2			12.821	10.5948	8.4787
	Padé		11.4160	10.333	9.8510	

TABLE VI. The discontinuity m_c of m at the critical point for gauge models.

Group	Approx.	$d=3$	$d=4$	$d=5$	$d=6$	$d=\infty$
Z_2	0	0.8643	0.9416	0.9616	0.9704	
	2	0.9921	0.9868	0.9865	0.9869	0.990 61
	3	0.9913	0.9858	0.9859	0.9865	
U(1)	0	0.6558	0.7913	0.8296	0.8479	
	2		0.8949	0.8793	0.8707	0.9004
	Padé	0.7829	0.8272	0.8469	0.8582	
SU(2)	0	0.4983	0.7042	0.7619	0.7892	
	2			0.8683	0.8400	0.8682
	Padé		0.7726	0.7938	0.8081	

defect of mean-field theory, which may predict spurious transitions.

Table V also exhibits the convergence of the expansion. For large d, β_c^* should be close to the mean field value (exact for d infinite), and convergence is expected to be rapid. Values obtained for β_c^*, either by expanding b up to second order in β [as in Eq. (4.22)] or by using the Padé approximant for b to the same order, are indeed close to each other. Furthermore, inclusion of third-order terms does not seem to improve the second-order results significantly. Such a fact already appeared for the scalar field model of Sec. III; a possible explanation is the occurrence of new irreducible skeletons at even orders only.

The discontinuity of the parameter m, which jumps from 0 to m_c at β_c is given in Table VI. Since m is the average value of the field A which varies on a unit sphere, it is bounded by 1. It is quite striking that m_c is close to this maximum. Consider for instance the case of the group Z_2 in the limit $d\to\infty$. Using the expression (4.21), in the equations (2.18), one is led to

$$\ln\cosh(\beta_c^* m_c^3) - \tfrac{3}{4}\beta_c^* m_c^4 = 0\,, \qquad \tanh(\beta_c^* m_c^3) - m_c = 0\,, \tag{4.23}$$

which yield $m_c = 0.9906$, $\beta_c^* = 2.755$.

In the case of U(1), the transition disappears for dimension 3 when we include second-order corrections. This is similar to the disappearance of the transition for the group Z_2 in 2 dimensions. It may be that the transition, which is a first-order one for high dimensionalities, becomes a second-order transition around $d=4$, then disappears. It is also possible that, for a continuous group, the low-lying excitations destroy the order at low dimension. A subtle kind of phase transition might remain at the limiting dimension. This phenomenon is known[8] to occur for the scalar models of Sec. III: whereas the two-dimensional Ising model ($d=2$, $n=1$) exhibits a usual second-order transition (like the Yang-Mills model for $d=3$, $n=1$), a transition without ordering takes place for $d=2$, $n\geq 2$.

For gauge fields, the limiting dimension is not unambiguously ascertained. In particular, we see in Table V that for SU(2), the predictions of the plain expansion and the Padé expansion differ for $d=4$. Note finally that in this case m is replaced by a 2×2 matrix, equal to the average value $\langle A_{ij}\rangle$ of spacelike field variables. Whereas A_{ij} is unitary, its average belongs to the algebra of SU(2), and has the form $u_0 + i\vec{u}\cdot\vec{\sigma}$. Hence $(u_0+i\vec{u}\cdot\vec{\sigma})(u_0+i\vec{u}\cdot\vec{\sigma})^\dagger = m^2$ is a c number bounded by 1. When looking for the stationary value of F, one finds $\vec{u}=0$. The corresponding numbers for m_c are listed in Table VI.

V. CONCLUSION

A diagrammatic approach, combining perturbative and variational techniques, allows one to compute various quantities of interest, such as critical couplings. A good numerical convergence seems to be achieved even in low dimension, and a check is provided by some known results. In particular, for the case of a Z_2 gauge group for $d=4$, we have shown in paper II that a transition occurs at $\beta_c^* = 3.5254$, to be compared with the value 3.5257 obtained with the present techniques developed to third order (Table V).

The method is based on a self-consistent determination of m, the average value of the field for a vanishing source. For the scalar model of Sec. III, with its second-order transition, m is a natural physical order parameter. It is somehow surprising to obtain quite satisfactory results for gauge fields, using the same techniques. Indeed, the analysis given in paper II shows that m has not the usual interpretation of an order parameter, and that no local order parameter exists in the Yang-Mills models. It was therefore important to test the validity of the method.

We have taken care to perform the calculations in the Coulomb gauge, suppressing the timelike field variables A_{ij}, and taking m as the (common) average value of the spacelike ones. This procedure was essential. We have tried to perform the same calculations in a straightforward fashion, without breaking formal gauge invariance, m being then the average of all variables A_{ij}. Although both approaches agree qualitatively to order zero, higher-order terms yield nonsensical results when no gauge condition is imposed. On the contrary, Tables V and VI show the success of second-order calculations in the Coulomb gauge, even in low dimensions.

The nature of the transition remains to some extent an open question. Since terms beyond second order yield $1/d$ corrections, the mean field approximation becomes very likely exact for $d\to\infty$. The transition of the Yang-Mills model is therefore a first-order one in high enough dimension, the function F having a discontinuous derivative. Not only is the transition a first-order one, but m jumps from 0 to a value close to its maximum, equal to 1. Ordering thus appears bluntly if we characterize it by m. It may, however be that the order parameter associated with the average $\langle \prod A_{ij} \rangle$ of the product of fields along a large loop has a smoother behavior.

Finally, as seen from the variations of β_c^* and m_c exhibited in Tables II, V, and VI, the transition sets in with more and more difficulty when the dimension decreases (and also when the number of components of the field increases). A limiting dimension exists, below which no ordering takes place. For this limiting dimension, a phase transition may still exist, possibly of second order (as for the group Z_2 in 3 dimensions), with a critical behavior.

[1]R. Balian, J. M. Drouffe, and C. Itzykson, Phys. Rev. D 10, 3376 (1974); preceding paper, 11, 2098 (1975). These papers will be referred to as paper I and paper II, respectively.
[2]K. Wilson, Phys. Rev. D 10, 2445 (1974).
[3]F. Englert, Phys. Rev. 129, 567 (1963).
[4]C. Domb, Adv. Phys. 9, 149 (1960); 19, 339 (1970).
[5]C. Bloch, *Studies in Statistical Mechanics* (North-Holland, Amsterdam, 1965), Vol. 3, Secs. 21–24.
[6]M. E. Fisher and D. S. Gaunt, Phys. Rev. 133, A224 (1964); R. Abe, Prog. Theor. Phys. 47, 62 (1972).
[7]H. E. Stanley, Phys. Rev. 176, 718 (1968).
[8]B. Jancovici, J. Phys. (Paris) 32, C-185 (1971).

PHYSICAL REVIEW D VOLUME 19, NUMBER 8 15 APRIL 1979

Errata

Erratum: Gauge fields on a lattice. III. Strong-coupling expansions and transition points [Phys. Rev. D 11, 2104 (1975)]

R. Balian, J. M. Drouffe, and C. Itzykson

We apologize for some erroneous coefficients in formulas (4.17), (4.18), and (4.19). We display here the correct free energy for the groups Z_2, U(1), and SU(2) and give also the corresponding results for Z_3 and SU(3) gauge groups.

For Z_2,

$$\begin{aligned}\frac{F}{d(d-1)} =& \tfrac{1}{4}\beta^2 - \tfrac{1}{24}\beta^4 + (\tfrac{1}{6}d - \tfrac{29}{90})\beta^6 + (-\tfrac{1}{3}d + \tfrac{3343}{5040})\beta^8 \\ &+ (d^2 - \tfrac{184}{45}d + \tfrac{118\,471}{28\,350})\beta^{10} + (-\tfrac{8}{3}d^2 + \tfrac{121\,153}{11\,340}d - \tfrac{20\,022\,781}{1\,871\,100})\beta^{12} \\ &+ (10\,d^3 - \tfrac{208}{3}d^2 + \tfrac{935\,561}{5\,670}d - \tfrac{5\,647\,451\,354}{42\,567\,525})\beta^{14} \\ &+ (-26\,d^3 + \tfrac{129\,161}{840}d^2 - \tfrac{376\,639\,121}{1\,247\,400}d + \tfrac{4\,021\,634\,721\,191}{20\,432\,412\,000})\beta^{16} + O(\beta^{18})\,.\end{aligned}$$

For Z_3,

$$\begin{aligned}\frac{F}{d(d-1)} =& \tfrac{1}{8}\beta^2 + \tfrac{1}{48}\beta^3 - \tfrac{1}{128}\beta^4 - \tfrac{1}{256}\beta^5 + (\tfrac{1}{192}d - \tfrac{51}{5120})\beta^6 + (\tfrac{1}{128}d - \tfrac{153}{10\,240})\beta^7 \\ &+ (\tfrac{1}{1024}d - \tfrac{2\,187}{1\,146\,880})\beta^8 + (-\tfrac{35}{6\,144}d + \tfrac{46\,597}{4\,128\,768})\beta^9 \\ &+ (\tfrac{1}{512}d^2 - \tfrac{1\,017}{81\,920}d + \tfrac{779\,381}{45\,875\,200})\beta^{10} + (\tfrac{3}{512}d^2 - \tfrac{4\,047}{163\,840}d + \tfrac{2\,383\,531}{91\,750\,400})\beta^{11} \\ &+ (\tfrac{137}{24\,576}d^2 - \tfrac{52\,709}{2\,293\,760}d + \tfrac{191\,096\,159}{8\,074\,035\,200})\beta^{12} \\ &+ (-\tfrac{15}{8\,192}d^2 + \tfrac{57}{8\,192}d - \tfrac{3\,041\,827}{461\,373\,440})\beta^{13} \\ &+ (\tfrac{5}{4\,096}d^3 - \tfrac{533}{32\,768}d^2 + \tfrac{3\,826\,173}{73\,400\,320}d - \tfrac{287\,774\,341\,033}{5\,877\,897\,625\,600})\beta^{14} \\ &+ (\tfrac{17}{3\,072}d^3 - \tfrac{7\,027}{163\,840}d^2 + \tfrac{34\,256\,119}{314\,572\,800}d - \tfrac{15\,959\,874\,120\,733}{176\,336\,928\,768\,000})\beta^{15} \\ &+ (\tfrac{603}{65\,536}d^3 - \tfrac{4\,553\,361}{73\,400\,320}d^2 + \tfrac{18\,321\,594\,271}{129\,184\,563\,200}d - \tfrac{205\,245\,882\,159\,867}{1\,880\,927\,240\,192\,000})\beta^{16} + O(\beta^{17})\,.\end{aligned}$$

For U(1),

$$\begin{aligned}\frac{F}{d(d-1)} =& \tfrac{1}{8}\beta^2 - \tfrac{1}{128}\beta^4 + (\tfrac{1}{192}d - \tfrac{11}{1\,152})\beta^6 + (-\tfrac{1}{256}d + \tfrac{757}{98\,304})\beta^8 \\ &+ (\tfrac{1}{512}d^2 - \tfrac{85}{12\,288}d + \tfrac{2\,473}{409\,600})\beta^{10} + (-\tfrac{29}{12\,288}d^2 + \tfrac{2\,467}{262\,144}d - \tfrac{1\,992\,533}{212\,336\,640})\beta^{12} \\ &+ (\tfrac{5}{4\,096}d^3 - \tfrac{237}{32\,768}d^2 + \tfrac{178\,003}{11\,796\,480}d - \tfrac{38\,197\,099}{3\,468\,165\,120})\beta^{14} \\ &+ (-\tfrac{15}{8192}d^3 + \tfrac{1\,485}{131\,072}d^2 - \tfrac{53\,956\,913}{2\,264\,924\,160}d + \tfrac{11\,483\,169\,709}{676\,457\,349\,120})\beta^{16} + O(\beta^{18})\,.\end{aligned}$$

For SU(2),

$$\begin{aligned}\frac{F}{d(d-1)} =& \tfrac{1}{4}\beta^2 - \tfrac{1}{48}\beta^4 + (\tfrac{1}{96}d - \tfrac{5}{288})\beta^6 + (-\tfrac{1}{96}d + \tfrac{29}{1\,440})\beta^8 \\ &+ (\tfrac{1}{256}d^2 - \tfrac{49}{4608}d + \tfrac{1\,001}{172\,800})\beta^{10} + (-\tfrac{7}{1024}d^2 + \tfrac{32\,131}{1\,244\,160}d - \tfrac{211\,991}{8\,709\,120})\beta^{12} \\ &+ (\tfrac{5}{2048}d^3 - \tfrac{43}{4096}d^2 + \tfrac{5\,341}{368\,640}d - \tfrac{264\,497}{40\,642\,560})\beta^{14} \\ &+ (-\tfrac{47}{8192}d^3 + \tfrac{7\,030\,933}{212\,336\,640}d^2 - \tfrac{97\,100\,911}{1\,486\,356\,480}d + \tfrac{1\,474\,972\,157}{33\,443\,020\,800})\beta^{16} + O(\beta^{18})\,.\end{aligned}$$

PHYSICAL REVIEW D VOLUME 19, NUMBER 8 15 APRIL 1979

For SU(3),

$$\begin{aligned}\frac{F}{d(d-1)} =& \tfrac{1}{2}\beta^2+\tfrac{1}{6}\beta^3-\tfrac{1}{24}\beta^5+(\tfrac{1}{243}d-\tfrac{113}{3\,888})\beta^6+(\tfrac{1}{81}d-\tfrac{133}{6\,480})\beta^7+(\tfrac{5}{324}d-\tfrac{1\,069}{51\,840})\beta^8\\
&+(\tfrac{5}{972}d-\tfrac{509}{77\,760})\beta^9+(\tfrac{2}{6561}d^2-\tfrac{157}{11\,664}d+\tfrac{490\,757}{20\,995\,200})\beta^{10}+(\tfrac{4}{2\,187}d^2-\tfrac{59}{2160}d+\tfrac{435\,299}{9\,797\,760})\beta^{11}\\
&+(\tfrac{1\,775}{354\,294}d^2-\tfrac{218\,824\,907}{7\,255\,941\,120}d+\tfrac{1\,682\,010\,779}{42\,326\,323\,200})\beta^{12}\\
&+(\tfrac{440}{59\,049}d^2-\tfrac{13\,919\,677}{604\,661\,760}d+\tfrac{7\,603\,159}{440\,899\,200})\beta^{13}\\
&+(\tfrac{20}{531\,441}d^3+\tfrac{8\,377}{2\,125\,764}d^2-\tfrac{12\,469\,727}{5\,441\,955\,840}d-\tfrac{14\,239\,256\,399}{1\,333\,279\,180\,800})\beta^{14}\\
&+(\tfrac{544}{1\,594\,323}d^3-\tfrac{69\,331}{7\,971\,615}d^2+\tfrac{106\,962\,409}{2\,821\,754\,880}d-\tfrac{3\,474\,317\,893}{79\,361\,856\,000})\beta^{15}\\
&+(\tfrac{2\,323}{1\,594\,323}d^3-\tfrac{5\,838\,272\,899}{220\,399\,211\,520}d^2+\tfrac{10\,597\,782\,658\,021}{123\,423\,558\,451\,200}d-\tfrac{6\,402\,970\,751\,747}{82\,282\,372\,300\,800})\beta^{16}\\
&+O(\beta^{17})\,.\end{aligned}$$

Hamiltonian formulation of Wilson's lattice gauge theories

John Kogut*
Laboratory of Nuclear Studies, Cornell University, Ithaca, New York 14853

Leonard Susskind†
Belfer Graduate School of Science, Yeshiva University, New York, New York
and Tel Aviv University, Ramat Aviv, Israel
and Laboratory of Nuclear Studies, Cornell University, Ithaca, New York
(Received 9 July 1974)

Wilson's lattice gauge model is presented as a canonical Hamiltonian theory. The structure of the model is reduced to the interactions of an infinite collection of coupled rigid rotators. The gauge-invariant configuration space consists of a collection of strings with quarks at their ends. The strings are lines of non-Abelian electric flux. In the strong-coupling limit the dynamics is best described in terms of these strings. Quark confinement is a result of the inability to break a string without producing a pair.

I. INTRODUCTION

The quark model has systematized a very large amount of information concerning the hadron spectrum. However, free isolated quarks do not appear to exist. In order to confine quarks into baryons and mesons, one is then led to suppose that the field-theoretic coupling between quarks becomes strong at large distances. This explanation is, however, somewhat perplexing because the forces between quarks at small distances appear to be weak. Such behavior can in principle be found in renormalizable field theories in which effective coupling constants can change from one size scale to the next. Clearly, in order to understand the successes of the quarkless quark model we need a theory in which weak short-distance forces give rise to strong long-range forces. The only theory in which this behavior appears possible is one containing non-Abelian (Yang-Mills) gauge fields.

It is instructive to recall why this behavior does not occur in conventional formulations of Abelian vector-gluon theories (electrodynamics, for example). Consider a static free charge of magnitude e inserted into the vacuum of quantum electrodynamics. As is well known, the electrodynamic vacuum is an ordinary dielectric,[1] so the free charge creates a polarization charge of *opposite* sign. The polarization charge is distributed in the vicinity of the free charge. Therefore, the total charge contained within a sphere of radius r is $eZ(r)$, where $Z(r)$ is a fraction less than 1 which decreases as r increases. The factor $Z(r)$ causes the intensity of electromagnetic interactions to be dependent on the distance scales involved.[2] In fact, if we are only interested in long-wavelength phenomena in electrodynamics, we can ignore all the short-distance fluctuations of the theory and replace the bare electric charge e by the screened or renormalized charge. More precisely, long-wavelength phenomena are insensitive to a cutoff at length λ if the bare charge is replaced by $eZ(\lambda)$. Since $Z(\lambda)$ decreases as λ increases, this theory has just the reverse behavior of what we want.

In theories with Yang-Mills fields the interaction between a pair of static charges is also governed by an effective coupling constant $gZ(r)$. As in electrodynamics, a cutoff version of Yang-Mills theory must replace g by $gZ(\lambda)$. This time it is found, however, that $Z(\lambda)$ can be an increasing function of λ.[3] The implication is that the effective couplings between the low-momentum modes of the theory may become very strong although the shorter-distance behavior may not involve strong coupling.

In this paper we shall be interested in the large-distance properties of a non-Abelian theory assuming that the effective coupling $g(\lambda)$ is sufficiently large to use Wilson's strong-coupling methods.[4] An ultraviolet cutoff is introduced into the theory through a spatial lattice. This construction destroys most of the space-time symmetries of relativistic field theories. For this reason the theory discussed here is not a realistic Yang-Mills theory. However, following Wilson,[4] we are mainly interested in determining the special effects of exact gauge invariance in strongly coupled gauge theories. As a result of this study, we find that quarks can be confined in locally gauge-invariant theories. The confining mechanism is the appearance of one-dimensional electric flux tubes which must link separated quarks.[5] The appropriate description of the strongly coupled limit consists of a theory of interacting, propagating strings.

This paper is organized into eight sections. In Sec. II we describe the field theory of fermions on a spatial lattice. The theory has global but not local non-Abelian symmetry. In Sec. III we develop the principle of local gauge invariance. In the weak-coupling limit the theory reduces to a standard theory when the spatial cutoff is taken to zero. In Sec. IV we develop the canonical formalism for an SU(2) gauge theory. The fundamental gauge-field degree of freedom which links adjacent lattice points is mathematically equivalent to a rigid rotator. The theory of rigid rotators is a helpful guide in discussing the gauge theory. In Sec. V the physical space of states of the strong-coupling theory is constructed and is described in terms of stringlike excitations of a gauge-invariant vacuum. The strings are the non-Abelian analogs of electric flux lines. In Sec. VI we include the dynamics of the gauge field into the lattice model and obtain a Hamiltonian for a discrete theory of fermions and gauge fields. In the weak-coupling continuum limit the usual Yang-Mills theory with fermions is retrieved.[6] In Sec. VII we consider the dynamics of the stringlike excitations and calculate the energy in various configurations. In particular, we show that the energy of a well-separated quark pair increases linearly with the distance between them.[7] Then we develop a perturbation theory around the strong-coupling limit. It expresses the solution of the theory as a series expansion in inverse powers of the coupling constant. Physically, the higher-order effects cause quantum fluctuations in the string configurations. If these fluctuations grow too large, they could invalidate the quark binding mechanism. In the final section we discuss and summarize several of our conclusions.

II. FERMION FIELDS ON A LATTICE

We begin by formulating the Dirac equation on a spatial cubic lattice. An arbitrary point on the lattice is denoted by a triplet of integers $\vec{r} = (r_x, r_y, r_z)$. The unit lattice vectors are denoted $\hat{m}_x$, $\hat{m}_y$, $\hat{m}_z$, $\hat{m}_{-x}$, $\hat{m}_{-y}$, and $\hat{m}_{-z}$ pointing along the x, y, z, $-x$, $-y$, and $-z$ axes, respectively (Fig. 1). By definition $\hat{m}_{-x} = -\hat{m}_x$, etc. We use the convention that summations over the lattice vectors include the six directions. The spaces between neighboring lattice points (*links*) are denoted by a position and a lattice vector $(\vec{r}, \hat{m}_i)$. The same interval can be denoted $(\vec{r} + \hat{m}_i, -\hat{m}_i)$. It will prove convenient to label lattice sites as odd or even by the following prescription: A site r is called even (odd) if $(-1)^{r_x+r_y+r_z} \equiv (-1)^r$ is even (odd). On each lattice site we define a *two*-component spinor $\psi(r)$. A discrete Hamiltonian can easily be constructed such that it yields the conventional Dirac theory in the continuum limit. It reads

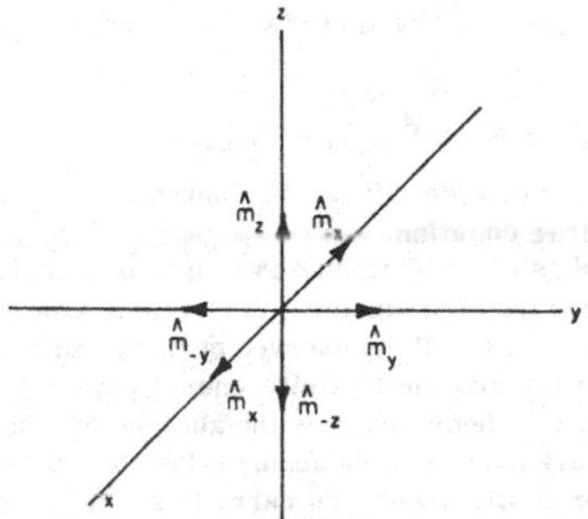

FIG. 1. Definition of unit lattice vectors.

$$H = a^{-1} \sum_{r,n} \psi^\dagger(r) \frac{\vec{\sigma}\cdot\vec{n}}{i} \psi(r+n) + m_0 \sum_r (-1)^r \psi^\dagger(r)\psi(r), \tag{2.1}$$

where a is the lattice spacing. If we postulate the canonical anticommutation relations

$$\{\psi_\alpha(r), \psi_\beta^\dagger(r')\} = \delta_{\alpha\beta}\delta_{r,r'}, \quad \text{etc.}, \tag{2.2}$$

the equation of motion of the spinor field becomes

$$i\dot{\psi}(r) = [\psi(r), H]$$
$$= a^{-1} \sum_n \frac{\vec{\sigma}\cdot\vec{n}}{i} \psi(r+n) + m_0(-1)^r \psi(r). \tag{2.3a}$$

This equation may be rewritten

$$i\dot{\psi}(r) = \frac{1}{2a} \sum_n \frac{\vec{\sigma}\cdot\vec{n}}{i} [\psi(r+n) - \psi(r-n)] + m_0(-1)^r \psi(r). \tag{2.3b}$$

Consider the continuity properties of the solutions to this equation as $a \to 0$. The finite-energy solutions ($\dot{\psi}$ finite) require that

$$\psi(r+n) - \psi(r-n) \sim a \tag{2.4}$$

as $a \to 0$. However, $\psi(r) - \psi(r+m)$ is not constrained by the discrete Dirac equation. Thus, in order to define fields with finite derivatives, we must introduce two separate fields for even and odd lattice sites. In the continuum limit we represent these two fields as upper and lower components of a four-component Dirac spinor, the upper (lower) components being the fields on the even (odd) lattice points. The discrete Dirac equation may then be approximated,

$$i\dot{\psi}_{\rm upper} = -i\vec{\sigma}\cdot\vec{\nabla}\psi_{\rm lower} + m_0\psi_{\rm upper}\,,$$
$$i\dot{\psi}_{\rm lower} = -i\vec{\sigma}\cdot\vec{\nabla}\psi_{\rm upper} - m_0\psi_{\rm lower}\,, \tag{2.5}$$

which one can identify as the conventional continuum Dirac equation.

For physical applications ψ will have both "color"[8] and ordinary SU(3) indices. The role of color is to provide a locally conserved quantum number whose vanishing in the finite-energy physical spectrum of the theory implies the absence of triality. The quark-confining mechanism then becomes one of color confinement. To carry this scheme out, the color degrees of freedom [not the ordinary SU(3)] will be coupled to colored Yang-Mills fields. For illustrative purposes we will ignore ordinary SU(3) and replace the color group by an SU(2) group.

The transformation of the fermion field under *global* gauge transformations reads

$$\hat{\psi}(r) = e^{i\vec{\tau}\cdot\vec{\omega}/2}\psi(r) \equiv V\psi(r)\,. \tag{2.6}$$

The Hamiltonian Eq. (2.1) is clearly invariant under such transformations. Since the global transformation rotates the fermion field identically over all points of space, this global invariance of the theory still permits color to be compared at separated points. This freedom will be lost when local gauge invariance is built into the theory.[6]

III. PRINCIPLE OF LOCAL GAUGE INVARIANCE

A local gauge transformation on the fermion field is written

$$\hat{\psi}(r) = e^{i\vec{\tau}\cdot\vec{\omega}(r)/2}\psi(r) \equiv V(r)\psi(r)\,, \tag{3.1}$$

where $\vec{\omega}(r)$ can now depend on the position r. In general the full gauge group consists of transformations which depend upon time as well as position. The canonical formalism is significantly more difficult when the full time-dependent gauge transformations are considered. We will therefore only discuss the invariance for spatially dependent gauge functions. This will then allow us to set the time component of the vector potential to zero when the gauge field enters the theory. There is, in fact, no loss of generality in this procedure.[9]

The Hamiltonian in Eq. (2.1) is not locally gauge-invariant since it involves the product of fermion fields at separated points. To compensate this lack of local invariance, we introduce a gauge field. This is done as follows.[4] On each link (r,m) we place a gauge field $\vec{B}(r,m)$ and a unitary transformation,

$$U_{1/2}(r,m) = \exp[i\tfrac{1}{2}\vec{\tau}\cdot\vec{B}(r,m)]\,. \tag{3.2}$$

The subscript $\frac{1}{2}$ on U denotes the fundamental representation of the SU(2) color group. We make the convention $B(r,m) = -B(r+m,-m)$. We note that the Fermi field is associated with the lattice points themselves, but the gauge fields are associated with *links* between points. This is so because the gauge field *transports* color information between lattice points. The two indices of the matrix

$$U_{1/2}(r,m)^i{}_j$$

are identified with the two ends of the link m. The upper (lower) index is associated with the beginning (end) of the link as depicted in Fig. 2. The gauge transformation acts on $U(r,m)$ according to,

$$U'(r,m) = V(r)U(r,m)V^{-1}(r+m)\,. \tag{3.3}$$

Since in general $V(r)$ and $V(r+m)$ are different, our gauge-invariant equations will require invariance under right multiplication and left multiplication separately. The matrices $U(r,m)$ can now be used to convert gauge-noninvariant products of spatially separated fields to gauge-invariant products. For example, an operator such as,

$$\psi^{\dagger}(r)\psi(r+m)$$

transforms under gauge transformations to

$$\psi^{\dagger}(r)V^{-1}(r)V(r+m)\psi(r+m)\,.$$

However, the operator

$$\psi^{\dagger}(r)U(r,m)\psi(r+m) \tag{3.4}$$

is gauge-invariant. The gauge transformations $V(r)$ and $V(r+m)$ acting on the ends of the link [indices of $U(r,m)$] undo the gauge transformations of the fermion fields. We can now apply this procedure to the Hamiltonian to render it gauge-invariant:

$$H = a^{-1}\sum_{r,m}\psi^{\dagger}(r)\frac{\vec{\sigma}\cdot\vec{m}}{i}U(r,m)\psi(r+m) + m_0\sum(-1)^r\psi^{\dagger}(r)\psi(r)\,. \tag{3.5}$$

Let us now consider the continuum limit $(a\to 0)$ of this Hamiltonian. To do this we write

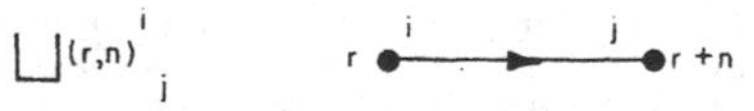

FIG. 2. The gauge field U is defined on the links. The two indices of U refer to the ends of the links.

$$H = a^3 \sum \psi'^\dagger(r) \frac{\vec{\sigma}\cdot\vec{m}}{ia} e^{i\vec{\tau}\cdot\vec{B}(r,m)/2}(1 - e^{-i\vec{\tau}\cdot\vec{B}(r,m)/2})\psi'(r+m) + a^3 \sum \psi'^\dagger(r) \frac{\vec{\sigma}\cdot\vec{m}}{ia}\psi'(r+m) + m_0 a^3 \sum (-1)^r \psi'^\dagger(r)\psi'(r), \tag{3.6}$$

where $\psi'(r) = a^{-3/2}\psi(r)$. In order to take the continuum limit smoothly, it is essential to assume that the operator $1 - \exp[-i\frac{1}{2}\vec{\tau}\cdot\vec{B}(r,m)]$ tends to zero as the lattice spacing goes to zero. That is, a small lattice spacing is only compatible with dynamics in which the magnitude of the operator $\vec{B}(r,m)$ is small, $\sim a$. Then the exponential in Eq. (3.6) can be expanded and the Hamiltonian becomes

$$H = a^3 \sum_{r,m} \psi'^\dagger(r)\left(\frac{\vec{\sigma}\cdot\vec{m}}{ia}\right)[i\tfrac{1}{2}\vec{\tau}\cdot\vec{B}(r,m)]\psi'(r+m) + a^3 \sum \psi'^\dagger(r)\frac{\vec{\sigma}\cdot\vec{m}}{ia}\psi'(r+m) + m_0 a^3 \sum (-1)^r \psi'^\dagger(r)\psi'(r). \tag{3.7}$$

In the continuum limit the quantity

$$\psi'^\dagger(r)\frac{\vec{\sigma}\cdot\vec{m}}{ia}\psi'(r+m)$$

becomes the kinetic-energy term

$$\bar{\psi}'(r)\gamma_i\partial_i\psi'(r).$$

Then the full Hamiltonian becomes

$$H = \int_r \left[\bar{\psi}'(r)\, i\gamma_i\partial_i\psi'(r) - \bar{\psi}'(r)\gamma_i \frac{\vec{B}_i(r)}{a}\cdot\frac{\vec{\tau}}{2}\psi'(r) + m_0\bar{\psi}'(r)\psi'(r)\right]. \tag{3.8}$$

This is the usual Yang-Mills gauge theory with fermions if we identify

$$B_i(r) = agA_i(r),$$

where $A_i(r)$ is the vector potential and g is the coupling constant. At this stage the gauge field does not enter the dynamics as a bona fide degree of freedom. This limitation will be remedied in Sec. VI.

IV. THE RIGID ROTATOR

In this section we shall consider the nature of the gauge-field degree of freedom on a single link. In ordinary scalar-particle field theory, the field degree of freedom at a point is an anharmonic oscillator. The derivative terms in the Hamiltonian couple adjacent oscillators. In Yang-Mills theory the local degree of freedom $U(r,m)$ is an element of a group. In our example the group is O(3). Since this is a non-Abelian compact group, the topology of the configuration space at a link is closed and nontrivial. This leads to complexities in the canonical formalism. Fortunately, the configuration space of a well-known mechanical system, the rigid rotator, is identical to these degrees of freedom.

We shall first review the kinematics of the quantum rigid rotator. A configuration of the rigid rotator is specified by a rotation from the space-fixed to a set of body-fixed axes.[10] The rotation may be represented in the form

$$U_j = \exp(i\vec{T}_j\cdot\vec{\Omega}),$$

where $T_{j\alpha}$ ($\alpha = 1, 2, 3$) are representation matrices of the generators of the rotation group for angular momentum j. In the spinor representation the elements of the matrix $\exp(i\vec{T}_{1/2}\cdot\vec{\Omega})$ are Cayley-Klein parameters. We introduce a notation for matrices in which lower (upper) components refer to space (body) axes. For example, if V_i are components of a vector in the space-fixed frame, then $(U_1)^I{}_i V_i = V^I$ are the corresponding body components.

The relationship between body and space axes for the rigid rotator is the same as the relation between the indices on the two ends of a link in the Yang-Mills theory. The action of a rotation of space axes on U is given by left multiplication by the appropriate rotation matrix, V, say. Similarly, a rotation of the body axes relative to the body is given by right multiplication. The requirement of local gauge invariance in Yang-Mills theory translates into invariance under *separate* space and body rotations. This invariance requires that the rotator be spherical, since only the spherical rotator has invariance under rotations of the body axes.

The angular velocity vector of the rigid rotator is defined as the time derivative of $\vec{\Omega}$,

$$\vec{\omega} = \frac{d}{dt}\vec{\Omega}.$$

The angular momentum $\vec{J}$ (generator of space rotations) of the rotator is given by $I\vec{\omega}$, where I is the moment of inertia of the rigid body. The Hamiltonian reads

$$H = J^2/(2I) = \tfrac{1}{2}I\omega^2 = \tfrac{1}{2}I\Omega^2. \tag{4.1}$$

Since the moment of inertia tensor is diagonal, the Hamiltonian is invariant under individual body and space rotations as required by gauge invariance. In fact, if $\vec{\mathcal{J}} = U_1\vec{J}$ is the generator of *body* rotations, the Hamiltonian may be rewritten

$$H = \mathcal{J}^2/(2I). \tag{4.2}$$

The fundamental canonical commutation relations are most easily given in terms of $\vec{J}$, $\vec{\mathcal{J}}$, and U. The commutator of J with U follows from the fact that $\vec{J}$ generates spatial (lower indices) rotations,

$$[J_i, (U_j)^a{}_b] = (T_{ji})_{bc}(U_j)^a{}_c \tag{4.3a}$$

or

$$[J_i, U_j] = T_{ji} U_j \quad \text{(no sum on } j\text{)}, \tag{4.3b}$$

where j labels the representation of the rotation group. Similarly, the commutator of the generator of body rotations with U reads

$$[\mathcal{J}^i, U_j] = U_j T_j{}^i \quad \text{(no sum on } j\text{)}. \tag{4.4}$$

In the Yang-Mills theory a *global* color rotation is given by rotating the degrees of freedom over all space equally, i.e., $V(r,m)$ is independent of r and m. Thus, each link transforms according to $U \to VUV^{-1}$. Therefore, the rigid rotator analog is a simultaneous and equal rotation of body and space axes. This transformation is generated by the difference of the body and space angular momenta.[11]

In the quantum theory of the spherical rotator, the eigenvectors are classified as simultaneous eigenvectors of the operators J^2, J_z, $\mathcal{J}^2$, and $\mathcal{J}_z$. They transform as multiplets under body and space rotations. Since $J^2 = \mathcal{J}^2$ as operators, we label states with quantum numbers J^2, J_z, and $\mathcal{J}_z$. The energy of a state is

$$E = j(j+1)/(2I) \tag{4.5}$$

and its multiplicity is $(2j+1)^2$. The space of states may be constructed from the singlet ground state $|0\rangle$ by using the matrix elements of U in the spin-$\frac{1}{2}$ representation as ladder operators. Alternatively one may apply $\exp(i\,\vec{T}_j \cdot \vec{\Omega})$ to the ground state. Consider first the quantities

$$(U_j)^i{}_l|0\rangle = [\exp(i\,\vec{T}_j \cdot \vec{\Omega})]^i{}_l|0\rangle\,. \tag{4.6}$$

Using Eqs. (4.1) and (4.3b) and the fact that the ground state satisfies

$$\vec{J}|0\rangle = \vec{\mathcal{J}}|0\rangle = 0\,,$$

we compute

$$\begin{aligned} H(U_j)^i{}_l|0\rangle &= \frac{1}{2I} J^2 (U_j)^i{}_l|0\rangle \\ &= \frac{1}{2I}[J^2, (U_j)^i{}_l]|0\rangle \\ &= \frac{1}{2I} J_k [J_k, (U_j)^i{}_l]|0\rangle \\ &= \frac{1}{2I}[J_k, (T_k)_{lm}(U_j)^i{}_m]|0\rangle \\ &= \frac{1}{2I} T^2{}_{ln}(U_j)^i{}_n|0\rangle \\ &= \frac{1}{2I} j(j+1)(U_j)^i{}_l|0\rangle \end{aligned} \tag{4.7}$$

as claimed. In particular, U_j generates states of definite energy $j(j+1)/(2I)$.

To use spin-$\frac{1}{2}$ representation matrices as ladder operators, we note that the representations $\vec{T}_j$ can can be constructed from $\vec{T}_{1/2}$. For example, in the case $j = 1$,

$$(U_1)^\alpha{}_\beta = \mathrm{tr} U_{1/2}{}^{-1} \sigma^\beta U_{1/2} \sigma^\alpha\,. \tag{4.8}$$

In summary, we can make the following set of correspondences between the Yang-Mills theory and the rigid rotator:

Simultaneous body and space rotation

$\to$ global color rotation,

separate body and space rotations

$\to$ local gauge transformation,

body index $\to$ final end of link,

space index $\to$ beginning end of link,

$\vec{\Omega} \to \vec{B}$,

$\vec{\omega} = \vec{J}/I \to \dfrac{d\vec{B}}{dl}$.

The body- (space-) fixed angular momenta correspond to the generators of gauge transformations which rotate one end of a link and do not affect the other end. We denote these operators in the Yang-Mills theory Q_- and Q_+, where the $-$ $(+)$ indicates the beginning (end) of a link. The total color Q carried by a link is the difference $Q_+ - Q_-$.[12] Since the operators $\mathcal{J}$ and J are related by $\mathcal{J}^\alpha = (U_1)^\alpha{}_\beta J_\beta$ for a rigid rotator, it follows by analogy that

$$Q_+^\alpha = (U_1)^\alpha{}_\beta (Q_-)_\beta \tag{4.9}$$

and

$$\begin{aligned} Q &= Q_+ - Q_- \\ &= [U(r,m) - 1]Q_- \\ &= [1 - U^{-1}(r,m)]Q_+\,. \end{aligned} \tag{4.10}$$

V. GAUGE-INVARIANT SPACE OF STATES

A. Pure Yang-Mills field

Evidently the space of states of the Yang-Mills field is the product of an infinite number of rigid-rotator spaces. However, not all of the states are physically relevant. The physical states are drawn from the space of gauge-invariant states. Let us first consider the generator of gauge transformations. An arbitrary gauge transformation can be built from individual gauge transformations at the points of the lattice. Therefore, it suffices to consider just a gauge transformation at the lattice site i. Six links emanate from the point and each is effected by the gauge transformation

$$[G^\alpha, U] = [Q_+^\alpha, U],$$

where G is the generator of gauge transformations at position r. Thus, the generator must be equal to the sum of the Q_+ over the six links,

$$G(r) = \sum_m Q_+(r, m) \tag{5.1}$$

We have seen in Sec. IV that Q_+ is proportional to $\dot{B}$ (in analogy with the space-fixed angular momentum). Accordingly, the generator $G(r)$ may be written as

$$G(r) = \text{const} \times \sum_m \dot{B}(r, m). \tag{5.2a}$$

The time derivative of the vector potential can be identified with the component of the non-Abelian electric field at position r in the direction m. The sum over electric fields emanating from a single site is the lattice analog of $\vec{\nabla}\cdot\vec{E}$ at the lattice site r. Because the electric field itself varies from $Q_-(r, m)$ to $Q_+(r, m)$ along a link, there is an additional contribution to the lattice analog of $\vec{\nabla}\cdot\vec{E}$ which is associated with the links. This additional source is just $Q_+(r, m) - Q_-(r, m)$, or the charge carried by the link. Thus the generator may be rewritten

$$G(r) = \vec{\nabla}\cdot\vec{E}(r) - \tfrac{1}{2}\sum_m Q(r, m). \tag{5.2b}$$

The gauge invariance of the physical sector is defined by $G(r)|\psi\rangle = 0$. Identifying $\frac{1}{2}\sum_m Q(r, m)$ as the local color density ρ_G, this constraint becomes the familiar condition $\vec{\nabla}\cdot\vec{E} = \rho_G$.

The gauge-invariant space of states may be constructed by starting with the gauge-invariant state $|0\rangle_G$, which is defined as the product over lattice sites of the individual gauge-field ground states. The full space of states is given by acting with any product of components of the $U_{1/2}(r, m)$,

$$\prod_{r, m \in \{s\}} U_{1/2}(r, m)^i{}_j |0\rangle_G, \tag{5.3}$$

where the product goes over all r and m belonging to some set $\{s\}$. The set $\{s\}$ may include any link any number of times. In general Eq. (5.3) describes a gauge-invariant state only if the color indices at each point are contracted to form a local singlet. Indices associated with different lattice sites may not be contracted since they do not transform identically under *local* gauge transformations. For example, the state,

$$U_{1/2}(r, m)^i{}_j |0\rangle$$

is not gauge-invariant since it has uncontracted indices. The state

$$U_{1/2}(1)^i{}_j U_{1/2}(2)^j{}_k U_{1/2}(3)^k{}_l U_{1/2}(4)^l{}_i |0\rangle,$$

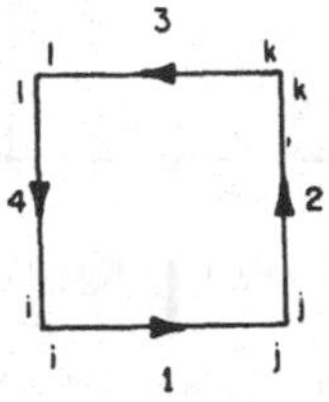

FIG. 3. Graphical representation for the gauge-invariant operator $\mathrm{tr}U_{1/2}(1)U_{1/2}(2)U_{1/2}(3)U_{1/2}(4)$.

where 1, 2, 3, and 4 refer to the links shown in Fig. 3, is gauge-invariant. However, if the contraction of indices did not involve the same lattice site, a similar object such as

$$U_{1/2}(1)^i{}_j U_{1/2}(3)^j{}_k U_{1/2}(2)^k{}_l U_{1/2}(4)^l{}_i |0\rangle$$

would not be gauge-invariant.

A simple pictorial representation of the construction of gauge-invariant states can be given. We begin with an oriented closed path of links Γ. For each link on Γ we associate an operator

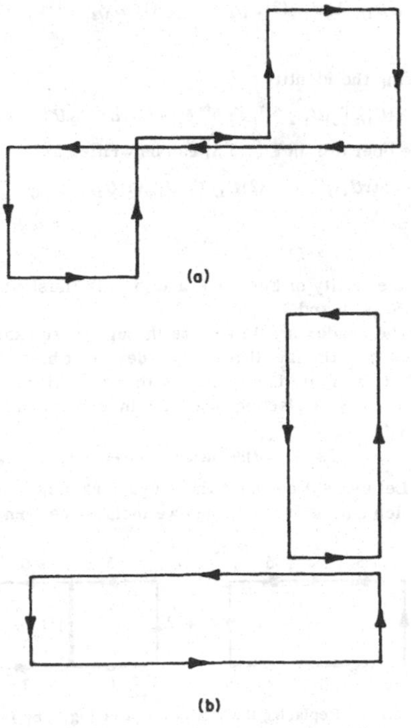

FIG. 4. The operators (a) $U_{1/2}(\Gamma)$ and (b) $U_{1/2}(\Gamma_1)U_{1/2}(\Gamma_2)$.

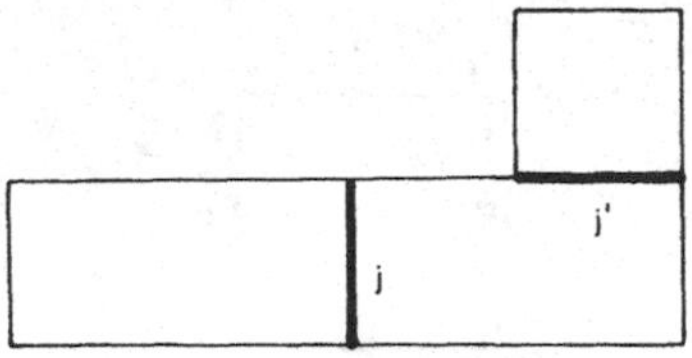

FIG. 5. Replacing overlapping "spin"-$\frac{1}{2}$ flux lines by lines of "spin" j.

$U_{1/2}(r, m)$. We construct the matrix trace,

$$U(\Gamma) = \mathrm{tr}U_{1/2}(r, n)U_{1/2}(r+n, m) \times U_{1/2}(r+n+m, l)\cdots U_{1/2}(r-s, -s),$$

where the factors occur in the order indicated by the path Γ. Obviously $U(\Gamma)$ is gauge-invariant. One can now apply any product of $U(\Gamma)$'s to the ground state $|0\rangle_G$ to produce the gauge-invariant subspace. Individual links may be covered more than once either by an individual Γ as in Fig. 4(a) or by two or more Γ's as in Fig. 4(b). Thus an alternative to the field description is provided by a configuration space in which a set of closed flux lines or strings are specified.

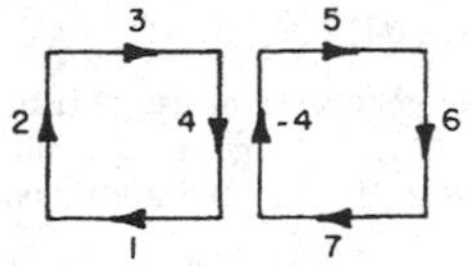

FIG. 6. Product of boxes with an overlapping link.

Another way to characterize the gauge-invariant states is to consider products of the $U_j(r, m)$ where j is not necessarily $\frac{1}{2}$. Again all indices must be contracted as before, but no link is covered more than once. The pictorial representation associated with this construction involves closed paths which can branch as in Fig. 5. Each segment is labeled by a value of j and the vertices involve the proper Clebsch-Gordan coefficients. We will construct an example of the equivalence of the two procedures. Consider the paths shown in Fig. 6. We associate with them the operator

$$\mathrm{tr}U_{1/2}(1)U_{1/2}(2)U_{1/2}(3)U_{1/2}(4)\,\mathrm{tr}U_{1/2}(-4)U_{1/2}(5)U_{1/2}(6)U_{1/2}(7) = \mathrm{tr}U_{1/2}(1)U_{1/2}(2)U_{1/2}(3)U_{1/2}(4) \times \mathrm{tr}U_{1/2}{}^{-1}(4)U_{1/2}(5)U_{1/2}(6)U_{1/2}(7). \tag{5.4}$$

Using the identity

$$(U_{1/2})^i{}_j(U_{1/2}{}^{-1})^k{}_l = \tfrac{1}{2}\delta^{ik}\delta_{jl} + \tfrac{1}{4}(\mathrm{tr}U^{-1}\tau_\beta U\tau_\alpha)(\tau_\beta)_{ik}(\tau_\alpha)_{jl}, \tag{5.5}$$

the operator in Eq. (5.4) can be written as

$$\tfrac{1}{2}\mathrm{tr}U_{1/2}(1)U_{1/2}(2)U_{1/2}(3)U_{1/2}(5)U_{1/2}(6)U_{1/2}(7) + \tfrac{1}{4}[\mathrm{tr}U_{1/2}(1)U_{1/2}(2)U_{1/2}(3)\tau_\alpha U_{1/2}(5)U_{1/2}(6)U_{1/2}(7)\tau_\beta]U_1(4)^\alpha{}_\beta. \tag{5.6}$$

The equality of Eqs. (5.4) and (5.6) is illustrated in Figs. 6 and 7.

The reader should realize through these examples that the fact that every index must be matched at each site is the non-Abelian equivalent of the continuity of electric flux lines in an Abelian theory.[13]

B. Yang-Mills theory with fermions

Let us next consider the gauge-invariant states which can be formed when we include the fermion

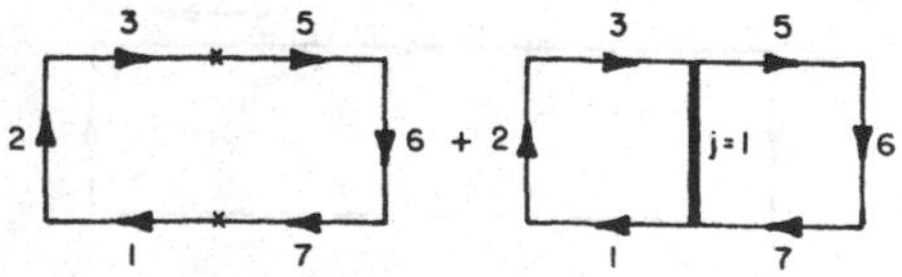

FIG. 7. Replacing the flux in link 4 of Fig. 6 by $j=0$ and $j=1$ flux lines.

field ψ. We can construct a gauge-invariant state by considering the lowest eigenstate of the gauge-invariant charge-conjugation-invariant operator

$$\sum_r (-1)^r \psi^\dagger(r)\psi(r)|0\rangle_F. \tag{5.7}$$

The state $|0\rangle_F$ is a product of fermion vacua over all the lattice sites. The product state $|0\rangle = |0\rangle_F|0\rangle_G$ is gauge-invariant.

Now, in addition to the operators $U(\Gamma)$ formed from closed paths, we can form gauge-invariant operators from paths with ends (Fig. 8). For ex-

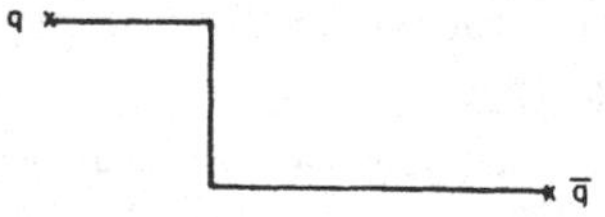

FIG. 8. A $q\bar{q}$ state with its accompanying flux line.

ample, consider a path Γ beginning at r and ending at site s. One can form the gauge-invariant operators

$$U(\Gamma,\Sigma)=\psi^{\dagger}(r)\Sigma U_{1/2}(r,n)U_{1/2}(r+n,m)\cdots$$
$$\times U_{1/2}(s-l,l)\psi(s), \quad (5.8)$$

where Σ is any 2×2 spin matrix.

The physical significance of the lines between occupied sites is interesting. They represent lines of electric flux. To see this observe that the operator $Q_+(r,m)$ is proportional to the electric field at site r and points in the direction m. At all links where there is no $U_{1/2}(r,m)$, $Q_+(r,m)$ gives zero. On the links through which a single Γ_j has passed, $Q_+{}^2(r,m)$ gives $j(j+1)$. In this sense one can think of these lines as containing electric flux of magnitude $[j(j+1)]^{1/2}$.

Now that the fermions have been added to the theory, the generator of gauge transformations at point r must include the additional operator $\psi^{\dagger}(r)\frac{1}{2}\vec{\tau}\psi(r)$, which generates color rotations of ψ. The full gauge-invariance condition on the space of states becomes

$$\left[\sum_m Q_+^{\alpha}(r,m)-\psi^{\dagger}(r)\tfrac{1}{2}\tau^{\alpha}\psi(r)\right]\Big|\Big\rangle=0. \quad (5.9)$$

This is analogous to the condition

$$\vec{\nabla}\cdot\vec{E}=\rho_G+\rho_F,$$

where ρ_G and ρ_F are the color densities of the gauge and Fermi fields.

VI. THE GAUGE-FIELD HAMILTONIAN

We must add a pure gauge-field term to the Hamiltonian describing fermions in order to give the field B some nontrivial dynamics. Since we are requiring local gauge invariance, the Hamiltonian must be built from gauge-invariant operators. Thus the Hamiltonian may contain objects like $U(\Gamma)$. However, since all components of the U's commute with one another, such terms will not be enough to produce nontrivial dynamics.

In addition to the U's, gauge-invariant operators can be built from the $Q_{\pm}(r,m)$. In particular $Q_+{}^2$ $(=Q_-{}^2)$ is the analog of the total angular momentum J^2 of the rigid rotator. Since J^2 commutes with both space and body rotations of the rigid rotator, $Q_+{}^2$ commutes with left and right gauge transformations. $Q_+{}^2$ is therefore gauge-invariant. Furthermore, since $Q_+{}^2$ does not commute with U, its appearance in the Hamiltonian will generate nontrivial dynamics. Accordingly, we include in the Hamiltonian a term

$$\sum_{r,m} Q_+{}^2(r,m)/(2I), \quad (6.1)$$

where I is a constant. This expression is, of course, analogous to the energy of an assembly of uncoupled rotators.

Clearly, from the rotator analogy ($J=I\omega$) we may deduce the equation of motion

$$\dot{B}=-i[B,H]=Q_+/I. \quad (6.2)$$

This equation states that the operator $Q_+(r,m)$ is proportional to the electric flux emanating from the lattice point r in the direction m. This fact was noted previously in Secs. IV and V.

In order that the pure Yang-Mills theory be nontrivial we must introduce terms which couple different links. To do this we make use of the operators $U(\Gamma)$. There is a great deal of arbitrariness in choosing the additional term(s). Following Wilson[4] we pick the simplest object which reproduces continuum Yang-Mills theory when $a\to 0$. Accordingly, let us consider the continuum limit of $U(\Gamma)$, where Γ is shown in Fig. 9. As $a\to 0$ we require as in Sec. III that the field B tend to zero $\sim a$. Thus we write the expansion,

$$U_{1/2}\cong 1+iag\vec{A}\cdot\tfrac{1}{2}\vec{\tau}-\tfrac{1}{2}a^2g^2\vec{A}\cdot\tfrac{1}{2}\vec{\tau}\vec{A}\cdot\tfrac{1}{2}\vec{\tau}+\cdots. \quad (6.3)$$

We assume that in the limit $a\to 0$ the field A becomes sufficiently smooth so that

$$A(r,n)-A(r+m,n)\cong a\vec{\nabla}A_n(r)\cdot\hat{m}.$$

Now we can expand $U_{1/2}(\Gamma)$ in powers of a. After some algebra we obtain

$$\mathrm{tr}U(\Gamma)=-\tfrac{3}{2}a^4g^4F^2, \quad (6.4)$$

where

$$F^{\alpha}_{mn}=\partial_m A^{\alpha}_n-\partial_n A^{\alpha}_m+g\epsilon^{\alpha\beta\gamma}A^{\beta}_n A^{\gamma}_m. \quad (6.5)$$

This quantity in Eq. (6.4) is familiar from the conventional continuum form of the Yang-Mills Hamiltonian.

Now we can collect together the various pieces of the Hamiltonian defining the discrete theory:

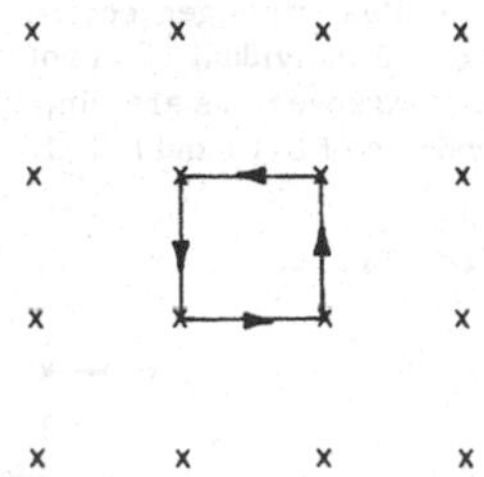

FIG. 9. The term U (box) in the gauge-field Hamiltonian.

$$H=\frac{a}{2g^2}\sum_{r,m}\dot{B}^2(r,m)+\frac{4}{ag^2}\sum \mathrm{tr}U_{1/2}(r,n)U_{1/2}(r+n,m)U_{1/2}(r+n+m,-n)U_{1/2}(r+m,-m)$$
$$+a^{-1}\sum\psi^+(r)\frac{\vec{\sigma}\cdot\vec{n}}{i}U(r,n)\psi(r+n)+m_0\sum(-1)^r\psi^+(r)\psi(r)\,. \tag{6.6}$$

The coefficients of the various terms are determined by requiring that H must have the usual continuum ($a\to 0$) limit.

Comparing the first term of Eq. (6.6) with the energy of a rigid rotator and recalling the correspondence $\dot{B}\to\omega$, we note the correspondence

$$I\to a/g^2\,.$$

It then follows from Eq. (6.1) that

$$J\to Q_+(r,n)=a\dot{B}(r+n,\,-n)/g^2\,.$$

Finally, recall that the color carried by a link is

$$Q=[U_1(r,n)-1]\,Q_-\,, \tag{4.10'}$$

which becomes in the continuum limit

$$Q^i\cong(iagA\cdot T_1)^i{}_j\,\frac{a^2}{g}\,\dot{A}_j\,,$$
$$Q^i/a^3\cong(\vec{A}\times\vec{E})_i\,. \tag{6.7}$$

This is the familiar expression for the color carried by the gauge field in the continuum Yang-Mills theory.

VII. ENERGY CONSIDERATIONS AND PERTURBATION THEORY

The qualitative features of a solution to the Hamiltonian in Eq. (6.6) depend upon which term dominates. For large g, the first term

$$H_0=\frac{g^2}{2a}\sum_{r,n}Q_+^2(r,n)$$
$$=\frac{a}{2g^2}\sum\dot{B}^2(r,n) \tag{7.1}$$

dominates and the remaining terms may be treated as perturbations. Since the first term does not couple adjacent lattice sites, its eigenvectors are product eigenvectors of the individual rigid rotators. The gauge-invariant eigenvectors are simply the states defined by products of $U(\Gamma)$ and $U(\Gamma,\Sigma)$ applied to $|0\rangle=|0\rangle_G|0\rangle_F$. Evidently in this strong-coupling approximation there are no fluctuations in the stringlike flux-line states. The energy of a state is most easily described in a representation utilizing the $U_j(r,m)$ operators introduced earlier. A typical state is pictured in Fig. 5. The energy of that state receives a contribution from each link equal to $j(j+1)/(2I)$.

This picture of the strongly coupled Yang-Mills theory in terms of a collection of stringlike flux lines is the central result of our analysis. It should be compared with the phenomenological use of stringlike degrees of freedom which has been widely used in describing hadrons.[14]

An important element of the Yang-Mills theory is that the electric flux is quantized. Since electric flux (Q_+) satisfies the commutation relations of the non-Abelian generators of gauge transformations, it cannot be indefinitely subdivided. This means that a unit of electric flux of magnitude $Q_+^2=\frac{1}{2}(\frac{1}{2}+1)$ cannot split. This is to be contrasted with a conventional Abelian gauge theory in which both charge and flux can be arbitrarily subdivided.[15]

Now consider some examples. The ground state is the state $|0\rangle$ with no excited flux lines. The next-lowest-energy gauge-invariant states involve a single excited link with a fermion-antifermion pair at its ends. All such states have energy $\frac{1}{2}(\frac{1}{2}+1)/(2I)$ in the strong-coupling limit in which only the term in Eq. (7.1) is considered in H. In the pure Yang-Mills theory the lowest-energy gauge-invariant excitation involves four links as in Fig. 9. It has energy $4\times\frac{1}{2}(\frac{1}{2}+1)/(2I)$.

The states involving fermions can have nontrivial SU(3) quantum numbers and may be identified as mesons. The Yang-Mills gauge-invariant excitations are SU(3) singlets. Excited states of these objects can be identified with states in which more than the minimum number of links

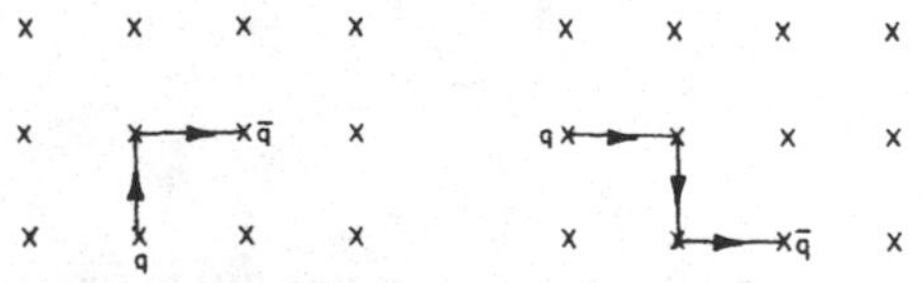

FIG. 10. Two excited states of a $q\bar{q}$ system.

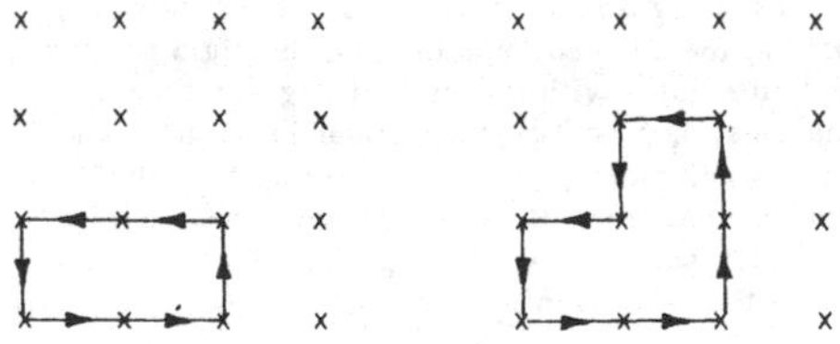

FIG. 11. Two excited states of a gauge-invariant excitation.

are involved. For example, in Fig. 10 we show two possible $q\bar{q}$ states, and in Fig. 11 we show two possible Yang-Mills gauge-invariant excitations. The open-ended $q\bar{q}$ flux lines are similar to the original "dual string"[14] used phenomenologically to describe ordinary mesons. The excitations along trajectories are analogous to excited $q\bar{q}$ states occupying more than one link. The Yang-Mills gauge-invariant excitation also has a counterpart in the dual model. It is closed SU(3)-singlet dual Pomeron string.[16]

Let us consider the force law between widely separated quarks in the strongly coupled limit. The potential energy is defined as the lowest energy compatible with the presence of a quark at site r and an antiquark at site s. For simplicity choose the sites r and s to lie in a given row as depicted in Fig. 12. The minimum-energy gauge-invariant state is obviously given by exciting the shortest path of links connecting the $q\bar{q}$ pair. The energy associated with the configuration shown in Fig. 12 is

$$(L/a)\,\tfrac{1}{2}(\tfrac{1}{2}+1)/(2I)\,, \tag{7.2}$$

where L is the separation distance between the quarks. Since the potential energy increases linearly with distance, the force between the quarks is independent of their separation. Such a force is clearly sufficient to confine quarks. As discussed in Refs. 5 and 7, this force law between unscreened charges is identical to the classical force laws in one-dimensional gauge theories.[17] It is also clear that the energy of confinement is stored on the line (flux tube) between the quarks (Fig. 12).

The force law between $j=\frac{1}{2}$ (color triplet) objects contrasts sharply with the force law between hypothetical $j=1$ (color octet) static objects. Physically the reason is that the low-energy state of a distance pair of $j=1$ objects can be constructed by screening the color of the static objects by the color gauge field. To see this explicitly, define the field $\phi_\alpha(r)$, which creates the $j=1$ particle at site r. The minimum-energy gauge-invariant state compatible with a single such object is

$$\phi_\alpha(r)\operatorname{tr} U_{1/2}(r,n)\,U_{1/2}(r+n,m) \times U_{1/2}(r+n+m,-n)\,U_{1/2}(r+m,-m)\tau^\alpha|0\rangle\,.$$

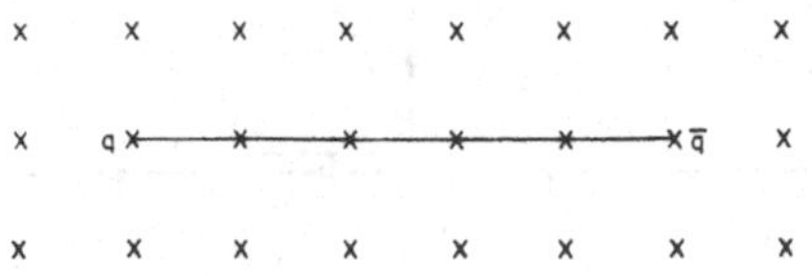

FIG. 12. Minimum-energy configuration for a separated $q\bar{q}$ system in the strong-coupling limit.

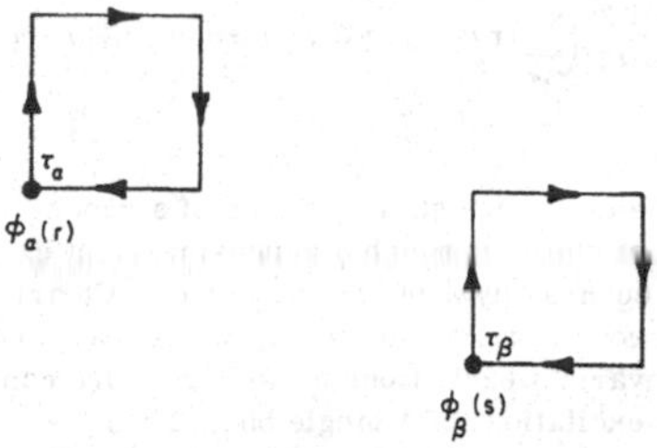

FIG. 13. Minimum-energy configuration for a separated pair of color-1 objects.

This state clearly has finite energy. The state of minimum energy of two such particles at large distance is obtained by independently applying such operators at distant sites (Fig. 13). The resultant energy is independent of the distance between the objects, so the force is entirely screened. The resulting well-separated objects are each colorless.

In summary, the strong-coupling limit is characterized by nonfluctuating flux-line configurations. This means that the fluctuations of the vector potentials A are as large as possible. This corresponds to the fact that the relative orientation of the body- and space-fixed axes in the rigid rotator's ground state is maximally uncertain.

Now we shall consider perturbations around the strong-coupling limit. The effects of the second term in H,

$$\begin{aligned} V &= \sum_{r,n} V(r,n) \\ &= \frac{4}{ag^2}\sum \operatorname{tr} U_{1/2}(r,n)\,U_{1/2}(r+n,m) \\ &\qquad \times U_{1/2}(r+n+m,-n)\,U_{1/2}(r+m,-m) \\ &\quad + \text{H.c.} \end{aligned} \tag{7.3}$$

correct the eigenstates of H_0 to $O(g^{-1})$ and correct the energy of these states to $O(g^{-2})$. In fact the term V plays three roles. The first is to diminish the fluctuations of the magnetic field.[18] The second is to create fluctuations of the string configurations. And the third is to propagate excitations through the lattice.

The first role is very simple. This follows because in the continuum limit V becomes the square of the magnetic field.

The second two roles can be studied in perturbation theory. For example, consider the corrections to the vacuum state of pure Yang-Mills theory. The first-order correction to a state Ψ_0 is

$$\Psi_1 = \frac{1}{E_0 - H_0}\,V\Psi_0\,. \tag{7.4}$$

Therefore, the correction to the vacuum state is

$$\Psi_1 = \frac{8I}{3ag^2}\Big(\sum_{r,n} \mathrm{tr}\, U_{1/2}(r,n)U_{1/2}(r+n,m)U_{1/2}(r+n+m,-n)U_{1/2}(r+m,-m) + \mathrm{H.c.}\Big)\,|0\rangle\,. \tag{7.5}$$

Thus, the corrected state consists of a superposition of states in which a gauge-invariant excitation occurs anywhere on the lattice. Clearly higher-order perturbation theory will create many gauge-invariant excitations as well as more complicated excitations of a single box. Thus the vacuum becomes a fluctuating sea of closed flux loops. The first-order correction to the vacuum energy is given by

$$\Delta E(\text{vacuum}) = \left\langle 0 \left| V \frac{-1}{H_0} V \right| 0 \right\rangle, \tag{7.6}$$

and an explicit calculation shows that

$$\Delta E(\text{vacuum})/\text{vol.} = -32/a^4g^6\,. \tag{7.7}$$

Next let us consider the corrections to the energy and state of a separated $q\bar{q}$ pair in lowest-order strong-coupling perturbation theory. The correction to the state is

$$\Delta\Psi(q\bar{q}) = \frac{1}{E_0 - H_0} V\psi^\dagger(r)U_{1/2}(r,n)U_{1/2}(r+n,m)\cdots \times U_{1/2}(s-l,l)\psi(s)|0\rangle \tag{7.8}$$

and is best analyzed graphically. First there are terms in which the boxes of V do not overlap with the original string connecting the $q\bar{q}$ pair. They are shown in Fig. 14 and may be regarded as corrections to the vacuum. Next there are corrections in which one of the lines in V overlaps with the original string. There are two possibilities as indicated in Figs. 15(a) and 15(b). Namely, the overlapping flux lines may be parallel or antiparallel. The two diagrams of Fig. 15 may be rewritten as in Fig. 16 in terms of $j=0$ and $j=1$ flux lines. Clearly these corrections cause both the position and structure of the string to fluctuate.

The change in energy of the $q\bar{q}$ configuration can be computed by standard perturbation theory. After subtracting off terms contributing to the vacuum energy, we find that the correction is a sum over the original links of the flux line. The contribution of each link is $\sim\text{const}/g^6$. Therefore the new potential energy is

$$\frac{L}{a}\left[\frac{\frac{1}{2}(\frac{1}{2}+1)}{(2I)} - \frac{\text{const}}{g^6}\right]. \tag{7.9}$$

It is important to note that the energy is still proportional to the distance between the fermions. The fermions are still confined.

When higher orders in the perturbation are considered, the energy continues to grow linearly with the length of the string for large interquark separation. However, deviations occur for short strings. To see why this is so, consider an intermediate state in which two adjacent boxes are involved as in Fig. 17. The contribution from such diagrams will again be summed along the length of the flux line and therefore lead to a linear force law for large distances. However, when the original flux line is less than two lattice sites long, this term becomes inoperative.

We now turn to a perturbation theory description of the propagation of gauge-invariant excitations through the lattice. To illustrate the effects of V, consider the propagation of the symmetric gauge-invariant excitation.[19] Recall that any localized box is an eigenvector of H_0 with eigenvalue $4\times\frac{1}{2}(\frac{1}{2}+1)/(2I)$. So, in the strong-coupling limit the momentum eigenstates (which are linear superpositions with weighting factors e^{ikr} of localized boxes at position r) exhibit no momentum dependence in their energy spectrum. Now consider the effects of V acting upon a localized gauge-invaraint excitation. The interesting terms occur when one of the sides of the gauge-invariant excitation coincide with a link in V. This gives rise to the states shown in Fig. 18. Allowing V to act twice, it may act on the original site of the gauge-invariant excitation and annihilate it, leaving a displaced gauge-invariant excitation. In this way

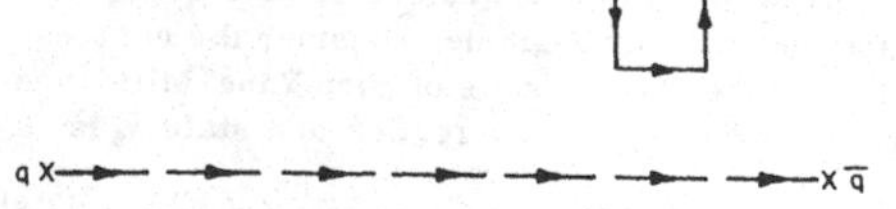

FIG. 14. A disconnected contribution to V acting on a $q\bar{q}$ state.

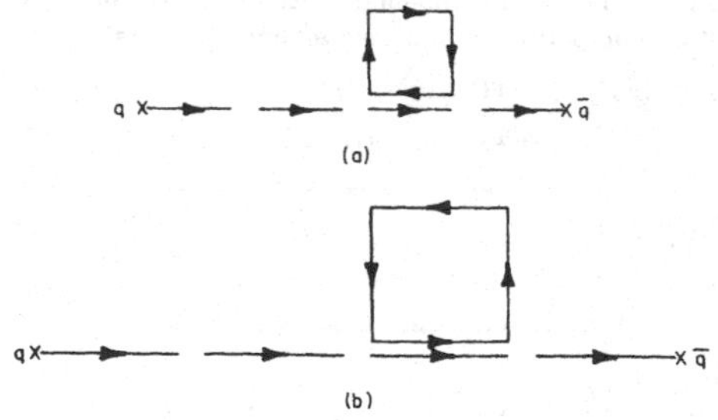

FIG. 15. Connected contributions to V acting on a $q\bar{q}$ state.

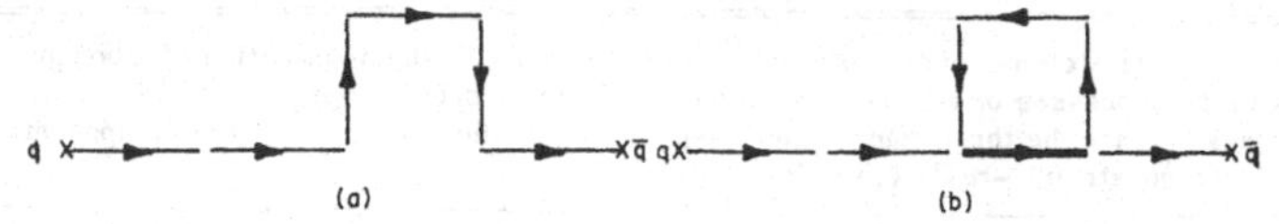

FIG. 16. Replacing the doubly excited link in Fig. 15 by an unexcited link or a $j=1$ link.

the perturbation can cause the gauge-invariant excitation to propagate through the lattice.

In order to study the propagation properties, we introduce the following notation. We denote $|r, m\rangle$ as a gauge-invariant excitation at r with polarization m. To define this terminology consider the three box states at r: $|r, x\rangle$, $|r, y\rangle$, and $|r, z\rangle$. The state $|r, x\rangle$ is bounded by the links (r, m_y), (r, m_z), $(r+m_y, m_z)$, and $(r+m_z, m_y)$ as depicted in Fig. 19.

Now consider a momentum eigenstate

$$|k, n\rangle = \sum_r e^{ik\cdot r}|r, n\rangle \,. \tag{7.10}$$

We will calculate the lowest-order nonvanishing correction to the energy of the state $|k, m\rangle$ and thereby exhibit a nontrivial dispersion law for these excitations. The first-order matrix element of the perturbation vanishes identically,

$$\langle k, l|V|k, l\rangle = 0\,.$$

To see this, note that if the box created by $V(r, m)$ has no line in common with $\langle r', l'|$ or $|r, l\rangle$, then $\langle r', l'|V(r, m)|r, l\rangle$ is zero. If it has one or more line in common with the states $|r, l\rangle$ or $|r', l'\rangle$, then one has two possible situations. Either $V(r, m)$ has one line in common with $|r, l\rangle$ or $|r', l'\rangle$, in which case a rectangle having six links is made (such a rectangle cannot project back onto the final state) or if $V(r, m)$ completely overlaps with $|r, l\rangle$ or $|r', l'\rangle$, then each side of the box is excited to a state of color $j=1$ or 0, and does not project back onto the final state which is a gauge-invariant excitation with $j=\frac{1}{2}$ sides.

The next order of perturbation theory is described with the matrix elements

$$\left\langle r', l' \middle| V \frac{1}{E-H_0} V \middle| r, l \right\rangle .$$

We must compute the eigenvalues of this matrix. The possible nonvanishing matrix elements are as follows. If $|r, l\rangle = |r', l'\rangle$, V may create a nonadjacent box, a box with one line in common or a completely overlapping box with $|r, l\rangle$. The nonoverlapping box contributes

$$\left\langle 0 \middle| V \frac{1}{H_0} V \middle| 0 \right\rangle ,$$

which we identify as the correction to the vacuum energy per box to this order. Since only the difference of the excitation energy with the vacuum is physically significant, we must subtract off this vacuum energy. This leaves over

$$\sum_{\substack{\text{boxes having no lines}\\ \text{in common with } |r,l\rangle}} = \sum_{\text{all boxes}} - \sum_{\substack{\text{boxes with a line}\\ \text{in common}}} .$$

So, subtracting out the vacuum energy gives us -13 times the energy of a vacuum box. The counting factor 13 comes from the fact that there are 12 boxes with one line in common with a given box and one box which completely overlaps with that box. Another contribution to the matrix element has an intermediate state with an adjacent box (Figs. 6 and 7). Finally, there is the case in which V creates a box completely overlapping $|r, l\rangle$. Call the magnitude of this contribution A. The other matrix elements which can exist involve $|r, l\rangle$ and $|r', l'\rangle$ with a single line in common. There are four possible contributions to this matrix element which are illustrated in Fig. 18 and

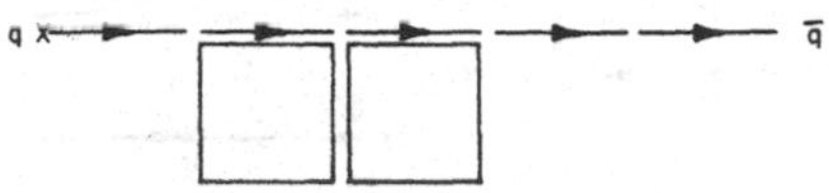

FIG. 17. A higher-order correction to a separated $q\bar{q}$ state.

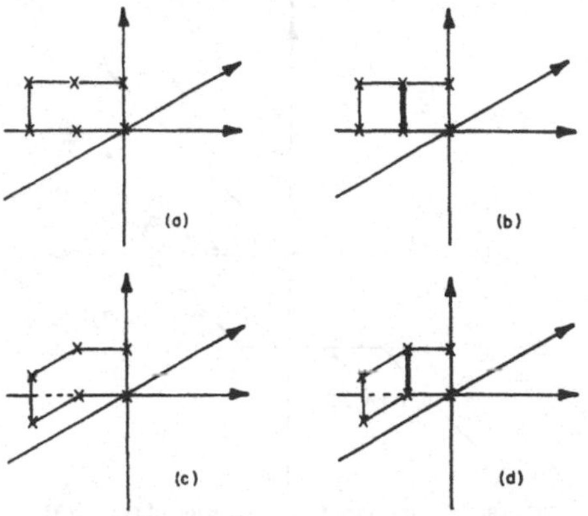

FIG. 18. V applied to a gauge-invariant excitation.

TABLE I. Matrix elements of $V(E_0-H)^{-1}V$ corresponding to the propagation of a box according to the processes of Fig. 18. The values in the table should be multiplied by a factor $-B$. Here $k_{x,y,z}$ are the three components of spatial momentum of the box. The components k_i satisfy the constraint $-\pi a^{-1} < k_i < \pi a^{-1}$.

	$\|k, m_z\rangle$	$\|k, m_y\rangle$	$\|k, m_x\rangle$
$\|k, m_z\rangle$	$2(\cos k_x + \cos k_y)$	$1+e^{ikz}+e^{-iky}+e^{ik(z-y)}$	$1+e^{ikz}+e^{-ikx}+e^{ik(z-x)}$
$\|k, m_y\rangle$		$2(\cos k_z + \cos k_x)$	$1+e^{iky}+e^{-ikx}+e^{ik(y-x)}$
$\|k, m_x\rangle$			$2(\cos k_z + \cos k_y)$

have been discussed above. These terms contribute a coefficient $-B$ to the matrix element.

Now we can collect together the matrix elements of $V(E_0-H)^{-1}V$ in the states $|k, l\rangle$ and $|k', l'\rangle$. They are tabulated in Table I. The energy eigenvalues are given by $4\times\frac{1}{2}(\frac{1}{2}+1)/(2I)+A$ plus the eigenvalues of the 3×3 matrix of Table I. For small momentum in the z direction the eigenvalues and eigenvectors read

$$\tfrac{3}{2}I + A + Bk_z^2 \qquad (0, 1, -1) \quad \text{spin } 2$$
$$\tfrac{3}{2}I + A + \tfrac{2}{3}Bk_z^2 \qquad (1, a, a)$$
$$\tfrac{3}{2}I + A - 6B + \tfrac{1}{3}Bk_z^2 \qquad (1, a', a') \quad \text{spin } 0\ .$$

At rest the three eigenvectors are $(0, 1, -1)$, which is a spin-2 object, $(2, 1, -1)$, and $(1, 1, 1)$, which is a rotational singlet. In general it is not possible to classify these states according to the rotation group because the lattice has only cubic symmetry. However, if we consider a rotation by 90° about the z axis, we see that the states $(1, 1, 1)$ and $(2, -1, -1)$ are invariant. The state $(0, -1, 1)$ changes sign under this rotation and is therefore classified as spin 2.

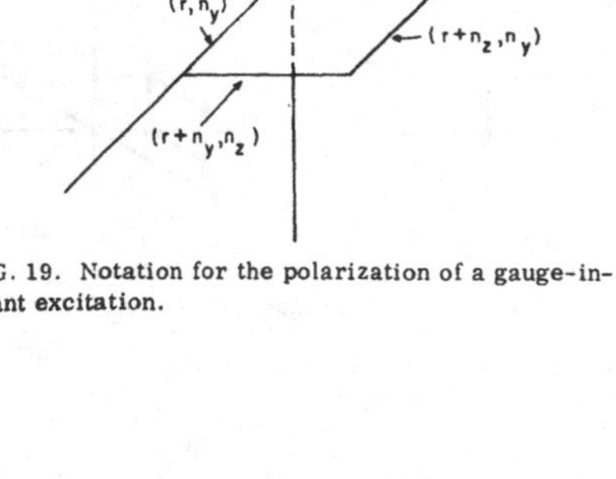

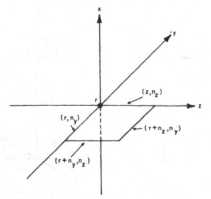

FIG. 19. Notation for the polarization of a gauge-invariant excitation.

Let us now return to the problem of quark confinement when the fermion piece of the Hamiltonian Eq. (6.6) is accounted for. The state with a single fluctuating flux line connecting two quarks is no longer an energy eigenstate. The fermion term considered as a perturbation describes processes in which a $q\bar{q}$ pair and its flux line is created. If that flux line overlaps with a link of the original flux line connecting the initial quarks, it may leave that link in an unexcited state (Fig. 20). Thus these processes allow the original string to break, i.e., they screen the long-range interquark forces. However, they do not allow free quarks to escape since each segment in Fig. 20 must be colorless. This situation is clearly very closely analogous to the phenomenon of vacuum polarization and screening in one-dimensional quantum electrodynamics which also confines quarks[17] and eliminates long-range forces.

Nevertheless, quark confinement can fail in the four-dimensional lattice theory if the very high-order terms in the perturbation-series expansion become important. If the terms in the Hamiltonian which cause fluctuations in the flux line become dominant, then electric flux will fail to be collimated along a line between quarks. Then the long-range force which would permanently bind quarks may disappear.

VIII. CONCLUSIONS AND DISCUSSION

The main result of this paper is that strongly coupled Yang-Mills theory on a lattice describes interacting propagating stringlike disturbances. The strings are elementary quantized lines of electric flux which can only end on charges.

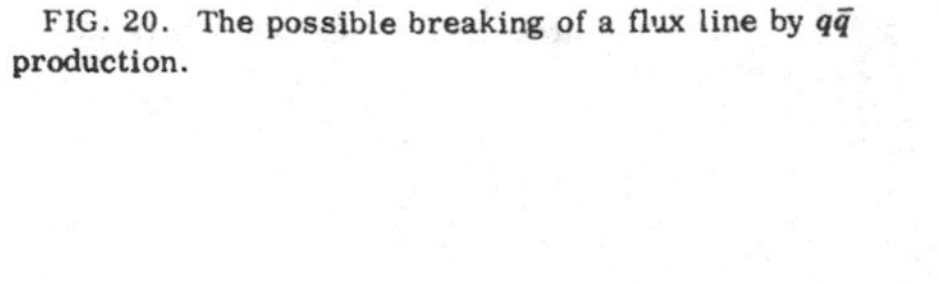

FIG. 20. The possible breaking of a flux line by $q\bar{q}$ production.

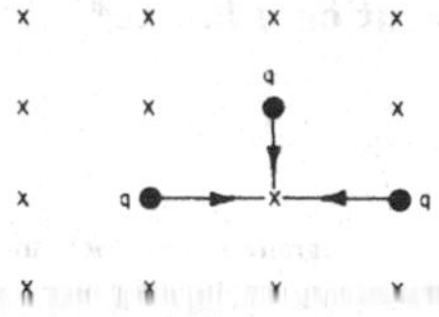

FIG. 21. A typical baryon configuration in a SU(3)-color theory.

For physical applications the gauge group should be SU(3) (color) rather than SU(2) (color). The main new feature of this generalization is that three quark indices can be contracted at a point with the antisymmetric coupling ϵ_{ijk}. This allows baryons to be formed from three flux lines joined at the center. It would be interesting to study these objects (see Fig. 21.)

We have not discussed in this article a number of difficult theoretical questions which remain unsolved. The most important is to show that renormalization effects really do lead to a strongly coupled theory at large distances. This will presumably require a renormalization-group approach to the theory in which it is first formulated on a lattice with a small spacing and a small coupling constant. The degrees of freedom may be "thinned out" according to Wilson's method[20] until an effective description with large lattice spacing is found. We hope such an analysis will justify our use of strong-coupling methods. In addition, one must hope that the effective description will define a theory which is not sensitive to the initial lattice of small spacing (except through certain coupling and wave-function renormalization constants). Then covariance will be restored and the theory will be (potentially) realistic. If this can not be done, our approach will be invalid.

ACKNOWLEDGMENT

The authors thank K. G. Wilson for introducing them to gauge theories formulated on a lattice and for many stimulating conversations. We also thank K. Gottfried for discussions concerning the quantum rigid rotator.

*Work supported in part by the National Science Foundation.

†Work supported in part by the National Science Foundation under Grant No. GP-38863.

[1]The earliest discussion of the phenomenon in a field-theoretic framework appears in V. F. Weisskopf, Phys. Rev. 56, 72 (1939).

[2]M. Gell-Mann and F. E. Low, Phys. Rev. 95, 1300 (1954); K. Symanzik, Commun. Math. Phys. 23, 49 (1971); C. G. Callan, Phys. Rev. D 5, 3202 (1972).

[3]G. 't Hooft, in Proceedings of the Marseilles Conference on Gauge Theories, 1972 (unpublished); H. D. Politzer, Phys. Rev. Lett. 30, 1346 (1973); D. J. Gross and F. Wilczek, Phys. Rev. Lett. 30, 1343 (1973).

[4]Kenneth G. Wilson, Phys. Rev. D 10, 2445 (1974).

[5]J. Kogut and Leonard Susskind, Phys. Rev. D 9, 3501 (1974).

[6]C. N. Yang and R. L. Mills, Phys. Rev. 96, 191 (1954).

[7]Linear force laws have been suggested by E. P. Tryon [Phys. Rev. Lett. 28, 1605 (1972)] and more recently by K. Kaufman [private communication from R. P. Feynman to one of us (L.S.)]. The theories discussed in Ref. 5 also have this behavior.

[8]For a recent discussion of color quark models, see H. Fritzsch and M. Gell-Mann, in *Proceedings of the XVI International Conference on High Energy Physics, Chicago-Batavia, Ill., 1972*, edited by J. D. Jackson and A. Roberts (NAL, Batavia, Ill., 1973), Vol. 2, p. 135.

[9]To prove this one can derive the equations of motion and explicitly verify that they can be written in a manifestly gauge-invariant form.

[10]The rigid rotator is discussed in H. Goldstein, *Classical Mechanics* (Addison-Wesley, Reading, Mass., 1958).

[11]The reader may be confused by the simultaneous use of two linear vector spaces. Each component of the $(2j+1)^2$-dimensional matrices U_j and the $(2j+1)$-dimensional vector J are operators in the Hilbert space of states of the quantum system. The matrices T_j are c numbers in this space. The commutation relations refer to this space and not to the $(2j+1)$-dimensional vector space on which T_j acts.

[12]It is the *difference* because the commutation relations of $\mathcal{J}$ differ in sign from those of J. See, for example, L. D. Landau and E. M. Lifshitz, *Quantum Mechanics* (Addison-Wesley, Reading, Mass., 1958).

[13]The notion of continuous flux lines was invented and used by Faraday.

[14]The string model of hadrons was originally suggested by L. Susskind, Phys. Rev. Lett. 23, 545 (1969). See Y. Nambu, in *Symmetries and Quark Models*, edited by R. Chand (Gordon and Breach, New York, 1970); H. B. Nielsen, Nordita report, 1969 (unpublished).

[15]However, Wilson's Abelian gauge theory formulated on a lattice has quantized flux since the gauge field B is again treated as an angular variable. See Ref. 4 for details.

[16]G. Frye and L. Susskind, Phys. Lett. 31B, 589 (1970).

[17]A. Casher, J. Kogut, and Leonard Susskind, Phys. Rev. Lett. 31, 792 (1973).

[18]The term "magnetic field" refers to $F_{ij}^{\alpha} = \partial_i A_j^{\alpha} - \partial_j A_i^{\alpha} + g\epsilon^{\alpha\beta\gamma} A_i^{\beta} A_j^{\gamma}$, $i, j = 1, 2, 3$.

[19]The term "symmetric gauge-invariant excitation" indicates the operator $\frac{1}{2}(\mathrm{tr}\,U_1U_2U_3U_4 + \mathrm{tr}\,U_4U_3U_2U_1)$, where the links 1, 2, 3 and 4 bound a box.

[20]For a discussion of and references to the Kadanoff-Wilson block-spin analysis see the review article, K. G. Wilson and J. Kogut, Phys. Rep. 12C, 75 (1974).

PHYSICAL REVIEW D VOLUME 15, NUMBER 4 15 FEBRUARY 1977

Gauge fixing, the transfer matrix, and confinement on a lattice*

Michael Creutz

Brookhaven National Laboratory, Upton, New York 11973

(Received 29 October 1976)

We use the transfer-matrix formalism of statistical mechanics to relate Wilson's Lagrangian approach and the Kogut-Susskind Hamiltonian approach to gauge theories on a lattice. As a preliminary we discuss gauge fixing in Wilson's theory. This process leaves invariant Green's functions of gauge singlet operators. Taking the timelike lattice spacing to zero, we extract the Kogut-Susskind Hamiltonian from the transfer matrix in the gauge $A_0 = 0$.

I. INTRODUCTION

A theory of quarks interacting with non-Abelian gauge mesons has recently attracted considerable attention as a potential theory of the strong interaction.[1] The esthetic appeal of such a theory is heightened by the successes of gauge theory elsewhere in particle physics, first and most dramatically in quantum electrodynamics and more recently in the renormalizable theories of the weak interaction.

The strong-interaction theory is conjectured to differ in one fundamental respect from previous applications of gauge theory. The physical spectrum of states should not contain isolated particle states corresponding to the fundamental fields, i.e., quarks and vector "gluons," but rather should include only bound states which are singlets under the gauge group. This conjectured "quark confinement" would provide a most elegant resolution of the successes of the quark model with the lack of observation of free quarks.

Unfortunately there is a dearth of theoretical evidence supporting confinement. Renormalization-group arguments indicate that, for low momenta, Green's functions reflect a large effective coupling constant.[1] This only means that perturbation theory is not a reliable tool for investigating widely separated quarks. Indeed, low orders of perturbation theory give no clear signal of a nascent confinement.[2]

To circumvent conventional perturbation theory, Wilson proposed placing the theory on a discrete space-time lattice and then perturbing in the kinetic term for the gauge field.[3] In this formulation gauge invariance remains an exact local symmetry of the action and naturally leads to confinement. The hope is that the artifice of going to a lattice is nothing but an ultraviolet cutoff. Unfortunately it is not known if the continuum limit exists and is Poincaré invariant.

Balian, Drouffe, and Itzykson have presented a further analysis of this theory.[4] One important feature of their work is a mean field-theory calculation indicating that in a sufficiently high number of space-time dimensions a phase transition will occur as the coupling constant is varied. For strong coupling one obtains the confined phase studied by Wilson whereas for small coupling one reverts to a quantum-electrodynamics–like theory of free quarks and gluons. Migdal has presented approximate arguments that for non-Abelian gauge theories the transition to the unconfined phase requires more than four space-time dimensions.[5] Presumably the transition can occur for Abelian theories in four dimensions and has occurred for quantum electrodynamics, where we certainly do not want confinement. From this point of view, the photon is the Goldstone boson of the ordered phase.[6]

Kogut and Susskind have pursued an alternative approach to lattice gauge theory.[7] Keeping time as a continuous variable, they make three-dimensional space discrete. Working in the gauge $A_0 = 0$, they define a Hamiltonian for quark- and gauge-field degrees of freedom on the space lattice. Confinement arises from the invariance of this Hamiltonian under time-independent gauge transformations. Recently, in conjunction with several other authors, they have investigated the spectrum of this Hamiltonian in a strong-coupling approximation.[8]

In this paper we investigate the relation between the Lagrangian approach of Wilson and the Hamiltonian approach of Kogut and Susskind. Our main tool is the transfer-matrix method of statistical mechanics. The use of the transfer matrix was suggested by Wilson[3] as a method of relating his Euclidean path integral with quantum mechanics in Minkowski space-time.

One of the virtues of Wilson's formulation is that it does not require a gauge selection for quantization; however, the gauge invariance does allow gauge fixing without altering the physics of gauge-invariant quantities. We will discuss this point and then go to the gauge $A_0 = 0$, as used by Kogut and Susskind. In this gauge the transfer matrix assumes a particularly simple form. Once the transfer matrix and the Hilbert space in which it acts are obtained, we find the Hamiltonian for the continuous-time theory by taking the timelike lattice spacing to zero.

The plan of this paper is as follows. In the next section we review Wilson's theory and discuss gauge invariance and its relation to confinement. In Sec. III we discuss the possibility of gauge fixing in the theory. Section IV reviews the use of the transfer matrix to obtain conventional quantum mechanics from a discrete-time path integral. In Sec. V we find the transfer matrix and the continuous-time Hamiltonian for pure gauge fields. We study the Fermi fields in Sec. VI and give the complete coupled Hamiltonian in the concluding Sec. VII.

II. REVIEW OF WILSON'S THEORY

Electrodynamics can be formulated in terms of a path-dependent phase acquired by the wave function of a charged particle moving through an electromagnetic field.[9] For a non-Abelian gauge theory this phase is replaced by a rotation in the internal-symmetry space of the theory. This notion of a nonintegrable phase factor forms the basis of Wilson's prescription for formulating gauge fields on a lattice. Representing the gauge-field degrees of freedom, an internal-symmetry rotation is associated with each pair of nearest-neighbor sites on the lattice. The fermion degrees of freedom reside on the sites. Whenever a fermion moves from one site to a neighboring one, its wave function undergoes the corresponding internal-symmetry rotation.

We now formulate the theory precisely. Working on a hypercubical lattice of spacing a, we label sites on the corners of the hypercubes by an index i. We consider a four-component spinor field $\psi_i{}^\alpha$ for each site i and for every value of the internal-symmetry index α. The internal-symmetry group is a connected unitary group $\mathcal{G}$ of matrices $g^{\alpha\beta}$. For simplicity we assume that the field $\psi_i{}^\alpha$ transforms under the fundamental representation

$$\psi_i{}^\alpha \to g^{\alpha\beta}\psi_i{}^\beta , \tag{2.1}$$

where a sum over β is understood. For every pair of nearest-neighbor sites $\{i,j\}$ we introduce the gauge field $U_{ij}{}^{\alpha\beta}$ which is a matrix in the group $\mathcal{G}$. Under interchange of i and j we do not obtain a new degree of freedom but rather require that U become its inverse in the group

$$(U_{ji})^{\alpha\beta} = (U_{ij}{}^{-1})^{\alpha\beta} . \tag{2.2}$$

In terms of these degrees of freedom the Euclidean action defining the theory is

$$\begin{aligned} S = &-a^4 \sum_{i,j} \frac{A_{ij}}{2a}\bar{\psi}_i(1-\gamma_\mu e_{ij}{}^\mu)\, U_{ij}\psi_j \\ &+ a^4 \sum_i \left(\frac{4}{a} - m\right)\bar{\psi}_i\psi_i \\ &- \frac{1}{8g^2}\sum_{ijkl} P_{ijkl}\, \mathrm{Tr}(U_{ij}U_{jk}U_{kl}U_{li}) , \end{aligned} \tag{2.3}$$

where we suppress internal-symmetry indices,

$$A_{ij} = \begin{cases} 1 & \text{if } i \text{ and } j \text{ are nearest neighbors} \\ 0 & \text{otherwise,} \end{cases} \tag{2.4}$$

$e_{ij}{}^\mu$ is a unit vector pointing from site i to site j, γ_μ are *Euclidean* Dirac matrices satisfying

$$\begin{aligned} [\gamma_\mu , \gamma_\nu]_+ &= 2\delta_{\mu\nu} , \\ \gamma_\mu^\dagger &= \gamma_\mu , \end{aligned} \tag{2.5}$$

a Euclidean sum is understood in $\gamma_\mu e^\mu$, g is the bare gauge field coupling, and

$$P_{ijkl} = \begin{cases} 1 & \text{if } i,j,k,l \text{ run around a "plaquette," i.e., a square of side } a \text{ in the lattice} \\ 0 & \text{otherwise.} \end{cases} \tag{2.6}$$

A discussion of the reduction of Eq. (2.3) in the limit $a \to 0$ to the usual classical-gauge-theory action is contained in Refs. 3 and 4. The projection matrices $(1-\gamma_\mu e_\mu)$ inserted in Eq. (2.3) ensure that free fermions will have low energy only for low momenta. Without these factors, one will have low-energy fermions with momenta of order π/a.

From the action in Eq. (2.3), Wilson defines Euclidean Green's functions by the path-integral formula

$$G(\psi_i, \psi_j, \ldots, \bar{\psi}_l ; U_{mn}, \ldots, U_{yz}) = \frac{1}{Z}\int [d\psi dU]\, e^{-S}\, \psi_i \cdots \bar{\psi}_l U_{mn} \cdots U_{yz} , \tag{2.7}$$

where

$$Z = \int [d\psi dU]\, e^{-S} . \tag{2.8}$$

The integral over fermion degrees of freedom is standard and will be defined precisely in Sec. VI. The measure $[dU]$ means that each independent U_{ij} is integrated over the internal-symmetry group with the Haar measure. This measure for compact groups has the properties

$$\int dg f(g) = \int dg f(gg_0) = \int dg f(g_0 g) = \int dg f(g^{-1}) \tag{2.9}$$

for any function $f(g)$ over the group and for any group element g_0. The measure is normalized

$$\int dg = 1. \tag{2.10}$$

The formal argument for confinement is based on the local gauge symmetry of the action in Eq. (2.3). Given an arbitrary group element g_i for each site i, the action is invariant under

$$\begin{aligned} \psi_i{}^{\alpha} &\to g_i{}^{\alpha\beta}\psi_i{}^{\beta}, \\ U_{ij}{}^{\alpha\beta} &\to (g_i U_{ij} g_j{}^{-1})^{\alpha\beta}. \end{aligned} \tag{2.11}$$

The integration measure in Eq. (2.7) is also invariant under this transformation; consequently, if boundary conditions on the lattice can be ignored, we have

$$\begin{aligned} G(\psi_i, \ldots; U_{mn}, \ldots) \\ = G(g_i\psi_i, \ldots; g_m U_{mn} g_n{}^{-1}, \ldots). \end{aligned} \tag{2.12}$$

This relation is true for arbitrary g_i; thus G can only depend on locally singlet combinations of the ψ_i and U_{ij}. In particular we conclude

$$G(\bar{\psi}_i, \psi_j) = 0 \quad \text{for } i \neq j, \tag{2.13}$$

which means that free quarks cannot propagate.

This argument can break down if boundary conditions imposed on the U_{ij} at the lattice edges affect the U_{ij} deep in the interior. In other words, for infinite lattices a phase transition to an ordered state may occur. This transition has been discussed in Refs. 4 and 5 and is expected to occur for the lattice version of quantum electrodynamics if this type of theory is to describe unconfined electrons and photons.

III. GAUGE FIXING AND THE LATTICE THEORY

A virtue of the Wilson approach is that we need not choose a gauge before quantizing. The integrals over the U_{ij} are finite because the group G is compact; consequently, no infinities arise from integrating over all gauges. On the other hand, gauge invariance of the action still permits working in a fixed gauge without affecting Green's functions of gauge-invariant operators. In this section we define gauge fixing in the lattice theory.

Let $P(\psi, \bar{\psi}, U)$ be some polynomial in the fields which is invariant under the general gauge transformation of Eq. (2.11). Associated with this gauge-invariant polynomial is a Green's function

$$G(P) = \frac{1}{Z}\int [d\psi dU]\, e^{-S} P(\psi, \bar{\psi}, U). \tag{3.1}$$

We begin the gauge-fixing process by concentrating on a single link from site i to site j. A δ function $\delta(g', g)$ on the group $\mathcal{G}$ has the properties

$$\begin{aligned} \int dg\, \delta(g', g) f(g) &= \int dg\, \delta(g, g') f(g) = f(g'), \\ \delta(g, g') &= \delta(g_0 g g_1, g_0 g' g_1) \end{aligned} \tag{3.2}$$

for arbitrary g_0 and g_1. Consider the integral

$$I(P, g_0) = \frac{1}{Z}\int [d\psi dU]\, \delta(U_{ij}, g_0)\, e^{-S} P(\psi, \bar{\psi}, U), \tag{3.3}$$

where g_0 is some group element. We have

$$\int dg_0 I(P, g_0) = G(P). \tag{3.4}$$

If we now perform the gauge transformation of Eq. (2.11) on Eq. (3.3), we obtain

$$I(P, g_0) = I(P, g_i{}^{-1} g_0 g_j). \tag{3.5}$$

Since g_i and g_j are arbitrary, $I(P, g_0)$ must be independent of g_0. Equation (3.4) then implies

$$\begin{aligned} G(P) &= I(P, g_0) \\ &= \frac{1}{Z}\int [d\psi dU]\, \delta(U_{ij}, g_0)\, e^{-S} P(\psi, \bar{\psi}, U). \end{aligned} \tag{3.6}$$

Thus, to calculate a gauge-invariant Green's function we can set any particular U_{ij} to an arbitrary group element and only integrate over the remaining U's.

This process can be repeated to fix more U's. The final result is that we can arbitrarily fix any set of U's as long as this set contains no closed loops; i.e., the fixed U's form a tree (possibly disconnected). A gauge is completely determined by first choosing a maximal tree T, a tree to which no more links can be added without forming a loop. Then the U_{ij} on the tree are set to arbitrary elements g_{ij}. The Green's functions of gauge-invariant operators are then found by integrating over the remaining U's. The general formula is

$$\begin{aligned} G(P) = \frac{1}{Z}\int [d\psi dU]\Big[\prod_{\{i,j\}\ T} \delta(U_{ij}, g_{ij})\Big] \\ \times e^{-S} P(\psi, \bar{\psi}, U). \end{aligned} \tag{3.7}$$

The notation $\{i,j\}$ means the link connecting sites i and j with arbitrary orientation. An example of a maximal tree is shown in Fig. 1.

The most natural element of G to use for fixing U's is the identity element. The simplest trees are those with many straight branches. The gauge we shall use for the rest of this paper corresponds to taking $A_0=0$ and was discussed briefly in Ref. 4. The tree for this case includes all timelike links, upon which the U_{ij} are set equal to the unit matrix. We are still free to make time-independent gauge transformations, which corresponds to the possibility of fixing a set of spacelike U's that link together the timelike branches of the tree. Such a tree is illustrated in Fig. 2.

The utility of the gauge $A_0=0$ is that the trace of four U's in a timelike plaquette becomes a trace of only two U's at subsequent times. We relabel the lattice sites with two indices i and t, where i represents the space coordinates and $a_0 t$ the time. Here we allow the timelike lattice spacing a_0 to differ from the spacelike spacing a. The action for the lattice gauge theory in the gauge $A_0=0$ then becomes

$$\begin{aligned} S = & -a^3 a_0 \sum_{i,j,t} \frac{A_{ij}}{2a} \bar{\psi}_{i,t}(1-\vec{\gamma}\cdot\hat{e}_{ij})U_{ij,t}\psi_{j,t} - a^3 a_0 \sum_{i,t} \frac{1}{2a_0}[\bar{\psi}_{i,t+1}(1+\gamma_0)\psi_{i,t} + \bar{\psi}_{i,t}(1-\gamma_0)\psi_{i,t+1}] \\ & + a^3 a_0 \sum_{i,t}\left(\frac{3}{a}+\frac{1}{a_0}-m\right)\bar{\psi}_{i,t}\psi_{i,t} - \frac{a}{2g^2 a_0}\sum_{i,j,t} A_{ij}\,\mathrm{Tr}(U_{ij,t+1}{}^{-1}U_{ij,t}) \\ & - \frac{a_0}{8g^2 a}\sum_{ijkl,t} P_{ijkl}\,\mathrm{Tr}(U_{ij,t}U_{jk,t}U_{kl,t}U_{li,t}) . \end{aligned} \tag{3.8}$$

Here the definitions of A_{ij} and P_{ijkl} in Eqs. (2.4) and (2.6) are restricted to the spacelike lattice. Note that the term coupling U's at different times resembles the Hamiltonian for the statistical mechanics of a set of one-dimensional classical spin chains with nearest-neighbor interactions. The final term in Eq. (3.8) represents a 4-spin coupling between the chains. Note that in two-dimensional space-time there is no interchain coupling and the pure-gauge part of the theory is a trivial one-dimensional statistical-mechanics problem.

IV. TRANSFER MATRIX AND QUANTUM MECHANICS

In this section we review the use of the transfer matrix to relate a Feynman path integral to the conventional operator formulation of quantum mechanics. The discussion is similar to that given in Feynman's original paper.[10] A treatment of a conventional scalar field theory is in Ref. 11. We illustrate the method on the one-degree-of-freedom harmonic oscillator with Lagrangian

$$L=\tfrac{1}{2}\dot{x}^2-\tfrac{1}{2}\omega^2 x^2 . \tag{4.1}$$

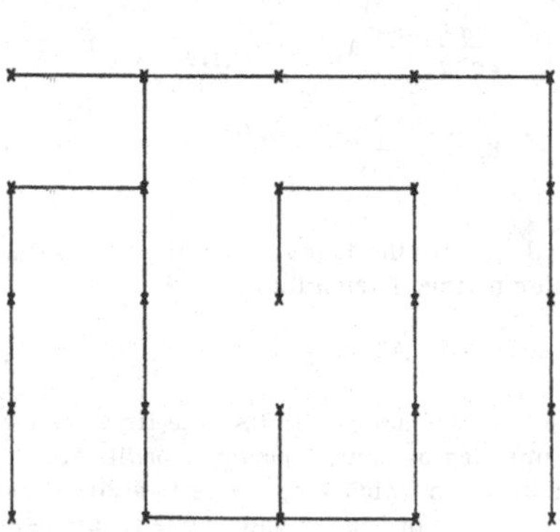

FIG. 1. An example of a maximal tree on a two-dimensional lattice. The U's corresponding to all links on the tree can be set to arbitrary group elements by the gauge-fixing process.

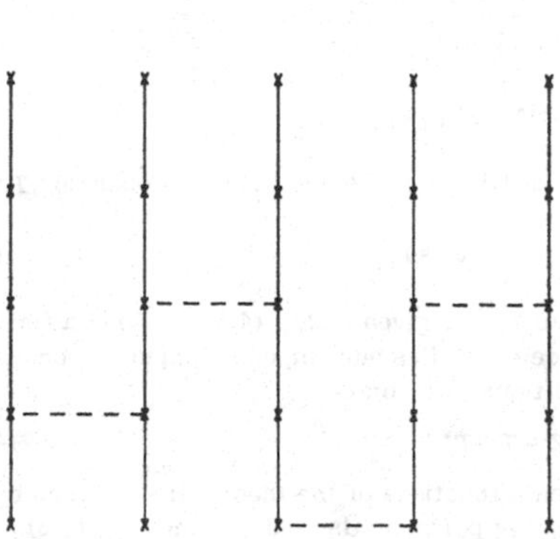

FIG. 2. A tree corresponding to the gauge $A_0=0$. Here the vertical direction represents time. The dashed links can be fixed by time-independent gauge transformations.

Going to an imaginary time lattice of spacing a leads to the action

$$S = a \sum_i \left[\tfrac{1}{2} \left(\frac{x_{i+1} - x_i}{a} \right)^2 + \frac{\omega^2}{2} x_i^2 \right]. \quad (4.2)$$

The path-integral prescription involves integrals such as

$$Z = \int [dx] e^{-S}. \quad (4.3)$$

The key to the transfer-matrix method is that the local nature of (4.2) allows Z to be written

$$Z = \int \prod_i (dx_i T_{x_{i+1} x_i}), \quad (4.4)$$

where

$$T_{x',x} = \exp\left[-\frac{1}{2a}(x' - x)^2 - \frac{a\omega^2}{2} x^2 \right]. \quad (4.5)$$

Consider the Hilbert space of functions of fast decrease with inner product

$$\langle \psi' | \psi \rangle = \int dx\, \psi'^*(x) \psi(x). \quad (4.6)$$

For notational convenience we expand states in this space in terms of the nonnormalizable basis $\{|x\rangle\}$ so that

$$|\psi\rangle = \int dx\, \psi(x) |x\rangle,$$

$$\langle x' | x \rangle = \delta(x' - x), \quad (4.7)$$

$$1 = \int dx\, |x\rangle \langle x|.$$

Introduce the operators $\hat{p}$ and $\hat{x}$ with properties

$$\hat{x} |x\rangle = x |x\rangle,$$

$$[\hat{p}, \hat{x}] = -i, \quad (4.8)$$

$$e^{-i\hat{p}a} |x\rangle |x + a\rangle.$$

In this Hilbert space we define the operator T by

$$\langle x' | T | x \rangle = T_{x'x}, \quad (4.9)$$

where $T_{x'x}$ is given in Eq. (4.5). Taking a finite lattice of N sites and imposing periodic boundary conditions, we obtain

$$Z = \mathrm{Tr}(T^N). \quad (4.10)$$

Green's functions of the theory are obtained by inserting polynomials of the x_i into Eq. (4.3). This corresponds to inserting the operator $\hat{x}$ in the appropriate places in the trace of Eq. (4.10).

Connection with the usual Hamiltonian is made by taking the lattice spacing a to zero. Using Eqs. (4.8), we write T in terms of $\hat{p}$ and $\hat{x}$

$$\begin{aligned} T &= \int d\Delta\, e^{-(1/2a)\Delta^2} e^{-i\hat{p}\Delta} e^{-a\omega^2 \hat{x}^2/2} \\ &= (2\pi a)^{1/2} e^{-a\hat{p}^2/2} e^{-a\omega^2\hat{x}^2/2} \\ &= (2\pi a)^{1/2} \exp[-aH + O(a^2)], \end{aligned} \quad (4.11)$$

where

$$H = \tfrac{1}{2}(\hat{p}^2 + \omega^2 \hat{x}^2) \quad (4.12)$$

is the usual harmonic-oscillator Hamiltonian.

The procedure for going from a path-integral to a Hilbert-space formulation of quantum mechanics consists of two steps. First, construct the transfer matrix T and the space on which it acts. Second, take the logarithm of the transfer matrix and identify the coefficient of the linear term in the lattice spacing as the Hamiltonian. Physically, the transfer matrix propagates the system from one time to the next, and this is the role played by the exponentiated Hamiltonian in the Hilbert-space formulation. In statistical mechanics the thermodynamic properties of a system are determined by the largest eigenvalue of the transfer matrix. In ordinary quantum mechanics the corresponding eigenvector has the lowest eigenvalue of the Hamiltonian, i.e., it is the vacuum or ground state of the system. In the remainder of this paper we will construct the transfer matrix for the lattice gauge theory in the gauge $A_0 = 0$.

V. PURE GAUGE FIELDS

We postpone discussion of the quark fields to the next section and consider here only the pure-gauge part of the action in Eq. (3.8)

$$\begin{aligned} S = &-\frac{a}{2g^2 a_0} \sum_{ij,t} A_{ij} \mathrm{Tr}(U_{ij,t+1}^{-1} U_{ij,t}) \\ &- \frac{a_0}{8g^2 a} \sum_{ijkl,t} P_{ijkl} \mathrm{Tr}(U_{ij,t} U_{jk,t} U_{kl,t} U_{li,t}). \end{aligned} \quad (5.1)$$

By analogy with the last section we wish to find a transfer matrix T such that

$$Z = \int [dU] e^{-S} = \mathrm{Tr}(T^N), \quad (5.2)$$

where N is the number of discrete times and we have imposed periodic boundary conditions.

The space in which T operates is a direct product of spaces of square-integrable functions over the group G. A state $|\psi\rangle$ in this space is specified by a set of square-integrable functions $\psi_{ij}(g)$ over the group, one such function corresponding to each different pair of nearest-neighbor sites $\{i,j\}$ on a spacelike lattice. A pair is not considered as being different if i and j are interchanged. The inner product in this space is

$$\langle\psi'|\psi\rangle = \prod_{\{i,j\}}\left[\int dg\, \psi_{ij}'^{*}(g)\psi_{ij}(g)\right], \tag{5.3}$$

where the product runs over all different nearest-neighbor pairs $\{i,j\}$. We expand the general state in the nonnormalizable basis $\{|U\rangle\}$, where a state in this basis is determined by a group element U_{ij} for each nearest-neighbor pair $\{i,j\}$. The U_{ij} satisfy a condition similar to Eq. (2.2),

$$U_{ij} = U_{ji}^{-1}. \tag{5.4}$$

The overlap in this basis is

$$\langle U'|U\rangle = \prod_{\{i,j\}}\delta(U_{ij}', U_{ij}). \tag{5.5}$$

Completeness is written

$$1 = \int [dU]|U\rangle\langle U|. \tag{5.6}$$

The general state is then expanded

$$|\psi\rangle = \int [dU]|U\rangle \prod_{\{i,j\}}\psi_{ij}(U_{ij}). \tag{5.7}$$

Working in this Hilbert space, we can immediately write down an operator T satisfying Eq. (5.2)

$$\langle U'|T|U\rangle = \exp\left[\frac{a}{2g^2a_0}\sum_{ij}A_{ij}\,\mathrm{Tr}U_{ij}'^{-1}U_{ij}\right]\exp\left[\frac{a_0}{8g^2a}\sum_{i,j,k,l}P_{ijkl}\,\mathrm{Tr}(U_{ij}U_{jk}U_{kl}U_{li})\right]. \tag{5.8}$$

Just as we expressed T for the harmonic oscillator in terms of the operators $\hat{p}$ and $\hat{x}$, we would like to write this T in terms of some simple operators in the present Hilbert space. We begin by defining a set of matrix operators $\hat{U}_{ij}^{\alpha\beta}$ and unitary operators $R_{ij}(g)$,

$$\hat{U}_{ij}^{\alpha\beta}|U\rangle = U_{ij}^{\alpha\beta}|U\rangle, \tag{5.9}$$

$$R_{ij}(g)|U\rangle = |U'\rangle, \tag{5.10}$$

where

$$U_{kl}' = U_{kl},\ \text{for } \{k,l\}\neq\{i,j\}$$
$$U_{ij}' = gU_{ij}. \tag{5.11}$$

The operators $R_{ij}(g)$ satisfy the group-representation property

$$R_{ij}(g)R_{ij}(g') = R_{ij}(gg'). \tag{5.12}$$

In terms of these operators we express T as

$$T = \left(\prod_{\{i,j\}}\left\{\int dg\, R_{ij}(g)\exp\left[\frac{a}{2g^2a_0}\mathrm{Tr}(g+g^{-1})\right]\right\}\right)$$
$$\times\exp\left(\frac{a_0}{8g^2a}\sum_{ijkl}P_{ijkl}\hat{U}_{ij}\hat{U}_{jk}\hat{U}_{kl}\hat{U}_{li}\right). \tag{5.13}$$

This form is not yet simple enough to discuss the limit $a_0\to 0$. To proceed we must review some group theory. We parametrize the elements of our unitary group in the standard form

$$g = g(x) = e^{i\Sigma_m x_m\Lambda_m} \equiv e^{ix\cdot\Lambda}, \tag{5.14}$$

where $\{\Lambda_m\}$ is a set of Hermitian matrices that generate the group. We orthonormalize them such that

$$\mathrm{Tr}(\Lambda_m\Lambda_n) = \delta_{mn}. \tag{5.15}$$

The Λ_α satisfy an algebra

$$[\Lambda_l, \Lambda_m] = if_{lmn}\Lambda_n, \tag{5.16}$$

where $f_{\alpha\beta\gamma}$ is totally antisymmetric in its indices. The group integration measure takes the form

$$d(g(x)) = J(x)\prod_m dx_m, \tag{5.17}$$

where the Jacobian function $J(x)$ is determined by the group multiplication law. It satisfies

$$J(x) = J(-x) \tag{5.18}$$

and in a neighborhood of $x_m = 0$ it is regular and nonvanishing.

Because of the representation property in Eq. (5.12), $R_{ij}(g)$ can be written

$$R_{ij}(g(x)) = e^{i\Sigma_m x_m l_{ij}^m} = e^{ix\cdot l_{ij}}, \tag{5.19}$$

where the l_{ij}^m are Hermitian operators with the following properties:

$$[l_{ij}^l, l_{ij}^m] = if_{lmn}l_{ij}^n, \tag{5.20}$$

$$[l_{ij}^m, \hat{U}_{ij}] = -\Lambda^m\hat{U}_{ij}, \tag{5.21}$$

$$[l_{ij}^m, \hat{U}_{ji}] = \hat{U}_{ji}\Lambda^m, \tag{5.22}$$

$$[l_{ij}^2, l_{ij}^m] = 0 = [l_{ij}^2, R_{ij}(g)], \tag{5.23}$$

where

$$l_{ij}^2 = l_{ji}^2 = \sum_m l_{ij}^m l_{ij}^m \tag{5.24}$$

is the quadratic Casimir operator for the group. In our Hilbert space of square-integrable functions over the group, the l_{ij}^m represent differential operators in the group parameters.

Using Eqs. (5.14), (5.17), and (5.19), we write

$$T=\prod_{\{i,j\}}\left\{\int\left(\prod_m dx^m\right)J(x)\,e^{il_{ij}\cdot x}\exp\left[\frac{a}{2g^2a_0}\operatorname{Tr}(2\cos\Lambda\cdot x)\right]\right\}\exp\left[\frac{a_0}{8g^2a}\sum_{ijkl}P_{ijkl}\operatorname{Tr}(\hat{U}_{ij}\hat{U}_{jk}\hat{U}_{kl}\hat{U}_{li})\right]. \tag{5.25}$$

When $a_0\to 0$, the integral over x is dominated by x near the maximum of $\operatorname{Tr}(2\cos\Lambda\cdot x)$. For a unitary group this maximum always occurs near $x=0$; consequently, we have

$$\operatorname{Tr}(2\cos\Lambda\cdot x)=2n-x^2+O(x^4), \tag{5.26}$$

where n is the dimension of the group matrices. Inserting Eq. (5.26) into Eq. (5.25), we do the Gaussian x integrations with the result

$$T=N\exp[-a_0H+O(a_0^2)], \tag{5.27}$$

where

$$H=\frac{g^2}{2a}\sum_{\{i,j\}}l_{ij}^2-\frac{1}{8g^2a}\sum_{ijkl}P_{ijkl}\operatorname{Tr}(\hat{U}_{ij}\hat{U}_{jk}\hat{U}_{kl}\hat{U}_{li}) \tag{5.28}$$

and N is an irrelevant constant factor. This is the gauge-field part of the Hamiltonian used by Kogut and Susskind.

We close this section with a brief discussion of the remaining gauge freedom of the theory. As we have only specified $A_0=0$, we can still do time-independent gauge transformations. An operator that performs such a transformation is $\prod_i J_i(g_i)$, where g_i are arbitrary group elements and

$$J_i(g(x))=\exp\left(ix\cdot\sum_j A_{ij}\,l_{ij}\right) \tag{5.29}$$

generates a local gauge transformation at the site i. Using the invariance of the group integration measure and the cyclic properties of the trace, one can show

$$[T,J_i(g)]=0=[H,J_i(g)]; \tag{5.30}$$

consequently, $J_i(g)$ generates a symmetry of the theory.

VI. QUARK FIELDS

In this section we apply the transfer-matrix formalism to the quark fields. This discussion does not exactly parallel the treatments in Secs. IV and V for two reasons: (1) The quark Lagrangian is only linear in the time derivatives of the quark field, and (2) the quark fields obey anticommutation relations. In addition, technical differences in Wilson's approach to fermions preclude our obtaining exactly the Kogut-Susskind quark Hamiltonian. Wilson uses a four-component field at each site, whereas in Ref. 8 a one-component field suffices.

In the gauge $A_0=0$ the coupling term between the quarks and gauge fields only involves fields at equal times; consequently, it will be trivial to find the coupled Hamiltonian from the free-fermion Hamiltonian. Thus, for simplicity we extract from Eq. (3.8) the action for free lattice fermions

$$S=-a^3a_0\sum_{i,t}\frac{1}{2a_0}[\bar{\psi}_{i,t+1}(1+\gamma_0)\psi_{i,t}+\bar{\psi}_{it}(1-\gamma_0)\psi_{i,t+1}]-a^3a_0\sum_{i,j,t}\frac{A_{ij}}{2a}\bar{\psi}_{i,t}(1-\vec{\gamma}\cdot\vec{e}_{ij})\psi_{j,t}$$
$$+a^3a_0\sum_{i,t}\left(\frac{3}{a}+\frac{1}{a_0}-m\right)\bar{\psi}_{i,t}\psi_{i,t}. \tag{6.1}$$

The transfer matrix will act on a Hilbert space of fermions on a three-space lattice. The fundamental spinor operators χ_i^α in this space satisfy

$$[\chi_i^\alpha,\chi_j^{\beta\dagger}]_+=a^3\delta_{ij}\delta_{\alpha\beta},\qquad [\chi_i^\alpha,\chi_j^\beta]_+=0. \tag{6.2}$$

To simplify the following formulas, we introduce the four-by-four projection matrices

$$P_\pm=\tfrac{1}{2}(1\pm\gamma_0). \tag{6.3}$$

We define a "bare vacuum" state $|0\rangle$ by

$$P_+\chi_i^\alpha|0\rangle=0,\qquad \bar{\chi}_i^\alpha P_-|0\rangle=0,\qquad \langle 0|0\rangle=1. \tag{6.4}$$

The general state in the Hilbert space is generated by application of polynomials in χ_i^α and $\bar{\chi}_i^\alpha$ to this vacuum.

We now consider a set of anticommuting objects

$\psi_i{}^\alpha$ and $\bar{\psi}_i{}^\alpha$ and form a "Fermi-coherent" state[12]

$$|\psi\rangle = \exp\left[a^3 \sum_i (\bar{\chi}_i P_+ \psi_i + \bar{\psi}_i P_- \chi_i)\right] |0\rangle . \quad (6.5)$$

This state has the properties

$$\langle 0|\psi\rangle = 1, \quad P_+\chi_i{}^\alpha|\psi\rangle = P_+\psi_i{}^\alpha|\psi\rangle , \quad \bar{\chi}_i{}^\alpha P_-|\psi\rangle = -\bar{\psi}_i{}^\alpha P_-|\psi\rangle . \quad (6.6)$$

The overlap between two of these states is

$$\langle\psi'|\psi\rangle = \exp\left[a^3\sum_i (\bar{\psi}_i' P_+\psi_i + \bar{\psi}_i P_- \psi_i')\right], \quad (6.7)$$

where we have assumed that ψ_i and ψ_i' anticommute. Note that if we set $\psi_i = \psi_{i,t}$ and $\psi_i' = \psi_{i,t+1}$, then the argument of this exponential is just the time-changing term in Eq. (6.1). The states $|\psi\rangle$ form an overcomplete set just as the coherent states for boson fields do.[13] In analogy with the boson case, we define a completeness relation

$$1 = \int [d\psi d\psi^\dagger] |\psi\rangle\langle\psi| e^{-a^3\Sigma_i \bar{\psi}_i \psi_i} . \quad (6.8)$$

This equation can be considered as defining the integral over the anticommuting objects ψ and $\psi^\dagger$ at one time. Integrals of various polynomials in ψ and $\psi^\dagger$ can be defined from this equation using Eq. (6.6). Note that the argument of the exponential in Eq. (6.8) is the remaining term in the action that does not contain a factor of a_0.

In this fermion Hilbert space we define the operator

$$T = \exp(-a_0 H) , \quad (6.9)$$

where

$$H = a^3 \sum_{i,j} \left[\frac{1}{2a} A_{ij} \bar{\chi}_i (1 - \vec{\gamma}\cdot\vec{e}_{ij})\chi_j\right] - a^3\sum_i \left(\frac{3}{a} - m\right) : \bar{\chi}_i \chi_i : . \quad (6.10)$$

Here the normal-ordering symbol : : means with respect to the vacuum of Eq. (6.3). Matrix elements of T between the Fermi-coherent states are

$$\langle\psi_{t+1}|T|\psi_t\rangle = \exp\left[+a_0 a^3 \sum_{ij} \frac{1}{2a} A_{ij} (\bar{\psi}_{i,t+1} P_+ + \bar{\psi}_{i,t} P_-)(1 - \vec{\gamma}\cdot\vec{e}_{ij})(P_+\psi_{j,t} + P_-\psi_{j,t+1}) - a_0 a^3 \sum_i \left(\frac{3}{a} - m\right)(\bar{\psi}_{i,t+1}P_+ + \bar{\psi}_{i,t}P_-)(P_+\psi_{i,t} + P_-\psi_{i,t+1})\right]\langle\psi_{t+1}|\psi_t\rangle . \quad (6.11)$$

To proceed we consider small a_0 and assume

$$\psi_{i,t+1} - \psi_{i,t} = O(a_0) . \quad (6.12)$$

This implies

$$P_+\psi_{j,t} + P_-\psi_{j,t+1} = \psi_{j,t} + P_-(\psi_{j,t+1} - \psi_{j,t}) = \psi_{j,t} + O(a_0) . \quad (6.13)$$

In Eq. (6.11) this gives

$$\langle\psi_{t+1}|T|\psi_t\rangle = \exp\left[+a_0 a^3 \sum_{i,j} \frac{1}{2a} A_{ij} \bar{\psi}_i (1 - \vec{\gamma}\cdot\vec{e}_{ij})\psi_j - a_0 a^3 \sum_i \left(\frac{3}{a} - m\right)\bar{\psi}_i\psi_i + O(a_0{}^2)\right]\langle\psi_{t+1}|\psi_t\rangle . \quad (6.14)$$

Up to the $O(a_0{}^2)$ term, the argument of this exponential gives the remaining terms in the action of Eq. (6.1). Combining Eqs. (6.7), (6.8), and (6.14), we obtain

$$Z = \int [d\psi d\psi^\dagger] e^{-S} = \mathrm{Tr}(\{\exp[-a_0 H + O(a_0{}^2)]\}^N) \quad (6.15)$$

with H given by Eq. (6.10).

VII. CONCLUSIONS

Combining the results of Secs. V and VI, we can easily construct the full interacting Hamiltonian

$$H = \frac{g^2}{2a}\sum_{\{i,j\}} l_{ij}{}^2 - \frac{1}{8g^2 a}\sum_{ijkl} P_{ijkl}\,\mathrm{Tr}(\hat{U}_{ij}\hat{U}_{jk}\hat{U}_{kl}\hat{U}_{li}) + a^3\sum_{i,j}\left[\frac{1}{2a} A_{ij}\bar{\chi}_i (1 - \vec{\gamma}\cdot\vec{e}_{ij})\hat{U}_{ij}\chi_j\right] - a^3\sum_i\left(\frac{3}{a} - m\right) : \bar{\chi}_i\chi_i : . \quad (7.1)$$

This differs from the Hamiltonian of Kogut and Susskind only in that we treat the fermions with Wilson's projection-operator technique. We regard this as a technical point, although a careful discussion of the chiral invariance of the theory for $m=0$ may reveal something more.[14] Indeed, the complications in placing fermions on a lattice are intimately related to the Adler-Bell-Jackiw anomaly in current algebra[15] and deserve further study.

In summary, modulo these technical points in the treatment of fermions, we have derived the Kogut-Susskind Hamiltonian for lattice gauge theories from Wilson's Lagrangian formulation.

This required addition of a gauge-fixing term to Wilson's theory. In the gauge $A_0=0$, the gauge-field part of the theory is equivalent to a statistical-mechanical system of coupled one-dimensional spin chains. The nature of the phase transition to the unconfined phase needs further study. In particular, ordinary quantum electrodynamics should be obtained in a continuum limit from the ordered phase.

If the Wilson and the Kogut-Susskind approaches are equivalent, which is preferable? The answer is clearly a matter of taste. In the Wilson form, space-time symmetry is more apparent, as is the relation of the possible phase transition to a statistical-mechanics problem. The particle spectrum is given by the singularity structure of Green's functions. In the Kogut-Susskind approach one deals with a generalization of conventional quantum mechanics with a well-defined Hamiltonian operator. The spectrum of the theory is the spectrum of this Hamiltonian, and the phase transition should be related to a level crossing in the infinite-volume limit.

ACKNOWLEDGMENT

I am grateful to I. Muzinich and F. Paige for discussions on the meaning of group integration.

*Work supported by Energy Research and Development Administration.

[1]H. D. Politzer, Phys. Rep. 14C, 129 (1974).
[2]T. Appelquist, J. Carazzone, H. Kluberg-Stern, and M. Roth, Phys. Rev. Lett. 36, 768 (1976); 36, 1161(E) (1976); Y.-P. Yao, *ibid.* 36, 653 (1976).
[3]K. Wilson, Phys. Rev. D 10, 2445 (1974); lectures at 1975 International School of Physics "Ettore Majorana" (unpublished).
[4]R. Balian, J. M. Drouffe, and C. Itzykson, Phys. Rev. D 10, 3376 (1974); 11, 2098 (1975); 11, 2104 (1975).
[5]A. A. Migdal, Zh. Eksp. Teor. Fiz. 69, 810 (1975) [Sov. Phys.-JETP 42, 413 (1975)]; *ibid.* 69, 1457 (1975) [42, 743 (1975)].
[6]E. A. Ivanov and V. I. Ogievetsky, Dubna Report No. JINR E2-9822, 1976 (unpublished).
[7]J. Kogut and L. Susskind, Phys. Rev. D 11, 395 (1975).
[8]T. Banks, S. Raby, L. Susskind, J. Kogut, D. R. T. Jones, P. N. Scharbach, and D. K. Sinclair, Phys. Rev. D 15, 1111 (1977).
[9]S. Mandelstam, Ann. Phys. (N.Y.) 19, 1 (1962); C. N. Yang, Phys. Rev. Lett. 33, 445 (1974).
[10]R. P. Feynman, Rev. Mod. Phys. 20, 367 (1948).
[11]K. G. Wilson and J. Kogut, Phys. Rep. 12C, 75 (1974).
[12]For a theory of bosons with a Lagrangian linear in time derivatives, one would use conventional coherent states.
[13]R. J. Glauber, Phys. Rev. 131, 2766 (1963).
[14]Another approach to treating fermions on a lattice with emphasis on chiral invariance is given in S. D. Drell, M. Weinstein, and S. Yankielowicz, Phys. Rev. D 14, 1627 (1976).
[15]J. S. Bell and R. Jackiw, Nuovo Cimento 60A, 47 (1969); S. L. Adler, Phys. Rev. 117, 2426 (1969).

Nuclear Physics B210[FS6] (1982) 310–336

GAUGE THEORY ON A RANDOM LATTICE*

N.H. CHRIST, R. FRIEDBERG and T.D. LEE

Columbia University, New York, N.Y. 10027, USA

Received 7 June 1982

A general formulation of gauge theory on a random lattice is developed and the strong coupling limit of the Wilson string tension worked out. The confining force found in this strong coupling limit is identical to that predicted by the relativistic string model. In particular, the force between two color-triplet charges is a constant for large separation and the tube of electric flux joining the charges fluctuates, giving it a net thickness proportional to the logarithm of its length.

1. Introduction

In an earlier paper** [1] we developed a discrete approximation to continuum field theory based on the introduction of a random lattice. A particularly attractive definition of a random lattice was given and some properties of a scalar field theory defined on such a lattice were investigated. In this paper we carry these ideas further, considering a gauge theory defined on our random lattice. The construction follows closely the regular lattice gauge theory defined by Wilson [2] with the triangles which naturally occur in the random lattice taking the place of the elementary square plaquettes of the regular lattice. Just as for the theory defined on a regular lattice, this random lattice gauge theory has a strong coupling limit which can be analyzed analytically and which confines charges in the fundamental representation of the gauge group. However, in contrast to the strong coupling limit on the regular lattice, the confining potential is rotationally symmetric and the tube of electric flux connecting the separated charges acquires a thickness which grows as the logarithm of its length. The properties of this flux tube are identical to those of a relativistic string.

In order to obtain these analytical results, it is necessary to perform an average over an ensemble of random lattices. This averaging process can be viewed in two ways: The first perspective approximates the space-time continuum by a particular random lattice chosen to satisfy some conditions of randomness and isotropy. Quantum mechanical expectation values are then to be computed for this single lattice. For example, a numerical Monte Carlo calculation might be performed with a single random lattice. In order to determine such an expectation value analytically, one might attempt to compute an average of that expectation value over an ensemble

* This research was supported in part by the US Department of Energy.

** Hereafter referred to as I.

of such lattices, which corresponds to a "quenched" average in the usual terminology.

The second point of view treats the positions of the lattice sites themselves as dynamical variables, so that we may add to the usual path integration an additional integration over the positions of random lattice sites. This can be called an "annealed" average. As we shall see, these two perspectives agree in the strong coupling limit, Outside the strong coupling region these two averages may differ, each having its own advantages.

In sect. 2 we define a gauge theory on our random lattice, and in sect. 3 we discuss the issues associated with these alternative averages in the context of computing a Wilson loop expectation value. The strong coupling limit of this Wilson loop expectation value is evaluated in sect. 4 and the string tension deduced. Finally, in sect. 5 we compute the thickness of the confining string implied by this strong coupling calculation and compare it with the calculation of a similar quantity in the relativistic string model.

We note that the approach to the strong coupling limit developed here can be applied in a straightforward way to the calculation of other quantities, such as the glueball mass and that the analogy with the string model persists. This will be discussed in detail in a separate paper.

2. Lattice gauge theory action

As in the conventional treatment of lattice gauge theory, we introduce gauge variables by assigning a group element $U(i,j)$ to a link connecting the points (i.e., sites) i and j in the lattice. We will consider an SU(N) gauge theory so that $U(i,j)$ can be viewed as an arbitrary $N\times N$ unitary matrix with determinant 1. It is convenient to define both $U(i,j)$ and $U(j,i)$ with $U(j,i)=U(i,j)^{-1}$. Next, with each elementary triangle of vertices i, j and k we associate a group element $U_{ijk}=U(i,j)U(j,k)U(k,i)$. Recall that in our construction of the random lattice we identify a "cluster" of $D+1$ vertices with the property that the $D-1$ dimensional sphere, on whose surface the cluster of $D+1$ points lies, has no lattice points in its interior. Each pair of points in such a cluster is joined by a link. An elementary triangle Δ_{ijk} is then formed of any three points i, j and k in the same cluster and the corresponding links.

The lattice action is defined using these group variables U_{ijk}:

$$\mathscr{A}_{\mathrm{L}}=\frac{1}{g^2}\sum_{\Delta_{ijk}}\kappa_{ijk}f(U_{ijk})\,, \tag{2.1}$$

where g is the coupling constant, and the sum extends over all triangles Δ_{ijk}. The coefficients κ_{ijk} obey a condition, specified below, which assures the proper continuum limit*. The action (2.1) will be invariant under the gauge transformation

$$U(i,j)\to u_i U(i,j) u_j^{-1}, \tag{2.2}$$

for an assignment of group elements u_i to the ith lattice site if the function $f(U)$ obeys

$$f(uUu^{-1})=f(U) \tag{2.3}$$

for all elements u and U. Finally we require that when the SU(N) matrix U is near the identity matrix I,

$$U=I+i\lambda^a\varepsilon^a+\mathrm{O}(\varepsilon^2), \tag{2.4}$$

then

$$f(U)\sim\sum_a(\varepsilon^a)^2+\mathrm{O}(\varepsilon^3), \tag{2.5}$$

where ε^a are small, the repeated index a is summed over from 1 to N^2-1, and the $N\times N$ hermitian generators λ^a obey the usual normalization condition

$$\mathrm{tr}\,(\lambda^a\lambda^b)=2\delta^{ab}. \tag{2.6}$$

The simple choice for the function $f(U)$,

$$f(U)=\tfrac{1}{2}\,\mathrm{tr}\,(I-U)+\mathrm{c.c.}=\mathrm{Re}\,[\mathrm{tr}\,(I-U)], \tag{2.7}$$

corresponds to the conventional Wilson action.

Next let us determine the condition that the coefficient κ_{ijk} must satisfy in order that our lattice action approach the usual continuum gauge theory action

$$\mathcal{A}_{\mathrm{C}}=\tfrac{1}{4}\int_\Omega F^a_{\mu\nu}F^a_{\mu\nu}\,\mathrm{d}^Dx \tag{2.8}$$

in the continuum limit. Here the Yang-Mills field strength is given by

$$F^a_{\mu\nu}=\partial_\mu A^a_\nu-\partial_\nu A^a_\mu+gf^{abc}A^b_\mu A^c_\nu, \tag{2.9}$$

where A^a_μ is the gauge potential and f^{abc} the SU(N) structure constants with

$$[\lambda^a,\lambda^b]=2i\lambda^c f^{abc}. \tag{2.10}$$

Begin with a continuum gauge field $A^a_\mu(x)$ defined in the volume Ω. We can introduce a group element $U(i,j)$ for linked sites i and j by

$$U(i,j)=\mathrm{P}\left[\exp\left(-ig\int_{x_j}^{x_i}A^a_\mu\lambda^a\,\mathrm{d}x_\mu\right)\right], \tag{2.11}$$

where the symbol P indicates the usual "path ordering" of the exponential. If $\mathcal{A}_{\mathrm{L}}$

* A particularly attractive choice of κ_{ijk} will be given in the companion paper.

is then evaluated for this assignment of $U(i,j)$, we require that $\mathscr{A}_L$ should approach the continuum action $\mathscr{A}_C$ of the gauge field A^a_μ in the limit that the density ρ of lattice sites becomes large for a fixed volume Ω:

$$\lim_{\rho\to\infty} \mathscr{A}_L = \mathscr{A}_C . \tag{2.12}$$

In the limit of large density, the range of integration in the exponent of (2.11) becomes small and we can approximate $U(i,j)$ by the power series

$$U(i,j) = I - ig\int_{x_j}^{x_i} A^a_\mu \lambda^a \, dx_\mu + \tfrac{1}{2}\mathrm{P}\left[-ig\int_{x_j}^{x_i} A^a_\mu \lambda^a \, dx_\mu\right]^2 + \mathrm{O}(\bar{l}^3) , \tag{2.13}$$

where $\bar{l}$ is the average length of a link. Thus the large density limit in (2.12) is also a "weak field" limit in the sense that our lattice variables $U(i,j)$ approach I. If the approximation (2.13) is used for the group elements defining the product U_{ijk}, then we find

$$U_{ijk} = I - \tfrac{1}{2} ig\lambda^a F^a_{\mu\nu} \Delta_{\mu\nu}(ijk) + \mathrm{O}(\bar{l}^3) . \tag{2.14}$$

Throughout, the subscripts μ and ν denote the euclidean-space index which can be $1, 2, \ldots, D$, the superscripts a, b and c are internal symmetry indices, each varying from 1 to N^2-1 and i, j and k (whether as subscripts or arguments) refer to the lattice sites $1, 2, \ldots, \mathcal{N}$. Here $\Delta_{\mu\nu}(ijk)$ is an antisymmetric tensor in μ and ν, parallel to the plane of the i, j, k triangle with a magnitude related to the triangle's area, A_{ijk}:

$$\Delta_{\mu\nu}(ijk)\Delta_{\mu\nu}(ijk) = 2(A_{ijk})^2 ; \tag{2.15}$$

more specifically,

$$\Delta_{\mu\nu}(ijk) = \tfrac{1}{6}[l_\mu(ij)l_\nu(jk) + l_\mu(jk)l_\nu(ki) + l_\mu(ki)l_\nu(ij) - (\mu \rightleftarrows \nu)] , \tag{2.16}$$

where $l_\mu(ij)$ is the μth component of the relative coordinate vector from j to i. Those terms linear in g in (2.14) follow from an application of Stokes' theorem while the order g^2 terms can be determined by gauge invariance; both are $\mathrm{O}(\bar{l}^2)$. If this form for U_{ijk} is substituted into the lattice action (2.1) and the condition (2.5) used to evaluate $f(U_{ijk})$, we find in the limit of $\rho\to\infty$

$$\mathscr{A}_L \to \tfrac{1}{4} \sum_{\Delta_{ijk}} \kappa_{ijk} \Delta_{\mu\nu}(ijk)\Delta_{\rho\sigma}(ijk) F^a_{\mu\nu} F^a_{\rho\sigma} . \tag{2.17}$$

If we assume that κ_{ijk} is a rotationally invariant function of the triangle i, j, k (i.e., κ_{ijk} depends only on the length of the three links ij, jk and ki), then (2.17) becomes

$$\mathscr{A}_L \to \int d^D x F^a_{\mu\nu} F^a_{\mu\nu} \frac{\rho_2}{D(D-1)} \langle \kappa_{ijk} (A_{ijk})^2 \rangle . \tag{2.18}$$

Here ρ_2 is the number of triangles (2-simplices) per unit volume and $\langle\ \rangle$ represents an average over the distribution of triangles appearing in the random lattice. In obtaining the result (2.18) we have used the following formula for the angular average of the product $\Delta_{\mu\nu}(ijk)\Delta_{\rho\sigma}(ijk)$:

$$\langle \Delta_{\mu\nu}(ijk)\Delta_{\rho\sigma}(ijk)\rangle_{\text{ang}} = \frac{2(A_{ijk})^2}{D(D-1)}[\delta_{\mu\rho}\delta_{\nu\sigma} - \delta_{\mu\sigma}\delta_{\nu\rho}], \tag{2.19}$$

which follows from the antisymmetry of $\Delta_{\mu\nu}(ijk)$ in μ and ν and the normalization condition (2.15). Thus, if the coefficients κ_{ijk} are related to the area A_{ijk} of the triangle by the condition

$$\langle \kappa_{ijk}(A_{ijk})^2\rangle \frac{\rho_2}{D(D-1)} = \frac{1}{4}, \tag{2.20}$$

then (2.1) defines a lattice gauge theory action with the proper weak-field, continuum limit.

If we add to the lattice action (2.1) a linear coupling between the group element $U(i,j)$ and an external $N\times N$ matrix source $J(i,j)$,

$$\sum_{i,j} \operatorname{tr}[U(i,j)J(i,j) + \text{h.c.}], \tag{2.21}$$

where the sum extends over all pairs of sites i and j connected by a link, we can define the generating functional for a quantum gauge theory on the random lattice as the Feynman path integral

$$\mathcal{G}_{\text{L}}(J) = \prod_{i,j} \int \mathrm{d}[U(i,j)] \exp\left\{-\mathcal{A}_{\text{L}} - \sum_{i,j} \operatorname{tr}[U(i,j)J(i,j) + \text{h.c.}]\right\}. \tag{2.22}$$

This equation defines the quantum gauge theory for a particular random lattice. Later we will find it convenient to generalize (2.22) by adding to the functional integral an average over an ensemble of random lattices.

3. Wilson loop and ensemble averages

The question of confinement is closely connected with the expectation value of the Wilson loop operator, which in a continuum theory is given by

$$\langle 0|\mathrm{P}\left[\exp\left(-ig\oint_{\text{C}} A^a_\mu \lambda^a \,\mathrm{d}x_\mu\right)\right]|0\rangle, \tag{3.1}$$

where C denotes a smooth closed loop lying in a two-dimensional plane. For a given random lattice L we can find a second continuous curve C_L made up of links belonging to the lattice and approximating the smooth curve C as follows: Recall (see sect. 5 of I) that in the dual lattice we divide space into cells of convex polyhedra ω_i composed of those points in space nearest to the site i. A lattice link then joins the sites i and i' if the corresponding polyhedra ω_i and $\omega_{i'}$ share a common face. Thus, by following the smooth curve C as it passes through a sequence of bordering polyhedra $\omega_i, \omega_j, \omega_k, \ldots$, we can map out a corresponding sequence of sites $i, j, k, \ldots$. The chain of links connecting these sites forms our curve C_L. We define the random lattice approximation to (3.1) as

$$W_L(C) = \frac{1}{Q_L} \mathcal{W}_L(C), \tag{3.2}$$

where

$$Q_L \equiv \int \prod_{k,l} d[U(k,l)] e^{-\mathcal{A}_L}, \tag{3.3}$$

$$\mathcal{W}_L(C) \equiv \int \prod_{k,l} d[U(k,l)] e^{-\mathcal{A}_L} \operatorname{tr}\left[\prod_{i,j \in C_L} U(i,j)\right], \tag{3.4}$$

in which the first product extends over all linked pairs of sites k and l, and the second product over $i, j \in C_L$ is understood to be path-ordered with the point i following j as one moves around the curve.

In order to give $W_L(C)$ a unique definition and to render it manifestly invariant under both translations and rotations, it is necessary to perform an average such as one of the following:

(i) Consider a particular random lattice L in a large volume Ω which contains $\mathcal{N} = \rho\Omega$ random sites. Draw a Wilson loop C along any smooth closed path. Evaluate $W_L(C)$ according to (3.2). The loop C can of course be placed anywhere in Ω. The average of $W_L(C)$ over all translations of C (called C-average and indicated by a subscript C),

$$\langle W_L(C)\rangle_C = \left\langle \frac{1}{Q_L} \mathcal{W}_L(C)\right\rangle_C, \tag{3.5}$$

in the limit $\Omega \to \infty$ at a constant density ρ then gives the physical result. Since Q_L is independent of C, it can be taken outside the average; hence (3.5) can also be written as

$$\langle W_L(C)\rangle_C = \frac{1}{Q_L} \langle \mathcal{W}_L(C)\rangle_C. \tag{3.6}$$

(ii) Alternatively, we may perform an average over an ensemble of lattices, each having $\mathcal{N}$ points randomly distributed in a volume Ω. If we use (2.22) to define a generating functional $\mathcal{G}_L(J)$ for a particular random lattice L with sites at $r_1, \ldots, r_{\mathcal{N}}$ in Ω, then the corresponding ensemble average over different L is*

$$\langle \mathcal{G}_L(J)\rangle_L = \Omega^{-\mathcal{N}} \int \prod_{i=1}^{\mathcal{N}} d^D r_i \mathcal{G}_L(J), \tag{3.7}$$

* In order that $\mathcal{G}_L(J)$ can be averaged for fixed J, it is necessary that the matrix source $J(i,j)$ be obtained from a continuous function $J(r, r')$ with $J(i,j) = J(r_i, r_j)$.

where the subscript L outside the bracket indicates that the average is over an ensemble of lattices.

The corresponding ensemble average of the Wilson loop functional may be done in two different ways:

(a) the quenched average

$$\langle W_L(C)\rangle_L \equiv \Omega^{-\mathcal{N}} \int \prod_{i=1}^{\mathcal{N}} d^D r_i W_L(C)$$

$$= \Omega^{-\mathcal{N}} \int \prod_{i=1}^{\mathcal{N}} d^D r_i \frac{1}{Q_L} \mathcal{W}_L(C) , \tag{3.8}$$

and (b) the annealed average

$$W(C) \equiv \frac{\langle \mathcal{W}_L(C)\rangle_L}{\langle Q_L\rangle_L} , \tag{3.9}$$

where

$$\langle \mathcal{W}_L(C)\rangle_L = \Omega^{-\mathcal{N}} \int \prod_{i=1}^{\mathcal{N}} d^D r_i \mathcal{W}_L(C) , \tag{3.10}$$

$$\langle Q_L\rangle_L = \Omega^{-\mathcal{N}} \int \prod_{i=1}^{\mathcal{N}} d^D r_i Q_L . \tag{3.11}$$

In the limit of infinite Ω, these averages are all manifestly invariant under translation and rotation.

We shall now show that the C-average in (i) is equivalent to the quenched average in (ii). Let us label the lattice and its volume used in (i) by L and Ω, those used in (ii) by L$'$ and Ω'. We take

$$\Omega = N\Omega' , \tag{3.12}$$

where N is a large number. The much larger random lattice L can be viewed as a composite of N different random lattices L$'(1)$, L$'(2)$, . . . , L$'(N)$, each in a region of volume Ω' which is assumed to be of a sufficient size so that effects due to surfaces between different L$'(n)$ can be neglected. Hence,

$$Q_L = \prod_{n=1}^{N} Q_{L'(n)} . \tag{3.13}$$

Consider now a specific term $\mathcal{W}_L(C)$ in the C-average (3.6). When $\Omega \to \infty$ with N held fixed, we may assume that the Wilson loop C lies within one of the smaller lattices, say L$'(a)$. From (3.3) and (3.4), we see that

$$\mathcal{W}_L(C) = \mathcal{W}_{L'(a)}(C) \prod_{n \neq a} Q_{L'(n)} , \tag{3.14}$$

and therefore

$$W_L(C) = \frac{1}{Q_L} \mathscr{W}_L(C) = \frac{1}{Q_{L'(a)}} \mathscr{W}_{L'(a)}(C) = W_{L'(a)}(C) . \tag{3.15}$$

The average over the position of C is then the same as the average over different L′(a); i.e.,

$$\langle W_L(C)\rangle_C = \langle W_L(C)\rangle_L . \tag{3.16}$$

Consequently, the C-average in (i) is equivalent to the quenched average in (ii).

In the sections that follow, we are mostly concerned with the evaluation of the Wilson loop functional in the strong coupling limit. As we shall see (eq. (4.6) in the next section), in that limit $Q_L \to$ constant, which makes the quenched average (3.8) the *same* as the annealed average (3.9). Away from the strong coupling limit, these two averages are in general different. We view them as alternative possibilities. In the quenched average, the lattice sites r_i appear as purely kinematic variables; each random lattice L is weighted entirely according to the kinematic factor $\prod d^D r_i$. However, in the annealed average, the lattice sites are treated on a par with other dynamic variables. In (3.9)–(3.11), the degrees of freedom describing our random lattice take on an added physical significance; therefore, this particular approach appears to us to be of a deeper nature. It generalizes the usual rigid lattice field theory in a fundamental way, and gives a formulation of the random lattice field theory that is perhaps more far-reaching.

Remark. There exist also other possibilities which we do *not* adopt. For example, we may consider a sequence of larger and larger loops C of a given shape, all passing through the same arbitrarily selected site in a particular random lattice L in a volume Ω. The string tension T can be obtained by taking the logarithm of $W_L(C)$ in the limits first $\Omega \to \infty$ and then the loop-size $\to \infty$. The result can be quite different from that obtained from the logarithm of either (3.8) or (3.9). [This difference remains even if we eventually also carry out an average of T; i.e., an average of ln $W_L(C)$, not the logarithm of an average of $W_L(C)$.]

As an illustration of this difference, let us consider an example of simplified roulette games, each consisting of spinning the wheel N times. The croupier always wins, and his take per game is $M = e^{n\lambda}$ where λ is a constant and n is the number of times that, say, 00 appears. Since the probability is $C^N_n p^n q^{N-n}$ where $p = \frac{1}{38}$ and $q = \frac{37}{38}$, the averages of M and $\ln M$ are $\langle M\rangle = (pe^\lambda + q)^N$ and $\langle \ln M\rangle = Np\lambda$. For a single game of very large N, $\lim_{N\to\infty} (1/N) \ln M$ is $p\lambda = (1/N)\langle \ln M\rangle$ which is *different* from $(1/N) \ln \langle M\rangle = \ln (pe^\lambda + q)$. This difference disappears when $\lambda \to 0$ (weak coupling), but is largest when $\lambda \to \infty$ (strong coupling). If we regard M as the analog of $W_L(C)$ and N as that of the loop area, then $(1/N) \ln \langle M\rangle$ corresponds to the string tension evaluated according to either the quenched average (3.8) or the annealed average (3.9). In this example since the analog of Q_L is

$$Q = \sum_{n=1}^{N} C^N_n p^n q^{N-n} = (p+q)^N = 1 ,$$

these two averages (quenched and annealed) are the same. Notice that when $\lambda \to \infty$, $(1/N)\ln\langle M\rangle \to \lambda + \ln p$ which seeks out the single configuration $n = N$, similar to the strong coupling calculation to be discussed in the next section.

To avoid ambiguity, throughout this paper we shall adhere only to the approaches described in (i) or (ii); i.e., either the quenched average (3.8) or the annealed average (3.9) of $W_L(C)$ is always taken at the beginning, and then we let the volume $\Omega \to \infty$. In the evaluation of the string tension, we take the limit of an infinite loop-size only in the very end.

4. String tension in the strong coupling limit

One of the attractive features of the conventional regular lattice gauge theory is the possibility of an analytic analysis of the strong coupling limit. Of particular interest is the Wilson loop functional, defined by (3.1). If the closed loop C lies in a lattice plane and is the border of an area made up of n elementary plaquettes, then, in the strong coupling limit, the Wilson loop functional behaves as

$$\sim (1/g^2)^n . \tag{4.1}$$

This result follows if we use the Wilson action, based on the function (2.7), and indicates a constant force between color triplet charges with an energy per unit separation or "string tension" proportional to $\ln g^2$. For a regular lattice, the Wilson loop functional varies when we change the orientation of the loop. In this section, we shall carry out a similar strong coupling calculation for the random lattice gauge theory, and shall obtain a rotationally symmetric result.

Let us take an arbitrary random lattice and begin with the expression $W_L(C)$ given by (3.2). For simplicity, we will study the strong coupling limit of $W_L(C)$ using the action

$$\mathscr{A}_L = \frac{1}{g^2} \sum_{\Delta_{ijk}} \kappa_{ijk} \operatorname{Re}[\operatorname{tr}(1 - U_{ijk})] \tag{4.2}$$

in accordance with (2.1) and (2.7). For large coupling we can expand the exponential of the action, keeping only the zeroth and first order terms for each of the fundamental triangles:

$$W_L(C) = \frac{1}{Q_L} \mathscr{W}_L(C) ,$$

where

$$Q_L = \int \prod_{k,l} d[U(k,l)] \prod_{\Delta_{ijk}} \left\{ 1 - \frac{1}{g^2} \kappa_{ijk} \operatorname{Re}[\operatorname{tr}(1 - U_{ijk})] \right\} , \tag{4.3}$$

$$W_L(C) = \int \prod_{k,l} d[U(k,l)] \, \mathrm{tr}\left[\prod_{i,j\in C_L} U(i,j)\right]$$
$$\times \prod_{\Delta_{ijk}} \left\{1 - \frac{1}{g^2}\kappa_{ijk}\, \mathrm{Re}\,[\mathrm{tr}\,(1-U_{ijk})]\right\}. \quad (4.4)$$

Since the differential $d[U]$ and the matrix elements of U and U^{-1} satisfy the identities

$$\int d[U] = N,$$
$$\int d[U] U_{\alpha\beta} = 0, \quad (4.5)$$
$$\int d[U] U_{\alpha\beta}(U^{-1})_{\gamma\rho} = \delta_{\alpha\rho}\delta_{\beta\gamma},$$

the leading contribution to Q_L when $g^2 \to \infty$ is

$$Q_L = N^{\mathcal{N}_1} = \text{constant}, \quad (4.6)$$

where $\mathcal{N}_1$ is the total number of links (1-simplices) in the lattice. Therefore,

$$W_L = N^{-\mathcal{N}_1} \mathcal{W}_L(C) \quad (4.7)$$

in the strong coupling limit. The limiting value can be obtained by taking only the smallest number of $1/g^2$ terms in the third product in (4.4) that is necessary for a non-zero result. We observe that for each $U(i,j)$ appearing in the second product in (4.4), its inverse $U(j,i)$ must be present in the expansion of $\exp(-\mathcal{A}_L)$, i.e., in the last product in (4.4). Similarly for each factor tr U_{ijk} there must appear elsewhere in the integrand the variables $U(j,i)$, $U(k,j)$ and $U(i,k)$.

Thus the leading order term in the strong coupling expansion of $W_L(C)$ corresponds to a continuous surface $\mathcal{S}_C$ formed of triangles Δ_{ijk} and bounded by the links in C_L. In fact, it must correspond to that surface containing the smallest number of triangles for the given curve C. If we assign each triangle an orientation given by the ordering of the Wilson loop which forms the boundary of the surface $\mathcal{S}_C$, then for SU(N) with $N \neq 2$, each triangle Δ_{ijk} corresponds to that term in the expression

$$\mathrm{Re}\,\mathrm{tr}\,(U_{ijk}) = \tfrac{1}{2}[\mathrm{tr}\, U_{ijk} + \mathrm{tr}\, U_{kji}] \quad (4.8)$$

in which the order of the vertices i, j, k is opposite to the orientation of the loop product $\prod_{i,j\in C_L} U(i,j)$ in (4.4). For SU(2) these two terms in (4.8) are equal and we obtain an extra factor of 2. Hence, substituting (4.5) into (4.4) we obtain the strong coupling limit of the Wilson loop expectation value for a particular random lattice L:

$$W_L(C) \cong g^{-2n} N^{1-\mathcal{N}_\ell} \prod_{\Delta_{ijk}\in\mathcal{S}_C} [\tfrac{1}{2}(1+\delta_{N2})\kappa_{ijk}], \quad (4.9)$$

where

$$n = \text{number of triangles contained in } \mathcal{S}_C \tag{4.10}$$

and $\mathcal{N}_\ell$ is the number of links contained in these n triangles. Next, we evaluate the ensemble average of $W_L(C)$. From (4.9) and (3.8), or (3.9), it follows that in the strong coupling limit either the quenched or the annealed average gives (omitting the subscript L outside the bracket in (3.8) or (3.9))

$$\langle W_L(C)\rangle = \Omega^{-\mathcal{N}} \int \prod_{i=1}^{\mathcal{N}} d^D r_i N^{1-\mathcal{N}_\ell} g^{-2n} \prod_{\Delta_{ijk}\in\mathcal{S}_C} [\tfrac{1}{2}(1+\delta_{N2})\kappa_{ijk}]. \tag{4.11}$$

Both n and $\mathcal{S}_C$ are functions of the sites $1, 2, \ldots, \mathcal{N}$ which define the lattice. In this calculation the factor g^{-2n} weights very strongly those lattice configurations which contain a minimal surface $\mathcal{S}_C$ made up of a small number n of triangles.

To evaluate the lattice average (4.9), we first relabel the lattice sites so that

$$1, 2, \ldots, \nu \text{ are the sites defining the surface } \mathcal{S}_C, \tag{4.12}$$

and $\nu+1, \ldots, \mathcal{N}$ the remaining ones. In order that a particular triangle, say Δ_{ijk}, that is present in $\mathcal{S}_C$ actually be realized as a plaquette in the lattice, there must be a cluster of $D+1$ sites including the three sites i, j, k. We recall from sect. 2 of I that $D+1$ sites form a cluster if and only if the $(D-1)$-dimensional sphere passing through them has an empty interior. Thus it is necessary, in order that Δ_{ijk} be a plaquette, that *some* $(D-1)$-dimensional sphere passing through i, j, k have an empty interior. But this condition is also sufficient since the largest such sphere must pass through $D+1$ sites, otherwise it could be made larger. It follows that for any plaquette Δ_{ijk} there is a *smallest* $(D-1)$-dimensional sphere S_{ijk} that passes through i, j, k and has an empty interior. This sphere may pass through any number

$$m_{ijk} = 0, 1, \ldots, D-2 \tag{4.13}$$

of additional sites besides i, j and k. [We can ignore the case when S_{ijk} passes through more than $D+1$ sites (i.e., $m_{ijk} > D-2$), since it is of zero measure.]

For notational clarity, we shall refer to those Δ_{ijk} on the surface $\mathcal{S}_C$ and their associated smallest empty spheres S_{ijk} by Δ_α and S_α where $\alpha = 1, 2, \ldots, n$. Define

$$V \equiv \text{union of volumes contained inside } S_1, S_2, \ldots, S_n. \tag{4.14}$$

By definition, there is no lattice site inside V. Given the positions of sites $1,2,\ldots,\nu$ which determine $\mathcal{S}_C$, the volume V has a minimum value $V_{\min}(1,\ldots,\nu)$.

In the limit $\Omega\to\infty$ and $\mathcal{N}\to\infty$ at a constant site density $\rho = \mathcal{N}/\Omega$, the integration over the positions of the lattice sites in (4.11) can be written as a sum of integrals over ν-sites given by (4.12) for various values of ν:

$$\lim_{\Omega\to\infty} \Omega^{-\mathcal{N}} \int \prod_{i=1}^{\mathcal{N}} d^D r_i = \sum_{\nu=1}^{\mathcal{N}} \int \prod_{j=1}^{\nu} d^D r_j \mathcal{P}(1,\ldots,\nu). \tag{4.15}$$

The weighting factor $\mathscr{P}(1, \ldots, \nu)$ thus defined is due to the integration over the remaining $\nu+1, \nu+2, \ldots, \mathcal{N}$ sites. In terms of $\mathscr{P}(1, \ldots, \nu)$, (4.11) becomes

$$\langle W_L(C)\rangle = \sum_{\nu=1}^{\mathcal{N}} \int \prod_{i=1}^{\nu} d^D r_i N^{1-\mathcal{N}} g^{-2n} \times \prod_{\Delta_{ijk}\in\mathscr{S}_C} [\tfrac{1}{2}(1+\delta_{N2})\kappa_{ijk}]\mathscr{P}(1, \ldots, \nu)\,. \tag{4.16}$$

It is useful to express $\mathscr{P}(1, \ldots, \nu)$ as an integral over the volume V defined by (4.14):

$$\mathscr{P}(1, \ldots, \nu) = \int_{V_{\min}}^{\infty} J(V)\,dV\, e^{-\rho V}\,, \tag{4.17}$$

where $e^{-\rho V}$ is the probability that the inside of V is empty of lattice sites and $J(V)$ is the jacobian.

To evaluate $\langle W_L(C)\rangle$ in the limit of large C and large g^2, we shall use the saddle-point approximation. The leading large g^2 behavior of $\langle W_L(C)\rangle$ is determined by the factors g^{-2n} in (4.16) and $e^{-\rho V}$ in (4.17). As g^2 becomes large, the factor g^{-2n} enhances those configurations with a small number n of triangles. However, as n decreases, the volume V, defined by (4.14), necessarily grows, making $e^{-\rho V}$ small. Hence at given site-positions $1, 2, \ldots, \nu$ (therefore, also at fixed n), the largest contribution to $\langle W_L(C)\rangle$ comes from regions near the lower limit $V = V_{\min}$ in the integral (4.17), i.e. the maximum of $e^{-\rho V}$; in particular, as we shall see, the deviation of $\mathscr{P}(1, \ldots, \nu)$ from its saddle-point value $e^{-\rho V_{\min}}$ can be estimated to be

$$\mathscr{P}(1, \ldots, \nu) \sim \frac{\rho^{\nu}}{\nu!} \prod_{\alpha=1}^{n} (\rho^{2/D} A_\alpha)^{D(D-2)/4}\, e^{-\rho V_{\min}}\,, \tag{4.18}$$

where A_α is the area of the plaquette Δ_α contained in $\mathscr{S}_C$. In the strong coupling limit it will turn out that* $V_{\min}$ is $\sim A(\ln g^2)^{(D-2)/D}$, A_α is $\sim(\ln g^2)^{2/D}$, and ν and n are $\sim A(\ln g^2)^{-2/D}$, where

$$A = \sum_{\alpha=1}^{n} A_\alpha \tag{4.19}$$

is the area of the Wilson loop. Thus, if we neglect $\ln \ln g^2$ as compared to $\ln g^2$, then the approximation

$$\mathscr{P}(1, \ldots, \nu) \cong e^{-\rho V_{\min}} \tag{4.20}$$

suffices for the strong coupling limit calculation. In what follows we shall first discuss how the estimate (4.18) is obtained, and then evaluate the string tension by using the approximation (4.20).

* See (4.47)–(4.49) below.

To understand the estimate (4.18) it is useful to discuss further the previously introduced notion of the smallest empty sphere S_{ijk} that is associated with each plaquette Δ_{ijk}. Consider the simple case when the space-dimension D is 3. From (4.13), we see that m_{ijk} can only be 0 or 1. The probability $P(i, j, k)$ that sites i, j and k are linked to form a plaquette may be written as a sum according to $m_{ijk} = 0$ or 1:

$$P(i, j, k) = P_0(i, j, k) + P_1(i, j, k) . \tag{4.21}$$

By $P_0(i, j, k)$ we mean the probability that the smallest empty sphere S_{ijk} passing through i, j and k passes through *no* other point (i.e., $m_{ijk} = 0$). But this would mean that i, j and k must lie on a great circle of S_{ijk}. [Otherwise, we could make the sphere smaller by moving its center towards the plaquette Δ_{ijk}, keeping of course i, j and k on its surface. Because we are considering the case $m_{ijk} = 0$, the surface S_{ijk} cannot encounter any other site during this move; thus eventually i, j and k must lie on a great circle of S_{ijk}.] In the plane that contains Δ_{ijk}, let r be the radius of the circumscribed circle of Δ_{ijk}. $P_0(i, j, k)$ is then simply the probability that a sphere of radius $R = r$ has empty interior, or

$$P_0(i, j, k) = e^{-4\pi r^3 \rho/3} , \tag{4.22}$$

where, as before, ρ is the site density. Likewise, by $P_1(i, j, k)$ we mean the probability that the smallest empty sphere passing through i, j and k also passes through one other site, say p. [Hence, it corresponds to $m_{ijk} = 1$.] The plane that contains Δ_{ijk} cuts the sphere S_{ijk} into two caps; p must lie on the smaller cap. Let x be the perpendicular distance from the center of the sphere to the plaquette Δ_{ijk}. As x increases by dx, the smaller cap sweeps out an infinitesimal volume $d\tau(x)$, which is the volume available to the site p. The whole sphere, whose interior we require to be empty, has a radius $R = (r^2 + x^2)^{1/2}$. Therefore

$$P_1(i, j, k) = 2 \int_0^\infty e^{-4\pi(r^2+x^2)^{3/2}\rho/3} \rho \, d\tau(x) , \tag{4.23}$$

where the factor 2 allows for displacement in either direction.

In (4.23), the exponent for small x is

$$-\tfrac{1}{3}4\pi(r^2+x^2)^{3/2}\rho = -\tfrac{1}{3}4\pi r^3 \rho - 2\pi\rho r x^2 + O(x^4) ,$$

which implies that only those x for which $\rho r x^2 = O(1)$ are significant. In the strong coupling limit, as we shall see, $\rho r^3 \sim \ln g^2 \gg 1$; therefore we need only consider $x \ll r$. In this approximation one has $d\tau(x) \cong \pi r^2 \, dx$, and so (4.23) becomes

$$\begin{aligned} P_1(i, j, k) &\cong 2\, e^{-4\pi r^3 \rho/3} \int_0^\infty e^{-2\pi \rho r x^2} \pi r^2 \rho \, dx \\ &= \pi (\tfrac{1}{2}\rho r^3)^{1/2} e^{-4\pi r^3 \rho/3} . \end{aligned} \tag{4.24}$$

In terms of the area A_{ijk} of the triangle Δ_{ijk}, we see that (4.24) can be written as

$$P_1(i,j,k) = \text{constant} \times (\rho A_{ijk}^{3/2})^{1/2}\, e^{-4\pi r^3 \rho/3}\,, \tag{4.25}$$

where the constant depends on the shape of Δ_{ijk}, but is O(1), while A_{ijk} is $O(r^2) = O(\ln^{2/3} g^2)$. Note that from (4.22) and (4.24)

$$P_1(i,j,k) \gg P_0(i,j,k) \tag{4.26}$$

in the strong coupling limit.

The above discussion can be generalized to any dimension D. Just as in (4.21), the probability $P(i,j,k)$ that sites i, j and k are linked to form a plaquette can be written as

$$P(i,j,k) = \sum_{m=0}^{D-2} P_m(i,j,k)\,, \tag{4.27}$$

where $P_m(i,j,k)$ corresponds to the case $m = m_{ijk}$ of (4.13). The Poisson formula gives

$$P_0(i,j,k) = \exp(-\rho v_{ijk}) \tag{4.28}$$

where v_{ijk} is the volume of a sphere of radius r in D-dimensions. Similarly to (4.24), we have (shown in the appendix)

$$P_m(i,j,k) = \text{constant} \times (\rho A_{ijk}^{D/2})^{m/2} \exp(-\rho v_{ijk})\,. \tag{4.29}$$

When $D = 3$ and $m = 1$, (4.29) reduces to (4.25).

Next, we examine the more complicated probability that the ν sites $1, 2, \ldots, \nu$ of (4.12) should be linked to form the n plaquettes $\Delta_1, \Delta_2, \ldots, \Delta_n$ which define the surface $\mathcal{S}_C$. To each Δ_α ($\alpha = 1, 2, \ldots, n$) there is a smallest $(D-1)$-dimensional sphere S_α with empty interior. The surface S_α passes through i, j, k and m_α additional sites, where $m_\alpha = 0, 1, \ldots, D-2$. In accord with (4.12), we may label these additional sites as

$$\nu+1, \ldots, \nu+m_1;\quad \nu+m_1+1, \ldots, \nu+m_1+m_2;\quad \ldots, \nu+\sum_{\alpha=1}^{n} m_\alpha\,, \tag{4.30}$$

so that $\nu+1, \ldots, \nu+m_1$ refer to sites on S_1, $\nu+m_1+1, \ldots, \nu+m_1+m_2$ to sites on S_2, etc. (The possibility that two or more of these $\nu+\sum_{\alpha=1}^{n} m_\alpha$ sites are identical is discussed in the appendix.)

As we shall see, this probability is closely related to the evaluation of $\mathcal{P}(1, \ldots, \nu)$ defined earlier. From the definition (4.15), $\mathcal{P}(1, \ldots, \nu)$ is determined by the integration over the positions of all the sites $\nu+1, \nu+2, \ldots, \mathcal{N}$ not on $\mathcal{S}_C$. We have

$$\mathcal{P}(1,2,\ldots,\nu)=\Omega^{-\mathcal{N}}\sum_{m_\alpha}\frac{\mathcal{N}!}{(\mathcal{N}-\nu-M)!\nu!\prod_\alpha m_\alpha!}$$

$$\times\int\prod_{i=1}^{M}\mathrm{d}^D r_{\nu+i}\int\prod_{j=\nu+M+1}^{\mathcal{N}}\mathrm{d}^D r_j\,, \tag{4.31}$$

where

$$M=\sum_{\alpha=1}^{n} m_\alpha\,, \tag{4.32}$$

and as before each m_α can vary from 0 to $D-2$. The sites $1, 2, \ldots, \nu$ are on $\mathcal{S}_C$, $\nu+1, \ldots, \nu+M$ are on the surface of V given by (4.14), but $\nu+M+1, \ldots, \mathcal{N}$ are outside V. Therefore, the last integration in (4.31) can be readily carried out:

$$\int\prod_{j=\nu+M+1}^{\mathcal{N}}\mathrm{d}^D r_j=(\Omega-V)^{\mathcal{N}-\nu-M}. \tag{4.33}$$

Substituting this expression into (4.31), we find in the limit $\Omega\to\infty$ and $\mathcal{N}\to\infty$ at a constant $\rho=\mathcal{N}/\infty$:

$$\mathcal{P}(1,\ldots,\nu)=\frac{\rho^\nu}{\nu!}\sum_{m_\alpha}\frac{\rho^M}{\prod_\alpha m_\alpha!}\int \mathrm{e}^{-\rho V}\prod_{i=1}^{M}\mathrm{d}^D r_{\nu+i}\,. \tag{4.34}$$

The integration over $\prod_{i=1}^{M}\mathrm{d}^D r_{\nu+i}$ can be carried out by following arguments similar to those used in the derivation of (4.24) and (4.29). The details are given in the appendix; for each distribution $\{m_\alpha\}$, the result is, similar to (4.29),

$$\frac{\rho^M}{\prod_\alpha m_\alpha!}\int \mathrm{e}^{-\rho V}\prod_{i=1}^{M}\mathrm{d}^D r_{\nu+i}\cong \mathrm{e}^{-\rho V_{\min}}\prod_{\alpha=1}^{n}\left[\sum_{m=0}^{D-2}C_\alpha(m)\rho^{m/2}A_\alpha^{mD/4}\right], \tag{4.35}$$

where $V_{\min}$ is the minimum of V and $C_\alpha(m)$, although dependent on $1, 2, \ldots, \nu$, remains O(1) when $g\to\infty$. On the other hand, as we shall see, the area A_α of the triangle Δ_α is

$$A_\alpha\sim(\ln g^2)^{2/D}\,. \tag{4.36}$$

Substituting (4.35) into (4.34) we have

$$\mathcal{P}(1,\ldots,\nu)\cong\frac{\rho^\nu}{\nu!}\mathrm{e}^{-\rho V_{\min}}\prod_{\alpha=1}^{n}\left[\sum_{m=0}^{D-2}C_\alpha(m)\rho^{m/2}A_\alpha^{mD/4}\right]. \tag{4.37}$$

Because of (4.36), the dominant term inside the square bracket is when $m=D-2$; (4.37) then leads to (4.18). For the leading strong coupling calculation, approximation (4.20) is sufficient; i.e., we can set

$$\mathcal{P}(1,\ldots,\nu)\cong e^{-\rho V_{min}}\,.$$

To derive V_{min}, we need only consider the case when the smallest empty sphere S_{ijk} of Δ_{ijk} is the one that has sites i, j and k on its great circle. Since the surface $\mathcal{S}_C$ is made up of n triangles Δ_{ijk}, the volume V_{min} is the union of the corresponding n spheres S_{ijk} determined in this way. If we consider a term in the expression (4.16) for $\langle W_L(C)\rangle$ with a fixed number of triangles n, the largest contribution to it will come from that arrangement of the n triangles with the maximum value of $\mathcal{P}$ or, referring to (4.20), with the smallest minimum empty volume V_{min}. Clearly V_{min} will be smallest if the n triangles and their ν vertices $1,\ldots,\nu$ all lie in the two-dimensional plane containing the original curve C. The dependence of V_{min} on the size and shape of a particular triangle, say Δ_{ijk}, is complicated by the overlap between the $D-1$ dimensional sphere S_{ijk} determined by that triangle and the spheres determined by neighboring triangles. However, we can uniquely associate with a triangle Δ_{ijk} a D-dimensional volume V_{ijk} consisting of that portion of the interior of the sphere S_{ijk} which is swept out by the triangle if it is displaced upward and downward in directions perpendicular to the surface $\mathcal{S}_C$ (see fig. 1).

For a given triangle this corresponding swept-out volume V_{ijk} is straightforward to compute. If the triangle has sides with lengths l_1, l_2 and l_3, the radius of the circumscribing circle is given by

$$r=\frac{l_1l_2l_3}{[(l_1^2+l_2^2+l_3^2)^2-2(l_1^4+l_2^4+l_3^4)]^{1/2}}\,. \tag{4.38}$$

The volume V_{ijk} for such a triangle is then the volume of the $D-1$ dimensional

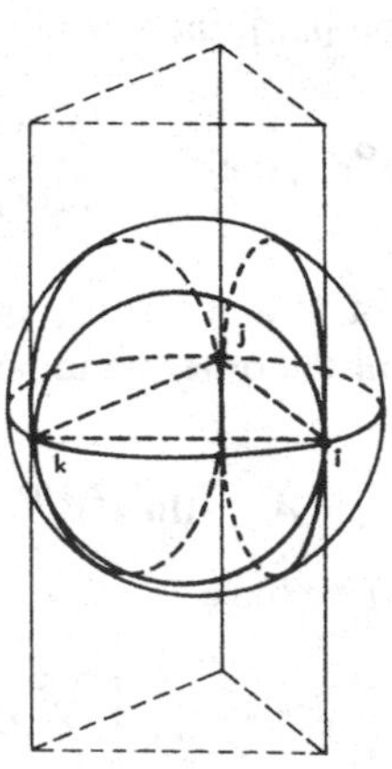

Fig. 1. The sphere S_{ijk} of minimum volume that must be empty if the triangle Δ_{ijk} is to occur in our lattice. That portion of the sphere lying inside the vertical, right triangular prism contains the volume V_{ijk}, uniquely associated with that triangle.

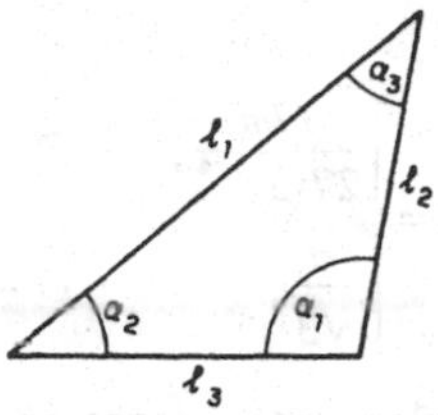

Fig. 2. One of the triangles making up the surface bounded by the Wilson loop C. Here we show the sides l_i and the opposite angles α_i ($i = 1, 2, 3$).

sphere with radius r minus the three spherical caps visible in fig. 1. The cap corresponding to the side l_i has a base which is a $D-2$ dimensional sphere of diameter l_i. The volume for such a cap is

$$\tfrac{1}{3}\pi r^3(2-3\cos\alpha_i+\cos^3\alpha_i)\,, \qquad \text{for } D=3\,, \tag{4.39}$$

$$\tfrac{1}{6}\pi r^4(3\alpha_i-5\sin\alpha_i\cos\alpha_i+2\sin\alpha_i\cos^3\alpha_i)\,, \qquad \text{for } D=4\,. \tag{4.40}$$

Here α_i is the interior angle opposite the side l_i (see fig. 2). Thus for such a triangle Δ_{ijk} in three or four dimensions, the corresponding volume V_{ijk} is

$$V_{ijk}=\tfrac{1}{3}\pi r^3\left[-2+\sum_{i=1}^{3}(3\cos\alpha_i-\cos^3\alpha_i)\right], \qquad \text{for } D=3\,, \tag{4.41}$$

$$V_{ijk}=\tfrac{1}{6}\pi r^4\left[3\pi-\sum_{i=1}^{3}(3\alpha_i-5\sin\alpha_i\cos\alpha_i+2\sin\alpha_i\cos^3\alpha_i)\right], \qquad \text{for } D=4\,. \tag{4.42}$$

The sum of V_{ijk} over all triangles Δ_{ijk} in $\mathcal{S}_C$ gives the corresponding minimum empty volume V_{min} in (4.20).

Next we will show that the arrangement of n triangles which covers a given large area A and corresponds to the smallest V_{min} is a regular array of equilateral triangles. Consider a triangle Δ_{ijk} of a particular shape specified by the angles α_1, α_2, α_3. We can relate the triangle's area A_{ijk} and the corresponding volume V_{ijk} by

$$V_{ijk}=\gamma_{ijk}[A_{ijk}]^{D/2}\,. \tag{4.43}$$

Thus $\gamma_{ijk}=\gamma_{ijk}(\alpha_1,\alpha_2,\alpha_3)$ serves as a scale factor; it is determined by the ratio of the volume V_{ijk}, given by (4.41) or (4.42), divided by the $\tfrac{1}{2}D$ power of the triangle's area

$$A_{ijk}=r^2\sum_{i=1}^{3}\sin\alpha_i\cos\alpha_i\,.$$

We observe that the function γ_{ijk} is minimum for an equilateral triangle, $\alpha_1=\alpha_2=$

$\alpha_3 = \frac{1}{3}\pi$, and this minimum is

$$\gamma_{\min} = \begin{cases} \dfrac{17\pi}{27(3)^{1/4}}, & D=3, \\ \dfrac{\pi}{\sqrt{3}}, & D=4. \end{cases} \tag{4.44}$$

Hence if we consider a particular arrangement of n triangles covering our large area A, each with individual area A_α and scale factor γ_α, where α (as before, replacing the subscripts ijk) varies from 1 to n, then the minimum empty volume that appears in (4.20) for this arrangement is

$$V_{\min} = \sum_{\alpha=1}^{n} \gamma_\alpha (A_\alpha)^{D/2} \geqslant \gamma_{\min} \sum_{\alpha} (A_\alpha)^{D/2} \geqslant \gamma_{\min} n (A/n)^{D/2}. \tag{4.45}$$

Here the last inequality states that if an area A is separated into a sum of n pieces $A = \sum_{\alpha=1}^{n} A_\alpha$, it is the separation into equal areas $A_\alpha = A/n$ which makes the sum $\sum_{\alpha=1}^{n} (A_\alpha)^{D/2}$ minimum for $D>2$. The inequality (4.45) implies that the smallest value of $V_{\min}$ is obtained for an arrangement of equilateral triangles of equal area.

With this result and the coefficient $\gamma_{\min}$ given above, we can write down the leading contributions to $\langle W_L(C)\rangle$ coming from those lattices in our average which contain a planar surface $\mathcal{S}_C$ bounded by the curve C, and consisting of n equilateral triangles of equal area $A_\alpha = A/n$:

$$\langle W_L(C)\rangle \sim \sum_n \exp\left[-n \ln g^2 - n\rho\gamma_{\min}(A/n)^{D/2}\right]. \tag{4.46}$$

The largest term in the sum corresponds to that value of n for which the exponent is minimum:

$$n = A\left[\frac{(D-2)\rho\gamma_{\min}}{2\ln g^2}\right]^{2/D}, \tag{4.47}$$

which gives for each triangle an area

$$\bar{A}_\Delta = \left[\frac{2\ln g^2}{(D-2)\rho\gamma_{\min}}\right]^{2/D} \tag{4.48}$$

and a side-length

$$\bar{l} = 2(\tfrac{1}{3})^{1/4}\left[\frac{2\ln g^2}{(D-2)\rho\gamma_{\min}}\right]^{1/D}. \tag{4.49}$$

Hence,

$$\langle W_L(C)\rangle \sim \exp\left(-AD[\tfrac{1}{2}\rho\gamma_{\mathrm{mim}}]^{2/D}[\ln g^2/(D-2)]^{(D-2)/D}\right). \tag{4.50}$$

Thus the Wilson-loop expectation value $\langle W_L(C)\rangle$ has a computable area-law behavior in the strong-coupling limit with the string tension $T \equiv -(1/A)\ln\langle W_L(C)\rangle$ given by

$$T = \begin{cases} \dfrac{1}{\sqrt{3}}\left(\dfrac{17\pi}{6}\right)^{2/3} (\rho^2 \ln g^2)^{1/3} = 2.4797(\rho^2 \ln g^2)^{1/3}\,, & D=3\,, \\ 2\left(\dfrac{\pi}{\sqrt{3}}\right)^{1/2} (\rho \ln g^2)^{1/2} = 2.6936(\rho \ln g^2)^{1/2}\,, & D=4\,. \end{cases} \tag{4.51}$$

5. String thickness

In the preceding section we analyzed the strong coupling limit of the Wilson-loop expectation value. That limit was determined in a saddle-point approximation with the saddle point corresponding to a regular array of equilateral triangles lying in the plane of the Wilson loop. The area of each of these triangles was proportional to $\ln^{2/D} g^2$ as implied by (4.48). Of course, in this saddle-point approximation the major contribution to the integral comes from a continuum of such triangular arrays corresponding to fluctuations about the saddle point. Because of these fluctuations, the typical surface of triangles does not lie precisely in the plane of the Wilson loop but fluctuates about that plane with an amplitude which grows logarithmically with the loop area. As we shall see, the average square of the distance between the fluctuating surface of triangles and the plane of the Wilson loop is given by

$$d^2 = \frac{1}{2\pi T}\ln A\,, \tag{5.1}$$

where T is the strong coupling string tension found in the previous section and the distance d is measured near the center of the Wilson loop. We will refer to d as the "string thickness". The relation (5.1) between string thickness and string tension is identical to that for a relativistic string [3].

This rather simple result is to be contrasted with the behavior of the string thickness obtained in the strong coupling limit for a gauge theory on a regular, hypercubic lattice [2]. For such a regular lattice we should distinguish two cases: (i) If the Wilson loop lies in a lattice plane, then the strong coupling limit comes from a single set of plaquettes, those lying in the plane of the Wilson loop and enclosed by it. Nearby configurations involving plaquettes outside this plane require an extra factor $(1/g^2)^4$ and thus have an action separated by a non-zero gap from the action of the unique leading order term. The strong-coupling string thickness for such a loop is zero. As g^2 is decreased from a "strong-coupling" value, it has been argued [4, 5] that the string thickness changes discontinuously, the "roughening transition". Below the roughening transition the string thickness is expected to increase with area as in (5.1). (ii) If the Wilson loop does not lie in a lattice plane, then there will be a number of different "staircase"-like surfaces of plaquettes that make equal contributions to the Wilson-loop expectation value. The thickness corresponding to such a configuration grows as the square root of the area of the Wilson loop in the limit of large loop area provided that the shape and orientation of the loop are held fixed. Thus in the strong-coupling limit for a regular lattice, both the string tension and thickness depend strongly on the orientation of the Wilson loop from which they were derived*.

In order to discuss the string thickness for our random lattice, we begin with a large, fixed Wilson loop C, chosen to lie in the x, y plane. Let the vector $\boldsymbol{r}_{l,m}$ locate the vertices of that regular array of triangles which is the saddle point of the expectation-value integration. The two integers l, m are the coordinates $\boldsymbol{r}_{l,m}$ with respect to axes meeting at a 60° angle:

$$\boldsymbol{r}_{l,m} = l\hat{\alpha} + m\hat{\beta}\,. \tag{5.2}$$

Here $\hat{\alpha}$ and $\hat{\beta}$ are unit vectors which lie in the x, y plane wth $\hat{\alpha}\cdot\hat{\beta} = \frac{1}{2}$. We will choose the vertex $\boldsymbol{r}_{0,0}$ to lie near the center of the Wilson loop. Next consider a second random lattice in our ensemble which lies near the saddle point referred to above. It also contains a surface of triangles whose vertices $\boldsymbol{r}'_{l,m}$ lie near the positions $\boldsymbol{r}_{l,m}$:

$$\boldsymbol{r}'_{l,m} = \boldsymbol{r}_{l,m} + \delta\boldsymbol{r}_{l,m}\,. \tag{5.3}$$

As will be shown below, the major contribution to the expectation-value integration comes from the region $|\delta\boldsymbol{r}_{l,m}|^2 \sim 1/\ln^{2/D} g^2$ so the fluctuations $\delta\boldsymbol{r}_{l,m}$ about the saddle point can be treated as small.

To leading order the $\delta\boldsymbol{r}_{l,m}$ dependence in the integrand of (4.15) appears only in the excluded volume V. Again to leading order, we can replace V by $V_{\min}(\delta\boldsymbol{r}_{l,m})$, the smallest volume which must be free of all lattice sites if the vertices given by (5.3) are to be linked into the specified triangular array. Let us decompose the fluctuations $\delta\boldsymbol{r}_{l,m}$ into two components, $\boldsymbol{\xi}_{l,m}$ parallel to the x, y plane and $\boldsymbol{\zeta}_{l,m}$ perpendicular to that plane:

$$\delta\boldsymbol{r}_{l,m} = \boldsymbol{\xi}_{l,m} + \boldsymbol{\zeta}_{l,m}\,. \tag{5.4}$$

Since $V_{\min}$ is symmetric under reflection in the x, y plane, $\boldsymbol{\zeta}_{l,m} \to -\boldsymbol{\zeta}_{l,m}$, the variables $\boldsymbol{\xi}_{l,m}$ and $\boldsymbol{\zeta}_{l,m}$ .decouple in $V_{\min}$ through the quadratic approximation. Thus we can compute the string thickness, which involves only the fluctuations $\boldsymbol{\zeta}_{l,m}$, by considering only configurations with $\boldsymbol{\xi}_{l,m} = 0$. The string thickness d is then given by

$$d^2 = \frac{\int \prod_{l,m} \mathrm{d}^2\zeta_{l,m} [\boldsymbol{\zeta}_{0,0}]^2\, \mathrm{e}^{-\rho V_{\min}(\boldsymbol{\zeta}_{l,m})}}{\int \prod_{l,m} \mathrm{d}^2\zeta_{l,m}\, \mathrm{e}^{-V_{\min}(\boldsymbol{\zeta}_{l,m})}}\,, \tag{5.5}$$

where in $V_{\min}$ we can neglect all powers of $\boldsymbol{\zeta}_{l,m}$ higher than quadratic.

* Kogut, Sinclair, Pearson, Richardson and Shigemitsu [6] have shown that higher-order terms in the strong-coupling expansion tend to restore the rotational symmetry at least for the string tension.

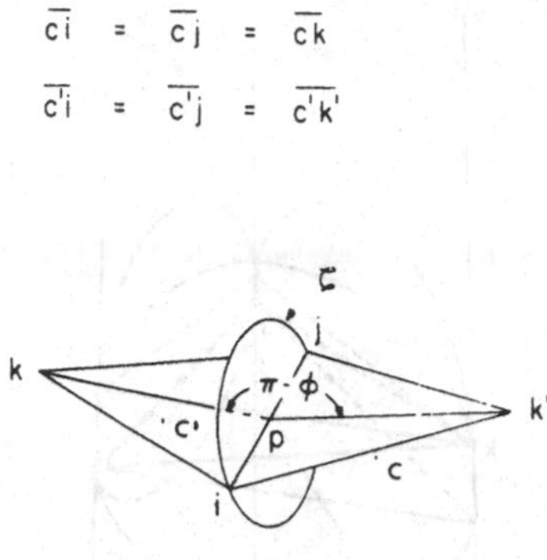

Fig. 3. Circle $\mathscr{C}$ is the intersection of the circumspheres S and S′, whose centers are c and c′. Link $\overline{ij}$ is perpendicular to $\overline{kp}$ and $\overline{k'p}$. When $\xi_{l,m} = 0$, p is the midpoint of $\overline{ij}$.

We will now compute $V_{\min}(\zeta_{l,m})$. Consider two adjacent triangles, say Δ_{ijk} and $\Delta_{ijk'}$. Let S and S′ be the corresponding circumspheres which contribute to $V_{\min}(\zeta_{l,m})$ and which must be empty if the triangles Δ_{ijk} and $\Delta_{ijk'}$ are to occur in the lattice. The vertices of Δ_{ijk} lie on the surface of S and those of $\Delta_{ijk'}$ on S′. If all the $\zeta_{l,m}$ are zero, the triangle Δ_{ijk} lies on a great circle of S, and $\Delta_{ijk'}$ on that of S′; however for non-zero $\zeta_{l,m}$ this is no longer true. For example, if the angle ϕ between the planes defined by the triangles Δ_{ijk} and $\Delta_{ijk'}$ is non-zero, the centers c and c′ of the spheres S and S′ will be drawn out of the planes of the respective triangles as shown in fig. 3. We can divide the overlapping volumes of S and S′ into two pieces, each associated with one of the two triangles, if we erect a plane passing through their common edge $\overline{ij}$ which contains the $D-2$ dimensional sphere (the circle $\mathscr{C}$ in fig. 3) of points common to the surfaces of S and S′. Thus to the triangle Δ_{ijk} we associate the volume of the sphere S minus the volume of the cap defined by the plane just described. If this process is continued to eliminate overlaps with the other two triangles which border Δ_{ijk}, the resulting volume V_{ijk} will be that of S minus the volume of three caps, defined by the three planes passing through the three sides of Δ_{ijk}. In contrast to the planar array of triangles discussed in sect. 4, these caps may overlap, leaving a volume V_{ijk} with faces of various sorts: segments of spheres, disks and cylinders (fig. 4). The complete volume $V_{\min}(\zeta_{l,m})$ is the union of the V_{ijk} corresponding to all the triangles Δ_{ijk} in the surface.

For a particular triangle Δ_{ijk}, the volume V_{ijk} depends on the lengths of the three sides of Δ_{ijk} and on the locations c, c′, c″ and c‴ of the centers of the sphere S and the spheres S′, S″ and S‴ for the three triangles adjacent to Δ_{ijk}. Since each of the lengths of the sides of Δ_{ijk} depends quadratically on $\zeta_{l,m}$, we can separate these two influences on V_{ijk} through second order

$$\delta V_{ijk} = \delta V^{\mathrm{I}}_{ijk} + \delta V^{\mathrm{II}}_{ijk}\,. \tag{5.6}$$

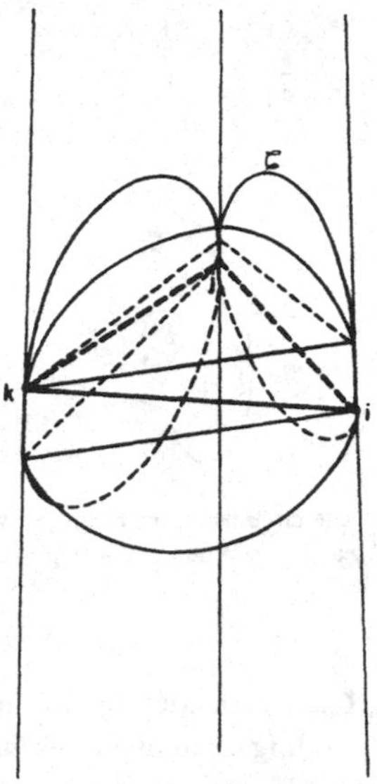

Fig. 4. Points i, j, k and the circle $\mathscr{C}$ are the same as those shown in fig. 3. Likewise the three vertical lines form a distorted version of the right triangular prism of fig. 1. The three circles represent the intersection of the sphere determined by the triangle ijk with the sphere defined by each of the neighboring triangles. The three nearly vertical lines are formed by the intersections of the planes defined by these circles.

Here δV_{ijk} is the change in V_{ijk} from the value of (4.41) and (4.42) for the planar, equilateral array. $\delta V^{\rm I}_{ijk}$ is the change in V_{ijk} that comes from the variation in the lengths of the triangle's sides, treating the centers c, c′, c″ and c‴ as each lying in the plane of Δ_{ijk}. Likewise, $\delta V^{\rm II}_{ijk}$ is the variation in V_{ijk} arising from the displacement of the centers c, c′, c″ and c‴ away from the plane of Δ_{ijk}. The corresponding change in $V_{\rm min}$ from its saddle-point value is

$$\delta V_{\rm min} = \sum_{\Delta_{ijk}} \delta V_{ijk} \,. \tag{5.7}$$

If we denote by δl_1, δl_2 and δl_3 the deviation of the length of each of the sides of the triangle Δ_{ijk} from the saddle-point values, then dimensional argument and the threefold symmetry of the equilateral triangle imply

$$\delta V^{\rm I}_{ijk} = \tfrac{1}{3}(\delta l_1 + \delta l_2 + \delta l_3) D \bar{V}_{ijk} / \bar{l} \,, \tag{5.8}$$

where $\bar{V}_{ijk} = \gamma_{\rm min}(\bar{A}_\Delta)^{D/2}$ and $\bar{l}$ are the saddle-point values of the excluded volume and triangle-side length, given by (4.44), (4.48) and (4.49). The sum of all the $\delta V^{\rm I}_{ijk}$

in terms of our orignal variables $\zeta_{l,m}$ is

$$\delta V^{\mathrm{I}}_{\min} = \sum_{\Delta_{ijk}} \delta V^{\mathrm{I}}_{ijk}$$

$$= \tfrac{1}{3}[D\gamma_{\min}(\bar{A}_\Delta)^{D/2}/\bar{l}^2] \sum_{l,m} \{(\zeta_{l+1,m} - \zeta_{l,m})^2 + (\zeta_{l,m+1} - \zeta_{l,m})^2 + (\zeta_{l-1,m+1} - \zeta_{l,m})^2\}\,. \tag{5.9}$$

In deriving this result we have used the expression

$$\tfrac{1}{2}(\zeta_{l,m} - \zeta_{l',m'})^2/\bar{l} \tag{5.10}$$

for the change in the length of the link connecting the vertices l, m and l', m', and also the fact that each link appears in the $\delta V^{\mathrm{I}}_{ijk}$ of two triangles.

We may regard $\zeta_{l,m} = \zeta(r_{l,m})$ where $r_{l,m}$ is given by (5.2). In the long-wavelength limit, $\zeta_{l+1,m} - \zeta_{l,m}$ can be approximated by $(\hat{a} \cdot \nabla)\zeta$; likewise, the sum inside the curly bracket in (5.9) becomes $\tfrac{3}{2}\bar{l}^2[(\partial\zeta/\partial x)^2 + (\partial\zeta/\partial y)^2]$. Since the number of lattice sites equals $\tfrac{1}{2}$ times that of triangles, we have

$$\delta V^{\mathrm{I}}_{\min} = \tfrac{1}{4} D\gamma_{\min}(\bar{A}_\Delta)^{D/2-1} \int_A \mathrm{d}x\, \mathrm{d}y \left[\left(\frac{\partial\zeta}{\partial x}\right)^2 + \left(\frac{\partial\zeta}{\partial y}\right)^2\right] + \mathrm{O}((\nabla\zeta)^4)\,, \tag{5.11}$$

where the integral extends over the total area A of the Wilson loop.

The second contribution $\delta V^{\mathrm{II}}_{ijk}$ in (5.6) is in general considerably more complicated to evaluate. However, by definition, it vanishes when the angle ϕ between the planes of the neighboring triangles Δ_{ijk} and $\Delta_{ijk'}$ in fig. 3 is zero. Hence, $\delta V^{\mathrm{II}}_{ijk}$ is proportional to the curvature of the surface of triangles. As will be shown below, this makes it quartic in the gradient of ζ. Therefore, in the long-wavelength limit, we can neglect $\delta V^{\mathrm{II}}_{ijk}$ and set $\delta V_{\min} = \delta V^{\mathrm{I}}_{\min}$. By using (4.48)–(4.51) and neglecting $\mathrm{O}(\nabla\zeta)^4$, we see that

$$\rho\delta V_{\min} = \tfrac{1}{2} T \int_A \mathrm{d}x\, \mathrm{d}y \left[\left(\frac{\partial\zeta}{\partial x}\right)^2 + \left(\frac{\partial\zeta}{\partial y}\right)^2\right], \tag{5.12}$$

which is identical to the action in the string model [7, 8], with T as the string tension.

To exhibit more explicitly the behavior of $\delta V^{\mathrm{II}}_{ijk}$, we introduce the Fourier transform

$$\zeta(x, y) = \sum_{q_1, q_2} \frac{1}{\sqrt{A}} \mathrm{e}^{i(q_1 x + q_2 y)} \tilde{\zeta}(q_1 q_2)\,. \tag{5.13}$$

The q-dependence of $\delta V^{\mathrm{II}}_{ijk}$ can be derived if we first express ϕ in terms of the displacements ζ_i, ζ_j, ζ_k, $\zeta_{k'}$ of the vertices of Δ_{ijk} and $\Delta_{ijk'}$. From fig. 3, we see that

$$\cos(\pi - \phi) = \frac{\boldsymbol{k'p} \cdot \boldsymbol{kp}}{|\boldsymbol{k'p}||\boldsymbol{kp}|}\,, \tag{5.14}$$

where

$$\begin{aligned} \boldsymbol{kp} &= -\boldsymbol{h} + \zeta_k - \tfrac{1}{2}(\zeta_i + \zeta_j)\,, \\ \boldsymbol{k'p} &= +\boldsymbol{h} + \zeta_{k'} - \tfrac{1}{2}(\zeta_i + \zeta_j)\,, \end{aligned} \tag{5.15}$$

with $\boldsymbol{h}$ the vector connecting the midpoint p of $\overline{ij}$ with k' for the saddle-point configuration. If (5.15) is substituted into (5.14), one obtains

$$\phi^2 = \frac{1}{h^2}(\zeta_k + \zeta_{k'} - \zeta_i - \zeta_j)^2\,, \tag{5.16}$$

where $h = \frac{1}{2}\sqrt{3}\,\bar{l}$. Next, we replace the function $\zeta_{l,m}$ in (5.16) with a single Fourier amplitude $\tilde{\zeta}(q_1, q_2)\exp(i\boldsymbol{q}\cdot\boldsymbol{r}_{l,m})$ from (5.13). Let us choose the line $\overline{ij}$ to be along the y-axis and parallel to $\hat{\beta}$ in (5.2); ϕ^2 and consequently also $\delta V^{\mathrm{II}}_{ijk}$ are proportional to

$$[\mathrm{e}^{i(q_\beta - q_\alpha)} + \mathrm{e}^{iq_\alpha} - \mathrm{e}^{iq_\beta} - 1]^2 = O(q^4)\,, \tag{5.17}$$

where $q_\beta = \boldsymbol{q}\cdot\hat{\beta} = q_2$ and $q_\alpha = \boldsymbol{q}\cdot\hat{\alpha} = \frac{1}{2}\sqrt{3}q_1 + \frac{1}{2}q_2$. Thus, in the long-wavelength limit (5.12) is valid. In terms of the Fourier components of ζ, it becomes, when the area A of the Wilson loop is large,

$$\rho\,\delta V_{\min} = \int \frac{T\,\mathrm{d}^2q}{2(2\pi)^2}\,|\tilde{\zeta}(q_1, q_2)|^2[q_1^2 + q_2^2]\,. \tag{5.18}$$

Substituting this expression into (5.5) and recalling that ζ is a $(D-2)$-dimensional vector, we obtain

$$d^2 = (D-2)\int \frac{\mathrm{d}^2q}{(2\pi)^2}\,\frac{1}{Tq^2} = \frac{D-2}{4\pi T}\ln A\,, \tag{5.19}$$

since the lower limit of the integral is $q^2 \sim A^{-1}$. When $D = 4$, this becomes (5.1).

We thank Professor J.M. Luttinger for a discussion of the quenched and annealed averages as used in condensed matter physics and Professor A.H. Mueller for a very helpful explanation of the physics of the relativistic string.

Appendix

To pass from (4.34) to (4.37), we recall from the text that each triangle (indexed by $\alpha = 1, \ldots, n$) is associated with m_α additional sites lying on the smallest empty sphere that passes through the three vertices of the triangle. We shall first ignore the fact that neighboring spheres may overlap. In that case the integral in (4.34) yields a product of independent factors similar to (4.29), one for each triangle. The result is of the form (4.37), provided that we can show that the integral over m sites associated with a triangle Δ of area A_Δ (where $\rho A_\Delta^{D/2} \gg 1$) yields a result proportional to $A_\Delta^{mD/4}$. This was done in the text for the case $D = 3$, in which m can take only the values 0 and 1. For $m = 1$, the extra factor is given by (4.24):

$$\pi(\tfrac{1}{2}\rho r^3)^{1/2} \sim A_\Delta^{3/4} = A_\Delta^{mD/4}\,, \tag{A.1}$$

since $A_\Delta \sim r^2$.

To show how a similar result arises for higher D, we shall first consider a simpler problem, in which the triangle Δ_{ijk} is replaced by a link l_{ij} of length $2r$. The number m now ranges from 0 to $D-1$. For $D=2$, the reasoning parallels the derivation of (4.22) and (4.24) in the text, and gives

$$P_0(i,j)=e^{-\pi r^2\rho}, \tag{A.2}$$

$$P_1(i,j)\cong 2\,e^{-\pi r^2}\int_0^\infty e^{-\pi\rho x^2}2r\rho\,dx$$

$$=2(\rho r^2)^{1/2}\,e^{-\pi r^2\rho}. \tag{A.3}$$

For $D=3$, there are three possibilities, $m=0, 1, 2$. To see how they arise, imagine that we have found an empty sphere passing through i and j. We move the center of the sphere toward the midpoint of $\overline{ij}$, shrinking the radius so that the sphere always passes through i and j. If we can reach the midpoint with the sphere still empty, then $m=0$. If not, we must have encountered a third site p_1 that prevented us from moving the sphere further in the desired direction. Now the sites i, j, p_1 lie on a small circle of the sphere. We move the center of the sphere toward the center of this circle, shrinking the radius so that the sphere always passes through i, j and p_1. If we can reach our goal, then $m=1$. If not, we encountered a fourth point p_2. Once the sphere passes through i, j, p_1, p_2 it cannot be moved any more, and $m=2$.

For the case $m=0$ the probability is given by an argument like that leading to (A.2):

$$P_0=e^{-4\pi r^3\rho/3}. \tag{A.4}$$

To find P_1, we let x denote the displacement of the center of the sphere (in any direction normal to l_{ij}) from the midpoint of l_{ij}. Then (4.23) holds, but without the factor 2, and with a different meaning for $d\tau(x)$. For a fixed x, the site p_1 lies on the surface of revolution generated by the smaller arc through i and j with radius $\sqrt{r^2+x^2}$. As x increases by dx, this surface sweeps out a volume

$$d\tau(x)=\int_0^\pi (2\pi r\sin\theta)r\sin\theta\,d\theta\cdot dx$$

$$=\pi^2r^2\,dx, \tag{A.5}$$

where $r\sin\theta$ is the perpendicular distance from p to $\overline{ij}$ in the approximation $x\ll r$, as explained before (4.24). Therefore

$$P_1(i,j)\cong e^{-4\pi r^3\rho/3}\int_0^\infty e^{-2\pi\rho r x^2}\pi^2r^2\rho\,dx$$

$$=\tfrac{1}{2}\pi^2(\tfrac{1}{2}\rho r^3)^{1/2}e^{-4\pi r^3\rho/3}. \tag{A.6}$$

Comparing (A.6) with (A.3), we see that in each case we obtained a factor $(\rho r^D)^{1/2} = \rho r^{Dm/2}$ for $m = 1$.

To evaluate P_2, we remember that i, j, p_1 lie on a small circle of the sphere passing through i, j, p_1, p_2. We denote by x_1 the displacement of the center of this circle from the midpoint of l_{ij}, and by x_2 the displacement of the center of the sphere from the center of the circle. The integral over x_2 is exactly like the one evaluated in (4.24), with x replaced by x_2, r by $\sqrt{r^2 + x_1^2}$, and p_1 playing the role of k. Thus

$$P_2(i, j) = \tfrac{1}{2} \int_0^\infty F(x_1)\, d\tau(x_1)\,, \tag{A.7}$$

where

$$\begin{aligned} F(x_1) &\cong 2\, e^{-\pi(r^2+x_1^2)^{3/2}\rho/3} \int_0^\infty e^{-2\pi\rho\sqrt{r^2+x_1^2}\,x_2^2}\, \pi(r^2 + x_1^2)\rho\, dx_2 \\ &= \pi[\tfrac{1}{2}\rho(r^2 + x_1^2)^{3/2}]^{1/2}\, e^{-4\pi(r^2+x_1^2)^{3/2}\rho/3}\,, \end{aligned} \tag{A.8}$$

and $d\tau(x_1)$ is given by exactly the same reasoning that led to (A.5). The factor $\frac{1}{2}$ in (A.7) compensates for overcounting $p_1 \leftrightarrow p_2$; it corresponds to the factor $1/m_\alpha!$ in (4.34).

Substituting (A.5) and (A.8) into (A.7), and making the approximation $x_1 \ll r$ because of the exponential in (A.8), we find

$$\begin{aligned} P_2(i, j) &\cong \tfrac{1}{2}\pi(\tfrac{1}{2}\rho r^3)^{1/2}\, e^{-4\pi r^3\rho/3} \int_0^\infty e^{-2\pi\rho x_1^2} \cdot \pi^2 r^2 \rho\, dx_1 \\ &= \tfrac{1}{2}\pi(\tfrac{1}{2}\rho r^3)^{1/2}\, e^{-4\pi r^3\rho/3}\tfrac{1}{2}\pi^2(\tfrac{1}{2}\rho r^3)^{1/2} \\ &= \tfrac{1}{4}\pi^3(\tfrac{1}{2}\rho r^3)\, e^{-4\pi r^3\rho/3}\,. \end{aligned} \tag{A.9}$$

The significant factor is now $\rho r^3 = \rho r^D = \rho r^{Dm/2}$.

The preceding examples illustrate the pattern which holds for higher D as well. Each new site restricting the sphere brings in a factor $(\rho r^D)^{1/2}$. Therefore, in general we have

$$P_m(i, j) = b_m(\rho r^D)^{m/2}\, e^{-\rho v_{ij}}\,, \tag{A.10}$$

where b_m is O(1) and v_{ij} is the volume of the sphere of radius r, in D dimensions.

For a triangle Δ_{ijk} instead of a link l_{ij}, the calculations are similar, only everything is raised by one dimension. Again one obtains

$$P_m(i, j, k) = b'_m(\rho r^D)^{m/2}\, e^{-\rho v_{ijk}}\,, \tag{A.11}$$

where b'_m is O(1), r is the radius of the circle drawn through i, j, k, and v_{ijk} is the volume in D dimensions of the sphere of radius r. Eq. (A.11) is identical to (4.29) in the text. Since the area $A_{ijk} \sim r^2$, we can write (A.11) as

$$P_m(i, j, k) = c'_m(\rho A_{ijk}^{D/2})^{m/2}\, e^{-\rho v_{ijk}}, \qquad (A.12)$$

where c'_m depends on m, D and the shape of Δ_{ijk}, but is O(1).

To pass from (A.12) to (4.37), we must consider the overlap of neighboring spheres. The main effect of this overlap is that the final formula contains not $e^{-\rho \Sigma v_{ijk}}$ but $e^{-\rho V_{min}}$ where V_{min} is the volume of the union of spheres as discussed in the text. In addition, the factors c'_m are replaced by smaller factors c_m, also of O(1), since the extra m_α sites for Δ_α are not allowed to penetrate neighboring spheres.

There is also the possibility that one additional site may do double duty, lying on the intersection of two neighboring spheres. Letting $\tilde{x}$ stand for a typical displacement and $\tilde{r}$ for a typical radius, such a point would have an available volume $\sim \tilde{r}^{D-2}\tilde{x}^2$ (as compared to a single site on one sphere which has available volume $\sim \tilde{r}^{D-1}x$). Since $\rho\tilde{r}^{D-2}\tilde{x}^2 \sim 1$, the order of magnitude of the corresponding term of (4.37) is the same as if the double-duty site were absent. (Triple-duty sites actually reduce the order of magnitude.) The effect is that the leading term of (4.37), $m_\alpha = D-2$ for all α, is unaffected by multiple-duty sites, although the lower-order terms are not quite correctly represented by the product form given there. In any case, the estimate (4.18) is established.

References

[1] N.H. Christ, R. Friedberg and T.D. Lee, Nucl. Phys. B202 (1982) 89
[2] K. Wilson, Phys. Rev. D10 (1974) 2455
[3] M. Luscher, Nucl. Phys. B180 [FS2] (1981) 1
[4] A. Hasenfratz, E. Hasenfratz and P. Hasenfratz, Nucl. Phys. B180 [FS2] (1981) 353
[5] C. Itzykson, M.E. Peskin and J.B. Zuber, Phys. Lett. 95B (1980) 259
[6] J.B. Kogut, D.R. Sinclair, R.B. Pearson, I.L. Richardson and J. Shigemitsu, Phys. Rev. D23 (1981) 2945
[7] Y. Nambu, Proc. Conf. on Symmetries and quark models, Detroit, 1969 (Gordon and Breach, New York, 1970), p. 269
[8] C. Rebbi, Phys. Reports 12 (1974) 1

Nuclear Physics B210[FS6] (1982) 337–346

WEIGHTS OF LINKS AND PLAQUETTES IN A RANDOM LATTICE*

N.H. CHRIST, R. FRIEDBERG and T.D. LEE

Columbia University, New York, N.Y. 10027, USA

Received 7 June 1982

In a random lattice, links and plaquettes have varying sizes and orientations. Therefore, they should enter the action function with different weights. A particularly attractive choice of these weights is presented; its consequences for spin-0, $\frac{1}{2}$ and 1 fields on such a lattice are discussed.

1. Introduction

In two preceding papers** [1, 2], we gave the construction of a random lattice and a discussion of the spin-0 and spin-1 fields on such a lattice. The usual gradient or curl operator in a continuum theory is replaced by the appropriate difference of operators on linked points or their product around a plaquette. These links and plaquettes have varying sizes and orientations, and can carry different weights in the lattice action. A single constraint is placed on these weights by the requirement that the lattice action should approach the continuum limit when the site density approaches infinity. In I and II, the weights are left arbitrary except for this constraint. In this paper, we focus on a particularly natural choice of these weights which, as we shall see, has several very attractive consequences. With this choice of weights, the spin-0 and spin-1 field equations are solved exactly by potentials which depend linearly on the position of the lattice sites (provided the weak coupling limit is taken for the spin-1 case). Likewise, the action for such fields can be evaluated analytically.

2. Spin-0 field

In the notation of I, the lattice action for a zero-mass free field is

$$A_L(\Omega)=\frac{1}{4}\sum_{i,j}\lambda_{ij}(\phi_i-\phi_j)^2\,, \tag{1}$$

where i and j denote the random lattice sites 1, 2, . . . , $\mathcal{N}$ and λ_{ij} are the weights:

$$\lambda_{ij}=0\,,\qquad \text{if } i \text{ and } j \text{ are not linked}\,. \tag{2}$$

* This research was supported in part by the US Department of Energy.

** Hereafter, they will be referred to as I and II.

The whole system is contained in a D-dimensional euclidean volume Ω. We recall that to each site i in the random lattice there corresponds a convex D-dimensional polyhedron ω_i in the dual lattice. To each link l_{ij} connecting sites i and j there corresponds a $(D-1)$-dimensional polyhedral surface s_{ij} of ω_i, and to each plaquette Δ_{ijk} with vertices at i, j and k there corresponds a $(D-2)$-dimensional polyhedral surface τ_{ijk} of s_{ij}, etc. We choose the weight λ_{ij} between any linked pair i and j to be

$$\lambda_{ij} = s_{ij}/l_{ij}\,, \tag{3}$$

where l_{ij} is the length of the link and s_{ij} the "volume" of the corresponding $(D-1)$-dimensional surface in the dual lattice. [If $D=2$, then s_{ij} denotes the border length in the dual lattice. See sect. 5 of I.]

Theorem 1.

$$\sum_j l^{\mu}_{ij}\lambda_{ij} = 0 \tag{4}$$

of all lattice sites i, and

$$L^{\mu\nu} \equiv \sum_{i,j} l^{\mu}_{ij} l^{\nu}_{ij}\lambda_{ij} = 2\Omega\delta^{\mu\nu}\,, \tag{5}$$

where λ_{ij} is given by (2) and (3), and l^{μ}_{ij} is the μ-component of $\boldsymbol{l}_{ij} = \boldsymbol{r}_j - \boldsymbol{r}_i$ with $\mu = 1, 2, \ldots, D$. Throughout this paper, the position vector of any point p is denoted by $\boldsymbol{r}_{\mathrm{p}}$.

Proof. Take any constant vector $\boldsymbol{V}$. From Gauss' theorem we have

$$0 = \int_{\omega_i} \boldsymbol{\nabla}\cdot\boldsymbol{V}\,\mathrm{d}^D r = \int \boldsymbol{V}\cdot\mathrm{d}\boldsymbol{S}\,. \tag{6}$$

Because the $(D-1)$-dimensional polyhedra s_{ij} (with i fixed but j varying) form the boundary of ω_i, and because $\hat{l}_{ij} \equiv \boldsymbol{l}_{ij}/l_{ij}$ is the normal vector of s_{ij}, (6) becomes

$$0 = \boldsymbol{V}\cdot\sum_j \hat{l}_{ij} s_{ij}$$

which implies the first part, (4), of the theorem.

To prove (5), we again apply Gauss' theorem, but to the following integral:

$$\omega_i\delta^{\mu\nu} = \int_{\omega_i} \frac{\partial}{\partial x^{\nu}}(\boldsymbol{r}-\boldsymbol{r}_i)^{\mu}\,\mathrm{d}^D r = \int (\boldsymbol{r}-\boldsymbol{r}_i)^{\mu}\,\mathrm{d}S^{\nu}$$

$$= \sum_j r^{\mu}_{ic}\hat{l}^{\nu}_{ij} s_{ij}\,, \tag{7}$$

where x^{ν} is the ν-component of $\boldsymbol{r}$ and r^{μ}_{ic} is the μ-component of $\boldsymbol{r}_{ic} = \boldsymbol{r}_c - \boldsymbol{r}_i$. The vector $\boldsymbol{r}_c$ is given by

$$\boldsymbol{r}_c \equiv \frac{1}{s_{ij}}\int_{s_{ij}} \boldsymbol{r}\,\mathrm{d}^{D-1} r\,, \tag{8}$$

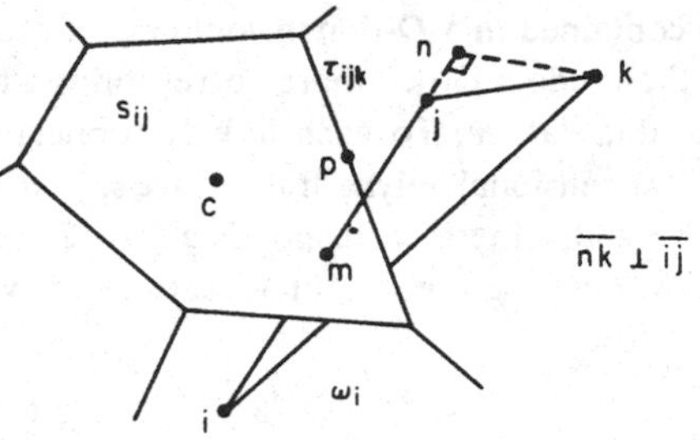

Fig. 1. The duals of i, l_{ij} and Δ_{ijk} are ω_i, s_{ij} and τ_{ijk}. The center of mass of s_{ij} is c, and midpoint of l_{ij} is m. Hence, $\boldsymbol{im} = \boldsymbol{mj} = \frac{1}{2}\boldsymbol{l}_{ij}$ and $\mathbf{mc} = \frac{1}{2}\boldsymbol{d}_{ij}$ where $\boldsymbol{d}_{ij}$ is defined by (10).

and is the position vector of the center of mass c of s_{ij}. (See fig. 1.) The link-vector l_{ij} is related to $\boldsymbol{r}_{ic}$ and $\boldsymbol{r}_{jc}$ by

$$\boldsymbol{l}_{ij} = \boldsymbol{r}_{ic} - \boldsymbol{r}_{jc}\,. \tag{9}$$

Define

$$\boldsymbol{d}_{ij} \equiv \boldsymbol{r}_{ic} + \boldsymbol{r}_{jc}\,. \tag{10}$$

Eq. (7) can then be written as

$$\omega_i \delta^{\mu\nu} = \tfrac{1}{2} \sum_j (\boldsymbol{l}_{ij} + \boldsymbol{d}_{ij})^{\mu} \hat{l}^{\nu}_{ij} s_{ij}\,.$$

Summing over all dual cells ω_i and noting that $\boldsymbol{d}_{ij}$ and s_{ij} are even under the exchange of i and j but $\boldsymbol{l}_{ij}$ and $\hat{l}_{ij}$ are odd, we derive

$$\Omega \delta^{\mu\nu} = \sum_i \omega_i \delta^{\mu\nu} = \tfrac{1}{2} \sum_{i,j} l^{\mu}_{ij} \hat{l}^{\nu}_{ij} s_{ij} \tag{11}$$

which establishes (5) and completes the proof of theorem 1.

By using the action (1), we see that the massless free field equation is

$$\sum_j \lambda_{ij} (\phi_i - \phi_j) = 0\,. \tag{12}$$

From theorem 1 it follows that a field ϕ_i which depends linearly on the site position $\boldsymbol{r}_i$ solves this field equation. Let

$$\phi_i = \phi_0 + \boldsymbol{V} \cdot \boldsymbol{r}_i\,, \tag{13}$$

where ϕ_0 and $\boldsymbol{V}$ are constants. Substituting (13) into (12) and applying (4), we find

$$\sum_j \lambda_{ij} (\phi_i - \phi_j) = -\boldsymbol{V} \cdot \sum_j \lambda_{ij} \boldsymbol{l}_{ij} = 0\,, \tag{14}$$

at all sites i. Furthermore, because of (5) the corresponding action is

$$A_L(\Omega) = \tfrac{1}{4}\sum_{i,j} \lambda_{ij}(\mathbf{V}\cdot \mathbf{l}_{ij})^2 = \tfrac{1}{2}\mathbf{V}^2\Omega\,. \tag{15}$$

Another consequence of (4) is that the bulk resistivity Z_η given by eq (6.46) of I, is unity; this is, of course, related to the statement that (13) is a solution of the massless free field equation (12).

For $D=2$, a stronger form of (15) exists (which, however, does not generalize to higher D or to spin 1). This is discussed in the appendix.

3. Spin-½ field

The continuum action in a D-dimensional euclidean volume Ω is a hermitian bilinear form given by*

$$A_C(\Omega) = \int_\Omega \left(\psi^\dagger\gamma^0\gamma^\mu \frac{\partial}{\partial x^\mu}\psi + m\psi^\dagger\gamma^0\psi\right) \mathrm{d}^D r\,, \tag{16}$$

where the dagger denotes hermitian conjugation and $\gamma^0, \gamma^1, \ldots, \gamma^D$ are anticommuting hermitian matrices whose squares are the unit matrix. The corresponding lattice action is

$$A_L(\Omega) - \tfrac{1}{2}\sum_{i,j} \psi_i^\dagger\gamma^0\gamma^\mu l_{ij}^\mu \lambda_{ij}\psi_j + m\sum_i \omega_i\psi_i^\dagger\gamma^0\psi_i\,, \tag{17}$$

where λ_{ij} is given by (2) and (3) and ω_i is the volume of the D-dimensional polyhedron dual to i. Because of (4), it follows that

$$\psi_i = \text{a constant spinor} \tag{18}$$

satisfies the massless field equation:

$$\sum_j \gamma^\mu l_{ij}^\mu \lambda_{ij}\psi_j = 0 \tag{19}$$

at all sites i.

Next, we consider the function

$$\psi_i = \chi\, \mathrm{e}^{i\mathbf{p}\cdot\mathbf{r}_i} \tag{20}$$

where χ is a constant spinor; the corresponding lattice action is

$$A_L(\Omega) = \tfrac{1}{2}(\chi^\dagger\gamma^0\gamma^\mu\chi)\sum_{i,j} l_{ij}^\mu\lambda_{ij}\,\mathrm{e}^{i\mathbf{p}\cdot\mathbf{l}_{ij}} + m\chi^\dagger\gamma^0\chi\Omega\,.$$

* We exclude other hermitian choices such as $A_C(\Omega) = \int_\Omega (i\psi^\dagger\gamma^\mu(\partial/\partial x^\mu)\psi + m\psi^\dagger\psi)\,\mathrm{d}^D r$ by requiring the field equation to be $(\gamma^\mu(\partial/\partial x^\mu) + m)\psi = 0$. The index μ is summed from 1 to D.

For small $\boldsymbol{p}$, $e^{i\boldsymbol{p}\cdot\boldsymbol{l}_{ij}} = 1 + i\boldsymbol{p}\cdot\boldsymbol{l}_{ij} - \frac{1}{2}(\boldsymbol{p}\cdot\boldsymbol{l}_{ij})^2 + \cdots$. By using (5) and neglecting $O(\boldsymbol{p}^3)$, we see that

$$A_L(\Omega) = [i(\chi^\dagger\gamma^0\gamma^\mu\chi)p^\mu + m\chi^\dagger\gamma^0\chi]\Omega\,,$$

which has the same form as in the continuum case.

The lattice action (17) may be resolved in terms of its eigenvectors $\psi_i(n)$ and eigenvalues $E(n)$:

$$\tfrac{1}{2}\sum_j \gamma^\mu l^\mu_{ij}\lambda_{ij}\psi_j(n) + m\omega_i\psi_i(n) = E(n)\omega_i\gamma^0\psi_i(n)\,. \tag{21}$$

We write

$$A_L(\Omega) = \sum_{n,i} E(n)\omega_i\psi_i(n)^\dagger\psi_i(n)\,.$$

It is easy to verify that (20) is not a solution of the eigenvector equation (21). Hence, the low-lying spectrum $E(n)$ of the lattice action is more complicated than the continuum case; its structure will be analyzed in a later paper.

4. Spin-1 field

In the notation of II, the action of a lattice gauge theory is

$$A_L(\Omega) = \frac{1}{g^2}\sum_{\Delta_{ijk}} \kappa_{ijk} f(U_{ijk})\,, \tag{22}$$

where κ_{ijk} is the weight of the plaquette Δ_{ijk}, g is the coupling constant and in the weak field approximation

$$f(U_{ijk}) = \tfrac{1}{4}g^2(\Delta^{\mu\nu}_{ijk}F^{\mu\nu})(\Delta^{\rho\sigma}_{ijk}F^{\rho\sigma})\,, \tag{23}$$

where*

$$\Delta^{\mu\nu}_{ijk} = \tfrac{1}{6}[l^\mu_{ij}l^\nu_{jk} + l^\mu_{jk}l^\nu_{ki} + l^\mu_{ki}l^\nu_{ij} - (\mu \rightleftarrows \nu)]\,, \tag{24}$$

which makes

$$A_L(\Omega) = \tfrac{1}{4}\sum_{\Delta_{ijk}} \kappa_{ijk}\Delta^{\mu\nu}_{ijk}\Delta^{\rho\sigma}_{ijk}F^{\mu\nu}F^{\rho\sigma}\,, \tag{25}$$

where, as in (22), the sum extends over all plaquettes Δ_{ijk}. [See (2.17) of II.]

As mentioned earlier, each plaquette Δ_{ijk} in the random lattice has a dual which is a $(D-2)$-dimensional convex polyhedron τ_{ijk}. These τ_{ijk}, with i and j fixed but k varying, form the surface of the $(D-1)$-dimensional polyhedron s_{ij}, which in

* $\Delta^{\mu\nu}_{ijk}$, $F^{\mu\nu}$ here are the same tensors as $\Delta_{\mu\nu}(ijk)$ and $F^a_{\mu\nu}$ of II, but with the internal symmetry index a suppressed.

turn is the dual of the link l_{ij}. We choose the weight κ_{ijk} of the plaquette Δ_{ijk} to be

$$\kappa_{ijk} = \frac{\tau_{ijk}}{2\Delta_{ijk}}, \tag{26}$$

where Δ_{ijk} is the area of the plaquette and τ_{ijk} the $(D-2)$-dimensional volume of its dual.

Theorem 2.

$$\sum_k \kappa_{ijk}\Delta_{ijk}^{\mu\nu} = 0 \tag{27}$$

for all sites i and j that are linked, and

$$T^{\mu\nu\rho\sigma} \equiv \tfrac{1}{4}\sum_{\Delta_{ijk}} \kappa_{ijk}\Delta_{ijk}^{\mu\nu}\Delta_{ijk}^{\rho\sigma} = \tfrac{1}{8}(\delta^{\mu\rho}\delta^{\nu\sigma} - \delta^{\mu\sigma}\delta^{\nu\rho})\Omega\,. \tag{28}$$

Proof. Consider the triangle Δ_{ijk}. Choose the point n lying on $\overline{ij}$ (or its extension) so that $\overline{nk}$ is perpendicular to $\overline{ij}$, as shown in fig. 1. Let $\hat{r}_{nk}$ be the unit vector along $r_{nk} \equiv r_k - r_n$. The tensor $\Delta_{ijk}^{\mu\nu}$, (24), can also be expressed in terms of the components of $\hat{l}_{ij}$ and $\hat{r}_{nk}$:

$$\Delta_{ijk}^{\mu\nu} = (\hat{l}_{ij}^{\mu}\hat{r}_{nk}^{\nu} - \hat{l}_{ij}^{\nu}\hat{r}_{nk}^{\mu})\Delta_{ijk}\,. \tag{29}$$

To derive (27), take any constant vector V lying in the $(D-1)$-dimensional space that contains s_{ij}. Similarly to (6), we apply Gauss' theorem to the volume integral of the derivative of the μ-component of V over s_{ij}:

$$0 = \int_{s_{ij}} \frac{\partial}{\partial x^{\nu}} V^{\mu}\, \mathrm{d}^{D-1}r = \sum_k \int_{\tau_{ijk}} \hat{e}_{ijk} V^{\mu}\, \mathrm{d}^{D-2}r\,, \tag{30}$$

where as before x^{ν} is the ν-component of r; $\hat{e}_{ijk}$ denotes the unit vector normal to τ_{ijk} and lying in s_{ij}, the dual of l_{ij}. Hence $\hat{e}_{ijk}$ is also normal to l_{ij}. Recall that $\hat{r}_{nk} \perp l_{ij}$ and is also normal to τ_{ijk} (since $\hat{r}_{nk}$ is on Δ_{ijk} which is normal to its dual τ_{ijk}). Thus, $\hat{e}_{ijk} = \hat{r}_{nk}$ and (30) becomes

$$0 = \sum_k \hat{r}_{nk}^{\nu}\tau_{ijk}V^{\mu}\,. \tag{31}$$

Therefore we have

$$\sum_k \tau_{ijk}\hat{r}_{nk}^{\nu} = 0\,, \tag{32}$$

which, when multiplied by $\hat{l}_{ij}^{\mu}$, gives

$$\sum_k \tau_{ijk}\hat{l}_{ij}^{\mu}\hat{r}_{nk}^{\nu} = 0\,. \tag{33}$$

Combining (26) and (29) with (33), we see that (27) holds.

Next, we turn to $T^{\mu\nu\rho\sigma}$, defined by (28). We may write

$$T^{\mu\nu\rho\sigma} = \tfrac{1}{48} \sum_{i,j,k} \frac{1}{\Delta_{ijk}} \Delta^{\mu\nu}_{ijk} \Delta^{\rho\sigma}_{ijk} \tau_{ijk}$$

$$= \tfrac{1}{48} \sum_{i,j,k} \frac{1}{\Delta_{ijk}} \Delta^{\mu\nu}_{ijk} \int_{\tau_{ijk}} \Delta^{\rho\sigma}_{ijk}\, \mathrm{d}^{D-2} r_{\mathrm{p}} \,, \tag{34}$$

where p is a point inside τ_{ijk} as shown in fig. 1, and the sum extends over all i, j, k which counts each Δ_{ijk} 3! times. Draw lines $\overline{i\mathrm{p}}$, $\overline{j\mathrm{p}}$ and $\overline{k\mathrm{p}}$. By considering the line integrals $\oint (x^\rho\, \mathrm{d}x^\sigma - x^\sigma\, \mathrm{d}x^\rho)$ around the triangles ijk, $ij\mathrm{p}$, $jk\mathrm{p}$ and $ki\mathrm{p}$, we have

$$\Delta^{\rho\sigma}_{ijk} = \Delta^{\rho\sigma}_{ij\mathrm{p}} + \Delta^{\rho\sigma}_{jk\mathrm{p}} + \Delta^{\rho\sigma}_{ki\mathrm{p}} \,. \tag{35}$$

Because $\Delta^{\mu\nu}_{ijk}$ is symmetric under a cyclic permutation of ijk, the three terms on the right-hand side of (35) give equal contributions to (34), which can then be written as

$$T^{\mu\nu\rho\sigma} = \tfrac{1}{16} \sum_{i,j,k} \frac{1}{\Delta_{ijk}} \Delta^{\mu\nu}_{ijk} \int_{\tau_{ijk}} \Delta^{\rho\sigma}_{ij\mathrm{p}}\, \mathrm{d}^{D-2} r_{\mathrm{p}} \,. \tag{36}$$

Let m be the midpoint of l_{ij} and $\boldsymbol{r}_{\mathrm{mp}} = \boldsymbol{r}_{\mathrm{p}} - \boldsymbol{r}_{\mathrm{m}}$. Because s_{ij} is the dual of l_{ij}, the $(D-1)$-dimensional space that contains s_{ij} bisects l_{ij}; therefore m is inside that space and $\boldsymbol{r}_{\mathrm{mp}} \perp \boldsymbol{l}_{ij}$. In terms of the components of $\boldsymbol{l}_{ij}$ and $\boldsymbol{r}_{\mathrm{mp}}$, we write

$$\Delta^{\rho\sigma}_{ij\mathrm{p}} = \tfrac{1}{2}(l^{\rho}_{ij} r^{\sigma}_{\mathrm{mp}} - l^{\sigma}_{ij} r^{\rho}_{\mathrm{mp}}) \,. \tag{37}$$

Substituting this expression and (29) into (36), we have

$$T^{\mu\nu\rho\sigma} = \tfrac{1}{32} \sum_{i,j,k} (\hat{l}^{\mu}_{ij} \hat{r}^{\nu}_{nk} - \hat{l}^{\nu}_{ij} \hat{r}^{\mu}_{nk}) \int_{\tau_{ijk}} (l^{\rho}_{ij} r^{\sigma}_{\mathrm{mp}} - l^{\sigma}_{ij} r^{\rho}_{\mathrm{mp}})\, \mathrm{d}^{D-2} r_{\mathrm{p}}$$

$$= t^{\mu\nu\rho\sigma} - t^{\nu\mu\rho\sigma} - t^{\mu\nu\sigma\rho} + t^{\nu\mu\sigma\rho} , \tag{38}$$

where

$$t^{\mu\nu\rho\sigma} = \tfrac{1}{32} \sum_{i,j} \hat{l}^{\mu}_{ij} l^{\rho}_{ij} \sum_{k} \int_{\tau_{ijk}} \hat{r}^{\nu}_{nk} r^{\sigma}_{\mathrm{mp}}\, \mathrm{d}^{D-2} r_{\mathrm{p}} \,. \tag{39}$$

Apply Gauss' theorem to the following integral over the $(D-1)$-dimensional polyhedron s_{ij}. Remembering that the surface of s_{ij} is τ_{ijk} and $\hat{r}_{nk}$ is the normal to τ_{ijk} that lies in s_{ij}, we obtain

$$\delta^{\nu\sigma} \mathrm{s}_{ij} = \int_{\mathrm{s}_{ij}} \frac{\partial}{\partial x^{\nu}} (\boldsymbol{r} - \boldsymbol{r}_{\mathrm{m}})^{\sigma}\, \mathrm{d}^{D-1} r = \sum_{k} \int_{\tau_{ijk}} \hat{r}^{\nu}_{nk} r^{\sigma}_{\mathrm{mp}}\, \mathrm{d}^{D-2} r_{\mathrm{p}} \,, \tag{40}$$

which, when substituted into (39), leads to

$$t^{\mu\nu\rho\sigma} = \tfrac{1}{32} \sum_{i,j} \hat{l}^{\mu}_{ij} l^{\rho}_{ij} \mathrm{s}_{ij} \delta^{\nu\sigma}$$

$$= \tfrac{1}{16} \Omega \delta^{\mu\rho} \delta^{\nu\sigma} , \tag{41}$$

where the last equality follows from (5). Combining (41) with (38), we establish (28). Theorem 2 is then proved.

Consider now the case when the field $F^{\mu\nu}$ is a constant; i.e., in (23), it is independent of the plaquette Δ_{ijk}. On account of (28), the action $A_L(\Omega)$ in the weak field approximation (25) becomes simply

$$A_L(\Omega) = \tfrac{1}{4} F^{\mu\nu} F^{\mu\nu} \Omega \,. \tag{42}$$

In addition, because of (27), the corresponding field equation

$$\sum_k \kappa_{ijk} \Delta_{ijk}^{\mu\nu} F^{\mu\nu} = 0 \tag{43}$$

is valid for any linked pair of sites i and j.

We thank Mr. H.-C. Ren for his considerable help in recognizing the properties of this definition of the action for a scalar field on the random lattice.

Appendix

We present here a special result for the spin-zero field in two dimensions. Consider any triangle Δ_{ijk}. The border* b_{ij} (dual to the link l_{ij}) can be cut into an internal and an external part:

$$b_{ij} = b_{ij}^{\text{int}\,(ijk)} + b_{ij}^{\text{ext}\,(ijk)} \,. \tag{A.1}$$

If the border does not cut the link, (A.1) still holds with one of the terms on the right negative.

Since λ_{ij} is proportional to b_{ij}, it can be similarly decomposed:

$$\lambda_{ij} = \lambda_{ij}^{\text{int}\,(ijk)} + \lambda_{ij}^{\text{ext}\,(ijk)} \,. \tag{A.2}$$

Then the action (1) can be written

$$A_L(\Omega) = \tfrac{1}{6} \sum_{i,j,k} A_L(ijk) \,, \tag{A.3}$$

where

$$\begin{aligned} A_L(ijk) &= \tfrac{1}{2} \lambda_{ij}^{\text{int}\,(ijk)} (\phi_i - \phi_j)^2 + \text{cyclic permutations} \\ &= \tfrac{1}{2} \frac{b_{ij}^{\text{int}\,(ijk)}}{l_{ij}} (\phi_i - \phi_j)^2 + \text{cyclic permutations} \end{aligned} \tag{A.4}$$

and the sum in (A.3) is over triples i, j, k forming a plaquette.

The foregoing can be generalized to any dimension, but the following theorem holds only for $D = 2$.

* The s_{ij} in the text becomes b_{ij} for $D = 2$.

Theorem 3. Suppose sites i, j, k form a plaquette. Let ϕ_i, ϕ_j, ϕ_k be given any values whatever, and let $\boldsymbol{E}$ be the unique vector satisfying

$$\boldsymbol{E}\cdot\boldsymbol{r}_i-\phi_i=\boldsymbol{E}\cdot\boldsymbol{r}_j-\phi_j=\boldsymbol{E}\cdot\boldsymbol{r}_k-\phi_k\,. \tag{A.5}$$

Then

$$A_{\mathrm{L}}(ijk)=\tfrac{1}{2}E^2\Delta_{ijk}\,, \tag{A.6}$$

where Δ_{ijk} is the area of the plaquette.

Proof. Without loss of generality we may take the circumcenter of the triangle as the origin of coordinates. Moreover, we may rotate the coordinate system so that $\boldsymbol{E}$ points in the y-direction. Then (writing β_{ij} for $b_{ij}^{\mathrm{int}\,(ijk)}$)

$$\begin{aligned}A_{\mathrm{L}}&=\frac{1}{2}\frac{\beta_{ij}}{l_{ij}}(l_{ij}^y)^2E^2+\text{cyclic permutations}\\&=\tfrac{1}{2}\beta_{ij}^x l_{ij}^y E^2+\text{cyclic permutations}\,.\end{aligned} \tag{A.7}$$

Here we regard $\boldsymbol{\beta}_{ij}$ as a vector pointing from the circumcenter to the midpoint of l_{ij}, and assume that the vertices i, j, k appear in counterclockwise order.

Since the origin is at the circumcenter, we have

$$\boldsymbol{\beta}_{ij}=\tfrac{1}{2}(\boldsymbol{r}_j+\boldsymbol{r}_i)\,, \tag{A.8}$$

and so

$$\begin{aligned}A_{\mathrm{L}}&=\tfrac{1}{4}(r_j^x+r_i^x)(r_j^y-r_i^y)E^2+\text{cyclic permutations}\\&=\tfrac{1}{4}E^2(r_j^xr_j^y-r_i^xr_i^y+r_i^xr_j^y-r_j^xr_i^y)+\text{cyclic permutations}\\&=\tfrac{1}{4}E^2(r_i^xr_j^y-r_j^xr_i^y)+\text{cyclic permutations}\,,\end{aligned} \tag{A.9}$$

since the first two terms of the second line cancel in the sum over permutations. But $\tfrac{1}{2}(r_i^xr_j^y-r_j^xr_i^y)$ is the area of the triangle made by i, j and the circumcenter; therefore (A.9) is identical to (A.6). The same result is obtained if the circumcenter lies outside the triangle, when due allowance is made for negative signs in the intermediate steps.

In three dimensions we have the following counter-example: let ijk form an equilateral triangle, with $\phi_i=\phi_j=\phi_k$. Let p be equidistant from i, j, k, but a distance h from their plane. Let $\phi_{\mathrm{p}}-\phi_i=Eh$. Let s be the distance from p to the midpoint of l_{ij}, and $s\gamma$ the distance from p to the circumcenter of i, j, p. Then the analog to (A.4) gives

$$A_{\mathrm{L}}(ijkl)=\tfrac{1}{4}\gamma h\Delta_{ijk}E^2\,, \tag{A.10}$$

whereas the volume of the tetrahedron is

$$\Omega_{ijkl}=\tfrac{1}{3}h\Delta_{ijk}\,. \tag{A.11}$$

The analog to (A.6) holds only if $\gamma = \frac{2}{3}$, which makes the circumcenter of i, j, l coincide with the intersection of medians. This happens only if the tetrahedron is regular.

References

[1] N.H. Christ, R. Friedberg and T.D. Lee, Nucl. Phys. B202 (1982) 89
[2] N.H. Christ, R. Friedberg and T.D. Lee, Nucl. Phys. B210[FS6] (1982) 310

VOLUME 42, NUMBER 21 PHYSICAL REVIEW LETTERS 21 MAY 1979

Experiments with a Gauge-Invariant Ising System

Michael Creutz, Laurence Jacobs, and Claudio Rebbi
Physics Department, Brookhaven National Laboratory, Upton, New York 11973
(Received 19 March 1979)

Using Monte Carlo techniques, we evaluate the path integral for the four-dimensional lattice gauge theory with a Z_2 gauge group. The system exhibits a first-order transition. This is contrary to the implications of the approximate Migdal recursion relations but consistant with mean-field-theory arguments. Our "data" agree well with a low-temperature expansion and the exact duality between the high- and low-temperature phases.

Based on a non-Abelian gauge theory, the standard model of hadronic dynamics may simultaneously confine quarks in physical hadrons and possess asymptotic freedom, a vanishing effective coupling at short distances. Central to an understanding of this picture is the study of phase transitions in lattice gauge theory.

Proposed by Wilson as a nonperturbative regularization procedure, lattice gauge theory allows ˙ong-coupling expansion which demonstrates k confinement for sufficiently large bare coupling.[1] Nevertheless, conventional weak-coupling perturbation theory suggests a possible electrodynamicslike nonconfining phase. Using mean-field arguments, Balian, Drouffe, and Itzykson have found evidence that in enough space-time dimensions lattice gauge theories will indeed posses two distinct phases depending on the coupling strength.[2] It is essential for the standard model that four space-time dimensions be insufficient for such a transition to occur with an SU(3) gauge group.

Using renormalization-group transformations with approximations based on bond moving, Migdal has argued that four dimensions represent a critical case for lattice gauge theory, just as two dimensions are critical for phase transitions in conventional spin systems with nearest-neighbor interactions.[3] Indeed, Migdal's relations are identical for gauge theory in d dimensions and spin systems in $d/2$ dimensions. Thus, the nonexistence of a phase transition in the O(3) Heisenberg model in two dimensions is touted as evidence for the absence of a nonconfining phase in non-Abelian gauge theories. Further, the interesting and rather complicated phase structure of the X-Y model in two dimensions has been correlated with the possibility of avoiding confinement in a lattice version of electrodynamics based on a U(1) gauge group.

With Wilson's lattice cutoff, one can go beyond the usual continuous Lie groups and consider theories based on discrete groups. The simplest such group is Z_2, the addition of integers modulo 2. As discussed by Balian, Drouffe, and Itzykson, this group provides a gauge-invariant version of the Ising model.[4] The Migdal recursion relation suggests an analogous phase structure between this model in four dimensions and the conventional Ising model in two dimensions. The latter model is exactly solvable and exhibits a second-order phase transition between a disordered and a ferromagnetic state. The purpose of this Letter is to present results, obtained by a Monte Carlo simulation, which strongly indicate that the phase transition in the four-dimensional Z_2 gauge theory is of the first order. Thus, we find evidence of a breakdown of the analogy between this model and the two-dimensional Ising model.

Monte Carlo simulations have provided a useful tool for studying statistical systems of lower dimensionality.[5] In applying this method, one constructs by an iterative procedure a sequence of configurations, Σ_1, Σ_2, $\Sigma_3, \ldots$, which eventually simulates statistical equilibrium. Given any configuration Σ_i, a new configuration Σ_i' is obtained from Σ_i by changing one of the statistical variables (spins) of the lattice. Σ_{i+1} is set equal to Σ_i or to Σ_i' with a definite conditional probability, P, which depends on the actions (or internal energies) of Σ_i and Σ_i'. This probability is chosen so as to ensure that, when equilibrium is reached, the states occur in the sequence with density proportional to the Boltzmann factor. The procedure is continued until all the spins of the lattice have been tested many times and it has become clear that equilibrium has been attained. The states occurring in the sequence then provide a good sample of the correct statistical sum.

A difficulty in the application of the method to four-dimensional systems resides in the large number of spins one has to consider if one wants to incorporate a reasonable number of lattice sites in each linear dimension. To overcome this problem we have developed a technique for processing simultaneously in a high-speed computer all the lattice variables situated along a definite direction, thus effectively reducing a four-dimensional lattice to a three-dimensional lattice of the same linear size. We shall refer to this technique as multispin coding (MSC).

While we shall present details of the technique elsewhere, we mention here the basic underlying idea. MSC takes advantage of the fact that the memory locations, or words, of high-speed computers are designed to store numbers with a high degree of precision, and therefore contain large numbers of binary digits (bits). In a computation where each of the variables can take only the values 0 and 1, representable by a single bit, it is a waste to allocate one word of memory for each spin. By MSC we denote the use of the individual bits of a single memory word to record the values of many different spins. Typically, one may place the spins associated with a fixed value of x, y, z but all values of t in a single word. Besides reducing the amount of memory storage required, MSC allows the simultaneous execution of computations which would otherwise be done sequentially. For instance, suppose that in a Z_2 lattice gauge the bits of the memory words A and B code as ones and zeros the spins located on the links emanating from sites with fixed x, y, z, variable t, and directed along the x and y directions respectively. The "exclusive or" instruction, $C = ([A \text{ and } (\text{not } B)] \text{ or } [(\text{not } A) \text{ and } B])$, a simple and fast computer operation, will perform the group multiplication of all the spins in A and B. In other words, MSC turns the computer into a fast array processor.

The Z_2 lattice gauge theory is formulated in terms of spin variables s_{ij} defined over the links of a hypercubical lattice. Each spin $s_{ij} = s_{ji}$ can be 0 or 1; i and j are indices which denote neighboring lattice sites. The action describing the interaction of these spins is

$$S = 2\sum_{\square}\Big(\sum_{s_{ij}\in\square} s_{ij}\Big)_{\text{mod } 2}, \tag{1}$$

where $\square$ represents a plaquette or elementary square in the lattice, and the outer sum is over all such squares. The factor of 2 is introduced in the definition of S so as to conform with the conventions of Ref. 4.

Working on a lattice of N sites, we define a "free energy"

$$F = N^{-1}\ln Z, \tag{2}$$

where the partition function, Z, is given by

$$Z = \sum_{s_{ij}} e^{-\beta S}. \tag{3}$$

The average action per plaquette is

$$E \equiv (6N)^{-1}\langle S\rangle = -\tfrac{1}{6}dF/d\beta. \tag{4}$$

A phase transition is signaled by a singularity in F or its derivatives. In particular, E should exhibit a discontinuity at a first-order phase transition and should be continuous (but with a discontinuous or singular derivative) at a second-order one.

Some exact results for this model have been obtained. In particular, the model is self-dual,[4,6] the theory at β being related to the theory at $\beta^* = -\frac{1}{2}\ln\tanh\beta$. The self-duality relations for E read

$$E(\beta) = 1 - \tanh\beta - (\sinh 2\beta)^{-1}E(\beta^*). \tag{5}$$

If we assume a unique critical point β_c, self-duality gives

$$\beta_c = \tfrac{1}{2}\ln(1+\sqrt{2}) = 0.4407. \tag{6}$$

We have used a Monte Carlo simulation to calculate $E(\beta)$ for a lattice extending eight sites in the x, y, and z directions and twenty sites in the t direction. To minimize surface effects, we imposed periodic boundary conditions.

We now describe our results. Figure 1 shows the values of E as a function of β obtained in a Monte Carlo simulation where all the spins were probed once and then β was varied slightly before proceeding to another sweep of the lattice. Starting from a completely ordered lattice and $\beta = 1.2$, we reduced β to $\beta = 0$ in steps of 0.0006. At $\beta = 0$ (infinite temperature) the simulation randomizes the lattice. We then increased β back to $\beta = 1.2$

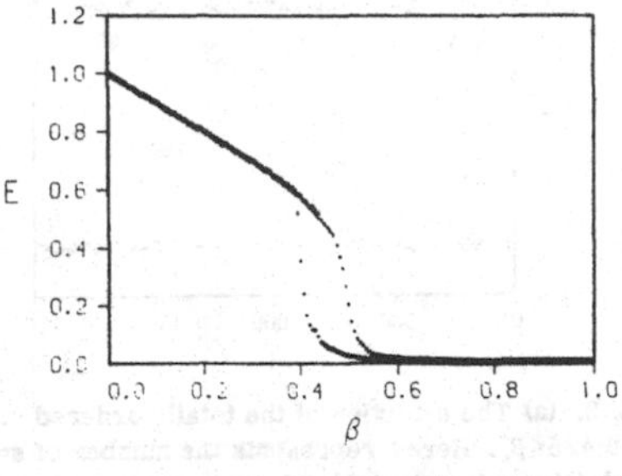

FIG. 1. The average energy per plaquette as a function of β. The system was heated (+) from $\beta = 1.2$ to $\beta = 0$ and then cooled (×) back to $\beta = 1.2$ in steps of 0.0006. Points are plotted every tenth step. The solid curves represent the low- and high-temperature expansions given by Eq. (8) and its dual equation.

with the same step size. In this "experiment" statistical equilibrium is never actually reached; rather, the lattice is heated across the transition temperature first and then cooled back. The clear hysteresis apparent in Fig. 1 is strongly indicative of a first-order phase transition. This hysteresis is due to the metastability of the ordered phase for $\beta<\beta_c$ and of the disordered phase for $\beta>\beta_c$ (superheating and supercooling).

Starting with extreme initial data represented by either a totally disordered configuration (spins chosen randomly) or a completely ordered state (all spins equal to zero), we perform fixed-temperature simulations ranging from a few hundred to several thousand sweeps of the lattice. The length of these simulations was dictated by the time it took the system to reach thermal equilibrium at a given temperature. In Fig. 2(a) we display a typical result of this procedure for $\beta \neq \beta_c$. The graph shows the evolution of the totally ordered state at $\beta=0.425$. After a rapid initial relaxation (on the order of ten sweeps) the system remains metastable for a few hundred sweeps and then abruptly decays into the (high energy) stable phase. In contrast, starting from either extreme initial state, two distinct phases appear to be stable at the critical temperature, with no sign of drift in the value of E, as is shown in Fig. 2(b). This behavior is unique to systems with first-order transitions and, apart from the observed hysteresis, forms the main basis for our conclusions.

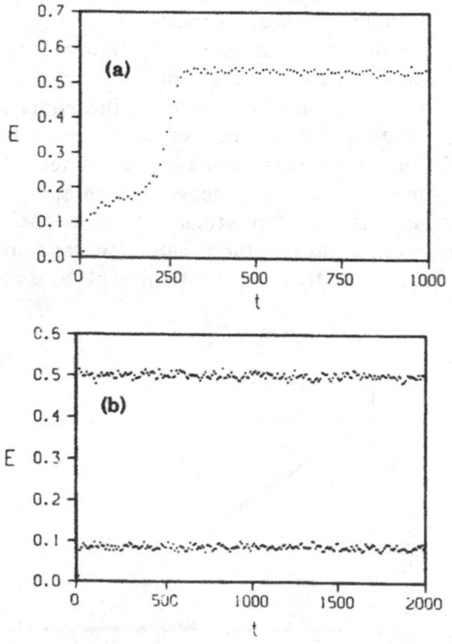

FIG. 2. (a) The evolution of the totally ordered state at $\beta=0.425<\beta_c$. Here t represents the number of sweeps of the lattice. A metastable phase is apparent for $t \lesssim 250$, followed by a transition to the stable phase. (b) The evolution of the totally ordered and the completely disordered states at the critical temperature. One notices the coexistence of two apparently stable phases.

For this system we are fortunate to know the temperature of the transition beforehand. Because of the strong tendency to superheat and supercool, initial conditions entirely in one phase or the other would not permit an accurate determination of an unknown critical temperature. As we wish to study other groups, we need a technique for locating β_c. Therefore we have studied initial conditions with half the lattice randomized and the other half in a ground-state configuration. After a rapid initial relaxation of the two halves into stable or metastable configurations, we observe a linear approach to a single phase. This behavior is shown in Fig. 3. The linear region is suggestive of the unstable phase "dissolving" into the stable phase at the boundary. The direction of this linear behavior determines immediately on which side of the critical temperature one is working.

As a test of our methods, we have carried out simulations on the three-dimensional Z_2 gauge theory. This system is known to undergo a sec-

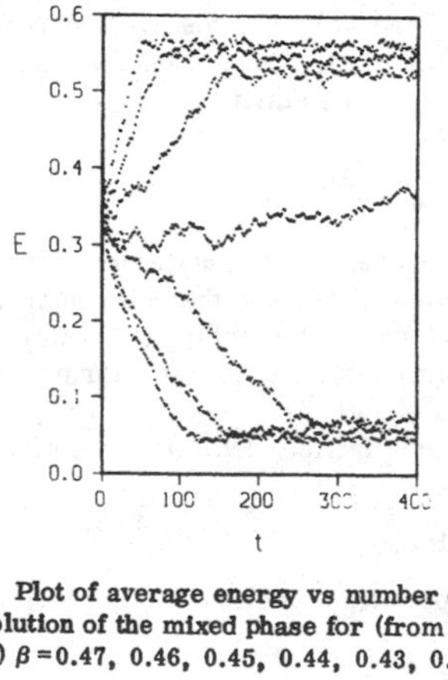

FIG. 3. Plot of average energy vs number of sweeps in the evolution of the mixed phase for (from the lowest curve) $\beta=0.47$, 0.46, 0.45, 0.44, 0.43, 0.42, and 0.41.

VOLUME 42, NUMBER 21 PHYSICAL REVIEW LETTERS 21 MAY 1979

ond-order transition at $\beta_c = 0.7613$.[4] The signals alluded to in the preceding paragraphs are absent in this case: given any initial configuration, the approach to equilibrium was gradual; moreover, at β_c both extreme initial configurations evolve smoothly and are seen to merge after a short relaxation time (on the order of 500 sweeps for a $16 \times 16 \times 30$ lattice). With use of the mixed phase technique described above, $E(\beta)$ is found to be continuous at β_c, as expected at a second-order transition.

Finally, we wish to remark that a low-temperature expansion can be derived for the system. For the energy we obtain

$$E = 8e^{-12\beta}[1 + 15x^2 + x^3 + 273x^4 + O(x^5)],$$
$$x = e^{-4\beta}. \qquad (7)$$

The dual of this expression gives a high-temperature expansion with corrections of order (β^{15}). Equation (7) may be Padé approximated,

$$E = 8e^{-12\beta}\left\{\frac{1 - \frac{1}{15}x - \frac{719}{225}x^2}{1 - \frac{1}{15}x - \frac{4094}{225}x^2} + O(x^5)\right\}. \qquad (8)$$

This function and its dual from Eq. (5) are plotted along with the "data" in Fig. 1, with excellent agreement. Note that the expression in Eq. (8) is singular at $\beta \approx 0.36 < \beta_c$. This is presumably an estimate of the maximum temperature for metastability of the superheated phase.

We wish to thank V. Emery, B. McCoy, F. Paige, R. Peierls, and R. Swendsen for many interesting discussions. This manuscript was submitted under U. S. Department of Energy Contract No. EY-76-C-02-0016.

[1]K. G. Wilson, Phys. Rev. D 10, 2445 (1974).

[2]R. Balian, J. M. Drouffe, and C. Itzykson, Phys. Rev. D 10, 3376 (1974).

[3]A. A. Migdal, Zh. Eksp. Teor. Fiz. 69, 457, 810 (1975), [Sov. Phys. JETP 42, 413, 743 (1975)]; L. P. Kadanoff, Rev. Mod. Phys. 49, 267 (1977).

[4]R. Balian, J. M. Drouffe, and C. Itzykson, Phys. Rev. D 11, 2098, 2104 (1975).

[5]For a review see K. Binder, in *Phase Transitions and Critical Phenomena*, edited by C. Domb and M. S. Green (Academic, New York, 1976), Vol. 5B.

[6]F. J. Wegner, J. Math. Phys. (N.Y.) 12, 2259 (1971).

PHYSICAL REVIEW D VOLUME 20, NUMBER 8 15 OCTOBER 1979

Monte Carlo study of Abelian lattice gauge theories

Michael Creutz, Laurence Jacobs,* and Claudio Rebbi
Physics Department, Brookhaven National Laboratory, Upton, New York 11973
(Received 21 June 1979)

Using Monte Carlo techniques, we study the thermodynamics of four-dimensional Euclidean lattice gauge theories, with gauge groups Z_N and U(1). For $N \leq 4$ the models exhibit a single first-order phase transition, while for $N \geq 5$ we observe two transitions of higher order. As N increases, one of these transitions moves toward zero temperature, whereas the other remains at finite temperature and survives in the U(1) limit. The behavior of the Wilson loop factor is also analyzed for the Z_2 and Z_6 models.

I. INTRODUCTION

Lattice gauge theories currently provide one of the most promising approaches toward a demonstration of quark confinement through an interaction with non-Abelian gauge fields.[1] The lattice formulation introduces a nonperturbative ultraviolet cutoff rendering the theory well defined. Wilson's expansion in terms of strings shows confinement for strong coupling[1]; however, ordinary perturbation theory via a spin-wave expansion suggests a possible unconfined phase for weak coupling. According to conventional lore,[2] four space-time dimensions represent a critical case where the spin-wave phase never appears for non-Abelian continuous gauge groups, but does appear for the Abelian group U(1) describing conventional electrodynamics. This parallels the critical nature of two dimensions for systems of spins interacting through nearest-neighbor couplings; the Heisenberg model based on the non-Abelian symmetry O(3) has only a disordered phase in two dimensions,[3] whereas with the Abelian group U(1) there is also a low-temperature phase with correlation functions behaving as a power of separation at large distances.[4]

Balian, Drouffe, and Itzykson suggested the study of discrete gauge groups as a practice ground for understanding the phase structure of lattice gauge theories.[5] Particularly interesting is the group Z_N, the set of complex Nth roots of unity with ordinary multiplication as the group operation. For $N=2$ we have a gauge-invariant generalization of the Ising model,[6] while when N goes to infinity we obtain U(1), the group of relevance to electrodynamics. In an earlier paper we used Monte Carlo techniques to argue that the Z_2 model has a first-order phase transition at the temperature where the system is self-dual.[7] In this article we extend our investigation to the groups Z_N and U(1).

Our results give evidence for a single, first-order phase transition in the Z_3 and Z_4 theories, at the self-dual temperatures.[8] For larger values of N, the Z_N model appears instead to undergo two phase transitions of higher order, at two temperatures, the higher of which shows little N dependence, whereas the other decreases for increasing N as $[1-\cos(2\pi/N)]=O(1/N^2)$. The former transition survives the U(1) limit. This pattern of phase transitions agrees nicely with recent theoretical arguments that for N large enough the Z_N theory should possess three phases, a disordered one at high temperature, an ordered one similar to that seen in the Z_2 model at low temperature, and a third intermediate phase mimicking the unconfined phase of the U(1) theory.[9] For the Z_6 model, where the two transitions are separated enough to delineate a clear intermediate phase and yet the ordered phase extends to a temperature sufficiently high to allow for an efficient Monte Carlo procedure, we have also studied the behavior of Wilson loops.[1] They appear to decrease exponentially with the area of the enclosed region in the disordered phase and with the perimeter in the intermediate- and low-temperature phases.

In Sec. II we define the models under consideration. Section III reviews the Monte Carlo technique used to evaluate the statistical sums. Section IV presents our results and Sec. V contains a few closing remarks.

II. THE MODELS

We formulate the theory on a four-dimensional hypercubical lattice. Associated with each link joining a pair of nearest-neighbor sites i and j is an element U_{ij} of the group Z_N defined by

$$Z_N=\{e^{2\pi i n/N}\,|\,n=0,1,\ldots,N-1\}. \tag{2.1}$$

This set forms an Abelian group under ordinary multiplication. As N goes to infinity we obtain the group U(1). The U_{ij} are oriented on the links of the lattice in the sense that we require

$$U_{ji}=U_{ij}^{*}. \tag{2.2}$$

The action describing the interaction of these

spins is

$$S=\frac{1}{8}\sum_{i,j,k,l} P_{ijkl}(1-U_{ij}U_{jk}U_{kl}U_{li}), \qquad (2.3)$$

where

$$P_{ijkl}=\begin{cases} 1 \text{ if } i,j,k,l \text{ label vertices circulating around an elementary square or "plaquette" of the lattice}, \\ 0 \text{ otherwise.}\end{cases} \qquad (2.4)$$

The factor of $\frac{1}{8}$ in Eq. (2.3) is inserted because each plaquette is counted eight times in the sum, four times in each of the two orientations. The action is real because the contribution from a definite orientation of a plaquette is the complex conjugate of that for the opposite one. The total contribution from any plaquette to S lies in the interval $[0, 2]$.

We insert this action into a path integral to define a partition function at temperature $T=1/\beta$,

$$Z=\sum_{\{U_{ij}\}} e^{-\beta S}, \qquad (2.5)$$

where the sum runs over all possible values of the link variables U_{ij}. The "free energy" of the system is an intensive quantity defined by

$$F=\frac{1}{N_s}\ln Z, \qquad (2.6)$$

where N_s is the number of sites on the lattice. A phase transition is defined as a singularity in the infinite-volume limit of F considered as a function of β. Differentiating the free energy gives the average action per plaquette

$$E\equiv\langle 1-\mathrm{Re}\,U_{ij}U_{jk}U_{kl}U_{li}\rangle=-\frac{1}{6}\frac{\partial}{\partial\beta}F, \qquad (2.7)$$

where the sites i, j, k, and l circulate around an elementary square. The factor $\frac{1}{6}$ is the ratio of the number of sites to the number of plaquettes in a four-dimensional lattice. This quantity E is the order parameter we concentrate upon in our numerical work. At a first-order phase transition E is discontinuous in β, while for higher-order transitions it is continuous.

III. THE MONTE CARLO TECHNIQUE

The partition function represents a sum over all configurations of the system, i.e., over all possible values for the statistical variables U_{ij}. Although for a bounded system this is a finite sum, the number of terms involved is so large even for systems of rather small size that it cannot be evaluated directly by numerical methods. In a Monte Carlo computation one generates a sequence of configurations which simulates an ensemble of states in thermal equilibrium at inverse temperature β. A sum over the states in this sequence approximates the full statistical sum.[10]

More specifically, the Monte Carlo technique consists of setting up a Markovian process. At each step a state of the system Σ_i is transformed into a new state Σ_{i+1} (which may coincide with Σ_i) according to a probability matrix $P(\Sigma_i, \Sigma_{i+1})$. This matrix must have the Boltzmann distribution as an eigenvector, i.e., the Markovian process must transform an ensemble in equilibrium into itself. A sufficient condition for this is an equation of detailed balance

$$\frac{P(\Sigma,\Sigma')}{P(\Sigma',\Sigma)}=\exp\{-\beta[S(\Sigma')-S(\Sigma)]\}, \qquad (3.1)$$

where $S(\Sigma)$ represents the action of the state Σ.

This is quite general and does not determine the probability matrix uniquely. Considerations of efficiency and computational feasibility eventually dictate its form. In our work we have followed the rather common procedure of probing the spins of the lattice one at a time, so that Σ_i and Σ_{i+1} differ at most in the value of a single statistical variable U_{ij}. Properly speaking, then, we are not dealing with a single probability matrix, but rather a collection of matrices $P^{(ij)}(\Sigma,\Sigma')$, one for each statistical variable. The entries of $P^{(ij)}(\Sigma,\Sigma')$ are zero unless Σ and Σ' differ at most in the value of U_{ij}. A single Monte Carlo step consists in acting with one of the $P^{(ij)}$ on the state vector. All the statistical variables are probed in succession, and therefore the matrix defining the Markovian process is

$$P=P^{(i_Nj_N)}\cdots P^{(i_2j_2)}P^{(i_1j_1)}, \qquad (3.2)$$

where the product runs over all links in the lattice. We shall refer to the application of P to the state vector as one Monte Carlo iteration, or a sweep of the lattice. In other words, the spins are probed in orderly succession and stochastically set to new values. When all the spins have been analyzed, one proceeds to a new iteration.

This still leaves open the detailed form of $P^{(ij)}(\Sigma,\Sigma')$. We have performed computations with two distinct probability matrices, corresponding to different algorithms for the stochastic changing of spins. One method, introduced by Metropolis *et al.*,[11] begins by choosing randomly a new value $U'_{ij}\neq U_{ij}$ from the group elements with uniform probability. If the action is lowered by the replacement of U_{ij} with U'_{ij}, the spin is set to this new value; if $\Delta S\geq 0$, a random number r with uniform distribution between 0 and 1 is generated and the spin is changed to U'_{ij} only if $r<\exp(-\beta\Delta S)$. This simple algorithm ensures that Eq. (3.1) is satisfied.

We have used this method in conjunction with a

computational technique allowing the storage of many spin variables in a single memory word of a computer and involving many spins in parallel computations with logic operations. This procedure makes the actual computations very fast, effectively reducing the dimensionality of the lattice from four to three. We have briefly described this technique in a previous communication[7] and shall not elaborate on it further here. With this method we work on a periodic lattice of 8 sites in the x, y, and z directions, and 20 sites in the t direction. The total number of spin variables is thus 40 960.[12]

As an alternative procedure, we have used an algorithm which selects a new value U'_{ij} for the spin variable in a stochastic manner with probability distribution proportional to the Boltzmann factor $\exp(-\beta S)$. The previous value of U_{ij} plays no direct role in the selection procedure. This method corresponds to placing successively each spin of the lattice in contact with a heat bath and will be referred to by this name. The corresponding probability matrix also obeys Eq. (3.1). The heat-bath method lends itself less well to the multiple storage of spins and the computations involved are slower; however, this is often offset by a faster convergence to equilibrium. In what follows, we shall explicitly indicate those results obtained with the heat-bath technique, leaving it understood that the others were derived by the procedure of Metropolis *et al.*[11] Frequently we have used the two methods independently to evaluate similar quantities as a check on the consistency of the results.

IV. RESULTS

A thermal cycle of a statistical system can provide a general overview of its phase structure. The temperature T is gradually varied while the internal energy is measured. By making the variation of the temperature sufficiently slow, the system can be kept generally close to thermal equilibrium, even though this is in principle never quite reached. In the neighborhood of a phase transition, however, the relaxation time increases and so does the departure from equilibrium. A hysteresis effect will then signal the presence of the transition.

These features, which would be observed in a true experiment, are also apparent in a Monte Carlo simulation. We have therefore evaluated the behavior of E in such a thermal cycle with gauge groups Z_2, Z_3, Z_4, Z_5, Z_6, and Z_8. The results are displayed in Figs. 1(a)–1(f). Starting with a definite value β_0 for the inverse temperature and with a completely ordered configuration where all U_{ij} are set to the group identity, we have performed Monte Carlo computations while changing β slightly after each iteration. Thus β is reduced to zero (infinite temperature) and then returned to β_0. We have chosen β_0 of 1.2 for Z_2, 2.0 for Z_3, Z_4, and Z_5, 2.5 for Z_6, and 3.5 for Z_8. The total number of iterations in the thermal cycles is 4000 for the models Z_2–Z_5, 5000 for Z_6, and 7000 for Z_8. The points are plotted every 16 iterations for Z_2 and every 10 for the other systems.

Hysteresis effects are apparent in all the dia-

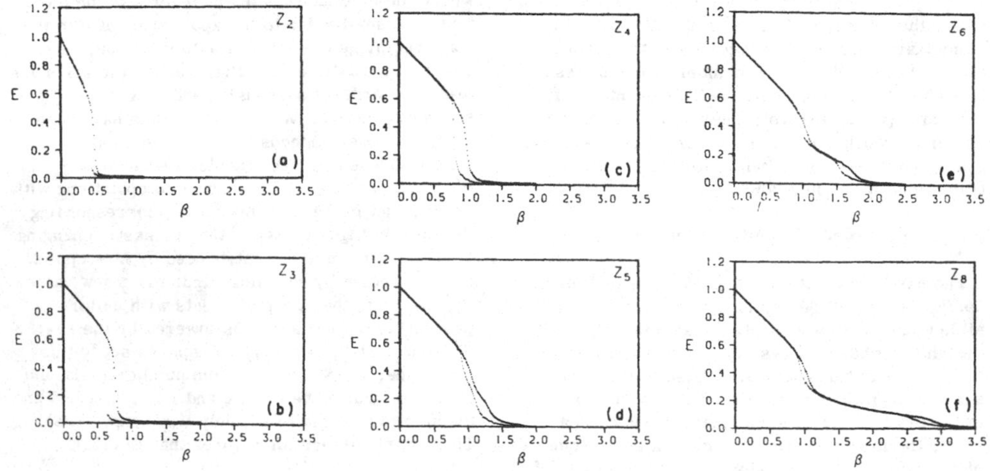

FIG. 1. Thermal cycles on the models Z_2, Z_3, Z_4, Z_5, Z_6, and Z_8.

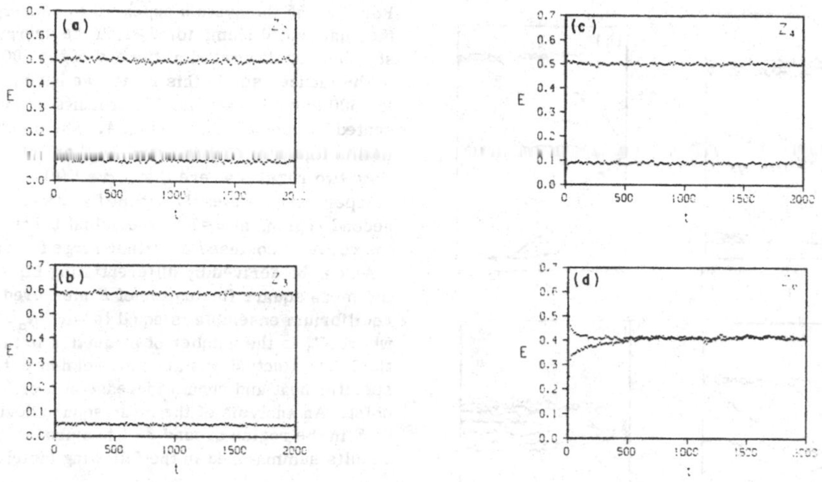

FIG. 2. Long runs at a critical temperature for the models Z_2, Z_2, Z_4, and Z_6.

grams; however, their qualitative features change drastically in the passage from Z_4 to higher groups. For Z_2, Z_3, and Z_4 one observes a single, rather steep hysteresis cycle. With Z_5 the cycle begins to separate into two less steep loops. This doubling becomes manifest with Z_6 and Z_8. Thus, the thermal cycles suggest a single transition with Z_2, Z_3, and Z_4, and two separate ones with Z_6, Z_8, and, most likely, Z_5.

The pronounced jumps in the cycles with Z_2, Z_3, and Z_4 suggest first-order transitions. At the temperature of a first-order phase change, a system can exist in either or two distinct stable phases, an ordered one with smaller internal energy and a disordered one with greater internal energy. Both phases become metastable on the "wrong" side of the transition. At a temperature slightly higher, for instance, than the critical value, the ordered phase is metastable and can exhibit an extremely long relaxation time to equilibrium. This property of superheating or supercooling, which represents a hindrance to reaching equilibrium from an inappropriate starting phase, can be exploited to verify the nature of the transition. Figures 2(a)–2(d) illustrate the results of long (2000 iterations) Monte Carlo simulations with β fixed near a critical point. The values selected for β are 0.4407, 0.67, 0.88, and 1.0 for the models Z_2, Z_3, Z_4, and Z_6, respectively. The critical values $\beta_c=\frac{1}{2}\ln(1+\sqrt{2})$ for Z_2, $\frac{2}{3}\ln[2/(\sqrt{3}-1)]$ for Z_3, and $\ln(1+\sqrt{2})$ for Z_4 are known from the self-duality of these models.[5,8] The value 1.0 for Z_6 was selected on the basis of further analysis presented below. The two series of points in each of these figures correspond to the values of E observed every 10 iterations. The starting configuration was completely ordered for the lower set of points and totally random for the upper set. For the groups Z_2, Z_3, and Z_4 the systems appear to converge in remarkably few iterations to one of the two definite phases, both of which then remain stable. The behavior of the Z_6 model is drastically different; irrespective of the initial configuration the internal energy converges rather slowly to a unique value. This is indicative of a phase transition of higher than first order.

The fact that Monte Carlo simulations starting from a totally ordered and a totally disordered lattice produce internal energies converging to a common value is a good sign that thermal equilibrium is being reached. Unfortunately, in the neighborhood of a first-order phase transition, the convergence of the metastable phase to thermal equilibrium becomes too slow to make this check feasible. To investigate the properties of the system in these regions and to determine the critical value of β more accurately than from a simple inspection of the hysteresis cycles, we have resorted to the following stratagem. A mixed configuration, in which all the U_{ij} with time coordinate between 1 and 10 are set equal to the identity and the remaining half are chosen randomly, has been used as initial configuration. 400 Monte Carlo iterations have then been performed for a few values of β selected in the regions of the hysteresis cycles. The results are displayed in Figs. 3(a)–3(d) for the groups Z_2, Z_3, Z_4, and Z_6. All diagrams show the results obtained with 7

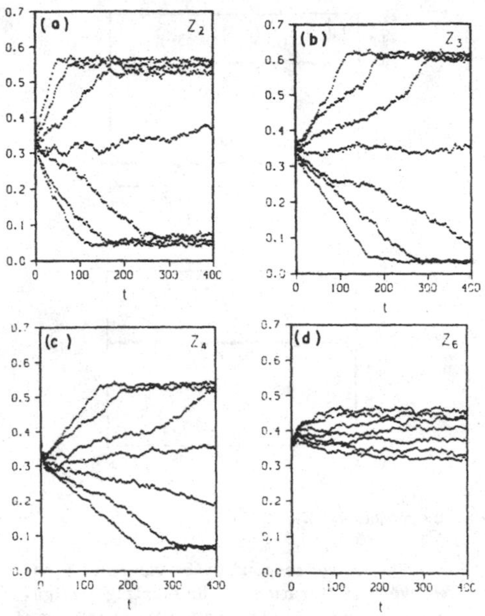

FIG. 3. Mixed phase running with groups Z_2, Z_3, Z_4, and Z_6.

values of β, in steps of 0.01, starting from 0.41, 0.64, 0.85, and 0.97 for the groups Z_2, Z_3, Z_4, and Z_6, respectively, the lower series of points corresponding to larger values of β. For Z_2, Z_3, and Z_4 one notices a quick initial relaxation of the value of E, corresponding to a rapid approach of the two halves of the system to the two phases which can coexist at the critical point, followed by a rather linear drift as the boundary between the two phases shifts until the stable phase overtakes the whole lattice. At the critical temperature both phases are stable and the drift is absent. From Figs. 3(a)–3(c) one can read off the self-dual points $\beta = 0.44$, 0.67, and 0.88 for the models Z_2, Z_3, and Z_4, respectively. The behavior of the mixed phase for Z_6 is strikingly different. One notices a convergence to an equilibrium value of E varying smoothly with β, suggesting a continous transition.

The model Z_6 represents the first case where the three-phase structure is well separated. We have tried to determine E as a function of β in more detail by a series of Monte Carlo simulations where the initial configuration consisted of layers of ordered and disordered spins. Then 500 iterations were performed for values of β set equal to 0.5, 0.9, 0.95, 0.98, 1, 1.02, 1.05, 1.1, 1.3, 1.5, 1.55, 1.58, 1.6, 1.62, 1.65, 1.7, and 2. For $\beta \leq 1.55$ the system appeared to converge to thermal equilibrium; for $\beta > 1.55$ the energy was still moving downward at the end of the 500 sweeps of the lattice; so, in this region we have proceeded for 500 more iterations. The results are represented by open circles in Fig. 4. The other points in this figure are taken from the thermal cycle. Note two regions where the curve $E(\beta)$ becomes steeper. One is clearly outlined at $\beta \approx 1$. In the second region, at $\beta \approx 1.6$, the actual behavior of the curve is confused by rather large fluctuations.

As can be derived by differentiating Eq. (2.7), the mean square fluctuation of E measured in an equilibrium ensemble is equal to $-(1/N_\square)(\partial E/\partial\beta)$ where $N_\square$ is the number of plaquettes in the lattice. The fluctuations are thus related to the specific heat and should increase at a critical point. An analysis of the mean square deviation of E in the region around $\beta = 1$ has produced the results summarized in the following table:

β	$\langle E\rangle$	$N_\square \times (\langle E^2\rangle - \langle E\rangle^2)$
0.9	0.529	0.810
0.95	0.487	0.778
0.98	0.447	0.997
1	0.411	1.28
1.02	0.337	0.586
1.05	0.310	0.441
1.1	0.285	0.560

The increase in $\langle E^2\rangle - \langle E\rangle^2$ at $\beta \approx 1$ is apparent, but the limited statistics does not allow quantitative comparisons.

Diagrams of E versus β have been obtained with the heat-bath method for the Z_{20} and U(1) models [with lattice size of 8^4 for Z_{20} and 5^4 for U(1)] and are reproduced in Figs. 5 and 6. The first phase transition occurs always at $\beta \approx 1$, the second moves to $\beta \approx 16$ in Z_{20} and is not present in the explored range of β in U(1).

For the marginal Z_5 theory, we used the heat-bath method on an 8^4 lattice at $\beta = 1.0$, 1.1, and 1.2. At $\beta = 1.1$ with either a random or an or-

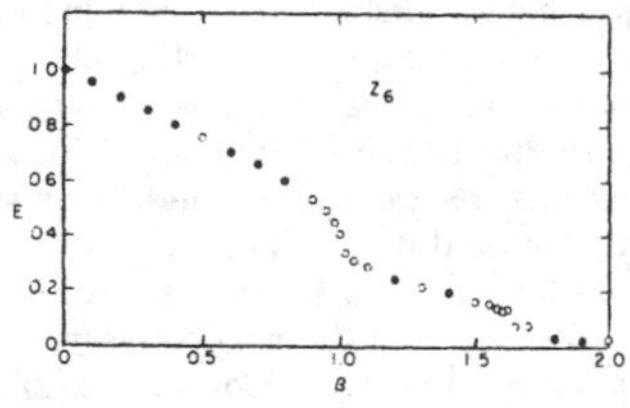

FIG. 4. The internal energy as a function of β for Z_6. The open circles represent the results of the long runs described in the text. The other points are from the thermal cycles.

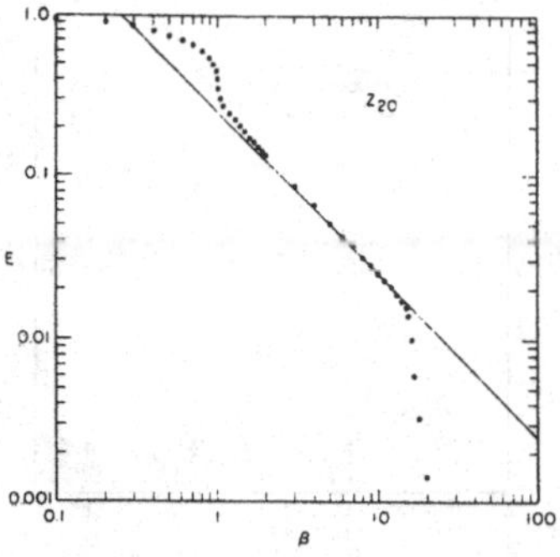

FIG. 5. The internal energy as a function of β for Z_{20}. The straight line represents the asymptotic spin-wave result $1/4\beta$.

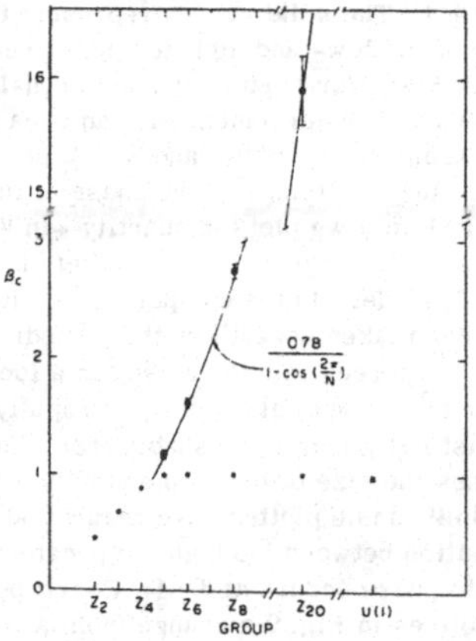

FIG. 7. Critical points as a function of symmetry group.

dered start the system reached equilibrium within 100 iterations. However, at either $\beta=1.0$, or $\beta=1.2$ the convergence rate was substantially reduced and similar to that seen with Z_6 at $\beta=1.0$. Thus the Z_5 model possesses the three-phase structure of the higher-order groups.

The values obtained for the critical β's are plotted versus the order of the group in Fig. 7. These points are obtained using the heat-bath method and repeatedly adjusting β so that E is maintained in the middle of the hysteresis cycles. The solid line represents the curves $\beta_c=0.78/[1-\cos(2\pi/N)]$. The fit suggests that the relevant quantity in determining the position of the low-temperature phase transition is the magnitude of the gap between the action of an unexcited plaquette and the action of a plaquette in its lowest state of excitation.

Finally, we have studied the behavior of the Wilson loop factor, defined as

$$W=\langle\mathrm{Re}(U_{i_1i_2}U_{i_2i_3}\cdots U_{i_ni_1})\rangle, \tag{4.1}$$

where the group elements are associated with links forming a closed loop in the lattice. For an elementary square W equals $1-E$ and ranges in value between 1, for a totally ordered lattice, and 0, for a totally disordered lattice. If one increases the size of the loop, W always approaches zero (in the limit of ∞ lattice size), but the modality of the approach depends on the phase of the system. It has been argued[1] that in a confining, disordered phase, W decays following an area law

$$W\approx\exp(-cA), \tag{4.2}$$

A being the area enclosed by the loop, whereas in an ordered, nonconfining phase, W should decay following a perimeter law

$$W\approx\exp(-c'L), \tag{4.3}$$

L being the length of the perimeter of the loop. In Fig. 8 we compare the values of $1-W$ for a single plaquette and a square loop of side 2 in the Z_2 model (results obtained with the heat-bath method,

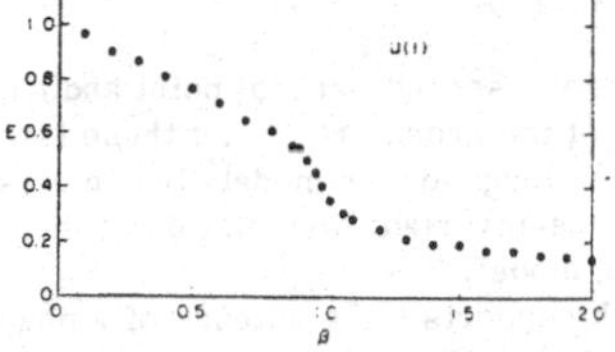

FIG. 6. The internal energy as a function for β for U(1).

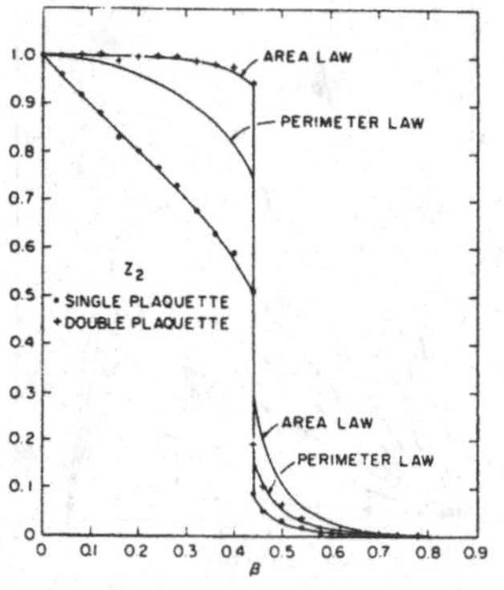

FIG. 8. Square Wilson loops of sides one and two for the group Z_2.

lattice size 8^4). The solid curves represent the predictions of the low- and high-temperature expansions for E and for perimeter- and area-law behaviors of W. The agreement with an area law in the high-temperature phase and with a perimeter law in the low-temperature phase is impressive. In Fig. 9 we plot the quantity $-\ln W$ versus the side of a square loop for a set of values of β in the Z_6 model. In the computations, the loops have been taken spacelike; the periodicity of the lattice then forces $\ln W$ to vanish for a loop of side 8. For the lowest values of β, W rapidly becomes statistically indistinguishable from 0 as one increases the size of the loop and only a few points for $\ln W$ can be plotted. We recall that the phase transition between the high temperature and intermediate phase occurs at $\beta \approx 1$. Correspondingly one notices in Fig. 9 a change from a concave to a rather linear behavior in the curves, suggesting indeed the transition from an area to a perimeter law for W.

V. DISCUSSION

Our work has many features of an experiment. One has a small piece of a four-dimensional crystal, heats it up, cools it, and measures its internal energy. It is, admittedly, a very small crystal, but still one hopes the statistics will be sufficiently good to provide relevant information on the properties of the medium. We have studied the effect of lattice size by performing simulations with the Z_2 gauge theory on hypercubical lattices ranging from 2^4 to 8^4 lattice sites. In Fig. 10 we show the average plaquette and root-mean-square fluctuation at the critical temperature with both ordered and disordered initial conditions. The

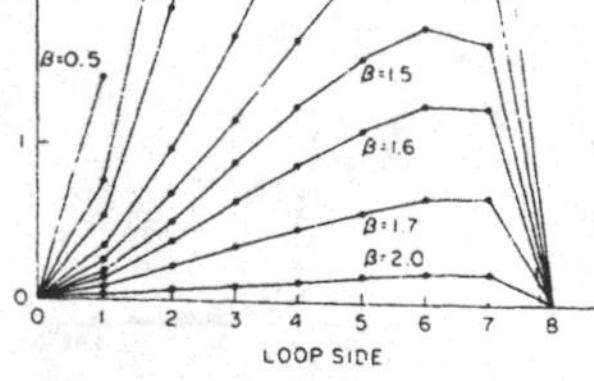

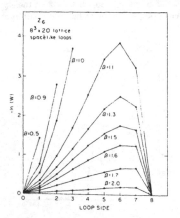

FIG. 9. Square Wilson loops for Z_6 as a function of loop side.

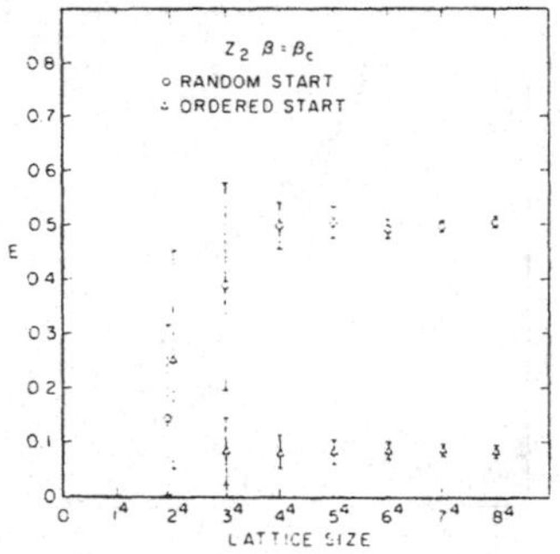

FIG. 10. Size dependence of the two-phase structure with Z_2.

plotted points and error bars are the average over the last 100 of 110 heat-bath iterations. Note that the two-phase structure is already clear on a 4^4 lattice; increasing the size only decreases the fluctuations.

From our work emerges a rather clear phase structure for the abelian lattice gauge theories. The appearance of three phases when the order of the group exceeds 4 is particularly noticeable. It is remarkable that such a structure has been predicted using entirely theoretical arguments based on the approach to the U(1) limit.[9] Our results confirm nicely their analysis.

To a qualitative overview of the thermal properties of the Z_N and U(1) models, we add elements of quantitative information. The curves giving E as a function of β are tantamount to an experimental determination of the partition function. The critical values of β, at which the phase transitions occur, when not known from self-duality, are determined to a reasonable degree of accuracy.

As already noted, the low-temperature transition for the high-order groups scales well with the action of the first excited state above the ground state. For β in the vicinity of this second transition the energy scales approximately as

$$E(\beta) = \frac{1}{\beta} f\left(\frac{\beta}{\beta_2}\right), \tag{5.1}$$

where β_2 is the second critical point and f is independent of the group order. At these low temperatures the large-order models become essentially a gauge-invariant formulation of the surface-roughening model.[13]

Our work supports the existence of a phase transition for U(1) lattice gauge theory. Such a transition is essential if Wilson's formulation is to describe the prototype of all gauge theories, quantum electrodynamics. Indeed, if the approach could not avoid the confinement of electrons, its use for any gauge theory would be suspect. We still have not shown that the low-temperature phase of this system contains a massless excitation, the photon. As some evidence for this, we note that a spin-wave expansion at large β would begin by approximating

$$\mathrm{Re}\Big(\prod_{\square} U_{ij}\Big) = \cos\Big(\sum_{\square}\theta_{ij}\Big) \approx 1 - \tfrac{1}{2}\Big(\sum_{\square}\theta_{ij}\Big)^2, \quad (5.2)$$

where the group elements are represented as

$$U_{ij} = \exp(i\theta_{ij}) . \quad (5.3)$$

The above replacement makes the path integral Gaussian and solvable, and one finds

$$E \xrightarrow[\beta\to\infty]{} \frac{1}{4\beta} . \quad (5.4)$$

This function is plotted with the results for Z_{20} in Fig. 5. Our results for the U(1) model and the Z_N models in the intermediate phase approach this behavior at large β and thus support the existence of spin-wave excitations.

ACKNOWLEDGMENT

This work was supported by the U.S. Department of Energy under Contract No. EY-76-C-02-0016.

*Present address: Instituto de Fisica, Universidad Nacional de México, Apdo. Postal 20-364, Mexico 20, D. F., Mexico.

[1]K. G. Wilson, Phys. Rev. D 10, 2445 (1974).

[2]L. P. Kadanoff, Rev. Mod. Phys. 49, 267 (1977).

[3]R. E. Peierls, Ann. Inst. Henri Poincaré 5, 177 (1935); N. D. Mermin and H. Wagner, Phys. Rev. Lett. 22, 1133 (1966).

[4]J. M. Kosterlitz and D. J. Thouless, J. Phys. C 6, 1181 (1973).

[5]R. Balian, J. M. Drouffe, and C. Itzykson, Phys. Rev. D 10, 3376 (1974); 11, 2098 (1975); 11, 2104 (1975).

[6]F. J. Wegner, J. Math. Phys. 12, 2259 (1971).

[7]M. Creutz, L. Jacobs, and C. Rebbi, Phys. Rev. Lett. 42, 1390 (1979).

[8]C. P. Korthals-Altes, Nucl. Phys. B142, 315 (1978); T. Yoneya, *ibid*. B144, 195 (1978).

[9]S. Elitzur, R. B. Pearson, and J. Shigemitsu, Phys. Rev. D 19, 3698 (1979); D. Horn, M. Weinstein, and S. Yankielowicz, *ibid*. 19, 3715 (1979); A. Guth, A. Ukawa, and P. Windey (unpublished).

[10]K. Binder, in *Phase Transitions and Critical Phenomena*, edited by C. Domb and M. S. Green (Academic, New York, 1976), Vol. 5B.

[11]N. Metropolis, A. W. Rosenbluth, M. N. Rosenbluth, A. H. Teller, and E. Teller, J. Chem. Phys. 21, 1087 (1953).

[12]While in principle it is possible to reduce the number of independent spin variables by a choice of gauge, imposing such a constraint renders the Monte Carlo computation less efficient. Indeed, many excitations involving only a few spins acquire a nonlocal aspect when a gauge is imposed and are reproduced only slowly with a procedure where spins are changed locally. This is particularly true at low temperature, where the acceptance factor accompanying an increase in the action is small. Consequently, we have not imposed a gauge constraint for the results presented here.

[13]S. T. Chui and J. D. Weeks, Phys. Rev. B 14, 4798 (1976).

Volume 95B, number 1 PHYSICS LETTERS 8 September 1980

PHASE TRANSITION IN FOUR-DIMENSIONAL COMPACT QED

B. LAUTRUP [1] and M. NAUENBERG [2]
CERN, Geneva, Switzerland

Received 10 June 1980

The energy and the specific heat of the four-dimensional U(1) lattice gauge model is evaluated by Monte Carlo simulations on lattices of size L^4, where L = 4, 5 and 6, and evidence is presented for the occurrence of a second-order phase transition. A finite size scaling analysis of our results gives the critical value of the coupling constant $e_c^2 = 0.995$ and a correlation length exponent $\nu \approx 1/3$.

It is known that compact quantum electrodynamics [1] [discrete U(1) gauge theory] on a four-dimensional lattice confines static electric charges in the strong coupling limit [2], while for weak coupling the interaction obeys the familiar Coulomb law [3]. These properties imply the occurrence of a phase transition at some intermediate value of the coupling constant [1], but in spite of considerable theoretical work on this model [4] there exist at present only conjectures on the order of this phase transition. Recently Creutz, Rebbi and their collaborators [5–8] have carried out a series of Monte Carlo studies for abelian and non-abelian lattice gauge models. For the U(1) model in four dimensions, Creutz found that the plaquette energy exhibits a hysteresis loop indicative of a phase transition [‡1]. He suggested that this is a higher order transition because of the occurrence of large fluctuations and slow convergence of the energy as a function of the number of Monte Carlo iterations. However, to determine the properties of this phase transition it is essential to evaluate also the specific heat or possibly higher order derivatives of the energy as a function of the coupling constant e^2, and the size of the lattice. These quantities depend on the correlations between plaquettes which are expected to increase at large distances in the neighbourhood of a higher order phase transition. The correlation length ζ diverges at the phase transition point e_c^2 as $\zeta \sim |e^2 - e_c^2|^{-\nu}$, where ν is a critical exponent. For a finite system of linear dimension L, $\zeta \lesssim L$, and therefore it is important to apply finite size scaling theory [9] to obtain the critical properties of the infinite system. We present in this letter Monte Carlo results obtained by computer calculations for the energy and the specific heat of the discrete U(1) gauge model on four-dimensional lattices of size L^4, where L = 4, 5 and 6. The analysis of our data with the finite size scaling theory [9] described below indicates clearly the occurrence of a second-order phase transition in this model at $e_c^2 = 0.995$ with a correlation exponent $\nu \approx 1/3$.

The Monte Carlo procedure used in our work is the heat bath method of Creutz et al. [5]. We show in fig. 1 the mean plaquette energy E_p as a function of $\beta = 1/e^2$ for a lattice of size L = 5. To achieve convergence to around 1% starting from cold and hot initial lattice configurations, we have found it necessary to sample up to 7000 iterations near the critical region, $0.90 < \beta < 1.0$. For comparison we have also plotted in fig. 1 our calculations for E_p obtained from low and high temperature series expansions [10]. These agree very well with our Monte Carlo re-

[1] Permanent address: The Niels Bohr Institute, 2100 Copenhagen Ø, Denmark.
[2] Permanent address: University of California, Santa Cruz, CA 95060, USA.
[‡1] Evidence for this hysteresis loop was not shown in ref. [5], and therefore the statements made in this paper concerning a phase transition in the U(1) lattice gauge model were not supported by the data.

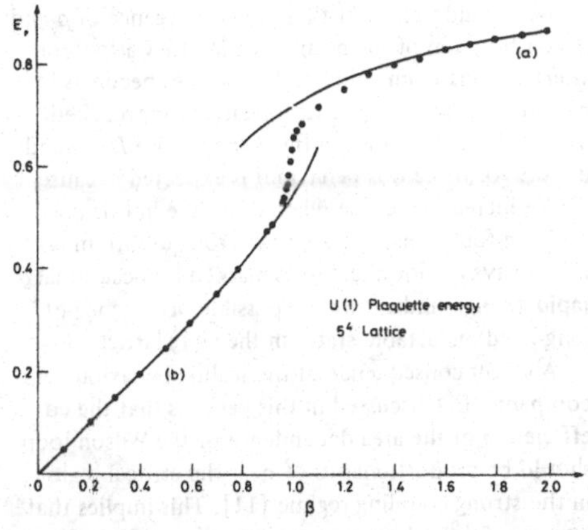

Fig. 1. The U(1) plaquette energy E_P as a function of $\beta = 1/e^2$ for a lattice of size 5^4. The curves labelled (a) and (b) are obtained from low and high temperature series expansions, respectively, where [10]

(a) $E_P = 1 - 1/4\beta - 1/32\beta^2$,

(b) $E_P = \dfrac{\beta/2 + (1/2)(\beta/2)^3 + (49/12)(\beta/2)^5}{1 + (\beta/2)^2 + (1/4)(\beta/2)^4 + (1/36)(\beta/2)^6}$.

sults except in the critical region.

We have evaluated also the specific heat of the U(1) lattice gauge model as a function of β and the lattice size L. We present in fig. 2 our results for the specific heat per plaquette divided by the single plaquette energy fluctuation ‡2, for values of β near the critical region and L = 4, 5 and 6. This normalized specific heat ρ is given by

$$\rho = 1 + \sum_{p' \neq p} [\langle E_p E_{p'} \rangle - \langle E_p \rangle^2] / [\langle E_p^2 \rangle - \langle E_p \rangle^2] , \quad (1)$$

where E_p is the energy (action) of the pth plaquette, and ⟨ ⟩ denotes the thermal average computed by Monte Carlo simulations. It is evident from fig. 2 that ρ has a maximum as a function of β which increases rapidly as a function of L indicating the occurrence of a second-order phase transition in this model.

In order to obtain the critical parameters for the

‡2 We use this normalized specific heat ρ because of its simple asymptotic properties: (a) $\rho \to 1$ as $\beta \to 0$, and (b) $\rho \to d/2$ as $\beta \to \infty$.

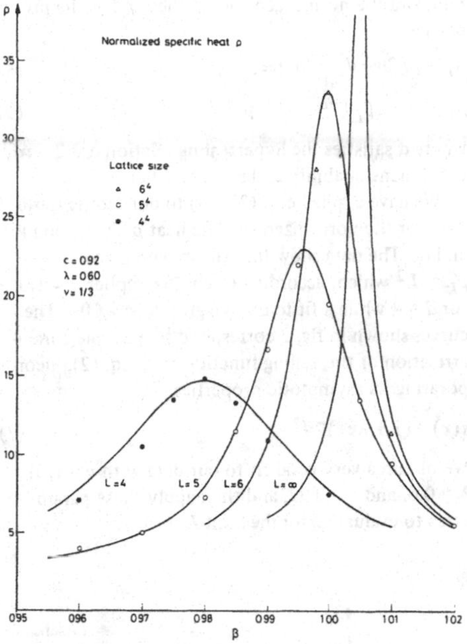

Fig. 2. The normalized specific heat ρ as a function β near the critical region for lattices of size L^4 where L = 4, 5 and 6. The curves are obtained from the scaling function $s(x)$, eq. (7), with $\nu = 1/3$, $\lambda = 0.6$, and $c = 0.92$

infinite lattice we apply a finite size scaling analysis to our data. Assume that the correlation length $\zeta_L(\beta)$ for a system of linear dimensions L takes the form

$$\zeta_L(\beta) = L s(x) , \quad (2)$$

where

$$x = |\beta - \beta_L| L^{1/\nu} \quad (3)$$

and $s(x)$ is a scaling function with the properties that: (a) $s(x)$ is analytic for small x and $s(0) = 1$; (b) as $x \to \infty$, $s(x) \sim x^{-\nu}$, where ν is the correlation length critical exponent. Then $\zeta_L \sim |\beta - \beta_L|^{-\nu}$ for $\beta \sim \beta_L$, and ζ_L has a maximum value $\zeta_L = L$ at $\beta = \beta_L$. It can be readily shown that eqs. (2), (3) imply that for large L

$$\beta_L - \beta_c \sim L^{-1/\nu} , \quad (4)$$

where β_c is the critical value of β for the infinite system, and the normalized specific heat ρ has the properties

$$\rho_L = cL^{2/\nu - d}, \quad \text{at } \beta = \beta_L , \tag{5}$$

$$\rho_L \sim |\beta - \beta_L|^{-\alpha}, \quad \text{for } \beta \sim \beta_L , \tag{6}$$

where α satisfies the hyperscaling relation $\alpha = 2 - \nu d$, d = dimensionability and c is a constant.

We have applied eqs. (3)–(5) to our Monte Carlo data for the normalized specific heat ρ_L to obtain ν and β_c. The data show that the maximum value of $\rho_L \sim L^2$ which, according to eq. (5) implies $\nu \approx 1/3$ for $d = 4$ while a fit to eq. (4) gives $\beta_c = 1.005$. The curves shown in fig. 2 correspond to a simple parametrization of the scaling function $s(x)$, eq. (2), incorporating its asymptotic properties,

$$s(x) = (1 + \lambda x^2)^{-\nu/2} . \tag{7}$$

We obtain a very good fit to our data with $\nu = 1/3$, $\lambda = 0.6$, and $c = 0.92$, and then apply these parameters to evaluate ρ for the limit $L \to \infty$.

We should point out that the convergence of ρ and E as a function of the number of Monte Carlo iterations starting from cold and hot lattices becomes increasingly slow as the critical region is approached. This is illustrated for ρ in fig. 3 in the case $L = 5$ and $\beta = 0.990$. While this behaviour is expected because of the increasing lattice relaxation time in this region, we have found that the approach to equilibrium is not always monotonic, but is marked by occasional rapid changes indicating the possible occurrence of long-lived metastable states in the U(1) lattice.

A direct consequence of the scaling behaviour of compact QED discussed in this paper is that the coefficient a of the area dependence of the Wilson loop should be proportional to ζ^2 near the critical region in the strong coupling regime [11]. This implies that $a \sim |\beta - \beta_c|^{2\nu}$ where we have found $\nu \approx 1/3$. A Monte Carlo study of the Wilson loops for this system should verify this behaviour and provide another determination for β_L and ν.

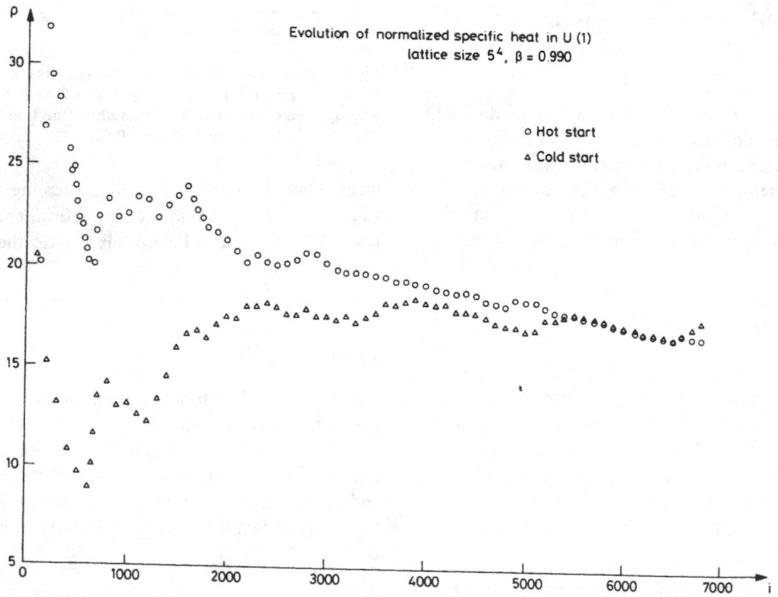

Fig. 3. The normalized specific heat ρ as a function of the number of Monte Carlo iterations for $\beta = 0.990$ and $L = 5$ starting from a cold and a hot lattice.

Volume 95B, number 1 PHYSICS LETTERS 8 September 1980

We would like to thank P. Hasenfratz and M. Virasoro for many valuable and stimulating discussions. One of us (M.N.) is indebted to M. Peskin for enlightening comments on U(1) lattice gauge theory. We appreciate also the special support of L. Van Hove, J. Prentki and V. Soergel to carry out our computer calculations, and we wish to thank them for the hospitality extended to us at CERN.

References

[1] K.G. Wilson, Phys. Rev. 10 (1974) 2245.
[2] K. Osterwalder and E. Seiler, Ann. Phys. (NY) 110 (1978) 440.
[3] A.H. Guth, Cornell preprint (1980).
[4] For a recent review on lattice gauge theory, see: J.B. Kogut, Rev. Math. Phys. 51 (1979) 659.
[5] M. Creutz, L. Jacobs and C. Rebbi, Phys. Rev. D20 (1979) 1915.
[6] M. Creutz, Phys. Rev. Lett. 43 (1979) 553.
[7] M. Creutz, Brookhaven preprints (1979, 1980).
[8] C. Rebbi, Brookhaven preprint (December 1979).
[9] M.E. Fisher, in: Proc. Intern. School of Physics Enrico Fermi (Varenna, 1970) Course No. 51, ed. M.S. Green (Academic Press, New York, 1971).
[10] B. Lautrup and M. Nauenberg, in preparation.
[11] M. Peskin, Ann. Phys. 113 (1978) 122.

PHYSICAL REVIEW D VOLUME 22, NUMBER 10 15 NOVEMBER 1980

Topological excitations and Monte Carlo simulation of Abelian gauge theory

T. A. DeGrand
Department of Physics, University of California, Santa Barbara, California 93106

Doug Toussaint
Department of Physics, and Institute for Theoretical Physics, University of California, Santa Barbara, California 93106
(Received 24 June 1980)

We study the phase structure of lattice electrodynamics in three and four dimensions using Monte Carlo simulation, with special emphasis on the topological excitations of the theory. We formulate an operational definition of a monopole and measure the density of monopoles as a function of coupling constant. In three dimensions and for strong coupling in four dimensions monopoles screen external magnetic fields. Below a critical coupling in four dimensions the external field penetrates into the bulk of the medium; this long-range correlation essentially shows that the lattice theory in weak coupling is characterized by a massless photon.

I. INTRODUCTION

Lattice gauge theories provide an attractive way of studying the strong-coupling behavior of gauge theories.[1] Even the simplest such theory, where the internal symmetry group is U(1), is of theoretical interest. In three dimensions this theory shows confinement for all values of the coupling. This behavior can be understood as the result of topological excitations in the theory.[2,3] In four dimensions the theory must have two phases.[4] The high-temperature phase, common to all lattice gauge theories, is a confining phase. The low-temperature phase provides the continuum limit of the theory. If the continuum limit is to be free electrodynamics, then the low-temperature phase must be somewhat special; it must contain a massless photon.

The recent work of Creutz, Jacobs, and Rebbi[5] and of other authors[6] has shown that Monte Carlo simulation is a valuable tool for investigating gauge theories. In this paper we describe a Monte Carlo study of Abelian lattice gauge theory from the point of view of the topological excitations.

In three dimensions the topological excitations are pointlike magnetic monopoles. Using an operational definition of a magnetic monopole, we find that our Monte Carlo simulations do contain monopoles. We then impose boundary conditions corresponding to immersing the system in an external magnetic field and find that the monopoles move in such a way as to screen the field. All this is exactly as expected; the strong-coupling phase is like a magnetic superconductor where electric flux is confined to flux tubes, and magnetic fields are screened.

In four dimensions the topological excitations are strings of monopole current. It is thought that the phase transition in the four-dimensional theory arises from the unbinding of closed loops of monopole string.[3] We find that when an external magnetic field is applied to a four-dimensional system there is a dramatic change in its behavior at $1/e^2 = \beta \approx 0.99-1.00$. For smaller β the field is shielded, while for larger β a finite field penetrates the bulk of the material. Thus, by observing a long-range field, we obtain a direct verification of the transition to a Coulomb phase.

Most of our calculations have used the Wilson form of the action. In this form the partition function is

$$Z_{\text{Wilson}} = \int [d\Theta_\mu(r)] \exp\left[\beta \sum_{r,\mu<\nu} \cos\Theta_{\mu\nu}(r)\right], \quad (1.1)$$

where Θ_μ is the angular variable associated with the link at r in the μ direction and

$$\Theta_{\mu\nu}(r) = \Theta_\mu(r) + \Theta_\nu(r+\hat{\mu}) - \Theta_\mu(r+\hat{\nu}) - \Theta_\nu(r)$$

is the sum of the four angles around the unit plaquette. In this form the theory has a phase transition at $\beta \approx 0.99$. We have also carried out some simulations using the Villain form of the partition function,

$$Z_{\text{Villain}} = \int [d\Theta] \sum_{r,\mu<\nu} \sum_{n=-\infty}^{\infty} \exp[-\tfrac{1}{2}\beta(\Theta_{\mu\nu}(r) - 2n\pi)^2] \quad (1.2)$$

We find that this action yields the same qualitative behavior as the Wilson action [Eq. (1.1)], but the transition point is at $\beta_c \approx 0.62$.

The internal symmetry group under which Θ transforms is chosen to be $Z(N)$ with N large—50–200, typically. $Z(N)$ theories for $N \geq 6$ are known to have three phases—a low-β electrically confining phase, a medium-β phase which is presumed to be Coulombic, and a very-large-β phase which is magnetically confining.[5,7] The high-β critical point is known to be proportional to N^2 so that for the large N's studied here it occurs at β much too high to be seen. So the $Z(N)$ theories are equivalent to U(1) for all practical purposes, and we shall often refer to them as such.

Our interest in $Z(N)$ rather than U(1) symmetry is purely technical. The time required for performing computations is considerably shortened, allowing us to study larger systems than would otherwise be possible. With this preparation we turn to the actual computations.

In Sec. II we briefly discuss Monte Carlo simulation of U(1) gauge theory. In Sec. III we describe how to find topological excitations in Monte Carlo data, and Sec. IV is an analysis of the behavior of these systems in external magnetic fields. Our conclusions are summarized in Sec. V. Some technical details of our Monte Carlo methods and a review of some simple theoretical ideas about monopoles are relegated to the appendices.

II. MONTE CARLO SIMULATION

The aim of Monte Carlo simulation is to generate a sequence of configurations for the system in such a way that the probability of producing a given configuration is given by Boltzmann weighting. That is, if two configurations have actions S_1 and S_2, the ratio of the probabilities of finding the configurations should be

$$\frac{P_1}{P_2} = \exp(S_1 - S_2) . \tag{2.1}$$

Monte Carlo simulations typically generate new configurations from old configurations by changing one variable at a time according to an algorithm designed to reproduce the configuration weighting (2.1) after many cycles. This process is repeated for every variable in the system. (Application of the algorithm to every variable in the system will be called "a pass through the system.") We use the standard Metropolis method for generating configurations.[8] After a sufficient number of passes for the system to each equilibrium (i.e., to become essentially independent of the starting configuration) expectation values of physical quantities may be measured by averaging their values over many successive configurations. Because configurations on successive passes are highly correlated, it is convenient to make several passes through the system between measurements. Further comments on our Monte Carlo program will be deferred until Appendix A.

Perhaps the most fundamental quantity that can be measured is the average energy (action). This quantity was extensively studied in the original work of Creutz, Jacobs, and Rebbi.[5] In three dimensions the energy is an entirely smooth function of the inverse temperature β, consistent with a one-phase structure. In four dimensions the energy shows pronounced hysteresis for β near 1, as expected if there is a phase transition there. One can test whether the putative phase transition is first order or continuous by beginning with a completely random or completely ordered initial configuration, presumably typical of the large- and small-β phases, and watching the energy as a function of the number of passes through the lattice. As shown in Fig. 1 and in the results of Creutz *et al.*,[5] the internal energy in the two cases converges to a common value, indicating that the phase transition is continuous.

The standard probe of the physics of a lattice gauge theory is the Wilson loop,

$$W(C) = \left\langle \mathrm{Tr} \sum_{i\,C} U_i \right\rangle . \tag{2.2}$$

Expectation values of small Wilson loops can easily be evaluated in a Monte Carlo simulation. We expect that in a confining phase large Wilson loops will be proportional to exp(−area), while in a phase with free charge large loops behave as exp(−perimeter). The expectation values of small loops in four dimensions are graphed in Fig. 2. For small β, these small loops show an area-law behavior. However, for $\beta > 1$ the behavior is unclear. Our data for loops up to 3×3 for $\beta > 1$ do not fit well to any simple form. We expect small loops to show large corrections to the asymptotic form in a phase with massless particles.[9] Clearly, we want better indicators of the nature of the phases than small Wilson loops. To find such indicators we turn to a study of the topological excitations of the theory, monopoles in three dimensions, and monopole strings in four dimensions.

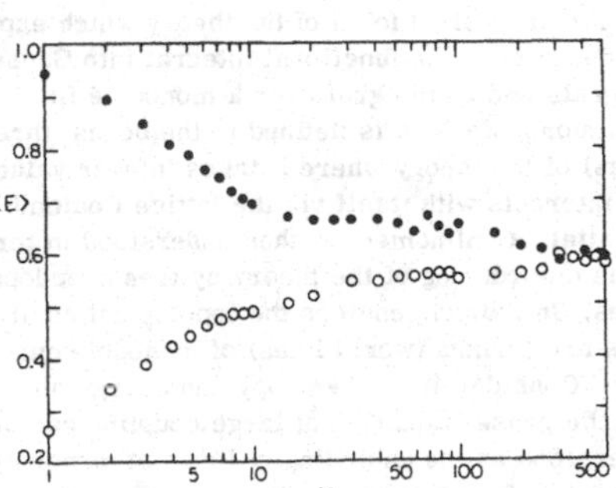

FIG. 1. Internal energy $\langle E\rangle$ vs N, the pass number, for a 6^4 $Z(100)$ system at $\beta = 0.99$, very close to the critical temperature.

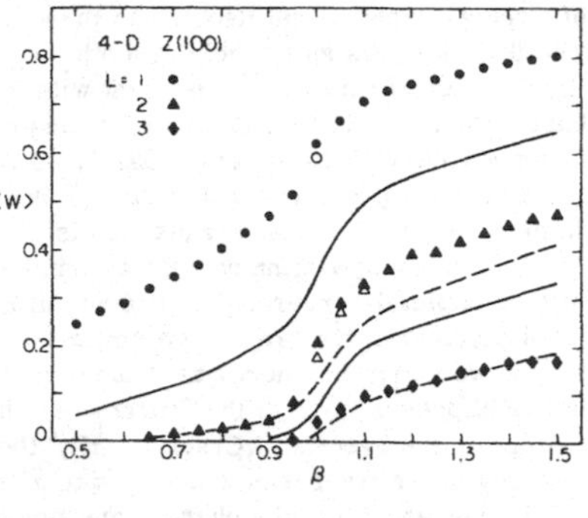

FIG. 2. Wilson loops of size $L \times L$ in $Z(100)$ on a 5^4 lattice, for $L = 1,2,3$. The smooth lines are direct extrapolations of a perimeter law from one L to $L+1$; the broken lines are extrapolations of an area law. Closed symbols represent data taken in runs where β decreased; open symbols show data taken in runs where β increased, for β near β_c, where hysteresis is seen.

III. MONOPOLES IN LATTICE THEORIES

It is well known that confinement in three-dimensional Abelian gauge theories can be understood in terms of magnetic monopoles. This is true both in lattice gauge theory with a compact gauge group[3] and in a continuum theory where the U(1) symmetry results from the breaking of a compact group via the Higgs mechanism.[2] This mechanism may be analogous to the effects of instantons or merons in four-dimensional non-Abelian theories.[10] The analysis of three-dimensional lattice theory by Banks, Myerson, and Kogut,[3] which follows the analysis of the two-dimensional XY model by José, Kadanoff, Kirkpatrick, and Nelson,[11] proceeds by performing a series of transformations on the partition function for the Villain form of the theory which explicitly decompose the functional integral into Gaussian integrals and an integral over a monopole field. This monopole field is defined in the boxes (three-cubes) of the theory where it takes integer values, and interacts with itself via the lattice Coulomb potential. Confinement is then understood in terms of the disordering of the theory by free pseudoparticles. In four dimensions the topological excitations are strings (world lines) of monopole current. Consideration of entropy and energy suggests that the phase transition at large coupling can be understood as the unbinding of loops of monopole current to form a condensate.

We are naturally led to ask whether monopoles are present in Monte Carlo simulations and whether the observed behavior of the theory can be understood in terms of these monopoles. We begin with the simplest question: Are there monopoles in three dimensions?

We search for monopoles by using Gauss's law. By measuring the total magnetic flux emanating from a closed surface in the lattice we can determine whether or not the surface encloses a monopole. For small angles Θ, the flux is defined by

$$ds \cdot B = \sum_{\text{surface}} ds_a \epsilon_{abc} \tfrac{1}{2} (\nabla_b \Theta_c - \nabla_c \Theta_b) = \sum_{\text{surface}} \Theta_p \,, \tag{3.1}$$

where Θ_p is the oriented plaquette angle, a gauge-invariant quantity. Clearly, if we used exactly this definition, we would find a net flux of zero for any closed surface since each link would be included twice, once with each sign. However, our definition of flux should be periodic in the Θ_p. In particular, if the plaquette angle is 2π, the plaquette carries zero energy. Such a configuration should be regarded as a Dirac string passing through the plaquette. Our algorithm for evaluating the flux is as follows. We assume that the plaquette angle $\Theta_{\mu\nu}$ consists of two pieces: physical fluctuations which lie in the range $-\pi$ to π, and Dirac strings which carry 2π units of flux. Defining $\overline{\Theta}_{\mu\nu} = \Theta_{\mu\nu} - 2\pi n_{\mu\nu}$, where $n_{\mu\nu}$ is the number of strings through the plaquette, we measure the monopole number M inside a surface:

$$2\pi M = \sum_{\text{surface}} \overline{\Theta}_{\mu\nu} \tag{3.2}$$

$$= \sum_{\substack{\text{boxes}\\ \text{inside surface}}} \nabla_\mu \epsilon_{\mu\nu\lambda} \overline{\Theta}_{\nu\lambda} \tag{3.3}$$

$$= 2\pi \sum_{\text{boxes}} \epsilon_{\mu\nu\lambda} \nabla_\mu n_{\nu\lambda} \,. \tag{3.4}$$

Equation (3.3) is the lattice equivalent of $M(x) = \vec{\nabla} \cdot \vec{B}$ and (3.4) says that the monopole number in a volume is given by the net number of Dirac strings entering the volume. It should be clear that adding multiples of 2π to any of the link variables in a configuration can move the Dirac strings around but cannot change the net number entering a volume, and that the monopole number in any volume is the algebraic sum of the monopole numbers of its subvolumes. Therefore, to count the monopoles in our system we examine cubes of unit volume. Examples taken directly from computer data of a monopole configuration and a monopole-antimonopole pair are shown in Fig. 3. By looking at successive Monte Carlo configurations, one can observe the movement of monopoles and the creation and annihilation of monopole-antimonopole pairs.

Because monopoles are collective excitations of many link variables, they move slowly in Monte

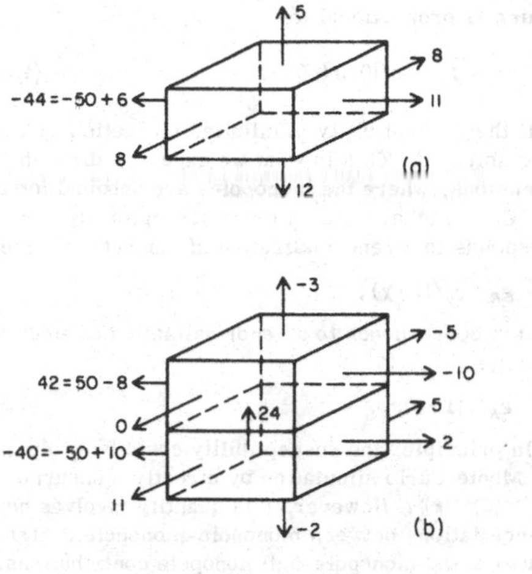

FIG. 3. A monopole (a) and a monopole-antimonopole pair (b) in $Z(50)$. Arrows label the flux out of each face. The strings, carrying ± 50 units of flux, flow through the left-hand faces.

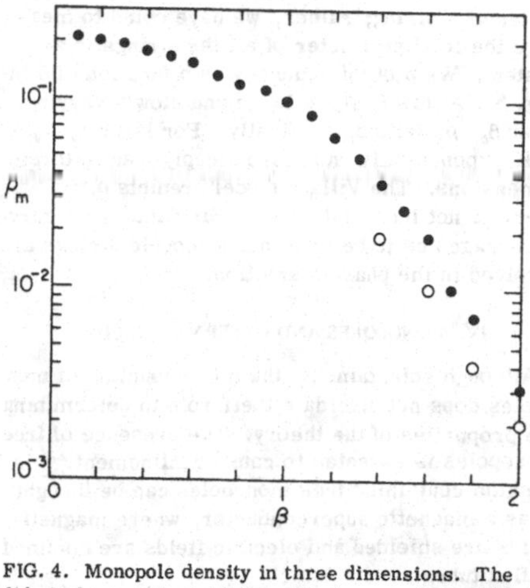

FIG. 4. Monopole density in three dimensions. The solid circles are the total density of monopoles, while the open circles are the density of "isolated" monopoles, for which no adjacent box contains an antimonopole.

Carlo simulations. To change the monopole number inside a surface, or to move the monopole around, at least one plaquette must pass through its maximum energy.

The definition of monopoles used here is similar to a definition of vortices in the XY model used in Monte Carlo calculations by Chester and Tobochnik.[12] It should be emphasized that our decomposition of a particular configuration into monopoles and fluctuations is not exactly the same as Banks, Myerson, and Kogut's decomposition of the functional integral into integrations over Gaussian variables and monopole variables. It is only at very large β (low temperature), where monopoles are heavy and Gaussian fluctuations really are small, that the two decompositions should coincide. Among other differences, the monopole field defined by Banks, Myerson, and Kogut can take on all integral values, while the monopole number used here is kinematically limited to be no larger than two in a unit cube.

The simplest measurement we can make is the density of monopoles. The average density of monopoles or antimonopoles $\rho_m = \frac{1}{2}(\langle N_{M^+}\rangle + \langle N_{M^-}\rangle)/V$ as a function of β is displayed in Fig. 4. It is seen to be a smoothly falling function of β. For the range of β in which Monte Carlo calculations are practical, most of the monopoles and antimonopoles are in pairs, with a monopole and antimonopole in adjacent boxes. Figure 4 also shows the density of "isolated" monopoles—those for which none of the neighboring cubes contains an antimonopole. It can be seen that as β increases, the density of close pairs falls off more rapidly than the density of isolated monopoles. Both the density of pairs and the density of isolated monopoles appear to be falling off exponentially with slopes in reasonable agreement with theoretical estimates using the Villain form or the dilute-gas approximation (DGA). These estimates are discussed in Appendices B and C. Finally we remark that for $\beta > 1$, nearly all our monopoles have $M = \pm 1$. At lower β, occasional $M = \pm 2$ monopoles are seen. Simple arguments suggest $\rho_{M=2} \approx \rho_{M=1}{}^4$, so it is not surprising that double monopoles are scarce objects.

The pointlike typological excitations in three dimensions become one-dimensional objects in four dimensions. These objects are the world lines of three-dimensional monopoles, or continuous strings of monopole current. The presence of a monopole current in a given direction at any point in space may be found by the same method used in three dimensions, simply measuring the flux (or counting Dirac strings) flowing through a three-dimensional surface oriented normally to the current. That is, we measure

$$2\pi M_\mu(x) = \epsilon_{\mu\nu\alpha\beta}\nabla_\nu \bar{\Theta}_{\alpha\beta}(x) \tag{3.5}$$

as a test for a monopole current at x. Conservation of topological charge demands that $\partial_\mu M_\mu(x) = 0$, so monopole strings form continuous loops. The study of monopole strings is somewhat more complicated than the study of monopoles. We have not found it profitable to measure the average perimeter of a string; rather, we have opted to measure the total perimeter of all the strings in a system. We plot this quantity as a function of β in Fig. 5. At low β, ρ_M is large and slowly varying. Near β_c, ρ_M falls dramatically. For large β, ρ_M falls exponentially and more steeply than in three dimensions. The Villain model predicts $\rho_s \sim e^{-\pi^2\beta}$, which is not inconsistent with our data. This curve encourages us to believe that monopole strings are involved in the phase transition.

IV. MONOPOLES AND EXTERNAL FIELDS

Although entertaining, the mere counting of monopoles does not elucidate their role in determining the properties of the theory. The presence of free monopoles is expected to cause confinement.[2,3] A medium containing free monopoles can be thought of as a magnetic superconductor, where magnetic fields are shielded and electric fields are confined to flux tubes.

The quantity of interest is the polarizability of the monopole gas[13]

$$\chi_{ij} \approx \frac{\partial^2}{\partial B_i \partial B_j} \ln Z\,, \tag{4.1}$$

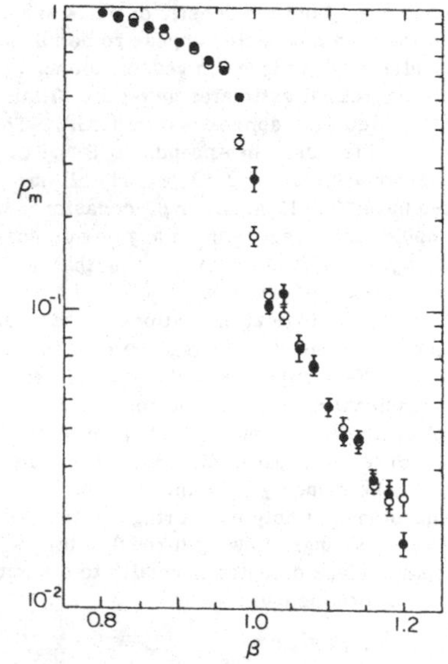

FIG. 5. Expectation value of the monopole string density vs β in $Z(100)$ for a 5^4 lattice. Open circles are found by stepping down in β, closed circles by stepping up in β. At each point the first 50 passes are discarded, then 40 measurements, each separated by five passes, are carried out.

which is proportional to

$$\chi \approx -\sum_r \langle r^2 M(0) M(r) \rangle\,. \tag{4.2}$$

If the susceptibility is infinite, magnetic fields are shielded. This is what we expect in three dimensions, where the monopoles are unbound for all β. On the other hand, a finite susceptibility corresponds to a renormalization of magnetic charge

$$g_R = g_0/(1+\chi)\,,$$

which corresponds to a renormalization of electric charge

$$e_R = (1+\chi) e_0\,. \tag{4.3}$$

In principle, the susceptibility could be evaluated in Monte Carlo simulation by directly measuring $\langle r^2 M(0) M(r)\rangle$. However, this quantity involves near cancellations between monopole-monopole contributions and monopole-antimonopole contributions. As a result, it is difficult to evaluate accurately and we experienced little success in our attempts to do so.

A better way to proceed is to introduce an external magnetic field and study the response of the monopoles to it by measuring the field deep inside the system. The magnetic field inside the system will be a sum of the external field and a field generated by the monopoles in terms of a permeability μ, where $\mu = 1+\chi$, $\vec{H}_{\text{external}} = \mu \vec{H}_{\text{internal}}$. For parallel-plate geometry we measure the magnetic flux Φ through planes perpendicular to the external field. Then H is proportional to Φ and a susceptibility may be measured directly:

$$\chi = \frac{\Phi_{\text{external}}}{\Phi_{\text{internal}}} - 1\,.$$

We begin by considering three dimensions. Consider a finite cube of a lattice gauge theory, illustrated in Fig. 6, and fix all the plaquette angles on the xy faces to some small value. This amounts to introducing a uniformly distributed magnetic flux through the face. We then put in periodic boundary conditions on the other faces of the cube. However, unless the flux is a multiple of 2π, we must introduce a twist in the boundary conditions.[13-15] That is, when leaving the top (xz) boundary of the cube and reentering the bottom, we add some amount to the links in the x direction. Clearly, the sum of the twist angles on the boundary of a face must equal the magnetic flux through the face modulo 2π. In essence, by fixing the flux at the boundary we have put our medium in a magnetic capacitor. We can now use the Monte Carlo algorithm on this system and measure the location of the monopoles. Equivalently, we may measure the total flux (as defined in Sec. III) through each xy plane.

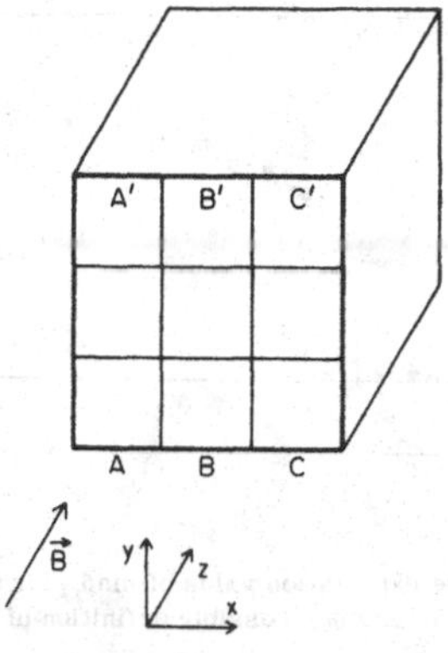

FIG. 6. A finite three-dimensional system. An evenly distributed magnetic flux in the Z direction is imposed by fixing all the plaquette angles on the front face to some value. A "twist" in the boundary conditions involves setting link A' equal to link A plus some increment. Periodic boundary conditions are used in the other two directions.

Some of the results of this three-dimensional experiment are shown in Fig. 7. We see that for β of order 1 the magnetic field is screened over a range of one or two lattice spacings. Thus we can actually observe the screening of magnetic charge by the free monopoles in the medium.

In four dimensions the experiment is more interesting. It is easiest to imagine the experimental situation by thinking of three dimensions plus time. For each value of time, we fix plaquettes exactly as in three dimensions. This amounts to embedding the system in an external field oriented in the z direction which is time independent and constant in x and y. We measure the time-averaged flux through planes of constant z. The quantity $\Phi(z)/\Phi(z=0)$ is plotted for two values of β in Fig. 8, using data taken on a $5\times5\times9\times5$ system. At small β the external field is again shielded after a small penetration. However, for $\beta>1$ the situation is completely different. The field in the center of the system is a constant fraction of its value at the edge. Thus the susceptibility is finite, and magnetic charge is renormalized by the monopole strings.

By testing the ability of the medium to support a long-range magnetic field, we are essentially testing for a massless photon. Thus our measurements give a direct verification of the Coulomb phase of U(1) lattice theory.

We have also measured the expectation value of $\sin\Theta_p$ in planes perpendicular to the flux. This is another possible definition of the physical flux and

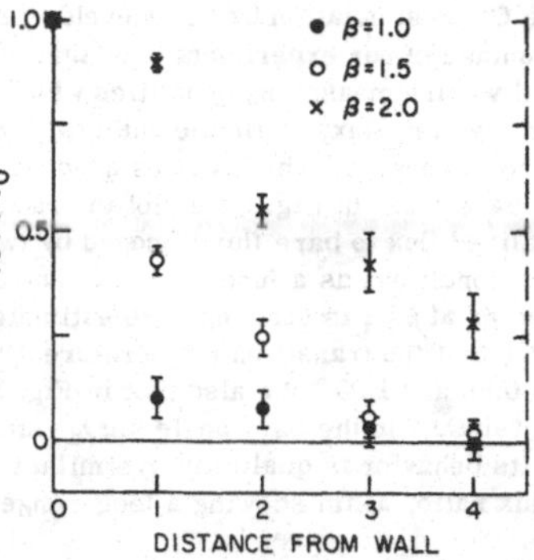

FIG. 7. Screening of a magnetic field in three dimensions in a 10^3 system. The plaquettes in one plane (at the left) are fixed to θ_0, and we plot the expectation value of the reduced angle $\bar{\theta}$ as a function of distance from the fixed plane.

behaves qualitatively like our usual definition.

Clearly the quantity in which we are most interested is the expectation value of the flux at an infinite distance from the plane of fixed plaquettes. Imagine extending our system a long way in the z direction and chopping a segment out of the middle. Information about the bare flux through the system is carried by the twist in the boundary conditions, so we may still carry out the experiment. However, the twist only defines the flux modulo 2π, so we are restricted to fluxes between $-\pi$ and π. We have now arrived at precisely the twisted boundary conditions introduced by 't Hooft[14] and used

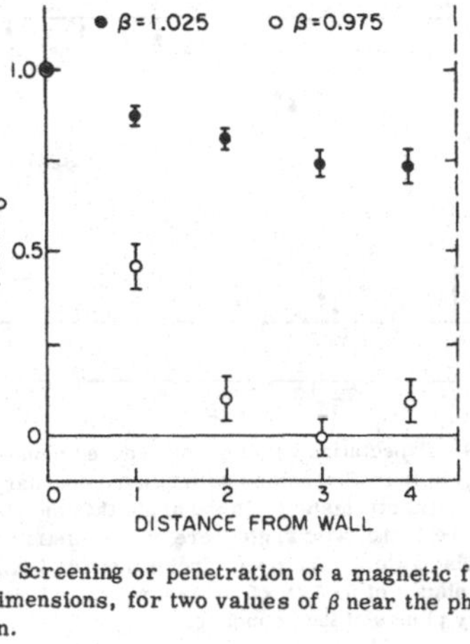

FIG. 8. Screening or penetration of a magnetic field in four dimensions, for two values of β near the phase transition.

in Monte Carlo simulation by Groeneveld *et al.*[15] The advantage of our experiment over that of Ref. 15 is that we are measuring quantities which are linear in the bare flux, while the authors of Ref. 15 studied the energy, which varies quadratically with the bare flux. In Fig. 9 we plot the ratio of renormalized flux to bare flux imposed by twisted boundary conditions as a function of β. The change in character at $\beta \approx 1$ is striking. We estimate from this curve that the transition temperature β_c lies between 0.98 and 1.00. We also plot in Fig. 10 the ratio of $\langle \sin\Theta_{xy}\rangle$ to the bare angle $\sin\Theta_0 = \sin(\text{flux}/N_x N_y)$. Its behavior is qualitatively similar to that of the flux ratio, again showing a long-range field for $\beta > 1$.

Theoretical arguments suggest that the phase transition may occur when the renormalized coupling reaches a universal critical value.[13] This would mean that as β is lowered toward β_c the susceptibility increases to a finite value and then suddenly becomes infinite. In this case we would expect the "renormalized field" that we are measuring to have a discontinuity at β_c. Our data are insufficient to decide whether the flux measured here has such a discontinuity. We repeat that the operational definitions of renormalized flux used here are not quantitatively the same as the definition used in the theoretical arguments, but we can expect them to have qualitatively similar behavior.

We have also carried out some experiments on the Villain form of the theory, where the exponential of the plaquette action is given by

$$\exp[S(\Theta_p)] = \sum_{n=-\infty}^{+\infty} \exp[-\tfrac{1}{2}\beta(\Theta - 2\pi n)^2]\,. \qquad (4.4)$$

This form of the action is of interest because it

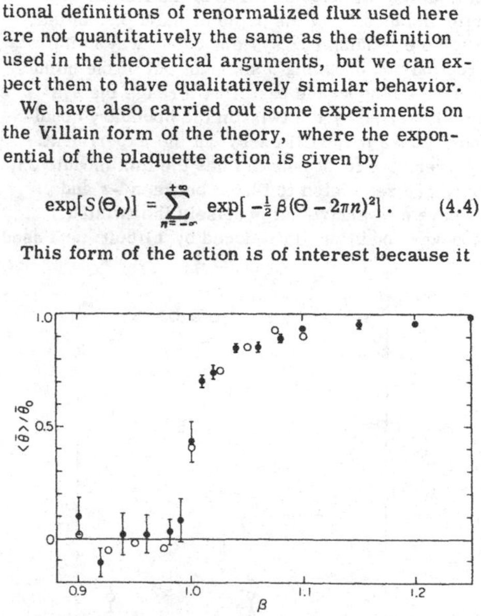

FIG. 9. Expectation value of the reduced plaquette angle $\bar{\theta}_{xy}$ in four dimensions with twisted boundary conditions (solid circles). θ_0 is the angle that the plaquettes would have if the twist angle were evenly distributed over all the plaquettes. The open circles are $\langle\bar{\theta}\rangle$ in the four central planes of $5\times5\times9\times5$ system with fixed plaquettes on the xy planes of the boundary.

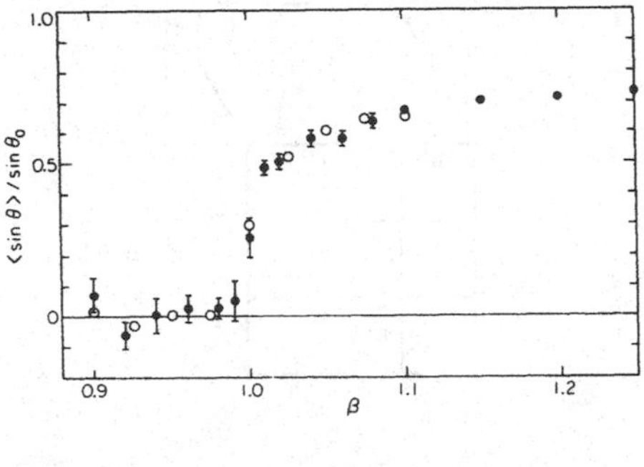

FIG. 10. The expectation value of $\sin\theta_{xy}$ for the same sample. This is another possible definition of the magnetic field.

lends itself to theoretical analysis. It is in this form of the theory that the partition function can be explicitly decomposed into Gaussian fluctuations and monopole excitations,[3] and it is the Villain form of the action that provides a strikingly good approximation to a fixed line under the Migdal approximate renormalization-group transformation.[11,16] In Fig. 11 we plot $\mu^{-1} = \Phi_{\text{renormalized}}/\Phi_{\text{bare}}$ for this theory, showing a phase transition at $\beta_c \sim 0.62$ or 0.63 and the same sharp behavior of μ as we saw before.

V. CONCLUSION

In this study of lattice U(1) gauge theory we have demonstrated that it is possible to formulate a description of a monopole which is accessible to measurement in Monte Carlo simulation. We have

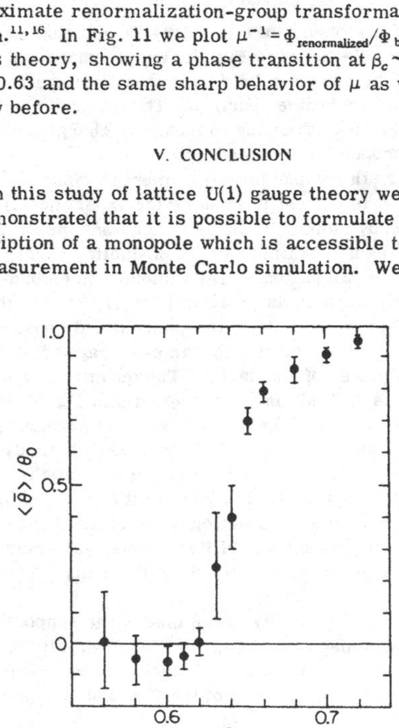

FIG. 11. $\langle\bar{\theta}_{xy}\rangle$ in the Villain form of the theory with twisted boundary conditions.

shown that these monopoles screen external magnetic fields when the theory is in an electrically confining phase. The long-distance penetration of the external field in the high-temperature phase of the four-dimensional theory is evidence that this phase possesses a massless photon. In addition, we are able to measure the renormalization of the external field in the Coulomb phase due to the monopole currents.

Although we have studied systems of only modest size, the nature of the effects we have measured are such that they should be equally valid in systems of infinite extent. In that respect, monopoles and external fields are much better signals for the nature of the phases of a system than are measurements of small Wilson loops whose area and perimeter are of comparable magnitude.

Two extensions of our work suggest themselves. It should be a simple matter to study the Abelian Higgs model in an external field and map the phase boundary between the Higgs/confinement sector and the massless photon sector. It may also be possible to extend the techniques of this paper to studies of topological excitations in non-Abelian gauge or spin theories, monopoles, or instantons. Work on these problems is in progress.

We believe that the results of our studies provide an extremely physical picture of the interplay of topological excitations and critical behavior in lattice photondynamics. Our results also indicate the profitability of using Monte Carlo techniques to study collective excitations of the degrees of freedom of gauge and spin systems.

ACKNOWLEDGMENTS

We thank John Cardy, Sudip Chakravarty, Mike Creutz, Lawrence Jacobs, John Kogut, and Bob Pearson for discussions. This work was supported by the National Science Foundation under Grants Nos. PHY78-08439 and PHY77-27084.

APPENDIX A

This appendix contains some technical comments on our Monte Carlo procedure and on our error estimation procedure. The basic principles of Monte Carlo simulation have been much discussed in the recent literature,[5,6,17] so we concentrate here on some points special to our approach. As noted in the Introduction, we approximated U(1) by $Z(N)$ for N reasonably large. This allowed us to use integer arithmetic in evaluating the plaquette angles. Also, because in $Z(N)$ a plaquette angle has roughly $4N$ possible values, one can simply tabulate all the possible Boltzmann weights at the beginning of the program. Hence no trigonometric functions or exponentials need to be evaluated in the inner loops of the program. We used values of N ranging from 50 to 200 and found no detectable dependence on N.

We used boundary conditions which differ slightly from the usual periodic boundary conditions. Figure 12 illustrates our "skewed" boundary conditions in two dimensions. The blocks represent a finite two-dimensional system replicated an infinite number of times. Ordinary periodic boundary conditions amount to connecting the blocks (really all the same block) as illustrated in Fig. 12(a), while skewed boundary conditions involve connecting the blocks as in Fig. 12(b). Operationally the skewed boundary conditions require that every time one passes out of the right-hand side of the lattice, he reenters the lattice from the left-hand side, but one row higher. In general, when we leave our finite system in the direction of the ith coordinate, we imagine reentering the system with the ith coordinate reset to zero, but the $(i+1)$th coordinate increased by one. The advantage of this method is that all the variables may be arranged in a one-dimensional array $x(j)$. The structure of a d-dimensional lattice is then described by d numbers $J(i)$, $i=1$ to d, which tell how to move one lattice spacing in any direction. The variable which is one unit away from a given variable $x(k_0)$ in the ith direction is simply $x(k_0+J(i))$. The only remaining problem is that this rule may sometimes lead to an index slightly outside the array. This is easily handled by appending a "virtual" copy of the first part of the array onto the end of the array, and appending a copy of the end of the array to the beginning of the array (see Fig. 13). If the action is

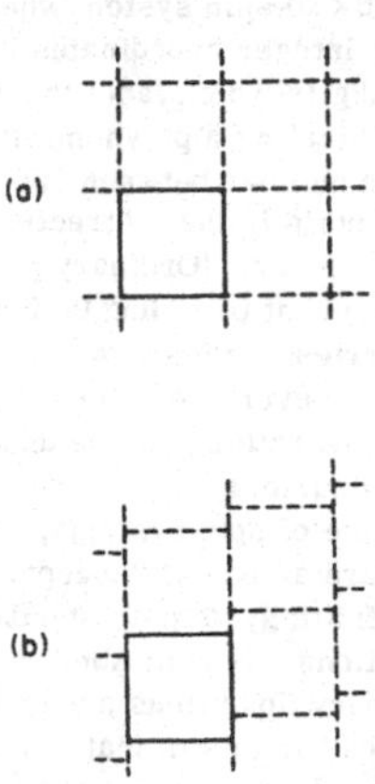

FIG. 12. Periodic boundary conditions. Each square represents a copy of a finite system. (a) Ordinary periodic boundary conditions; (b) skewed boundary conditions.

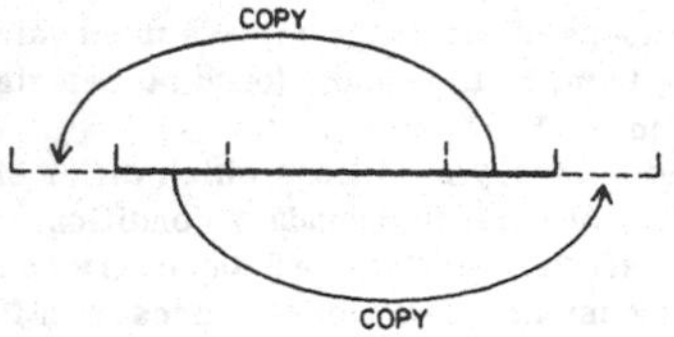

FIG. 13. The virtual lattice. The heavy line indicates the real lattice, stored as a one-dimensional array. The dashed lines indicate the virtual lattice, which is a copy of part of the real lattice.

defined on an elementary plaquette, the fraction of the array that must be duplicated is $1/N$, where N is the size in the most slowly varying direction. Of course, we do not apply the Monte Carlo algorithm to the spins in the virtual array, but simply update them to duplicate the "real" array as necessary.

The advantage of these boundary conditions is that to find a link one site away in the ith direction from the current link we simply add an offset to the index. In contrast, in the most naive implementation of periodic boundary conditions it is necessary to increment one coordinate, reduce it to the correct range (e.g., by the modulo function), and then evaluate a $(d-1)$-order polynomial to locate it anyway, since computer memory is a one-dimensional array. Clearly, the speed advantages of skewed boundary conditions increase rapidly as the number of dimensions increases.

An analogy may help the reader who is confused. Imagine a $10\times 10\times 10$ spin system where a site is labeled by three integer coordinates between 0 and 9. Placing the digits which label the site side by side (that is, evaluating a polynomial), we find each site labeled by a number between 000 and 999. To translate by n_y units in the y direction, we simply add $0n_y0$ to the number. (Ordinary periodic boundary conditions amount to adding $0n_y0$ with the proviso that all carries are discarded.) In the actual computation, one never needs the three coordinates of a point; everything can be done using the one (three digit) number.

Even in the case of spin systems, skewed boundary conditions are an equally acceptable approximation to an infinite system as standard periodic boundary conditions. For an Abelian gauge theory these two boundary conditions are gauge equivalent. The reason for this is that any configuration that is skewed periodic can be gauge transformed into a configuration which is periodic in the ordinary sense.[13] The only gauge-invariant quantity which can characterize boundary conditions is the Wilson loop going around the entire system.[13]

Clearly, for skewed periodic or ordinary periodic boundary conditions, this Wilson loop is equal to the identity element. However, this loop may be given a nontrivial value by introducing a "twist" in the boundary conditions.[14,15] This can be accomplished, for example, by setting the links along the top edge of the square in Fig. 12 equal to the links on the bottom edge plus some increment. The sum of all these increments, the twist, is gauge invariant and is equal modulo 2π to our usual definition of the magnetic flux through the loop in any particular configuration. With our representation of the lattice, such a twist is easily imposed by adding the increments to the appropriate elements on the virtual lattice illustrated in Fig. 13.

The alert reader may have noted that with skewed boundary conditions a plane that naively divides the lattice—e.g., the plane enclosed by the Wilson loop made up of N steps in the x direction, N steps in the y direction, N steps in the $-x$ direction, and N steps in the $-y$ direction—may not be quite closed. We avoid this problem by choosing the plane defined by the two most slowly varying coordinates when imposing nonzero twists or measuring fluxes. In that plane these annoyances do not occur.

We use the standard Metropolis method for generating configurations.[8,17] When examining a particular variable, we choose a trial value for that variable which lies within a fixed range of the old value. The size of this range is adjusted to provide the desired acceptance ratio (fraction of variables changed in one pass). We used acceptance ratios of 0.4 or 0.5 for most of our work. The efficiency of the simulation appeared largely insensitive to the exact value of the acceptance ratio.

Our programs were written in FORTRAN and run on a VAX 11-780 computer. On this machine one pass through a 5^4 lattice (2500 variables) required 1.5 sec.

Because lattice configurations separated by one Monte Carlo pass are highly correlated, it is advantageous to make several passes between measurements of the quantities to be evaluated. We typically made three to five passes between measurements. Even so, very close to β_c in four dimensions we sometimes observed correlation coefficients of successive measurements as high as 0.8. A typical run used 100 to 200 passes to thermalize the lattice, followed by 200 to 500 passes during which measurements were made.

The statistical uncertainty of a Monte Carlo measurement may be estimated in several ways. The simplest method, valid when correlations are small, is to take the standard deviation of the mean corrected for correlations

$$\Delta x^2 = \frac{1}{N-1}\left(\frac{1}{N}\sum x_i{}^2 - \bar{x}^2\right)(1+2c_1+\cdots), \quad \text{(A1)}$$

where the x_i are the individual measurements and c_1 is the correlation coefficient of successive measurements

$$c_1 = \frac{\langle x_i x_{i+1}\rangle - \bar{x}^2}{\langle x^2\rangle - \bar{x}^2}. \quad \text{(A2)}$$

The error bars in Fig. 5 were computed in this way. A method which avoids the effect of short-time correlations is to group the sequence of measurements into subsequences, average the subsequences, and compute the standard deviation of the partial averages. When the correlation between successive measurements was small, the two formulas agreed well.

For β very close to β_c in the four-dimensional theory, equilibration takes a very long time. This is especially true for quantities involving collective excitations. Therefore, for the measurements of flux in Figs. 7–11, we made a number of runs (typically 4 to 12) from different starting configurations. The error bars on these figures are obtained from the standard deviations of the averages of the separate runs.

APPENDIX B: MONOPOLES IN THE VILLAIN APPROXIMATION

This appendix recapitulates the transformation of U(1) lattice theory in Villain approximation. It contains no new physics but does serve as a repository for various Villain formulas. The formulas are taken from Banks, Meyerson, and Kogut.[3]

Let us compute the partition function $Z(J)$ in the presence of an external current loop $J_\mu(r)$ for a three-dimensional U(1) theory:

$$Z(J) = \int [d\Theta_\mu(r)] \exp\left[\beta \sum_{r,\mu<\nu} \cos\Theta_{\mu\nu}(r) + i\sum_{r,\mu}\Theta_\mu(r)J_\mu(r)\right], \quad \text{(B1)}$$

where $\Theta_\mu(r)$ is the angle on the μ-directed link at r and

$$\Theta^{(r)}_{\mu\nu} = \partial_\mu\Theta_\nu(r) - \partial_\nu\Theta_\mu(r) \quad \text{(B2)}$$

(all derivatives being finite-difference operators). The Villain approximation consists of making a character expansion on the action

$$e^{\beta\cos\Theta_{\mu\nu}(r)} = \sum_{l_{\mu\nu}=-\infty}^{\infty} e^{i l_{\mu\nu}\Theta_{\mu\nu}} I_{l_{\mu\nu}}(\beta), \quad \text{(B3)}$$

combined with the large-β expansion of the modified Bessel function

$$I_l(\beta) \sim e^{-l^2/2\beta}/(2\pi\beta)^{1/2}. \quad \text{(B4)}$$

Inserting (B3) and (B4) into (B1), we evaluate all the Θ_μ integrals and obtain

$$Z(J) \simeq \sum_{r,\mu,\nu,l_{\mu\nu}} \delta_{\partial_\nu l_{\mu\nu}+J_\mu(r),0} \exp\left[-\frac{1}{2\beta}\sum_{r,\mu<\nu} l_{\mu\nu}{}^2(r)\right]. \quad \text{(B5)}$$

The constraint equation is solved by taking

$$l_{\mu\nu}(r) = n^\mu(n\cdot\partial)^{-1}J^\nu - n^\nu(n\cdot\partial)^{-1}J^\mu + \epsilon_{\mu\nu\lambda}\partial_\lambda l(r), \quad \text{(B6)}$$

where n is a unit vector. Equation (B5) is poorly convergent so we perform a Poisson resummation

$$\sum_{l=-\infty}^{\infty} g(l) = \sum_{m=-\infty}^{\infty}\int d\phi\, g(\phi)e^{il m\phi}, \quad \text{(B7)}$$

perform the Gaussian integration over ϕ, and find

$$Z(J) = Z_{\text{monopole}} Z_{\text{spin-wave}} Z_{\text{ext}}, \quad \text{(B8)}$$

where

$$Z_{\text{spin-wave}} = \int [d\phi] \exp\left(-\frac{1}{2\beta}\sum_{r,\lambda}[\partial_\lambda\phi(r)]^2\right), \quad \text{(B9a)}$$

$$Z_{\text{monopole}} = \sum_r \sum_{m(r)=-\infty}^{\infty} \exp\left(-2\pi^2\beta\sum_{r\,\partial r'} m(r)V(r-r')m(r') + 2\pi i\sum_{r,r'}\partial_\nu B^\nu(r)V(r-r')m(r')\right), \quad \text{(B9b)}$$

$$Z_{\text{ext}} = \exp\left(-\frac{1}{2\beta}\sum_{r,r',\mu} J_\mu(r)V(r-r')J_\mu(r')\right), \quad \text{(B9c)}$$

and $\nabla^2 V(r-r') = \delta_{r,r'}$; i.e., V is the lattice Coulomb Green's function. $B_\nu(r)$ is the magnetic field generated by J_μ,

$$B_\nu(r) = \epsilon_{\nu\mu\lambda}n_\mu(n\cdot\partial)^{-1}J_\lambda(r) \quad \text{or} \quad \epsilon_{\alpha\beta\gamma}\partial^\beta B^\gamma(r) = J_\alpha(r). \quad \text{(B10)}$$

A Villain monopole is an integer-valued scalar field whose values range from $-\infty$ to ∞ and whose self-interactions and interactions with the external magnetic field are Coulombic.

At large β the density of monopoles is low: Monopoles are widely separated and the density of monopoles depends in lowest approximation only on their self-interactions. Then the density of isolated monopoles is

$$\rho_M \sim \exp[-2\pi^2 \beta V(0)\beta] \quad \text{(B11)}$$

and the density of isolated monopole-antimonopole pairs in adjacent lattice sites is

$$\rho_{M\bar{M}} \sim \exp\{-2\pi^2[2V(0) - 2V(1)]\beta\}. \quad \text{(B12)}$$

These densities are numerically $\rho_M \sim \exp(-5\beta)$ and $\rho_{M\bar{M}} \sim \exp(-6.58\beta)$, using $V(0) \sim 0.253$ and $V(0) - V(1) = \frac{1}{6}$.

The derivation of a monopole partition function for the four-dimensional theory is essentially identical. Equation (B8) is recovered, with

$$Z_{\text{spin wave}} = \int [dA_\mu] \exp\left(-\frac{1}{2\beta} F_{\mu\nu} F^{\mu\nu}\right) \quad \text{(B13)}$$

and

$$Z_{\text{monopole}} = \sum_{r,\mu} \sum_{m_\mu(r)} \delta_{\partial_\mu m_\mu(r),0} \exp\left(-2\pi^2\beta \sum_{\mu,r,r'} m_\mu(r) m^\mu(r') V(r-r') + 2\pi i \sum_{\mu,r,r'} m_\mu(r) V(r-r') \epsilon_{\mu\nu\alpha\beta} n_\nu \partial_\alpha (n\cdot\bar{\partial})^{-1} J_\rho(r')\right) \quad \text{(B14)}$$

describing the interaction of closed current loops m_μ with each other and with external fields. At large β the only energetically allowed topological excitations are closed loops of length four. Then

$$\rho_{\text{string}} \sim \exp\{-2\pi^2\beta[4\nu(0) - 4\nu(1)]\}, \quad \text{(B15)}$$

but $\nu(0) - \nu(1) = \frac{1}{8}$ so $\rho \sim \exp(-\pi^2\beta)$.

APPENDIX C: SIMPLE DILUTE-GAS APPROXIMATIONS

We can make some simple dilute-gas approximations for monopole densities in three dimensions for β large. The partition function for a system of N boxes

$$Z = \int [d\theta_\mu] \exp\left(\sum_{\text{plaq}} \beta \cos\theta_p\right) \quad \text{(C1)}$$

is a function of about $2N$ link variables after gauge fixing. It may be (formally) integrated to give

$$Z(\beta) = \left(\frac{1}{\sqrt{\beta}}\right)^{2N} e^{2N\beta} e^{-S_{\min}} (\det M)^{-1/2}, \quad \text{(C2)}$$

where $-S_{\min}$ is the exponent of Eq. (C1) evaluated at some minimum-action configuration and

$$M_{ij} = -\frac{\partial}{\partial\theta_i} \frac{\partial}{\partial\theta_j} \sum_p \cos\theta_p \Big|_{\theta_p = \min} \quad \text{(C3)}$$

is the matrix of second derivatives evaluated at the minimum. At large β an absolute minimum is just all $\theta_p = 0$. We may, however, imagine (local) minima of the action which are topologically stable monopole configurations. The density of monopoles in the system will then just be given by the ratio of partition functions evaluated about monopole and vacuum configurations

$$\rho_{\text{monopole}}(\beta) = \frac{Z(\text{monopole})}{Z(\text{vacuum})} = \left(\frac{\det M\,(\text{mono})}{\det M(\text{vac})}\right)^{-1/2} e^{S(\text{vac}) - S(\text{mono})}. \quad \text{(C4)}$$

The action S is proportional to β while the determinants are independent of β. Therefore, at large β the monopole density has the form

$$\rho_M = \exp\left[-\left(m\beta - \tfrac{1}{2} \ln\det \frac{M(\text{mono})}{M(\text{vac})}\right)\right], \quad \text{(C5)}$$

where $m\beta$ is a bare monopole mass and the determinants provide the first quantum correction. The determinants are independent of β because at large β the monopole is essentially restricted to lie in the center of a box—i.e., the plaquette angles of the box must all be nearly equal. Therefore, even the putative translation modes become Gaussian.

This approximation to the monopole density becomes valid when the motion of a monopole from one box to another requires tunneling through a high barrier. In this limit Monte Carlo calculations become impractical because the fluctuations in the positions of monopoles from pass to pass become tiny and the system thermalizes slowly. Therefore, the dilute-gas approximation really becomes good only for larger β than used in our experiments. Nevertheless, it is interesting to use this approximation to make rough estimates for the density of monopoles or monopole-antimonopole pairs at large β.

First, let us imagine a system consisting of a single cube of unit size. After gauge fixing, it has five independent link variables. The ground state of the system is $\theta_p = 0$ for each of the six plaquettes, while a one-monopole configuration can be generated by placing 2π units of (outgoing) flux over the six plaquettes, along with -2π units of Dirac string in one face. The minimum-action configuration is symmetric: each face carries $\pi/3$ units of flux (plus a string somewhere). Since all $\cos\theta_p$ are equal to $\frac{1}{2}$,

$$M_{ij}(\text{monopole}) = \tfrac{1}{2} M_{ij}(\text{vacuum}) \tag{C6}$$

and

$$\rho_M = \sqrt{32}\, e^{-3\beta}\,. \tag{C7}$$

One may perform similar analyses for larger systems. For example, a monopole centered in a $3\times3\times3$ cube has a mass $m\beta$ [$m\beta \equiv S(\text{vac}) - S(\text{monopole})$] $\simeq (3+0.96)\beta$. In bigger systems one rapidly builds up the usual Coulomb self-energy of an isolated charge. For comparison with our Monte Carlo simulation, however, these calculations are probably unreliable: If the monopole density is too dilute, monopoles move too slowly for practical Monte Carlo calculation, and if the density is higher, interactions become important at long distance. A value $m\sim 3$ to 4 which takes into account only the short-distance part of the monopole field is probably a reasonable estimation for a monopole density when that density is such that monopole separation is typically only a few lattice spacings.

The density of monopole-antimonopole pairs is somewhat harder to estimate. In a universe of two cubes we have

$$M \equiv S(\text{vac}) - S(\text{pair}) = \left[\sum_{p_1} (1-\cos\theta_{p_1}) + \sum_{p_2} (1-\cos\theta_{p_2}) - (1-\cos\theta_{p_{12}})\right]\beta\,, \tag{C8}$$

where $p_{1,2}$ are plaquettes on the two cubes which share plaquette p_{12}. While each cube separately has a minimum configuration at $\theta_p=\pi/3$, the system may lower M by increasing the flux through the shared face until the pair decays. For example, the symmetric combination ($\theta_p=\pi/3$) has $M=\frac{11}{2}\beta$, but a configuration with $\theta_{p_{12}}=\pi$, where all other $\theta_p=\pi/5$, has $M=3.91$. This last figure represents a lower limit on the mass of an $M\bar{M}$ pair.

---

[1]K. G. Wilson, Phys. Rev. D 10, 2445 (1974); J. Kogut and L. Susskind, *ibid.* 11, 395 (1975).

[2]A. M. Polyakov, Nucl. Phys. B120, 429 (1977).

[3]T. Banks, R. Myerson, and J. Kogut, Nucl. Phys. B129, 493 (1977); R. Savit, Phys. Rev. Lett. 39, 55 (1977).

[4]K. Osterwalder and E. Seiler, Ann. Phys. (N.Y.) 110, 440 (1978); A. Guth, Phys. Rev. D 21, 2291 (1980).

[5]M. Creutz, L. Jacobs, and C. Rebbi, Phys. Rev. Lett. 42, 1390 (1979); Phys. Rev. D 20, 1915 (1979).

[6]M. Creutz, Phys. Rev. Lett. 43, 553 (1979); Phys. Rev. D 21, 1006 (1978); 21, 2308 (1978); K. G. Wilson, Cornell report, 1979 (unpublished); C. Rebbi, Phys. Rev. D 21, 3350 (1980); G. Bhanot and B. Freedman, Brookhaven report, 1980 (unpublished); G. A. Jongeward and J. D. Stack, Phys. Rev. D 21, 3360 (1980); S. Shenker and J. Tobochnik, Phys. Rev. B 22, 4462 (1980).

[7]S. Elitzur, R. B. Pearson, and J. Shigemitsu, Phys. Rev. D 19, 3698 (1979); D. Horn, M. Weinstein, and S. Yankielowicz, *ibid.* 19, 3715 (1979); A. Ukawa, P. Windey, and A. Guth, *ibid.* 21, 1013 (1980).

[8]N. Metropolis, A. Rosenbluth, M. Rosenbluth, A. Teller, and E. Teller, J. Chem. Phys. 21, 1087 (1953).

[9]Logarithmic corrections to the perimeter law were first noticed by E. Fradkin and L. Susskind, Phys. Rev. D 17, 2637 (1978). We thank John Cardy for a discussion on this point.

[10]A. Belavin, A. Polyakov, P. Schwartz, and Y. Tyupkin, Phys. Lett. 59B, 851 (1975); G. 't Hooft, Phys. Rev. D 14, 3432 (1976); C. Callan, R. Dashen, and D. Gross, *ibid.* 20, 3279 (1979), plus references therein.

[11]J. José, L. P. Kadanoff, S. Kirkpatrick, and D. R. Nelson, Phys. Rev. B 16, 1217 (1977).

[12]J. Tobochnik and G. Chester, Phys. Rev. B 20, 3761 (1979).

[13]J. Cardy, Nucl. Phys. B (to be published).

[14]G. 't Hooft, Nucl. Phys. B153, 141 (1979).

[15]J. Groeneveld, J. Jurkiewicz, and C. P. Korthals Altes, Utrecht report, 1980 (unpublished).

[16]A. A. Migdal, Zh. Eksp. Teor. Fiz. 69, 810 (1975); 69, 1457 (1975) [Sov. Phys. JETP 42, 413 (1976); 42, 743 (1976)]; L. P. Kadanoff, Ann. Phys. (N.Y.) 100, 359 (1976).

[17]For a review of Monte Carlo methods applied to spin systems, see K. Binder, in *Phase Transitions and Critical Phenomena*, edited by C. Domb and M. S. Green (Academic, New York, 1976), Vol. 5B.

VOLUME 43, NUMBER 8 PHYSICAL REVIEW LETTERS 20 AUGUST 1979

Confinement and the Critical Dimensionality of Space-Time

Michael Creutz
Department of Physics, Brookhaven National Laboratory, Upton, New York 11973
(Received 11 June 1979)

Using Monte Carlo techniques, we study pure SU(2) gauge fields in four and five space-time dimensions and a compact SO(2) gauge field in four dimensions. Ultraviolet divergences are regulated with Wilson's lattice prescription. Both SU(2) in five dimensions and SO(2) in four dimensions show clear phase transitions between the confining regime at strong coupling and a spin-wave phase at weak coupling. No phase change is seen for the four-dimensional SU(2) theory.

The standard theory of hadronic interactions is based on quarks interacting with non-Abelian gauge fields. The viability of this picture depends on the conjectured phenomenon of confinement, wherein the only physically observable particles are invariant under the gauge group. Thus far, the only demonstration of this property is in the strong-coupling limit and with a space-time lattice regulating ultraviolet divergences.[1] Approximate renormalization-group arguments[2] suggest that four space-time dimensions represent a critical case where confinement persists for all couplings when the gauge group is non-Abelian. In contrast, Abelian groups should exhibit a phase transition to a nonconfining weak-coupling phase containing massless gauge bosons. Thus arises the conjecture that in our four-dimensional (4D) world, the lattice formulation of electrodynamics can avoid confinement of electrons, while the continuum limit of the strong-interaction gauge theory can exhibit asymptotic freedom, a vanishing coupling at short distances.

Recent Monte Carlo results have given mixed support for these arguments. For the four-dimensional gauge-invariant Ising model, the observed transition is first order, contrary to the approximate renormalization-group prediction of a second-order transition analogous to that in the conventional two-dimensional Ising model.[3] However, for Z_n with $n \geq 5$ and SO(2) symmetries, the predicted similarities between the four-dimensional gauge models and the two-dimensional

VOLUME 43, NUMBER 8 PHYSICAL REVIEW LETTERS 20 AUGUST 1979

spin systems are confirmed.[4]

In this note I report results on Monte Carlo studies of SU(2) gauge theory. To show the critical nature of four dimensions I ran these simulations for both four- and five-dimensional lattices. I also make comparisons with the Abelian group SO(2) [isomorphic to U(1)]. I work with pure gauge fields on the assumption that the addition of a few fermion species represents a perturbation that will not spoil confinement. Although the group of physical interest is SU(3), I study SU(2) because of its simpler structure. As confinement is connected with disorder in the lattice formulation, and as adding more degrees of freedom should increase disorder, confinement with SU(2) gauge fields should imply confinement with SU(3).

The system is formulated on a hypercubical lattice. Associated with the link joining any pair of nearest-neighbor sites i and j is an element U_{ij} of the gauge group (i and j label sites and should not be confused with the implicit matrix indices on the group elements). The wave function of a particle traversing the respective link undergoes an internal-symmetry rotation corresponding to U_{ij}. The reverse path gives the conjugate rotation

$$U_{ji} = (U_{ij})^{-1}, \tag{1}$$

where the inverse is in the group sense. The quantum theory is defined via the path integral

$$Z = \int \Big(\prod_{\{i,j\}} dU_{ij} \Big) e^{-\beta S(U)}, \tag{2}$$

where the integral includes all links and uses the invariant group measure. The action is that defined by Wilson,

$$S(U) = \sum_{\Box} S_{\Box}, \tag{3}$$

where the sum extends over all elementary squares or "plaquettes" $\Box$ and

$$S_{\Box} = 1 - \tfrac{1}{2} \operatorname{Tr}(U_{ij} U_{jk} U_{kl} U_{li}). \tag{4}$$

Here i, j, k, and l are some labeling of the sites going around the square $\Box$. The normalization is such that for the groups SU(2) and SO(2) any plaquette contributes a number between zero and two to the action. As shown by Wilson,[1] this action reduces in the classical continuum limit to the usual gauge-theory action with β proportional to the inverse square of the coupling constant.

Equation (1) is formally identical to the partition function of a statistical mechanical system with Hamiltonian S and at inverse temperature β.

The Monte Carlo algorithm consists of successively touching a heat bath to each link of the lattice while holding fixed the group elements on the remaining links. Repeating this procedure will eventually produce a sequence of states which simulates an ensemble of such systems in thermal equilibrium.[5] Green's functions for the quantum theory follow from correlation functions in the states of the ensemble.

Beginning in some initial configuration, we pass through the entire lattice varying one link at a time. At each link's turn, a new group element g is selected to occupy that position. This choice is made randomly from the entire gauge group with weighting proportional to the Boltz-

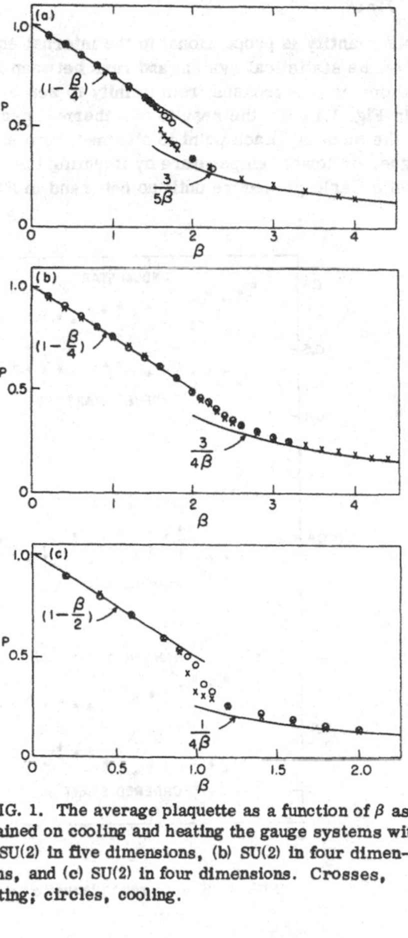

FIG. 1. The average plaquette as a function of β as obtained on cooling and heating the gauge systems with (a) SU(2) in five dimensions, (b) SU(2) in four dimensions, and (c) SU(2) in four dimensions. Crosses, heating; circles, cooling.

mann factor

$$B(g) = \exp[-\beta S(g)], \quad (5)$$

where $S(g)$ is the action evaluated with the given link having group element g and all other links fixed with their previous values. The old value for the current link plays no direct role in this procedure. In what follows an iteration is defined as one application of this algorithm to each link in the lattice.

For the four-dimensional models we use a $5\times 5\times 5\times 5$ lattice while for the five-dimensional simulation we work on a $4\times 4\times 4\times 4\times 4$ lattice. To minimize surface effects we impose periodic boundary conditions. As an order parameter we use the average action per plaquette,

$$P = \langle S_{\square} \rangle. \quad (6)$$

This quantity is proportional to the internal energy of the statistical system and runs between zero and one as β decreases from infinity to zero.

In Fig. 1 I show the results of a thermal cycle on the models. Each point is obtained from either higher or lower temperature by iterating the Monte Carlo procedure until no net trend in P is observed over six iterations. Plotted is the average value of P over these six iterations. The heating runs are initiated with a totally ordered lattice while the cooling runs start with all link variables chosen randomly. Note that the four-dimensional SO(2) and the five-dimensional SU(2) models show clear hysteresis effects indicative of phase transitions. The four-dimensional SU(2) model shows no similar gross structure, although convergence appears to be slightly reduced in the region $2.2 \lesssim \beta \lesssim 2.5$. I further discuss this region below.

In order to quote critical temperatures for the observed transitions, I select a value P_0 for P in the middle of the hysteresis loops and then iterate while adjusting β until P fluctuates around P_0. Choosing $P_0 = 0.52$ for SU(2) in five dimensions and $P_0 = 0.4$ for SO(2) in four dimensions, I obtain

$$\beta_c = 0.987 \pm 0.023, \quad \text{SO(2) in 4D}, \quad (7)$$

$$\beta_c = 1.642 \pm 0.015, \quad \text{SU(2) in 5D}. \quad (8)$$

To investigate whether the transitions are of first or higher order, I made extended runs at

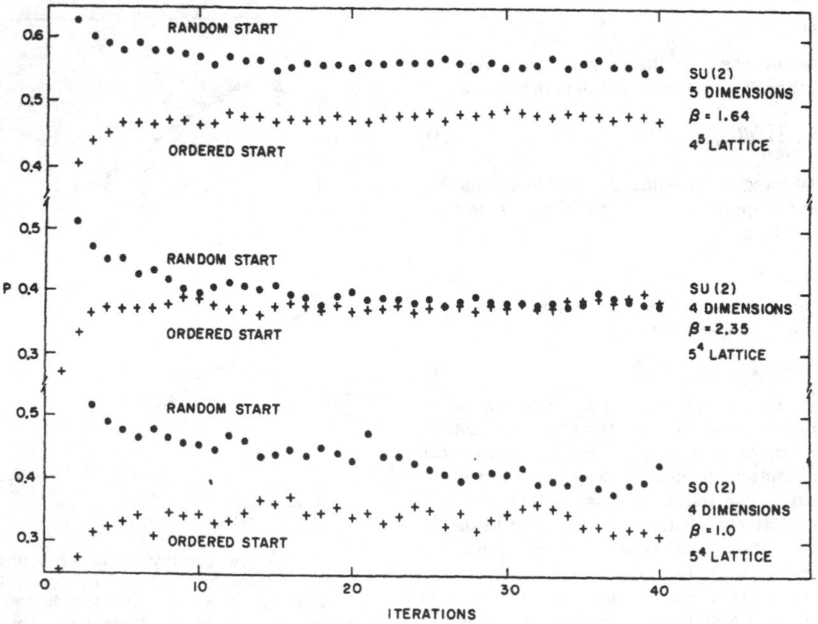

FIG. 2. The average plaquette as a function of number of iterations at a fixed β.

the critical temperatures with both ordered and disordered starts. In Fig. 2 we see that for SU(2) in five dimensions the two runs stabilize at different values, indicative of a first-order transition. In contrast, for SO(2) in four dimensions these runs show large fluctuations and continue to converge slowly, suggesting a higher-order transition.

In Fig. 2 I also show the results of similar runs with the four-dimensional SU(2) model at $\beta=2.35$. This corresponds to a temperature in the middle of the slow-convergence region alluded to above. The two runs converge after about thirty iterations, while fluctuations are considerably controlled relative to those seen in the SO(2) model. I feel that the reduced convergence in this region is not evidence for a phase transition, but rather a consequence of the critical nature of four dimensions.

At low β (high temperature) the points follow the strong-coupling limit

$$P=1-\tfrac{1}{2}\beta+O(\beta^3) \text{ for SO(2)}, \tag{9}$$

$$P=1-\tfrac{1}{4}\beta+O(\beta^3) \text{ for SU(2)}. \tag{10}$$

The large-β behavior can be estimated by keeping only those terms in the action which are quadratic in parameters describing the group manifold. This yields a Gaussian path integral and implies

$$P \underset{\beta\to\infty}{\longrightarrow} n/\beta d, \tag{11}$$

where n is the number of group generators and d is the dimensionality of space-time. The functions in Eqs. (9)–(11) are plotted along with the "data" in Fig. 1. Note that this inverse-β behavior at large β is approached in all the models. I do not expect any further phase transitions for β above the onset of this spin-wave behavior.

In conclusion, I have presented Monte Carlo evidence that the confinement phase of SU(2) lattice gauge theory in four dimensions extends to all values of coupling. This means that the continuum limit of this theory simultaneously exhibits confinement and asymptotic freedom. Of course this is not an analytic proof; indeed, there could exist a subtle transition not readily observable in the average plaquette. I regard this as unlikely in the light of the extreme clarity of the transitions seen with the other models.

This work was supported by the U. S. Department of Energy under Contract No. EY-76-C-02-0016.

[1]K. G. Wilson, Phys. Rev. D 10, 2445 (1974).
[2]A. A. Migdal, Zh. Eksp. Teor. Fiz. 69, 810 (1975), and 69, 457 (1975) [Sov. Phys. JETP 42, 413 (1975), and 42, 473 (1975)]; L. P. Kadanoff, Rev. Mod. Phys. 49, 267 (1977).
[3]M. Creutz, L. Jacobs, and C. Rebbi, Phys. Rev. Lett. 42, 1390 (1979).
[4]M. Creutz, L. Jacobs, and C. Rebbi, to be published.
[5]K. Binder, in *Phase Transitions and Critical Phenomena*, edited by C. Domb and M. S. Green (Academic, New York, 1976), Vol. 5B.

PHYSICAL REVIEW D VOLUME 21, NUMBER 12 15 JUNE 1980

Phase structure of non-Abelian lattice gauge theories

Claudio Rebbi
Department of Physics, Brookhaven National Laboratory, Upton, New York 11973
(Received 23 January 1980)

The phase structure of four-dimensional lattice gauge theories based on finite non-Abelian groups is studied by Monte Carlo computations. All models examined exhibit a two-phase structure with a first-order phase transition. In three systems where the gauge group is a discrete subgroup of SU(2) the critical temperature moves toward zero as the order of the group increases and the high-temperature phase has confining properties.

I. INTRODUCTION

The lattice formulation of a gauge field theory offers a very powerful technique to study its quantum properties.[1] It provides a regularization of the ultraviolet divergences and allows strong-coupling expansions. The continuum theory is recovered in the limit where the correlation length becomes infinite: it is therefore quite crucial to have a knowledge of the possible phase transitions. Thus, the charge-confining properties of non-Abelian gauge models are related to the absence of any phase transition: The confinement observed in the strong-coupling regime is believed to extend all the way to the zero-temperature limit, where one recovers the continuum system. On the contrary, the existence of free charges in quantum electrodynamics requires a phase transition, separating a strong-coupling, confining phase from a low-temperature, spin-wave phase in the corresponding lattice theory.

Very recently numerical methods based on the Monte Carlo technique have been used to obtain information about the phase structure of a variety of gauge models.[2-5] The results have proven quite encouraging and agree nicely with the conclusions of other analyses, based on perturbative or semiclassical expansions.[6-8] More specifically, in Refs. 2 and 3 Abelian gauge theories have been investigated, while in Refs. 4 and 5 the non-Abelian system with gauge group SU(2) has been studied.

A remarkable feature of lattice gauge theories is that discrete gauge groups may also be considered.[9] Thus, together with the model with U(1) gauge group, one may study the whole category of systems with the finite, Abelian groups Z_N. In the limit $N\to\infty$ one expects to recover the properties of the U(1) theory. Indeed, one of the main results of the numerical analysis of Ref. 3 consists in the observation, for N large enough, of a three-phase structure in the Z_N models, with two phase transitions, one of which disappears at zero temperature, while the other survives in the U(1) limit. Considerations about this limit have also formed the main ingredient in the study of the Z_N models of Ref. 6.

The interrelations between the properties of lattice gauge theories with discrete and continuum groups motivated this work, where we present Monte Carlo results obtained for a variety of gauge systems with finite, non-Abelian groups. The main emphasis will be placed on models where the gauge group is a subgroup of SU(2). Three of these systems, with gauge groups of 8, 24, and 48 elements, respectively, have been analyzed: All exhibit a single, very clear phase transition, which definitely moves toward zero temperature as the order of the group increases. Internal energy and disorder parameters (Wilson loop factors) of the high-temperature phase agree almost up to the transition point with those already determined for SU(2).[4]

Contrary to the case of U(1), the manifold of SU(2) cannot be filled with points of discrete subgroups which become dense in a suitable limit. Only a finite number of nontrivial subgroups of SU(2), related to the symmetries of the regular polyhedra, exists. But this is a limitation only in principle. We recall from Ref. 3 that the two phase transitions in the Z_N models are well separated already for $N=8$, with one transition essentially where it is observed in the U(1) theory, the other at a temperature low enough to approach the limit of reliability of the computation. The model with a 48-element group considered here has the same energy (or action) gap as Z_8 (which is contained as a subgroup) and the only phase transition has already moved to a temperature lower than that for the Z_8 theory. Thus our results, we believe, corroborate strongly the notion that four-dimensional lattice gauge theories with non-Abelian continuum groups posses a single, confining phase.

Section II contains a brief description of the models considered and of the computational technique used. Section III presents the actual numerical results. Section IV is devoted to a few words of conclusion.

II. DESCRIPTION OF THE MODELS AND OUTLINE OF THE COMPUTATION

A lattice gauge theory with group $\mathcal{G}$ is defined by associating an element $U_{ij} \in \mathcal{G}$ to each link joining neighboring sites i and j. $U_{ji} = U^{-1}{}_{ij}$ and, in a gauge transformation, $U_{ij} \to U'_{ij} = G^{-1}{}_i U_{ij} G_j$, with the elements $G_i \in \mathcal{G}$ defined locally at each site i. The quantities (Wilson loop factors)

$$W_\gamma = \mathrm{Tr}\{U_{i_1 i_2} U_{i_2 i_3} \cdot \cdot U_{i_N i_1}\}, \tag{2.1}$$

where the sites $i_1, \ldots, i_N$ form a closed loop γ and Tr denotes a class function (i.e., $\mathrm{Tr}G^{-1}UG = \mathrm{Tr}U$), are gauge invariant.

In the applications to quantum field theory the lattice is usually taken to be a four-dimensional hypercubical lattice. Quantum averages are defined with a weight $e^{-\beta S}$, where the action S is given by a sum of suitable functions of loop factors extended to all elementary squares of the lattice (plaquettes)[10]:

$$S = \sum_{\Box} f(W_\Box). \tag{2.2}$$

Of particular interest are the normalization factor itself, or partition function,

$$Z = \sum_{\{U_{ij}\}} e^{-\beta S}, \tag{2.3}$$

the free energy, defined as

$$F = \frac{1}{N_s} \ln Z, \tag{2.4}$$

N_s being the number of sites in the lattice, which becomes independent of the lattice size as $N_s \to \infty$, and the internal energy

$$E = \langle f(W_\Box)\rangle = \frac{1}{6}\frac{\partial}{\partial\beta} F. \tag{2.5}$$

In this article we shall study the models obtained with the following choices for $\mathcal{G}$:

(i) The 8-element group of quaternions, denoted by Q, generated for instance by the matrices $i\sigma_x$ and $i\sigma_y$ (σ_i being the Pauli matrices).

(ii) The 24-element group $\bar{T}$ generated by the matrices $\frac{1}{2} + \frac{1}{2} i\sqrt{3}\sigma_z$ and $\frac{1}{2} + (i\sqrt{2}/\sqrt{3})\sigma_y - (i/2\sqrt{3})\sigma_z$. T contains Z_2 as invariant subgroup and the factor group $T = \bar{T}/Z_2$ is the rotation group of the tetrahedron.

(iii) The 48-element group $\bar{O}$ generated by the matrices $1/\sqrt{2} + (i/\sqrt{2})\sigma_x$ and $1/\sqrt{2} + (i/\sqrt{2})\sigma_y$. $\bar{O}$ also contains Z_2 as invariant subgroup and the factor group $O = \bar{O}/Z_2$ is the rotation group of the octahedron.

(iv) The 24-element group O.

(v) The permutation group of three elements S_3.

The groups Q, $\bar{T}$, and $\bar{O}$ are subgroups of SU(2). Their elements are represented by matrices of the form

$$u = \cos\vartheta + i \sin\vartheta \vec{\sigma} \cdot \hat{n}, \tag{2.6}$$

$\hat{n}$ being a unit vector, and we choose $\cos\vartheta$ as the class function Tr appearing in Eq. (2.1). The action is then defined by

$$f(W_\Box) = 1 - W_\Box. \tag{2.7}$$

This agrees, in particular, with the normalization used in Refs. 2–4 and allows a direct comparison of results, without rescalings.

The elements of O may also be represented by matrices of the form (2.6), identifying however u with $-u$. W is then defined by $W = \cos^2\vartheta$, as appropriate for the rotation group O(3). $f(W_\Box)$ is again set equal to $1 - W_\Box$.

The six elements of S_3 fall into three classes, one containing the identity I, another the two permutations of all 3 elements, C and C^2, the third one the permutations P_i which leave the ith element fixed. I, C, and C^2 together form the invariant subgroup Z_3 and, to achieve the same normalization as in Ref. 3, we assign $f(W_\Box) = 0$ to I, $f(W_\Box) = \frac{3}{2}$ to C and C^2. The choice of action for the remaining class is quite arbitrary and we have performed computations with the three values $f(\{P_i\}) = \frac{1}{2}$, 1, and $\frac{3}{2}$. Averages of loop factors have not been evaluated for this group, so the choice of $W_\Box$ itself is irrelevant.

The groups Q, $\bar{T}$, and $\bar{O}$ are all subgroups of SU(2), and one of the main purposes of this work is to study what happens to the phases of the corresponding gauge theories as the points representing the group elements become denser within the manifold of SU(2). For comparison with the work of Ref. 3, we notice that $\bar{O}$ has subgroups isomorphic to Z_8 and, in particular, the gap between the action of an unexcited plaquette and the action of a plaquette in the lowest state of excitation is the same in both cases. O has been studied to see how factoring out the center of the group alters the properties of the model. S_3 has been considered for its own sake, also because the numerical analysis could be extended to this system with a minimal cost in computing.

Quantum averages are evaluated numerically by the Monte Carlo technique: One generates a sequence of states Σ_i in such a way that a definite configuration of the spin variables U_{ij} appears with a probability proportional to the Boltzmann factor:

$$P(\Sigma) \propto e^{-\beta S(\Sigma)}. \tag{2.8}$$

The quantum average of an operator $A(\Sigma)$ is then approximated by an average over configurations in the sequence

$$\langle A \rangle \approx \frac{1}{N} \sum_{i=i_0}^{i_0+N-1} A(\Sigma_i). \tag{2.9}$$

A number of states encountered at the beginning of the sequence is excluded to ensure that statistical equilibrium has been reached.

The configuration Σ_{i+1} is obtained from Σ_i by a stochastic process, whereby one of the spins of the lattice U_{ij} is set to a new value U'_{ij} (possibly equal to U_{ij}) according to a definite probability matrix $p(U_{ij} \to U'_{ij})$. p is defined so that in statistical equilibrium Eq. (2.8) is satisfied. After U_{ij}, a new spin $U_{i'j'}$ is reset according to p, and so on until all the spins of the lattice are probed in succession. This completes one Monte Carlo iteration. The whole process is then repeated and many iterations are used to construct the sequence of states appearing in Eq. (2.9). We refer the reader to Refs. 11 and 2–5 for a more detailed discussion of the method.

The matrix $p(U_{ij} \to U'_{ij})$ used in this analysis is constructed as follows. For all possible choices U'_{ij} the total action $S'(U'_{ij})$ of the plaquettes containing the link ij is evaluated. Then p is chosen proportional to $e^{-\beta S'(U'_{ij})}$:

$$p(U'_{ij}) = e^{-\beta S'(U'_{ij})} \Big/ \sum_{U'_{ij}} e^{-\beta S'(U'_{ij})}. \tag{2.10}$$

The statistical process thus corresponds to touching the spin U_{ij} with a heat reservoir at inverse temperature β, all other spins being held fixed. The procedure generated by this choice of p has been called the heat-bath algorithm in Refs. 2–4.

An alternative possibility for p, originally introduced by Metropolis *et al.*[12] and widely used, consists of selecting only one new candidate value U'_{ij} for the spin. If the new choice lowers the action, U_{ij} is changed to the new value. If not, the change is made with conditional probability $\exp\{-\beta[S(U'_{ij}) - S(U_{ij})]\}$. The computer time needed to probe one spin with this algorithm is shorter than in the heat-bath method, but the heat-bath algorithm converges to statistical equilibrium faster (the relative efficiency depending on the number of possible values for the spins) and thus fewer Monte Carlo iterations are needed. In the context of the present analysis we found that the gain in convergence outweighs the loss in computer time and the heat-bath algorithm has been used throughout.

A technique of storing many spins in a single memory word of the computer (multi-spin-coding: see Ref. 2) to reduce memory requirements and processing time has been utilized. The computations have been performed with a lattice extending for 8 sites in each of the three spatial directions and 10 sites in the temporal one, subject to periodic boundary conditions but without any gauge constraint. The total number of spin variables is then 20 480, which, on the basis of previous results,[2-4] should be sufficient to produce reliable averages without excessive statistical fluctuations.

III. THE PHASE STRUCTURE

While the basic feature of the Monte Carlo technique consists always in the stochastic readjustment of the spins, there are a variety of computations that can be done for a definite model. There is arbitrariness in the choice of the initial configuration; there is also the option of varying the parameter β every iteration or every few iterations, thus subjecting the system to a change in temperature. In a sense, the Monte Carlo algorithm creates a small specimen of the material inside the computer. But the choice of what experiment to do with it is left open.

In this study we have done the following three types of computation:

(i) Starting at a definite initial value β_0 of β (determined on the basis of trial computations) and with the system completely ordered (all U_{ij} set equal to the identity), 10 iterations are performed, the internal energy is evaluated, and β is lowered by an amount $\Delta\beta$. The procedure is repeated until β becomes zero (infinite-temperature limit) and then β is raised again, in steps of $\Delta\beta$, up to the original value β_0. This computation, which we shall refer to as "simulation of a thermal cycle," provides a general overview of the phase structure of the system. The internal energy E is indeed a one-valued function of β (with the possible exception of a first-order critical point β_c where E can take two values, E_+ and E_-) so that in equilibrium a single value for the internal energy should be measured at any definite temperature. By changing β, however, the system is constantly kept slightly off-equilibrium. If the variation of β is slow, the departure from equilibrium is generally very small and effectively a single value for E is determined. But near a phase transition the relaxation time becomes large and the plot of E versus β displays the typical shape of a hysteresis loop.

(ii) In the initial configuration half of the spins (those for instance emanating from sites with temporal coordinate ≤ 5) are set equal to the identity, the other half are chosen at random. A definite number of Monte Carlo iterations is then performed for a few fixed values of β, selected in the vicinity of a phase transition, and the internal energy E is plotted versus the number t of iterations. These computations, which we shall call "mixed-phase runs," provide information about the nature of the transition. At a critical point of the first order there are two different, stable phases. Slightly off the critical temperature one of the two phases remains stable, the other becomes metastable: In a mixed-phase run the system approaches rather quickly a configuration in which half the lattice is in the stable phase and the other half is in the metastable phase, but then the boundary of the stable phase expands until this phase overtakes the whole lattice and one observes a drift in E towards either E_+ or E_-, according to the value of β. This behavior is qualitatively different from what is seen in a phase transition of higher order, where the curves $E(t)$ tend to equilibrium values which vary continuously when β is changed.

(iii) The loop factors W for rectangular paths of sides m and n are evaluated after equilibrium is reached at a definite temperature, averaging over all spacelike loops and also over a few Monte Carlo iterations. W plays the role of order parameter. Increasing the size of the loop the correlations among the spins decrease, W tends to zero and $-\ln W$ therefore to infinity.[13] It has been argued that $-\ln W$ should increase like the area of the loop in a disordered, confining phase, like the perimeter in a nonconfining phase.[1] Thus, the behavior of $-\ln W$ versus m and n can be taken as a measure of the order of the system.

The thermal cycles of the models with groups Q, $\tilde{T}$, and $\tilde{O}$ are reproduced in Figs. 1(a)–1(c). The range of temperatures has been covered in

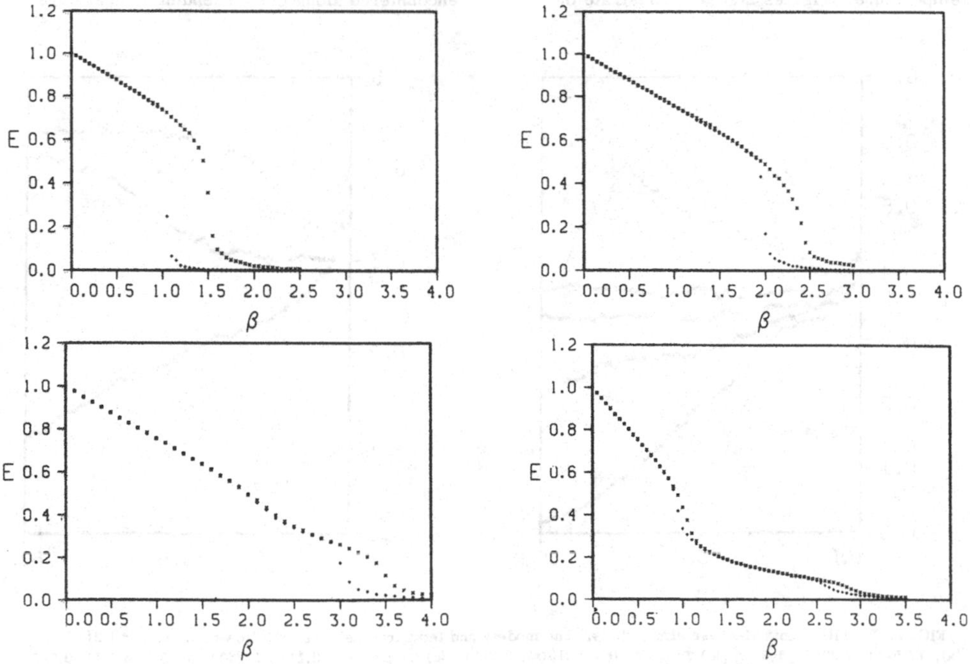

FIG. 1. Thermal cycles for the models with groups Q, $\tilde{T}$, $\tilde{O}$, and Z_8. + (×) denotes the values of E measured while increasing (decreasing) the temperature.

steps $\Delta\beta = 0.05$ with the Q and $\tilde{T}$ systems, $\Delta\beta = 0.1$ with the $\tilde{O}$ system. A hysteresis loop signaling a phase transition is apparent in all diagrams and it is also evident that the critical point moves toward zero temperature as the order of group increases: The groups Q, $\tilde{T}$, and $\tilde{O}$, we recall, are subgroups of SU(2) with 8, 24, and 48 elements, respectively. The thermal cycle of the Z_8 model (from Ref. 3) is presented in Fig. 1(d) for comparison. The Abelian group Z_8 is contained as a subgroup in $\tilde{O}$ and the two models have the same energy (or action) gap. But the difference between their thermal cycles is impressive. The Abelian model gives clear evidence of two phase transitions and therefore of three phases: a high-temperature, confining phase,[14] an ordered phase, likely to disappear at zero temperature when the order of the group increases, and an intermediate, nonconfining spin-wave phase. There is no sign of a spin-wave phase in the non-Abelian systems, and only the high-temperature, confining phase[14] and the ordered, low-temperature phase appear to be present.

Mixed-phase runs have been made to determine the order of the phase transition and the critical temperature. Figures 2(a)–2(c) illustrate the results of runs of 80 iterations with the Q, $\tilde{T}$, and $\tilde{O}$ models, respectively. The values of β are indicated in the figure captions. Figure 2(d), presented for comparison, shows the result of a mixed-phase simulation with the Z_6 model near a critical point of higher order.[3] In Figs. 2(b) and 2(c) a divergence of the curves $E(t)$ to limiting values E_+ and E_- is apparent. This trend is not so noticeable for the intermediate values of β in Fig. 2(a), but the results of longer runs, presented in Fig. 2(e), suggest a first-order transition for the system with gauge group Q as well. A longer run [see Fig. 2(f)] has also been done for the $\tilde{O}$ model to locate more precisely the critical point. These computations indicate that the systems with gauge group Q, $\tilde{T}$, and $\tilde{O}$ all undergo first-order transitions, at critical temperatures given by $\beta_c = 1.23 \pm 0.02$, $\beta_c = 2.175 \pm 0.025$, and $\beta_c = 3.21 \pm 0.01$, respectively.

The behavior of the Wilson loop factors has been studied in the model with gauge group $\tilde{O}$. Starting from an initial configuration where the system is in statistical equilibrium, averages of the loop factors have been taken over 12 Monte Carlo iterations and over all spacelike loops. For initial configurations we have chosen those encountered along the descending temperature

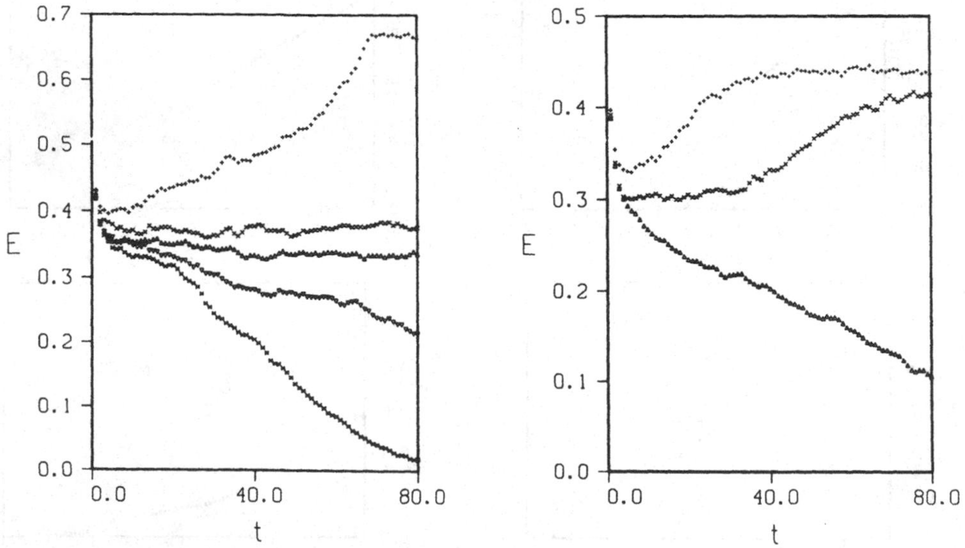

FIG. 2. Results of mixed-phase simulations. The models and temperatures are as follows: (a) Q; $\beta = 1.2(+)$, 1.23 (×), 1.25(△), 1.27(▽), 1.3(*). (b) $\tilde{T}$; $\beta = 2.1(+)$, 2.15(×), 2.2(△). (c) $\tilde{O}$; $\beta = 3(+)$, 3.1(×), 3.2(△), 3.3(▽), 3.4(*), 3.5(□), 3.6(◇). (d) Z_6; β+ 0.97(+), 0.98(×), 0.99(△), 1(▽), 1.01(*), 1.02(□), 1.03(◇). (e) Q; $\beta = 1.23(+)$, 1.25(×), 1.27(△). (f) $\tilde{O}$; $\beta = 3.2(+)$, 3.22(×), 3.24(△).

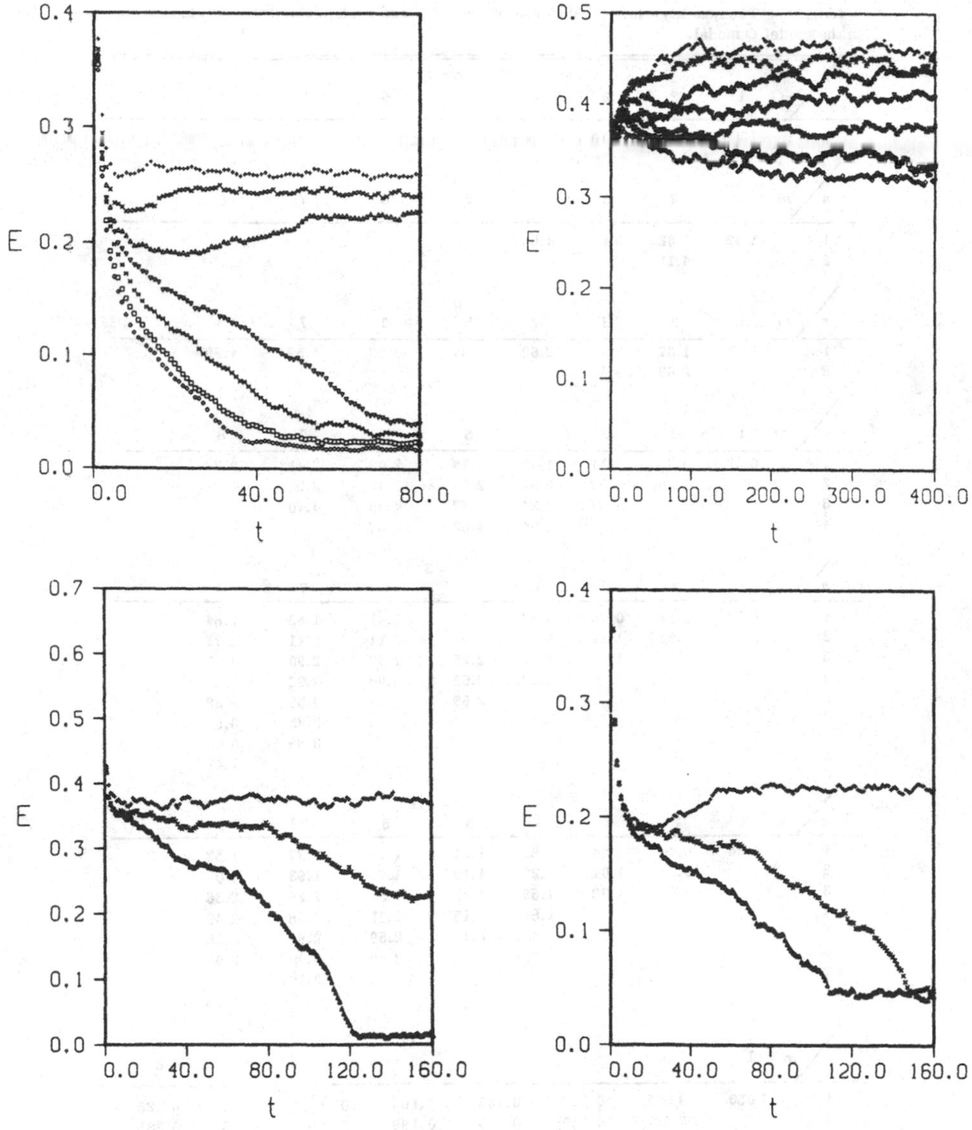

FIG. 2. (*Continued*)

branch of the thermal cycle for $\beta=1$, 1.5, 2, 2.5, and 3; those encountered at the end of the long mixed-phase runs (160 iterations) for $\beta=3.2$ and 3.22. In Table I we reproduce the values of $-\ln W$ thus determined for all the loops which give $-\ln W < 5$. Larger values of $-\ln W$ correspond to loop factors so close to zero that they cannot be distinguished from statistical fluctuations. These statistical fluctuations, not reported in the table, are of the order of 10^{-2}, with the exception of the

TABLE I. The quantity $-\ln W$ for different values of β and rectangular loops of sides m, n in the model $\bar{O}$ model.

$\beta = 1$

n \ m	1	2	3	4	5	6	7	8
1	1.42	2.79	4.10					

$\beta = 1.5$

n \ m	1	2	3	4	5	6	7	8
1	1.02	2.02	3.01	4.06				
2		4.11						

$\beta = 2$

n \ m	1	2	3	4	5	6	7	8
1	0.70	1.37	2.03	2.69	3.34	4.03	4.59	4.99
2		2.69	4.17					

$\beta = 2.5$

n \ m	1	2	3	4	5	6	7	8
1	0.42	0.78	1.11	1.43	1.75	2.08	2.41	2.75
2		1.35	1.87	2.37	2.90	3.40	3.93	4.34
3			2.57	3.14	3.77	4.39	4.70	
4				3.94	4.52	4.87		

$\beta = 3$

n \ m	1	2	3	4	5	6	7	8
1	0.30	0.54	0.76	0.98	1.20	1.41	1.63	1.84
2		0.90	1.21	1.51	1.81	2.11	2.41	2.71
3			1.56	1.91	2.25	2.62	2.90	3.07
4				2.27	2.62	2.98	3.26	3.23
5					2.99	3.28	3.59	3.38
6						3.65	3.88	3.51
7							3.89	3.05
8								1.41

$\beta = 3.2$

n \ m	1	2	3	4	5	6	7	8
1	0.26	0.46	0.64	0.82	1.01	1.19	1.37	1.53
2		0.75	1.01	1.25	1.49	1.71	1.93	2.07
3			1.32	1.59	1.83	2.05	2.26	2.36
4				1.89	2.15	2.35	2.58	2.49
5					2.40	2.59	2.81	2.58
6						2.89	3.00	2.56
7							3.10	2.39
8								1.28

$\beta = 3.22$

n \ m	1	2	3	4	5	6	7	8
1	0.050	0.079	0.107	0.135	0.163	0.192	0.219	0.225
2		0.111	0.140	0.169	0.199	0.229	0.255	0.235
3			0.170	0.199	0.229	0.259	0.284	0.236
4				0.228	0.259	0.289	0.313	0.236
5					0.290	0.319	0.341	0.236
6						0.349	0.369	0.235
7							0.390	0.228
8								0.045

lowest temperature ($\beta = 3.22$), where they are smaller.

From the table it is apparent that $-\ln W$ increases with the area of the loop in the high-temperature phase, whereas it is area independent in the low-temperature phase. This behavior becomes particularly manifest if one compares the values of $-\ln W$ for loops of the same perimeter and different areas, such as are found along the diagonals.

A quantitative determination of the coefficient T in the area term, or string tension, may also be attempted, but finite-size effects introduce some degree of ambiguity. Indeed, loops of size one and two are very likely too small for a measurement of the size dependence of the loops (they tend to give a larger area term, when inserted in a fit); on the other hand, as the loops approach the size of the lattice, increased correlations due to the periodic boundary conditions tend to make $-\ln W$ smaller ($-\ln W$ would be exactly zero for square loops of side eight in an Abelian system). Thus, for instance, for $\beta = 2.5$ comparison of loops measuring 3×3 and 4×2 gives $T = 0.20$, comparison of loops measuring 4×4 and 5×3 gives $T = 0.17$, a fit to the square loops of sides 1, 2, and 3 gives $T = 0.145$, a fit to those of sides 2, 3, and 4 gives $T = 0.075$. In spite of this degree of uncertainty, the computation reveals clearly a confining behavior of the high-temperature phase, with a disorder parameter T which decreases for increasing β, has a discontinuity at the phase transitions, and vanishes in the low-temperature phase.

Simulations of thermal cycles and mixed-phase runs have been done for the models with gauge groups O and S_3. The results are displayed in Figs. 3 and 4. $\Delta\beta = 0.05$ for all thermal cycles. The three cycles for the S_3 model correspond to different choices of the action to be associated with the odd permutations P_i.

All thermal cycles show hysteresis loops. In the S_3 model, as one might expect on general scaling considerations, lowering the action $f(\{P_i\})$ has the effect of increasing the value of β where the loop appears. The mixed-phase runs

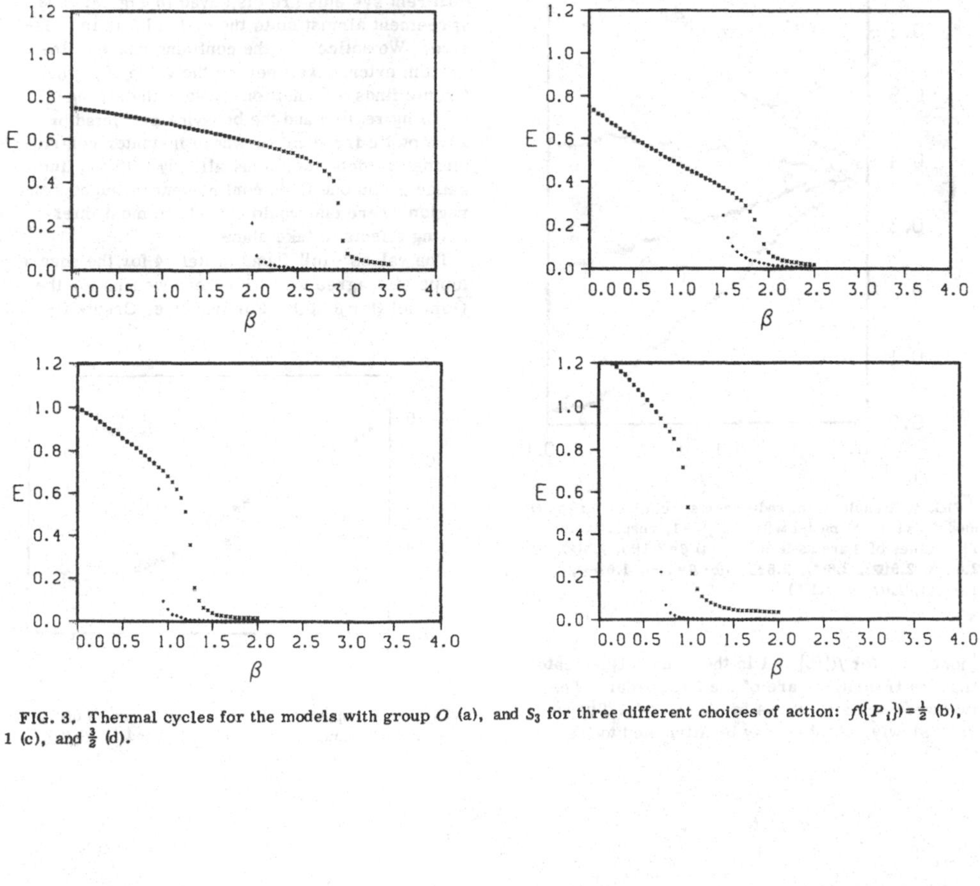

FIG. 3. Thermal cycles for the models with group O (a), and S_3 for three different choices of action: $f(\{P_i\}) = \frac{1}{2}$ (b), 1 (c), and $\frac{3}{2}$ (d).

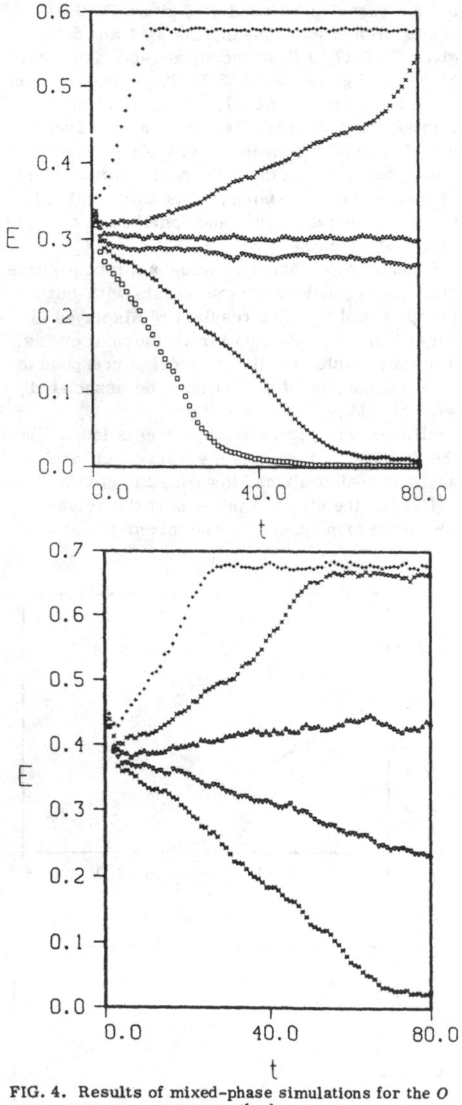

FIG. 4. Results of mixed-phase simulations for the O model and the S_3 model with $f(\{P_i\})=1$, respectively. The values of β are as follows: (a) $\beta=2.1(+)$, $2.3(\times)$, $2.4(\triangle)$, $2.5(\nabla)$, $2.6(*)$, $2.8(\square)$. (b) $\beta=1(+)$, $1.03(\times)$, $1.05(\triangle)$, $1.07(\nabla)$, $1.1(*)$.

[done only for $f(\{P_i\})=1$ in the S_3 model] indicate that the transitions are of the first order. The two intermediate curves in the O model diverge very slowly, but this may be attributed to the temperatures being very close to the critical temperature. The behavior of the other curves is rather typical of a first-order phase transition.

IV. CONCLUSIONS

Our results give numerical evidence that all non-Abelian models considered have a two-phase structure with a first-order phase transition. In the three systems where the gauge group is a subgroup of SU(2) the critical point definitely moves toward zero temperature as the order of the group increases. The analysis of the Wilson loop factors done for the $\tilde{O}$ model, moreover, shows that the high-temperature phase is confining, with a string tension that becomes discontinuously zero at the critical point.

Very much as the phase structure of the Z_N models is suggestive of that of the system with gauge group U(1), this study strongly supports the notion that a single, confining phase should be present in the model with gauge group SU(2). We have compared our results for the $\tilde{O}$ model with those obtained by Creutz[4] for the SU(2) model. The values found for the internal energy in the two different systems are displayed in Fig. 5: The agreement almost up to the critical β is impressive. We notice that the confining phase of the $\tilde{O}$ system extends well beyond the value of β where Creutz finds a transition between the strong-coupling regime and the behavior predicted by asymptotic freedom[4] or where instanton contributions are detected;[8] thus already with the finite gauge group one finds confinement throughout the region where one would expect the most interesting effects to take place.

The values $-\ln W$ found in Ref. 4 for the square loops also agree well with those we find for the $\tilde{O}$ model (for $\beta<\beta_c$). For instance, Creutz finds

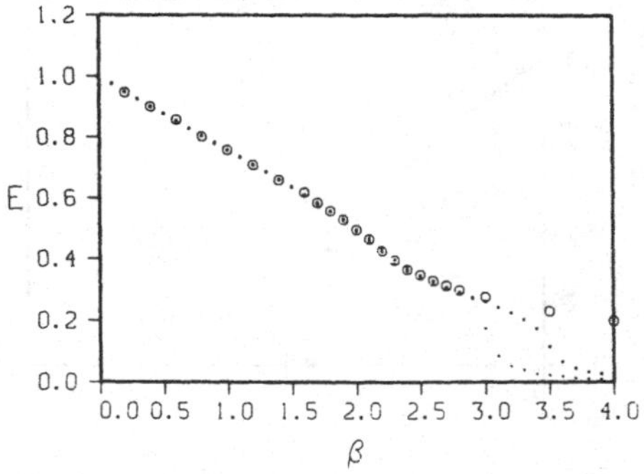

FIG. 5. Comparison of the internal energies of the systems with gauge group $\tilde{O}$ (+ and $\times$) and SU(2) ($\bigcirc$).

values 0.43, 1.36, 2.46, and 3.50 for square loops of side 1, 2, 3, and 4 at $\beta = 2.5$ to be compared with our values 0.42, 1.35, 2.57, and 3.94. This remarkable agreement between quantities measured in the two models suggests the interesting possibility of using the $\tilde{O}$ model, or the model based on the 120-element subgroup of SU(2), for reliable approximate computations of observables of the SU(2) system itself. The saving in memory requirements and computing time could allow the study of larger lattices.

Our understanding of non-Abelian gauge systems has certainly progressed during the last few years. The hypothesis of confinement, that not so long ago was a mere conjecture, is now strongly supported by a variety of numerical, perturbative, and semiclassical computations. The time may be ripe to promote the hypothesis into a theorem, with the rigor of analytical proof.

ACKNOWLEDGMENTS

I gratefully acknowledge the participation of Laurence Jacobs in early computations done for the system with gauge group $\tilde{Q}$. This research was performed under Contract No. DE-AC02-76CH00016 with the U.S. Department of Energy.

[1]K. G. Wilson, Phys. Rev. D 10, 2445 (1974).

[2]M. Creutz, L. Jacobs, and C. Rebbi, Phys. Rev. Lett. 42, 1390 (1979).

[3]M. Creutz, L. Jacobs, and C. Rebbi, Phys. Rev. D 20, 1915 (1979).

[4]M. Creutz, Phys. Rev. Lett. 43, 553 (1979); Phys. Rev. D 21, 2308 (1980).

[5]K. G. Wilson, Cornell University report, 1979 (unpublished).

[6]S. Elitzur, R. B. Pearson, and J. Shigemitsu, Phys. Rev. D 19, 3698 (1979); D. Horn, M. Weinstein, and S. Yankielowicz, *ibid.* 19, 3715 (1979); A. Ukawa, P. Windey, and A. Guth, *ibid.* 21, 1013 (1980).

[7]J. Kogut, R. B. Pearson, and J. Shigemitsu, Phys. Rev. Lett. 43, 484 (1979).

[8]C. G. Callan, R. Dashen, and D. J. Gross, Phys. Rev. D 20, 3279 (1979); Phys. Rev. Lett. 44, 435 (1980).

[9]F. J. Wegner, J. Math. Phys. 12, 2259 (1971); R. Balian, J. M. Drouffe, and C. Itzykson, Phys. Rev. D 10, 3376 (1974); 11, 2098 (1975); 11, 2104 (1975).

[10]While with our definition $f(W_\square)$ is again a loop factor with different class function Tr, later it will be convenient to assume a specific form for Tr in Eq. (2.1) and the elementary action is denoted $f(W_\square)$ for consistency.

[11]K. Binder, in *Phase Transitions and Critical Phenomena*, edited by C. Domb and M. S. Green (Academic, New York, 1976), Vol. 5B.

[12]N. Metropolis, A. W. Rosenbluth, M. N. Rosenbluth, A. H. Teller, and E. Teller, J. Chem. Phys. 21, 1087 (1953).

[13]Up to boundary effects. The periodic boundary conditions introduce correlations when the size of the loop becomes close to extent of the system in the spatial directions and eventually a decrease in $-\ln W$ is observed. For a square loop of side 8, W can still be different from 1 (the value it would take in an Abelian system) because of the noncommutativity of the U_{ij}.

[14]The evidence that the high-temperature phase has confining properties is obtained from the analysis of the loop factors, to be presented later for the $\tilde{O}$ model and discussed in Ref. 3 for the Z_N models.

Computer Physics Communications 25 (1982) 275–287
North-Holland Publishing Company

A FAST ALGORITHM FOR MONTE CARLO SIMULATIONS OF 4-d LATTICE GAUGE THEORIES WITH FINITE GROUPS

Gyan BHANOT
Institute for Advanced Study, Princeton, NJ 08540, USA

Christian B. LANG
CERN, Geneva, Switzerland

and

Claudio REBBI
Physics Department, Brookhaven National Laboratory, Upon, NY 11973, USA

Received 16 October 1981

PROGRAM SUMMARY

Title of program: LATGAUGEMC

Catalogue number: AAVI

Program obtainable from: CPC Program Library, Queen's University of Belfast, N. Ireland (see application form in this issue)

Computer: CDC 7600; *Installation*: Brookhaven National Laboratory, Upton, NY, 11973 and CERN, Geneva, Switzerland

The program is operable on any computer which has a word structured memory with a wordlength of at least 28 bits, the possibility to implement mask operations (bit by bit logical product) and circular left-shift, and FORTRAN IV

Operating system: CDC Scope

Programming language used: FORTRAN IV

High speed storage required: 19634 D (46262_8) for the example

No. of bits in a word: 60

Overlay structure: none

Peripherals used: mass storage (amount depends on the application, 15741 D computer words in the example)

No. of cards in combined program and test deck: 393

Card punching code: 029 code

Keywords: lattice gauge theory, discrete group, 4-dimensions, variable lattice size, periodic boundary conditions, Monte Carlo, Metropolis method

Nature of the physical problem
The program LATGAUGEMC generates configurations of the link-gauge variables of a 4-dimensional hypercubic lattice of linear extent 2, 4 or 8 with periodic boundary conditions. The link variables take values from any group where the number of group elements plus the number of elements equal to their own inverse is less than 128. The Monte Carlo method used to update the link variables is the Metropolis method [1]. The configurations generated by the program are representative of the ensemble of configurations on a finite lattice that defines the partition function of a lattice gauge field theory. They may be used to evaluate expectation values of functions of the gauge variables. The program has been used by the authors [2] to study the 4-dimensional SU(2) lattice gauge theory. It could be used for Monte Carlo simulations of other problems in statistical mechanics and particle physics.

Typical running time
The actual running time depends on the number NUPG of times a single link variable is upgraded before proceeding to

the next one. With NUPG=6, the running time is ≈50 μs/variable per MC iteration on a CDC 7600.

Unusual features of the program

a) The group elements are labelled so that the operation of finding the inverse of any element is a single FORTRAN command.

b) The periodic boundary conditions on the lattice are implemented using shift and mask operations to perform the necessary modulo arithmetic.

These features result in a considerable increase in the speed of the Monte Carlo over other conventional methods.

References

[1] N. Metropolis, A.W. Rosenbluth, A.H. Teller and E. Teller, J. Chem. Phys. 21 (1953) 1087.

[2] C. Rebbi, Phys. Rev. D 21 (1980) 3350.
D. Petcher and D.H. Weingarten, Phys. Rev. D 22 (1980) 2465.
G. Bhanot and C. Rebbi, Nucl. Phys. B180 (1981) 469.
C.B. Lang, C. Rebbi, P. Salomonson and B.S. Skagerstam, Phys. Lett. 101B (1981) 173.

LONG WRITE-UP

1. Introduction

This paper describes a program to perform Monte Carlo simulations of 4-dimensional lattice gauge theories [1,2] defined on discrete gauge groups [3]. The number of elements in the gauge group plus the number of elements that are equal to their inverse must not exceed 127. The program as written contains the parameter set for a lattice of 4^4 sites and the 120 element icosahedral subgroup Y of SU(2). The lattice size can be changed to 2^4 sites or 8^4 sites and the group Y may be changed to any other group using information supplied in the form of comment statements in the program.

The properties of the gauge group that are used in the program have to be supplied as data in tables MT, ET and IG. Detailed instructions on how to create these tables for an arbitrary gauge group are given in section 7.

2. The physical model

The lattice gauge theory on a group G is defined by the action [1],

$$S(\{U\},\beta)=\beta\sum_{(i\mu\nu)}\left[1-\frac{1}{2N}\mathrm{Re}\left\{\mathrm{Tr}\left(V(i\mu\nu)+V^{-1}(i\mu\nu)\right)\right\}\right], \tag{2.1a}$$

$$V(i\mu\nu)=U_{i,\mu}U_{i+\mu,\nu}U^{-1}_{i+\nu,\mu}U^{-1}_{i,\nu}. \tag{2.1b}$$

The U's are the link-gauge variables (represented by complex $N\times N$ matrices). They are elements of G and are defined on the links of a 4-dimensional hypercubic lattice; i labels the lattice sites and $\mu,\nu=1,2,3,4$ label the four directions. The sum in (2.1a) is over all plaquettes (unit squares) of the lattice, each plaquette being considered once in the sum. Each term of the sum represents the product (2.1b) of the link variables around a plaquette $(i\mu\nu)$. The U variables are oriented in the sense that

$$U^{-1}_{i,\mu}=U_{i+\mu,-\mu}. \tag{2.1c}$$

The dynamics is defined by the partition function $Z(\beta)$ and the free energy $F(\beta)$.

$$Z(\beta)=e^{F(\beta)}=\sum_{\{U\}}\exp\left[-S(\{U\},\beta)\right] \tag{2.2}$$

where the sum runs over all configurations of the U variables and the parameter β is the inverse temperature.

The expectation value $\bar{A}(\beta)$ of any function $A(\{U\})$ is given by,

$$\bar{A}(\beta)=\frac{1}{Z(\beta)}\sum_{\{U\}}A(\{U\})\exp\left[-S(\{U\},\beta)\right]. \tag{2.3}$$

The formulation described above was used by Wilson [1] to study the confinement of quarks in strong-interaction particle physics. Since then, the lattice gauge theory has found applications in many diverse areas of particle and statistical physics [2,3].

3. The Metropolis method

Eqs. (2.1)–(2.3) define a problem of statistical mechanics. It is well known that statistical systems can be analysed by computer simulations. This is done by generating a sequence of configurations of the link-gauge variables such that the probability of having in the sequence a configuration with action S is $\exp(-S)$. A widely followed procedure to obtain such a sequence is based on an algorithm, originally introduced by Metropolis et al. [4] which we now describe.

Starting from an arbitrary configuration (with the link-variables assigned any set of values from the group G), one steps through the lattice link by link and updates the old value U of the field on the link to a new value U' using the following algorithm:

a) The contribution $S(U)$ of the link being updated to the total action $S(\{U\},\beta)$ is calculated. This involves taking the product of U's on the links bordering the plaquettes which are connected to the link being updated. A trial element U' is constructed by multiplying U with a randomly chosen group element (in our case a neighbour element of the identity). Then $S(U')$, the contribution with U replaced by U' is computed.

b) If $S(U') \leqslant S(U)$ then U is replaced by U'.

c) If $S(U') > S(U)$ a (pseudo) random number R with uniform distribution in (0, 1) is generated. If $R < \exp[S(U) - S(U')]$ then U is replaced by U'. Otherwise, the old value U on the link is retained.

While a particular link variable is being updated, all other variables are held fixed. Steps a), b) and c) may be repeated $n_u \geqslant 1$ times at each link before moving to another variable, provided that the number n_u is kept fixed until all the link variables on the lattice are updated.

One application of the above algorithm to every link of the lattice is called an iteration. It can be shown that the algorithm method described above asymptotically generates configurations with weight $\exp[-S(\{U\},\beta)]$ and thus can be used for all approximate evaluations of the statistical averages.

4. Labelling the group elements

Since the group G has a finite number of elements, the possible values that the field U can take may be labelled by integers. It is convenient to set up the labelling as follows:

Suppose G has m elements, of which n are equal to their inverse. We enlarge G to a set G' of $m+n-1$ elements with a definite order. Thus

$$G': \{U_p; p = 1, 2, \ldots, m+n-1\},$$

(i) the element $U_{(m+n)/2}$ is the identity *;

(ii) the elements $U_1, U_2, \ldots, U_{n-1}$ are the remaining self-inverse elements;

(iii) the set $S_1 = \{U_n, U_{n+1}, U_{n+2}, \ldots, U_{[(n+m)/2]-1}\}$ has elements which have their inverse elements in the set $S_2 = \{U_{[(n+m)/2]+1}, U_{[(n+m/2]+2}, \ldots, U_m\}$. The ordering of elements in S_1 and S_2 is such that the inverse of $U_{[(m+n)/2]-i}$ in S_1 is $U_{[(m+n)/2]+i}$ in S_2 ($i = 1, 2, \ldots, (m-n)/2$).

(iv) the element U_{m+i} is identical to the element U_{n-i} ($i = 1, 2, \ldots, (n-1)$).

This labelling assigns one or two integers p ($p = 1, 2, \ldots, m+n-1$) to each element U_p of the set G'. The advantage of this method of labelling the group elements is that the inverse of an element $U_i \in G'$ is the element $U_{m+n-i} \in G'$. Thus

$$U_i^{-1} = U_{m+n-i} \; (i = 1, 2, \ldots, (m+n-1)), \qquad (4.1a)$$

or, symbolically

$$(i)^{-1} \equiv (m+n-i). \qquad (4.1b)$$

5. The program

The updating of the lattice (iteration) is performed by calling a subroutine UPDATE (IDIR) which updates all links pointing along IDIR.

We now describe the most relevant features of LATGAUGEMC.

* Note that $m+n$ is always even.

5.1. Locating a link on the lattice

The variables $U_{i,\mu}$ of the theory are defined on links. The location L of each variable is therefore given by five integers, four coordinates x, y, z, t of the site to which the link is connected and the direction μ along which it points. x, y, z and t are assumed to be integers ranging from 1 to the lattice size. For a lattice of linear extent 2^n, define

$$L=\mu+(x-1)2^2+(y-1)2^{n+2}+(z-1)2^{2n+2}+(t-1)s^{3n+2}. \quad (5.1)$$

Clearly, L ranges from 1 to $4(2^n)^4$. This assigns one integer L to each link on the 4-dimensional lattice.

The gauge-variables are stored in an array LV of size $4(2^n)^4$. LV(L) contains the integer $p(L)$ labelling the value of the field on the link L (see section 4 for the labelling convention). Thus,

$$\mathrm{LV}(L)=p(L). \quad (5.2)$$

5.2. Locating neighbouring links with periodic boundary conditions

Given a link L [eq. (5.1)], the Metropolis method (section 3) requires that we locate all other links on plaquettes connected to L. This is accomplished in the program by using forward shifts FX, FY, FZ, FT and backward shifts BX, BY, BZ, BT. These, when appropriately added to L, give the coordinates of the neighbouring links. Periodic boundary conditions are imposed by the use of masks M1, M2 and M3. We shall illustrate the procedure by describing how it works on the lattice of 4^4 sites.

The number S representing the site to which L is connected is $S=L-\mu$.

On a 4^4 lattice, S is a number of up to 10 bits. Counting from right to left, the first two bits contain zero, the next two contain $(x-1)$, the next two $(y-1)$ and so on. Also,

$$S_{\max}=1111111100\ (\text{binary}). \quad (5.3)$$

To find the coordinates of a site neighbouring S, we perform the following steps:
Define

$$S'=\mathrm{AND}(S+\mathrm{SHIFT}(S,15),\mathrm{M3}), \quad (5.4)$$

where

$$\mathrm{M3}=1100110000000110011000\ (\text{binary}). \quad (5.5)$$

This operation has the effect of putting zeros in the original location of $(x-1)$ and $(z-1)$ and of shifting these to the bit positions 18, 19 and 22, 23, respectively. $(y-1)$ and $(t-1)$ are in bit positions 5, 6 and 9, 10, respectively. Now define the forward and backward shifts:

$$\mathrm{FX}=2^{17},\ \mathrm{BX}=3\times2^{17},\quad \mathrm{FY}=2^4,\ \mathrm{BY}=3\times2^4,$$
$$\mathrm{FZ}=2^{21},\ \mathrm{BZ}=3\times2^{21},\quad \mathrm{FT}=2^8,\ \mathrm{BT}=3\times2^8. \quad (5.6)$$

To shift the site index S forward or backward along a given direction, we add the appropriate shift from (5.6) to S'. Suppose we want to move forward along the x direction and then want the index for the link pointing along μ' from this new site. Then, define

$$L'=S'+\mathrm{FX}+\mu. \quad (5.7)$$

This adds 1 to the location for $(x-1)$ in S' and puts μ' in the bit positions 1, 2, 3. Now we impose the periodic boundary condition using the mask M1:

$$\mathrm{M1}=1100110000000110011011\ (\text{binary}). \quad (5.8)$$

The logical operation

$$L''=\mathrm{AND}(L',\mathrm{M1}) \quad (5.9)$$

removes any carry-over bit in the addition of 1 to $x-1$ and thus effectively makes the addition a sum modulo 4, enforcing the periodicity of the lattice.

Finally, we bring L'' into the standard form [eq. (5.5)] by the operation

$$L'''=\mathrm{AND}(\mathrm{SHIFT}(L'',-15)+L'',\mathrm{M2}), \quad (5.10)$$

with

$$\mathrm{M2}=1111111111\ (\text{binary}). \quad (5.11)$$

L''' is then the index for the link pointing along μ' shifted from L by one site forward along x.

The appropriate shifts and masks for other lattice sites are defined in the program as comments. The subroutine UPDATE upgrades all the link variables in the x direction first, then the directions are permuted in the main program and the link variables with the other orientations are successively upgraded.

5.3. The Monte Carlo method

Once all the link labels for the link to be updated and all links on plaquettes connected to it are recovered, we read the values of the fields on these links from the array LV. Next, we use the multiplication table MT to find $S(U_L)$ and $S(U'_L)$ where U'_L is obtained by multiplying U_L by one of the elements nearest the identity (chosen randomly from the array IG of all such elements). Then, the Metropolis algorithm (described in section 3), is used to decide whether or not to change U_L to U'_L. Each link is updated $n_u = [(q+1)/2]$ times before moving to the next link on the lattice.

This updating procedure is performed once on each lattice link to complete one iteration of the lattice. As the lattice links are being updated, we measure the order parameter E defined by,

$$E = \left\langle 1 - \frac{1}{2N} \mathrm{Re}\left(\mathrm{Tr}\left(U_\Box + U_\Box^{-1}\right)\right)\right\rangle = -\frac{\partial F}{\partial \beta} \tag{5.12}$$

where $\Box$ stands for a lattice plaquette; E is the average value of the action. E is measured by initializing it to zero before each iteration, incrementing it by $S(U)$ or $S(U')$ (depending on whether or not U was changed to U') as each link is updated, and finally, normalizing it by dividing by $6N_L$ at the end of the iteration.

E can be used as an indicator of the degree to which the lattice has approached equilibrium after a certain number of iterations. For instance, if we initialize the lattice links to the unit element, the initial lattice has $E = 0$. As we perform iterations at a given β, E will change until it stabilizes around some mean value $\bar{E}(\beta)$ characteristic of the temperature $1/\beta$. At this point, the lattice configuration is near thermal equilibrium, at least for the order parameter E. It should be noticed, however, that although the average value of E agrees with the correct thermal average when equilibrium is reached, the quantity E, as defined in the program, does not represent the mean value of the plaquette action of any definite configuration. Thus, in studies where the actual distribution of the action is relevant, such as analyses of fluctuations, E should be evaluated separately, stripping for instance the loop 2 in the subroutine UPDATE.

6. Variables

6.1. List of variables, their names in LATGAUGEMC and symbols used for them in this write-up

Variable	Definition	Symbol in write-up
NELE	number of group elements	m
NSCE	number of elements equal to their inverse	n
NEAR	number of elements closest to the unit element	q
NLVA	actual number of links (equal to four times the number of lattice sites) or higher	N_L
E	average action per plaquette	E
FX, FY, FZ, FT	forward shifts	FX, FY, FZ, FT
BX, BY, BZ, BT	backward shifts	BX, BY, BZ, BT
M1, M2, M3	masks	M1, M2, M3
NUPG	number of Monte Carlo upgrades performed at a given link before moving to the next	n_u

6.2. Common blocks

The program contains four labelled common blocks.

Common block name and length	Variable list	Definition *
COM1 length=23 words	IS11, IS12 IS13, IS21, IS22, IS23, IS31, IS32, IS33, IS41, IS42, IS43 IS51, IS52, IS53, IS61, IS62, IS63, M1, M2, M3, MI, NUPG	IS11 through IS63 are integer increments that find the coordinates of links on plaquettes connected to the link being updated. M1, M2 and M3 are masks, MI=NELE+NSCE
COM2 length=2+NLVA words	LLEN1, LV(NLVA), E	LLEN1 is the number of lattice sites. LV(I) contains the value of the group element at link location I
COM3 length=(NMUL+NELE +NEAR) words	MT(NMUL), ET(NELE), IG(NEAR)	MT is the multiplication table which is used to find the group product of two elements. ET(I) contains the action of the Ith group element. IG contains the identifiers of the elements closest to the identity
COM4 length=NELE	BFA(NELE)	BFA(I) contains the Boltzman weight exp($-\beta\cdot$ET(I)) of the Ith group element

7. Procedure to construct MT, ET and IG

Storage requirements:
MT: $(2^7+1)(m+n-1)$ words,
ET: m words,
IG: q words.

7.1. Constructing MT

(i) Construct the m, $N\times N$ matrices of the group G.
(ii) Construct the set G' of $(m+n-1)$ matrices $\{U_p;\ p=1,2,\ldots,(m+n-1)\}$ ordered as described in section 5.
(iii) Consider a pair of matrices U_i and $U_j\in G'$. Let $U_iU_j=U_k$ $(k=1,2,\ldots,m)$.

Now define,

$$l=i+\mathrm{SHIFT}(j,7), \tag{7.1a}$$

and set

$$\mathrm{MT}(l)=k. \tag{7.1b}$$

Repeat this for all possible values of i and j $(i,j=1,2,\ldots,m+n-1)$ to create the multiplication table.

7.2. Constructing the energy table ET

For Wilson's definition [1] of the action [eq. (2.1a)] set

$$\mathrm{ET}(p)=1-\frac{1}{2N}\mathrm{Re}(\mathrm{Tr}(U_p+U_{m+p-1})),\quad (p=1,2,\ldots,m). \tag{7.2}$$

* See section 5 for definitions not given below.

7.3. Constructing IG, the table of elements nearest the identity

Find the set of integers $\{p_i, i=1,2,\ldots,q\}$ for which ET(p_i) has a minimum non-vanishing value. Set

$$IG(i)=p_i, \quad (i=1,2,\ldots,q). \tag{7.3}$$

Remarks

The program makes use of intrinsic functions such as AND or SHIFT which implement simple machine operations but are not standard FORTRAN functions. However, the availability of such functions in a wide range of compilers justifies in our opinion their usage. In any case, the corresponding operations can always be implemented through machine language subroutines or by equivalent arithmetic operations. The functions are defined as follows:

AND(I, J) is the bit by bit logical AND of I with J.

SHIFT(I, N) is the content of I circularly shifted to the left by N bits.

Wherever SHIFT(I, −N) is used, it may be either implemented by a right-shift (circular or with sign extension) or substituted by SHIFT(I, $n-$N) where n denotes the number of bits of a computer word.

The program utilizes the CDC in-line random number generator function RANF(DUMMY). Any fast generator of random numbers with uniform distribution in (0, 1) may be used instead.

References

[1] K.G. Wilson, Phys. Rev. D 10 (1974) 2445.

[2] R. Balian, J.M. Drouffe and C. Itzykson, Phys. Rev. D 10 (1974) 3376, D 11 (1975) 2098, 2104, D 19 (1979) 2514.
M. Creutz, L. Jacobs and C. Rebbi, Phys. Rev. Lett. 42 (1979) 1390; Phys. Rev. D 20 (1979) 1915.
M. Creutz, Phys. Rev. Lett. 43 (1979) 553; Phys. Rev. D 21 (1980) 2308; Phys Rev. Lett. 45 (1980) 331.
G. Toulouse and P. Vannimenus, Phys. Rep. 67 (1980) 47.
For a more complete discussion and detailed listing of references, see e.g. M. Creutz, Lectures given at the Ettore Majorana School in Erice, BNL preprint 1981, or C. Rebbi, Lectures given at the GIFT School in San Felin the Guixols and at the ICPT, Trieste, ICTP preprint (1981).

[3] C. Rebbi, Phys. Rev. D 21 (1980) 3350.
D. Petcher and D.H. Weingarten, Phys. Rev. D 22 (1980) 2465.
G. Bhanot and C. Rebbi, Nucl. Phys. B180 (1981) 469.
C.B. Lang, C. Rebbi, P. Salomonson and B.S. Skagerstam, Phys. Lett 101B (1981) 173.

[4] N. Metropolis, A.W. Rosenbluth, A.H. Teller and E. Teller, J. Chem. Phys. 21 (1953) 1087.

Appendix

Listing of the program card-deck

```
      PROGRAM MAIN(INPUT,OUTPUT,TAPE1)
C
C            --------------------------------------------------------------
C            PROGRAM [illegible] FOR MONTE CARLO SIMULATIONS
C            OF 4-DIMENSIONAL LATTICE GAUGE SYSTEMS WITH A FINITE
C            GAUGE GROUP.
C            --------------------------------------------------------------
C            AUTHORS: G. BHANOT, C. B. LANG, C. REBBI
C            --------------------------------------------------------------
C
C            THE PROGRAM GENERATES CONFIGURATIONS OF THE LINK-GAUGE
C            VARIABLES OF A 4-DIMENSIONAL HYPER-CUBIC LATTICE OF SIZE
C            2**4, 4**4, 8**4 TO BE SPECIFIED. THE NUMBER OF ELEMENTS
C            IN THE GAUGE GROUP PLUS THE NUMBER OF ELEMENTS EQUAL TO
C            THEIR INVERSE MUST NOT EXCEED 127. APART FROM THIS,
C            THE GAUGE GROUP MAY HAVE AN ARBITRARY STRUCTURE, WHICH
C            IS GIVEN IN A MULTIPLICATION TABLE MT. THIS TABLE
C            TOGETHER WITH AN ACTION TABLE (GIVING THE VALUE OF THE
C            ACTION FOR EACH GROUP ELEMENT) AND AN ARRAY GIVING THE
C            INDICES OF THE GROUP ELEMENTS CLOSEST TO THE IDENTITY
C            HAS TO BE CREATED BEFORE EXECUTION OF THE PROGRAM.
C            USUALLY THIS IS DONE ONLY ONCE AND THE TABLES ARE
C            PERMANENTLY KEPT ON MASS STRORAGE (TAPE1).
C            THE MONTE-CARLO METHOD IS THE MODIFIED METROPOLIS METHOD.
C
C            THIS PROGRAM HAS THE PARAMETERS SET FOR A 4**4 LATTICE
C            AND THE 120-ELEMENT ICOSAHEDRAL SUBGROUP OF SU(2).
C            IN ORDER TO VARY THESE SPECIFICATIONS ALL NECESSARY
C            CHANGES ARE INDICATED IN COMMENTS. THE FOLLOWING
C            ABBREVIATIONS ARE USED:                *)
C                 NELE = ACTUAL NUMBER OF GROUP ELEMENTS
C                 NSCE = NUMBER OF ELEMENTS EQUAL TO THEIR INVERSE
C                 NEAR = NUMBER OF GROUP ELEMENTS CLOSEST TO THE UNIT
C                        ELEMENT
C                 NMUL = MAXIMAL NUMBER OF NECESSARY ENTRIES IN THE
C                        MULTIPLICATION TABLE.(THIS NUMBER IS EASILY
C                        OBTAINED BY COMPUTING THE DECIMAL EQUIVALENT
C                        OF A 14-BIT BINARY NUMBER WHICH IS OBTAINED
C                        BY WRITING (NELE+NSCE-1) TWO TIMES IN ITS
C                        BINARY REPRESENTATION.)
C                 NLVA = ACTUAL NUMBER OF LINKS (FOUR TIMES THE
C                        NUMBER OF LATTICE POINTS) OR HIGHER.
C
C            THE PROGRAM USES THREE NON-STANDARD FUNCTIONS (CF. LONG
C            WRITEUP):
C                 RANF(DUMMY) PROVIDES A REAL RANDOM NUMBER UNIFORMLY
C                             DISTRIBUTED IN THE INTERVAL (0.,1.),
C                             DUMMY IS A DUMMY VARIABLE.
C                 AND(I,J)    GIVES THE BIT BY BIT LOGICAL PRODUCT OF I
C                             WITH J.
C                 SHIFT(I,N)  GIVES THE CONTENT OF I CIRCULARLY SHIFTED
C                             TO THE LEFT BY N BITS. IF N IS NEGATIVE
C                             IT DENOTES A SHIFT TO THE RIGHT WHICH
C                             MAY OR MAY NOT BE WITH SIGN EXTENSION.
C
C            --------------------------------------------------------------
C
```

```
      COMMON/COM1/IS11,IS12,IS13,IS21,IS22,IS23,IS31,IS32,IS33,       0065
     1            IS41,IS42,IS43,IS51,IS52,IS53,IS61,IS62,IS63,       0066
     2            M1,M2,M3,MI,NUPG                                     0067
C                                                                     0068
C             FOR DIFFERENT LATTICE SIZE LV(1024) SHOULD BE           0069
C             CHANGED TO LV(NLVA).              *)                     0070
C                                                                     0071
      COMMON/COM2/LLEN1,LV(1024),E                                    0072
C                                                                     0073
C             FOR OTHER GROUPS MT(15609),ET(120),IG(12), AND BFA(120) 0074
C             SHOULD BE CHANGED TO MT(NMUL),ET(NELE),IG(NEAR) AND     0075
C             BFA(NELE).                       *)                     0076
C                                                                     0077
      COMMON/COM3/MT(15609),ET(120),IG(12)                           0078
      COMMON/COM4/BFA(120)                                            0079
      INTEGER FX,FY,FZ,FT,BX,BY,BZ,BT                                 0080
      INTEGER FXA,FYA,FZA,FTA,BXA,BYA,BZA,BTA                         0081
C                                                                     0082
C             THESE ARE THE INCREMENTS FOR A 4**4 LATTICE:            0083
C                                                                     0084
      DATA FX,FXA,BX,BXA/2*400000B,2*1400000B/                        0085
      DATA FY,FYA,BY,BYA/2*20B,2*60B/                                 0086
      DATA FZ,FZA,BZ,BZA/2*10000000B,2*30000000B/                    0087
      DATA FT,FTA,BT,BTA/2*400B,2*1400B/                              0088
      DATA M1,M2,M3/31401467B,3777B,31401460B/                        0089
C                                                                     0090
C             FOR A 2**4 LATTICE USE INSTEAD THE FOLLOWING STATEMENTS:0091
C                                                                     0092
C      DATA FX,FXA,BX,BXA/4*400000B/                                  0093
C      DATA FY,FYA,BY,BYA/4*10B/                                      0094
C      DATA FZ,FZA,BZ,BZA/4*20000000B/                                0095
C      DATA FT,FTA,BT,BTA/4*40B/                                      0096
C      DATA M1,M2,M3/2400057B,177B,2400050B/                          0097
C                                                                     0098
C             FOR A 8**4 LATTICE USE INSTEAD THE FOLLOWING STATEMENTS:0099
C                                                                     0100
C      DATA FX,FXA,BX,BXA/2*400000B,2*3400000B/                       0101
C      DATA FY,FYA,BY,BYA/2*40B,2*340B/                               0102
C      DATA FZ,FZA,BZ,BZA/2*40000000B,2*340000000B/                   0103
C      DATA FT,FTA,BT,BTA/2*4000B,2*34000B/                           0104
C      DATA M1,M2,M3/343434347B,77777B,343434340B/                    0105
C                                                                     0106
      DATA MX,MXA,MY,MYA,MZ,MZA,MW,MWA/2*1,2*2,2*3,2*4/              0107
C                                                                     0108
C             FOR 2**4 :   LSIZE=1                                    0109
C             FOR 4**4 :   LSIZE=2                                    0110
C             FOR 8**4 :   LSIZE=3                                    0111
C                                                                     0112
      LSIZE=2                                                         0113
C                                                                     0114
C             FOR ANOTHER GROUP CHANGE 120,2 AND 12 TO THE APPROPRIATE0115
C             VALUES OF NELE, NSCE AND NEAR.                          0116
C                                                                     0117
      NELE=120                                                        0118
      NSCE=2                                                          0119
      NEAR=12                                                         0120
C                                                                     0121
```

```
C              INITIALIZATION                                         0122
C                                                                     0123
      ELE=NELE                                                        0124
      MI=NELE+NSCE                                                    0125
      MIH=MI/2                                                        0126
      NUPG=(NEAR+1)/2                                                 0127
      NSIZE=2**LSIZE                                                  0128
      LLEN1=NSIZE**4                                                  0129
      LLEN2=4*LLEN1                                                   0130
      PL4=24.*LLEN1                                                   0131
      PRINT 100                                                       0132
100   FORMAT(1H1,48HTEST RUN FOR A NON-ABELIAN GAUGE THEORY WITH THE, 0133
     1    /,49H ICOSAHEDRAL SUBGROUP OF SU(2) AS GAUGE GROUP       ,  0134
     2    /,52H -------------------------------------------------,/)  0135
      PRINT 101,NSIZE,NELE                                            0136
101   FORMAT(36H PROGRAM FOR LATTICE GAUGE THEORY                    0137
     1    /,52H -------------------------------------------------,//, 0138
     2         16H LATTICE SIZE : ,I1,3H**4,/,                        0139
     3         28H DISCRETE GAUGE GROUP WITH ,I3,10H ELEMENTS ,//,    0140
     4         38H B DENOTES THE COUPLING PARAMETER BETA,/,           0141
     5         46H E DENOTES THE AVERAGE ACTION PER PLAQUETTE    ,    0142
     6    //,52H -------------------------------------------------,/) 0143
C                                                                     0144
C              READ THE MULTIPLICATION TABLE, ACTION TABLE, AND LIST OF 0145
C              ELEMENTS CLOSEST TO THE UNIT ELEMENTS.                 0146
C                                                                     0147
      REWIND 1                                                        0148
      READ(1) MT,ET,IG                                                0149
C                                                                     0150
C              INITIALIZE THE LATTICE:                                0151
C                    ISTART=0     ALL ELEMENTS FROZEN TO UNITY        0152
C                    ISTART=1     RANDOM START (HOT START)            0153
C                                                                     0154
      ISTART=1                                                        0155
      DO 1 I=1,LLEN2                                                  0156
1     LV(I)=MIH*(1-ISTART)+ISTART*(INT(ELE*RANF(DUMMY))+1)           0157
C                                                                     0158
C              START OF B LOOP                                        0159
C                                                                     0160
      DO 2 IB=1,100                                                   0161
      B=(IB-1)*0.05                                                   0162
C                                                                     0163
C              COMPUTE ALL BOLTZMANN FACTORS                          0164
C                                                                     0165
      DO 3 I=1,NELE                                                   0166
3     BFA(I)=EXP(-B*ET(I))                                           0167
      E=0.                                                            0168
C                                                                     0169
C              UPDATE LINKS IN ALL FOUR DIRECTIONS.                   0170
C              FOR EACH CALL OF THE SUBROUTINE UPDATE ALL LINKS       0171
C              IN THE CORRESPONDING DIRECTION ARE UPDATED.            0172
C              THE INCREMENTS ARE PERMUTED CORRESPONDINGLY.           0173
C                                                                     0174
      DO 4 IDIR=1,4                                                   0175
      IS11=MY                                                         0176
      IS12=FY+MX                                                      0177
      IS13=FX+MY                                                      0178
      IS21=BY+MY                                                      0179
      IS22=BY+MX                                                      0180
      IS23=FX+BY+MY                                                   0181
      IS31=MZ                                                         0182
      IS32=FZ+MX                                                      0183
      IS33=FX+MZ                                                      0184
      IS41=BZ+MZ                                                      0185
      IS42=BZ+MX                                                      0186
      IS43=FX+BZ+MZ                                                   0187
      IS51=MW                                                         0188
      IS52=FT+MX                                                      0189
      IS53=FX+MW                                                      0190
      IS61=BT+MW                                                      0191
      IS62=BT+MX                                                      0192
      IS63=FX+BT+MW                                                   0193
      CALL UPDATE(IDIR)                                               0194
      MX=MYA                                                          0195
      FX=FYA                                                          0196
      BX=BYA                                                          0197
      MY=MZA                                                          0198
      FY=FZA                                                          0199
      BY=BZA                                                          0200
      MZ=MWA                                                          0201
      FZ=FTA                                                          0202
      BZ=BTA                                                          0203
      MW=MXA                                                          0204
      FT=FXA                                                          0205
      BT=BXA                                                          0206
      MXA=MX                                                          0207
      FXA=FX                                                          0208
      BXA=BX                                                          0209
```

```
        MYA=MY                                                          0210
        FYA=FY                                                          0211
        BYA=BY                                                          0212
        MZA=MZ                                                          0213
        FZA=FZ                                                          0214
        BZA=BZ                                                          0215
        MWA=MW                                                          0216
        FTA=FT                                                          0217
        BTA=BT                                                          0218
4       CONTINUE                                                        0219
C                                                                       0220
C               DIVIDE THE TOTAL ACTION E BY THE NUMBER OF PLAQUETTES AND 0221
C               BY FOUR (DUE TO THE OVER COUNTING IN THE SUBROUTINE).  0222
C               ATTENTION: THIS E IS THEN THE AVERAGE OVER THE LATTICE 0223
C               WHILE IT IS UPDATED, THUS ALSO AN AVERAGE IN "TIME", AND 0224
C               NOT AN AVERAGE OVER THE ACTUAL LATTICE CONFIGURATION AT 0225
C               THIS MOMENT.                                           0226
C               CONSIDERING DETERMINATIONS OF FLUCTUATIONS THE USE OF  0227
C               THESE NUMBERS WOULD INDUCE AN ERROR. IN THIS CASE      0228
C               ONE HAS TO REALLY CALCULATE THE AVERAGE OVER THE ACTUAL 0229
C               CONFIGURATION BY E.G. SKIPPING THE UPDATE LOOP 2 IN THE 0230
C               SUBROUTINE UPDATE.                                     0231
C                                                                       0232
        E=E/PL4                                                        0233
        PRINT 102,B,E                                                 0234
102     FORMAT(10H        B=  ,F10.5,8H       E=  ,F10.5)             0235
2       CONTINUE                                                      0236
        PRINT 103                                                     0237
103     FORMAT(/,52H ----------------------------------------------->) 0238
        STOP                                                          0239
        END                                                          0240
        SUBROUTINE UPDATE(IDIR)                                      0241
C                                                                       0242
C               SUBROUTINE TO UPDATE ALL LINKS VARIABLE IN A DIRECTION. 0243
C                                                                       0244
C               FOR CHANGES OF THE COMMONS /COM2/ AND /COM3/ DUE TO OTHER 0245
C               LATTICE SIZE OR ANOTHER GROUP CONSULT THE MAIN PROGRAM. 0246
C                                                                       0247
        COMMON/COM1/IS11,IS12,IS13,IS21,IS22,IS23,IS31,IS32,IS33,    0248
     1          IS41,IS42,IS43,IS51,IS52,IS53,IS61,IS62,IS63,        0249
     2          M1,M2,M3,MI,NUPG                                     0250
        COMMON/COM2/LLEN1,LV(1024),E                                 0251
        COMMON/COM3/MT(15609),ET(120),IG(12)                         0252
        COMMON/COM4/BFA(120)                                         0253
C                                                                       0254
C               GO THROUGH THE WHOLE LATTICE AND UPDATE THE          0255
C               LINK POINTING FROM THE LATTICE POINT IN THE          0256
C               DIRECTION IDIR.                                      0257
C                                                                       0258
        DO 1 IND1=1,LLEN1                                            0259
        IND=SHIFT(IND1-1,2)                                          0260
        N=IND+IDIR                                                   0261
        NA=AND(IND+SHIFT(IND,15),M3)                                 0262
C                                                                       0263
C               PLAQUETTE NUMBER 1                                   0264
C               COMPUTE ADDRESSES OF THE OTHER THREE LINKS OF THE    0265
C               PLAQUETTE.                                           0266
C                                                                       0267
        N11=AND(NA+IS11,M1)                                          0268
        N11=AND(N11+SHIFT(N11,-15),M2)                               0269
        N12=AND(NA+IS12,M1)                                          0270
        N12=AND(N12+SHIFT(N12,-15),M2)                               0271
        N13=AND(NA+IS13,M1)                                          0272
        N13=AND(N13+SHIFT(N13,-15),M2)                               0273
        L11=LV(N11)                                                  0274
        L12=SHIFT(LV(N12),7)                                         0275
        L13=SHIFT(MI-LV(N13),7)                                      0276
        L1=MT(L13+MT(L11+L12))                                       0277
C                                                                       0278
C               NOW L1 CONTAINS THE ELEMENT NUMBER OF THE ORDERED PRODUCT 0279
C               OF THE GROUP ELEMENTS OF LINKS 1,2, AND 3 OF THIS FIRST 0280
C               PLAQUETTE.                                           0281
C               THE OTHER PLAQUETTES ARE TREATED CORRESPONDINGLY.    0282
C                                                                       0283
C               PLAQUETTE NUMBER 2                                   0284
C                                                                       0285
        N21=AND(NA+IS21,M1)                                          0286
        N21=AND(N21+SHIFT(N21,-15),M2)                               0287
        N22=AND(NA+IS22,M1)                                          0288
        N22=AND(N22+SHIFT(N22,-15),M2)                               0289
        N23=AND(NA+IS23,M1)                                          0290
        N23=AND(N23+SHIFT(N23,-15),M2)                               0291
        L21=MI-LV(N21)                                               0292
        L22=SHIFT(LV(N22),7)                                         0293
        L23=SHIFT(LV(N23),7)                                         0294
        L2=MT(L23+MT(L21+L22))                                       0295
C                                                                       0296
C               PLAQUETTE NUMBER 3                                   0297
```

```
C                                                                          0298
        N31=AND(NA+IS31,M1)                                                0299
        N31=AND(N31+SHIFT(N31,-15),M2)                                     0300
        N32=AND(NA+IS32,M1)                                                0301
        N32=AND(N32+SHIFT(N32,-15),M2)                                     0302
        N33=AND(NA+IS33,M1)                                                0303
        N33=AND(N33+SHIFT(N33,-15),M2)                                     0304
        L31=LV(N31)                                                        0305
        L32=SHIFT(LV(N32),7)                                               0306
        L33=SHIFT(MI-LV(N33),7)                                            0307
        L3=MT(L33+MT(L31+L32))                                             0308
C                                                                          0309
C                   PLAQUETTE NUMBER 4                                     0310
C                                                                          0311
        N41=AND(NA+IS41,M1)                                                0312
        N41=AND(N41+SHIFT(N41,-15),M2)                                     0313
        N42=AND(NA+IS42,M1)                                                0314
        N42=AND(N42+SHIFT(N42,-15),M2)                                     0315
        N43=AND(NA+IS43,M1)                                                0316
        N43=AND(N43+SHIFT(N43,-15),M2)                                     0317
        L41=MI-LV(N41)                                                     0318
        L42=SHIFT(LV(N42),7)                                               0319
        L43=SHIFT(LV(N43),7)                                               0320
        L4=MT(L43+MT(L41+L42))                                             0321
C                                                                          0322
C                   PLAQUETTE NUMBER 5                                     0323
C                                                                          0324
        N51=AND(NA+IS51,M1)                                                0325
        N51=AND(N51+SHIFT(N51,-15),M2)                                     0326
        N52=AND(NA+IS52,M1)                                                0327
        N52=AND(N52+SHIFT(N52,-15),M2)                                     0328
        N53=AND(NA+IS53,M1)                                                0329
        N53=AND(N53+SHIFT(N53,-15),M2)                                     0330
        L51=LV(N51)                                                        0331
        L52=SHIFT(LV(N52),7)                                               0332
        L53=SHIFT(MI-LV(N53),7)                                            0333
        L5=MT(L53+MT(L51+L52))                                             0334
C                                                                          0335
C                   PLAQUETTE NUMBER 6                                     0336
C                                                                          0337
        N61=AND(NA+IS61,M1)                                                0338
        N61=AND(N61+SHIFT(N61,-15),M2)                                     0339
        N62=AND(NA+IS62,M1)                                                0340
        N62=AND(N62+SHIFT(N62,-15),M2)                                     0341
        N63=AND(NA+IS63,M1)                                                0342
        N63=AND(N63+SHIFT(N63,-15),M2)                                     0343
        L61=MI-LV(N61)                                                     0344
        L62=SHIFT(LV(N62),7)                                               0345
        L63=SHIFT(LV(N63),7)                                               0346
        L6=MT(L63+MT(L61+L62))                                             0347
C                                                                          0348
C                   OLD LINK ELEMENT:                                      0349
C                                                                          0350
        LO=SHIFT(MI-LV(N),7)                                               0351
C                                                                          0352
C                   COMPUTE THE OLD BOLTZMANN FACTOR:                      0353
C                                                                          0354
        BFO=BFA(MT(LO+L1))*BFA(MT(LO+L2))*BFA(MT(LO+L3))*                  0355
     1      BFA(MT(LO+L4))*BFA(MT(LO+L5))*BFA(MT(LO+L6))                   0356
        DO 2 IU=1,NUPG                                                     0357
C                                                                          0358
C                   CHOOSE RANDOMLY ONE OF THE NEIGHBOURS OF THE UNIT ELEMENT. 0359
C                   LN CONTAINS THE POSSIBLE NEW LINK VARIABLE.            0360
C                                                                          0361
        IR=IG(1+INT(12*RANF(DUMMY)))                                       0362
        LN=SHIFT(MT(LO+IR),7)                                              0363
C                                                                          0364
C                   COMPUTE THE NEW BOLTZMANNFACTOR:                       0365
C                                                                          0366
        BFN=BFA(MT(LN+L1))*BFA(MT(LN+L2))*BFA(MT(LN+L3))*                  0367
     1      BFA(MT(LN+L4))*BFA(MT(LN+L5))*BFA(MT(LN+L6))                   0368
C                                                                          0369
C                   TEST FOR ACCEPTANCE:                                   0370
C                                                                          0371
        IF(BFO*RANF(DUMMY).GT.BFN)GOTO 2                                   0372
C                                                                          0373
C                   ACCEPT THE CHANGE.                                     0374
C                                                                          0375
        LO=LN                                                              0376
        BFO=BFN                                                            0377
      2 CONTINUE                                                           0378
C                                                                          0379
C                   FIX THE LINK VARIABLE.                                 0380
C                                                                          0381
        LV(N)=MI-SHIFT(LO,-7)                                             0382
C                                                                          0383
C                   COMPUTE THE CONTRIBUTION OF THE SIX BORDERING PLAQUETTES 0384
C                   TO THE TOTAL ACTION                                    0385
C                                                                          0386
        E=E+ET(MT(L1+LO))+ET(MT(L2+LO))+ET(MT(L3+LO))+                    0387
     1      ET(MT(L4+LO))+ET(MT(L5+LO))+ET(MT(L6+LO))                      0388
      1 CONTINUE                                                           0389
        RETURN                                                             0390
        END                                                                0391
```

TEST RUN OUTPUT

```
TEST RUN FOR A NON-ABELIAN GAUGE THEORY WITH THE
ICOSAHEDRAL SUBGROUP OF SU(2) AS GAUGE GROUP
-----------------------------------------------

PROGRAM FOR LATTICE GAUGE THEORY
-----------------------------------------------

LATTICE SIZE :  4**4
DISCRETE GAUGE GROUP WITH  120 ELEMENTS

B DENOTES THE COUPLING PARAMETER BETA
E DENOTES THE AVERAGE ACTION PER PLAQUETTE

-----------------------------------------------

     B=     0.00000     E=     1.00154
     B=      .05000     E=      .99158
     B=      .10000     E=      .97713
     B=      .15000     E=      .96684
     B=      .20000     E=      .95440
     B=      .25000     E=      .93991
     B=      .30000     E=      .92763
     B=      .35000     E=      .90839
     B=      .40000     E=      .89939
     B=      .45000     E=      .90119
     B=      .50000     E=      .88449
     B=      .55000     E=      .86908
     B=      .60000     E=      .86234
     B=      .65000     E=      .82684
     B=      .70000     E=      .81862
     B=      .75000     E=      .81227
     B=      .80000     E=      .79628
     B=      .85000     E=      .79790
     B=      .90000     E=      .77754
     B=      .95000     E=      .77825
     B=     1.00000     E=      .76095
     B=     1.05000     E=      .74486
     B=     1.10000     E=      .75206
     B=     1.15000     E=      .73343
     B=     1.20000     E=      .70469
     B=     1.25000     E=      .70575
     B=     1.30000     E=      .68589
     B=     1.35000     E=      .67793
     B=     1.40000     E=      .67553
     B=     1.45000     E=      .66852
     B=     1.50000     E=      .63665
     B=     1.55000     E=      .62285
     B=     1.60000     E=      .61611
     B=     1.65000     E=      .60681
     B=     1.70000     E=      .60638
     B=     1.75000     E=      .59425
     B=     1.80000     E=      .59513
     B=     1.85000     E=      .55680
     B=     1.90000     E=      .54129
     B=     1.95000     E=      .54129
     B=     2.00000     E=      .52153
     B=     2.05000     E=      .49989
     B=     2.10000     E=      .49692
     B=     2.15000     E=      .48754
     B=     2.20000     E=      .45910
     B=     2.25000     E=      .46525
     B=     2.30000     E=      .44965
     B=     2.35000     E=      .43360
     B=     2.40000     E=      .41680
     B=     2.45000     E=      .41328
     B=     2.50000     E=      .41739
     B=     2.55000     E=      .39862
     B=     2.60000     E=      .37693
     B=     2.65000     E=      .37853
     B=     2.70000     E=      .36458
     B=     2.75000     E=      .36321
     B=     2.80000     E=      .34357
     B=     2.85000     E=      .33281
     B=     2.90000     E=      .31477
     B=     2.95000     E=      .31708
     B=     3.00000     E=      .30388
     B=     3.05000     E=      .29180
     B=     3.10000     E=      .28286
     B=     3.15000     E=      .27765
     B=     3.20000     E=      .27507
     B=     3.25000     E=      .26173
     B=     3.30000     E=      .26001
     B=     3.35000     E=      .25274
     B=     3.40000     E=      .24571
     B=     3.45000     E=      .25348
     B=     3.50000     E=      .24970
     B=     3.55000     E=      .24695
     B=     3.60000     E=      .23071
     B=     3.65000     E=      .23195
     B=     3.70000     E=      .23223
     B=     3.75000     E=      .21793
     B=     3.80000     E=      .21806
     B=     3.85000     E=      .21429
     B=     3.90000     E=      .20256
     B=     3.95000     E=      .20058
     B=     4.00000     E=      .20842
     B=     4.05000     E=      .19756
     B=     4.10000     E=      .19714
     B=     4.15000     E=      .19503
     B=     4.20000     E=      .19310
     B=     4.25000     E=      .19043
     B=     4.30000     E=      .18324
     B=     4.35000     E=      .18111
     B=     4.40000     E=      .17758
     B=     4.45000     E=      .17383
     B=     4.50000     E=      .17586
     B=     4.55000     E=      .17482
     B=     4.60000     E=      .17769
     B=     4.65000     E=      .17597
     B=     4.70000     E=      .16569
     B=     4.75000     E=      .16400
     B=     4.80000     E=      .16537
     B=     4.85000     E=      .16445
     B=     4.90000     E=      .16584
     B=     4.95000     E=      .16292

-----------------------------------------------
```

VOLUME 30, NUMBER 26 PHYSICAL REVIEW LETTERS 25 JUNE 1973

Ultraviolet Behavior of Non-Abelian Gauge Theories*

David J. Gross† and Frank Wilczek
Joseph Henry Laboratories, Princeton University, Princeton, New Jersey 08540
(Received 27 April 1973)

It is shown that a wide class of non-Abelian gauge theories have, up to calculable logarithmic corrections, free-field–theory asymptotic behavior. It is suggested that Bjorken scaling may be obtained from strong-interaction dynamics based on non-Abelian gauge symmetry.

Non-Abelian gauge theories have received much attention recently as a means of constructing unified and renormalizable theories of the weak and electromagnetic interactions.[1] In this note we report on an investigation of the ultraviolet (UV) asymptotic behavior of such theories. We have found that they possess the remarkable feature, perhaps unique among renormalizable theories, of asymptotically approaching free-field theory. Such asymptotically free theories will exhibit, for matrix elements of currents between on-mass-shell states, Bjorken scaling. We therefore suggest that one should look to a non-Abelian gauge theory of the strong interactions to provide the explanation for Bjorken scaling, which has so far eluded field-theoretic understanding.

The UV behavior of renormalizable field theories can be discussed using the renormalization-group equations,[2,3] which for a theory involving one field (say $g\varphi^4$) are

$$[m\partial/\partial m+\beta(g)\,\partial/\partial g-n\gamma(g)]\Gamma_{\mathrm{asy}}^{(n)}(g;P_1,\ldots,P_n)=0. \tag{1}$$

$\Gamma_{\mathrm{asy}}^{(n)}$ is the asymptotic part of the one-particle–irreducible renormalized n-particle Green's function, $\beta(g)$ and $\gamma(g)$ are finite functions of the renormalized coupling constant g, and m is either the renormalized mass or, in the case of massless particles, the Euclidean momentum at which the theory is renormalized.[4] If we set $P_i=\lambda q_i^{\,0}$, where $q_i^{\,0}$ are (nonexceptional) Euclidean momenta, then (1) determines the λ dependence of $\Gamma^{(n)}$:

$$\Gamma^{(n)}(g;P_i)=\lambda^D\Gamma^{(n)}(\bar g(g,t);q_i)\exp\Big[-n\int_0^t\gamma(\bar g(g,t'))\,dt'\Big], \tag{2}$$

where $t=\ln\lambda$, D is the dimension (in mass units) of $\Gamma^{(n)}$, and $\bar g$, the invariant coupling constant, is the solution of

$$d\bar g/dt=\beta(\bar g),\quad \bar g(g,0)=g. \tag{3}$$

The UV behavior of $\Gamma^{(n)}$ ($\lambda\to+\infty$) is determined by the large-t behavior of $\bar g$ which in turn is controlled by the zeros of β: $\beta(g_f)=0$. These fixed points of the renormalization-group equations are said to be UV stable [infrared (IR) stable] if $\bar g\to g_f$ as $t\to+\infty$ ($-\infty$) for $\bar g(0)$ near g_f. If the physical coupling constant is in the domain of attraction of a UV-stable fixed point, then

$$\Gamma^{(n)}(g;P_i)\underset{\lambda\to\infty}{\simeq}\lambda^{D-n\gamma(g_f)}\Gamma^{(n)}(g_f;q_i)\exp\Big\{-n\int_0^\infty[\gamma(\bar g(g,t))-\gamma(g_f)]\,dt\Big\}, \tag{4}$$

so that $\gamma(g_f)$ is the anomalous dimension of the field. As Wilson has stressed, the UV behavior is determined by the theory at the fixed point ($g=g_f$).[5]

In general, the dimensions of operators at a fixed point are not canonical, i.e., $\gamma(g_f)\neq 0$. If we wish to explain Bjorken scaling, we must assume the existence of a tower of operators with canonical dimensions. Recently, it has been argued for all but gauge theories, that this can only occur if the fixed point is at the origin, $g_f=0$, so that the theory is asymptotically free.[6,7] In that case the anomalous dimensions of all operators vanish, one obtains naive scaling up to finite and calculable powers of $\ln\lambda$, and the structure of operator products at short distances is that of free-field theory.[7] Therefore, the existence of such a fixed point, for a theory of the strong interactions, might explain Bjorken scaling and the success of naive light-cone or parton-model relations. Unfortunately, it appears that the fixed point at the origin, which is common to all theories, is not UV stable.[8,9] The only exception would seem to be non-Abelian gauge theories, which hitherto have not been explored in this regard.

Let us consider a Yang-Mills theory given by the Lagrangian

$$\mathcal{L} = -\tfrac{1}{4}F_{\mu\nu}{}^a F_a{}^{\mu\nu},$$
$$F_{\mu\nu}{}^a = \partial_\mu A_\nu{}^a - \partial_\nu A_\mu{}^a - gC_{abc}A_\mu{}^b A_\nu{}^c, \tag{5}$$

where the C_{abc} are the structure constants of some (semisimple) Lie group G. Since the theory is massless, the renormalization is performed at an (arbitrary) Euclidean point. For example, the wave-function renormalization constant $Z_3(g, \Lambda/m)$ will be defined in terms of the unrenormalized vector-meson propagator $D_{\mu\nu}{}^{ab}$ (in the Landau gauge),

$$D_{\mu\nu}{}^{ab}(P)\big|_{P^2=-m^2} = \left(g_{\mu\nu}+\frac{P_\mu P_\nu}{m^2}\right)\frac{iZ_3}{m^2}\delta_{ab}. \tag{6}$$

(For a thorough discussion of the renormalization see the work of Lee and Zinn-Justin.[10]) The renormalization-group equations for this theory are easily derived.[11] In the Landau gauge they are identical with (1). In order to investigate the stability of the origin, it is sufficient to calculate β to lowest order in perturbation theory. To this order we have

$$\beta(g) = \frac{\partial g}{\partial \ln m}\bigg|_{\Lambda, g_0} = -g\frac{\partial}{\partial \ln\Lambda}\left(\frac{Z_3^{3/2}}{Z_1}\right), \tag{7}$$

where Λ is a UV cutoff, and Z_1 the charge-renormalization constant. In Abelian gauge theories $Z_3 = Z_1 = 1 - g^2 C \ln\Lambda$ ($C>0$), as a consequence of gauge invariance and the Källén-Lehman representation, and thus $\beta(g) \cong g^3$ which leads to IR stability at $g=0$. Non-Abelian theories have no such requirement; Z_3 and Z_1 are gauge dependent and can be greater than 1. Thus $\beta(g)$, which must be gauge independent in lowest order, could have any sign at $g=0$. We have calculated Z_1 and Z_3 for the above Lagrangian, and we find that[12]

$$\beta_V = -(g^3/16\pi^2)\tfrac{11}{3}C_2(G) + O(g^5), \tag{8}$$

where $C_2(G)$ is the quadratic Casimir operator of the adjoint representation of the group G: $\sum_{b,c} c_{abc} \times c_{dbc} = C_2(G)\delta_{ad}$ [e.g., $C_2(\mathrm{SU}(N)) = N$]. The solution of (3) is then $\bar{g}^2(t) = g^2/(1-2\beta_V g^{-1}t)$, and $\bar{g} \to 0$ as $t\to\infty$ as long as the physical coupling constant g is in the domain of attraction of the origin.[13]

We have thus established that for all non-Abelian gauge theories based on semisimple Lie groups the origin is UV stable. It is easy to incorporate fermions into such a theory without destroying the UV stability. The fermion interaction is given by $L_F = \bar{\psi}(i\gamma\cdot\partial - g\gamma\cdot B^a M^a)\psi +$ mass terms, where M^a are the matrices of some representation R of the gauge group G. The only effect of the fermions is to change the value of $\beta(g)$ by the amount[11]

$$\beta_F(R) = (g^3/16\pi^2)\tfrac{4}{3}T(R), \tag{9}$$

where $\mathrm{Tr}(M^a M^b) = T(R)\delta_{ab}$, $T(R) = C_2(R)d(R)/r$, $d(R)$ is the dimension of the representation R, and r is the order of the group, i.e., the number of generators, and $C_2(R)$ is the quadratic Casimir operator of the representation. Although the fermions tend to destabilize the origin, there is room to spare. For example, in the case of SU(3): $\beta_V = -11$, whereas $\beta_F(3) = \frac{2}{3}$, $\beta_F(8) = 4$, etc., so that one could accomodate as many as sixteen triplets. One can therefore construct many asymptotically free theories with fermions. The vector mesons, however, will remain massless until the gauge symmetry is spontaneously broken. One might hope that this would be a consequence of the dynamics,[14] but at the present the only known way of achieving this is to introduce scalar Higgs mesons, whose nonvanishing vacuum expectation values break the symmetry.

The introduction of scalar mesons has a very destabilizing effect on the UV stability of the origin. Their contribution to $\beta(g)$ is small; a scalar meson transforming under a complex (real) representation R of the gauge group adds to β a term equal to $\frac{1}{4}$ ($\frac{1}{8}$) of Eq. (9). The problem with scalar mesons is that they necessarily have their own quartic couplings, and one must deal with a new coupling constant. Consider the Lagrangian for the coupling of scalars belonging to a representation R of G:

$$\mathcal{L} = \tfrac{1}{2}[(\partial_\mu - igB_\mu{}^a M^a)\vec{\varphi}]^2 - \lambda(\vec{\varphi}\cdot\vec{\varphi})^2 + V(\vec{\varphi}). \tag{10}$$

where $V(\varphi)$ contains cubic, quadratic, and linear terms in φ (which have no effect on the UV behavior of the theory) plus, perhaps, additional quartic terms invariant under G. The renormalization-group equations have an additional term, $\beta_\lambda(g,\lambda)\partial/\partial\lambda$, and one must investigate the UV stability of the origin ($g=\lambda=0$) with respect to both g and λ [if there are other quartic invariants in $V(\varphi)$ there will be additional coupling constants to consider]. The structure of the renormalization-group equation for g is unchanged to lowest order, whereas for the coupling constant $\Gamma \equiv \lambda/g^2$ we have[11]

$$d\bar{\Gamma}(\Gamma, t, g^2)/dt = \bar{g}^2[A\bar{\Gamma}^2 + B\bar{\Gamma} + C] \tag{11}$$

(where we have neglected terms of order g^4, $g^4\Gamma$,

VOLUME 30, NUMBER 26 PHYSICAL REVIEW LETTERS 25 JUNE 1973

$g^4\Gamma^2$, and $g^4\Gamma^3$). In the absence of vector mesons ($g=0$) this equation is UV unstable at $\lambda=0$, since A is strictly positive and λ must be positive.[15] The vector mesons contribute to B and C and tend to stabilize the origin. If the right-hand side of (11) has positive zeros ($C>0$, $B<0$, and $B^2>4AC$), then for Γ less than the larger zero of (11) we will have that $\lambda\to+0$ as $t\to\infty$. We have investigated the structure of these equations for a large class of gauge theories and representations of the scalar mesons. We have found many examples of theories which contain scalar mesons and are UV stable.[11] These include (a) SU(N) if the scalar mesons belong to the adjoint representation for $N\geqslant 6$; (b) SU(N)$\otimes$SU(N) if the scalars belong to the $(N,\bar{N})$ representation for $N\geqslant 5$; (c) SU(N) with the scalars transforming as a symmetric tensor for $N\geqslant 9$; and many others. In all of these models it is necessary for the theory to contain a large number of fermions in order to make β_g small; otherwise $\bar{g}$ approaches zero too rapidly for the vector mesons to stabilize the scalar couplings.

Unfortunately, in none of these models can the gauge symmetry be totally broken by the Higgs mechanism. The requirement that the interactions of the scalar mesons be renormalizable so severely constrains the form of Lagrangian that the ground state invariably is invariant under some non-Abelian subgroup of the gauge group. If one tries to overcome this by larger representations for the scalar mesons, UV instability inevitably occurs.

It thus appears to be very difficult to retain UV stability and break the gauge symmetry by explicitly introducing Higgs mesons. Since the Higgs mesons are so restrictive, we would prefer to believe that spontaneous symmetry breaking would arise dynamically.[14] This is suggested by the IR instability of the theories, which assures us that perturbation theory is not trustworthy with respect to the stability of the symmetric theory nor to its particle content.

With this hope in mind one can construct many interesting models of the strong interactions. One particularly appealing model is based on three triplets[16] of fermions, with Gell-Mann's SU(3)$\otimes$SU(3) as a global symmetry and an SU(3) "color" gauge group to provide the strong interactions. That is, the generators of the strong-interaction gauge group commute with ordinary SU(3)$\otimes$SU(3) currents and mix quarks with the same isospin and hypercharge but different "color." In such a model the vector mesons are neutral, and the structure of the operator product expansion of electromagnetic or weak currents is (assuming that the strong coupling constant is in the domain of attraction of the origin!) essentially that of the free quark model (up to calculable logarithmic corrections).[11]

Finally, we note that theories of the weak and electromagnetic interactions, built on semisimple Lie groups,[17] will be asymptotically free if we again ignore the complications due to the Higgs particles. This suggests that the program of Baker, Johnson, Willey, and Adler[18] to calculate the fine-structure constant as the value of the UV-stable fixed point in quantum electrodynamics might fail for such theories.

---

*Research supported by the U.S. Air Force Office of Scientific Research under Contract No. F-44620-71-C-0180.

†Alfred P. Sloan Foundation Research Fellow.

[1]S. Weinberg, Phys. Rev. Lett. 19, 1264 (1967). For an extensive review as well as a list of references, see B. W. Lee, in Proceedings of the Sixteenth International Conference on High Energy Physics, National Accelerator Laboratory, Batavia, Illinois, 1972 (to be published).

[2]M. Gell-Mann and F. E. Low, Phys. Rev. 95, 1300 (1954).

[3]C. G. Callan, Phys. Rev. D 2, 1541 (1970); K. Symanzik, Commun. Math. Phys. 18, 227 (1970).

[4]The basic assumption underlying the derivation and utilization of the renormalization group equations is that the large Euclidean momentum behavior of the theory is the same as the sum, to all orders, of the leading powers in perturbation theory.

[5]K. Wilson, Phys. Rev. D 3, 1818 (1971).

[6]G. Parisi, to be published.

[7]C. G. Callan and D. J. Gross, to be published.

[8]A. Zee, to be published.

[9]S. Coleman and D. J. Gross, to be published.

[10]B. W. Lee and J. Zinn-Justin, Phys. Rev. D 5, 3121 (1972).

[11]Full details will be given in a forthcoming publication: D. J. Gross and F. Wilczek, to be published.

[12]After completion of this calculation we were informed of an independent calculation of β for gauge theories coupled to fermions by H. D. Politzer [private communication, and following Letter, Phys. Rev. Lett. 30, 1346 (1973)].

[13]K. Wilson has suggested that the coupling constants of the strong interactions are determined to be IR-stable fixed points. For nongauge theories the IR stability of the origin in four-dimensional field theories implies that theories so constructed are trivial, at least in a domain about the origin. Our results suggest that non-Abelian gauge theories might possess IR-stable fixed points at nonvanishing values of the coupling constants.

VOLUME 30, NUMBER 26 PHYSICAL REVIEW LETTERS 25 JUNE 1973

[14]Y. Nambu and G. Jona-Lasino, Phys. Rev. 122, 345 (1961); S. Coleman and E. Weinberg, Phys. Rev. D 7, 1888 (1973).

[15]K. Symanzik (to be published) has recently suggested that one consider a $\lambda\varphi^4$ theory with a negative λ to achieve UV stability at $\lambda = 0$. However, one can show, using the renormalization-group equations, that in such theory the ground-state energy is unbounded from below (S. Coleman, private communication).

[16]W. A. Bardeen, H. Fritzsch, and M. Gell-Mann, CERN Report No. CERN-TH-1538, 1972 (to be published).

[17]H. Georgi and S. L. Glashow, Phys. Rev. Lett. 28, 1494 (1972); S. Weinberg, Phys. Rev. D 5, 1962 (1972).

[18]For a review of this program, see S. L. Adler, in Proceedings of the Sixteenth International Conference on High Energy Physics, National Accelerator Laboratory, Batavia, Illinois, 1972 (to be published).

VOLUME 30, NUMBER 26 PHYSICAL REVIEW LETTERS 25 JUNE 1973

Reliable Perturbative Results for Strong Interactions?*

H. David Politzer
Jefferson Physical Laboratories, Harvard University, Cambridge, Massachusetts 02138
(Received 3 May 1973)

An explicit calculation shows perturbation theory to be arbitrarily good for the deep Euclidean Green's functions of any Yang-Mills theory and of many Yang-Mills theories with fermions. Under the hypothesis that spontaneous symmetry breakdown is of dynamical origin, these symmetric Green's functions are the asymptotic forms of the physically significant spontaneously broken solution, whose coupling could be strong.

Renormalization-group techniques hold great promise for studying short-distance and strong-coupling problems in field theory.[1,2] Symanzik[2] has emphasized the role that perturbation theory might play in approximating the otherwise unknown functions that occur in these discussions. But specific models in four dimensions that had been investigated yielded (in this context) disappointing results.[3] This note reports an intriguing contrary finding for any generalized Yang-Mills theory and theories including a wide class of fermion representations. For these one-coupling-constant theories (or generalizations involving product groups) the coefficient function in the Callan-Symanzik equations commonly called $\beta(g)$ is negative near $g=0$.

The constrast with quantum electrodynamics (QED) might be illuminating. Renormalization of QED must be carried out at off-mass-shell points because of infrared divergences. For small e^2, we expect perturbation theory to be good in some neighborhood of the normalization point. But what about the inevitable logarithms of momenta that grow as we approach the mass shell or as some momenta go to infinity? In QED, the mass-shell divergences do not occur in observable predictions, when we take due account of the experimental situation. The renormalization-group technique[4] provides a somewhat opaque analysis of this situation. Loosely speaking,[5] the effective coupling of soft photons goes to zero, compensating for the fact that there are more and more of them. But the large-p^2 divergence represents a real breakdown of perturbation theory. It is commonly said that for momenta such that $e^2\ln(p^2/m^2)\sim 1$, higher orders become comparable, and hence a calculation to any finite order is meaningless in this domain. The renormalization group technique shows that the effective coupling grows with momenta.

The behavior in the two momentum regimes is reversed in a Yang-Mills theory. The effective coupling goes to zero for large momenta, but as p^2's approach zero, higher-order corrections become comparable. Thus perturbation theory tells *nothing* about the mass-shell structure of the symmetric theory. Even for arbitrarily small g^2, there is no sense in which the interacting theory is a small perturbation on a free multiplet of massless vector mesons. The truly catastrophic infrared problem makes a symmetric particle interpretation impossible. Thus, though one can well approximate asymptotic Green's functions, to what particle states do they refer?

Consider theories defined by the Lagrangian

$$\mathcal{L}=-\tfrac{1}{4}F_{\mu\nu}{}^{a}F^{a\mu\nu}+i\bar{\psi}_i\gamma\cdot D_{ij}\psi_j, \tag{1}$$

where

$$F_{\mu\nu}{}^{a}=\partial_\mu A_\nu{}^{a}-\partial_\nu A_\mu{}^{a}+gf^{abc}A_\mu{}^{b}A_\nu{}^{c},$$

and

$$D_{ij}^{\mu}=\partial^{\mu}\delta_{ij}-igA^{a\mu}T_{ij}^{a},$$

the f^{abc} are the group structure constants, and the T^a are representation matrices corresponding to the fermion multiplet. (One may be interested in models with massless fermions because of their group structure or because they have the same asymptotic forms[6] as massive theories.) The normalizations of the conventionally defined irreducible vertices for n mesons and n' fermions, $\Gamma^{n,n'}$, must refer to some mass M. The renormalization-group equation reads

$$\left(M\frac{\partial}{\partial M}+\beta(g)\frac{\partial}{\partial g}+n\gamma_A(g)+n'\gamma_\psi(g)\right)\Gamma^{n,n'}=0. \qquad (2)$$

Putting it in this form makes use of the first available simplification, proper choice of gauge. Equation (2) describes how finite renormalizations accompanied by a change in g and a rescaling of the fields leave the $\Gamma^{n,n'}$ unchanged. Consider gauges defined by α in the zeroth-order propagator

$$\Delta_{\mu\nu}(p^2)=\frac{-g_{\mu\nu}+p_\mu p_\nu/p^2}{p^2}+\alpha\frac{p_\mu p_\nu}{p^4}.$$

The generalized Ward identities[7] imply that there are no higher-order corrections to the longitudinal part. But if the fields are rescaled as in Eq. (2), α must be changed to leave Γ^2 invariant. Hence α should occur in Eq. (2) much as g does, and one would have to study the $\Gamma^{n,n'}$ for arbitrary α to determine the coefficient functions perturbatively. But for $\alpha=0$ initially, it remains zero under finite renormalizations; so it suffices to study the theory in a Landau gauge.

To first order, the meson inverse propagator is

$$\Gamma_{\mu\nu}^{2ab}(p,-p)=\delta^{ab}(-g_{\mu\nu}p^2+p_\mu p_\nu)[1+(\tfrac{13}{3}c_1-\tfrac{8}{3}c_2)(g/4\pi)^2\ln(-p^2/M^2)], \qquad (3)$$

where

$$f_{acd}f_{bcd}=2c_1\delta_{ab},\quad \mathrm{tr}(T^aT^b)=2c_2\delta_{ab},$$

and $c_1>_0$ and $c_2\geq 0$. [For SU(2), $c_1=1$, c_2(isodoublet) $=\frac{1}{4}$, and c_2(isotriplet) $=1$.] To first order (only), the fermion self-energy is proportional to the self-energy in massless QED, which vanishes in the Landau gauge. Similarly, the contribution to the fermion-vector three-point vertex correction proportional to the first-order QED correction needs no subtractions and contains no reference to M. Calculation of the remaining correction, involving the meson self-coupling, yields

$$\Gamma_{\mu ij}^{1,2a}(0,p,-p)=gT_{ij}^{a}\gamma_\mu[1-\tfrac{3}{2}c_1(g/4\pi)^2\ln(-p^2/M^2)]. \qquad (4)$$

Applying Eq. (2) to these functions at their normalization points yields

$$\gamma_\psi(g)=0+O(g^4),\quad \gamma_A(g)=(\tfrac{13}{3}c_1-\tfrac{8}{3}c_2)(g/4\pi)^2+O(g^4),\quad \beta(g)=-(\tfrac{22}{3}c_1-\tfrac{8}{3}c_2)g(g/4\pi)^2+O(g^5).$$

It is also apparent, by inspecting the graphs, that to this order the coupling constants of product groups do not enter into each other's β functions.

For the case where there are no fermions, the coefficient functions can be obtained by setting $c_2=0$. (Even though the fermion-vector vertex, which had been used implicitly to define g, is no longer present, it can be simulated by introducing two multiplets of spinor fields with the same group transformations but opposite statistics. The physical effects of internal fermions are canceled by the ghosts —spinor fields with Bose statistics.) Alternatively, one can study the corrections to the three-meson vertex. Define F by

$$\Gamma_{\lambda\mu\nu}^{3abc}(p,-p,0)=f^{abc}(p_\lambda g_{\mu\nu}+p_\mu g_{\nu\lambda}-2p_\nu g_{\lambda\mu})gF(p^2/M^2,g^2). \qquad (5)$$

The normalization condition is $F(-1,g^2)=1$ (up to a phase convention.) To first order

$$F=1+\tfrac{17}{8}c_1(g/4\pi)^2\ln(-p^2/M^2) \qquad (6)$$

which yields the same β as described above.

The renormalization-group "improved" perturbation theory[4,5] extends results valid near the normalization point by effectively moving that point. The improved vertex functions are constructed from the straightforward perturbative ones, involving a momentum-scale-dependent effective coupling $g'(g,t)$, where $t=\frac{1}{2}\ln(s/M^2)$ and s sets the scale, e.g., $s=\sum(-p_i^2)$. $g'(g,t)$ is defined by

$$\partial g'/\partial t=\beta(g'),\qquad g'(g,0)=g. \qquad (7)$$

For the approximate β's derived above, $\beta = -bg^3$,

$$g'^2 \approx g^2/(1+2bg^2t). \qquad (8)$$

Thus for a pure meson theory or for theories including not too many fermions (in the sense that $c_2 < \frac{11}{4}c_1$), g' goes to zero for asymptotic momenta, i.e., $t\to\infty$. The $\Gamma^{n,n'}$ show a well-defined slow approach to quasifree field values.

It is worth remembering that successive orders of perturbation theory give the behavior of β for infinitesimal g and, strictly speaking, say nothing about finite g. Making a polynomial fit to a perturbative result for β is pure conjecture.

Hypothesizing that β stays negative (at least into the domain of strong coupling constant) relates all theories defined by Eq. (1) [with g less than the first zero of $\beta(g)$] to the model with g arbitrarily small by a change in mass scale. They all share the same asymptotic Green's functions, differing only by how large is asymptotic.

To utilize this result, we make the following hypothesis: The gauge symmetry breaks down spontaneously as a result of the dynamics. Consequently, the fields obtain (in general massive) particle interpretation—the Higgs phenomenon. As yet, nothing is known about the particle spectrum, the low-energy dynamics, or particles describable only by composite fields. But the Callan-Symanzik analysis says that the asymptotic Green's functions for the "dressed" fundamental fermion and vector fields are the symmetric functions discussed above.[8]

[An alternative is to introduce fundamental scalar fields, in terms of which the group transformation properties of the vacuum can be studied.[9] But these theories are not in general ultraviolet stable in terms of the additional coupling constants that must be introduced. Particular models which are ultraviolet stable as well as spontaneously asymmetric have been found.[10] But gauge theories of fermions (only) have aesthetic attractions, including the possibility of a dynamical determination of the dimensionless coupling constant.[9]]

Hypotheses of this type go back to the work of Nambu and collaborators.[11] In the renormalizable massless theories including scalars that have been studied,[9] infrared instability is a necessary condition for spontaneous symmetry breakdown.[12] The model of Nambu and Jona-Lasinio can be treated by the methods of Coleman and Weinberg.[9] The model is defined by

$$\mathcal{L} = i\bar{\psi}\gamma\cdot\partial\psi + g_0[(\bar{\psi}\psi)^2 - (\bar{\psi}\gamma_5\psi)^2] \qquad (9)$$

and the stipulation that the momentum integrals are cut off at some Euclidean mass squared Λ^2. Define a scalar

$$\varphi(x) = g_0\bar{\psi}(x)\psi(x)$$

and an analogous pseudoscalar, which one can do because of the cutoff. A study of the Green's functions in the one-loop approximation yields all the original results. But the existence of the vacuum-degenerate solution requires the dimensionless parameter characterizing the theory to satisfy $g_0\Lambda^2 > 2\pi^2$. But this is the condition that the one-loop correction to fermion-fermion scattering be at least as important as the tree approximation.

The situation is similar in the renormalizable models. $\lambda\varphi^4$ is stable for small λ because the one-loop corrections are small. But in massless scalar QED, photon-loop corrections of order e^4 can dominate over the lowest order φ-φ scattering (order λ) for both λ and e arbitrarily small. The requirement is just that $\lambda < e^4$. In this light, the problem with the Nambu–Jona-Lasinio model is not its nonrenormalizability but that in the domain of large $g_0\Lambda^2$, where spontaneous breakdown is alleged to occur, higher loop corrections are likely to dominate. (In the framework of the original solution, more complex infinite chains and self-energy graphs dominate over the ones studied.) In theories defined by Eq. (1), composite scalar densities can also be defined and studied in perturbation theory. But the condition that the one-loop approximation imply vacuum degeneracy requires that the expansion parameter be large, rendering the application of perturbation theory suspect.

The author thanks Sidney Coleman and Erick Weinberg, who have offered insights and advice freely, and the latter especially for his help in the computations.

*Work supported in part by the U. S. Air Force Office of Scientific Research under Contract No. F44620-70-C-0030.

[1]Of central importance is the work reviewed in K. Wilson and J. Kogut, "The Renormalization Group and the ϵ Expansion" (to be published); K. Johnson and M. Baker, "Some Speculations on High Energy Quantum Electrodynamics" (to be published); S. Adler, Phys. Rev. D 5, 3021 (1972).

[2]K. Symanzik, DESY Report No. DESY 72/73, 1972

(to be published), and references therein.

[3]A. Zee, "Study of the Renormalization Group for Small Coupling Constants" (to be published).

[4]N. N. Bogoliubov and D. V. Shirkov, *Introduction to the Theory of Quantized Fields* (Interscience, New York, 1959).

[5]Definitions of the relevant quantities will be given, but for the general theory and derivations see Refs. 1, 2, and 4; S. Coleman, in the Proceedings of the 1971 International Summer School "Ettore Majorana" (Academic, New York, to be published); S. Coleman and E. Weinberg, Phys. Rev. D 7, 1888 (1973), whose conventions we follow.

[6]Asymptotic refers to a particular set of Euclidean momenta as they are collectively scaled upward.

[7]E. g., B. W. Lee and J. Zinn-Justin, Phys. Rev. D 5, 3121 (1972); G. 't Hooft, Nucl. Phys. B33, 173 (1971). which also include details of Feynman rules. regularization, etc.

[8]Configurations where the symmetric theory has infrared singularities not present in the massive case are discussed in detail by Symanzik (Ref. 2).

[9]Coleman and Weinberg, Ref. 5.

[10]D. Gross and F. Wilczek, preceding Letter [Phys. Rev. Lett. 30, 1343 (1973)].

[11]Y. Nambu and G. Jona-Lasinio, Phys. Rev. 122, 345 (1961).

[12]$\lambda\varphi^4$ theory with $\lambda < 0$ is ultraviolet stable (Ref. 2) and hence infrared unstable but cannot be physically interpreted in perturbation theory. Using the computations of Ref. 9, for $\lambda < 0$ "improved" perturbation theory is arbitrarily good for large field strengths. In particular, the potential whose minimum determines the vacuum decreases without bound for large field.

Volume 93B, number 1,2 PHYSICS LETTERS 2 June 1980

THE CONNECTION BETWEEN THE Λ PARAMETERS OF LATTICE AND CONTINUUM QCD

Anna HASENFRATZ
L. Eötvös University, Budapest, Hungary
and
Peter HASENFRATZ [1]
CERN, Geneva, Switzerland

Received 21 March 1980

By calculating the two- and three-point functions at the one-loop level in weak-coupling lattice QCD we have found the numerical relation between the Λ parameter of the lattice theory and that of the continuum theory:

$\Lambda^{\mathrm{MOM}}_{(\alpha=1)} = 83.5\,\Lambda^{\mathrm{lattice}}$ for SU(3), $\Lambda^{\mathrm{MOM}}_{(\alpha=1)} = 57.5\,\Lambda^{\mathrm{lattice}}$ for SU(2).

This relation sets the scale for all lattice calculations in QCD.

The correspondence between euclidean quantum field theory and classical statistical mechanics provided new insights into the problem of quark confinement. In this context it is natural to use a lattice to regularize the theory. Wilson's way of defining a gauge theory on a lattice [1,2] preserves exact gauge symmetry and at the same time techniques used in statistical physics become applicable through this regularization method.

For finite lattice distance a there is no euclidean invariance. Though the remaining symmetries (parity, rotations 90°, gauge invariance) put strong constraints on the possible form of Green's functions, there is no general proof that euclidean invariance will be recovered in the $a \to 0$ continuum limit. Nevertheless it is a general hope, supported by increasing evidence [3–8], that in the $a \to 0$, $g_0 \to 0$ limit a euclidean invariant, asymptotically free, confining theory will be obtained, which in the perturbative weak region is completely equivalent to perturbative continuum QCD.

In this limit dimensional physical quantities are expected to have a well defined, unique dependence on g_0, dictated by the requirement of cut-off independence. A physical mass should behave as

$$m = ca^{-1}e^{-1/(2\beta_0 g_0^2)}(\beta_0 g_0^2)^{-\beta_1/2\beta_0^2}[1 + O(g_0^3)] \equiv c\Lambda^{\mathrm{lattice}}, \tag{1}$$

where β_0, β_1 are the first coefficients of the lattice β function

$$-a(\mathrm{d}/\mathrm{d}a)g_0 \equiv \beta(g_0) = -\beta_0 g_0^3 - \beta_1 g_0^5 + \ldots, \tag{2}$$

while $\Lambda^{\mathrm{lattice}}$ is a cut-off independent mass parameter, defined analogously to that of the Λ parameter of the continuum theory [9–12]:

$$\Lambda = M\,e^{-1/[2\beta_0 g^2(M)]}[\beta_0 g^2(M)]^{-\beta_1/2\beta_0^2}[1 + O(g^2(M))]\,. \tag{3}$$

Here M is a mass introduced in the renormalization (or regularization) process, and $g^2(M)$ is the corresponding coupling constant. Different renormalization schemes differ in what is meant by M and $g^2(M)$, therefore Λ, as defined in eq. (3), is scheme dependent [12].

[1] On leave from the Central Research Institute for Physics, Budapest, Hungary.

In principle, Λ is measurable in deep inelastic processes. However, the necessary two-loop calculations are available only in the continuum theory [13,14]. In order to find $\Lambda^{\rm lattice}$, that is, to set a scale for lattice calculations, $\Lambda^{\rm lattice}$ should be connected with one of the Λ parameters of the continuum theory, with $\Lambda^{\rm MOM}$ for instance.

Monte-Carlo calculations [3] and other considerations [15] seem to require $\Lambda^{\rm lattice}$ to be of the order of a few MeV. On the other hand, $\Lambda^{\rm MOM}$ is measured to be of the order of a few hundred MeV, therefore a large, dimensionless connecting number seems to be required between $\Lambda^{\rm MOM}$ and $\Lambda^{\rm lattice}$. At first sight it is rather embarrassing, since the scheme dependence of the Λ parameter in the continuum theory is not so dramatic [12].

In this letter we summarize the results of a rather long calculation aimed at calculating the connection between $\Lambda^{\rm lattice}$ and $\Lambda^{\rm MOM}$. We have calculated the two- and three-point functions at the one-loop level in lattice QCD. We have explicitly demonstrated that there are no unwanted divergences and that all of the non-covariant terms cancel in the $a \to 0$ continuum limit. The remaining logarithmic divergences and therefore the first term of the β function are identical to that of the continuum theory. By also calculating the finite terms, we obtained

$$\Lambda^{\rm MOM}_{\rm (Feynman\ gauge)} = 83.5\ \Lambda^{\rm lattice} \quad \text{for SU(3)}, \qquad \Lambda^{\rm MOM}_{\rm (Feynman\ gauge)} = 57.5\ \Lambda^{\rm lattice} \quad \text{for SU(2)}. \tag{4}$$

These surprisingly large numbers are due to the specific regularization of momentum space integrals and to the new type of vertices of the lattice theory. As certain parts of our work may be useful for future weak coupling calculations, we intend to publish the details soon [16].

In a weak coupling expansion the vector potentials $A^a_{n\mu}$ are used as basic variables. The total action is a sum of Wilson's action, the integration measure, the gauge fixing term and the ghost terms. The one-loop corrections to the two- and three-point functions are of the order g_0^2 and g_0^3, respectively. Since $U_{n\mu} = \exp(iag_0)A^a_{n\mu}T^a$, every $A^a_{n\mu}$ carries a factor g_0, therefore Wilson's action should be expanded up to the fifth order in $A^a_{n\mu}$ (there is a factor $1/g_0^2$ in front), while the integration measure and the ghost terms are expanded up to the third order in the vector potentials. We worked in Feynman gauge, the gauge fixing term has the form $A_{\rm gf} = a^4 \Sigma_n \Sigma_{\mu,\nu} - \frac{1}{2}(\Delta_\mu A^a_{n-\hat\mu,\mu})^2$, where $\Delta_\mu f_n = 1/a(f_{n+\hat\mu} - f_n)$. The ghost terms can be obtained by using the Faddeev–Popov trick (for a review see Abers and Lee [17]: the procedure on the lattice has been discussed by Baaquie [8]). We found it advantageous not to introduce explicit ghost fields, but to work with the original non-local expressions.

In the following we consider the case of SU(3). The total relevant action has the following structure:

$$A_{\rm tot} = a^4 \sum_n \sum_{\mu,\nu} [A_2 + A_{\rm gf} + g_0 A_3 + g_0^2 A_4 + g_0^3 A_5] + A_{\rm measure} + A_{\rm ghost},$$

where

$$A_2 = -\tfrac{1}{4}(\Delta_\mu A^a_{n\nu} - \Delta_\nu A^a_{n\nu})^2,$$

$$A_3 = f^{abc}[A^a_{n\mu} A^b_{n\nu} \Delta_\mu A^c_{n\nu} + \tfrac{1}{2} a A^a_{n\mu} \Delta_\nu A^b_{n\mu} \Delta_\mu A^c_{n\nu}],$$

$$A_4 = -\tfrac{1}{4} f^{abt} f^{cdt} A^a_{n\mu} A^b_{n\nu} A^c_{n\mu} A^d_{n\nu} + \tfrac{1}{2} f^{abt} f^{cdt} a \Delta_\mu A^a_{n\nu} A^b_{n\mu} A^c_{n\mu} A^d_{n\nu} + \dots$$

(9 different fourth-order terms),

$$A_5 = -\tfrac{1}{6} f^{abt} f^{tcq} f^{qde} a^2 A^a_{n\mu} A^b_{n\nu} A^c_{n\mu} \Delta_\mu A^d_{n\nu} A^e_{n\mu} + \dots \tag{5}$$

(14 different fifth-order terms),

$$A_{\rm measure} = -\tfrac{1}{8} a^2 g_0^2 \sum_{n,\mu} A^a_{n\mu} A^a_{n\mu}$$

(there are no third-order terms in the measure),

Volume 93B, number 1,2 PHYSICS LETTERS 2 June 1980

$$A_{\text{ghost}} = \tfrac{3}{2} g_0^2 \int \frac{d^4k}{(2\pi)^4} \int \frac{d^4p}{(2\pi)^4} \frac{1}{(\hat{k}\hat{k}^*)(\hat{p}\hat{p}^*)} \sum_{\mu,\nu} \hat{k}_\mu (1 + \tfrac{1}{2} a\hat{p}^*_\mu) \hat{p}_\nu (1 + \tfrac{1}{2} a\hat{k}^*_\nu) A^a_\mu(k-p) A^a_\nu(p-k) + \dots$$

(4 different non-local expressions).

Here A_{ghost} is expressed in terms of momentum space integrals, where

$$\int d^4k \equiv \iiiint_{-\pi/a}^{\pi/a} d^4k, \quad \hat{k}_\mu = \frac{1}{a}(e^{-ik_\mu a} - 1), \quad (\hat{k}\hat{k}^*) = \sum_\mu \hat{k}_\mu \hat{k}^*_\mu . \tag{6}$$

After fixing the gauge, the integration region for $A^a_{n\mu}$ can be extended to $(-\infty, \infty)$. The Feynman rules can be straightforwardly derived. The propagator has the simple form

$$\underset{k}{\overset{a,\mu \qquad b,\mu'}{\overline{\qquad\qquad}}} \quad \frac{1}{(2\pi)^4}\, \delta_{ab} \frac{\delta_{\mu\mu'}}{(\hat{k}\hat{k}^*)}, \tag{7}$$

but there is a large number of three-, four- and five-point vertices.

At the one-loop level $\Gamma^{(2),ab}_{\mu\mu'}(p)$ and $\Gamma^{(3)abc}_{\mu\nu\rho}(p,k,q)$ are determined by the following type of graphs:

$\Gamma^{(2),ab}_{\mu\mu'}(p) = \frac{1}{2}$  $+ \frac{1}{2}$  + (contribution from the ghost and the measure),

$\Gamma^{(3)abc}_{\mu\nu\rho}(p,k,q) =$  $+ \frac{1}{2}$  + crossed $+ \frac{1}{2}$ 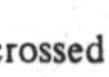 + (ghost contribution) . (8)

Due to the large number of vertices with a complicated, non-covariant structure, and due to the unpleasant form of momentum conservation on the lattice [if $q_\mu = -(p_\mu + k_\mu)$, then $\hat{q}_\mu = \hat{p}^*_\mu + \hat{k}^*_\mu + \hat{p}^*_\mu \hat{k}^*_\mu$], an enormous number of terms have been produced at intermediate stages of the calculation. To avoid algebraic errors none of the restrictions of the symmetries of the lattice theory have been used during the calculation, therefore the requirement of parity invariance, gauge invariance and invariance under rotations of 90° provided checks on our final results.

The momentum space integrals are over a four-dimensional hypercube. Several integrals had to be calculated numerically. In calculating $\Gamma^{(2),ab}_{\mu\mu'}(p)$ we have explicitly demonstrated that all the quadratic and linear divergences cancel, and that there are no non-covariant terms left in the $a \to 0$ continuum limit. The remaining terms have the form:

$$\Gamma^{(2)ab}_{\mu\mu'}(p) \underset{a\to 0}{=} \delta_{ab} g_0^2 \left[\delta_{\mu\mu'} p^2 \left(\frac{1}{16\pi^2} \cdot 5 \ln \frac{\pi^2}{a^2p^2} + 0.297694 \dots \right) - p_\mu p_{\mu'} \left(\frac{1}{16\pi^2} \cdot 5 \ln \frac{\pi^2}{a^2p^2} + 0.297721 \dots \right) \right]. \tag{9}$$

The difference between the finite terms is of the order of 10^{-5}, which is the precision of our numerical integrations. There is no correction to the longitudinal part of the propagator in accordance with gauge invariance.

For the three-point function our strategy was the following. First we have shown explicitly that there are no linear divergences and that all the non-covariant terms cancel in the $a \to 0$ limit. Next, we kept only those terms which are linear in the external momenta, because only those are relevant in determining the connection between the Λ parameters.

On the lattice it is much easier to calculate the trigluon vertex at an asymmetric point – say at $p^2 = q^2 = M^2$, $k = 0$ – than at the symmetric point $p^2 = k^2 = q^2 = M^2$. Let us denote the corresponding momentum space subtraction procedure by $\widetilde{\text{MOM}}$ and MOM, respectively. In the first step we found the connection between Λ^{lattice} and $\Lambda^{\widetilde{\text{MOM}}}$, and then the relation between $\Lambda^{\widetilde{\text{MOM}}}$ and Λ^{MOM} has been calculated. This last step can be done easily because $\Lambda^{\widetilde{\text{MOM}}}$ and Λ^{MOM} can be connected within the continuum theory. On the lattice we obtained:

$$\Gamma^{(3)abc}_{\mu\nu\rho}(p,k,q)|_{k=0,p^2=q^2=M^2} = if^{abc}g_0^3\Big[(\delta_{\mu\nu}p_\rho + \delta_{\nu\rho}p_\mu - 2\delta_{\mu\rho}p_\nu)\left(\frac{1}{16\pi^2}\cdot(-2)\ln\frac{\pi^2}{a^2M^2} - 0.23096\right)$$

$$- \delta_{\mu\rho}p_\nu\cdot 0.02528\Big] + \text{(terms of higher order in the external momenta)}. \tag{10}$$

The first term is proportional to the bare trigluon vertex taken at $k = 0$. As the counterterm is proportional to this vertex, consistent renormalizability requires that the second term $-\delta_{\mu\rho}p_\nu \cdot 0.02528$ must be independent of the regularization procedure. To check this point we calculated this term directly in the continuum theory using dimensional regularization. We obtained $-\delta_{\mu\nu}p_\nu(1/4\pi^2) = -\delta_{\mu\rho}p_\nu\cdot 0.02533$. The difference is again of the order of 10^{-5}, which is the precision of our numerical integrations.

$\widetilde{\mathrm{MOM}}$ is defined by requiring the propagator to be equal to the bare propagator at $p^2 = M^2$, and the three point vertex to have no correction proportional to the bare vertex at the point $k = 0, p^2 = q^2 = M^2$. This condition gives immediately

$$Z_3^{\widetilde{\mathrm{MOM}}} = 1 + g_0^2\left(\frac{1}{16\pi^2}\cdot 5\ln\frac{\pi^2}{a^2M^2} + 0.2977\right), \quad Z_1^{\widetilde{\mathrm{MOM}}} = 1 + g_0^2\left(\frac{1}{16\pi^2}\cdot 2\ln\frac{\pi^2}{a^2M^2} + 0.23096\right). \tag{11}$$

The relation between the renormalized and bare coupling constants is given by

$$[g(M)^{\widetilde{\mathrm{MOM}}}]^2 = g_0^2\frac{(Z_3)^3}{(Z_1)^2} = g_0^2\left[1 + g_0^2\left(\frac{11}{16\pi^2}\ln\frac{\pi^2}{a^2M^2} + 0.4312\right)\right], \tag{12}$$

which gives $\beta_0 = 11/16\pi^2$, as was expected. On the other hand, the finite term gives the relation between the Λ parameters. In the limit $g_0 \to 0, a \to 0$, eqs. (1), (3) and (12) give immediately

$$\Lambda^{\widetilde{\mathrm{MOM}}} = \pi e^{0.4312/2\beta_0}\Lambda^{\text{lattice}}, \quad \Lambda^{\widetilde{\mathrm{MOM}}}_{\text{(Feynman gauge)}} = 69.4\,\Lambda^{\text{lattice}}, \quad \text{for SU(3)}. \tag{13}$$

The last step is to connect $\Lambda^{\widetilde{\mathrm{MOM}}}$ with Λ^{MOM}. There is nothing distriguished about Λ^{MOM}, except that this is the definition which is used widely in deep inelastic phenomenology. To find this connection we have calculated the three-point vertex at the asymmetric point in the continuum theory, while at the symmetric point it has been calculated previously by Celmaster and Gonzalves [12] using dimensional regularization. From these results we obtained

$$\Lambda^{\mathrm{MOM}}_{\text{(Feynman gauge)}} = 1.2025\,\Lambda^{\widetilde{\mathrm{MOM}}}_{\text{(Feynman gauge)}}. \tag{14}$$

Eqs. (13) and (14) give our final result quoted in eq. (4). In the continuum theory the ratio between different Λ's is the same for SU(2) and SU(3); the reason is that both the constant terms in the two- and three-point functions and β_0 are proportional to the Casimir operator $C_2(G)$, and so it is cancelled ‡1. On the lattice, however, there are terms spoiling this simple result, and for SU(2) one obtains a somewhat smaller ratio [eq. (14)].

In his Monte-Carlo calculation Creutz used the lowest order definition for Λ, without the extra factor of $(\beta_0 g_0^2)^{-\beta_1/2\beta_0^2}$ [3]. This factor is changing slowly compared to the first factor, but it gives a normalizing factor of the order of three in the relevant coupling constant region (which is $2 < \beta < 3$ in Creutz notation). Therefore the effective connecting number in this region is $\Lambda^{\mathrm{MOM}} \approx 57.5\cdot 3\Lambda^{\text{Creutz}}$. Comparing this relation to the Monte-Carlo result (tension)$^{1/2} \approx 200\,\Lambda^{\text{Creutz}}$ and remembering that (tension)$^{1/2} \approx \Lambda^{\mathrm{MOM}}$, the numerical consistency is better than what we could expect in this "light-quarkless, 10^4 steps large, computer laboratory".

One of us (A.H.) is indebted to P. Hrasko for his interest and help and to M. Huszar for his help in finding

‡1 We are indebted to E. Witten for a discussion on this point.

the integration measure. P.H. is indebted to J. Kuti, M. Nauenberg, W. van Neerven, G. Parisi, C. Sachrajda, R. Stora M. Virasoro, J. Zinn-Justin, K. Wilson and E. Witten for useful discussions. We are indebted to M. Veltman for his help by introducing a new command in SCHOONSCHIP which helped to find and to check the expanded action. At certain stages we used also the symbolic program REDUCE.

References

[1] K.G. Wilson, Phys. Rev. D10 (1974) 2445.
[2] R. Balian, J.M. Drouffe and C. Itzykson, Phys. Rev. D10 (1974) 3374.
[3] M. Creutz, BNL preprint (1979).
[4] M. Creutz, Phys. Rev. Lett. 43 (1979) 553.
[5] K.Wilson, Cornell preprint (1979).
[6] C. Rebbi, BNL preprint (1980).
[7] J. Kogut, R.B. Pearson and J. Shigemitsu, Phys. Rev. Lett. 43 (1979) 484.
[8] B.E. Baaquie, Phys. Rev. D16 (1977) 2612.
[9] For a recent comprehensive summary see: J. Ellis and C.T. Sachrajda, Ref. TH. 2782, CERN preprint (1979).
[10] M. Bace, Phys. Lett. 78B (1978) 132.
[11] A.J. Buras, Fermilab-Conf-79/65-THY (1979).
[12] W. Celmaster and R.J. Gonzalves, Phys. Rev. Lett. 42 (1979) 1435; Phys. Rev. D20 (1979) 1420.
[13] E.G. Floratos, D.A. Ross and C.T. Sachrajda, Nucl. Phys. B129 (1977) 66; Erratum B139 (1978) 545; B152 (1975) 493.
[14] W.A. Bardeen, A.J. Buras, D.W. Duke and T. Muta, Phys. Rev. D18 (1978) 3998.
[15] C.G. Callan, R. Dashen and D.J. Gross, Princeton Univ. preprint (1980).
[16] A. Hasenfratz and P. Hasenfratz, in preparation.
[17] E.S. Abers and B.W. Lee, Phys. Rep. 9C (1973) 1.

PHYSICAL REVIEW D VOLUME 23, NUMBER 10 15 MAY 1981

Relationship between lattice and continuum definitions of the gauge-theory coupling

Roger Dashen
The Institute for Advanced Study, Princeton, New Jersey 08540

David J. Gross
Princeton University, Princeton, New Jersey 08544
(Received 21 July 1980)

We generalize the background-field method to lattice gauge theories. By evaluating the lattice partition function in a weak background lattice field for weak coupling we determine the relationship between the lattice definition of the coupling and that of dimensional regularization. This corresponds to a determination of the ratio of the renormalization scale parameters for these definitions. We find, for an SU(N) lattice gauge theory with the Wilson action, that $(\Lambda_{\overline{MS}}/\Lambda_L)_{SU(N)} = 38.852\,704 \exp(-3\pi^2/11N^2)$.

I. INTRODUCTION

In recent months significant progress has been made in understanding the dynamics of the simplest approximation to quantum chromodynamics (QCD)—namely, quarkless non-Abelian gauge theories. Calculations performed using a variety of different methods, including semiclassical techniques,[1] direct Monte Carlo integration for a Euclidean lattice theory,[2,3] and strong-coupling lattice Hamiltonian expansions,[3] all yield an extremely simple picture of the transition from weak coupling at short distances to strong coupling at large distances. These calculations strongly support the contention that QCD confines color and yield a relation between the slope of the linear heavy-quark potential (the "string tension") and the renormalization scale parameter Λ of the theory. If not for the fact that quarks are absent in these calculations this relation would provide a crucial quantitative test of QCD.

The strongest evidence for confinement comes from the Monte Carlo calculations of Creutz and Wilson.[2,3] Here one directly evaluates, using Monte Carlo integration, the expectation value of Wilson loops for a Euclidean lattice gauge theory on a periodic lattice of spacing a. One then looks for an area-law behavior of large Wilson loops and determines the value of the string tension σ, in units of a^{-2}, as a function of the lattice coupling g. In order to establish that the continuum theory has linear confinement one must show that σ remains constant as g^2 approaches zero (proportional to $1/\ln a\Lambda$) as it must in the continuum limit. The relation between the bare coupling g and the lattice spacing a is determined by asymptotically free perturbation theory,[5-7]

$$x = 8\pi^2/g^2N = \tfrac{11}{6}\ln[1/(\Lambda a)^2] + \tfrac{17}{22}\ln\{\ln[1/(\Lambda a)^2]\}. \tag{1.1}$$

This formula can be regarded as specifying the meaning of the renormalization-group scale parameter Λ. A change in the value of Λ shifts x by a finite amount to this order and different renormalization schemes may yield different values of this parameter. We shall refer to the renormalization scale parameter that appears in the lattice theory as Λ_L.

Now since, for small coupling, g satisfies Eq. (1.1) then σa^2 must behave, for weak coupling, as

$$\sigma a^2 = \sigma/\Lambda_L^2(a\Lambda_L)^2 = (\sigma/\Lambda_L^2)\left(\frac{48\pi^2}{11g^2N}\right)^{51/121} \exp(-24\pi^2/11g^2N)\,. \tag{1.2}$$

If one can show that σa^2 as determined by Monte Carlo integration has the above dependence on g for sufficiently small g, then this is an indication of linear confinement. Furthermore, one can calculate the numerical value of σ/Λ_L^2.

The calculations of Creutz and Wilson show that, for SU(2), (a) the behavior of large Wilson loops is consistent with an area law, (b) the behavior of σa^2 exhibits an abrupt change from strong-coupling behavior ($\sigma a^2 = \ln g^2$) to the weak-coupling behavior of Eq. (1.2) at $g^2 \approx 2$, and (c) the value of σ/Λ_L^2 is 80 ± 20.[8]

A similar abrupt transition from strong-coupling to weak-coupling behavior at $g^2 \approx 2$ was found by Kogut, Pearson, and Shigemitsu, who calculated the string tension in a strong-coupling expansion for a Hamiltonian lattice gauge theory and extrapolated their results, using Padé approximants to weak coupling.[4]

The semiclassical results of Ref. 1 are based on a totally different approach. Here one employs semiclassical techniques to construct an effective lattice gauge theory. The basic assumption is that one can obtain an adequate representation of the physics of certain observables (e.g., planar Wilson loops) in terms of a simple Wilson action characterized by one coupling constant $g(a)$. The coupling-constant renormalization which yields $g(a)$ is calculated by considering the response of the theory to weakly varying external fields, and consists of the perturbative renormalization given in Eq. (1.2) and a further renormalization due to instantons. One then tests whether the a dependence of the coupling is such as to yield a constant string tension $\sigma \equiv \ln g^2(a)/a^2$ once g^2 is large enough. If the basic assumption is valid this should occur once $g^2 \gtrsim 2$, where strong coupling sets in.

The results of this calculation were as follows: (a) An abrupt transition from weak-coupling to strong-coupling behavior occurred at $g^2 \approx 2$ [for SU(2)] at which point the instanton renormalization suddenly turned on. (b) Over a range of distances for which $2 \leq g^2 \leq 16$, where the semiclassical approximation was reasonable, the string tension was independent of g^2 or a, lending support to the basic assumption. (c) The value of σ was determined to be $\sqrt{\sigma} \approx (70 \pm 30)\,\Lambda_L$,[9] in good agreement with Ref. 2 considering the uncertainties in both calculations.

In all of these calculations it is essential to know the value of Λ_L in terms of more familiar renormalization-scheme scale parameters. For example, if the lattice gauge theorists are to compare their calculation of σ with experiment (ignoring the absence of quarks) it is necessary to know the value of Λ_L in MeV units. The standard asymptotic-freedom predictions, however, are performed using dimensional regularization and there is some evidence from scaling deviations that $\Lambda_{\overline{MS}}$, for the modified-minimal-subtraction scheme, is in the range of 300–600 MeV. The string tension can also be estimated (again ignoring light quarks) from the slope of the Regge trajectory or the heavy-quark potential to be of order (450 MeV)^2. To compare these it is necessary to know the ratio $\Lambda_L/\Lambda_{\overline{MS}}$.

In the semiclassical calculation it is even more essential to know Λ_L, since the calculation rests on 't Hooft's evaluation of the density of instantons using a continuum Pauli-Villars regularization scheme. In order to evaluate the effective lattice coupling it is necessary to convert to the lattice coupling—and this can (and does) have a big effect on the instanton density.

In Ref. 1 the relation between Λ_L and $\Lambda_{\overline{MS}}$ was estimated to be ≈ 6.8. This estimate was based on a comparison of a one-loop calculation involving scalar particles in the continuum theory (using Pauli-Villars or dimensional regularization) with the analogous lattice calculation. Equating these one-loop calculations yields the above value of $\Lambda_L/\Lambda_{\overline{MS}}$, which differs from one due to the difference between lattice and continuum propagators. The effects of the change in the structure of vertices were not taken into account.

Recently Hasenfratz and Hasenfratz have performed a more careful calculation of Λ_L.[10] They explicitly evaluated the two- and three-point functions to the one-loop level for the weak-coupling SU(N) lattice gauge theory, thereby precisely determining the relation between Λ_L and $\Lambda_{\overline{MS}}$. Having heard of their calculation we undertook to independently calculate this ratio. This calculation is presented below. Our results agree with those of Ref. 10.[11] In any case we thought it useful to present them since our method of calculation is quite different. We generalize the well-known background-field method[12] to the lattice and calculate the effective Lagrangian for weak coupling in the presence of a slowly varying background lattice field. The actual calculations required to relate the lattice and the continuum theories are then quite simple. In fact a simple intuitive explanation of the difference between lattice and continuum perturbation theory can be given which yields, with almost no work, the ratio of $\Lambda_L/\Lambda_{\overline{MS}}$ to an accuracy of 25%.

II. THE LATTICE BACKGROUND-FIELD METHOD

In this section we shall adapt the familiar background-field technique[12] to lattice gauge theories, and show how it can be used to evaluate Λ_L. First we shall define our notation and then we shall outline the method. The explicit calculation will be presented in Sec. III.

A. Notation

Consider Wilson's[13] formulation of the SU(N) Euclidean lattice gauge theory.[14] We label the sites of the lattice by an integer-valued four-vector x. Links are then labeled by pairs $(x, x+\mu)$, $\mu = 1, \ldots, 4$, where the four vector μ_α is equal to $\delta_{\mu\alpha}$. The basic variables of the lattice gauge theory are $N\times N$ unitary matrices defined on the links and denoted by $U_{x,x+\mu} \equiv U^\dagger_{x+\mu,x}$. The lattice action involves the plaquette variables, defined on elementary squares. At the point x in the μ-ν plane,

$$W_{\mu\nu}(x) \equiv U_{x,x+\mu} U_{x+\mu,x+\mu+\nu} U^\dagger_{x+\nu,x+\mu+\nu} U^\dagger_{x,x+\nu} = W^\dagger_{\nu\mu}(x)$$

$$\equiv \tfrac{1}{2}[G_{\mu\nu}(x) + i H_{\mu\nu}(x)], \quad (2.1)$$

where $G_{\mu\nu}(x)$ and $H_{\mu\nu}(x)$ are Hermitian. If λ^α, $\alpha = 1, \ldots, N^2-1$, are the $N\times N$ matrix representations of the generators of SU(N), normalized so that $\mathrm{Tr}\,\lambda^\alpha\lambda^\beta = 2\delta_{\alpha\beta}$, then we define $H^\alpha_{\mu\nu}$ by

$$H_{\mu\nu}=H^{\alpha}_{\mu\nu}\tfrac{1}{2}\lambda^{\alpha}\ . \tag{2.2}$$

In the continuum limit, i.e., as the lattice spacing $a\to 0$,

$$U_{x,x+\mu} \doteq P\exp\left(i\int_{xa}^{(x+\mu)a} A\cdot dx\right),$$

$$W_{\mu\nu}(x)\approx I+iF_{\mu\nu}(x)a^2,$$

and

$$H^{\alpha}_{\mu\nu}=2a^2F^{\alpha}_{\mu\nu},$$

where $F_{\mu\nu}=F^{\alpha}_{\mu\nu}\lambda_{\alpha}/2$ is the continuum field strength.

The Wilson action is

$$S\equiv\frac{1}{g^2(a)}\sum_x\sum_{\nu>\mu}\mathcal{L}_{\mu\nu}(x)\ , \tag{2.3}$$

$$\begin{aligned}\mathcal{L}_{\mu\nu}(x)&\equiv \mathrm{tr}[1-W_{\mu\nu}(x)]+\mathrm{H.c.}\\ &=\mathrm{tr}[2-G_{\mu\nu}(x)],\end{aligned} \tag{2.4}$$

and reduces in the continuum limit to the Yang-Mills action

$$S=\int d^4x\tfrac{1}{4}\sum_{\alpha,\mu,\nu}(F^{\alpha}_{\mu\nu})^2\ . \tag{2.5}$$

Finally we shall define lattice covariant derivatives. Given a set of link variables U (gauge fields) and a matrix-valued function defined on lattice sites (i.e., matter fields) the lattice covariant derivatives are defined as follows:

$$\begin{aligned}D_{\mu}f(x)&\equiv U_{x,x+\mu}f(x+\mu)U^{\dagger}_{x,x+\mu}-f(x)\ ,\\ \overline{D}_{\mu}f(x)&\equiv U^{\dagger}_{x,x+\mu}f(x-\mu)U_{x+\mu,x}-f(x)\ .\end{aligned} \tag{2.6}$$

Note that under gauge transformations

$$\begin{aligned}U_{x,x+\mu}&\to V(x)U_{x,x+\mu}V^{\dagger}(x+\mu)\ ,\\ f(x)&\to V(x)f(x)V^{\dagger}(x)\ ,\\ D_{\mu}f(x)&\to V(x)D_{\mu}f(x)V^{\dagger}(x)\ .\end{aligned} \tag{2.7}$$

Also when all $U=1$ (vanishing gauge field), D_{μ} ($\overline{D}_{\mu}$) reduces to the ordinary lattice derivative Δ_{μ} ($\overline{\Delta}_{\mu}$), where

$$\begin{aligned}\Delta_{\mu}f(x)&=f(x+\mu)-f(x),\\ \overline{\Delta}_{\mu}f(x)&=f(x-\mu)-f(x)\ .\end{aligned} \tag{2.8}$$

The analog of integration by parts on the lattice is

$$\begin{aligned}&\sum_x \mathrm{tr}[g(x)(D_{\mu}f(x))]=\sum_x\mathrm{tr}[(\overline{D}_{\mu}g(x))f(x)],\\ &\sum_{x,\mu}\mathrm{tr}[(D_{\mu}g(x))(D^{\mu}f(x))]=\sum_{x,\mu}\mathrm{tr}[(\overline{D}^{\mu}g(x))(\overline{D}_{\mu}f(x))],\end{aligned} \tag{2.9}$$

as can easily be verified.

B. The strategy

Our strategy will be as follows. Let U^0 denote a set of link variables which represent a solution of the classical lattice equations of motion on a lattice of spacing a.[15] Then the lattice action can be expanded about $U=U^0$, which is an extremum of the action

$$S=S_0+S_2+\text{cubic terms}+\cdots\ , \tag{2.10}$$

where $S_0=S(U^0)$ and S_2 is quadratic in the fluctuations of U about U^0. We now perform a saddle-point approximation to the functional integral representation of the partition function (the Euclidean vacuum-to-vacuum transition amplitude)

$$Z_a\equiv\int[dU]e^{-S(U)}\ , \tag{2.11}$$

where $[dU]=\prod_x\prod_{\mu}dU_{x,x+\mu}$ and $dU_{x,x+\mu}$ is the Haar measure on SU(N). It is convenient to introduce, as in the continuum theory, gauge-fixing terms S_{gf}, and the compensating ghost terms S_{gh}. The precise form of these terms will be given in the following section. We can then expand about $U=U^0$ as follows:

$$Z_a\approx e^{-S^0}\int[dU]e^{-(S_2+S_{gf}+S_{gh})}[1+O(g(a)^2)], \tag{2.12}$$

where $g(a)$ is the (bare) coupling appropriate for a lattice of spacing a. Note that S^0, which is simply the vacuum energy in the background field U^0, is of order $1/g^2(a)$ and since we will calculate $\ln Z_a$ up to order $(g^2)^0$ the $O(g^2(a))$ terms in (2.12) can be ignored.

A similar calculation of Z can be performed in the continuum theory, and compared with the lattice calculation. This is most easily done for weak, slowly varying fields, i.e., $aA^0_{\mu}\ll 1$ and $a\partial_{\nu}A^0_{\mu}\ll A^0_{\mu}$, or $U_{x,x+\mu}\approx 1+i\int A^0\cdot dx$. To perform the continuum calculation we must regularize the theory in one way or another. We shall choose a Pauli-Villars regularization scheme, but dimensional regularization would be equally good. If one compares these two calculations, for weak slowly varying background fields one should find that

$$-\ln\left(\frac{Z_a}{Z_m}\right)=\left[\frac{1}{4g^2(a)}-\frac{1}{4g^2(m)}+d(ma)\right]\int d^4x(F^{\alpha}_{\mu\nu})^2\ , \tag{2.13}$$

where m is the Pauli-Villars regulator mass, $g(m)$ the bare coupling for the continuum theory, and $d(ma)$ is an (infrared finite) number. Given our knowledge of the Yang-Mills β function we know that $g^2(a)$ (for $a\to 0$) is given by (1.1) and that analogously $g^2(m)$ is given (for $m\to\infty$) by

$$x_{PV}(m)=\frac{8\pi^2}{g^2(m)N}=\tfrac{11}{6}\ln\left(\frac{m}{\Lambda_{PV}}\right)^2+\tfrac{17}{22}\ln\left[\ln\left(\frac{m}{\Lambda_{PV}}\right)^2\right], \tag{2.14}$$

thus defining the Pauli-Villars renormalization scale parameter Λ_{PV}. The difference between x_{PV}/N and x_L/N, when $a\to 0$ and $m\to\infty$, with ma kept fixed, is thus

$$x_L-x_{PV}=\tfrac{11}{3}\ln\left(ma\frac{\Lambda_{PV}}{\Lambda_L}\right)+O\left(\frac{1}{\ln m}\right). \tag{2.15}$$

One expects (and this expectation is upheld as we shall see by explicit calculation) that the difference between the lattice and continuum calculation is simply a finite renormalization (or redefinition) of the coupling and that therefore $\ln(Z_a/Z_m)$ is independent of m or $1/a$ as $m\to\infty$. Therefore it must be that

$$d(ma)=\frac{11N}{96\pi^2}(\ln ma+C(N)), \tag{2.16}$$

where $C(N)$ is a pure number (perhaps N dependent). In that case

$$\ln\frac{Z_a}{Z_m}=\frac{11N}{96\pi^2}\left(\ln\frac{\Lambda_L}{\Lambda_{PV}}-C(N)\right)\int d^4x\,(F_{\mu\nu})^2, \tag{2.17}$$

and if we require that the two calculations yield the same physics, i.e., that Z_a/Z_m, then Λ_L/Λ_{PV} is determined to be

$$\frac{\Lambda_L}{\Lambda_{PV}}=\exp(C(N)). \tag{2.18}$$

Let us note some important points:

(i) The ratio Λ_L/Λ_{PV} can be computed by performing a one-loop calculation in the presence of a background field.

(ii) By considering only weak fields that are slowly varying we can greatly simplify our calculation. Any contribution to S_2 that produces, when we carry out the Gaussian lattice integrals, terms such as $(F_{\mu\nu})^4$ or $(D_\lambda F_{\mu\nu})^2$ that are small compared to $(F_{\mu\nu})^2$ can be dropped.

(iii) The calculation can be performed in a way that is manifestly infrared finite. Thus, in evaluating $\ln(Z_a/Z_m)$, all integrals that occur can be evaluated at zero external momentum. This enormously simplifies the numerical labor.

III. THE CALCULATION

A. The measure

Let us parametrize the fluctuations of the link variables U by four traceless Hermitian matrices $\alpha^\mu(x)$ defined by

$$U_{x,x+\mu}\equiv e^{ig\alpha^\mu(x)}U^0_{x,x+\mu}. \tag{3.1}$$

Owing to the invariance of the measure $[dU]$ under right (or left) multiplication,

$$[dU]=[d(e^{ig\alpha}U^0)]$$
$$=[de^{ig\alpha}]=[d\alpha][1+O(g^2\alpha^2)]. \tag{3.2}$$

For our purposes, terms of order $g^2\alpha^2$ can be ignored and thus we can take

$$[dU]=[d\alpha]=\prod_{x,\mu}d\alpha^\mu(x), \tag{3.3}$$

and, for the same reason, allow $\alpha^\mu(x)$ to range over an infinite range.

B. The quadratic action

Next we need to compute S_2. We begin with the plaquette variable

$$W_{\mu\nu}(x)=\exp[ig\alpha^\mu(x)]U^0_{x,x+\mu}\exp[ig\alpha_\nu(x+\mu)]U^0_{x+\mu,x+\mu+\nu}U^{0\dagger}_{x+\mu+\nu,x+\nu}\exp[-ig\alpha^\mu(x+\nu)]U^\dagger_{x,x+\nu}\exp[-ig\alpha^\nu(x)], \tag{3.4}$$

whose trace can be written using $Ve^{i\alpha}=\exp(iV\alpha V^{-1})V$ as

$$\mathrm{tr}\,W_{\mu\nu}(x)=\mathrm{tr}\{\exp[-ig(D^0_\nu\alpha^\mu(x)+\alpha^\mu(x))]\exp[ig\alpha^\nu(x)]\exp[ig\alpha^\mu(x)]\exp[ig(D^0_\mu\alpha^\nu(x)+\alpha^\nu(x))]W^0_{\mu\nu}\}, \tag{3.5}$$

where D_μ^0 ($W_{\mu\nu}^0$) is the lattice covariant derivative (plaquette variable) with $U=U^0$. Next we combine terms using

$$e^{gA}e^{gB}=\exp\{g(A+B)+(g^2/2)[A,B]\}(1+O(g^3))$$

to obtain

$$\operatorname{tr} W_{\mu\nu}(x)=\operatorname{tr}\{[\exp(igD_\mu^0\alpha^\nu(x)-igD_\nu^0\alpha^\mu(x)) + (g^2E_{\mu\nu}(x))]W_{\mu\nu}^0(x)\}+O(g^3)\ , \tag{3.6}$$

where

$$E_{\mu\nu}(x)=[\alpha^\nu(x),\ \alpha^\mu(x)]+\tfrac{1}{2}[\alpha^\nu(x),D_\mu^0\alpha^\nu(x)] +\tfrac{1}{2}[D_\nu^0\alpha^\mu(x),\ \alpha^\mu(x)] +\tfrac{1}{2}[D_\nu^0\alpha^\mu(x),\ D_\mu^0\alpha^\nu(x)]\ . \tag{3.7}$$

Now we expand and keep only the quadratic terms in α to obtain

$$2\operatorname{Re}\operatorname{tr}[W_{\mu\nu}(x)-W_{\mu\nu}^0(x)] =-\tfrac{1}{2}g^2\operatorname{tr}[(D_\mu^0\alpha^\nu(x)-D_\nu^0\alpha^\mu(x))^2G_{\mu\nu}^0(x)] +ig^2\operatorname{tr}[E_{\mu\nu}(x),H_{\mu\nu}^0(x)]+O(g^3)\ , \tag{3.8}$$

where $G_{\mu\nu}$ and $H_{\mu\nu}$ and defined in Eq. (2.1).

For weak background fields $G_{\mu\nu}^0=2+O((F_{\mu\nu})^2)$, so that the first term can be simplified to equal

$$-\operatorname{tr}[(D_\mu^0\alpha^\nu(x)-D_\nu^0\alpha^\mu(x))^2] -\tfrac{1}{2}\operatorname{tr}[(\Delta_\mu\alpha^\nu(x)-\Delta_\nu\alpha^\mu(x))^2]\frac{\operatorname{tr}(G_{\mu\nu}^0-2)}{N}\ , \tag{3.9}$$

where we have thrown away terms of higher order than $O(F^2)$ and used the fact that when the α's are integrated out only $\operatorname{tr}G_{\mu\nu}^0$ will appear so that $G_{\mu\nu}^0$ can be replaced by $\operatorname{tr}G_{\mu\nu}^0/N$.

C. Gauge fixing

In perturbing about a background field it is useful to work in a background-field gauge. The lattice analog of this gauge condition is

$$\sum_{\mu=1}^{4}(U_{x,x+\mu}U_{x,x+\mu}^{0\dagger}-U_{x-\mu,x}^\dagger U_{x-\mu,x}^0)=0\ , \tag{3.10}$$

which is trivially satisfied when $U^0=U$. This, when expanded, yields the condition on the α's

$$\sum_{\mu=1}^{4}\bar{D}_\mu^0\alpha^\mu(x)=0\ . \tag{3.11}$$

We shall work (for convenience) in the background-field Feynman gauge, by adding to the action the gauge-fixing term

$$S_{gf}=\frac{1}{g^2}\sum_x\operatorname{tr}\left[\left(\sum_{\mu=1}^{4}D^\mu\alpha_\mu(x)\right)^2\right]\ . \tag{3.12}$$

A straightforward computation shows that the corresponding Faddeev-Popov determinant is $\operatorname{Det}[\sum_{\mu=1}^4\bar{D}_\mu^0D_\mu]$. In the usual way it can be represented by a Gaussian integral over a set of complex matrix ghost fields $\phi(x)$ whose action is

$$S_{gh}=\frac{1}{g^2}\sum_x\sum_{\mu=1}^{4}\operatorname{tr}[(D^\mu\phi(x))^\dagger(D^\mu\phi(x))]\ . \tag{3.13}$$

D. The full quadratic action

The full quadratic action involving the gauge fields is given by $S=S_2+S_{gf}$. We find it useful to collect the various terms into four groups. The $(D_\mu^0\alpha^\nu-D_\nu^0\alpha^\mu)^2$ term in S_2 combines with S_{gf} to yield

$$S_{sc}+\frac{1}{2g^2}\sum_x\sum_{\nu>\mu}\operatorname{tr}\{\alpha^\nu([\bar{D}_\mu^0,D_\nu^0]\alpha^\mu)+\alpha^\mu([\bar{D}_\nu^0,D_\mu^0]\alpha^\nu)\}\ , \tag{3.14}$$

where S_{sc} is the action of a collection of scalar matter fields α^λ ($\lambda=1,\ldots,4$),

$$S_{sc}=\tfrac{1}{2}\sum_x\sum_{\lambda,\mu}\operatorname{tr}[(D_\mu^0\alpha^\lambda)(D_\mu^0\alpha^\lambda)]\ . \tag{3.15}$$

The second term in Eq. (3.9) can be written as

$$S_T=-\frac{a^4}{48Ng^2}\sum_{x,\alpha,\beta}(F^{\alpha\beta}(x))^2\sum_{\nu>\mu}\operatorname{tr}(\Delta_\mu\alpha^\nu(x)-\Delta_\nu\alpha^\mu(x))^2\ , \tag{3.16}$$

where we have used the fact that when the α's are integrated out $\langle(F_{\alpha\beta})^2\rangle$ will be independent of α and β and have replaced $(F_{\alpha\beta})^2$ by $\frac{1}{6}\sum_{\alpha>\beta}(F_{\alpha\beta})^2$.

Finally the two terms S_A and S_B are combinations of the last term in Eq. (2.26) and the last term in Eq. (3.14),

$$S_A=\frac{a^2}{g^2}\sum_x\sum_{\nu>\mu}\operatorname{tr}[A_{\mu\nu}(x)F_{\mu\nu}(x)]\ ,$$
$$A_{\mu\nu}(x)=-2i\{2[\alpha^\nu,\alpha^\mu]+[\alpha^\nu,D_\nu^0\alpha^\nu] +[D_\mu^0\alpha^\nu,\alpha^\mu]-\tfrac{1}{2}[D_\nu^0\alpha^\mu,D_\mu^0\alpha^\nu]\}\ , \tag{3.17}$$

and

$$S_B=\frac{a^2}{g^2}\sum_x\sum_{\nu>\mu}\operatorname{tr}[B_{\mu\nu}(x)F_{\mu\nu}]\ ,$$
$$B_{\mu\nu}(x)=-i\{[\alpha^\nu(x),D_\mu^0\alpha^\nu(x)] +[D_\nu^0\alpha^\mu(x),\alpha^\mu(x)]\}\ . \tag{3.18}$$

Collecting everything together we have

$$S_2+S_{gf}=S_{sc}+S_T+S_A+S_B\ . \tag{3.19}$$

In the continuum limit only S_{sc} and S_A survive and these become

$$S_{sc}=\tfrac{1}{2}\int d^4x\operatorname{Tr}(D_\mu^0\delta A^\lambda)^2 \tag{3.20}$$

and

$$S_A = -2i \int d^4x \, \mathrm{tr}([\delta A^\nu, \delta A^\mu] F^{\mu\nu}) \tag{3.21}$$

when we expand A_μ about the background gauge field $A^0_\mu, A_\mu = A^0_\mu + \delta A_\mu$.

In doing perturbation theory about the background field the free piece of the action will determine the propagator of α^μ or δA^μ. For the lattice theory this is

$$S_f = \tfrac{1}{2} \sum_x \sum_{\lambda,\mu} \mathrm{tr}[(\Delta_\mu \alpha_\lambda(x))^2] , \tag{3.22}$$

whereas for the continuum theory it is simply

$$S_f = \tfrac{1}{2} \int d^4x \, \mathrm{tr}(\partial_\mu \delta A_\lambda)^2 , \tag{3.23}$$

corresponding to the momentum-space propagators in the Feynman gauge,

$$\frac{\delta_{\alpha\beta}}{\sum_{\mu=1}^4 \Delta_\mu \bar{\Delta}_\mu} = \frac{\delta_{\alpha\beta} a^2}{\sum_{\mu=1}^4 4\sin^2(p_\mu a/2)}$$

for the lattice theory, and $\delta_{\alpha\beta}/\partial^2 = \delta_{\alpha\beta} a^2/p^2$ for the continuum theory.

We now wish to calculate the quantity $d(ma)$ defined in Eq. (2.13). Denote by $\langle\theta\rangle_a$ the expectation value of an observable θ in the lattice theory using the free action S_f:

$$\langle\theta\rangle_a = \mathrm{const} \times \int [d\alpha] e^{-S_f} \theta , \tag{3.24}$$

where the normalization is fixed by $\langle 1\rangle = 1$. Let $\langle\theta\rangle_m$ denote the corresponding average in the continuum theory using Pauli-Villars regulators (m denotes the Pauli-Villars mass). Then it is easy to see that

$$-d(ma) \int d^4x (F^\alpha_{\mu\nu})^2 = \tfrac{1}{2} \ln(Z^{sc}_a / Z^{sc}_m) + I_1 + I_2 + I_3 , \tag{3.25}$$

where Z^{sc}_a or Z^{sc}_m are the respective partition functions computed using the action S_{sc} [Eqs. (3.15) and (3.20)], with no additional ghost terms. The complex ghost fields simply cancel half of the contribution of the four scalar particles in S_{sc} and thus account for the factor of $\frac{1}{2}$ in the above equation. Finally

$$I_1 = -\langle S_T\rangle_a ,$$
$$I_2 = \tfrac{1}{2} \langle (S_B)^2\rangle_a , \tag{3.26}$$
$$I_3 = \tfrac{1}{2}[\langle (S_A)^2\rangle_a - \langle (S_A)^2\rangle_m] .$$

S_T appears only to lowest order since it is already of order $(F_{\mu\nu})^2$. It is trivial to see that there can be no linear terms arising from S_A or S_B. The only nontrivial point is that the cross terms between S_A, S_B, or S_{sc} is a straightforward consequence of the fact that S_A is odd in the vector indices of α_μ while S_B and S_{sc} are even. There is probably an equally simple reason for the vanishing of the S_B and S_{sc} cross terms; however, we have been unable to find it, and can only rely on explicit calculation. The numerical values of these four terms are calculated in the next section.

IV. THE RESULTS

We now proceed to evaluate the separate contributions to the coupling-constant renormalization. First we consider I_1, which equals $-\langle S_T\rangle_a$. This term, which is responsible in factor for 75% of the ratio Λ_L/Λ_{PV}, has a simple origin and can be easily analytically calculated. It arises because the Wilson action is, unlike the continuum action, a bounded function of the gauge field. Thus when we expand $\mathcal{L}_{\mu\nu}$ in powers of $F_{\mu\nu}$ we obtain $\mathcal{L}_{\mu\nu} = \frac{1}{2}(F_{\mu\nu})^2 - \mathrm{const} \times (F_{\mu\nu})^4$. The quartic term is of course negative, thus effectively reducing the value of $1/g^2(a)$. This yields a negative contribution to $C(N)$ thus reducing the value of Λ_L/Λ_{PV}. Clearly such a term has no classical analog. Indeed S_T is a dimension-eight (irrelevant) operator, which is multiplied by a^4 and whose only contribution as $a \to 0$ is in its contribution to dimension-four operators which diverge as a^{-4}. Thus its only effect is to yield a finite coupling renormalization. It is trivial to evaluate this term since the only Feynman graph that contributes to $\langle S_T\rangle_a$ is a tadpole graph [see Fig. 1(a)] with one propagator, and a vertex which equals the inverse propagator. These therefore cancel yielding

$$I_1 = -\langle S_T\rangle_a = \frac{N^2-1}{32N} \int d^4x (F^\alpha_{\mu\nu})^2 \int (dp)_a \frac{D(p)}{D(p)}$$
$$= \frac{N^2-1}{32N} \int d^4x (F^\alpha_{\mu\nu})^2 , \tag{4.1}$$

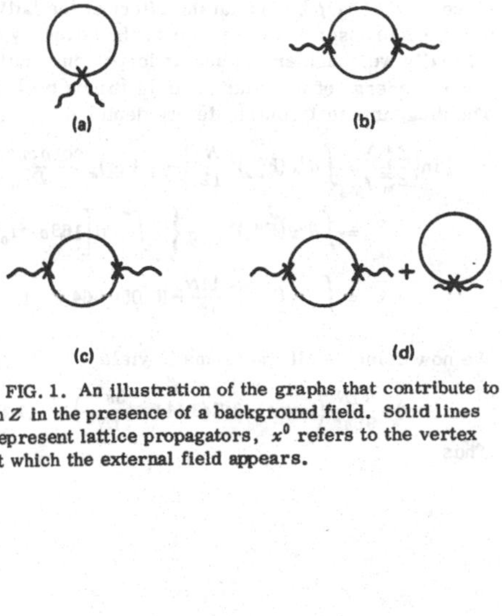

FIG. 1. An illustration of the graphs that contribute to ln Z in the presence of a background field. Solid lines represent lattice propagators, x^0 refers to the vertex at which the external field appears.

where we denote the lattice momentum-space integral by

$$\int (dp)_a \equiv \prod_{\mu=1}^{4} \int_{-\pi/a}^{+\pi/a} \frac{dp_\mu a}{(2\pi)^4} . \tag{4.2}$$

Note that this term, unlike all the other terms that we encounter, is not proportional to N (the factor of N^2-1 is simply the number of gauge modes in the loop), and is solely responsible for the N dependence of $C(N)$ and Λ_L.

I_2 also has no classical limit, since it arises from S_B a dimension-five operator. The only graph that contributes is depicted in Fig. 1(b) and yields

$$I_2 = \tfrac{1}{2}\langle (S_B)^2 \rangle_a$$
$$= \int d^4x (F^\alpha_{\mu\nu})^2 N \int \frac{(dp)_a 16\pi^2 \sin^2 p_1 a}{D^2(p)} . \tag{4.3}$$

Here we have used the fact that we can replace D_μ, in Eq. (3.18), by Δ_μ since we are working to $O(F^2)$ only, and that in momentum space

$$\Delta_\mu(p) = e^{ip_\mu a} - 1, \quad \bar{\Delta}_\mu(p) = e^{-ip_\mu a} - 1 . \tag{4.4}$$

This is an example of a nontrivial lattice momentum-space integral which must be evaluated numerically. However, all such one-loop integrals can be reduced to one-dimensional integrals by expressing $D^{-k}(p)$ as

$$D^{-k}(p) = \int_0^\infty d\beta \frac{\beta^{k-1}}{\Gamma(k)} \prod_{\mu=1}^{4} e^{-2\beta(1-\cos p_\mu a)} . \tag{4.5}$$

Then all $\int_{-\pi/a}^{+\pi/a} dp_\mu$ integrals can be evaluated in terms of Bessel functions. After such an exercise we find

$$I_1 = \int d^4x (F^\alpha_{\mu\nu})^2 \frac{N}{8} \int_0^\infty d\beta\, e^{-8\beta} I_0^3(2\beta) I_1(2\beta)$$
$$= \int d^4x (F^\alpha_{\mu\nu})^2 \frac{11N}{96\pi^2} (0.322\,288) , \tag{4.6}$$

where the number in the parentheses is the contribution to $-C(N) = \ln(\Lambda_{PV}/\Lambda_L)$. All numerical integrations were performed with an accuracy of one part in 10^{-6}.

The term I_3 does have a corresponding continuum limit and contains logarithmic divergences as $a \to 0$. The corresponding continuum calculation is done with a Pauli-Villars regulator for the δA_μ field. Again, in Eq. (3.17), the D_μ's can be replaced by Δ_μ's to this order, and the only graphs which contribute are depicted in Fig. 1(c). They yield

$$I_3 = \int d^4x (F^\alpha_{\mu\nu})^2 N \left\{ \int (dp)_a \frac{[(1+\cos p_1 a)/2][(1+\cos p_2 a)/2]}{D^2(p)} - \int \frac{d^4p}{(2\pi)^4} \left[\frac{1}{p^4} - \frac{1}{(p^2+m^2)^2} \right] \right\}$$
$$= \int d^4x (F^\alpha_{\mu\nu})^2 \frac{N}{4} \int_0^\infty d\beta \left[\beta e^{-8\beta} I_0^2(2\beta)(I_0(2\beta)+I_1(2\beta))^2 - \frac{(1-e^{-\beta m^2 a^2})}{4\pi^2 \beta} \right] \tag{4.7}$$
$$= \int d^4x (F^\alpha_{\mu\nu})^2 \frac{11N}{96\pi^2} (0.678\,249 - \tfrac{12}{11} \ln ma) . \tag{4.8}$$

Note that both integrals in Eq. (4.7) have infrared divergences; however, their difference is finite. In subtracting these two integrals we have been careful to first introduce infrared regulators (e.g., a gluon mass) in order to avoid error. This term is essentially the only one considered in Ref. 1. However, there we only took into account the effects of the lattice propagator, which yields a positive contribution since $1/D(p) > 1/p^2$, and not the effect of the lattice derivatives which produce the term $(1+\cos p_1 a)/2$ which suppresses the integrand on the boundary of the Brillouin zone ($p_1 a = \pm\pi$).

Finally we consider the scalar loops contribution $\frac{1}{2}\ln(Z^{sc}_a/Z^{sc}_m)$. There is simply the difference of the vacuum energy of a scalar particle (times two) in the external field for the lattice and continuum theories. The diagrams that contribute are depicted in Fig. 1(d) and, after some work, yield

$$\tfrac{1}{2}\ln\left(\frac{Z^{sc}_a}{Z^{sc}_m}\right) = \int d^4x (F^\alpha_{\mu\nu})^2 \frac{N}{12} \left\{ -\int (dp)_a \frac{\cos p_1 a \cos p_2 a}{D^2(p)} + \int \frac{d^4p}{(2\pi)^4} \left[\frac{1}{p^4} - \frac{m^4}{(p^2+m^2)^4} \right] \right\}$$
$$= \int d^4x (F^\alpha_{\mu\nu})^2 \frac{N}{192} \left\{ -\int_0^\infty d\beta \left[16\beta e^{-8\beta} I_0^2(2\beta) I_1^2(2\beta) + \frac{1-e^{-\beta m^2 a^2}}{\beta\pi^2} \right] + \frac{5}{6} \right\}$$
$$= \int d^4x (F^\alpha_{\mu\nu})^2 \frac{11N}{96\pi^2} (0.050\,864 + \tfrac{1}{12} \ln ma) . \tag{4.9}$$

We now combine all the terms to yield

$$d(ma) = \frac{11N}{96\pi^2} \left(\ln ma - 3.743\,111 + \frac{3\pi^2}{11N^2} \right) . \tag{4.10}$$

Thus

$$C(N) = \frac{3\pi^2}{11N^2} - 3.743\,111 \tag{4.11}$$

and

$$\left(\frac{\Lambda_L}{\Lambda_{PV}}\right)_{SU(N)} = 0.023\,680 e^{3\pi^2/11N^2}.$$

This yields

$$\left(\frac{\Lambda_L}{\Lambda_{PV}}\right)_{SU(2)} = 0.046\,413, \quad \left(\frac{\Lambda_{PV}}{\Lambda_L}\right)_{SU(2)} = 21.545\,854,$$

$$\left(\frac{\Lambda_L}{\Lambda_{PV}}\right)_{SU(3)} = 0.031\,936, \quad \left(\frac{\Lambda_{PV}}{\Lambda_L}\right)_{SU(3)} = 31.321\,2960,$$

$$\left(\frac{\Lambda_L}{\Lambda_{PV}}\right)_{SU(\infty)} = 0.023\,68, \quad \left(\frac{\Lambda_{PV}}{\Lambda_L}\right)_{SU(\infty)} = 42.229\,161.$$

We can now compare the lattice definition of the coupling to other more familiar continuum definitions, i.e., dimensional regularization, $\overline{\text{MS}}$, or momentum-space regularization. To do this it suffices to establish the relation between Λ_{PV} and Λ's corresponding to these definitions. This can be done quite easily in the continuum theory using, for example, the background-field method described above.

The relation of Λ_{PV} to Λ_{DR} is determined by comparing calculations using Pauli-Villars regulators to those using dimensional regularization and minimal subtraction. This calculation was performed by 't Hooft[16] with the result (γ = Euler's constant)

$$\ln\left(\frac{\Lambda_{PV}}{\Lambda_{DR}}\right) = \tfrac{1}{2}\ln 4\pi - \gamma/2 + \tfrac{1}{12} = 1.060\,238,$$
$$\left(\frac{\Lambda_{PV}}{\Lambda_{DR}}\right) = 2.887\,057. \tag{4.12}$$

In the so-called $\overline{\text{MS}}$ scheme[17] the coupling is defined with a finite renormalization which corresponds to removing the factor $\frac{1}{2}(\ln 4\pi - \gamma)$ from the above equation. Thus the Pauli-Villars and $\overline{\text{MS}}$ Λ's are almost identical,

$$\left(\frac{\Lambda_{PV}}{\Lambda_{\overline{MS}}}\right) = e^{1/12} = 1.086\,904. \tag{4.13}$$

Finally we can use the relation, established by Celmaster and Gonsalves,[18] between Λ_{DR} and Λ_{MOM}, where Λ_{MOM} corresponds to a (symmetric) momentum-space subtraction procedure. This yields a gauge-dependent definition of g^2 and Λ_{MOM}, which is, in the absence of matter fields,

$$\left(\frac{\Lambda_{MOM}}{\Lambda_{DR}}\right)_{\text{Landau gauge}} = 8.86,$$
$$\left(\frac{\Lambda_{MOM}}{\Lambda_{DR}}\right)_{\text{Feynman gauge}} = 7.69, \tag{4.14}$$
$$\left(\frac{\Lambda_{MOM}}{\Lambda_{PV}}\right)_{\text{Feynman gauge}} = 2.66.$$

Using these relations we can compare Λ_L to Λ_{DR}, $\Lambda_{\overline{MS}}$, or Λ_{MOM}. Thus, for example,

$$\left(\frac{\Lambda_{\overline{MS}}}{\Lambda_L}\right)_{SU(N)} = 38.852\,704\, e^{-3\pi^2/11N^2},$$
$$\left(\frac{\Lambda_{MOM,\ \text{Feynman gauge}}}{\Lambda_L}\right)_{SU(N)} = 112.33\, e^{-3\pi^2/11N^2}. \tag{4.15}$$

We can now compare with Hasenfratz and Hasenfratz's calculation, in which they find (in the Feynman gauge)[10]

$$\left(\frac{\Lambda_{MOM}}{\Lambda_L}\right)_{SU(2)} = 57.5, \quad \left(\frac{\Lambda_{MOM}}{\Lambda_L}\right)_{SU(3)} = 83.5,$$

whereas we find

$$\left(\frac{\Lambda_{MOM}}{\Lambda_L}\right)_{SU(2)} = 57.3, \quad \left(\frac{\Lambda_{MOM}}{\Lambda_L}\right)_{SU(3)} = 83.29. \tag{4.16}$$

The slight disagreement is probably not significant.

Finally we note that the large value of $\Lambda_{\overline{MS}}/\Lambda_L$ or Λ_{MOM}/Λ_L implies that the calculated values of $\sqrt{\sigma}/\Lambda_L$ are much closer to those one would expect in the real world. Thus in the case of SU(3) the values determined by semiclassical or Monte Carlo techniques lie in the range $\sqrt{\sigma} = (150-250)\Lambda_L$. When translated to $\Lambda_{\overline{MS}}$ this means that $\sqrt{\sigma} = (7-12)\Lambda_{\overline{MS}}$. In the real world $\sqrt{\sigma} \sim 450$ MeV and $\Lambda_{\overline{MS}} \sim (200-600)$ MeV. Some authors have tried to decrease the remaining discrepancy by using Λ_{MOM} to obtain $\sqrt{\sigma} = (1.5-2.5)\Lambda_{MOM}$. However, this is problematic since the value of Λ_{MOM} is highly dependent on the number of fermion fields and Λ_{MOM} is, in any case, determined by fits to experiment to be larger than $\Lambda_{\overline{MS}}$. We instead ascribe the discrepancy to the absence of quarks in these treatments. In the semiclassical approach it is clear that the introduction of quarks will lead to a decrease in $\sqrt{\sigma}/\Lambda_L$ by an amount which we can crudely estimate to be about 2–5.

ACKNOWLEDGMENTS

The work of R.D. was sponsored by the Department of Energy under Grant No. EY-76-S-02-2220. The work of D.J.G. was sponsored by the National Science Foundation under Grant No. PHY78-01221.

[1]C. Callan, R. Dashen, and D. Gross, Phys. Rev. D 19, 1826 (1979); Phys. Rev. Lett. 44, 435 (1980).
[2]M. Creutz, Phys. Rev. D 21, 2308 (1980).
[3]K. Wilson, report, 1980 (unpublished).
[4]J. Kogut, R. Pearson, and J. Shigemitsu, Phys. Rev. Lett. 43, 484 (1979); J. Kogut and J. Shigemitsu, Phys. Rev. Lett. 45, 410 (1980).
[5]D. Gross and F. Wilczek, Phys. Rev. Lett. 30, 1343 (1973).
[6]H. Politzer, Phys. Rev. Lett. 30, 1346 (1973).
[7]W. Caswell, Phys. Rev. Lett. 33, 344 (1974); D. R. Jones, Nucl. Phys. B75, 531 (1974).
[8]The value quoted by Creutz in Ref. 2 was based on a one-loop approximation to Eq. (1.1). The value quoted in the text is that recently determined by Creutz using Eq. (1.1); M. Creutz, Phys. Rev. Lett. 45, 313 (1980).
[9]This value differs from that presented in Ref. 1 due largely to the revised value of $\Lambda_L/\Lambda_{\overline{MS}}$.
[10]A. Hasenfratz and P. Hasenfratz, Phys. Lett. 93B, 165 (1980).
[11]It has recently been pointed out by P. Weisz (unpublished that 't Hooft's calculation of $\Lambda_{PV}/\Lambda_{DR}$ is slightly wrong. This correction has been taken into account in this paper, and removes the discrepancy between our calculation and that of Hasenfratz and Hasenfratz that previously existed.
[12]B. S. DeWitt, Phys. Rev. 162, 1195 (1967); 162, 1239 (1967); J. Honerkamp, Nucl. Phys. B48, 269 (1972); G. 't Hooft, *ibid.* B62, 444 (1973).
[13]K. Wilson, Phys. Rev. 10, 2445 (1974).
[14]Our methods would be equally applicable to any lattice gauge theory which reduced to the ordinary Yang-Mills theory in the continuum limit. These would only differ by a finite renormalization of g and thus by their value of Λ_L.
[15]As in the continuum background-field method we shall never need to know the precise form of the background field.
[16]G. 't Hooft, Phys. Rev. D 14, 3432 (1976). See Ref. 11.
[17]W. Bardeen, A. Buras, D. Duke, and T. Muta, Phys. Rev. D 18, 3998 (1978).
[18]W. Celmaster and R. Gonsalves, Phys. Rev. D 20, 1420 (1979).

VOLUME 43, NUMBER 7 PHYSICAL REVIEW LETTERS 13 AUGUST 1979

Quantum-Chromodynamic β Function at Intermediate and Strong Coupling

John B. Kogut,[a] Robert B. Pearson, and Junko Shigemitsu
The Institute for Advanced Study, Princeton, New Jersey 08540
(Received 5 June 1979)

With use of strong coupling methods, the energy of a long flux string is computed and this quantity is used to renormalize the theory. This gives an expansion for the renormalization-group beta function which smoothly extrapolates from the strong-coupling limit to the asymptotic-freedom value. The theory is described by three qualitatively distinct regions over this range.

The introduction of gauge-invariant lattice theories in Euclidean[1] and Hamiltonian[2] form has made it possible to study non-Abelian gauge theories directly in the phase in which quarks are confined thus achieving one of the major roles which quantum chromodynamics (QCD) is expected to play in strong-interaction physics. The other role is at short distances where the theory's asymptotic freedom[3] predicts almost free-particle behavior for quarks, with great phenomenological success.[4] A major obstacle to date has been the inability of either lattice or continuum versions to bridge the enormous qualitative gap in quark dynamics between the two. There is, however, one way in which the two kinds of quark behavior seem naturally related, and that is from the point of view of the renormalization-group beta function. Thus it is this quantity that we study in this Letter.

The lattice gauge theory is defined in terms of the partition function[1]

$$Z(\beta=1/g^2) = \int\Big(\prod_l dg\Big)\prod_p \exp\{\beta \operatorname{tr}[U(p)+\text{H.c.}]\}, \quad (1)$$

where l runs over all links and p over all plaquettes of a regular hypercubical lattice, dg is the normalized Haar measure over SU(3) for each link of the lattice, and $U(p)$ is the unitary $[\underline{3}]$ representation of the product of the group elements on the boundary of the plaquette p. If the transfer matrix for this partition function is constructed in timelike axial gauge and the limit that the timelike lattice spacing vanishes is taken,[5] one obtains a Hamiltonian corresponding[2] to (1)

$$H=(g^2/2a)\Big\{\sum_l \vec{E}_l^{\,2} - x\sum_p \operatorname{tr}[U(p)+\text{H.c.}]\Big\}, \qquad x=2/g^4, \quad (2)$$

where $\vec{E}_l^{\,2}$ is the quadratic Casimir operator C_2, l runs over all links and p all plaquettes of a regular cubical lattice. Strong-coupling expansions, i.e., expansions in inverse powers of g, may be obtained from (1) and (2). For the Euclidean theory, one may use the Fourier decomposition over characters,

$$\exp\{\beta[\chi_3(g)+\chi_{3^*}(g)]\} = I(\beta)\sum_\nu \omega_\nu(\beta)\chi_\nu(g), \quad (3)$$

where ν runs over the representations of SU(3) and

$$I(\beta)=\int dg \exp[\beta(\chi_3+\chi_{3^*})]$$
$$\omega_\nu(\beta)=\int dg\, \chi_{\nu^*}(g)\exp[\beta(\chi_3+\chi_{3^*})]/I(\beta), \quad (4)$$

and group theory results on the integration of products of characters together with conventional graphical methods for high-temperature series expansions to obtain expansions for quantities of interest. For the Hamiltonian theory, one uses standard Rayleigh-Schrödinger perturbation theory in a Fock space of irreducible representations of SU(3) on each link diagonalizing $\vec{E}_l^{\,2}$. The action of products of U's on the ground state may be computed with use of SU(3) Clebsch-Gordan coefficients to decompose them. Matrix elements of gauge-invariant operators may be expressed in terms of group invariants e.g., dimensionalities, and 3-j and 6-j symbols, etc.[6]

In order to renormalize the strong-coupling expansion, we must select some dimensionful property of the theory to be held fixed. The coupling constant then becomes a function of this quantity and the cutoff, which in this case is the lattice spacing. There are no more free parameters in the theory. The quantity which we hold fixed is the coefficient of the linear term in the potential between two widely separated quarks. This is equivalent to the coefficient of the area in the expectation value of the Wilson loop, or the energy per unit length of an infinitely long string in the Hamiltonian theory. The matrix elements are to be computed in the gauge sector of the theory without the effects of quark loops which can screen the flux. There are several important advantages to this choice of normalization condition: (1) it is a natural and straightforward quantity to compute in the strong coupling expansion, (2) this condition assures that the theory always remains in the confined phase, and (3) this parameter has direct phenomenological consequences on Regge behavior and "quarkonium" spectrum. Having fixed the surface tension (or string tension, depending on circumstances), the manner in which the coupling constant varies with the cutoff is described in the usual way by the beta function:

$$\beta(g)/g = -d\ln g/d\ln a. \tag{5}$$

We report here a calculation of the surface tension in both Euclidean and Hamiltonian SU(3) lattice gauge theories to order $1/g^{20}$ in the strong-coupling expansion. The Hamiltonian result is[7]

$$\mu=(g^2/2a^2)[4/3-(11/153)x^2-(61/1632)x^3-0.012\,711\,501\,8x^4-0.003\,067\,187\,52x^5-\ldots],\quad x=2/g^4. \tag{6}$$

The Euclidean result is

$$\mu=(-1/a^2)[\ln\omega+4\omega^4+12\omega^5-10\omega^6-36\omega^7+391\omega^8/2+1131\omega^9/10+2\,550\,837\omega^{10}/512+\ldots], \tag{7}$$

where $\omega=\omega_3/3$ is the natural expansion parameter for the Euclidean theory. Differentiating (6) and (7) with respect to a gives, respectively, the results for $\beta(g)/g$ for Hamiltonian theory,

$$-\beta(g)/g=1-(11/51)x^2-(183/1088)x^3-0.041\,378\,583\,3x^4+0.034\,436\,440\,6x^5+\ldots, \tag{8}$$

and, for Euclidean theory,

$$-\beta(g)/g=(2d\ln g/d\ln\omega)[\ln\omega(1-16\omega^4-60\omega^5+60\omega^6+252\omega^7+121\omega^8/2+9021\omega^9/10-1\,690\,677\omega^{10}/512+\ldots)+4\omega^4+12\omega^5-10\omega^6-36\omega^7+263\omega^8/2-1989\omega^9/10+298\,037\omega^{10}/5120+\ldots]. \tag{9}$$

In Fig. 1, we present a plot of Padé approximates to the Hamiltonian beta function. The Euclidean beta function differs in details but has the same qualitative features. The asymptotic behavior for large g differs reflecting the different way in which they treat the cutoff of short distances. For $g \gtrsim 1.5$ the effects of nonleading terms in the strong-coupling expansion are small, giving a 10% or smaller correction to the beta function. For g between 1.5 and about 0.8, the higher-order strong-coupling effects become more important, driving the beta function down to the region of the weak-coupling value. Near $g=0.9$, there is clear evidence that the Hamiltonian strong-coupling beta function is "trying to match" onto the weak-coupling function, with a sharp break from the decreasing behavior in the region 1.5 to 0.9. Below this region, where the two functions overlap, the extrapolations of the strong-coupling expansion become unreliable. At these small values of the coupling constant the perturbative expansion to the beta function is valid, and in fact the higher-order corrections to it are quite small.

Experience with previous applications of the strong-coupling expansion to the calculation of beta functions[8] leads us to expect that when further terms of the strong-coupling expansion become available the accuracy with which the extrapolated strong-coupling beta function agrees with the weak-coupling result will improve. Nevertheless, the current degree of agreement is remarkable.

These results have several implications for physics. For the first time, there is clear evidence that strong-coupling methods may be capable of reproducing the asymptotic freedom which is necessary to describe the short-distance properties of hadrons. By combining the weak- and strong-coupling results which have an overlap region where they are both valid, one obtains

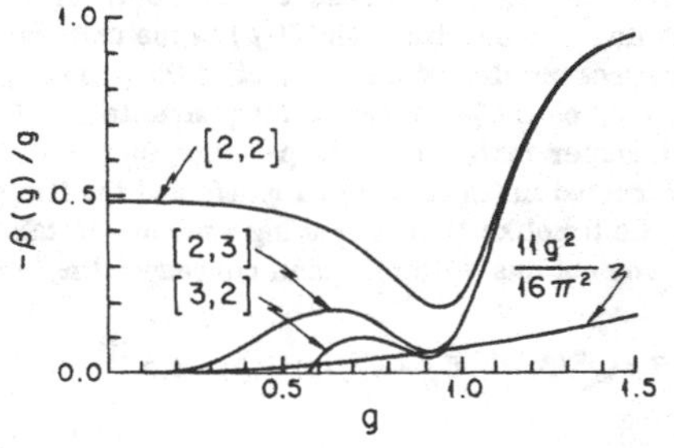

FIG. 1. Padé approximates to the Hamiltonian beta function.

for the first time a quantitative description of the theory over the entire range of coupling constants. In fact, we may use the weak-coupling result for the beta function to extrapolate the string tension to zero lattice spacing, which gives, after renormalization, a value for $g(a)$ for all a. If we choose a physical value for the string tension, we obtain a zero-parameter approximation to $\alpha = g^2/4\pi$ since $g(a)$ is an approximation to the renormalized coupling constant at a length scale of a.

There is a clear separation of the physics into three regions. As previously mentioned, for $g \lesssim 0.8$ weak coupling prevails; for $0.8 \lesssim g \lesssim 1.5$, there is a transition region between weak and strong coupling; and for $g \gtrsim 1.5$, one is in a strong-coupling regime. From the point of the strong-coupling expansion, the onset of the transition is caused by the thawing of fluctuations in the string, allowing the flux to spread out lowering the energy. From the weak-coupling end, the onset of the transition cannot be perturbative in origin since the higher-order corrections to the beta function[9] are far too small to account for it. However, estimates of the effects of instantons on the weak-coupling beta function made by Callan, Dashen, and Gross[10] give a correction which has all the desired qualitative features. Thus it appears that the origin of the transition from the weak-coupling point of view is the effect of vacuum tunneling, i.e., gauge fluctuations which begin to trap the flux. Just the opposite of the effect which causes the transition from the strong-coupling end. So we see that the main characteristic of the transition region is the smooth transition from flux confined to strings to unconfined flux over a narrow range of coupling constant.

A lattice theory is of course only an approximation at any finite lattice spacing and in order to recover a complete physical description one must be able to take the limit that the lattice spacing vanishes. During most of this process the coupling will be quite small and essentially one may use weak-coupling methods. However, one begins in the strong-coupling regime and must be able to safely make the transition to the weak-coupling regime. The importance of the present result is that it demonstrates that this is possible with only a few orders in the strong-coupling expansion. Several further areas of exploration are suggested. A combination of strong-coupling Padé methods with the renormalization group at weak coupling as suggested by Wilson[11] may prove to be very powerful. It is also of interest to study the strong-coupling expansion for a modified Hamiltonian with the leading irrelevant operators removed by adding six link terms. And, of course, efforts are under way to extend the strong-coupling series to higher orders. We feel that it should be possible to eventually compute corrections to order $1/g^{36}$ using present methods. (The calculation of the g^{-24} coefficient is nearing completion.)

We wish to extend heartfelt thanks to those people who have aided this work in one way or another. Special thanks go to Roger Dashen for explaining the role of instantons in the weak-coupling beta function, to D. Sinclair who helped in the early stages of this calculation, and to Rich Fredrickson whose extensive study of the representations of SU(3) make this calculation possible. We also wish to acknowledge the aid of MACSYMA, which is supported in part by the Department of Energy. This work was supported in part by Department of Energy Contract No. EY76-02-2220, and in part by the National Science Foundation under Contract No. PHY-77-25279.

[a] Permanent address: Department of Physics, University of Illinois, Urbana, Ill. 61801.

[1] K. G. Wilson, Phys. Rev. D 10, 2445 (1974); R. Balian, J. M. Drouffe, and C. Itzykson, Phys. Rev. D 11, 395 (1975).

[2] J. B. Kogut and L. Susskind, Phys. Rev. D 11, 395 (1975).

[3] G. 't Hooft, in *Proceedings of the Conference on Gauge Theories, Marseilles, 1972* (unpublished); H. D. Politzer, Phys. Rev. Lett. 30, 1346 (1973); D. J. Gross and F. Wilczek, Phys. Rev. Lett. 30, 1343 (1973).

[4] J. Ellis, SLAC Report No. SLAC-PUB-2121, 1978 (to be published).

[5] M. Creutz, Phys. Rev. D 15, 1128 (1977).

[6] R. Frederickson, to be published. A table of 6-j symbols involving up to 42-dimensional representations was used here.

[7] The exact coefficients are

$$-\frac{737\,327\,120\,374\,220\,449}{58\,004\,722\,308\,819\,686\,400}$$

$$-\frac{98\,631\,094\,843\,173\,218\,126\,309}{32\,156\,851\,302\,637\,820\,478\,720\,000}.$$

[8] C. J. Hamer, J. B. Kogut, and L. Susskind, Phys. Rev. Lett. 41, 1337 (1978).

[9] W. Caswell, Phys. Rev. Lett. 33, 244 (1974); D. R. T. Jones, Nucl. Phys. B75, 531 (1974).

[10] C. G. Callan, Jr., R. Dashen, and D. J. Gross, to be published.

[11] K. G. Wilson, private communication.

PHYSICAL REVIEW D VOLUME 21, NUMBER 8 15 APRIL 1980

Monte Carlo study of quantized SU(2) gauge theory

Michael Creutz
Department of Physics, Brookhaven National Laboratory, Upton, New York 11973
(Received 24 October 1979)

Using Monte Carlo techniques, we evaluate path integrals for pure SU(2) gauge fields. Wilson's regularization procedure on a lattice of up to 10^4 sites controls ultraviolet divergences. Our renormalization prescription, based on confinement, is to hold fixed the string tension, the coefficient of the asymptotic linear potential between sources in the fundamental representation of the gauge group. Upon reducing the cutoff, we observe a logarithmic decrease of the bare coupling constant in a manner consistent with the perturbative renormalization-group prediction. This supports the coexistence of confinement and asymptotic freedom for quantized non-Abelian gauge fields.

I. INTRODUCTION

Gauge theories currently dominate our understanding of elementary particle physics. Indeed, we now conceive that all interactions represent ramifications of underlying local symmetries. The elegant inclusion of the strong nuclear force into this picture demands the phenomenon of confinement; indeed, physical hadrons should be gauge-singlet bound states of the fundamental quark and gluon degrees of freedom. At our present level of knowledge, confinement appears to play a role solely for the unbroken non-Abelian gauge theory of strong interactions.

Theoretical evidence for quark confinement by gauge fields is remarkably sparse. Renormalization-group arguments imply that perturbation theory may be inapplicable at large distances,[1] thus dismissing the lack of perturbative evidence for confinement of quarks.[2] Studies of large orders in the weak-coupling expansion[3] as well as semiclassical treatments[4,5] all suggest important nonperturbative effects in non-Abelian gauge theories.

Any true nonperturbative analysis requires a means of controlling the ultraviolet divergences of field theory in a manner independent of Feynman diagrams. Wilson's formulation of gauge theory on a lattice provides such a cutoff scheme.[6] This particular regulator also preserves an exact local symmetry. With the cutoff in place, Wilson derived a strong-coupling expansion in terms of quarks connected by strings. In this picture confinement arises naturally; however, to take the continuum limit one must leave the strong-coupling domain and the expansion could fail to converge. Balian, Drouffe, and Itzykson have presented arguments that in a sufficient number of space-time dimensions, the lattice theory will exhibit a phase transition between the strong-coupling region of confinement and the weak-coupling perturbative regime.[7] Such a transition is essential for the lattice formulation of conventional electrodynamics where photons and electrons exist as free particles.

Renormalization-group analysis implies that for short-distance phenomena the effective coupling of non-Abelian gauge theories becomes small and perturbative results become valid.[2] If this "asymptotic freedom" is to arise in the confining phase of Wilson's formulation, then four space-time dimensions must be inadequate to support the transition of Ref. 7. As evidence for this, Migdal has presented an approximate nonperturbative recursion relation between different values of the cutoff parameter.[8] He finds a close analogy between d-dimensional gauge theories and $(d/2)$-dimensional nearest-neighbor spin systems of statistical mechanics. On this basis he concludes that four dimensions represents a critical case where gauge theories based on non-Abelian groups only possess the confining phase, whereas the Abelian group U(1) of electrodynamics possesses a peculiar transition similar to that occurring in the two-dimensional "XY" model.[9]

Recently, Monte Carlo techniques have proven to be a powerful nonperturbative tool for analysis of quantized gauge fields.[10,11] We have seen clear confinement–spin-wave phase transitions for U(1) lattice gauge theory in four space-time dimensions and for the SU(2) theory in five dimensions.[10] In contrast, this transition appears to be absent for the four-dimensional SU(2) model. These results support the Migdal arguments on the existence of phase transitions; however, the observation of first-order transitions in Z_2, Z_3, and Z_4 lattice gauge theories rather than the predicted second-order critical points shows that Migdal's approximate recursion relations may misidentify the nature of the transition.[11]

In this paper we extend our analysis of the four-dimensional SU(2) theory. Working on lattices of up to 10^4 sites, we emphasize the continuum limit of the theory. Renormalizing such that the string tension, the asymptotic linear part of the quark-antiquark potential, remains fixed,[12] we obtain the cutoff dependence of the bare coupling constant. Passing out of the strong-coupling regime at large lattice spacing, we observe the onset of a logarithmic decrease of the bare change as the continuum limit is approached. This decrease is at a rate consistent with the prediction of asymptotic freedom. As the renormalization scheme is based on confinement, this provides strong evidence that quantized SU(2) Yang-Mills fields simultaneously exhibit confinement and asymptotic freedom. This unifies the nonperturbative lattice formulation and the more conventional perturbative treatments of continuum non-Abelian gauge theory.

As discussed in Ref. 10, our Monte Carlo algorithm consists of successively touching a heat bath to each gauge variable in the system. After setting up the theory, we describe in Sec. II how special features of the group SU(2) allow the efficient execution of this procedure. Section III contains a review of the asymptotic-freedom prediction for the cutoff dependence of the bare charge. Section IV presents our numerical results connecting the bare coupling, the string tension, and the cutoff. In Sec. V we draw some further conclusions from the analysis.

II. THE ALGORITHM

The model is Wilson's lattice gauge theory with gauge group SU(2).[6] A link variable U_{ij}, which is an element of SU(2), is associated with each nearest-neighbor pair of sites i and j on a four-dimensional simple hypercubic lattice. The reversed link is associated with the inverse element

$$U_{ij}=(U_{ji})^{-1}. \tag{2.1}$$

To eliminate surface effects we impose periodic boundary conditions. Regarding the connection with Minkowski-space formulation as well established, we work entirely in Euclidian space.[13,14] The path integral

$$Z=\int\Big(\prod_{(i,j)} dU_{ij}\Big)e^{-\beta S(U)} \tag{2.2}$$

defines the quantum theory. Here the integral includes all independent link variables and uses the invariant group measure. The action S is a sum over all elementary squares or plaquettes □ in the lattice,

$$S(U)=\sum_{\Box} S_{\Box}, \tag{2.3}$$

where

$$S_{\Box}=1-\tfrac{1}{2}\operatorname{Tr}(U_{ij}U_{jk}U_{kl}U_{li}). \tag{2.4}$$

Here i, j, k, and l label the sites circulating about the square □. Our normalization is such that any plaquette contributes a number between zero and two to the action. In the next section we review the classical continuum limit of this theory and show that it reduces to the conventional Yang-Mills system with coupling e_0 given by

$$e_0^2=\frac{4}{\beta}. \tag{2.5}$$

Equation (2.2) represents the partition function of a statistical system at temperature $T=1/\beta$. We treat the system by obtaining an ensemble of configurations which simulates an ensemble in equilibrium at this temperature. Starting in some initial configuration, we successively touch a heat bath to each link variable. Each U_{ij} in turn is replaced with a new element U'_{ij} chosen randomly from the entire group with probability density proportional to the Boltzmann factor,

$$dP(U')\sim\exp[-\beta S(U')]dU', \tag{2.6}$$

where $S(U')$ is the action evaluated with the given link having value U'_{ij} and all other links fixed at their previous values. In what follows, one iteration refers to one application of this procedure to every link in the lattice.

This algorithm satisfies the detailed-balance requirements which ensure that any ensemble will eventually be brought to equilibrium.[15] Of all possible Monte Carlo algorithms which vary only a single spin at a time, the above method leads to equilibrium in the least number of iterations. This follows because repeated application of any valid algorithm to a single spin eventually simulates the heat bath.

The action of lattice gauge theory carries an exact local gauge symmetry. The action is unchanged by the replacement

$$U_{ij}\to g_iU_{ij}g_j^{-1}, \tag{2.7}$$

where the g_i represent arbitrary group elements associated with each site i. This symmetry allows gauge fixing; thus, in an axial gauge all U_{ij} on links parallel to some particular axis would be set to the identity and held fixed in the path integral.[14] This procedure does not affect gauge-invariant correlation functions. Gauge fixing is an essential first step in the conventional perturbative expansion about the classical ground state.[16] Nonetheless, with Wilson's cutoff the theory is well defined without going into a particular gauge. Furthermore, the process of gauge fixing introduces long-range interactions which are not well simulated by a local Monte Carlo algorithm. Although the final results are the same, we have found that a fixed gauge increases the convergence time for a simulation. Thus, the numerical results of this paper include an integral over gauges.

We now discuss in detail the algorithm for generation of group elements with weighting given in Eq. (2.6). We parametrize SU(2) elements in the form

$$U=a_0 I+i\vec{a}\cdot\vec{\sigma}, \quad (2.8)$$

where a_μ is a real four-vector of unit length

$$a_0^2+\vec{a}^2=1. \quad (2.9)$$

The 2×2 identity matrix is denoted I and the three Pauli matrices satisfy

$$\sigma^\alpha\sigma^\beta=i\epsilon^{\alpha\beta\gamma}\sigma^\gamma+\delta^{\alpha\beta}I, \quad \mathrm{Tr}(\sigma^\alpha)=0. \quad (2.10)$$

Here, $\epsilon^{\alpha\beta\gamma}$ is totally antisymmetric with $\epsilon^{123}=1$ and $\sigma^{\alpha\beta}$ is the Kronecker function

$$\delta^{\alpha\beta}=\begin{cases}1, & \alpha=\beta\\ 0, & \alpha\neq\beta.\end{cases}$$

Our group elements are stored in the computer memory as the four numbers a_μ for each link. In this notation the invariant group measure takes the simple form

$$dU=\frac{1}{2\pi^2}\,\delta(a^2-1)d^4a. \quad (2.11)$$

While working on a particular link $\{i,j\}$, we need consider only the contribution to the action coming from the six plaquettes containing that link. If we denote by $\tilde{U}_\alpha$, $\alpha=1,\dots,6$, the six products of three link variables which interact with the link in question, then Eq. (2.6) assumes the form

$$dP(U)\sim dg\exp\left[\tfrac{1}{2}\beta\,\mathrm{Tr}\left(U\sum_{\alpha=1}^{6}\tilde{U}_\alpha\right)\right]. \quad (2.12)$$

A useful property of elements of the group SU(2) is that any sum of them is proportional to another SU(2) element. In particular, it follows from representation (2.8) that

$$\sum_{\alpha=1}^{6}\tilde{U}_\alpha=k\bar{U}, \quad (2.13)$$

where k is given by the determinant

$$k=\left|\sum_{\alpha=1}^{6}\tilde{U}\right|^{1/2}, \quad (2.14)$$

and $\bar{U}$ is an element of SU(2). The utility of this observation appears when we use the invariance of the group to measure to write

$$dP(U\bar{U}^{-1})\sim dU\exp(\tfrac{1}{2}\beta k\,\mathrm{Tr}U) = \frac{1}{2\pi^2}\,\delta(a^2-1)d^4a\exp(\beta k a_0). \quad (2.15)$$

The problem reduces to generating points randomly on the surface of the unit sphere in four dimensions with exponential weighting along the a_0 direction. Generating an element U in this manner, we replace the link variable on the lattice with the product

$$U'_{ij}=U\bar{U}^{-1}. \quad (2.16)$$

To generate the appropriately weighted points on the four-dimensional sphere, first note that the integration over $|\vec{a}|$ can be done using the δ function

$$\delta(a^2-1)d^4a\exp(\beta k a_0)=\tfrac{1}{2}da_0 d\Omega(1-a_0^2)^{1/2}\times\exp(\beta k a_0), \quad (2.17)$$

where $d\Omega$ is the differential solid angle of $\vec{a}$ and $\vec{a}$ has length $(1-a_0^2)^{1/2}$. Thus we need to generate a_0 stochastically in the interval $[-1, +1]$ with probability

$$P(a_0)\sim(1-a_0^2)^{1/2}\exp(\beta k a_0), \quad (2.18)$$

and the direction of $\vec{a}$ is chosen totally randomly. Our algorithm for the a_0 selection begins with a trial a_0,

$$a_0=1+\frac{1}{\beta k}\ln(x), \quad (2.19)$$

where x is a random number uniformly distributed in the region

$$e^{-2\beta k}<x<1. \quad (2.20)$$

This generates a_0 distributed with exponential weight $e^{\beta k a_0}$. To correct for the factor $(1-a_0^2)^{1/2}$ in Eq. (2.18), reject this a_0 with probability $1-(1-a_0^2)^{1/2}$ and select a new trial a_0. Repeat this until an a_0 is accepted. We leave it to the interested reader to design his own scheme for randomly selecting the direction for $\vec{a}$. This completes the algorithm.

Once the lattice is in equilibrium, one can measure any desired correlation function. One simple quantity we shall use extensively is the average action per plaquette:

$$P=\langle S_\Box\rangle. \quad (2.21)$$

This quantity is equivalent to the "internal energy" of the equivalent statistical-mechanical system. As β runs from zero to infinity, P falls from one to zero. In addition to P we will study Wilson loops, which we define in the next section.

For initial conditions, we either select all elements randomly from the group or we set them all to the identity. Agreement of simulations from these independent starting configurations is a test of convergence. Other "mixed" starting configurations as described in Refs. 8–10 are of no particular value for this system which appears to lack any phase transition at finite β.

III. THE CONTINUUM LIMIT AND THE RENORMALIZATION GROUP

We begin this section with a review of the classical continuum limit of the lattice theory.[6,7] We do this in order to establish the relation of β to the charge of conventional perturbation theory, a low-temperature expansion about this limit. We then present the renormalization-group prediction for the cutoff dependence of the bare coupling for the quantized theory. Finally we define our renormalization prescription for the continuum limit of the quantum theory.

The classical SU(2) Yang-Mills theory follows from the Lagrangian density[17]

$$\mathcal{L} = \tfrac{1}{4} F^{\alpha}_{\mu\nu} F^{\alpha}_{\mu\nu}, \tag{3.1}$$

where the internal-symmetry index α runs from one to three, and $F^{\alpha}_{\mu\nu}$ is defined in terms of potentials

$$F^{\alpha}_{\mu\nu} = \partial_\mu A^{\alpha}_{\nu} - \partial_\nu A^{\alpha}_{\mu} - e_0 \epsilon^{\alpha\beta\gamma} A^{\beta}_{\mu} A^{\gamma}_{\nu}. \tag{3.2}$$

Using the Pauli matrices of Eq. (2.10) gives a convenient matrix formulation

$$A_\mu = \tfrac{1}{2} \sigma^\alpha A^\alpha_\mu, \tag{3.3}$$

$$A^\alpha_\mu = \mathrm{Tr}(\sigma^\alpha A_\mu), \tag{3.4}$$

$$F_{\mu\nu} = \tfrac{1}{2} \sigma^\alpha F^\alpha_{\mu\nu} = \partial_\mu A_\nu - \partial_\nu A_\mu + i e_0 [A_\mu, A_\nu], \tag{3.5}$$

$$\mathcal{L} = \tfrac{1}{2} \mathrm{Tr}(F_{\mu\nu} F_{\mu\nu}). \tag{3.6}$$

To connect the lattice theory with this Lagrangian, identify

$$U_{ij} = \exp\left[-i e_0 (x^j - x^i)_\mu A_\mu \left(\frac{x^i + x^j}{2} \right) \right], \tag{3.7}$$

where x^i_μ are the coordinates of the site i. Considering a plaquette in the (μ, ν) plane and Taylor expanding A_μ about the center of this plaquette, we find with a little suppressed algebra

$$\begin{aligned} S_\Box &= 1 - \tfrac{1}{2} \mathrm{Tr}(U_{ij} U_{jk} U_{kl} U_{li}) \\ &= \tfrac{1}{8} e_0^2 a^4 \mathrm{Tr}(F_{\mu\nu} F_{\mu\nu} + F_{\nu\mu} F_{\nu\mu}) + O(a^5), \end{aligned} \tag{3.8}$$

with no implied sum over μ and ν. Here, a is the lattice spacing. Combining the a^4 with the sum over plaquettes gives a four-dimensional integral over space-time; consequently, we conclude that

$$\tfrac{1}{4} \beta e_0^2 = 1 \tag{3.9}$$

reproduces the classical theory when the lattice spacing goes to zero. This justifies Eq. (2.5).

We now turn to the renormalization group. In a conventional perturbation treatment of the quantized theory one defines a renormalized charge e_R in terms of some physical correlation function at a scale of mass μ. The precise definition is a matter of convention; we merely assume it is made such that when an ultraviolet cutoff of length scale a is in effect

$$e_R(e_0, \mu, a) = e_0 + O(e_0^3). \tag{3.10}$$

The continuum limit of the quantum theory follows by taking a to zero while adjusting e_0 to keep the physically defined e_R fixed. Of course, in a theory with more parameters such as bare masses, additional physical quantities need fixing. The Gell-Mann–Low function[18] is defined as

$$\gamma(e_R) = \mu \frac{\partial}{\partial \mu} e_R(e_0, \mu, a). \tag{3.11}$$

If the continuum limit of the theory is physically sensible $\gamma(e_R)$ should remain finite as a is taken to zero. For SU(2) the perturbative expansion of $\gamma(e_R)$ begins[1]

$$\gamma(e_R) = -\frac{11 e_R^3}{24\pi^2} + O(e_R^5). \tag{3.12}$$

Through terms of order e_R^5, the function $\gamma(e_R)$ is independent of the details of the definition of e_R.

Remarkably, if a perturbative analysis is ever valid for SU(2) gauge theory, then Eq. (3.12) tells us how e_0 must vary in the continuum limit. Because e_R is being held fixed, we have

$$0 = a \frac{d}{da} e_R(e_0, \mu, a) = \frac{\partial e_R}{\partial e_0} a \frac{d e_0}{da} + a \frac{\partial e_R}{\partial a}. \tag{3.13}$$

Simple dimensional analysis gives

$$a \frac{\partial e_R}{\partial a} = \mu \frac{\partial e_R}{\partial \mu} = \gamma(e_R). \tag{3.14}$$

Combining Eqs. (3.10)–(3.14), we obtain

$$a \frac{d e_0}{da} = \frac{11 e_0^3}{24\pi^2} + O(e_0^5). \tag{3.15}$$

We now assume that e_0 does not get hung up on a fixed point away from the origin. If e_0 eventually is small enough that the $O(e_0^5)$ term in Eq. (3.15) is negligible, then we can integrate to obtain

$$e_0^2 = \frac{12\pi^2}{11 \ln(\hat{a}/a)} + O((\ln a)^{-2}). \tag{3.16}$$

This analysis does not determine the integration constant $\hat{a}$. For later comparison with our Monte Carlo results, we rewrite Eq. (3.16) in the form

$$a^2 \underset{e_0 \to 0}{\sim} \hat{a}^2 \exp\left(-\frac{24\pi^2}{11 e_0^2} \right) = \hat{a}^2 \exp\left(-\frac{6\pi^2 \beta}{11} \right). \tag{3.17}$$

This is the prediction of asymptotic freedom.

In the lattice formulation a natural physical quantity to use for renormalization purposes is the string tension.[12] This quantity can be extracted from a study of Wilson loops. For a closed contour C comprised of links in the lattice, the Wilson loop is defined

$$W(C)=\left\langle \tfrac{1}{2}\,\mathrm{Tr}\left(\prod_C U\right)_{\mathrm{PO}}\right\rangle . \tag{3.18}$$

Here, PO represents "path ordering"; that is, the U_{ij} are ordered and oriented as they are encountered in circulating around the contour. The simplest Wilson loop arises when C is a single plaquette, in which case

$$W(\square)=1-\langle S_\square\rangle . \tag{3.19}$$

If for large separations the interaction energy of two static sources in the fundamental representation of the gauge group increases linearly with distance, then one expects for large contours[6]

$$\ln W(C)=-KA(C)+O(p(C)), \tag{3.20}$$

where $A(C)$ is the minimum area enclosed by C and $p(C)$ is the contour perimeter. The string tension K is the coefficient of the linear part of the static quark-antiquark potential. Measuring A in physical units, it equals the lattice spacing squared times $N_\square(C)$, the minimum number of plaquettes forming a surface bounded by C. Thus we write

$$\ln W(C)=-(a^2K)N_\square(C)+O(p(C)). \tag{3.21}$$

By measuring the exponential falloff of $W(C)$ with $N_\square$, we effectively measure the dimensionless quantity a^2K as a function of the bare coupling e_0.

As the string tension is a physical quantity, we can use it to define the continuum limit of the theory. Thus we consider the renormalization scheme of holding K fixed by adjusting e_0 while the lattice spacing a is taken to zero. As K is a nonperturbative quantity, this scheme does not directly use an analog of Eq. (3.10). However, the asymptotic-freedom result in Eq. (3.17) is independent of renormalization method, and thus we expect to find for small e_0, i.e., large β,

$$a^2K\sim\exp\left(\frac{-6\pi^2\beta}{11}\right). \tag{3.22}$$

Indeed, the left-hand side of Eq. (3.15) represents a definition of a Gell-Mann–Low function for this renormalization prescription.

Although the classical theory contains only dimensionless parameters, the quantum theory requires introduction of a dimensional quantity such as K. This is an example of the phenomenon of dimensional transmutation as discussed by Coleman and Weinberg.[19] Note also the essential singularity at vanishing coupling in Eq. (3.22). Thus the string tension is a strictly nonperturbative quantity.

In the small-β region Wilson's strong-coupling expansion is valid. The first nonvanishing term in this expansion is

$$W(C)\sim(\tfrac{1}{4}\beta)^{N_\square} . \tag{3.23}$$

Consequently, we expect at high temperatures

$$a^2K\sim-\ln(\tfrac{1}{4}\beta) . \tag{3.24}$$

The main goal of the calculations of the next section is to join the behaviors in Eqs. (3.22) and (3.24).

IV. RESULTS

In Fig. 1, we illustrate the convergence of the Monte Carlo procedure. Working at $\beta=2.3$, we show the average plaquette P, defined in Eq. (2.21), as a function of number of iterations for a total of 30 iterations.[20] We show runs begun both randomly as well as ordered and on lattices of dimensions $4\times4\times4\times4$, $6\times6\times6\times6$, $8\times8\times8\times8$, and $10\times10\times10\times10$. Note that the convergence rate is essentially independent of lattice size; only the fluctuations grow on the smaller lattices. This supports the absence of a phase transition in this region.

In Fig. 2 we study the evolution from the zero-temperature ordered state on an 8^4 lattice for several values of β. Note that the convergence time is not strongly β dependent; a slight increase appears in the range $\beta\approx2$–2.4. At all β, equilibrium is essentially complete after 20 iterations. Convergence becomes extremely good in the high- and low-temperature regions; consequently, the method is not tied to either strong or weak coupling.

In Fig. 3 we show the expectation values of

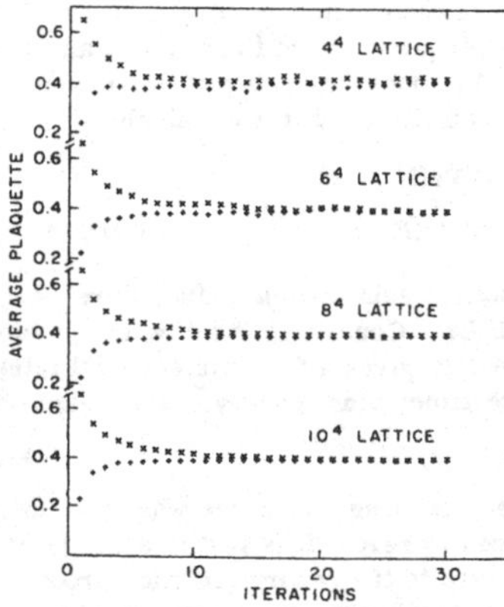

FIG. 1. The average plaquette as a function of number of iterations at $\beta=2.3$.

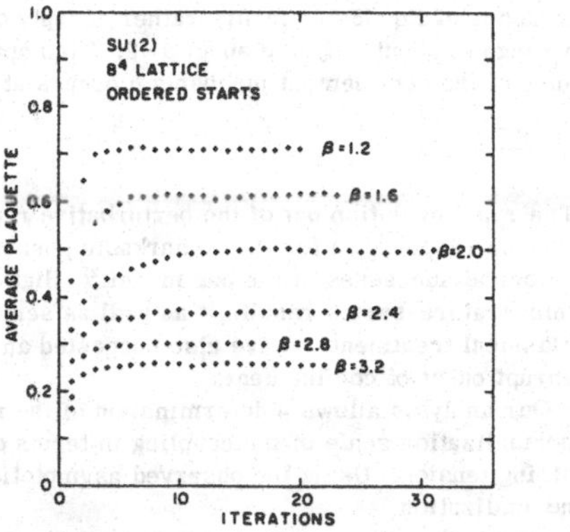

FIG. 2. The evolution of the average plaquette at several values of β.

square Wilson loops at $\beta=3$ as a function of lattice size. These loops are taken to lie in a fundamental plane of the lattice and are up to six links on a side. Each measurement is an average over all similar loops in the lattice and the error bars represent the standard deviation for the fluctuations over five iterations after attaining equilibrium. As intuitively expected, larger loops show the finite-size effects most strongly. On a 10^4 lattice, loops of up to five sites on a side appear to have stabilized. This represents the largest loop used in the subsequent analysis.

In Fig. 4 we summarize the values for these square Wilson loops as a function of β. In the important region from $\beta=2.1$ to $\beta=3$, these numbers came from a 10^4 lattice whereas elsewhere we used an 8^4 lattice. In Fig. 4 we also plot the

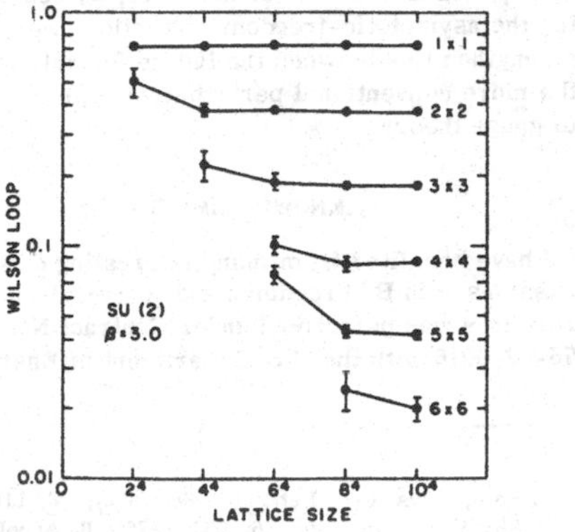

FIG. 3. Wilson loops at $\beta=3$ as a function of lattice size.

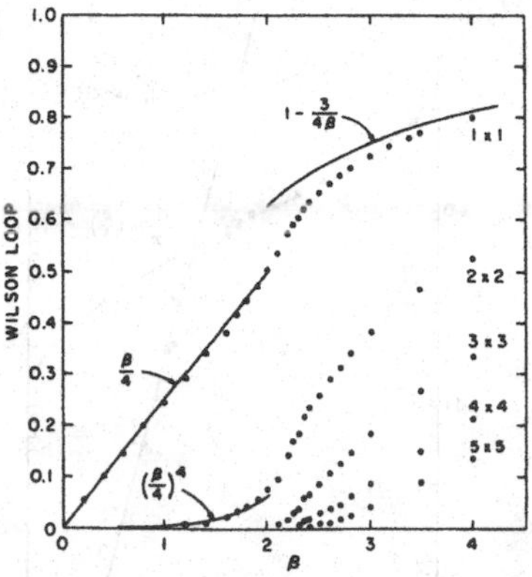

FIG. 4. Wilson loops as a function of β.

strong-coupling result

$$W(\square) \underset{\beta\to 0}{\sim} \tfrac{1}{4}\beta,$$
$$W(2\times 2) \underset{\beta\to 0}{\sim} (\tfrac{1}{4}\beta)^4, \tag{4.1}$$

and the weak-coupling limit[10]

$$W(\square) \underset{\beta\to\infty}{\sim} 1-\frac{3}{4\beta}. \tag{4.2}$$

All loops approach unity inversely with β at low temperatures.

To extract a string tension at a given value of β, we fit these loops to the form

$$W(S)=\exp[-(A+BS+CS^2)], \tag{4.3}$$

where S is the loop side. We adjusted the parameters A, B, and C to minimize the mean square deviation of this fit from the "measured" loops. In Fig. 5 we show some representative fits. Be-

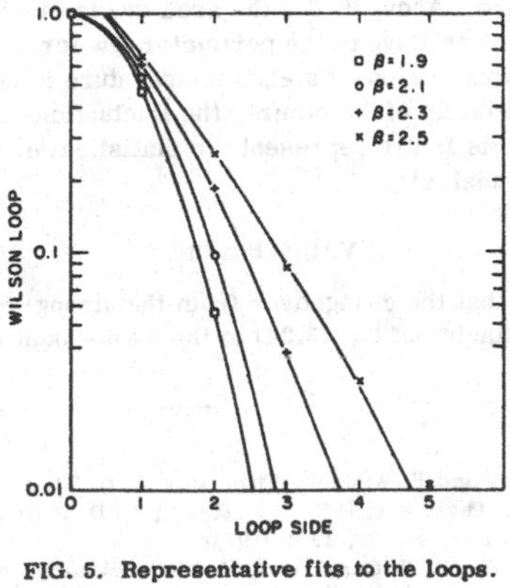

FIG. 5. Representative fits to the loops.

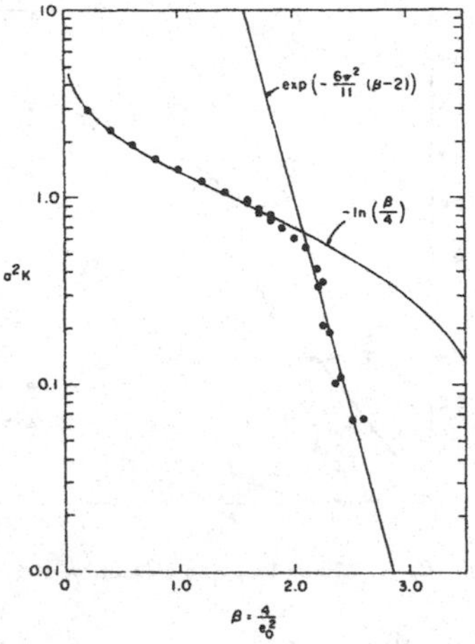

FIG. 6. The cutoff squared times the string tension as a function of β. The solid lines are the strong- and weak-coupling limits.

low $\beta = 2.1$ only loops of side 1 and 2 are significantly different from zero so we must include the loop of side 0 in the fit. Below $\beta = 1.6$ only the loop of side 1 is significant and we assume the area term C dominates. From Eq. (3.21) we identify

$$C = a^2K. \tag{4.4}$$

In Fig. 6 we summarize these results by plotting a^2K versus β. Here we also plot the strong-coupling result of Eq. (3.24) and the weak-coupling conclusion of Eq. (3.22) with an arbitrarily chosen normalization. From $\beta = 1.6$ to 1.8 we plot both the least-square fit and the result of assuming pure area-law behavior. For $\beta = 2.2$ and 2.25 we plot fits including and not including the loop of side zero. Above $\beta = 2.5$ the area law is too subdominant relative to the perimeter law for accurate determination. As each temperature is treated independently of the others, the fluctuations apparent in this figure represent the statistical error of this analysis.

V. DISCUSSION

Note that the changeover from the strong-coupling behavior of Eq. (3.24) to the weak-coupling behavior of Eq. (3.22) occurs rather sharply over a range of about 10% in β about $\beta = 2$. This appearance of the confinement mechanism occurs at

$$\frac{e_0^2}{4\pi} \approx 0.16\,. \tag{5.1}$$

The rapid evolution out of the perturbative regime may be responsible for the remarkable phenomenological successes of the bag model.[21] High-temperature-series results,[12] as well as semiclassical treatments,[5] have also suggested an abrupt onset of confinement.

Our analysis allows a determination of the renormalization scale of the coupling in terms of the string tension. Using the observed asymptotic normalization

$$a^2K \underset{\beta\to\infty}{\sim} \exp\left(-\frac{6\pi^2}{11}(\beta-2)\right), \tag{5.2}$$

we can solve for e_0^2 to give

$$\frac{e_0^2}{4\pi} \underset{a\to 0}{\sim} \frac{3\pi}{11\ln(1/a\Lambda)}, \tag{5.3}$$

where the renormalization scale is

$$\Lambda \approx \sqrt{K}\exp\left(-\frac{6\pi^2}{11}\right) \approx \frac{1}{200}\sqrt{K}\,. \tag{5.4}$$

Thus we see the appearance of a rather large dimensionless number. The uncertainty in this coefficient is roughly a factor of two because of the large coefficient in the exponential. The renormalization mass should be strongly dependent on both the gauge group and addition of quarks.

We have shown the onset of asymptotic freedom for the bare coupling constant in a renormalization scheme based on confinement. This is strongly suggestive that SU(2) non-Abelian gauge theory simultaneously exhibits confinement and asymptotic freedom. Furthermore, by reproducing the asymptotic-freedom prediction, we strengthen ties between the lattice formulation and the more conventional perturbative approaches to gauge theory.

ACKNOWLEDGMENTS

I have benefited from many interesting discussions with B. Freedman and R. Swendsen. This research was performed under Contract No. EY-76-02-0016 with the U.S. Department of Energy.

[1]D. Gross and F. Wilczek, Phys. Rev. Lett. 30, 1346 (1973); Phys. Rev. D 8, 3633 (1973); H. D. Politzer, Phys. Rev. Lett. 30, 1343 (1973).
[2]T. Appelquist, J. Carazzone, H. Kluberg-Stern, and M. Roth, Phys. Rev. Lett. 36, 768 (1976); 36, 1161(E) (1976); Y. P. Yao, *ibid.* 36, 653 (1976); T. Appelquist, M. Dine, and I. Muzinich, Phys. Lett. 69B, 231 (1977).
[3]L. N. Lipatov, Zh. Eksp. Teor. Fiz. Pis'ma Red. 25,

116 (1977) [JETP Lett. 25, 104 (1977)]; E. Brezin, J. C. Le Guillou, and J. Zinn-Justin, Phys. Rev. D 15, 1544 (1977); 15, 1588 (1977).

[4]A. M. Polyakov, Phys. Lett. 59B, 82 (1975); G. 't Hooft, Phys. Rev. D 14, 3432 (1976).

[5]C. G. Callan, R. F. Dashen, and D. J. Gross, Phys. Rev. D 10, 1000 (1070).

[6]K. Wilson, Phys. Rev. D 10, 2445 (1974).

[7]R. Balian, J. M. Drouffe, and C. Itzykson, Phys. Rev. D 10, 3376 (1974); 11, 2098 (1975); 11, 2104 (1975).

[8]A. A. Migdal, Zh. Eksp. Teor. Fiz. 69, 810 (1975); 69, 1457 (1975) [Sov. Phys.—JETP 42, 413 (1975); 42, 743 (1975)]; L. P. Kadanoff, Rev. Mod. Phys. 49, 267 (1977).

[9]J. M. Kosterlitz and D. J. Thouless, J. Phys. C 6, 1181 (1973).

[10]M. Creutz, Phys. Rev. Lett. 43, 553 (1979).

[11]M. Creutz, L. Jacobs, and C. Rebbi, Phys. Rev. Lett. 42, 1390 (1979); Phys. Rev. D 20, 1915 (1979).

[12]J. B. Kogut, R. B. Pearson, and J. Shigemitsu, Phys. Rev. Lett. 43, 484 (1979).

[13]K. Osterwalder and R. Schrader, Commun. Math. Phys. 42, 281 (1975).

[14]M. Creutz, Phys. Rev. D 15, 1128 (1977).

[15]K. Binder, in *Phase Transitions and Critical Phenomena*, edited by C. Domb and M. S. Green (Academic, New York, 1976), Vol. 5B.

[16]S. Coleman, in *Laws of Hadronic Matter*, edited by A. Zichichi (Academic, New York, 1975).

[17]C. N. Yang and R. Mills, Phys. Rev. 96, 191 (1954).

[18]M. Gell-Mann and F. E. Low, Phys. Rev. 95, 1300 (1954).

[19]S. Coleman and E. Weinberg, Phys. Rev. D 7, 1888 (1973).

[20]The plotted values represent an average obtained during the respective iteration. After touching each link with the heat bath, we insert the values of all six plaquettes containing that link into this average. Thus every plaquette enters four times, once for each link it contains.

[21]A. Chodos, R. L. Jaffe, K. Johnson, C. B. Thorn, and V. F. Weisskopf, Phys. Rev. D 9, 3471 (1974); T. Degrand, R. L. Jaffe, K. Johnson, and J. Kiskis, *ibid.* 12, 2060 (1975).

VOLUME 45, NUMBER 5 PHYSICAL REVIEW LETTERS 4 AUGUST 1980

Asymptotic-Freedom Scales

Michael Creutz
Physics Department, Brookhaven National Laboratory, Upton, New York 11974
(Received 12 May 1980)

Using Monte Carlo methods with Wilson's lattice cutoff, the asymptotic-freedom scales of SU(2) and SU(3) gauge theories without quarks are calculated.

PACS numbers: 11.10.Np, 11.10.Jj, 11.30.Jw

The standard SU(3) Yang-Mills theory of the strong interaction is asymptotically free.[1] The effective coupling constant decreases logarithmically at short distances. This ultraviolet behavior permits perturbative analysis of high-momentum-transfer processes. In this paper, I present a calculation of the parameters setting the scale of this phenomenon in pure SU(2) and SU(3) gauge theories.

In a non-Abelian gauge theory with an ultraviolet cutoff, the bare charge g_0 goes to zero with the logarithm of the cutoff parameter[2]

$$g_0^2 = \frac{1}{\gamma_0 \ln(1/\Lambda_0^2 a^2) + (\gamma_1/\gamma_0)\ln[\ln(1/\Lambda_0^2 a^2)] + O(g_0^2)}. \quad (1)$$

Here γ_0 and γ_1 are the first two coefficients in the perturbative expansion of the Gell-Mann–Low function[3]

$$\gamma(g_0) = a\,dg_0/da = \gamma_0 g_0^3 + \gamma_1 g_0^5 + O(g_0^7). \quad (2)$$

The cutoff length a is the lattice spacing in a lattice formulation. The parameter Λ_0 is the asymptotic-freedom scale associated with the renormalization scheme being used. For SU(N) gauge groups the coefficients in Eq. (2) are[2,4]

$$\gamma_0 = \tfrac{11}{3}(N/16\pi^2), \tag{3}$$

$$\gamma_1 = \tfrac{34}{3}(N/16\pi^2)^2. \tag{4}$$

These first two coefficients are independent of renormalization prescription.

Equation (1) defines the scale Λ_0 and can be rewritten

$$\Lambda_0 = \lim_{a\to 0}\frac{1}{a}[(\gamma_0 g_0^2(a))]^{(-\gamma_1/2\gamma_0^2)}\exp\left(\frac{-1}{2\gamma_0 g_0^2(a)}\right). \tag{5}$$

My Monte Carlo results for these scales in the pure gauge theories (no quarks) are

$$\Lambda_0 = (1.3\pm 0.2)\times 10^{-2}\sqrt{K}, \quad \mathrm{SU}(2); \tag{6}$$

$$\Lambda_0 = (5.0\pm 1.5)\times 10^{-3}\sqrt{K}, \quad \mathrm{SU}(3). \tag{7}$$

Here I have used Wilson's lattice regulator[5] and K is the string tension, the coefficient of the linear potential between widely separated sources in the fundamental representation of the gauge group. I will return later to the method of calculating these numbers.

At first sight these small numbers are rather surprising, coming as they do from a theory with no small dimensionless parameters. However the value of Λ_0 is not independent of renormalization scheme.[6] Since it is defined in the weak-coupling limit, perturbative calculations to one-loop order can relate different definitions. Hasenfratz and Hasenfratz have recently done a lengthy analysis relating this Λ_0 to the more conventional scale Λ^{MOM} defined by the three-point vertex in Feynman gauge and at a given scale in momentum space. Their results are

$$\Lambda^{\mathrm{MOM}} = 57.5\Lambda_0, \quad \mathrm{SU}(2); \tag{8}$$

$$\Lambda^{\mathrm{MOM}} = 83.5\Lambda_0, \quad \mathrm{SU}(3). \tag{9}$$

These large factors partially cancel the small numbers in Eqs. (6) and (7); combining them gives

$$\Lambda^{\mathrm{MOM}} = (0.75\pm 0.12)\sqrt{K}, \quad \mathrm{SU}(2); \tag{10}$$

$$\Lambda^{\mathrm{MOM}} = (0.42\pm 0.13)\sqrt{K}, \quad \mathrm{SU}(3). \tag{11}$$

If we accept the string model[8] connection between K and the Regge slope α'

$$K = 1/2\pi\alpha' \tag{12}$$

and use $\alpha' = 1.0\ (\mathrm{GeV})^{-2}$, then we conclude for SU(3)

$$\Lambda^{\mathrm{MOM}} = 170\pm 50\ \mathrm{MeV}. \tag{13}$$

Some caution may be necessary in the phenomenological interpretation of this number because I have not included effects of virtual quark loops.

I now turn to the method of calculation. For SU(2) gauge theory I worked with a lattice of 10^4 sites except at strong coupling where 8^4 sufficed. I used the heat bath Monte Carlo algorithm of Creutz.[9] For SU(3) I used a Metropolis[10] scheme similar in spirit to that employed by Wilson.[11] This is inherently less efficient than the SU(2) procedure; so most SU(3) running was on a 4^4 lattice. One value of coupling, $g_0^2 = 0.902$, was studied on a 6^4-site system. As the SU(3) lattices are rather small, the conclusions depend heavily on an assumed similarity with the SU(2) case.

After bringing the lattice into a typical equilibrium state at some value of the bare coupling constant, I measured the expectation values of rectangular Wilson loops $W(I,J)$ where I and J are the dimensions of the loop in fundamental lattice units. From these loops, I construct the quantities

$$\chi(I,J) = -\ln\left(\frac{W(I,J)W(I-1,J-1)}{W(I,J-1)W(I-1,J)}\right). \tag{14}$$

In this combination overall constant factors and perimeter behaviors of the loops will cancel out. For $I \gg J \gg 1$, $\chi(I,J)$ is proportional to the force between a quark and an antiquark separated by distance Ja. The motivation for introducing χ is that in a region where the loops are dominated by an area law

$$W(I,J) \sim e^{-KA}, \tag{15}$$

where $A = a^2IJ$ is the loop area, it directly measures the string tension K

$$\chi \sim a^2K. \tag{16}$$

This happens both when I and J are large and when the bare coupling is large. However in the weak-coupling limit with I and J held fixed, χ should have a perturbative expansion

$$\chi(I,J) = a_1 g_0^2 + O(g_0^4). \tag{17}$$

For example, a simple calculation gives

$$\chi(1,1) \underset{g_0^2\to 0}{\sim} \begin{cases} \tfrac{3}{16} g_0^2, & \mathrm{SU}(2) \\ \tfrac{1}{3} g_0^2, & \mathrm{SU}(3). \end{cases} \tag{18}$$

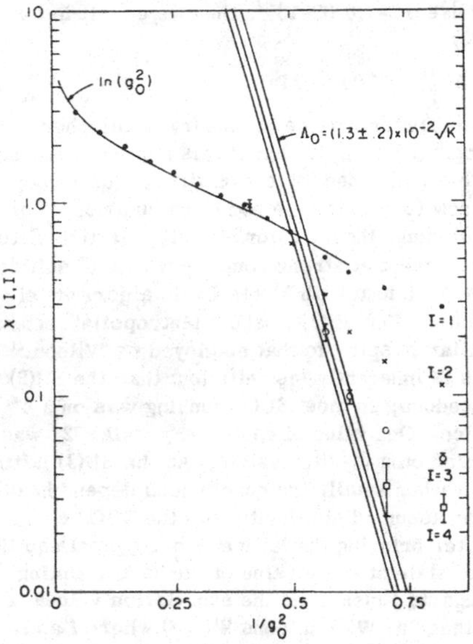

FIG. 1. The quantities $\chi(I, I)$ for SU(2) gauge theory as a function of $1/g_0^2$.

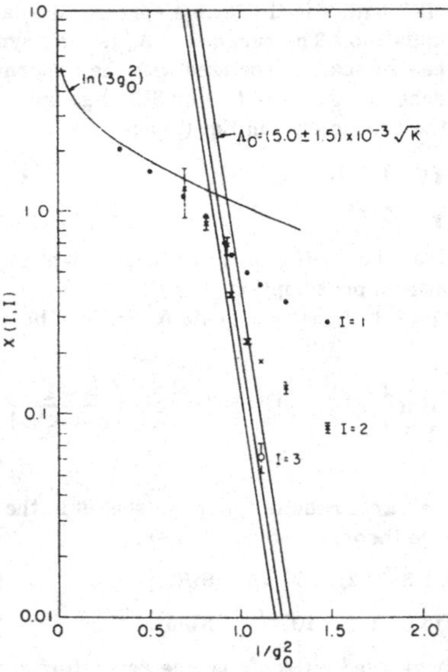

FIG. 2. The quantities $\chi(I, I)$ for SU(3) gauge theory.

This power behavior is radically different than the essential singularity expected for the right-hand side of Eq. (16)

$$a^2K \underset{g_0^2\to 0}{\sim} \frac{K}{\Lambda_0^2}(\gamma_0 g_0^2)^{(-\gamma_1/\gamma_0^2)}\exp\left(-\frac{1}{\gamma_0 g_0^2}\right). \quad (19)$$

This is a consequence of Eq. (5) coupled with the renormalization prescription of holding the physical string tension K fixed as the cutoff is removed.[9,12] In summary, for strong coupling, i.e., $g_0^2>1$, we expect all $\chi(I,J)$ to be equal to the coefficient of the area law, as in Eq. (16), but as g_0^2 is reduced, small I and J should give a χ deviating from a^2K. Thus the envelope of curves of $\chi(I,J)$ for all I and J plotted versus the coupling should be the true value of a^2K as a function of g_0^2. Use of this envelope avoids ambiguities in the fitting procedure used in Ref. 9.

In Fig. 1, I plot the value of $\chi(I,I)$ for $I=1-4$ versus $1/g_0^2$ for the gauge group SU(2). The error bars on the points represent the standard deviation of the mean taken from an ensemble of five configurations. At stronger couplings, the large loops have large relative errors but are consistent with χ approaching the values from smaller loops. On this graph, I show the strong-coupling limit for all $\chi(I,J)$,

$$\chi(I,J)=\ln(g_0^2)+O(1/g_0^4). \quad (20)$$

The weak-coupling limit for a^2K as given in Eq. (19) appears in the graph as a band corresponding to the value of Λ_0 in Eq. (6). The error in Λ_0 is a purely subjective estimate.

Figure 2 is essentially the same as Fig. 1 except the gauge group is now the physically interesting SU(3). As most of the points are from a 4^4 lattice, only $\chi(1,1)$ and $\chi(2,2)$ are plotted. At $g_0^2=0.902$ the run on a 6^4 lattice gave one point for $\chi(3,3)$. The strong-coupling behavior is now

$$\chi(I,J)=\ln(3g_0^2)+O(1/g_0^2). \quad (21)$$

Note that the corrections for SU(3) begin in a lower order of the strong-coupling expansion than for the SU(2) case in Eq. (20). These small lattices in and of themselves do not allow any precise conclusions; however, assuming a similarity with the SU(2) picture, I have plotted the band from Eq. (19) giving the Λ_0 in Eq. (7). Note that the strong-coupling deviation from the asymptotic-freedom behavior sets in rather abruptly at g_0^2

VOLUME 45, NUMBER 5 PHYSICAL REVIEW LETTERS 4 AUGUST 1980

~1. This is in excellent agreement with the series results of Ref. 12.

This research has been performed under Contract No. DE-AC02-76CH00016 with the U. S. Department of Energy.

[1]C. N. Yang and R. Mills, Phys. Rev. 96, 191 (1954); D. Gross and F. Wilczek, Phys. Rev. Lett. 30, 1346 (1973), and Phys. Rev. D 8, 3633 (1973); H. D. Politzer, Phys. Rev. Lett. 30, 1343 (1973).

[2]W. E. Caswell, Phys. Rev. Lett. 33, 244 (1974).

[3]M. Gell-Mann and F. E. Low, Phys. Rev. 95, 1300 (1954).

[4]D. R. T. Jones, Nucl. Phys. B75, 531 (1974).

[5]K. G. Wilson, Phys. Rev. D 10, 2445 (1974).

[6]W. Celmaster and R. J. Gonsalves, Phys. Rev. D 20, 1420 (1979).

[7]A. Hansenfratz and P. Hansenfratz, CERN Report No. TH. 2727-CERN, 1980 (unpublished).

[8]P. Goddard, J. Goldstone, C. Rebbi, and C. B. Thorn, Phys. Rev. D 20, 2096 (1979).

[9]M. Creutz, Phys. Rev. D 21, 2308 (1980).

[10]N. Metropolis, A. W. Rosenbluth, M. N. Rosenbluth, A. H. Teller, and E. Teller, J. Chem. Phys. 21, 1087 (1953).

[11]K. G. Wilson, to be published.

[12]J. Kogut, R. B. Pearson, and J. Shigemitsu, Phys. Rev. Lett. 43, 484 (1979).

Nuclear Physics B190 [FS3] (1981) 349–356

STRING TENSION IN SU(3) LATTICE GAUGE THEORY

E. PIETARINEN[1]

Research Institute for Theoretical Physics, University of Helsinki, Finland

Received 3 February 1981

Wilson loop expectation values have been determined in SU(3) latttice gauge theory without fermions using Monte Carlo methods and considering lattices of up to 10^4 sites. A heat bath technique has been developed in order to enhance the statistical independence of successive lattice configurations.

1. Introduction

Monte Carlo methods provide a convenient tool for studying lattice gauge theories numerically. Exact gauge symmetry is preserved in the lattice formulation [1, 2], while certain other symmetry properties of the continuum theory are supposed to be recovered in the limit when the lattice spacing goes to zero, which also corresponds to the vanishing of the bare coupling constant. For finite lattices one hopes to observe the behaviour implied by the renormalization prescription, where some physical quantity is kept fixed, when the lattice spacing goes to zero. Such an observation cannot, of course, be regarded as any proof of the existence and properties of the continuum limit, since the lattice size is severely restricted in a numerical calculation. The simplest hypothesis, supported by numerical evidence for the absence of phase transitions [3, 4], is to join the numerical results with the known renormalization group behaviour. In this way Creutz determined the string tension (coefficient of static quark–antiquark linear potential) for SU(2) [3] and recently for SU(3) [4] gauge groups in the euclidean formulation of a lattice gauge theory without fermions. In the present work the string tension is determined by an independent technique and using larger lattices and higher statistics than in ref. [4]. For this purpose a heat bath method was developed for the SU(3) gauge group as described in sect. 2. Numerical results are presented and discussed in sect. 3.

2. The heat bath method for the SU(3) gauge group

Let us consider a hypercubic 4-dimensional lattice of size d^4 with periodic boundary conditions. At each link b there is an element $U(\mathrm{b}) \in \mathrm{SU}(3)$. The problem

[1] Supported by the Academy of Finland.

is to calculate averages using the partition function

$$Z = \int \prod_{b} dU(b) \exp\left\{\frac{2}{g^2} \mathrm{Re} \sum_{p} \mathrm{Tr}\, U(p)\right\}, \tag{1}$$

where, for each plaquette p, $U(p)$ is the ordered product of the four link matrices surrounding the plaquette and dU is the Haar measure for SU(3).

The heat bath method (for SU(2) see ref. [3]) consists of selecting each link in turn in a definite order and replacing the group element by a new one chosen randomly according to the distribution (1), where other link variables are kept fixed at their momentary values. This procedure is repeated to generate a large number of lattice configurations, over which the required quantities are averaged.

In practice we therefore have to generate $U \in SU(3)$ according to the distribution

$$I = \int dU \exp\{\mathrm{Re}\,\mathrm{Tr}\,(T^+U)\}, \tag{2}$$

where the 3×3 matrix T is obtained as a sum of 6 SU(3) matrices multiplied with $2/g^2$. If T and U are represented in terms of complex 3-component vectors

$$T = (\boldsymbol{A}, \boldsymbol{B}, \boldsymbol{C}), \qquad U = (\boldsymbol{X}, \boldsymbol{Y}, \boldsymbol{Z}), \tag{3}$$

the distribution can be written as

$$I = \int d^6X\delta(\boldsymbol{X}^* \cdot \boldsymbol{X} - 1) \int d^6Y\delta(\boldsymbol{Y}^* \cdot \boldsymbol{Y} - 1)\delta^2(\boldsymbol{X}^* \cdot \boldsymbol{Y})|_{\boldsymbol{Z}=(\boldsymbol{X}\times\boldsymbol{Y})^*}$$

$$\times \exp\{\mathrm{Re}\,[\boldsymbol{A}^* \cdot \boldsymbol{X} + \boldsymbol{B}^* \cdot \boldsymbol{Y} + \boldsymbol{C}^* \cdot \boldsymbol{Z}]\}, \tag{4}$$

where $\int d^6X = \int d\,\mathrm{Re}\, X_0\, d\,\mathrm{Im}\, X_0 \ldots d\,\mathrm{Im}\, X_2$ and $\delta^2(Z) = \delta(\mathrm{Re}\, Z)\delta(\mathrm{Im}\, Z)$ etc. It is easy to show that this representation is a Haar measure for SU(3). Let us write $\boldsymbol{X}$, $\boldsymbol{Y}$ in terms of orthonormal unit vectors to be specified later in detail:

$$\boldsymbol{X} = X_0\boldsymbol{a}_0 + X_1\boldsymbol{a}_1 + X_2\boldsymbol{a}_2, \qquad \boldsymbol{Y} = Y_0\boldsymbol{b}_0 + Y_1\boldsymbol{b}_1 + Y_2\boldsymbol{b}_2. \tag{5}$$

Choosing $\boldsymbol{b}_0 \equiv \boldsymbol{X}/|\boldsymbol{X}|$ we can integrate over Y_0 using the δ-function. Next let us substitute

$$\boldsymbol{b}_1 = (\boldsymbol{B} - (\boldsymbol{B} \cdot \boldsymbol{b}_0^*)\boldsymbol{b}_0 + (\boldsymbol{C} \times \boldsymbol{X})^*)/f(\boldsymbol{X}), \tag{6}$$

where $f(\boldsymbol{X})$ is a real normalization factor. Then $\boldsymbol{b}_2$ is determined from $\boldsymbol{b}_2 = (\boldsymbol{b}_0 \times \boldsymbol{b}_1)^*$. Using these definitions the distribution (4) has been transformed into

$$I = \int d^6X\delta(\boldsymbol{X}^* \cdot \boldsymbol{X} - 1)\, e^{\mathrm{Re}\,\boldsymbol{A}^* \cdot \boldsymbol{X}} \int_0^1 d\alpha_6 \int_0^1 d\alpha_7 \int_0^1 d\alpha_8 \sqrt{1 - \mathrm{Re}^2\, Y_1}\, \frac{\sinh f(\boldsymbol{X})}{f(\boldsymbol{X})}\, 4\pi, \tag{7}$$

where

$$\alpha_6 = (e^{f\,\mathrm{Re}\,Y_1} - e^{-f})/(e^f - e^{-f}), \tag{8}$$

$$\alpha_7 = \tfrac{1}{2}(\text{Im } Y_1/\sqrt{1-\text{Re}^2 Y_1}+1)\,, \tag{9}$$

$$e^{2\pi i\alpha_8} = Y_2/|Y_2|\,. \tag{10}$$

Next let us take

$$\boldsymbol{a}_0 = [\boldsymbol{A}+\lambda(\boldsymbol{B}\times\boldsymbol{C})^*]/h\,, \tag{11}$$

where (real) λ will be fixed later and h is a real normalization factor. Dropping unessential numerical factors the whole integral can now be written as

$$I = \prod_{k=1}^{8}\left(\int_0^1 d\alpha_k\right)[1-\text{Re}^2 X_0]^{3/2}[1-\text{Re}^2 Y_1]^{1/2}\frac{\sinh f(\boldsymbol{X})}{f(\boldsymbol{X})}\, e^{-\lambda \text{ Re}(\boldsymbol{B}\times\boldsymbol{C}\cdot\boldsymbol{X})}\,, \tag{12}$$

where

$$\alpha_1 = (e^{h \text{ Re} X_0} - e^{-h})/(e^h - e^{-h})\,, \tag{13}$$

$$\alpha_2 = [\text{Im } X_0(1-\text{Re}^2 X_0) - \tfrac{1}{3}\text{Im}^3 X_0 + \tfrac{2}{3}(1-\text{Re}^2 X_0)^{3/2}]/\tfrac{4}{3}(1-\text{Re}^2 X_0)^{3/2}\,, \tag{14}$$

$$\alpha_3 = |X_1|^2/(1-|X_0|^2)\,, \tag{15}$$

$$e^{2\pi i\alpha_4} = X_1/|X_1|\,, \tag{16}$$

$$e^{2\pi i\alpha_5} = X_2/|X_2|\,. \tag{17}$$

The parameter λ should be chosen such that the remaining distribution is as flat as possible. A simple choice is

$$\lambda = (\coth\xi - \xi^{-1})\xi^{-1}\,, \qquad \xi = \sqrt{|\boldsymbol{B}|^2+|\boldsymbol{C}|^2}\,. \tag{18}$$

In practice $\alpha_1, \ldots \alpha_8$ in eqs. (8)–(10) and (13)–(17) are generated uniformly in the interval (0, 1) and the corresponding U is calculated. The remaining weight factor (integrand) in eq. (12) is then compared with the product of a random number and the maximum possible weight in eq. (12). If the actual integrand is smaller, the process of generating $\alpha_1, \ldots \alpha_8$ has to be repeated. In the average only about 5 trials are needed. The whole procedure of generating SU(3) link matrices is nearly an order of magnitude slower than a corresponding calculation with the SU(2) gauge group. The main advantage of the heat bath method over the method of Metropolis et al. [5] used by Creutz [4] is that each link matrix is changed with certainty, thus correlations between successive lattice configurations are minimized.

3. Results for the string tension

Calculations were performed for several values of the coupling constant $0.9 \leq 1/g^2 \leq 1.2$ and lattice with 6^4, 8^4 and 10^4 sites. At smaller values of $1/g^2$ one may study smaller lattices while $1/g^2 \simeq 1.1-1.2$ requires the largest 8^4 and 10^4 lattices in order to extract the string tension.

Following Creutz [4], the quantity

$$\chi(I,J)=\log[W(I,J)W(I-1,J-1)/W(I,J-1)W(I-1,J)] \tag{19}$$

was calculated from the measured values of Wilson loops $W(I,J)$ with side lengths of I and J lattice spacings. This quantity gives directly the string tension provided the area law dominates:

$$\chi(I,J)=a^2K. \tag{20}$$

A contribution proportional to the perimeter of the Wilson loops is cancelled in eq. (19). The renormalization group gives the following dependence between the lattice spacing a and the coupling constant g:

$$g^2(a)=\left\{\frac{11}{16\pi^2}\log \Lambda^{-2}a^{-2}+\frac{51}{88\pi^2}\log\log \Lambda^{-2}a^{-2}+O(g^2)\right\}^{-1}. \tag{21}$$

Solving for Λ, the scale parameter of QCD, and multiplying both sides of the equation with K, one obtains [4]

$$a^2K=\frac{K}{\Lambda^2}\left(\frac{11g^2}{16\pi^2}\right)^{-102/121} e^{-16\pi^2/11g^2}. \tag{22}$$

On the left-hand side of eq. (22) we can use the numerical Monte Carlo results, while the right-hand side indicates how the result is expected to vary as a function of g^2 provided $1/g^2$ is large enough for eq. (21) to be accurate. Extracting the coefficient in front of the right-hand side of eq. (22), one obtains the square of the QCD scale parameter Λ in units of the physical string tension K, which can be (model dependently) extracted e.g. from the Regge slope α'; $K\sim 1/2\pi\alpha'$.

The expectation values are plotted in fig. 1 as functions of $1/g^2$. The expectation values grow quite rapidly in a narrow region for values of $1/g^2$ slightly below 1. The smallest 1×1 Wilson loop also gives the expectation value of the action directly.

The calculated values of the string tension are displayed in fig. 2, which is similar to the plot made by Creutz [4]. The present values correspond to lattices with 6^4, 8^4 and 10^4 sites as compared with 4^4 and 6^4 by Creutz. The agreement is quite good, and the result strengthens the interpretation that the dependence in accordance with eq. (22) is present for string tension values of an order of magnitude smaller than observed in ref. [4]. The indicated QCD scale parameter values in fig. 2 already take into account the difference between lattice and momentum space definitions [6], [7]. More important than the actual numerical value of the scale parameter is the fact that the expected renormalization group behaviour is seen to continue from $\chi(3,3)$ by Creutz to $\chi(4,4)$ and $\chi(5,5)$. Despite this apparent consistency in the results there exists the possibility that the results are misleading due to the possibility of a roughening transition [13–15], which involves fluctuations difficult to accommodate in a computer simulation.

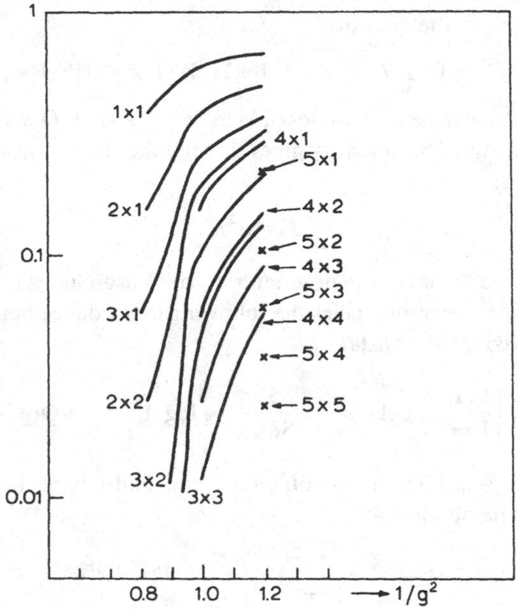

Fig. 1. Expectation values of Wilson loops of different sizes as functions of $1/g^2$. The results are normalized such that they grow from 0 at $g^2 = \infty$ to 1 at $g^2 = 0$.

Lower statistics computer calculations, not shown here in detail, indicate that the Wilson loop expectation values smoothly follow the continuation of the curves in fig. 1 as $1/g^2$ grows. Similarly $\chi(I, J)$ values bend away from the enveloping curve as observed in ref. [4]. An extension of the calculation of $\chi(I, J)$ for much larger values of I and J than in the present work is hardly possible. The calculations by Wilson [8] are designed for a direct verification of the renormalization group behaviour by a completely different method. A related result was obtained also in ref. [9] where in the case of the SU(2) gauge group another physical quantity was found to obey a similar renormalization behaviour as a function of size and the coupling constant.

An alternative method of calculating the string tension is to consider correlations between line operators defined in the following way: Consider a line through a periodic lattice of linear length d and define

$$L_{x,\mu} = \mathrm{Tr}\,[U(x, \mu)U(x+\hat{\mu}, \mu) \cdots U(x+(d-1)\hat{\mu}, \mu)] \tag{23}$$

where $U(x, \mu)$ is a link matrix at point x to direction μ and $\hat{\mu}$ is a unit vector along the μ-axis. Correlations between parallel lines with end points at x and y correspond to expectation values of rectangular Wilson loops of sides $|x-y|$ and d. Formally the calculation of $\langle L_{x,\mu}\rangle$ and $\langle L_{x,\mu}L_{y,\mu}\rangle$ is the same problem as in refs. [10, 11], where finite temperature lattice gauge theories were considered, except that the SU(2)

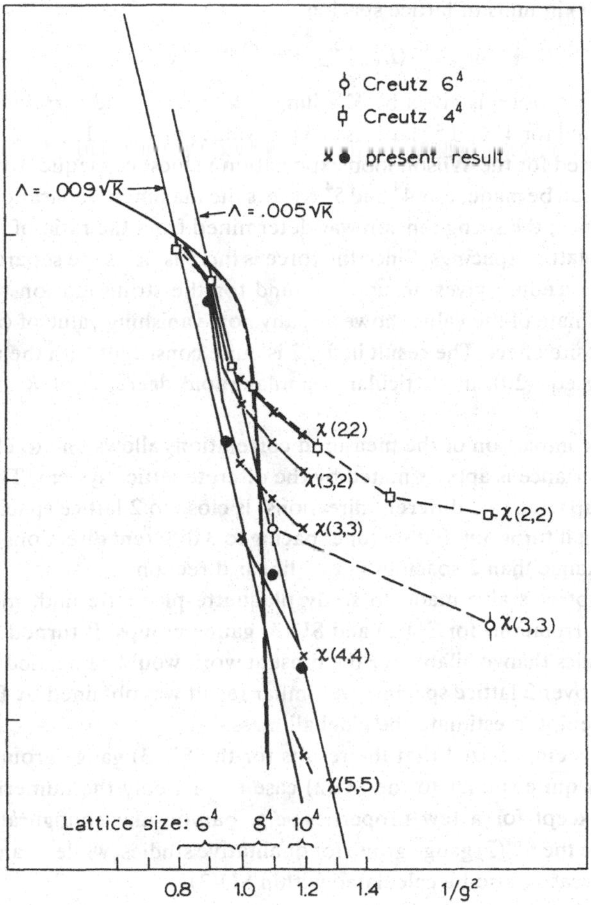

Fig. 2. Values of the string tension [$\chi(I, J)$ from eq. (19)] plotted versus $1/g^2$. Also shown are the results by Creutz [4] as well as the strong coupling expansion from ref. [15] (solid curve). The black dots show the values from the line correlations, eq. (24). The lines describe the renormalization group behaviour, eq. (22) for $\Lambda = 0.005\sqrt{K}$ ($\Lambda_{mom} \simeq 170$ MeV) and $\Lambda = 0.009\sqrt{K}$ ($\Lambda_{mom} \simeq 300$ MeV).

gauge group was used. In refs. [10, 11] the expectation value $\langle L_{x,\mu} \rangle$ was used as an order parameter; in particular, a non-zero value of $\langle L_{x,\mu} \rangle$ signifies a finite self-energy for a quark source suggesting Coulomb-type forces and a possibility of quark liberation at finite temperature. Choosing a lattice of at least 5^4 sites, we have found that $\langle L_{x,\mu} \rangle$ vanishes to the required precision and therefore the results below correspond to zero temperature apart from the technical limitation of using finite (small) lattices. From the line correlations we can extract the static quark-antiquark potential $V(r)$ in units of lattice spacing

$$\langle L_{x,\mu} L^*_{y,\mu} \rangle \propto e^{-dV(|x-y|)}, \tag{24}$$

and the string tension is given by $K = \lim_{r\to\infty} V(r)/a^2 r$. The correlations $\langle L_{x,\mu} L^*_{y,\mu}\rangle$ were calculated for 4^4 and 5^4 lattices at a few values of $1/g^2$. The statistics was much lower than used for the Wilson loop expectation values; consequently, only a rough comparison can be made. For 4^4 and 5^4 lattices the maximum separation is two lattice spacings; hence, the string tension was determined from the ratio of correlations at one and two lattice spacings. Since the force is increasing as the separation becomes small, this procedure gives an upper bound for the string tension rather than an accurate estimate of the value: however, any non-vanishing value of $\langle L_{x,\mu}\rangle$ probably has the opposite effect. The result in fig. 2 is quite consistent with the more accurate results using eq. (20), in particular a similar rapid decrease of K is detected for $1/g^2 > 1$.

A simple comparison of the measured correlations allows one to check how well rotation invariance is approximated by the discrete lattice system. The distance $\sqrt{2}$ (one lattice spacing to 2 different directions) is close to 2 lattice spacings in a single direction, but it turns out that $\sqrt{3}$ (one spacing to 3 different directions) appears to be a longer distance than 2 spacings in one linear direction.

An attempt was also made to study plaquette-plaquette and, more generally, loop-loop correlations for SU(2) and SU(3) gauge groups. It turned out that much higher statistics than available for the present work would be needed for separation distances of over 2 lattice spacings. A similar result was obtained by Berg [12], who made an attempt to estimate the glueball mass.

Finally it is emphasized that the results for the SU(3) gauge group appear to be qualitatively quite similar to the SU(2) case so far; only the numerical values are different. Except for a few properties, e.g. baryon-like configurations, one can therefore use the SU(2) gauge group for qualitative studies, while quantitative results can, in most cases, also be calculated within SU(3).

I am grateful to Prof. H. Joos for hospitality and support at DESY, where part of this work was done. I am also grateful for discussions with M. Creutz, G. Mack, G. Münster and P. Weisz.

References

[1] K. Wilson, Phys. Rev. D10 (1974) 2445
[2] J. Kogut, L. Susskind, Phys. Rev. D11 (1975) 395
[3] M. Creutz, Phys. Rev. D21 (1980) 2308
[4] M. Creutz, Phys. Rev. Lett. 45 (1980) 313
[5] N. Metropolis et al., J. Chem. Phys. 21 (1953) 1087
[6] A. Hasenfratz and P. Hasenfratz, Phys. Lett. 93B (1980) 165
[7] R. Dashen and D. Gross, Princeton Univ. preprint (1980)

[8] K. Wilson, Cornell Univ. preprint CLNS/80/442 (1980)
[9] G. Mack and E. Pietarinen, Phys. Lett. 48B (1980) 397
[10] J. Kuti, J. Polónyi and K. Szlachányi, Phys. Lett. 98B (1981) 199
[11] L.D. McLerran and B. Svetitsky, Phys. Lett. 98B (1981) 195
[12] B. Berg, DESY preprint 80/82 (1980)
[13] A. Hasenfratz, E. Hasenfratz and P. Hasenfratz, Nucl. Phys. B180 [FS2] (1981) 353
[14] M. Lüscher, G. Münster and P. Weisz, Nucl. Phys. B180 [FS2] (1981) 1
[15] J.M. Drouffe and J.B. Zuber, Nucl. Phys. B180 [FS2] (1981) 264

Volume 96B, number 1,2 PHYSICS LETTERS 20 October 1980

ESTIMATE OF THE RELATION BETWEEN SCALE PARAMETERS AND THE STRING TENSION BY STRONG COUPLING METHODS

Gernot MÜNSTER and Peter WEISZ
II. Institut für Theoretische Physik der Universität Hamburg, Hamburg, Fed. Rep. Germany

Received 24 June 1980

We estimate the relation between the scale parameter Λ_L and the string tension α in pure SU(3) gauge theory using strong coupling expansions for euclidean lattice gauge theory up to 12th order. The result is $\Lambda_L = 3.7 \times 10^{-3}\sqrt{\alpha}$. For the more conventional scale parameter Λ^{MOM} this gives $\Lambda^{MOM} = 0.31\sqrt{\alpha}$ by use of the proportionality factor of Hasenfratz. Results for the string tension of Z_3 lattice gauge theory are also discussed.

It is generally believed that in non-abelian pure gauge theories far separated static quarks are confined by a linear potential $V(r) = \alpha \cdot r$. The string tension α is not computable by perturbation theory. Due to dimensional transmutation [1] there exists no free dimensionless parameter in asymptotically free gauge theory. The only parameter is a scale parameter Λ which has the dimension of a mass. Every other physical quantity such as α is proportional to some power of Λ with a numerical factor which is fixed by the theory and can in principle be calculated by non-perturbative methods.

The natural framework for such problems is lattice gauge theory [2,3]. It provides us with a non-perturbative cut-off which respects local gauge symmetry. Strong coupling pure lattice gauge theories are known to confine static quarks [4]. The main question is whether the theory possesses a continuum limit with persisting confinement property. Due to asymptotic freedom [5] the continuum limit of SU(2) or SU(3) lattice gauge theory is supposed to be a weak coupling limit [6,7], where the lattice spacing a and the bare coupling g simultaneously go to zero. Recent investigations suggest that this picture is correct [8–11].

In this paper we use strong coupling expansions to calculate the relation between α and Λ. We consider euclidean lattice gauge theories on a hypercubical lattice. Our main interest is the case of the gauge group SU(3) in $\nu = 4$ dimensions, but we also investigate the gauge group Z_3 and dimension $\nu = 3$. The gauge field $U(b) \in SU(3)$ is attached to links b of the lattice. The ordered product of the variables $U(b)$ on the boundary of an elementary plaquette p is called $U(p)$. The action is

$$L = \tfrac{1}{3}\beta \sum_p \mathrm{Re\,Tr}\, U(p). \qquad (1)$$

The sum extends over all unoriented plaquettes of the lattice and β is related to the usual coupling constant g by

$$\beta = 6/g^2. \qquad (2)$$

We use the methods described in detail by one of us in ref. [12], where SU(2) was considered, to calculate the cluster expansion for the string tension of pure SU(3) and Z_3 [13] lattice gauge theories up to 12th order. The series for SU(3) has been calculated up to 10th order previously by Kogut et al. [8].

The natural expansion parameters are the Fourier coefficients $0 \leqslant a_r(\beta) < 1$ in the series

$$\exp(\tfrac{1}{3}\beta\ \mathrm{Re\,Tr}\, U) = N(\beta)\Big[1 + \sum_{r\neq 1} d_r a_r(\beta)\chi_r(U)\Big], \qquad (3)$$

where the sum extends over all inequivalent nontrivial irreducible representations r of SU(3) with dimension d_r, and χ_r are the corresponding primitive characters. Representations are denoted in the usual way: r = 1,

Volume 96B, number 1,2 PHYSICS LETTERS 20 October 1980

$3, \bar{3}, \ldots$. N is an irrelevant normalization factor. For Z_3 the natural expansion parameter is [13]:

$$x = [1 - \exp(-\tfrac{3}{2}\beta)]/[1 + 2\exp(-\tfrac{3}{2}\beta)]. \tag{4}$$

Our results for the string tension are given in table 1 for $\nu = 3$ and $\nu = 4$ dimensions. Expanding the series for SU(3), $\nu = 4$ in terms of $u = a_3$ the result is

$$a^2\alpha = -\ln u - 4u^4 - 12u^5 + 10u^6 + 36u^7 - \tfrac{391}{2}u^8 - \tfrac{1131}{10}u^9 - \tfrac{2550837}{5120}u^{10} + \tfrac{5218287}{2048}u^{11} - \tfrac{273755099}{61440}u^{12}. \tag{5}$$

This is in agreement with the 10th order result of Kogut et al. [8].

The relation between α and Λ in the continuum limit can be estimated in the following way. If asymptotic freedom in the form predicted by perturbation theory is true for the physical theory and confinement persists in the continuum limit, the string tension should behave like [14,15]

$$\alpha \approx Ca^{-2}(\beta_0 g^2)^{-(\beta_1/\beta_0^2)} \exp(-\beta_0^{-1} g^{-2}), \tag{6}$$

in the weak coupling limit of the lattice theory. β_0 and β_1 are the lowest coefficients in the Gell-Mann–Low function with

$$\beta_0 = 11/(16\pi^2), \quad \beta_1 = 102/(16\pi^2)^2, \quad \text{for SU(3)}, \tag{7}$$

and C is a constant. The theory can be renormalized by holding the string tension fixed, while a and g go to zero [8]. Then a renormalization scale parameter Λ_L is given by

$$\Lambda_L^2 = \alpha \cdot C^{-1}. \tag{8}$$

Monte Carlo calculations [9,11,15] indicate that the above picture is correct. They show a rather small intermediate coupling region in which a changeover from strong coupling to weak coupling asymptotic freedom behaviour takes place. High temperature expansions for the SU(2) string tension up to 12th order [16,12] also show such an abrupt breakaway. They yield values for the string tension, which are in good agreement with the Monte Carlo data at strong and intermediate coupling. If one adjusts the constant C such that the weak coupling function (6) touches the high temperature curve, the obtained value for C agrees with the result of the Monte Carlo computation.

This experience suggests applying the same procedure to the case of the gauge group SU(3). In fig. 1 we have plotted the expansion of α according to table 1 up to 10th, 11th and 12th order. Fitting the weak coupling function to the 12th order curve we get

$$\Lambda_L = 3.7 \times 10^{-3}\sqrt{\alpha}. \tag{9}$$

This is in agreement with Creutz's result for SU(3) [15]:

$$\Lambda_L = (5.0 \pm 1.5) \times 10^{-3}\sqrt{\alpha} \quad \text{(Creutz)}. \tag{10}$$

We would like to add some critical remarks concerning our procedure. Above $\beta \approx 6$ the contributions of the higher orders are relatively large. In particular the 11th order curve differs significantly from the curves representing the series up to 10th and 12th or-

Table 1

SU(3) $\quad u = a_3, v_1 = a_6, v_2 = a_8, w_1 = a_{10}, w_2 = a_{15m}$

$\nu = 3$:

$$\begin{aligned} a^2\alpha = {} & -\ln u - 2u^4 - 6u^5 + 32u^6 - 12u^4v_1 - 16u^4v_2 \\ & - 46u^8 + 21u^9 + 48u^{10} + 48u^8v_1 - 8u^8v_2 - 72u^6v_1^2 \\ & - 96u^6v_2^2 - 12v_1^5 - 16v_2^5 + 414u^{11} + 90u^9v_1 \\ & + 144u^9v_2 - 432u^7v_1v_2 - 32v_1^5v_2u^{-1} - 32v_2^5v_1u^{-1} \\ & - \tfrac{8552}{3}u^{12} + 1164u^{10}v_1 + 1552u^{10}v_2 - 30u^9w_1 \\ & - 135u^9w_2 + 64u^8v_2^2 + 384u^8v_1v_2 - 360u^6v_1^2v_2 \\ & - 224u^6v_2^3 - 48u^4v_1^4 - 64u^4v_2^4 + 144v_1^6 + 128v_2^6 \\ & - 40v_1^5w_1u^{-1} - 80v_2^5w_2u^{-1} - 60v_1^5w_2u^{-1} \end{aligned}$$

$\nu = 4$:

$$\begin{aligned} a^2\alpha = {} & -\ln u - 4u^4 - 12u^5 + 64u^6 - 24u^4v_1 - 32u^4v_2 \\ & - 128u^8 - 186u^9 + 876u^{10} - 336u^8v_1 - 592u^8v_2 \\ & - 144u^6v_1^2 - 192u^6v_2^2 - 24v_1^5 - 32v_2^5 + 4836u^{11} \\ & - 1332u^9v_1 - 1632u^9v_2 - 864u^7v_1v_2 - 64v_1^5v_2u^{-1} \\ & - 64v_2^5v_1u^{-1} - \tfrac{59656}{3}u^{12} + 10344u^{10}v_1 + 13792u^{10}v_2 \\ & - 180u^9w_1 - 810u^9w_2 - 1296u^8v_1^2 - 2176u^8v_2^2 \\ & - 2688u^8v_1v_2 - 720u^6v_1^2v_2 - 448u^6v_2^3 - 96u^4v_1^4 \\ & - 128u^4v_2^4 + 288v_1^6 + 256v_2^6 - 80v_1^5w_1u^{-1} \\ & - 160v_2^5w_2u^{-1} - 120v_1^5w_2u^{-1} \end{aligned}$$

$Z_3 \quad x = [1 - \exp(-\tfrac{3}{2}\beta)]/[1 + 2\exp(-\tfrac{3}{2}\beta)]$

$\nu = 3$:

$$a^2\alpha = -\ln x - 2x^4 - 2x^5 - 22x^8 + 7x^9 - 29x^{10} - 6x^{11} - \tfrac{428}{3}x^{12}$$

$\nu = 4$:

$$a^2\alpha = -\ln x - 4x^4 - 4x^5 - 80x^8 - 62x^9 - 130x^{10} - 20x^{11} - \tfrac{5776}{3}x^{12}$$

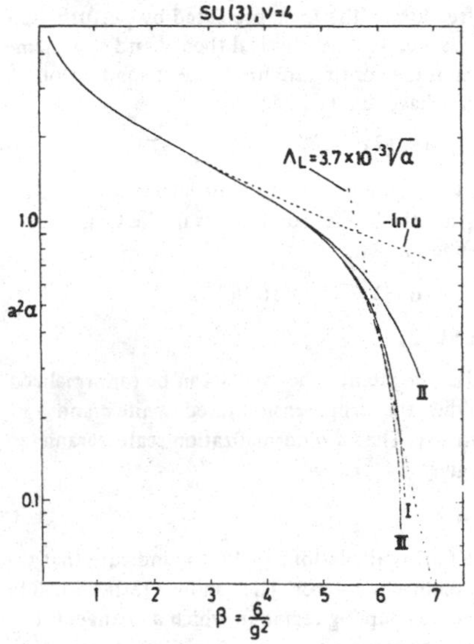

Fig. 1. The string tension α times the lattice spacing squared as a function of $\beta = 6/g^2$ for SU(3) lattice gauge theory in $\nu = 4$ dimensions. The solid lines represent results of strong coupling expansions up to 10th (I), 11th (II), and 12th (III) order. The dashed lines are the lowest order strong coupling curve and the fitted weak coupling function (6).

der and is situated even above the 8th and 9th order curves. Therefore one has to be cautious in using the series at these values of β. On the other hand the sequence of even order curves moves down uniformly and appears to converge even in the intermediate region. Remembering that for SU(2) only even orders contribute, one gets the impression that the sequence of even order curves behaves better than the odd order ones, and we propose to use the even order curves to extract numerical results. For comparison we quote the numbers extracted from some lower order curves:

$$\begin{aligned} \Lambda_L \cdot \alpha^{-1/2} &= 2.4 \times 10^{-3}, \quad \text{up to 8th order}, \\ &= 3.4 \times 10^{-3}, \quad \text{up to 10th order}, \\ &(= 1.7 \times 10^{-3}, \quad \text{up to 11th order}). \end{aligned} \tag{11}$$

One might think of an extrapolation of the strong coupling series for α, e.g. by Padé methods. But the expected weak coupling behaviour is qualitatively very different from what Padé approximants produce. Furthermore from consideration of graphs we expect that in higher orders in the expansion there are relatively more contributions with positive sign than in the low orders, while Padé approximants merely extrapolate the low order terms. Therefore we do not see any reason why Padé approximants should be closer to the truth.

The relation between Λ_L and the better known scale parameter Λ^{MOM} [17] has been investigated by Hasenfratz and Hasenfratz [14]. They find

$$\Lambda^{\text{MOM}} = 83.5\, \Lambda_L\,. \tag{12}$$

Using this relation one gets with formula (9)

$$\Lambda^{\text{MOM}} = 0.31\sqrt{\alpha}\,. \tag{13}$$

Inserting the "experimental" value $\sqrt{\alpha} \approx 450$ MeV [18] one obtains

$$\Lambda^{\text{MOM}} \approx 140 \text{ MeV}\,. \tag{14}$$

This number cannot yet be taken too seriously, because dynamical quarks have been neglected.

Finally we would like to discuss the expansion for the string tension of Z_3 lattice gauge theory in 4 dimensions. The theory is self-dual [13] and is supposed to undergo a first order phase transition at the point

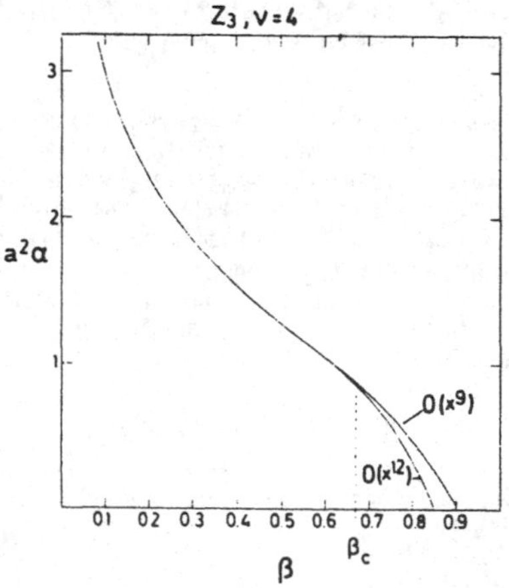

Fig. 2. The string tension times the lattice spacing squared for Z_3 lattice gauge theory in $\nu = 4$ dimensions. The lines represent results of strong coupling expansions up to 9th and 12th order. The critical coupling β_c is indicated.

$\beta_c = 0.670$ [19]. For $\beta > \beta_c$ the string tension is zero. Because of the first order nature of the transition α may have a discontinuity at β_c. In fig. 2 we plot the high temperature series for α up to 9th and 12th order. At β_c it appears to converge to some finite value around 0.83. This supports the expected first order nature of the transition.

We would like to thank G. Mack, M. Lüscher and B. Berg for discussions. One of us (P.W.) thanks the Deutsche Forschungsgemeinschaft for financial support.

Note added. After completion of this work, Drouffe sent us an unpublished paper (Stony Brook preprint ITP-SB-78-35) in which he calculates the high temperature series for the string tension of several three-dimensional models up to 16th order, including SU(3) and Z_3. Our results agree with his.

References

[1] S. Coleman and E. Weinberg, Phys. Rev. D7 (1973) 1888; D. Gross and A. Neveu, Phys. Rev. D10 (1974) 3235.
[2] K. Wilson, Phys. Rev. D10 (1974) 2445.
[3] J. Kogut and L. Susskind, Phys. Rev. D11 (1975) 395.
[4] K. Osterwalder and E. Seiler, Ann. Phys. (NY) 110 (1978) 440.
[5] G. 't Hooft (1972), unpublished; H. Politzer, Phys. Rev. Lett. 30 (1973) 1346; D. Gross and F. Wilczek, Phys. Rev. Lett. 30 (1973) 1343.
[6] J. Kogut, Rev. Mod. Phys. 51 (1979) 659.
[7] G. Mack, Properties of lattice gauge theory models at low temperatures, DESY-report 80/03 (Jan. 1980), to be published in: Recent developments in gauge theories, eds. G. 't Hooft et al. (Plenum, New York, 1980).
[8] J. Kogut, R. Pearson and J. Shigemitsu, Phys. Rev. Lett. 43 (1979) 484.
[9] M. Creutz, Solving quantized SU(2) gauge theory, Brookhaven preprint BNL-26847 (Sep. 1979).
[10] G. Mack, Predictions of a theory of quark confinement DESY 80/21 (March 1980).
[11] G. Mack and E. Pietarinen, Phys. Lett. 94B (1980) 397.
[12] G. Münster, High temperature expansions for the free energy of vortices respectively the string tension in lattice gauge theories, DESY 80/44 (June 1980).
[13] T. Yoneya, Nucl. Phys. B144 (1978) 195; C. Korthals-Altes, Nucl. Phys. B142 (1978) 315.
[14] A. Hasenfratz and P. Hasenfratz, Phys. Lett. 93B (1980) 165.
[15] M. Creutz, Asymptotic freedom scales, Brookhaven preprint (April 1980).
[16] G. Münster, Phys. Lett. 95B (1980) 59.
[17] See A. Buras, Rev. Mod. Phys. 52 (1980) 199.
[18] E. Eichten, K. Gottfried, T. Kinoshita, K.D. Lane and T.M. Yan, Phys. Rev. D21 (1980) 203; P. Goddard, J. Goldstone, C. Rebbi and C.B. Thorn, Nucl. Phys. B56 (1973) 109.
[19] S. Elitzur, R. Pearson and J. Shigemitsu, Phys. Rev. D19 (1979) 3698; A. Ukawa, P. Windey and A. Guth, Phys. Rev. D21 (1980) 1013; M. Creutz, L. Jacobs and C. Rebbi, Phys. Rev. D20 (1979) 1915.

Volume 100B, number 6

ERRATA

G. Münster and P. Weisz, Estimate of the relation between scale parameters and the string tension by strong coupling methods, *Phys. Lett.* **96**B(1980) 119.

To the string tension series for SU(3) and Z_3 lattice gauge theory $\nu = 4$ dimensions [table 1 and eq. (5) on page 120] contributions $-192u^{12}$ and $-64x^{12}$, respectively, have to be added, as has been pointed out to us by J.M. Prouffe and J.Z. Zuber. Any figures and numerical estimates remain unchanged.

Volume 98B, number 3 PHYSICS LETTERS 8 January 1981

A MONTE CARLO STUDY OF SU(2) YANG–MILLS THEORY AT FINITE TEMPERATURE ☆

Larry D. McLERRAN [1] and Benjamin SVETITSKY [2]
Stanford Linear Accelerator Center, Stanford University, Stanford, CA 94305, USA

Received 4 August 1980
Revised manuscript received 6 November 1980

We employ Monte Carlo methods to study SU(2) Yang–Mills theory in the absence of fermions, on a lattice, at finite temperature. We determine the temperature at which a second-order transition takes place between confined and unconfined phases, in terms of the string tension.

Systems of quarks and gluons at finite temperature and density have been the subject of much recent interest [1–3]. Assuming the existence of a quark-freeing transition, perturbative calculations, and non-perturbative calculations involving dilute instanton gases, can be and have been performed [1–4]. At high densities and temperatures, the QCD coupling strength is small and such calculations are self-consistent.

At lower temperatures and densities, non-perturbative confinement effects are important and perturbative calculations may lose their validity. A non-perturbative estimate of the transition temperature and the order of the transition would be useful, as would calculations of the entropy density, the energy density, and the quark–quark potential both in the confined phase and in the quark–gluon plasma. These results could be employed in the study of such plasmas as they may occur in neutron stars, heavy ion collisions, or the early universe [2,3,5,6].

In this letter, we present some results of a Monte Carlo lattice calculation [7] for a gluon plasma at finite temperature. We have treated an SU(2) gauge theory in the absence of fermions. The existence of a quark-freeing transition in this system has been demonstrated, in the strong coupling limit of the hamiltonian theory, by Susskind and by Polyakov [8]. An extrapolation to the continuum limit is however not possible without information about the intermediate coupling regime, which the Monte Carlo calculation provides.

In this note we shall display only our results for the critical transition temperature and the order of the transition. The energy density, entropy, and quark–quark potential will be discussed at length elsewhere.

The partition function of the SU(2) gauge theory at temperature $T = 1/\beta$ may be written as a euclidean path integral

$$Z = \int \mathcal{D}A^\mu \exp\left[-\int_0^\beta \mathrm{d}t \int \mathrm{d}^3x\, \mathcal{L}_\mathrm{E}\right], \tag{1}$$

where we have indicated in the exponent the integral of the euclidean action density

$$\mathcal{L}_\mathrm{E} = g^{-2} F^a_{\mu\nu} F^a_{\mu\nu}, \tag{2}$$

over a finite interval of imaginary time $0 \leqslant \tau \leqslant \beta$; we impose periodic boundary conditions

$$A_\mu(x, \tau) = A_\mu(x, \tau + \beta). \tag{3}$$

An ultraviolet cutoff may be introduced into (1) by formulating the theory on a space–time lattice of points (x, τ) with the degrees of freedom

☆ Work supported by the Department of Energy, contract DE-AC03-76SF00515.
[1] Present address: Physics Department, University of Washington, Seattle, WA 98195, USA.
[2] Present address: Institute for Theoretical Physics, University of California, Santa Barbara, CA 93106, USA.

$U_\mu(x, \tau) \equiv \exp ia\sigma^a A^a_\mu(x, \tau) \in SU(2)$,

defined on the links $\{\mu\}$ at each site (a is the lattice spacing). Then (1) is replaced by

$$Z = \int \Big[\prod_{\text{links}} dU \Big] \exp \Big[-\frac{1}{g^2} \sum_p \Big(1 - \tfrac{1}{2}\mathrm{Tr}(UUUU)_p \Big) \Big], \tag{4}$$

where we have written the lattice action as a sum over plaquettes $\{p\}$ of the trace of the product of U's around each plaquette. dU represents the invariant integration measure on SU(2) at each link. Our lattice is comprised of $N_t \times N_x^3$ sites: N_x, the number of sites along any spatial axis, is finite only as a matter of convenience, while the finiteness of N_t, the number of sites along the time axis, is supposed to introduce effects of nonzero temperature when $N_t \ll N_x$.

A convenient order parameter with which to study the phase structure of (4) is provided by an adaptation of the Wilson loop integral. Consider

$$L(x) = \mathrm{Tr} \prod_{\tau=1}^{N_t} U_0(x, \tau), \tag{5}$$

which is the trace of the product of U's along a time-oriented string running the temporal length of the lattice [8]. By virtue of the periodic boundary condition, the string is a closed loop, and $L(x)$ is therefore gauge invariant.

The expectation value of $L(x)$ in the ensemble defined by (4) yields the free energy F_q of an isolated quark (relative to the volume) via

$$e^{-\beta F_q} = \langle L(x) \rangle ; \tag{6}$$

a review of the elementary derivation of the path integral (1) makes this apparent. Further, the two-point function of $L(x)$ is related to the free energy $V(R)$ of a $q\bar{q}$ pair according to

$$e^{-\beta V(R)} = \langle L(x) L(x+R) \rangle . \tag{7}$$

We will refer to $V(R)$ as the $q\bar{q}$ potential.

There is a global symmetry operation on the system (4) which reverses the sign of L. To display it, it is convenient first to do a partial gauge-fixing. The gauge invariance of (5) shows that it is impossible in general to fix $A_0 = 0$ (i.e., $U_0 = 1$) everywhere; the best we can do is $U_0(x, \tau) = 1$ for $\tau \neq 1$. Then $L(x) = \mathrm{Tr}\, U_0(x, 1)$. Now it is apparent that the action is invariant under the global transformation $U_0 \to -U_0$, which transforms $L \to -L$.

Thus $\langle L \rangle$ is reminiscent of the magnetization in a

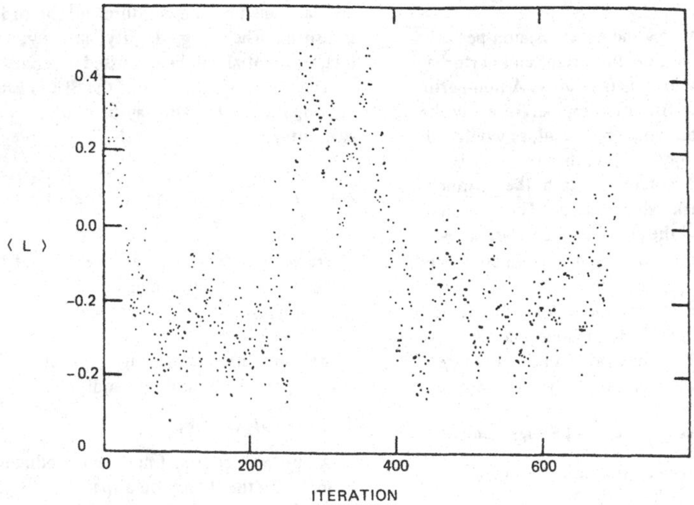

Fig. 1. Fluctuations in magnetization for $N_t = 1$, $1/g^2 = 0.85$. In this case $1/g^2_{cr} \approx 0.75$.

three-dimensional Ising system. If $\langle L\rangle = 0$, meaning $F_q = \infty$, we expect

$$\langle L(x)L(x+R)\rangle \xrightarrow[|R|\to\infty]{} e^{-|R|/\xi},$$

showing that an isolated quark has infinite free energy and the $q\bar{q}$ potential is linear. On the other hand, a spontaneous symmetry-breaking magnetization $\langle L\rangle = M$ will lead to

$$\langle L(x)L(x+R)\rangle \xrightarrow[|R|\to\infty]{} M^2,$$

meaning that $V(R) \to$ constant and free quarks exist.

We have studied (4) on a small lattice, with Monte Carlo methods. These techniques have been described adequately elsewhere [7]. Choosing N_t, N_x, and g^2, we would either start with a magnetized lattice or an unmagnetized lattice and wait until $\langle L\rangle$ stabilized. Sometimes an otherwise stable nonzero magnetization would flip sign in a small number of iterations, showing nucleation and growth of a bubble (see fig. 1). This means, of course, that one must check for other finite volume effects.

We have found that, for various values of N_x and $N_t < N_x$, there is indeed a transition between an unmagnetized (confined) strong coupling phase and a magnetized (liberated) weak coupling phase as we vary g^2. Fig. 2 shows this behavior for $N_t = 3$. The curves for $N_x = 5, 6$ and 7 almost coincide, showing that effects of finite spatial volume have almost disappeared

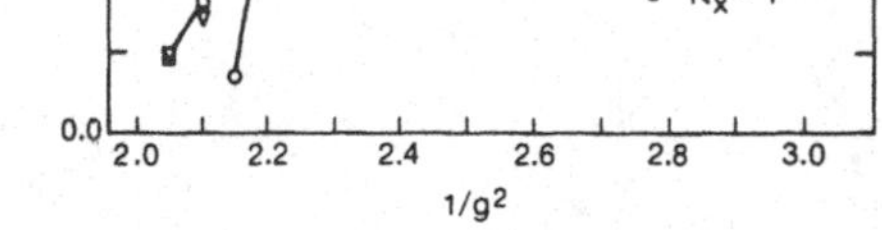

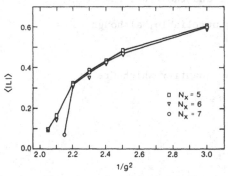

Fig. 2. Magnetization curves for $N_t = 3$. We display $\langle|L|\rangle$ rather than $\langle L\rangle$ to remove effects of domain nucleation as shown in fig. 1. Points for $N_x = 5$ and for $N_x = 7$ are joined to guide the eye.

and that, on a scale set by $\beta \propto N_t$, we have reached the thermodynamic limit. For the smaller values of N_x, there is a slight distortion of the critical behavior.

In order to understand the implications of our results for the continuum theory, we must adopt a renormalization scheme. As the lattice space goes to zero, a sensible physical parameter to keep fixed is the string tension at zero temperature. For all intents and purposes, zero temperature is reached when $N_t = N_x$, and we may look to Creutz's work [7] for the desired variation of bare coupling with lattice spacing. We write the continuum inverse temperature as a limit as $a, g \to 0$ and $N_t, N_x \to \infty$:

$$\beta = \lim a(g^2)N_t\,. \tag{8}$$

In particular, the critical β depends on the critical g^2, which in turn depends on N_t:

$$\beta_{cr} = \lim_{N_t\to\infty} a(g^2_{cr}(N_t))N_t\,. \tag{9}$$

Creutz's work shows that lowest-order strong and weak coupling perturbation theory account well for the behavior of $a(g^2)$ except in a small interval around $1/g^2 = 2$:

$$a^2 = \frac{1}{K}\left(-\log\frac{1}{4g^2}\right), \qquad \frac{1}{g^2} \lesssim 2\,, \tag{10}$$

$$a^2 = \frac{1}{K}\exp\left[-\frac{6\pi^2}{11}\left(\frac{1}{g^2}-2\right)\right], \qquad \frac{1}{g^2} \gtrsim 2\,. \tag{11}$$

K is the string tension, related to the Regge slope $\alpha' \approx 1\ \mathrm{GeV}^{-2}$ by $K = 1/2\pi\alpha'$ (for definiteness we use this number from the real world to discuss the SU(2) Yang–Mills theory). In table 1 we display values of $1/g^2_{cr}$, $a(g^2_{cr})$, and aN_t for several (finite) values of N_t. Note that the approximants to β_{cr} for $N_t = 2$ and for

Table 1
Approximants to β_{cr}. Typical error in $1/g^2_{cr}$ may be estimated from fig. 2; most of the error in T_{cr} will come from Creutz's determination of the renormalization curve.

Lattice size ($N_t \times N_x^3$)	$1/g^2_{cr}$	$a(g^2_{cr})$ (GeV^{-1})	aN_t (GeV^{-1})	T_{cr} (MeV)
1×5^3	0.75	3.2	3.2	310
2×5^3	1.8	2.2	4.4	230
3×6^3	2.15	1.66	5.0	200

$N_t = 3$ are close together, showing that the continuum limit is well represented by $N_t = 3$.

Finally, we note that the continuity of the order parameter across the transition, as shown in fig. 2, is characteristic of a second-order transition.

We thank M. Creutz for providing us with a copy of his Monte Carlo computer program. We are grateful for useful conversations with him, with T. DeGrand, R. Giles, G. Lasher, S. Shenker and N. Weiss, and with our colleagues at SLAC. Similar work by Kuti et al. [9] has come to our attention. We thank J. Kuti and J. Polonyi for pointing out an error in fig. 2 as it appeared in the original manuscript. This work was supported by the Department of Energy under contract DE-AC03-76SF00515.

References

[1] J.C. Collins and M.J. Perry, Phys. Rev. Lett. 34 (1975) 1353.
[2] P.D. Morley and M.B. Kislinger, Phys. Rep. 51 (1979) 63, and references therein.
[3] E.V. Shuryak, Phys. Rep. 61 (1980) 71, and references therein.
[4] D.J. Gross, R.D. Pisarski and L.G. Yaffe, Princeton preprint (1980), unpublished, and references therein.
[5] G. Steigman, Lectures at Summer School in Theoretical Physics: Physical cosmology (Les Houches, France, 1978), and references therein.
[6] R. Anishetty, P. Koehler and L. McLerran, SLAC-PUB-2565 (1980), and references therein.
[7] M. Creutz, Phys. Rev. D21 (1980) 2308, and references therein.
[8] A.M. Polyakov, Phys, Lett. 72B (1978) 477; L. Susskind, Phys. Rev. D20 (1979) 2610.
[9] J. Kuti, talk presented XXth Intern. Conf. on High energy physics (Madison, WI, July 1980); J. Kuti, J. Polonyi and K. Szlachanyi, Phys. Lett. 98B (1981) 199.

Notes Added:

1) In an $N_t = 1$ lattice, since a given spacelike link appears twice in a timelike plaquette, the heat-bath algorithm for that link should differ from that used for $N_t > 1$ lattices. This was neglected in our calculation. Fig. 1 and the first row of Table 1 are therefore quantitatively unreliable, although qualitatively correct.

2) We use the notation $1/g^2$ for the same quantity which Creutz (Ref. 7) and other authors label as $\beta = 4/e_o^2$.

Volume 98B, number 3 PHYSICS LETTERS 8 January 1981

MONTE CARLO STUDY OF SU(2) GAUGE THEORY AT FINITE TEMPERATURE

J. KUTI, J. POLÓNYI and K. SZLACHÁNYI

Central Research Institute for Physics, H-1525 Budapest, Hungary

Received 9 September 1980

We find numerical evidence for the phase transition between the confinement phase and free Coulomb phase of SU(2) Yang–Mills theory with lattice cut-off. The search for the critical temperature is based on a Monte Carlo study of the string tension between a heavy $Q\bar{Q}$-pair in a heat bath. The arbitrary normalization 0.2 GeV^2 is used for the string tension at zero temperature when a smooth extrapolation of the lattice theory to the continuum limit is carried out. Our numerical estimate for the critical temperature is $T_C \approx 160 \pm 30$ MeV in the absence of quark degrees of freedom. It is suggested that the phase transition is of second-order.

A few weeks ago we announced [1] the first numerical evidence for the existence of a phase transition between the confinement phase and free Coulomb phase of the SU(2) Yang–Mills theory with lattice cut-off.

There has been a long-standing conjecture that a phase transition must take place between the high temperature and low temperature phases of a non-abelian gauge field theory [2]. Polyakov [3] and Susskind [4] gave convincing arguments for the presence of this phase transition in the strong coupling limit of the lattice model.

Our Monte Carlo calculation confirms the existence of a phase transition between confined and liberated phases in the strong coupling limit of SU(2) lattice gauge theory. Besides, we find the two phases and the critical point in the region of intermediate coupling where a smooth extrapolation of the lattice model to its continuum limit exists.

It did not escape our attention that the results presented here may be useful for the early universe, for quark matter search in heavy ion collisions, and for a broader view and better understanding of the confinement problem in quantum chromodynamics. It is remarkable that an environment can be simulated in the computer which corresponds to thermal quark liberation at a temperature of a few hundred MeV.

The physical properties of a quantum field theory at finite temperature can be calculated in terms of the partition function,

$$Z = \mathrm{Tr}[\exp(-\beta H)] , \qquad (1)$$

and the thermal averages of physical observables,

$$\langle O \rangle = Z^{-1}\, \mathrm{Tr}[O \exp(-\beta H)] , \qquad (2)$$

where $\beta = 1/T$ is the inverse temperature with $k_B = 1$.

The partition function of the SU(2) gauge theory in the absence of quarks may be written as a euclidean path integral

$$Z = \int \mathcal{D}A_\mu \exp\left[-(1/2g^2) \int_0^\beta \mathrm{d}t \int \mathrm{d}^3x\, \mathrm{tr}\, F_{\mu\nu}F_{\mu\nu}\right], \qquad (3)$$

over periodic gauge fields,

$$A_\mu(\beta, x) = A_\mu(0, x) , \qquad (4)$$

with period β in the fictive imaginary time direction. The standard notation $A_\mu = A_\mu^a \sigma^a/2$ is used throughout the paper. The index a runs from 1 to 3 in SU(2) and σ^a denotes the standard Pauli matrices. The trace tr operates on SU(2) matrices.

In order to study the string tension in a heat bath we have to introduce an external color source Q at location R and a color sink $\bar{Q}$ at the origin. The free energy $V(\beta, R)$ of this heavy $Q\bar{Q}$-pair is related to the

0 031–9163/81/0000–0000/\$ 02.50

Volume 98B, number 3 PHYSICS LETTERS 8 January 1981

correlation function of thermal Wilson loops by the formula [‡1]

$$\langle \mathrm{tr}\, W(0)\, \mathrm{tr}\, W^{+}(R)\rangle = \exp[-\beta V(\beta, R)] \,. \quad (5)$$

The thermal Wilson loop $W(R)$ is defined as a closed path in the fictive imaginary time direction,

$$W(R) = \mathcal{P} \exp\left[\mathrm{i} \int_0^\beta \mathrm{d}t\, A_0(t, R)\right] , \quad (6)$$

where $\mathcal{P}$ denotes path ordering in the standard fashion.

The free energy $V(\beta, R)$ is a measure of the potential energy between the heavy Q$\bar{\mathrm{Q}}$-pair at finite temperature. In the confinement phase we expect the behavior

$$\langle \mathrm{tr}\, W(0)\, \mathrm{tr}\, W^{+}(R)\rangle \sim \mathrm{const} \cdot \exp[-\beta\sigma(\beta)R] \quad (7)$$

at large distances. The free Coulomb phase is characterized by a screened Coulomb potential, and accordingly,

$$\langle \mathrm{tr}\, W(0)\, \mathrm{tr}\, W^{+}(R)\rangle \underset{R\to\infty}{\sim} \mathrm{const} \cdot [1 + \beta(\tfrac{3}{16} g^2/\pi R) \exp(-\kappa R)] \,. \quad (8)$$

The effective string tension $\sigma(\beta)$ at finite temperature is defined by eq. (7). The Debye screening length κ^{-1} in eq. (8) is known to be $\kappa^2 = \frac{2}{3} g^2 T^2$ in the lowest order of perturbation theory.

The two phases are clearly distinguished by the different behaviors of the correlation function in eqs. (7) and (8). The trace of the gauge invariant thermal Wilson loop may be regarded as an order parameter in SU(2) gauge theory at finite temperature. The thermal average of tr $W(R)$ measures the free energy of an isolated quark with respect to the vacuum. It vanishes in the confinement phase, since the infinite free energy of the isolated quark is in the exponent of eq. (5). The order parameter tr $W(R)$ is some non-vanishing constant in the Coulomb phase where it is related to the finite self-energy of an isolated free quark on the lattice.

For the numerical evaluation of the functional integral on the left-hand side of eq. (5) of a lattice cutoff is introduced in the model following the standard procedure [5]. The periodic boundary condition in the imaginary time direction is required by the finite temperature of the heat bath. To eliminate surface effects in the three-dimensional physical space we also impose periodic boundary conditions in the three spatial directions.

The lagrangian formulation of the continuum theory requires a symmetric lattice with equal lattice spacing in the spatial and imaginary time directions. The hamiltonian method starts directly from eq. (1) and operates with the transfer matrix [8]. A dense slicing is then required operationally in the imaginary time direction for fixed spatial cut-off a [1]. The two methods yield compatible results for the phase transition.

First, we study the symmetric lattice. The inverse temperature β is given in lattice spacing units a by the relation $\beta = n_t a$ where n_t is the number of lattice sites in the imaginary time direction. The spatial volume $n_s^3 a^3$ must be reasonably large for the calculation of thermodynamical quantities. There is no other restriction on the spatial size and the number of sites, n_s, in the three spatial directions, is limited only by the performance of the computer.

The partition function

$$Z = \int \prod_{\{i,j\}} \mathrm{d}U_{ij} \exp[-4g^{-2} S(U)] \,, \quad (9)$$

defines now the thermodynamics of SU(2) gauge theory. The finite dimensional integral in eq. (9) includes all independent link variables U_{ij}, and $\mathrm{d}U_{ij}$ designates the invariant group measure of SU(2). The action S is a sum over all elementary plaquettes,

$$S = \sum_{\mathrm{plaquettes}} [1 - \tfrac{1}{2}\mathrm{tr}(U_{ij} U_{jk} U_{kl} U_{li})] \,,$$

where i, j, k and l represent the labeling of the sites around a plaquette. The connection between the link variable U_{ij} and the exponentiated gauge field variable $\exp(\mathrm{i}aA_\mu)$ is well known [5].

The Monte Carlo method [6,7] was applied for the calculation of the order parameter tr $W(R)$ and for the evaluation of the correlation function tr $W(0)$ tr $W(R)$. In our program the heat bath method of Creutz [7] was implemented for sweeping through all lattice sites in each step of the iteration towards thermal equilibrium.

The order parameter is shown in fig. 1a for the inverse temperature $\beta = 3a$ at different values of the

‡1 The details of our calculation will be published elsewhere.

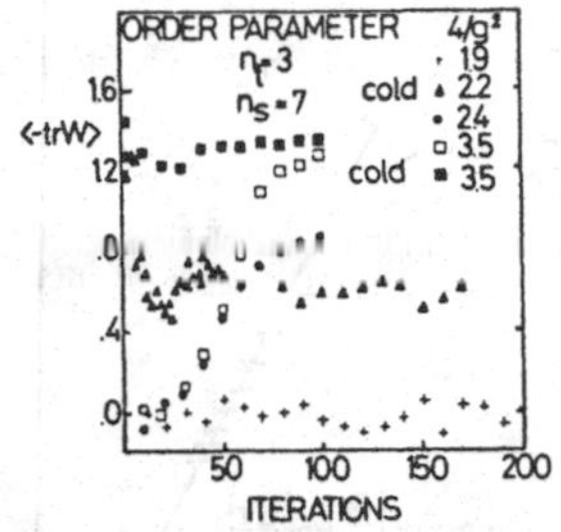

(a)

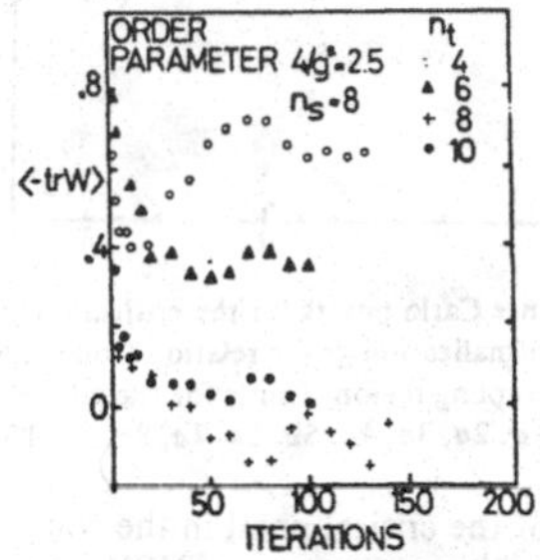

(b)

Fig. 1. (a) The evolution of the order parameter ⟨tr W⟩ is shown at $\beta = 3a$ for various values of the coupling constant as the number of iteration steps increases. The value of the order parameter averaged over 10 iterations is plotted. Cold starts from ⟨tr W⟩ = −2 are indicated and the corresponding runs are more densely populated for the first few iterations. For the black triangle points the plot is interrupted between iteration steps 50 and 80 for clarity of the figure. (b) Runs are shown for fixed coupling as the temperature varies. All runs are selected with cold starts. The order parameter disappears in the noise for $\beta > 10a$.

coupling constant g^2. At $4/g^2 = 1.9$ the order parameter drops to zero and a phase transition occurs in the system. The same behavior of the order parameter is seen in fig. 1b for a fixed value of the coupling constant at $4/g^2 = 2.5$ as the inverse temperature β varies. The thermal average of tr W drops to zero at about $\beta = 10a$.

It is easy to show that the action S is invariant under the global symmetry transformation $U \to -U$ in a selected spacelike hyperplane on each link in the imaginary time direction. The transformation flips the order parameter tr $W \to -$tr W.

In the symmetric disordered phase ⟨tr W⟩ vanishes and the free energy of an isolated quark is infinite. Therefore, this phase confines quarks and the asymptotic form

$$\langle \mathrm{tr}\, W(0)\, \mathrm{tr}\, W(R)\rangle \xrightarrow[R\to\infty]{} \mathrm{const}\cdot \exp(-\beta\sigma(\beta)R)$$

is observed. In the free Coulomb phase a spontaneous symmetry breaking occurs and the behavior

$$\langle \mathrm{tr}\, W(0)\, \mathrm{tr}\, W(R)\rangle \xrightarrow[R\to\infty]{} \langle \mathrm{tr}\, W\rangle^2 \neq 0$$

is expected. This implies finite free energy for isolated free quarks on the lattice.

The observation of symmetry breaking in the Coulomb phase is influenced by the finite size of the spatial lattice. The order parameter develops a constant expectation value for long time periods and the probability that it flips sign gradually decreases with growing n_s.

In fig. 1b only cold starts are shown where all U's are set to the same constant matrix at the beginning of the iteration. The run with $\beta = 6a$ actually starts at the positive value ⟨tr W⟩ = +2 and we had to change the sign of the points on the plot throughout this particular run for convenient comparison with other values of β. Similarly, in fig. 1a we changed the sign of ⟨tr W⟩ for the run at $4/g^2 = 3.5$ with a hot start. The U variables are set to random matrices at hot starts and the runs begin with ⟨tr W⟩ = 0. The runs at $4/g^2 = 3.5$ with hot and cold starts reach the same thermodynamical limit within one hundred iteration steps.

The behavior of the correlation function is shown in fig. 2a at $\beta = 4a$ and $g^2 = 2$ in the confinement phase. We find the numerical value $\sigma(\beta)a^2 = 0.54$ for the tension as extracted from the exponential shape of the correlation function. Creutz measures $\sigma(0)a^2 = 0.6$ at $g^2 = 2$. His result corresponds to zero temperature within some technical limitations.

Fig. 2b depicts the correlation function in the free Coulomb phase at $\beta = 4a$ and $4/g^2 = 2.3$. The arrow marks the value of ⟨tr W⟩^{2} which is the asymptotic limit of the correlation function.

In our search for the phase transition point the order parameter never exhibits a discontinuous jump at T_c and a second-order transition is suggested. It is also supported by the observation of large fluctuations near the critical point.

The critical coupling constant is shown in fig. 3 at

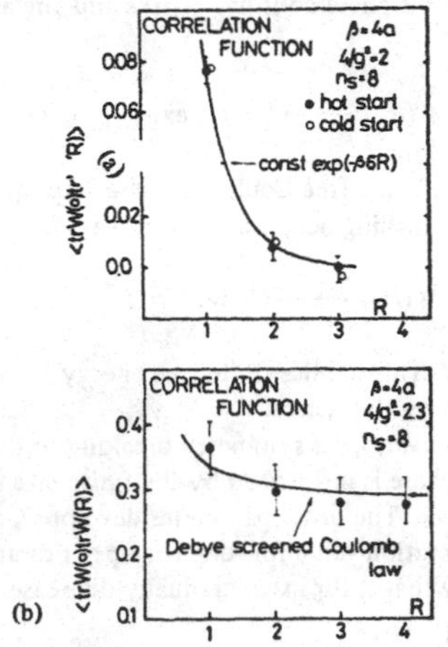

Fig. 2. (a) The correlation function with exponential decay at $g^2 = 2$ determines the tension $\sigma(\beta)$ in the confinement phase at $\beta = 4a$ near the critical point. R is given in lattice spacing units. (b) The correlation function is shown at the same temperature but for weaker coupling $4/g^2 = 2.3$. The points follow the Debye screened Coulomb law with screening length κ^{-1} determined by the temperature and coupling.

various temperatures. The interpretation of these results requires a smooth extrapolation to the continuum limit of the theory. The scale is set on the lattice by the lattice Λ parameter in lattice spacing units as

$$\Lambda = \lim_{a\to 0} a^{-1}\,[\gamma_0 g^2(a)]^{-\gamma_1/2\gamma_0^2} \exp[-1/2\gamma_0 g^2(a)] \qquad (10)$$

in the continuum limit. The coupling constant $g(a)$ is used throughout the Monte Carlo calculations. For SU(N) gauge groups the coefficients in eq. (10) are $\gamma_0 = \frac{11}{3}N/16\pi^2$ and $\gamma_1 = \frac{34}{3}(N/16\pi^2)^2$.

The lattice Λ parameter is related to Λ^{MOM} in the continuum limit theory by a recent calculation of Hasenfratz and Hasenfratz [9]. Creutz calculated in his Monte Carlo program the tension at zero temperature in terms of the Λ parameter in the SU(2) gauge theory. He finds [10]:

$$\Lambda = (1.3 \pm 0.2) \times 10^{-2}\, \sigma^{1/2}(0)\,. \qquad (11)$$

The arbitrary normalization 0.2 GeV^2 is used for the string tension at zero temperature in our numeri-

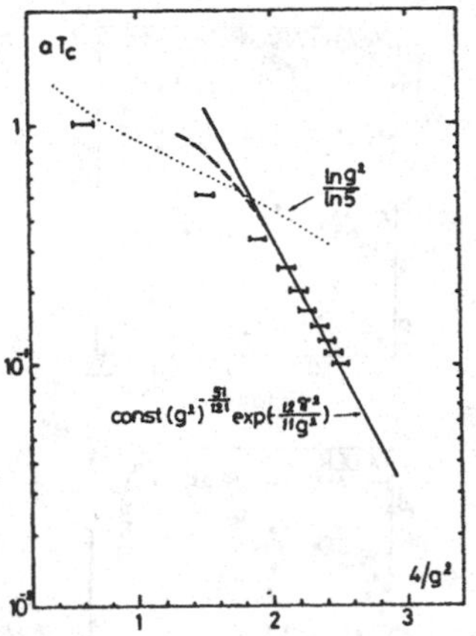

Fig. 3. Our Monte Carlo points for the critical temperature follow the renormalization group relation (solid line) in the intermediate coupling region. Points for the critical coupling are given at $\beta = a, 2a, 3a, 4a, 5a, 6a, 7a, 8a, 9a, 10a$.

cal estimate of the critical point in the continuum limit. This would correspond to Λ^{MOM} = 330 MeV in the continuum SU(2) gauge theory [9].

Our calculated Monte Carlo points in fig. 3 follow the renormalization group relation

$$T_c a = \text{const} \cdot [g^2(a)]^{-51/121} \exp[-\tfrac{12}{11}\pi^2/g^2(a)] \quad (12)$$

for $4/g^2 \geqslant 2$. The otherwise arbitrary constant in eq. (12) is determined by the Monte Carlo points. The best estimate of the critical temperature with the presented extrapolation to the continuum limit is

$$T_c = (0.35 \pm 0.05)\,\sigma(0)^{1/2},$$

$$\text{or } T_c = 160 \pm 30 \text{ MeV}. \qquad (13)$$

The error bar on the critical point in eq. (13) is twofold. There is a statistical error in Creutz's relation of eq. (11). Our statistical inaccuracy is represented in fig. 3 by the horizontal error bars of the critical coupling for a given value of Ta.

There are finite temperature corrections to the relation between the lattice spacing a and coupling consta[nt] g which we calculated in Coulomb gauge on the one-loop level. These corrections are small for $4/g^2 > 2$ where $T_c \ll a^{-1}$.

Volume 98B, number 3 PHYSICS LETTERS 8 January 1981

Our estimate for the renormalization of the coupling at finite temperature is as follows. We calculate the Coulomb force on the scale a at temperature T on a superfine lattice a_0 as

$$\left.\frac{g^2(a,T)}{p^2}\right|_{p^2=a^{-2}} = \frac{g^2(a_0)}{p^2 - \Pi_{00}(p_0=0,p^2,g^2(a_0),T)}, \tag{14}$$

where the right-hand side is also taken at $p^2 = 1/a^2$. The polarization tensor Π_{00} is given by

$\Pi_{00} = \frac{1}{2}$ 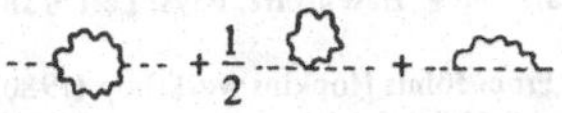 (15)

$$= \Pi_{00}(p_0=0,p^2,g^2(a_0),T=0) + g^2(a_0)\,\tfrac{1}{18}N_c T^2,$$

for SU(N) gauge theory, in the limit $p^2 \gg T^2$.

Eq. (14) is obtained by the bubble summation of the one-loop diagrams in Coulomb gauge (footnote 1).

Eqs. (14) and (15) relate the coupling constant $g^2(a,T)$ on two different scales and different temperatures as

$$g^2(a_2,T_2) = g^2(a_1,T_1) \times \{1 + g^2(a_1,T_1)[\tfrac{11}{24}\pi^{-2}N_c \ln(a_1/a_2) + \tfrac{1}{18}N_c(T_1^2a_1^2 - T_2^2a_2^2)]\}^{-1}.$$

The dashed line in fig. 3 includes this finite temperature correction.

The dotted line in fig. 3 is an estimate of the critical temperature (footnote 1) in the strong coupling limit:

$$T_c \approx a\sigma(0)/\ln 5 .$$

Its derivation is based on the observation that the partition function develops a singularity at the critical point by the condensation of chromo-electric vortices.

At about $4/g^2 = 1.9$ the Monte Carlo points break away from the corrected curve (dashed line in fig. 3) and gradually approach the strong coupling estimate.

It is interesting to search for the critical point of the phase transition in the hamiltonian formulation of the theory. It also provides a consistency check of the overall picture.

Our starting point is eq. (1) with the lattice hamiltonian of Kogut and Susskind [11]. The partition function can be approximated through the transfer matrix [8] by

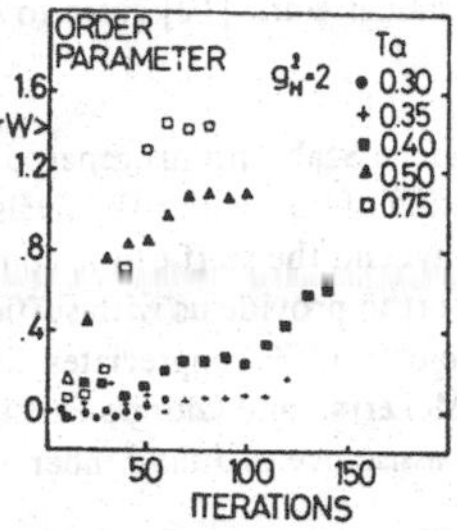

Fig. 4. The order parameter is shown for runs at $g_H^2 = 2$ in the hamiltonian formulation. Typically, twelve to sixteen time slices were adequate for inverse temperatures around $\beta = 3a$. The order parameter disappears in the noise at about $Ta = 0.3$.

$$Z = \int \prod_{\{i,j\}} dU_{ij} \times \exp\Bigg\{ -\frac{4}{g_H^2}\frac{\tau}{a} \sum_{\substack{\text{spacelike}\\ \text{plaquettes}}} [1 - \tfrac{1}{2}\,\mathrm{tr}(U_{ij}U_{jk}U_{kl}U_{li})] - \frac{4}{g_H^2}\frac{a}{\tau} \sum_{\substack{\text{timelike}\\ \text{plaquettes}}} [1 - \tfrac{1}{2}\,\mathrm{tr}(U_{ij}U_{jk}U_{kl}U_{li})]\Bigg\}, \tag{16}$$

where τ is a single slice of the interval β. Eq. (16) becomes exact in the limit $\tau/a \to 0$ for fixed lattice spacing a of the three-dimensional space. The coupling constant $g_H(a)$ appears in the original lattice hamiltonian [11]. The expectation values in eq. (2) can be calculated in a similar way.

The phase transition is shown in fig. 4 for a fixed coupling $4/g_H^2 = 2$. The critical temperature was found at $T_c a = 0.32$. For a smaller coupling $4/g_H^2 = 2.15$ we find $T_c a = 0.22$. The two points fit the continuum limit renormalization scheme with $T_c \approx 0.35[\sigma(0)]^{1/2}$ provided that the lattice scale parameter in the hamiltonian formulation is approximately the same as in the lagrangian method.

Our numerical calculation may be regarded as the first direct determination of the hamiltonian scale parameter Λ_H in terms of the lagrangian Λ parameter. If the applied transfer matrix approximation with our dense slicing of the finite time interval β is adequate, we predict $\Lambda_H \approx \Lambda$.

After the work reported here was complete and

presented [1], similar work [12] came to our attention.

We thank Ferenc Szabó for his generous help during the preparation of our work. We are also indebted to Magda Zimányi and the staff of the computer center for their effort to provide us with sufficient computer time. One of us (J.K.) appreciates discussions with Larry D. McLerran and Giorgio Parisi. He also appreciates the assistance of Julia Ember.

References

[1] J. Kuti, talk presented at the XXth Intern. Conf. on High energy physics (Madison, July 18, 1980);
J. Kuti, J. Polónyi and K. Szlachányi, to be published in the conference proceedings.
[2] N. Cabibbo and G. Parisi, Phys. Lett. 59B (1975) 67;
J.C. Collins and M.J. Perry, Phys. Rev. Lett. 34 (1975) 1353;
P.D. Morley and M.B. Kislinger, Phys. Rep. 51C (1979) 63, and references therein;
E.V. Shuryak, Phys. Rep. 61C (1980) 71, and references therein;
for a recent interesting work, see D.J. Gross, R.D. Pisarski and L.G. Yaffe, Princeton preprint (1980).
[3] A.M. Polyakov, Phys. Lett. 72B (1978) 477.
[4] L. Susskind, Phys. Rev. D20 (1979) 2610.
[5] For a review, see, e.g., J.M. Drouffe and C. Itzykson, Phys. Rep. 38C (1978) 133.
[6] K.G. Wilson, Cornell preprint (1979).
[7] M. Creutz, Brookhaven preprint (1979).
[8] M. Creutz, Phys. Rev. D15 (1977) 1128.
[9] A. Hasenfratz and P. Hasenfratz, Phys. Lett. 93B (1980) 165.
[10] M. Creutz, Proc. Johns Hopkins workshop (1980) p. 85.
[11] J. Kogut and L. Susskind, Phys. Rev. D11 (1975) 395.
[12] L.D. McLerran and B. Svetitsky, SLAC-PUB-2572 (1980).

Volume 101B, number 1,2 PHYSICS LETTERS 30 April 1981

HIGH TEMPERATURE SU(2) GLUON MATTER ON THE LATTICE

J. ENGELS, F. KARSCH and H. SATZ
Department of Theoretical Physics, University of Bielefeld, Germany
and
I. MONTVAY
II. Institut für Theoretische Physik der Universität Hamburg [1], *Germany*

Received 21 January 1981

We calculate by Monte Carlo simulation on the lattice the energy density ϵ of an SU(2) Yang–Mills system at finite physical temperature. First, we study the high temperature form of ϵ, showing that the conventional euclidean lattice formulation converges to the parameter-free Stefan–Boltzmann limit of a free gluon gas in the continuum. Secondly, we show that the specific heat of gluon matter exhibits a sharp peak at the transition point from the confined phase to the color-screened gluon gas. The resulting transition temperature is found to be 210 ± 10 MeV.

Recent Monte Carlo studies of SU(2) Yang–Mills systems [1,2] have provided strong indications that gluon matter experiences a phase transition at a critical temperature T_c around 160–230 MeV, changing from a confined phase below T_c to one in which color Debye screening renders the gluons effectively free, above T_c. Such a transition was expected from strong coupling considerations [3] as well as from a variety of phenomenological approaches [4]; nevertheless, Monte Carlo calculations on the lattice so far constitute the only way of treating in one approach the entire range from confinement to "free" gluons.

The aim of this note is to further investigate by Monte Carlo methods the thermodynamics of finite temperature gluon matter. First, we want to connect at high temperature the lattice formulation with the results of perturbative QCD [5–7] in the continuum, where the Stefan–Boltzmann form of the energy density provides a completely determined parameter-free limit. As second point, we shall show that the specific heart of gluon matter exhibits a singularity-like peak at the deconfinement transition – giving us a very physical and clear-cut way to determine the transition temperature T_c.

[1] Supported by the Bundesministerium für Forschung und Technologie, Bonn, Germany.

We start, as in refs. [1,2], from the lattice partition function

$$Z(N, N_\beta, g^2) \equiv \int \prod_{\{i,j\}} dU_{ij} \exp[-(4/g^2)S(U)] , \quad (1)$$

where dU_{ij} is the invariant SU(2) measure and U_{ij} the corresponding link variable; g denotes the bare coupling constant. The action

$$S = \sum_{\{P\}} (1 - \tfrac{1}{2} \mathrm{tr}\, U_{ij} U_{jk} U_{kl} U_{li}) , \quad (2)$$

is a sum over all plaquettes $\{P\}$, where i, j, k, l label the sites defining the plaquette. The lattice underlying eqs. (1) and (2) has N sites in each spatial direction and N_β sites in the temperature (= imaginary time) direction.

Eq. (2) holds for equal lattice spacing in all directions. Choosing the lattice spacing in the temperature direction, a_β, to be different from the lattice spacing in the spatial directions, a, eq. (2) goes over into

$$S = \left(\frac{a_\beta}{a} \sum_{\{P_s\}} (1 - \tfrac{1}{2} \mathrm{tr}\, U_{ij} \ldots U_{li}) + \frac{a}{a_\beta} \sum_{\{P_\beta\}} (1 - \tfrac{1}{2} \mathrm{tr}\, U_{ij} \ldots U_{li}) \right). \quad (3)$$

Here the "space-like" plaquettes $\{P_s\}$ have only links in spatial directions and the "temperature-like" plaquettes $\{P_\beta\}$ have two (opposite) links in the temperature direction.

Imposing periodic boundary conditions in the temperature direction, for sufficiently large N and N_β but fixed $\beta = N_\beta a_\beta$, this lattice describes a system of temperature $T = \beta^{-1}$. Periodic boundary conditions in the spatial directions (although not necessary) are usually also assumed. In the thermodynamic limit the overall spatial volume $V = (Na)^3$ has to be large enough for the spatial surface effects to become negligible. Therefore, generally speaking, in Monte Carlo simulations of finite temperature systems, $N_\beta \ll N$ is required (at least for $a_\beta = a$) in order to minimize the finite volume effects in comparison with the effects of periodicity in the temperature direction. The temperature $T = 0$ is obtained by holding a and a_β fixed as N and N_β become "infinite". Note that there is no particular connection between $T = 0$ and a finite symmetric lattice, just as there is no reason, other than simplicity, for spatially symmetric lattices. N_μ must just be large enough in each direction μ to leave the results insensitive to a change of N_μ. (For a discussion of finite temperature effects in previous Monte Carlo lattice calculations see ref. [8].)

In the continuum, the energy density ϵ is given by

$$\epsilon = -V^{-1}\,[\partial \ln Z(\beta, V)/\partial\beta]_V\,, \tag{4}$$

and the pressure p by

$$p = \beta^{-1}\,[\partial \ln Z(\beta, V)/\partial V]_\beta\,. \tag{5}$$

On the lattice, with fixed N and N_β, we write

$$\partial/\partial\beta = N_\beta^{-1}\,\partial/\partial a_\beta\,, \tag{6}$$

$$\partial/\partial V = (3N^3a^2)^{-1}\,\partial/\partial a\,, \tag{7}$$

for the derivatives in eqs. (4) and (5). With eqs. (1) and (3), this yields

$$\epsilon = \epsilon_a + \epsilon_g\,, \quad p = p_a + p_g \tag{8}$$

for energy density and pressure, where

$$\epsilon_a \equiv 4(N^3N_\beta a^3 a_\beta g^2)^{-1}\Big\langle \frac{a_\beta}{a}\sum_{\{P_s\}}[1 - \tfrac{1}{2}\,\mathrm{tr}\,U_{ij}\ldots U_{li}] - \frac{a}{a_\beta}\sum_{\{P_\beta\}}[1 - \tfrac{1}{2}\,\mathrm{tr}\,U_{ij}\ldots U_{li}]\Big\rangle\,, \tag{9}$$

and, with eq. (3),

$$\epsilon_g \equiv 4(N^3N_\beta a^3)^{-1}(\partial g^{-2}/\partial a_\beta)_a\,\langle S\rangle\,,$$

$$p_g \equiv -\tfrac{4}{3}(N^3N_\beta a^2 a_\beta)^{-1}(\partial g^{-2}/\partial a)_{a_\beta}\,\langle S\rangle\,. \tag{10}$$

Here $\langle\ \rangle$ denotes, as usual, the thermodynamic average over the partition function (1).

The pressure p_a and energy density ϵ_a "at constant g" are found to satisfy the zero mass relation

$$p_a = \tfrac{1}{3}\epsilon_a\,, \tag{11}$$

already on the lattice.

In the continuum, perturbation theory gives as high temperature limit [5–7]

$$\epsilon = (N_g\pi^2/15)T^4\,[1 - \alpha_s 5N_c/\pi + \alpha_s^{3/2}\,80(N_c/\pi)^{3/2}/\sqrt{3} \pm \ldots]\,, \tag{12}$$

where $N_g = N_c^2 - 1$ denotes the number of gluons [3 for SU(2)] and N_c the number of colors [2 for SU(2)]; α_s is the running coupling constant of QCD

$$\alpha_s = g_s^2/16\pi = 3\pi/22N_c \ln(4T/\Lambda)\,, \tag{13}$$

with Λ as the continuum normalization parameter. For sufficiently high temperatures, the continuum limit of the SU(2) lattice system must thus attain the Stefan–Boltzmann form

$$\epsilon_{SB} = \tfrac{1}{5}\pi^2T^4\,, \tag{14}$$

independently of the choice of continuum or lattice normalization parameters Λ and Λ_L. One of our aims is to check if eq. (8) indeed converges to this parameter-free limit. In view of the complex relationship between euclidean lattice and hamiltonian continuum formulations, such a convergence is not a priori evident.

The lattice spacing a, the lattice scale parameter Λ_L and the coupling g are in the connected limit ($a \to 0$) connected by the renormalization group relation

$$\Lambda_L a = (11g^2/24\pi^2)^{-51/121}\exp(-12\pi^2/11g^2)\,, \tag{15}$$

Monte Carlo calculations have yielded [9]

$$\Lambda_L = (1.3 \pm 0.2)\times 10^{-2}[\sigma(0)]^{1/2} \tag{16}$$

for Λ_L in terms of the string tension $\sigma(0)$ at $T = 0$. With the relation $\sigma(0) = (2\pi\alpha')^{-1}$ and a Regge slope $\alpha' = 1$ (GeV)$^{-2}$, this gives

$$\Lambda_L = 5.2 \pm 0.8\ \text{MeV}\,. \tag{17}$$

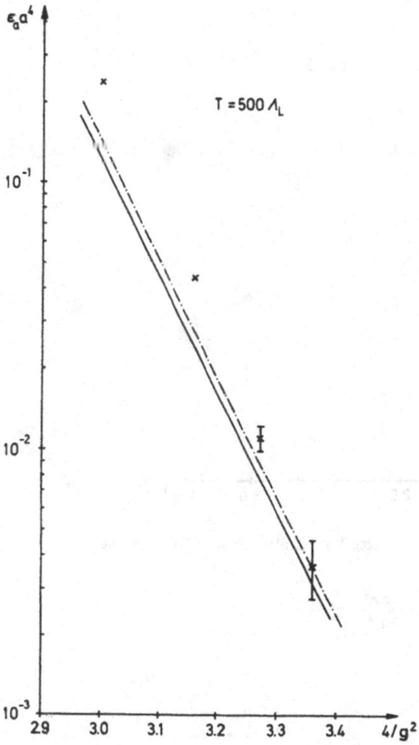

Fig. 1. Energy density of gluon matter versus $4/g^2$, at fixed temperature $T = 500\ \Lambda_L$, after about 600 iterations. The solid line gives the Stefan–Boltzmann limit; the dashed line includes the perturbative corrections of eq. (12).

This value of Λ_L, when inserted in the SU(2) relation [10]

$$\Lambda \approx 57.5\ \Lambda_L \tag{18}$$

leads to $\Lambda \approx 300 \pm 50$ MeV, in reasonable consistency with what one would expect from deep inelastic scattering or e^+e^- annihilation data. – These connections being given, we shall from now on, for convenience, measure T in units of Λ_L.

To calculate the "constant g" part (9) of the lattice energy density, we consider a $10^3 \times N_\beta$ lattice with isotropic spacing $a = a_\beta$. Holding the temperature $T = (N_\beta a)^{-1}$ fixed, we let N_β run from $N_\beta = 2$ to $N_\beta = 5$

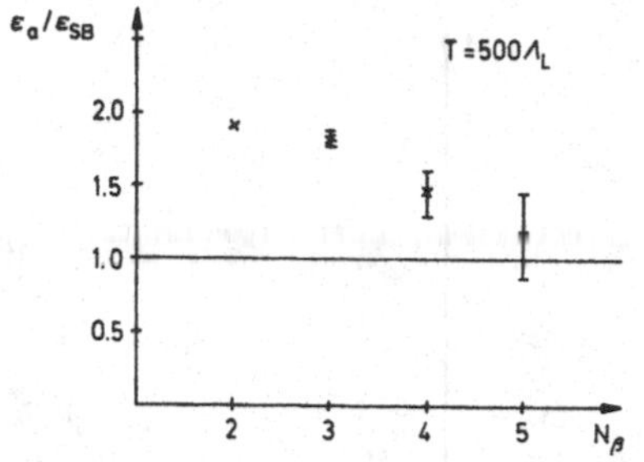

Fig. 2. Energy density of gluon matter versus lattice size, at $T = 500\ \Lambda_L$, in comparison with the Stefan–Boltzmann limit ϵ_{SB}; the dashed line lincludes the perturbative corrections of eq. (12).

(varying a with N_β accordingly). We can now ask two questions: Has the energy density ϵ_a at the spacing a corresponding to a given T already reached the continuum limit? If this is the case, then the relation between g and a is given by eq. (15). Does ϵ_a at high temperature approach the Stefan–Boltzmann form? This need not be the case, since we have so far ignored the second term in eq. (8).

If ϵ_a does satisfy the Stefan–Boltzmann relation

$$\epsilon_a a^4 = \pi^2/(5N_\beta^4), \tag{19}$$

at a given temperature $T = \tau\Lambda_L$, then we have

$$\epsilon_a a^4 = \tfrac{1}{5}\pi^2\tau^4 \times [(11g^2/24\pi^2)^{-51/121} \exp(-12\pi^2/11g^2)]^4 \tag{20}$$

if at the lattice spacing $a = (N_\beta \tau \Lambda_L)^{-1}$, which corresponds to g, the energy density ϵ_a is in the continuum limit. Relation (20) thus gives us a parameter-free $g \to 0$ and high temperature limit.

In fig. 1 we show for $\tau = 500$ (corresponding to $T \approx 2.6$ GeV) the result of our Monte Carlo calculations, using the same methods as ref. [9]. Besides the Stefan–Boltzmann limit (20), we also display the limiting curve obtained by including the higher order corrections of eq. (12), using $\Lambda/\Lambda_L = 57.5$. In spite of the relatively small lattice size (manageable for computers), both the absolute value and the g-dependence of our Monte Carlo results are seen to agree quite well with the free gluon gas limit. To illustrate the dependence of our results on the lattice size, we shown in fig. 2 a

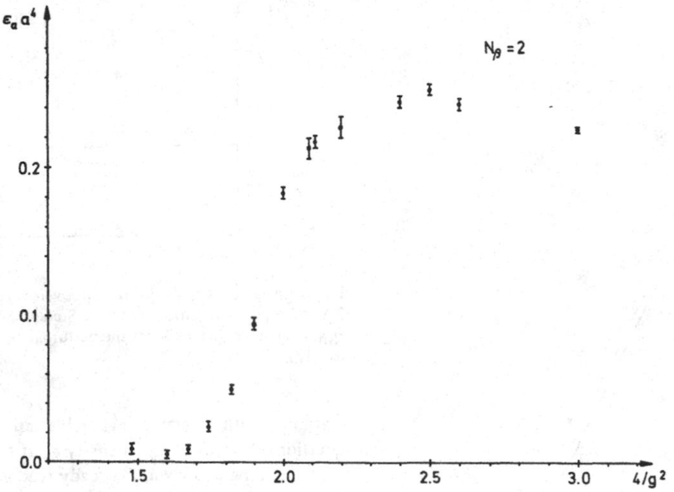

Fig. 3. Energy density of gluon matter versus $4/g^2$, at fixed lattice size $N_\beta = 2$, after about 500 iterations.

plot of ϵ_a/ϵ_{SB} as function of N_β; the convergence to the free gluon gas appears to be fairly rapid.

So far, we have considered only the first term ϵ_a of eq. (8), and that alone was seen to yield the correct Stefan–Boltzmann limit. To check that the second term, $\langle S\rangle\,(\partial g^{-2}/\partial a_\beta)_a$, is indeed negligible, we have numerically calculated the variation of g as a function of a_β at fixed a.

The best way to define the lattice coupling constant g in a finite temperature calculation is to choose a dimensional thermodynamic quantity and fix its physical value in terms of g. (This is analogous to the usual procedure of fixing in field theory at $T = 0$, say, the physical value of the string tension in order to define the dependence of g on a [9,11].) An obvious choice in the gluon system is to fix the physical value of the critical temperature for the deconfining phase transition. It is convenient to introduce, instead of a and a_β, the variables t_c and α:

$$t_c \equiv aT_c, \quad \alpha \equiv a_\beta/a \tag{21}$$

implying

$$a(\partial g^{-2}/\partial a)_{a_\beta} = t_c(\partial g^{-2}/\partial t_c)_\alpha - \alpha(\partial g^{-2}/\partial\alpha)_{t_c}$$

$$a(\partial g^{-2}/\partial a_\beta)_a = (\partial g^{-2}/\partial\alpha)_{t_c}\,. \tag{22}$$

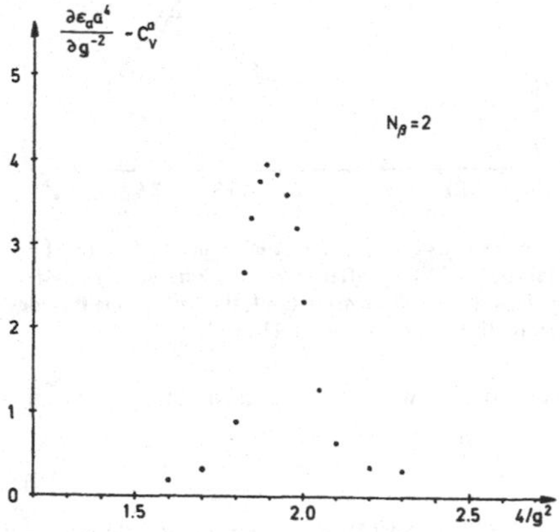

Fig. 4. Specific heat of gluon matter versus $4/g^2$, at fixed lattice size $N_\beta = 2$.

From the connection of the lattice spacing a with the lattice scale parameter Λ_L in eq. (15), there follows for small g, independently of the value of Λ_L:

$$t_c(\partial g^{-2}/\partial t_c)_{\alpha=1} \to -11/(12\pi^2)\,. \tag{23}$$

The value of $(\partial g^{-2}/\partial\alpha)_{t_c}$ can be determined numeri-

Volume 101B, number 1,2 PHYSICS LETTERS 30 April 1981

cally by looking for the variation of the critical value of g^{-2} which corresponds to T_c. When we go, for instance, from $N_\beta = 4$ and $\alpha = 1$ to $N_\beta = 3$ and $\alpha = 4/3$ the value of a and t_c is unchanged. The change in the critical value of g^{-2} gives a numerical estimate for $(\partial g^{-2}/\partial\alpha)_{t_c}$. The resulting value in our range of g^2 (about $g^2 = 1-2$) is small (of the order of a few percent of g^{-2}).

In ϵ_g, the derivative $(\partial g^{-2}/\partial\alpha)_{t_c}$ is multiplied by the average action $\langle S\rangle$, which is not small and does not vanish exponentially for $g \to 0$, as would be required by the renormalization group. However, ϵ still contains the zero point term ("vacuum energy density") inherent in the euclidean formulation [12].

This term is infinite for $g \to 0$, but it does not depend on the temperature. Hence it can be removed by calculating the differences of $\langle S\rangle$ between two temperatures at fixed g. These differences, multiplied by $(\partial g^{-2}/\partial\alpha)_{t_c}$, are less than 1% of ϵ_a in the temperature range we checked ($T \gtrsim 100\ \Lambda_L$). Hence ϵ_a is for $a = a_\beta$ indeed a good approximation of the energy density at these temperatures.

Having established the connection between the lattice energy density and its high temperature continuum limit, we now turn to the temperature region around the deconfinement transition. In fig. 3, we show the behaviour of ϵ_a for $N_\beta = 2$ as function of g^2. Apart from a scale change, this also gives us the temperature dependence of ϵ_a, since at fixed N_β the coupling g and the lattice spacing a are related either by the renormalization group relation (15), or, in the non-asymptotic regime, by numerical [9] and/or strong coupling [11,13] results. The corresponding specific heat

$$C_V^a = \partial\epsilon_a/\partial T \approx \partial\epsilon_a a^4/\partial g^{-2}\,, \qquad (24)$$

is shown in fig. 4. At $4/g^2 \approx 1.90$ (for $N_\beta = 2$) we have a clear peak as the signal of the deconfinement transition from gluonium matter to gluon gas. (Due to the sharpness of the peak, its position is not changed noticeably if we differentiate with respect to g^{-2} instead of T.)

To show that the transition observed here in terms of the specific heat is indeed the deconfinement transition defined in refs. [1,2] through the average Wilson loop $|\langle L\rangle|$, we display in fig. 5 our results for $|\langle L\rangle|$ at $N_\beta = 2$. The change from $\langle L\rangle \approx 0$ in the confined phase to $\langle L\rangle \neq 0$ in the gluon gas is seen to occur at the

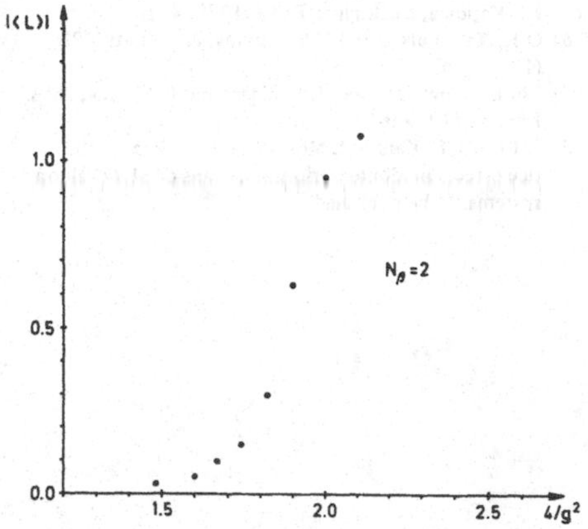

Fig. 5. Average Wilson loop versus $4/g^2$, at fixed lattice size $N_\beta = 2$.

same g^2 as the rise of ϵ_a in fig. 3. We note that a determination of T_c through the peak of the specific heat, which seems physicallly the most clear-cut, leads to a higher T_c than what one obtains by placing the transition at that g^2 where $\langle L\rangle$ becomes non-zero. This may account for the lower T_c found in ref. [1].

To assure that the transition occurs indeed at fixed temperature, we have also calculated C_V^a for $N_\beta = 4$. The peak then lies at $4/g^2 \approx 2.28$; using eq. (15), this gives $T_c = 40 \pm 2\ \Lambda_L$, which by eq. (17) yields $T_c = 210 \pm 10$ MeV. This is in agreement with the value obtained for $N_\beta = 2$, if there the non-asymptotic part of the curve of Creutz [9] is used.

Finally we note that at lower temperatures, in the region of T_c, both energy density and specific heat will in addition have contributions from ϵ_g; these, however, are not expected to modify significantly the location of the transition.

References

[1] J. Kuti, J. Polónyi and K. Szlachányi, Phys. Lett. 98B (1981) 199.
[2] L.D. McLerran and B. Svetitsky, Phys. Lett. 98B (1981) 195.
[3] L. Susskind, Phys. Rev. D20 (1979) 2610.
[4] For a review, see e.g. H. Satz, On critical phenomena in strong interaction physics, Bielefeld preprint BI-TP 80/33 (1980); to be published in: Proc. 17th Winter School of Theoretical physica (Karpacz, Poland).

[5] J.I. Kapusta, Nucl. Phys. B148 (1979) 461.
[6] O.K. Kalashnikov and V.V. Klimov, Phys. Lett. 88B (1979) 328.
[7] For a review, see e.g.: M. Kislinger and P. Morley, Phys. Rep. 51 (1979) 64.
[8] J. Engels, F. Karsch, I. Montvay and H. Satz, Finite lattice effects in Monte Carlo simulations of SU(2) gluon systems, to be published.
[9] M. Creutz, Phys. Rev. D21 (1980) 2308.
[10] A. Hasenfratz and P. Hasenfratz, Phys. Lett. 93B (1980) 165.
[11] J.B. Kogut, R.B. Pearson and J. Shigemitsu, Phys. Rev. Lett. 43 (1979) 484.
[12] C. Bernard, Phys. Rev. D9 (1974) 3312.
[13] G. Münster, Phys. Lett. 95B (1980) 59.

Nuclear Physics B180[FS2] (1981) 469–482

SU(2) STRING TENSION, GLUEBALL MASS AND INTERQUARK POTENTIAL BY MONTE CARLO COMPUTATIONS

G. BHANOT
Brookhaven National Laboratory, Upton, NY 11973, USA

C. REBBI[1]
CERN, Geneva, Switzerland

Received 3 December 1980

Monte Carlo simulations of the SU(2) gauge system on a large (16^4) lattice and with high statistics are performed to determine several quantities of physical interest. Previous evaluations of the ratio between string tension, σ, and scale constant, Λ_0, are confirmed. The mass of the glueball is found to be approximately $3\sqrt{\sigma}$ and the potential between static charges at very small separation is measured.

1. Introduction

Monte Carlo simulations have provided us with extremely important results about lattice gauge theories [1–4]. Yet the fact of working in four dimensions imposes severe constraints on the computation: the size of the lattice is necessarily small, one runs into boundary effects and the statistics are often poor. We have tried to refine the analyses done so far and to obtain new results by performing Monte Carlo simulations on a fairly large lattice (lattice size 16, with periodic boundary conditions; number of gauge variables $4 \times 16^4 = 262144$), for a non-abelian model defined by the 120-element icosahedral subgroup $\tilde{Y}$ of SU(2). In this article, we present the results of our investigation.

It has been shown already that models using a finite subgroup of SU(2) as gauge group can give a very good approximation to the SU(2) theory itself throughout a range of values of the parameter $\beta = 4/g^2$ [3,4]. In particular, the $\tilde{Y}$ system undergoes the phase transition to the broken symmetry phase, characteristic of the finite nature of the gauge group, at $\beta_c \approx 5.99$. This value far exceeds the value $\beta \sim 2$, where the transition from the strong coupling to the weak coupling behaviour takes place, and one expects the $\tilde{Y}$ model to reproduce quite well the properties of the

[1] On leave from Brookhaven National Laboratory, Upton, NY 11973, USA.

SU(2) gauge theory throughout the interesting physical region. The possibility of using a multiplication table for the group operation and of storing many group elements (spins) in the same word of the computer permits a substantial reduction in processing time and memory requirements [5]. This has allowed us to perform Monte Carlo simulations on a 16^4 lattice, still keeping the time needed for one sweep of the entire lattice (on a CDC 7600) to the acceptable value of ~ 15 seconds.

Working with a large lattice presents some definite advantages. Finite size effects are less likely to occur. For the determination of the string tension, for instance, we have measured the expectation value of Wilson loops of size up to six by six. Since the loops are embedded in a lattice extending for 16 sites in each direction, the distortion of their expectation values due to boundary effects is almost certainly negligible. A more subtle advantage has to do with the possible occurrence of a deconfining phase transition in finite systems. A study of temperature effects reveals that on a lattice, in principle infinite in its spatial extent, but periodic over d lattice sites in the time direction, the system undergoes a transition to a Debye-screened Coulomb phase as β increases beyond some definite $\beta_c(d)$ [6]. d acts like an inverse temperature ($da = 1/T$, where a is the lattice spacing), and the larger d is, the greater is the value of β_c. The point is that a signal of this transition is also observed with lattices of finite spatial extent (even of the order of d), so, to play it safe, one should not rely on Monte Carlo computations at a value of β which exceeds β_c to derive properties of the confining phase. With a larger system, higher values of β are allowed.

A large lattice size implies a very large number of spin variables: so, even averaging over a single lattice configuration in thermodynamical equilibrium, the statistics are quite good. We have made definite use of this improvement in statistics to evaluate quantities which have previously escaped measurement. Of course, there is really no true gain from working with a larger lattice. The statistics are improved if averages are taken over a fixed number of Monte Carlo iterations, but, with a smaller lattice, one could do more iterations during the same processing time. Smaller systems (as long as one does not run into finite size effects) may be more convenient if large-scale fluctuations are present: the relaxation time becomes longer when the size of the system increases and more steps become necessary to obtain reliable results. This typically happens near a phase transition. It has been recently argued [7] that a phase transition (of the roughening type) may occur in a variety of lattice gauge theories, including the one under consideration. This transition, however, would not destroy confinement or otherwise alter the basic properties of the theory, and, as we have not observed long-range fluctuations in the quantities we have measured, we believe that the possible presence of a roughening transition would not spoil the accuracy of our results. Finally, a lattice measuring 16^4 may be large enough to permit two or three iterations of the renormalization transformation in real space [8]. However, this interesting application is left to future investigations.

2. Outline of the computation

We have simulated the dynamics of the $\tilde{Y}$ model using the so-called improved Metropolis method [9]. Given a definite group element U_{ij} on the link between the sites i and j, a new, trial value, $\hat{U}_{ij}$, is obtained by multiplying U_{ij} by one of the 12 neighbours G_i of the identity chosen at random. By "neighbours of the identity" we mean the following. Writing any element of $\tilde{Y}$ as $U=\cos\frac{1}{2}\theta+\mathrm{i}\boldsymbol{\sigma}\cdot\hat{\boldsymbol{n}}\sin\frac{1}{2}\theta$, the identity is, of course, obtained with $\theta=0$, and the G_i's are those elements which correspond to the lowest non-zero value of $|\theta|$ (namely $|\theta|=\frac{2}{5}\pi$). They correspond to the minimal non-trivial rotations which leave the icosahedron invariant in the factor group $Y=\tilde{Y}/Z_2$, and any two of them (provided they are not the inverses of each other) generate the whole group.

$\hat{U}_{ij}$ is accepted to replace U_{ij} if the change lowers the action S; otherwise, if S is increased by ΔS, it is accepted with conditional probability $e^{-\beta\Delta S}$. The action is given by

$$S=\sum_{\square}(1-W_{\square}), \tag{2.1}$$

where the sum is extended over all elementary squares of the lattice (plaquettes) and the Wilson factor for any closed loop $\gamma(i_1i_2i_3\ldots i_Ni_1)$ is defined as

$$W_{\gamma}=\tfrac{1}{2}\,\mathrm{tr}\left(U_{i_1i_2}U_{i_2i_3}\ldots U_{i_Ni_1}\right). \tag{2.2}$$

To optimize the rate of convergence to statistical equilibrium, six upgrading steps are performed on any definite link, keeping the values of all the other U_{ij}'s fixed. Then a new spin is upgraded, and so on, until all the U_{ij}'s have been analyzed, thus completing a Monte Carlo iteration. The sequence of MC iterations defines a markovian chain, where eventually the probability of encountering any configuration Σ of the gauge variables becomes proportional to the Boltzmann factor $e^{-\beta S(\Sigma)}$. Quantum mechanical expectation values of observables $A(\Sigma)$ may then be approximated by

$$\langle A\rangle\approx\frac{1}{N}\sum_{i=i_0+1}^{i_0+N}A(\Sigma_i), \tag{2.3}$$

where the Σ_i's are the configurations occurring in the sequence.

We have started from a completely random system at $\beta=0$ and have gradually increased β in steps of $\Delta\beta=0.01$ up to $\beta=4.5$, performing, in general, one iteration at each value of β. The procedure corresponds to a slow cooling of the whole system, which should maintain it reasonably close to statistical equilibrium at each step. Measurements of the internal energy, $E\equiv\langle S_{\square}\rangle$, produced results in agreement with independent determinations of $E(\beta)$ from longer MC runs or MC simulations with different initial conditions, thus giving assurance that any departure from equilibrium cannot be too drastic.

27 values of β have been singled out ($\beta = 0.25, 0.5, \ldots, 1.5, 1.6, 1.7, \ldots, 3, 3.25, 3.5, \ldots, 4.5$). At these values of β, two additional MC iterations have been performed and then the whole set of gauge variables was recorded on magnetic tape. Thus we obtained 27 configurations of the system in approximate statistical equilibrium at the corresponding values of β. Some of the measurements have then been made averaging the observables over these configurations. For other measurements, aimed at greater statistical accuracy, some of the 27 configurations have been taken as initial states and averages have been made over several further MC iterations, performed now at fixed β. The quantities which we have measured are the following:

(i) the expectation value $E \equiv \langle S \rangle$ of the action of an individual plaquette;

(ii) the expectation values of Wilson factors $\overline{W}(m,n) \equiv \langle W_\gamma \rangle$ for m by n rectangular loops, extending up to $m = n = 6$;

(iii) the plaquette-plaquette correlation function

$$C(d) \equiv \langle W_\Box W_{\Box'} \rangle - \langle W_\Box \rangle^2 , \tag{2.4}$$

where $\Box$ and $\Box'$ stand for two plaquettes facing each other and separated by d lattice links;

(iv) the expectation value

$$Q(d) \equiv \langle W_\lambda W_{\lambda'} \rangle \tag{2.5}$$

of the product of loop factors associated with parallel straight segments λ and λ' running through the lattice and separated by d lattice links. These segments define closed loops by virtue of the periodic boundary conditions. The topology of the lattice is indeed that of a torus, and the paths λ and λ' wind around the torus once.

Whenever we say that an average has been taken over a lattice, we mean that the quantity has been averaged over all possible locations and orientations within the lattice. Thus, for instance, a measurement of $C(d)$ over a single lattice configuration involves an average over 6 (orientations of the plaquettes) $\times\ 16^4$ (locations within the lattice) $= 393\,216$ numbers. In the graphs we shall frequently reproduce error bars. When the quantity has been measured averaging over many MC iterations, the error has been inferred from the fluctuations in the averages taken over the individual lattices. When the averages have been taken over a single configuration, the errors have been estimated partioning the lattice into four sublattices and evaluating the fluctuation of the averages over the sublattices.

In sect. 3, we describe our results and discuss their physical implications.

3. Results and interpretation

3.1. THE EXPECTATION VALUE OF THE ACTION

Fig. 1 gives an over-all view of the function $E(\beta)$ for the gauge model under consideration. The numbers have been obtained with a rather small lattice (4^4), doing a simulation of a thermal cycle, where β is first decreased from 6 to 0 ($\times$), and then increased back to 6 ($\cdot$) in small steps after each MC iteration, and from mixed phase runs. In these mixed phase runs, we start from a lattice which is half ordered ($U_{ij}=1$) and half disordered (U_{ij} chosen at random), and perform several iterations at fixed β until E appears to converge to a definite value. Mixed phase runs overcome some of the problems associated with metastabilities, especially near first-order phase transitions. The purpose of fig. 1 is to show that the system undergoes a discontinuous transition at $\beta_c \approx 5.9$ (see also ref. [4]), without noticeable critical points for lower values of β.

A more refined measurement of $E(\beta)$ is presented in fig. 2. The values are those obtained averaging over the individual lattices of size 16^4 encountered while increasing β from 0 to 4.5, as described in sect. 2. The continuous lines show Padé approximants of the strong coupling expansion carried to order J^{20}, where

$$J=\int \mathrm{d}U(\tfrac{1}{2}\operatorname{tr}U)\mathrm{e}^{\beta \operatorname{tr}U/2} \Big/ \int \mathrm{d}U\mathrm{e}^{\beta \operatorname{tr}U/2}$$

$$=\frac{\partial}{\partial\beta}\ln\left[\int_0^{2\pi}\mathrm{e}^{\beta\cos\theta}\sin^2\theta\,\mathrm{d}\theta\right]. \tag{3.1}$$

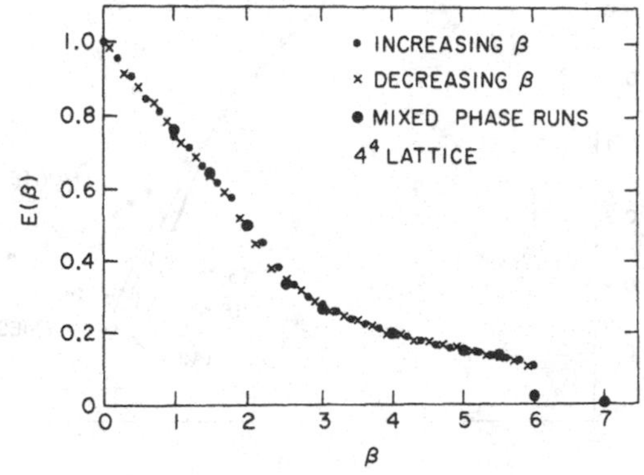

Fig. 1. Relation between internal energy $E=\langle S_{\square}\rangle$ and inverse temperature β in the $\tilde{Y}$ gauge model (determined from a 4^4 lattice).

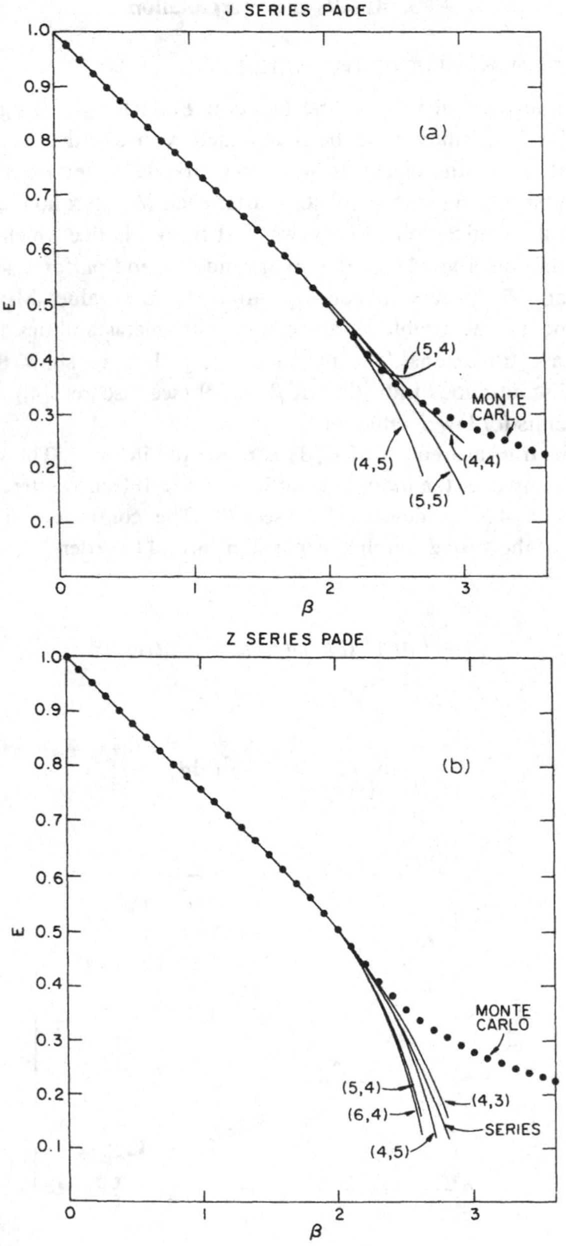

Fig. 2. The relation $E = E(\beta)$ determined with better statistical accuracy on a 16^4 lattice ($\cdot$) and fits (solid lines) obtained by Padé approximants of the strong coupling series.

The strong coupling series

$$E = 1 - J\{1 + 4J^4 - 8J^6 + \tfrac{188}{3}J^8 - 196.9975309J^{10} + 1349.23786J^{12}$$
$$-5939.062747J^{14} + 34280.32143J^{16}$$
$$-175554.005J^{18} + 977457.232J^{20} - \cdots\}, \tag{3.2}$$

has been derived by K. Wilson, who has kindly allowed us to reproduce his results. Fig. 2a illustrates the Padé approximants obtained directly from eq. (3.2), while in fig. 2b the approximants after the conformal mapping $J^2 = Z/(1-2Z)$ are displayed. Series and numerical results agree very well up to $\beta \approx 2$, after which value the results of the MC computation exhibit a rather swift departure from the predictions of the strong coupling expansion.

3.2. WILSON LOOP FACTORS

We have measured $\overline{W}(m,n)$ for rectangular loops of size up to 6 by 6. The measurements have been done over the individual lattices recorded on magnetic tape at the 27 preselected values of β and also, for $\beta = 2.25$, 2.5, 2.75, 3 and 3.25, over the last 30 lattices of long MC runs (50 iterations) at fixed β (starting from the configurations on tape at the same or the closest value of β). These last runs were made to obtain results with good statistical accuracy.

A signal of confinement is the presence in $-\ln \overline{W}$ of a term growing like the area enclosed by the loop. To isolate such a term, it is convenient to compare the expectation values of W_γ for square loops and rectangular loops with the same perimeter, and differing only by one lattice unit of area. Thus we measured the quantities

$$K(I,\beta) = -\ln \frac{\overline{W}(I,I)}{\overline{W}(I+1,I-1)}. \tag{3.3}$$

In so far as $\overline{W}$ is dominated by a term growing like the area, $K(I,\beta)$ gives the coefficient of this term and is therefore directly related to the physical string tension:

$$K = a^2\sigma, \tag{3.4}$$

a being the lattice spacing. In the scaling regime, the whole dependence of K on β should be in the relation between β and a, as predicted by the renormalization group:

$$a^2(\beta) = \frac{1}{\Lambda_0^2}\left(\tfrac{6}{11}\pi^2\beta\right)^{102/121} \exp\{-\tfrac{6}{11}\pi^2\beta\}, \tag{3.5}$$

Λ_0^2 being the lattice regularization parameter.

Indeed, one expects eq. (3.4) to be true only for sufficiently large I at any given β. If we keep I fixed and increase β, thus decreasing the lattice spacing, eventually the physical size of the loop becomes small and the expectation value of W_γ can be computed in perturbation theory. One then finds a large β behaviour given by a power series in $1/\beta$. Thus we expect the scaling behaviour expressed by eqs. (3.4) and (3.5) to appear only as an envelope of the curves $K(I,\beta)$ determined for different values of I.

In fig. 3a we plot all the values we have measured for $K(I,\beta)$ up to $I=5$, whereas in fig. 3b we reproduce the points (and errors) determined through the longer MC runs. From fig. 3a (where the first term in the strong coupling expansion of the string tension is also reproduced) it is apparent that all the curves $K(I,\beta)$ agree, for low β, with the strong coupling expansion, then follow for a while the scaling behaviour until the weak coupling perturbative regime takes over. Of the points with better statistics (fig. 3b) we have selected those which appear to lie closer to the envelope of the curves $K(I,\beta)$ (i.e., the points with $I=2,3,4$ at $\beta=2.25, I=4,5$ at $\beta=2.5, I=5$ at $\beta=2.75$) and have determined the ratio σ/Λ_0^2 fitting the scaling curve [eqs. (3.4) and (3.5)] through these points. We find

$$\Lambda_0=(0.011\pm0.002)\sqrt{\sigma}\,, \tag{3.6}$$

which agrees with a previous measurement by Creutz [2].

3.3. THE GLUEBALL MASS

On general grounds, the plaquette-plaquette correlation function $C(d)$ [see eq. (2.4)] is expected to behave as

$$C(d)\sim C_0\exp\{-\mu(\beta)d\} \tag{3.7}$$

for large separation.

In the scaling limit ($\beta\to\infty$) the dimensionless mass gap $\mu(\beta)$ should go to zero with $a(\beta)$, in such a way that the physical mass

$$M=\mu a^{-1} \tag{3.8}$$

approaches a finite limit. This is interpreted as the mass of the lowest quantum state (the glueball) coupled to the $F_{\mu\nu}^{\ a}F^{\mu\nu}_{\ \ a}$ gauge-invariant operator.

In practice, we are limited to measurements with finite values of d. This is partly because of the finite extent of the lattice (d cannot be larger than 8 on a 16^4 lattice), but especially because the number to be measured decreases very rapidly as d increases and the signal soon becomes buried in the noise of statistical fluctuations. Thus we have measured a d-dependent mass, defined according to the equation

$$\mu(d,\beta)=-\ln\frac{C(d)}{C(d-1)}\,. \tag{3.9}$$

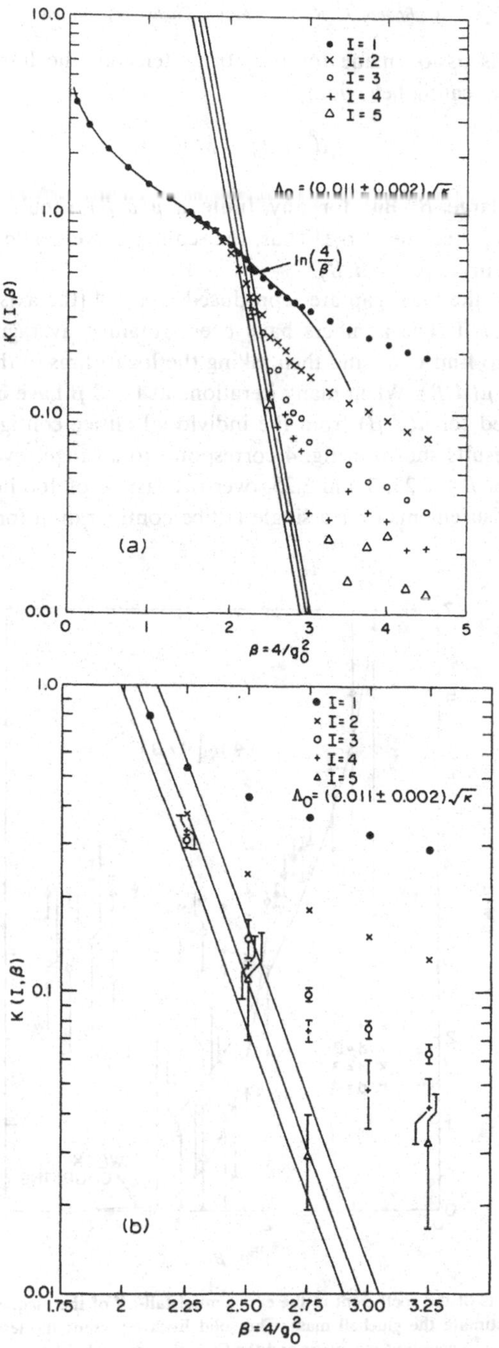

Fig. 3. Increments of the negative logarithm of $\langle W_\gamma \rangle$ as the area enclosed by the loop γ increases by one lattice unit, based on the comparison of $I \times I$ square loops with $(I+1) \times (I-1)$ rectangular loops. (a) reproduces all data, (b) only those with better statistical accuracy. The solid lines represent the leading term in the strong coupling expansion and the fit to the scaling behaviour.

Repeating the discussion made for the string tension, the limit of μ for $d \to \infty$ should follow the scaling behaviour

$$\mu(\infty, \beta) \sim Ma(\beta) \tag{3.10}$$

for sufficiently large β. But for any finite $d, \mu(d, \beta)$ should eventually behave perturbatively as β becomes large. Thus, the scaling curve would appear only as the envelope of the curves $\mu = \mu(d, \beta)$.

Our results for the mass gap are reproduced in fig. 4 [the axis is labelled $aM(d)$, rather than $\mu(d, \beta)$]. The numbers have been obtained averaging the correlations over one lattice to find $C(d)$ and then taking the logarithms of the ratios $C(d)/C(d-1)$ to evaluate $\mu(d, \beta)$. When many iterations at fixed β have been performed, the numbers obtained for $\mu(d, \beta)$ from the individual lattice configurations have been averaged. The results shown in fig. 4 correspond to averages over the last 30 of 50 MC iterations for $\beta = 2.25$, 3 and 3.25, over the last 70 of 100 iterations for $\beta = 2.5$ and 2.75, to measurements over a single lattice configuration for all other values of

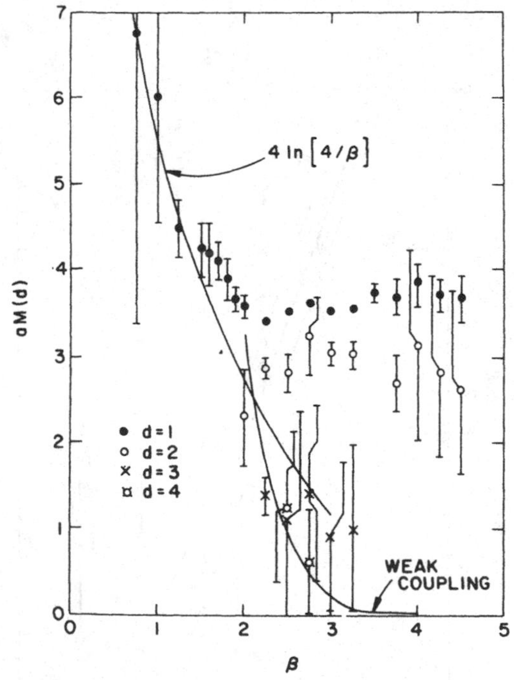

Fig. 4. Measurements of the coefficient in the exponential fall-off of the plaquette-plaquette correlation function, used to estimate the glueball mass. The solid lines represent the leading term in the strong coupling expansion and the fit to the scaling behaviour.

β. The difficulty of the measurements, due to the statistical fluctuations, is indicated by the large errors, and by the fact that averages over 30 MC iterations have been necessary in order to obtain meaningful signals for $d=3$ and over 70 iterations for $d=4$.

The solid lines in fig. 4 represent the lowest order strong coupling prediction and the (weak coupling) scaling behaviour of eq. (3.10). Assuming that the points with $d=2$ for $\beta=2, d=3$ for $\beta=2.25$ and 2.5, $d=4$ for $\beta=2.5$ and 2.75, belong to the envelope of the functions $\mu(d,\beta)$, a fit of the scaling curve through these points gives:

$$M=(3\pm1)\sqrt{\sigma}\,. \tag{3.11}$$

3.4. THE INTERQUARK POTENTIAL

The expectation value $\overline{W}(d,d')$ of Wilson factors for rectangular loops is expected to behave as

$$\overline{W}(d,d')\sim\exp\{-d'a\phi(d)\} \tag{3.12}$$

in the limit $d'\gg d, \phi(d)$ being the energy of two static sources (quark-antiquark system) at a distance $r=da$. If we take $d'=16$, the maximum length permissible with our lattice, $\overline{W}(d,16)$ will not differ in an appreciable way from the expectation value $Q(d)$ of the product of Wilson factors for parallel loops at separation d running through the lattice. Hence one also expects

$$Q(d)\approx\exp\{-16a\phi(d)\} \tag{3.13}$$

and a measurement of $Q(d)$ ought to allow a determination of the interquark potential at distances $r=da$.

The correct interpretation of $Q(d)$ is, however, more subtle. With a lattice extending for d' sites along the time direction and periodic boundary conditions, and of infinite extent along the spatial directions, one can demonstrate the relation

$$Q(d)=\exp\{-V(d)/T\}, \tag{3.14}$$

where $T=1/ad'$ is the physical temperature and $V(d)$ is the free energy of the system. Neglecting the entropy term in $V(d), \phi(d)$ may still be interpreted as an interquark potential, but not at vanishing temperature. In particular, if T is large enough, the system may be in the Debye screened phase [6, 10] and a measurement of $Q(d)$ would not reveal a confining force between static sources. An infinite extent of the lattice along the spatial directions is, in principle, necessary to produce a critical behaviour, but, as is often the case with Monte Carlo computations, a signal of the deconfining phase transition is observed even with a lattice of rather limited spatial extent.

Monte Carlo simulations give $T_c \simeq 27\Lambda_0$ for the temperature of the decoupling phase transition in the SU(2) model [6]. Combining the requirement $T = 1/a(\beta)d' < T_c$ with the functional dependence of the lattice spacing a on β [eq. (3.5)], one finds, for each value of d', the maximum β (and therefore the lowest value of the bare coupling constant $g_0 = 2/\sqrt{\beta}$) which still avoids the phase transition. (In particular, the larger d' is, the lower is the minimum safe value of g_0, which shows the advantage of working with a large lattice, as discussed in sect. 1.) On the other hand, for low β, the expectation value of the product of Wilson factors for loops running through the lattice is very small and, as soon as the loops are separated, it becomes indistinguishable from statistical fluctuations. To obtain meaningful numbers for $Q(d)$, we measured this quantity at the largest value of β (= 4.5) met in our computation and the expectation values were determined averaging over 100 lattice configurations in a long MC run (at fixed β, with initial configuration the one recorded on magnetic tape, obtained, we recall, at the end of a cooling process of $450 + 2 \times 27 = 504$ MC iterations). The results for the interquark potential are reproduced in fig. 5, where we plot the shifted potential $a\tilde{\phi}(d) = a\phi(d) - c$ ($c = 0.161$).

$\beta = 4.5$ is larger than the value of β obtained from $16a(\beta) = 1/T_c$; thus we may be measuring the Debye screened potential rather than the confining potential among static quarks. But the Debye screening length is then much larger than the maximum separation allowed by our computation, and the string tension, on the

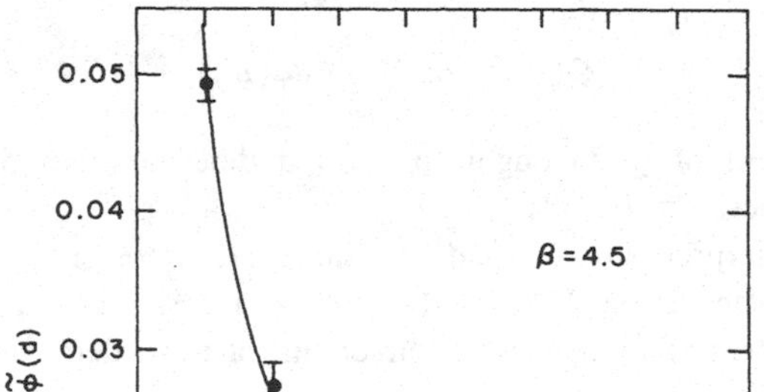

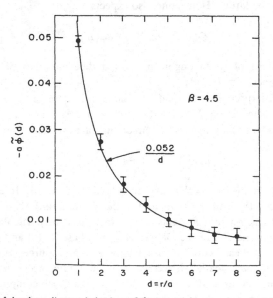

Fig. 5. Estimate of the short-distance behaviour of the potential between quark and antiquark, based on the analysis of parallel loops winding through the lattice.

other hand, has become so small that a confining part of the potential would also appear only at much larger distances. Indeed, we have fit our results for $a\phi(d)$ both to a Debye screened form

$$a\phi(d)=-\frac{\alpha e^{-\mu d}}{d}+c, \tag{3.15}$$

and to a linear plus Coulomb form

$$a\phi(d)=-\frac{\alpha}{d}+a^2\sigma d+c, \tag{3.16}$$

obtaining

$$\alpha=0.053, \quad \mu=-0.052, \quad c=0.164, \tag{3.17}$$

and

$$\alpha=0.052, \quad c=0.161 \tag{3.18}$$

in the two cases. Thus, whether the system is in the deconfined phase or not, we are effectively measuring the Coulomb part of the potential at very small separation ($a(\beta)\approx 0.0019\sigma^{-1/2}$, for $\beta=4.5$), obtaining the result

$$\phi_{\text{Coulomb}}(r)\approx\frac{0.052}{r}, \quad r\approx(0.002\text{–}0.016)\sigma^{-1/2}. \tag{3.19}$$

The smallness of the coefficient is not surprising, in view of the very short scale of distances.

4. Conclusions

The results presented in this article illustrate once again, we believe, the power of Monte Carlo computations. The confining properties of the SU(2) lattice gauge theory and its scaling behaviour toward the continuum limit emerge clearly. Moreover, various numbers of physical significance can be determined. We have confirmed previous measurements of the string tension and have evaluated the mass of the glueball and the parameters of the interquark potential at short distances.

But the limitations of the method should also be apparent. One computes averages over quantum mechanical fluctuations, suitably simulated. The presence of large errors of statistical nature is often unavoidable, especially when the quantities to be measured are small. The errors can be reduced only by averaging over longer Monte Carlo runs, but a reduction of the error by one order of magnitude would typically require a hundred times more data and would make the computation prohibitively costly.

There are many other quantities of physical interest which can be measured, and refinement of the results obtained up to now is still possible. But perhaps the most interesting future applications of the Monte Carlo method will consist of investigations where the power of computer simulations coupled with suitable analytical techniques will allow some insight into the mechanism of confinement.

Many interesting discussions with Mike Creutz and Kenneth Wilson are gratefully acknowledged. We also wish to thank Kenneth Wilson for allowing us to reproduce his strong coupling results.

Note added in proof

After completion of the manuscript we have learnt of a work by B. Berg [11], where results on the glueball mass in agreement with ours are presented.

References

[1] K. Wilson, Cargèse Lecture Notes (1979);
M. Creutz, L. Jacobs and C. Rebbi, Phys. Rev. Lett. 42 (1979) 1390; Phys. Rev. D20 (1979) 1915
[2] M. Creutz, Phys. Rev. Lett. 43 (1979) 553; Phys. Rev. D21 (1980) 2308; Phys. Rev. Lett. 45 (1980) 313
[3] C. Rebbi, Phys. Rev. D21 (1980) 3350
[4] D. Petcher and D. Weingarten, Indiana University preprint (1980)
[5] L. Jacobs and C. Rebbi, J. Comp. Phys., to be published
[6] J. Kuti, J. Polonyi and K. Szlachanyi, Central Research Institute for Physics, Budapest, preprint (1980);
L.D. McLerran and B. Svetitsky, SLAC preprint (1980)
[7] A. Hasenfratz, E. Hasenfratz and P. Hasenfratz, CERN preprint (1980);
C. Itzykson, M.E. Peskin and J.B. Zuber, Saclay preprint (1980)
[8] K. Ma, Phys. Rev. Lett. 37 (1976) 461;
R.H. Swendsen, Phys. Rev. Lett. 42 (1979) 859; Phys. Rev. B20 (1979) 2080;
H.W. Blöte and R.H. Swendsen, Phys. Rev. Lett. 43 (1979) 799; Phys. Rev. B20 (1979) 2077;
C. Rebbi and R.H. Swendsen, Phys. Rev. B21 (1980) 4094
[9] N. Metropolis, A.W. Rosenbluth, A.H. Teller and E. Teller, J. Chem. Phys. 21 (1953) 1087
[10] A.M. Polyakov, Phys. Lett. 72B (1978) 477;
L. Susskind, Phys. Rev. D20 (1979) 2610
[11] B. Berg, Phys. Lett. 97B (1980) 401

PHYSICAL REVIEW D VOLUME 27, NUMBER 2 15 JANUARY 1983

Heavy quark potential in SU(2) lattice gauge theory

John D. Stack
Department of Physics, University of Illinois at Urbana-Champaign, 1110 West Green Street, Urbana, Illinois 61801
(Received 7 September 1982)

Using Monte Carlo methods and the icosahedral approximation to SU(2), the heavy quark potential has been determined for SU(2) lattice gauge theory. The range of distances covered is 0.05 fm to 1 fm for a string tension of 400 MeV. The results are compared to perturbation theory, string theory, and the phenomenological potential of Martin. A simple Coulomb plus linear form gives a good overall representation of the potential.

I. INTRODUCTION

The chromodynamic potential between a heavy quark and antiquark arises from a fundamental force and as such certainly ranks in importance with the Coulomb potential between electric charges and Newton's gravitational potential between bodies with mass. Yet in contrast to its two more famous predecessors, the quark potential does not admit a simple theoretical treatment. Even with infinitely heavy sources, the calculation of the potential is highly nontrivial, due to the nonlinear dynamics of gauge fields. Computer simulations of lattice gauge theories are currently playing an important role in understanding this dynamics. This paper reports on our Monte Carlo calculations of the heavy quark potential for SU(2) lattice gauge theory without fermions. This theory is believed to have a heavy quark potential which is similar in all its features to that for lattice QCD, but is obviously simpler to treat computationally. Actually, in order to work on relatively large lattices, we simplified further and replaced SU(2) by its 120-element icosahedral subgroup $\bar{I}$. However, the physics remains that of SU(2), since in the region of interest to us here, Petcher and Weingarten,[1] and Bhanot and Rebbi[2] have previously shown that the 120-element subgroup produces results indistinguishable from full SU(2).

Our results, which build on previous calculations of the string tension K, determine the continuum heavy quark potential up to a single overall constant for SU(2) pure gauge theory, over the range $0.1 \leq x \leq 2.0$, where $x = R/\xi$, R is the interquark separation, and ξ is the correlation length which is essentially $1/\sqrt{K}$. The precise definition of ξ is given in (1) and (2) below. In the units of hadron physics, this range of separations is equivalent to $0.05\ \text{fm} \leq R \leq 1\ \text{fm}$, if $\sqrt{K}$ is given the realistic value of 400 MeV. This is the important range of distances for ψ and Υ spectroscopy and is also a range where analytic methods are ineffective at present, giving a strong motivation for extending the results obtained here to QCD itself.

The x values we probed in this calculation were determined by our computer resources. In any calculation on a finite lattice, there are three characteristic lengths on which a physical quantity can depend in principle; the lattice spacing a, the correlation length ξ, and the lattice size L. For the continuum limit, we want the system to have "forgotten" a and L, leaving ξ as the only relevant length. Formally, this requires $\xi/a \to \infty$ and $L/\xi \to \infty$. However, as first shown by Creutz,[3] Monte Carlo calculations on a finite lattice in $d=4$ show continuum renormalization-group behavior for $\xi/a \gtrsim 2$, so the system loses its dependence on a extremely rapidly. As ξ grows, finite-lattice-size effects will eventually cause distortions in any physical quantity when ξ becomes comparable to the lattice size L. The two requirements $\xi/a \gtrsim 2$ and $\xi/L \lesssim 1$ determine a range of couplings for which continuum information can be obtained in a given calculation. The present work was done on lattices $16a \times L^3$ in size, at various L from $L = 8a$ to $16a$. To determine the corresponding β values, we need a measure of ξ. The string tension K sets the physical length scale in considerations of the quark potential and it is natural to define ξ to be proportional to $1/\sqrt{K}$, with a coefficient of ~ 1. To be specific, we define

$$\xi = 0.012/\Lambda_L , \tag{1}$$

where Λ_L is the usual two-loop expression for the Λ parameter on a Euclidean lattice,

$$\Lambda_L = (1/a)(6\pi^2\beta/11)^{51/121}\exp(-3\pi^2\beta/11) . \tag{2}$$

We chose the coefficient of $1/\Lambda_L$ in (1) to be close to previous Monte Carlo estimates of $\Lambda_L/\sqrt{K}$.[1-4]

Figure 1 shows the plot of ξ vs β, using (1). As can be seen from a glance at this figure, if we require that $2 \lesssim \xi/a \lesssim 16$, the present calculation can make contact with the continuum limit for the approximate interval $2.3 \lesssim \beta \lesssim 3.1$.

Allowing ξ/L to range up to ~ 1 may be overly generous. However, by studying various values of L/a from 8 to 16, we have found that distortions due to finite lattice size are less severe in SU(2) lattice gauge theory than in a typical Abelian gauge theory, where the more stringent requirement $\xi/L \lesssim \frac{1}{3}$ is reasonable.[5] Even allowing $\xi \sim L$, Fig. 1 shows that it will be difficult with present-day computers to study the continuum limit for β significantly larger than 3.0, due to the exponential growth of ξ with β.

We deduced the quark potential from Monte Carlo data for rectangular Wilson loops, $W(T,R)$. As described in more detail in Sec. II, runs were made at a number of different β values in the interval $2.2 \lesssim \beta \lesssim 3.1$. (The values $\beta=2.2$ and 2.25 violate somewhat our criterion that ξ be at least two lattice spacings. They were included to try to gain more information on the large-R potential.) In this interval, the quark potential $V(R)$, which has had the quark self-energy removed, hopefully has negligible dependence on the lattice size and lattice spacing, so we can regard ξV as a scaling function, i.e., a function solely of $x=R/\xi$. We calculated Wilson loops only for $R/a=1,2,3,4$ so each value of β gives the potential at just four values of x. However, by varying β, we can vary x continuously, even though R/a is discrete. The number of different x values at

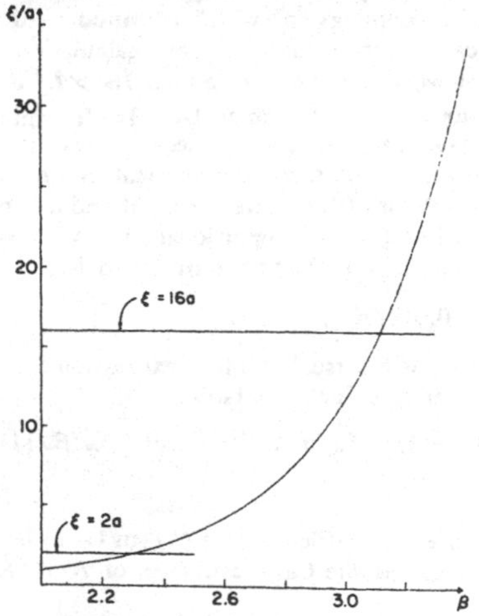

FIG. 1. ξ vs β, using (1) and (2).

which we can calculate the potential and the spacing between them is then limited only by the available computer time. The largest x values ($x_{max} \sim 2$) come from the smallest β values, while the smallest ($x_{min} \sim 0.1$) come from the largest β values. In this way, we can map out ξV as a function of x, in principle as accurately as desired, over the interval $x_{min} \leq x \leq x_{max}$. The assumption that the results do not depend on L can be tested by working on lattices of different size, and the assumption that the results are independent of the lattice spacing a can be tested by seeing how well scaling works.

Even in the scaling region, the potential deduced directly from Wilson loops is not the continuum potential $V(R)$, but rather a quantity we call the lattice potential and denote by $V_l(R)$. For T sufficiently large compared to R we have

$$W(T,R) \sim \exp[-TV_l(R)] \ . \tag{3}$$

The relation between the continuum and lattice potentials is

$$V_l(R) = V(R) + V_0 \ , \tag{4}$$

where V_0 is lattice-spacing dependent and R independent, i.e., $V_0 = g(\beta)/a$, for some function $g(\beta)$. Physically, V_0 represents the self-energy of the infinitely heavy quarks which sit on the perimeter of the Wilson loop. It contributes a term proportional to the perimeter of the loop in the exponent of $W(T,R)$ and means that $W(T,R)$ is not a scaling function, but has dependence on R/a and T/a in addition to R/ξ and T/ξ. If we were using only one value of β, V_0 would be a harmless constant. But in order to merge results at different β values together and use scaling, it is essential to eliminate V_0. Our procedure for doing this is described in detail in Sec. III. In essence, it is very simple. Since V_0 is independent of R, it does not appear in the force between two quarks. The force at a set of discrete points in x can be gotten by forming the differences $[V_l(R+a) - V_l(R)]/a$. (In practice, we use only $R=a$. See Sec. III for further discussion.) Putting a smooth curve through these points and integrating numerically, we get a determination of the continuum potential up to a constant of integration. After this arbitrary constant is chosen, a value of V_0 can be determined at each β by applying (4) at, for example, $R=a$. We can then test the self-consistency of the whole procedure. Removing $V_0(\beta)$ from the lattice potentials $V_l(R)$, multiplying by ξ, and plotting versus x should map all the Monte Carlo data onto a smooth curve which represents the continuum potential. This works quite well and is a nontrivial test, since only half the Monte Carlo data are used in the numerical integration.

In Sec. IV, we discuss various fits to the potential in both large- and small-distance regions, and using a phenomenological potential applied very successfully to heavy quark spectroscopy by Martin. We find that a good overall description of the potential can be obtained with a simple linear plus Coulomb fit.

II. MONTE CARLO METHOD AND WILSON LOOPS

A. Monte Carlo parameters

We used the Wilson action, namely,

$$S=\frac{\beta}{2}\sum_{\bar{k}\,\bar{l}\,\bar{x}} \mathrm{tr}[1-U_k(\bar{x})U_l(\bar{x}+\bar{k}a)U_k^\dagger(\bar{x}+\bar{l}a)U_l^\dagger(\bar{x})]\,. \tag{5}$$

The link variable $U_k(\bar{x})$ is an element of the group $\bar{I}$, in the representation which is the restriction to $\bar{I}$ of the fundamental representation of SU(2). [If we refer to the SU(2) quantum number as isospin, then the first five irreducible representations of SU(2) with isospin $\frac{1}{2}$, 1, $\frac{3}{2}$, 2, and $\frac{5}{2}$ are also irreducible when restricted to $\bar{I}$.]

Given the action, the goal of the Monte Carlo process is to generate a sequence of configurations [sets of $U_k(\bar{x})$] with weight proportional to $\exp(-S)$. The basic step is the "link update," which begins by generating a trial link $U_k'(\bar{x})$, by multiplying the present link with a group element randomly chosen from the 12 nearest neighbors of the group identity. The decision to accept the trial link $U_k'(\bar{x})$ or retain $U_k(\bar{x})$ is then made by applying the standard Metropolis algorithm.[6] In our calculations, the rate at which trial links were accepted varied from $\sim 40\%$ near $\beta=2.3$ to $\sim 25\%$ near $\beta=3.1$.

The restriction of the gauge group to $\bar{I}$ has a number of computational advantages.[2] From the viewpoint of memory requirements, a link variable in $\bar{I}$ can be represented by a 7-bit integer and 8 links can then be stored in one 60-bit computer word. In addition, the use of Boltzmann factor tables, group character and multiplication tables, plus standard Boolean operations (shift, mask, etc.), are useful in reducing execution time. By applying these techniques, plus tinkering with innermost loops, the method of indexing links, etc., we arrived at a final program in which a link update took approximately 50 μsec on a CDC Cyber 175.

A typical run consisted of $\sim 10^8$ link updates. The number of sweeps through the lattice varied, depending on lattice size. For example, at $\beta=2.3$, where $\xi\sim 2a$, we ran 350 sweeps through a 16×8^3 lattice, updating each link 10 times. On the other hand, at $\beta=3.1$, $\xi\sim 16a$ and it was important to run on a large lattice. Here we ran 80 sweeps through a 16^4-lattice, updating each link five times. All of our runs began with a completely ordered configuration which had the group identity on every link of the lattice. Periodic boundary conditions were used for three directions (those corresponding to L^3), while helical boundary conditions were used for the remaining direction which always had side $16a$. Data were taken at β values from $\beta=2.2$ to 2.9 in steps of 0.05, as well as at the isolated values $\beta=2.325$, 3.0, 3.1.

Wilson loops $W(T,R)$ with $T/a=2$–8 and $R/a=1$–4 were measured every sweep through the lattice. The time spent measuring Wilson loops varied from 10 to 20% of the total execution time. The loops measured were those oriented the "long" way on our $16a\times L^3$ lattices, where $L=8a$ and $10a$ was used for $\beta<2.6$ (with the exception of $\beta=2.325$, where $L=6a$ was used), and $L=16a$ for $\beta\geq 2.6$. For $\beta\geq 2.6$, some runs were made with $L=10a$ and $13a$ to check for finite-lattice-size effects, which were found to be small in all cases. In early runs, we recorded the characters of the loops, and then averaged these over all eight nontrivial irreducible representations of $\bar{I}$. However, the resulting numbers were ridiculously small for all but the first few representations, so in later runs we recorded only the loops averaged over isospins $\frac{1}{2}$, 1, $\frac{3}{2}$, and 2. Here, we concentrate on the fundamental representation, which determines the potential between two heavy quarks with isospin $\frac{1}{2}$. Where comparison is possible, our results for fundamental representation loops are in excellent agreement with the high-statistics results obtained for full SU(2) by Berg and Stehr.[4]

B. The lattice potential

There is no hard and fast rule for judging when the lattice is in equilibrium and configurations weighted with $\exp(-S)$ are actually being generated. The trial link acceptance rate and the geometrically small Wilson loops stabilize after only a few sweeps of the lattice. However, the larger loops, which have numerically smaller values, can take somewhat longer. We generally took a conservative approach and omitted the first $\frac{1}{4}$ to $\frac{1}{3}$ of the run to ensure equilibrium, and computed the average of the loops over the rest of the run. For our runs of $\sim 10^8$ link updates, $\Delta W(T,R) \lesssim 0.007$, where $\Delta W(T,R)$ is the uncertainty in the loop mean value. The mean value of all measured loops is several times larger than $\Delta W(T,R)$ for $\beta > 2.45$, but as β decreases below 2.45, the mean value of loops with $R/a \geq 3$ and $T/a \geq 6$ becomes increasingly comparable to $\Delta W(T,R)$.

The lattice potential $V_l(R)$ measures the ground-state energy of the system of heavy quark, heavy antiquark, and gauge field. On a lattice infinite in the Euclidean-time direction, $V_l(R)$ is precisely defined by

$$V_l(R) = -\lim_{T\to\infty} \ln[W(R,T)]/T \,. \tag{6}$$

This formal limit is clearly not possible on a finite lattice. Instead we kept $T > R$, and tested for exponential behavior in T. Fortunately, in all cases, there is strong evidence for exponential behavior of $W(T,R)$ as soon as $T > R$. We got the value of the lattice potential by fitting $\ln W(T,R)$ to a straight line in T for $T > R$, the slope giving $V_l(R)$. The fit was standard least squares, with the points weighted with their inverse statistical errors. In Fig. 2, we show $\ln[W(T,R)]$ vs T for $\beta = 2.55$. The plots for other values of β look very similiar, except that the statistical errors become steadily more visible as β decreases. For 2.2 and 2.25 we were only able to determine the potential for $R/a = 1$ and 2, due to the very small values of $W(T,R)$ for $R/a = 3$ and 4.

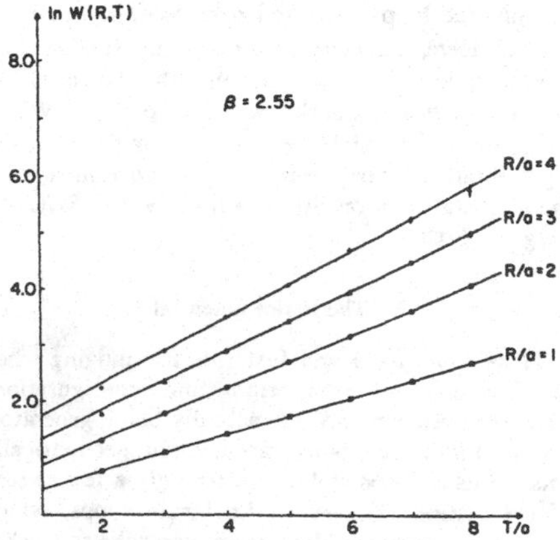

FIG. 2. $\ln W(T,R)$ vs T, at $\beta = 2.55$.

III. THE FORCE AND CONTINUUM POTENTIAL

As mentioned in the Introduction, the lattice potential does not scale, since it contains the self-energy $V_0(\beta)$. This is illustrated in Fig. 3, where we plot the dimensionless lattice potential ξV_l vs x.[7] Our goal is to get a relative determination of $V_0(\beta)$, so that removing it from $V_l(R)$ will move the sets of points which lie on different curves in Fig. 3 onto a single curve which then will determine the continuum potential.

Since $V_0(\beta)$ is independent of R, it does not affect the force between two quarks,

$$F(R) = \frac{\partial V}{\partial R} = \frac{\partial V_l}{\partial R} \,. \tag{7}$$

The approach we took was to eliminate $V_0(\beta)$ by integrating the force between two quarks with respect to x. Our values for $V_l(R)$ at $R/a = 1,2$ are statistically very accurate for all measured values of β, so these values of R/a were used to determine F. The two values $R/a = 1,2$ define x values $x_2 = 2a/\xi$, $x_1 = a/\xi$, and their difference $\Delta x = x_2 - x_1$ at each β. In terms of these quantities, we may write the dimensionless force $\xi^2 F$ as

$$\xi^2 F(x_i) = \xi[V(x_2) - V(x_1)]/\Delta x \,, \tag{8}$$

where x_i lies between x_1 and x_2 and will be specified

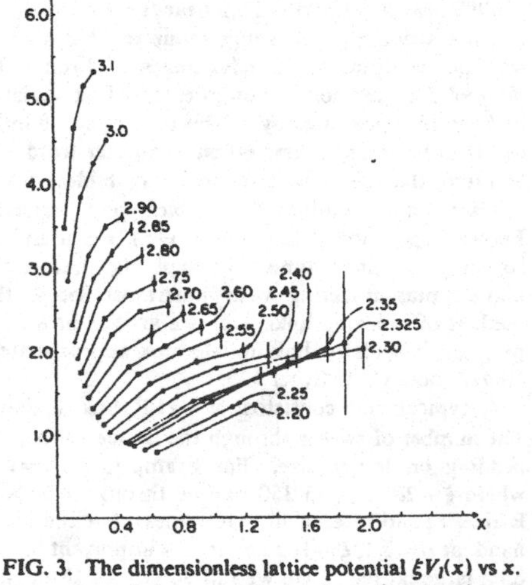

FIG. 3. The dimensionless lattice potential $\xi V_l(x)$ vs x. The same symbol represents a given β value in all succeeding plots. The solid lines connect points with the same β.

shortly. Before doing that, it is important to discuss one feature of definition (8), namely, that it tacitly assumes that the continuum limit applies at the small distances $R/a=1,2$. This cannot literally be true; for the system to "forget" the lattice spacing, it is necessary to be in the regime $R/a \gg 1$, R/ξ arbitrary. However, it is part of the lore of spin systems that once ξ is in the scaling region, spin-spin correlation functions assume their scaling forms quite accurately for small values of R/a, even $R/a=1,2$.[8,9] (Here R refers to the distance between spins.) Since Wilson loops are the analog for lattice gauge theories of spin-spin correlation functions for spin systems, it is very natural to apply a similar assumption here, with the expectation that it will introduce some small errors ($\sim 5\%$).

Although $R/a \gg 1$ is clearly inaccessible in our calculations, it might be objected that using $R/a=2,3$ or $R/a=3,4$ to determine the force would at least be a step in the right direction. We have checked that it makes a quantitatively negligible difference to use $R/a=2,3$ rather than $R/a=1,2$ to determine $\xi^2 F$. On the other hand, using $R/a=3,4$ suffers from the difficulty that the statistical accuracy of $V_l(R)$ at $R/a=4$ decreases rapidly as β falls below 2.45. Thus using $R/a=1,2$ seems to us to be the best procedure.

Now, let us turn to the assignment of the intermediate distance x_i. Since V_0 has dropped out of (8), the right-hand side of this equation is the difference of the dimensionless continuum potential $\xi V(x)$ at the points x_1 and x_2, divided by their separation Δx. (There is a small penalty to be paid for violating the condition $R/a \gg 1$, as discussed above.) From (7), $\xi^2 F(x_i)$ is $\xi(\partial V/\partial x)(x_i)$. If we were really able to take the limit $\Delta x \to 0$, then x_i could be any point in the interval $[x_1,x_2]$. However, Δx is finite, ranging from 0.06 near $\beta=3$ to 0.5 near $\beta=2.2$. Nevertheless, the mean-value theorem of elementary calculus still states that for some x_i in the interval $[x_1,x_2]$, the right side of (8) gives the exact value of $\xi^2 F$. To get a definite result for x_i, we make the assumption that the potential can locally be represented as a linear term plus a Coulomb term, i.e., $\xi V(x)=Ax-B/x$, where A and B are either constants or functions which vary slowly compared to x and $1/x$. As discussed further in the next section, this general sort of form for the potential represents the leading behavior in both large- and small-distance regions. In between, analytic methods say nothing, but is is reasonable to apply it there also, since we expect the potential to be a smooth, simple function despite the difficulty of calculating it.

Under the assumption that A and B are either constants or slowly varying compared to x and $1/x$, the right-hand side of (8) becomes $A+B/2(x_1)^2$. But this is the same as the derivative $\xi\partial V/\partial x$ (again ignoring the variation of A and B over the interval Δx), evaluated at $x_i=\sqrt{2}x_1$. This then fixes x_i. In Fig. 4, we show $\xi^2 F$ as determined by (8) vs x.

The next step is to numerically integrate $\xi^2 F$ to get ξV. This was done by putting a smooth curve through the points in Fig. 4, using a cubic spline interpolation. Cubic splines are widely used in interpolation and have the desirable properties of introducing minimal extraneous curvature and converging to the interpolated function as the separation between points goes to zero.[10] From our point of view, using interpolation here rather than some form of least-squares fit allows us to determine V_0 locally in β and does not require a commitment on the overall shape of the potential. The interval over which the interpolation is carried out ranges from $x=\sqrt{2}a/\xi(3.1)=0.092$ to $x=\sqrt{2}a/\xi(2.2)=0.895$.

Once the spline interpolation to $\xi^2 F$ is made, we numerically integrate, fixing the constant of integration by assigning ξV the arbitrary value -2.0, at $x=\sqrt{2}a/\xi(3.1)$. We can then determine the (relative) self-energy at each β by computing the difference between ξV_l and ξV as found by numerical integration. This difference is computed at the x value which corresponds to $R/a=1$ for all β values for which this point falls in the spline interpolation interval, $R/a=2$ otherwise. Having in this way determined a V_0 for each β, we can now move the lattice potentials of Fig. 3 relative to each other to

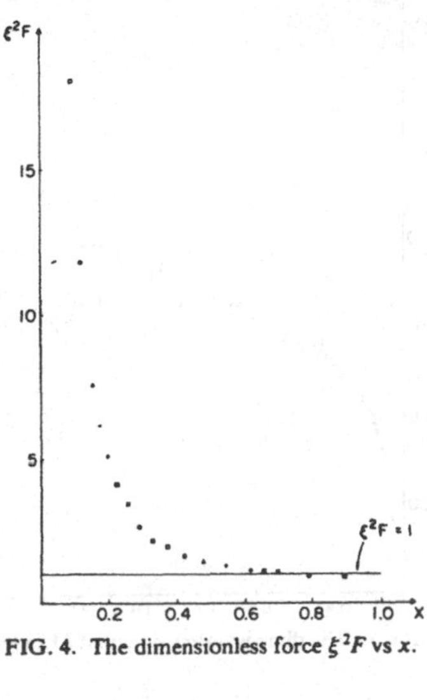

FIG. 4. The dimensionless force $\xi^2 F$ vs x.

get values which represent the continuum potential. If our procedure is self-consistent, the $R/a=3,4$ points which have not been used so far, should fall on the same curve as the $R/a=1,2$ points. The result is shown in Fig. 5. Although there are some minor violations, the overall quality of the scaling appears to be excellent.

We close this section by briefly discussing the assumptions underlying our results so far. To restate them, we have made the following assumptions. (i) The correlation length is given by (1) and (2). (ii) In the scaling region, there is only a small error introduced by working at $R/a=1,2$ rather than $R/a \gg 1$. (iii) The intermediate distance x_i is given by $x_i=\sqrt{2}x_1$, or equivalently, the potential is locally linear plus Coulomb. Note that no free, adjustable parameters have been introduced. With regard to assumption (i), it is the functional form of Eq. (2) which is important, the value for the coefficient of $1/\Lambda_L$ in (1) has been chosen merely for convenience. We have done some rough tests of this functional form by varying the exponent in (2) away from $3\pi^2/11$. The results do not scale as well as those in Fig. 4, although it would be hard to show convincingly that $3\pi^2/11$ was the correct value with our data alone. The same holds true for the $(\beta)^{51/121}$ term multiplying $\exp(-3\pi^2\beta/11)$. The form of ξ is definitely a hypothesis here from which we hope to learn about V. We have also checked assumption (iii) by varying x_i/x_1 away from $\sqrt{2}$. Here only minor variations wreck scaling, and although we have not tried to do it, our data could probably be used to show that in order for ξV to scale, x_i/x_1 must take a value near $\sqrt{2}$. Other than the fact that the results scale well, we have no way to directly check assumption (ii).

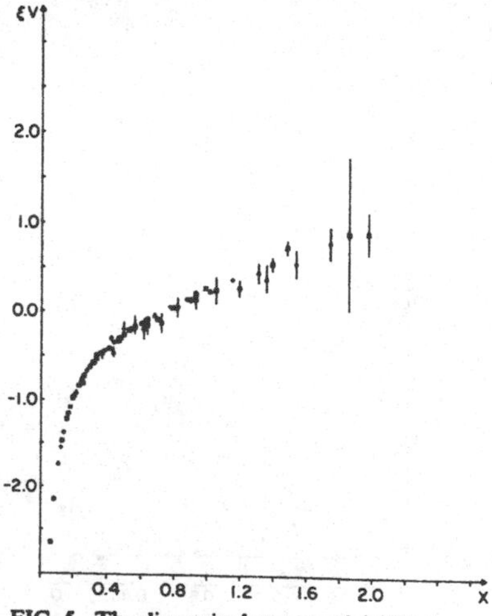

FIG. 5. The dimensionless potential $\xi V(x)$ vs x.

IV. FITS TO THE POTENTIAL

A. Theory

Before discussing the various fits we have made to our Monte Carlo potential, let us review what is known theoretically, where there is understanding only in the extreme large- and small-distance regions. For $\sqrt{K}R \gg 1$, the potential is predicted[11] to take the limiting form

$$V(R) \rightarrow KR - \pi/12R\,, \tag{9}$$

where the first term is the energy of the static chromoelectric flux tube connecting the heavy quarks, and the second term arises from the fluctuations of this flux tube or string. The $1/R$ term in (9) has not yet been derived from gauge field theory, but follows from a treatment analogous to a particle path-integral description of a propagator in field theory.[12] In the analogy, K plays the same role in the string case that the mass does in the particle case. Since a particle path integral does give the correct large-distance behavior of a massive field-theory propagator, it seems quite likely that (9) is correct. The analogy would suggest that there are further correction terms to (9), involving higher inverse powers of R, but nothing is known about this at present.

At short distances, $\sqrt{K}R \ll 1$, renormalized perturbation theory makes a precise prediction for the leading behavior;

$$V(R) \rightarrow \frac{-(9\pi/22)}{RL(R)}\left[1-\frac{102}{121}\frac{\ln[L(R)]}{L(R)}\right], \tag{10}$$

where $L(R)=2\ln(1/\Lambda' R)$. Correction terms to (10) can be of two types; (i) perturbative corrections (three loops and beyond), in which the basic structure remains a Coulomb term, modulated by logarithms and (ii) nonperturbative corrections, e.g., a term $\Lambda' R$, with modulating logarithms. The latter sort of corrections would be analogous to higher twist terms in electroproduction. It should be noted that if as assumed by some authors,[13,14] the small-R behavior of $V(R)$ is controlled by the vacuum matrix elements of local operators, then the leading nonperturbative correction to ξV is of the form $(\Lambda' R)^3$, rather than $\Lambda' R$. This difference would have a negligible effect on the values of ξV at small x which were determined in the last section. The reason is that the Coulomb term dominates at small x, and therefore the value $x_i/x_1=\sqrt{2}$ which we used is still quite accurate.

As has been discussed extensively in the literature, Λ parameters, although renormalization-group invariants, are not fundamental physical quantities like K, since they depend on the method of renormalization. Equation (10) is written in a scheme, natural in discussing the potential, in which there is no $[L(R)]^{-2}$ term on the right side of (10). The Λ parameter in this scheme, denoted by Λ', is close in magnitude to $\Lambda_{\overline{MS}}$ and Λ_{MOM} and is related to Λ_L in SU(2) pure gauge theory by $\Lambda' = 56.49\Lambda_L$.[15,16]

B. Fits to the potential

We have made a number of least-squares fits to the data for ξV determined in the last section. Most of them are motivated by our basic picture of the potential, that it can be represented locally by a Coulomb term plus a linear term. All are simple, involving at most two parameters. An attempt has been made to explore fairly thoroughly the effects of various cuts on the data; in x, the couplings included, the R/a values included, etc. Our linear plus Coulomb fits are parametrized in terms of constants $K'\xi^2$ and α as follows:

$$\xi V = (K'\xi^2)x - \alpha/x \ , \tag{11}$$

where ξ is always given by (1), and we will quote either $K'\xi^2$ or $\Lambda_L/\sqrt{K}$, along with α. (In the last section, $K'\xi^2$ and α were denoted as A and B.) The prime on K denotes the fact that unless the fit is restricted to large x, the linear term only determines an effective string tension. The prime on K will be dropped when discussing the large-x region of the data. Since it is extremely difficult to gather accurate Monte Carlo data for appreciable values of x, we will be very generous in deciding what constitutes a "large" value of x.

We begin by approaching the data from the perspective of large-distance physics. The main standard of comparison here is the previous Monte Carlo calculations of the string tension.[1-4] All of these calculations use the method introduced by Creutz,[3] in which the string tension is extracted from ratios of nearly square Wilson loops. The basic assumption of the method is that, apart from a perimeter term, the Wilson loops obey an area law. The analogous assumption for us is that the potential is purely linear. In Fig. 6, we show the linear fit to all the data with $x \geq x_{min} = 0.5$. The value of $K\xi^2$ is 0.94, and varies only slightly when we perform cuts such as eliminating the points with $\beta = 2.2$ and 2.25, which lie at the edge of the scaling region. However, fits of this type are not stable under an increase of x_{min}. The value of $K\xi^2$ steadily decreases, reaching 0.83 for a linear fit to all the data with $x \geq 0.9$. Beyond $x_{min} = 0.9$, we gradually run out of accurate

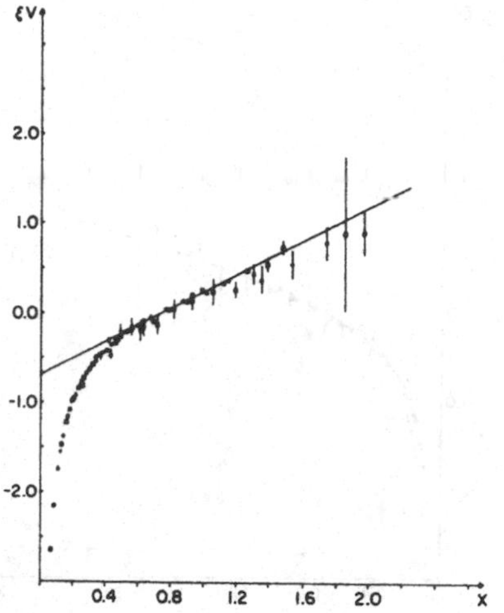

FIG. 6. Straight-line fit to ξV for $x \geq 0.5$.

data points, so the parameters for fits with $x_{min} > 0.9$ are not very well determined. A value of $K\xi^2 = 0.83$ corresponds, using (1), to $\Lambda_L/\sqrt{K} = 0.13$, which is consistent with results obtained using Creutz ratios.[1-4]

The fits just described do not really constitute an objective measure of K, since the value of $K\xi^2$ obtained was not stable. The fact that $K\xi^2$ was decreasing with increasing x_{min} suggests the presence of an attractive Coulomb term, in addition to the linear term. We have explored the possibility of a Coulomb term for $x \geq 0.5$ in two different fits. The first constrains $\alpha = \pi/12$ as in (9). The resulting fit to all data with $x \geq 0.5$ is shown in Fig. 7. The statistical quality of this fit is about the same as the purely linear fit of Fig. 6. The value of $K\xi^2$ is now 0.50, considerably smaller than for the purely linear case. As we increase x_{min}, we again find that the results change, but this time $K\xi^2$ *increases*, reaching 0.57 for $x_{min} = 0.90$. Although the behavior of the purely linear fit suggested the presence of a Coulomb term, the value $\alpha = \pi/12$ is clearly too large, and the correct value of α apparently lies between 0 and $\pi/12$. This is confirmed by a two-parameter fit to the large-x region, where we get a best value for $K\xi^2$ of 0.72, corresponding to $\Lambda_L/\sqrt{K} = 0.014$, while $\alpha = 0.11$, less than half $\pi/12$. The errors are fairly large, $\sim 15\%$ for $\Lambda_L/\sqrt{K}$, $\sim 20\%$ for α. Thus we favor a string tension somewhat smaller than that determined with Creutz ratios. This is not surprising, since as mentioned earlier, the Creutz ratio method implicitly as-

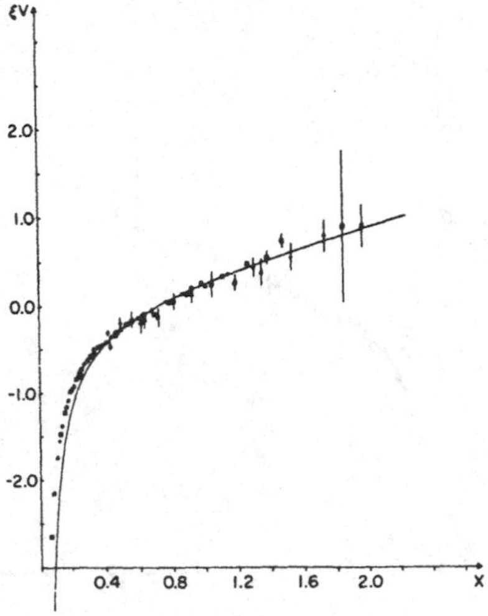

FIG. 7. Fit to ξV with $\alpha = \pi/12$, $x \geq 0.5$.

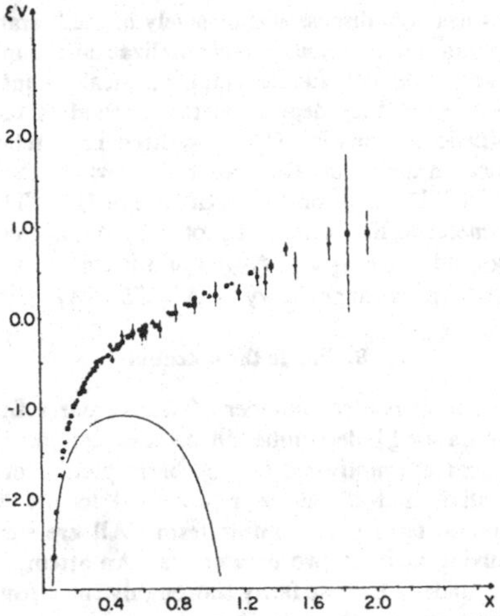

FIG. 8. Two-loop perturbation theory compared to ξV.

sumes a linear potential, whereas we claim that a Coulomb term is also present. The Coulomb term accounts for a small ($\sim 10\%$ at $x = 1$), but measurable amount of the variation of the potential over the range of x values we are probing. Since we are not yet in the region $\sqrt{K} R \gg 1$, there is no direct conflict with the prediction (9) for α.

We now turn attention to the small-x region. The obvious first question is whether perturbation theory applies. In Fig. 8, we show two-loop perturbation theory from (10), plotted versus our data. A constant has been adjusted so that (10) goes through $x = 0.065$ which is the smallest x data point. While there is brief contact between the data and perturbation theory, the data soon rise sharply above perturbation theory. As is clear from Fig. 8, perturbation theory eventually breaks down completely, but its quantitative validity probably disappears much sooner. From (10), the effective Coulomb term of two-loop perturbation theory has $\alpha = 0.155$ at our smallest x value. The one-loop approximation is approximately 25% larger. There are of course no positive powers of x in straight perturbation theory.

To gain further insight into the small-x region, we performed local two-parameter fits to the data using a Coulomb term plus a linear term over intervals $x_{\min}$ to $x_{\min} + 0.20$, with $x_{\min} = 0.0$, 0.10, 0.20, 0.30, and 0.40. In this way, we can determine an effective local value of $K'\xi^2$ and α. The results are shown in Table I. There are several features worth noting. The value of α generally decreases as smaller x values are sampled. This is in qualitative accord with asymptotic freedom. Quantitatively, α is smaller than the value obtained from (10), but this is only a 10% effect at $x = 0.065$ and could easily be accounted for there by higher orders in perturbation theory. It should also be remembered that although the Monte Carlo calculation is nonperturbative, in the short-distance region it is also approximate in that we are working on lattices with $\xi \sim L$ and using the icosahedral approximation to SU(2) which must eventually break down at short distances. Finally, from Table I there is clear evidence for a linear term at all the small-x values we can study. A large value of $K'\xi^2$ is found in the interval $x = 0.0$ to $x = 0.2$, but after that $K'\xi^2$ settles down to values of the same order of magnitude that we find at large x.

We have also studied fits using a form of the potential applied very successfully to ψ and γ spectroscopy by Martin.[17] Here the potential is parametrized by

$$\xi V = A + Bx^{\gamma} . \tag{12}$$

TABLE I. Effective Coulomb and linear terms at short distance.

α	$K'\xi^2$	$x_{\min}$	$x_{\max}$
0.135	2.132	0.0	0.2
0.158	0.962	0.1	0.3
0.176	0.700	0.2	0.4
0.170	0.635	0.3	0.5
0.180	0.674	0.4	0.6

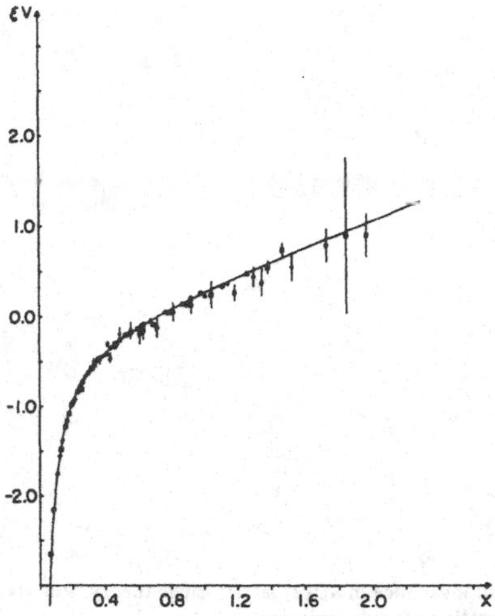

FIG. 9. Linear+Coulomb fit to all the data.

These fits do not work as well as our linear plus Coulomb fits. Further, the value of γ is very sensitive to the x interval chosen. To give three examples, a fit over the interval $x=0.15$ to $x=0.8$ gives $\gamma=-0.43$, the interval $x=0.3$ to 1.4 gives $\gamma=0.07$, while $x\geq 0.5$ gives $\gamma=0.25$.

Finally, although the effective strengths of linear and Coulomb terms both vary with x, a simple linear plus Coulomb fit to all the data actually works quite well as shown in Fig. 9. The parameters of the fit are well determined; $K'\xi^2=0.71\pm0.02$ and $\alpha=0.164\pm0.008$. This fit is very stable against various cuts, and if there were a hadron spectroscopy for SU(2) pure gauge theory, would represent the best simple parametrization of the potential to use.

V. Conclusion

Using standard Monte Carlo methods and strong but reasonable assumptions on scaling, we have shown that the spin-independent part of the heavy quark potential can be determined for non-Abelian lattice gauge theory. Two extensions of the present work are needed in order to confront directly the rich spectroscopy of heavy quark mesons; the inclusion of spin-dependent terms in the potential and the extension to SU(3) as the gauge group. Both appear to be feasible with present-day computers, although it will clearly be necessary to work on smaller lattices than those used here for the case of an SU(3) gauge group. This may not be a difficulty, since we found that finite-lattice-size effects were small for SU(2), and recent work has suggested that they become completely negligible for SU(N) as $N\rightarrow\infty$.[18–20] Thus lattices considerably smaller than 16^4 may suffice for SU(3). We hope to report on the spin-dependent part of the potential and the extension to SU(3) in future work.

ACKNOWLEDGMENTS

I would like to thank John Kogut for discussions and help with some of the calculations, and the University of Illinois Research Board for a grant of computer time. This research was supported by the U. S. National Science Foundation, Grant No. NSF-PHY-81-09494.

[1]D. Petcher and D. H. Weingarten, Phys. Rev. D 22, 2465 (1980).

[2]G. Bhanot and C. Rebbi, Nucl. Phys. B180, 469 (1981).

[3]M. Creutz, Phys. Rev. Lett. 45, 313 (1980); Phys. Rev. D 21, 2308 (1980).

[4]B. Berg and J. Stehr, Z. Phys. C 9, 349 (1981).

[5]G. Jongeward and J. Stack (unpublished).

[6]N. Metropolis, A. W. Rosenbluth, M. N. Rosenbluth, A. H. Teller, and E. Teller, J. Chem. Phys. 21, 1087 (1953).

[7]The error bars shown in Fig. 3 are naive statistical errors, multiplied by 2. The values of Wilson loops are correlated from sweep to sweep, so treating them as independent tends to underestimate statistical errors. See Berg and Stehr, Ref. 4, for further discussion.

[8]M. E. Fisher and R. J. Burford, Phys. Rev. 156, 583 (1967).

[9]M. Ferer, M. A. Moore, and M. Wortiz, Phys. Rev. Lett. 22, 1382 (1969).

[10]See, for example, J. Stoer and R. Bulirsch, *Introduction to Numerical Analysis* (Springer, New York, 1980), pp. 93–106.

[11]M. Lüscher, K. Symanzik, and P. Weisz, Nucl. Phys. B173, 365 (1980).

[12]J. Stack and M. Stone, Phys. Lett. 100B, 476 (1981).

[13]C. A. Flory, Phys. Lett. 101B, 98 (1981).

[14]A. Soni and M. D. Tran, Phys. Lett. 109B, 393 (1982).

[15]A. Billoire, Phys. Lett. 104B, 472 (1981).

[16]E. Kovacs, Phys. Rev. D 25, 3312 (1982).

[17]A. Martin, Phys. Lett. 100B, 511 (1981); 93B, 338 (1980).

[18]T. Eguchi and H. Kawai, Phys. Rev. Lett. 48, 1063 (1982).

[19]G. Bhanot, U. Heller, and H. Neuberger, Phys. Lett. 113B, 47 (1982).

[20]D. Gross and Y. Kitazawa, Princeton report, 1982 (unpublished).

Volume 115B, number 2 PHYSICS LETTERS 26 August 1982

POTENTIAL AND RESTORATION OF ROTATIONAL SYMMETRY IN SU(2) LATTICE GAUGE THEORY

C.B. LANG[1]
CERN, Geneva, Switzerland

and

C. REBBI
Brookhaven National Laboratory, Upton, Ny, USA

Received 15 April 1982

We study by numerical techniques the restoration of rotational invariance in SU(2) lattice gauge theory. Results on the potential between two static sources in the fundamental representation are also represented.

During the last few years numerical simulations of lattice gauge theories [1–3] have emerged as a very powerful technique for obtaining information about the properties of the corresponding continuum quantum systems. Although the lattice introduces a distortion of space–time, one expects that by a renormalization procedure whereby the lattice spacing is sent to zero while the bare coupling constant is simultaneously readjusted, a continuum limit can be obtained where the theory exhibits full Poincaré invariance. In the all-important non-abelian models, where the continuum limit corresponds to a vanishing value of the bare coupling constant g, numerical computations of several physical quantities [4–10] as functions of g reveal a rather sudden transition (crossover) from a strong coupling behaviour to the scaling behaviour predicted by the renormalization group. One expects then that the symmetry properties of the models should also undergo a rapid change in the crossover region, from the symmetries of the lattice to an approximate invariance under the full O(4) group of euclidean rotations. This recovery of rotational invariance has been related to the occurrence of a roughening transition [11,12] in the expectation value of some observables.

In this letter we wish to present numerical results, obtained by Monte Carlo simulations, which show how rotational invariance is re-established as one proceeds through the crossover region in the SU(2) lattice gauge theory. The quantity we have studied is the potential $V(r)$ between two static sources in the fundamental representation, placed at arbitrary space separation vector $r = (r_x, r_y, r_z)$ on the lattice. $V(r)$ is measured considering the expectation value of the product of two Wilson factors

$$W = \mathrm{tr}(U_{i_1 i_2} U_{i_2 i_3} \dots U_{i_N i_1}) , \tag{1}$$

associated with the loop extending through the lattice in the time direction and closed by virtue of the periodic boundary conditions.

$$V(r) = -T \ln\langle W(r)\, W(0)\rangle , \tag{2}$$

T is the physical temperature of the system, determined by the relation

$$T = 1/n_t a(g, \Lambda) , \tag{3}$$

where n_t is the extension of the system along the time direction (in lattice units), a the lattice spacing, g the bare coupling constant and Λ the lattice scale param-

[1] Address after May 1st, 1982: Institut f. Theor. Physik, Universität Graz, A-8010 Graz, Austria.

eter. V is, strictly speaking, the free energy of the quantum system in the presence of two static sources at separation r.

In the course of our analysis we have measured $\langle W\rangle$ and $\langle W(r)\, W(0)\rangle$ for a wide range of the parameters and thus obtained detailed information on the behaviour of the potential as a function of physical separation and temperature. We also present our results on the potential, although they constitute to some extent a verification of earlier analyses by Kuti et al. [6] and by McLerran and Svetitsky [7]. An investigation of the recovery of rotational invariance in the abelian U(1) model, based on an approach similar to ours, has been carried out by De Grand and Toussaint [13].

The model is defined in euclidean space–time with Wilson's action. The Boltzmann factor is then

$$\exp\left(-\beta \sum_{\square} \tfrac{1}{2}\mathrm{tr}(1 - U_{\square})\right), \tag{4}$$

where the sum extends over all plaquettes of the lattice, U is the parallel transporter around the plaquette and the coupling parameter β is related to the nonrenormalized coupling constant g by $\beta = 4/g^2$. We have considered lattices extending for n_s sites in each of the space directions and n_t sites in the time direction, with periodic boundary conditions. We approximated the full gauge group by its discrete icosahedral subgroup $\tilde{Y}$ to improve the efficiency of the computation. It is well known that the $\tilde{Y}$ model gives a very good approximation to the SU(2) theory throughout the physically relevant region [14,5]. The details of the program used have been published in ref. [15].

We have made measurements of expectation values of thermal loop factors [i.e., the Wilson factors for transport along the time axis at fixed space coordinates, see eq. (1)] and of products of loop factors at arbitrary space-like lattice separation $\boldsymbol{n} = (n_x, n_y, n_z)$. We denote these quantum averages by

$$w_0 = \langle W\rangle , \tag{5}$$

$$w(\boldsymbol{n}) = \langle W(\boldsymbol{n})\, W(0)\rangle . \tag{6}$$

We studied lattices with $n_t = 4,6,8$, $n_s = 8$ and $n_t = 6$, $n_s = 16$. For a proper interpretation of w_0 and $w(\boldsymbol{n})$ in terms of a free energy, i.e.

$$w_0 = \exp[-V_0/T] , \tag{7}$$

$$w(\boldsymbol{n}) = \exp[-V(r)/T] \tag{8}$$

[$r = an$ and $T = 1/an_t$, see eq. (3)], n_s should be much larger than n_t. In practice one observes that the numerical results remain reasonably stable even if n_s becomes comparable to n_t. For the systems with $n_s = 8$ the averages were taken over several hundred (typically 1000) Monte Carlo iterations starting from configurations at approximate equilibrium, which were obtained by adiabatically cooling the system from $\beta = 0$ in a few hundred iterations. For the simulations with $n_s = 16$, eight replicas of the configuration with $n_s = 8$ obtained at the end of the simulation were combined into a single, larger lattice and then 300 further iterations were performed. We have measured the correlations $w(\boldsymbol{n})$ at all separations $\boldsymbol{n}$ with $|\boldsymbol{n}| < 5$, folding the data obtained with separation vectors related by symmetries of the lattice to ameliorate the statistics. However, only those values of $\boldsymbol{n}$ for which $w(\boldsymbol{n})$ was larger than the error induced by fluctuations were kept for further analysis.

The first question we want to address is the recovery of rotational invariance. To estimate the degree of distortion introduced in the potential by the lattice we proceeded as follows. The values of the potential $V(r)$ were obtained through eq. (2). One dimensional arrays of aligned, nearby spatial lattice sites

$$\boldsymbol{n}_j = \boldsymbol{n}_0 + j\Delta\boldsymbol{n} , \tag{9}$$

were then considered. By "nearby" we mean that only small steps $\Delta\boldsymbol{n}$ [for instance, $\Delta\boldsymbol{n} = (1,1,0)$] were taken into account; the lines defined by eq. (9), however, were not constrained to pass through the origin. The values

$$V_j = V(r_j = an_j) . \tag{10}$$

were fit in terms of a Coulomb plus linear form

$$V(r) = c_0 + c_1/r + c_2 r . \tag{11}$$

separately for each of these lines. The vectors $\boldsymbol{r}^{(k)}$ where the fitted potentials took a definite value V_0 were finally determined. Clearly, the points $\boldsymbol{r}^{(k)}$ (k labels the different rectilinear arrays of lattice sites) give a representation of the equipotential surfaces of the interpolated potential function.

The results of the above analysis, for $\beta = 2$, $n_s = 8$, $n_t = 4$ and for $\beta = 2.25$, $n_s = 16$, $n_t = 6$, are displayed in figs. 1a and 1b, respectively. Points corresponding

Volume 115B, number 2 PHYSICS LETTERS 26 August 1982

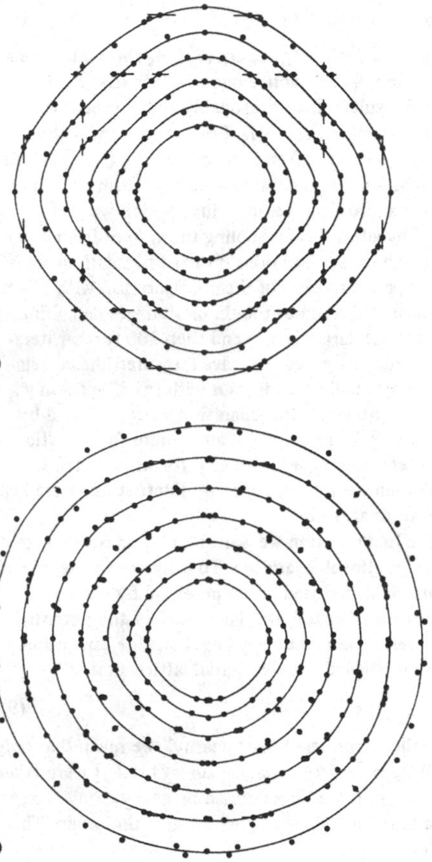

(a)

(b)

Fig. 1. Restoration of rotational invariance from (a) $\beta = 2$, $n_s = 8$, $n_t = 4$ to (b) $\beta = 2.25$, $n_s = 16$, $n_t = 6$; the equipotential curves are obtained through fits as described in the text.

to several different constant values of the potential are shown in the graphs and contours defined by the equation

$$r(\vartheta) = r_0 + r_1 \cos 4\vartheta\,, \qquad (12)$$

are fitted through them. These lines have been added to the figures both to provide a guide to the eye (otherwise points obtained with different V_0 should be labelled or plotted separately to avoid confusion) and to give a more objective indication of the departure from rotational symmetry. Notice that the crossover from the strong coupling regime to the scaling regime occurs between $\beta \approx 2$ and $\beta \approx 2.2$. The fact that the transition is accompanied by a restoration of rotational invariance is manifest from the figures.

We turn now to the behaviour of the potential as a function of separation and temperature. We have used the results of refs. [4] and [5] to establish a relation between string tension κ, lattice spacing a and coupling parameter β. The asymptotic form

$$a(\beta) = (11/6\pi^2\beta)^{-51/121} \exp(-3\pi^2\beta/11)\,, \qquad (13)$$

has been used down to $\beta = 2.15$ and the functional form deduced directly from the measurement of the string tension has been used for lower values of β. The slight discrepancy between the results of refs. [4] and [5] ($\Lambda = 0.013\sqrt{\kappa}$ versus $\Lambda = 0.011\sqrt{\kappa}$) is probably due to the different methods used to extract the string tension from the expectation value of rectangular loop factors. For definiteness we assume $\Lambda = 0.013\sqrt{\kappa}$ and, in the following, all dimensional quantities will be thus expressed in units of $\sqrt{\kappa}$.

As the physical temperature increases, the system undergoes a transition from the confining phase to a Debye screened, unconfined phase. In refs. [6] and [7] the critical temperature for the transition was estimated to be $T_c \approx 0.39$ (in units of $\sqrt{\kappa}$). The occurrence of the phase transition is characterized by the fact that the free energy of a single source becomes finite. Thus one has (for a lattice with $n_s = \infty$) $w_0 = 0$ for $T < T_c$, w_0 finite for $T > T_c$. (To be precise, w_0 should be defined through the limit $w(r) \xrightarrow[|r|\to\infty]{} |w_0|^2$ or by introducing a source coupled to W which is then sent to zero.) Our measurements of w_0 are reproduced in fig. 2. The results are consistent with $T_c \approx 0.39$. In refs. [16] a slightly larger value for T_c is quoted, but the discrepancy may be explained by the different procedure used to obtain T_c. Taking for the critical temperature the point where the curve giving $|w_0|$ has maximal slope in fig. 2 would indeed produce a higher T_c. The discrepancy should disappear in the infinite volume limit.

Assuming a critical temperature of $0.39\sqrt{\kappa}$, we fitted out data for the on-axis and off-axis potential $V(r)$ in terms of the coulombic plus linear form, i.e.,

$$V(r) = c_0 + c_1 V_c(r) + c_2 r\,, \qquad (14)$$

Volume 115B, number 2 PHYSICS LETTERS 26 August 1982

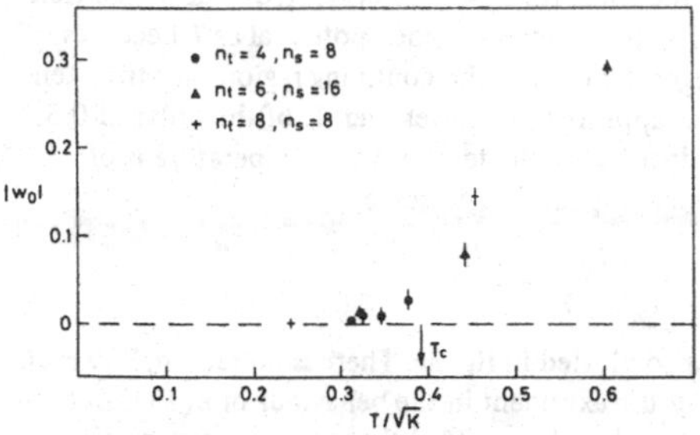

Fig. 2. Determination of the critical temperature.

for $T < T_c$, and in terms of a Debye-screened form, i.e.,

$$V(r) = c_0 + c_1 V_c(r) \exp(-c_2 r) , \tag{15}$$

for $T > T_c$. In the continuum limit $V_c(r)$ would be given by $1/\pi r$, but one cannot expect, even in the scaling region, an amount of rotational symmetry larger than that present in the lattice Coulomb potential

$$V_c(r) = \frac{1}{(2\pi)^3} \int_{-\pi/a}^{\pi/a} d^3k \, \frac{\exp(-ik \cdot r)}{\Sigma_i \sin^2(k_i a/2)} . \tag{16}$$

Thus we have used the lattice form of eq. (16) in the fit. [We have verified that replacing the integration in eq. (16) with a finite sum, as would be appropriate for a finite size lattice, to a high degree of approximation only affects the potential through the addition of a constant term. Since the constant in eqs. (14) and (15) is not physically meaningful in the continuum limit (it corresponds to a self-energy which must be renormalized) we have disregarded finite volume corrections to $V_c(r)$.]

Our results for the potential are summarized in table 1, where we reproduce the values of β at which the measurements have been made, the size of the lattice, the range r_1 to r_2 over which $V(r)$ has been evaluated (r_0 is also the lattice spacing), the physical temperature T, and the values found for the parameters c_1 and c_2. All dimension-full quantities (r_0, r_1, T, c_2) are given in units of corresponding powers of $\sqrt{\kappa}$, the square-root of the physical string tension.

Errors have been included in this table when the statistical fluctuations in the determination of $V(r)$ appeared to induce an uncertainty in the fit. The discrepancies between the values found for the parameters of the potential at identical or very close values of the temperature may be explained in part by this degree of uncertainty. But they are more likely due to finite size effects, to insufficient thermalization for the larger lattices and, especially, to the fact that the

Table 1
The fits to the potential.

	β	$n_s^3 \times n_t$	r_0	r_1	T	c_1	c_2
confining	2.25	$8^3 \times 8$	0.516	1.153	0.242	−0.909	0.447 ± 0.034
	2	$8^3 \times 4$	0.800	2.400	0.312	−1.767	0.500 ± 0.036
	2.25	$8^3 \times 6$	0.516	1.548	0.323	−1.016	0.439 ± 0.056
	2.25	$16^3 \times 6$	0.516	2.248	0.323	−0.923	0.518 ± 0.026
	2.05	$8^3 \times 4$	0.768	2.304	0.326	−1.682	0.395 ± 0.008
	2.10	$8^3 \times 4$	0.721	2.163	0.347	−1.156	0.537 ± 0.035
	2.15	$8^3 \times 4$	0.662	1.986	0.378	−1.218	0.456 ± 0.030
screened	2.375	$8^3 \times 6$	0.377	1.131	0.442	−0.870	0.401 ± 0.501
	2.375	$16^3 \times 6$	0.377	1.846	0.442	−1.295	0.034 ± 0.056
	2.5	$8^3 \times 8$	0.275	0.825	0.454	−0.714	1.116 ± 0.738
	2.5	$8^3 \times 6$	0.275	0.825	0.606	−0.598	2.916 ± 0.073
	2.5	$16^3 \times 6$	0.275	1.348	0.606	−0.558	1.695 ± 0.225
	2.75	$8^3 \times 8$	0.146	0.438	0.856	−0.472	0 ± 8.9
	2.75	$8^3 \times 6$	0.146	0.438	1.141	−0.424 ± 1.061	8.7 ± 7.4
	3	$8^3 \times 6$	0.077	0.231	2.156	−0.263 ± 0.165	12.9 ± 8.8

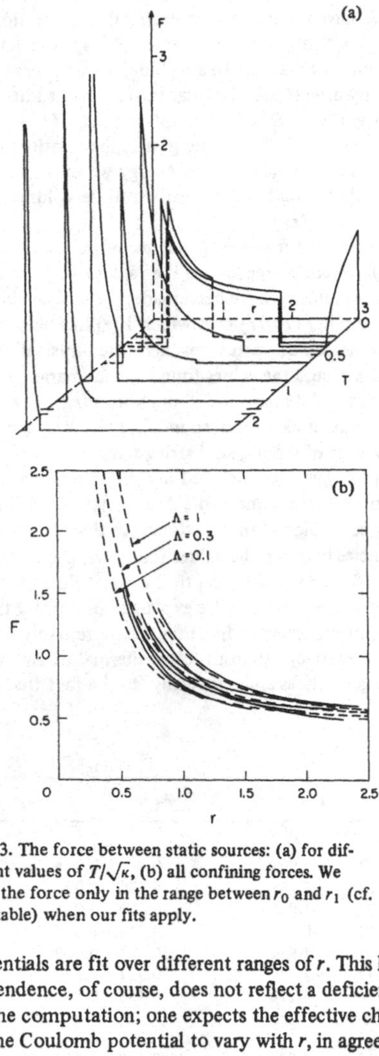

Fig. 3. The force between static sources: (a) for different values of $T/\sqrt{\kappa}$, (b) all confining forces. We plot the force only in the range between r_0 and r_1 (cf. the table) when our fits apply.

potentials are fit over different ranges of r. This last dependence, of course, does not reflect a deficiency of the computation; one expects the effective charge in the Coulomb potential to vary with r, in agreement with the renormalization group.

Altogether, the pattern which emerges for the behaviour of the potential is rather consistent. In fig. 3a we display the force between static sources derived from the potentials we found [after replacement of $V_c(r)$ with its continuum form $1/\pi r$] for a few values of the temperature. The confining nature of the force for $T < T_c$ is evident, as well as the transition to a Debye-screened potential as T becomes larger than T_c. In the confining region the string tension appears to be lower than 1, of the order of 0.5. A decrease in the tension with temperature is of course expected. The curve

$$\kappa(T) = \left(\frac{T_c - T}{T}\right)^{1/2},$$

is also plotted in fig. 3a. There is no reason, however, why the exponent in the behaviour of $\kappa(T)$ should be 1/2, and we have added that specific curve to the graph only to exemplify a possible temperature dependence.

In fig. 3b we plot all the confining forces (solid lines), corresponding to the first seven entries of the table. Onto these data we superimpose graphs of the function

$$F(r) = \frac{9\pi}{22}\left(\frac{1}{r^2 \ln(1 + 1/\Lambda_c^2 r^2)} - \Lambda_c^2\right) + k, \qquad (17)$$

with $k = 0.46$ and $\Lambda_c = 0.1$, 0.3 and 1 (broken lines). The expression in eq. (17) reproduces the expected asymptotic freedom behaviour for small r and the constant behaviour ($F \to k$) for large r. Apart from that it is not singled out by any theoretical consideration. The figure does show however that the potentials are consistent with the expected asymptotic behaviour and also that the parameter Λ_c determining the strength of the running charge is much larger than the lattice parameter ($\Lambda_{\text{lattice}} \sim 0.013$), in agreement with the results of ref. [4] (cf. also ref. [17]).

A refinement of our analysis, based on longer MC simulations and larger lattices, could be used to obtain a rather precise expression for the potential between static sources. This could then serve as an input for phenomenological calculations of the mass spectrum, although it appears today that the spectrum can be computed directly by Monte Carlo simulations [18], bypassing the evaluation of the potential.

It is gratifying, however, to see that the transition to the scaling domain is accompanied by a restoration of rotational symmetry and that the behaviour of the potential with separation and temperature conforms with theoretical expectations. Our analysis, we believe, provides yet another piece of evidence for the correctness of the lattice formulation of quantum gauge theories.

Volume 115B, number 2 PHYSICS LETTERS 26 August 1982

References

[1] M. Creutz, L. Jacobs and C. Rebbi, Phys. Rev. Lett. 42 (1979) 1390; Phys. Rev. D20 (1979) 1915.
[2] M. Creutz, Phys. Rev. Lett. 43 (1979) 553; Phys. Rev. D21 (1980) 2308.
[3] K. Wilson, Cargèse Lecture Notes, (1979), Vol. 59 (Plenum, New York, 1980).
[4] M. Creutz, Phys. Rev. Lett. 45 (1980) 313.
[5] G. Bhanot and C. Rebbi, Nucl. Phys. B180 (1981) 469.
[6] J. Kuti, J. Polonyi and K. Szlachanyi, Phys. Lett. 98B (1981) 199.
[7] L.D. McLerran and B. Svetitsky, Phys. Rev. D24 (1981) 450.
[8] C.B. Lang, C. Rebbi, P. Salomonson and B.S. Skagerstam, Phys. Lett. 101B (1981) 173; and to be published in Phys. Rev.
[9] B. Berg, A. Billoire and C. Rebbi, Brookhaven preprint (1982), to be published in Ann. Phys.
[10] B. Berg and A. Billoire, CERN preprint TH. 3230 (1982).
[11] A. Hasenfratz, E. Hasenfratz and P. Hasenfratz, Nucl. Phys. B180 [FS2] (1981) 353;
C. Itzykson, M. Peskin and J.B. Zuber, Phys. Lett. 95B (1980) 259;
M. Lüscher, G. Münster and P. Weisz, Nucl. Phys. B180 [FS2] (1981) 1.
[12] J.B. Kogut, D.K. Sinclair, R.B. Pearson, J.L. Richardson and J. Shigemitsu, Phys. Rev. D23 (1981) 2945.
[13] T.A. DeGrand and D. Toussaint, Phys. Rev. D24 (1981) 466.
[14] D. Petcher and D. Weingarten, Phys. Rev. D22 (1980) 2465.
[15] G. Bhanot, C.B. Lang and C. Rebbi, CERN preprint TH. 3178 (1981), to be published in: Comput. Phys. Commun. 25 (1982).
[16] J. Engels, F. Karsch, I. Montvay and H. Satz, Phys. Lett. 101B (1980) 89;
J. Engels, F. Karsch and H. Satz, Bielefeld preprint BI-TP 81/28 (1981).
[17] A. Billoire, Phys. Lett. 104B (1981) 472.
[18] H. Hamber and G. Parisi, Phys. Rev. Lett. 47 (1981) 1792;
E. Marinari, G. Parisi and C. Rebbi, Phys. Rev. Lett. 47 (1981) 1795;
D. Weingarten, Phys. Lett. 109B (1982) 57;
H. Hamber, E. Marinari, G. Parisi and C. Rebbi, Phys. Lett. 108B (1982) 314;
A. Hasenfratz, P. Hasenfratz, Z. Kunszt and C.B. Lang, Phys. Lett. 110B (1982) 289.

Reprinted from ANNALS OF PHYSICS
All Rights Reserved by Academic Press, New York and London

Vol. 142, No. 1, August 1982

Monte Carlo Estimates of the $SU(2)$ Mass Gap

B. BERG* AND A. BILLOIRE

CERN, Geneva, Switzerland

AND

C. REBBI

Physics Department, Brookhaven National Laboratory, Upton, New York 11973

Received February 8, 1982

Combining Monte Carlo and variational techniques, we compute the mass gap (or glueball mass) in the $SU(2)$ lattice gauge theory. We obtain the estimate $m_g = (2.4 \pm 0.6)\sqrt{K}$, K being the string tension. We also discuss previous results derived by numerical or strong coupling methods and present Monte Carlo data on the second moment of the correlation length.

I. INTRODUCTION AND SUMMARY OF RESULTS

In this article we present results for the mass gap (the glueball mass) in the $SU(2)$ lattice gauge theory, obtained by means of Monte Carlo simulations. Previous numerical and strong coupling reults are also reviewed and discussed.

We consider the Euclidean $SU(2)$ lattice gauge theory in $d = 4$ dimensions with Wilson's action [1]. The expectation values of gauge invariant observables $O\{U\}$ are given by

$$\langle O \rangle = \frac{1}{Z} \int \prod_b dU(b)\, O\{U\} \cdot \exp\left\{-\frac{1}{2}\beta \sum_p \mathrm{tr}[1 - U(\dot{p})]\right\}. \tag{1.1}$$

The dynamical variables $U(b) \in SU(2)$ are attached to the links b of a four-dimensional hypercubical lattice with lattice spacing a. The summation is over all elementary plaquettes p of the lattice and $U(\dot{p})$ is the parallel transporter around the plaquette p.

In the weak coupling limit, $\beta = 4/g^2 \to \infty$, the lattice theory is conjectured to define a continuum quantum field theory. The determination of the mass spectrum requires non-perturbative computations. According to weak coupling renormalization group

* On leave from II. Institut für Theoretische Physik der Universität Hamburg, West Germany.

analysis, any physical mass has to be proportional to the lattice mass scale, given by

$$\Lambda_L = \frac{1}{a}\left(\frac{11}{24\pi^2}g^2\right)^{(-51/121)} \exp\left(-\frac{12\pi^2}{11g^2}\right), \tag{1.2}$$

and vanishes therefore to all orders of weak coupling perturbation theory. In the following we shall use units with $a = 1$. The lattice mass scale has been related [2] to more conventional mass scales of perturbation theory, for instance, $\Lambda_{\text{MOM}} = 57.5\,\Lambda_L$.

Monte Carlo (MC) methods, adapted to lattice gauge theories [3, 4], provide a powerful tool for calculating non-perturbative quantities. The $SU(2)$ string tension K was the first such observable determined by a MC calculation (Creutz [5, 6]). The computation has been repeated by several authors [7–9], and all results are consistent with a value

$$\Lambda_L = (0.0119 \pm 0.0015)\sqrt{K}. \tag{1.3}$$

For the (subjective) estimate of the error, we have taken Refs. [8, 9] as independent investigations. Less accurate results exist for the $SU(3)$ string tension [6, 10], and recently results on the meson spectrum have also been obtained [11].

In this article we present a MC study of the mass gap m_g in pure $SU(2)$ quantum gauge theory. The mass gap is defined to be the mass of the lowest lying quantum state above the vacuum (frequently called the glueball). One believes that by dynamical mass generation $m_g > 0$ in the $SU(2)$ Yang–Mills theory, and therefore

$$m_g = \text{const} \cdot \Lambda_L. \tag{1.4}$$

The problem is to determine the constant. Using the mean value of Eq. (1.3), we will always express m_g in units of $\sqrt{K}$. Table I summarizes previous estimates of the glueball mass from MC [8, 12, 13] and strong coupling (SC) [14] calculations. (An estimate of m_g based on a rather short strong coupling expension was also presented in Ref. [15].)

MC studies of the glueball mass turned out to be much more difficult than corresponding estimates of the string tension. The first estimates [8, 12] relied on an investigation of plaquette–plaquette correlations,

$$\rho(\mathbf{x}, t) = \rho(x) = \tfrac{1}{4}\langle \text{Tr}\, U(\dot{p}_1)\, \text{Tr}\, U(\dot{p}_2)\rangle - \tfrac{1}{4}\langle \text{Tr}\, U(\dot{p}_1)\rangle\langle \text{Tr}\, U(\dot{p}_2)\rangle, \tag{1.5}$$

TABLE I

Previous Results on the Glueball Mass m_g in Units $\sqrt{K}$

m_g	Method	Reference
3.5 ± 1.2	MC	Berg [12]
3.2 ± 1.1	MC	Bhanot and Rebbi [8]
1.7 ± 0.8	SC	Münster [14]
1.8 ± 0.6	MC	Engels *et al.* [13]

where p_1 and p_2 are two plaquettes at a distance $x = (x, t)$. On general grounds (see, e.g., Ref. [14], and references therein) one expects a leading behavior

$$\rho(x) \propto |x|^{\text{power}} \exp(-|x/\xi|) \qquad \text{for} \quad |x| \gg \xi. \tag{1.6}$$

Here ξ is the correlation length and the mass gap is given by

$$m_g = \xi^{-1}.$$

One is interested in the scaling region, i.e., the domain of values of β, where ξ starts to grow to infinity following the behavior of Eqs. (1.2)–(1.4). Due to statistical fluctuations, reliable MC results for $\rho(x)$ can be obtained only up to $|x| = 3$. At first sight it would seem hopeless to determine m_g on the basis of correlations at such short range. On the other hand the analysis of the string tension indicates that the scaling regime sets in when the correlation length is still very small. To extract the physical string tension K, one forms ratios of expectation values $W_{I,J}$ of $I \times J$ rectangular Wilson loop factors in such a way that unwanted parameter dependences factor out. A typical quantity being considered is [5, 6]

$$X(I, J) = -\ln \left(\frac{W_{I,J} W_{I-1,J-1}}{W_{I-1,J} W_{I,J-1}} \right) \tag{1.7}$$

which should approach K as I and J grow large. The fact that $X(3, 3)$ already appears to scale with a value close to the actual string tension in the region $2.1 < \beta < 2.5$ indicates that $\xi < 2$ at $\beta = 2.5$. This gives the bound

$$m_g > 1.7 \sqrt{K}. \tag{1.8}$$

If the scaling region sets in when the correlation length extends over less than one or two lattice sites, the possibility of determining m_g from a MC study of plaquette–plaquette correlations is not ruled out. A difficulty arises though because of the power dependence in Eq. (1.6), which at small distances may obscure the exponential behavior. The results of Refs. [8, 12] were essentially obtained by matching a signal for the onset of scaling of the correlation length on the first order SC expansion. This implied a rather large uncertainty, although the MC result for the glueball mass (from correlations at short distances) followed the first order SC result up to an amazingly large value of β.

MC results for the string tension are self-consistent in the sense that a region of β exists where the data follow unambiguously the expected scaling behavior. The above MC results for the glueball mass are not self-consistent in that sense, because an estimate of the glueball mass was only obtained under the assumption of scaling, after observing a signal. Additional evidence that the scaling region is reached at $\beta \approx 2.1$ comes from consistence [16] with finite size scaling predictions.

Münster [14] has estimated the glueball mass by means of a SC expansion. Due to singularities in the complex coupling constant plane, the series is clearly not convergent in the region where the weak coupling extrapolation is done. The results depend on the choice of expansion variable and on the extrapolation procedure. One has to assume rapid crossover from strong to weak coupling, where the series (written in a chosen expansion variable) breaks down. Such a procedure is again not self-consistent (in the sense explained above). In combination with MC results SC expansions are, however, extremely valuable. In Section II we comment in more detail on existing SC expansions, and compare with MC results, where possible. We also present some new SC results on plaquette correlations at finite distance.

In the present paper we refine the analysis of the glueball mass by using a variational principle. (The possibility of using a variational technique to determine m_g was pointed out by K. Wilson [Communication at the Abingdon Meeting on Lattice Gauge Theories, March, 1981].) It is well known that

$$m_g = \min_{|\psi\rangle} \left(-\frac{1}{t} \ln \frac{\langle \psi | e^{-Ht} | \psi \rangle}{\langle \psi | \psi \rangle} \right), \qquad \langle 0 | \psi \rangle = 0. \tag{1.9}$$

We have assumed a discrete spectrum and the variation goes over all states $|\psi\rangle$ which are orthogonal to the vacuum. In our investigation we truncate the set of states to

$$|\psi\rangle = \sum_{i=1}^{12} c_i O_i |0\rangle,$$

where the sum runs over a definite set of operators O_i chosen to excite only states of vanishing spin and lattice momentum. With MC simulations we calculate expectation values

$$\langle 0 | O_i(0) \, e^{-Ht} O_j(0) \, | 0 \rangle = \langle 0 | O_i(0) \, O_j(t) \, | 0 \rangle, \tag{1.10}$$

and after assembling significant statistics the minimization is done. Advantages of the procedure are

(1) Even for a large correlation length, glueball estimates are (in principle) possible from correlations at small "time" distances, by including enough operators. (In the MC investigation typically $t = 1, 2$.)

(2) The procedure allows checks from the consistency of the results obtained with different time separations.

(3) One obtains information about the glueball wave function.

The main disadvantage is that minimization (1.9) requires increasing accuracy of the MC results for the correlations if the number of operators is increased. In particular there are operators which contribute little to the glueball wave function but add noise. We have accurate results for $t = 1$ and reasonable results for $t = 2$. Both are consistent and give the estimate

$$m_g = (2.4 \pm 0.6) \sqrt{K}. \tag{1.11}$$

We can also see the onset of scaling. The MC results have been obtained on a $L^3 \times T = 4^3 \times 16$ lattice. The 120-element icosahedral subgroup of $SU(2)$ has been used as a gauge group to speed up the computation. It is well known that models with finite subgroups of $SU(2)$ of sufficiently high order provide very good approximations to the $SU(2)$ lattice gauge theory itself [7, 8, 17]. The details of the calculation are presented in Section III.

In Section IV we reproduce MC results on the second moment definition [18] of the correlation length, obtained on 8^4 lattices which are imbedded in a 16^4 lattice. For reasons discussed in that section the results remain, however, rather inconclusive. Final remarks are presented in Section V.

II. Strong Coupling Expansions Versus Monte Carlo Calculations

Strong coupling (SC) calculations try to extract information about the continuum theory ($\beta \to \infty$) by means of a systematic expansion for β small. The aim is to obtain reliable results which reach into the scaling region and can then be continued using the known weak coupling renormalization group behavior. The program has been put forward by Kogut *et al.* [19] in the Hamiltonian formulation (for $SU(3)$) and by Münster [14, 20] in the Euclidean formulation.

Due to singularities in the complex coupling constant plane, it is, however, not obvious that the scaling regime can actually be reached by SC expansions. The results and the stability of Pade approximants depend crucially on the choice of expansion variable.

From the formulation of the theory (Eq. (1.1)) β seems to be a natural expansion variable. The calculations, however, are more conveniently carried out in terms of characters of the gauge group. For $SU(2)$ this leads one to consider

$$u = I_2(\beta)/I_1(\beta) \tag{2.1}$$

(I_n modified Bessel function; see Eq. 8.445 of Ref. [21]) as expansion variable. As a matter of fact the results improve considerably under the switch from β to u. This is due to (conjectured) singularities on the negative u^2-axis, which may be avoided by introducing a new variable z by means of a conformal transformation

$$z = u^2/(1 + xu^2). \tag{2.2}$$

The choice [8] $x = 2$ gives an excellent fit to the internal nergy far into the scaling region (up tp $\beta \approx 2.3$), and Pade approximants in this variable turn out to be remarkably stable. The choice $x = 6$ and the possibility of expanding in terms of $E_p(\beta) = 1 - E(\beta)$ (E = internal energy) has been discussed elsewhere [22]. It has also been remarked in Ref. [22] that a reliable analysis of the singularity structure in the complex β-plane would necessarily require more orders in the SC expansion than are at present (and presumably also in the future) available.

In summary there seems to be a rather large amount of subjective freedom in extrapolating predictions from the existing SC series expansions into the scaling domain. In the next two subsections we illustrate this uncertainty by comparing SC expansions in the variables β, u, and z $(x = 2)$ derived for several observables. Without relying on further (Pade) extrapolation methods, we consider the last two available orders of the SC series and wherever possible we estimate their range of validity by comparing with MC data.

II.(a) *Internal Energy and Specific Heat*

The internal energy E and specific heat C are defined by

$$E = 1 - \left\langle \frac{1}{2} \operatorname{Tr} U(\dot{p}) \right\rangle, \qquad C = -\beta^2 \frac{dE}{d\beta}. \tag{2.3}$$

For the internal energy the SC series has been derived by Wilson [23] up to order u^{21}. One easily obtains the corresponding series (up to order u^{22}) for the specific heat. The coefficients for both expansions are given in Table II.

In Fig. 1 we consider the internal energy and compare the last two orders of the SC expansions in the variables β, u, and z $(x = 2)$ (only u^2-terms are expanded in powers of z; the results are stable if one also expands the leftover factor of u) with MC data from the models with the icosahedral gauge group [8] and the $SU(2)$ gauge group [9]. It is clear that the expansion in β fails very early, whereas the expansion in u remains reasonably accurate up to the onset of the scaling region and the expansion in z follows the data far into it. At $\beta \approx 3$ the MC data approach the lowest order spin-wave result

$$E = 3/4\beta. \tag{2.4}$$

TABLE II

Coefficients for the SC Expansion of Internal Energy and Specific Heat

	$E \cdot u^{-1} - u^{-1}$	C
u^0	−1.	0.
u^2	0.	4.
u^4	−4.	$-2.66\bar{6}$
u^6	8.	$79.11\bar{1}$
u^8	$-62.66\bar{6}$	$-277.33\bar{3}$
u^{10}	196.9975309	2,388.06914
u^{12}	−1,349.23786	−10,121.3648
u^{14}	5,939.062747	75,448.627
u^{16}	−34,280.32143	−401,204.466
u^{18}	175,554.005	2,553,295.44
u^{20}	−977,457.232	−14,817,680.6
u^{22}	/.	90,490,898.4

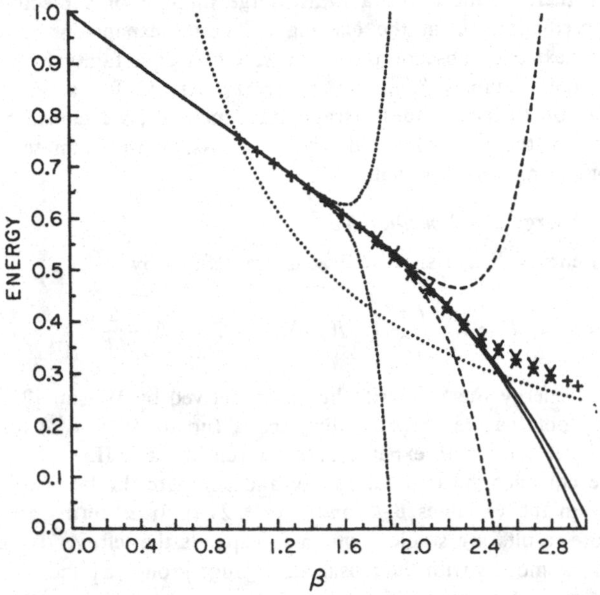

FIG. 1. SC expansion for the internal energy. Expansion variables: β (– · –), u (– – –) and z (——). The lower curve corresponds always to the highest order. The crosses (×) represent MC data for the full group (obtained on a 8^4 lattice) and the symbols (+) MC data for the icosahedral subgroup (obtained on a 16^4 lattice [8], cf. also Table VII in Section IV). The dotted line is the spin-wave result.

Without incorporating the asymptotic spin-wave behavior, the SC expansion in z with the choice $x = 6$ [22] does not give a better fit to the data than with $x = 2$.

In Fig. 2 the corresponding comparisons are done for the specific heat. In all the variables the expansions fail earlier than for the internal energy. Again the expansion variable β clearly gives the worst results and the expansion in z is the most stable one. In the region where the expansion is supposed to be valid, the z variable is not really better than the u variable. This also applies to the choice of z of Ref. [22]. Several MC data are included in Fig. 2. The data with the (large) error bars are from a direct calculation of the specific heat with high statistics on a 16^4 lattice (see Table VII, Section IV). The data without error bars are obtained by numerical differentiation from high statistics data for the internal energy. With this procedure one gets rather accurate numbers, but it is difficult to estimate the systematic error because statistical fluctuations prevent taking $\Delta\beta$ very small. We have used a step $\Delta\beta = 0.1$. From the investigation of Lautrup and Nauenberg [24] one knows that the peak around $\beta \approx 2.25$ is even more pronounced on a 4^4 lattice.

II.(b) *String Tension and Glueball Mass*

The SC expansion for the $SU(2)$ string tension has been carried out [20] to order

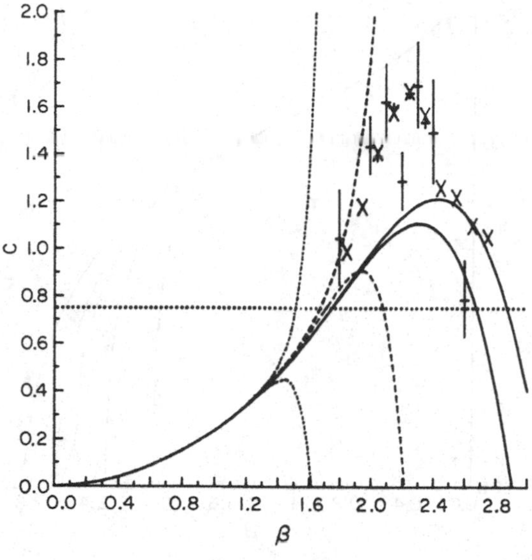

FIG. 2. SC expansion for the specific heat. Expansion variables: β (– · –), u (– – –) and z (——). The upper curve corresponds always to the highest order. The MC data for the full group (×) and the isosahedral subgroup (+) are obtained by numerical differentiation of the internal energy (Fig. 1). The MC data with error bars are direct measurements (cf. Table VII in Section IV). The dotted line is the spin-wave result.

u^{12}. The coefficients are given in Table III. In Fig. 3 we compare, again in the variables β, u, and z ($x = 2$), the last two orders of the SC series with the asymptotic behavior obtained using the MC result of Eq. (1.3). An estimate of the string tension from the SC expansion in the variable u would come closest to the MC estimate. Since corrections from the last orders all have the same trend and the change from $O(u^{10})$ to $O(u^{12})$ is significant, it is not clear that u remains a better variable than z in higher orders.

TABLE III

SC Coefficients for String Tension, Glueball Mass, and Finite Distance Plaquette–Plaquette Correlations

	$K + \ln u$	$m_g + 4\ln u$	$m_g(1) + 4\ln u$	$\rho(0) - 1$	$\rho(1)$
u^2	0.	2.	2.	−2.	0.
u^4	−4.	$-32.66\bar{6}$	$-32.66\bar{6}$	$10.66\bar{6}$	1.
u^6	0.	−51.8123457	−27.8123457	$-55.644\bar{4}$	−4.
u^8	$-58.66\bar{6}$	−719.440329	—	$290.177\bar{7}$	$49.33\bar{3}$
u^{10}	−27.0024691	—	—	—	−185.165432
u^{12}	−1211.95391	—	—	—	—

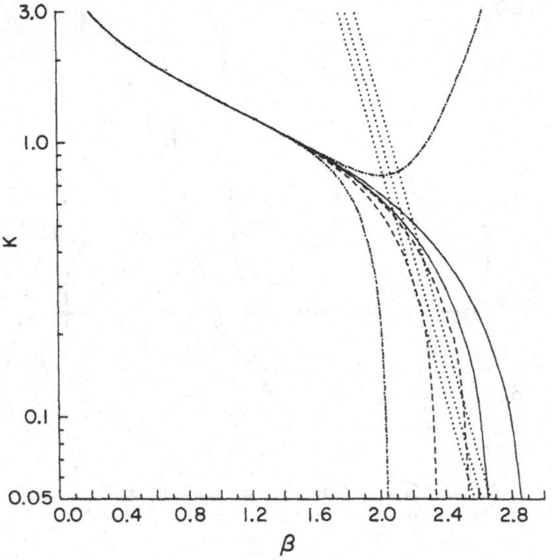

FIG. 3. SC expansions for the string tension. Expansion variables: β (– · –), u (– – –) and z (——). The lower curve always corresponds to the highest order. The dotted lines represent the expected asymptotic behavior (including errors) with the value of string tension used in this paper.

As the MC estimate for the string tension is obtained from rather small finite-size Wilson loops, it is quite instructive to compare the MC data directly with SC results for finite-size Wilson loops. This has been done in Ref. [25] and it has been found that the SC expansion fails before the scaling region is reached.

Finally we discuss SC results for the glueball mass. We consider plaquette–plaquette correlations $\rho(\mathbf{x}, t)$ (see Eq. (1.5)) ar a fixed distance t. Following Ref. [14], we sum over all space-like orientations of p_1 and p_2 separately and over all spatial locations of p_1 and p_2 with fixed time difference t. We call $\rho(t)$ the correlation function obtained in this way with a suitable normalization fixed by $\rho(0) = 1$ at $\beta = 0$. From finite distance correlations we define

$$m_g(t) = -\ln \frac{\rho(t)}{\rho(t-1)} \qquad (t = 1, 2, \ldots), \tag{2.5}$$

and the glueball mass is $m_g = \lim_{t \to \infty} m_g(t)$.

For a SC expansion truncated to a finite order u^n there always exists a finite t such that $m_g = m_g(t)$ to this order. In Table III the results of Ref. [14] for m_g to order u^8 and for $m_g(1)$ to order u^6 are given. ($m_g(1)$ is calculated from $\rho(1)$ and $\rho(0)$, also given in Table III. Partial results were previously obtained by Münster [26].) It is remarkable that to order u^4, $m_g(1)$ and m_g already agree. Thus the presently available

SC series does not contain information beyond correlations at a rather small distance, presumably $t = 3$.

In Fig. 4 the last two orders of the SC expansion for the glueball mass are given. The MC data are for $m_g(2)$. They come close to the last order for m_g if the variable z is used, and close to m_g in order u^6 if the variable u is used. The available orders for m_g are rather low and the expansion certainly breaks down before the scaling region is reached. The dotted lines represent the estimate obtained in this paper (cf. next section). It is not difficult to imagine a cross-over from the SC results to these lines, i.e., the result of this paper and the SC results are consistent.

If we took the last order of the SC series in the variable u seriously up to $1.7 < \beta < 1.8$ and arranged immediate scaling afterwards, we would get a glueball mass $m_g \approx 1.4 \sqrt{K}$. This should be taken as a lower bound. We regard the variable β already unreliable in this region. The use of the variable z would give a slightly higer value $m_g \approx 1.65 \sqrt{K}$, which is close to the bound of Eq. (1.8). Higher orders in the

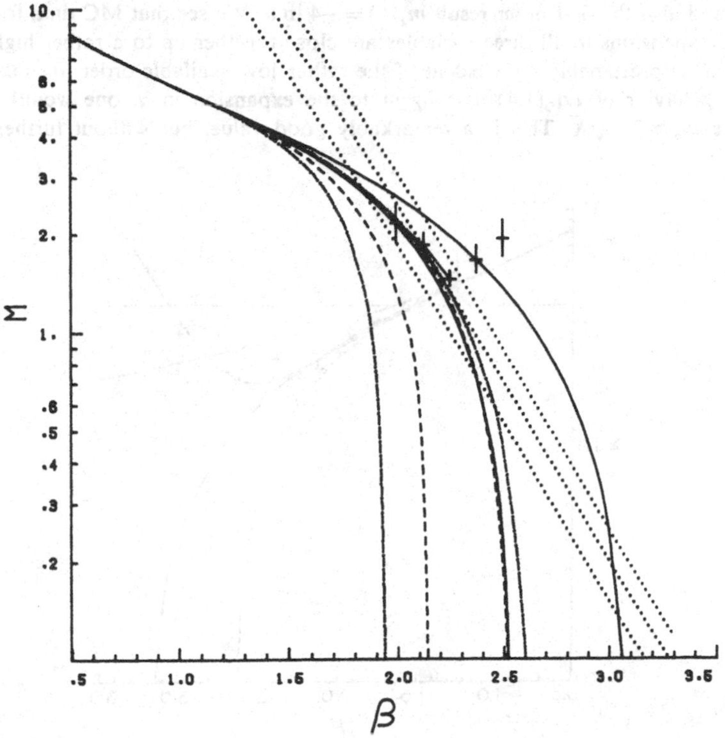

FIG. 4. SC expansion for the glueball mass. Expansion variables: β (– · –), u (– – –) and z (——). The lower curve always corresponds to the highest order. The dotted lines correspond to the glueball mass estimate (and error) of the present paper. MC data are for $m_g(2)$ on a $4^3 \times 16$ lattice.

SC expansion should improve the reliability of the series in the region $1.7 < \beta < 1.8$.

An estimate of the glueball mass from the MC data for $m_g(2)$ would give $m_g \approx 2.7\sqrt{K}$, which should be taken as an upper bound and is within the error bar of our improved estimate. The MC results for $2 < \beta < 2.25$ appear to lie on a line with a slope remarkably close to the one predicted by scaling, the deviation presumably due to the rather small distance. However, because of the lowest order spin-wave result

$$m_g(2) \xrightarrow[\beta \to \infty]{} 3.49 \tag{2.6}$$

(on a $4^3 \times 16$ lattice), the data must eventually go up again. This explains why the observed scaling of the correlation length (as determined from measurements at small distances) extends over a much shorter range than in the case of the string tension (from small size Wilson loops).

Figure 5 presents the last order of the SC expansion for $m_g(1)$ in the variables β, u, and z, and also the first order result $m_g(1) = -4 \ln u$. We see that MC data for $m_g(1)$ and SC expansions in all three variables are close together up to a rather high value of β. This is presumably an accident of the rather low available order. By taking the scaling behavior of Eq. (1.4) as tangent to the expansion in u, one would get the estimate $m_g \approx 2.7\sqrt{K}$. This is a remarkably good value, but without further infor-

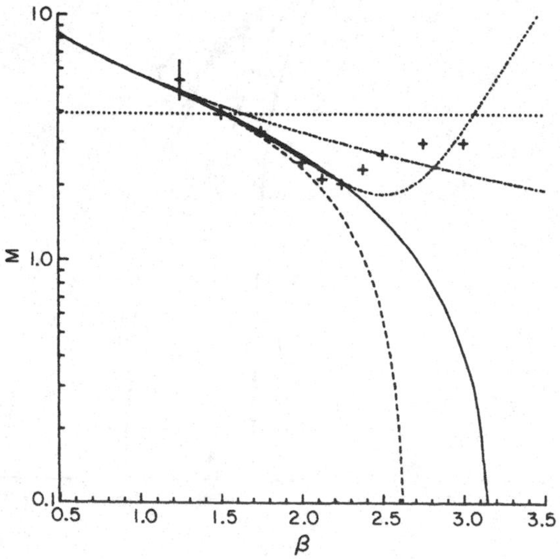

FIG. 5. SC expansion for $m_g(1)$. Last order in the variables β (– · · –), u (– – –) and z (——), also first order (– · · –, lower curve). The dotted line is the spin-wave result. The crosses represent MC data (and error bars) for $m_g(1)$ on a $4^3 \times 16$ lattice.

mation there would be no reason to trust it. (The variable z would give a much higher value.)

The lowest order spin-wave approximation on a $4^3 \times 16$ lattice is also displayed in Fig. 5. The value

$$m_g(1) = 3.96 \tag{2.7}$$

is much higher than MC data for $m_g(1)$ in the scaling region.

III. Monte Carlo Investigation of the Glueball Mass

Previous attempts [8, 12] to determine the value of the glueball mass through MC simulations were based on measurements of correlations between facing plaquettes. However, as shown by Münster [14], it is better to consider correlation functions of the operator

$$O_1(t) = \frac{1}{N_p} \sum_{\mathbf{x},\,\text{orientations}} \frac{1}{2} \operatorname{Tr} U(p_{\mathbf{x},t}). \tag{3.1}$$

(N_p is the number of space-like plaquettes.) The summation over all space positions projects onto the zero momentum subspace, leading to a behavior without power law corrections:

$$\langle O_1(t)\, O_1(0)\rangle_c = \langle O_1(t)\, O_1(0)\rangle - \langle O_1(0)\rangle^2 \underset{t\to\infty}{\sim} e^{-m_g t}. \tag{3.2}$$

A further improvement of the glueball mass estimate can be obtained in the following way. Rather than considering a single operator O_1 as in Eq. (3.1), one forms linear combinations

$$O = \sum_i c_i O_i, \tag{3.3}$$

where the O_i operators are given by formulae analogous to Eq. (3.1) but with the traces taken over path ordered operators of a different kind. If the linear combinations could be adjusted so that the operator O excited *only* the glueball state from the vacuum, then of course the correlation function

$$\langle O(t)\, O(0)\rangle_c$$

would be given by const $\times\, e^{-m_g t}$ at all separations and the mass gap could be determined comparing the values for $t = 0$ and 1. While in general it is not possible to find the operator which projects exactly over the first excited level, the coefficients in Eq. (3.3) may be adjusted so as to minimize the rate of decay of the correlation function. This is the procedure we have followed in our analysis. It is analogous to the Hamiltonian variational technique and leads to better upper bounds for the value of m_g.

As operators O_i we have considered all the path ordered operators corresponding to paths of length 4 and 6, the trace being taken in the $\frac{1}{2}$, 1, or $\frac{3}{2}$ representations of $SU(2)$. Thus O_2, O_3, and O_4 are defined by a formula identical to Eq. (3.1), but with the transport along the paths indicated in Fig. 6 rather than along p. Operators O_5, O_6, O_7, O_8 (O_9, O_{10}, O_{11}, O_{12}) are the analogues of operators O_1, O_2, O_3, and O_4, respectively, in the $j=1$ ($j=\frac{3}{2}$) representations. For instance,

$$O_5(t)=\frac{1}{N_p}\sum_{\mathbf{x},\mathrm{or.}}\frac{1}{2}\left((\mathrm{Tr}\, U(\dot{p}_{\mathbf{x},t}))^2-1\right), \tag{3.4}$$

$$O_9(t)=\frac{1}{N_p}\sum_{\mathbf{x},\mathrm{or.}}\frac{1}{2}\left((\mathrm{Tr}\, U(\dot{p}_{\mathbf{x},t}))^3-2\,\mathrm{Tr}\, U(\dot{p}_{\mathbf{x},t})\right). \tag{3.5}$$

We have measured the 78 correlations

$$C_{ij}(t)=N_p(\langle O_i(t)\, O_j(0)\rangle-\langle O_i(0)\rangle\langle O_j(0)>) \tag{3.6}$$

for $t=0$, 1, 2, and 3 by Monte Carlo simulations on a $4^3\times 16$ lattice. The rather long length ($N_t=16$) in the time direction eliminates boundary effects due periodicity in time. The choice of the space size N_s is rather critical. As the signal-to-noise ratio goes like $N_s^{-3/2}$, it is better to take N_s as small as possible. However, as we will show at the end of this section, strong finite-size effects are present in a $2^3\times N_t$ lattice. Spurious boundary effects are presumably also introduced if the separation t in time become comparable to N_s. A 4^3 space lattice seems thus optimal. (Lattice lengths given by powers of 2 allow more efficient implementations of periodicity in the computer program.)

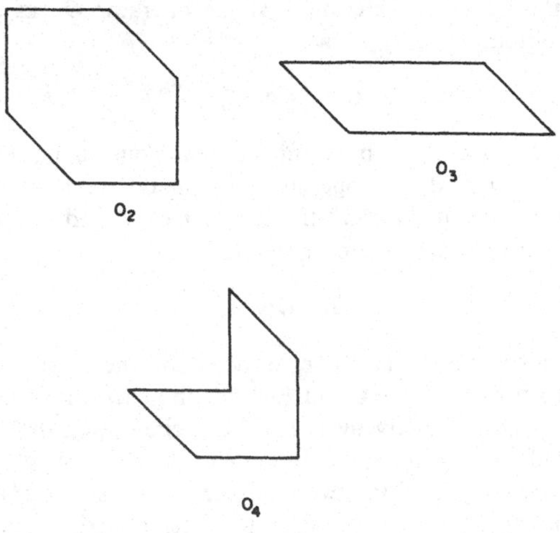

FIG. 6. The paths used to define the operators considered in this paper.

TABLE IV

MC Data for the Mean Values of the 12 Operators Considered in this Paper, on a $4^3 \times 16$ Lattice

β	Sweeps	1	2	3	4	5	6	7	8	9	10	11	12
0.25	4,000	0.06231	0.00019	0.00379	0.00395	0.00401	0.00016	−0.00007	0.00006	0.00009	0.00005	0.00004	0.00001
		0.00014	0.00012	0.00009	0.00008	0.00015	0.00012	0.00012	0.00007	0.00013	0.00016	0.00009	0.00007
0.50	1,000	0.12428	0.00400	0.01521	0.01576	0.01522	−0.00032	0.00008	0.00032	0.00130	−0.00040	0.00019	−0.00021
		0.00033	0.00015	0.00020	0.00016	0.00032	0.00032	0.00021	0.00009	0.00039	0.00025	0.00016	0.00017
0.75	1,000	0.18351	0.01219	0.03340	0.03492	0.03394	0.00033	0.00103	0.00085	0.00447	−0.00006	−0.00006	−0.00013
		0.00039	0.00014	0.00017	0.00014	0.00029	0.00016	0.00016	0.00014	0.00037	0.00032	0.00028	0.00011
1.00	1,000	0.24282	0.02880	0.05906	0.06230	0.06017	0.00103	0.00216	0.00282	0.00964	0.00012	−0.00039	0.00004
		0.00020	0.00030	0.00020	0.00028	0.00039	0.00041	0.00010	0.00015	0.00028	0.00015	0.00016	0.00016
1.25	5,000	0.30190	0.05396	0.09147	0.09793	0.09387	0.00232	0.00584	0.00729	0.01951	−0.00001	0.00013	0.00026
		0.00016	0.00018	0.00016	0.00015	0.00013	0.00012	0.00011	0.00008	0.00014	0.00010	0.00009	0.00008
1.50	5,000	0.36250	0.09244	0.13213	0.14459	0.13733	0.00649	0.01276	0.01641	0.03525	0.00011	0.00076	0.00096
		0.00018	0.00019	0.00015	0.00018	0.00017	0.00010	0.00011	0.00008	0.00016	0.00008	0.00009	0.00007
1.75	10,000	0.42703	0.14786	0.18498	0.20514	0.19424	0.01700	0.02621	0.03439	0.06045	0.00097	0.00188	0.00316
		0.00015	0.00019	0.00016	0.00017	0.00016	0.00011	0.00008	0.00008	0.00012	0.00009	0.00006	0.00005
2.00	30,000	0.50138	0.23188	0.25928	0.28902	0.27396	0.04334	0.05431	0.07178	0.10337	0.00430	0.00623	0.01008
		0.00011	0.00017	0.00014	0.00015	0.0013	0.00008	0.00007	0.00008	0.00010	0.00005	0.00004	0.00003
2.125	30,000	0.54427	0.28930	0.30991	0.34342	0.32759	0.06916	0.08041	0.10424	0.13680	0.00880	0.01150	0.01811
		0.00014	0.00022	0.00019	0.00020	0.00018	0.00011	0.00012	0.00013	0.00013	0.00005	0.00005	0.00004
2.250	40,000	0.58992	0.35665	0.37063	0.40530	0.39173	0.10861	0.11983	0.15011	0.18173	0.01795	0.02174	0.03227
		0.00015	0.00026	0.00024	0.00022	0.00022	0.00017	0.00017	0.00018	0.00016	0.00006	0.00006	0.00006
2.375	40,000	0.62702	0.41291	0.42291	0.45668	0.45058	0.15138	0.16226	0.19723	0.22810	0.03080	0.03562	0.05029
		0.00010	0.00018	0.00017	0.00015	0.00016	0.00014	0.00014	0.00014	0.00013	0.00006	0.00006	0.00006
2.50	20,000	0.65394	0.45320	0.46115	0.49379	0.49803	0.18935	0.19971	0.23793	0.26967	0.04499	0.05051	0.06901
		0.00008	0.00014	0.00014	0.00013	0.00015	0.00015	0.00015	0.00014	0.00014	0.00009	0.00009	0.00008
2.75	10,000	0.69370	0.51239	0.51773	0.54867	0.57602	0.25756	0.26626	0.30944	0.34643	0.07722	0.08320	0.10920
		0.00011	0.00019	0.00019	0.00016	0.00022	0.00024	0.00024	0.00023	0.00024	0.00015	0.00015	0.00015
3.00	10,000	0.72407	0.55796	0.56176	0.59101	0.64177	0.32048	0.32775	0.37410	0.41898	0.11410	0.12027	0.15305
		0.00007	0.00013	0.00015	0.00011	0.00017	0.00019	0.00022	0.00018	0.00021	0.00015	0.00016	0.00014
3.25	2,000	0.74870	0.59525	0.59830	0.62564	0.69918	0.37920	0.38574	0.43357	0.48845	0.15501	0.16145	0.19985
		0.00017	0.00033	0.00030	0.00027	0.00043	0.00052	0.00056	0.00047	0.00059	0.00041	0.00049	0.00041
3.50	2,000	0.76877	0.62557	0.62773	0.65398	0.74888	0.43200	0.43703	0.48675	0.55364	0.19694	0.20256	0.24692
		0.00014	0.00024	0.00029	0.00020	0.00035	0.00042	0.00053	0.00039	0.00048	0.00036	0.00046	0.00036
3.75	2,000	0.78573	0.65184	0.65361	0.67832	0.79275	0.48141	0.48594	0.53581	0.61447	0.24037	0.24631	0.29465
		0.00013	0.00024	0.00022	0.00020	0.00036	0.00048	0.00045	0.00042	0.00056	0.00047	0.00045	0.00044
4.00	2,000	0.80099	0.67546	0.67701	0.70025	0.83388	0.52887	0.53313	0.58265	0.67508	0.28619	0.29248	0.34410
		0.00013	0.00024	0.00024	0.00019	0.00035	0.00052	0.00047	0.00043	0.00051	0.00056	0.00046	0.00049

Note. For each entry, the estimated error is placed just under the mean value.

Our results for the averages $\langle O_i \rangle$, $i = 1, 12$, are reproduced in Table IV, together with the number of sweeps performed. In this section all error bars are computed averaging our results inside bins of 100 (or more if explicitly stated) consecutive sweeps and treating these averages as independent measurements. Results are stable upon variations of the bin size. The first 10% of the sweeps are used for thermalization and are not included in the analysis.

In Figs. 7 and 8 we plot $C_{11}(0)$ and $C_{44}(0)$ as representative examples of $t = 0$ correlations. The shape of these curves is qualitatively determined by the low β-behavior

$$C_{ii}(0) \sim \tfrac{1}{4} + A_i u^2 \qquad (i = 1, 12), \tag{3.7}$$

(all A_i coefficients except for A_1 are positive; their values can be found in Table V), and the high β-behavior

$$C_{ii}(0) \sim B_i/\beta^2. \tag{3.8}$$

(We have calculated explicitly only B_1; see Appendix).

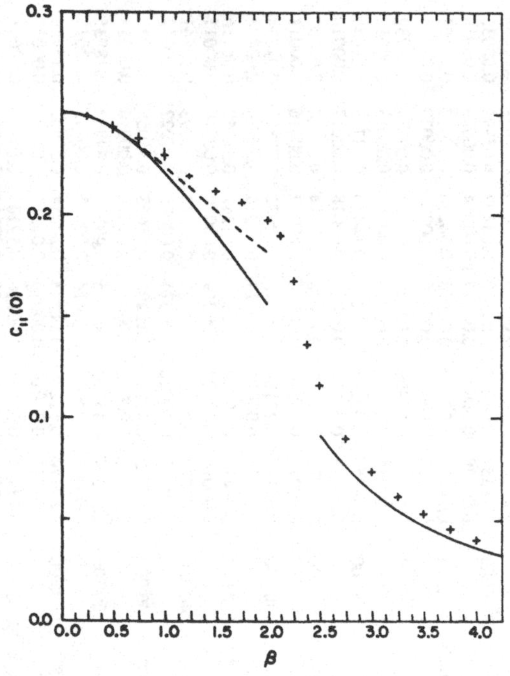

FIG. 7. MC data for $C_{11}(0)$ compared with first order, low β, expansion in the variables $u(\beta)$ (straihgt line) and $z(\beta)$ (dashed line), and with the spin-wave result for high β.

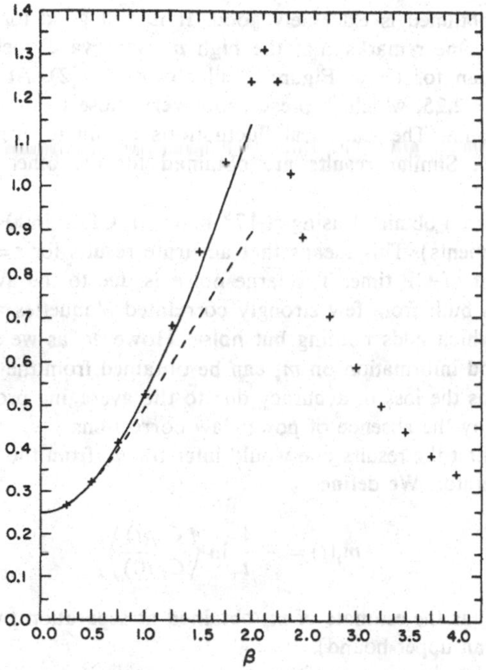

FIG. 8. MC data for $C_{44}(0)$ compared with the first order expansion for low β.

TABLE V

Low β Expansion for $C_{ii}(0)$, $C_{ii}(1)$ and the Corresponding Glueball Mass Estimate, for $i = 1$–5

i	$C_{ii}(0)$	$C_{ii}(1)$	$m_i(t)$
1	$\frac{1}{4}-\frac{u^2}{2}$	$\frac{u^4}{4}$	$-4\ln u$
2	$\frac{1}{4}+\frac{3u^2}{2}$	$\frac{u^6}{4}$	$-6\ln u$
3	$\frac{1}{4}+\frac{3u^2}{2}$	$\frac{9u^6}{4}$	$-6\ln u-\ln 9$
4	$\frac{1}{4}+\frac{19u^2}{4}$	$\frac{17u^6}{4}$	$-6\ln u-\ln 17$
5	$\frac{1}{4}+\frac{u^2}{2}$	$\frac{4u^8}{81}$	$-8\ln u-\ln\frac{16}{81}$

The precision obtained is extremely good. It is still good for $t=1$ as shown in Fig. 9 for $C_{11}(1)$. One remarks that the high β "spin-wave" behavior sets in later (in β) for $t=1$ than for $t=0$. Figure 10 illustrates $C_{11}(2)$. At $t=3$, we obtain a signal only for $\beta=2.25$, which is presumably very close to the point where $C_{11}(3)$ reaches its maximum. The statistical fluctuations dominate over the signal for all other values of β. Similar results are obtained for the other correlations $C_{ii}(t)$, $t=1, 2, 3$.

Our data have been obtained using $\simeq 17^H$ hours of CDC 7600 (60% for upgrades, 40% for measurements). This means that accurate results for $t=3$ would require a prohibitive amount of CP time. The large noise is due to the averaging over space locations. $C_{ij}(t)$ is built from few strongly correlated plaquettes and a lot of weakly correlated ones, which adds nothing but noise. However, as we shall see later, with our procedure good information on m_g can be obtained from the data with $t=1$ and $t=2$ already. Thus the loss of accuracy due to the averaging procedure seems to be overcompensated by the absence of power law corrections.

Let us present first the results one would infer for m_g from the correlation function of a single O_i operator. We define

$$m_i(t) = -\frac{1}{t}\ln\left\langle\frac{C_{ii}(t)}{C_{ii}(0)}\right\rangle \tag{3.9}$$

and consider $m_i(t)$ as the estimate of m_g obtained at separation t with the operator O_i ($m_i(t)$ is actually an upper bound).

As explained already, we first form averages over 100 consecutive sweeps. The final means and standard derivations are taken over these averages. Note that the alternative definition

$$\tilde{m}_i(t) = -\ln\left\langle\frac{C_{ii}(t)}{C_{ii}(t-1)}\right\rangle \tag{3.10}$$

gives better (lower) estimates but with t times bigger error bars. If the error bars on $m_i(t)$ are already rather large, then the alternative definition is not good for our purposes.

The expected behavior of the $m_i(1)$'s for low β is given in Table V. It is clear that as β goes to zero, the glueball reduces to a pure one-plaquette state. When β grows, all length 6 operators become important, as can be appreciated from Figs. 11 and 12, where the MC results for $m_1(t)$ and $m_4(t)$ are displayed. The $j=1$, $j=\frac{3}{2}$ operators remain, however, of negligible importance for the glueball wave function, as one can deduce from their rate of decay (see Figs. 13 and 14, for $m_5(t)$ and $m_9(t)$).

Going back to $m_1(t)$, one notices that in the spin-wave limit this quantity decreases very slowly with t (see Table VI), taking for low t values comparable to those assumed at $\beta \simeq 1.5$. As a consequence, for small t, $m_1(t)$ cannot follow the (rapidly decreasing) scaling behavior over an extended domain, since it has to ultimately reach the spin-wave value. The situation from this point of view is very different from the case of the string tension, where the spin-wave results decrease rather rapidly with

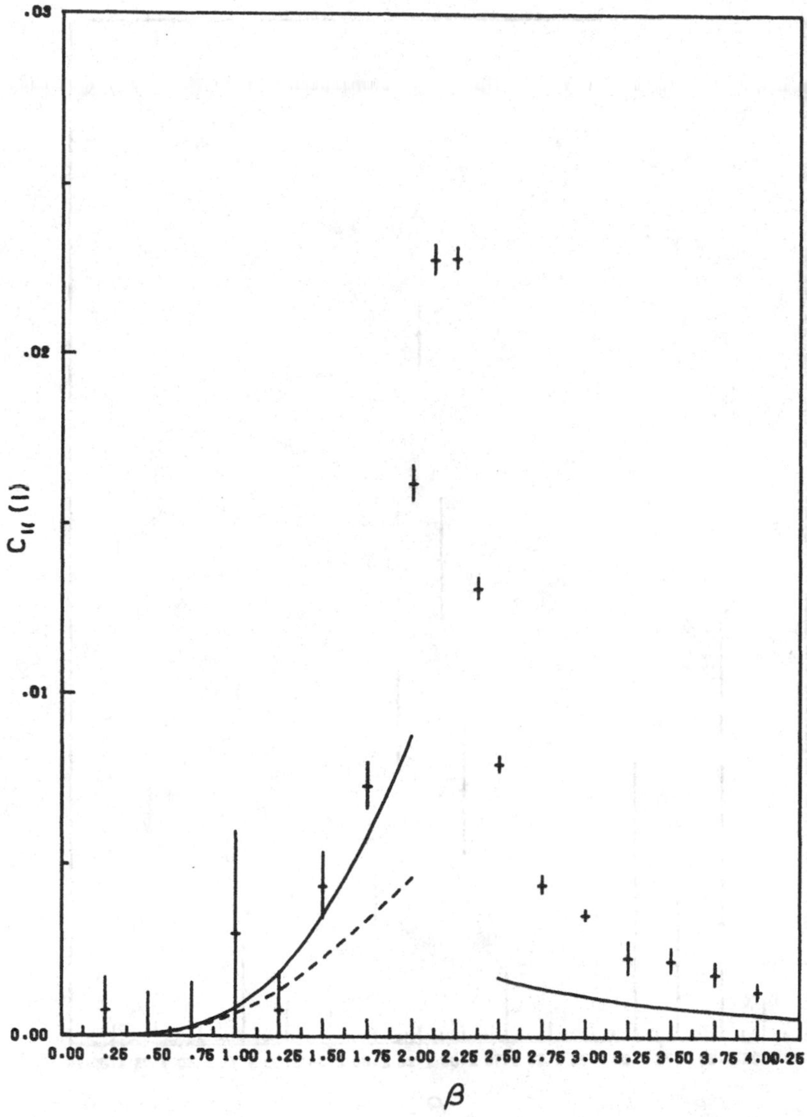

FIG. 9. MC data for $C_{11}(1)$ compared with expected behavior for low and high β.

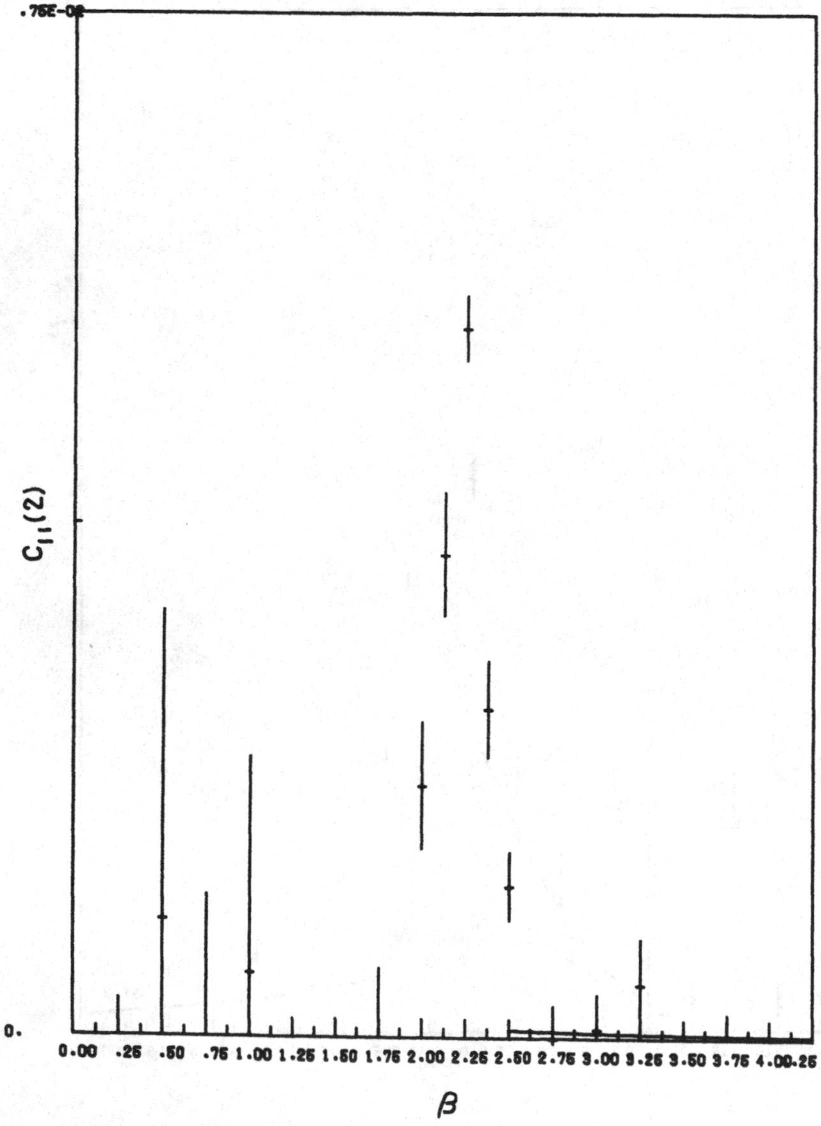

FIG. 10. MC data for $C_{11}(2)$.

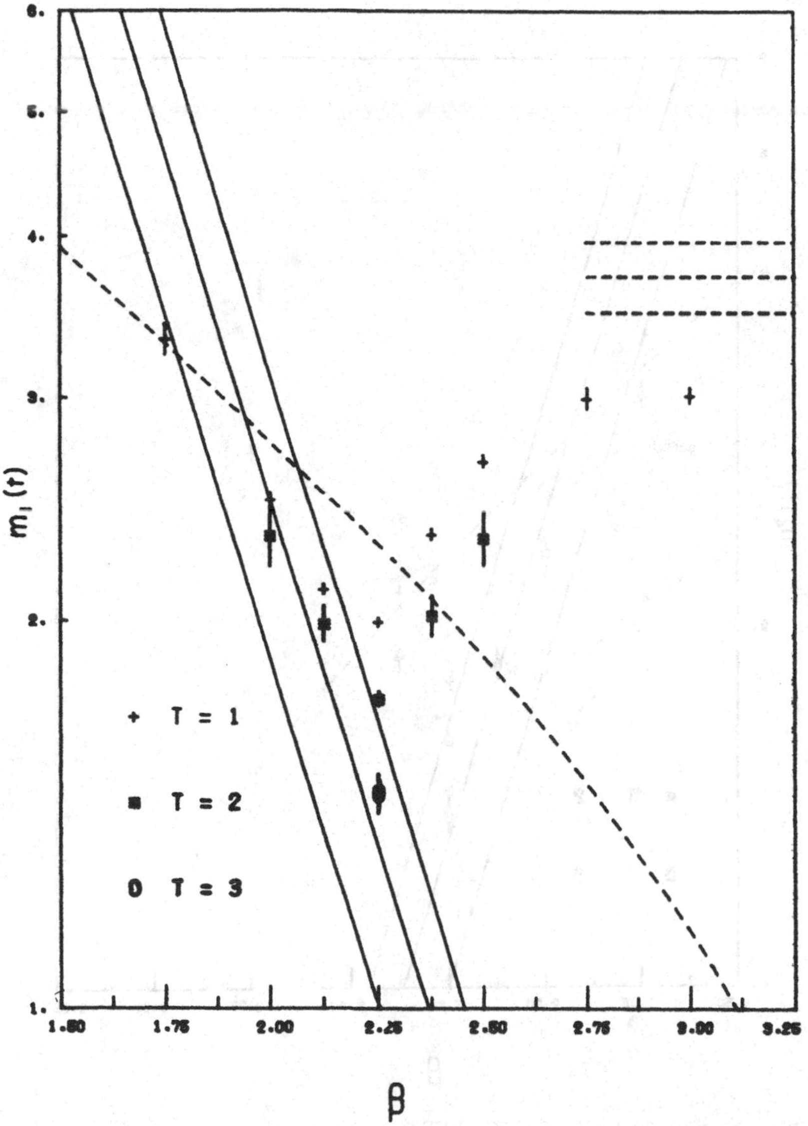

FIG. 11. Estimate of the glueball mass from the correlation function of the operator O_1. The three horizontal lines on the right of the figure are high β limits obtained with $t = 1$, 2 and 3. The scaling behavior corresponding to $m_g = 2.4 \pm 0.6$ is also shown.

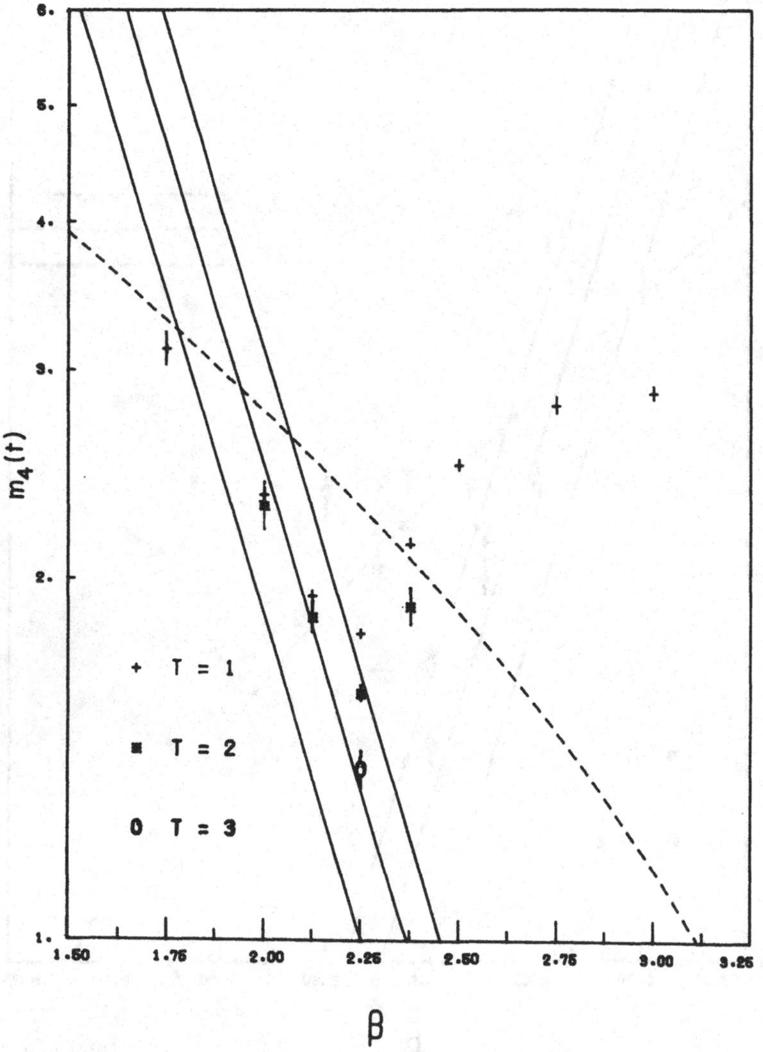

FIG. 12. Estimate of the glueball mass from the correlation function of the operator O_4.

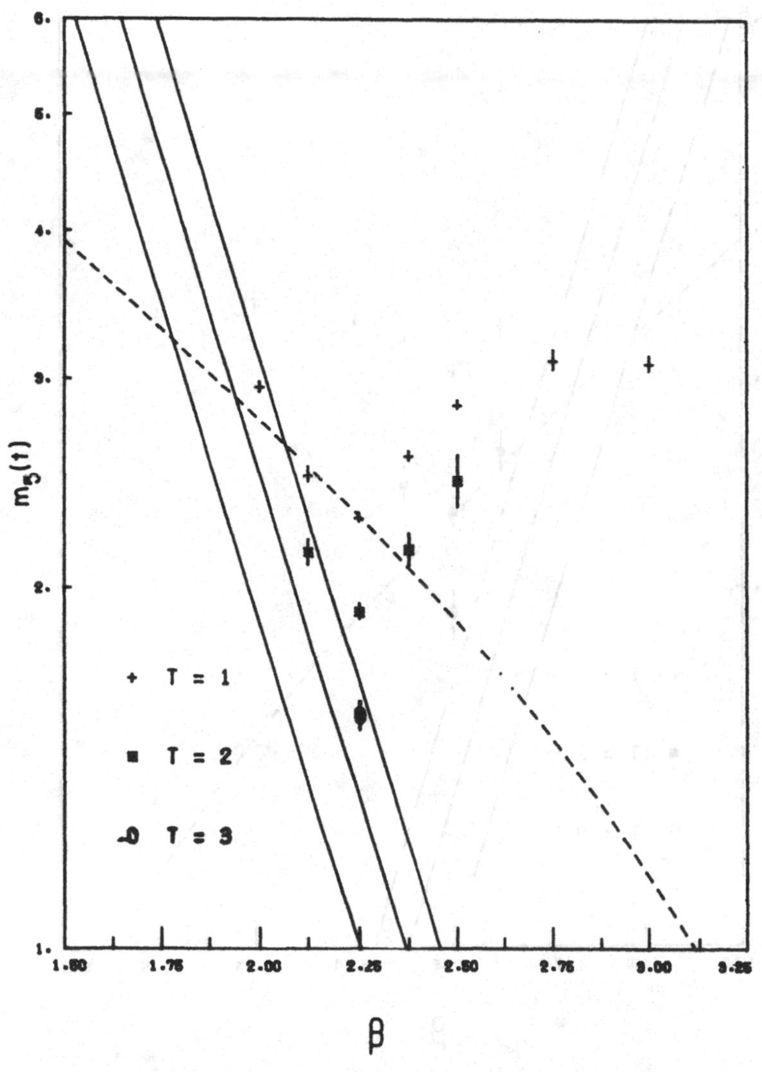

FIG. 13. Estimate of the glueball mass from the correlation functions of O_5.

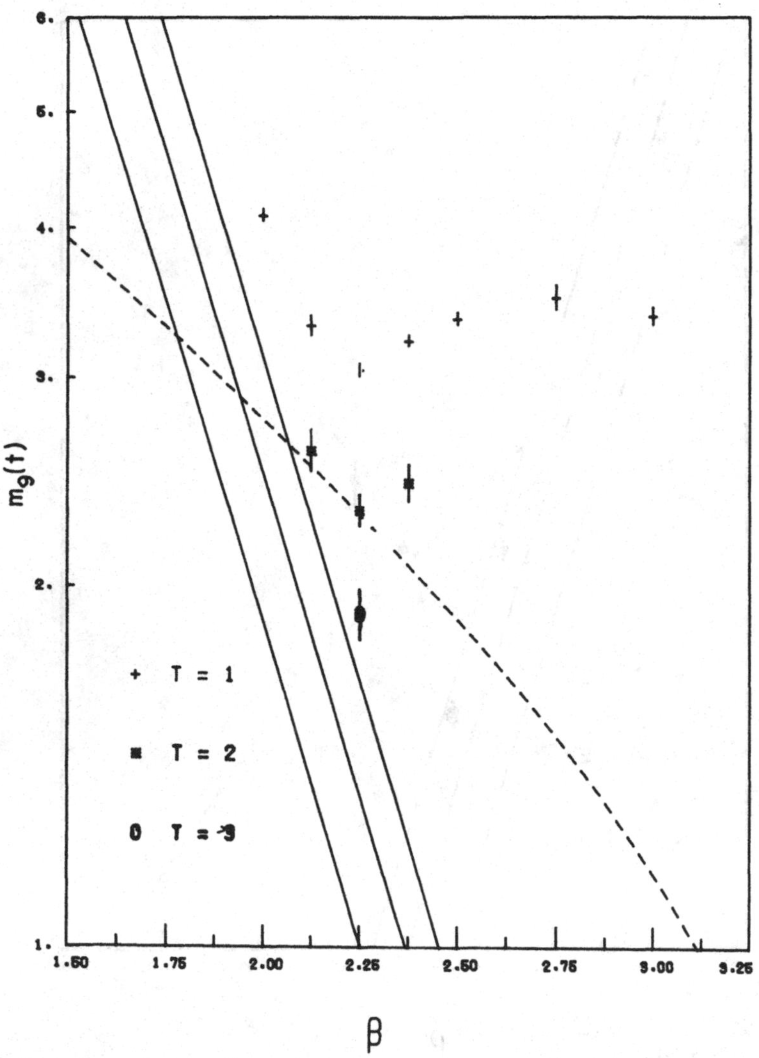

FIG. 14. Estimate of the glueball mass from the correlation functions of O_9.

TABLE VI

Glueball Mass Estimate from the Finite Distance Correlation Function C_{11} in the Spin-Wave Approximation

	$2^3 \times 16$ Lattice		$4^3 \times 16$ Lattice		$8^3 \times 16$ Lattice	
t	$m(t)$	$\bar{m}(t)$	$m(t)$	$\bar{m}(t)$	$m(t)$	$\bar{m}(t)$
1	4.13	4.13	3.96	3.96	3.92	3.92
2	3.97	3.80	3.73	3.49	3.58	3.24
3	3.85	3.63	3.50	3.06	3.20	2.43
4	3.78	3.56	3.33	2.81	2.88	1.92
5	3.73	3.54	3.20	2.70	2.64	1.68
6	3.70	3.53	3.11	2.65	2.45	1.52

the size of the loop [25, 27]. There is, however, a small range of β where approximate scaling behavior is observed.

The variational principle leads to the equation

$$m_g = -\frac{1}{t} \ln \left\langle \text{Max} \frac{\sum_{ij} c_i c_j C_{ij}(t)}{\sum_{ij} c_i c_j C_{ij}(0)} \right\rangle. \tag{3.11}$$

The actual use of the formula requires some care. Before minimization, the result is a rather smooth function of the coefficients c_i. So, if inside one bin one of the correlations $C_{ii}(t)$, due to some fluctuation, happens to be particularly large, the minimization procedure will give a big weight to the operator i. m_g is thus systematically underestimated.

We have done two things in order to avoid this bias. The first is to exclude the operators which, on the basis of the numerical results, are found to contribute only little to the glueball wave function, but basically add noise. This turned out to be the case for all non-spin $\frac{1}{2}$ operators and for the operator O_2. Only the operators O_1, O_3, and O_4 were thus eventually kept in the analysis.

The second way of reducing the bias on m_g is to analyze what happens as one considers bigger and bigger bin sizes. The estimate of Eq. (3.11) can only be trusted if there exists a value for the bin size above which it becomes constant (inside error bars). We have found indeed that such a constant estimate is obtained for ($\beta = 2 - 2.50$, $\Delta t = 1, 2$) as the bin size exceeds 5000 sweeps.

Our final results are displayed in Fig. 15. The effect of the minimization on the estimate of the glueball mass is quite appreciable. As already stated in the Introduction, it leads us to the value $m_g = 2.4 \pm 0.6 \sqrt{K}$.

We close this section presenting data for $m_1(t)$ on a $2^3 \times 16$ lattice (Fig. 16). These show considerable finite-size effects when compared to Fig. 11. Although this small lattice allows simulations faster by an order of magnitude than with the $4^3 \times 16$ lattice, it cannot be used to determine the glueball mass.

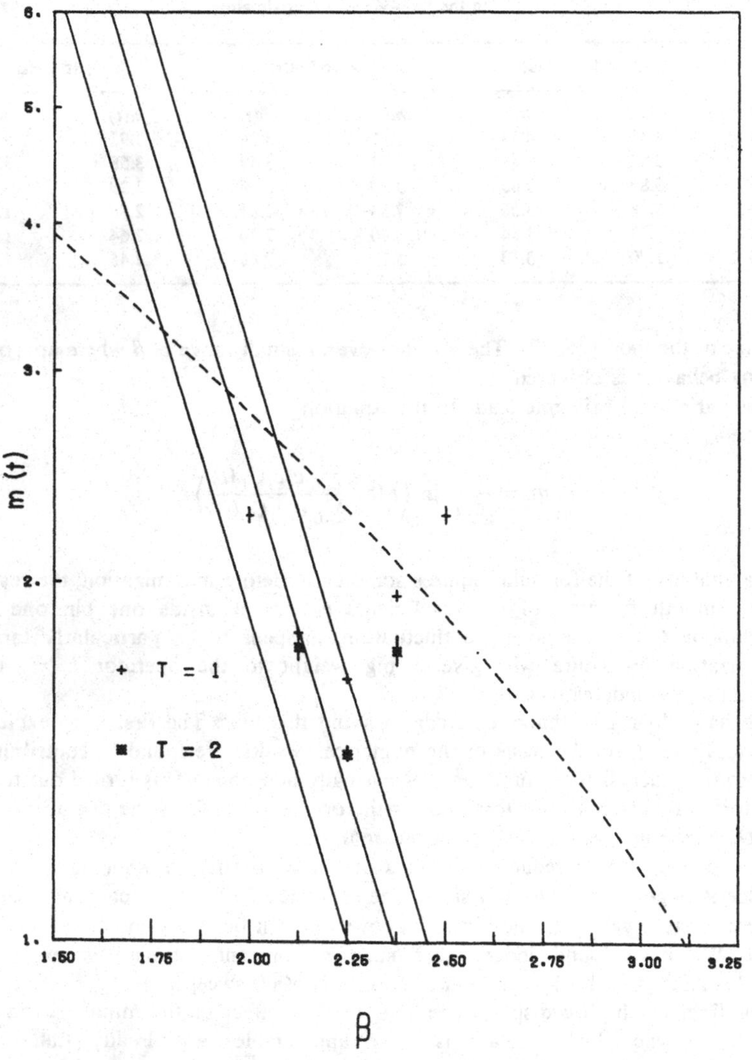

FIG. 15. Refined estimate of the glueball mass using the minimization procedure described in the text. We have only plotted those results which remain stable as one increases the bin size.

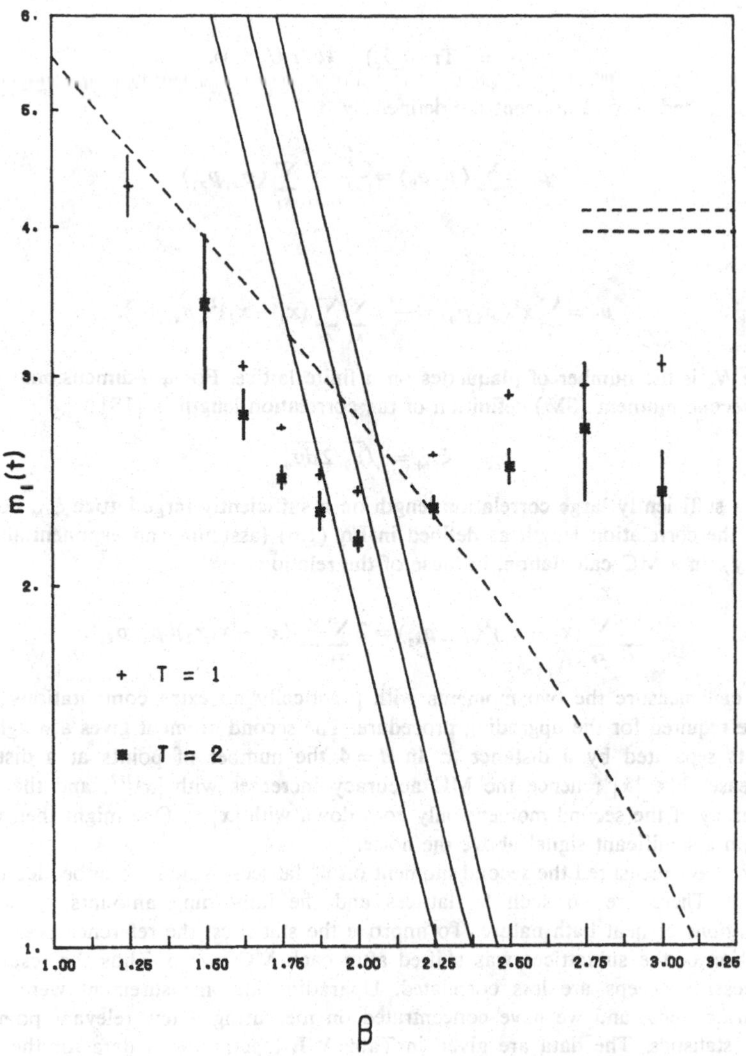

FIG. 16. Estimate of the glueball mass from the correlation function of O_1 on a small ($2^3 \times 16$) lattice.

IV. Second Moment Definition of the Correlation Length

Let

$$p_x = \tfrac{1}{2}\,\mathrm{Tr}\, U(\dot{p}_x) - \tfrac{1}{2}\langle \mathrm{Tr}\, U(\dot{p}_x)\rangle. \tag{4.1}$$

The zero and second moment are defined by

$$\mu_0 = \sum_x \langle p_x p_0 \rangle = \frac{1}{N_p} \sum_{x_1} \sum_{x_2} \langle p_{x_1} p_{x_2} \rangle \tag{4.2}$$

and

$$\mu_2 = \sum_x x^2 \langle p_x p_0 \rangle = \frac{1}{N_p} \sum_{x_1} \sum_{x_2} (x_1 - x_2)^2 \langle p_{x_1} p_{x_2} \rangle. \tag{4.3}$$

Here N_p is the number of plaquettes on a finite lattice. For a d-dimensional system, the second moment (SM) definition of the correlation length is [18]

$$\xi_{SM} = \sqrt{\mu_2 / 2d\mu_0}. \tag{4.4}$$

For a sufficiently large correlation length on a sufficiently large lattice ξ_{SM} coincides with the correlation length as defined in Eq. (1.6) (assuming an exponential rate of decay). In a MC calculation, because of the relation

$$\sum_{x_1} \sum_{x_2} (x_1 - x_2)^2 \langle p_{x_1} p_{x_2} \rangle = 2 \sum_{x_1} \sum_{x_2} (x_1^2 - x_1 x_2) \langle p_{x_1} p_{x_2} \rangle,$$

one can measure the two moments with practically no extra computations beyond those required for the upgrading procedure. The second moment gives a weight x^2 to points separated by a distance x. In $d = 4$ the number of points at a distance x increases like $|x|^3$; hence the MC accuracy increases with $|x|^{3/2}$, and the relative accuracy of the second moment only goes down with $|x|^{1/2}$. One might then hope to obtain a significant signal above the noise.

We have measured the second moment on 8^4 lattices, which are imbedded in a 16^4 lattice. There are 16 such 8^4 lattices and the imbedding amounts to boundary conditions of heat bath nature. To improve the statistics, the reference point for the position of the sublattices was shifted after each MC sweep. Thus the results from successive sweeps are less correlated. Upgrading and measurement were done in separate steps, and we have concentrated on measuring a few relevant points with high statistics. The data are given in Table VII, together with data for the specific heat $C = \beta^2 \mu_0$ on the full 16^4 lattice and on the 8^4 sublattices, and data for the internal energy on the full 16^4 lattice.

The result is rather disappointing. Although the statistics are sufficient to achieve error bars <1 for the correlation length, nevertheless no increase of ξ_{SM} with β is observed. This may be due to the following two reasons:

TABLE VII

MC Data for: Internal Energy and Specific Heat on a 16^4 Lattice; Specific Heat, Second Moment and Second Moment Result for the Correlation Length from 16 8^4 Sublattices; and Number of Sweeps and Number of Sweeps for Equilibrium (Eq.)

β	E, 16^4	C, 16^4	C, 8^4	μ_2	ξ_{SM}	Sweeps (Eq.)
1.8	0.55896 ± 0.00019	1.05 ± 0.20	0.85 ± 0.12	0.09 ± 0.35	0.3 ± 0.3	100 (98)
2.0	0.49883 ± 0.00011	1.44 ± 0.12	1.10 ± 0.02	0.02 ± 0.14	0.1 ± 0.2	480(250)
2.1	0.46557 ± 0.00010	1.63 ± 0.16	1.24 ± 0.04	0.21 ± 0.14	0.3 ± 0.1	480(200)
2.2	0.43078 ± 0.00008	1.29 ± 0.12	1.27 ± 0.02	0.00 ± 0.10	0.0 ± 0.1	480(200)
2.3	0.39785 ± 0.00014	1.70 ± 0.18	1.24 ± 0.03	0.16 ± 0.11	0.3 ± 0.1	480(200)
2.4	0.36993 ± 0.00009	1.50 ± 0.22	1.12 ± 0.03	0.07 ± 0.08	0.2 ± 0.1	480(200)
2.6	0.33007 ± 0.00009	0.79 ± 0.16	0.92 ± 0.03	-0.05 ± 0.08	/	200(100)

(1) Through the imbedding procedure, significant correlations are lost because the boundary of an 8^4 lattice contains more sites than the lattice itself. The reader may observe from Table VII that the peak in the specific heat is flattened on the 8^4 sublattices.

(2) There can be a disturbing, rather large power behavior at small distances.

We tried to take into account both points, assuming a behavior

$$\rho(x) = \frac{K_1(m(r + r_0))}{(r + r_0)^\alpha}, \qquad (r = |x|) \tag{4.5}$$

(K_1 modified Bessel function; see Eq. (24) of Ref. [21]) for the correlation function. For a smooth enough behavior at small distances (e.g., $\alpha = 1$, $r_0 = 0.1$), one should clearly see an increase of ξ_{SM} in the region where the correlation length ξ becomes >1 (i.e., $\beta > 2.3$). The situation changes drastically for a large power law at short distances (e.g., $\alpha = 2.5$, $r_0 = 0.1$): then the second moment definition gives much too small values, in agreement with the results from Table VII.

In conclusion the data presented in this section supports a rather short correlation length in the scaling region. However, because of the circumstances explained above, a precise measurement of the correlation length on the basis of the second moment definition cannot be obtained.

V. Conclusions

The mass gap in four-dimensional non-Abelian lattice gauge theory is a quantity of fundamental importance, but at the same time quite difficult to determine, be it by expansion methods or numerical techniques. With our analysis we hope to have obtained a reliable estimate of its value. The results previously presented in the literature, given the various factors of uncertainty discussed in this article, are consistent with ours.

The mass gap, beyond being a quantity of theoretical relevance, may be experimentally observable as the mass of a possible resonance with the quantum numbers of the vacuum. However, both the passage to an $SU(3)$ gauge group and, especially, the inclusion of fermionic (quark) degrees of freedom would most likely alter the result obtained in this paper, which in a pure $SU(2)$ gauge theory places the glueball at a mass of 1.2 GeV.

APPENDIX: SPIN-WAVE ANALYSIS

When the correlation length is larger than the other relevant scales, a perturbative spin-wave analysis becomes possible. Expanding in terms of the field A_μ defined by

$$U_\mu(x) = e^{i\sum_a A_\mu^a(x)(\sigma^a/2)}$$

and keeping only terms of lowest order the action reduces to

$$S + S_{GF} = \sum_x \frac{\beta}{8} \sum_a \left(\sum_\mu \Delta_\mu A_\mu^a(x) \right)^2,$$

where Δ_μ is the lattice differential in the μ direction

$$\Delta_\mu f(x) = f(x + e_\mu) - f(x)$$

and a suitable gauge fixing term has been added.

TABLE VIII
Glueball Mass Estimate from Finite Distance Plaquette–Plaquette Correlation in the Spin–Wave Approximation

	8^4 Lattice		16^4 Lattice		32^4 Lattice	
t	$m(t)$	$\tilde{m}(t)$	$m(t)$	$\tilde{m}(t)$	$m(t)$	$\tilde{m}(t)$
1	3.97	3.97	3.96	3.96	3.96	3.96
2	3.81	3.65	3.80	3.64	3.80	3.64
3	3.58	3.12	3.60	3.19	3.60	3.19
4			3.36	2.64	3.36	2.64
5			3.11	2.09	3.10	2.09
6			2.86	1.59	2.86	1.66
7			2.60	1.08	2.64	1.35
8					2.46	1.14
9					2.30	0.99
10					2.15	0.87

This quadratic form is diagonalized by going to Fourier space:

$$A_\mu^a(x) = \frac{(2\pi)^2}{N_s^3 N_t} \sum_k \varepsilon_\mu^a(k)\, e^{-ikx}.$$

This gives

$$S + S_{GF} = \frac{4\pi^4 \beta}{N_s^3 N_t} \sum_{h,\nu} \Pi(h)\, \varepsilon_\nu(h)\, \varepsilon_\nu^*(h),$$

where $\Pi(h) = \sum_\mu (1 - \cos h_\mu)$. Green's functions are then obtained by straihgtforward Gaussian integration. One gets, in particular,

$$C_{ii}(t) = \frac{1}{N_s^6 N_t^2 \beta^2} \sum_{\substack{\mathbf{h}, h_4, k_4 \\ (\mathbf{h}, h_4) \neq 0 \\ (\mathbf{h}, k_4) \neq 0}} \frac{\cos(h_4 t) \cos(k_4 t)\, \Pi(\mathbf{h}, 0)^2}{\Pi(\mathbf{h}, h_4)\, \Pi(\mathbf{h}, k_4)},$$

$$\rho(\mathbf{0}, t) = \frac{2}{3 N_s^6 N_t^2 \beta^2} \left(\sum_{h \neq 0} \frac{\cos(h_4 t)\, \Pi(\mathbf{h}, 0)}{\Pi(\mathbf{h}, \mathrm{h}_4)} \right)^2.$$

Numerical values of the qualities $m_1(t)$ and $\tilde{m}_1(t)$ (see Eqs. (3.9), (3.10)) in the spin-wave approximation can be found in Table VI. Similar quantities, derived from simple plaquette–plaquette correlations, are given in Table VIII.

Acknowledgments

We would like to thank G. Münster for correspondence and unpublished notes [26].

Note added in Proof. Independently analogous studies, albeit focusing on different sets of operators, have been carried out by Falcioni *et al.* [28] and by Ishikawa *et al.* [29]. The final results are compatible. For further developments see Ref. [30].

References

1. K. Wilson, *Phys. Rev. D* **10** (1974), 2445.
2. A. Hasenfratz and P. Hasenfratz, *Phys. Lett. B* **93** (1980), 165.
3. K. Wilson, Cargese Lectures, 1979.
4. M. Creutz, L. Jacobs, and C. Rebbi, *Phys. Rev. D* **20** (1979), 1915.
5. M. Creutz, *Phys. Rev. D* **21** (1980), 2308.
6. M. Creutz, *Phys. Rev. Lett.* **45** (1980), 313.
7. D. Petcher and D. H. Weingarten, *Phys. Rev. D* **22** (1980), 2465.
8. G. Bhanot and C. Rebbi, *Nucl. Phys. B* **180** (1981), 469.
9. B. Berg and J. Stehr, *Z. Phys. C* **9** (1981), 33.
10. E. Pietarinen, *Nucl. Phys. B* **90** (1981), 349.

11. H. Hamber and G. Parisi, *Phys. Rev. Lett.* **47** (1981), 1792; E. Marinari, G. Parisi, and C. Rebbi, *Phys. Rev. Lett.* **47** (1981), 1795; D. Weingarten, *Phys. Lett. B* **109** (1982), 57; H. Hamber, E. Marinari, G. Parisi, and C. Rebbi, *Phys. Lett. B* **108** (1982), 314.
12. B. Berg, *Phys. Lett. B* **97** (1980), 401.
13. J. Engels, F. Karsch, H. Satz, and I. Montvay, *Phys. Lett. B* **102** (1981), 332.
14. G. Münster, *Nucl. Phys B* **190** (1981), 454; erratum, *Nucl. Phys. B* **200** (1982), 536.
15. G. Bhanot, *Phys. Lett. B* **101** (1981), 95.
16. R. Brower, M. Nauenberg, and T. Schalk, *Phys. Rev. D* **24** (1981), 548; R. Brower, M. Creutz, and M. Nauenberg, University of California, Santa Cruz, preprint, 1981.
17. C. Rebbi, *Phys. Rev. D* **21** (1980), 3350.
18. M. E. Fischer and R. F. Burford, *Phys. Rev.* **156** (1967), 583.
19. J. B. Kogut, D. Sinclair, and L. Susskind, *Nucl. Phys. B* **144** (1976), 199; J. B. Kogut, R. B. Pearson, and J. Shigemitsu, *Phys. Rev. Lett.* **43** (1979), 484; *Phys. Lett. B* **98** (1981), 63.
20. G. Münster, *Phys. Lett. B* **95** (1980), 59; *Nucl. Phys. B* **180** (1981), 23.
21. I. S. Gradshteyn and I. M. Ryshik, "Table of Integrals, Series and Products," Academic Press, New York, 1965.
22. M. Falcioni, E. Marinari, M. L. Paciello, G. Parisi, and B. Taglienti, *Phys. Lett. B* **102** (1981), 270; *Nucl. Phys. B* **190** (1981), 782.
23. K. Wilson, unpublished.
24. B. Lautrup and M. Nauenberg, *Phys. Rev. Lett.* **45** (1980), 1755.
25. J. P. Kovall and H. Neuberger, *Nucl. Phys. B* **189** (1981), 535.
26. G. Münster, unpublished.
27. T. Hattori and H. Kawai, *Phys. Lett. B* **105** (1981), 43.
28. M. Falcioni, E. Marinari, M. L. Paciello, G. Parisi, F. Rapuano, B. Taglienti, and Zhang Yi-cheng, *Phys. Lett. B* **110** (1982), 295.
29. K. Ishikawa, M. Teper, and G. Schierholz, *Phys. Lett. B* **110** (1982), 399.
30. B. Berg, Lecture notes presented at the John Hopkins workshop in Florence, CERN preprint, 1982.

Volume 110B, number 3,4 PHYSICS LETTERS 1 April 1982

ON THE MASSES OF THE GLUEBALLS IN PURE SU(2) LATTICE GAUGE THEORY

M. FALCIONI [a,b], E. MARINARI [a,b], M.L. PACIELLO [a], G. PARISI [c],
F. RAPUANO [a,b], B. TAGLIENTI [a] and ZHANG Yi-Cheng [d]
[a] *Istituto Nazionale di Fisica Nucleare, Sezione di Roma, Rome, Italy*
[b] *Istituto di Fisica "G. Marconi", Università di Roma, Rome, Italy*
[c] *Istituto di Fisica, Facoltà di Ingegneria, Università di Roma, Rome, Italy and INFN, Laboratori Nazionali di Frascati, Italy*
[d] *Scuola Internazionale Superiore di Studi Avanzati, Trieste, Italy*

Received 6 January 1982

Using a Monte Carlo method to measure the correlation functions of different quantities, the mass spectrum of the glueball is investigated for an SU(2) gauge model in an 8^4 lattice.

One of the main purposes of Monte Carlo like simulations of field theories is to compute the mass spectrum of the theory. This is normally done by measuring the correlation function G of an appropriate local operator:

$$G(x) = \langle O(x)O(0)\rangle - \langle O(x)\rangle\langle O(0)\rangle . \quad (1)$$

Indeed we know that when in four-dimensional euclidean theories $|x| \to \infty$

$$G(x) \sim (m^{+1/2}/|x|^{+3/2}) \exp(-m|x|) , \quad (2)$$

where m is the mass of the lowest state $|s\rangle$ such that $\langle 0|O|s\rangle \neq 0$, i.e. the lightest state with the same quantum numbers of O (if we neglect accidental cancellations). If the correlation function is measured in the conventional way, i.e. using its definition (1), the measurement of the mass may be rather problematic, especially for composite states: we need to know the correlation function in the large-x region where G is very small: unfortunately the statistical error on $G(x)$ is roughly speaking independent of the distance and proportional at the best to $\langle O^2(0)\rangle/N^{+1/2}$, N being the number of Monte Carlo steps.

In spite of these difficulties some estimates have been obtained for the glueball mass in four dimensions, either by measuring the correlation functions with high statistics [1], or by using a slightly modified version of this method (the finite scaling [2, 3]) which is unfortunately unable (at least in the presently used version with untwisted boundary conditions) to distinguish among particles with different spin–parity. However, it is difficult to make further progress in this direction.

A possible alternative would be to use an indirect algorithm to compute the correlation functions in order to decrease the statistical errors of many orders of magnitude. This has been accomplished in a few cases [3–6], but it is not evident if the method can be successfully applied to glueballs.

An other alternative, which has been suggested by Wilson [7], consists in trying to extract the masses using the correlation functions of the theory at small and not at large distance. This can be done by measuring correlations of many operators. Let us explain Wilson's proposal.

We introduce on our lattice a time axis t, and we construct a list of operators, $O_\alpha(t)$ having the same quantum numbers. These operators are local in time, but non local in space; they are constructed by summing and multiplying fields defined at the same time slice. In lattice gauge theories a convenient form for the operators would be the trace of a Wilson loop of different shapes. Using the transfer matrix formalism the validity of the Källen–Lehmann representation has been proved:

$$\langle O_\alpha(t)O_\beta(0)\rangle = \sum_n \langle 0|O_\alpha|n\rangle\langle n|O_\beta|0\rangle \exp(-E_n t), \tag{3}$$

where the sum over n runs overall the possible states of the theory; however if the operator carries zero spatial momentum [i.e. $O_\alpha(t) = \Sigma_{x,y,z} \widetilde{O}_\alpha(x.y,z,t)$], only states at rest contribute to the right-hand side of (3). If we suppose that the number of states N contributing to eq. (3) is finite, the masses of all these states can be extracted from the correlation functions of N operators at distance n and $1 + n$. Indeed let us define

$$\mu_{\alpha\beta} \equiv \langle O_\alpha(n)O_\beta(0)\rangle = \langle\alpha|\beta\rangle \equiv C^{(n)}_{\alpha\beta},$$

$$|\alpha\rangle \equiv T^{n/2}O_\alpha(0)|0\rangle,$$

$$C^{(n+1)}_{\alpha\beta} = \langle O_\alpha(n+1)O_\beta(0)\rangle \equiv \langle\alpha|T|\beta\rangle,$$

$$T \equiv \sum_n |n\rangle\langle n| \exp(-m_n), \tag{4}$$

where T is the transfer matrix.

It is evident that in the generic case the states $|\alpha\rangle$ form a nonorthonormal basis of an N-dimensional Hilbert space. It is convenient to consider the matrix $\widetilde{T} = M^{-1/2}C^{(n+1)}M^{-1/2}$, i.e. the matrix T in an orthonormal basis. If t_n are the eigenvalues of T we get

$$m_n = -\ln t_n. \tag{5}$$

A consistency check of the approach could be done by computing the correlation functions at higher distances, e.g.:

$$C^{(n+2)}_{\alpha\beta} \equiv \langle O_\alpha(n+2)O_\beta(0)\rangle = \langle\alpha|T^2|\beta\rangle$$
$$= \sum_{\gamma,\delta} C^{(n+1)}_{\alpha\gamma} M^{-1}_{\gamma\delta} C^{(n+1)}_{\delta\beta}. \tag{6}$$

It is clear that we do not need too big lattices in the time direction, provided the condition $\exp(-Lm_0) \ll 1$ is satisfied. (m_0 is the mass we are looking for). In space direction it is convenient to consider lattices which are not much bigger than the actual size of the particles one is considering.

In this note we report some results from a modified form of this method for the lower masses of the glueballs of different spins in a pure SU(2) lattice gauge theory at one value of the bare coupling constant.

We have considered a 8^4 lattice with periodic boundary conditions; the action has the Manton form. We have chosen this action because it does not give rise to the pronounced peak in the specific heat versus β [8]. So the correlation functions in the transition region from the strong coupling to the scaling regime, is expected to exhibit a smoother behaviour than with the Wilson action.

After we verified that $\beta \geqslant 1.4$ is in the scaling region by preliminary runs we have taken most of our data at $\beta = 1.55$. Following the analysis of ref. [8] this value of β corresponds to a lattice spacing $a = \sigma^{-1/2}$ (0.57 –0.67) ~ 1.24 GeV^{-1} if $\sqrt{\sigma} = 0.5$ GeV, σ being the value of the string tension, as extracted from the ψ spectrum. We approximate the SU(2) group with its maximal finite subgroup [1,9]: the symmetry group of the icosahedron.

The gauge invariant operators we have considered are:

$$O_\alpha(t) = \sum_{\substack{xy,xz,yz\\ \text{planes}}} [1 - \tfrac{1}{2}\mathrm{tr}(\Pi_i^{(\alpha)}U_i)],$$

where $\Pi_i^{(\alpha)}U_i$ is the product of SU(2) matrices defined on links belonging to the paths P_α listed in fig. 1, and t runs from 1 to the lattice size.

The results for the connected correlation functions are shown in table 1. The standard deviations for the values of table 1 are all ~ 1.0. These values have been obtained using 9 000 equilibrium configurations; each configuration gives $32 = 8 \times 4$ contributions to the correlation functions, the factor 8 comes from the different choices of the origin, the factor 4 comes from different choices of the time axis. The total CPU time

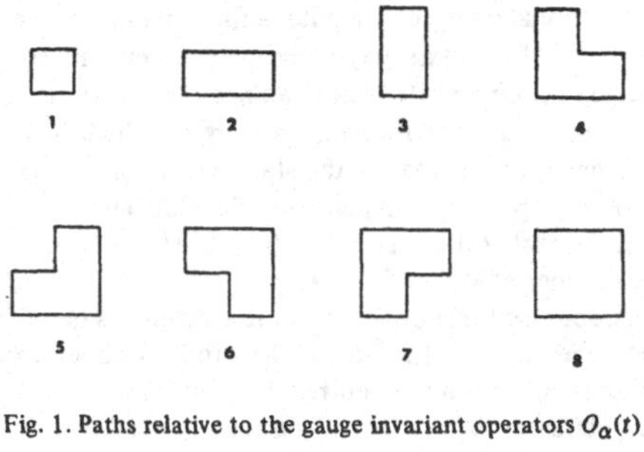

Fig. 1. Paths relative to the gauge invariant operators $O_\alpha(t)$

Volume 110B, number 3,4 PHYSICS LETTERS 1 April 1982

Table 1
Connected correlation functions at distance 0, 1, 2.

232.52							
260.78	515.12						
260.99	297.95	515.30					
223.33	370.35	370.48	539.97				
222.53	370.24	369.76	338.67	538.97			
223.23	370.14	370.11	403.00	338.61	539.41		
223.25	370.52	370.24	339.71	402.92	338.76	540.67	
179.69	339.79	339.86	424.96	423.91	423.99	424.97	599.07
18.36							
25.82	39.11						
26.05	37.22	40.26					
25.61	38.32	38.94	39.50				
25.15	37.45	38.42	37.98	38.81			
25.85	38.71	39.21	39.53	39.06	39.73		
24.80	37.19	38.61	38.25	38.26	37.68	38.74	
24.83	38.32	39.09	40.83	40.69	39.95	40.01	42.66
2.18							
2.98	4.96						
3.42	5.39	5.74					
3.00	5.09	4.52	4.26				
3.18	6.14	5.75	5.58	6.12			
3.63	5.76	5.98	5.72	5.91	6.62		
2.83	4.71	5.43	4.80	5.54	5.05	5.60	
3.95	6.81	6.84	6.39	7.48	7.28	7.29	8.57

used on the INFN VAX 11/780 of Rome is approximately 150 hours.

A first analysis on the correlation functions at distance 0 and 1 and then at distance 2, as indicated from eq. (6) gives us rather inconsistent results. That would mean the mass spectrum contributing to the correlation functions at distance 0 is very different from that at distance 1 or 2. This appears also roughly looking at the decrease of the correlation functions versus the distance for fixed β.

Then we performed the analysis previously described using correlation functions at distance 1, 2 and with some care, 3. In the analysis of spin 0 states we found it convenient to reduce the original 8×8 correlation matrices to a 4×4 matrix by averaging the 8×8 matrix elements corresponding to indistinguishable configurations of the paths, for example we averaged the elements 14, 15, 16, 17 of the 8×8 matrices, and so on.

We chose for the matrix elements of T the basis in which $C^{(1)}_{\alpha\beta}$ is diagonal; namely: $\widetilde{T} = (C^{(1)})^{-1/2} C^{(2)} \times (C^{(1)})^{-1/2}$. The eigenvalues $C^{(1)}_{n}$ are the following:

$$135.34, \quad 363, \quad 0.87, \quad 0.07 \, .$$

As there is one eigenvalue definitely bigger than the others, we project $\widetilde{T}$ only on the corresponding eigenvector. This is equivalent to consider the correlation function of a definite combination of the operators O_α.

The result is:

$$t_0 = 0.157 \, .$$

We have consistently verified that the eigenvectors corresponding to the largest eigenvalue of $C^{(1)}$ and $C^{(2)}$ are approximately the same.

From eq. (5) we estimate the following values for the spin zero mass:

$$m_0 = (1.85 \pm 0.23) a^{-1} \, .$$

The error we quoted is the standard deviation relative to the values of m_0 obtained from six samples of 1500 events.

The correlation functions at distance 3 have large statistical errors but they are consistent with the value of t_0; we expect for t_0^2 the value 2.5×10^{-2} while we measure $(0.6 \pm 2.5) \times 10^{-2}$.

To select different spin states [10,11] we have con-

tation and perform an MC calculation for the six correlation functions between these operators. The MC calculation is carried out on a $4^3 \times 8$ lattice with cyclic boundary conditions. 8 is idei.ufied with the time direction. Correlations are measured between different time planes at distance $t = 0, 1, 2$. We sum over all space positions to project onto the zero-momentum states and sum over all space-like orientations to select out the $J = 0$ state. This means our correlation functions are

$$\rho_{ij}(t) = \langle \Box_i(0) \Box_j(t) \rangle ,$$

with

$$\Box_i(t) = \sum_{x,\,\mathrm{orientations}} (O_i(x,t) - \langle O_i(x,t) \rangle). \qquad (4)$$

The normalization of the operators O_i is taken such that $\rho_{ii}(0) = 1/18$ for $\beta = 0$.

For the upgrading of the lattice we have used the SU(3) heat bath method of Pietarinen [12]. On the calculation we have spent 17 h CDC 7600 computer time, roughly 2/3 for upgrading and 1/3 for measurements. The whole procedure has turned out to be slower by a factor of 30 than in the similar calculation for SU(2) [3]. We had spent about the same time on the SU(2) calculation, hence the computer time used for the SU(3) calculation is rather small.

Our main results are given in figs. 2 and 3 and in the tables. They are more encouraging than even an optimist might have hoped. Fig. 2 represents MC data for the glueball mass from plaquette–plaquette correlations at a distance $t = 1$, and fig. 3 gives the glueball mass after minimization over contributions from all the six considered correlations (again at a distance $t = 1$) has been carried out. The three solid lines are parallel to the weak-coupling renormalization group (RG) prediction (2), and represent the glueball estimates from figs. 2 and 3.

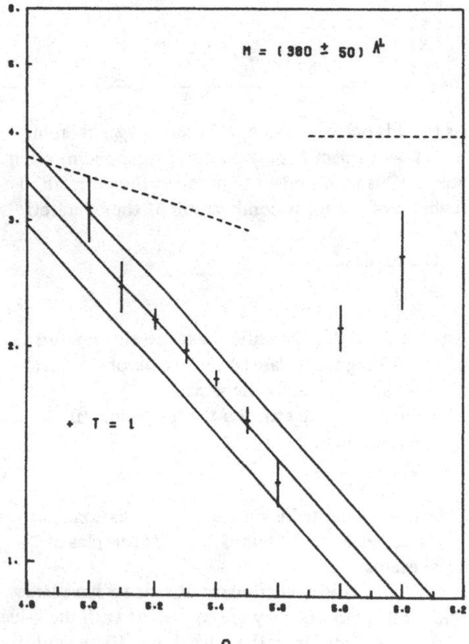

Fig. 2. Glueball results from plaquette–plaquette correlations.

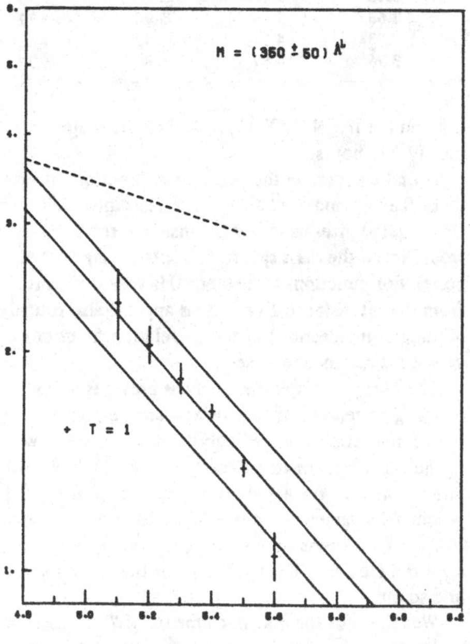

Fig. 3. Glueball results after minimization.

Volume 113B, number 1 PHYSICS LETTERS 3 June 1982

Table 1
Correlations of the operations O_i ($i = 1, ..., 3$) of fig. 1 with themselves and the number of sweeps for measurement.

β	$C_{11}(0)$	$C_{22}(0)$	$C_{33}(0)$	$C_{11}(1)$	$C_{22}(1)$	$C_{33}(1)$	Sweeps
5.0	0.095 ± 0.002	0.178 ± 0.005	0.408 ± 0.011	0.004 ± 0.001	0.009 ± 0.003	0.015 ± 0.007	800
5.1	0.093 ± 0.002	0.186 ± 0.004	0.444 ± 0.006	0.008 ± 0.002	0.015 ± 0.004	0.040 ± 0.007	800
5.2	0.099 ± 0.002	0.210 ± 0.003	0.495 ± 0.011	0.011 ± 0.001	0.024 ± 0.002	0.058 ± 0.003	2000
5.3	0.100 ± 0.001	0.228 ± 0.004	0.530 ± 0.011	0.014 ± 0.001	0.033 ± 0.003	0.081 ± 0.009	2000
5.4	0.102 ± 0.001	0.257 ± 0.003	0.583 ± 0.009	0.017 ± 0.001	0.045 ± 0.002	0.106 ± 0.006	2000
5.5	0.104 ± 0.002	0.298 ± 0.007	0.640 ± 0.013	0.021 ± 0.002	0.070 ± 0.005	0.150 ± 0.010	2000
5.6	0.106 ± 0.004	0.339 ± 0.013	0.693 ± 0.027	0.029 ± 0.003	0.107 ± 0.012	0.216 ± 0.025	2000
5.8	0.069 ± 0.001	0.247 ± 0.007	0.464 ± 0.010	0.008 ± 0.001	0.037 ± 0.005	0.062 ± 0.008	400
6.0	0.061 ± 0.001	0.222 ± 0.006	0.406 ± 0.013	0.004 ± 0.002	0.022 ± 0.006	0.030 ± 0.012	200

Table 2
Mixed correlations of the operators O_i of fig. 1.

β	$C_{12}(0)$	$C_{13}(0)$	$C_{23}(0)$	$C_{12}(1)$	$C_{13}(1)$	$C_{23}(1)$
5.0	0.110 ± 0.003	0.178 ± 0.005	0.214 ± 0.006	0.007 ± 0.002	0.007 ± 0.003	0.010 ± 0.005
5.1	0.112 ± 0.002	0.186 ± 0.003	0.231 ± 0.005	0.012 ± 0.002	0.019 ± 0.003	0.024 ± 0.005
5.2	0.125 ± 0.002	0.205 ± 0.004	0.268 ± 0.006	0.016 ± 0.001	0.025 ± 0.002	0.038 ± 0.003
5.3	0.133 ± 0.002	0.215 ± 0.004	0.297 ± 0.006	0.022 ± 0.002	0.033 ± 0.003	0.050 ± 0.004
5.4	0.145 ± 0.002	0.230 ± 0.003	0.339 ± 0.005	0.027 ± 0.001	0.042 ± 0.002	0.068 ± 0.004
5.5	0.160 ± 0.003	0.245 ± 0.005	0.391 ± 0.009	0.039 ± 0.003	0.057 ± 0.004	0.101 ± 0.006
5.6	0.175 ± 0.007	0.260 ± 0.010	0.444 ± 0.019	0.055 ± 0.006	0.078 ± 0.010	0.151 ± 0.018
5.8	0.118 ± 0.003	0.171 ± 0.003	0.302 ± 0.008	0.016 ± 0.003	0.022 ± 0.003	0.049 ± 0.007
6.0	0.104 ± 0.003	0.151 ± 0.004	0.266 ± 0.008	0.009 ± 0.003	0.011 ± 0.005	0.027 ± 0.008

In fig. 2 we have also included the spin wave result (tree approximation for $\beta \to \infty$ on the finite lattice), and (as in fig. 3) the first two orders of the SC expansion ($\beta \to 0$). See the broken lines. From the tables the interested reader may read off the correlations of the operators and the number of sweeps used for measurements. We have first used 600 sweeps to bring one lattice "adiabatically" into the considered scaling region, and then spent at each value of β an additional number of 200 sweeps for equilibrium. For plaquette–plaquette correlations at distance $t = 0$ we have checked our MC data in the SC limit, and for the high value $\beta = 10$ we have checked our MC results on the spin wave result of fig. 2.

In figs. 2 and 3 and in the tables the mean values are from all data and the error bars are obtained by comparison relative to 10 samples of consecutive events. To prevent statistical fluctuations of the error bars, the number of subsamples has to be reasonably large. On the other hand, as explained in ref. [3], subsamples have to be rather big to give a good wave function and to minimize the bias brought about by the minimization procedure. Our way of calculating the error bar in some sense also accounts for the bias. The bias of mean values from averages over rather large samples of size 400 lowers for $5.2 \leqslant \beta \leqslant 5.6$ the result by 5% in the worst case ($\beta = 5.2, 5.6$). We have also obtained results at distance $t = 2$. They are still afflicted by rather large error bars and do not yet allow a reliable minimization. For the plaquette–plaquette correlations alone the mean values for the glueball mass from distance $t = 2$ are (as expected) systematically below the mean values from distance $t = 1$, and the result is well compatible with eq. (5) below.

The results of fig. 3, as well as of fig. 2, fit very well the weak-coupling RG prediction and allow the rather accurate relation

$$m(0^+) = (350 \pm 50)\, \Lambda^L . \qquad (5)$$

The error bar in eq. (5) comes from a trajectory of

points which follow the weak-coupling RG behaviour, and tries to estimate (in a somewhat subjective manner) the order of magnitude of possible corrections in the approach to the continuum limit. This kind of error bar should not be confused with the statistical error bar coming from a single point, as used in the final results of refs. [4–8].

There are several notable points. First of all, contrary to our SU(2) results, the glueball mass from correlations of 0^+ plaquette eigenstates at a distance $t = 1$ already exhibits scaling in a certain region of β. An estimate of the glueball mass from these correlations alone (fig. 2) would give a value $m(0^+) = (380 \pm 50)\,\Lambda^L$, which is already close to the value (5). That the one-plaquette operator gives a reasonable result in a certain range of the scaling region comes presumably from the higher complexity of the SU(3) gauge group. Before starting this investigation we had conjectured this, but we did not expect the effect to be so drastic. After minimization the glueball wave functions turn out to be rather flat in the involved operator. Our SU(3) results are self-consistent in the sense that the data follow the predicted scaling behaviour. With respect to this they are even better than our SU(2) results [3].

Our glueball mass data undershoot (in the region where we observe scaling) the strong-coupling series of ref. [11] (except the PADE [2,5]). Therefore the value (5) is smaller than the SC extrapolation (3). There is, however, an overlap within two error bars, and it is notable that the lowest value within Münster's error bars gives a tangent to the highest order of the SC expansion.

Our value (5) is in disagreement (i.e., no overlap within two error bars) with a value $M = (720 \pm 100)\Lambda^L$ quoted in ref. [7]. No details are given there. We would like to remark that reliable results cannot be obtained from finite-size scaling effects without having a detailed understanding of the degeneracy and the excited levels.

A subtle question is to extract a GeV value for the glueball mass from our result (5). If SU(3) hadron spectroscopy [7,9] continues to make rapid progress, the best attitude is certainly to make a comparison with experimentally well-known quark bound states. We here rely on the MC estimates of the SU(3) string tension. The values are (ref. [13])

$$\Lambda^L = (5 \pm 1.5) \times 10^{-3} \sqrt{K},$$

and (ref. [12])

$$\Lambda^L = (7 \pm 2) \times 10^{-3} \sqrt{K}.$$

As these values are less accurate than our estimate of the glueball $m(0^+)$, we take the mean value of eq. (5) for the glueball mass, and $\Lambda^L = (6 \pm 2) \times 10^{-3} \sqrt{K}$ for the string tension. With $\sqrt{K} = 440$ MeV we thus obtain

$$m(0^+) = (920 \pm 310)\ \text{MeV}. \qquad (6)$$

Summary and outlook: we have found scaling behaviour for the SU(3) glueball mass $m(0^+)$ in a certain region of the coupling β. With eq. (5) we have obtained an accurate value for the glueball mass in units of Λ^L. In a forthcoming paper we will present results for the excited SU(3) glueball states. In the more distant future one may include quark corrections.

Note added in proof. Gernot Münster kindly informed us about a further correction to his calculation [11]. The glueball estimate from SC expansion becomes now

$$m_g(0^+) = (430 \pm 60)\,\Lambda^L,$$

in agreement with our result.

References

[1] B. Berg, Phys. Lett. 97B (1980) 401.
[2] G. Bhanot and C. Rebbi, Nucl. Phys. B180 [FS2] (1981) 469.
[3] B. Berg, A Billoire and C. Rebbi, Brookhaven National Laboratory preprint (1981) submitted to Ann. Phys. (NY).
[4] M. Falcioni et al., Phys. Lett. 110B (1982) 295.
[5] K. Ishikawa, M. Teper and G. Schierholz, Phys. Lett. 110B (1982) 399.
[6] K.H. Mütter and K. Schilling, to be published (private communication).
[7] H. Hamber and G. Parisi, Phys. Rev. Lett. 47 (1981) 1792.
[8] E. Marinari, G. Parisi and C. Rebbi, Phys. Rev. Lett. 47 (1981) 1795.
[9] A. Hasenfratz, Z. Kunszt, P. Hasenfratz and C. Lang, Phys. Lett. 110B (1982) 289.
[10] A. Hasenfratz and P. Hasenfratz, Phys. Lett. 93B (1980) 165.
[11] G. Münster, Nucl. Phys. B190 [FS3] (1981) 439; and Erratum, to be published.
[12] E. Pietarinen, Nucl. Phys. B190 [FS3] (1981) 349.
[13] M. Creutz, Phys. Rev. Lett. 45 (1980) 313.

Volume 116B, number 6 PHYSICS LETTERS 28 October 1982

SU(3) LATTICE MONTE CARLO CALCULATION OF THE GLUEBALL MASS SPECTRUM

K. ISHIKAWA, M. TEPER
Deutsches Elektronen-Synchrotron DESY, Hamburg, Fed. Rep. Germany

and

G. SCHIERHOLZ
II. Institut für Theoretische Physik der Universität Hamburg, Hamburg, Fed. Rep. Germany

Received 5 May 1982
Revised manuscript received 24 June 1982

We have calculated the glueball masses of various spins and parities in SU(3) gauge theory. Our first results give $m_M(0^{++}) = (3.6 \pm 0.2)\Lambda_{mom}$, $m_E(0^{++}) = (4.3 \pm 0.3)\Lambda_{mom}$, $m(0^{-+}) = (7.2^{+1.6}_{-0.9})\Lambda_{mom}$, $m_M(2^{++}) = (8.1 \pm 1.1)\Lambda_{mom}$ and $m_E(2^{++}) = (8.3^{+1.6}_{-1.0})\Lambda_{mom}$ as well as information on the glueball wave functions.

In this letter we extend our recent calculation of the glueball mass spectrum [1] to the gauge group SU(3) (i.e., QCD).

We follow the calculational procedure outlined in our previous letter [1]. That is, we perform a variational calculation by choosing some à priori reasonable class of wave functions $\{\phi\}$ and vary ϕ within that class to maximize the expectation value

$$\frac{\Gamma_a}{\Gamma_0} = \frac{\langle\phi_{p=0}(t+a)\phi_{p=0}(t)\rangle}{\langle\phi_{p=0}(t)\phi_{p=0}(t)\rangle} = \frac{\langle\phi_{p=0}(t)e^{-Ha}\phi_{p=0}(t)\rangle}{\langle\phi_{p=0}(t)\phi_{p=0}(t)\rangle} = \frac{\sum_{n=0}^{\infty} e^{-m_n a}|\langle n|\phi_{p=0}(t)|\Omega\rangle|^2}{\sum_{n=0}^{\infty}|\langle n|\phi_{p=0}(t)|\Omega\rangle|^2}, \tag{1}$$

where a is the lattice spacing and t the euclidean time. The resulting wave function is then used to calculate

$$\frac{\Gamma_{2a}}{\Gamma_a} = \frac{\langle\phi_{p=0}(t+2a)\phi_{p=0}(t)\rangle}{\langle\phi_{p=0}(t+a)\phi_{p=0}(t)\rangle} = \frac{\langle\phi_{p=0}(t)e^{-H2a}\phi_{p=0}(t)\rangle}{\langle\phi_{p=0}(t)e^{-Ha}\phi_{p=0}(t)\rangle} = \frac{\sum_{n=0}^{\infty} e^{-m_n 2a}|\langle n|\phi_{p=0}(t)|\Omega\rangle|^2}{\sum_{n=0}^{\infty} e^{-m_n a}|\langle n|\phi_{p=0}(t)|\Omega\rangle|^2}, \tag{2}$$

the logarithm of which will give us the lowest-lying glueball mass ($m = m_0$):

$$m = -a^{-1}\ln(\Gamma_{2a}/\Gamma_a) . \tag{3}$$

The idea behind this procedure is that (i) increasing the theoretical accuracy involves searching for a larger signal and (ii) the higher mass states in (2) are double suppressed by the choice of the wave function *and* the exponential damping factor

It follows that (1) provides an *upper* bound on the mass:

$$-a^{-1}\ln(\Gamma_a/\Gamma_0) \geq m . \tag{4}$$

The *pure* variational calculations of the glueball mass(es) [2,3] assume that this upper bound will be close to the actual mass(es). How much this is actually so for the rather small lattices and the limited class of wave functions considered can be checked [1] and will be discussed further along with the results.

We work on a $4^3 \cdot 8$ lattice with 8 lattice points in the time direction. From our previous experience [1], where we have considered 4^4, $4^3 \cdot 8$, 6^4 and 8^4 lattices, we infer that the $4^3 \cdot 8$ lattice is large enough for our purposes provided that the coupling parameter β is chosen appropriately. Working on a "rectangular" lattice has the further advantage that it lowers the physical temperature at nearly the same computer

Volume 116B, number 6 PHYSICS LETTERS 28 October 1982

cost and, hence, better approximates zero temperature QCD.

An appropriate value of β is one for which

$$a < \text{glueball size} < 2a\,. \qquad (5)$$

To two loops the lattice spacing is given by ($\beta = 6/g^2$)

$$a = (83.5/\Lambda_{\text{mom}})(\tfrac{8}{33}\pi^2\beta)^{51/121}\exp(-\tfrac{4}{33}\pi^2\beta)\,, \qquad (6)$$

where the overall scale is set by the string tension ‡1 [we "measure" $\Lambda_{\text{mom}} = (0.48 \pm 0.05)\sqrt{K}$ at $\beta = 5.7$].

$$\Lambda_{\text{mom}} \approx 0.5\sqrt{K}\,, \qquad (7)$$

which we believe is $\sqrt{K} \approx 400$ MeV, corresponding to unit Regge slope. Assuming the the glueball size to be of the order of 0.5 fermi (which proves to be correct in the actual calculation), this constrains β to the region $5 \lesssim \beta \lesssim 6$. We also require that β lies in the continuum region. Leaning on the results of ref. [4] we infer that $5 \lesssim \beta$ is a suitable value.

The occurence of a sharp peak in the specific heat at ‡2 $\beta = 5.45 \pm 0.1$ and a rapid increase in the correlation length near this peak, which has been attributed to nearby complex β plane singularities [6], cuts the range of admissible values of β further down to $5.6 \lesssim \beta$. We have taken "data" at $\beta = 5.7$ and $\beta = 5.9$ which we have checked lie in the scaling region. Incidentally, the authors in ref. [3] have taken most of their "data" below and at the peak in the specific heat which casts some doubt on their results.

We have constructed states ϕ with $J^{PC} = 0^{++}, 0^{-+}, 1^{--}, 1^{+-}$ and 2^{++} which we expect to be low-lying. Except for 0^{-+} they can be composed of planar loops with purely space-like links ($0^{++}, 1^{+-}, 2^{++}$) or space- and time-like links ($0^{++}, 1^{--}, 2^{++}$) as shown schematically in fig. 1. To project onto zero momentum states the sum is to be taken over all spacial lattice points. The loops that we have considered in our variational calculation are shown in fig. 2. We have confined ourselves to loops that extend only one lattice spacing in time ($\mathcal{O}_4$ and $\mathcal{O}_5$, cf. fig. 2). That is to reduce the amount of time-like overlap in Γ_a and to avoid time-like overlaps in Γ_{2a} at all which, otherwise, could lead us into conflict with physical positivity [7]. Note that for 0^{++} and 2^{++} two types of different operators arise: magnetic ($\sim B^2$) with exclusively space-like extensions and electric ($\sim E^2$) extending also in time

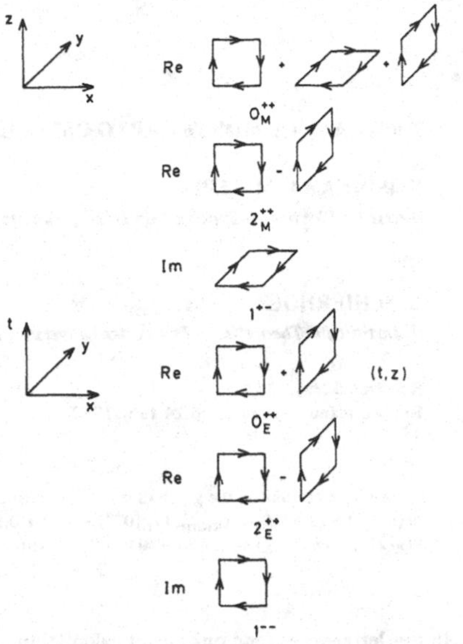

Fig. 1. 0^{++}, 2^{++}, 1^{+-} and 1^{--} wave functions composed of planar loops. The loops are understood to be summed over all space-like translations. The subscripts M and E stand for magnetic and electric, respectively.

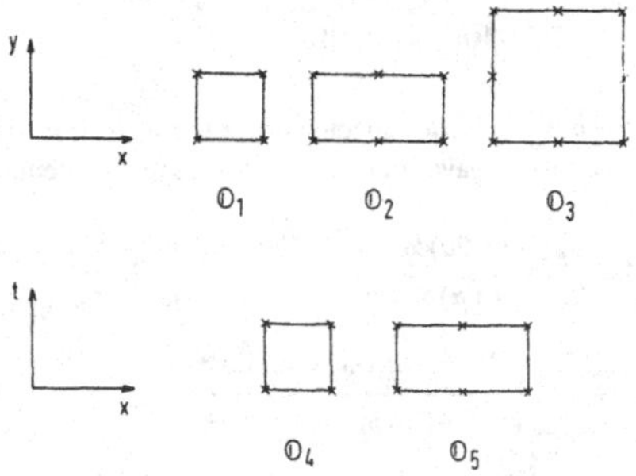

Fig. 2. Loops considered in our calculation (x, y stand for any space direction).

‡1 We haven taken the mean value of the string tensions "measured" by the authors in ref. [4].

‡2 This has been obtained by a straight-line interpolation $1/g^2 \sim N$, N being the number of colours, of the SU(2), SU(4), SU(5) and SU(6) peak/phase transition values reported in ref. [5]. Note that they fall very nicely on a straight line.

Volume 116B, number 6 PHYSICS LETTERS 28 October 1982

(cf. fig. 1). This doubling of glueball states has also been found in the bag model [8]. The most natural trial wave function for 0^{-+} is $F\widetilde{F}$, the topological charge density, which is of the type $\boldsymbol{E} \cdot \boldsymbol{B}$. We have taken $\Sigma_{i,j,k=\pm 1}^{\pm 3} \widetilde{\epsilon}_{0ijk} F^{0i} F^{jk}$ instead which extends only one lattice spacing in the time direction in contrast to $F\widetilde{F}$ (which does two) but else has the same quantum numbers and topology. The reason is to secure physical positivity [7] as mentioned before. There are various ways of implementing this operator on the lattice. We use the definition employed in ref. [9].

We have performed several independent lines of iterations of the lattice configuration using the heat-bath method [10]. Before taking any "data" we have heated up the lattice starting from cold ($\beta = \infty$) and hot ($\beta = 0$) starts. So far we have collected $\gtrsim$24000 "events" at $\beta = 5.7$ and $\gtrsim$ 7000 "events" at $\beta = 5.9$. Our first results for SU(3) glueball masses and wave functions are summarized below. A more detailed presentation will be given elsewhere.

0^{++}. As 0^{++} has vacuum quantum numbers, it is necessary to replace ϕ by $\phi - \langle\phi\rangle$ in (1) and (2) to ensure that the vacuum state does not appear as intermediate state.

We first treat the magnetic and electric 0^{++} glueball states independently, i.e., not accounting for any mixing between them.

(a) Magnetic. After maximization we obtain at $\beta = 5.7$

$$\Gamma_a/\Gamma_0 = 0.22 \pm 0.01\,, \tag{8}$$

which, according to (4), (6), corresponds to the upper bound

$$m_{\mathrm{M}}(0^{++}) \leqslant (5.52 \pm 0.18)\Lambda_{\mathrm{mom}} \tag{9}$$

(where the subscript M stands for magnetic). The accompanying best wave function is (cf. fig. 2)

$$\phi_{\mathrm{M}}(0^{++}) \approx 0.33\, \mathcal{O}_1 + 0.40\, \mathcal{O}_2 + 0.85\, \mathcal{O}_3\,, \tag{10}$$

where the operators on the right-hand side are understood to be summed over all orientations in accord with fig. 1 and over the spacial lattice for any fixed t. For ease of writing we have dropped the subscript $\boldsymbol{P} = 0$.

In the next step we compute Γ_{2a}/Γ_a now using the wave function (10). We find

$$\Gamma_{2a}/\Gamma_a = 0.37 \pm 0.02\,, \tag{11}$$

which gives the mass

$$m_{\mathrm{M}}(0^{++}) = (3.6 \pm 0.2)\Lambda_{\mathrm{mom}}\,. \tag{12}$$

We observe that the straightforward variational calculation would have yielded a considerably higher 0^{++} glueball mass (9) which disqualifies the pure variational approach. This is not surprising as the class of possible wave functions is rather limited for such a small lattice.

(b) Electric. Our best wave function is ($\beta = 5.7$; cf. fig. 2)

$$\phi_{\mathrm{E}}(0^{++}) \approx 0.42\, \mathcal{O}_4 + 0.9\, \mathcal{O}_5 \tag{13}$$

(where the subscript E stands for electric) which corresponds to

$$\Gamma_a/\Gamma_0 = 0.16 \pm 0.01 \tag{14}$$

(note that Γ_a is positive in line with physical positivity [7]). This results in the upper bound

$$m_{\mathrm{E}}(0^{++}) \lesssim (6.68 \pm 0.22)\Lambda_{\mathrm{mom}}\,. \tag{15}$$

Using the optimized wave function (13), we then obtain

$$\Gamma_{2a}/\Gamma_a = 0.31 \pm 0.02\,, \tag{16}$$

which gives the mass

$$m_{\mathrm{E}}(0^{++}) = (4.3 \pm 0.3)\Lambda_{\mathrm{mom}}\,. \tag{17}$$

Note that also in this case the pure variational calculation cannot be trusted.

(c) Mixing. In reality the magnetic and electric glueball operators will mix. Accordingly, we write

$$\phi = \cos\Theta\, \phi_{\mathrm{M}} + \sin\Theta\, \phi_{\mathrm{E}}\,. \tag{18}$$

The physical eigenstates and mass eigenvalues of the transfer matrix are then obtained by searching for two orthogonal wave functions $\phi_{1,2}$ that maximize and minimize, respectively, Γ_{2a}/Γ_a. We have attempted this. But, due to the relatively large errors, we have no reliable numbers yet and even cannot tell with certainty that there are two states (or just one), not to mention the additional mixing with quark–antiquark states. We postpone the details to a forthcoming publication.

At $\beta = 5.9$ our best wave functions have a rather small projection onto the lowest lying state which results in a large error of Γ_{2a}/Γ_a.

Volume 116B, number 6 PHYSICS LETTERS 28 October 1982

2^{++}. We now go through the same analysis for 2^{++}.

(a) Magnetic. At $\beta = 5.7$ the best wave function is (cf. fig. 2)

$$\phi_M(2^{++}) \approx 0.18\, \mathcal{O}_1 + 0.46\, \mathcal{O}_2 + 0.87\, \mathcal{O}_3 \tag{19}$$

with

$$\Gamma_a/\Gamma_0 = 0.058 \pm 0.002\,, \tag{20}$$

which gives the upper bound

$$m_M(2^{++}) \leqslant (10.3 \pm 0.2)\Lambda_{mom}\,. \tag{21}$$

Using (19) we then obtain

$$\Gamma_{2a}/\Gamma_a = 0.11 \pm 0.03\,, \tag{22}$$

which gives

$$m_M(2^{++}) = (8.0 \pm 1.2)\Lambda_{mom}\,. \tag{23}$$

Our results at $\beta = 5.9$ (with $\sim 1/4$ the statistics) agree very well with that. The optimal wave function is

$$\phi_M(2^{++}) \approx 0.14\, \mathcal{O}_1 + 0.34\, \mathcal{O}_2 + 0.93\, \mathcal{O}_3 \tag{24}$$

and gives only a slightly higher bound on m while

$$\Gamma_{2a}/\Gamma_a = 0.16 \pm 0.03\,, \tag{25}$$

which results in

$$m_M(2^{++}) = (8.4 \pm 1.0)\Lambda_{mom}\,. \tag{26}$$

This is to say that Γ_{2a}/Γ_a follows nicely the weak coupling renormalization group trajectory.

By taking the statistical average of (23) and (26) we finally obtain the mass

$$m_M(2^{++}) = (8.1 \pm 1.1)\Lambda_{mom}\,. \tag{27}$$

(b) Electric. At $\beta = 5.7$ the best wave function is (cf. fig. 2)

$$\phi_E(2^{++}) \approx \mathcal{O}_5 \tag{28}$$

corresponding to

$$\Gamma_a/\Gamma_0 = 0.022 \pm 0.002 \tag{29}$$

(note that Γ_a is positive in line with physical positivity [7]) which leads to the bound

$$m_E(2^{++}) \leqslant (13.9 \pm 1.0)\Lambda_{mom}\,. \tag{30}$$

From (28) we then obtain

$$\Gamma_{2a}/\Gamma_a = 0.11 \pm 0.06 \tag{31}$$

and

$$m_E(2^{++}) = (8.0\,^{+2.9}_{-1.6})\Lambda_{mom}\,. \tag{32}$$

At $\beta = 5.9$ (with roughly the same statistics) we find the optimal wave function to be the same and

$$\Gamma_{2a}/\Gamma_a = 0.15 \pm 0.04\,, \tag{33}$$

which leads us to

$$m_E(2^{++}) = (8.7\,^{+1.4}_{-0.9})\Lambda_{mom} \tag{34}$$

in good agreement with scaling.

The statistical average of (32) and (34) gives finally the mass

$$m_E(2^{++}) = (8.3\,^{+1.6}_{-1.0})\Lambda_{mom}\,, \tag{35}$$

(c) Mixing. Our data do not allow any conclusions yet as to the mixing between the magnetic and electric 2^{++} glueball operators.

0^{-+}. In this case physical positivity [7] demands that the (naive) correlation functions Γ_a, Γ_{2a}, etc. are negative because the proper positive metric involves reflecting the time in one of the $\Sigma^{\pm 3}_{i,j,k=\pm 1}\tilde{\epsilon}_{0ijk} \times F^{0i}F^{jk}$. We have found that Γ_a (and Γ_{2a}, Γ_{3a}, though these have no time-like overlaps) is (are) indeed negative.

Since the calculation of $\Sigma^{\pm 3}_{i,j,k=\pm 1}\tilde{\epsilon}_{0ijk}F^{0i}F^{jk}$ on the lattice is very slow and the time required to obtain a reasonable signal/error ratio for $\boldsymbol{P} = 0$ wave functions is outside the present bounds of possibility, we have used momentum smeared wave functions here {as already before in our previous letter [1] on SU(2)}. That is, instead of summing the wave functions individually over all spacial lattice points to project onto zero momentum states, only nearest neighbour correlations are taken into account. We have calculated 0^{++} and 2^{++} energies in exactly the same way and, by comparing with the 0^{++} glueball mass stated above (2^{++} is less suited because of the relatively large errors), corrected for the effect of momentum smearing. We found consistently $\langle \boldsymbol{P}^2\rangle \approx 4/a^2$.

All our "measurements" are at $\beta = 5.7$. From $|\Gamma_a/\Gamma_0|$ we obtain the mass upper bound

$$m(0^{-+}) \leqslant (8.46 \pm 0.20)\Lambda_{mom}\,, \tag{36}$$

while Γ_{2a}/Γ_a yields the mass estimate

$$m(0^{-+}) = (7.2\,^{+1.6}_{-0.9})\Lambda_{mom}\,. \tag{37}$$

The proximity of (36) and (37), which is a reflection

of $\Gamma_{2a}/\Gamma_a \approx |\Gamma_a/\Gamma_0|$, suggests that $\Sigma^{\pm 3}_{i,j,k=\pm 1} \tilde{\epsilon}_{0ijk} \times F^{0i}F^{jk}$ is a good 0^{-+} wave function.

1^{+-}, 1^{--}. We have no useful bounds and numbers yet in this case.

So far we have expressed the glueball masses in units of Λ_{mom}. To make contact to experiment [11] we may use (7) now to fix the absolute scale (but note the obvious caveats). If we do so, we obtain the mass spectrum

$$m_{\mathrm{M}}(0^{++}) = (720 \pm 40)\ \mathrm{MeV}\,,$$
$$m_{\mathrm{E}}(0^{++}) = (850 \pm 50)\ \mathrm{MeV}\,,$$
$$m(0^{-+}) = (1430^{+320}_{-180})\ \mathrm{MeV}\,,$$
$$m_{\mathrm{M}}(2^{++}) = (1620 \pm 220)\ \mathrm{MeV}\,,$$
$$m_{\mathrm{E}}(2^{++}) = (1670^{+320}_{-200})\ \mathrm{MeV} \qquad (38)$$

[where we have ignored the uncertainties in (7)].

To summarize, we have reported first results of a calculation of the low-lying glueball masses based on altogether $\gtrsim$31 000 Monte Carlo "events". Our aim for the future is to reduce the errors and to make a more systematic survey of the low-lying glueball spectrum, in particular: to look for "oddballs".

We like to thank H. Joos, H.S. Sharatchandra, K. Symanzik and P. Weisz for useful discussions. We also thank E. Pietarinen for the use of his heat-bath and iteration routines.

References

[1] K. Ishikawa, G. Schierholz and M. Teper, Phys. Lett. 110B (1982) 399.
[2] M. Falcioni et al., Phys. Lett. 110B (1982) 295.
[3] B. Berg and A. Billoire, Phys. Lett. 113B (1982) 65.
[4] M. Creutz, Phys. Rev. Lett. 45 (1980) 313; E. Pietarinen, Nucl. Phys. B190 [FS3] (1981) 349.
[5] B. Lautrup and M. Nauenberg, Phys. Rev. Lett. 45 (1980) 1755; M. Creutz, Brookhaven preprint BNL 29301 (1981); M. Creutz and K.J.M. Moriarty, Brookhaven preprint BNL 29782 (1981).
[6] M. Falcioni, E. Marinari, M.L. Paciello, G. Parisi and B. Taglienti, Phys. Lett. 102B (1981) 270.
[7] M. Lüscher, Common. Math. Phys. 54 (1977) 283; K. Osterwalder and E. Seilder, Ann. Phys. 110 (1978) 440.
[8] R.L. Jaffe and K. Johnson, Phys. Lett. 60B (1976) 201.
[9] P. Di Vecchia, K. Fabricius, G.C. Rossi and G. Veneziano, Nucl. Phys. B192 (1981) 392.
[10] E. Pietarinen, Nucl. Phys. B190 [FS3] (1981) 349.
[11] E.g., D.L. Scharre, in: Proc. 10th Intern. Symp. on Lepton and photon interactions at high energies (Bonn, August 1981), ed. W. Pfeil (Physikalisches Institut, University of Bonn, 1981) p. 163.

Volume 117B, number 1, 2 PHYSICS LETTERS 4 November 1982

DETERMINATION OF THE CORRELATION LENGTH BY VARYING THE BOUNDARY CONDITIONS IN LATTICE GAUGE THEORIES

K.H. MÜTTER
Physics Department, University of Wuppertal, Germany
and
K. SCHILLING [1]
CERN, Geneva, Switzerland

Received 23 February 1982

By Monte Carlo techniques we study the response of the expectation values of Wilson loops (sampled from the interior of a lattice) to the change of boundary conditions. In the case of the pure SU(2) gauge theory, we find, within reasonable computer time, a clear numerical signal for the correlation between the boundary and the interior plaquettes over a distance of up to four lattice spacings. As a result of our computations on 8^4 and 10^4 lattices, we verify the asymptotic freedom behaviour of the correlation length for β values up to 2.3–2.4 and estimate the glueball mass to be $m_g = (2.35 \pm 0.30)\sqrt{\sigma}$.

Ever since the exciting discovery of asymptotic freedom behaviour of the string tension calculated from the area decrease of large Wilson loops in SU(2) and SU(3) lattice gauge theories [1] it has been a challenge to establish a scaling region of the correlation length $\xi(\beta)$ as well. The practical problem turned out to be that Monte Carlo simulations would need extremely long computer time before yielding any significant signal for the correlation function $\rho(r)$ between plaquettes at distances r more than two lattice spacings apart [2]. The short distance behaviour of $\rho(r)$ is, however, rapidly dominated by perturbative effects as the correlation length increases with β. Therefore, the β-region over which the asymptotic freedom behaviour

$$\xi(\beta) = (\Lambda_L/m_g)(\tfrac{6}{11}\pi^2\beta)^{-51/121} \exp(\tfrac{3}{11}\pi^2\beta)$$

$$\text{(for the SU(2) case)} \qquad (1)$$

can be tested, is very limited ($2.0 \lesssim \beta \lesssim 2.2$) so far.

There appear to exist two possibilities to obtain a reasonably large scaling domain in Monte Carlo calculations for the correlation functions [‡1]:

(1) Concentrate on short distances $r = 1$ and $r = 2$, but include the correlations of more complicated Wilson loops than simple plaquettes and disentangle the true long range behaviour due to the lowest lying 0^+ glueball state. The method is to expand the glueball state in terms of certain Wilson loop operators acting on the vacuum and varying the coefficients in this expansion until the state with lowest energy is reached. This variational method was proposed by Wilson a year ago and performed most recently in ref. [6].

(2) Search for a situation where the long range ($r = 3$ or 4) correlation is large enough to be detectable in Monte Carlo simulation.

[1] On leave of absence from Physics Department, University of Wuppertal till March 1st, 1982.

[‡1] A more speculative third method has been recently advanced by Brower et al. [3]. They propose to use finite size scaling a la Fischer [4] to determine the correlation length. Engels et al. [5] proposed to extract the glueball mass (without verification of the asymptotic freedom behaviour of ξ) from finite temperature lattice calculations by describing energy density and specific heat in terms of an ideal gas of resonances.

Volume 117B, number 1, 2 PHYSICS LETTERS 4 November 1982

In this paper we pursue the second option by exploiting the sensitivity of a Wilson loop average to variations of the boundary conditions in finite lattices. The maximal signal is achieved by changing from free to fixed boundary conditions. As we showed in a recent paper [7] the Wilson loop average in the thermodynamical limit ($V \to \infty$) is bounded from above and below by its expectation value calculated in a finite lattice V with fixed and free boundary conditions, respectively:

$$L(r) = \langle W(C)\rangle_{0,V} \leqslant \langle W(C)\rangle_{V\to\infty} \leqslant \langle W(C)\rangle_{1,V} = U(r). \tag{2}$$

Here the index 0 and 1 denotes free and fixed boundary conditions respectively The quality of these bounds of course crucially depends on the distance r between the loop C and the boundary ∂V.

It will be shown below that up to third and higher order correlations, the difference $\Delta = U - L$ between the two bounds has the following structure in terms of second order correlation functions:

$$\frac{\Delta}{L} = \Big(1 + \sum_{P\neq P'\in\partial V} \tilde{\rho}(P, P')\Big)^{-1} \sum_{P\in\partial V} \rho(P, C), \tag{3}$$

where

$$\rho(P, C) = \langle O(P) W(C)\rangle_0 / \langle O(P)\rangle_0 \langle W(C)\rangle_0 - 1, \tag{4a}$$

and

$$\tilde{\rho}(P, P') = \langle O(P) O(P')\rangle_0 / \langle O(P)\rangle_0 \langle O(P')\rangle_0 - 1. \tag{4b}$$

In eq. (3) the sums extend over the plaquettes P on the boundary ∂V. $O(P)$ is a local operator and the averages in eq. (4) have to be performed with free boundary conditions. In this paper we shall demonstrate that Monte Carlo calculations yield a significant signal for the quantity Δ even if all the distances $r(P, C)$ in the correlation functions $\rho(P, C)$ [cf. eq. (3)] exceed three lattice spacings.

As an example we have chosen the pure SU(2) gauge theory – approximated by its 120 element icosaheder subgroup T [8] – on a hypercubical lattice $V = N^4$ ($N = 8, 10$) with the Wilson action [9]:

$$S = \frac{1}{2} \sum_{P\in V} \mathrm{tr}(U(P) + U^+(P)), \tag{5}$$

where

$$U(P) = \prod_{\ell\in P} U(\ell)$$

stands for the ordered product of the four link variables $U(\ell)$, $\ell \in P$. Expectation values are calculated with respect to the partition function

$$Z(\beta) = \prod_{\ell\in V} \int \mathrm{d}U(\ell) \exp(\beta S). \tag{6}$$

For the subgroup T, the integral in eq. (6) turns into a sum which is calculated by the standard Metropolis process. Free boundary conditions are defined by simply integrating over all links inside the boundary ∂V. Fixed boundary conditions are realized by freezing all plaquette variables on the boundary ∂V to the unit element

$$U(P) = 1 \quad \text{for } P \in \partial V. \tag{7}$$

Let us now sketch the derivation of eqs. (3) and (4): the condition (7) for the SU(2) case is equivalent to the vanishing of the plaquette angles $\theta(P)$, defined by

$$\cos\theta(P) = \mathrm{tr}(U(P) + U^+(P))/(2\ \mathrm{tr}\ 1). \tag{8}$$

It is obvious that any expectation value with fixed boundary conditions can be converted into another such quantity with free boundary conditions. In particular one derives

$$\frac{\Delta}{L} = \frac{\langle W(C) \Pi_{P\in\partial V} O(P)\rangle_0}{\langle W(C)\rangle_0 \langle \Pi_{P\in\partial V} O(P)\rangle_0} - 1, \tag{9}$$

where

$$O(P) = \delta(\theta(P)). \tag{10}$$

Now we perform a cluster expansion of the expectation values encountered in eq. (9):

$$\Big\langle \prod_{P\in\partial V} O(P)\Big\rangle_0 = \prod_{P\in\partial V} \langle O(P)\rangle_0 + \sum_{P_1\neq P_2\in\partial V} [\langle O(P_1) O(P_2)\rangle_0 - \langle O(P_1)\rangle_0 \langle O(P_2)\rangle_0] \times \prod_{P_3\neq P_1,P_2} \langle O(P_3)\rangle_0 + \ldots, \tag{11}$$

$$\Big\langle W(C) \prod_{P\in\partial V} O(P)\Big\rangle_0 - \langle W(C)\rangle_0 \Big\langle \prod_{P\in\partial V} O(P)\Big\rangle_0 = \sum_{P_1\in\partial V} [\langle O(P_1) W(C)\rangle_0 - \langle O(P_1)\rangle_0 \langle W(C)\rangle_0] \times \prod_{P_2\neq P_1\in\partial V} \langle O(P_2)\rangle + \ldots . \tag{12}$$

Volume 117B, number 1, 2 PHYSICS LETTERS 4 November 1982

In view of the fact that in the β region of interest the correlation length ξ is of order $\lesssim 1$ we drop the higher order correlations in eq. (9). As a result eq. (3) is easily recovered.

In our computations, we have chosen the loops C to be simple plaquettes Q sampled from concentric cubes

$$K_d = (N - 2d)^4, \tag{13}$$

with distance d to the boundary ∂V. In a given lattice $V = N^4$ the bounds $U(d)$ and $L(d)$ are expected to become tighter with increasing d. The numerical situation is exemplified for $N = 8$ in table 1, where the most precise previous values for the mean plaquette obtained with periodic boundary conditions are also given for reference. Comparing these results one observes that the averages performed with free and fixed boundary conditions are always below and above those with periodic boundary conditions, which is a numerical support of the inequalities eq. (2).

In fig. 1 we have plotted for an 8^4 lattice the difference $\Delta(d) = U(d) - L(d)$ as a function of β and for distances $d = 1, 2, 3$ (connected by full lines to guide the eye) and the corresponding results for the 10^4 lattice (connected by dotted lines). This figure illustrates that our method indeed extends the accessible range for correlation measurements from $d = 2$ to $d = 4$.

In order to extract the correlation length from these data, we assume an exponential behaviour for

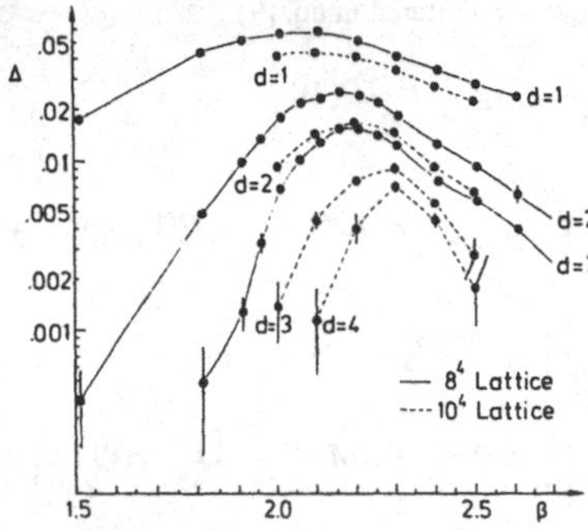

Fig. 1. The quantities $\Delta = U - L$ versus β. Full lines connect the results obtained in an 8^4 lattice, for various distances d of the interior cube to the boundary. Dashed lines connect respective results in a 10^4 lattice.

the correlation function appearing in eq. (3):

$$\rho(\mathrm{P, Q}) \approx \rho_0 \exp[-r(\mathrm{P, Q})/\xi(\beta)] \tag{14}$$

where $r(\mathrm{P, Q})$ denotes the euclidean distance between the plaquettes P in the boundary ∂V and the measured plaquette Q in the concentric cube K_d [cf. eq. (13)]. With this ansatz the quantity $\Delta(d)$ is related to the correlation length $\xi(\beta)$ in the following way:

$$\Delta(d) = n(d)^{-1} \rho_0 \frac{\Sigma_{\mathrm{Q} \in K_d} \Sigma_{\mathrm{P} \in \partial V} \exp[-r(\mathrm{P, Q})/\xi(\mathrm{P})]}{1 + \Sigma_{\mathrm{P}_1 \neq \mathrm{P}_2 \in \partial V} \tilde{\rho}(\mathrm{P}_1, \mathrm{P}_2)}, \tag{15}$$

where $n(d)$ is the number of plaquettes Q in the cube K_d. By means of eq. (10) we extract the inverse correlation length $m(\beta) = \xi(\beta)^{-1}$ from the measured ratios $\Delta(d)/\Delta(d-1)$ in a 8^4 lattice with $d = 2, 3$ and in a 10^4 lattice with $d = 2, 3, 4$. The results in figs. 2a and 2b display the pattern familiar since Creutz's famous results [1] for the string tension: a high temperature regime fairly well accounted for by the high temperature expansion in lowest order [10] [‡2], a rapid turnover into the asymptotic freedom domain and a breakoff into a spin wave dominated region, whose position depends strongly on the distance d involved in the measurement. We emphasize that in both lattices the scaling domain – where the correlation length $\xi(\beta)$ follows the prediction of asymptotic freedom [cf. eq. (1)] – is now clearly visible. While the data taken at a short distance ($d = 2$) start deviating from the asymptotic freedom behaviour near $\beta = 2.1$, the long distance measurements ($d = 4$) provide us with a considerably larger scaling domain, extending up to $\beta = 2.4$.

In figs. 2a and 2b the full straight lines – with slopes predicted by asymptotic freedom – mark the bandwidths of our data. According to eq. (1) we obtain for the glueball mass:

$$m_g = (2.2 \pm 0.3)\sqrt{\sigma} \quad \text{from the } 8^4 \text{ lattice} \tag{16}$$

and

$$m_g = (2.5 \pm 0.3)\sqrt{\sigma} \quad \text{from the } 10^4 \text{ lattice} \tag{17}$$

in terms of the string tension σ, using $\Lambda_L = 0.0119\sqrt{\sigma}$ [11]. Our best estimate for the glueball mass is thus

‡2 A thorough discussion of the numerical situation of high temperature series expansion results for m will be given by Berg et al. [11].

Table 1
Upper and lower bounds $U(d)$ and $L(d)$ in the 8^4 lattice. The row labelled BNS refers to the mean plaquette in the infinite lattice calculated in ref. [3] by extrapolating periodic boundary results in various finite lattices to $N \to \infty$. BS quotes the periodic boundary condition results for the mean plaquette obtained by Berg and Stehr [2] in an 8^4 lattice. Our errors are calculated from the fluctuations between subsamples over a 200 sweeps each over the whole lattice $K_0 = 8^4$. The accuracy for $L(2)$, $U(2)$, $L(3)$, $U(3)$ was reached by updating and measuring over K_2, K_3 more frequently in between each sweep over $K_0 = V$. The relative frequencies for these sweeps were adjusted to be $f_0 : f_2 : f_3 = 1:5:60$.

β	number of free boundary conditions	$L(1) \times 10^5$	$L(2) \times 10^5$	$L(3) \times 10^5$	BNS $\times 10^5$	BS $\times 10^5$	$U(3) \times 10^5$	$U(2) \times 10^5$	$U(1) \times 10^5$	number of fixed boundary conditions
1.5	4600	36375 ± 59	36263 ± 13	36235 ± 16			36258 ± 16	36301 ± 14	38106 ± 56	5400
1.8	7200	44050 ± 40	44094 ± 12	44097 ± 19		44075 ± 10	44145 ± 24	44573 ± 14	48350 ± 40	6400
1.9	8000	46834 ± 60	47005 ± 14	47031 ± 21		46980 ± 12	47156 ± 19	47975 ± 15	51972 ± 29	7200
1.95	4600		48524 ± 27	48518 ± 20			48844 ± 29	49845 ± 16		5600
2.0	5200	49862 ± 32	50057 ± 18	50083 ± 27		50095 ± 25	50764 ± 35	51809 ± 32	55447 ± 65	5600
2.05	4200		51702 ± 22	51735 ± 36	517400		52731 ± 32	53820 ± 21		5400
2.10	7000	53060 ± 60	53395 ± 20	53409 ± 25	534500	53461 ± 16	54670 ± 35	55686 ± 30	58760 ± 50	4200
2.15	4600		55021 ± 15	55113 ± 27	551800		56635 ± 40	57507 ± 25		3600
2.2	2800	56329 ± 77	56683 ± 30	56836 ± 53	569200	56877 ± 13	58374 ± 36	59061 ± 23	61390 ± 120	3400
2.25	2200		58327 ± 31	58480 ± 53	586200		59892 ± 34	60445 ± 22		2800
2.30	3000	59476 ± 60	59903 ± 32	60033 ± 44	602000	60183 ± 14	61227 ± 23	61691 ± 24	63690 ± 50	2600
2.4	2800	62150 ± 30	62686 ± 23	62819 ± 38	630000	63032 ± 12	63553 ± 23	63905 ± 15	65530 ± 20	2800
2.5	3000	64400 ± 30	64880 ± 18	64969 ± 21	651900	65135 ± 34	65536 ± 24	65779 ± 16	67130 ± 20	3000

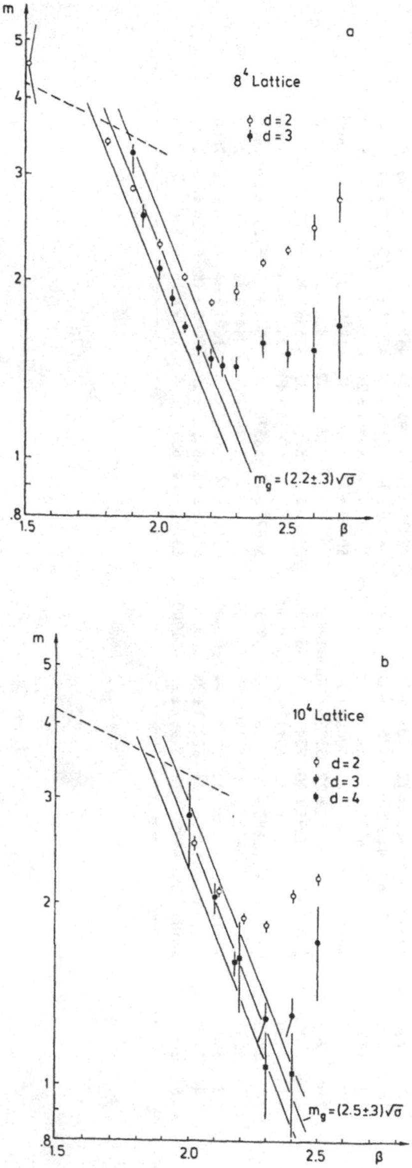

◀ Fig. 2. (a) Inverse correlation length $m(\beta)$ obtained in an 8^4 lattice. Open points are obtained from $\Delta(2)/\Delta(1)$, full points from $\Delta(3)/\Delta(2)$ according to eq. (10). The dotted line represents $m = -4 \ln I_2(\beta)/I_1(\beta)$, the lowest order strong coupling result. The full lines have the asymptotic freedom slope $-3\pi^2/11$ and normalization $m_g = (2.2 \pm 0.3)\sqrt{\sigma}$ (with $\Lambda_L = 0.0119\sqrt{\sigma}$), according to eq. (1). (b) Inverse correlation length $m(\beta)$ obtained in a 10^4 lattice. Open and full points as in fig. 2a, squares obtained from $\Delta(4)/\Delta(3)$. Full lines present asymptotic freedom predictions normalized to $m_g = (2.5 \pm 0.3)\sqrt{\sigma}$ (with $\Lambda_L = 0.0119\sqrt{\sigma}$).

$m_g = (2.35 \pm 0.3)\sqrt{\sigma}$, which is in beautiful agreement with the most advanced calculation using the variational method by Berg et al. [6]. They quote a value of $m_g = (2.4 \pm 0.6)\sqrt{\sigma}$. Let us mention that the computer effort to obtain correlation signals over four lattice spacings by our method is quite reasonable; the 10^4 lattice results presented here were obtained with a total of 21 000 sweeps in 20 h of CDC 7600.

In conclusion, the method of varying the boundary conditions appears to be very useful in the study of lattice gauge theories. It would be very desirable to try it in the SU(3) case as well. Another open question is how the results obtained here depend on the specific choice of the action. Work along this line is in progress.

We enjoyed useful discussions with Dr. B. Berg, Dr. C.B. Lang and Dr. I. Montvay. Thanks in particular to P. Hasenfratz for helping to improve the manuscript by constructive criticism. We are grateful to the authors of ref. [8] for permission to use their group multiplication table programme. B. Berg kindly communicated to us the draft of ref. [11] prior to publication. K.S. appreciates the hospitality of the CERN Theory Division. The 10^4 lattice results were obtained on the CERN CDC 7600 computer.

References

[1] M. Creutz, Phys. Rev. D21 (1980) 2308; Phys. Rev. Lett. 45 (1980) 313;
E. Pietarinen, Nucl. Phys. B190 [FS3] (1981) 349.
[2] B. Berg, Phys. Lett. 97B (1980) 401;
B. Berg and I. Stehr, Z. Phys. C9 (1981) 333;
G. Bhanot and C. Rebbi, Nucl. Phys. B180 [FS2] (1981) 469.
[3] R. Brower, M. Nauenberg and T. Schalk, Phys. Rev. D24 (1981) 548.

[4] M.E. Fisher, in Proc. Int. School of Physics, Enrico Fermi, Varenna (1970) Course No. 51, ed. M.S. Green, (Academic Press, New York).
[5] J. Engels, F. Karsch, I. Montvay and H. Satz, Phys. Lett. 102B (1981) 332 and Bielefeld preprint BI TP 81/29.
[6] K.G. Wilson, talk given at the Abingdon meeting on Lattice Gauge Theories (March 1981);
B. Berg, A. Billoire and C. Rebbi, Brookhaven preprint, Monte Carlo estimate of the SU(2) massgap (1982);
K. Ishikawa, G. Schierholz and M. Teper, DESY preprint 81-089 (1981);
M. Falcioni et al., Phys. Lett. 110B (1982) 295;
B. Berg and A. Billoire, Phys. Lett. 113B (1982) 65.
[7] K.H. Mütter and K. Schilling, Nucl. Phys. B200 [FS4] (1982) 362.
[8] G. Bhanot, C.B. Lang and C. Rebbi, CERN preprint TH-3178 (1980).
[9] K. Wilson, Phys. Rev. D19 (1975) 2445.
[10] G. Münster, Nucl. Phys. B190 [FS3] (1981) 548.
[11] B. Berg et al., Brookhaven preprint (1982).

Nuclear Physics B153 (1979) 141–160

A PROPERTY OF ELECTRIC AND MAGNETIC FLUX IN NON-ABELIAN GAUGE THEORIES

G. 't HOOFT
Institute for Theoretical Physics, University of Utrecht, The Netherlands*

Received 19 January 1979

Pure non-Abelian gauge models with gauge group SU(N) are considered in a box with periodic boundary conditions at various temperatures β^{-1}. Electric and magnetic flux are defined in a gauge-invariant way. The free energy of the system satisfies an exact duality equation, following from Euclidean invariance. The equation relates properties of the electric and the magnetic fields. Conclusions that can be drawn for instance are that for $N \leq 3$ one cannot have both electric and magnetic confinement, and that the infrared structure of the Georgi–Glashow model is self-dual.

1. Introduction

The forces between quarks in a hadron are most likely described by a non-Abelian gauge theory without scalar fields. No precise perturbative schemes are known to compute mass spectra and scattering matrix elements in this theory. Nevertheless it is understood [1, 2] that non-Abelian local gauge symmetry can be realized in Nature in several ways: either some scalar field combination undergoes an explicit or dynamical Higgs phenomenon causing the vector bosons to become massive and quarks to become "liberated", or a disordered phase causes permanent "confinement" of quarks and absence of any reminiscence of gauge (color) symmetry. As is stressed by Mandelstam [3], local gauge invariance is not a symmetry in Hilbert space such as the usual global symmetries. Hilbert space can be set up entirely using only gauge-invariant operators acting on the vacuum. This is why local gauge invariance is sometimes obscured in the long-distance structure of a theory, and why neither of the two modes mentioned above (Higgs *versus* confinement) necessarily contains massless particles.

The confinement mode on the one hand and the complete Higgs "symmetry-breakdown" mode on the other hand are found to be dual to each other in the sense of electric-magnetic duality or Kramers–Wannier [4, 5] duality.

* Address: Princetonplein 5, P.B. 80006, 3508 TA Utrecht.

However, there are also self-dual modes such as the "Georgi–Glashow mode" (to be discussed in sect. 9) and also the critical point between Higgs and confinement. As argued in ref. [1], such modes must contain massless particles.

Here we consider pure SU(N) gauge theories in four dimensions. For any closed curve C in 3-dimensional space we have operators $A(\mathrm{C})$ and $B(\mathrm{C})$ satisfying the commutation rule [1]

$$A(\mathrm{C})B(\mathrm{C}') = B(\mathrm{C}')A(\mathrm{C}) \exp((2\pi i n/N), \tag{1.1}$$

where n is the number of times the curve C′ winds around C in a certain direction.

The operator

$$A(\mathrm{C}) = \frac{1}{N} \mathrm{Tr}\, \mathrm{P} \exp \oint_{\mathrm{C}} igA_k(x)\, \mathrm{d}x^k \equiv \mathrm{e}^{i\Phi_\mathrm{B}}, \tag{1.2}$$

measures, in a certain sense, the total magnetic flux going through C. Here P is the symbol for path ordering along the curve, and A_k are the space components of the vector field in matrix notation. In fact we might consider Φ_B as a definition of magnetic flux*. In the Higgs mode, Φ_B tends to be quantized in units of $2\pi/N$, whereas it is only defined modulo 2π. On the other hand, $A(\mathrm{C})$ can also be considered to be the creation operator of a "bare" (i.e., infinitely narrow) electric flux line along the curve C.

The operator $B(\mathrm{C})$, defined in ref. [1], satisfies formally

$$(B(\mathrm{C}))^N = 1, \tag{1.3}$$

and we can set $B(\mathrm{C}) = \mathrm{e}^{i\Phi_\mathrm{E}}$ where Φ_E is the electric flux going through C. It is quantized also in units $2\pi/N$ and also defined modulo 2π. Conversely, $B(\mathrm{C})$ is the creation operator of a "bare" magnetic flux line along C.

We notice a striking but not complete resemblance between the A- and B-type operators. Mandelstam [3] also attempts to write $B(\mathrm{C})$ in terms of an electric non-Abelian vector potential $B^a_\mu(x)$. We will not need such a potential for our considerations.

The definition of A and B is gauge invariant so our concepts of quantized electric and magnetic flux also have a gauge-invariant interpretation. This is in contrast with the usual definition of the electromagnetic fields in terms of the covariant curls $G^a_{\mu\nu}$ which are not quantized, but not gauge-invariant either. Note however, that $B(\mathrm{C})$ is only uniquely defined in theories *without* quarks or any other particles whose fields are not invariant under the center Z(N) of SU(N). This is because such particles have gauge-invariant electric charge corresponding to a total flux $\Phi_\mathrm{E} = 2\pi/N$ and thus spoil electric flux conservation.

* However, it must be borne in mind that, defined this way, flux is a not strictly additive quantity. Later, another definition will be given.

It is the purpose of this paper to study some properties of pure gauge theories before quarks are added to them. It is generally believed [6] that in such theories electric flux lines behave as unbreakable strings with universal thickness and a universal tension force (to be fitted with the experimentally measured value of 14 tons). We will investigate however the various possibilities that may arise in general in pure SU(N) gauge theories. In particular we will show that the energy of an electric flux is related to that of a magnetic flux by a dually symmetric formula. We will spell out in detail why, in the case of $N \leq 3$, only electric flux lines *or* magnetic flux lines, but not both, may behave as quantized Nambu strings [7].

One can imagine some internal parameters in a theory which we can vary. Then a transition point between the two modes discussed above might be found. Our dual equation (6.3) will then require long-range interactions associated with massless particles. But the transition can also take place *via* a mode that has the long-distance structure of the Georgi–Glashow model [8], a possibility not considered in ref. [1]. This model is characterized by ordinary photons, electrically charged bosons and magnetically charged [9] particles, and satisfies eq. (6.3) in a dually symmetric way. In that case there will be at least two critical points.

We consider a rectangular box with sides a_1, a_2, a_3 and with periodic boundary conditions. In the box is a pure SU(N) gauge system at a certain temperature. As is well-known, field theories at finite temperature $1/\beta$ can be regarded as statistical systems in a space with Euclidean metric, bounded by periodic boundary conditions in the imaginary time direction [10], with period equal to β. If we set $\beta = a_4$ then we have a box in Euclidean 4-dimensional space with sides a_μ and periodic boundary conditions in all four directions. We will then modify the boundary conditions such that we have a certain number of electric and magnetic flux quanta going in various directions through the box and consider the free energy as a_k and $\beta \to \infty$. It will turn out that the total free energy at temperature $1/\beta$ of a system with given electric and magnetic flux configuration can be expressed in terms of functional integrals over twisted bundles of gauge potentials in the box, and our relation between the electric and the magnetic energy is obtained *via* a rotation over 90° in Euclidean space. The relationship is symmetric and, as we said, rules out simultaneous electric and magnetic string formation (for $N \leq 3$). Not only will we reobtain the result of ref. [1], but also a quantitatively more precise result: the energy of magnetic flux in QCD drops exponentially with the area through which it goes, and the exponent is expressed in terms of the string constant (sect. 8). The choice of gauge in the Euclidean box must be done with some care. We elaborate on that in the appendix.

2. The twisted gauge field

If all fields are invariant under the center elements of the gauge group then we actually have an SU(N)/Z(N) theory. For such a theory, when put in a box, various different classes of periodic boundary conditions may be considered. Let us first concentrate on the 1,2 direction explicitly. The most general periodic boundary condition is

$$A_\mu(a_1, x_2) = \Omega_1(x_2) A_\mu(0, x_2)\,, \tag{2.1a}$$

$$A_\mu(x_1, a_2) = \Omega_2(x_1) A_\mu(x_1, 0)\,, \tag{2.1b}$$

where $A_\mu(x_1, x_2)$ is the vector potential in matrix notation and $\Omega_{1,2}$ are gauge rotations. Here ΩA_μ is short for $\Omega A_\mu \Omega^{-1} + (1/gi)\Omega\partial_\mu\Omega^{-1}$. How to perform the functional integrations over the values of $A_\mu(x)$ and $\Omega_{1,2}$ is explained in the appendix. To get no contradiction in the corners we must have

$$\Omega_1(a_2)\Omega_2(0) = \Omega_2(a_1)\Omega_1(0) Z\,, \tag{2.2}$$

where Z is an element of the centre Z(N) of SU(N). It is possible to find a gauge rotation $\Omega(x_1, x_2)$ such that

$$\Omega(x_1, 0) = I\,, \qquad \Omega(x_1, a_2) = \Omega_2(x_1)\,,$$

and $\Omega(x_1, x_2)$ must be continuous and differentiable on the rectangle, but not necessarily periodic. Then the transformation

$$A \to \Omega(x_1, x_2)^{-1} A\,,$$

brings the boundary conditions (2.1) into

$$A_\mu(a_1, x_2) = \Omega(x_2) A_\mu(0, x_2)\,, \tag{2.3a}$$

$$A_\mu(x_1, a_2) = A_\mu(x_1, 0)\,, \tag{2.3b}$$

where $\Omega(x_2)$ is a new gauge rotation satisfying

$$\Omega(a_2) = \Omega(0) Z\,. \tag{2.4}$$

Here Z is the same as in eq. (2.2). Further transformations on $\Omega(x_2)$ are possible, but since Z is an element of a discrete class there are no transformations that remove Z from expression (2.4). We conclude that there exist N non-gauge equivalent choices for the periodic boundary conditions on the gauge vector potential, when considered on a two-dimensional plane (torus).

This observation can be seen to hold independently for all pairs (μ, ν) of directions in (Euclidean) 4-space. Therefore, all together, there are N^6 distinct non-gauge equivalent choices for the boundary conditions [11]. We can label these by giving the six integers $n_{\mu\nu}$ for $\mu \neq \nu$; $0 \leq n_{\mu\nu} < N$. For later use, we define

$$n_{4i} = n_i\,, \qquad (i = 1, 2, 3)\,,$$

$$n_{ij} = m_k\,, \qquad (i, j, k \text{ even permutation of } 1, 2, 3)\,. \tag{2.5}$$

Functional integrals will be performed under these twisted boundary conditions:

$$W\{\boldsymbol{n}, \boldsymbol{m}; a_\mu\} = C \int_{\{\boldsymbol{n},\boldsymbol{m}\}} \mathrm{D}A \exp S(A) , \tag{2.6}$$

where $\int_{\{\boldsymbol{n},\boldsymbol{m}\}}\mathrm{D}A$ stands for integration only over those fields A that are twisted according to the integers n_i, m_i. It is important that the divergent renormalization effects are independent of the choice made at the boundaries. Therefore C is a common normalization factor, independent of n_i and m_i. Clearly $W\{\boldsymbol{n}, \boldsymbol{m}; a\}/W\{\boldsymbol{0}, \boldsymbol{0}; a\}$ are relevant quantities, dependent on the sizes a_μ of the box. We will study these functions.

One thing will be obvious here: W must be invariant under those simultaneous permutations of $n_{\mu\nu}$ and a_μ that correspond to orthogonal rotations in Euclidean 4-space. Nevertheless, identities between W-functions thus obtained will have non-trivial consequences, as we shall see. It is important to note here that also the gauge-fixing procedure can be made Euclidean invariant (see appendix).

3. Magnetic flux in a box

Let us formally consider quantization in the $A_0 = 0$ gauge. Then the magnetic field operators commute with the vector potentials $A_i(\boldsymbol{x})$. Therefore, all classical field configurations carry a well-specified amount of magnetic flux. At first sight it might seem to be a good idea to use the operator $A(\mathrm{C})$ as defined in sect. 1 to define total magnetic flux in a certain direction in the box, by choosing C to be a loop orthogonal to that direction. However, the quantity Φ_B obtained this way is not strictly additive (except when C runs in a certain type of vacuum) and therefore the periodic boundary conditions will not guarantee its being conserved. Indeed, Φ_B would not be conserved.

A more convenient way to define magnetic flux in the 3-direction of the box can be found by first considering some curve C defined by

$$\mathrm{C} = \{\boldsymbol{x}(\sigma)\}, \qquad 0 \leq \sigma < 1 ,$$

$$\boldsymbol{x}(0) = \begin{pmatrix} x_1 \\ x_2 \\ 0 \end{pmatrix}, \qquad \boldsymbol{x}(1) = \begin{pmatrix} x_1 \\ x_2 \\ a_3 \end{pmatrix}. \tag{3.1}$$

The operator $B(\mathrm{C})$ as defined in ref. [1] then creates one unit of flux in the 3-direction. Nearly everywhere $B(\mathrm{C})$ is a pure gauge rotation $\Omega(\boldsymbol{x})$, singular for $\boldsymbol{x} \in \mathrm{C}$. But $\Omega(\boldsymbol{x})$ is prescribed to make a jump by a factor $\mathrm{e}^{2\pi i/N}$ at a certain angle with respect to the curve C. This cut in Ω could be located for instance at all points

$$\{\boldsymbol{x}(\sigma) + \lambda \boldsymbol{a}_1 , \lambda > 0\} . \tag{3.2}$$

What distinguishes $B(C)$ from a pure gauge rotation is that in the transformation law for $A_i(\boldsymbol{x})$, the (singular) derivative across this cut is replaced by zero. In other words: the "string" (3.2) is unphysical, and can be gauge transformed to other positions. The new field $A_i(\boldsymbol{x})$, after the operation $B(C)$, can be made to satisfy the same boundary conditions (2.3) and (2.4), but Z in (2.4) has jumped by one unit. Clearly, the integer m_3, as defined in (2.5) counts how many times an operator such as $B(C)$ has acted. In other words, m_3 counts a conserved variety of magnetic flux in the 3-direction. In general, $(m_1, m_2, m_3) = \boldsymbol{m}$ will be considered to be the (integer valued) magnetic flux vector. It is the direct analogue of the ordinary (continuously valued) magnetic flux for the Abelian case.

4. Electric flux in a box

Electric field operators do not commute with the vector potentials. Therefore we must examine the quantized theory (in the $A_0 = 0$ gauge) before establishing the concept of electric flux.

The gauge restriction $A_0 = 0$ leaves invariance under time-independent gauge rotations $\Omega(\boldsymbol{x})$. States in Hilbert space must then be representations of this invariance group. For all those $\Omega(\boldsymbol{x})$ which can be continuously and uniformly connected to the identity element we must choose the trivial representation (see, also, the appendix):

$$\{\Omega(\boldsymbol{x})\}|\psi\rangle = |\psi\rangle . \tag{4.1}$$

This is necessary if we want a theory that approaches the usual theory when $a_i \to \infty$.

But there are other homotopy classes of $\Omega(\boldsymbol{x})$, which cannot be deformed continuously towards the identity. First there is the familiar 2nd Chern class of mappings $\Omega(\boldsymbol{x})$ that are related to instanton effects [12] and have representations described by an arbitrary angle θ. These cannot be directly related to electric or magnetic flux. Now the gauge transformations in a box with periodic boundary conditions (in 3-space) form other homotopy classes besides those that gave the θ vacuum. Consider namely an $\Omega(\boldsymbol{x})$ with the properties

$$\Omega\begin{pmatrix} a_1 \\ x_2 \\ x_3 \end{pmatrix} = \Omega\begin{pmatrix} 0 \\ x_2 \\ x_3 \end{pmatrix} Z_1 , \qquad \Omega\begin{pmatrix} x_1 \\ a_2 \\ x_3 \end{pmatrix} = \Omega\begin{pmatrix} x_1 \\ 0 \\ x_3 \end{pmatrix} Z_2 ,$$

$$\Omega\begin{pmatrix} x_1 \\ x_2 \\ a_3 \end{pmatrix} = \Omega\begin{pmatrix} x_1 \\ x_2 \\ 0 \end{pmatrix} Z_3 , \tag{4.2}$$

where at least one of the $Z_{1,2,3}$ is a non-trivial element of the center of SU(N). This $\Omega(\boldsymbol{x})$ is a gauge rotation that leaves the boundary conditions (2.3) and (2.4) invariant. Any choice of $Z_{1,2,3}$ forms a distinct homotopy class, characterized by three integers (k_1, k_2, k_3). Since they are an invariance of the Hamiltonian, we must have

$$\Omega|\psi\rangle = \mathrm{e}^{i\omega(k_1,k_2,k_3)}|\psi\rangle\,. \tag{4.3}$$

These angles ω need not vanish, but we do have, if we indicate elements of the homotopy class (k_1, k_2, k_3) by $\Omega[\boldsymbol{k}]$:

$$\Omega[\boldsymbol{k}_1]\Omega[\boldsymbol{k}_2] = \Omega[\boldsymbol{k}_1 + \boldsymbol{k}_2(\mathrm{mod}\ N)]\,, \tag{4.4}$$

and $(\Omega[\boldsymbol{k}])^N$ is homotopically equivalent with the identity. Therefore, ω must satisfy

$$\omega(\boldsymbol{k}) = \frac{2\pi}{N}\sum_i e_i k_i\,, \tag{4.5}$$

where $e_{1,2,3}$ are again three integers, defined modulo N. They represent new conserved quantities in the system.

Let us again consider a curve C in the 3-direction, as defined in eq. (3.1). Now we construct the operator $A(\mathrm{C})$:

$$A(\mathrm{C}) = \frac{1}{N}\,\mathrm{Tr}\,\mathrm{P}\exp ig\int_0^1 \mathrm{d}\sigma \frac{\mathrm{d}\boldsymbol{x}}{\mathrm{d}\sigma}\cdot \boldsymbol{A}(\boldsymbol{x}(\sigma))\,, \tag{4.6}$$

and consider

$$|\psi'\rangle = A(\mathrm{C})|\psi\rangle\,. \tag{4.7}$$

The quantity $A(\mathrm{C})$ is not invariant under $\Omega[k_1, k_2, k_3]$ if $k_3 \neq 0$:

$$A(\mathrm{C}) \to \frac{1}{N}\,\mathrm{Tr}\,\Omega(\boldsymbol{x}(0))\{\mathrm{P}\exp\ldots\}\Omega^{-1}(\boldsymbol{x}(1)) = \mathrm{e}^{-2\pi i k_3/N}A(\mathrm{C})\,. \tag{4.8}$$

Therefore,

$$A(\mathrm{C})\Omega[\boldsymbol{k}]|\psi\rangle = \Omega[\boldsymbol{k}]\,\mathrm{e}^{-2\pi i k_3/N}A(\mathrm{C})|\psi\rangle\,. \tag{4.9}$$

If $|\psi\rangle$ satisfies eq. (4.3), then

$$\Omega[\boldsymbol{k}]|\psi'\rangle = \mathrm{e}^{i\omega(\boldsymbol{k})+2\pi i k_3/N}|\psi'\rangle\,, \tag{4.10}$$

so that $|\psi'\rangle$ has e_3 replaced by e_3+1. Clearly, the integers e_i count how many times an operator of the form $A(\mathrm{C})$ acted in the various directions. Now $A(\mathrm{C})$ can be considered to be the creation operator of an electric flux line [1, 2, 6]. The conserved integers e_i therefore indicate total amount of electric flux in the three directions. Again this definition of electrix flux corresponds to the usual one in an Abelian model, apart from normalization (the total flux of a quark is normalized to be equal to one).

5. The functional integral for a fixed flux configuration

As one can conclude from the previous sections, a state $|\psi\rangle$ can be restricted to have a fixed magnetic flux (m_1, m_2, m_3) and a fixed electric flux (e_1, e_2, e_3). These six restrictions do not interfere with each other. We now wish to consider the vacuum functional integral with a given fixed $(\boldsymbol{m}, \boldsymbol{e})$ configuration.

It will be clear from the preceding that the integers m_i in eq. (2.5) must be chosen to have the values of the required magnetic flux. How do we fix the electric flux? We wish to compute the free energy F defined by

$$e^{-\beta F} = \mathrm{Tr} P(\boldsymbol{e}, \boldsymbol{m})\, e^{-\beta H}, \tag{5.1}$$

where H is the Hamiltonian of the theory and P is a projection operator that selects the required electric flux $\boldsymbol{e}$ and magnetic flux $\boldsymbol{m}$. We saw how to select the magnetic flux $\boldsymbol{m}$. The electric flux $\boldsymbol{e}$ is selected by using the operators $\Omega[\boldsymbol{k}]$ defined in sect. 4. Since $|\psi\rangle$ must satisfy eqs. (4.3) and (4.5) we may put

$$P(\boldsymbol{e}, \boldsymbol{m}) = \frac{1}{N^3} \sum_{\boldsymbol{k}} c^{-2\pi i (\boldsymbol{k}\cdot\boldsymbol{e})/N} \Omega[\boldsymbol{k}]. \tag{5.2}$$

Therefore

$$e^{-\beta F(\boldsymbol{e}, \boldsymbol{m}; \boldsymbol{a}, \beta)} = \frac{1}{N^3} \sum_{\boldsymbol{k}} e^{-2\pi i (\boldsymbol{k}\cdot\boldsymbol{e})/N}\, \mathrm{Tr}\, \Omega[\boldsymbol{k}]\, e^{-\beta H}, \tag{5.3}$$

The operator $e^{-\beta H}$ is the evolution operator in Euclidean space over a distance β in the 4-direction. Taking the trace in Hilbert space implies a periodicity condition in the 4-direction [10]. But the operator $\Omega[\boldsymbol{k}]$ is also inserted. It implies a twist in the periodicity. Indeed, as can be read off immediately from eq. (4.2), we have, for each choice of $\boldsymbol{k}$, a boundary condition such as eqs. (2.3) and (2.4) both with the directions 1, 2 replaced by 1, 4 or 2, 4 or 3, 4. Thus we find

$$e^{-\beta F(\boldsymbol{e}, \boldsymbol{m}; \boldsymbol{a}, \beta)} = \frac{1}{N^3} \sum_{\boldsymbol{k}} e^{-2\pi i (\boldsymbol{k}\cdot\boldsymbol{e})/N}\, W\{\boldsymbol{k}, \boldsymbol{m}; a_\mu\}, \tag{5.4}$$

with $a_4 = \beta$, and W defined as in eq. (2.6).

The author has actually made an attempt to find a formulation of the Feynman rules to compute W perturbatively. But in particular $W\{0, 0; a_\mu\}$, needed for normalization, contains a quite complicated multidimensional integral, even at lowest order, that so far kept him from doing explicit numerical calculations.

In sect. 6 we will see that even without doing any numerical calculations, some conclusions can be drawn from Euclidean symmetry on W.

In the limit $\beta \to \infty$ (or $T \to 0$), F becomes the energy of the lowest state with the given flux configuration. We are interested in the behaviour of this energy as a_i become large.

6. Duality

Clearly, W will be invariant under joint rotations of a_μ and $n_{\mu\nu}$ in Euclidean space. In particular, let us perform the SO(4) rotation as given by the matrix

$$\begin{pmatrix} 0 & -1 & & \\ 1 & 0 & & \\ & & 0 & 1 \\ & & -1 & 0 \end{pmatrix} \tag{6.1}$$

(keeping in mind that $k_i = n_{4i}$ and $n_{ij} = \varepsilon_{ijk} m_k$). Let the first two components of a vector $\boldsymbol{x}$ be denoted by $\tilde{x}$, and let $\hat{\tilde{a}}$ be $\tilde{a}$ with its two components interchanged. Then we find

$$W\{\tilde{k}, k_3, \tilde{m}, m_3; \tilde{a}, a_3, \beta\} = W\{\tilde{m}, k_3, \tilde{k}, m_3; \hat{\tilde{a}}, \beta, a_3\}. \tag{6.2}$$

The consequence of this is

$$\exp\{-\beta F(\tilde{e}, e_3, \tilde{m}, m_3; \tilde{a}, a_3, \beta)\}$$
$$= N^{-2} \sum_{\tilde{k}, \tilde{l}} \exp\left\{\frac{2\pi i}{N}[-(\tilde{k}\cdot\tilde{e}) + (\tilde{l}\cdot\tilde{m})] - a_3 F(\tilde{l}, e_3, \tilde{k}, m_3; \hat{\tilde{a}}, \beta, a_3)\right\}. \tag{6.3}$$

Here N^{-2} normalizes the Fourier transforms. Notice the complete "dual" symmetry under interchange of all $\boldsymbol{e}$ with $\boldsymbol{m}$ and *vice versa* in eq. (6.3). Right- and left-hand side of eq. (6.3) differ by a Fourier transformation and dual interchange with respect to only $\tilde{e}$ and $\tilde{m}$. Later, we will frequently put e_3 and m_3 equal to zero. Eq. (6.3) will be referred to as the "duality equation". It must be stressed that so far no approximation has been made. Our duality equation (6.3) for pure non-Abelian gauge theories is exact.

7. Condensation

7.1. Light fluxes

We now may ask which asymptotic structure of F as $\boldsymbol{a}$ and β tend to infinity is compatible with duality, eq. (6.3). First let us assume that no massless physical particles occur (what may happen in the presence of massless particles, which is the most difficult case, is briefly discussed in sect. 9). The asymptotic region (for large a_μ) will then be approached exponentially. We immediately notice that not all F are allowed to tend to zero. Then, namely, βF would tend to zero too and contradiction arises with eq. (6.3). If a_3 is sufficiently large then the major contributions to the sum in eq. (6.3) come from those values of $\tilde{l}$ and $\tilde{k}$ for which F vanishes. One can then easily deduce from eq. (6.3) that of all N^4 values of $(\tilde{e}, \tilde{m})$

exactly N^2 combinations must give vanishing F. Let us call these the "light" fluxes. The others send F to infinity. These we call "heavy" fluxes. And for any pair $(\boldsymbol{e}_{(1)}, \boldsymbol{m}_{(1)}), (\boldsymbol{e}_{(2)}, \boldsymbol{m}_{(2)})$ of light fluxes with $e_3(1) = e_3(2)$ and $m_3(1) = m_3(2)$, we must have:

$$\tilde{e}_{(1)} \cdot \tilde{m}_{(2)} = \tilde{e}_{(2)} \cdot \tilde{m}_{(1)}, \qquad (\text{modulo } N) . \tag{7.1}$$

The number N^2 is necessary to cancel N^{-2} in eq. (6.3), and the condition (7.1) is necessary to cancel the imaginary part of the exponent in eq. (6.3). A further restriction follows from the requirement that W in eq. (5.4) must be positive: if $(\boldsymbol{e}, \boldsymbol{m})$ is a light flux, then $(0, \boldsymbol{m})$ must also be a light flux. This excludes some exotic solutions of eq. (7.1). In SU(2) and SU(3) it follows that either all electric or all magnetic fluxes are light. In SU(4) there is a third possibility: it could be that only the even electric and even magnetic fluxes are light, and so on.

7.2. *Heavy fluxes*

For all other values of $(\boldsymbol{e}, \boldsymbol{m})$ the free energy F must not tend to zero as $\boldsymbol{a}, \beta$ become large. This is different from the Abelian case, where the energy of any given flux, say in the x direction, behaves as

$$E \to \frac{Ca_1}{a_2 a_3}, \tag{7.2}$$

where C is a constant. Thus, unlike Abelian fluxes, the heavy fluxes cannot spread out in space and produce a lower and lower energy density as the box becomes large. Physical intuition then tells us that the only possible alternative is string formation: the fluxes form narrow flux tubes with constant energy per unit of length. If the Higgs mode is realized then these heavy fluxes are the magnetic ones, known as Nielsen–Olesen flux tubes [7, 13]. But if the magnetic fluxes are the light ones then the heavy electric fluxes behave like strings as is argued in the literature [1–3, 6]. We now see how our duality equation, (6.3), forces us to accept the possibility of such an electric string mode, and tells us that the transition towards such a string mode from the Higgs mode must be through one or more phase transition(s) with massless particles at the critical point(s) [1].

From now on we will assume that the magnetic fluxes are light and the electric ones heavy (confinement mode) unless stated otherwise (the other case can always be obtained by the trivial replacement $e \leftrightarrow m$, and we will not discuss any further the exceptional situations possible in SU(4) and larger groups).

Thus we now have a clear idea about the energy of the heavy fluxes: at $\beta \to \infty$ we must have

$$F(\beta) \to E \to \rho a_1 , \tag{7.3}$$

if the flux is in the x-direction. Here E is the free energy at zero temperature and ρ is the string constant. In sect. 8 we will derive the precise asymptotic form of the energy of a (light) magnetic flux. We need, however, one more piece of information.

7.3. Factorization

We will assume absence of interference between electric and magnetic fluxes in the limit $a_\mu \to \infty$. Physically this is quite acceptable. Not only does this hold for Abelian fields; it holds as soon as we assume that strings occupy only a negligible portion of total space whereas magnetic fields fill the whole space. We will refer to this property as "factorization":

$$F(\boldsymbol{e}, \boldsymbol{m}; a_\mu) \Rightarrow F_e(\boldsymbol{e}; a_\mu) + F_m(\boldsymbol{m}; a_\mu) . \tag{7.4}$$

Note that $F_e(\boldsymbol{e}; a_\mu)$ will not always factorize with respect to the three components of $\boldsymbol{e}$. In the case of a square box, $\boldsymbol{e} = (1, 1, 0)$ will correspond to a string running in the diagonal direction, so for sufficiently large β:

$$\begin{aligned} F_e(1, 1, 0; a_\mu) &\to \sqrt{2} F_e(1, 0, 0; a_\mu) \\ &\to \sqrt{2} F_e(0, 1, 0; a_\mu) . \end{aligned} \tag{7.5}$$

Factorization can only hold for $\beta \gg a_i$ otherwise contradictions would arise between eqs. (6.3) and (7.5). If $\beta \gg a_{1,2}$ only, then factorization is still possible provided e_3 and m_3 are kept zero. This restriction is physically understandable. It means that when very long electric and magnetic flux lines are forced to stay close together then some interference will occur. If $e_3 = m_3 = 0$, eq. (6.3) implies:

$$\exp\{-\beta F_m(\tilde{m}, 0; \boldsymbol{a}, \beta)\} = N^{-1} \sum_{\tilde{l}} \exp\left\{\frac{2\pi i}{N}(\tilde{l} \cdot \tilde{m}) - a_3 F_e(\tilde{l}, 0; \hat{\tilde{a}}, \beta, a_3)\right\} , \tag{7.6}$$

$$\beta, a_3 \gg a_1, a_2 .$$

8. Computation of the free energy

We would like to know the behavior of the energy of a magnetic field:

$$E_m(\tilde{m}, 0; \boldsymbol{a}) = \lim_{\beta \to \infty} F_m(\tilde{m}, 0; \boldsymbol{a}, \beta) \tag{8.1}$$

(where $\beta \to \infty$ first, a_i is taken large afterwards). We can use eq. (7, 6), provided we know how F_e behaves at finite β, when $a_3 \to \infty$. Expression (7.3) only holds for large β, but we can use it as a starting point.

What is F_e at finite β? Clearly this is a problem of statistical physics since β can be interpreted as an inverse temperature. As $a_3 \to \infty$ our box becomes very large and we must allow for the possibility that thermal oscillations produce more than one string, even if $\boldsymbol{e} = (1, 0, 0)$. Consider fig. 1. For the time being we ignore strings

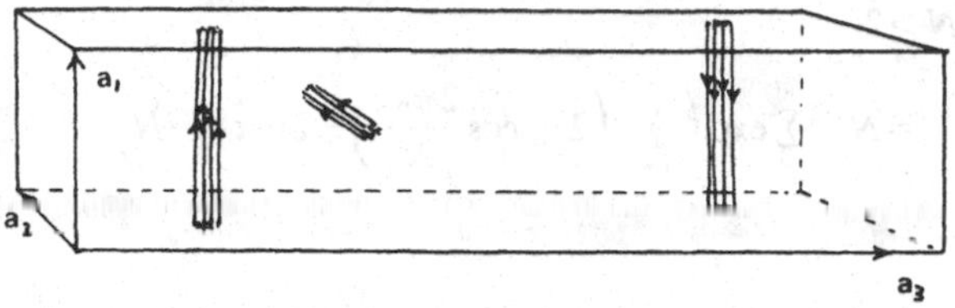

Fig. 1. Elongated box at finite temperature. Several strings may be produced by thermal oscillations. Total flux in the 1- and 2-directions is fixed (modulo N). Flux in the 3-direction is chosen to be zero.

that run diagonally. Later in this section it will be confirmed that they indeed may be neglected.

So let the total electric flux $\boldsymbol{e}$ be fixed to be $(e_1, e_2, 0)$. The Boltzmann factor for one string in the 1-direction is $e^{-\beta\rho a_1}$. It may pierce the 2-3 surface at any point. Therefore, the contribution of a single 1-string to the partition sum would be

$$\lambda a_2 a_3 \,e^{-\beta\rho a_1} \equiv \gamma_1 \,, \tag{8.2}$$

where λ is some elementary constant, ρ is the string constant. If two strings go in the positive 1-direction then their contribution is

$$\frac{1}{2!}\gamma_1^2 \,, \tag{8.3}$$

(a combinatorial factor for interchange symmetry is included) and so on. If β is large enough then the strings will be far apart on the average, and interactions may be neglected. If the group parameter $N>2$ then the two string orientations (up and down) must be distinguished.

To get the total partition sum we must add all possible multi-string configurations in the 1 and 2 directions with the restriction that total flux, modulo N, is given by (e_1, e_2):

$$\begin{aligned} &\exp[-\beta(F_e(e_1, e_2; \boldsymbol{a}, \beta) + C(\boldsymbol{a}, \beta))] \\ &\quad = \sum_{n_1^\pm, n_2^\pm} \frac{1}{(n_1^+)!(n_1^-)!(n_2^+)!(n_2^-)!} \gamma_1^{n_1^+ + n_1^-} \gamma_2^{n_2^+ + n_2^-} \\ &\quad \times \delta_N(n_1^+ - n_1^- - e_1)\,\delta_N(n_2^+ - n_2^- - e_2) \,, \end{aligned} \tag{8.4}$$

where $C(\boldsymbol{a}, \beta)$ is an as yet to be determined normalization term. Here we took $N>2$. If $N=2$ we must put $n_{1,2}^- = 0$. The functions $\delta_N(x)$ are defined to be one, if x is a multiple of N, and zero otherwise. We have

$$\delta_N(x) = \frac{1}{N}\sum_{k=0}^{N-1} e^{2\pi i k x/N} \,. \tag{8.5}$$

Therefore, if $N>2$:

$$e^{-\beta(F_e+C)}=N^{-2}\sum_{\vec{k}}\exp\left(\sum_{a=1}^{2}\left(2\gamma_a\cos\frac{2\pi k_a}{N}\right)-2\pi i\vec{k}\cdot\vec{e}/N\right), \tag{8.6}$$

and if $N=2$:

$$e^{-\beta(F_e+C)}=\tfrac{1}{4}\sum_{\vec{k}}(-1)^{\vec{k}\cdot\vec{e}}\exp\left(\sum_{a=1}^{2}\gamma_a(-1)^{k_a}\right). \tag{8.7}$$

Usually, C will be adapted so that $F(0;\boldsymbol{a},\beta)=0$. The asymptotic behavior for large a_3 can be read off from eqs. (8.2) and (8.6), (8.7). By some remarkable accident, eqs. (8.6) and (8.7) have again the form of a Fourier transform so they are easy to insert into the duality equation (7.6) in order to obtain F_m. We find:

$$\exp\left[-\beta(F_m(\tilde{m},0;\boldsymbol{a},\beta)+C(\boldsymbol{a},\beta))\right]=N^{-1}\exp\left[(2-\delta_{N2})\sum_a\hat{\gamma}_a\cos\frac{2\pi m_a}{N}\right], \tag{8.8}$$

with

$$\hat{\gamma}_1=\lambda a_1\beta\, e^{-\rho a_2 a_3},\qquad \hat{\gamma}_2=\lambda a_2\beta\, e^{-\rho a_1 a_3}. \tag{8.9}$$

Now we can take the limit $\beta\to\infty$ to obtain the energy of a magnetic flux:

$$e^{-\beta F_m(\tilde{m},0;\boldsymbol{a},\beta)}\to e^{-\beta E_m(\tilde{m},0;\boldsymbol{a})}, \tag{8.10}$$

with

$$E_m(\tilde{m},0;\boldsymbol{a})=\sum_i E_i(m_i,\boldsymbol{a}), \tag{8.11a}$$

$$E_1(m_1,\boldsymbol{a})=R(m_1)a_1\, e^{-\rho a_2 a_3}, \tag{8.11b}$$

$$E_2(m_2,\boldsymbol{a})=R(m_2)a_2\, e^{-\rho a_1 a_3}, \tag{8.11c}$$

$$R(m)=\lambda(2-\delta_{N2})\left(1-\cos\frac{2\pi m}{N}\right). \tag{8.11d}$$

The 1 in eq. (8.11d) arises from the normalization condition. $C(\boldsymbol{a},\beta)$ in eq. (8.8) must be such that $F(0;\boldsymbol{a},\beta)\to 0$. Of course eq. (8.11) is expected to hold only for sufficiently large a_i.

In deriving the asymptotic behaviour of $E_m(\boldsymbol{m};\boldsymbol{a})$ we have neglected contributions in eq. (8.4) from strings that run diagonally, besides the ones sketched in fig. 1. As promised we can now easily justify that. They namely would have an associated Boltzmann factor

$$\gamma_{12}=\frac{\lambda a_1 a_2 a_3}{\sqrt{a_1^2+a_2^2}}\exp\{-\beta\rho\sqrt{a_1^2+a_2^2}\}, \tag{8.12}$$

etc. In eq. (8.8) this would give extra terms with

$$\hat{\gamma}_{12}=\frac{\lambda a_1 a_2 \beta}{\sqrt{a_1^2+a_2^2}}\exp\{-\rho a_3\sqrt{a_1^2+a_2^2}\}\,, \tag{8.13}$$

etc. And in eq. (8.11) we would get extra terms going like

$$\exp\{-\rho a_3\sqrt{a_1^2+a_2^2}\}\,, \tag{8.14}$$

which decrease faster than eqs. (8.11b, c) and therefore can be neglected. Eqs. (8.11) give the asymptotic behaviour of the energy of magnetic fluxes exactly. Observe that the flux energy is proportional to the length of the flux lines, and decreases exponentially with the *area* through which the flux lines go. Of course, by extrapolation, we expect also

$$E(\boldsymbol{m},\boldsymbol{a})=\sum_i E_i(m_i,\boldsymbol{a})\,, \tag{8.15}$$

for non-vanishing $m_{1,2,3}$ (if one m_i vanishes this follows directly from eq. (8.11a)). Obviously, this behaviour of the magnetic fluxes is quite opposite to that of the confining electric fluxes. The coefficient ρ in the exponent must be precisely the string constant. It multiplies an area, not a distance and therefore cannot directly be linked to the mass of a physical particle. A consequence of this rapidly decreasing energy of magnetic fields is that objects with color-magnetic charge are not confined in Quantum Chromodynamics. To the contrary, they only show short-range interactions. This was conjectured but not proved by several authors [3, 14].

9. The massless particle phase

When massless particles are present the separation between light and heavy fluxes becomes impossible. The dependence on $\boldsymbol{a}$ and β of the flux energies could be quite complicated and we have not yet succeeded in classifying the various possibilities consistent with eq. (6.3). As indicated in ref. [1], a possible phase-transition point between Higgs phase and confinement phase must show massless excitations which one could study. However, there is another realization of eq. (6.3) through massless particles that probably does not correspond to a critical transition point but may occupy a finite region in parameter space. It is when one or more unbroken U(1) groups survive the Higgs mechanism.

Let us consider the case SU(2) with an isospin-one Higgs field [8] leaving as an apparent local symmetry group the subgroup U(1). As we will see this realization which we will refer to as the "Georgi–Glashow mode" is self-dual. Indeed, Montonen and Olive [15] observed a dual resemblance between magnetic monopoles and charged vector particles in this model. Let us, by way of exercise, estimate the free energy of electric and magnetic fluxes in this case.

If all components of $\boldsymbol{a}$ and $\boldsymbol{\beta}$ are sufficiently large, then only the U(1) Maxwell fields determine the free energy. None of the massive particles will give a noticeable direct contribution to the free energy, because their Boltzmann factors $e^{-\beta m}$ are too small. But they give an indirect contribution, in the following way. By rare thermal fluctuations (or by quantum tunneling) a pair of oppositely charged vector bosons may be created. One member may separate, go through one of the periodic walls (i.e., wind around the torus) and meet its companion from the other side after which they annihilate. We then obtain a Maxwell field configuration in the box where the U(1) electric flux in one direction has increased by 2 units (to keep our original notation, the charged vector bosons have electric charge g = two units, and an elementary doublet would have charges $\pm\frac{1}{2}g = \pm$ one unit). Indeed, since we have $N=2$ in our example, electric flux was only defined modulo 2. Thus, when we say that the electric flux is one unit in a certain direction, we really have to take into account all possible fluctuations that add an even number (positive or negative) to this flux, together with their Boltzmann factors. Clearly, the same must be done with the magnetic fields, because magnetic monopoles exist in this model and can be pair-created [9].

Since we are dealing with Maxwell fields, factorization (in the sense of sect. 7) holds:

$$F = F_e + F_m, \qquad F_e(\boldsymbol{e}; \boldsymbol{a}, \boldsymbol{\beta}) = \sum_i F_{ei}(e_i; \boldsymbol{a}, \boldsymbol{\beta}),$$
$$F_m(\boldsymbol{m}; \boldsymbol{a}, \boldsymbol{\beta}) = \sum_i F_{mi}(m_i; \boldsymbol{a}, \boldsymbol{\beta}). \tag{9.1}$$

Let us first compute the free energy of an electric flux. If the flux were completely fixed in the U(1) sense, we would have

$$F_{e1}(k; \boldsymbol{a}, \boldsymbol{\beta}) = \frac{g^2 a_1 k^2}{8 a_2 a_3}, \tag{9.2}$$

where g is the gauge coupling constant, and k is the integer that indicates the flux in our natural units. Because of the above explained tunneling phenomenon, in our SU(2) theory we have only $k=0$ or 1. The other values are reached by thermal excitations. If we normalize

$$F_{e1}(0; \boldsymbol{a}, \boldsymbol{\beta}) = 0, \tag{9.3}$$

then

$$\exp[-\beta F_{e1}(1; \boldsymbol{a}, \boldsymbol{\beta})] = \frac{\sum_{k=-\infty}^{\infty} \exp\left[-\beta \frac{g^2 a_1}{2 a_2 a_3}(k+\frac{1}{2})^2\right]}{\sum_{k=-\infty}^{\infty} \exp\left[-\beta \frac{g^2 a_1}{2 a_2 a_3} k^2\right]}, \tag{9.4}$$

and similarly for F_{e2}, F_{e3}.

The magnetic monopoles in the theory have magnetic charge $4\pi/g$, which is two elementary flux units. Therefore, replacing, in eq. (9.4), g by $4\pi/g$ we find the formula for the magnetic energy:

$$\exp\left[-\beta F_{m1}(1; \boldsymbol{a}, \beta)\right] = \frac{\sum_{k=-\infty}^{\infty} \exp\left[-\beta \frac{8\pi^2 a_1}{g^2 a_2 a_3}(k+\tfrac{1}{2})^2\right]}{\sum_{k=-\infty}^{\infty} \exp\left[-\beta \frac{8\pi^2 a_1}{g^2 a_2 a_3} k^2\right]} . \tag{9.5}$$

Using the formulas

$$\sum_{k=-\infty}^{\infty} e^{-\lambda k^2} = \sqrt{\{\pi/\lambda\}} \sum_{k=-\infty}^{\infty} e^{-\pi^2 k^2/\lambda} , \tag{9.6a}$$

and

$$\sum_{k=-\infty}^{\infty} (-1)^k e^{-\lambda k^2} = \sqrt{\{\pi/\lambda\}} \sum_{k=-\infty}^{\infty} e^{-\pi^2 (k+\frac{1}{2})^2/\lambda} , \tag{9.6b}$$

we find that eq. (7.6) is satisfied, up to an irrelevant normalization factor.

When g is varied, eqs. (9.4) and (9.5) go over into each other continuously. This is why we say that this phase is self-dual. The fact that in this "Georgi–Glashow phase" eq. (6.3) is realized in a self-dual way is in our opinion a non-trivial observation.

10. The twisted functional integral

We now know how to compute the energy of electric and magnetic fluxes, if one knows the functional integrals

$$W\{n_{\mu\nu}; a_\mu\} = C \int_{\{n_{\mu\nu}\}} DA \exp S(A) \tag{10.1}$$

in a Euclidean box with sides a_μ and twisted boundary conditions given by the integers $n_{\mu\nu}$. So we ask: what is the asymptotic form of W in the various phases (Higgs/Confinement/Georgi-Glashow)? To be specific we consider the case $N = 3$ and

$$W \equiv W\{n_{12} = 1, \text{rest} = 0; a_\mu\} / W\{0; a_\mu\} . \tag{10.2}$$

When all a_μ are larger than the "hadronic" mass scale we put

$$a_1 a_2 = \Sigma_1 , \qquad a_3 a_4 = \Sigma_2 , \tag{10.3}$$

We find that W essentially depends only on Σ_1 and Σ_2.

(*a*) *The Higgs phase.* When the Higgs mechanism removes the local symmetry completely in the usual way then the magnetic flux quantizes into Nielsen–Olesen tubes. Eq. (8.6) holds if F_e is replaced by F_m and $N=3$. Since factorization is assumed and all n_{4i} are kept zero, we find

$$W = e^{-\beta(F_m(m_3=1)-F_m(0))} = \frac{1-e^{-3\gamma}}{1+2e^{-3\gamma}}\,; \tag{10.4}$$

$$\gamma = \lambda\Sigma_1\, e^{-\rho\Sigma_2}\,, \tag{10.5}$$

where λ and ρ are defined in sect. 8. The latter is the string constant for the Nielsen–Olesen string.

(*b*) *The confinement phase.* The absolute confinement phase is described by eqs. (8.6)–(8.9) directly. We find

$$W = e^{-\lambda\Sigma_2}\exp[-\rho\Sigma_1] \tag{10.6}$$

(*c*) *The Georgi–Glashow phase.* We take the Georgi-Glashow phase as an example of a self-dual phase with massless particles. Eq. (9.5) applies to the case $N=2$. We can easily* extend it to $N=3$, and we find

$$W = \frac{\sum_{k=-\infty}^{\infty}\exp\{-\lambda(k+\frac{1}{3})^2\Sigma_2/\Sigma_1\}}{\sum_{k=-\infty}^{\infty}\exp\{-\lambda k^2\Sigma_2/\Sigma_1\}} = \frac{\sum_k \exp\left\{-\frac{\pi^2}{\lambda}k^2\Sigma_1/\Sigma_2+\frac{2k\pi i}{3}\right\}}{\sum_k \exp\left\{-\frac{\pi^2}{\lambda}k^2\Sigma_1/\Sigma_2\right\}}\,. \tag{10.7}$$

Here λ is some charge parameter.

Now let us consider the limit $\Sigma_1\to\infty$ first, then Σ_2 large, for the three cases a, b and c. We find

$$W \to 1 - \text{const}\cdot e^{-\Sigma_1 f(\Sigma_2)} \tag{10.8}$$

with in case a:

$$f(\Sigma_2) = 3\lambda\, e^{-\rho\Sigma_2}\,, \tag{10.9}$$

case b:

$$f(\Sigma_2) = \rho\,, \tag{10.10}$$

case c:

$$f(\Sigma_2) = \pi^2/\lambda\Sigma_2\,. \tag{10.11}$$

If on the other hand $\Sigma_2\to\infty$ first then

$$W \to \text{const}\cdot e^{-\Sigma_2 g(\Sigma_1)} \tag{10.12}$$

* There are various ways to describe this phase as a partial Higgs phase but the final result in terms of electrically and magnetically charged particles is independent of this description.

with in case a:

$$g(\Sigma_1) = \rho\,, \tag{10.13}$$

case b:

$$g(\Sigma_1) = \lambda\, e^{-\rho\Sigma_1}\,, \tag{10.14}$$

case c:

$$g(\Sigma_1) = \lambda/9\Sigma_1\,. \tag{10.15}$$

In all three cases the convergence towards the limit form is exponential. If W can be computed for reasonably large Σ_1, Σ_2 then it can be found out with some confidence which of the various phases is realized.

11. Conclusion

When this investigation was started it was with a view to finding a scheme for quantitative calculations for the hadron spectrum in QCD. In particular we want to express the string constant ρ in terms of the distance scales set by the renormalized coupling constant. Our starting point was eq. (5.4), in which the quantity W has a perturbation expansion that, term by term, is free of infrared divergences. Borel resummation procedures [16], corrected for instanton effects [17] could perhaps give a fairly reliable result. And then it would be easy to check which of the asymptotic forms of the previous section apply.

However, Euclidean invariance is reduced to rotations over 90° and artifacts due to this mutilation of the continuous Euclidean symmetry turned out to be formidable. (The reader is invited to compute the *zeroth* order term. He will then understand our problem, which is technical, not fundamental.)

On the other hand, qualitative study of eq. (5.4) gave us much insight in the long-distance structure of gauge theories. For instance, we found the proof of a conjecture made several times in the literature; simultaneous electric and magnetic confinement is impossible. If electric confinement is assumed, then the energy of a magnetic flux can be computed and is found to vanish exponentially as the size of the box increases. And we found that the Georgi–Glashow model, in which magnetic monopoles exist, has a long-distance structure which is self-dual. This is a statement for which we needed no such details as spin or mass spectrum of magnetic monopoles and dyons [15]. It follows solely from the gauge group structure of the model.

Finally, we hope that understanding of the asymptotic form for "twisted functional integrals" will be helpful in finding reliable calculational schemes for quantum chromodynamics.

The author acknowledges fruitful discussions with Dr. C. P. Korthals Altes.

Appendix

We should formulate precisely the prescription for gauge fixing in the finite Euclidean box. In this appendix we will concentrate on the continuous parts of the gauge group. How to treat the various homotopy classes of gauge field configurations ("twisted fibre bundles") is explained at length in the text.

Consider the gauge condition $A_0 = 0$. We then have an Hamiltonian H depending on $\boldsymbol{A}(\boldsymbol{x})$ and $\pi(\boldsymbol{x})$, the latter being the momenta conjugate to $\boldsymbol{A}(\boldsymbol{x})$. Now H is still invariant under (time independent) gauge transformations

$$\boldsymbol{A}(\boldsymbol{x}) \to \Omega(\boldsymbol{x})\boldsymbol{A}(\boldsymbol{x}), \tag{A.1}$$

where we use the same notation as in eqs. (2.1). In a Hilbert space of all field configurations $\boldsymbol{A}(\boldsymbol{x})$, the Hamiltonian commutes with the gauge rotation Ω defined by

$$\Omega\psi\{\boldsymbol{A}(\boldsymbol{x})\} \equiv \psi\{\Omega^{-1}(\boldsymbol{x})\boldsymbol{A}(\boldsymbol{x})\}. \tag{A.2}$$

So states in Hilbert space may be chosen to be representations of Ω. If Ω is in the same homotopy class as the identity, then we must choose the trivial representation:

$$\Omega|\psi\rangle = |\psi\rangle. \tag{A.3}$$

States that satisfy eq. (A.3) form the physical subspace of the above Hilbert space. To characterize those physical states it is sufficient to specify $\psi\{\boldsymbol{A}(\boldsymbol{x})\}$ for those $\boldsymbol{A}(\boldsymbol{x})$ that satisfy a gauge condition, such as

$$A_3(\boldsymbol{x}) = A_2(x_1, x_2, 0) = A_1(x_1, 0, 0) = 0. \tag{A.4}$$

However, the condition (A.4) is only compatible with the periodic boundary conditions if the functions Ω in eqs. (2.1) are allowed to be physical degrees of freedom: in a functional integral we must integrate over the values of $\Omega_{1,2}$.

Now we wish to express $\mathrm{Tr}\, e^{-\beta H}$ in terms of functional integrals:

$$\mathrm{Tr}\, e^{-\beta H} = \sum_{\psi(t=o)} \langle\psi(t=-i\beta)|\psi(t=0)\rangle. \tag{A.5}$$

The sum is over a basis set of physical states only. We write

$$|\psi(t=0)\rangle = \int \mathrm{D}\Omega\,\Omega|\boldsymbol{A}(\boldsymbol{x})\rangle, \tag{A.6}$$

where $\boldsymbol{A}(\boldsymbol{x})$ satisfies eq. (A.4). We see that

$$\mathrm{Tr}\, e^{-\beta H} = \int \mathrm{D}A\, e^{S(A)}, \tag{A.7}$$

in Euclidean space, if: (i) the gauge is completely specified, for instance by choosing $A_4 = 0$ and eq. (A.4) at $t = 0$ and (ii) the functions Ω that describe the periodic boundary conditions in all four directions are integrated over. Comparing eq. (A.7) with eq. (5.4) we see that $e = 0$ and the values of m are summed over. The other cases can be obtained by inserting P as is done in sect. 5.

Our formulation of the periodic boundary conditions in 3-space corresponds to the requirement that only gauge-invariant expressions must be periodic. An alternative would be to require also $A_\mu(x, t)$ to be periodic but that would exclude the possibility to define magnetic flux.

We notice that the gauge restrictions on the functional integral (A.7) imply invariance of eq. (A.7) under permutations of the Euclidean coordinates, a property we needed to derive eq. (6.3).

References

[1] G. 't Hooft, Nucl. Phys. B138 (1978) 1.
[2] A.M. Polyakov, Phys. Lett. 59B (1975) 82; 72B (1978) 477; Nucl. Phys. B120 (1977) 429.
[3] S. Mandelstam, Berkeley preprint (1978).
[4] S. Mandelstam, Phys. Rev. D11 (1975) 3026.
[5] H.A. Kramers and G.H. Wannier, Phys. Rev. 60 (1941) 252;
L.P. Kadanoff and H. Ceva, Phys. Rev. B3 (1971) 3918.
[6] S. Weinberg, Phys. Rev. Lett. 31 (1973) 494;
H. Fritzsch, M. Gell-Mann, and H. Leutwyler, Phys. Lett. 47B (1973) 365;
J. Kogut and L. Susskind, Phys. Rev. D9 (1974) 3501;
K.G. Wilson, Phys. Rev. D10 (1974) 2445.
[7] Y. Nambu, Phys. Rev. D10 (1974) 4262.
[8] M. Georgi and S.L. Glashow, Phys. Rev. Lett. 28 (1972) 1494.
[9] G. 't Hooft, Nucl. Phys. B79 (1974) 276;
A.M. Polyakov, JETP Lett. 20 (1974) 194.
[10] C.W. Bernard, Phys. Rev. D9 (1974) 3312;
L. Dolan and R. Jackiw, Phys. Rev. D9 (1974) 3320.
[11] C. Isham, private communication.
[12] R. Jackiw and C. Rebbi, Phys. Rev. Lett. 37 (1976) 172;
C. Callan, R. Dashen and D. Gross, Phys. Lett. 63B (1976) 334.
[13] H.B. Nielsen and P. Olesen, Nucl. Phys. B61 (1973) 45.
[14] G. 't Hooft, Nucl. Phys. B105 (1976) 538.
[15] C. Montonen and D. Olive, Phys. Lett. 72B (1977) 117.
[16] L.N. Lipatov, JETP Pisma 24 (1976) 179, and Leningrad preprints;
E. Brézin et al., Phys. Rev. D15 (1977) 1544, 1558;
G. 't Hooft, Erice lecture notes 1977;
G. Parisi, Phys. Lett. 66B (1977) 167.
[17] W.Y. Crutchfield, SUNY, Stony Brook preprint ITP-SB-78-59 (1978).

Note added by author.

In the reprinted paper one simplification was tacitly assumed: the instanton angle θ was mentioned but subsequently ignored. So we should read the paper with the understanding that $\theta = o$ is taken throughout. What happens when $\theta \neq o$ however is quite interesting. It is discussed in my Schladming lectures[1]). The positivity condition mentioned in Sect. (7.1) then does not necessarily hold and new "oblique" confinement modes will be possible. They are discussed in[2]), from a slightly different viewpoint.

Utrecht, January 1983

References.

1) G. 't Hooft, Confinement and Topology in Non-Abelian Gauge Theories. Lectures given at the Schladming Winterschool. Feb . 20-29, 1980. Acta Physica Austriaca Suppl. XXII, 531 (1980).

2) G. 't Hooft, Nucl. Phys. B190 (1981) 455.

Nuclear Physics B205[FS5] (1982) 141–167

MONOPOLES, VORTICES AND CONFINEMENT

G. MACK

II. Institut für Theoretische Physik der Universität Hamburg, Germany

E. PIETARINEN

Deutsches Elektronen-Synchrotron, Hamburg, Germany
and
Research Institute for Theoretical Physics, University of Helsinki, Finland

Received 19 October 1981

An exact relation is established between an SO(3) lattice gauge theory model without monopoles, and a corresponding SU(2) model. Elimination of the monopoles (and their strings) leads to a substantial lowering of the entropy of thin vortices and a corresponding decrease of the string tension for low β. This is revealed by approximate calculations of the vortex free energy and is confirmed by Monte Carlo data. The value of the physical transition temperature to "hot gluon soup" is also lowered considerably.

1. Introduction and summary of Monte Carlo results

Four-dimensional pure SU(2) lattice gauge theory can be interpreted as a Z_2 (gauge) theory with monopoles and fluctuating coupling constants [1, 2]. Condensation of these monopoles and/or the associated Z_2 strings can lead to phase transitions [1, 3–8]. In the present paper we investigate what happens if both the monopoles and Z_2 strings are eliminated from the model (by giving infinite energy to the strings). Such a modification does not affect the *formal* continuum limit.

We consider a 4-dimensional hypercubic lattice Λ made of sites x, links b, plaquettes p, cubes c, and hypercubes h. The boundary of a plaquette is sum of four links, etc. Let $\overline{U}=\{\overline{U}(\mathrm{b})\in \mathrm{SO}(3)\}$ be an SO(3) lattice gauge field on Λ and choose representatives $\mathrm{U(b)}\in \mathrm{SU}(2)$ of the cosets $\overline{\mathrm{U}}(\mathrm{b})\in \mathrm{SO}(3)=\mathrm{SU}(2)/Z_2$ in an arbitrary way. A gauge group SO(3) admits monopoles whose magnetic charge is added mod (2) because the fundamental group $\pi_1(\mathrm{SO}(3))=Z_2$. Translating the well known definition of monopoles in the continuum [9] to the lattice (see appendix A) one defines [1]*

$$\rho_c(\bar{U}) = \prod_{p \in \partial c} \operatorname{sign} \operatorname{tr} U(\partial p) = \pm 1 \in Z_2 \tag{1.1a}$$

for every 3-dimensional cube c. The product runs over the six plaquettes in the boundary of c. $\rho_c(\bar{U})$ is independent of the choice of representatives $U(b)$, and depends therefore only on $\bar{U}$, because it is invariant under the substitution $U(b) \to U(b)\gamma(b)$ with $\gamma(b) = \pm 1$. A conserved magnetic current $j_{\mu\nu\lambda}(x) \in F_2 = \{0, 1\}$ is now defined by

$$\rho_c = \exp i\pi j_{\mu\nu\lambda}(x). \tag{1.1b}$$

c = cube with corners $x, x + e_\mu, x + e_\mu + e_\nu, \ldots, x + e_\lambda$. At a given time the monopoles are in the spacelike cubes c where $\rho_c(\bar{U}) = -1$. (The definitions generalize to $SU(N)/Z_N$ in the obvious way. Orientation of plaquettes has to be watched if $N \geqslant 3$.) The world lines of the monopoles are closed loops on the dual lattice. In an SU(2) theory the same definition (1.1) is used (i.e. the monopoles are the monopoles of the SO(3) gauge field that is obtained from the SU(2) gauge field by forming cosets.) The monopoles are now attached to Z_2 *strings* that carry energy [1, 10]. They consist of plaquettes where

$$\operatorname{sign} \operatorname{tr} U(\partial p) = -1. \tag{1.2}$$

The world sheets of the Z_2 strings are 2-dimensional surfaces on the dual lattice. They are either closed or bordered by monopole loops.

In order to put our work into perspective, we briefly review what is known about the above-mentioned phase transitions.

The prototype of a phase transition associated with Z_2 *string* condensation occurs in Wegner's pure Z_2 gauge theory [11, 12]. Such a transition was also proven to exist in an SU(2) model in which the monopoles were eliminated by a constraint (MP model) [13]. Its transition point was determined by Monte Carlo computations by Brower, Kessler and Levine [5]. Our Monte Carlo data presented in fig. 1 confirm that it is associated with condensation of Z_2 strings. The result for $1 - \langle \operatorname{sign} \operatorname{tr} U(\partial p) \rangle$ behaves very much like the internal energy of Wegner's Z_2 model.

Monte Carlo evidence for phase transitions associated with *monopole* condensation in SO(3) models was found by Halliday and Schwimmer and by Greensite and Lautrup [3, 4]. Brower, Kessler and Levine [5] have shown that the monopole density in the standard SU(2) model with Wilson action rises rapidly in the vicinity of its

* Notation: If C is a path composed of links $b_1 \cdots b_n$ then $U(C) = U(b_n) \cdots U(b_1)$. In particular $U(\partial p) = U(b_4) \cdots U(b_1)$ for a plaquette p with boundary $\partial p \equiv \dot{p}$ consisting of oriented links $b_1 \cdots b_4$.

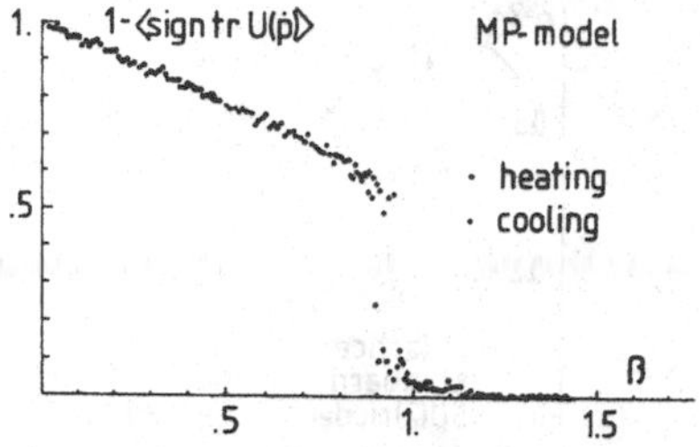

Fig. 1. Monte Carlo data for 1-⟨sign tr $U(\partial p)$⟩ in the SU(2) MP model (3.11). The calculations were done on a 3^4 lattice.

"rapid transition" from weak to strong coupling, and they proposed that this transition is due to condensation of monopoles together with their strings.

One can suppress monopoles by adding a suitable second term L_2 to the standard SU(2) action of Wilson,

$$L_1 = \tfrac{1}{2}\beta \sum_{\mathrm{p}} \operatorname{tr} U(\partial \mathrm{p}). \tag{1.3}$$

One can either choose $L_2 = \Sigma_c \lambda \rho_c(\overline{U})$ as in ref. [5], or $L_2 =$ action for an SO(3) theory [6–8]. The phase diagrams of such theories with two coupling parameters were studied with the Monte Carlo method by Brower, Kessler and Levine [5], Bhanot and Creutz [6], and Caneschi, Halliday and Schwimmer [7]. First-order phase-transition lines are found that are associated with either condensation of monopoles, or of Z_2 strings, or of both. Condensation of monopoles enhances the order parameters (monopole density · 2)

$$\overline{M} = 1 - \langle \rho_c(\overline{U}) \rangle, \tag{1.4}$$

while condensation of Z_2 strings in an SU(2) theory enhances the order parameter [7]

$$\overline{E} = 1 - \langle \varepsilon_b(U) \rangle, \qquad \varepsilon_b = \prod_{\mathrm{p} \in \partial^* \mathrm{b}} \operatorname{sign} \operatorname{tr} U(\partial \mathrm{p}). \tag{1.5}$$

The product runs over the six plaquettes p that have b in their boundary. These order parameters were computed by Caneschi, Halliday and Schwimmer [7]. In this way they were able to elucidate the nature of the observed phase transitions.

One of the observed first-order phase-transition lines has a jump of both $\overline{E}$ and $\overline{M}$ and projects towards the point in the two parameter plane which corresponds to the "rapid transition" in the standard SU(2) model with action L_1, $\beta \approx 2.2$, but it stops before reaching it [6,7]. A peak in the specific heat C_V of the standard SU(2) model at $\beta = 2.2$ was observed by Lautrup and Nauenberg [14]. Lang et al. [15] pointed out

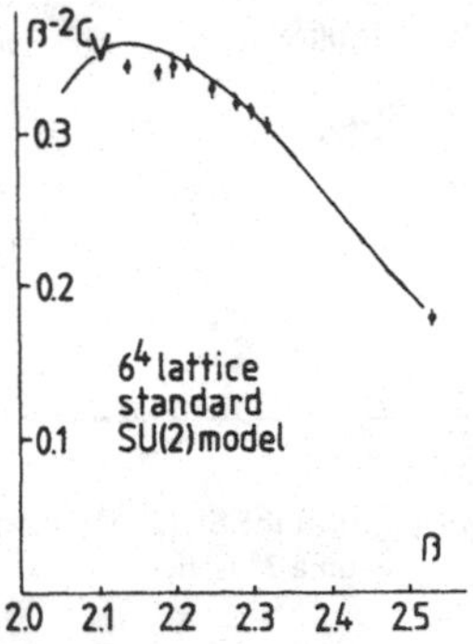

Fig. 2. The specific heat C_V of the standard SU(2) model with Wilson action (1.3), computed on a 6^4 lattice. The line represents the finite size scaling polynomial fit of Nauenberg, Schalk and Brower to their data for the internal energy. For further explanation see text.

that it may represent a 3rd order phase transition (a change in slope of $\beta^{-2}C_V$). In fig. 2 we present Monte Carlo data of our own for $\beta^{-2}C_V$. They are more accurate and from a larger lattice than data available before. Each data point is determined from the fluctuations in internal energy during 20 000 or more sweeps through the 6^4 lattice, using the heat bath method. The statistical errors were determined by dividing into M subsamples of about 400 sweeps each. They represent the mean square deviation of the averages over individual subsamples divided by $\sqrt{M}$. The result is consistent with a third-order phase transition but it does not establish it since a smooth curve could also be drawn through the points. Moderate agreement is found with the fit of Nauenberg, Schalk and Brower [16] to their data for the internal energy. This fit obeys finite size scaling and reduces to a polynomial fit for infinite lattice size. High-temperature series appear to favor the absence of any phase transition in the standard model [17]. A tentative explanation of the absence of a (first-order) phase transition is that small monopole loops become abundant before the monopoles are liberated, and their liberation (= condensation of Z_2 strings) becomes thereby thermodynamically insignificant. The new theory of first-order phase transitions of Dobrushin, Shlosman and Kotecký [18] and the work of refs. [7, 19] offer some promise that a better understanding will soon be reached.

In the formal continuum limit, the SU(2) lattice action L_1 becomes equal to the Yang-Mills action for a gauge theory in continuous euclidean space-time [20] because $U(\partial p) \to 1$ as a random variable when $\beta \to \infty$ [1]. We see from this that the monopoles and Z_2 strings can be eliminated by a constraint $\operatorname{sign} \operatorname{tr} U(\partial p) \equiv +1$ without affecting the formal continuum limit. We call the resulting model a "*positive plaquette model*".

Similarly, monopoles in the SO(3) theory with Wilson action can be eliminated by a constraint $\rho_c \equiv 1$ without affecting the formal continuum limit because $\rho_c(\overline{U}) = 1$ if

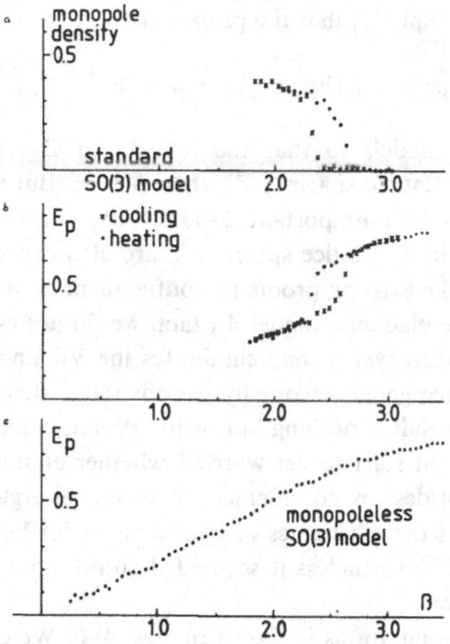

Fig. 3. (a) Monopole density and (b) internal energy in the SO(3) model with lagrangian $\frac{1}{3}\beta \operatorname{tr} \overline{U}(\partial p)$, compared with (c) the internal energy of the corresponding monopoleless SO(3) model (2.1).

$\overline{U}$ is locally close to a pure gauge. The formal continuum limit of the theories with gauge group SU(2) and SO(3) is the same.

Neither of these constraints violates gauge invariance of the models in question. On the basis of the above discussion of the nature of observed phase transitions we expect that elimination of the monopoles in the SO(3) model eliminates its phase transition, and the elimination of Z_2 strings and monopoles in the SU(2) models eliminates the "glitch" in its internal energy at the "rapid transition point" $\beta \approx 2.2$ [12]. This is indeed the case. In fig. 3 we compare Monte Carlo data for the internal energy of an SO(3) model with and without monopoles. The monopole density $\frac{1}{2}\overline{M}$ is also shown. Calculations for the SU(2) case were done by Brower, Kessler and Levine [5].

It is, in fact, not necessary to consider the constraint SU(2) and SO(3) models separately. In sect. 2 we establish an exact and explicit relation between such models. All observables, including in particular the expectation value of the Wilson loop [20] for fractionally charged static quarks can be transcribed from the positive plaquette SU(2) model to the corresponding monopoleless SO(3) model. The relation (2.14) for the Wilson loop is a simplified and more explicit version of a formula that was obtained by one of us in ref. [21]. It is very instructive and will elucidate both the topological origin of confinement and its basis in the peculiar locality properties of (classical) gauge field theories [33].

Chessboard estimates imply [1] that the probability

$$\text{prob}(\text{tr}\, U(\partial p) < 0) < \text{const} \cdot e^{-\beta/13} \tag{1.6}$$

in the standard SU(2) model, so that the density of Z_2 strings goes to zero exponentially in units of (lattice spacing)$^{-3}$, when $\beta \to \infty$. But this is insufficient to guarantee that they become unimportant dynamically, because the string tension and mass squared in units of (lattice spacing)$^{-2}$ are also expected to tend to zero exponentially. None of the existing proofs of confinement at strong coupling (small β) applies to the positive plaquette model. In fact, we do not even have a proof for the special case $\beta = 0$. Moreover, if one eliminates the Villain monopoles [22] in a 3-dimensional U(1) lattice gauge theory by a constraint, then linear confinement goes away (because the result is nothing but ordinary non-compact electrodynamics [23]). Therefore, one might start to get worried whether elimination of monopoles and Z_2 strings might not destroy confinement. Theoretical arguments based on our present understanding of the mechanics of confinement for large β [24] imply that this should not happen. Nevertheless it seemed desirable to study the question by Monte Carlo computation.

The result of such computations is shown in figs. 4–6. We consider lattices Λ of size $n_t \times n_s^3$ with periodic boundary conditions, and loops C_x that wind through Λ as shown in fig. 7. Let

$$L(x) = \text{tr}\, U(C_x). \tag{1.7}$$

n_t^{-1} can be interpreted as a physical temperature in units of (lattice spacing)$^{-1}$. A deconfining phase transition to "hot gluon soup" [25] with $L(x) \neq 0$ is expected to occur when n_t is lowered. Fig. 4 shows that such a transition does indeed occur, but the transition temperature n_t^{-1} is lowered dramatically compared to the standard

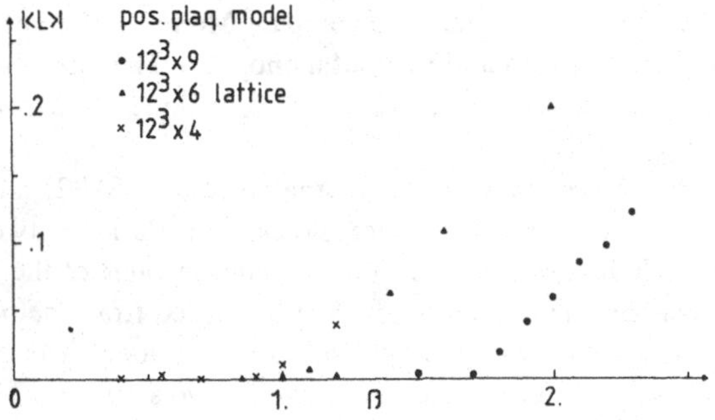

Fig. 4. The order parameter $\langle L(\mathbf{0}) \rangle$ in the positive plaquette model (2.2a, b) for different sizes $n_t \times n_s^3$ of the lattice. The value of n_t^{-1} where $\langle L(\mathbf{0}) \rangle$ reaches zero can be interpreted as a physical transition temperature to "hot gluon soup" (in units of lattice spacing a^{-1}).

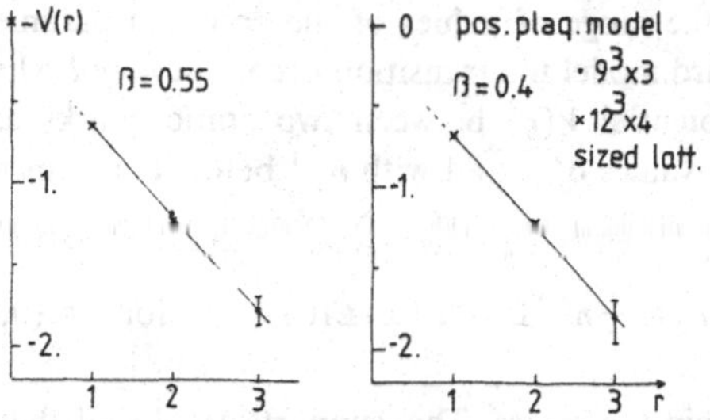

Fig. 5. The potential $V(r)$ between static quarks as defined by eq. (1.8) at two values of β, for n_t^{-1} below the transition temperature.

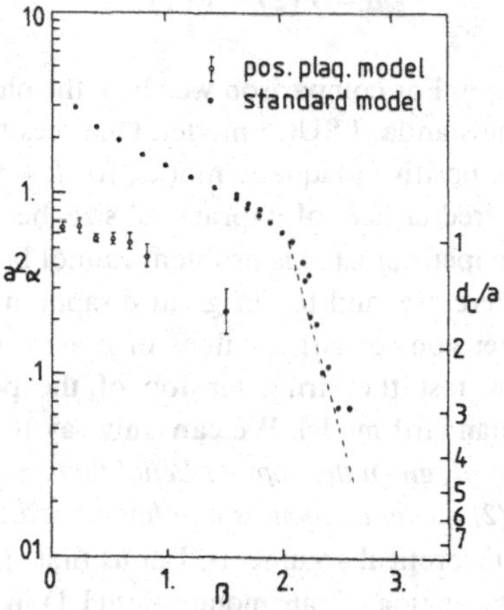

Fig. 6. The string tension for the positive plaquette model (2.2a,b) compared with Creutz' Monte Carlo data for the string tension of the standard SU(2) model with Wilson action (1.3). The line represents the fit by eqs. (4.2), (4.3a), with the values of α_0, β_1 which give the best fit to the Monte Carlo data for $\partial\nu/\partial\beta$ (fig. 8) in the standard model. Raising β_1 to 2.09 would give a perfect fit. We attribute this small discrepancy to systematic errors in the determination of α in [27].

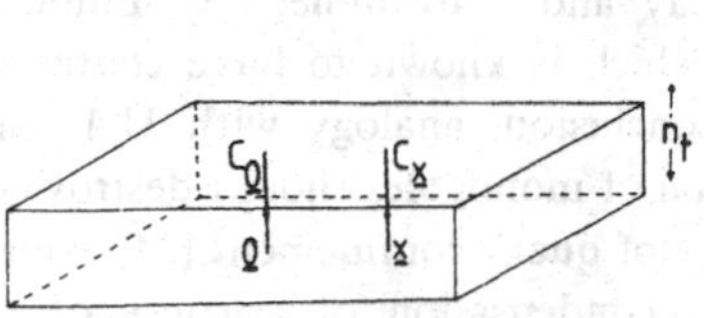

Fig. 7. The paths used in defining the order parameters and correlation functions $\langle L(0)\rangle$ and $\langle L(x)L(0)\rangle$. See text before eq. (1.7).

SU(2) model, for the range of values of the coupling parameter β where we have data. In the standard model the transition occurs at $n_t \approx 2$ when $\beta = 1.8$ [26]. In fig. 5 we show the potential $V(r)$ between two static quarks as a function of their distance r, for two values of β and with n_t^{-1} below the transition temperature. It is defined by [26]

$$V(r) = -n_t^{-1} \ln 4 \langle L(x) L(0) \rangle , \qquad \text{for } x = (0,0,r) . \tag{1.8}$$

It looks linear within the errors. This supports the belief that the confinement has not been destroyed. Assuming that $V(r)$ can be fit by a linear function of r down to $r = 1$ one can determine the string tension α by

$$\alpha = V(2) - V(1) . \tag{1.9}$$

The result is shown in fig. 6. For comparison we show the old data of Creutz [27] for the string tension α in the standard SU(2) model. One sees that the string tension is considerably lower in the positive plaquette model, for $\beta \leqslant 1.5$. Calculation of α for larger β would have required lattices of impractical size, because of the low value of the physical transition temperature. (This problem cannot be avoided by considering Wilson loops, either.) Therefore, and to our great disappointment, we were not able to determine the string tension for larger values of β as would have been necessary to see whether and how fast the string tension of the positive plaquette model approaches that of the standard model. We can only say that *the approach is not as fast as one might have hoped, given the popular belief that one is close to the continuum limit of the standard SU(2) model as soon as one has passed the "rapid transition"*.

Now, we come to the theoretical arguments. Let us first straighten out the analogy with the U(1) theory. The analog of our monopoles (1.1) in a U(1) theory would be Glimm-Jaffe monopoles [28] rather than Villain monopoles [22]. The Villain monopoles are defined in terms of auxiliary variables that exist in some models. Eliminating them destroys the U(1) gauge invariance and leads to a gauge theory with gauge group $\mathbb{R}$ = universal covering of U(1). Villain monopoles can also be defined in the SO(3) model of Halliday and Schwimmer [3]. Eliminating them produces the standard SU(2) model which is known to have confinement at strong and intermediate coupling. In conclusion, analogy with U(1) theory produces no sound argument that elimination of monopoles should destroy confinement.

The effective Z_2 theory of quark confinement [24] explains confinement for large β as a consequence of condensation of vortices of thickness $d \geqslant d_c(\beta)$, with $d_c(\beta)/a \to \infty$ exponentially as $\beta \to \infty$ (a = lattice spacing), whereas thinner vortices freeze out. The Z_2 strings are a special kind of thin vortex of thickness 1 lattice spacing. Using the effective Z_2 theory, and the simplest of approximations to compute the free energy of thin vortices, the suppression of the string tension α at low β can be explained quantitatively. This will be shown in sect. 3. The calculations suggest that there are important *correlations between vortices and monopoles*, because elimination of monopoles cuts down the entropy associated with internal structure of the vortices*. It is tempting to speculate that the same kind of correlation exists between fat vortices and fat monopoles. The fat monopoles are defined by the same formula (1.1a) except that $U(\partial p)$ is replaced by the parallel transporter around the boundary of a square P of larger side length (= a plaquette of a block lattice) see [1].

In the last section of this paper we take the opportunity to present some new Monte Carlo data for the vortex free energy [29]. They add to those presented in our first paper [30]. Its conclusions are unchanged.

2. Exact relation between models with gauge group SO(3) and SU(2)

In this section we will establish an exact relation between monopoleless SO(3) models and corresponding "positive plaquette" SU(2) models.

Our SO(3) models without monopoles have Gibbs measure

$$d\bar{\mu}(\bar{U}) = \frac{1}{\bar{Z}} \prod_b d\bar{U}(b) \exp\left[\sum_p \bar{\mathcal{L}}(\bar{U}(\partial p))\right] \prod_c \theta(\rho_c(\bar{U})), \qquad (2.1a)$$

with

$$\bar{\mathcal{L}}(\bar{V}) = \tfrac{1}{2}\beta\left[(\mathrm{tr}\,\bar{V}+1)^{1/2} - 2\right], \quad \text{(pos. square root)}, \qquad (2.1b)$$

or

$$\bar{\mathcal{L}}(\bar{V}) = \tfrac{1}{3}\beta[\mathrm{tr}\,\bar{V} - 3], \quad \text{for } \bar{V} \in SO(3). \qquad (2.1c)$$

$d\bar{U}$ is normalized Haar measure on SO(3), and tr is here a trace of real orthogonal 3×3 matrices. Evidently the models are SO(3) gauge invariant. Expectation values of observables $\bar{F}(U)$ are defined by

$$\langle \bar{F} \rangle_{SO(3)} = \int d\bar{\mu}(\bar{U}) F(\bar{U}).$$

The corresponding SU(2) models involve variables which are 2×2 matrices $U(b) \in$ SU(2). Their Gibbs measure is given by

$$d\mu(U) = \frac{1}{Z} \prod_b dU(b) \exp\left[\sum_p \mathcal{L}(U(\partial p))\right], \qquad (2.2a)$$

* Samuel has argued before that monopoles are important for confinement, especially for gauge group SU(3) [42]. But his monopoles are defined differently and we are at present unable to say what the relation might be.

with

$$\mathfrak{L}(V)=\begin{cases}\frac{1}{2}\beta[\operatorname{tr}V-2], & \text{if } \operatorname{tr}V>0,\\ -\infty, & \text{otherwise,}\end{cases} \tag{2.2b}$$

or

$$\mathfrak{L}(V)=\begin{cases}\frac{1}{3}\beta\left[(\operatorname{tr}V)^2-4\right], & \text{if } \operatorname{tr}V>0,\\ -\infty, & \text{otherwise}.\end{cases} \tag{2.2c}$$

Expectation values of observables $F(U)$ are defined by

$$\langle F\rangle_{SU(2)}=\int d\mu(U)F(U). \tag{2.2d}$$

Validity of the exact relation between these SU(2) models and the SO(3) models (2.1) will require matching boundary conditions. For the SU(2) model it is required that the boundary conditions are invariant under substitutions $U(b)\to U(b)\gamma(b)$ with $\gamma(b)=\pm 1$, for $b\in\partial\Lambda$. Thus we may impose either free boundary conditions, or periodic boundary conditions for cosets $\bar{U}(b)$ (but not for $U(b)$ themselves), and the same boundary conditions for the SO(3) model.

Now we will transcribe expectation values of observables from this SU(2) model to the SO(3) model. The relation is simple for (gauge-invariant, local*) observables $F(U)=\bar{F}(\bar{U})$ which depend on U only through cosets $\bar{U}\in SO(3)=SU(2)/Z_2$, so that they can be regarded as observables of the SO(3) theory in a natural way:

$$\langle F\rangle_{SU(2)}=\langle\bar{F}\rangle_{SO(3)}. \tag{2.3}$$

We proceed to the proof of this relation. From the Kronecker decomposition of the Kronecker product of two 2-dimensional representations of SU(2) it follows that

$$(\operatorname{tr}V)^2=\operatorname{tr}\bar{V}+1.$$

Consequently

$$\bar{\mathfrak{L}}(\bar{V})=\mathfrak{L}(V), \qquad \text{if } \operatorname{tr}V\geqslant 0. \tag{2.4}$$

* Local means here that F should not depend on $U(b)$ attached to links b in the boundary $\partial\Lambda$ of the (finite) lattice Λ.

We note also that the SU(2) models (2.2) have no monopoles. By definition, the monopoles are the monopoles of the SO(3) gauge field $\bar{U}$ that is obtained from U by forming cosets. Since $U(\mathrm{b})$ is a representative of $\bar{U}(\mathrm{b})$ it follows that

$$\rho_c(\bar{U}) = \prod_{\mathrm{p}\in\partial c} \operatorname{sign} \operatorname{tr} U(\partial \mathrm{p}) \equiv 1, \tag{2.5}$$

because the measure (2.2) has in its support only gauge fields U with $\operatorname{tr} U(\partial \mathrm{p}) > 0$. Because of relation (2.4) the expectation value $\langle F\rangle_{\mathrm{SU}(2)}$ takes the form

$$\langle F\rangle_{\mathrm{SU}(2)} = Z^{-1}\int \prod \mathrm{d}U(\mathrm{b}) \exp\left[\sum_{\mathrm{p}} \bar{\mathfrak{L}}(\bar{U}(\partial \mathrm{p}))\right] \prod_{\mathrm{p}} \theta(\operatorname{tr} U(\partial \mathrm{p})) \bar{F}(\bar{U}). \tag{2.6}$$

We introduce the normalized Haar measure on Z_2

$$\int \mathrm{d}\gamma(\cdots) = \tfrac{1}{2} \sum_{\gamma=\pm 1} (\cdots). \tag{2.7}$$

The corresponding δ-function is $\delta(\gamma) = 1 + \gamma$. It satisfies $\int \mathrm{d}\gamma \delta(\gamma) f(\gamma) = f(1)$. We will make use of identities of the form

$$\int_{\mathrm{SU}(2)} \mathrm{d}V f(V) = \int_{\mathrm{SO}(3)} \mathrm{d}\bar{V} \int_{Z_2} \mathrm{d}\gamma f(V\gamma). \tag{2.8}$$

They are obtained by making a variable substitution $V \to V\gamma$, and averaging over γ. The relation (2.8) then follows from invariance of the Haar measure; $\int \mathrm{d}\gamma f(V\gamma)$ depends on V only through the coset $\bar{V}$. Our choice of boundary conditions admits a variable substitution $U(\mathrm{b}) \to U(\mathrm{b})\gamma(\mathrm{b})$ in (2.6). Averaging over $\gamma(\mathrm{b})$ we obtain

$$\langle F\rangle_{\mathrm{SU}(2)} = Z^{-1}\int \prod_{\mathrm{b}} \mathrm{d}\bar{U}(\mathrm{b}) \exp\left[\sum_{\mathrm{p}} \bar{\mathfrak{L}}(\bar{U}(\partial \mathrm{p}))\right] \bar{F}(\bar{U})$$

$$\times 2^{-N_\mathrm{p}} \int \prod \mathrm{d}\gamma(\mathrm{b}) \prod_{\mathrm{b}} \delta(\gamma(\partial \mathrm{p}) \operatorname{sign} \operatorname{tr} U(\partial \mathrm{p})).$$

N_p is the number of plaquettes in Λ. Integration over Z_2 gauge fields $\gamma(\mathrm{b})$ is equivalent to integration over Z_2 field strengths $\sigma(\mathrm{p})$, subject to the constraints imposed by the 2nd Maxwell equations. Thus

$$\langle F\rangle_{SU(2)}=2^{-N_p}Z^{-1}\int\prod_b d\bar{U}(b)\exp\Big[\sum_p \bar{\mathfrak{L}}(\bar{U}(\partial p))\Big]\bar{F}(\bar{U})$$

$$\times\int\prod_p d\sigma(p)\delta(\sigma(p)\,\mathrm{sign\,tr}\,U(\partial p))\prod_c\delta\Big(\prod_{p\in\partial c}\sigma(p)\Big)$$

$$=2^{-N_p}Z^{-1}\int\prod d\bar{U}(b)\exp\Big[\sum_p \bar{\mathfrak{L}}(\bar{U}(\partial p))\Big]\bar{F}(\bar{U})\prod_c\delta(\rho_c(\bar{U}))$$

$$=2^{-N_p+N_c}(\bar{Z}/Z)\int d\bar{\mu}(\bar{U})F(\bar{U}). \tag{2.9}$$

The partition functions are defined so that $\langle 1\rangle=1$. Specializing to $F=\bar{F}=1$ we conclude that $2^{-N_p+N_c}(\bar{Z}/Z)=1$. Relation (2.9) therefore reduces to (2.3). Proof completed.

We wish to extend our relation to observables which are not of the simple form $F=\bar{F}(\bar{U})$. It will suffice to consider the Wilson loop expectation value for fractional charge.

Given a closed loop C which consists of links $b_1,\dots,b_n$, the parallel transporter $U(C)$ around C is the ordered product $U(b_n)\cdots U(b_1)$. Let χ_l be the character of the $2l+1$ dimensional representation of SU(2). Then $\chi_l(U(C))$ is the Wilson loop observable for static quarks of charge l. It depends only on $\bar{U}$ if and only if l is integer. It will be instructive to consider fractional and integer l at the same time. Up to eq. (2.12) below our discussion will be general, without reference to particular models.

We will introduce new variables $\sigma(b)=\pm 1$ and $W(b)\in SU(2)$ with $\mathrm{tr}\,W(b)\geq 0$. They are functions of the original variables $U(b')$ with the following properties [21,31].

(1) *Locality*: $W(b)$ and $\sigma(b)$ depend only on $U(b')$ on links b′ in a neighborhood of one lattice spacing of b, and $W(b)$ depends in fact only on cosets $\bar{U}(b)\in SO(3)=SU(2)/Z_2$.

(2) *Gauge invariance*: $W(b)$ are gauge invariant, whereas a gauge transformation of the variables $U(b)$ induces a Z_2 gauge transformation $\sigma(b)\to\omega(x)\sigma(b)\omega(y)^{-1}$ for $b=(x,y)$, with $\omega(\cdot)=\pm 1$. Therefore the field strength $\sigma(\partial p)=\sigma(b_4)\cdots\sigma(b_1)$ (product over links on the boundary of a plaquette) is gauge invariant.

(3) *Completeness*: There exists a gauge transformation $S(\cdot)$ (dependent on U) such that

$$W(b)\sigma(b)=S(x)U(b)S(y)^{-1},\qquad \text{for } b=(x,y), \tag{2.10}$$

and $\mathrm{tr}\,W(b)\geq 0$.

Properties (1) and (2) will be summarized by saying that $W(\mathrm{b})$ and $\sigma(\partial \mathrm{p})$ are *local gauge invariants*. One says that a *vortex soul* passes through the plaquette p if $\sigma(\partial\mathrm{p}) = -1$. Vortex souls form closed 2-dimensional surfaces on the dual lattice, in 4 dimensions. For weak coupling β^{-1} these vortex souls have a topological interpretation analogous to the zero of the Higgs field in a Nielsen-Olesen vortex [32], see sect. 10 of ref. [21].

Properties (1)–(3) above do not fix the variables W, σ uniquely. To get an explicit definition one must choose a local gauge [33] (a generalization of what is called a unitary gauge in Higgs models). An example will be given at the end of this section. It will therefore in general depend on the choice of local gauge where a vortex has its soul. (This explains the name "soul": one does not know very precisely where it is, but vortices can be counted by counting souls.) The formulae below are, however, valid for any choice since they only depend on the above properties.

Consider a loop C which is boundary of a surface Ξ. From eq. (2.10) and the property of characters $\chi_l(\gamma V) = \gamma^{2l}\chi_l(V)$ for $\gamma = \pm 1$ it follows that

$$\chi_l(V) = \begin{cases} \chi_l(W(\mathrm{C})) \prod\limits_{\mathrm{p}\in\Xi} \sigma(\partial\mathrm{p}) & \text{for } l = \frac{1}{2}, \frac{3}{2}, \ldots, \\ \chi_l(W(\mathrm{C})), & \text{for } l = 0, 1, 2, \ldots. \end{cases} \tag{2.11}$$

$\chi_l(W(\mathrm{C}))$ is a sum of products of local gauge invariants that are localized near the path C. For instance,

$$\chi_{1/2}(W(\mathrm{C})) = \mathrm{tr}\, W(\mathrm{C}) = \sum_{\alpha_1 \cdots \alpha_n} W_{\alpha_1\alpha_n}(\mathrm{b}_n) \cdots W_{\alpha_2\alpha_1}(\mathrm{b}_1) \tag{2.12}$$

if C is composed of links $\mathrm{b}_1 \cdots \mathrm{b}_n$. In the case of fractional charge there appears another factor which is not of this form. It counts the number of vortex souls that wind around C. It has been verified by Göpfert [34] to all orders of the high-temperature expansion in the standard SU(2) model that $\langle \chi_{1/2}(W(\mathrm{C}))\rangle$ is perimeter-law behaved, and that $\langle \chi_{1/2}(U(\mathrm{C}))\rangle \sim \langle \Pi\sigma(\partial\mathrm{p})\rangle$ up to a perimeter-law behaved factor, so that the probability distribution of vortex souls determines the string tension. For those local gauges that were considered in [34], $\langle \Pi\sigma(\partial\mathrm{p})\rangle$ depends on the choice of local gauge only through a (in some cases oscillatory) perimeter-law behaved factor.

Now we return to our special models (2.2). The Gibbs measure (2.2) is supported on configurations U with $\mathrm{sign}\,\mathrm{tr}\, U(\partial\mathrm{p}) \equiv 1$. Therefore it follows from eq. (2.10) that

$$\sigma(\partial\mathrm{p}) = \mathrm{sign}\,\mathrm{tr}\, W(\partial\mathrm{p}), \qquad \text{for models (2.2)}. \tag{2.13}$$

As a result

$$\langle \chi_l(U(\mathrm{C}))\rangle_{\mathrm{SU}(2)} = \left\langle \chi_l(W(\mathrm{C})) \left[\prod_{\mathrm{p}\in\Xi} \mathrm{sign}\,\mathrm{tr}\, W(\partial\mathrm{p}) \right]^{2l} \right\rangle_{\mathrm{SU}(2)} .$$

By property (1), W depends only on cosets $\bar{U}$. Therefore we may now apply eq. (2.3) to obtain our *final result*. If l is integer, χ_l is a character of an SO(3) representation and we may identify $\chi_l(V)=\chi_l(\bar{V})$. Thus finally, for any surface Ξ with boundary C,

$$\langle\chi_l(U(\mathrm{C}))\rangle_{\mathrm{SU}(2)}=\begin{cases}\langle\chi_l(W(\mathrm{C}))\prod\limits_{\mathrm{p}\in\Xi}\operatorname{sign}\operatorname{tr}W(\partial\mathrm{p})\rangle_{\mathrm{SO}(3)}, & \text{for } l=\frac{1}{2},\frac{3}{2},\ldots,\\ \langle\chi_l(W(\mathrm{C}))\rangle_{\mathrm{SO}(3)}=\langle\chi_l(\bar{U}(\mathrm{C}))\rangle_{\mathrm{SO}(3)}, & \text{for } l=0,1,2,\ldots.\end{cases} \tag{2.14}$$

$\prod\operatorname{sign}\operatorname{tr}W(\partial\mathrm{p})$ is independent of the choice of Ξ because of the constraints of the model which eliminate monopoles. This follows from eq. (2.13), since $\prod_{\mathrm{p}\in\partial\mathrm{c}}\sigma(\partial\mathrm{p})\equiv 1$ for $\sigma(\partial\mathrm{p})=\prod_{\mathrm{b}\in\partial\mathrm{p}}\sigma(\mathrm{b})$ (2nd Maxwell equation).

Finally we exhibit an example of a local gauge. To construct W, it suffices to specify the gauge transformation $S(x)$ in (2.10). Since $\operatorname{tr}W(\mathrm{b})\geqslant 0$ this fixes $\sigma(\mathrm{b})=\operatorname{sign}\operatorname{tr}S(x)U(\mathrm{b})S(y)^{-1}$. Let $\mathrm{p}_{ij}(x)$ be the plaquette protruding from corner point x in positive ij direction. Then one may define the magnetic field matrix $\boldsymbol{B}(x)=(B^a{}_k(x))$ by

$$U(\mathrm{p}_{ij}(x))=A+i\sum_a\tau^aB^a{}_k(x),\qquad (ijk=123\text{ or cyclic}). \tag{2.15}$$

τ^a are Pauli matrices. One can now define $\bar{S}(x)\in\mathrm{SO}(3)$ by the decomposition

$$\boldsymbol{B}(x)=\bar{S}(x)\boldsymbol{P}(x), \tag{2.16}$$

where the 3×3 matrix $\boldsymbol{P}$ is required to be either positive or negative semidefinite. $S(x)$ is chosen as the representative of the coset $\bar{S}(x)$ with $\operatorname{tr}S(x)\geqslant 0$. $S(x)$ is uniquely defined for almost all gauge fields U, and W,σ are therefore well-defined random variables (on a finite lattice, and also in general by an argument based on the Markov property, compare [35]). It is easy to verify that properties (1)–(3) above are satisfied.

3. Predictions of the effective Z_2 theory of confinement

The model (2.2a, b) differs from the SU(2) MP model by the absence of Z_2 strings, and from the standard SU(2) model by the absence of monopoles as well as Z_2 strings. The world sheets of these Z_2 strings consist of plaquettes p in Λ where $\operatorname{sign}\operatorname{tr}U(\partial\mathrm{p})=-1$. They are closed surfaces on the dual lattice Λ^* in the MP model, whereas in the standard model these surfaces may be either closed or have a boundary which is world line of a monopole. It consists of cubes c in Λ (= links in Λ^*) with $\rho_{\mathrm{c}}=-1$.

The closed Z_2 strings are a special kind of "thin" vortices. They have a thickness of only one lattice spacing. At small β they condense in both models (MP and standard) and this can be exploited to prove confinement. In the SU(2) model (2.2) the Z_2 strings are eliminated by a constraint. Therefore, we expect a reduction of the string tension at small β. But we do not expect that confinement will be destroyed completely, because "fat" vortices can take over and confine static quarks in the same way as they do in the standard model at low β where the Z_2 strings are frozen (because they cost too much energy which cannot be compensated by their configurational entropy).

To obtain somewhat more quantitative information we appeal to the effective Z_2 theory of confinement which was described in ref. [24]. In this theory, an effective Z_2 coupling constant $\beta_{\text{eff}}(d)$ is introduced which determines the chemical potential of a vortex of thickness d. More precisely it determines its energy, or internal free energy (which includes the entropy due to fluctuations in internal structure) whereas the configurational entropy is (approximately) the same as in a Z_2 theory on a lattice of lattice spacing d. Vortices of thickness d will condense if $\beta_{\text{eff}}(d) \leqslant \beta_c \approx 0.44$ (the transition point in Wegner's Z_2 gauge theory model). This requires that $d \geqslant d_c(\beta)$. A distinction is made between a "high-temperature" region where $\beta_{\text{eff}}(a) < \beta_c$ ($a =$ lattice spacing) so that $d_c = a$ and vortices of thickness a condense, and a "low-temperature" region where $d_c > a$. The high-temperature region ends at the value $\beta = \beta_1$ of the coupling parameter where $\beta_{\text{eff}}(a) = \beta_c$. In the low-temperature region d_c is determined from the equation $\beta_{\text{eff}}(d_c) = \beta_c$, and the string tension is given by $\alpha = \alpha_0/d_c^2$, where α_0 is the string tension in Wegner's Z_2 model just below its phase-transition point. The value $\alpha_0 = 0.54$ is obtained by extrapolating the result of high temperature expansions to order $2k = 8, 10, 12, 14$ using a linear function of $1/k$ [36].

For the following calculations we need an approximate expression for $\beta_{\text{eff}}(d)$. It is given by 't Hooft's version of a vortex free energy $\nu(d)$ [29]. To obtain this approximation one imagines cutting a piece of area d'^2 out of a vortex sheet of thickness d and simulating the effect of its environment by imposing periodic boundary conditions [24]. This gives

$$\beta_{\text{eff}}(d) \approx \nu(d) = -\tfrac{1}{2}(d/d')^2 \ln[Z(\text{block, t.b.c.})/Z(\text{block, p.b.c.})], \quad (3.1)$$

and is supposed to be reasonably accurate for β not too large, depending on d, so that $d_c(\beta) \leqslant d$. $Z(\cdots)$ are partition functions for a block of size $d \times d \times d' \times d'$ ($d' \geqslant d$) with periodic boundary conditions (p.b.c.), and twisted periodic boundary conditions (t.b.c.) with one twist in the 12-plane, respectively. The result should be independent of d' for $d' \gg d$.

The periodic (twisted) boundary conditions assure that there is an even (odd) number of vortex souls passing through the intersection Ξ of every plane $x_3, x_4 =$ const. with the block. This is easily seen as follows [31]. Periodic boundary conditions imply in particular periodicity of the Z_2 variables $\sigma(b)$ that were introduced in sect 2. It follows that the Z_2 parallel transporters satisfy

$$1=\sigma(\partial\Xi)=\prod_{p\in\Xi}\sigma(\partial p), \qquad \text{for p.b.c.}$$

This says that an even number of vortex souls passes through Ξ. The singular gauge transformation on $\partial\Lambda$ (or a neighborhood of it) which changes periodic into twisted boundary configuration takes parallel transporters $\sigma(\partial\Xi)\to-\sigma(\partial\Xi)$. Therefore the number of vortex souls through Ξ is changed from even to odd.

We wish to obtain an estimate for the string tension of our positive plaquette model near $\beta=0$. To get it we try to determine whether vortices of thickness 1 lattice spacing can condense at all in this model. The answer is not trivial, in spite of the constraint $\operatorname{tr}U(\partial p)>0$, if we count vortices by counting their souls as is suggested by the consideration of sect. 2. In ref. [30] we pointed out that twisted boundary conditions can be fulfilled by a pure gauge. This means that *creation of vortex souls costs entropy but no energy*. It follows that t.b.c. can be fulfilled without making $\operatorname{tr}U(\partial p)\leqslant 0$ for any plaquette p, even in the case $d=a$. For instance $U(\partial p)=-U_1U_2U_1^{-1}U_2^{-2}=1$ for $U_1=i\sigma_1, U_2=i\sigma_2$ (quaternions).

The effective Z_2 theory of confinement has so far given good results while using only the simplest of approximations. Encouraged by this we will be bold and use approximation (3.1) for $d=a$. Moreover, we will drop the plaquettes in the boundary of the blocks from the action. (Consider them as part of the environment.) To check that these approximations are not unreasonable we will first try them out on the standard SU(2) model at low β. The partition function of that model is

$$Z=\int\prod dU(b)\exp\left[\sum_p\tfrac{1}{2}\beta\operatorname{tr}U(\partial p)\right]. \tag{3.2}$$

Therefore, we obtain

$$Z(\text{block},\pm)=\left\{\int dU dV\exp\left[\pm\tfrac{1}{2}\beta\operatorname{tr}UVU^{-1}V^{-1}\right]\right\}^{d'^2/a^2}.$$

$+(-)$ stands for p.b.c. (t.b.c.). We evaluate the integral by noting that $UVU^{-1}V^{-1}\in SU(2)$ for any U,V. Therefore there must exist a measure $\rho(W)dW$ on SU(2) such that

$$\int dUdVf(UVU^{-1}V^{-1})=\int\rho(W)dWf(W). \tag{3.3}$$

Explicitly (see appendix B),

$$\rho(W)=\frac{2\pi-\theta}{4\sin\frac{1}{2}\theta},\qquad \text{for } W=Se^{i\sigma_3\theta/2}S^{-1},\qquad \theta=0\cdots 2\pi. \tag{3.4}$$

θ is the angle of rotation. For the purpose of integrating class functions we may use Weyl's integration formula to substitute

$$\mathrm{d}W=\frac{1}{\pi}\sin^2\tfrac{1}{2}\theta\,\mathrm{d}\theta,\quad(\theta=0\cdots 2\pi). \tag{3.5}$$

As a result we obtain (I_0 is the modified Bessel function)

$$\int \mathrm{d}U\mathrm{d}V\exp[\tfrac{1}{2}\beta\operatorname{tr}UVU^{-1}V^{-1}]=\beta^{-1}(\mathrm{e}^{\beta}-I_0(\beta)). \tag{3.6}$$

Thus, finally

$$\nu(a)=\tfrac{1}{2}\ln\{(\mathrm{e}^{\beta}-I_0(\beta))/(I_0(\beta)-\mathrm{e}^{-\beta})\}=\tfrac{1}{4}\beta+\cdots, \tag{3.7}$$

for the standard model. We can use this to determine the end point β_1 of the high-temperature region where vortices of thickness a cease to condense. Eq. (3.7) gives $\nu(a)=0.44$ at $\beta=\beta_1=2.05$. This is in very good agreement with Monte Carlo data (see sect. 4). Moreover, for $\beta<\beta_1$ the string tension α of the model should equal the string tension α_{Z_2} in Wegner's Z_2 model at coupling parameter $\beta_{\mathrm{eff}}(a)$. Since $\alpha_{Z_2}(\beta)=-\ln\beta+\ldots$, we find, using eqs. (3.1) and (3.7)

$$\alpha=-\ln\tfrac{1}{4}\beta+\cdots. \tag{3.8}$$

This is the correct result, to order β^0.

Now that we have seen how well our simple approximation works for the standard model, we apply it to the positive plaquette model (2.2a, b). We obtain

$$Z(\text{block},\pm)=\left\{\int \mathrm{d}U\mathrm{d}V\theta(\pm\operatorname{tr}UVU^{-1}V^{-1})\exp[\pm\tfrac{1}{2}\beta\operatorname{tr}UVU^{-1}V^{-1}]\right\}^{d'^2/a^2}.$$

The integral can be evaluated in the same manner as before. It is expressible in terms of the modified Bessel and Struve functions I_0 and L_0 [37]. As a result

$$\begin{aligned}\nu(a)&=\tfrac{1}{2}\ln\{(2\mathrm{e}^{\beta}-1-I_0(\beta)-L_0(\beta))/(I_0(\beta)+L_0(\beta)-1)\}\\&=\tfrac{1}{2}\ln(\pi-1)+\frac{\pi(4-\pi)}{16(\pi-1)}\beta+\cdots\\&=0.38+0.08\beta+\cdots.\end{aligned} \tag{3.9}$$

This is close to and slightly below the critical value $\beta_c = 0.44$ for $\beta = 0$ and has a very small slope there. Therefore, the string tension α in the model (2.2a, b) should be $(a = 1)$

$$\alpha \gtrsim \alpha_0 \approx 0.54, \qquad \text{at } \beta = 0 \tag{3.10}$$

(equal to the string tension of Wegner's Z_2 model at $\beta \approx 0.38$) and it should change little with β near $\beta = 0$. Ignoring the errors inherent in our approximation, vortices of thickness a are found to just barely condense at $\beta = 0$, since $\nu(a) \leqslant 0.44$ for $\beta \leqslant 0.82$. In fig. 6 we present some rough Monte Carlo data for the string tension of the model. They were obtained in the manner described in the introduction. We see that they are in agreement with the result of our theoretical calculations.

Finally we will now also consider the MP model. It has Z_2 strings but no monopoles. It was proven in ref. [13] that this model has a phase transition that is associated with condensation of Z_2 strings. Partition functions are defined by the formula

$$Z = \int \prod \mathrm{d}U(\mathrm{b}) \exp\left[\tfrac{1}{2}\beta \operatorname{tr} U(\partial \mathrm{p})\right] \prod_{\mathrm{c}} \theta\left(\rho_{\mathrm{c}}(\bar{U})\right). \tag{3.11}$$

We compute the vortex free energy $\nu(a)$ using the same approximations as before. Because of the absence of monopoles and the periodic boundary conditions, we must either have $\operatorname{tr} U(\partial \mathrm{p}) \geqslant 0$ for all plaquettes in the interior of the block, or $\operatorname{tr} U(\partial \mathrm{p}) \leqslant 0$ for all of them. $Z(\text{block}, \pm)$ are therefore sums of two contributions:

$$Z(\text{block}, \pm) = \left[\int \mathrm{d}U \mathrm{d}V \theta(\pm \operatorname{tr} UVU^{-1}V^{-1}) \exp\left(\pm\tfrac{1}{2}\beta \operatorname{tr} UVU^{-1}V^{-1}\right)\right]^{d'^2/a^2}$$
$$+ \left[\int \mathrm{d}U \mathrm{d}V \theta(\mp \operatorname{tr} UVU^{-1}V^{-1}) \exp\left(\pm\tfrac{1}{2}\beta \operatorname{tr} UVU^{-1}V^{-1}\right)\right]^{d'^2/a^2}.$$

The integrals can be evaluated in the same manner as before with the result

$$Z(\text{block}, +) = \left[\frac{1}{2\beta}\left(2\mathrm{e}^{\beta} - 1 - I_0(\beta) - L_0(\beta)\right)\right]^{d'^2/a^2}$$
$$+ \left[\frac{1}{2\beta}\left(1 - I_0(\beta) + L_0(\beta)\right)\right]^{d'^2/a^2},$$

$$Z(\text{block}, -) = \left[\frac{1}{2\beta}\left(-1 + I_0(\beta) + L_0(\beta)\right)\right]^{d'^2/a^2}$$
$$+ \left[\frac{1}{2\beta}\left(-2\mathrm{e}^{-\beta} + 1 + I_0(\beta) - L_0(\beta)\right)\right]^{d'^2/a^2}. \tag{3.12}$$

We are interested in the behavior for $d'/a \gg 1$. In Z(block, +) the first term dominates in this limit for all β. In Z(block, −) the second term, which comes from the contributions of Z_2 strings, dominates for small β. Therefore,

$$\nu(a) - \frac{3\pi}{8(\pi-1)}\beta + , \qquad \text{for small } \beta, \qquad (3.13)$$

and the string tension behaves like

$$\alpha = -\ln\frac{3\pi}{8(\pi-1)}\beta + \cdots, \qquad \text{for small } \beta. \qquad (3.14)$$

For large enough β the first term, which comes from vortices that are not Z_2 strings, dominates Z(block, −). The transition between the two possibilities occurs at the value of $\beta = \beta_{c,MP} = 0.82$ where both terms are equal. At this value of β the vortex free energy assumes the value $\nu(a) = 0.45$ which happens to be very nearly equal to the critical value $\beta_c = 0.44$ for the Z_2 coupling constant. From this we deduce that the string tension should behave as follows. For small β condensation of Z_2 strings leads to a large value of the string tension as given by eq. (3.14). With increasing β the string tension falls. At $\beta = \beta_{c,MP} \approx 0.82$ it reaches a value around $\alpha_0 \approx 0.54$ which agrees with the prediction for the string tension for the positive plaquette model at that value of β. At $\beta > \beta_{c,MP}$ the Z_2 strings are no longer able to condense, and vortices which are not Z_2 strings take over. Their thickness has to grow with β.

We do not have Monte Carlo data for the string tension of the MP model. But there are indirect indications that the above predictions are essentially correct. First, the result (3.14) is consistent with the rigorous inequality of ref. [13] which implies that $\alpha \geqslant -\ln\beta$ for $\beta \to 0$. Second, the predicted value of the position $\beta_{c,MP}$ of the phase transition agrees well with the value $\beta_{c,MP} \approx 0.9$ that was obtained from Monte Carlo computations by Brower, Kessler and Levine [5], compare fig. 1. Third, we have also Monte Carlo evidence that the probability of Z_2 strings in the MP model becomes very small for β above $\beta_{c,MP}$. This follow from the results of fig. 1 for $\langle \operatorname{sign} \operatorname{tr} U(\partial p)\rangle$.

Let us summarize our conclusions.

(i) The effective Z_2 theory appears to work well with very simple approximations.

(ii) There is a strong correlation between thin vortices and monopoles. If the monopoles are suppressed by a constraint, the entropy of these vortices is lowered substantially. As a result, the string tension is also lowered substantially, for $\beta < \beta_1$ (the range of β where thin vortices condense in the standard model).The simplicity of our approximations makes it very clear how the loss of entropy comes about because of a loss of possibilities in internal structure of the vortices.

Let us now turn to a discussion of what should happen at large β. Formally, the standard model, the MP model, and the positive plaquette model (2.2) all have the same continuum limit. The example of the 3-dimensional U(1) lattice gauge theory [23] shows that arguments based on such formal properties need not be reliable. There is, however, a partial result. In ref. [1] it was shown that the monopoles in the standard SU(2) model are confined for large β. Moreover, it was predicted in ref. [24] on the basis of the effective Z_2 theory of confinement that the string tension for the standard model and for the MP model should behave in the same way for large β. By the same argument the same should also hold for the positive plaquette model (2.2). One expects therefore that the correlation between monopoles and the fat vortices (which are needed to produce confinement for large β) will become less and less pronounced with increasing thickness d_c of these vortices. At the moment we can only hope that future Monte Carlo data will eventually confirm these predictions.

So far we have only considered monopoles of size 1 lattice spacing. In a sense such small monopoles in an SO(3) theory are the only ones which really deserve to be called monopoles (compare appendix A). From the point of view of a block spin or renormalization group picture [38] it is, however, natural to consider also "fat monopoles" as have been introduced in ref. [1]. It would be interesting to know how they are correlated with fat vortices. An investigation in this direction was suggested in refs. [5, 7]. A specific conjecture has been advanced by Iwasaki* [39]. He believes that the most important fat monopoles are those that are hidden inside instantons (in 4 dimensions). A stringent lower bound on the cost of energy of fat monopoles and their strings was established in [1].

4. Monte Carlo data for the vortex free energy

In our first paper [30] we presented Monte Carlo data for the derivative $\partial\nu/\partial\beta$ of the vortex free energy ν as defined in eq. (3.1), for cubic lattices of side length $d' = d \leqslant 5a$, and we also compared them to predictions of the effective Z_2 theory of confinement. Since then we have collected more such data, and we have extended the computations to a lattice of side length $d' = d = 6a$. We take the opportunity to present these data in fig. 8. The conclusions are the same as in ref. [30]. The fits represent the following predictions of the effective Z_2 theory for intermediate β [24]:

$$\nu(d) = [d/d_c(\beta)]^2 \exp[-\alpha d^2], \qquad \text{for } d \gg d_c, \tag{4.1}$$

where the string tension is given by

$$\alpha = \alpha_0 \exp[-\tfrac{6}{11}\pi^2(\beta - \beta_1)], \qquad \text{for } \beta > \beta_1, \tag{4.2}$$

* In Iwasaki's work only thin vortices are considered. The spreading mechanism, by which fat vortices can lower their free energy and confine static quarks for large β, is not considered there.

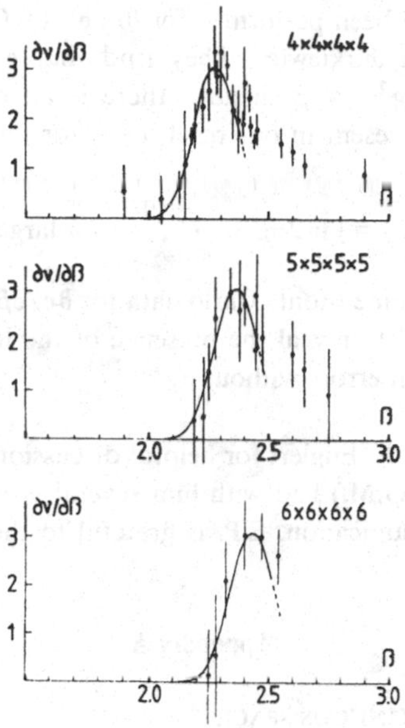

Fig. 8. Monte Carlo data for the derivative $\partial\nu/\partial\beta$ of the vortex free energy $\nu(d)$ of the standard SU(2) model with Wilson action (1.3). The lines represent a fit by eqs. (4.1)–(4.3a), with $\alpha_0 = 0.54a^{-2}$ and $\beta_1 = 2.06$.

and

$$(d_c/a)^2 = \begin{cases} \alpha_0/\alpha, & \text{for } \beta \geq \beta_c, \quad (4.3a) \\ 1, & \text{for } \beta \leq \beta_c. \quad (4.3b) \end{cases}$$

a = lattice spacing, and α_0 and β_1 have the meaning described in sect. 3. In the fits we put $\alpha_0 = 0.54a^{-2}$ and $\beta_1 = 2.06$. All data are for $\beta > \beta_1$. (Below $\beta_1, \partial\nu/\partial\beta$ is too small to be computable on a reasonably big lattice.) Very good agreement is seen. Münster has shown [40] that the asymptotic behavior of ν is $\nu = (d^2/a^2)\exp[-\alpha d^2]$ to all orders of the *high-temperature* expansion. If we were allowed to combine with eq. (4.2) for the string tension, which fits Creutz' Monte Carlo data [27] for $\beta > 2.15$ or so, we would get a ratio 16:25:36 of the height of the maxima of $\partial\nu/\partial\beta$ for lattice size $4a$, $5a$ and $6a$. In contrast, the predictions (4.1), (4.3a) of the effective Z_2 theory for $\beta > \beta_1$ differ from this expression by a factor (d^2/d_c^2) which depends exponentially on β. It predicts equal height of the maxima. The data decide in favor of this alternative.

Note added. The asymptotic behavior of $\nu(d)$ for large β and fixed d/a can, in principle, be calculated by the saddle-point method. In practise this is very difficult, but a calculation has now been performed for $\partial\nu/\partial\beta$ by Gonzalez-Arroyo, Groeneveld, Korthals Altes and Jurkiewicz. They find that ν does not admit a series expansion in powers of $g^2=4/\beta$; instead there is an extra logarithmic term. A logarithmic term is also present in our result (3.7) for the special case $d=a$ which gives

$$\nu(a)=\tfrac{1}{4}\ln 2\pi\beta+\cdots, \qquad \text{for large } \beta.$$

In ref. [30] we presented some Monte Carlo data for $\partial\nu/\partial\beta$ on a 3^4 lattice for large β (up to $\beta=9$). They failed to reveal the presence of the logarithmic term in ν – the data could be fitted within errors without it.

We would like to thank F. Englert for helpful discussions on monopoles and their strings which one of us (G.M.) had with him several years ago, and C.P. Korthals Altes for a private communication. E.P. is grateful to the DESY theory group for their kind hospitality.

Appendix A

SO(3) MONOPOLES IN CONTINUOUS SPACE

We will adopt the fibre bundle point of view which has been advocated in particular by Wu and Yang [9].

Consider the 3-dimensional space $\mathrm{M}=\mathbb{R}^3-\{x_0\}$ which is obtained from euclidean space by removing a single point x_0. x_0 will be the monopoles site. Since the 2nd Maxwell equation (Bianchi identity) is not true at the site of a monopole, a gauge field (vector potential) will only exist away from x_0.

One imagines that a 3-dimensional real vector space V_x is attached to every point x of M. One may envisage introducing matter fields ϕ eventually which take their values $\phi(x)\in\mathrm{V}_x$. To formulate differential equations for them one needs a topology (and differentiable structure) on the space E of pairs (x,v); $x\in\mathrm{M}, v\in\mathrm{V}_x$. It should satisfy certain requirements (the scalar product $\langle\ ,\ \rangle$ should be continuous, and E should be a euclidean space locally). When E is equipped with such a topology it is called a vector bundle with structure group SO(3).

The *Naheinformationsprinzip* [33] asserts that there is no *a priori* way of comparing directions in vector spaces V_x and V_y that are attached to different points $x=y$. Instead, a map

$$\mathcal{U}(\mathrm{C}):\mathrm{V}_x\to\mathrm{V}_y$$

is assumed to be given for every path C from x to y. It should depend smoothly on C, preserve the scalar product so that

$$\langle v_1,v_2\rangle=\langle\mathcal{U}(\mathrm{C})v_1,\mathcal{U}(\mathrm{C})v_2\rangle, \qquad \text{for } v_1,v_2\in\mathrm{V}_x,$$

and obey the composition rule

$$\mathcal{U}(C_2 \circ C_1) = \mathcal{U}(C_2)\mathcal{U}(C_1) : V_x \to V_z,$$

if $C_2 \circ C_1$ is obtained by juxtaposition of a path C_1 from x to y and a path C_2 from y to z. $\mathcal{U}(C)$ are called parallel transporters. They are said to specify a connection.

Our space M is topologically a sphere S^2. The classification of vector bundles over spheres has been described in Steenrod's book [41]. The result is that they are specified by homotopy classes of maps from the structure group G of the vector bundle into the equator of the sphere. Here $G = SO(3)$ and the equator of S^2 is a circle S^1. Since $\pi_1(SO(3)) = Z_2$ there are two inequivalent bundles [42]. One is trivial, i.e. isomorphic to $M \times V_x$. This vector bundle can be extended to a vector bundle over $\mathbb{R}^3$, while the other one cannot. One says that there is a monopole at x_0 if the vector bundle E is non-trivial.

To obtain a vector potential from $\mathcal{U}$ one needs to specify a moving frame first. A moving frame $\boldsymbol{e}$ specifies an orthonormal basis $\boldsymbol{e}(x) = (e_1(x), e_2(x), e_3(x))$ of vectors $e_i(x) \in V_x$ for every $x \in M$. One would like to have them depend smoothly on x; the problem caused by this requirement will be discusssed below. A moving frame specifies a coordinate system on E: A point $(x, v) \in E$ can be specified by real numbers (x^μ, v^i), $v^i = \langle v, e_i(x)\rangle$.

Given a moving frame and a connection, i.e. parallel transporters $\mathcal{U}(C)$, one can define parallel transport matrices $\bar{U}(C) = (\bar{U}^i{}_j(C)) \in SO(3)$ by

$$\mathcal{U}(C)e_i(x) = e_j(y)\bar{U}^j{}_i(C).$$

x and y are initial and final points of C. The vector potential $A_\mu(x) = (A^i{}_{j\mu}(x))$ is a matrix in the Lie algebra so(3) and is defined in terms of the parallel transport matrices by the familiar relation

$$\bar{U}(C) = \mathrm{T}\exp\left[-\int_C A_\mu \mathrm{d}x^\mu\right].$$

One of the fundamental results in the theory of fibre bundles asserts that it is impossible to find an everywhere smooth moving frame in a non-trivial vector bundle (= a smooth section in the associated principal fibre bundle) [41]. It follows that the presence of a monopole at $x = x_0$ enforces the presence of singularities in any moving frame over $\mathbb{R}^3 - \{x_0\}$. In other words, a non-singular global coordinate system does not exist. Let C_D be any path from x_0 to infinity. Since $\mathbb{R}^3 - C_D$ is topologically trivial, the fibre bundle E reduces to a trivial fibre bundle over $\mathbb{R}^3 - C_D$. Therefore, the moving frame can be chosen so that it has singularities only on C_D. C_D is called a Dirac string. The singularities of the moving frame cause singularities of the vector potential on C_D as well. These are not singularities of the

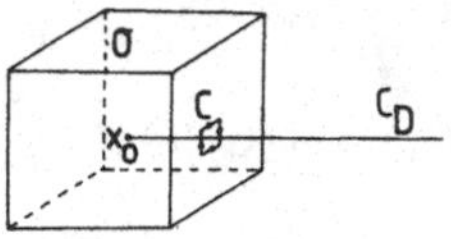

Fig. 9. A monopole at x_0 and its Dirac string C_D.

connection $\mathcal{U}$ but are merely due to the singularities of the coordinate system – the Dirac string is unphysical.

Given a vector potential $A_\mu(x)$, parallel transport matrices $U(C)$ in SU(2) – rather than SO(3) – can also be defined by

$$U(C) = \exp\left[-\tfrac{1}{2}\int \varepsilon_{ikj}\frac{\tau^i}{2}A^j{}_{k\mu}(x)\,dx^\mu\right].$$

τ^i are Pauli matrices. The singularities of the vector potential on the Dirac string have the consequence that

$$\tfrac{1}{2}\operatorname{tr}U(C) \approx -1,$$

if C is a closed path of infinitesimal length which winds around the Dirac string C_D, see fig. 9. In contrast, $\frac{1}{2}\operatorname{tr}U(C) \approx 1$ for any boundary of an infinitesimal area that is not crossed by C_D (and lies away from x_0).

We will now consider a more general situation with several monopoles at some positions x_i. Given an open set O, we may define the magnetic charge $Q = 0, 1$ in O to be equal to the number of monopoles in O modulo 2. There is a formula for Q. Suppose that O is a cube, and none of the monopoles is on its boundary ∂O. We may superimpose a lattice of small lattice spacing a on the continuum $\mathbb{R}^3$ so that ∂O is a union of plaquettes p of this lattice. We can count monopole charge by counting Dirac strings. Therefore

$$e^{i\pi Q} = \prod_{p \in \partial O} \operatorname{sign}\operatorname{tr}U(\partial p)$$

for sufficiently small lattice spacing a (depending on O and U). This motivates the definition of monopoles in an SO(3) lattice gauge theory that is described in the introduction.

The world line of a monopole in space time $\mathbb{R}^4$ is a line without end points – either closed, or extending to infinity.

Let us emphasize once more that monopoles such as are discussed here can only exist when the gauge group G has a non-trivial first homotopy group $\pi_1(G)$, and that the Dirac string attached to such monopoles is unphysical and unobservable. If we had started with 2-dimensional complex vector spaces V_x in which (classical) fields or Schrödinger wave functions for quarks could take their values, and with a corresponding gauge group SU(2), then any vector bundle over $\mathbb{R}^3 - \{x_0\}$ would be trivial and there would be no monopoles. In an SU(2) *lattice* gauge theory one can interpret the gauge field as an SO(3) gauge field and use this to define monopoles as discussed in sect. 1, but the string attached to such monopoles is not unphysical – it costs energy [10].

Appendix B

PROOF OF FORMULA (3.4)

We wish to determine the measure $\rho(W)\mathrm{d}W$ on SU(2) which satisfies

$$\int \mathrm{d}U\mathrm{d}V f(UVU^{-1}V^{-1}) = \int \rho(W)\mathrm{d}W f(W). \tag{B.1}$$

$\mathrm{d}U$ is the normalized Haar measure on SU(2). By making a substitution $U \to SUS^{-1}, V \to SVS^{-1}$ with $S \in \mathrm{SU}(2)$, we see that ρ is a class function. Therefore it admits a character expansion

$$\rho(W) = \sum_{j=0,1/2,1,\ldots} a_j^{-1}\chi_j(W). \tag{B.2}$$

χ_j is the character of the $2j+1$ dimensional irreducible representation of SU(2). It remains to determine the coefficients a_j^{-1}. We use orthogonality and the defining non-linear integral equation of irreducible characters,

$$\int \mathrm{d}U\chi_j(U)\chi_k(U^{-1}) = \delta_{jk}, \qquad \int \mathrm{d}U\chi_j(UV_1U^{-1}V_2) = \frac{1}{2j+1}\chi_j(V_1)\chi_j(V_2). \tag{B.3}$$

From eq. (B.1) we find that $a_j = 2j+1$. Thus

$$\rho(W) = \sum_{j=0,1/2,\ldots} \frac{1}{2j+1}\frac{\sin(j+\frac{1}{2})\theta}{\sin\frac{1}{2}\theta} = \frac{2\pi-\theta}{4\sin\frac{1}{2}\theta}, \qquad \text{for } \theta = 0\cdots 2\pi. \tag{B.4}$$

The last equation is well known, especially to electrical engineers.

References

[1] G. Mack and V.B. Petkova, Z_2 monopoles in the standard SU(2) lattice gauge theory model, DESY 79/22 (April, 1979), Z. Phys. C, to be published
[2] E. Tomboulis, Phys. Rev. D23 (1981) 2371

[3] I.G. Halliday and A. Schwimmer, Phys. Lett. 101B (1981) 327;
J. Greensite and B. Lautrup, Phys. Rev. Lett. 47 (1981) 9
[4] I.G. Halliday and A. Schwimmer, Phys. Lett. 102B (1981) 337
[5] R.C. Brower, D.A. Kessler, and H. Levine, Phys. Rev. Lett. 47 (1981) 621
[6] G. Bhanot and M. Creutz, preprint Brookhaven (May, 1981)
[7] L. Caneshi, I.G. Halliday and A. Schwimmer, Nucl. Phys. B200[FS4] (1982) 409
[8] S.B. Khochlachev and Yu.M. Makeenko, Moscow preprint ITEP-126 (1979); ZhETF (USSR) 80 (1981) 448
[9] T.T. Wu and C.N. Yang, Phys. Rev. D12 (1975) 3845;
S. Mandelstam, Phys. Rev. D19 (1979) 2391
[10] F. Englert, Electric and magnetic confinement schemes, Cargése lectures 1977;
F. Englert and P. Windey, Dynamical and topological considerations on quark confinement, Les Houches Lectures, 1978
[11] F. Wegner, J. Math. Phys. 12 (1971) 2259;
B. Balian, J.M. Drouffe and C. Itzykson, Phys. Rev. D10 (1974) 3376
[12] M. Creutz, L. Jacobs and C. Rebbi, Phys. Rev. Lett. 42 (1979) 1390; Phys. Rev. D20 (1979) 1915;
M. Creutz, Phys. Rev. Lett. 43 (1979) 553
[13] G. Mack and V.B. Petkova, Ann. of Phys. 123 (1979) 442
[14] B. Lautrup and M. Nauenberg, Phys. Rev. Lett. 45 (1980) 1755
[15] C.B. Lang, C. Rebbi, P. Salomonson and S.B. Skagerstam, Phys. Lett. 101B (1981) 173
[16] M. Nauenberg, T. Schalk and R. Brower, Phys. Rev. D24 (1981) 548
[17] M. Falcioni, E. Marinari, M.L. Paciello, G. Parisi and B. Taglienti, Phys. Lett. to be published; Nucl. Phys. B190[FS3] (1981) 782
[18] R.L. Dobrushin and S.B. Shlosman, Phases corresponding to the local energy minima, to be published;
R. Kotecký and S.B. Shlosman, First order transitions in large entropy lattice models, preprint IHES/P/81/37, Bures-sur-Yvette (July, 1981);
R. Kotecký, Comm. Math. Phys. 82 (1981) 391
[19] P. Cvitanović, J. Greensite and B. Lautrup, Phys. Lett. 105B (1981) 197
[20] K. Wilson, Phys. Rev. D10 (1974) 2445
[21] G. Mack, Properties of lattice gauge theory models at low temperature. DESY-report 80/03, *in* Recent developments in gauge theories, ed. G. 't Hooft et al. (Plenum Press, New York, 1980)
[22] T. Banks, R. Myerson and J. Kogut, Nucl. Phys. B129 (1977) 493;
J Villain, J. de Phys. 36 (1975) 581
[23] M. Göpfert and G. Mack, Comm. Math. Phys. 82 (1982) 545
[24] G. Mack, Phys. Rev. Lett. 45 (1980) 1378
[25] A.M. Polyakov, Phys. Lett. 72B (1978) 477;
L. Susskind, Phys. Rev. D20 (1979) 2610
[26] L. McLerran and B. Svetitsky, Phys. Lett. 98B (1981) 195; Phys. Rev. D24 (1981) 450;
L. Kuti, J. Polónyi and K. Szlachanyi, Phys. Lett. 98B (1981) 1991;
J. Engels, F. Karsch, H. Satz and I. Montvay, Phys. Lett. 101B (1981) 89; 102B (1981) 332
[27] M. Creutz, Phys. Rev. D21 (1980) 2309
[28] J. Glimm and A. Jaffe, Comm. Math. Phys. 56 (1977) 195; Phys. Lett. 66B (1977) 67
[29] G. 't Hooft, Nucl. Phys. B153 (1979) 141;
G. Mack and V.B. Petkova, Ann. of Phys. 125 (1980) 117
[30] G. Mack and E. Pietarinen, Phys. Lett. 94B (1980) 397
[31] G. Mack, Acta Phys. Austriaca Suppl. XXII (1980) 509
[32] H.B. Nielsen and P. Olesen, Nucl. Phys. B61 (1973) 45
[33] G. Mack, Fortschr. Phys. 29 (1981) 135
[34] M. Göpfert, Nucl. Phys. B190[FS3] (1981) 151
[35] J. Glimm, Cargése lectures 1979, *in* Recent developments in gauge theories, ed. G. 't Hooft et al. (Plenum Press, New York, 1980)
[36] N. Kimura, Prog. Theor. Phys. 64 (1980) 310;
J.M. Drouffe and J.B. Zuber, Nucl. Phys. B180[FS2] (1981) 253

[37] I.S. Gradshteyn and I.M. Ryzhik, Table of integrals, series and products (Academic Press, New York, 1965)
[38] L.P. Kadanoff, Physics 2 (1965) 263;
K. Wilson, Phys. Rev. D2 (1970) 1473
[39] Y. Iwasaki, The structure of the vacuum II – non-abelian gauge models, Tsukuba-University preprint UTHEP-82 (1981)
[40] G. Münster, Nucl. Phys. B180[FS2] (1981) 23
[41] N. Steenrod, The topology of fibre bundles, Princeton University Press (1951)
[42] H.C. Tze and Z.F. Ezawa, Phys. Rev. D14 (1976) 1006
[43] S. Samuel, Nucl. Phys. B154 (1979) 62
[44] C.P. Korthals Altes, private communication

Nuclear Physics B192 (1981) 392–408

PRELIMINARY EVIDENCE FOR $U_A(1)$ BREAKING IN QCD FROM LATTICE CALCULATIONS

P. DI VECCHIA and K. FABRICIUS

Physics Department, University of Wuppertal, Wuppertal, FRG

G.C. ROSSI

Istituto di Fisica dell'Università Rome and INFN, Sezione di Roma, Italy

G. VENEZIANO

CERN, Geneva, Switzerland

Received 15 June 1981

We suggest a simple definition of the topological charge density $Q(x)$ in the lattice Yang-Mills theory and evaluate $A \equiv \int d^4x \langle Q(x)Q(0)\rangle$ in SU(2) by Monte Carlo simulation. The "data" interpolate well between the strong and weak coupling expansions, which we compute to order g^{-12} and g^6, respectively. After subtraction of the perturbative tail, our points exhibit the expected asymptotic freedom behaviour giving $A^{1/4} \simeq (0.11 \pm 0.02)K^{1/2}$, K being the SU(2) quarkless string tension. Although a larger value for $A^{1/4}K^{-1/2}$ would be preferable, we are led to conclude (at least tentatively) that the $U_A(1)$ problem of QCD is indeed solved perturbatively in the quark loop expansion.

1. Introduction

In the last couple of years the idea [1] that a non-trivial topological charge in QCD can solve its U(1) problem [i.e., explicitly break the U(1) axial symmetry] has been made more precise and quantitative [2, 3] through the use of expansions of the $1/N$ type. The situation, which can be summarized in terms of a simple effective lagrangian [4], is as follows:

(i) In massless QCD, at the one-quark loop level (or, alternatively, at leading order in $1/N_{\text{colour}}$), a set of L^2 unmixed Goldstone bosons $\pi_{ij}(i, j = \text{u,d,s} \cdots)$ occurs as a result of the spontaneous breaking* of $U(L) \otimes U(L)$ chiral symmetry down to $U(L)_{\text{vector}}$ (L is the number of massless flavours).

* Several arguments for the necessity of such breaking have been given under different (reasonable) assumptions [5, 6].

(ii) The strong axial anomaly equation,

$$\partial_\mu J_{\mu 5}(x) \equiv \partial_\mu \sum_{i=1}^{L} \bar{\psi}_i \gamma_\mu \gamma_5 \psi_i = 2LQ(x), \tag{1.1}$$

$$Q(x) = \frac{g^2}{64\pi^2} \varepsilon_{\mu\nu\rho\sigma} F^a_{\mu\nu} F^a_{\rho\sigma}, \tag{1.2}$$

implies that, at the two-loop (OZI violating) level, the pseudoscalar mass matrix in the flavourless sector is non-zero. The matrix element between π_{ii} and π_{jj} is given by

$$\mathcal{M}^2_{ij} = 2F_\pi^{-2}A, \qquad i, j = 1 \cdots L, \tag{1.3}$$

with $F_\pi \simeq 95$ MeV the pion decay constant and, in euclidean space,

$$A = \int d^4x \langle 0|Q(x)Q(0)|0\rangle \Big|_{\text{no quarks}}. \tag{1.4}$$

In order to estimate A from the actual pseudoscalar spectrum, one has to add to M^2_{ij} the standard current algebra contribution coming from non-zero (but small) quarks masses. One thus obtains [3, 7] the relation (for $L = 3$):

$$\frac{2LA}{F_\pi^2} = \left(m^2_\eta + m^2_{\eta'} - 2m^2_K\right)_{\text{exp}} = 0.726\ \text{GeV}^2, \qquad A \simeq (180\ \text{MeV})^4. \tag{1.5}$$

The resolution of the $U_A(1)$ problem, as well as a successful description of the dynamics of the pseudoscalar nonet, is thus reduced to explaining why $A \neq 0$ in the quarkless theory*, although it vanishes to all orders in perturbation theory.

From general arguments [8], $Q(x)$ and A are expected to be renormalization group invariant, in agreement with their relationship with physical masses. This implies that the dependence of A upon the subtraction point μ and the coupling constant g is of the form

$$A = c\left[\mu \exp\left(\int_g dg'/\beta(g')\right)\right]^4 \approx c\left[\mu\left(-\beta_0 g^2\right)^{\beta_1/2\beta_0^2} \exp\left(\frac{1}{2\beta_0 g^2}\right)\right]^4, \tag{1.6}$$

where $\beta(g) \simeq \beta_0 g^3 + \beta_1 g^5 + O(g^7)$ and

$$\beta_0 = -\frac{11}{3}\frac{N}{16\pi^2}, \qquad \beta_1 = -\frac{34}{3}\left(\frac{N}{16\pi^2}\right)^2. \tag{1.7}$$

* It was argued [6] by one of us, (G.V.), that even the spontaneous breaking of chiral symmetry in QCD follows from $A \neq 0$.

in the quarkless SU(N) gauge theory. The dimensionless quantity c is what we would like to compute. At present the only techniques available for computing dimensional quantities like A are those provided by first regulating the theory on a lattice and then by performing either strong coupling expansions [9–11] or numerical Monte Carlo simulations [12, 13].

In this paper we present, for the case of SU(2), a first indication that the renormalization group behaviour (1.6) with a non-zero (positive) value of c emerges from lattice calculations. The result is obtained using a simple minded (but, we think, attractive) definition of $Q(x)$ on the lattice. The idea is, of course, that in the physical continuum limit ($a \to 0$), any definition with the correct naive $a \to 0$ limit should give the same result.

What is relevant is, as usual, the ratio of $A^{1/4}$ to another physical parameter with dimensions of a mass. Comparing our curves with those of Creutz [13], we extract

$$A^{1/4} = (0.11 \pm 0.02) K^{1/2}, \tag{1.8}$$

where K is the SU(2) quarkless string tension. If one takes for K the value $(2\pi\alpha')^{-1}$ of the dual string model with $\alpha' = 0.9\ \mathrm{GeV}^{-2}$, then one obtains

$$A^{\mathrm{SU(2)}} \simeq (45 \pm 10\ \mathrm{MeV})^4, \tag{1.9}$$

i.e., a rather small number compared to the desired value [eq. (1.5)]. On the other hand, assuming that $A^{1/4} K^{-1/2}_{\text{no quarks}}$ varies little from SU(2) to SU(3) (as suggested by $1/N$ expansion considerations) and using the known relation [11] between $K^{1/2}$ and Λ_{MOM} in SU(3), one would get

$$A^{\mathrm{SU(3)}} \simeq \left(0.26 \Lambda^{\mathrm{SU(3)}}_{\mathrm{MOM}}\right)^4 \tag{1.10}$$

which is much more acceptable. A discussion of this numerology will be given at the end of sect. 5.

The outline of the paper is as follows. In sect. 2 we discuss some possible definitions of $Q(x)$ in the lattice, showing that they all suffer from a non-zero perturbative expansion in g (even at zero momentum). In sect. 3 we compute the first non-trivial term in the perturbative expansion of A, adopting for $Q(x)$ a definition which appears most suitable for joining the strong coupling behaviour with the asymptotic freedom regime. In sect. 4 we describe our strong coupling calculations up to order $(1/g^2)^6$, improved by character expansion techniques [10]. In sect. 5 we present the output of our Monte Carlo simulation and we discuss the subtraction of the perturbative tail. We then show the emergence of the expected behaviour (1.6) in a region of the weak coupling regime and, finally, we extract and discuss the numerical relations (1.8) and (1.10). An outline of future investigations is given in sect. 6.

2. Definitions of $Q(x)$ on the lattice

A rather natural definition of $Q(x)$ on the lattice was given sometime ago by Peskin [15] on the basis of the well-known property [9] of the product of four link variables around one plaquette of side a:

$$U_{n,\mu\nu} \equiv U_{n,\mu} U_{n+\mu,\nu} U^{+}_{n+\nu,\mu} U^{+}_{n,\nu} \equiv \exp\left\{ i \sum_b \mathcal{F}^b_{\mu\nu,n} T^b \right\},$$

$$\mathcal{F}^b_{\mu\nu,n} \underset{a\to 0}{\to} g a^2 F^b_{\mu\nu}(x_n) + \mathrm{O}(a^3), \tag{2.1}$$

where T^b are the generators of the gauge group and $F^b_{\mu\nu}$ is the usual field strength tensor. In the same way as one obtains from (2.1) the Yang-Mills action as

$$\sum_{n,\mu\nu} \mathrm{Tr}(U_{n,\mu\nu} - 1) \underset{a\to 0}{\to} -\tfrac{1}{4} g^2 a^4 F^b_{\mu\nu} F^b_{\mu\nu} + \mathrm{O}(a^5), \tag{2.2}$$

one also finds

$$Q^{(\mathrm{P})}_n \equiv -\frac{\varepsilon_{\mu\nu\rho\sigma}}{32\pi^2} \mathrm{Tr}\left[U_{n,\mu\nu} U_{n,\rho\sigma}\right] \underset{a\to 0}{\to} a^4 Q(x_n) + \mathrm{O}(a^5). \tag{2.3}$$

We see that $Q^{(\mathrm{P})}_n$ is a sum of eight-link Wilson loops of the type shown in fig. 1. Moreover, because of the $\varepsilon_{\mu\nu\rho\sigma}$ tensor, $Q^{(\mathrm{P})}$ contains links in every one of the four space-time directions. On the other hand, this definition does not symmetrically treat the positive and negative directions of each axis and this prevents $Q^{(\mathrm{P})}$ from having definite parity. The problem is easily overcome by the symmetrized definition*

$$Q^{(\mathrm{S})}_n \equiv -\sum_{(\mu,\nu,\rho,\sigma)=\pm 1}^{\pm 4} \frac{\tilde{\varepsilon}_{\mu\nu\rho\sigma}}{2^4 32\pi^2} \mathrm{Tr}\left[U_{n,\mu\nu} U_{n,\rho\sigma}\right], \tag{2.4}$$

with $1 = \tilde{\varepsilon}_{1234} = -\tilde{\varepsilon}_{2134} = -\tilde{\varepsilon}_{-1234}$, etc., which has the same naive $a \to 0$ limit as $Q^{(\mathrm{P})}_n$. Eqs. (2.3) and (2.4) then imply that our quantity A is given by

$$A \underset{a\to 0}{\to} \frac{1}{a^4} \sum_n \langle 0 | Q^{(\mathrm{S})}_n Q^{(\mathrm{S})}_0 | 0 \rangle \equiv A^{(\mathrm{S})}_{\mathrm{L}} \tag{2.5}$$

in the naive continuum limit.

An alternative, more natural, definition of Q on the lattice can be obtained by first realizing that the Wilson loop of fig. 1 consists of two orthogonal plaquettes lying on the elementary hypercube of fig. 2. The latter contains eight three-

* We are grateful to P. Hasenfratz for very constructive remarks on this point.

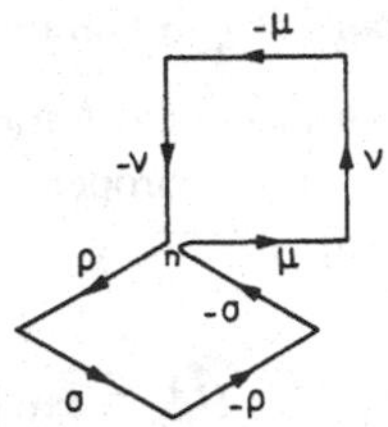

Fig. 1. Wilson loop appearing in Peskin's original definition [15] of the topological charge density.

dimensional cubes, 24 plaquettes, 32 links and 16 vertices, labelled in fig. 2, so that labels of opposite vertices differ by eight. A typical loop in the definition (2.5) is then (1,2,3,4,1,5,6,2,1) with the point 1 assuming a privileged role with respect to the other six points.

A definition of Q treating all the 16 vertices symmetrically is

$$Q_n = -\frac{1}{64\pi^2}\sum_{i=1}^{48} W_n(\mathrm{P}_i)(-1)^{\pi_i}, \tag{2.6}$$

where n is now a hypercube (not a site) label and $W_n(\mathrm{P}_i)$ is the Wilson loop operator for the ith closed path P_i on the nth hypercube. A generic P_i starts from one of the 16 vertices, goes in four steps to the opposite vertex in any possible way and comes back in the same (now unique) order. There are $4! \times 16/8 = 48$ inequivalent oriented paths. Examples of P_i are $\mathrm{P}_0 = (1,2,3,7,9,10,11,15)$, $\mathrm{P}_1 = (1,2,6,7,9,10,14,15)$, etc. The 48 paths will be classified better in section 4. Obviously, all the eight vertices of P_i are treated symmetrically and the same is true for positive and negative directions along each axis. Hence the definition (2.6) has automatically the correct parity property provided the signs $(-1)^{\pi_i}$ are defined in agreement with the $\tilde{\varepsilon}$ symbol of eq.

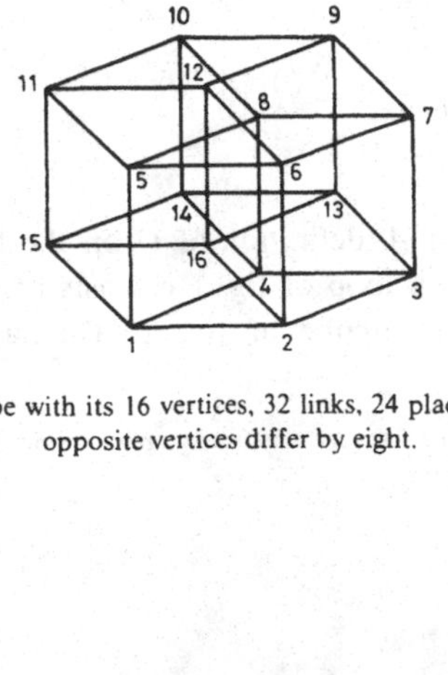

Fig. 2. The elementary hypercube with its 16 vertices, 32 links, 24 plaquettes and eight cubes. Labels of opposite vertices differ by eight.

(2.5) (+1 for P_0, −1 for P_1, etc). An alternative way to write (2.6) is then

$$Q_n = -\frac{1}{2^4 32\pi^2} \sum_{(\mu,\nu,\rho,\sigma)=\pm 1}^{\pm 4}$$

$$\times \tilde{\varepsilon}_{\mu\nu\rho\sigma} \mathrm{Tr}\left[U_{n,\mu} U_{n+\mu,\nu} U_{n+\mu+\nu,\rho} U_{n+\mu+\nu+\rho,\sigma} U^{+}_{n+\nu+\rho+\sigma,\mu} U^{+}_{n+\rho+\sigma,\nu} U^{+}_{n+\sigma,\rho} U^{+}_{n,\sigma}\right]. \tag{2.7}$$

With either definition one still has

$$A \underset{a\to 0}{\to} \frac{1}{a^4} \sum_n \langle 0|Q_n Q_0|0\rangle \equiv A_L. \tag{2.8}$$

We have found the second definition [(2.6) or (2.7)] to be preferable to (2.5) on several accounts. (i) The strong coupling expansion of $A_L^{(S)}$ is more complicated than the one of A_L because it contains, from the start, plaquettes in the definition of $Q_n^{(S)}$. (ii) While the strong coupling expansion of A_L starts alternating in sign (see sect. 4) and shows a clear tendency to bend down at values of $\beta = 4/g^2$ around 2, $A_L^{(S)}$ starts from a lower value at $\beta = 0$ and appears to be flattish at small β. This presumably means the need for many terms in the strong coupling series before the expected turnover begins to appear. (iii) Monte Carlo calculations are more time consuming with the first definition because of its smaller symmetry. For these reasons we stick to the second definition (2.6) in this first exploratory study.

A common feature of the lattice versions of local operators is that their expectation values are non-zero in perturbation theory. A typical case [16] is the expectation value of the plaquette which, in the naive continuum limit, is related to the expectation value of $g^2 F^a_{\mu\nu} F^a_{\mu\nu}$ [see eq. (2.2)]. This latter quantity is set to zero in continuum perturbation theory as any other physical dimensional quantity, while one finds [16] that the perturbative ($\beta \to \infty$) tail of the plaquette is quite sizeable with respect to the physically relevant non-perturbative part even near the turnover point ($\beta \sim 2$). One could have hoped that this problem would not arise for the operator $\int d^4x Q(x)$ which vanishes automatically in perturbation theory, $Q(x)$ being a total divergence. This is not the case, however, for either one of our definitions of Q_L, since, in a small g expansion of Q_L, operators of arbitrarily high dimension which are not total divergences do occur. Although they appear in the expansion multiplied by correspondingly high powers of a, their matrix elements diverge for $a \to 0$ precisely as required by naive dimensional counting so that their perturbative contributions are all of the same order in a.

As in ref. [16], the attitude that we shall take here is that, for each operator of dimension higher than 4, once the perturbative contribution has been subtracted out, the remainder should behave as $\exp(-\gamma/g^2)$ as $g^2 \to 0$ with a coefficient γ larger than that of the (lowest dimension) operator which survives in the naive $a \to 0$ limit. Unless this is the case, the possibility of extracting renormalization group invariant quantities such as A or $\langle g^2F^2\rangle$ from lattice calculations appears to be hopeless. In this respect, the results obtained for $\langle g^2F^2\rangle$ in refs. [16, 17] are certainly encouraging.

3. The weak coupling expansion of A_{L}

We have already stated that the weak coupling expansion of A_{L}, as given by eq. (2.8), is non-vanishing. In order to see how this happens, consider, for instance, the weak coupling expansion* of the quantity $Q_n^{(\mathrm{P})}$ defined in eq. (2.3). Defining as usual

$$U_{n,\mu}\equiv\exp\left(igaA^b_{n,\mu}T^b\right), \tag{3.1}$$

one finds

$$Q_n^{(\mathrm{P})}=\frac{a^4g^2}{64\pi^2}\varepsilon_{\mu\nu\rho\sigma}\Delta_\mu A^b_{n,\nu}\Delta_\rho A^b_{n,\sigma}+\mathrm{O}(g^3), \tag{3.2}$$

where Δ_μ is the finite difference operator

$$\Delta_\mu f_n=(f_{n+\mu}-f_n)/a. \tag{3.3}$$

Going over to momentum space, one obviously has:

$$\Delta_\mu\underset{\mathrm{FT}}{\rightarrow}(\mathrm{e}^{-ip_\mu a}-1)/a\underset{a\to 0}{\rightarrow}-ip_\mu+\mathrm{O}\left(ap_\mu^2\right). \tag{3.4}$$

One can then check that, to this order, $Q_n^{(\mathrm{P})}$ acts as a two-gluon vertex of the form

$$\frac{Q^{\mathrm{P}}(k)}{a^4}\sim g^2\varepsilon_{\mu\nu\rho\sigma}\int_{-\pi/a}^{\pi/a}\mathrm{d}^4p\,\mathrm{d}^4q\,\delta^{(4)}(k+p+q)\frac{(\mathrm{e}^{-ip_\mu a}-1)(\mathrm{e}^{-iq_\rho a}-1)}{a^2}$$
$$\times A^c_\nu(-p)A^c_\sigma(-q). \tag{3.5}$$

Clearly this expression is not vanishing at $k=0$ ($p=-q$) except in the limit $a\to 0$. Inserting the expression (3.5) in the definition of $A_{\mathrm{L}}^{(\mathrm{P})}$, eq. (2.4), one immediately finds a non-zero contribution to this quantity at order g^4. One checks, however, that with the symmetrized expression (2.6) the two-gluon vertex itself goes to zero at $k=0$. Unfortunately, this result does not persist to higher orders in g. In this respect, the definition (2.6) of Q_n, which we adopt in the following, behaves in the same way as the definition (2.4). A straightforward, but lengthy and tedious calculation gives the following result for the first non-trivial contribution to A_{L} in

* We are very grateful to P. Hasenfratz for having initiated us to the techniques of weak coupling lattice calculations.

the SU(2) case:

$$\pi^4 2^{18} a^4 A_{\mathrm{L}} = C_3 \beta^{-3} + \mathrm{O}(\beta^{-4}) = C_3(\tfrac{1}{4}g^2)^3 + \mathrm{O}(g^8), \tag{3.6a}$$

$$C_3 = 3 \cdot 2^9 \int_{-\pi}^{+\pi} \frac{\mathrm{d}^4 k}{(2\pi)^4} \int_{-\pi}^{+\pi} \frac{\mathrm{d}^4 p}{(2\pi)^4} \int_{-\pi}^{+\pi} \mathrm{d}^4 q\, \delta^{(4)}(k+p+q) \sum_{\mu\nu\rho\sigma} \varepsilon_{\mu\nu\rho\sigma}\varepsilon_{\mu\nu\rho\sigma}$$

$$\times \left[\sum_\alpha (1-\cos k_\alpha) \cdot \sum_\beta (1-\cos q_\beta) \cdot \sum_\alpha (1-\cos p_\gamma) \right]^{-1}$$

$$\times \Big\{ \tfrac{1}{3} \big| - C_\mu (\mathrm{e}^{-ik_\sigma} + \mathrm{e}^{-iq_\sigma})(\mathrm{e}^{-iq_\nu} + \mathrm{e}^{-ip_\nu})(\mathrm{e}^{-ik_\rho} + \mathrm{e}^{-ip_\rho})$$

$$+ \sin k_\mu (1+\mathrm{e}^{ik_\nu}) \big[(1+\mathrm{e}^{ip_\sigma})(\mathrm{e}^{-ik_\rho} + \mathrm{e}^{-ip_\rho}) + (1+\mathrm{e}^{iq_\rho})(\mathrm{e}^{-iq_\sigma} + \mathrm{e}^{-ik_\sigma}) \big]$$

$$+ \sin q_\mu (1+\mathrm{e}^{iq_\rho}) \big[(1+\mathrm{e}^{ik_\nu})(\mathrm{e}^{-ik_\sigma} + \mathrm{e}^{-iq_\sigma}) + (1+\mathrm{e}^{ip_\sigma})(\mathrm{e}^{-ip_\nu} + \mathrm{e}^{-iq_\nu}) \big]$$

$$+ \sin p_\mu (1+\mathrm{e}^{ip_\sigma}) \big[(1+\mathrm{e}^{iq_\rho})(\mathrm{e}^{-iq_\nu} + \mathrm{e}^{-ip_\nu}) + (1+\mathrm{e}^{ik_\nu})(\mathrm{e}^{-ik_\rho} + \mathrm{e}^{-ip_\rho}) \big] \big|^2$$

$$+ 16(1-\cos k_\sigma)\big[1+\cos(q_\nu - p_\nu)\big] \sin p_\mu \sin q_\rho$$

$$\times \big[\sin p_\mu \sin q_\rho - \sin p_\rho \sin q_\mu \big] \Big\}, \tag{3.6b}$$

where

$$C_\mu = \sin k_\mu + \sin p_\mu + \sin q_\mu . \tag{3.7}$$

The length of eq. (3.6) is a clear indication that going beyond this order may prove to be an impossible task. In table 1 we give the numerical result for C_3 for the case of a finite lattice of sizes 3^4 to 5^4 as well as for the case of an infinite lattice. Notice that

TABLE 1
Values of C_3 [eq. (3.6)] for increasing lattice size

	C_3
3^4	12539.6
4^4	14321.4
5^4	14760.7
∞^4	15034 ± 92

The integral in the last line has been computed by a Monte Carlo method using $1.6 \cdot 10^6$ points.

the finite lattice results do not differ by more than 15% from that of the infinite lattice. For the case of SU(N), the values of table 1 have to be multiplied by the factor $\frac{1}{6}N(N^2-1)$. As already discussed in sect. 2, we shall use this result to make a subtraction on our Monte Carlo data for A_L (see sect. 5).

4. Strong coupling expansion of A_L

Following standard techniques [9–11], we have performed a strong coupling expansion of A_L, as defined by eqs. (2.6) and (2.8) up to order g^{-12}. The lowest order terms in $1/g^2$ come from picking up the term with $n=0$ in the sum over n of eq. (2.8) so that the paths appearing in Q_0 and Q_n through eq. (2.6) lie on the *same* hypercube. Since we can arbitrarily choose the path in Q_0 to be $P_0 = (1,2,3,7,9,10,11,15)$ (and gain a factor 48), we shall have

$$a^4A_L = 48\left(\frac{1}{64\pi^2}\right)^2 \sum_{i=1}^{48} (-1)^{\pi_0+\pi_i}\langle W_0(P_i)W_0(P_0)\rangle + (n\neq 0). \tag{4.1}$$

Evaluation of (4.1) can be done by first classifying each of the 48 paths P_i with respect to P_0. One finds that the 24 distinct (unoriented) paths can be subdivided as follows:

(i) P_0 itself;

(ii) four paths like (1,2,6,7,9,10,14,15) which share with P_0 six vertices and four links;

(iii) eight paths like (1,2,6,12,9, etc.) and

(iv) four paths like (1,2,16,12,9, etc.) both sharing with P_0 four vertices and two links;

(v) two paths like (1,4,14,13,9, etc.) with four vertices in common with P_0;

(vi) four paths like (1,4,14,13,9, etc.) with just two vertices in common with P_0 and finally

(vii) one path with (4,8,5,6,12,16,13,14) with no vertex in common with P_0.

Each one of the above 24 paths can be taken in either direction. In the SU(2) case the orientation of the path does not matter and from now on we shall restrict ourselves to this case.

The systematics of the strong coupling expansion is made easier by the use of the character expansion technique [10]. For SU(2), and up to the order β^6 at which we work, it is enough to use the formula

$$\exp\left(\tfrac{1}{4}\beta \operatorname{Tr}(U+U^+)\right) = 1 + \tfrac{1}{4}\bar{\beta}\operatorname{Tr}(U+U^+) + \cdots, \tag{4.2}$$

where $\frac{1}{4}\bar{\beta}$ is the coefficient of the fundamental representation (spin $\frac{1}{2}$) character and the dots indicate characters in higher spin representations of SU(2). One finds

$$\bar{\beta}=4I_2(\beta)/I_1(\beta)=\beta(1-\tfrac{1}{24}\beta^2+\tfrac{1}{384}\beta^4+\cdots), \tag{4.3}$$

with $I_n(\beta)$ the usual modified Bessel function. We have checked that, to $O(\beta^6)$, higher spin characters do not contribute to A_L. Also, as far as we could check, the contributions $O(\beta^6)$ coming from terms with $n\neq 0$ in eq. (2.8) (surprisingly) cancel out leaving us with the evaluation of (4.1). Now, for P_i in each of the seven classes described above, we have different strong coupling contributions. The leading contribution as $\beta\to 0$ comes from $P_i=P_0$. Since the group integrals (contractions) give just one, we get

$$A_L \underset{\beta\to 0}{\to} \frac{1}{a^4}(48)\left(\frac{1}{64\pi^2}\right)^2\cdot 2=\frac{3}{2^7\pi^4}\frac{1}{a^4}\equiv A_L^\infty, \tag{4.4}$$

where the factor of two comes from the two orientations of P_0 and would be absent for SU(N), $N>2$. In a similar way, since $\mathrm{Tr}\,U=\mathrm{Tr}\,U^+$ for SU(2), the effective coefficient of the spin $\frac{1}{2}$ character in eq. (4.2) is $\frac{1}{2}\bar{\beta}$ rather than $\frac{1}{4}\bar{\beta}$. At order $\bar{\beta}^2$, we get contributions from P_i in class (ii) (described above) if we use twice the action expanded as in (4.2). Taking into account that there are four paths in this class, that the relative sign of P_i and P_0 is -1 and that the group integrals give $(\frac{1}{2})^2$, one gets a contribution

$$A_L^\infty\times 4\times(-1)\times(\tfrac{1}{2}\bar{\beta})^2\times(\tfrac{1}{2})^2=-\tfrac{1}{4}\bar{\beta}^2A_L^\infty. \tag{4.5}$$

In a similar way, the ten paths of classes (iii) and (v) give

$$A_L^\infty\times 10\times(+1)\times(\tfrac{1}{2}\bar{\beta})^4\times(\tfrac{1}{2})^4=\frac{5}{2^7}\bar{\beta}^4A_L^\infty. \tag{4.6}$$

For the contribution of order $\bar{\beta}^6$ the situation starts to get complicated. The paths in classes (iv) and (vi) are found to give a contribution

$$A_L^\infty\times 28\times(-1)\times(\tfrac{1}{2}\bar{\beta})^6\times(\tfrac{1}{2})^6=-\frac{7}{2^{10}}\bar{\beta}^6A_L^\infty. \tag{4.7}$$

There are, however, two sorts of new contributions to this order. One stems from the fact that previously found contributions of $O(\bar{\beta}^2)$ can be transformed into contributions of $O(\bar{\beta}^6)$ by replacing a plaquette with five others forming a cube together with the removed original one. Since there are four ways of building such a cube and the extra weight to pay for a cube is $(\frac{1}{4}\bar{\beta})^4$ one can effectively take this "plaquette excitation" contribution into account by the replacement

$$\bar{\beta}\to\bar{\beta}\,(1+\tfrac{1}{64}\bar{\beta}^4). \tag{4.8}$$

There is, however, a related effect due to the fact that one has to divide the strong coupling diagrams for $Q_n Q_0$ by the vacuum diagrams. The first non-trivial vacuum diagram is again a cube and gives

$$\langle 0|0\rangle = 1 + \sum_{\text{cubes}} \bar{\beta}^6/4^5 + \cdots . \tag{4.9}$$

Usually vacuum diagrams cancel against disconnected diagrams in the numerator. This does not happen however when the cube has one face in common with a preexisting plaquette of the connected diagrams. The effect (the so-called excluded volume effect) can be taken approximately into account by a further replacement, leading us to define the new strong coupling expansion variable:

$$\tilde{\beta} \equiv \bar{\beta}\left(1 + \tfrac{1}{64}\bar{\beta}^4(1 - \tfrac{1}{4}\bar{\beta}^2)\right). \tag{4.10}$$

In terms of $\tilde{\beta}$, collecting contributions (4.4) to (4.7) we get

$$A_L = A_L^\infty\left(1 - \tfrac{1}{4}\tilde{\beta}^2 + \frac{5}{2^7}\tilde{\beta}^4 - \frac{7}{2^{10}}\tilde{\beta}^6\right) + O(\tilde{\beta}^8), \qquad A_L^\infty \equiv \frac{1}{a^4}\frac{3}{2^7\pi^4}. \tag{4.11}$$

This is our final result to order β^6. It includes automatically, through the use of the new variable $\tilde{\beta}$, *some* higher order effects. In any case, the numerical results would not be much different if we had expressed everything in terms of β and had truncated the series at $O(\beta^6)$. We should point out however that, due to its complicated nature, the calculation of the $O(\beta^6)$ contribution should be rechecked by different techniques. We also feel that going systematically to higher orders already calls for the development and use of a computer programme.

Notice that the terms we computed alternate in sign and that all terms become roughly of the same order for $\beta \simeq 2$, i.e., around the point at which the famous turnover of the string tension occurs.

In the next sections the above strong coupling calculations will be compared with the Monte Carlo data.

5. Monte Carlo simulation and evaluation of A

Existing Monte Carlo techniques [12, 13] can be readily used for a numerical evaluation of A_L on a finite lattice. We have used the heat bath method on a 4^4 lattice with periodic boundary conditions (p.b.c.). The quantity actually computed is

$$a^4 A_L^{MC} = \frac{1}{V}\left\langle\left(\sum_{n\in V} Q_n\right)^2\right\rangle, \qquad V = 4^4, \tag{5.1}$$

which is equal to $\langle(\Sigma_{n\in V} Q_n)Q_0\rangle$ thanks to the discrete translational invariance of the periodic lattice. The use of p.b.c. on the link variables $U_{n,\mu}$ immediately raises the question of whether one can get at all a non-zero result for the volume integral of a quantity which, in the naive continuum limit, approaches a total divergence*. Upon integration by parts, opposite boundary conditions would seem to cancel against each other thanks to periodicity. The point is, however, that even for $a \to 0$, Q_n can only be written as a finite difference of a function of the $A_{n,\mu}$ variables defined via eq. (3.1). Clearly, $A_{n,\mu}$ is not uniquely determined by $U_{n,\mu}$ and periodicity in the U's does *not* imply periodicity in the A's. If we want to write the $a \to 0$ limit of Q_n as a derivative, we have to impose some sort of continuity on the A's [e.g., $ga(A_{n,\mu} - A_{n+\nu,\mu}) \ll 1$] or else have a singularity in Q_n. But then it is quite possible that p.b.c. in the U's plus continuity in the A's imply some non-periodicity of the A's and a non-zero value of $\Sigma_n Q_n$**. The above conclusion fits well with the intuitive expectation that the correlation function $\langle Q_n Q \rangle$ should vanish exponentially in the separation between the sites n and m. The characteristic scale is given by the inverse mass of the lightest pseudoscalar glueball: as long as this is substantially smaller than the size of our lattice, boundary effects on the integrated correlation $\langle \Sigma_n Q_n Q_0 \rangle$ should be negligible. The finiteness of our lattice then puts a lower limit on the value of g at which one can work. The hope is, of course, that the sought for asymptotic freedom behaviour sets in before this critical value of g is reached.

We took Monte Carlo data for a few strong coupling points (mainly for a check) and then mainly for values of β larger than 2. We ran up to 6000 sweeps for fixed β using, as link variables, elements of the full SU(2) group. We made hot and cold starts and checked that equilibrium was reached after a small number of sweeps. Nevertheless, fluctuations were rather large as compared, for instance, with those of the plaquette energy. In fig. 3, the data $\pi^4 2^{18} a^4 A_L^{MC}$ are presented as a function of $\beta = 4/g^2$ with their statistical errors. Strong and weak coupling curves are also given at various levels of approximation, i.e., to $O(\tilde\beta^2)$, $O(\tilde\beta^4)$ and $O(\tilde\beta^6)$ on the strong coupling side [eq. (4.11)] and to $O(\beta^{-3})$ on the weak coupling side. The agreement at small β, which persists up to $\beta \approx 1.5$–2, indicates the correct performance of our Monte Carlo programme. At the same time, the calculated $C_3\beta^{-3}$ term is seen to account for the "measured" values for $\beta \gtrsim 2.5$–3 (see, in particular, the point at $\beta = 4$).

The difference between Monte Carlo and the computed β^{-3} term is shown in fig. 4. We notice that the difference is small, systematically positive and decreasing at large β. A fit to a C_4/β^4 term in the region $\beta \gtrsim 2.8$ turns out to be possible with the result:

$$a^4 A_L^{MC} \underset{\beta \geq 2.8}{\approx} 5.608 \cdot 10^{-4} \beta^{-3} + 2.14 \cdot 10^{-4} \beta^{-4} \equiv A_L^{pert} . \qquad (5.2)$$

* Questions of this type were first raised in ref. [15].

** We are grateful to C. Rebbi and M.A. Virasoro for very useful discussions on this point and for communicating to us some unpublished work of theirs in two-dimensional compact QED, where they have found the above phenomenon to occur.

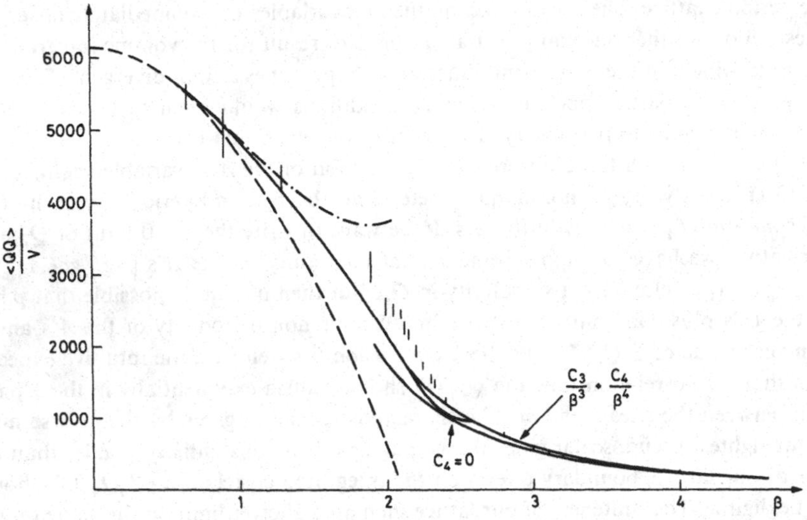

Fig. 3. Monte Carlo data for $\pi^4 2^{18} a^4 A_L$ with their statistical error bars are shown together with a partly computed, partly fitted, "perturbative tail". Strong coupling curves at order $\bar{\beta}^2$ (dashed line), $\bar{\beta}^4$ (dash-dotted line) and $\bar{\beta}^6$ (solid line) are also shown.

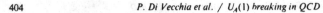

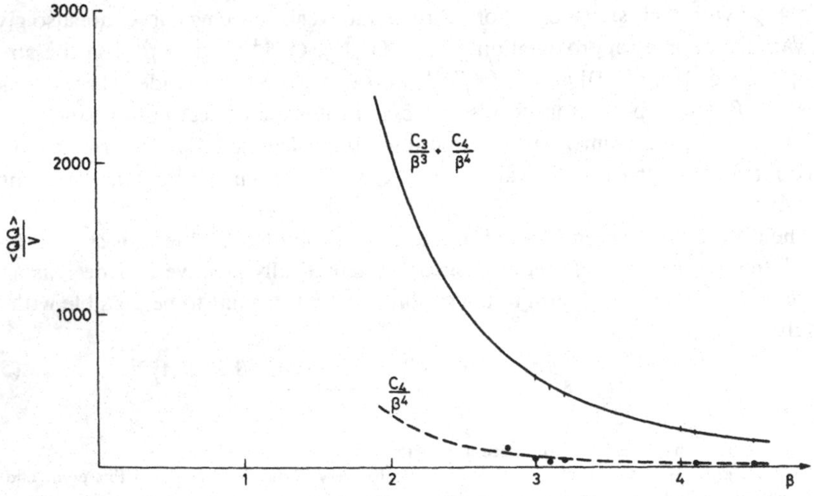

Fig. 4. Computed $C_3\beta^{-3}$ perturbative term. Fitted $C_4\beta^{-4}$ perturbative term.

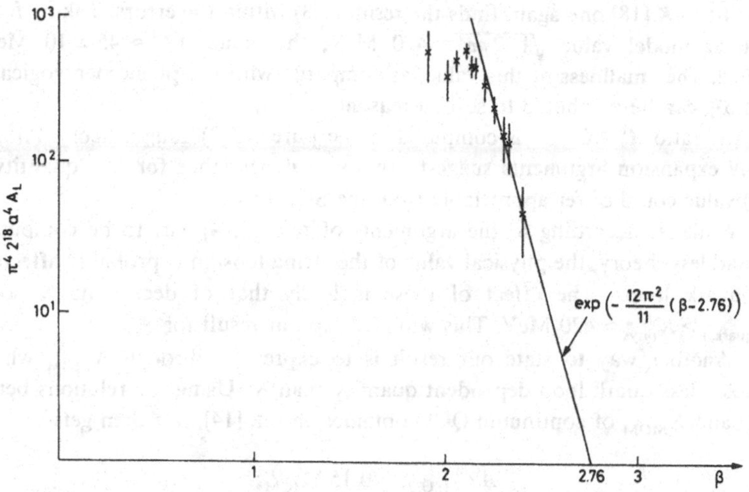

Fig. 5. Subtracted Monte Carlo data in the region $2.2 \leq \beta \leq 2.4$ and fit to the asymptotic freedom straight line $-\frac{12}{11}\pi^2(\beta-2.76)$.

Here, encouraged by the fact that the fit is indeed a good one and that the β^{-4} correction is small, we have assumed higher order perturbative corrections [$O(\beta^{-5})$, etc.] to be negligible in the region $2 \leq \beta \leq 3$, in which we are interested.

In fig. 5 we show, on a logarithmic scale, the difference between $a^4 A_L^{MC}$ and $a^4 A_L^{pert}$ [as defined by (5.2)] in the interval $2 \leq \beta \leq 2.8$. Amazingly, the points do lie on a straight line with the expected slope [eq. (1.6) is the one loop approximation $\beta_1 = 0$] giving

$$A^{\text{non-pert}} \equiv \left(A_L^{MC} - A_L^{pert}\right)$$

$$= a^{-4}\exp\left(-\frac{12\pi^2}{11}(\beta-\beta_A)\right), \quad \beta_A = 1.18 \pm 0.03, \quad (5.3)$$

allowing us to estimate the continuum limit value of A.

Eq. (5.3) can indeed be compared with the result of ref. [13] for the SU(2) string tension K:

$$K = a^{-2}\exp\left(-\frac{6\pi^2}{11}(\beta-\beta_K)\right), \quad \beta_K = 2.00 \pm 0.05. \quad (5.4)$$

Eqs. (5.3) and (5.4) give $A^{1/4} \approx (0.11 \pm 0.02)K^{1/2}$, the result anticipated in eq. (1.8).

Another fit to the subtracted data is possible (and even slightly better) with the two-loop asymptotic freedom formulae, eq. (1.6). Comparing that exercise to a similar fit to K [18] one again finds the result (1.8) within the errors. Taking for $\sqrt{K}$ the string model value $\sqrt{1/2\pi\alpha'} = 420$ MeV, the value $A^{1/4} \approx 45 \pm 10$ MeV is obtained. The smallness of this value, as compared with the phenomenological one [eq. (1.5)], can be attributed to several reasons:

(i) The ratio $A^{1/4}K^{-1/2}$ is computed in the pure SU(2) gauge theory. Although large-N expansion arguments suggest a weak N dependence for this quantity, the SU(3) value could differ appreciably from the SU(2) value.

(ii) While A, according to the arguments of refs. [3, 4], *has* to be computed in the quarkless theory, the physical value of the string tension is probably affected by light quark loops. The effect of these is likely that of decreasing K so that $K^{1/2}_{(\text{no quarks})} > K^{1/2}_{\text{phys}} \simeq 420$ MeV. This will push up our result for A.

(iii) Another way to state our result is to express A through Λ_{MOM} which is perhaps a less quark loop dependent quantity than K. Using the relations between Λ_{lattice} and Λ_{MOM} of continuum QCD obtained in ref. [14], one then gets

$$A^{1/4}|_{\text{SU(2)}} \simeq 0.15 \Lambda^{\text{SU(2)}}_{\text{MOM}}; \tag{5.5}$$

hence, $\Lambda^{\text{SU(2)}}_{\text{MOM}} \simeq 1.2$ GeV for $A^{1/4} = 180$ MeV.

(iv) A more favourable result is obtained by assuming $A/K^2_{\text{no q}}$ to be the same in SU(2) and SU(3). Using again ref. [14] one then gets

$$A^{1/4}|_{\text{SU(3)}} \simeq 0.26 \Lambda^{\text{SU(3)}}_{\text{MOM}}, \tag{5.6}$$

i.e., a good value of A for 500 MeV $< \Lambda_{\text{MOM}} <$ 900 MeV, which we can regard as very reasonable. But, of course, with $\Lambda_{\text{MOM}} = 700$ MeV one also predicts

$$K^{1/2}|_{\substack{\text{SU(3)}\\ \text{no quarks}}} \simeq 1.6 \text{ GeV}, \tag{5.7}$$

and light quarks have to intervene drastically to reduce $K^{1/2}$ to the "observed" string model value of 420 MeV.

(v) Another possible indication for a "large" value of $K^{1/2}|_{\text{no quarks}}$ is offered by a recent calculation of the mass of the lightest scalar glueball by Münster [19] giving

$$m_{\text{gl}} = (1.8 \pm 0.8) K^{1/2} \tag{5.8}$$

Since we might expect glueballs to be relatively heavy (say around 1.5 to 2 GeV) a larger value of $K^{1/2}$ (~ 1 GeV) would be preferred.

(vi) Finally, the recent calculations of $\langle g^2F^2\rangle|_{\text{no quarks}}$ fit the phenomenological value if a smaller value of Λ_{MOM} (~ 300 MeV) is used in apparent disagreement with our case. The discrepancy can be explained if one assumes, according to some theoretical arguments given in ref. [20], that $\langle g^2F^2\rangle|_{\text{no quarks}}$ is quite larger than the physical value in the presence of two (or three) light quarks.

In conclusion, although we would have preferred to find a larger value for A/K^2, we cannot say that we are, as yet, in contradiction with other results and/or with experiments also bearing in mind that several non-trivial current algebra extrapolations have been used in refs. [3, 4].

6. Outlook

Although we consider the results presented in this paper to be quite encouraging, much work is certainly still needed before one can claim that the non-perturbative value of A is different from zero and can reliably compute its magnitude.

Some of our limitations are of a theoretical nature, in particular our inability to define a Q_L free of perturbative contributions*. Others come from the limited statistics of the Monte Carlo data we have been able to collect so far.

For the future we plan to utilize faster programmes, possibly based on finite subgroups of SU(2) [22], in order to improve the statistics of our data and to extend our calculations. In particular we would like to explore:

(i) Effects of finite volume and/or of boundary conditions.

(ii) Dependence of the value of A from the lattice definition employed. In particular we plan to check whether the expected asymptotic freedom behaviour also emerges using the definition (2.5). Preliminary results seem to indicate that this is actually the case [23].

We are very much indebted to P. Hasenfratz for his important contributions during the early stages of this work. We also thank A. Di Giacomo, C. Lang and, in particular, C. Rebbi and M. Virasoro for very useful discussions.

Note added in proof

After completion of this work we received a paper by M. Lüscher (Bern preprint BUTP-10/1981) in which a new definition of the topological charge density on a four-dimensional lattice is given. Such a definition leads to a quantized topological charge and therefore does not suffer from perturbative contributions. We are presently considering its possible use.

References

[1] G. 't Hooft, Phys. Rev. Lett. 37 (1976) 8; Phys. Rev. D14 (1976) 3432
[2] E. Witten, Nucl. Phys. B156 (1979) 269
[3] G. Veneziano, Nucl. Phys. B159 (1979) 213

* A definition of topological charge which is free of perturbative contributions has been recently given for some two-dimensional models [21].

[4] P. Di Vecchia, Phys. Lett. 85B (1979) 357;
C. Rosenzweig, J. Schechter and G. Trahern, Phys. Rev. D21 (1980) 3388;
P. Di Vecchia and G. Veneziano, Nucl. Phys. B171 (1980) 253;
E. Witten, Ann. of Phys. 128 (1980) 363;
P. Di Vecchia, *in* Field theory and strong interactions, ed. P. Urban. (Springer-Verlag)
[5] A. Casher, Tel-Aviv University preprint, TAUP 734/79 (1979);
J.M. Cornwall, UCLA preprint TEP/9 (1980);
S. Drell, H. Quinn and M. Weinstein, SLAC preprint (1980);
G. 't Hooft, Lecture at 1979 Cargèse Summer Inst.;
S. Coleman and E. Witten, Phys. Rev. Lett. 45 (1980) 100
[6] G. Veneziano, Phys. Lett. 95B (1980) 90; CERN preprint TH-3042 (1981)
[7] P. Di Vecchia, F. Nicodemi, R. Pettorino and G. Veneziano, Nucl. Phys. B181 (1981) 318;
K. Kawarabayashi and N. Ohta, University of Tokyo preprint (May, 1980);
D.I. Dyakonov and M.J. Eides, Leningrad Nucl. Phys. Inst. preprint 639 (1981)
[8] R.J. Crewther, Riv. Nuovo Cim. 2 (1979) 63
[9] K. Wilson, Phys. Rev. D10 (1976) 2445
[10] P. Balian, J.M. Drouffe and C. Itzykson, Phys. Rev. D10 (1974) 3376; D11 (1975) 2098, 2104
[11] J.B. Kogut, R.B. Pearson and J. Shigemitsu, Phys. Rev. Lett. 43 (1979) 484
[12] K. Wilson, Lecture at 1979 Cargèse Summer Inst.;
M. Creutz, L. Jacobs and C. Rebbi, Phys. Rev. Lett. 42 (1979) 1390; Phys. Rev. D20 (1979) 1915
[13] M. Creutz, Phys. Rev. Lett. 43 (1979) 553; Phys. Rev. D21 (1980) 2308
[14] A. Hasenfratz and P. Hasenfratz, Phys. Lett. 93B (1980) 165
[15] M. Peskin, Cornell University preprint CLNS 395 (1978), Thesis
[16] A. Di Giacomo and G.C. Rossi, Phys. Lett. 100B (1981) 481;
T. Banks, R. Horsley, H.R. Rubinstein and U. Wolff, Nucl. Phys. B190[FS3] (1981) 692
[17] J. Kripfganz, Phys. Lett. 101B (1981) 169
[18] M. Creutz, Talk at Symp. on Topical questions in QCD, Niels Bohr Institute and Nordita, Copenhagen, June 1980; Brookhaven National Laboratory preprint BNL 27995 (1980)
[19] G. Münster, Nucl. Phys. B190[FS3] (1981) 439
[20] A.J. Vainshtein, V.I. Zakharov, V.A. Novikov and M.A. Shifman, ITEP preprint 87 (1980)
[21] B. Berg and M. Lüscher, Nucl. Phys. B190 (1981) 412;
M. Lüscher, Bern preprint BUTP-4/1981;
G. Martinelli, R. Petronzio and M.A. Virasoro, CERN preprint TH-3076 (1981)
[22] C. Rebbi, Phys. Rev. D21 (1980) 3350;
G. Bhanot and C. Rebbi, Nucl. Phys. B180[FS2] (1981) 469
[23] P. Di Vecchia, K. Fabricius, G.C. Rossi and G. Veneziano, in preparation

Volume 108B, number 4,5 PHYSICS LETTERS 28 January 1982

NUMERICAL CHECKS OF THE LATTICE DEFINITION INDEPENDENCE OF TOPOLOGICAL CHARGE FLUCTUATIONS

P. Di VECCHIA and K. FABRICIUS
Physics Department, University of Wuppertal, Wuppertal, Fed. Rep. Germany

G.C. ROSSI
Instituto di Fisica dell'Università, and INFN, Sezione di Roma, Rome, Italy

and

G. VENEZIANO
CERN, Geneva, Switzerland

Received 5 November 1981

The stability of our previous results on the topological charge fluctuation in SU(2) is checked against some variations of its lattice definition. The method consists of subtracting a partly computed perturbative tail to Monte Carlo data, whose high statistics are achieved by use of the icosahedral subgroup of SU(2).

In a previous paper [1] (referred to hereafter as I) we attempted a lattice calculation of the topological charge fluctuation:

$$A \equiv \int d^4x \, \langle 0|Q(x)Q(0)|0\rangle \,, \qquad (1)$$

in the quarkless SU(2) gauge theory, where

$Q(x) = (g^2/64\pi^2)\epsilon_{\mu\nu\rho\delta} F^a_{\mu\nu}(x) F^a_{\rho\sigma}(x)$

is the so-called topological charge density.

A non-zero value of A would provide a solution of the $U_A(1)$ problem of QCD [2]. Actually, for the agreement with the phenomenology of the pseudo-scalar nonet, one needs for colour SU(3):

$$A \overset{\text{phen}}{\approx} (180\ \text{MeV})^4 \,. \qquad (2)$$

In I, with a simple definition of $Q(x)$, and using Monte Carlo techniques for the full SU(2) group, we obtained:

$$A^{1/4} \approx (0.11 \pm 0.02) K^{1/2} \,, \qquad (3)$$

where K is the (quarkless) string tension.

In this paper we want so show that different lattice definitions of $Q(x)$ and A lead to results in agreement with eq. (3) (within errors), as required by the existence of a continuum limit of lattice QCD.

The defintion of $Q(x)$ on the lattice used in I can be written as

$$Q_n^{(1)} = - \sum_{(\mu,\nu,\rho,\sigma)=\pm 1}^{\pm 4} \frac{\tilde{\epsilon}_{\mu\nu\rho\sigma}}{2^4 \cdot 32\pi^4}$$

$$\times \operatorname{tr}(U_{n,\mu} U_{n+\mu,\nu} U_{n+\mu+\nu,\rho} U_{n+\mu+\nu+\rho,\sigma}$$

$$\times U^+_{n+\nu+\rho+\sigma,\mu} U^+_{n+\rho+\sigma,\nu} U^+_{n+\sigma,\rho} U^+_{n,\sigma}) \,, \qquad (4)$$

with $1 = \tilde{\epsilon}_{1234} = -\tilde{\epsilon}_{2134} = -\epsilon_{-1234}$ and $U_{n,\mu}$ the usual link variable. The link structure of eq. (4) is shown in fig. 1a.

In I we also mentioned another definition of $Q(x)$ that has the same naive continuum limit as $Q^{(1)}$. It reads [‡1] (fig. 1b):

$$Q_n^{(2)} = - \sum_{(\mu,\nu,\rho,\sigma)=\pm 1}^{\pm 4} \frac{\tilde{\epsilon}_{\mu\nu\rho\sigma}}{2^4 \cdot 32\pi^4} \operatorname{tr}(U_{n,\mu\nu} U_{n,\rho\sigma}) \,, \qquad (5)$$

‡1 This is the definition called $Q^{(s)}$ in I.

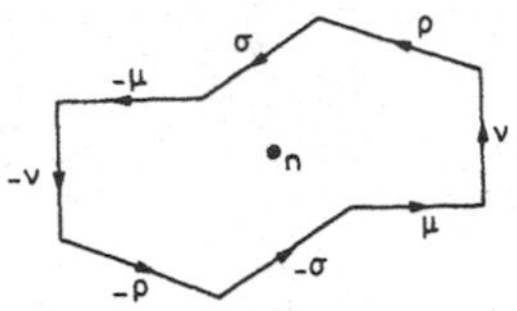

(a)

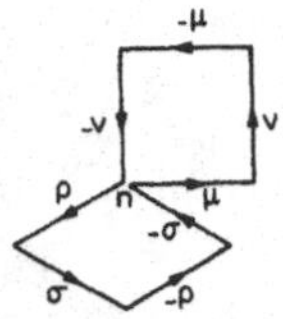

(b)

Fig. 1. (*a*) Link structure of the definition of Q_n given in eq. (4); (b) same for definition given in eq. (5).

where $U_{n,\mu\nu}$ is the usual plaquette variable constructed out of four links:

$$U_{n,\mu\nu} = U_{n,\mu}\, U_{n+\mu,\nu} U^+_{n+\nu,\mu}\, U^+_{n,\nu}\,. \tag{6}$$

In terms of the quantities (5) and (6), we can construct the following three lattice definitions of A:

$$A_{11} = \frac{1}{a^4}\sum_n \langle 0|Q_n^{(1)}Q_0^{(1)}|0\rangle\,, \tag{7a}$$

$$A_{22} = \frac{1}{a^4}\sum_n \langle 0|Q_n^{(2)}Q_0^{(2)}|0\rangle\,, \tag{7b}$$

$$A_{12} = \frac{1}{a^4}\sum_n \langle 0|\, Q_n^{(1)}Q_0^{(2)}|0\rangle\,. \tag{7c}$$

The results of I summarized by eq. (3) refer to A_{11}.

Here we shall present the results for A_{11}, A_{22} and A_{12} as obtained by Monte Carlo simulations on a 4^4 lattice with periodic boundary conditions. Our computer programme is based on that of ref. [3], which uses the 120 element (icosahedron) subgroup of SU(2). For each value of $\beta = 4/g^2$ we made 20 000 to 40 000 sweeps. We used hot and cold starts and checked that equilibrium was reached after a small number of sweeps. Consequently we have always ignored the first 500 sweeps.

The data for $\pi^4 \cdot 2^{18}\, a^4\, A$ are presented in figs. 2a, b, c for the three quantities defined respectively in eqs. (7a, b, c), with their statistical errors as a function of β. Note the large differences obtained in the data with the three definitions, which forced us to us different scales.

The data of fig. 2a agree well with those obtained in I, showing that the restriction to the icosahedral subgroup of SU(2) is an excellent approximation in the region of β considered. (We stay far away from the phase transition point of this finite group, $\beta_c \approx 6$.)

The strong coupling expansion for A_{11} has already been given in I [eqs. (4.4) and (4.11)] and we report it here as a function of β for completeness:

$$\pi^4 \cdot 2^{18} a^4 A_{11} = 6144\left(1 - \tfrac{1}{4}\beta^2 + \frac{23}{3\cdot 2^7}\beta^4 - \frac{139}{9\cdot 2^{10}}\beta^6 + O(\beta^8)\right). \tag{8}$$

For A_{22} and A_{12} one finds respectively

$$\pi^4 \cdot 2^{18} a^4 A_{22} = 2304[1 + O(\beta^4)]\,, \tag{9}$$

$$\pi^4 \cdot 2^{18} a^4 A_{12} = 72\beta^4 + O(\beta^6)\,. \tag{10}$$

This behaviour is in agreement with the shape of our Monte Carlo data for small β (see figs. 2a, b, c).

In order to extract the physical value of A we proceed as in I by first subtracting from each set of data the appropriate perturbative tail which we write as:

$$\pi^4 \cdot 2^{18} a^4 A^{\text{pert}} = C_3/\beta^3 + C_4/\beta^4\,. \tag{11}$$

The coefficients C_3 have been computed, giving for a 4^4 lattice $C_3 = 14321.2$, 6414 and 4690 for A_{11}, A_{22} and A_{12}, respectively. The computation of C_4 in (11) is very complicated and we have not attempted it. As in I, we have determined it by a fit of the data in the weak coupling region $2.8 < \beta < 4.5$. We obtain $C_4 = 4987$, 4897 and -884 for A_{11}, A_{22} and A_{12}, respectively. The perturbative tails, as defined in eq. (11), are also shown in figs. 2a, b, c on top of the data.

In fig. 3 we plot together, on a logarithmic scale, the differences between our monte Carlo data for $\pi^4 \cdot 2^{18}\, a^4\, A$ and the perturbative tail [eq. (11)] for the three definitions of A. The three sets of data agree well with the (two-loop) renormalization group behaviour shown by the solid lines [‡2]. Comparing these results with those on the string tension K [4], we get

[‡2] The results are practically unchanged if the one-loop renormalization group behaviour is used instead.

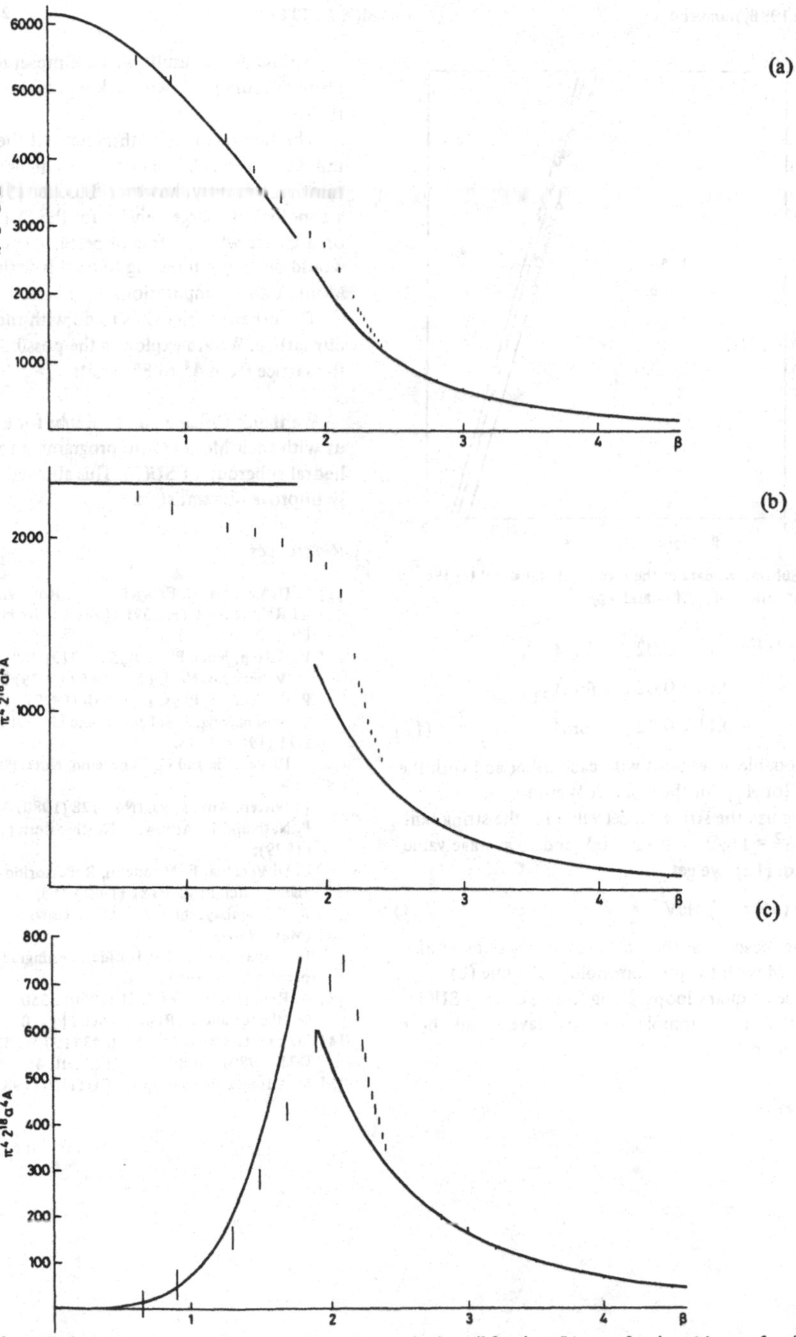

Fig. 2. (a) Monte Carlo data, strong coupling expansion and perturbative tail for A_{11}; (b) same for A_{22}; (c) same for A_{12}.

Volume 108B, number 4,5 PHYSICS LETTERS 28 January 1982

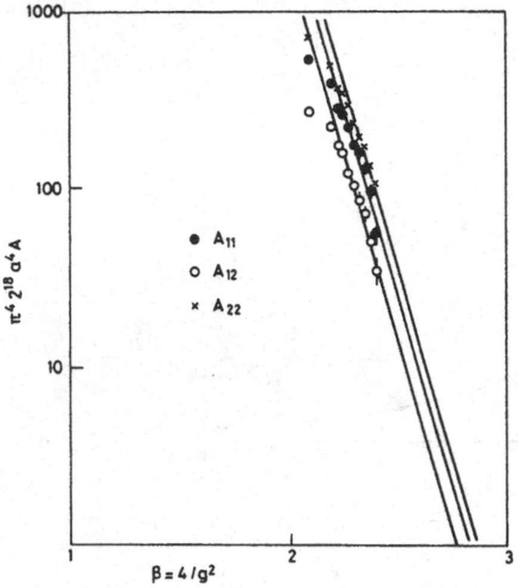

Fig. 3. Subtracted data in the region $2.1 \leqslant \beta \leqslant 2.4$ for the three definitions A_{11}, A_{22} and A_{12}.

$$A^{1/4}K^{-1/2} = 0.13 \pm 0.02\,, \quad \text{for } A_{11}\,,$$
$$= 0.14 \pm 0.02\,, \quad \text{for } A_{22}\,,$$
$$= 0.11 \pm 0.02\,, \quad \text{for } A_{12}\,, \qquad (12)$$

in reasonable agreement with each other and with the results for A_{11} for the full SU(2) group.

If we use the string model value for the string tension $K^{1/2} = 1/\sqrt{2\pi\alpha'} = 420$ MeV and an average value from eq. (12), we get

$$A^{1/4} = (55 \pm 10)\ \text{MeV}\,. \qquad (13)$$

Possible reasons for the smallness of this number as compared with the phenomenological value (2) [absence of quark loops, going from SU(2) to SU(3), current algebra extrapolation, etc.] have already been discussed in I.

Although the results we have presented here are quite encouraging, two weak points should be mentioned.

The first one is the subtraction of the perturbative tail which is a delicate one due to higher order uncertainties. Recently, however, Lüscher [5] constructed a topological charge density for the SU(2) gauge theory on a lattice which is free of perturbative corrections. It would be very interesting to use his definition in a Monte Carlo computation.

The other problem has to do with the small size of our lattice. We are exploring the possibility of enlarging the lattice from 4^4 to 8^4 points.

We thank C. Lang and C. Rebbi for kindly providing us with their Monte Carlo programme for the icosahedral subgroup of SU(2). This allowed us to drastically improve our statistics.

References

[1] P. Di Vecchia, K. Fabricius, G.C. Rossi and G. Veneziano, CERN preprint TH-3091 (1981), to be published in Nucl. Phys. B.
[2] E. Witten, Nucl. Phys. B156 (1979) 269;
G. Veneziano, Nucl. Phys. B159 (1979) 213;
P. Di Vecchia, Phys. Lett. 85B (1979) 213;
C. Rosenzweig, J. Schechter and G. Trahern, Phys. Rev. D21 (1980) 3388;
P. Di Vecchia and G. Veneziano, Nucl. Phys. B171 (1980) 253;
E. Witten, Ann. Phys. (NY) 128 (1980) 363;
P. Nath and R. Arnowitt, Northeastern preprint NUB-2417 (1979);
P. Di Vecchia, F. Nicodemi, R. Pettorino and G. Veneziano, Nucl. Phys. B181 (1981) 318;
K. Kawarabayashi and N. Ohta, Univ. of Tokyo preprint (May 1980);
D.I. Dyakonov and M.J. Eides, Leningrad Nucl. Phys. Inst. preprint 639 (1981).
[3] C. Rebbi, Phys. Rev. D21 (1980) 3350;
G. Bhanot and C. Rebbi, Nucl. Phys. B180 (1981) 469.
[4] M. Creutz, Phys. Rev. Lett. 43 (1979) 553; Phys. Rev. D21 (1980) 2308; Phys. Rev. Lett. 45 (1980) 313.
[5] M. Lüscher, Bern preprint BUTP-10 (1981).

Nuclear Physics B190[FS3] (1981) 288–300

THE ACTION OF SU(N) LATTICE GAUGE THEORY IN TERMS OF THE HEAT KERNEL ON THE GROUP MANIFOLD

P. MENOTTI

Istituto di Fisica dell' Università, Pisa and
INFN, Pisa, Italy

E. ONOFRI[1]

CERN, Geneva, Switzerland

Received 19 February 1981

We consider the heat kernel on the group manifold as an alternative to the Wilson action in lattice gauge theory, and we exhibit its strict analogy with the well-known Berezinski-Villain action. With the heat kernel action, the Gross-Witten singularity is rigorously absent in two dimensions. The similarity of the heat kernel action to the hamiltonian approach should provide a better convergence of the lagrangian strong coupling expansion, while its behaviour at weak coupling should simplify the analysis of the weak coupling perturbative expansion.

1. Introduction

There has been a great deal of effort recently to understand the transition from strong to weak coupling in lattice gauge theories. Numerical experiments [1,2] on small lattices with Wilson action show a rather sharp turnover from the strong coupling behaviour of the string tension to a weak coupling behaviour compatible with asymptotic freedom. In the leading $N \to \infty$ approximation of the U(N) gauge theory in two dimensions, Gross and Witten [3] have shown that the string tension σ is non-analytic in $\beta = (g^2N)^{-1}$ at some critical point β_c. Higher order corrections [4] show that a breakdown of the $1/N$ expansion occurs in some interval around β_c, where it should be necessary to take into account terms of order e^{-cN} (this is clear in the hamiltonian approach where the $N = \infty$ critical point corresponds to an aperiodic instanton-like classical trajectory in the equivalent Fermi gas [5]). It was conjectured that the $N = \infty$ singularity could manifest itself at finite N by a sharp bending of the function $\sigma(\beta)$ (which is what is actually observed) provided one can extend the result to four dimensions. On the other hand, the singularity would forbid a simple

[1] On leave from Istituto di Fisica della Università, Parma, Italy.

extrapolation from strong to weak coupling in the $1/N$ expansion. The question then arises: is this singularity unavoidable, i.e., is it a property of any gauge theory on the lattice, or rather is it a peculiarity of Wilson's action? A partial answer was found recently [6] by studying the alternative version of the action introduced by Manton [7]. With this choice*, $A_M(U_P) = g^{-2}\,\mathrm{dist}(U_P, I)^2$, the leading $N \to \infty$ contribution to the string tension in two dimensions is analytic in β (higher order corrections, however, are not known).

In our opinion, another much simpler action, which leads to a similar (actually stronger) result, i.e., no Gross-Witten singularity in two dimensions to *all* orders in $1/N$, was already encountered in various contexts** a couple of years ago, but perhaps overlooked from this point of view [8–11]. We want to consider it again here, because it appears to be the most natural non-abelian generalization of the so-called "Villain model"***. This too was noticed in refs. [8–11], but we are able to go one step further in making the analogy with Villain's model more explicit. The action we want to introduce is given in terms of the "heat kernel" on the gauge group, i.e., in terms of the matrix elements of $\exp\{\frac{1}{2}g^2\Delta\}$, where Δ is the Laplace-Beltrami operator (with respect to the invariant metric on the group). The precise form of the action will be introduced in sect. 2. For the moment, let us anticipate that the heat kernel on a compact Lie group can be given explicitly in terms of "periodic gaussians" in invariant angles (coordinates of the maximal abelian subgroup). This makes the connection with the U(1) Villain model more explicit, even if the non-abelian nature of the group does not lead to a direct interpretation of the theory in terms of vortices as happens in the abelian case [12]; we cannot exclude that this may be feasible, at least in some approximate sense.

Another virtue of the heat kernel action is that it leads to a straightforward computation of the renormalization transformation of the Migdal-Kadanoff type for all groups in the weak coupling limit (see appendix B). In two dimensions, the action is exactly self-reproducing, thus providing the exact renormalized two-dimensional action. In four dimensions, there are indications that the heat kernel could be very near to the stable distribution (notice that the Villain periodic gaussian used in ref. [13] is not exactly the heat kernel as it differs from it by a factor $\theta/\sin\theta$ for SU(2)—this factor favours our suggested action against the simple periodic gaussian).

The paper is organized as follows: In sect. 2 we derive the new form for the action, starting from the hamiltonian formulation of Kogut and Susskind, we discuss the Gross-Witten singularity and we give the Villain form of the action. Sect. 3 is

* $\mathrm{dist}(U, I)$ is the geodesic distance of U from the identity with respect to the invariant metric on the group.

** We are grateful to G. Veneziano for drawing our attention to ref. [9].

*** Manton's action, though completely acceptable from the point of view of gauge invariance and compactness, is singular at "conjugate" points. The heat kernel provides the most natural generalization to any riemannian manifolds of the usual gaussians in euclidean space.

devoted to a general discussion of the advantages of the heat kernel action, also keeping in mind the strong coupling expansion [14, 15] and the very recent numerical computations [16]. Since the mathematics of heat diffusion on riemannian manifolds is unfamiliar to most physicists, we give a self-contained account of it in appendix A where we derive the heat kernel on U(N) and SU(N) in the most elementary way.

2. The action defined by the heat kernel

The most natural way of introducing the "heat kernel" action is to start from the Kogut-Susskind [17] hamiltonian formalism and to try to recover a space-time (euclidean) formulation by the customary Feynman-Kac-Nelson formula. We have

$$H=\frac{g^2}{2a}\sum_{\ell} E_{\ell}^{a}E_{\ell}^{a}+\sum_{\mathrm{P}} V(U_{\mathrm{P}}), \tag{1}$$

where the summation extends over all links ℓ and over all plaquettes P contained in some finite volume $\Lambda \subset \mathbb{R}^3$; usually, $V(U_{\mathrm{P}})=\frac{1}{2}(g^2a)^{-1}\mathrm{tr}(2-U_{\mathrm{P}}-U_{\mathrm{P}}^{\dagger})$, but we are going to make a different choice very soon. The Casimir operator $\Sigma_{\alpha}E^{\alpha}E^{\alpha}$ coincides with the Laplace-Beltrami operator $-\Delta$ on the group with respect to the invariant metric [18] (see appendix A for details).

To calculate $\exp(-TH)$ one can apply Trotter's formula to obtain:

$$\langle U^{(0)}|\mathrm{e}^{-TH}|U\rangle = \lim_{\substack{n\to\infty\\ a_0=T/n}} \int[\mathrm{d}U^{(1)}]\dots\int[\mathrm{d}U^{(n-1)}]$$

$$\times\prod_{k=0}^{n-1}\left\langle U^{(k)}\left|\exp\left\{\frac{g^2a_0}{2a}\sum_{\ell}\Delta_{\ell}\right\}\right|U^{(k+1)}\right\rangle \exp\left\{-a_0\sum_{\mathrm{P}}V\left(U_{\mathrm{P}}^{(k+1)}\right)\right\}, \tag{2}$$

with $U^{(n)}\equiv U$. We have now to approximate this exact formula. First of all we take a large, but finite, n in such a way that $a_0=T/n=a$; second, one may approximate the matrix elements of $\exp\{\frac{1}{2}g^2\Sigma\Delta_{\ell}\}$ in such a way as to recover Wilson's action (in the temporal gauge). The choice we want to make instead is to keep the kernel $\langle U|\exp\Delta|U'\rangle$ exactly and to choose the potential $V(U_{\mathrm{P}})$ so as to have an isotropic interaction in the four-dimensional lattice. This means that we have to set

$$\mathrm{e}^{-aV(U_{\mathrm{P}})}=\left\langle I\left|\exp\{\tfrac{1}{2}g^2\Delta\}\right|U_{\mathrm{P}}\right\rangle$$

$$\equiv K\left(U_{\mathrm{P}},\tfrac{1}{2}g^2\right) \tag{3}$$

(I is the identity element of the group). In this way we have constructed an approximate solution for the dynamics of $\boldsymbol{H}$ in terms of a space-time lattice action defined by

$$e^{A} = \prod_{P} \frac{K\left(U_P, \frac{1}{2}g^2\right)}{K\left(I, \frac{1}{2}g^2\right)}. \tag{4}$$

(The normalization factor is chosen in order to have $A=0$ for constant field configurations.) This definition is consistent, because (i) the heat kernel $K(U, \frac{1}{2}g^2)$ is differentiable and strictly positive everywhere; (ii) the small g behaviour is the same as Wilson's (and Manton's) action after a suitable indentification of the coupling constant and (iii) the strong coupling confining properties are preserved [9] (as is expected from the character expansion eq. (6) which in the strong coupling limit $g \to \infty$ converges to Wilson's action). We now observe that in (1 + 1) dimensions, *no approximation whatsoever* is involved in truncating the Trotter formula at finite n, since there is no potential V. As a consequence, the space-time lattice will give the same string tension as the hamiltonian formalism, i.e.,

$$\sigma a^2 = \tfrac{1}{2}g^2 C_f = \tfrac{1}{4}\left(g^2 N\right)\left(1 - \frac{1}{N^2}\right) \tag{5}$$

where C_f is the quadratic Casimir of the fundamental representation of SU(N). This result can be easily checked from the character expansion [9], namely

$$K\left(U, \tfrac{1}{2}g^2\right) = \sum_{\mu} d_\mu \chi_\mu(U) e^{-g^2 C_\mu/2}, \tag{6}$$

where μ runs over all irreducible unitary representations $\mathcal{D}^{(\mu)}$ of SU(N), χ_μ is the character of $\mathcal{D}^{(\mu)}$, $d_\mu = \chi_\mu(I)$ its dimension and C_μ its quadratic Casimir invariant which represents the eigenvalue of $-\Delta$ in each invariant subspace of L_2[SU(N)].

The action we have defined was already considered from several points of view: as a weak coupling approximation to Wilson's action [9], as a strong coupling effective action in the hamiltonian approach [8,10], and as a generalized Villain model for CP^n models [11]. The main point we want to raise here is that $K(U, \frac{1}{2}g^2)$ admits, *in general*, a representation in terms of periodic gaussians in the invariant angles. Let $U = S\Phi S^\dagger$ where $\Phi = \mathrm{diag}[e^{i\phi_1}, \dots, e^{i\phi_N}]$ [$\Sigma\phi_i = 0$ for SU(N)]; K is a function of $(\phi_1, \dots, \phi_N)$ only and is given explicitly by

$$K\left(\phi, \tfrac{1}{2}g^2\right) = \mathfrak{N} \sum_{\{l\}=-\infty}^{\infty} \prod_{i<j} \frac{\phi_i - \phi_j + 2\pi(l_i - l_j)}{2\sin\frac{1}{2}\left[\phi_i - \phi_j + 2\pi(l_i - l_j)\right]} \exp\left\{-\frac{1}{g^2} \sum_j \left(\phi_j + 2\pi l_j\right)^2\right\}, \tag{7}$$

where $\mathcal{N}$ does not depend on ϕ [and factors out in eq. (4)] and the sum ranges over all integers $(l_1, \ldots, l_N)$ [subject to the constraint $\Sigma l_j = 0$ for SU(N)]. Eq. (7) is proved in appendix A.

In the weak coupling limit, all terms in the sum are exponentially small with respect to the dominant $l_k = 0$ term (in a neighbourhood of the identity $U = I$); the factors $\phi/\sin\phi$ appearing in eq. (7) have a convergent expansion near $\phi = 0$ and contribute to the second derivative of the action near $\phi = 0$. In fact we have

$$K \underset{\phi \to 0}{\simeq} \mathcal{N} \exp\left\{-\left(\frac{1}{g^2} - \tfrac{1}{24}N\right)\sum_j \phi_j^2\right\}, \tag{8}$$

where we have expanded $\phi/\sin\phi \sim \exp\{\frac{1}{6}\phi^2 + O(\phi^4)\}$, and we have taken into account the constraint $\Sigma\phi_k = 0$ which gives $\Sigma_{i<j}(\phi_i - \phi_j)^2 \equiv N\Sigma\phi_i^2$. As a simple application of eq. (8) we shall give in appendix B a straightforward derivation of the Migdal renormalization transformation. For SU(2) we have

$$K(\varphi, \tfrac{1}{2}g^2) = \mathcal{N} \sum_{-\infty}^{+\infty} \frac{\varphi + 2\pi n}{\sin\varphi} \exp\{-2g^{-2}(\varphi + 2\pi n)^2\}, \tag{9}$$

a formula which goes back to the fifties [19]. Eq. (7) is strongly reminiscent of the Berezinski-Villain action for abelian groups, where, through an exact transformation [12], the curl of the discrete field l_μ has been given the interpretation of vortex excitations, while the gradient part is used to *decompactify* the integration range. Here, the meaning of the integers $\{l_k\}$ is less clear due to the non-linear relationship between the plaquette invariant angles $\{\phi_j\}$ and the group integration variables U_ℓ.

3. Discussion

It is important to examine the potential implications of the heat kernel action which we have displayed in sect. 2. If one accepts universality, the actual continuum limit of the theory should be independent of the details of the interaction lagrangian; thus, in principle, Wilson's action is as good as any other action which preserves local gauge invariance and exhibits the compactness of the gauge group integration. On the other hand, from a practical point of view, the situation can be very different in the sense that one action may give rise to more quickly convergent algorithms (like strong coupling expansions or Monte Carlo calculations). The favourable points we want to stress about the heat kernel lagrangian are the following: (i) In two dimensions, the Gross-Witten singularity is rigorously absent to all orders in $1/N$; thus, in two dimensions such a singularity is an artefact of the Wilson's action which could reflect itself in a failure of the $1/N$ expansion in four dimensions. It is obvious that it is better to start from an interaction which is free from this disease, at least in two dimensions. (ii) In the weak coupling limit, Wilson's action contains [for SU(2)] terms of the type ϕ^4/g^2 which are washed away in crude weak coupling approximations and replaced by the smaller terms of order $O(1)\phi^4$ which are very well represented by the heat kernel up to non-negligible values ($\frac{1}{2}\beta = 2/g^2 \gtrsim 0.6$) of the coupling constant [13]. A smoother approach to the weak coupling region is also shown by the recent Monte Carlo calculation by Lang et al. [16] with the heat kernel action [eq. (9)]. The reduction of one order in $1/g^2$ of the ϕ^4 terms in the lagrangian, i.e., from $(1/g^2)\phi^4 = O(g^2)$ for Wilson's action to $\phi^4 = O(g^4)$ for the heat kernel action should also simplify the analytic comparison of lattice theory with the continuum theory [20]. (iii) At present, hamiltonian strong coupling expansions appear to be in better shape, as far as the radius of convergence is concerned, than lagrangian strong coupling expansions. In fact, if we define as the boundary β_w of the weak coupling region the point where the $\ln\sigma(\beta)$ curve is tangent to the asymptotic freedom line, then according to the analysis of Lüscher et al. [14] of the lagrangian series expansion, the string roughening point lies well below $\beta_w(\beta_R^{lag} \sim 1.9, \beta_w^{lag} \sim 2)$. According to the work of Kogut et al. [15], we have instead $\beta_R^{ham} > \beta_w^{ham}$, giving more confidence in the hamiltonian strong coupling expansion in the weak coupling region. If one were to perform Monte Carlo calculations in the hamiltonian formulation by an approximation of the evolution equation through Trotter's formula, preserving discrete rotational invariance (isotropy), one is led naturally to the heat kernel interaction as we have discussed in sect. 2.

We warmly thank G. Veneziano for his generous advice. We thank C.B. Lang, C. Rebbi, P. Salomonson and B.S. Skagerstam for interesting discussions and for communicating their results prior to publication. E.O. would like to thank his friends G. Marchesini for his continous encouragement and N.E. Hurt who introduced him to the beautiful mathematics of the heat diffusion on manifolds.

Appendix A

THE HEAT KERNEL ON U(N) AND SU(N) [21-24]

The heat kernel, i.e., the fundamental solution of the diffusion equation

$$\frac{\partial K}{\partial t} = \frac{1}{2} D \Delta K \tag{A.1}$$

on a Lie group G, or more generally, on a riemannian manifold, has a long history in the mathematical literature (see refs. [21,22] and references therein). In the physics literature, the diffusion equation was studied in detail in the case of the rotation group, in the context of the rotational brownian motion of molecules [19]. More recently [23,24], attention was focused on the free Schrödinger equation on Lie groups ($D = i\hbar/m$), the aim being to write a Feynman path integral for particles on riemannian manifolds. In this appendix, we calculate the heat kernel for U(N) and

SU(N). Let us first show that the Kogut-Susskind operator $\Sigma_\alpha E^\alpha E^\alpha$ is identical to the Laplace-Beltrami operator on the group. E^α is defined by the commutation property*

$$[E^\alpha, \hat{U}] = \lambda^\alpha \hat{U}, \tag{A.2}$$

where λ^α are the generators of the group in the fundamental representation and are normalized to $\operatorname{tr} \lambda^\alpha \lambda^\beta = \frac{1}{2}\delta^{\alpha\beta}$. Let us introduce canonical coordinates x_α, i.e., $U = \exp\{ix_\alpha \lambda_\alpha\}$: then we find

$$E^\alpha = -i\frac{\partial}{\partial x_\alpha} - \tfrac{1}{2} f_{\alpha\beta\nu} x_\beta \frac{\partial}{\partial x_\nu} + O(x^2), \tag{A.3}$$

$$\sum_\alpha E^\alpha E^\alpha = -\sum_\alpha \left(\frac{\partial}{\partial x_\alpha}\right)^2 + O(x^2). \tag{A.4}$$

Now observe that $\Sigma_\alpha E^\alpha E^\alpha$ is invariant under group translations; formally

$$(L_V \psi)(U) = \psi(V^{-1}U), \tag{A.5}$$

$$L_V E^\alpha E^\alpha = E^\alpha E^\alpha L_V. \tag{A.6}$$

Hence,

$$\begin{aligned}(E^\alpha E^\alpha \psi)(U) &= (L_V E^\alpha E^\alpha L_V^{-1}\psi)(U) \\ &= (E^\alpha E^\alpha L_U^{-1}\psi)(I) \\ &= \lim_{x\to 0} (E^\alpha E^\alpha L_U^{-1}\psi)(\exp\{ix_\beta\lambda_\beta\}) \\ &= -\sum_\alpha \left(\frac{\partial}{\partial x_\alpha}\right)^2 (L_U^{-1}\psi)(\exp\{ix_\beta\lambda_\beta\})\Big|_{x=0}.\end{aligned} \tag{A.7}$$

Now we define the metric $ds^2 = 2\operatorname{tr}(dU dU^\dagger)$ which is invariant under left and right translations. The Laplace-Beltrami operator Δ associated to ds^2 is also invariant and its action can be calculated as in eq. (A.7), namely

$$(\Delta\psi)(U) = \lim_{x\to 0} (\Delta L_U^{-1}\psi)(\exp\{ix_\beta\lambda_\beta\}) \tag{A.8}$$

* $\hat{U}$ is the Hilbert space operator, while U denotes its formal eigenvalues in a continous basis $|U\rangle$.

Since $ds^2 = 2\,\mathrm{tr}(\mathrm{d}\,e^{ix_\alpha\lambda_\alpha}\,\mathrm{d}\,e^{-ix_\beta\lambda_\beta}) = \Sigma_\alpha \mathrm{d}x^\alpha \mathrm{d}x^\alpha + O(x^2, \mathrm{d}x^2)$, we have $\Delta = -\Sigma_\alpha(\partial/\partial x_\alpha)^2 + O(x^2)$ near the identity. Hence $\Sigma_\alpha E^\alpha E^\alpha = -\Delta$ everywhere. It follows that the heat kernel has the well-known character expansion [eq. (6)].

To find the fundamental solution of the heat equation, let us now introduce polar coordinates $U = S\Phi S^\dagger$, $\Phi = \mathrm{diag}[e^{i\phi_1},\dots,e^{i\phi_N}]$, in terms of which we have

$$ds^2 = 2\sum_i \mathrm{d}\phi_i^2 + 2\sum_{i\neq j} \left|e^{i\phi_i} - e^{i\phi_j}\right|^2 \left|(S^\dagger \mathrm{d}S)_{ij}\right|^2. \tag{A.9}$$

Due to the invariance of Δ, the fundamental solution of eq. (A.1) (with $D = 2$)

$$K(U,t) = \langle I|e^{t\Delta}|U\rangle \tag{A.10}$$

is a class function, i.e., it is a function of $(\phi_1,\dots,\phi_N)$ only. It is sufficient then to consider the "radial" part of Δ, namely

$$\mathring{\Delta} = \tfrac{1}{2}\sum_i \frac{1}{J^2}\frac{\partial}{\partial\phi_i} J^2 \frac{\partial}{\partial\phi_i}, \tag{A.11}$$

where $J(\phi) = \Pi_{i<j}[2\sin\frac{1}{2}(\phi_i - \phi_j)]$ is the fourth root of the determinant of the metric tensor. A more convenient form for $\mathring{\Delta}$ is the following:

$$\mathring{\Delta} = \frac{1}{2J}\sum_i \left(\frac{\partial}{\partial\phi_i}\right)^2 J + \tfrac{1}{2}R_N, \tag{A.12}$$

with $R_N = \frac{1}{12}N(N^2 - 1)$.

Now we make use of the known solution of the corresponding problem for hermitian matrices. Namely, let $U = \exp iX$; the same transformation S which diagonalizes U brings X to its diagonal form $X = S\,\mathrm{diag}(\phi_1,\dots,\phi_N)S^\dagger$. The Laplace operator Δ_X on the space of hermitian matrices acting on invariant functions is well known to be

$$\Delta_X = \frac{1}{D(\phi_1,\dots,\phi_N)}\sum_k \left(\frac{\partial}{\partial\phi_k}\right)^2 D(\phi_1,\dots,\phi_N), \tag{A.13}$$

where

$$D(\phi) \equiv D(\phi_1,\dots,\phi_N) = \prod_{i<j}(\phi_i - \phi_j). \tag{A.14}$$

Hence we have

$$2\mathring{\Delta} = \frac{D(\phi)}{J(\phi)}\Delta_X \frac{J(\phi)}{D(\phi)} + R_N. \tag{A.15}$$

Since Δ_X is an ordinary Laplace operator in a euclidean space, the fundamental solution to the corresponding diffusion equation is a simple gaussian, namely

$$\frac{\partial}{\partial t}\mathcal{K}(X,t)=\tfrac{1}{2}\Delta_X\mathcal{K},$$

$$\mathcal{K}(X,t)=(2\pi t)^{-N^2/2}\exp\left\{-\frac{1}{2t}\operatorname{tr}X^2\right\} \tag{A.16}$$

Because of eq. (A.15),

$$K_0(U,t)=(2\pi t)^{-N^2/2}\frac{D(\phi)}{J(\phi)}\exp\left\{-\frac{1}{2t}\operatorname{tr}X^2+\tfrac{1}{2}R_N t\right\} \tag{A.17}$$

is a solution to eq. (A.1) which has the correct limit to $\delta(U)$ as $t\downarrow 0$ (up to a normalization); K_0, however, is not a regular solution everywhere (it is not periodic in ϕ_j). To obtain the true fundamental solution, we have to sum the values taken by K_0 on all points X such that $\exp iX=U$, i.e., over all values $(\phi_j+2\pi l_j)$ with integer $\{l_j\}$. This gives eq. (7) of the text*. The restriction to SU(N) is simply given by $\{\Sigma\phi_j=0,\Sigma l_j=0\}$. Notice that since $\mathcal{K}(X,t)$ is normalized with respect to the invariant volume $[\mathrm{d}X]=D(\phi_1,\dots,\phi_N)^2\mathrm{d}\phi_1,\dots,\mathrm{d}\phi_N[\mathrm{d}S]$, $K(U,t)$ is not automatically normalized according to $[\mathrm{d}U]=J(\phi)^2\mathrm{d}\phi_1,\dots,\mathrm{d}\phi_N[\mathrm{d}S]$; the proper normalization can be found in ref. [23], but it is irrelevant for our purposes. Notice also that the naive Villain form adopted in ref. [13] corresponds to the heat kernel of Δ_X summed over all inverse images $\exp^{-1}(U)$. There is nothing wrong with this choice, but the reproducing property of the heat kernel and the simple character expansion in terms of the Casimir invariant are lost.

We now want to show how the heat kernel on U(N) is related to the heat kernel on its subgroup U(1) and SU(N). Let us introduce the variables

$$\theta_j=\phi_j-\frac{1}{N}\sum_k\phi_k,$$

$$\theta_0=\frac{1}{N}\sum_k\phi_k. \tag{A.18}$$

$(\theta_0;\theta_1,\dots,\theta_N)$ are invariant coordinates on $\mathrm{U}(1)\otimes\mathrm{SU}(N)$. The correspondence $\mathrm{U}(N)\leftrightarrow\mathrm{U}(1)\otimes\mathrm{SU}(N)$ given by eq. (A.18) is multivalued, namely all the points

$$\theta_j^\nu=\theta_j-\frac{2\pi\nu}{N}+2\pi\nu\delta_{jN},\qquad(\nu=0,\dots,N-1),$$

$$\theta_0^\nu=\theta_0+\frac{2\pi\nu}{N}, \tag{A.19}$$

* This is essentially the method of "images" of the classical potential theory.

correspond to the same point in U(N). This is written formally as $\mathrm{U}(N)=\mathrm{SU}(N)\otimes\mathrm{U}(1)/Z_N$. It follows that

$$K_{\mathrm{U}(N)}(\phi,t)=\sum_{\nu}K_{\mathrm{SU}(N)}(\theta^{\nu},t)K_{\mathrm{U}(1)}(\theta_0^{\nu},t/N), \tag{A.20}$$

where $K_{\mathrm{U}(1)}$ is the usual Villain expression. Notice that since U(N) is not simple, the metric (A.9), i.e., $\mathrm{d}s^2=2\,\mathrm{tr}(\mathrm{d}U\mathrm{d}U^{\dagger})$ is not the only invariant metric. We are free to choose a relative weight between the U(1) and SU(N) content, i.e., the heat kernel contains two coupling constants, as should be expected from the theory on the continuum. The direct proof of eq. (A.20) runs as follows. The identity*

$$\sum\phi_i^2=\frac{1}{N}\sum_{i<j}(\phi_i-\phi_j)^2+\frac{1}{N}\Big(\sum\phi_i\Big)^2 \tag{A.21}$$

allows us to write eq. (7) as

$$K_{\mathrm{U}(N)}(\phi,t)=\sum_{\{l\}}\prod_{i<j}\Theta_t(\theta_i-\theta_j+2\pi(l_i-l_j))\exp\left[-\frac{N}{2t}\left(\theta_0+\frac{2\pi}{N}\sum l_j\right)^2\right], \tag{A.22}$$

where

$$\Theta_t(x)\equiv\frac{x}{2\sin\frac{1}{2}x}\exp\left\{-\frac{x^2}{2Nt}\right\}. \tag{A.23}$$

(Warning: we are ignoring numerical factors.)

Now let $\Sigma l_i=Nn+\nu$ $(0\leqslant\nu\leqslant N-1)$; we transform the sum over $l_1,\ldots,l_N$ as follows:

$$\sum_{\{l\}}=\sum_{\nu=0}^{N-1}\sum_{n=-\infty}^{+\infty}\sum_{\{l\}}\delta\Big(\sum_i l_i-nN-\nu\Big). \tag{A.24}$$

Since the arguments of Θ_t are invariant under a uniform translation $l_i\to l_i-n$, we find

$$K_{\mathrm{U}(N)}(\phi,t)=\sum_{\nu=0}^{N-1}\sum_{\{l\}}\delta\Big(\sum_i l_i-\nu\Big)\prod_{i<j}\Theta_t(\theta_i-\theta_j+2\pi(l_i-l_j))$$

$$\times\sum_{n=-\infty}^{+\infty}\exp\left\{-\frac{N}{2t}\left(\theta_0+\frac{2\pi\nu}{N}+2n\pi\right)^2\right\}. \tag{A.25}$$

* A special case of the identity $\Sigma_\mu(\alpha_\mu\phi)^2=C_2(\Sigma_i\gamma_i\cdot\phi_i)^2$ valid for any compact group, where α_μ are the positive roots, γ_i are simple roots and C_2 is the quadratic Casimir of the adjoint representation (see ref. [24]).

Now we set $\bar{l}_i = l_i - \nu\delta_{iN}$; the sum over $\{l\}$ in eq. (A.25) becomes

$$\sum_{\Sigma \bar{l}_i=0} \prod_{i<j} \Theta_t\left(\theta_i - \theta_j + 2\pi\left(\bar{l}_i - \bar{l}_j\right) + 2\pi\nu\left(\delta_{iN} - \delta_{jN}\right)\right)$$

$$\equiv \sum_{\Sigma l_i=0} \prod_{i<j} \Theta_t\left(\theta_i^\nu - \theta_j^\nu + 2\pi\left(l_i - l_j\right)\right) \equiv K_{SU(N)}(\theta^\nu, t), \tag{A.26}$$

where $\theta_i^\nu = \theta_i - 2\pi\nu/N + 2\pi\nu\delta_{iN}$ is precisely the Z_N transformation [eq. (A.19)].

Appendix B

MIGDAL-KADANOFF RECURSION RELATIONS

The mathematical structure of the heat kernel provides a straightforward technique for computing its power and its convolution in the weak coupling limit. Such operations appear, for example, in approximate renormalization transformations of the Migdal-Kadanoff type and can appear in other lattice calculations as well. We recall that one is confronted with two operations: the convolution $\mathfrak{B}_\lambda$

$$K(U, \tfrac{1}{2}g^2) \xrightarrow{\mathfrak{B}_\lambda} \underbrace{K \otimes K \otimes \ldots \otimes K}_{\lambda^2 \text{ times}}$$

$$\equiv \int [\mathrm{d}V_1] \ldots \int [\mathrm{d}V_{\lambda^2-1}] K(UV_1^\dagger, \tfrac{1}{2}g^2) K(V_1V_2^\dagger, \tfrac{1}{2}g^2) \cdots K(V_{\lambda^2-1}, \tfrac{1}{2}g^2), \tag{B.1}$$

and the superposition $\mathfrak{S}_\lambda$

$$K \xrightarrow{\mathfrak{S}_\lambda} K^{\lambda^{d-2}}, \tag{B.2}$$

where d is the space-time dimension. With the heat kernel, the first operation trivially gives

$$K(U, \tfrac{1}{2}g^2) \xrightarrow{\mathfrak{B}_\lambda} K(U, \tfrac{1}{2}\lambda^2 g^2). \tag{B.3}$$

In the weak coupling region, neglecting exponentially small terms, expanding

$\theta/\sin\theta \sim 1 + \frac{1}{6}\theta^2 + O(\theta^4)$ and using the identity (A.21) we get (for SU(N))

$$K\left(U, \tfrac{1}{2}g^2\right) = \mathfrak{N}\exp\left\{-\left(\frac{1}{g^2} - \tfrac{1}{24}N\right)\sum_i \phi_i^2 + O(\phi^4)\right\}, \tag{B.4}$$

where the $O(\phi^4)$ term is independent of g. Thus, for $g^2 \ll 1$

$$K\left(U, \tfrac{1}{2}g^2\right) \xrightarrow{\mathcal{S}_\lambda} K\left(U, \tfrac{1}{2}g'^2\right) \tag{B.5}$$

holds, where

$$\frac{1}{g'^2} - \tfrac{1}{24}N = \lambda^{d-2}\left(\frac{1}{g^2} - \tfrac{1}{24}N\right). \tag{B.6}$$

It follows that

$$\frac{1}{g^2} \xrightarrow{\mathcal{S}_\lambda \mathcal{B}_\lambda} \frac{\lambda^{d-4}}{g^2} - \tfrac{1}{24}N\left(\lambda^{d-2} - 1\right), \tag{B.7a}$$

$$\frac{1}{g^2} \xrightarrow{\mathcal{B}_\lambda \mathcal{S}_\lambda} \frac{\lambda^{d-4}}{g^2} - \tfrac{1}{24}N\left(\lambda^{d-2} - 1\right)\lambda^{-2}. \tag{B.7b}$$

Taking into account the identification

$$\frac{1}{g^2} - \tfrac{1}{24}N \equiv \frac{1}{g_0^2}, \tag{B.8}$$

where g_0 is the continuum bare coupling constant, we obtain [neglecting $O(g^6)$]

$$g_0^2 \xrightarrow{\mathcal{S}_\lambda \mathcal{B}_\lambda} \lambda^{4-d} g_0^2\left[1 + \tfrac{1}{24}N\left(\lambda^2 - 1\right)g_0^2\right], \tag{B.9a}$$

$$g_0^2 \xrightarrow{\mathcal{B}_\lambda \mathcal{S}_\lambda} \lambda^{4-d} g_0^2\left[1 + \tfrac{1}{24}N\lambda^{4-d}\left(1 - \lambda^{-2}\right)g_0^2\right], \tag{B.9b}$$

which are the well-known recursion relations [25]*.

* Eq. (B.9a) differs for $d \neq 4$ from eq. (52) of Migdal [25]. This is probably due to a misprint in Migdal's paper (see also ref. [26]).

A final comment about U(N). Since the Lie algebra u(N) of U(N) is the direct sum su(N) ⊕ u(1), the generators of su(N) and u(1) must decouple in the continuum limit. With the heat kernel action, we have shown in appendix A that $K_{U(N)} = \Sigma_{\gamma \in Z(N)} K^{\gamma}_{U(1)} K^{\gamma}_{SU(N)}$, but in the limit $g \to 0$ the terms corresponding to $\gamma \neq$ identity are exponentially small, i.e., the action decouples very early as opposed to Wilson's action which contains couplings of the order $g^{-2}\theta_0^2 \Sigma \theta_j^2$. Under the approximate recursion relation $\mathfrak{B}_\lambda \mathfrak{S}_\lambda$ and $\mathfrak{S}_\lambda \mathfrak{B}_\lambda$ one finds that the U(1) coupling is fixed, which means that under the renormalization transformation one has to modify the metric ds^2 in order to adjust the relative weight of U(1) and SU(N) actions.

References

[1] M. Creutz, Phys. Rev. D21 (1980) 2308
[2] C. Rebbi, Phys. Rev. D21 (1980) 3350;
G. Bhanot and C. Rebbi, Nucl. Phys. B180[FS2] (1981) 469
[3] D.J. Gross and E. Witten, Phys. Rev. D21 (1980) 446
[4] Y.Y. Goldschmidt, J. Math. Phys. 21 (1980) 1842
[5] H. Neuberger, Nucl. Phys. B179 (1981) 253
[6] C.B. Lang, P. Salomonson and B.S. Skagerstam, Nucl. Phys. B190[FS3] (1981) 337
[7] N.S. Manton, Phys. Lett. 96B (1980) 328
[8] A.M. Polyakov, Phys. Lett. 72B (1978) 447
[9] J.M. Drouffe, Phys. Rev. D18 (1978) 1174
[10] L. Susskind, Phys. Rev. D20 (1979) 2610
[11] M. Stone, Nucl. Phys. B152 (1979) 97
[12] A. M. Polyakov, unpublished lecture notes, ed. E. Gava, IAEA, Trieste IC/78/4;
J.V. Josè, L.P. Kadanoff, S. Kirkpatrick and D.R. Nelson, Phys. Rev. B16 (1977) 1217;
R. Savit, Phys. Rev. Lett. 39 (1977) 55; Phys. Rev. B17 (1978) 1340;
A.H. Guth, Phys. Rev. D21 (1980) 2291
[13] M. Nauenberg and D. Toussaint, Nucl. Phys. B190[FS3] (1981) 217
[14] M. Lüscher, G. Münster and P. Weisz, Nucl. Phys. B180[FS2] (1981) 1
[15] J.B. Kogut, R.P. Pearson and J. Shigemitsu, University of Illinois preprint ILL-TH-80-41 (1980)
[16] C.B. Lang, C. Rebbi, P. Salomonson and B.S. Skagerstam, CERN preprint TH-3021 (1981), Phys. Lett. B, to be published
[17] J. Kogut and L. Susskind, Phys. Rev. D11 (1975) 395;
J. Kogut, Phys. Reports 67 (1980) 67
[18] S.R. Wadia, Phys. Lett. 93B (1980) 403
[19] W.H. Furry, Phys. Rev. 107 (1957) 7;
M. Born and G.Z. Ludwig, Z. Phys. 150 (1958) 106
[20] A. Hasenfratz and P. Hasenfratz, Phys. Lett. 93B (1980) 165
[21] S.A. Molchanov, Russ. Math. Surv. 30:1 (1975) 1
[22] P.B. Gilkey, The index theorem and the heat equation (Publish or Perish, Boston, 1974);
N. Hurt and R. Hermann, Quantum statistical mechanics and Lie group harmonic analysis, part A (MATH SCI Press, Brookline, MA., 1980)
[23] J.S. Dowker, J. Phys. A3 (1970) 451; Ann. of Phys. 62 (1971) 361
[24] M.S. Marinov and M.V. Terentev, Sov. J. Nucl. Phys. 28 (1978) 729; Fortschr. Phys. 27 (1979) 511
[25] A.A. Migdal, JETP (Sov. Phys.) 42 (1976) 413, 743;
L.P. Kadanoff, Ann. of Phys. 100 (1976) 359; Rev. Mod. Phys. 49 (1977) 267;
S. Caracciolo and P. Menotti, Ann. of Phys. 122 (1979) 74
[26] S. Caracciolo, Tesi, Pisa (1979) unpublished

Volume 101B, number 3 PHYSICS LETTERS 7 May 1981

THE TRANSITION FROM STRONG COUPLING TO WEAK COUPLING IN THE SU(2) LATTICE GAUGE THEORY

C.B. LANG, C. REBBI [1], P. SALOMONSON and B.S. SKAGERSTAM
CERN, Geneva, Switzerland

Received 28 January 1981

The effects of modifications in the action for SU(2) lattice gauge theories are investigated by Monte Carlo computations. We find that Manton's action and a generalization of Villain's U(1) action lead to earlier approaches to scaling and evaluate the relevant parameters. We also find evidence of a higher-order phase transition with Manton's action.

General considerations. The lattice formulation of gauge field theories [1] is considered by particle physicists as a very convenient way of performing a gauge invariant regularization, which also allows a strong coupling expansion. Eventually, a continuum limit must be taken. From this point of view, the specific form chosen for the lattice action is not so relevant: it is expected that wide classes of actions lead to the same continuum theory when the coupling constant approaches some suitable scaling critical point. Yet much attention has also been paid to the properties of lattice systems away from their critical points: besides being interesting per se, it is believed that these investigations may provide a clue to the understanding of the corresponding continuum theories. It then becomes important to see how a change in the form of the action affects the features of the theory. It is clear that any property of the lattice system, such as the rapid cross-over from a strong coupling to a scaling regime observed with non-abelian gauge models [2], can be related to properties of the continuum theory [3] only if some degree of universality with respect to a modification of the lattice action exists.

These and other considerations have motivated us to perform a numerical analysis, by the Monte Carlo method, of the SU(2) lattice gauge theory defined with actions different from the one standardly used in these computations.

[1] On leave from Physics Department, Brookhaven National Laboratory, Upton, NY, USA.

The dynamical variables of a lattice gauge theory are, we recall, the finite group elements U_{ij} associated with the oriented links of the lattice. The ordered products $U_\square$ of the four U_{ij} associated with the links forming the sides of an elementary square constitute the plaquette variables. To form the action, one considers a class function (i.e., a function invariant under similarity transformations $U_\square \to G^{-1} U_\square G$, which guarantees gauge invariance) $f(U_\square)$ and then sums over all plaquettes [‡1]. The dependence on the coupling constant g, or on the parameter $\beta = 4/g^2$, is contained in f. Most often it is simply multiplicative: $f(U_\square, \beta) = \beta \hat{f}(U_\square)$, but this is not necessary.

If the gauge group is SU(2), the only one which will be considered in this paper, then $U_\square$ may be represented in the form:

$$U_\square = \cos\theta_\square + i\boldsymbol{\sigma}\cdot\hat{\boldsymbol{n}} \sin\theta_\square, \tag{1}$$

and f must be an even, periodic function of θ. The so-called Wilson form of the action is obtained choosing

$$f_W = \beta(1 - \tfrac{1}{2}\operatorname{tr} U_\square) = \beta(1 - \cos\theta_\square) \tag{2}$$

and, to our knowledge, it is the only one which has been used in Monte Carlo simulations of the SU(2) gauge theory. However, one would expect that any f,

[‡1] Effects of change in the form of the action, made to incorporate also variables extending to more than one plaquette, have been considered by Wilson [4]. This change is of a different nature than the one studied in the present paper.

with reasonable properties, could be used to define a lattice system with the same continuum limit. The action of a single plaquette should be minimum when $U_\square$ equals the identity; near the minimum it should have a quadratic behaviour which is convenient to normalize so that, at least for large β,

$$f(\theta,\beta) \underset{\theta\sim 0}{\sim} \tfrac{1}{2}\beta\theta^2 \qquad (3)$$

and it should have no local minima other than the values $\theta = 0 \pmod{2\pi}$.

In particular, interesting candidates which have been proposed and investigated in different contexts are Manton's form [5,6]:

$$f_M = \tfrac{1}{2}\beta\bar{\theta}_\square^2, \qquad (4)$$

where by $\bar{\theta}_\square$ we denote the value between $-\pi$ and π congruent to $\theta_\square$ modulus 2π, and a generalization of the form suggested by Villain for the XY model [7]. In this generalization one takes for $\exp(-f)$ the heat kernel over the group manifold [8,9] which is given by

$$\begin{aligned}\exp(-f_V) &= \left(\sum_{n=-\infty}^{\infty} (\sin\theta)^{-1}(\theta+2n\pi)\exp[-\tfrac{1}{2}\beta(\theta+2n\pi)^2]\right)\\ &\times\left(\sum_{n=-\infty}^{\infty}(1-4\beta n^2\pi^2)\exp[-2\beta n^2\pi^2]\right)^{-1}\\ &=\left(\sum_{l=0}^{\infty}(l+1)(\sin\theta)^{-1}\sin[(l+1)\theta]\right.\\ &\left.\times\exp[-(2\beta)^{-1}l(l+2)]\right)\\ &\times\left(\sum_{l=0}^{\infty}(l+1)^2\exp[-(2\beta)^{-1}l(l+2)]\right)^{-1}. \qquad (5)\end{aligned}$$

We have performed Monte Carlo simulations of the SU(2) lattice gauge system with these two actions, measuring the average value of the internal energy, its derivative, related to the specific heat of the system, and the string tension. We have found that using Manton's form of the action one obtains a smoother behaviour of the internal energy versus β than with Wilson's action [2,10–12]; in particular the inflexion point, which with Wilson's action leads to a pronounced peak [13] in the specific heat, is absent.

Measurements of the string tension, done both with Manton's action and with the generalized Villain's action, exhibit very clearly the scaling behaviour characteristic of the approach to the continuum limit and already observed with Wilson's form of the action [2,10, 11]. The transition from the strong coupling regime to the scaling regime occurs earlier in β, at $\beta \approx 1.5$ with Manton's action and at $\beta \approx 2.0$ with the generalized Villain's form. This gives larger values for the Λ_0 parameters than in Wilson's case, $\Lambda_0^{(M)} = (0.0616 \pm 0.0020)\sqrt{\sigma}$ and $\Lambda_0^{(V)} = (0.0206 \pm 0.0011)\sqrt{\sigma}$ ($\Lambda_0^{(W)} = (0.012 \pm 0.002)\sqrt{\sigma}$ [10,11]), which agree with theoretical expectations (see later).

Apart from this difference in the scales, the transition appears still rather abrupt and similar to what one observes with Wilson's action. But a most surprising result came up when we evaluated the derivative of the internal energy with Manton's action. The very good statistics of our simulation allowed us to determine this function very accurately, both from a direct evaluation of the derivative and from a computation of the energy fluctuations. The curve we obtained reveals a very marked, discontinuous change in slope at $\beta \approx 1.6$. Although a numerical computation can never distinguish between a real singularity and regular, but rapidly varying behaviour, the discontinuity in the second derivative of the internal energy, if indeed present, would signal a third-order phase transition.

In our computation we have approximated the full SU(2) gauge group with its 120 element icosahedral subgroup. That gauge system with finite subgroups of SU(2) may give very good approximations as has been shown already in the literature [12,14,11]. The technical advantage of working with a finite group is enormous: the possibility of using a multiplication table, stored in the computer memory, to implement the group operation and to perform other logical manipulations allows a very marked reduction in computer time. On the CDC 7600 at CERN it takes our algorithm $\approx$0.8 s to perform a full Monte Carlo iteration with an 8^4 lattice, i.e., to go through the upgrading procedure of all 16384 dynamical variables. The Monte Carlo method has been by now presented many times in the literature and we shall not dwell on it. We mention only that we have used a generalization of the algorithm of Metropolis et al. [15], in which the new trial value for a U_{ij} variable is obtained by multiplying U_{ij} by one of the 12 neighbours of the identity (i.e., the group elements of the form $\cos\theta + i\boldsymbol{\sigma}\cdot\hat{\boldsymbol{n}}\sin\theta$ with $|\theta| = 2\pi/10$, the lowest non-vanishing value) chosen at random;

Volume 101B, number 3 PHYSICS LETTERS 7 May 1981

this upgrading procedure is then repeated six times before going to another U_{ij}. We have done all computations on an 8^4 lattice, except for the analysis of the energy fluctuations, which has been done with long runs of 10 000 iterations on a 4^4 lattice.

The results. With Wilson's and Manton's actions we define the average plaquette contribution to the action of a certain lattice configuration with the factor of β removed:

$$S_P^W \equiv \frac{1}{N}\sum_\square (1 - \cos\theta_\square), \qquad (6)$$

$$S_P^M \equiv \frac{1}{N}\sum_\square \tfrac{1}{2}\bar{\theta}^2_\square, \qquad (7)$$

where N is the number of plaquettes; we call the expectation value $\langle S_P\rangle$ the "internal energy" E. In the generalized Villain's model the more complicated dependence of the action on β does not lend itself to a simple definition of internal energy and we have not considered this quantity.

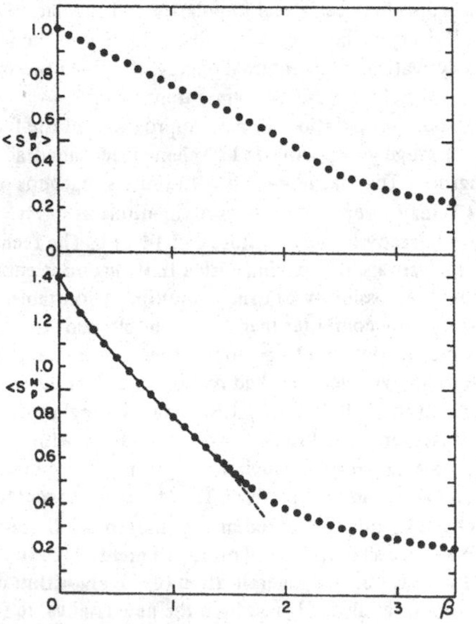

Fig. 1. The average plaquette contribution to the internal energy with Wilson's (upper part, from ref. [11]) and Manton's form of the action; the curve is the strong coupling expansion as described in the text.

In fig. 1a, taken from ref. [11], we reproduce the results of a Monte Carlo evaluation of the function $E_W(\beta)$. In fig. 1b we show the results of our determination of $E_M(\beta)$. The measurement has been done with a very large number of Monte Carlo iterations on an 8^4 lattice with periodic boundary conditions. The number of iterations performed to evaluate E varied from 120, for the values of β where less accuracy was sought (0.8–1), to 240 or 300 for $\beta = 1.1$–1.45, 1.55, 2.4 and 2.5, and to over 800 for the other points in the range 1.5–2 where we wanted to have a quite precise determination of E. The efficiency of our algorithm allowed us to perform all these iterations while still keeping the cost of the computing moderate; the errors inherent in the statistical fluctuations are thus sensibly reduced. A straightforward evaluation of these errors, according to the formula

$$\Delta E = (n-1)^{-1/2}\langle S_P^2 - \langle S_P\rangle^2\rangle^{1/2}$$

(n being the number of iterations), would indeed give $\Delta E \lesssim 1.3 \times 10^{-4}$ for the points between $\beta = 1.5$ and $\beta = 2$, but this number underestimates the possible effects of fluctuations with a range comparable with or longer than n, and therefore the true errors may be larger. Another detail that should be mentioned about the computational procedure is that we took as initial configurations for the runs used to evaluate E (and the string tension) those encountered along a very slow cooling process: starting from a completely disordered lattice several Monte Carlo iterations were performed to bring β to 0.8 and then 100 (or 200 for larger β) further iterations were done at each value of β before proceeding to the next one. This, of course, reduces any errors due to possible deviations from statistical equilibrium.

In fig. 1b we also reproduce the strong coupling expansion for E [carried uo to $O(\beta^{14})$]. In deriving it one utilizes the character expansion for Manton's action

$$\exp[-f_M(\theta,\beta)] = \alpha(\beta)\left[1 + \sum_{r>0} \nu_r c_r(\beta)\chi_r(U)\right], \qquad (8)$$

where the sum is over all non-trivial representations with dimensions ν_r. It should be noticed that, in distinction to Wilson's case, all $c_r(\beta)$ are of order β and in principle all representations have to be taken into account at every order. In practice, the contributions from higher representations become rapidly negligible. The strong coupling expansion for E is then obtained by standard methods (see, for instance, ref. [16]). (Incidentally, up to the order considered, insertion of the characters for the SU(2) group or its icosahedral subgroup makes very little difference in the curve for $E(\beta)$, thus supporting the validity of the discrete group approximation.) We notice in fig. 1b a rather marked difference in the behaviour of the function $E(\beta)$ with respect to Wilson's case. In particular, the inflection point, which leads to a pronounced peak in the specific heat, is absent. However, as with Wilson's action, the strong coupling expansion (to the order considered) appears to reproduce quite well the data up to some value of β, after which it departs from the numerical results rather abruptly.

Figs. 2a and 2b illustrate the results of our measurements of the string tension. If U_γ is the product of the U_{ij} variables along a closed path γ, the Wilson loop factor is defined as

$$W_\gamma = \langle \tfrac{1}{2} \operatorname{tr} U_\gamma \rangle. \tag{9}$$

The coefficient κ of an area term in the exponential fall-off

$$W_\gamma \sim \exp(-\kappa A_\gamma), \tag{10}$$

may be evaluated comparing loop factors for square and rectangular loops of the same perimeter and different areas [11,12] or, anyway, of suitably related sizes [10]. If γ is a rectangular path extending for m and n plaquettes and we denote W_γ by $W(m, n)$, some of the possible determinations of κ are given by

$$\kappa^{(A)}(\beta, m) = -\ln \frac{W(m,m)W(m-1,m-1)}{W(m,m-1)^2}, \tag{11}$$

$$\kappa^{(B)}(\beta, m) = -\ln \frac{W(m,m)}{W(m+1,m-1)} \tag{12}$$

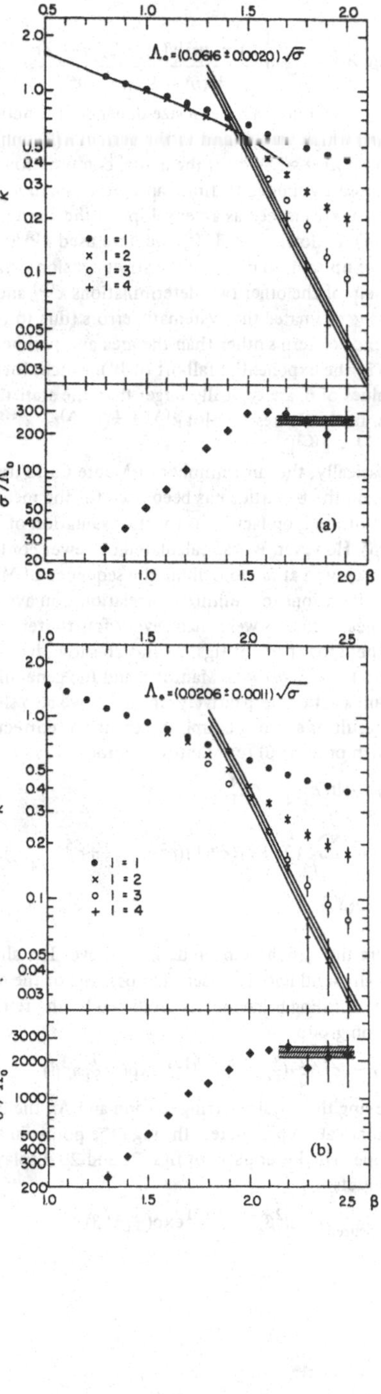

Fig. 2. (a) In the upper part we give the results for Wilson's loop averages W_γ (loop sizes up to 4 × 4) using Manton's action. In the lower part we display the points on the envelope again, but now the expected scaling behaviour has been divided out and the values should tend asymptotically to σ/Λ_0^2. The solid lines represent the strong coupling expansion and the fit to the scaling behaviour. (b) Like fig. 2(a), but using the generalization of Villain's form of the action.

and

$$\kappa^{(C)}(\beta, m) = -\tfrac{1}{2}\ln \frac{W(m,m)W(m-2,m-2)}{W(m-1,m-1)^2}. \qquad (13)$$

In any case, one evaluates a size dependent function $\kappa(\beta, m)$ which would tend to the correct $\kappa(\beta)$ only in the limit $m \to \infty$. Since in the actual computation one is necessarily limited to finite and rather small m, the function $\kappa(\beta)$ is seen as an envelope of the curves $\kappa(\beta, m)$. Following ref. [10], we have used $\kappa^{(A)}(\beta, m)$, with m up to 4, to measure the string tension; with the help of the other two determinations $\kappa^{(B)}$ and $\kappa^{(C)}$ we have estimated the systematic errors (due to the presence of terms other than the area and perimeter ones in the exponential fall-off of W), which, for higher values of β, are typically larger than the statistical ones. In the figures we plot $\kappa^{(A)} \pm \frac{1}{2}(|\kappa^{(A)} - \kappa^{(B)}| + |\kappa^{(A)} - \kappa^{(C)}|)$.

Basically, the same number of Monte Carlo iterations on the 8^4 lattice has been used for the measurement of the loop factors as for the evaluation of $E_M(\beta)$. However, W was calculated on fewer configurations, chosen at random along the sequence of Monte Carlo iterations to minimize correlations; on average the measurements were made every fourth iteration.

The upper parts of figs. 2a and 2b show the values found for $\kappa(\beta, m)$ with Manton's and the generalized Villain's action, respectively. In fig. 1a we also show the result of a strong coupling expansion (correct up to fifth order in β) for Manton's action

$$\kappa(\beta) = -\ln c_{1/2} - 4c_{1/2}^4$$
$$-4\sum_{r\geq 1/2}(2r+1)(r+1)(c_r^5 c_{r+1/2} + c_{r+1/2}^5 c_r)c_{1/2}^{-1}$$
$$+ O(\beta^6), \qquad (14)$$

where the $c_r(\beta)$ have been defined above. The almost straight solid lines represent the best fits of the expected scaling behaviour as predicted by the renormalization group

$$\kappa(\beta) \sim (\sigma/\Lambda_0^2)(\tfrac{6}{11}\pi^2\beta)^{102/121}\exp(-\tfrac{6}{11}\pi^2\beta) \qquad (15)$$

(σ being the physical string tension and Λ_0^2 the lattice regularization parameter) through the points in the envelope. The lower parts of figs. 2a and 2b display the products

$$\kappa_{\text{measured}}(\tfrac{6}{11}\pi^2\beta)^{-102/121}\exp(\tfrac{6}{11}\pi^2\beta), \qquad (16)$$

which should tend asymptotically to σ/Λ_0^2.

The numerical values we find for $\kappa(\beta)$ agree extremely well with the prediction of the strong coupling expansion for small β and with the expected scaling behaviour, when β exceeds some critical value. The Λ_0 parameters are determined to be

$$\Lambda_0^{(M)} = (0.0616 \pm 0.0020)\sqrt{\sigma}, \qquad (17)$$

$$\Lambda_0^{(V)} = (0.0206 \pm 0.0011)\sqrt{\sigma}, \qquad (18)$$

in the two cases. Both values are larger than the one determined with Wilson's form [10,11]

$$\Lambda_0^{(W)} = (0.012 \pm 0.002)\sqrt{\sigma}, \qquad (19)$$

as may be expected because of the absence or smallness of the term $O(\bar{\theta}^4)$ in the action [‡2]. The passage from the strong coupling behaviour to the scaling behaviour may appear slightly smoother in the graphs of figs. 2a and 2b than in the corresponding diagrams for the model with Wilson's action. However, in the present computation, the system had the possibility of relaxing toward statistical equilibrium for more iterations than in previous Monte Carlo work and, given the relative slowness of the approach to equilibrium in the transition region, this might justify slightly smoother curves. On the whole, the transition between the two regimes is still abrupt and this gives further support to the idea that some relevant physical effect may be occurring there [3,18].

The high accuracy in our determination of E_M on the 8^4 lattice has allowed us to obtain the derivative $-(\partial E_M/\partial\beta)$ just by forming finite differences. In the infinite volume limit this quantity is related to the specific heat of the (four-dimensional) system

$$C_M(\beta) = -\beta^2\, \partial E_M/\partial\beta. \qquad (20)$$

The results are represented by the crosses in fig. 3. One notices immediately that while $-(\partial E_M/\partial\beta)$ is continuous, there is an apparent break in the slope of the

‡2 This may be seen by examining the relevance of the various terms in the perturbative computation of $\Lambda_0^{(W)}/\Lambda_{\text{mom}}$ in ref. [17]. Adapting that computation to the derivation of $\Lambda_0^{(M)}/\Lambda_{\text{mom}}$ and $\Lambda_0^{(V)}/\Lambda_{\text{mom}}$ we find values for the ratios of scales which agree with $\Lambda_0^{(M)} > \Lambda_0^{(V)} > \Lambda_0^{(W)}$ but deviate somewhat from the numerical results. A more detailed investigation of perturbation theory for the ratios Λ_M/Λ_W and Λ_V/Λ_M is in progress.

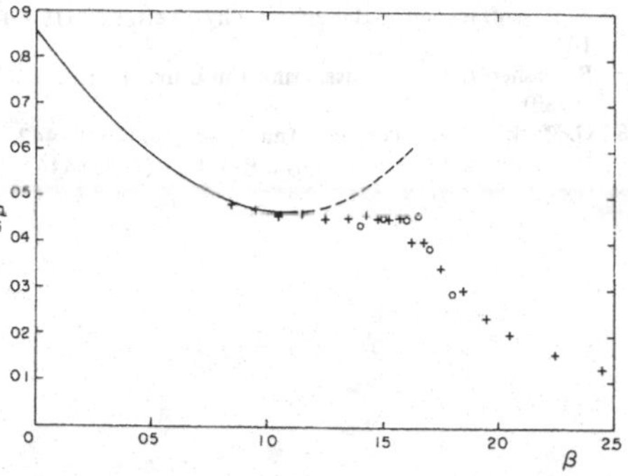

Fig. 3. The crosses denote $-\partial E/\partial\beta$ as obtained by taking the slopes between any two neighbour points of the internal energy (fig. 1) obtained on an 8^4 lattice with Manton's action. The open circles denote the same quantity, but derived from the fluctuations of the internal energy on a 4^4 lattice over 10 000 Monte Carlo iterations. The curve represents the strong coupling expansion as given in the text.

curve at $\beta = 1.6$. The point at $\beta = 1.625$ seems actually to lie outside any reasonable extrapolation of the curve. Assuming that the errors in E_M are entirely due to normally distributed statistical fluctuations, then the errors in $-(\partial E_M/\partial\beta)$ should be of size smaller than the crosses in the figure. However, as we have discussed already, long-range fluctuations may be responsible for larger errors in E, especially near a critical point, and this could justify the rather odd position of the point at $\beta = 1.625$.

The slope of the internal energy may also be evaluated from an analysis of the fluctuations in E itself, according to the formula

$$-\partial E/\partial\beta = N[\langle S_p^2\rangle - \langle S_p\rangle^2], \quad (21)$$

where the bracket denotes averages over the values measured after each Monte Carlo iteration and N is the number of plaquettes in the lattice. An accurate determination of $-(\partial E/\partial\beta)$ by this method requires averages over very large numbers of Monte Carlo iterations, precisely to allow possible long-range fluctuations to take place. We have therefore done the measurement on a 4^4 lattice with Monte Carlo runs of 10 000 iterations (required computing time $\approx$500 s on the CDC 7 600). The results are given by the open circles in fig. 3. The agreement with the determination of $-(\partial E/\partial\beta)$ by finite differences is very good and a break in the slope of the curve is again quite noticeable.

A real discontinuity in the second derivative of the internal energy would be a quite interesting feature of the model. It would correspond to a non-deconfining phase transition of a novel and intriguing nature (presumably third order). We notice that the results of Lautrup and Nauenberg [13], who studied the specific heat of the SU(2) system with Wilson's action, are not incompatible with a discontinuity in $-(\partial^2 E/\partial\beta^2)$. The non-scaling peak which they find could actually correspond to a sudden change of slope, which appears as a peak or a cusp because it occurs at the inflection point of $E(\beta)$ and thus is superimposed on a local maximum. It is also interesting that a discontinuity in the derivative of the specific heat is present in the $N \to \infty$ limit of the two-dimensional SU(N) model with Wilson's action [19], although the discontinuity disappears if Manton's action is used instead [6].

Unfortunately, from a numerical computation one can obtain only strong evidence for a discontinuity and the possibility that the variation be continuous, albeit very abrupt, is still open. In any case, we believe this unexpected behaviour of the specific heat to constitute a very interesting result and hope that further investigations will clarify its origin.

We are grateful to André Peterman for a critical reading of the manuscript. Useful conversations with Enrico Onofri are also gratefully acknowledged.

References

[1] K.G. Wilson, Phys. Rev. D10 (1974) 2445.
[2] M. Creutz, Phys. Rev. Lett. 43 (1979) 553; Phys. Rev. D21 (1980) 2308.
[3] C.G. Callan, R.F. Dashen and D.J. Gross, Phys. Rev. Lett. 44 (1980) 435.
[4] K.G. Wilson, Cargèse Lecture Notes (1979), Vol. 59 (Plenum, New York, 1980).
[5] N.S. Manton, Phys. Lett. 96B (1980) 328.
[6] C.B. Lang, P. Salomonson and B.S. Skagerstam, CERN preprint TH 2993 (1980); Phys. Lett. 100B (1981) 29.
[7] J. Villain, J. de Phys. 36 (1975) 581.
[8] J.M. Drouffe, Phys. Rev. D18 (1978) 1174.
[9] P. Menotti and E. Onofri, private communication.
[10] M. Creutz, Phys. Rev. Lett. 45 (1980) 313.
[11] G. Bhanot and C. Rebbi, CERN preprint TH 2979 (1980), to be published in Nucl. Phys. (FSI).
[12] C. Rebbi, Phys. Rev. D21 (1980) 3350.
[13] B. Lautrup and M. Nauenberg, Phys. Rev. Lett. 45 (1980) 755.

[14] D. Petcher and D.H. Weingarten, Phys. Rev. D22 (1980) 2465.
[15] V. Metropolis, A.W. Rosenbluth, A.H. Teller and E. Teller, J. Chem. Phys. 21 (1953) 1087.
[16] J.M. Drouffe, Nucl. Phys. B170 (1980) 91 (FSI).
[17] A. Hasenfratz and P. Hasenfratz, Phys. Lett. 93B (1980) 165;
R. Dashen and D.J. Gross, Princeton Univ. preprint (1980).
[18] G. Mack and V.B. Petkova, Ann. Phys. 123 (1979) 442.
[19] D.J. Gross and E. Witten, Phys. Rev. D21 (1980) 446.

Volume 101B, number 5 PHYSICS LETTERS 21 May 1981

THE PHASE STRUCTURE OF SU(N)/Z(N) LATTICE GAUGE THEORIES

I.G. HALLIDAY and A. SCHWIMMER [1]
Blackett Laboratory, Imperial College, London SW7 2BZ, UK

Received 9 March 1981

The phase structure of pure SU(N)/Z(N) lattice gauge theories in four dimensions is discussed. The presence of Z(N) monopoles plausibly leads to a phase transition. A Monte Carlo simulation of SO(3) shows the presence of a very strong, maybe first order, phase transition.

Pure gauge theories in the continuum are defined in terms of a Lie algebra. On the lattice the global structure of the group must be given. The lattice therefore clearly differentiates between theories with different groups and the same algebra. Such theories have the same naive continuum limit.

In the present note we investigate how SU(N)/Z(N) and SU(N) reach their joint continuum limit.

The general mechanism is that SU(N)/Z(N) where various elements of SU(N) are effectively identified possesses additional, topological, solitons. In the strong-coupling limit the solitons condense into the vacuum. When the coupling is decreased a phase transition may occur the vacuum becoming free of solitons. It is then very similar to the vacuum of the SU(N) theory. From this point on, as the coupling becomes weaker, the continuum limit is reached jointly by the two theories. An example of similar behaviour is compact QED in four dimensions [1].

The Wilson action for an SU(N)/Z(N) gauge theory is

$$S_{\mathrm{w}} = \sum_p [\chi_{\mathrm{a}}(U(\partial p)) - 1] , \tag{1}$$

where p are the plaquettes, ∂p the boundary of the plaquette, $U(\partial p) = U(b_4)U(b_3)\ldots U(b_1)$, U being SU(N) matrices defined on the bonds b_i. χ_{a} is the character in the adjoint representation.

The generating functional Z is obtained by integrating with the SU(N) invariant group measure $\mathrm{d}U(b)$.

$$Z_{\mathrm{w}} = \int \prod \mathrm{d}U(b) \exp(\beta S_{\mathrm{w}}) . \tag{2}$$

Clearly the action S_{w} gives equal weight to group elements which differ by an element of the center Z(N) i.e. such elements are identified. This theory has Z(N) monopoles with finite energy. A monopole configuration is built by choosing a set of plaquettes going to infinity which carry an element σ belonging to the center Z(N). They may be considered to be all the plaquettes pierced by a Dirac string. The energy of the string is zero. Due to the conservation of flux the same total amount of flux σ will be found spread on a big sphere surrounding the origin of the string [‡1]. Such a configuration is topologically stable since it is in a one-to-one correspondence with an element of $\Pi_1(\mathrm{SU}(N)/\mathrm{Z}(N)) = \mathrm{Z}(N)$ [‡2].

Due to the existence of these monopoles the above two phases are possible. A convenient way to make the above features explicit is to replace the Wilson generating functional (2) with the Villain form

$$Z_{\mathrm{v}} = \sum_{\sigma(p)} \int \mathrm{d}U(b) \exp\left(\beta \sum_p \mathrm{tr}(U(\partial p)) \cdot \sigma(p)\right) , \tag{3}$$

where $\sigma(p)$ is a plaquette variable $\in$ Z(N) and $U \in$ SU(N).

[1] On leave of absence from the Weizmann Institute, Israel.

[‡1] We stress that this monopole is different from a Z(N) monopole in a SU(N) theory whose Dirac string is visible i.e. it carries energy [2].

[‡2] In the continuum such monopoles can be explicitly constructed if one embeds SO(3) as an unbroken group in a bigger group e.g. SU(3) broken through a Higgs mechanism [3].

Volume 101B, number 5 PHYSICS LETTERS 21 May 1981

Z_v corresponds to an SU(N)/Z(N) theory since, on summing over σ, the character expansion of Z contains only zero N-ality representations of SU(N). The weight for two values of a bond variable differing by an element of the center is the same. This can be formally seen on remarking that (3) has a Kalb–Ramond [4] Z(N) gauge invariance

$$U(b) \to U(b)\,\xi(b)\,, \quad \sigma(p) \to \sigma(p)\,\xi(\partial p)\,, \tag{4}$$

for $\xi(b) \in Z(N)$ as well as the usual SU(N) gauge invariance.

The Z(N) monopoles are now exposed by writing

$$\sigma(p) = \exp[(2\pi i/N)\, n_{\mu\nu}]\,,$$

where $n_{\mu\nu}$ is a plaquette integer $0 \leqslant n_{\mu\nu} \leqslant N-1$. The most general expression for $n_{\mu\nu}$ is

$$n_{\mu\nu} = \Delta_\mu m_\nu - \Delta_\nu m_\mu + \epsilon_{\mu\nu\rho\sigma}\, a^\rho M^\sigma / a \cdot \Delta\,, \tag{5}$$

Here M^σ is a bond variable on the dual lattice

$$M^\sigma = \epsilon^{\sigma\rho\mu\nu} \Delta_\rho n_{\mu\nu} \tag{6}$$

and a^ρ is an arbitrary unit vector on the dual lattice. M^σ and m_μ fulfill the constraints

$$\Delta_\sigma M^\sigma = 0\,, \quad a^\mu m_\mu = 0\,. \tag{7}$$

Absorbing the m's in $U(b)$ we obtain

$$Z_v = \sum_{M^\sigma} \int dU(b) \exp\Big(\beta \sum_p \mathrm{tr}(U(\partial p)) \times \exp[(2\pi i/N)\, \epsilon_{\mu\nu\rho\sigma}\, a^\rho M^\sigma / a \cdot \Delta]\Big)\,. \tag{8}$$

Here M^σ is a conserved (mod N) monopole current. It represents a vacuum fluctuation caused by the creation and subsequent annihilation of a monopole–anti-monopole pair. The monopoles interact via the full SU(N) gauge theory.

To study the large distance behaviour of (3) we block-spin treating $\sigma(p)$ as a variable. We produce new interaction terms consistent with the symmetries of the problem. These are the SU(N) gauge invariance and the special Z(N) gauge invariance (4). The simplest new interaction term is

$$\gamma\sigma(\partial c) = \gamma\sigma(p_6) \ldots \sigma(p_1) = \gamma \exp[(2\pi i/N) M^\rho]\,, \tag{9}$$

where c is a cube and $p_1 \ldots p_6$ are the plaquettes of its boundary. This is the gauge interaction for a Kalb–Ramond theory or can be understood as a chemical potential for monopole loops. Thus we are led to study

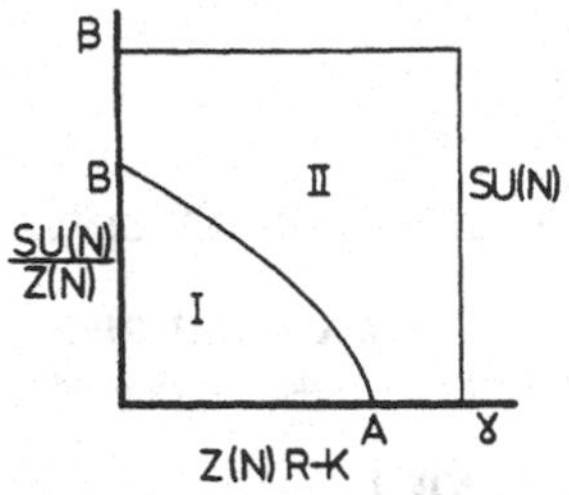

Fig. 1. The phase diagram for the Villain Model [eq. (10)].

the phase structure of a plaquette Z(N) interacting with an SU(N) gauge field through

$$\mathcal{L} = \beta\, \mathrm{tr}(U(\partial p))\sigma(p) + \gamma\sigma(\partial c)\,. \tag{10}$$

This theory has the following boundaries in the β, γ plane (fig. 1)

(a) For $\beta = 0$ it gives a Z(N) Kalb–Ramond gauge theory.

(b) For $\beta \to \infty$ $\sigma(p)$ becomes a pure gauge and so there is no γ dependence.

(c) For $\gamma = 0$ we have our SU(N)/Z(N) theory.

(d) For $\gamma \to \infty$ the $\sigma(p)$ become pure gauge and so can be absorbed in $U(b)$ giving the SU(N) gauge theory.

The plaquette Z(N) gauge theory, being dual to a Z(N) spin model [5], is known to have one second-order phase transition. (pt. A in fig. 1). This transition is a result of the competition between the chemical potential of the loops and their entropy which is proportional to their length. This transition is at least locally stable. The first term in a β-expansion produces a renormalisation

$$\gamma \to \gamma + \beta^6\,. \tag{11}$$

Then the phase boundary is

$$\gamma_c = \gamma + \beta^6\,. \tag{12}$$

If this line crosses the $\gamma = 0$ axis two phases for SU(N)/Z(N) are produced.

Region I where γ went to the $\gamma = 0$ fixed point of the plaquette theory and the monopole loops condensed.

Region II where γ went to the $\gamma = \infty$ fixed point i.e. at large distances $M^\rho = 0$ and the SU(N)/Z(N) vacuum is analytically connected to SU(N).

We remark that the phase structure is very different from what one would expect if the plaquettes were coupled to a $Z(N)$ bond gauge theory instead of $SU(N)$. In such a situation the phase boundary would join the $\gamma = \infty$ axis [6]. The entire Osterwalder–Seiler analyticity region [7] of such a model is mapped inside our region I.

We perform a duality transformation on the $\sigma(p)$ variables in (10) for the particular case $N = 2$ i.e. the SO(3) theory. Fixing the Z(2) gauge such that $\mathrm{Re}\, U_{11}(b) \geqslant 0$ we obtain after standard manipulations:

$$Z_v = \sum_{\xi(\dot{b}),\eta(\dot{s})} \int_{\mathrm{Re}\, U_{11} \geqslant 0} \mathrm{d}U(b) \exp\Big(f(\tilde{\gamma}) + \sum [f(\tilde{\beta}(p)) + \tilde{\beta}(p)\, \xi(\partial \dot{p})] + \sum_{\dot{b}} \tilde{\gamma} \cdot \xi(\dot{b})\, \eta(\partial \dot{b})\Big), \tag{13}$$

where for a variable x, $\tilde{x}$ and $f(\tilde{x})$ are defined as

$$\tilde{x} = \tfrac{1}{2} \ln \mathrm{cth}\, x, \quad f(x) = \tfrac{1}{2} \ln [\mathrm{ch}\, x\, \mathrm{sh}\, x], \tag{14}$$

$\beta(p) \equiv \beta\, \mathrm{tr}\, U(\partial p)$ and the variables ξ, η are Z(2) variables defined on dual bonds, $\dot{b}$, and sites $\dot{s}$ respectively.

From (13) we can deduce the degrees of freedom of the theory: $\eta(s)$ represents the phase of a fixed length Z(2) monopole field defined on the sites of the dual lattice. The monopoles interact with the SO(3) degrees of freedom through a Z(2) magnetic gauge field $\xi(\dot{b})$. The monopole–magnetic gauge field coupling is of the standard Higgs type i.e. $\gamma \xi(\dot{b})\, \eta(\partial \dot{b})$. Dual transforming the magnetic gauge field we obtain an electric Z(2) gauge field which with the SO(3) degrees of freedom produces the full SU(2). We believe that for any SO(3) lagrangian this mechanism represents correctly the degrees of freedom i.e. the full SU(2) gauge theory in addition to the monopoles. Therefore region I is a Z(2) magnetic Higgs phase in which the Z(2) monopoles are condensed screening the Z(2) magnetic loops.

In region II the screening by the monopoles does not occur but the exact behaviour of the magnetic loops depends on their interaction with the SO(3) degrees of freedom; i.e. on the structure of the SU(2) theory. In the simplest situation where big magnetic loops (of certain thickness) are always condensed [8] the transition between regions I and II can be understood as an interchange of the roles of monopoles and magnetic loops.

The simplest order parameters, therefore, would be the density of monopoles of big magnetic loops. The Wilson and 't Hooft loops defined in the fundamental representation take trivial values in an SO(3) theory (0 and 1, respectively). However, they can be modified to include information about the monopoles and the magnetic loops. The modified Wilson loop A depending explicitly on the plaquettes P on which it is defined is:

$$\langle \tilde{A}(P) \rangle \equiv \Big\langle \mathrm{tr}\Big(\prod_{b \in \partial P} U(b)\Big)\Big\rangle \prod_{p \in P} \sigma(p). \tag{15}$$

It is amusing to note that objects in the fundamental representation can be introduced in an $SU(N)/Z(N)$ theory when surrounded by the appropriate $Z(N)$ magnetic configuration.

Similarly by creating a $Z(N)$ magnetic loop and then making a duality transformation a modified 't Hooft loop $B(C^*)$ can be defined

$$\langle \tilde{B}(C^*) \rangle \equiv \Big\langle \exp\Big(\beta \sum_p \mathrm{tr}\, U(\partial p)\Big) \times \exp[(2\pi \mathrm{i}/N)\, \epsilon_{\mu\nu\rho\sigma}\, a^p \bar{M}^s / a \cdot \Delta]\Big\rangle, \tag{16}$$

where $\bar{M}^\sigma = 1$ for $\dot{b} \in C^*$, $\bar{M}^\sigma = 0$ otherwise.

In region II since $\gamma \to \infty$ at large distances, $\sigma(p) \to 1$, and the behaviour of A and B would be the same as the behaviour of A and B in an $SU(N)$ theory. In region I, A and B can be calculated in a strong-coupling expansion and they have an area and perimeter behaviour, respectively. Therefore, one does not expect a qualitative change in the behaviour of A and B across the transition line but just a change in the coefficient of the area or perimeter.

We substantiate the previous qualitative discussion with a Monte Carlo simulation of the SO(3) model. We work with the Wilson action:

$$Z_{\mathrm{W}} = \int \mathrm{d}U(b) \exp\Big(\frac{\beta}{2} \sum_p (\mathrm{tr}\, U(\partial p))^2 - 1\Big), \tag{17}$$

where $\mathrm{d}U$ is the invariant SU(2) measure. The normalization chosen leads in the naive continuum limit to the standard SU(2) Yang–Mills lagrangian.

We calculate the free energy per plaquette E:

$$E \equiv \frac{1}{\text{No. Plaquettes}} \frac{\partial \ln Z}{\partial \beta}. \tag{18}$$

We used the standard Metropolis method on lattices with sizes 3 and 4.

E as a function of β is plotted in fig. 2. The points at each temperature in the high and low temperature region represent typically 150 passes through the lat-

Volume 101B, number 5 PHYSICS LETTERS 21 May 1981

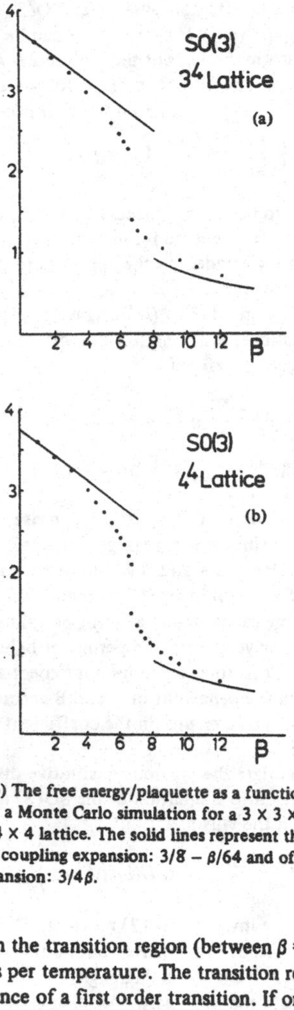

Fig. 2. (a), (b) The free energy/plaquette as a function of β calculated by a Monte Carlo simulation for a $3 \times 3 \times 3 \times 3$ and $4 \times 4 \times 4 \times 4$ lattice. The solid lines represent the results of the strong coupling expansion: $3/8 - \beta/64$ and of the weak coupling expansion: $3/4\beta$.

tice while in the transition region (between $\beta = 6-8$) 1500 passes per temperature. The transition region has the appearance of a first order transition. If one defines arbitrarily a relaxation time of a hundred passes per temperature a hysteresis curve is produced between $\beta \approx 6$ and $\beta \approx 8$ and there is a region of values for E ($\approx 0.15-0.20$) which we have not succeeded in reaching, as a final convergence point. They are of course crossed after a small number of interactions in converging from hot to cold values and vice versa. Obviously in order to establish clearly the order of the phase transition and its connection to the picture proposed above one needs much higher statistics on larger lattices and a study of the order parameters.

A phase transition in a two-dimensional identified σ-model which can be thought as the two-dimensional analogue of an SO(3) gauge theory was recently found through a strong-coupling hamiltonian method [9].

We comment finally on some implications of the structure seen in the SO(3) theory on the interpretation of SU(2) theories [10].

Since both SO(3) and SU(2) have instantons while their behaviour seems to be strikingly different it is very improbable that instantons play a dominant role in establishing the phase structure. If one normalizes arbitrarily the SO(3) theory in such a way that the values of E at $\beta = 0$ and ∞ coincide with the ones in SU(2) (this amounts to multiplying E and β with $\frac{8}{3}$ and $\frac{3}{8}$, respectively) the SO(3) transition point occurs at $\beta \approx 2.6$ i.e. in the region where a structure is seen in SU(2). Since the interaction of the Z(2) monopoles [2] of SU(2) is rather similar to the one we discussed here one wonders if the structure seen in SU(2) is not related to some activity of the monopoles.

We acknowledge very useful discussions and correspondence with T. Banks, E. Brezin, M. Creutz, G. Mack, D. Olive, M. Peskin, D.M. Scott and S. Yankielowicz. We thank M. Creutz for giving us a copy of his SU(2) Monte Carlo program. One of the authors (A.S.) acknowledges the hospitality of the Rutgers University High Energy Theory Group where part of this work was performed.

References

[1] T. Banks, R. Myerson and J. Kogut, Nucl. Phys. B129 (1977) 493;
M.E. Peskin, Ann. Phys. (NY) 113 (1978) 121;
M. Stone and P.R. Thomas, Phys. Rev. Lett. 41 (1978) 351.

[2] G. 't Hooft, Nucl. Phys. B138 (1978) 1;
S. Mandelstam, Phys. Rep. 23C (1976) 237;
G. Mack, in: Recent developments in gauge theories (Plenum, New York);
E. Tomboulis, Princeton Univ. preprint (1980);
L. Yaffe, Phys. Rev. D21 (1980) 1574.

[3] P. Goddard and D.I. Olive, Rep. Prog. Phys. 41 (1978) 91;
D.M. Scott, Imperial College Ph.D Thesis, unpublished;
see also: S. Samuel, Nucl. Phys. B154 (1979) 62.

[4] M. Kalb and P. Ramond, Phys. Rev. D9 (1974) 2273.

Volume 101B, number 5 PHYSICS LETTERS 21 May 1981

[5] R. Savit, Phys. Rev. Lett. 39 (1977) 55.
[6] E. Fradkin and S. Shenker, Phys. Rev. D19 (1979) 3682; T. Banks and E. Rabinovici, Nucl. Phys. B160 (1979) 349; A. Ukawa, P. Windey and A.H. Guth, Phys. Rev. D21 (1980) 1013.
[7] K. Osterwalder and E. Seiler, Ann. Phys. (NY) 110 (1978) 440.
[8] G. Mack, Phys. Rev. Lett. 45 (1980) 1378.
[9] S. Solomon, Phys. Lett. 100B (1981) 492.
[10] M. Creutz, Phys. Rev. Lett. 43 (1979) 553; B. Lautrup and M. Nauenberg, Phys. Rev. Lett. 45 (1980) 1755.

VOLUME 47, NUMBER 1 PHYSICAL REVIEW LETTERS 6 JULY 1981

First-Order Phase Transition in Four-Dimensional SO(3) Lattice Gauge Theory

J. Greensite and B. Lautrup
The Niels Bohr Institute, University of Copenhagen, DK-2100 Copenhagen Ø, Denmark
(Received 12 March 1981)

We present Monte Carlo evidence of a first-order phase transition in four-dimensional SO(3) lattice gauge theory. This result stands in sharp contrast to the known single-phase structure of the corresponding SU(2) theory, and suggests that the Z_2 center degrees of freedom may be important to the thermodynamics of the SU(2) gauge system.

PACS numbers: 11.10.Np

One of the major advantages of the lattice regularization of continuum gauge theory is that confinement can be understood as a property of the strong-coupling (or "high-temperature"), disordered phase of the gauge system. However, according to asymptotic freedom, $g^2 \to 0$ in the continuum limit; so the problem is to show that the "high-temperature" property of confinement persists in the "low-temperature" region down to $g^2 = 0$. The extrapolation of strong-coupling properties into the weak-coupling regime is possible if there is no phase transition separating the strong- and weak-coupling regimes, and this is the motivation for studying the thermodynamics of lattice gauge systems. The numerical results of Creutz[1] and Lautrup and Nauenberg[2] are strong evidence of the absence of any phase transition in four-dimensional SU(2) lattice theory.

It is natural to think that the absence of phase transitions in the SU(2) theory can be generalized to lattice theories with any continuous non-Abelian gauge group. However, the essential mechanism which frustrates a phase transition between the strong- and weak-coupling regimes in SU(2) is not really understood at present, while phase transitions have in fact been observed in certain other four-dimensional lattice gauge systems. Mean-field theory, for example, predicts the occurrence of first-order transitions in gauge systems,[3] and these are known to occur for four-dimensional lattice theories with some finite gauge groups.[4] Compact U(1) gauge theory seems to have a second-order transition in four dimensions.[5] Even in the SU(2) case, the specific heat has a sharp peak in the crossover region[2]; so it seems that there is "almost" a phase transition in the crossover region. One would like to somehow isolate the degrees of freedom in the SU(2) theory which prevent this transition from actually occuring. Now it is widely believed that topological configurations associated with the gauge-group center are responsible for confinement in the SU(2) weak-coupling regime[6]; so it might also be true that the Z_2 center degrees of freedom are crucial to the SU(2) thermodynamics.[7] It is this conjecture which has motivated us to study, by Monte Carlo techniques, the thermodynamics of a theory with a trivial group center, namely, four-dimensional SO(3) lattice gauge theory.

SO(3) lattice theory can be expressed in terms of integrals over the SU(2) group variables, however, the action is expressed as a trace in the adjoint, rather than the fundamental, representation of SU(2):

$$Z = \int \prod_l dU_l \, e^{\beta \sum_p S_p},$$

$$S_p = \chi_A(g_p)/\chi_A(1) = \tfrac{1}{3}[\chi_F^2(g_p) - 1] = \tfrac{1}{3}[\mathrm{tr}(UUU^\dagger U^\dagger)^2 - 1], \quad (1)$$

where the link variables U_l are SU(2) matrices, and the χ_A, χ_F are traces over the SU(2) group variables in the adjoint and fundamental representations, respectively. It should be noted that in SO(3) lattice gauge theory, in fact in any lattice gauge theory with a trivial center, quark color always can be screened by glue, i.e., a quark can bind with gluons to form a color singlet. Hence the Wilson loop must always follow a perimeter law even at strong couplings, and the strong-coupling (or "disordered") phase of such theories corresponds to color screening, rather than quark confinement *per se*.[8] The same color screening effect would occur in an SU(N) gauge theory with quarks in the adjoint representation. But the more interesting distinction between SU(2) and SO(3), apart from their available quark representations, is the fact that SO(3) cannot distinguish between link variables U and $-U$; this means, for example, that a "thin fluxon" configuration in SU(2) is indistinguishable from vacuum in the SO(3) theory. Conceivably, this loss of Z_2 degrees of freedom in the SO(3) theory could affect the thermodynamics so severely that a phase transition between strong and weak

VOLUME 47, NUMBER 1 PHYSICAL REVIEW LETTERS 6 JULY 1981

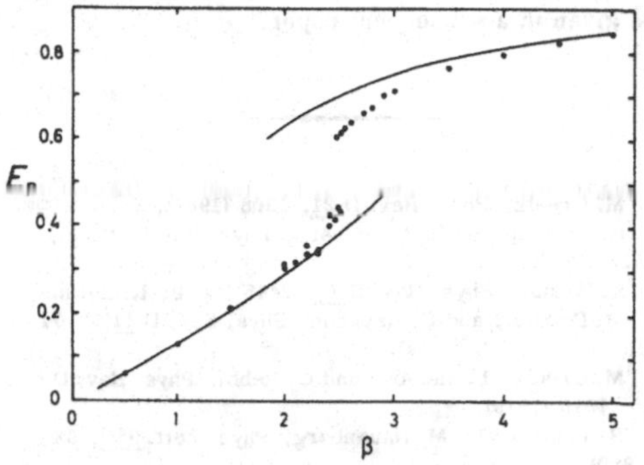

FIG. 1. The plaquette energy E_p as a function of β. The solid lower curve is a plot of the high-temperature expansion

$$E_p = \frac{1}{9}\beta + \frac{1}{54}\beta^2 - \frac{1}{1459}\beta^4 + \frac{199}{944784}\beta^5$$

and the upper solid curve is a plot of the low-temperature expansion $E_p = 1 - \frac{3}{4}\beta^{-1}$.

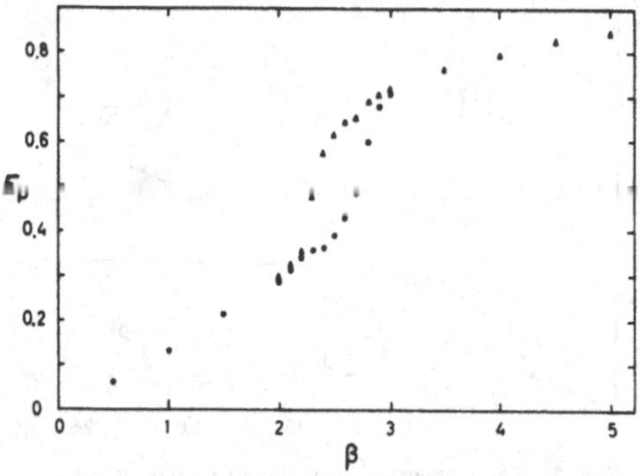

FIG. 2. SO(3) hysteresis curve in the plaquette energy E_p, taken from step-mode Monte Carlo runs with 100 iterations for each datum point. Points represented by triangles were taken from hot starts; circles represent cold starts.

couplings would occur. Using Monte Carlo methods on a 3^4 lattice, we have in fact found such a transition.

In Fig. 1 we show our results for the mean plaquette energy $E_p = \langle S_p \rangle$ as a function of $\beta = 3/2g^2$. This Monte Carlo data can be compared to the high- and low-temperature expansions for E_p, which are also plotted in Fig. 1. There is a very strong signal of a phase transition at $\beta = 2.48$, where E_p makes a sudden jump from $E_p = 0.43$ to $E_p = 0.60$ in a region of $\Delta\beta \simeq 0.01$. The only question is whether this is a first- or second-order transition. We conclude that it is a first-order transition because, to the limits of our computer data, there is a discontinuity between the upper and lower branches of the E_p curve. We were unable to resolve any intermediate points in the transition region. After 3000 iterations at the transition temperature $\beta = 2.48$, a Monte Carlo run with a cold start gives a data point on the upper branch of the curve, while a run at the same temperature with a hot start gives a data point on the lower branch. We have also run the Monte Carlo program in step mode with only 100 iterations per step and obtained the hysteresis curve of Fig. 2, which again shows a very pronounced two-phase structure. Practical constraints on computer time did not allow us to determine the normalized specific heat

$$\rho = \sigma^{-2}(\partial E_p/\partial\beta) = \sum_{p'} (\langle S_p S_{p'}\rangle - \langle S_p\rangle\langle S_{p'}\rangle)/(\langle S_p^2\rangle - \langle S_p\rangle^2) \quad (2)$$

with much reliability; we can only report that in the data available to us, in 3^4 as well as 4^4 lattices, the specific heat does not show any clear sign of forming a sharp peak, which would be characteristic of a second-order transition, and so this data seems consistent with a first-order transition. We have also seen, in the neighborhood of the transition point at $\beta = 2.48$, that the lattice system will remain in the metastable phase for hundreds of iterations, and then suddenly jump to the stable phase during the Monte Carlo run, which again is evidence of a well-separated two-phase structure. It is interesting to note that the first-order transition in SO(3) and the peak in the specific heat occurs at roughly the same value of β ($\beta = 2.48$ and 2.2, respectively). On the other hand, as a function of g^2, the SO(3) transition and SU(2) peak well separated. In Fig. 3 we have plotted the SU(2) and SO(3) plaquette action densities as a function of g^2, and on this scale the SO(3) transition seems very precocious.

We conclude that since the SO(3) gauge theory in four dimensions has a phase transition while the SU(2) theory does not, the center degrees of freedom could well be responsible for the difference in the thermodynamics of these very

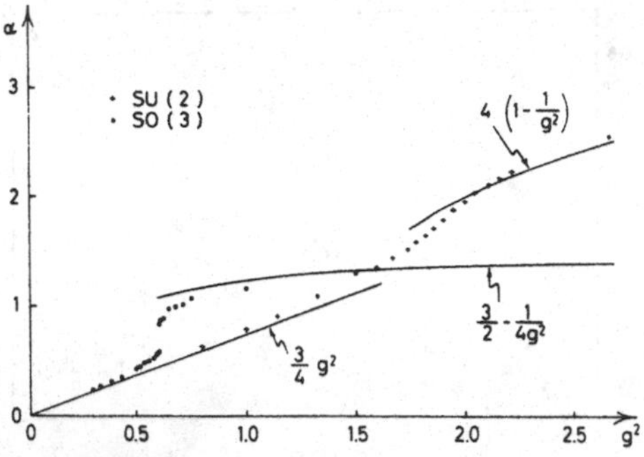

FIG. 3. The plaquette action densities plotted as a function of g^2. The action densities are $\alpha = 4 - 2\,\mathrm{tr}g$ for SU(2), and $\alpha = \frac{3}{2} - \frac{1}{2}\,\mathrm{tr}g$ for SU(3). In the weak-coupling limits the two groups must agree with each other for the case of the action density.

similar systems. It is also of interest that we now have an example of a continuous gauge group in four dimensions which, like some finite groups, undergoes a first-order phase transition as predicted by mean-field theory. In fact, a naive mean-field calculation by the Weiss self-consistent method gives a value $\beta_c = 2.59$ for the transition point, which is not far from our observed value $\beta_c = 2.48$. Further applications of mean-field methods to lattice gauge theories will be given in a subsequent paper.[9]

[1]M. Creutz, Phys. Rev. D 21, 1308 (1980).

[2]B. Lautrup and M. Nauenberg, Phys. Rev. Lett. 45, 1755 (1980).

[3]K. Wilson, Phys. Rev. D 10, 2445 (1974); R. Balian, J. M. Drouffe, and C. Itzykson, Phys. Rev. D 11, 2104 (1975).

[4]M. Creutz, L. Jacobs, and C. Rebbi, Phys. Rev. D 20, 1915 (1979).

[5]B. Lautrup and M. Nauenberg, Phys. Lett. 95B, 63 (1980).

[6]The fluxon confinement mechanism was suggested by G. 't Hooft, Nucl. Phys. B138, 1 (1978), and has been extensively discussed in the literature. See in particular the article by G. Mack, in *Recent Developments in Gauge Theories*, edited by G. 't Hooft *et al.* (Plenum, New York, 1980). The Z_N fluxon scheme has also been related to the "spaghetti" vacuum picture developed by the Copenhagen group; see J. Ambjørn and P. Olesen, Nucl. Phys. B170 [FS1], 265 (1980).

[7]In this connection, see also G. Mack and V. B. Petkova, Ann. Phys. (N.Y.) 123, 442 (1979).

[8]J. Kogut and L. Susskind, Phys. Rev. D 11, 395 (1975). Plaquette screening configurations which lead to a strong-coupling perimeter law can be found in the paper by J. M. Blairon, R. Brout, F. Englert, and J. Greensite, Nucl. Phys. B180 [FS2], 439 (1981).

[9]J. Greensite and B. Lautrup, Niels Bohr Institute Report No. NBI-HE-81-20 (to be published).

PHYSICAL REVIEW D VOLUME 24, NUMBER 12 15 DECEMBER 1981

Variant actions and phase structure in lattice gauge theory

Gyan Bhanot and Michael Creutz
Brookhaven National Laboratory, Upton, New York 11973
(Received 27 May 1981)

We study a simple generalization of Wilson's SU(2) lattice gauge theory. In various limits the model reduces to the usual SU(2), SO(3), or Z_2 models. Using Monte Carlo techniques on a four-dimensional lattice, we follow the known SO(3) and Z_2 first-order transitions into the phase diagram. They merge at a triple point and continue together to a critical end point. The peak in the specific heat of the SU(2) model is a shadow of this nearby singularity.

Monte Carlo studies with a lattice cutoff have given strong evidence for quark confinement in the asymptotically free SU(2) and SU(3) gauge theories. Until quite recently the lore was that lattice gauge theories based on non-Abelian gauge groups will not, in four dimensions, show any phase transitions separating a strong-coupling confining phase from the perturbative weak-coupling domain of the continuum limit. However, discoveries of unexpected transitions with the gauge groups SO(3), SU(4), and SU(5) have clouded this issue.[1,2]

The action used with a lattice cutoff (or indeed with any regulator) is highly nonunique. Wilson's formulation[3] is particularly elegant and thus has dominated research. As long as physics in the continuum limit is itself unique, the choice of lattice action is a matter of taste. However, when the lattice spacing is not small, variation of the action can modify the phase structure of the system. Indeed, the SO(3) and SU(5) transitions may be artifacts of the simple Wilson action. The mere existence of a phase transition does not necessarily imply a loss of confinement. A more general lattice Lagrangian may permit continuation around bothersome singularities.

With this motivation, we have studied a simple generalization of Wilson's model for the gauge group SU(2). We find that it is indeed possible to introduce a spurious transition in this theory. Our action has a two-parameter coupling space in which a critical point lies near to and is responsible for the rapid crossover from strong- to weak-coupling behavior in the conventional theory.

To stay as close as possible to the standard lattice gauge theory, we use the variables U_{ij} which are elements of the gauge group and are associated with the nearest-neighbor bonds $\{(i,j)\}$ of a four-dimensional simple hypercubic lattice. We also follow Wilson in keeping the action a function only of plaquette variables. Thus we restict

$$S=\sum_{\square} S_{\square}\,, \tag{1}$$

where the sum extends over all plaquettes $\square$ and $S_{\square}$ is a function only of $U_{\square}$, an ordered product of the four group elements about the square $\square$. Gauge invariance leads us to assume that $S_{\square}$ is a real class function of $U_{\square}$. Any such function can be expanded in characters

$$S_{\square}=\sum_R \beta_R \operatorname{Re}\operatorname{Tr}_R(U_{\square})\,, \tag{2}$$

where the sum is over all representations R of the group, Tr_R is the trace of $U_{\square}$ expressed in the given representation, and β_R are arbitrary coefficients. We throw away the imaginary part to keep the action real. The usual Wilson action keeps only the fundamental representation in this sum. Recently Manton[4] has suggested an alternative action where $S_{\square}$ is the square of the distance of $U_{\square}$ from the identity in the group manifold. This in general involves all representations and has been studied with Monte Carlo methods.[5]

In this paper we consider a two-parameter lattice action obtained by considering only the fundamental and adjoint representations in Eq. (2). For SU(2) we define our normalization such that

$$S_{\square}=\beta[1-\operatorname{Tr}(U_{\square})/2]+\beta_A[1-\operatorname{Tr}_A(U_{\square})/3]\,, \tag{3}$$

where Tr without a subscript is taken in the two-dimensional defining representation and Tr_A is in the three-dimensional adjoint representation. Note that for arbitrary SU(N) the adjoint trace is easily found using the identity

$$\mathrm{Tr}_A U = |\mathrm{Tr} U|^2 - 1 . \quad (4)$$

We insert this action into a path integral or partition function

$$Z = \int dU \exp\left[-\sum_{\Box} S_{\Box}\right] , \quad (5)$$

where the integration is over all bond variables. We shall see that this simple generalization of the Wilson action gives rise to a rich phase struture.

This model has several interesting limits. When $\beta_A = 0$ it reduces to the usual SU(2) theory. When $\beta = 0$ the action only depends on the adjoint representation, and corresponds to the standard Wilson theory based on the gauge group SO(3). This model is known[1] to have a first-order phase transition, which we confirm, at $\beta_A = 2.50 \pm 0.03$. Another interesting limit occurs on taking β_A to infinity. This drives all the plaquette variables to the identity in the adjoint representation. In the fundamental representation, each $U_{\Box}$ must then lie in the center of the group,

$$U_{\Box} \in \{\pm I\} . \quad (6)$$

This implies that under a gauge transformation all the bond variables can also be placed in the center, and we have the conventional Wegner Z_2 lattice gauge theory[6] at inverse temperature β. That model is known to exhibit a striking first-order phase transition at[7]

$$\beta = \tfrac{1}{2} \ln(1+\sqrt{2}) = 0.44 \ldots . \quad (7)$$

Thus at the outset we know that our model must possess a nontrivial phase diagram, with first-order lines entering from $\beta = 0.44$, $\beta_A = \infty$ and $\beta = 0$, $\beta_A = 2.5$.

We have used Monte Carlo simulation to follow these lines into the (β, β_A) plane. Our algorithm follows Metropolis *et al.*[8] Each link variable in turn is tentatively multiplied by a group element randomly selected from a table of twenty. This tentative change is accepted if a random number uniformly selected in the interval (0,1) is larger than the exponential of the change in the action. This satisfies the detailed balance requirement which ensures that an arbitrary ensemble of configurations will be brought closer to the Boltzmann distribution defined by the exponentiated action. One Monte Carlo iteration consists of applying this algorithm ten times to one link and then similarly touching all the other links of the lattice. The elements in the table were randomly selected with a coupling-dependent weighting towards the identity. A β_A-dependent fraction of the elements was selected near $-I$ to assist convergence when the Z_2 structure of the action is important. A new table was generated after each sweep through the lattice. Our boundary conditions were always periodic and no gauge fixing was imposed.

To check our results, we also studied our action for the discrete subgroup of SU(2) defined by the symmetries of an icosahedron.[9] For small β_A an extra transition due to the discrete nature of this subgroup was well separated from the interesting structures in the SU(2) model. As one expects from a simple two-state model for this "discreteness" transition, the critical couplings (β^C, β_A^C) for $\beta_A \lesssim 2$ lie on an approximately straight line well fit by

$$\beta^C = 6 - 2.4\beta_A^C . \quad (8)$$

It is clear from this that when β_A is of order 2-3, the discrete approximation strongly affects the phase structure. In this region we must treat the model using the full SU(2).

To monitor the behavior of this model, we measure separately the expectation of the two terms in the action. Thus we define

$$P = \langle 1 - \mathrm{Tr}(U_{\Box})/2 \rangle , \quad (9)$$

$$P_A = \langle 1 - \mathrm{Tr}_A(U_{\Box})/3 \rangle . \quad (10)$$

For a totally ordered lattice both these vanish whereas for random U_{ij} they both equal unity. Wherever $\beta = 0$ the Z_2 symmetry of the adjoint action gives $P = 1$ for any β_A. At a few values of β, β_A we also measured rectangular Wilson loops, the expectation of the trace of an ordered product of U_{ij} around a rectangle. These were measured in both the fundamental and adjoint representation.

In Fig. 1(a) we show a thermal cycle in β_A for $\beta = 0$. The lattice was 5^4 sites in size and each point represents P_A after thirty iterations at the given β_A. The crosses represent cooling of the lattice and the circles represent heating. Note the hysteresis effect due to the first-order phase transition in the SO(3) model. In Fig. 1(b) we show runs of 100 iterations starting both random and ordered at $\beta_A = 2.5$. Figure 1 represents an independent confirmation of the results in Ref. (1).

Figure 2 displays a similar thermal cycle in β with β_A fixed at 3.0. Here the order parameter P

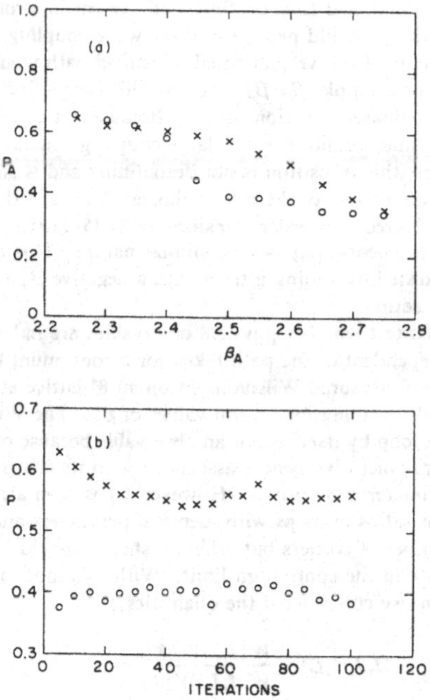

FIG. 1. (a) Thermal cycle in the SO(3) limit. Crosses represent cooling and circles represent heating. (b) Evolution of random and ordered configurations at $\beta_A=2.5$, $\beta=0$.

clearly shows the Z_2 transition. At this finite β_A, the transition has moved from the value in Eq. (7) to $\beta=0.50\pm0.02$. This shift to larger β is expected because decreasing β_A allows the system to be more disordered. Note that the transition is nearly unobservable in P_A.

By performing several similar thermal cycles on a 4^4 lattice we have followed both of these transitions further into the (β,β_A) plane and have obtained the phase diagram shown in Fig. 3. The Z_2 and SO(3) transitions meet at a triple point at $\beta=0.55\pm0.03$, $\beta_A=2.34\pm0.03$. Figure 4 shows three runs of one hundred iterations on a 5^4 lattice at the triple point. The three initial conditions were (1) ordered with every $U_{ij}=I$ (solid circles); (2) with each U_{ij} selected totally randomly from SU(2) (crosses); and (3) with each U_{ij} chosen randomly from $Z_2=\{\pm1\}$ (open circles). The system has three distinct stable phases at this point in coupling space.

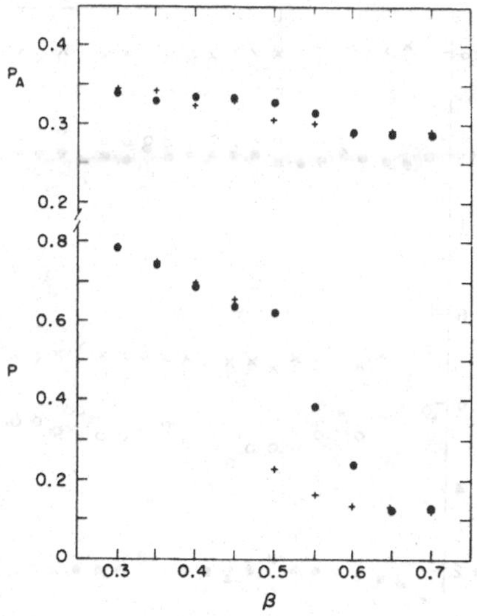

FIG. 2. Thermal cycle in β at $\beta_A=3.0$.

The third first-order line emerging from this triple point extends towards smaller β_A but terminates at a critical end point before reaching the β axis. By extrapolating the latent heats in P and in P_A to vanishing values, we quote $\beta=1.48\pm0.05$, $\beta_A=0.90\pm0.03$ as the coordinates of this new crit-

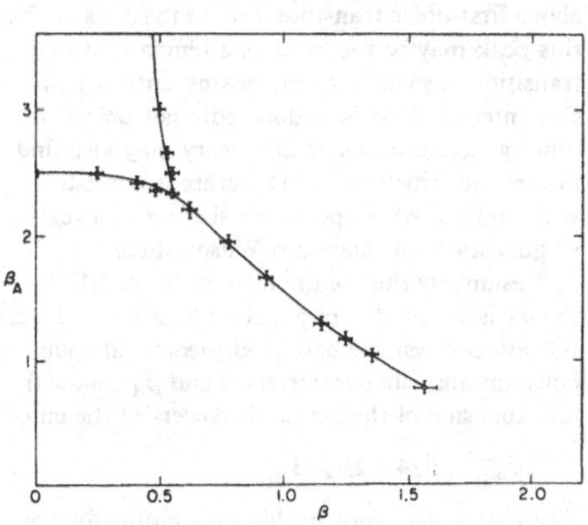

FIG. 3. The full phase diagram. The open circles represent the location of the triple point and the critical point. The solid circles trace out the first-order transition lines. The solid curves are drawn to guide the eye.

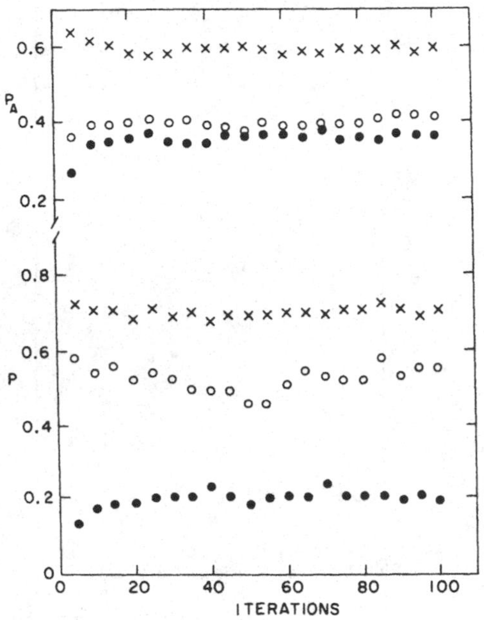

FIG. 4. Three Monte Carlo runs at the triple point.

ical point in SU(2) lattice gauge theory.

The conventional SU(2) theory exhibits a narrow but smooth peak in its specific heat[10] at $\beta=2.2$. This is directly at a naive extrapolation of the above first-order transition line to the β axis. Thus this peak may be regarded as a remnant of that transition, a shadow of the nearby critical point. Our interpretation is undoubtedly not unique as other generalizations of the theory may also find interesting structure. This picture is consistent with and indeed supports the absence of a real singularity in the standard Wilson theory.

Presumably the continum limit of the SU(2) theory is unique for physical observables. The connection between the bare field-theoretical coupling constant and our parameters β and β_A follow from an expansion of the action in powers of the cutoff

$$g_0^{-2}=\beta/4+2\beta_A/3\ . \tag{11}$$

The continuum limit in this asymptotically free theory requires taking g_0^2 to zero, but this can be done along many paths in the (β,β_A) plane. Previous work has concentrated on the trajectory $\beta_A=0$, $\beta\to\infty$. Along that path no singularities are encountered, and thus confinement, present in strong coupling, should persisit into the weak-coupling domain. However, an equally justified path would be, for example, $\beta=\beta_A\to\infty$. In this case a first-order phase transition occurs. Because one can continue around it in our large-coupling-constant space, this transition is not deconfining and is simply an artifact of the action choice. The recently discovered first-order transition in SU(5) lattice gauge theory may be of a similar nature. This can be tested by adding a term with a negative β_A to the action.

To test whether physical observables are indeed independent of the path taken for a continuum limit, we measured Wilson loops on an 8^4 lattice at weak coupling for several values of β_A. The Wilson loop by itself is not an observable because of ultraviolet divergences associated with its sharp perimeter and corners. However it has been argued that ratios of loops with identical perimeters and number of corners but different shapes should be finite in the continuum limit. With this motivation, we constructed the quantities

$$R(I,J,K,L)=\frac{W(I,J)W(K,L)}{W(I,L)W(J,K)}\ , \tag{12}$$

where $W(I,J)$ is the Wilson loop of dimensions I by J in lattice units. Wishing to compare points which give similar physics, we searched in β at fixed β_A for the points where $R(2,2,3,3,)$ had the fixed values 0.87 and 0.93. This gave rise to the points in the (β,β_A) plane shown in Fig. 5. In this figure we also show the terminating first-order line discussed above and the large-β transition from the discrete approximation. The dashed lines represent contours of constant bare charge from Eq. (11). If physics is indeed similar at all these points, all ratios R of Eq. (12) should match. In Fig. 6 we show various such ratios as functions of β_A at the $R(2,2,3,3)\approx0.87$ points from Fig. 5. To avoid clutter we have not included error bars, which are comparable to the scatter along the various curves. The comparison is quite good considering that finite cutoff corrections are ignored. We remark that if individual loops are compared without taking ratios as in Eq. (12), their values are not constant along these contours.

Note that in this comparison the bare charge is not a constant. In Fig. 5 we varied g_0^2 from less than unity to nearly 4 while holding $R(2,2,3,3)$ fixed. This variation is permissible as the bare charge is unobservable and depends on prescription. This dependence can be characterized with

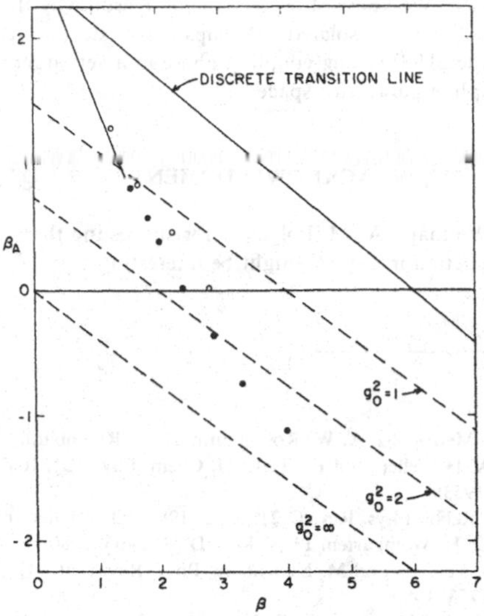

FIG. 5. Points of constant "physics." Solid circles represent $R(2,2,3,3)=0.87$, open circles represent $R(2,2,3,3)=0.93$.

an asymptotic freedom scale Λ_0 which depends on β_A. The quantity Λ_0 is an integration constant of the renormalization-group equation and is defined as the lattice spacing and g_0 became small by

$$\Lambda_0 = \frac{1}{a}\left[\gamma_0 g_0^2\right]^{-\gamma_1/2\gamma_0^2} \times \exp[-1/(2\gamma_0 g_0^2)][1+O(g_0^2)] \,. \quad (13)$$

Here γ_0 and γ_1 are the first two coefficients in the perturbative expansion of the Gell-Mann-Low function,[11]

$$\gamma(g_0)=a\,dg_0/da=\gamma_0 g_0^3+\gamma_1 g_0^5+O(g_0^7) \,. \quad (14)$$

For SU(2) we have

$$\gamma_0=\frac{11}{24\pi^2}\,, \quad (15)$$

$$\gamma_1=\frac{17}{96\pi^4}\,. \quad (16)$$

Assuming that the constant R contours follow a line of constant lattice spacing, we extract the β dependence for Λ_0 shown in Fig. 7. Note that the $R(2,2,3,3)=0.93$ and 0.87 results are consistent. Remarkably, the addition of β_A can change the lattice Λ_0 by several orders of magnitude.

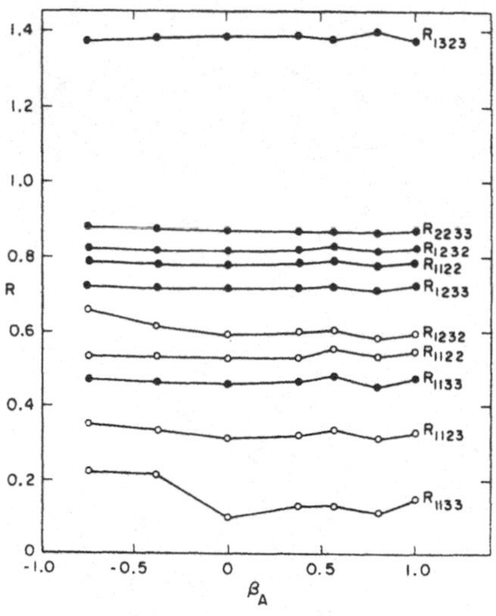

FIG. 6. Various loop ratios along the $R(2,2,3,3)=0.87$ contour. Solid circles are from loops in the fundamental representation, open circles from the adjoint.

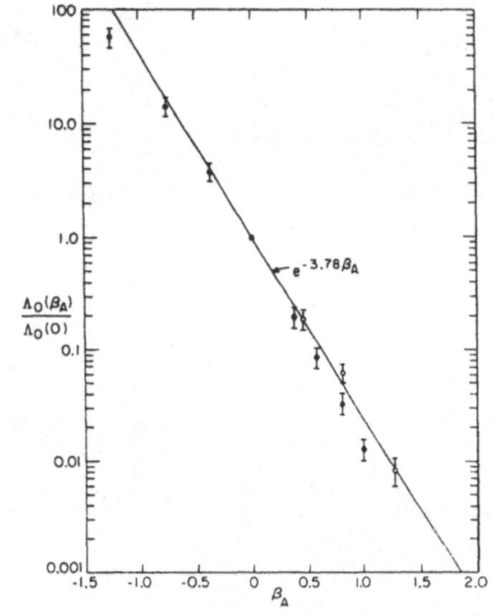

FIG. 7. The β_A dependence of the renormalization scale $\Lambda_0(\beta_A)$. The solid circles and open circles are from $R(2,2,3,3)=0.87$ and 0.93, respectively.

In conclusion, we have shown how a simple modification of the Wilson action can introduce a first-order phase transition separating the strong- and weak-coupling domains in SU(2) lattice gauge theory. The transition is not deconfining because it can be continued around in a larger coupling space. Our model has limits which select out Z_2, the center of the gauge group, and SO(3). A peculiar isolated phase at large β_A and small β is represented by the weak-coupling domain of the SO(3) model. We know of no convincing reason why this phase must be isolated. Perhaps it too is connected to the SU(2) strong-coupling phase in a yet larger coupling parameter space.[12]

ACKNOWLEDGMENT

We thank A. M. Polyakov for suggesting that the action in Eq. (3) might be interesting.

[1]I. G. Halliday and A. Schwimmer, Phys. Lett. 101B, 327 (1981); J. Greensite and B. Lautrup, Phys. Rev. Lett. 47, 9 (1981).
[2]M. Creutz, Phys. Rev. Lett. 46, 1441 (1981).
[3]K. Wilson, Phys. Rev. D 10, 2445 (1974).
[4]N. S. Manton, Phys. Lett. 96B, 328 (1980).
[5]C. B. Lang, C. Rebbi, P. Salomonson, and B. S. Skagerstam, Phys. Lett. 101B, 173 (1981).
[6]F. J. Wegner, J. Math. Phys. 12, 2259 (1971).
[7]M. Creutz, L. Jacobs, and C. Rebbi, Phys. Rev. Lett. 42, 1390 (1979).
[8]N. Metropolis, A. W. Rosenbluth, M. N. Rosenbluth, A. H. Teller, and E. Teller, J. Chem. Phys. 21, 1087 (1953).
[9]C. Rebbi, Phys. Rev. D 21, 3350 (1980); D. Petcher and D. H. Weingarten, Phys. Rev. D 22, 2465 (1980).
[10]B. Lautrup and M. Nauenberg, Phys. Rev. Lett. 45, 1755 (1980).
[11]M. Gell-Mann and F. E. Low, Phys. Rev. 95, 1300 (1954).
[12]I. G. Halliday and A. Schimmer, Phys. Lett. 102B, 337 (1981).

PHYSICAL REVIEW D VOLUME 21, NUMBER 4 15 FEBRUARY 1980

Phase diagrams for coupled spin-gauge systems

Michael Creutz
Department of Physics, Brookhaven National Laboratory, Upton, New York 11973
(Received 27 August 1979)

Using Monte Carlo techniques, we study Z_N lattice gauge theory coupled to a Higgs field represented by spins situated on the lattice sites. We present phase diagrams for the Z_2 and Z_6 theories with the Higgs field in the fundamental representation of the gauge group and for Z_6 gauge theory coupled to Z_3 Higgs fields.

I. INTRODUCTION

Recent results establish the utility of Monte Carlo procedures for the study of phase transitions in lattice gauge theory.[1-3] Here we extend those investigations to coupled spin-gauge systems. Placing spins on lattice sites provides a prototype for a matter field which can produce gauge-meson masses via the Higgs mechanism.[4] This system carries two coupling constants, β corresponding to the gauge-field self-interaction and β_H representing the strength of the nearest-neighbor gauge-invariant spin-spin interaction. For large β the gauge fields become ordered and the model reduces to a conventional nearest-neighbor spin system exhibiting the ferromagnetic transition responsible for the Higgs mechanism. For vanishing β_H the site spins disorder and the model reverts to the pure gauge theory. Depending on the gauge group and the dimensionality of space-time, there may be one or more phase transitions along this line.[1-3,5]

Several authors have discussed phase diagrams for these systems.[6] When the site spins are in the fundamental representation of the gauge group G, the ordered spin phase of the Higgs mechanism is analytically connected to the disordered phase of the pure gauge field. When the spins are in another representation R, the theory at large β_H reduces to a pure gauge theory with group G/R, i.e., the subgroup of G under which the matter fields are invariant. In this case the above-mentioned phases can be distinct. In this paper we use Monte Carlo techniques on a four-dimensional lattice to "experimentally" confirm this structure. We restrict our treatment to the discrete Abelian gauge groups Z_N.

In the next section we define the models and summarize the Monte Carlo procedure. In Sec. III we discuss the limiting regions bounding the phase diagram. Section IV contains the phase diagrams for Z_2 and Z_6 gauge fields. For Z_6 we consider matter fields in both Z_6 and Z_3. The latter case gives a nontrivial quotient group $Z_6/Z_3=Z_2$. Section V contains some discussion. Conclusions on U(1) are drawn from the Z_6 model at low and intermediate β.

II. THE MODELS AND THE METHOD

We work on a four-dimensional hypercubical lattice. On each site i we have a spin variable S_i taken from the group Z_M,

$$S_i \in Z_M=\{e^{2\pi i m/M} \mid m=1,\ldots,M\}. \tag{2.1}$$

For each pair of nearest-neighbor sites i and j we have a gauge or link variable U_{ij} in the group Z_N,

$$U_{ij}\in\{e^{2\pi i n/N}\mid n=1,\ldots,N\}. \tag{2.2}$$

We require that the quotient

$$l=N/M \tag{2.3}$$

be an integer so that Z_M is a subgroup of Z_N. The link variables are oriented in the sense that

$$U_{ij}=U_{ji}^*. \tag{2.4}$$

The dynamics of this system of spin and gauge degrees of freedom follows from the action

$$\mathcal{S}=\beta_H\sum_{(i,j)}\mathcal{S}_L(i,j)+\beta\sum_{\Box}\mathcal{S}_\Box. \tag{2.5}$$

The first sum is over all nearest-neighbor pairs of sites (i,j) where each such pair contributes

$$\mathcal{S}_L(i,j)=1-\mathrm{Re}(S_i U_{ij}^l S^{-1}{}_j), \tag{2.6}$$

where the power l is defined in Eq. (2.3). The second sum in Eq. (2.5) is over all elementary squares or plaquettes, each such square contributing

$$1-\mathrm{Re}(U_{ij}U_{jk}U_{kl}U_{li}), \tag{2.7}$$

where the sites i, j, k, and l circulate around the square $\Box$. A nearest-neighbor pair contributes to the action a number from the interval $[0,2\beta_H]$ and a plaquette contributes from $[0,2\beta]$.

We insert this action into a path integral or partition function

$$Z=\sum_{S_i,U_{ij}}e^{-\mathcal{S}}, \tag{2.8}$$

where the sum is over all allowed configurations of the link and site variables. The free energy of the system is defined as

$$F=\frac{1}{N_S}\ln Z, \tag{2.9}$$

where N_S is the number of sites in the lattice. The correlation functions we study are the average link and average plaquette defined by

$$L=\langle \mathcal{S}_L(i,j)\rangle=-\frac{1}{4}\frac{\partial}{\partial\beta_H}F(\beta,\beta_H), \tag{2.10}$$

$$P=\langle \mathcal{S}_\Box\rangle=-\frac{1}{6}\frac{\partial}{\partial\beta}F(\beta,\beta_H). \tag{2.11}$$

The factors $\frac{1}{4}$ and $\frac{1}{6}$ are the ratios of the number of sites to the number of links and plaquettes, respectively, in a four-dimensional lattice.

This system possesses a local gauge symmetry. Given an element g_i of Z_N associated with each site of the lattice, the action is unchanged by the replacement

$$\begin{aligned} U_{ij} &\to g_i U_{ij} g^{-1}{}_j, \\ S_i &\to S_i g^{-1}{}_i . \end{aligned} \tag{2.12}$$

Note that by selecting $g^{1}{}_i=S_i$ the spin variables all become unity. For gauge-invariant correlation functions the theory is thus equivalent to the pure gauge theory coupled to a non-gauge-invariant applied field of strength β_H. We call this choice the unitary gauge. It will be useful in the discussion of the small-β and the large-β_H limits of the theory.

Even for extremely modest lattice sizes, it is impractical to evaluate the sum in Eq. (2.8) directly. For Z_2 gauge and Higgs fields on a mere 2^4-site lattice this sum has already

$$2^{5\times 2^4}=1.2\times 10^{24} \tag{2.13}$$

terms. Indeed, this immoderate sum suggests a statistical treatment. The Monte Carlo method generates a sequence of states which simulates an ensemble of configurations in thermal equilibrium. In this ensemble the probability of finding a given configuration is proportional to the Boltzmann factor $e^{-\mathcal{S}}$. Expectation values in the states of this sequence should then fluctuate about the true correlation function of the full path integral.

We use a Monte Carlo algorithm which is equivalent to successively placing a heat bath in contact with the individual spins and links of the lattice. After touching any particular spin variable S_i, we replace it with a new value S'_i in a random manner weighted by the Boltzmann factor

$$B=\exp[-\mathcal{S}(S'_i)], \tag{2.14}$$

where $\mathcal{S}(S'_i)$ is the action evaluated with site i having spin S'_i and all other dynamical variables fixed at their previous values. We similarly treat the link variables. In the remainder of this paper, one Monte Carlo iteration refers to one application of this algorithm to every link and spin variable in the lattice.

For initial configurations we set all $S_i=1$ and set the links either randomly or to unity. (A situation where both site and link variables are random is gauge equivalent to random links and ordered spins.) These two initial states represent infinite and zero temperature. A measure of equilibrium is the agreement of correlation functions obtained with Monte Carlo iterations from these two initial conditions.

The phase diagrams of Sec. IV follow from simulations on a $5\times5\times5\times5$ lattice. We then check a few points in crucial regions on an 8^4 lattice. To minimize surface effects we always impose periodic boundary conditions. Although gauge invariance theoretically permits us to fix spins and only vary links, we have found that convergence of the Monte Carlo procedure is enhanced when the gauge is allowed to fluctuate. In addition to running at fixed β and β_H we have found it useful to adjust β and/or β_H after each iteration to search for values giving a desired average plaquette and/or average link. This allows a rapid determination of contours of constant correlation.

III. LIMITING REGIONS

We now discuss the four limits $\beta_H\to 0, \infty$ and $\beta\to 0, \infty$. For vanishing β_H the site spins randomize and the model reduces to pure Z_N gauge theory. For $N=2$ this model has a first-order phase transition at the self-dual[1] point $\beta=\frac{1}{2}\ln(1+\sqrt{2})=0.44\ldots$. For Z_6 we have two higher-order transitions occurring at[2]

$$\begin{aligned} \beta_1&=1.00\pm0.01, \\ \beta_2&=1.61\pm0.04. \end{aligned} \tag{3.1}$$

In Figs. 1 and 2 we show the average plaquette

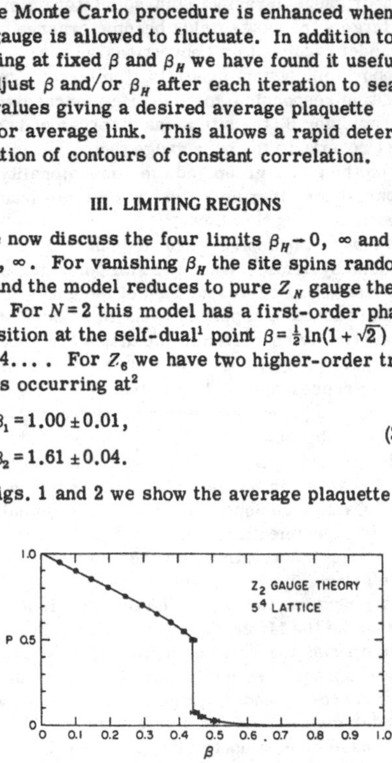

FIG. 1. The average plaquette as a function of β for pure Z_2 gauge theory. The solid line represents the series results in Ref. 1.

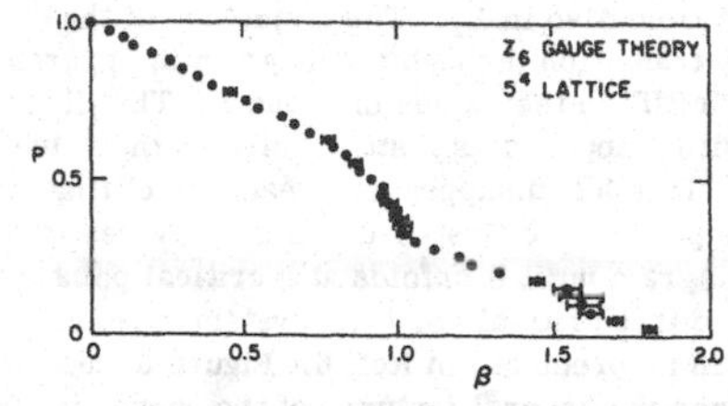

FIG. 2. The average plaquette as a function of β for pure Z_6 gauge theory.

as a function of β for the groups Z_2 and Z_6. Although in Refs. 1 and 2 we used larger lattices, this simulation shows the typical fluctuations expected on a 5^4 lattice. In Ref. 2 a study of lattice size showed no qualitative changes down to a 3^4 lattice; only the fluctuations grew as the size was reduced. In Fig. 1 we also show the series results for Z_2 quoted in Ref. 1.

When $\beta_H \to \infty$ the system must have vanishing average link L. In the unitary gauge this is equivalent to

$$U_{ij}^l = 1. \qquad (3.2)$$

If the Higgs field is in the fundamental representation, i.e., if $l=1$, the gauge fields must order and both P and L will vanish. However, if $l \neq 1$ then Eq. (3.2) only implies that all gauge variables lie in the quotient group $Z_N/Z_M = Z_l$. Thus as $\beta_H \to \infty$ the theory goes over to a pure Z_l gauge theory. Our treatment of Z_6 gauge fields coupled to Z_3 spins will reduce to Z_2 gauge theory in this limit.

The limit $\beta \to 0$ is trivial in the unitary gauge. Without the gauge interaction the link variables decouple and the average link is

$$L = -\frac{\partial}{\partial \beta_H} \ln\left(\sum_{U \in Z_N} \exp\{-\beta_H[1 - \mathrm{Re}(U^l)]\}\right). \qquad (3.3)$$

As each link is decoupled from the others, the average plaquette is

$$P = 1 - (1-L)^4 \delta_{l,1}. \qquad (3.4)$$

In Fig. 3 we plot the functions in Eqs. (3.3) and (3.4) as a function of β_H for the group Z_2. For comparison, we superpose Monte Carlo results obtained without gauge fixing on a 5^4 lattice.

Finally we come to the limit $\beta \to \infty$. Here all plaquettes must go to the identity. The gauge fields are then gauge equivalent to total ordering and the model reduces to a pure Z_M spin system with nearest-neighbor couplings. In Fig. 4 we show the results of simulations of this system with Z_2 Higgs fields. The Z_2 model is the Ising[7] model and the Z_3 model is equivalent to the three-state Potts[8] model, both in four dimensions. These

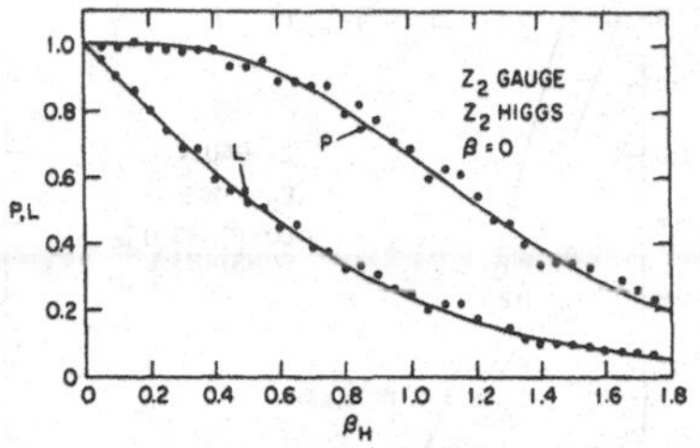

FIG. 3. The average plaquette and link as a function of β_H for the Z_2 system at $\beta = 0$. The solid lines are the exact result and the points are from Monte Carlo simulation.

systems all exhibit ferromagnetic phase transitions. Based on mean-field theory, conventional lore is that for Z_2 and Z_6 these transitions are second order. For Z_3, however, there is appreciable evidence that this is a first-order phase transition.[9] The inverse temperatures of these transitions are

$$\beta_H = \begin{cases} 0.150 \text{ for } Z_2 \\ 0.258 \text{ for } Z_3 \\ 0.34 \pm 0.01 \text{ for } Z_6 . \end{cases} \qquad (3.5)$$

The values for Z_2 and Z_3 are from Refs. 9 and 10, while for Z_6 we used our own analysis on an 8^4 lattice.

To summarize this section, the models under study have one or two transition lines entering the phase diagram from the axis $\beta_H = 0$, one transition at $\beta = \infty$ and zero or one transition at $\beta_H = \infty$. In the next section we will see how these lines connect in the interior of the diagram. The figures in the present section serve to indicate the typical fluctuations occurring in Monte Carlo simulations on a 5^4 lattice.

IV. THE DIAGRAMS

In Fig. 5 we show contours of constant L and P in the (β, β_H) plane for the gauge group Z_2 and

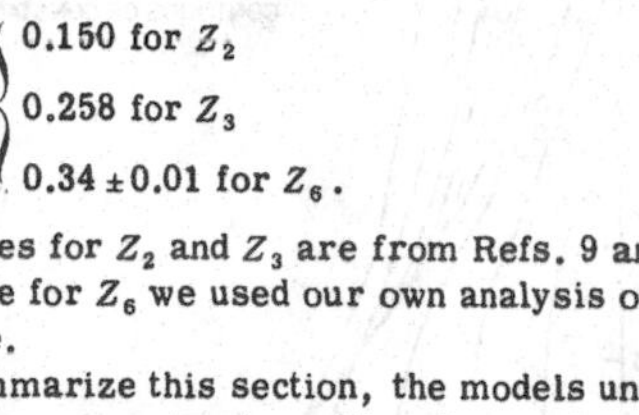

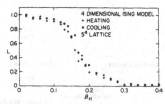

FIG. 4. A thermal cycle on the four-dimensional Ising model. Each point is the average link after 10 iterations at fixed β_H. At each β_H the lattice was started in the configuration obtained from an adjacent point, either hotter or cooler.

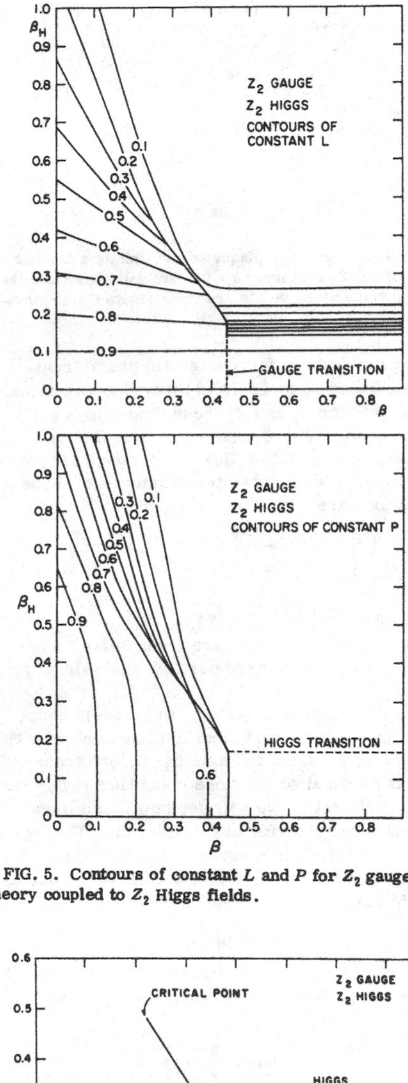

FIG. 5. Contours of constant L and P for Z_2 gauge theory coupled to Z_2 Higgs fields.

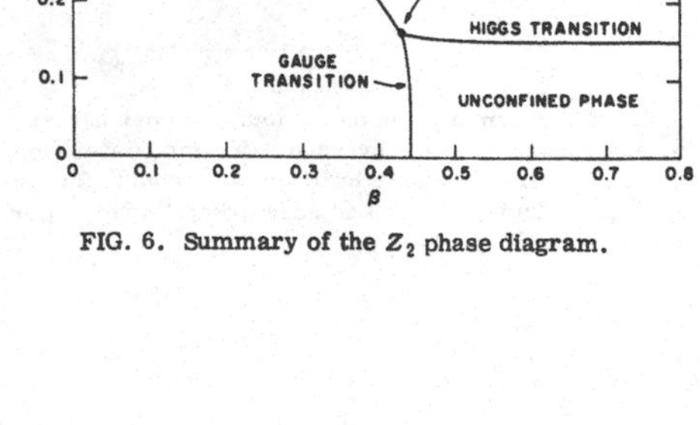

FIG. 6. Summary of the Z_2 phase diagram.

Higgs field also in Z_2. The trajectory of the gauge transition through the diagram is apparent as a "cliff" in the values of P and L. The Higgs transition appears as a steep "hill" in the value of L. This hill disappears beneath the cliff at a triple point. The first-order line continues into the diagram until it unfolds at a critical point. Beyond that critical point the system appears smooth as predicted in Ref. 6. Figure 6 summarizes the general features of the phase diagram.

As gauge fluctuations induce disorder, the Higgs transition should move to larger β_H as β is reduced.[6] However, the gauge field is so thoroughly

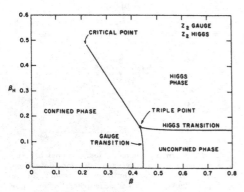

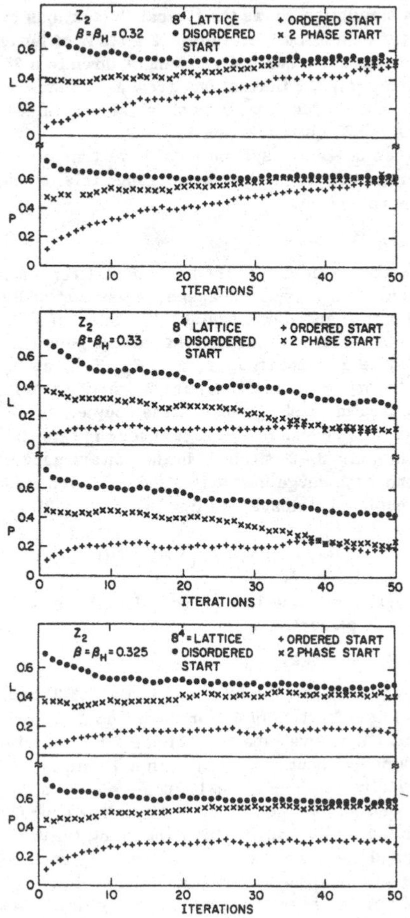

FIG. 7. Fifty iterations on an 8^4 lattice for the Z_2 system at (a) $\beta=\beta_H=0.32$, (b) $\beta=\beta_H=0.33$, and (c) $\beta=\beta_H=0.325$.

ordered for $\beta > 0.44$ that we cannot see this shift. Similarly the gauge transition should move to lower β as β_H is increased, but below the triple point this shift is also too small to appear in our analysis. Thus we place the triple point at $\beta = 0.43 \pm 0.02$, $\beta_H = 0.16 \pm 0.02$.

In Fig. 7 we show the average plaquette and link as a function of number of Monte Carlo iterations on an 8^4 lattice with β and β_H chosen near the transition line above the triple point. These runs were initiated with three distinct starting conditions, ordered, disordered, and a mixed phase. In the latter, half the lattice was ordered and half random. Such a state should evolve without being caught in a metastable phase associated with a first-order phase transition. Figure 7(a) shows that the point $\beta = \beta_H = 0.32$ lies on the disordered side of the transition line while Fig. 7(b) shows $\beta = \beta_H = 0.33$ is ordered. The first-order nature of the line is manifest in Fig. 7(c), where at $\beta = \beta_H = 0.325$ two stable phases appear and the mixed phase drifts rather slowly. The critical point where this first-order line terminates is difficult to locate precisely because of the steep behavior beyond it. We estimate $\beta = 0.22 \pm 0.03$ and $\beta_H = 0.48 \pm 0.03$ as its coordinates.

In Fig. 8 we show contours of P and L for the gauge group Z_6 with the Higgs field in the fundamental representation. The Higgs transition again appears as a steep slope in L. For β_H below this transition P is essentially independent of β_H.

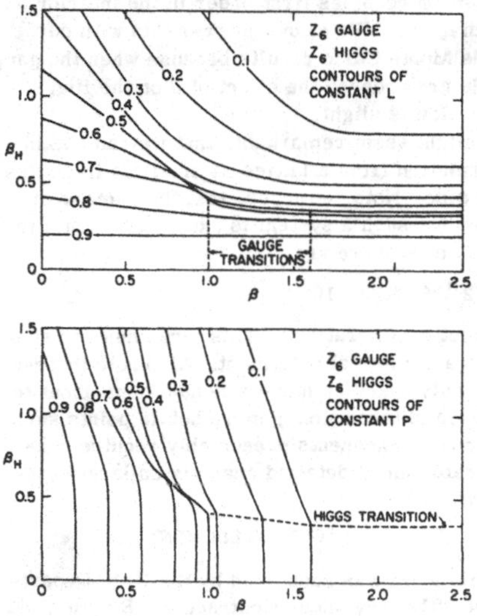

FIG. 8. Contours of constant L and P for the coupled Z_6 spin-gauge system.

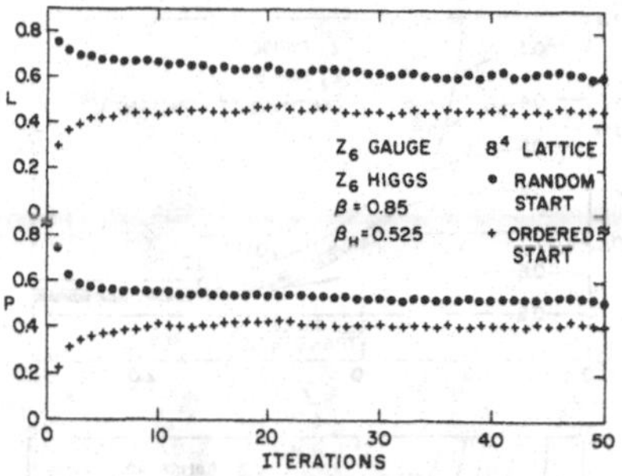

FIG. 9. Fifty iterations on an 8^4 lattice for the coupled Z_6 system at $\beta = 0.85$, $\beta_H = 0.525$.

The two gauge transitions thus proceed essentially at constant β until they join the Higgs transition at two separate triple points. For β below the junction of the Higgs and high-temperature gauge transitions, we have a single transition line terminating at a critical point similar to that seen with Z_2. This line appears to be first order, but this is not certain because the discontinuity across it is less substantial than in the Z_2 case. For the large-β triple point we quote

$$\beta = 1.60 \pm 0.05, \quad \beta_H = 0.35 \pm 0.02, \tag{4.1}$$

and for the low-β triple point

$$\beta = 0.98 \pm 0.03, \quad \beta_H = 0.42 \pm 0.03. \tag{4.2}$$

The location of the critical point is

$$\beta = 0.67 \pm 0.05, \quad \beta_H = 0.67 \pm 0.05. \tag{4.3}$$

The errors in the above numbers are subjective estimates. To argue that the line connecting the critical point with the low-β triple point is first order, we show in Fig. 9 runs on an 8^4 lattice at $\beta = 0.85$, $\beta_H = 0.525$ with both random and ordered

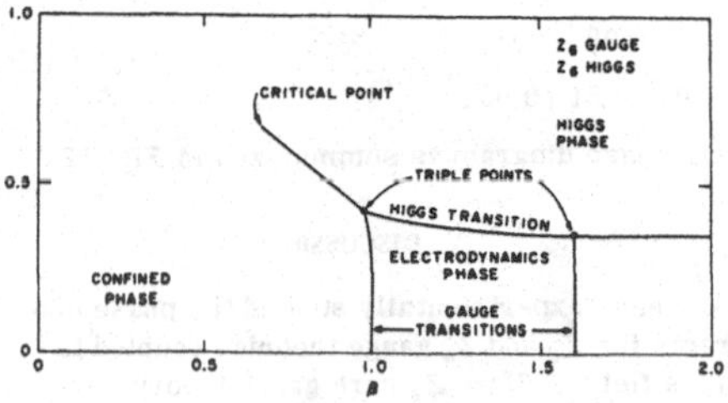

FIG. 10. Phase diagram for the Z_6 system.

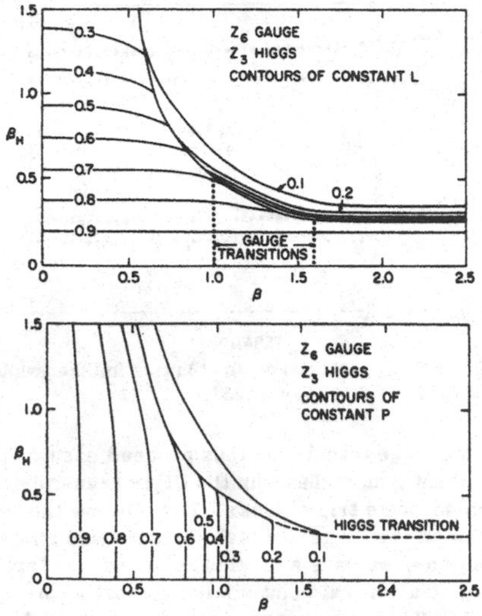

FIG. 11. Contours of constant L and P for Z_6 gauge fields coupled to Z_3 Higgs fields.

initial states. These runs appear to yield distinct phases, but the separation is small. The general features of the phase diagram for this system are summarized in Fig. 10.

Finally, in Fig. 11 we plot the P and L contours for Z_6 gauge fields with the Higgs field in the Z_3 representation. Qualitatively the diagram is similar to that seen in Fig. 8 except that the first-order line with β_H above the low-β triple point no longer unfolds. Rather it continues to large-β_H and becomes the first-order transition of the residual Z_2 theory. Both triple points are shifted to lower β_H because the Z_3 spin system is naturally more ordered than Z_6. The large-β triple point occurs at

$$\beta_H = 1.58 \pm 0.05\,, \qquad \beta_H = 0.28 \pm 0.02\,, \tag{4.4}$$

while the other occurs at

$$\beta = 0.98 \pm 0.03\,, \qquad \beta_H = 0.51 \pm 0.03\,. \tag{4.5}$$

This phase diagram is summarized in Fig. 12.

V. DISCUSSION

We have experimentally studied the phase diagrams for Z_2 and Z_6 gauge theories coupled to Higgs fields. Since Z_6 pure gauge theory behaves

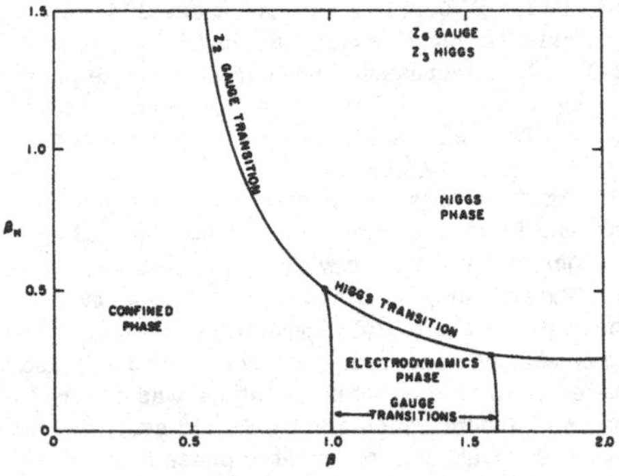

FIG. 12. Phase diagram for the Z_6 gauge, Z_3 spin system.

essentially as the U(1) model for β below the second transition,[2] the first triple point and the unfolding of a first-order line are likely properties of a similar phase diagram for the U(1) coupled Higgs-gauge system. The lack of unfolding for Z_3 site spins coupled to Z_6 gauge fields corresponds to the case of a doubly charged Higgs field in the U(1) system. This work confirms the basic structure of these phase diagrams as predicted in Ref. 6.

Except for the Z_3 case, when β is infinite the Higgs transition is second order. An interesting question is whether this critical point is part of a line of second-order transitions or if the Higgs transition becomes first order in the interior of the diagram. We cannot answer this with our crude Monte Carlo results because when the gauge fields are ordered the effect of β on the Higgs transition is slight.

It might seem remarkable that we can obtain information from a lattice as small as five sites on a side. Note, however, that the number of states for such a system is extremely large; for the Z_2 case there are

$$2^{5^4 \times 5} = 5.23 \times 10^{940}$$

distinct configurations. This large number supports a statistical treatment. Also note that we have only asked rather crude questions about the location of transitions; more subtle points such as critical exponents presumably require considerably more detailed analysis on larger lattices.

ACKNOWLEDGMENT

This work was supported by the U. S. Department of Energy under Contract No. EY-76-C-02-0016.

[1]M. Creutz, L. Jacobs, and C. Rebbi, Phys. Rev. Lett. 42, 1390 (1979).

[2]M. Creutz, L. Jacobs, and C. Rebbi, Phys. Rev. D 20, 1915 (1979).

[3]M. Creutz, Phys. Rev. Lett. 43, 553 (1979).

[4]P. Higgs, Phys. Rev. Lett. 13, 508 (1964).

[5]S. Elitzur, R. B. Pearson, and J. Shigemitsu, Phys. Rev. D 19, 3698 (1979); A. Ukawa, P. Windey, and A. H. Guth, Phys. Rev. D (to be published).

[6]R. Balian, J. Drouffe, and C. Itzykson, Phys. Rev. D 10, 3376 (1974); E. Fradkin and S. Shenker, *ibid.* 19, 3682 (1979); T. Banks and E. Rabinovici, Institute for Advanced Study report, 1979 (unpublished).

[7]E. Ising, Z. Phys. 31, 253 (1925).

[8]R. B. Potts, Proc. Cambridge Philos. Soc. 48, 106 (1952).

[9]H. Blöte and R. Swendsen, Phys. Rev. Lett. 43, 799 (1979).

[10]M. E. Fisher and D. S. Gaunt, Phys. Rev. 133, A224 (1964).

Volume 104B, number 4 PHYSICS LETTERS 3 September 1981

THE PHASE STRUCTURE OF A NON-ABELIAN GAUGE HIGGS FIELD SYSTEM

C.B. LANG, C. REBBI [1] and M. VIRASORO
CERN, Geneva, Switzerland

Received 22 June 1981

The phase structure of a non-abelian gauge Higgs field system is studied by Monte Carlo (MC) simulations. The analysis is done on a 4^4 lattice and approximating both the gauge group and the Higgs manifold by the icosahedral subgroup of SU(2). Results are presented for a Higgs field in the fundamental representation, in which case the symmetry is completely broken for sufficiently strong coupling, and for a Higgs field in the adjoint representation, in which case a U(1) symmetry (approximated by Z_{10}) survives. No evidence for a "Coulomb pocket" is found.

1. Introduction.

MC simulations have proven most valuable for investigating the phase structure of pure gauge field theories [1]. A step towards the final aim of understanding a coupled fermion gauge system consists of the introduction of scalar matter fields and the analysis of the resulting phase diagrams. For some abelian discrete gauge groups of low order (Z_6) this has been done by Creutz [2], to our knowledge no MC calculations for non-abelian gauge groups have been made. There are, however, distinct theoretical expectations [3–5], mainly based on arguments concerning the convergence of the cluster expansion for strong and/or weak couplings. As an independent check and in order to localize the critical lines, we have studied the phase structure of a coupled gauge Higgs system with gauge group SU(2) by MC methods.

To make this analysis feasible we had to comply with a few simplifications which we believe do not distort the physical results. We fix (compactify) the Higgs field to unit norm, i.e., we formally give the potential infinite weight. In the continuum limit, which will eventually be obtained at a critical point where the correlation length grows to infinity, the radial mode should not be relevant.

We work on a 4^4 lattice with periodic boundary conditions for both the gauge and the Higgs variables. Finite-size effects are small for bulk quantities, such as the average contribution to the action per plaquette or per link (internal energies). One MC iteration consists of an update of all gauge field variables without changing the Higgs variables and then vice versa. Most information has been derived from thermal cycles, where the coupling is changed by a small amount, five MC iterations are performed to get closer to equilibrium, then another ten MC iterations are used to obtain average values of the internal energies before proceeding to the next coupling. In this way the system is cooled to the highest values of the coupling constant and then heated again to starting value. The shape of the surfaces giving the internal energies as a function of the couplings, the presence of hysteresis loops or large fluctuations have all been used to locate the probable critical lines. As the thermal cycles are performed with "constant speed", a comparison of the shape of different hysteresis loops also allows us to draw inferences on the type of the phase transitions. Beyond the internal energies, we have also measured the expectation value of the covariant product of Higgs fields at different points. This quantity, together with its fluctuation, may be used to discriminate between the symmetric and Higgs phases of the theory.

Another simplification consisted of approximating the SU(2) gauge group with its 120 element icosahedral subgroup $\tilde{Y}$. The saving in computer time is very substantial and a previous analysis [6] has demonstrated

[1] On leave from Physics Department, Brookhaven National Laboratory, NY, USA.

Volume 104B, number 4 PHYSICS LETTERS 3 September 1981

that the icosahedral group gives, for the pure gauge system, results undistinguishable from those with gauge group SU(2) for values of the gauge coupling below $\beta_c = 6.05$. The latter value is the position of the phase transition specific to the discreteness of the group and (more or less) determined by the action gap. At this value, however, the correlation length in the pure gauge theory is already considerably larger than the lattice size and there will be no difference from a spin wave approximation. Any important structure (which is noticeable on lattices of this size) should have evolved already at smaller values of β and there should be no basic difference between the situation at $\beta = 6$ and at $\beta = \infty$ for the full group SU(2).

The action contains two terms:

$$S(U,\phi) = \beta \sum_P S_P(U_P) + \gamma \sum_{\substack{L \\ (i,j)}} S_L(\phi_i, \phi_j, U_{ij}) , \tag{1.1}$$

where the first sum is over all plaquettes and the second over all links of the lattice. The first part gives the pure gauge field action and we have used Wilson's [7] form

$$S_P(U_P) = \tfrac{1}{2}\,\mathrm{tr}(1 - U_P) , \tag{1.2}$$

normalized so that $0 \leqslant \langle S_P \rangle \leqslant 1$. The second sum gives the gauge invariant Higgs action and will have the general form

$$S_L(\phi_i, \phi_j, U_{ij}) = \mathrm{const} - [R(\phi_i)^+ R(U_{ij}) R(\phi_j) + \mathrm{h.c.}] , \tag{1.3}$$

where $R(\phi)$ and $R(U)$ denote vector and matrix representations of the Higgs- and the gauge-field elements of the group. We investigated the particularly interesting cases where the Higgs field is in (a) the fundamental or (b) the adjoint representation.

Although the discrete approximation could be restricted to the gauge variables, the major saving in computer time is achieved when the continuous Higgs manifold is also replaced by a finite collection of points. Moreover, we found it more elegant to treat both fields in the same way. When the Higgs field is in the fundamental representation, the isomorphism between the gauge and Higgs manifolds ($\approx S^3$) immediately suggests that the discretization used for the SU(2) group be applied also to the Higgs variables. Indeed, setting

$$V_i = \begin{pmatrix} \phi_i^1 & -\phi_i^{2*} \\ \phi_i^2 & \phi_i^{1*} \end{pmatrix} , \tag{1.4}$$

one establishes a one-to-one correspondence between the values of the Higgs field, normalized to one, and the unitary, unimodular matrices V_i. With a normalization such that $0 \leqslant \langle S_L \rangle \leqslant 1$, the Higgs action per link then takes the form

$$S_L^{(F)}(\phi_i, \phi_j, U_{ij}) = \tfrac{1}{2}\,\mathrm{tr}(1 - V_i U_{ij} V_j^+) . \tag{1.5}$$

The discrete approximation is obtained by restricting the matrices V_i, as well as the U_{ij}, to those belonging to the icosahedral subgroup $\widetilde{Y}$.

The case of the adjoint representation requires a little more consideration. In the continuous case the values of the Higgs field (belonging to S^2) may be represented by matrices $\boldsymbol{\sigma}\cdot\boldsymbol{\phi}_i, \boldsymbol{\phi}_i^2 = 1$, or equivalently, may again be represented by SU(2) matrices

$$V_i = \cos\theta + \mathrm{i}\boldsymbol{\sigma}\cdot\boldsymbol{\phi}_i \sin\theta , \tag{1.6}$$

but where the angle θ ($\neq 0, \pi$) is now kept fixed. This realizes the isomorphism

$$S^2 \approx S^3/S^1 \approx \mathrm{SU(2)/U(1)} . \tag{1.7}$$

Replacing V_i by the matrices of $\widetilde{Y}$ will thus produce a discretization of the Higgs manifold; the V_i cannot, however, range over the whole group, but must belong to one of the equivalence classes. Leaving aside the trivial one-element classes $\{I\}$ and $\{-I\}$, the icosahedral subgroup has four equivalence classes $n = 1-4$ of 12 elements [with $\theta = n\pi/5$ in eq. (1.6)], two classes of 20 elements (with $\theta = n\pi/3$) and one class of 30 elements ($\theta = \pi/2$). The rotations belonging to these various classes can be put into correspondence with the vertices, faces or sides of the icosahedron. Clearly the elements of larger class give the denser finite representation of S^2. However, they do not provide the discretization of the Higgs variables most useful for our purposes. Indeed, whereas the stability group of any Higgs vector in the continuous case is U(1), the stability group of the rotations belonging to the 12, 20 or 30 element classes are, respectively, Z_{10}, Z_6 and Z_4. These will be the surviving gauge groups when, for a sufficiently large Higgs coupling, the original gauge symmetry is maximally broken. Since we want this remnant gauge freedom to approximate U(1) as closely as possible, we have used one of the 12 element classes as a discrete version of the Higgs manifold. Specifically, we have represented the Higgs variables by the 12 matrices, $\epsilon\widetilde{\mathrm{Y}}$, of the form of eq. (1.6) with $\theta = \pi/5$. The Higgs action (normalized again so that its expectation value lies between 0 and 1) is then given by

$$S_L^{(A)} = (2\sin^2\tfrac{1}{5}\pi)^{-1}\,\mathrm{tr}(1 - V_i U_{ij} V_j^+ U_{ij}^+)\,. \qquad (1.8)$$

2. Expectation and results.

2.1. The Higgs field in the fundamental representation.

(a) Expectations. For $\gamma = 0$ the lagrangian reduces to that of the pure $\widetilde{\mathrm{Y}}$ gauge field theory, which has been investigated in detail [6]. To simplify the discussion for the cases $\gamma = \infty$ or $\beta = 0$ note that by a gauge transformation

$$V_i \to V_i' = V_i g_i\,, \qquad U_{ij} \to U_{ij}' = g_i^+ U_{ij} g_j\,,$$

one may transform the Higgs field away by choosing $g_i = V_i^+$ ("U gauge"). The lagrangian becomes one of a gauge system coupled to a non-gauge invariant field with strength γ. Thus for $\gamma = \infty$ all U's will be frozen to unify in order to minimize $\mathrm{tr}(1 - U)$; $\langle S_P\rangle$ and $\langle S_L\rangle$ vanish identically. For $\beta = 0$, only the link term contributes to the action and the integral factorizes. It may be solved explicitly and the MC results are in agreement with the analytical expression. For $\beta = \infty$ the U configuration becomes a gauge transform of the trivial (completely ordered) one; this leaves a spin system with global symmetry SU(2) (actually its icosahedral subgroup) and nearest neighbour coupling. Such a system will develop one phase transition of (presumably) second order. (As a matter of fact, some non-trivial features are associated with the finite size of the system; see later discussion.) From the clustering properties (strong/weak coupling expansion) one finds analyticity domains $\{\gamma, \beta \mid \text{both } \gamma, \beta \text{ small or } \gamma/\beta \text{ large enough}\}$ [3]. For $\gamma > 0$ there will be perimeter decay for both the Wilson loop and the 't Hooft loop [8] due to the screening of static charges through the Higgs field. For our discrete group there is also analyticity in $\{\gamma, \beta \mid \gamma \text{ small and } \beta \text{ large}\}$ (cf. Seiler [4]).

(b) Results. The results of several thermal cycles done at fixed β or fixed γ are all put together in fig. 1a. The upper graph illustrates the measured expectation value of the gauge action per plaquette $\langle S_P\rangle$, the lower graph the expectation value of the Higgs action per link $\langle S_L\rangle$. Hysteresis loops are evident in many of the cycles and are clear signs of the critical slowing down associated with phase transitions. A pronounced maximum in the slope of the curves (corresponding to a peak in the specific heat) can also be taken as a sign of a possible critical phenomenon. In the presence of a phase transition of higher order, one expects an increase in the fluctuations. This is clearly visible along some of the curves in fig. 1a. In particular, the line with $\gamma = 1$ in the graph for $\langle S_L\rangle$ runs very close to a line of phase transitions (indicated by the hysteresis loops in the simulations at $\beta = \text{const}$) and one notices a marked increase in the fluctuations relative, for instance, to the other lines with $\gamma = \text{const}$.

From these measurements we deduce the phase diagram exhibited in fig. 1b. The phase transition due to the discreteness of the group starts at $\beta = 6.05$, $\gamma = 0$ and moves to smaller β values with increasing γ. Above $\gamma \approx 5$ the hysteresis essentially disappears leaving a slope maximum; this indicates a change of the order of the phase transition. At $\gamma \gtrsim 6$ we were not able to detect a signal of this phase transition.

The other critical line is of much more interest – it should closely resemble the structure of the continuous group. From the comparison of the hysteresis we are led to attribute to that critical line a second-order phase transition everywhere. As expected this critical line ends in the γ, β plane at a position close to $\beta = 1.2$, $\gamma = 1.5$. Concerning the shape of the curve towards larger values of β we find no evidence of a splitting into two lines (which would lead to a pocket of Coulomb phase). It seems that the line of phase transition approaches asymptotically the value of γ where one observes a phase transition of the spin system.

The thermal cycle at $\beta = 6$ shows a wide hysteresis loop in the results for $\langle S_L\rangle$. This interesting effect deserves some discussion. When β becomes very large the gauge variables are frozen into a configuration with vanishing gauge action. Locally this must be a gauge transform of the identity and, in particular, by suitably arranging the Higgs variables, the minimum of the Higgs action may also locally be set to zero. However, because of the finite size and the periodicity of the boundary conditions, the gauge action may be zero even if the configuration of the gauge variables is not a global gauge transform of the identity. In such a configuration the minimum of the Higgs action is larger than zero. In principle, as γ increases, the configurations of the gauge field which are locally, but not globally ordered should become energetically unfavourable; but because the MC

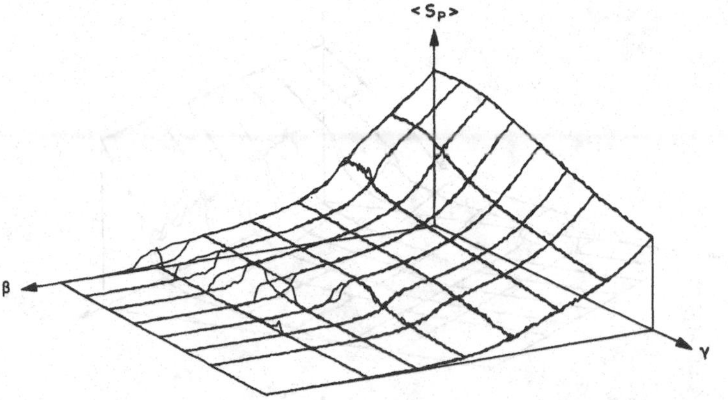

(a)

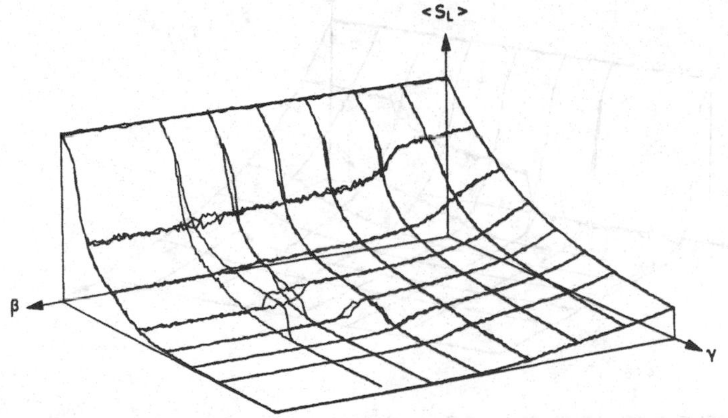

(b)

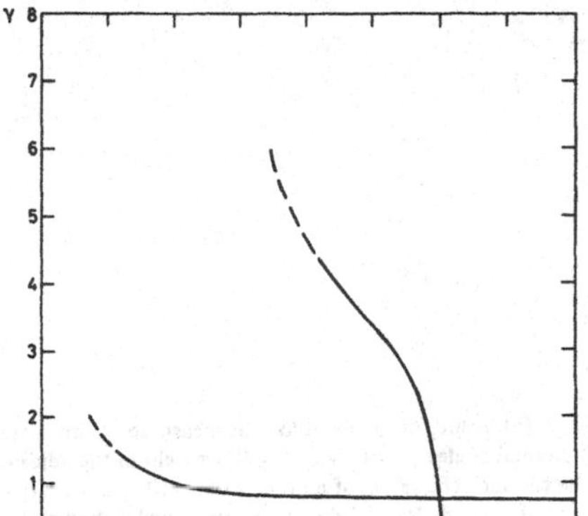

Fig. 1. (a) Values of $\langle S_P \rangle$ and $\langle S_L \rangle$ as measured in simulations of thermal cycles for case (a): the Higgs field in the fundamental representation. The values of β or γ are 0, 1, 2, 3, 4, 5, 6, 8. (b) The phase diagram deduced from the thermal cycles shown above.

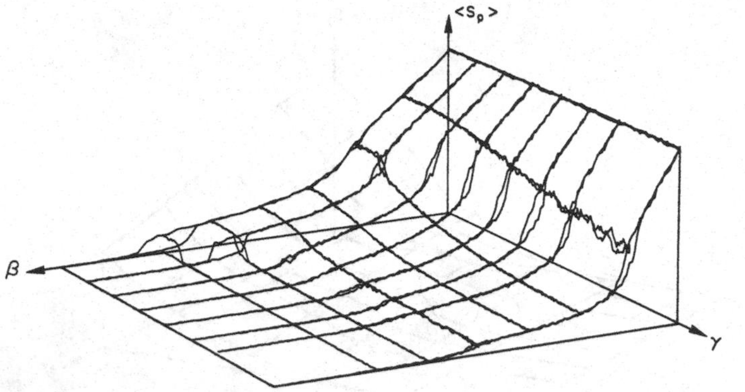

(a)

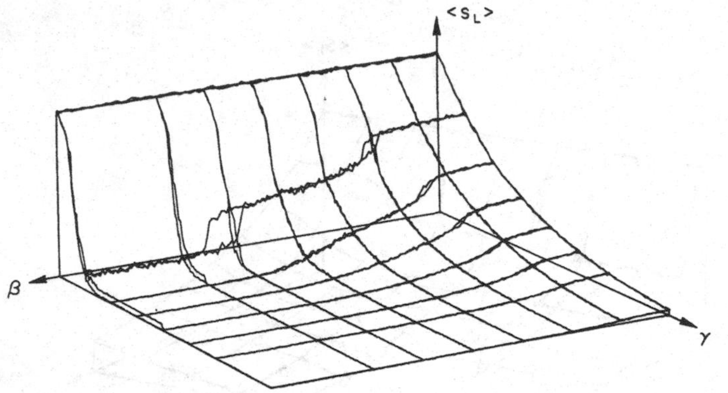

(b)

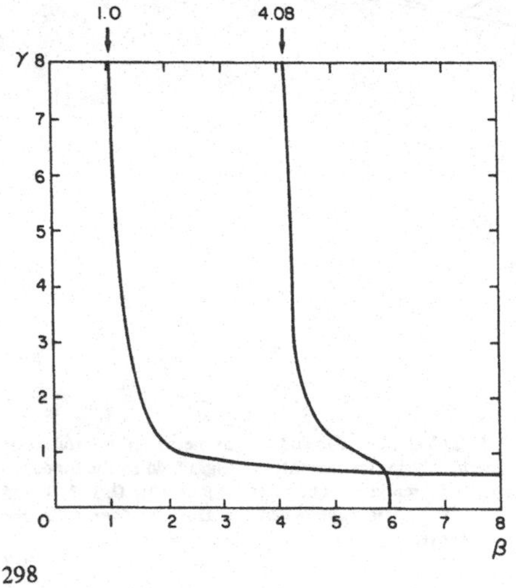

Fig. 2. (a) Values of $\langle S_P \rangle$ and $\langle S_L \rangle$ as measured in simulations of thermal cycles for case (b): the Higgs field in the adjoint representation. The values of β or γ are 0, 1, 2, 3, 4, 5, 6, 8. (b) The phase diagram deduced from the thermal cycles shown above.

Volume 104B, number 4 PHYSICS LETTERS 3 September 1981

upgrading is an essentially local process, the gauge field may remain locked into one of these configurations even at large γ. This is what is observed in the simulation at $\beta = 6$. The initial configuration taken from a run with $\gamma = 0$, still contains some disorder in the gauge variables because it is derived from the upper part of the hysteresis loop around the transition at $\beta_c \approx 6$ (see the diagram for $\langle S_P \rangle$). As the MC simulation proceeds and γ increases, both $\langle S_P \rangle$ and $\langle S_L \rangle$ decrease, but the Higgs action cannot get below some non-vanishing minimum value (although $\langle S_P \rangle$ is essentially zero), because of some global disorder frozen in. Eventually, for large γ (≈ 4), the ordering effect of the Higgs part of the action prevails and the whole system undergoes a transition to a more ordered configuration. It is interesting that this unfolding of the global disorder of the gauge field requires a temporary increase of $\langle S_P \rangle$, clearly evident as a small peak in the line with $\beta = 6$ in the upper part of the diagram. The simulation at $\beta = 8$ was done starting from the identity configuration of the gauge field and no effects attributable to global disorder are observed.

2.2. The Higgs field in the adjoint representation.

(a) Expectations. For $\beta = 0$ the integral again factorizes and may be solved explicitly (all ϕ_i may be transformed to some ϕ_0) confirming our MC results. For $\beta = \infty$ the system becomes a spin system with the icosahedral subgroup of SO(3) as a symmetry group, i.e., the classical Heisenberg model for a subgroup of O(3); one phase transition (presumably second order) is expected. For $\gamma = \infty$ the stability group is Z_{10} and we expect two phase transitions of second order at $\beta = 1$ and 4.08 [1]. Further results [3,4] are the evidence of a Higgs phase (perimeter decay of the Wilson loop, area decay of the 't Hooft loop) for both γ, β large and of a confinement phase (area decay for the Wilson loop, perimeter decay for the 't Hooft loop) for small β and γ.

(b) Results. The thermal cycles in this case are shown in fig. 2a. The analysis is done as for the previous case. The resulting phase diagram is shown in fig. 2b. We observe that both critical lines now continue until $\gamma = \infty$. Actually $\gamma = 8$ already provides a good approximation to a Z_{10} theory with its two phase transitions. Using the full continuous group SU(2) would lead to a U(1) theory at $\gamma = \infty$. This would not change the position of the first phase transition at $\beta = 1$ but the second phase transition now at $\beta = 4.08$ would move to infinity. At the same time the phase transition at $\gamma = 0, \beta = 6.05$ would move to infinite values of β. From the hysteresis we expect that the critical line from $\gamma = \infty, \beta = 1$ to $\gamma \approx 0.75, \beta = \infty$ indicates a second-order phase transition. The other one starts as a first order phase transition at $\gamma = 0, \beta = 6.05$ and appears to become second order above $\gamma \gtrsim 2$.

3. Looking at the Higgs confinement transition.

If one tries to characterize the transition from the Higgs region to the confinement region (with the matter field in the fundamental representation) one encounters the difficulty that there is no gauge-invariant order parameter [9]. In our model this is reflected in the phase diagram fig. 1 which shows just one single phase to the left of the $\beta = 6$ line. On the other hand, the naive picture just assumes that the Higgs field itself develops a vacuum expectation value. We would like to understand the degree of validity of this assumption.

We have therefore looked at the behaviour of the following operator:

$$X = \sum_{n_1} \tfrac{1}{2} \mathrm{tr}[V^{+}_{n_1+\Delta} W(n_1, \Delta) V_{n_1}] \qquad (3.1)$$

across the transition line at $\beta = 5.2$ ($\gamma = 0.6, 0.7, 0.8, 0.9$). Here n_1 and $n_1 + \Delta$ are the two antipodal points in a fundamental hypercube [i.e., $n_1 + \Delta = (n_{1x} + 1, n_{1y} + 1, n_{1z} + 1, n_{1t} + 1)$] and W is the average over the 24 paths of minimum length from n_1 to $n_1 + \Delta$ of the ordered products of the U field. In (3.1) the sum over n_1 sweeps the whole lattice. This is the only reasonable choice at $\beta = 5.2$ where the correlation length is larger than the lattice. In general we would choose a volume whose size is a definite fraction of the correlation length. Then X will develop a large expectation value (compared to its fluctuations) if $\phi^\dagger \phi$ does so.

We have measured X, at each value of γ, over 1000 successive configurations of the lattice. The histograms are shown in fig. 3. We observe that the average value is different from zero both below and above the transition at $\gamma = 0.75$. However, in the disordered phase the fluctuations and the average value are of the same order, while in the Higgs phase the average is much larger than the fluctuations. This is exactly what one expects to justify the semiclassical approximation around a new vacuum characterized (in a certain gauge) by $\phi \neq 0$.

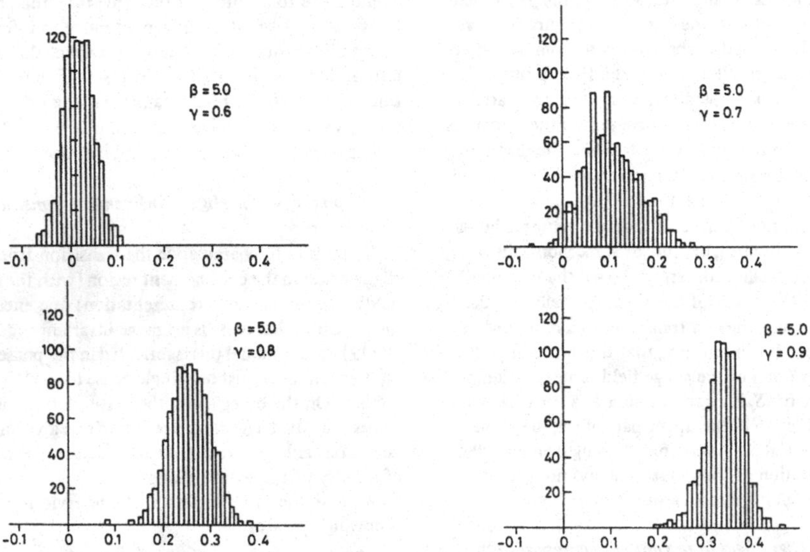

Fig. 3. Histograms showing the distribution of the instantaneous values of the operator X measured a total of 1000 times at $\gamma = 0.6, 0.7, 0.8, 0.9$.

4. Conclusions.

MC methods appear to be well suited to discussing coupled Higgs gauge systems. In the cases considered we find no surprises, i.e., no deviations from the theoretical expectations. However, contrary to the approaches based on series expansions, we are capable of determining the field phase structure quite efficiently. For a closer investigation of order parameters one will have to move to larger lattices.

One of us (C.L.) would like to acknowledge interesting discussions with S. Shenker.

References

[1] K. Wilson, Cargèse Lecture Notes (1979);
M. Creutz, L. Jacobs and C. Rebbi, Phys. Rev. Lett. 42 (1979) 1390; Phys. Rev. D20 (1979) 1915;
M. Creutz, Phys. Rev. Lett. 43 (1979) 553; Phys. Rev. D21 (1980) 2308; Phys. Rev. Lett. 45 (1980) 313;
C. Rebbi, Phys. Rev. D21 (1980) 3350.
[2] M. Creutz, Phys. Rev. D21 (1980) 1006.
[3] K. Osterwalder and E. Seiler, Ann. Phys. (NY) 110 (1978) 440;
E. Fradkin and S. Shenker, Phys. Rev. D19 (1979) 3682;
T. Banks and E. Rabinovici, Nucl. Phys. B160 (1979) 349.
[4] G. Münster, Z. Phys. C6 (1980) 175;
E. Seiler, Lausanne Lecture Notes (February 1981), unpublished.
[5] A. Ukawa, P. Windey and A.H. Guth, Phys. Rev. D21 (1980) 1013.
[6] G. Bhanot and C. Rebbi, Nucl. Phys., to be published.
[7] K. Wilson, Phys. Rev. D10 (1974) 2445.
[8] G. 't Hooft, Nucl. Phys. B138 (1978) 1; B153 (1979) 141.
[9] S. Elitzur, Phys. Rev. D12 (1975) 3978;
G.F. De Angelis, D. De Falco and F. Guerra, Lett. Nuovo Cimento 19 (1977) 75;
J. Fröhlich, G. Morchio and F. Strocchi, Phys. Lett. 97B (1980) 249.

Nuclear Physics B183 (1981) 103–140

LATTICE FERMIONS: SPECIES DOUBLING, CHIRAL INVARIANCE AND THE TRIANGLE ANOMALY

Luuk H. KARSTEN

Institute of Theoretical Physics, Physics Department, Stanford University, Stanford, CA 94305, USA

Jan SMIT

Instituut voor Theoretische Fysica, Universiteit van Amsterdam, Valckenierstraat 65, 1018 XE Amsterdam, The Netherlands

Received 28 October 1980

We investigate the formulation of fermions in lattice gauge theories, in weak coupling perturbation theory. It is shown that a lattice fermion formulation without species doubling and with explicit continuous chiral symmetry is impossible. We interpret this as a manifestation of the triangle anomaly. Green functions and (axial) vector flavor currents in lattice QCD with Wilson fermions are studied. We show that in the (weak coupling) continuum limit the theory is equivalent to perturbative continuum QCD: chiral invariant and with the usual anomaly in the flavor singlet axial current.

1. Introduction

The favorite theory for strong interactions, QCD, has its successes both at long and short distances. At short distances the effective coupling is small and one uses weak coupling perturbation theory (WCPT). At longer distances the coupling gets larger, WCPT ceases to be valid. For large distances one needs a non-perturbative regularization, like the lattice. On the lattice for large coupling quark confinement is explicit.

It is necessary for QCD that the connection between the theory formulated at large and at short distances is smooth and does not involve a phase transition. There has been progress, especially during the last year, showing that such a connection might exist, and in understanding the dynamics in this region [1–5]. These new developments, however, pertain to gauge theories without dynamical quarks.

An obstacle to the extension of such results to gauge theories with quarks is the uncertainty about the formulation of lattice fermions. The most straightforward fermion formulation produces too many fermions. Several ways of avoiding this multiplicity of fermions have been proposed in the literature [6–8]. One of these methods (SLAC fermions) has manifest chiral invariance, but has severe problems in reproducing continuum WCPT for weak coupling and small lattice distance [9]. The other two methods (Wilson's and Susskind's fermions) do not have these problems, but they have no explicit continuous chiral invariance on the lattice either. Chiral symmetry, realized in the Nambu–Goldstone mode, is supposed to be an important approximate symmetry of the strong interactions: one of the consequences is the smallness of the pion mass. Lack of chiral invariance of the continuum limit of lattice QCD will reflect directly on the value of the pion mass.

Indeed, strong coupling calculations with Susskind's fermions, with Padé extrapolation to weak coupling, give much too high an m_π/m_N ratio [10].

Calculations with Wilson's fermions [11, 12], however, indicate that with this method a low pion mass is no problem.

More recently it has been shown by one of us [13] that using Wilson's fermions in lattice QCD, at strong coupling within the $1/N$ expansion, a continuum limit can be taken in the pseudoscalar meson sector such that chiral symmetry emerges, realized in the Nambu–Goldstone mode. The results obtained fit in the Caldi–Pagels picture of spontaneous symmetry breaking: pseudoscalar mesons are Goldstone bosons, vector mesons "dormant" Goldstone bosons [14]; mass relations for mesons and (static) baryons were derived, identifying dynamical and current quark masses. Furthermore, axial and vector currents were discussed, resulting in relations for f_π, γ_ρ, and g_A.

The purpose of this paper is to investigate lattice fermions in gauge theories in WCPT. Although WCPT will not reveal the full content of the theory, we expect to gain valuable information. Studies in constructive field theory [15] suggest that the structure of counterterms remains valid beyond perturbation theory. Still more important, lattice QCD with fermions should in WCPT reduce to the usual continuum formulation. This is a necessary first step in showing that a smooth transition between large and small coupling exists.

In this paper we ask ourselves the following questions. First, why is it impossible to find a lattice fermion formulation without the multiplicity problem and with chiral symmetry? Second, is it indeed possible to use Wilson's lattice fermions in obtaining a chiral invariant QCD in the continuum limit? Third, what is the divergence structure of the axial vector currents?

The result of this investigation in WCPT extending previous work by us [9, 17] is as follows. In sect. 2 we present various lattice fermion formulations. It is shown that the naive fermion formulation produces 16 bona fide fermions. We demonstrate that one has in any lattice fermion formulation that makes sense in WCPT either a multiplicity of states or no explicit continuous chiral invariance. In sect. 3 the connection is made with the axial anomaly. In a good regularization, like the lattice, there is no place for anomalies. Either the axial anomaly has to be cancelled by extra fermions, the multiplicity of fermions in the naive fermion formulation serves this purpose, or the "anomaly" has to be introduced by explicit chiral symmetry breaking, which happens in Wilson's formulation. We illustrate this with an explicit calculation of the triangle diagram in a theory which interpolates between naive and

Wilson's fermions. We give an argument for why there are 16 fermions in the naive formulation and not two, as would be enough for anomaly cancelling.

In sects. 4, 5, lattice QCD with flavored Wilson fermions is investigated. In sect. 4 we calculate one-loop contributions to Green functions and discuss multi-loop contributions. The theory has a continuum limit identical to QCD in continuum perturbation theory, chiral invariant and independent of the extra parameter which is introduced by the chiral symmetry breaking "Wilson term" in Wilson's fermion formulation. Sect. 5 is devoted to (axial) vector currents, their definition, normalization and conservation. We calculate one-loop contributions to their matrix elements, and show that in the continuum limit the flavor non-singlet chiral currents are conserved, while the flavor singlet chiral current has the usual anomaly. It is explicitly shown that the anomaly is indeed generated by the Wilson term, but independent of the coefficient in front of this term (as long as it is non-zero). In sect. 6 we give our conclusions. In the appendix the Feynman rules for a gauge theory with Wilson fermions are listed.

2. Naive and sophisticated lattice fermions

We work on a four-dimensional euclidean hypercubic lattice. The lattice spacing is a, and the lattice points are labelled with

$$x_\mu = n_\mu a\,, \qquad n_\mu = 0, \pm 1, \pm 2, \ldots, \qquad \mu = 1, 2, 3, 4\,. \tag{2.1}$$

The range of the momenta is restricted to an interval of length $2\pi/a$, which can be chosen as

$$-\pi/a < k_\mu < \pi/a\,. \tag{2.2}$$

The natural way to find a lattice version of the continuum action for a free spin-$\frac{1}{2}$ fermion field ψ,

$$I = \int d^4x \left[-\sum_\mu \bar{\psi}(x) \frac{1}{i} \gamma_\mu \frac{\overleftrightarrow{\partial}_\mu}{2} \psi(x) - m\bar{\psi}(x)\psi(x) \right], \tag{2.3}$$

is to replace the differentials by differences. Then we find the naive lattice fermion action

$$I = +\sum_x \left(\sum_\mu \frac{-1}{2ia} [\bar{\psi}(x)\gamma_\mu \psi(x+a_\mu) - \bar{\psi}(x+a_\mu)\gamma_\mu \psi(x)] - m\bar{\psi}(x)\psi(x) \right). \tag{2.4}$$

Here $\sum_x = a^4 \sum_n$ and a_μ is a vector along the μ direction with length a.

The action (2.4) has a global U(1) invariance, $\hat{\psi}(x) = V\psi(x)$ and $\hat{\bar{\psi}}(x) = \bar{\psi}(x) V^{-1}$, $V \in \mathrm{U}(1)$, that we make local in the standard way by introducing a gauge field $U_\mu(x)$, defined on links $(x, x+a_\mu)$:

$$I = +\sum_x \left\{ \sum_\mu \frac{-1}{2ia} [\bar{\psi}(x)\gamma_\mu U_\mu(x)\psi(x+a_\mu) - \bar{\psi}(x+a_\mu)\gamma_\mu U^\dagger_\mu(x)\psi(x)] - m\bar{\psi}(x)\psi(x) \right\} + I(U)\,. \tag{2.5}$$

U transforms as follows:

$$\hat{U}_\mu(x) = V(x)U_\mu(x)V^{-1}(x+a_\mu)\,. \tag{2.6}$$

It will make no difference to our considerations in this section whether we take U(1) or a gauge group like SU(N).

For weak fields it is useful to introduce a vector potential $v_\mu(x)$ by

$$U_\mu(x) = \exp\,[igav_\mu(x)]\,. \tag{2.7}$$

Inserting this definition in (2.5), we find by Fourier transformation the (weak coupling) Feynman rules. The fermion propagator is

$$S(p) = \left(\sum_\mu \gamma_\mu \frac{1}{a}\sin p_\mu a + m\right)^{-1}. \tag{2.8}$$

The $\bar{\psi}\psi v_\mu$ vertex function is (cf. fig. 3b)

$$g\gamma_\mu \cos\tfrac{1}{2}(p-q)_\mu a\,, \qquad p+q+k=0\,. \tag{2.9}$$

Furthermore, one finds vertices with two or more v's, and the gauge field propagator, $D_{\mu\nu}(k)$. These are discussed in the appendix.

Next we analyse the classical continuum limit of the theory (2.5), i.e., the tree graphs for $a \to 0$. We will show that (2.5) in fact describes 16 fermions, with degenerate mass m and charge g. For this discussion it is convenient to shift the range of momenta to $\pi/2a < p_\mu < 3\pi/2a$, using periodicity over $2\pi/a$.

The fermion propagator $S(p)$, (2.8), does not vanish in the limit $a \to 0$ in 16 regions in momentum space: about the points with $p_\mu = 0$ or π/a. Call such a point generically $\bar{p}$. About these points $\bar{p}$ the propagator behaves in a similar way. First observe that the action (2.5) is invariant under a group of 16 symmetry transformations

$$\hat{\psi}(x) = T\psi(x)\,, \qquad \hat{\bar{\psi}}(x) = \bar{\psi}(x)T^{-1}\,, \tag{2.10}$$

with $T = 1$, $\gamma_\mu\gamma_5(-)^{x_\mu/a}$ or any product of these transformations. (The T form a discrete subgroup of the euclidean version of the U(4) symmetry group found recently [18]; in 2 lattice dimensions the discrete group was considered by Chodos and Healy [16].) In momentum space the T interrelate the points $\bar{p}$. For example $T = \gamma_1\gamma_5(-)^{x_1/a}$ shifts p_1 to $p_1 + \pi/a$ (modulo $2\pi/a$), and $\bar{p} = (0,0,0,0)$ is transformed into $\bar{p} = (\pi/a, 0, 0, 0)$ and vice versa, $\bar{p} = (0, \pi/a, 0, 0)$ into $\bar{p} = (\pi/a, \pi/a, 0, 0)$, etc. We denote the 16 T by the point $\bar{p}$ into which they transform $\bar{p} = (0,0,0,0)$: $T(\bar{p})$. $T(\bar{p})$ has a spin part $s(\bar{p})$ ($=1$ or $\gamma_\mu\gamma_5$, etc.).

Now we consider the propagator $S(p)$ about a point $\bar{p}$. Define $k=p-\bar{p}$, for $k=\mathrm{O}(a^0)$:

$$S(p)=\frac{m-\sum_\mu (1/a)\gamma_\mu \sin p_\mu a}{m^2+\sum_\mu (1/a^2)\sin^2 p_\mu a}$$

$$=\frac{m-\sum_\mu k_\mu \gamma_\mu \cos \bar{p}_\mu a}{m^2+\sum_\mu k_\mu^2}+\mathrm{O}(a)$$

$$=s(\bar{p})\frac{m-\sum_\mu \gamma_\mu k_\mu}{m^2+\sum_\mu k_\mu^2}s(\bar{p})^{-1}+\mathrm{O}(a)\,. \tag{2.11}$$

Next we rotate k_μ into Minkowski space, $\gamma^0=i\gamma_4$, $k^0=ik_4$:

$$S(k;\bar{p})=\frac{s(\bar{p})(\gamma^0 k^0-\boldsymbol{\gamma}\cdot\boldsymbol{k}+m)s^{-1}(\bar{p})}{-k^{02}+\boldsymbol{k}^2+m^2-i\varepsilon}+\mathrm{O}(a)\,. \tag{2.12}$$

As the $s(\bar{p})$ perform just a similarity transformation on the γ's, we have around each point $\bar{p}$ all the states of a free Dirac particle with mass m, 16 Dirac particles in total. Their wave functions can be found by considering the residue of the mass-shell pole of S: for $k^{02}=\boldsymbol{k}^2+m^2$, $k^0>0$,

$$s(\bar{p})(\gamma^0 k^0-\boldsymbol{\gamma}\cdot\boldsymbol{k}+m)s^{-1}(\bar{p})=\sum_\lambda [s(\bar{p})u_\lambda(\boldsymbol{k})][\bar{u}_\lambda(\boldsymbol{k})s^{-1}(\bar{p})]\,, \tag{2.13a}$$

and similarly for $k^0<0$:

$$=-\sum_\lambda [s(\bar{p})v_\lambda(-\boldsymbol{k})][\bar{v}_\lambda(-\boldsymbol{k})s^{-1}(\bar{p})]\,, \tag{2.13b}$$

with $^{(}\bar{u}^{)}_\lambda$ and $^{(}\bar{v}^{)}_\lambda$ the standard Dirac spinors [37]. The wave functions of the Dirac (anti-) particle near $\bar{p}$ are $s(\bar{p})u_\lambda(\boldsymbol{k})$, $s(\bar{p})v_\lambda(+\boldsymbol{k})$, $\bar{u}_\lambda(\boldsymbol{k})s^{-1}(\bar{p})$, and $\bar{v}_\lambda(+\boldsymbol{k})s^{-1}(\bar{p})$.

Even if we start with only particles with $p\approx 0$, we have to consider all 16 particles: they will be pair produced, they contribute to intermediate processes. In fig. 1, we start with a pair of $p\approx 0$ particles; in the final state we have a particle pair around general $\bar{p}$. The amplitude is, using the vertex function (2.9),

$$\sum_{\mu,\nu}\bar{v}(+\boldsymbol{p}_2)e\gamma^\mu \cos[\tfrac{1}{2}(p_1+p_2)_\mu a]u_1(\boldsymbol{p}_1)D_{\mu\nu}(k)$$
$$\times[\bar{u}(\boldsymbol{p}_3-\bar{\boldsymbol{p}})s^{-1}(\bar{p})e\gamma^\nu \cos[\tfrac{1}{2}(p_3+p_4)_\nu a][s(\bar{p})v(\boldsymbol{p}_4-\bar{\boldsymbol{p}})]$$
$$=\sum_{\mu,\nu}\bar{v}(\boldsymbol{p}_2)e\gamma^\mu u(\boldsymbol{p}_1)D_{\mu\nu}(k)\bar{u}(\boldsymbol{p}_3-\bar{\boldsymbol{p}})e\gamma^\nu v(\boldsymbol{p}_4-\bar{\boldsymbol{p}})+\mathrm{O}(a)\,. \tag{2.14}$$

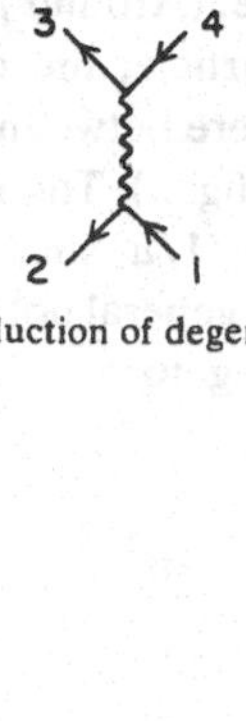

Fig. 1. Pair production of degenerate fermions.

From (2.14) we see that all 16 fermions can be produced in the tree approximation. Furthermore we see that their vector charge is the same, $+g$.

In sect. 3 we will show that the one-loop contribution to the vacuum polarization is multiplied with a factor 16. This, together with our tree-graph considerations, shows that the action (2.5) describes in weak coupling a theory with 16 fermions instead of one. (In general: if d dimensions are latticized, the multiplicity is 2^d.)

The most straightforward way to remove these extra particles is giving them a mass of the order of the cut-off. This method has been proposed by Wilson [6]. He adds an extra term

$$\frac{r}{2a}\sum_{x,\mu}[\bar{\psi}(x)U_\mu(x)\psi(x+a_\mu)+\bar{\psi}(x+a_\mu)U^\dagger_\mu(x)\psi(x)-2\bar{\psi}(x)\psi(x)] \qquad (2.15)$$

to the action (2.5). The $\bar{\psi}(x)U\psi(x\pm a_\mu)$ terms break the doubling symmetry (2.10). Wilson uses $r=1$, we take $0\leq r\leq 1$. The fermion propagator becomes

$$S(p)=\left[\sum_\mu \gamma_\mu \frac{1}{a}\sin p_\mu a+m+\frac{r}{a}\sum_\mu(1-\cos p_\mu a)\right]^{-1}. \qquad (2.16)$$

The propagator is $O(a^0)$ only for $p_\mu\approx 0$. The r term is a momentum-dependent mass term, and gives 15 fermions a mass $m+k2r/a$ ($k=1, 2, 3$ or 4). For small p_μ the extra term is $O(a)$. We have just one fermion with a relativistic spectrum in the continuum limit $a\to 0$.

An obvious disadvantage of Wilson's fermion formulation is that the extra term (2.15) breaks chiral symmetry: even for $m=0$ the fermion lattice action with the Wilson term is not chiral invariant. This complicates the use of Wilson fermions in a lattice formulation of field theories where chiral invariance is supposed to play an important role, as in quantum chromodynamics.

Is it possible to send the extra fermions away without losing chiral invariance? To answer this question we examine the general form of the propagator for lattice fermions compatible with continuous chiral invariance

$$S(p)=\left[\sum_\mu \gamma_\mu P_\mu(p)\right]^{-1}, \qquad P_\mu(p)\ \text{real}. \qquad (2.17)$$

To describe one Dirac fermion for $a\to 0$, $P_\mu(p)$, as a function of p_μ, should cross the $P_\mu=0$ axis once, say at $p_\mu=0$. By periodicity $P_\mu(p)$ crosses the axis again at $p_\mu=2\pi/a$, with the same derivative. (Around $p_\mu=0$, $P_\mu(p)=p_\mu+O(a)$ in order to get the propagator for a Dirac particle; the crossing at $p_\mu=2\pi/a$ represents of course the same particle.) Somewhere between 0 and $2\pi/a$ there has to be another crossing or a jump across the axis (fig. 2). The naive propagator (2.8) is an example with such an extra crossing $P_\mu(p)=(1/a)\sin ap_\mu$. Such behavior or a finite $[O(a^0)]$ jump means that there are extra (in general not relativistic) excitations in the theory, as we discussed for the naive propagator.

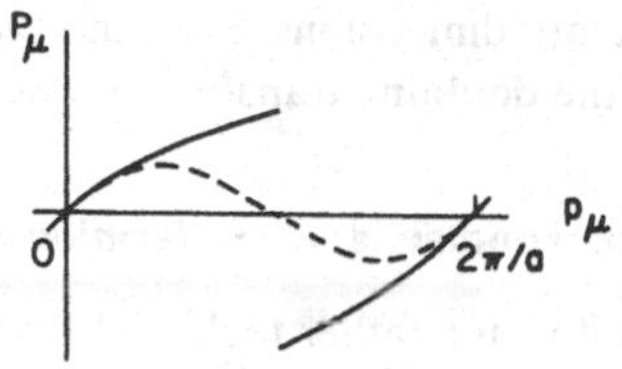

Fig. 2. The function $P_\mu(p)$ in (2.11): either a gap (full curve) or an extra axis crossing (dashed curve).

An infinite jump [O(1/a)] propagator seems to solve the fermion doubling. An example of such a propagator is the SLAC propagator [8]: $P_\mu = p_\mu$; for $-\pi/a < p_\mu < \pi/a$ and periodic over $2\pi/a$. There is a gap at $p_\mu = \pi/a$, with width $2\pi/a$. This gap implies a non-local lattice action, i.e., an action containing products of fields arbitrarily far apart. As we have shown elsewhere [9, 17] the vacuum polarization in straightforward perturbation theory in a gauge theory with SLAC fermions gets non-local not Lorentz covariant contributions, at the one-loop level $a \to 0$, because of this gap. A typical term is of the form

$$\frac{\delta_{\mu\nu} \sum_\rho |p_\rho||p_\nu| - |p_\nu||p_\mu|}{p_\nu p_\mu}. \tag{2.18}$$

This result can immediately be generalized, as inspection of that calculation shows, to all lattice gauge theories with fermions with an infinite gap in their propagator. This means that infinite-gap fermions do not make contact with covariant continuum perturbation theory for small g and $a \to 0$. (This is contrary to claims made in the literature [18]. In our opinion these claims have not been substantiated.)

It has been argued that straightforward perturbation theory is not applicable in the case of infinite-gap fermions because of infrared infinities [19]. While this is plausible, considering for example the extra term for $p \to 0$, any more careful definition of a perturbative expansion with chiral invariance will result in a propagator of the form (2.17) for which the above analysis is valid. Probably extra excitation will again arise. (Indeed, at strong coupling one has again the species doubling problem [18].)

In summary, it seems that if one insists on chiral invariance on the lattice one gets extra states and/or a non-covariant theory (for $a \to 0$), both of which one does not want.

In the foregoing we have tacitly assumed all four components of ψ defined on a site of the lattice. Susskind has proposed a lattice fermion formulation where $\psi(x)$ is a one-component field. This solves part of the naive degeneracy, and one has a discrete γ_5 invariance. But continuous chiral transformations cannot be defined. Probably because one has no Goldstone theorem for discrete symmetries, the pion mass comes out too high in strong coupling calculations [10]. (As Chodos and Healy have shown, in two lattice and one continuous dimensions, Susskind's method can be viewed as a maximal diagonalization of the doubling transformations (2.10), [16].)

3. The axial anomaly and the fermion doubling

In sect. 2 we showed that it is, for rather technical reasons, impossible to have a lattice gauge theory with just one fermion, with continuous chiral invariance and a covariant continuum limit of the weak coupling perturbation expansion. In this section we will show that there is a deeper reason for this fermion problem. The deeper reason is the axial anomaly, the Adler–Bell–Jackiw anomaly [31, 32].

Had we found a lattice fermion formulation with the above listed properties, then we would have a counter example to Adler's demonstration that such a regularization cannot be found [20]. Specifically, the one (fermion) loop contribution to the matrix element $\Gamma_{\alpha\mu\nu} = \langle 0|\mathrm{T}(A_\alpha v_\mu v_\nu)|0\rangle$, with A_α the axial current, cannot be regularized gauge invariantly such that A_α remains conserved.

In a lattice formulation one has to give up something: locality (SLAC fermions) with disasters in a straightforward weak coupling expansion, or explicit chiral invariance (Wilson), or one has more fermions cancelling the anomaly [23] (naive fermion method).

We will demonstrate that indeed the extra fermions in the naive method have axial charges such that the anomaly in $\Gamma_{\alpha\mu\nu}$ is cancelled and that the extra term in Wilson's formulation gives a contribution to the right-hand side of the axial Ward identity for $\Gamma_{\alpha\mu\nu}$ that is exactly the anomaly.

It is illuminating to partly lift the degeneracy of the naive fermion formulation by the addition of a very small Wilson term [17]. Consider the action

$$I = \sum_x \Big\{ -\sum_\mu \frac{1}{2a}\Big[\bar\psi(x)\frac{1}{i}\gamma_\mu U_\mu(x)\psi(x+a_\mu) - \bar\psi(x+a_\mu)\frac{1}{i}\gamma_\mu U^+_\mu(x)\psi(x)\Big] - M\bar\psi(x)\psi(x)$$
$$+\tfrac{1}{2}\tilde r \sum_\mu [\bar\psi(x)U_\mu(x)\psi(x+a_\mu) + \bar\psi(x+a_\mu)U^+_\mu(x)\psi(x)]\Big\} + I(U)\,, \qquad M = m + 4\tilde r\,. \tag{3.1}$$

For $\tilde r = 0$ we find the naive fermion action (2.5). Putting $\tilde r = r/a$, (3.1) becomes the Wilson fermion action (2.5)+(2.15) with gauge fields.

For $\tilde r = \mathrm{O}(a^0)$ the action (3.1) describes 16 fermions, but for $\tilde r \neq 0$ they are no longer mass degenerate. The fermion propagator is [cf. (2.16)]

$$S(p) = \Big[\sum_\mu \gamma_\mu \frac{1}{a}\sin p_\mu a + m + \tilde r \sum_\mu (1-\cos p_\mu a)\Big]^{-1}. \tag{3.2}$$

The behavior around the points $\bar p$ (components 0 or π/a) for $a \to 0$ is

$$S(p) = \Big[\sum_\mu \gamma_\mu \cos \bar p_\mu a (p-\bar p)_\mu + m + \tilde r \sum_\mu (1-\cos\bar p_\mu a)\Big]^{-1} + \mathrm{O}(a)\,,$$
$$(p-\bar p)_\mu = \mathrm{O}(a^0)\,. \tag{3.3}$$

The value of the $\tilde{r}$ term is 0, $2\tilde{r}$, $4\tilde{r}$, $6\tilde{r}$ or $8\tilde{r}$, if zero, one, two, three or four components of $\bar{p}$ are π/a. The 16 particles have masses m, $m+2\tilde{r}, \ldots, m+8\tilde{r}$.

We construct the axial current by introducing in the chiral invariant part of our action (3.1) (i.e., the part obtained putting $m=\tilde{r}=0$) an external axial gauge field $a_\mu(x)$, and calculating up to first order in $a_\mu(x)$. The action is

$$I=\sum_x\Big\{-\sum_\mu \frac{1}{2a}\Big[\bar{\psi}(x)\frac{1}{i}\gamma_\mu \exp\{ia[gv_\mu(x)+\gamma_5 a_\mu(x)]\}\psi(x+a_\mu)$$
$$-\bar{\psi}(x+a_\mu)\tfrac{1}{i}\gamma_\mu \exp\{-ia[gv_\mu(x)+\gamma_5 a_\mu(x)]\}\psi(x)\Big]$$
$$-M\bar{\psi}(x)\psi(x)+\tfrac{1}{2}\tilde{r}\sum_\mu\{\bar{\psi}(x)\exp[iagv_\mu(x)]\psi(x+a_\mu)$$
$$+\bar{\psi}(x+a_\mu)\exp[-iagv_\mu(x)]\psi(x)\}\Big\}+I(v)\,. \tag{3.4}$$

In this theory there are $\bar{\psi}\psi a(v)^n$ vertices ($n\geqslant 0$), the $\bar{\psi}\psi a$ vertex function is (fig. 3c):

$$\gamma_\mu\gamma_5\cos\tfrac{1}{2}(p-q)_\mu a\,,\qquad p+q+k=0\,. \tag{3.5}$$

To determine the axial charges of the 16 fermions we add wave functions to the fermion lines for a particle around $p=\bar{p}$ [cf. (2.14)]:

$$\bar{u}(p-\bar{p})s^{-1}(\bar{p})\cdot\gamma_\mu\gamma_5\cos[\tfrac{1}{2}(p+q)_\mu a]\cdot s(\bar{p})v(q-\bar{p})$$
$$=\bar{u}(p-\bar{p})\cdot s^{-1}(\bar{p})\gamma_\mu\gamma_5 s(\bar{p})\cos\bar{p}_\mu a\cdot v(q-\bar{p})+\mathrm{O}(a)\,. \tag{3.6}$$

If all the components of $\bar{p}$ are 0, then $s(\bar{p})=1$, and

$$s^{-1}(\bar{p})\gamma_\mu\gamma_5 s(\bar{p})\cos\bar{p}_\mu a=\gamma_\mu\gamma_5\,. \tag{3.7a}$$

We conclude that the $\bar{p}=0$ particle has axial charge +1. If one component of $\bar{p}$ is π/a, say $\bar{p}_\nu$, then $s(\bar{p})=\gamma_\nu\gamma_5$,

$$s^{-1}(\bar{p})\gamma_\mu\gamma_5 s(\bar{p})\cos\bar{p}_\mu a=-\gamma_\mu\gamma_5\,. \tag{3.7b}$$

The four particles around the points $\bar{p}$ with one component π/a have chiral charge -1. Similarly one proves that the six particles with two components of $\bar{p}$ equal to π/a have chiral charge +1, the four with three components of $\bar{p}=\pi/a$ have -1, and the one particle with $\bar{p}=(\pi/a, \pi/a, \pi/a, \pi/a)$ has +1.

In total there are 8 particles with +1, and 8 with -1: the fermion content of the theory is anomaly free.

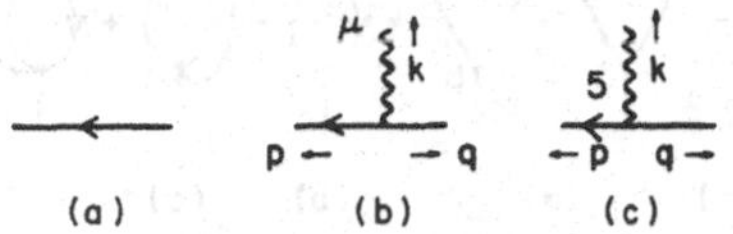

Fig. 3. Propagators and vertices: (a) fermion propagator, (b) $\bar{\psi}\psi v$ vertex, (c) $\bar{\psi}\psi a$-vertex.

TABLE 1
Properties of the 16 fermions in sects. 2, 3

$\bar{p}_\rho$ (units π/a)	$s(\bar{p})$	Multiplicity	Mass	Q	Q_5
0	1	1	m	$+g$	$+1$
$\delta_{\rho\mu}$	$\gamma_\mu\gamma_5$	4	$m+2\bar{r}$	$+g$	-1
$\delta_{\rho\mu}+\delta_{\rho\nu}$ $(\mu\neq\nu)$	$\gamma_\mu\gamma_5\gamma_\nu\gamma_5$	6	$m+4\bar{r}$	$+g$	$+1$
$\delta_{\rho\mu}+\delta_{\rho\nu}+\delta_{\rho\lambda}$ $(\mu\neq\nu\neq\lambda\neq\mu)$	$\gamma_\mu\gamma_5\gamma_\nu\gamma_5\gamma_\lambda\gamma_5$	4	$m+6\bar{r}$	$+g$	-1
1	γ_5	1	$m+8\bar{r}$	$+g$	$+1$

In table 1 we summarize the properties of the 16 fermions: about which $\bar{p}$, $s(\bar{p})$, multiplicity, mass, vector charge Q, and axial charge Q_5.

Next we calculate $\Gamma_{\mu\nu\alpha}=\langle 0|\mathrm{T}(v_\mu v_\nu a_\alpha)|0\rangle$ explicitly in one-loop order, for $a\to 0$. The contributing Feynman diagrams are listed in fig. 4. (The Feynman rules are discussed in the appendix.) The evaluation of $\Gamma_{\mu\nu\alpha}$ makes use of the vector Ward identity on the lattice,

$$\sum_\mu \frac{2}{a}\sin\tfrac{1}{2}ap_\mu\cdot\Gamma_{\mu\nu\alpha}(p,q)=0\,, \tag{3.8}$$

and of the Bose symmetry

$$\Gamma_{\mu\nu\alpha}(p,q)=\Gamma_{\nu\mu\alpha}(q,p)\,. \tag{3.9}$$

It is interesting to note that because of charge conjugation symmetry the two diagrams (a) are equal, and are therefore each by themselves Bose symmetric.

The Ward identity (3.8) can be derived either in the path integral formulation or directly from the integral representation of the diagrams in fig. 4. On the lattice the integrals over the internal momenta are well defined and over a finite interval. A technical detail that facilitates the derivation (which we do not give) is that shifts in the internal momenta leave the integrand invariant, because of periodicity.

We use for the evaluation an adaptation of the method Sharatchandra used for Wilson fermions [21]. The diagrams contributing to $\Gamma_{\mu\nu\alpha}$ are superficially linear divergent, $\mathrm{O}(a^{-1})$, if we count powers just as in continuum theory with one change. We have to take into account that anomalous vertices [21], i.e., vertices that vanish

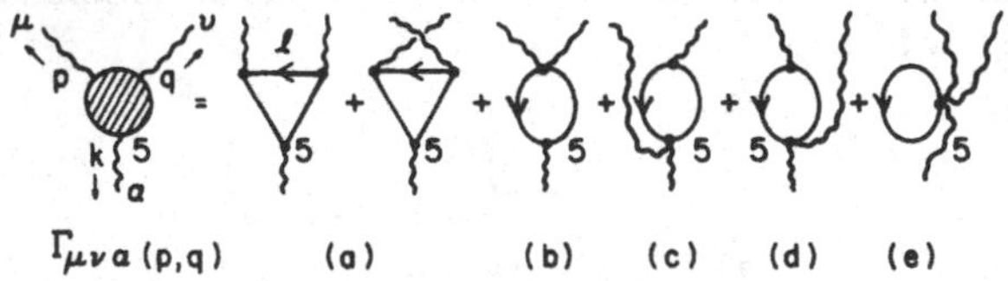

Fig. 4. One-loop contributions to $\Gamma_{\mu\nu\alpha}(p,q)$.

for $a \to 0$ at tree-graph level, like the $\bar{\psi}\psi vv$ vertex, have explicit powers of a. Diagram (b) is superficially $O(a^{-2})$ multiplied by an explicit a from the anomalous vertex, in total $O(a^{-1})$. We will show that this power counting makes sense in this theory on the lattice.

Make a Taylor expansion of $\Gamma_{\mu\nu\alpha}(p, q)$ in p and q to isolate the (possibly) infinite $O(a^{-1})$ contributions:

$$\Gamma_{\mu\nu\alpha}(p, q) = \Gamma^{(0)}_{\mu\nu\alpha} + \sum_{\rho} \Gamma^{(1)}_{\mu\nu\alpha\rho} p_\rho + \sum_{\rho} \Gamma^{(1)}_{\nu\mu\alpha\rho} q_\rho + \cdots . \tag{3.10}$$

Use was made of the Bose symmetry (3.9). The functions $\Gamma^{(0)}$ and $\Gamma^{(1)}$ only depend on m, $\tilde{r}$ and a.

It is crucial that propagator and vertex functions in this theory are analytic in the momenta, otherwise a Taylor expansion is not possible (as is the case with SLAC fermions [9]).

It will be shown that the terms indicated by dots in (3.10) are finite for $a \to 0$, and reduce to relativistic expressions. In this case the counterterms, $\Gamma^{(0)}$ and $\Gamma^{(1)}$, are zero, as can be seen using the vector Ward identity (3.8). To lowest order in p and q (3.8) reads

$$\sum_{\mu} p_\mu \Gamma^{(0)}_{\mu\nu\alpha} = 0 . \tag{3.11}$$

Immediately $\Gamma^{(0)}_{\mu\nu\alpha} = 0$ follows, as the components p_μ are independent. In first order in p and q, (3.8) is

$$\sum_{\mu,\rho} p_\mu \Gamma^{(1)}_{\mu\nu\alpha\rho} q_\rho = 0 . \tag{3.12}$$

Now $\Gamma^{(1)}_{\mu\nu\alpha\rho} = 0$ follows.

The next terms in the Taylor series converge if $a \to 0$. Choose as integration interval $(-\pi/2a, 3\pi/2a)$ for all components of the internal momentum l. Then divide the integration hypercube in 16 smaller hypercubes corresponding to $(-\pi/2a, \pi/2a)$ and $(\pi/2a, 3\pi/2a)$ for each l_ρ. First consider the hypercube $(-\pi/2a, \pi/2a)$ for all l_ρ. In this hypercube we can estimate sin la from the propagators: $|(1/a)(\sin la)| \geq 2/\pi |l|$. This allows us to use the result from renormalization theory, that power counting gives the highest degree of divergence. The terms indicated by dots in (3.10) are obtained by twice differentiating with respect to p and/or q. Consider first diagrams b, c, d, and e. After two differentiations their integral representations behave like $a^m O(a^{-n})$, with $m > 0$ (at least one a from an anomalous vertex) and $m - n > 0$: they vanish. If in diagrams a at least one of the differentiations works on a vertex function, the contribution vanishes by the same argument. Only if the differentiations work on propagators do we get a non-vanishing contribution, convergent for $a \to 0$.

Furthermore, it can be shown that one gets the same finite part if one replaces in the twice differentiated integrals for diagrams a, vertices and propagators by their

$a \to 0$ limits. The contribution from the first hypercube to $\Gamma_{\mu\nu\alpha}$ can be written as

$$\Gamma_{\mu\nu\alpha}(p,q) = \Gamma^{(0)}_{\mu\nu\alpha} + \sum_\rho \Gamma^{(1)}_{\mu\nu\alpha\rho} p_\rho + \sum_\rho \Gamma^{(1)}_{\nu\mu\alpha\rho} q_\rho$$

$$+ \Big(1 - |_0 - \sum_\rho p_\rho |_0 \, \partial_{p_\rho} - \sum_\rho q_\rho |_0 \, \partial_{q_\rho}\Big) \Gamma^{\text{cont}}_{\mu\nu\alpha}(p,q) + \mathrm{O}(a)\,, \qquad (3.13)$$

with $|_0 f(p,q) = f(0,0)$ and $\Gamma^{\text{cont}}_{\mu\nu\alpha}$ evaluated in some continuum regularization. Formula (3.13) is *mutatis mutandis* valid for a general set of diagrams. It allows us to isolate the terms specific for the lattice regularization, the terms $\Gamma^{(0)}$ and $\Gamma^{(1)}$ (which are zero here). This leads to the standard relativistic expression for the triangle graph for a particle with mass m [22].

A similar reasoning can be given for the other 15 integration hypercubes. We find

$$\Gamma_{\mu\nu\alpha}(p,q) = \sum_{i=1}^{16} g^2 Q_i^5 G_{\mu\nu\alpha}(p,q;m_i) + \mathrm{O}(a)\,. \qquad (3.14)$$

The index i is summed over the 16 particles (table 1), and the standard triangle result $G_{\mu\nu\alpha}$ is [22]

$$16\pi^2 G_{\mu\nu\alpha}(p,q;m)$$

$$= -\frac{1}{\pi^2}\Big\{\sum_\rho \varepsilon_{\rho\mu\nu\alpha} p_\rho A_1(p,q,m) + \sum_{\rho\sigma} \varepsilon_{\rho\sigma\mu\alpha} p_\rho q_\sigma p_\nu A_3(p,q,m)$$

$$+ \sum_{\rho\sigma} \varepsilon_{\rho\sigma\mu\alpha} p_\rho q_\sigma q_\nu A_4(p,q,m)\Big\} + (p,\mu \leftrightarrow q,\nu)\,. \qquad (3.15)$$

The invariant functions A_1, A_3 and A_4 are

$$I_{st}(p,q) = \int_0^1 \mathrm{d}x \int_0^{1-x} \mathrm{d}y \frac{x^s y^t}{m^2 + y(1-y)p^2 + 2xyp\cdot q + x(1-x)q^2}\,, \qquad (3.16)$$

$$A_1 = p\cdot q A_3 + q^2 A_4\,,$$
$$A_3 = -16\pi^2 I_{11}\,, \qquad (3.17)$$
$$A_4 = 16\pi^2 (I_{20} - I_{10})\,.$$

From (3.14) we directly calculate the axial Ward identity for $\Gamma_{\mu\nu\alpha}(p,q)$:

$$\sum_\alpha k_\alpha \Gamma_{\mu\nu\alpha} = +2 \sum_{i=1}^{16} m_i^2 \frac{g^2 Q_i^5}{2\pi^2} \sum_{\rho\sigma} \varepsilon_{\rho\sigma\mu\nu} p_\rho q_\sigma I_{00}(p,q;m_i) + \mathrm{O}(a)\,. \qquad (3.18)$$

The terms on the right-hand side of (3.18) are the usual mass terms of the fermions in the theory.

The one-loop contribution to $\langle 0|Ta_\alpha v_\mu v_\nu|0\rangle$ is anomaly free, satisfies the vector Ward identity, and is Lorentz covariant. The lattice theory (3.4) managed this by providing extra particles.

A more difficult question is: why did the lattice theory not just provide one extra particle which is enough to cancel the anomaly? To this question we only have an incomplete answer, which nevertheless gives some insight.

The argument is rather technical and starts with the observation that the anomaly has to do with the fact that in continuum theory the triangle graph is superficially linearly divergent. A consequence is that although the evaluation of the triangle gives without any regularization a finite answer, this answer depends on the choice of the loop variable l: a shift gives a different (finite) answer [24]. One choice of l leads to a gauge-invariant answer, a different choice to a chiral-invariant answer.

Suppose now that we discretize the continuum theory in only one dimension, say the 1-direction, keeping chiral invariance. The fermion propagator has the form

$$S(p)=\left[\gamma_1 P_1(p_1)+\sum_{\mu=2}^{4}\gamma_\mu p_\mu\right]^{-1}.$$

In this theory $\Gamma_{\mu\nu\alpha}$ is, for fixed a, only superficially logarithmically divergent; shifts in l give the same answer. The answer satisfies axial and vector Ward identities, also for $a\to 0$. The lattice has to provide at least one extra fermion for anomaly cancellation. This can only come from an extra zero of $P_1(p_1)$. As such a one lattice dimension theory can be obtained by taking the classical continuum limit in three lattice dimensions, the lattice fermion propagator on a four-dimensional lattice theory (with chiral invariance) has the form

$$S(p)=\left[\sum_\mu \gamma_\mu P_\mu(p)\right]^{-1}, \tag{3.19}$$

with (at least) one extra zero in each $P_\mu(p_\mu)$ beside $p_\mu=0$ for an anomaly cancelling particle: this gives $2^4=16$ particles.

From the answer (3.14) given for $\Gamma_{\mu\nu\alpha}$ in one loop in the above theory we can also heuristically deduct the answer for $\Gamma_{\mu\nu\alpha}$ for a Wilson fermion, i.e., the theory (3.4) with $\tilde{r}=r/a$, $r=\mathrm{O}(a^0)$. As $G_{\mu\nu\alpha}(p,q;M)$ behaves like $1/M^2$ for $M\to\infty$, we take the limit $\tilde{r}\to\infty$, the 15 extra particles with masses $\sim\tilde{r}$ decouple, and we get the Wilson $\Gamma_{\mu\nu\alpha}$:

$$\Gamma_{\mu\nu\alpha}(p,q)=g^2 G_{\mu\nu\alpha}(p,q;m)+\mathrm{O}(a). \tag{3.20}$$

The 15 mass terms in (3.18) transform in this limit into the anomaly, because $I_{00}(p,q,M)\sim 1/2M^2$ for $M\to\infty$. The axial Ward identity (3.18) becomes

$$\sum_\alpha k_\alpha \Gamma_{\mu\nu\alpha}=2m^2\frac{g^2}{2\pi^2}\sum_{\rho\sigma}\varepsilon_{\rho\sigma\mu\nu}p_\rho q_\sigma I_{00}(p,q;m)-\frac{g^2}{2\pi^2}\sum_{\rho\sigma}\varepsilon_{\rho\sigma\mu\nu}p_\rho q_\sigma+\mathrm{O}(a). \tag{3.21}$$

In the gauge theory with a Wilson fermion, $\Gamma_{\mu\nu\alpha}$ still satisfies the vector Ward identity and is Lorentz covariant ($a \to 0$), and $\Gamma_{\mu\nu\alpha}$ has the usual anomaly, which can be traced back to the explicit chiral symmetry breaking by the extra r term. (Actually we should be more careful: first taking the $a \to 0$ limit with $\tilde{r} = O(a^0)$, and then $\tilde{r} \to \infty$ need not be the same as the $a \to 0$ limit with $\tilde{r} = r/a = O(a^{-1})$. For $\Gamma_{\mu\nu\alpha}$ it is the same, as we will show in sect. 5.)

The same method used for calculating $\Gamma_{\mu\nu\alpha}$ in the one-loop approximation can be applied to calculate all Green functions in the $\tilde{r}$ theory (3.1) in one loop. The superficially convergent Green functions are immediately seen to be Lorentz covariant and equal to their counterpart in the continuum theory with 16 fermions and a gauge field. For divergent diagrams we have to prove that the infinite (for $a \to 0$) parts are Lorentz covariant.

We discuss one such Green function, the vacuum polarization. The fact that, as will come out, all 16 fermions contribute gives us, furthermore, the unitarity argument stressing that the theory contains 16 fermions.

Two diagrams contribute to the vacuum polarization $I_{\mu\nu} = \langle 0|\mathrm{T}(v_\mu v_\nu)|0\rangle$, $I^{(a)}_{\mu\nu}$ and $I^{(b)}_{\mu\nu}$ (fig. 5). The corresponding Feynman amplitudes are

$$I^{(a)}_{\mu\nu}(p) = -g^2 \int_l \mathrm{Tr}\,[V_\mu(l, l+p)S(l+p)V_\nu(l+p, l)S(l)]\,, \tag{3.22}$$

$$I^{(b)}_{\mu\nu}(p) = +g^2 \int_l \mathrm{Tr}\,[V_{\mu\nu}(p, p)S(l)]\,, \tag{3.23}$$

where the vertex functions are given in (A.8) ($r \to a\tilde{r}$).

$I_{\mu\nu}$ has the following properties:

$$I_{\mu\nu}(p) = I_{\nu\mu}(-p)\,, \qquad \text{(Bose symmetry)}\,, \tag{3.24}$$

$$\sum_\mu \frac{2}{a} \sin \tfrac{1}{2}p_\mu a \cdot I_{\mu\nu}(p) = 0, \qquad \text{(Ward identity)}\,, \tag{3.25}$$

$$I_{\mu\nu}(p) = I_{\nu\mu}(p)\,. \tag{3.26}$$

The last property is trivial for $I^{(b)}_{\mu\nu}$ and can be shown for $I^{(a)}_{\mu\nu}$ by evaluating the trace. The consequence of (3.24) and (3.26) is that only even terms in the Taylor expansion of $I_{\mu\nu}$ occur:

$$I_{\mu\nu}(p) = I^{(0)}_{\mu\nu} + \sum_{\rho,\sigma} I^{(2)}_{\mu\nu,\rho\sigma} p_\rho p_\sigma + \cdots\,. \tag{3.27}$$

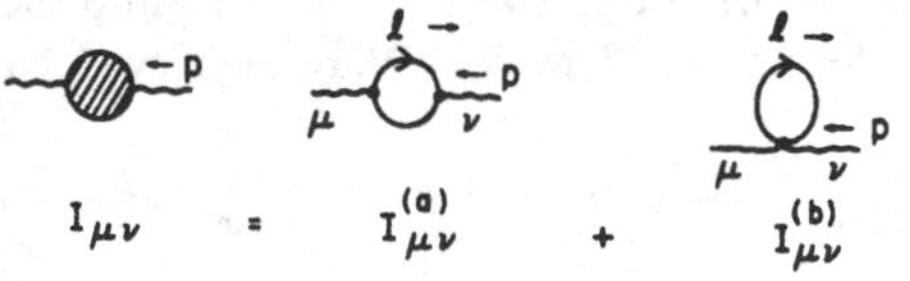

Fig. 5. One-loop contributions to $I_{\mu\nu}(p)$.

The Ward identity (3.25) gives $I^{(0)}_{\mu\nu}=0$ in lowest order in p. Actually, the fact that $I^{(0)}_{\mu\nu}$, which is $O(a^{-2})$, vanishes involves a crucial cancellation in the $p=0$ terms between diagrams a and b. This cancellation can be neatly demonstrated explicitly by using the differential forms (A.10) of the Ward identities and making a partial integration.

We have also explicitly evaluated $I^{(2)}$. The following less explicit procedure is easier, if one is interested in Lorentz covariance. One looks at the integral representation (3.22) for $I^{(a)}_{\mu\nu}$ to see what possible tensor structures can come out. One finds the following tensors, bilinear in p and symmetric in (μ, ν):

$$p_\nu^2+p_\mu^2\,,\qquad p_\nu p_\mu\,,\qquad \delta_{\mu\nu}p_\mu^2\,,\qquad \delta_{\mu\nu}p^2\,. \tag{3.28}$$

Only the linear combination $\delta_{\mu\nu}p^2-p_\mu p_\nu$ satisfies the Ward identity. Hence

$$\sum_{\rho\sigma} I^{(2)}_{\mu\nu\rho\sigma}p_\rho p_\sigma = I^{(2)}\cdot(\delta_{\mu\nu}p^2-p_\mu p_\nu)\,. \tag{3.29}$$

$I^{(2)}$ is a function of m, $\bar{r}$, and a.

By the same reasoning as for $\Gamma_{\mu\nu\alpha}$, one proves that the terms indicated by dots in (3.27), which are obtained by differentiating $I^{(a)}_{\mu\nu}(p)$ three or more times, are finite in the limit $a\to 0$. One makes the above described division of the integration interval in 16 hypercubes, and finds that one gets a sum of 16 terms, each corresponding to the relativistic contribution [25] of one of the 16 fermions (minus terms quadratic in p) [cf. (3.13)]:

$$I_{\mu\nu}(p)=I^{(2)}\cdot(\delta_{\mu\nu}p^2-p_\mu p_\nu)+\sum_{i=1}^{16}\Big(1-\sum_\rho p_\rho|_0\,\partial_{p_\rho}-\tfrac{1}{2}\sum_{\rho\sigma}p_\rho p_\sigma|_0\,\partial_{p_\rho}\partial_{p_\sigma}\Big)I^{\text{cont}}_{\mu\nu}(p, m_i)\,. \tag{3.30}$$

Indeed $I_{\mu\nu}(p)$ is Lorentz covariant, and has contributions from all 16 fermions to its finite part. In Minkowski space the imaginary part is then also a sum of 16 terms, and to have a unitary theory one needs to include the 16 fermions in the intermediate state, as is schematically depicted in fig. 6.

To summarize, we have argued in this section that the lattice generates the extra fermions needed to have an anomaly free axial current. This was illustrated by a tree-graph consideration to determine the axial charges of the particles and by a one-loop calculation of $\langle 0|T(v_\mu v_\nu a_\alpha)|0\rangle$.

Furthermore we discussed the vacuum polarization $\langle 0|T(v_\mu v_\nu)|0\rangle$ in order to show that all 16 fermions contribute in S-matrix elements and to show a general way of evaluating lattice Feynman diagrams in the limit $a\to 0$.

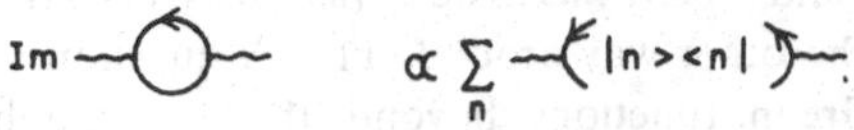

Fig. 6. Unitarity relation for $I_{\mu\nu}$.

The connection with the axial anomaly makes it clear that we can only have one fermion if we break chiral symmetry explicitly on the lattice. This is what happens in the Wilson lattice fermion formulation: a generalized (momentum-dependent) mass term breaks the chiral symmetry and removes the extra fermions. The axial vector current has the right anomaly, as we showed above.

In the following sections we discuss Wilson's fermions, and how one gets a chiral invariant continuum limit.

4. Wilson's lattice QCD

This section deals in more detail with Wilson's lattice quarks in the context of QCD. One-loop diagrams will be examined with emphasis on aspects of chiral symmetry.

The quark part of the action is given by (2.5), (2.15), where $U_\mu(x)$ is an element of SU(N) (N = number of colors). It is useful to change the notation: the mass terms in (2.5) and (2.15) can be regrouped into the form

$$I_{\text{mass}} = -\sum_x \bar{\psi}(x) M \psi(x) + \frac{r}{2a} \sum_{x,\mu} [\bar{\psi}(x) U_\mu(x) \psi(x+a_\mu) + \text{h.c.}] . \tag{4.1}$$

The parameter M is related to Wilson's [6] hopping parameter K by $2aM = 1/K$. Comparison of (2.5), (2.15) with (4.1) gives $M = m + 4r/a$. From now on, however, we shall reserve the notation m for the renormalized quark mass.

The mass action (4.1) has two adimensional mass parameters per flavor, aM and r. At the tree-graph level, however, the quark mass is given by the combination $m = M - 4r/a$ and requiring m independent of a determines aM to depend on a according to $aM = 4r + am$. So the continuum limit requires aM to approach the critical value $4r$ at which point the correlation length ξ in units of the lattice spacing, $\xi = 1/am$, diverges. The resulting physics depends only on the quark mass m and the gauge coupling g, and not on r. The r dependence is contained in so-called [21] anomalous vertices, defined as vertex functions which vanish in the tree graph continuum limit $a \to 0$.

In higher orders one may expect that aM has again to approach a critical value aM_c as $a \to 0$ in order that the physical masses remain finite in the continuum limit:

$$M = M_c + m_0 , \qquad am_0 \to 0 \text{ as } a \to 0 , \tag{4.2}$$

$$aM_c = 4r + O(g^2) = \text{function of } r \text{ and } g . \tag{4.3}$$

This assumes, as we shall do in the following, that r is assigned some value ($0 < r \leq 1$, flavor independent) and that M has to be adjusted as $a \to 0$ (flavor dependent), but it may also be done the other way around [11]. A question which arises is: are the continuum limit Green functions beyond the tree-graph approximation also independent of r? The answer is that the r dependence can be completely eliminated by the coupling constant, mass and wave-function renormalization. Sharatchandra has pointed out [21] that anomalous vertices (containing the r dependence) only lead to contact terms in primitively divergent diagrams and their effect can be absorbed in counterterms. This will be verified below.

Within the weak coupling expansion aM_c is defined as the value of aM where the quark mass vanishes. We define the latter as the pole mass in the theory with an infrared regulator, e.g., a gluon mass λ [cf. (A.5)]. However, we expect this mass to be infrared finite since it is known in the continuum theory that the pole mass is infrared infinite at least up to the two-loop approximation [38]. It is also gauge independent. Let $S'(p)$ be the complete unrenormalized quark propagator. On the lattice we may define the pole mass m by the condition $\det S'(p)^{-1} = 0$, for $p_4 = -im$, $p_1 = p_2 = p_3 = 0$. We want m to be much smaller than the cutoff π/a. For this it is assumed that aM has to be close to aM_c and am vanishes at $aM = aM_c$. For $a \to 0$ with m and p fixed, the inverse propagator becomes Lorentz covariant:

$$S'(p)^{-1} = A(p) + \sum_{\mu} \gamma_\mu B_\mu(p)\,, \tag{4.4a}$$

$$A(p) \sim S_0(p^2)\,, \qquad B_\mu(p) \sim p_\mu S_1(p^2)\,, \tag{4.4b}$$

as will be verified below to order g^2, and the pole mass m has to satisfy

$$S_0(-m^2) = m S_1(-m^2)\,. \tag{4.5}$$

To finite order in the weak coupling expansion, dynamical mass generation is absent and $mS_1(-m^2) \to 0$ as $m \to 0$, so at $m = 0$ the propagator is chiral invariant.

Intuitively, chiral symmetry will appear when there is a cancellation between the effects of the two mass terms in (4.1), for instance in the γ_μ independent part $A(p)$ of $S'(p)^{-1}$. Indeed, for free quarks $U \to 1$ in (4.1) and $A(p) = M - (r/a) \sum_\mu \cos ap_\mu = 0$ at $p = 0$, provided that $aM = 4r = aM_c(g = 0)$. For non-zero g this cancellation will take place at a smaller value of aM, as the effective strength of r is diminished by the fluctuating unitary U and $U^\dagger$ in (4.1), so we expect $A(0) = 0$ at $aM = aM_c < 4r$. For chiral symmetry it is not only necessary that $A(0) = 0$ but also that $A(p) \equiv 0$ in the continuum limit, as concluded above from (4.5), and this can be understood as follows. The function $A(p)$ is finite for gluon mass $\lambda = 0$ [as verified below to $O(g^2)$] and at $M = M_c$ it has the form $A(p) = f(y)/a$, with $f(y)$ an adimensional function of $y_\mu = ap_\mu$ depending parametrically on r and g, even in $y \to -y$, with $f(0) = 0$. At the tree-graph level $f(y) = r \sum_\mu (1 - \cos y_\mu) = \frac{1}{2} r y^2 + O(y^4)$. In higher orders this behavior is modified at most by logarithms, e.g., $f(y) \sim \text{const.} \times y^2 (\ln y)^k$. Then it follows that $A(0) = 0$ for $a \neq 0$ implies $A(p) \equiv 0$ in the limit $a \to 0$. So at $M = M_c$ chiral symmetry is expected in the perturbative Green functions, and the departure from chiral symmetry is measured by $m_0 = M - M_c$ in (4.2). We call m_0 the bare current quark mass.

A distinctive feature of Wilson's fermion method is the necessity of $O(a^{-1})$ mass counterterms. Writing

$$M - 4r/a = m - \delta m\,, \tag{4.6}$$

we find δm in perturbation theory in the usual way by requiring a independence of the renormalized Green functions. Then aM_c can be identified from

$$aM_c = 4r - \lim_{a\to 0} a\delta m\,. \tag{4.7}$$

Since we expect $aM_c < 4r$, δm will be positive with an a^{-1} dependence. Such counterterms reflect the strong breaking of chiral symmetry on the lattice which will be related in sect. 5 to the triangle anomaly. However, in common with Lorentz invariance, chiral invariance is to be recovered in the continuum limit.

A definition of M_c which is applicable to the strong coupling expansion is that at $M = M_c$ the pion mass vanishes [6, 12, 13]. Kawamoto [26] has given a definition of M_c which is independent of an expansion in g; namely, the expansion of $\langle\bar\psi(x)\psi(x)\rangle$ in powers of $K = (2aM)^{-1}$ at fixed r and g has $(2aM_c)^{-1}$ as its radius of convergence. This definition interpolates between the two given above, since at weak coupling the singularity in K of $\langle\bar\psi\psi\rangle$ comes from the vanishing quark mass and at strong coupling it comes from the vanishing pion mass. If confinement holds for all non-zero g, then the singularity is expected to correspond to a vanishing pion mass also at weak coupling. Defined in this way, M_c is manifestly gauge invariant. Note that $\langle\bar\psi\psi\rangle$ does not vanish for $r \neq 0$ even at $M = M_c$ and it can be analyzed with both the weak and the strong coupling expansions.

Consider the theory with two flavors, u and d, and take M flavor independent. Then there will be a "pion" (flavor triplet) and an "η" (flavor singlet). Define M_c such that at $M = M_c$ the mass gap vanishes, so either $m_\pi = 0$ or $m_\eta = 0$ at $M = M_c$; there is no reason why both should vanish, since singlet splitting is expected in the Green functions $\langle\bar\psi_a(x)i\gamma_5\psi^b(x)\bar\psi_c(y)i\gamma_5\psi^d(y)\rangle$ (a, b, ... = u, d). Fig. 7 shows the simplest non-trivial Feynman diagram leading to this splitting. Hence, there is no U(1) problem in Wilson's lattice QCD.

The same singlet splitting mechanism causes the anomaly in the flavor singlet axial current, relative to the non-singlet current. This can be analyzed in the weak coupling expansion and it raises the following interesting dilemma: the anomaly is somehow proportional to r since we have seen that at $r = 0$ there is no anomaly, yet the anomaly is to be r-independent (similarly it has been argued [13] that the "η"–"π" splitting is $\propto r$, yet it should not depend on r). It will be explicitly checked in sect. 5 that the anomaly is indeed r-independent at the one-loop level.

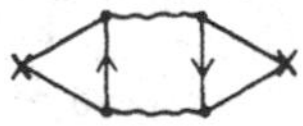

Fig. 7. Diagram causing "η"–"π" splitting.

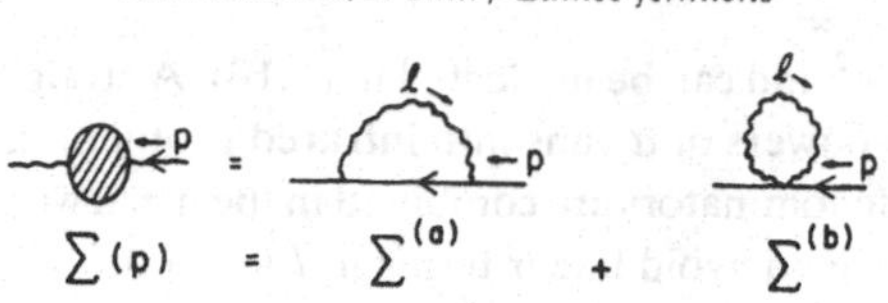

Fig. 8. One-loop contribution to $-\Sigma(p)$.

Turning next to the one-loop calculations, we note that all primitively convergent diagrams take the usual continuum theory form in the limit $a \to 0$. This can be shown with the reasoning of sect. 3 or by the scaling method used below. The regions near the cutoff are suppressed in this case by the Wilson mass term. So we concentrate on diagrams with primitive divergence $\geqslant 0$ involving fermions, in particular the quark self-energy function, the gluon-quark vertex function, and the gluon self-energy function.

The quark self-energy function described by the diagrams in fig. 8 has the integral representation

$$\Sigma(p) = \Sigma^{(a)}(p) + \Sigma^{(b)}(p)\,, \tag{4.8a}$$

$$\Sigma^{(a)}(p) = -\bar{g}^2 \int_l V_\mu(p, l+p) S(l+p) V_\nu(l+p, p) D_{\mu\nu}(l)\,,$$

$$\Sigma^{(b)}(p) = \bar{g}^2 \int_l V_{\mu\nu}(p, p) D_{\mu\nu}(l)\,, \tag{4.8b}$$

$$\bar{g}^2 = \tfrac{1}{2} g^2 (N^2 - 1)/N\,, \tag{4.9}$$

where the vertex functions and propagators are given in (A.5), (A.8) and (A.9). We shall now describe a direct method for evaluating such expressions, without recourse to Taylor expansions and comparison with the continuum theory. As an example, consider $\Sigma^{(a)}$ in the Feynman gauge. Rescaling the loop variable $l \to l/a$ leads to the form

$$\begin{aligned}\Sigma^{(a)}_{\mathrm{F}}(p) = -\bar{g}^2 a^{-1} \int_l & v_\mu(\tfrac{1}{2} l + ap)[\mathcal{M}(l+ap) - \gamma \cdot \sin(l+ap)] \\ & \times v_\mu(\tfrac{1}{2} l + ap)[\mathcal{M}^2(l+ap) + \sin^2(l+ap)]^{-1} \\ & \times [a^2 \lambda^2 + 4 \sin^2 \tfrac{1}{2} l]^{-1}\,,\end{aligned} \tag{4.10}$$

where the integrals over l now run from $-\pi$ to π and for generic k

$$v_\mu(k) = \gamma_\mu \cos k_\mu + r \sin k_\mu\,, \tag{4.11a}$$

$$\sin^2 k = \sum_\nu \sin^2 k_\nu\,, \qquad \gamma \cdot \sin k = \sum_\nu \gamma_\nu \sin k_\nu\,, \tag{4.11b}$$

$$\mathcal{M}(k) = aM - r \sum_\nu \cos k_\nu = am + r \sum_\nu (1 - \cos k_\nu) - a\delta m\,, \tag{4.11c}$$

where δm is of order g^2 and can be neglected in (4.10). A straightforward expansion of the integrand into powers of a runs into infrared problems at $l=0$. To deal with this, the propagator denominators are combined in the usual way, and the integration variable is shifted, so as to avoid linear terms in l as $l\to 0$:

$$\text{denominator} = \{l^2 + a^2[xm^2 + (1-x)\lambda^2 + x(1-x)p^2] + O(l^3)\}^{-2}, \qquad (4.12)$$

where x is the Feynman parameter. Then the integral is split into two regions: an inner region $l^2<\delta^2$ and the rest, the outer region. In the outer region $l^2>\delta^2$ a straightforward expansion in a is possible. The non-vanishing terms as $a\to 0$ constitute a covariant polynomial in the external momentum and in m, of degree 1, which is the primitive degree of divergence. The expansion coefficients are complicated integrals over the sines and cosines, sometimes diverging as $\delta\to 0$. Such a divergence is cancelled by a similar divergence coming from the integral over the inner region, since the complete expression is independent of δ. For the inner region integral, only the numerator is expanded in a leaving a covariant polynomial in both l and the external momenta and masses, since δ is chosen so small that the non-covariant higher orders in l do not matter. Using spherical symmetry the inner integral can be done explicitly in the limit $a\to 0$ ($a \ll \delta$) and only terms non-vanishing in δ are kept. A typical example is

$$\begin{aligned}
&\int_0^1 \mathrm{d}x \int_l (1-x) \sum_\mu p_\mu \gamma_\mu \cos l_\mu \cos^2 \tfrac{1}{2} l_\mu (x\{\mathcal{M}^2[l+(1-x)ap] \\
&\qquad + \sin^2[l+(1-x)ap]\} + (1-x)[a^2\lambda^2 + 4\sin^2 \tfrac{1}{2}(l-xap)])^{-2} \\
&= \tfrac{1}{2} \not{p} \lim_{\delta\to 0} \left[\int_{|l|>\delta} \cos l_4 \cos^2 \tfrac{1}{2} l_4 \{x[\mathcal{M}^2(l) + \sin^2 l] + (1-x) r \sin^2 \tfrac{1}{2} l\}^{-2} + \frac{\ln \delta^2}{16\pi^2} \right] \\
&\quad - \frac{\not{p}}{16\pi^2} \int_0^1 \mathrm{d}x (1-x) \{\ln a^2[xm^2 + (1-x)\lambda^2 + x(1-x)p^2] - 1\} + O(a).
\end{aligned} \qquad (4.13)$$

Note how the lattice permutational symmetry leads to the covariant contact term. Usually one is not interested in the value of the contact terms, and one can immediately proceed to the inner region integral, which is nothing but a rescaled covariant continuum theory integral with a euclidean invariant cutoff. The lattice takes care of Ward identities not being violated by shifts of integration variables. The evaluation of the fermion self-energy leads to the following form:

$$\begin{aligned}
\Sigma(p) &= \bar{g}^2 \left\{ \sigma_1(r) a^{-1} + m\sigma_2(r) + \not{p}\sigma_3(r) - \frac{1}{4\pi^2} \int_0^1 \mathrm{d}x [m + \tfrac{1}{2}(1-x)\not{p}] \ln a^2 \Delta \right\} \\
&\quad + \bar{g}^2 \zeta (m+\not{p}) \left\{ \sigma_4(a\lambda) + \int_0^1 \mathrm{d}x \frac{(1-x)[m-(1-x)\not{p}]}{16\pi^2 \Delta} (m+\not{p}) \right\},
\end{aligned} \qquad (4.14a)$$

$$\Delta = xm^2 + (1-x)\lambda^2 + x(1-x)p^2. \qquad (4.14b)$$

The contact terms contain all the r dependence. The non-trivial momentum dependence comes from the infrared divergence developing as $a \to 0$ at the origin $l = 0$, where the propagators and vertices take their continuum theory form. From the inverse quark propagator

$$S'(p)^{-1} = M - 4ra + \not{p} + \Sigma(p), \tag{4.15}$$

the critical M is found by letting $m \to 0$ and requiring $S'(0)^{-1} = 0$,

$$aM_c = 4r - \bar{g}^2 \sigma_1(r) + O(g^4). \tag{4.16}$$

Note that $S'(p)^{-1} \propto \not{p}$ at $M = M_c$. The expression for $\sigma_1(r)$ is easily identified from (4.10) by setting $p = 0$:

$$\sigma_1(r) = r \int_l \frac{\sin^2 l + 2\sin^2 \frac{1}{2}l[\cos^2 l + (1 - r^2)\sin^2 \frac{1}{2}l]}{[(2r\sin^2 \frac{1}{2}l)^2 + \sin^2 l]4\sin^2 \frac{1}{2}l}, \tag{4.17}$$

and $\sigma_1(r) > 0$ implies $aM_c < 4r$, as expected. We note that M_c is indeed gauge independent and infrared finite (there is a cancellation in the gauge-dependent a^{-1} terms of $\Sigma^{(a)}$ and $\Sigma^{(b)}$ which comes out easily by using Ward identities in the evaluation of the ζ terms). As usual in the continuum theory $\Sigma(p)$ is finite for gluon mass $\lambda \to 0$, but no longer analytic at $\not{p} = -m$. There is an infrared divergence in $\sigma_4(a\lambda)$ cancelling a similar divergence in the x-integral multiplying ζ:

$$\begin{aligned}\sigma_4(a\lambda) &= -\int_l [a^2\lambda^2 + 4\sin^2 \tfrac{1}{2}l]^{-2} \\ &= \frac{\ln a^2\lambda^2 + 1}{16\pi^2} - \lim_{\delta \to 0}\left[\int_{|l|>\delta} (4\sin^2 \tfrac{1}{2}l)^{-2} + \frac{\ln \delta^2}{16\pi^2}\right].\end{aligned} \tag{4.18}$$

The other σ's are given by similar complicated integrals over periodic functions. The renormalized inverse propagator $\tilde{S}'(p)^{-1}$ is obtained from $S'(p)^{-1}$ by multiplication by Z_2. Writing

$$\Sigma(p) = \Sigma_0(p^2) + \not{p}\Sigma_1(p^2), \tag{4.19}$$

we have [cf. (4.4b), (4.6)]

$$\tilde{S}_0(p^2) = Z_2[m - \delta m + \Sigma_0(p^2)], \tag{4.20a}$$

$$\tilde{S}_1(p^2) = Z_2[1 + \Sigma_1(p^2)], \tag{4.20b}$$

and requiring $\tilde{S}_{0,1}$ to be finite as $a \to 0$ and independent of r leads to

$$Z_2 = 1 + \bar{g}^2\left[\frac{(1-\zeta)}{16\pi^2}\ln a^2\mu^2 - \sigma_3(r) + C\right], \tag{4.21a}$$

$$\delta m = \bar{g}^2\sigma_1(r)a^{-1} + m(1 - Z_m), \tag{4.21b}$$

$$Z_m = 1 - (Z_2 - 1) + \bar{g}^2\left[\frac{\ln a^2\mu^2}{4\pi^2} - \sigma_2(r) + C'\right] + O(g^4), \tag{4.21c}$$

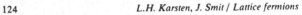

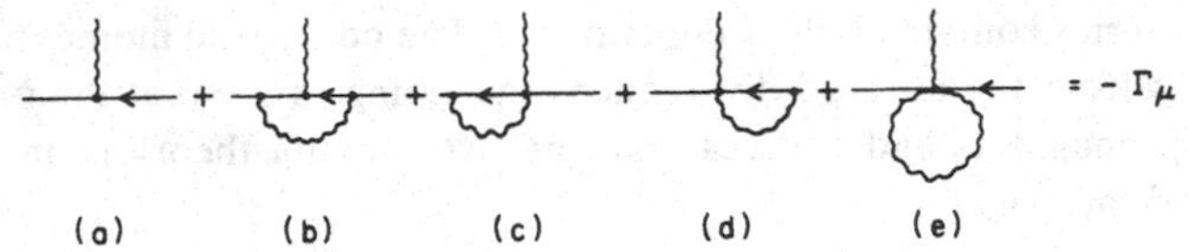

Fig. 9. One-loop approximation to $-\Gamma_\mu(p', p)$.

where μ is the renormalization scale parameter and the constants C and C' do not depend on a and r. If m is chosen to be the pole mass and if Z_2 is determined by the residue of the pole in $S'(p)$, then C and C' depend on m/μ, λ/μ and ζ such that Z_2 is independent of μ and δm independent of ζ, μ and λ for $\lambda \to 0$. Note that it is possible to write

$$m_0 \equiv M - M_c = mZ_m\,. \tag{4.21d}$$

The next example is the quark-gluon vertex function as described by the diagrams in fig. 9. The evaluation by the method described above is somewhat simpler than the previous example, since the divergence as $a \to 0$ is only logarithmic. Diagram b is easily brought into the standard covariant Feynman parametric form. The contact term coming from the expression [cf. (4.11)]

$$\int_{|l|>\delta} \frac{v_\rho(\frac{1}{2}l)[\mathcal{M}(l) - \gamma\cdot\sin l]v_\mu(l)[\mathcal{M}(l) - \gamma\cdot\sin l]v_\rho(\frac{1}{2}l)}{[\mathcal{M}^2(l) + \sin^2 l]^2 4\sin^2\frac{1}{2}l} \tag{4.22}$$

is seen to be proportional to γ_μ on grounds of reflection properties under $l \to -l$ of the various terms in the integrand. So the r-dependent terms in the vertex functions v_ρ, v_μ cause no chiral symmetry breaking. Expression (4.22) contains all the surviving r dependence as $a \to 0$. Diagrams c, d and e contain anomalous vertices and therefore lead to contact terms. Consider for example diagram d, which is proportional to

$$\int_l V_{\mu\rho}(p', l+p)S(l+p)V_\sigma(l+p, p)D_{\sigma\rho}(l)\,. \tag{4.23}$$

Scaling $l \to l/a$ and considering the inner region, leads to the integral (Feynman gauge, no summation over μ)

$$\int_0^1 \mathrm{d}x \int_{|l|<\delta} \frac{(-\gamma_\mu \frac{1}{2}l_\mu + r)(-\not{l})\gamma_\mu + \mathrm{O}(a, l^2)}{\{l^2 + a^2[xm^2 + (1-x)\lambda^2 + x(1-x)p^2]\}^2}\,. \tag{4.24}$$

As $a \to 0$ the possible non-trivial momentum dependence due to the logarithmic singularity at $l = 0$ does not contribute and the result is γ_μ (constant of order δ^2).

The third divergent one-loop diagram we discuss with Wilson's fermion is the vacuum polarization $I_{\mu\nu}(p)$. To duplicate as much as possible of the corresponding calculation with 16 fermions in sect. 3, we use the Taylor expansion method described there. Again the two contributing diagrams are in fig. 5. The amplitudes are the same as (3.22) and (3.23) with $\tilde{r}$ replaced by r/a, and color and flavor matrices added. Taylor expansion and the Ward identity give again

$$I_{\mu\nu}(p)=I^2(m,r,a)\cdot(\delta_{\mu\nu}p^2-p_\mu p_\nu)+\tilde{I}_{\mu\nu}(p)+\mathrm{O}(a)\,. \tag{4.25}$$

A division in 16 hypercubes is now, of course, not necessary to determine $\tilde{I}_{\mu\nu}(p)$. It is obtained from $I_{\mu\nu}$ in any (continuum) regularization scheme by subtracting up to second order in p [25]:

$$\tilde{I}_{\mu\nu}(p)=\left(1-\tfrac{1}{2}\sum_{\rho\sigma}p_\rho p_\sigma\partial_{p_\rho}\partial_{p_\sigma}|_0\right)I^{\text{cont}}_{\mu\nu}(p)\,, \tag{4.26}$$

$$I^{\text{cont}}_{\mu\nu}(p)=\frac{g^2}{2\pi^2}(\delta_{\mu\nu}p^2-p_\mu p_\nu)\int_0^1 \mathrm{d}z\, z(1-z)\ln[m^2+p^2z(1-z)]\,. \tag{4.27}$$

It is seen that $I_{\mu\nu}(p)$ is Lorentz covariant. The r dependence is again only present in the contact terms.

Concluding this section we make some comments about higher orders. Suppose we are dealing with the two-loop approximation. The counterterms needed for renormalization at the one-loop level must be written in their lattice gauge invariant form, such that they can be interpreted as absorbed in the parameters M and g. Then their insertion in a two-loop diagram has the proper periodicity, which allows for freely shifting the integration variables which, for example, secure the Ward identities. By again rescaling the loop integrations, the non-trivial momentum dependence can be seen to depend on the behavior of the integrand near $l=0$. It is the low momentum behavior of vertex insertions which determines the physics of the two-loop integral, and this is just what has been studied here for the one-loop integrals. Besides removing the a dependence, the counterterms also remove the r dependence from the low momentum behavior of the vertex insertions. Hence the non-trivial momentum dependence as $a\to 0$ will again be independent of r and the r dependence is in contact terms, which are absorbed by the two-loop level renormalization procedure.

5. Currents in Wilson's lattice QCD

In sect. 4 we verified by a few examples at the one-loop level that renormalized Green functions involving quarks in Wilson's lattice QCD are equal to the Green functions of the continuum theory, up to a finite renormalization. We see no reason why the lattice formulation should not be equivalent to the continuum formulation to any finite order in the weak coupling expansion. Experience of other authors [21, 28, 29] supports this assumption. In particular we found that at $M=M_c(r,g)$ the Green functions are chiral invariant and that the renormalized Green functions can be taken independent of r.

To make contact with current algebra and with the electroweak interactions it is desirable to have explicit expressions on the lattice of vector and axial vector currents, such that their matrix elements can be calculated independent of the weak coupling methods. We want these currents to be gauge invariant and local (i.e., $j_\mu(x)$ involves only fields at points x' which lie within a sphere about x of finite radius, in units of the lattice distance). Such currents are easily obtained by symmetry considerations leading to divergence equations or by coupling to external fields. Both these ways for obtaining currents will now be reviewed briefly.

In the first method one performs a transformation of variables on the quark fields in the partition function which has the form of a (γ_5 dependent) gauge transformation. Differentiation with respect to the parameters of this gauge transformation leads to the divergence equations

$$\langle \partial'_\mu V^\mu_\alpha(x) - D^V_\alpha(x) + i\bar{\eta}(x)\tfrac{1}{2}\lambda_\alpha\psi(x) - i\bar{\psi}(x)\tfrac{1}{2}\lambda_\alpha\eta(x)\rangle = 0\,, \tag{5.1}$$

$$\langle \partial'_\mu A^\mu_\alpha(x) - D^A_\alpha(x) + i\bar{\eta}(x)\tfrac{1}{2}\lambda_\alpha\gamma_5\psi(x) + i\bar{\psi}(x)\tfrac{1}{2}\lambda_\alpha\gamma_5\eta(x)\rangle = 0\,, \tag{5.2}$$

where the brackets denote the partition function average, $\eta(x)$ and $\bar{\eta}(x)$ are external quark sources, the matrices λ_α act on flavor $[\lambda_0 = (2/N_f)^{1/2}]$, and

$$\partial'_\mu f(x) = \frac{1}{a}[f(x) - f(x - a_\mu)]\,, \tag{5.3}$$

$$V^\mu_\alpha(x) = \tfrac{1}{2}\bar{\psi}(x)\gamma_\mu\,\tfrac{1}{2}\lambda_\alpha U_\mu(x)\psi(x + a_\mu) + \text{h.c.}\,, \tag{5.4}$$

$$A^\mu_\alpha(x) = \tfrac{1}{2}\bar{\psi}(x)\gamma_\mu\gamma_5\,\tfrac{1}{2}\lambda_\alpha U_\mu(x)\psi(x + a_\mu) + \text{h.c.}\,, \tag{5.5}$$

$$D^V_\alpha(x) = \bar{\psi}(x)i[\tfrac{1}{2}\lambda_\alpha, M]\psi(x) + \sum_\mu\left\{\left[\frac{ir}{2a}\bar{\psi}(x)\tfrac{1}{2}\lambda_\alpha U_\mu(x)\psi(x + a_\mu) + \text{h.c.}\right] - [x \to x - a_\mu]\right\}, \tag{5.6}$$

$$D^A_\alpha(x) = \bar{\psi}(x)i\gamma_5\{\tfrac{1}{2}\lambda_\alpha, M\}\psi(x) - \frac{r}{2a}\sum_\mu\{[\bar{\psi}(x)i\gamma_5\,\tfrac{1}{2}\lambda_\alpha U_\mu(x)\psi(x + a_\mu) + \text{h.c.}] + [x \to x - a_\mu]\}\,. \tag{5.7}$$

The summation over μ in (5.6) has the form of a divergence in the sense of (5.3), so these terms may be subtracted from $V^\mu_\alpha(x)$, thereby obtaining the current

$$\hat{V}^\mu_\alpha(x) = \tfrac{1}{2}\bar{\psi}(x)(\gamma_\mu - ir)\tfrac{1}{2}\lambda_\alpha U_\mu(x)\psi(x + a_\mu) + \text{h.c.}\,, \tag{5.8}$$

with the divergence equation

$$\left\langle \sum_\mu \partial'_\mu \hat{V}^\mu_\alpha - \bar{\psi}i[\tfrac{1}{2}\lambda_\alpha, m_0]\psi + i\bar{\eta}\,\tfrac{1}{2}\lambda_\alpha\psi - i\bar{\psi}\,\tfrac{1}{2}\lambda_\alpha\eta\right\rangle = 0\,, \tag{5.9}$$

($m_0 = M - M_c$). The current $\hat{V}$ is conserved in the case of flavor symmetry. A similar rearrangement is not suggested by the right-hand side of (5.7), which exhibits the chiral symmetry breaking as a sum of generalized mass terms.

The second method using external gauge fields has already been employed in sect. 3 [cf. (3.4)]. It consists of substituting into the action (3.1) in various places

$$\mathrm{U(color)} \to \mathrm{U(color)} \otimes \mathrm{U(flavor)}\,, \tag{5.10a}$$

with

$$U_\mu(x)(\text{flavor}) = \exp\{i\tfrac{1}{2}\lambda_\alpha[v^\alpha_\mu(x) + \gamma_5 a^\alpha_\mu(x)]\}\,, \tag{5.10b}$$

where v^α_μ and a^α_μ are the external gauge fields. Differentiating the action with respect to v^α_μ and a^α_μ gives the currents. If the substitution (5.10) is made only in the γ_μ part of the quark field gradient, then the currents V^μ_α and A^μ_α are obtained. Performing (5.10), $\gamma_5 \to 0$, also in the r dependent terms leads to $\hat{V}^\mu_\alpha$. Within a particular substitution scheme (5.10) (e.g., the $\hat{V}^\mu - A^\mu$ scheme), differentiation of the partition function with respect to the external fields generates Green functions involving currents.

In a more complete theory involving the electroweak interactions some linear combination of the external gauge fields will become dynamical fields, and $U_\mu(x)$(flavor) may appear in the Wilson mass term even with $\gamma_5 \nrightarrow 0$. The non-invariance of this mass term under γ_5 dependent transformations could be compensated, for example, by a Higgs field. Then extra terms would appear in the axial vector current in addition to $A^\mu_\alpha(x)$, such as the current $\hat{A}^\mu_\alpha$ introduced in ref. [13]:

$$\hat{A}^\mu_\alpha(x) = A^\mu_\alpha(x) + s\bar{\psi}(x+a_\mu)i\gamma_5\tfrac{1}{2}\lambda_\alpha\psi(x+a_\mu) - \tfrac{1}{2}r[\bar{\psi}(x)i\gamma_5\tfrac{1}{2}\lambda_\alpha U_\mu(x)\psi(x+a_\mu) + \text{h.c.}]\,, \tag{5.11}$$

where the s-term may be considered as added by hand. It is needed because otherwise $\hat{A}^\mu_\alpha$ would not have the properties of a current in the continuum limit. The tree-graph value of s is $s = r$.

So there is an ambiguity in the definition of currents on the lattice, which is similar to the ambiguity in the continuum theory where one has to give a meaning to the product of fields at the same space-time point. One way of dealing with this is to make the external fields dynamical, e.g., constructing a Weinberg–Salam–Glashow model on the lattice, generating the Wilson mass term by the Higgs mechanism. This can be done, but here we wish to stay within QCD and investigate the properties of the currents introduced above.

Consider the current $A^\mu_\alpha(x)$ and its divergence equation (5.2), (5.7). The axial symmetry breaker $D^\mathrm{A}_\alpha(x)$ consists of two terms which somehow should cancel in the chiral limit $M = M_\mathrm{c}$. Suppose this cancellation takes place in all matrix elements of the (flavor)non-singlet current, $\langle|D^\mathrm{A}_\alpha(x)|\rangle = 0$, $\alpha \neq 0$. Then there is no reason why it should cancel as well in all flavor singlet current matrix elements: $\langle|D^\mathrm{A}_0(x)|\rangle \neq 0$ for some matrix elements. This singlet splitting is due to closed fermion loops to which the non-singlet current cannot couple: it is the axial anomaly. There are essentially two types of matrix elements (in the weak coupling expansion) for testing the reasoning above, namely the one-quark matrix element $\langle p'|D^\mathrm{A}_\alpha|p\rangle$ which turns out to be zero at $M = M_c$ for both $\alpha \neq 0$ and $\alpha = 0$, and the transition element $\langle\text{gluons}|D^A_\alpha|0\rangle$, which is zero for $\alpha \neq 0$ and which for $\alpha = 0$ in the case of two gluons becomes the usual triangle anomaly at $M = M_c$. This will be shown below in the one-loop approximation.

However, there are other effects which have to do with the fact that the current A^μ_α is not conserved on the lattice (i.e., $\partial'_\mu A^\mu_\alpha \neq 0$). This causes a renormalization of the strength of the current by the color interaction. In the continuum theory this phenomenon is well known to happen for non-conserved currents. If the symmetry breaking is soft (dimension <4 operators) then the renormalization is finite (cutoff independent) and if the symmetry breaking is not soft (dimension ≥ 4 operators) then the renormalization is infinite. An example of the latter is the flavor singlet current which has the anomaly (dimension 4) and which needs an infinite ($\sim \ln \Lambda$) renormalization [20, 30, 41]. In the present case, $D^A_\alpha(x)$ has dimension 4 and the singlet current needs an infinite ($\sim \ln a$) renormalization but the non-singlet current A^μ_α suffers a finite renormalization depending on r. However, it is conserved in the chiral limit $M = M_c$.

Within the weak coupling expansion it is easily recognized how to identify this finite renormalization and dispose of it, but it could be a nuisance in a non-perturbative approach or in a strong coupling expansion. One needs a criterion for how to normalize the strength of the axial vector current, from first principles, which can be used in both the weak and strong coupling expansion. Current algebra comes to mind, but it seems somewhat *ad hoc*, and, moreover, it could be impractical at strong coupling, since one has first to calculate the states in a commutation relation.

The best normalization condition we can think of is based on the following argument. If the external gauge fields become dynamical such that they correspond to the electroweak bosons, then the anomaly structure of the vector and axial vector currents should be right, otherwise unitarity is violated in a covariant renormalizable gauge [23]. So we propose to use the vector currents $\hat{V}^\mu_\alpha(x)$, which suffer no renormalization, in the $\hat{V}\hat{V}A$ triangle anomaly as a criterion for normalizing $A^\mu_\alpha(x)$. Let A^μ_α multiplied by κ^α_A be the correctly normalized current, and

$$\sum_{x,y} e^{-ipx-iqy}\langle 0|T\hat{V}^\alpha_\mu(x)\hat{V}^\beta_\nu(y)\kappa^\gamma_A A^\gamma_\rho(0)|0\rangle = \tfrac{1}{2}d_{\alpha\beta\gamma}\Gamma_{\mu\nu\rho}(p,q)\,, \tag{5.12}$$

where $d_{\alpha\beta\gamma} = \frac{1}{4}\operatorname{Tr}\{\lambda_\alpha, \lambda_\beta\}\lambda_\gamma$ and $\gamma \neq 0$. Then κ_A is determined from

$$(p+q)_\rho\Gamma_{\mu\nu\rho}(p,q) = \varepsilon_{\mu\nu\sigma\tau}p_\sigma q_\tau\frac{1}{2\pi^2}[1 + O(p,q)]\,. \tag{5.13}$$

We shall refer to $\kappa_A A^\mu_\alpha$ (and $\kappa_V V^\mu_\alpha$) as the dynamical currents. The validity of this relation (5.13) to all orders in the weak coupling expansion of the continuum theory is usually referred to as the Adler–Bardeen theorem [30, 35, 36, 39, 41]. It will be demonstrated below that it is true on the lattice in lowest order with $\kappa_A = 1$ and we conjecture it to be true to all orders *if* κ_A is determined by the condition that the quark form factor F_1 for the dynamical current $\kappa_A A^\mu_\alpha$ satisfies $F_1 = 1$ at zero momentum transfer. With confined quarks we must impose (5.13) as a condition. It may be useful to extend (5.13) to the case $\gamma = 0$.

Before illustrating the above remarks by one-loop calculations, we want to say something about the currents V^μ_α and $\hat{A}^\mu_\alpha$. The current V^μ_α is not conserved on the lattice and it needs a κ_V to readjust its strength. All we can do with $\hat{A}^\mu_\alpha$ at weak coupling is adjust the constant s in (5.11) such that $\hat{A}^\mu_\alpha = A^\mu_\alpha$ in the continuum limit. So it seems we need not bother with V^μ_α and $\hat{A}^\mu_\alpha$, were it not that both currents have been used in a strong coupling calculation [13] within the hamiltonian method, giving interesting results: the relation $1 = \frac{3}{5} g_A \gamma_\rho (f_\pi / m_\rho) \sqrt{3}$ (satisfied to 12%) obtained by the use of V^μ_α and either A^μ_α or $\hat{A}^\mu_\alpha$. Furthermore in that continuous time formulation only the current $\hat{A}^\mu_\alpha$ with $s = \frac{1}{4} a M_c$ is consistent with both PCAC and Lorentz invariance (ref. [13] assumed $s = r$, which is incorrect). We leave this puzzling situation for what it's worth and shall use V^μ_α below only for illustrative purposes.

The one-loop calculations focus on the vector and axial vector quark form factors and on the triangle anomaly. First we determine how the one-quark matrix elements of the currents renormalize. We do this because we have more physical intuition for the quark form factors than for the off-shell Green functions. In defining on-shell quark states an infrared regulating gluon mass λ is used, defined in (A.5). The form factors are gauge independent in the limit $\lambda \to 0$, since to finite order in the weak coupling expansion the gauge dependence is proportional to λ^2 times some power of $\ln \lambda^2$. We fix κ by requiring the dynamical currents not to renormalize in the symmetry limit, i.e., $F_1(0)(\text{dyn}) = 1$, giving $\kappa(r, g) = 1 + \kappa(r) g^2 + O(g^4)$.

Next the axial vector–two-gluon vertex function is discussed and finally the triangle anomaly is verified to be given correctly by the two-gluon vertex function of $D^A_0(x)$ at $M = M_c$.

The Feynman rules involving the divergences $D^V_\alpha(x)$ and $D^A_\alpha(x)$ are conveniently obtained by adding to the action the source terms $\sum (d^V_\alpha D^V_\alpha + d^A_\alpha D^A_\alpha)$. Vertex functions for an external v^α_μ, $\hat{v}^\alpha_\mu$, a^α_μ, d^V_α or d^A_κ line are in fig. 14, with the corresponding functions given in the appendix. The one-loop approximation to the vector and axial vector vertex functions is again described by the diagrams in fig. 9, with the appropriate modifications. Taking, for simplicity, the flavor symmetric case, the vertex functions have the form

$$\Gamma^V_\mu(p', p) = \gamma_\mu [1 + \bar{g}^2 \Gamma^V_0(r)] + \text{conventional}, \tag{5.14a}$$

$$\Gamma^A_\mu(p', p) = \gamma_\mu \gamma_5 [1 + \bar{g}^2 \Gamma^A_0(r)] + \text{conventional}, \tag{5.14b}$$

where "conventional" means a conventional r-independent expression as in (4.14); the overall factor $\frac{1}{2}\lambda_\alpha$ is not written. All the r dependence is in the constants $\Gamma^V_0(r)$ and $\Gamma^A_0(r)$. For $\Gamma^{\hat{V}}_\mu(p', p)$ there is a corresponding $\Gamma^{\hat{V}}_0(r)$. Note that the vertex functions (5.14) are consistent with chiral symmetry.

The one-quark matrix elements of the currents are given by

$$\langle p' | j_\mu(0) | p \rangle = Z_2 \bar{u}(p') \Gamma_\mu(p', p) u(p), \tag{5.15}$$

where Z_2 is defined as the residue of the pole in the quark propagator. Defining Z_1^A by

$$\bar{u}(p)\Gamma_\mu^A(p,p)u(p) = \frac{1}{Z_1^A}\bar{u}(p)\gamma_\mu\gamma_5 u(p)\,, \tag{5.16}$$

and similarly for Z_1^V, $Z_1^{\hat{V}}$, the usual form factor $F_1[(p-p')^2]$ is found to have the normalization

$$F_1(0) = Z_2/Z_1\,. \tag{5.17}$$

The ratio Z_2/Z_1 is conveniently studied with the divergence equations (5.1) or (5.9) and (5.2), which lead to the Ward identities (as $a \to 0$, again omitting $\frac{1}{2}\lambda_\alpha$)

$$(p-p')_\mu\Gamma_\mu^V(p',p) = -iD^V(p',p) + S'(p)^{-1} - S'(p')^{-1}\,, \tag{5.18a}$$

$$(p-p')_\mu\Gamma_\mu^A(p',p) = -iD^A(p',p) - \gamma_5 S'(p)^{-1} - S'(p')^{-1}\gamma_5\,, \tag{5.18b}$$

and similarly for $\Gamma_\mu^{\hat{V}}$, with $D^{\hat{V}} = 0$ in the present flavor symmetric case. The diagrams for $D(p',p)$ are shown in fig. 10. Writing the inverse full fermion propagator as

$$S'(p)^{-1} = S_0(p^2) + pS_1(p^2)\,, \tag{5.19}$$

the constant Z_2 is given by

$$Z_2 = \{S_1(-m^2) + 2m[S_0'(-m^2) - mS_1'(-m^2)]\}^{-1}\,,$$

$$S_j'(p^2) = \frac{\partial}{\partial p^2}S_j(p^2)\,, \qquad j = 1, 2\,, \tag{5.20}$$

where the mass m satisfies eq. (4.5). Setting $p' = p$ in (5.18) gives

$$D^V(p,p) = 0\,, \tag{5.21}$$

$$D^A(p,p) = i\gamma_5 2S_0(p^2)\,, \tag{5.22}$$

and first differentiating with respect to p_μ, going on-shell with $p' = p$ leads to

$$F_1^V(0) = 1 + D_1^V\,, \qquad F_1^{\hat{V}}(0) = 1\,, \tag{5.23}$$

$$F_1^A(0) = Z_2 S_1(-m^2) + D_1^A\,, \tag{5.24a}$$

$$Z_2 S_1(-m^2) = 1 - 2m[S_0'(-m^2) - mS_1'(-m^2)] + O(g^4)\,, \tag{5.24b}$$

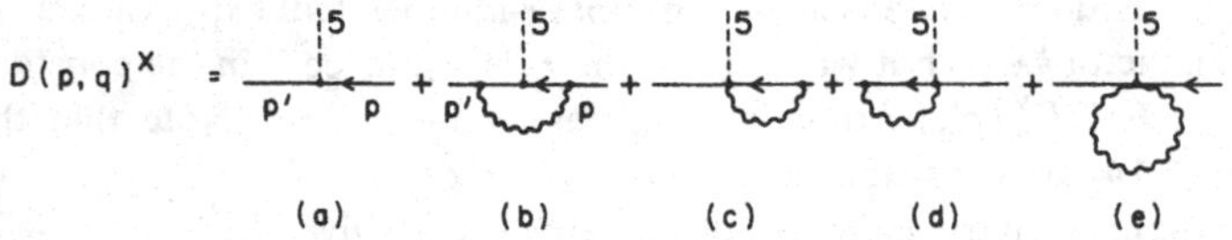

Fig. 10. One-loop approximation to $D^X(p',p)$; X = V or A.

where the constants $D_1^{A,V}$ are defined by, for example,

$$Z_2\bar{u}(p)\left[-i\frac{\partial}{\partial p_\mu}D^A(p',p)\right]_{p'=p}u(p)=D_1^A\bar{u}(p)\gamma_\mu\gamma_5 u(p)\,. \tag{5.25}$$

The current $\hat{V}^\mu$ is conserved on the lattice and its form factor at zero momentum transfer is unchanged from the tree-graph value $F_1^V=1$, as expected. The current V^μ is not conserved on the lattice and $F_1^V(0)\neq 1$. However, V^μ *is* conserved in the continuum limit. This is because the one-loop $D^V(p',p)$ has in the case of flavor symmetry the simple structure

$$D^V(p',p)=i\bar{g}^2 d^V(r)(\not p-\not p')+O(a)\,, \tag{5.26}$$

with the constant $d^V\propto r$. Hence $D^V(p',p)$ does not contribute when (5.18a) is sandwiched between equal mass spinors as in (5.15). The form (5.26) was to be expected from the remarks leading to eq. (5.8) and it is in accordance with the fact that the $D^V(x)$ vertices are anomalous in the case of flavor symmetry.

Similar remarks apply to A^μ and D^A: in the chiral limit ($M=M_c$, $m=0$)

$$D^A(p',p)=i\bar{g}^2 d^A(r)(\not p-\not p')\gamma_5+O(g^4)\,, \tag{5.27}$$

again with the constant $d^A\propto r$. In this case, (5.24) reduces to

$$F_1^A(0)=1+D_1^A\,, \tag{5.28}$$

with $D_1^A=\bar{g}^2 d^A(r)+O(g^4)$. The axial current matrix element is also conserved in the continuum limit, at $m=0$. Let us examine the one-loop approximation to $D^A(p',p)$ more closely (fig. 10). Diagram 10a represents the vertex function [cf. (A.13)]

$$a^{-1}[\mathcal{M}(ap')+\mathcal{M}(ap)]i\gamma_5=2(m-\delta m)i\gamma_5+O(a)\,. \tag{5.29}$$

Diagram 10b represents the expression

$$\bar{g}^2\int_l V_\mu(p',l+p')S(l+p')a^{-1}[\mathcal{M}(al+ap')+\mathcal{M}(al+ap)]$$

$$\times i\gamma_5 S(l+p)V_\nu(l+p,p)D_{\mu\nu}(l)\,, \tag{5.30}$$

where to this order in g^2 we may use $a^{-1}\mathcal{M}=m+a^{-1}\mathcal{M}_c$, with $\mathcal{M}_c$ the anomalous mass vertex function

$$\mathcal{M}_c(ak)=r\sum_\lambda(1-\cos ak_\lambda)+O(g^2)\,. \tag{5.31}$$

Denoting by $\mathcal{A}(p',p)$ the sum of the diagrams c–e plus the contribution of (5.31) to (5.30), we know that $\mathcal{A}(p',p)$ is a polynomial of degree 1 in the continuum limit, since it contains the anomalous vertices,

$$\mathcal{A}(p',p)=\bar{g}^2[2\mathcal{A}_0(r)a^{-1}+2m\mathcal{A}_1(r)]i\gamma_5+i\bar{g}^2 d^A(r)(\not p-\not p')\gamma_5+O(a)\,. \tag{5.32}$$

The a^{-1} terms in (5.32) and (5.29) cancel because of eq. (5.22), since M is chosen

($M=M_c+\cdots$) such that they are absent from the right-hand side of this equation:

$$-\delta m+\bar{g}^2a^{-1}\mathcal{A}_0(r)=m(Z_m-1)+O(g^4)\,, \tag{5.33}$$

where the form (4.21b) is used. Hence $D^A(p',p)$ can be written as

$$D^A(p',p)=2m\Gamma^P(p',p)+i\bar{g}^2d^A(r)(\not p-\not p')\gamma_5\,, \tag{5.34}$$

with

$$\Gamma^P(p',p)=i\gamma_5+[Z_m-1+\bar{g}^2\mathcal{A}_1(r)]i\gamma_5$$
$$+\bar{g}^2\int_l V_\mu(p',l+p')S(l+p')i\gamma_5S(l+p)V_\nu(l+p,p)D_{\mu\nu}(l)+O(g^4)\,. \tag{5.35}$$

For later use we give the result of evaluating the integral above in the Feynman gauge,

$$\bar{g}^2\int_l V_\mu Si\gamma_5SV_\nu D^F_{\mu\nu}=\bar{g}^2\Gamma^P_0(r)i\gamma_5+\frac{\bar{g}^2}{16\pi^2}\int_0^1 dx\int_0^{1-y}dy$$
$$\times\{-8i\gamma_5(\ln a^2\Delta-\tfrac{3}{2})+\Delta^{-1}\gamma_\mu[m-(1-x)\not p'+y\not p]i\gamma_5$$
$$\times[m+x\not p'-(1-y)\not p]\gamma_\mu\}+O(a)\,, \tag{5.36a}$$

$$\Delta=(x+y)m^2+(1-x-y)\lambda^2+x(1-x)p'^2+y(1-y)p^2-2xyp'p\,, \tag{5.36b}$$

$$\Gamma^P_0(r)=\lim_{\delta\to0}\left[\int_{|l|>\delta}\frac{\cos^2\frac12 l+r^2\sin^2\frac12 l}{[\mathcal{M}_c^2(l)+\sin^2 l]4\sin^2\frac12 l}+\frac{\ln\delta^2}{4\pi^2}\right]. \tag{5.36c}$$

The combination $Z_2D^A(p,p)$ is, to this order in g^2, independent of r and finite as $a\to0$, as follows from (5.22) since the right-hand side multiplied by Z_2 has the properties $Z_2S_0(p^2)=\tilde{S}_0(p^2)$ [cf. (4.20)]. Hence the r dependence in $Z_2+Z_m+\bar{g}^2\mathcal{A}_1(r)+\bar{g}^2\Gamma^P_0(r)$ cancels to $O(g^2)$ and $Z_2\Gamma^P(p',p)$ is r and a independent. The cancellation of the $\ln a$ terms can easily be checked explicitly. Hence, all the r dependence in $Z_2D^A(p',p)$ is in $d^A(r)$.

In the limit $m\to0$, (5.27) is indeed recovered from (5.34)–(5.36). Eq. (5.34) corresponds to the separation of $D^A(x)$ into a soft piece with scale dimension 3 (up to logarithms) and a hard piece with scale dimension 4, the $d^A(r)$ terms, which is absent in the continuum theory formulation. This hard piece is consistent with current conservation in the chiral limit, however. Crucial for obtaining chiral invariance was the cancellation of the a^{-1} terms in (5.33), which could be demonstrated with the help of the axial vector Ward identity.

The normalizing constants κ follow from (5.23)–(5.28):

$$\kappa_V=1-\bar{g}^2d^V(r)+O(g^4)\,,\qquad \kappa_{\hat{V}}=1\,, \tag{5.37a}$$

$$\kappa_A=1-\bar{g}^2d^A(r)+O(g^4)\,. \tag{5.37b}$$

With these κ's the dynamical currents have $F_1(0)(\mathrm{dyn}) = 1$ in the symmetry limit, where they are conserved as well. Their matrix elements are independent of r because the r dependence is to this order in g^2 only present in the constant part of F_1, not in its momentum dependence and neither in the other form factors $F_2, \ldots$ [cf. (5.14)]. This generalizes to the flavor non-symmetric situation since the mass dependence of F_1 is also independent of r. The κ's are flavor independent ($\alpha = 1, \ldots, N_f$) and to this order the same for the singlet ($\alpha = 0$) and the non-singlet ($\alpha \neq 0$) currents.

A comment is in order here with regard to the normalization condition $F_1^A(0)(\mathrm{dyn}) = 1$. It is sometimes argued that the axial vector form factor renormalizes at the quark level, which is interpreted as a possible explanation for $g_A < \frac{5}{3}$. To see how this comes about we compute $F_1^A(0)(\mathrm{dyn})$ from (5.24) using (5.19), (4.15), (4.14) and (5.25), (5.34)–(5.36) in the Feynman gauge (it is gauge independent) and find

$$F_1^A(0)(\mathrm{dyn}) = 1 - \frac{\bar{g}^2}{4\pi^2}\int_0^1 \mathrm{d}x \frac{x^3}{x^2 + (1-x)(\lambda^2/m^2)} + \mathrm{O}(g^4)\,, \tag{5.38}$$

in accordance with Langacker and Pagels [34]. Letting $m \to 0$ gives $F_1^A(0)(\mathrm{dyn}) \to 1$, letting $\lambda \to 0$ results in $F_1^A(0)(\mathrm{dyn}) \to 1 - \bar{g}^2/8\pi^2 + \mathrm{O}(g^4)$. We believe that keeping λ fixed as $m \to 0$ exhibits the physics of the chiral limit better than letting $\lambda \to 0$ in the first place, but it is clear that the criterion $F_1^A(0)(\mathrm{dyn}) = 1$ at $M = M_c$, $\lambda \neq 0$, is useful only in the framework of the weak coupling expansion, if quarks are confined.

Consider next the one-loop flavor singlet axial vector–two-gluon vertex function described by the diagrams in fig. 4, suitably reinterpreted for the present case. The dependence on the color indices simply factorizes in a Kronecker δ-function because of global color invariance. Then the reasoning of sect. 3 is applicable again, except that the contribution from the regions in momentum space near π/a is negligible here ($\tilde{r} \to r/a$) and the result of the $a \to 0$ limit is the standard continuum theory *VVA* vertex function with the standard triangle anomaly in its axial vector Ward identity. It is independent of r.

We want to relate $D_0^A(x)$ in (5.7) to the triangle anomaly when $M = M_c$. The lowest order two-gluon diagrams involving $D_0^A(x)$ are shown in fig. 11, which defines the vertex function $D_{\mu\nu}^{mn}(p, q)^A$. To this order the vertex renormalization (5.28) does not yet contribute and $M_c = 4r/a$, so that $D_0^A(x) \propto r$. Let us check that $D_{\mu\nu}^{mn}(p, q)^A$ becomes indeed, the customary anomaly in the continuum limit for $M = M_c$. Diagrams b–d in fig. 11 vanish because of the trace over Dirac matrices so

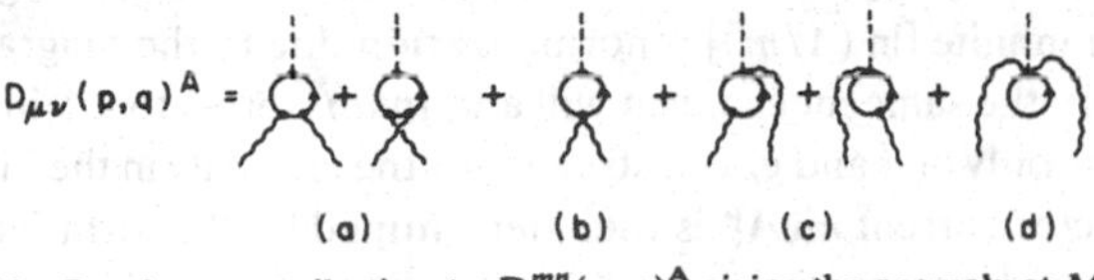

Fig. 11. One-loop contribution to $D_{\mu\nu}^{mn}(p, q)^A$ giving the anomaly at $M = M_c$.

$D^{mn}_{\mu\nu}(p,q)^{\mathrm{A}}$ is given by the two diagrams a which are equal by charge conjugation invariance:

$$D^{mn}_{\mu\nu}(p,q)^{\mathrm{A}} = \tfrac{1}{4}\delta_{mn}(2/N_{\mathrm{f}})^{1/2}D^{\mathrm{A}}_{\mu\nu}(p,q)\,, \tag{5.39a}$$

$$D^{\mathrm{A}}_{\mu\nu}(p,q) = -2g^2\int_l \mathrm{Tr}\,\{i\gamma_5 a^{-1}[\mathcal{M}(al+aq)+\mathcal{M}(al-ap)] \times S(l-p)V_\mu(l-p,l)S(l)V_\nu(l,l+q)S(l+q)\}\,. \tag{5.39b}$$

Scaling $l\to l/a$ brings out a factor a^{-2} and a straightforward expansion in a does not encounter an infrared divergence in the non-vanishing terms when $M=M_{\mathrm{c}}$, as then $\mathcal{M}$ becomes the anomalous vertex (5.31). The result can be written in the form

$$D^{\mathrm{A}}_{\mu\nu}(p,q)|_{M=M_{\mathrm{c}}} = ig^2\varepsilon_{\mu\nu\alpha\beta}p_\alpha q_\beta 16\int_l \cos l_\mu \cos l_\nu \cos l_\alpha \times\frac{[\mathcal{M}_{\mathrm{c}}^2(l)\cos l_\beta - 4r\mathcal{M}_{\mathrm{c}}(l)\sin^2 l_\beta]}{[\mathcal{M}_{\mathrm{c}}^2(l)+\sin^2 l]^3}\,, \tag{5.40}$$

where α and β are summed. The integrand is $\propto r^2$, yet the integral is independent of r: using the identity

$$\mathcal{M}_{\mathrm{c}}^2\cos l_\beta - 4r\mathcal{M}_{\mathrm{c}}\sin^2 l_\beta = \cos l_\beta(\mathcal{M}_{\mathrm{c}}^2+4\sin^2 l_\beta) + (\mathcal{M}_{\mathrm{c}}^2+\sin^2 l)^3\sin l_\beta\frac{\partial}{\partial l_\beta}(\mathcal{M}_{\mathrm{c}}^2+\sin^2 l)^{-2}\,, \tag{5.41}$$

and making a partial integration, with due regard to the behavior at the origin, leads to the standard "anomaly"

$$-iD^{\mathrm{A}}_{\mu\nu}(p,q)|_{M=M_{\mathrm{c}}} = -\frac{g^2}{2\pi^2}\varepsilon_{\mu\nu\alpha\beta}p_\alpha q_\beta\,. \tag{5.42}$$

For $M\neq M_{\mathrm{c}}$ the complete right-hand side of (3.21) is recovered.

In the two-loop approximation the strength of $D^{\mathrm{A}}_{\mu\nu}$ at $M=M_{\mathrm{c}}$ is affected by the vertex renormalization (5.28), but multiplication by κ_{A} will readjust this in accordance with the continuum theory. At the three-loop level for the $\hat{V}\hat{V}A$ vertex function the two-loop vertex correction of fig. 12 comes into play. This diagram, amongst others, causes a splitting between the singlet and non-singlet form factors. At this stage it is known [31] in the continuum theory that the singlet quark form factor needs an infinite [ln (Λ/m)] renormalization due to the diagram in fig. 12. In the lattice theory the same mechanism will add ln am terms to κ^0_{A} in addition to the terms depending only on r and g. The strength of the anomaly in the divergence of the singlet axial vector current $\kappa^0_{\mathrm{A}}A^\mu_0$ is then determined by the form factor normalization condition, and one may wonder how this compares with the Adler–Bardeen

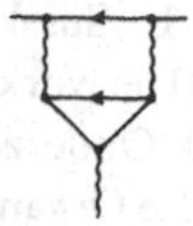

Fig. 12. Diagram causing singlet splitting in the quark form factors.

formula [31, 39, 40] in the continuum theory, which in unrenormalized form reads

$$\partial_\mu \bar{\psi}\gamma_\mu\gamma_5\psi = \bar{\psi}\{m_0, i\gamma_5\}\psi + \frac{g^2}{32\pi^2} G^m_{\mu\nu}\tilde{G}^m_{\mu\nu} . \tag{5.43}$$

The first term on the right-hand side does not need to be renormalized and one [30, 40, 41] renormalizes this equation by adding a term $(Z_A - 1)\partial_\mu \bar{\psi}\gamma_\mu\gamma_5\psi$ to both sides. There may also be other pseudoscalar densities mixing with $G\tilde{G}$ [30, 41]. Presumably the lattice and continuum formulations agree if κ^0_A and Z_A are fixed by the same normalization condition.

For the non-singlet currents we expect that the form factor determination of κ^α_A brings the divergence of $\kappa^\alpha_A A^\mu_\alpha$ into accordance with the Adler–Bardeen theorem (5.13).

6. Conclusions

We have shown that in a chiral-invariant lattice gauge theory with fermions the requirement of a Lorentz-invariant continuum limit at small but non-zero coupling implies species doubling. The simplest way to avoid species doubling is Wilson's method, which introduces explicit flavor-independent chiral symmetry breaking on the lattice. However, both weak coupling (this paper) and strong coupling investigations [13] indicate that the continuum limit of QCD with N_f flavored Wilson fermions has $SU(N_f) \times SU(N_f)$ chiral symmetry at zero current quark mass, but not U(1) chiral symmetry. There is no U(1) problem.

In the chiral continuum limit the flavor non-singlet axial vector currents are conserved and the singlet current has the usual "anomaly" as a residual effect of the chiral symmetry breaking on the lattice. The latter also affects the overall strength of the axial vector currents and this can be undone by multiplication by the constants κ_A. We have given a criterion for determining the κ_A independently of the quark form factors.

There are, fortunately, still many open questions. We did not discuss products of axial vector currents. We touched only very briefly the formulation of lattice gauge theories with fermions, where the dynamical gauge group is a chiral group. How does one put the Weinberg–Salam–Glashow model or any other unified gauge theory on a lattice?

While these and other questions deserve further work, the Wilson fermion method is a satisfactory formulation, at least for QCD.

One of us (LHK) wants to thank L. Susskind for the hospitality of Stanford University, and for useful discussions. The work of JS is part of the research program of the Stichting voor Fundamenteel Onderzoek der Materie (FOM), which is financially supported by the Nederlandse Organisatie voor Zuiver Wetenschappelijk Onderzoek (ZWO). LHK received a NATO Science Fellowship via ZWO.

Appendix

Our Minkowski-space γ-matrix conventions follow ref. [37]. In euclidean space $\gamma_4 = -i\gamma^0$, $\gamma_k = \gamma^k$, $k = 1, 2, 3$, such that $\{\gamma_\mu, \gamma_\nu\} = -2\delta_{\mu\nu}$. Occasionally the handy notation h.c. is used. In euclidean space, where ψ and $\bar{\psi}$ are independent variables, it is defined as

$$\sum_j c_j\bar{\psi}_1\Gamma_j\psi_2 + \text{h.c.} = \sum_j (c_j\bar{\psi}_1\Gamma_j\psi_2 + c_j^*\bar{\psi}_2\Gamma\psi_1)\,, \tag{A.1}$$

with

$$\Gamma_j = 1\,, \qquad i\gamma_5 = i\gamma_1\gamma_2\gamma_3\gamma_4,\ \ \gamma_\mu,\ \ \gamma_\mu\gamma_5\,, \qquad \sigma_{\mu\nu} = \tfrac{1}{2}i[\gamma_\mu, \gamma_\nu]\,. \tag{A.2}$$

The ε-tensor is fixed by $\varepsilon_{1234} = 1$.

For the weak coupling expansion it is necessary to use gauge fixing and a parametrization of $U_\mu(x)$ regular at $U_\mu(x) = 1$. We use the exponential parametrization; for example, in QCD

$$U_\mu(x) = \exp igat_m G_\mu^m(x)\,, \tag{A.3}$$

with $G_\mu^m(x)$ the gluon field and t_m the generators of SU(N) in the fundamental representation, normalized by Tr $t_m t_n = \frac{1}{2}\delta_{mn}$. Perturbative gauge fixing on the lattice in terms of $G_\mu^m(x)$ follows the continuum theory manipulations. The invariant measure $dU_\mu(x)$ on a link can be expressed in terms of the ordinary $dG_\mu^m(x)$ and a determinant depending on $G_\mu^m(x)$ (see for example [42], appendix E). We shall not need this determinant in this paper, which concentrates on the fermion aspects of the theory. For calculations testing the non-abelian nature of $U_\mu(x)$ at weak coupling see [28, 29]. The compact integration region for $G_\mu^m(x)$ of order $1/(ag)$ may be extended to $(-\infty, +\infty)$, since the difference this makes to the perturbative saddle-point evaluation of the partition function is of order $\exp(-1/g)$, hence non-perturbative.

The gauge field propagator is given by

$$\langle 0|\mathrm{T}G_\mu^m(x)G_\nu^n(0)|0\rangle = \int_p \mathrm{e}^{ipx}\, \mathrm{e}^{(ia/2)(p_\mu - p_\nu)}\delta_{mn}D_{\mu\nu}(p)\,, \tag{A.4}$$

$$D_{\mu\nu}(p) = \left[\delta_{\mu\nu} - \zeta\frac{S_\mu(p)S_\nu(p)}{\lambda^2 + S^2(p)}\right]\Big/[\lambda^2 + S^2(p)]\,, \tag{A.5}$$

with ζ specifying the gauge and

$$S_\mu(p) = (2/a)\sin\tfrac{1}{2}ap_\mu\,, \tag{A.6}$$

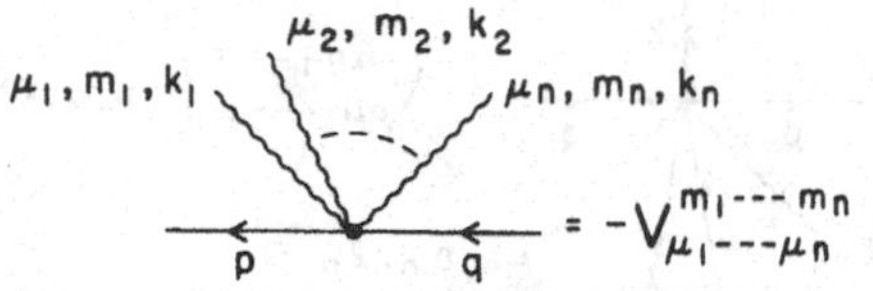

Fig. 13. Quark–gluon vertices.

$$\int_p = \int_{-\pi/a}^{\pi/a} \mathrm{d}^4p/(2\pi)^4 . \tag{A.7}$$

An infrared regulating mass λ has been inserted. Extracting the phase factor $\exp \frac{1}{2}ia(p_\mu - p_\nu)$ from $D_{\mu\nu}(p)$ in (A.4) leads to a real propagator and the corresponding real vertex functions given below. The quark-gauge field vertices are shown in fig. 13. We shall use as an illustrative example the QCD action (2.5), (2.15) together with (A.3) and the mass terms rewritten in the form (4.1). The bare real vertex functions depend only on $p+q$ and have the form

$$V^{m_1\cdots m_n}_{\mu_1\cdots\mu_n}(p,q) = g^n t_{m_1\cdots m_n} V_{\mu_1\cdots\mu_n}(p,q) , \tag{A.8a}$$

$$V_{\mu_1\cdots\mu_n}(p,q) = a^{n-1}\sum_\mu \delta_{\mu\mu_1}\cdots\delta_{\mu\mu_n} v^n_\mu(\tfrac{1}{2}ap+\tfrac{1}{2}aq) , \tag{A.8b}$$

$$t_{m_1\cdots m_n} = \frac{1}{n!}\sum_{\text{perm } P} t_{P_1}\cdots t_{P_n} , \tag{A.8c}$$

$$v^1_\mu(ap) \equiv v_\mu(ap) = \gamma_\mu \cos ap_\mu + r \sin ap_\mu , \tag{A.8d}$$

$$v^2_\mu(ap) = -\gamma_\mu \sin ap_\mu + r\cos ap_\mu , \tag{A.8e}$$

$$v^{n+2}_\mu(ap) = -v^n_\mu(ap) . \tag{A.8f}$$

The fermion propagator is given by

$$S(p) = a\left[\sum_\mu \gamma_\mu \sin ap_\mu + \mathcal{M}(ap)\right]^{-1} , \tag{A.9a}$$

$$\mathcal{M}(ap) = aM - r\sum_\mu \cos ap_\mu , \tag{A.9b}$$

and Ward's differential identity generalizes to

$$V_{\mu_1\cdots\mu_n}(p,p) = \frac{\partial}{\partial p_{\mu_1}}\cdots\frac{\partial}{\partial p_{\mu_n}} S(p)^{-1} . \tag{A.10}$$

Next the vertex functions involving the currents and their divergences as illustrated in fig. 14 will be given. Diagrams a represent the currents and their mathematical expression is similar to (A.8), as is evident from the method (5.10):

$$V^{\alpha m_1\cdots m_n}_{\rho\mu_1\cdots\mu_n}(p,q)^{\hat V} = \tfrac{1}{2}\lambda_\alpha g^n t_{m_1\cdots m_n} V_{\rho\mu_1\cdots\mu_n}(p,q) , \tag{A.11}$$

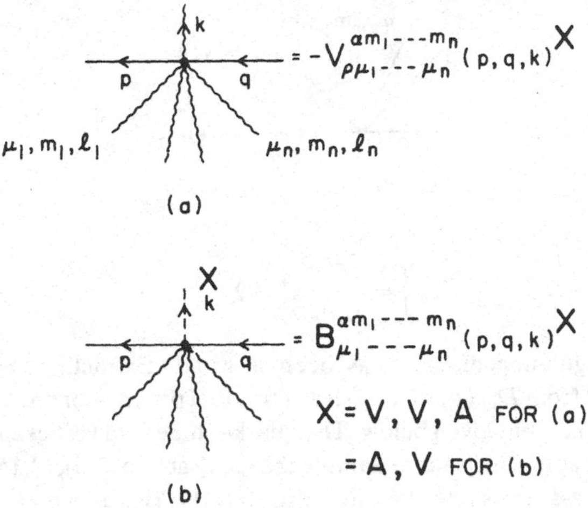

Fig. 14. Vertices involving currents and their divergences: (a) $= -V^{\alpha m_1\cdots m_n}_{\rho\mu_1\cdots\mu_n}(p, q, k)^X$; (b) $= B^{\alpha m_1\cdots m_n}_{\mu_1\cdots\mu_n}(p, q, k)^X$; X = $\hat{\text{V}}$, V or A.

from which the expression for X = V is obtained by setting $r = 0$ and from this result the expression for X = A by multiplication with γ_5. Vertex functions with multiple $\hat{v}^\alpha_\rho$ and/or a^α_ρ lines can also be derived easily. The X = A version of (A.11) with $\frac{1}{2}\lambda_\alpha \to 1$ is used in sect. 3, fig. 4.

For $n = 0$ the vector current divergence vertex in fig. 14b is given by

$$B^\alpha(p, q)^V = i[\tfrac{1}{2}\lambda_\alpha, m_0] + \tfrac{1}{2}i\lambda_\alpha[\mathcal{M}_c(aq) - \mathcal{M}_c(ap)]/a\,, \tag{A.12a}$$

where $m_0 = M - M_c$ and $\mathcal{M}_c$ is the mass function (A.9b) at $M = M_c$. For non-zero n the B's have the general form

$$B^{\alpha m_1\cdots m_n}_{\mu_1\cdots\mu_n}(p, q, k)^X = g^n a^{n-1} t_{m_1\cdots m_n}\, \tfrac{1}{2}\lambda_\alpha \sum_\mu \delta_{\mu\mu_1}\cdots\delta_{\mu\mu_n} B^n_\mu(p+q, k)^X\,, \tag{A.12b}$$

with $B^{n+2}_\mu = -B^n_\mu$, and

$$B^1_\mu(p, k)^V = i2\sin(\tfrac{1}{2}ak_\mu)[-r\cos\tfrac{1}{2}ap_\mu]\,, \tag{A.12c}$$

$$B^2_\mu(p, k)^V = i2\sin(\tfrac{1}{2}ak_\mu)[r\sin\tfrac{1}{2}ap_\mu]\,, \tag{A.12d}$$

$$B^1_\mu(p, k)^A = i\gamma_5 2\cos(\tfrac{1}{2}ak_\mu)[r\sin\tfrac{1}{2}ap_\mu]\,, \tag{A.13a}$$

$$B^2_\mu(p, k)^A = i\gamma_5 2\cos(\tfrac{1}{2}ak_\mu)[r\cos\tfrac{1}{2}ap_\mu]\,. \tag{A.13b}$$

For $n = 0$ the axial vector symmetry breaking vertex is

$$B^\alpha(p, k)^A = \{m_0, \tfrac{1}{2}\lambda_\alpha i\gamma_5\} + \tfrac{1}{2}\lambda_\alpha i\gamma_5[\mathcal{M}_c(ap) + \mathcal{M}_c(aq)]/a\,. \tag{A.13c}$$

Finally, note that all expressions on the lattice are periodic in momentum space with period $2\pi/a$. This makes it possible to replace the momentum conservation modulo

$2\pi/a$ ("Umklapp") by ordinary momentum conservation, and integration variables in loop integrals may be shifted freely.

References

[1] M. Creutz, Phys. Rev. Lett. 43 (1979) 553; Phys. Rev. D21 (1980) 2308
[2] J.B. Kogut, R.B. Pearson and J. Shigemitsu, Phys. Rev. Lett. 43 (1979) 484; J.B. Kogut and J. Shigemitsu, Phys. Rev. Lett. 45 (1980) 410
[3] C.G. Callan, R.F. Dashen and D.J. Gross, Phys. Rev. Lett. 44 (1980) 435
[4] G. Mack, DESY preprint 80/03 (1980), *in* Recent progress in gauge theories, ed. G. 't Hooft et al. (Cargese, 1979) (Plenum Press, New York, 1980)
[5] A., E. and P. Hasenfratz, Nucl. Phys. B180 [FS2] (1981) 353; C. Itzykson, M.E. Peskin and J.B. Zuber, Phys. Lett. 95B (1980) 259; M. Lüscher, G. Münster and P. Weisz, Nucl. Phys. B180 [FS2] (1981) 1
[6] K.G. Wilson, Phys. Rev. D10 (1974) 2445; *in* New phenomena in subnuclear physics, ed. A. Zichichi (Plenum, New York, 1977) (Erice, 1975)
[7] Y. Aharonov, A. Casher and L. Susskind, Phys. Rev. D5 (1972) 988; L. Susskind, Phys. Rev. D16 (1977) 3031
[8] S.D. Drell, M. Weinstein and S. Yankielowicz, Phys. Rev. D14 (1976) 487; 1627
[9] L.H. Karsten and J. Smit, Nucl. Phys. B144 (1978) 536; Phys. Lett. 85B (1979) 100
[10] T. Banks et al., Phys. Rev. D15 (1977) 1111
[11] J. Shigemitsu, Phys. Rev. D18 (1978) 1709
[12] B. de Wit and G. 't Hooft, Topological expansion in a lattice color gauge theory, Harvard preprint (1976), unpublished
[13] J. Smit, Nucl. Phys. B175 (1980) 307
[14] D.G. Caldi and H. Pagels, Phys. Rev. D14 (1976) 809; D15 (1977) 2668
[15] Constructive quantum field theory, ed. G. Velo and A.S. Wightman, Lecture notes in physics 25 (Springer, New York, 1973)
[16] A. Chodos and J.B. Healy, Nucl. Phys. B127 (1977) 426
[17] L.H. Karsten, Ph.D. thesis (Amsterdam, 1979); *in* Field theoretical methods in particle physics, ed. W. Rühl (Plenum Press, New York, 1980) (Kaiserslautern, 1979)
[18] B. Svetitsky, S.D. Drell, H.R. Quinn and M. Weinstein, Phys. Rev. D22 (1980) 490; 1190
[19] J.M. Rabin, Santa Barbara Lattice Gauge Theory Workshop (28/7–1/8 1980)
[20] S.L. Adler, *in* 1970 Brandeis University Summer Inst. in Theoretical physics, ed. S. Deser, M. Grisaru and H. Pendleton (MIT Press, Cambridge, Mass.)
[21] H.S. Sharatchandra, Phys. Rev. D18 (1978) 2042
[22] L. Rosenberg, Phys. Rev. 129 (1963) 2786
[23] C. Bouchiat, J. Iliopoulos and Ph. Meyer, Phys. Lett. 38B (1972) 519; D.J. Gross and R. Jackiw, Phys. Rev. D6 (1972) 477
[24] R. Jackiw, *in* Lectures on current algebra and its applications (Princeton University Press, Princeton NJ, 1972) (Brookhaven, 1970)
[25] R.P. Feynman, Phys. Rev. 76 (1949) 769
[26] N. Kawamoto, Towards the phase structure of euclidean lattice gauge theories with fermions, Amsterdam preprint (June, 1980)
[27] G. 't Hooft, Nucl. Phys. B61 (1973) 455
[28] A. and P. Hasenfratz, Phys. Lett. 93B (1980) 165
[29] B.E. Baaquie, Phys. Rev. D16 (1977) 2612
[30] T. Marinucci and M. Tonin, Nuovo Cim. 31A (1976) 381
[31] S.L. Adler, Phys. Rev. 177 (1969) 2426
[32] J.S. Bell and R. Jackiw, Nuovo Cim. 60A (1969) 47
[33] W.A. Bardeen, Proc. 16th Int. Conf. on High-energy physics, Batavia, Ill. (1972), ed. J.D. Jackson and A. Roberts (NAL, Batavia, Ill., 1973)

[34] P. Langacker and H. Pagels, Phys. Rev. D9 (1974) 3413
[35] S.-Y. Pi and S.-S. Shei, Phys. Rev. D11 (1975) 2946
[36] A. Zee, Phys. Rev. Lett. 29 (1972) 1198
[37] J.D. Bjorken and S.D. Drell, Relativistic quantum mechanics (McGraw-Hill, New York, London, 1964)
[38] R. Tarrach, The pole mass in perturbative QCD, Marseille preprint CNRS 80/P. 1215 (June, 1980)
[39] S.L. Adler and W.A. Bardeen, Phys. Rev. 182 (1969) 1517
[40] W.A. Bardeen, Nucl. Phys. B75 (1974) 246
[41] J.H. Lowenstein and B. Schroer, Phys. Rev. D7 (1973) 1929
[42] R.J. Finkelstein, Non-relativistic mechanics (Benjamin, Reading, Mass. 1973)

Nuclear Physics B185 (1981) 20–40

ABSENCE OF NEUTRINOS ON A LATTICE
(I). Proof by homotopy theory

H.B. NIELSEN

The Niels Bohr Institute and Nordita, Blegdamsvej 17, DK-2100 Copenhagen Ø, Denmark

M. NINOMIYA

Rutherford Laboratory, Chilton, Didcot, Oxon OX11 0QX, England

Received 20 November 1980
(Final version received 27 January 1981)

It is shown, by a homotopy theory argument, that for a general class of fermion theories on a Kogut–Susskind lattice an equal number of species (types) of left- and right-handed Weyl particles (neutrinos) necessarily appears in the continuum limit. We thus present a no-go theorem for putting theories of the weak interaction on a lattice. One of the most important consequences of our no-go theorem is that it is not possible, in strong interaction models, to solve the notorious species doubling problem of Dirac fermions on a lattice in a chirally invariant way.

1. Introduction

(a) It has been known for some time that incorporation of fermions on a lattice leads to further fermionic modes than those naively expected [1–3]. For example, a naive construction of a Weyl fermion (=neutrino) on a Kogut–Susskind lattice (i.e. discrete space and continuous time) yields eight Weyl fermions in the low energy regime. To be specific we consider the $3+1$ dimensional naive model of the Weyl equation on a lattice which is obtained by replacing $\boldsymbol{\partial}$ in the Weyl equation

$$i\frac{\partial}{\partial t}u(\boldsymbol{x})=\frac{1}{i}\boldsymbol{\sigma}\boldsymbol{\partial}u(\boldsymbol{x})$$

by differences

$$i\frac{\partial}{\partial t}u(\boldsymbol{x})=\sum_{i=1}^{3}\frac{1}{2i}\sigma_i\{u(\boldsymbol{x}+\boldsymbol{n}_i)-u(\boldsymbol{x}-\boldsymbol{n}_i)\}\,. \qquad (1.1)$$

Here $u(\boldsymbol{x})$ is a two-component spinor. We take the lattice constant to be unity, and $\boldsymbol{n}_i$ denotes the unit vector in the x_i direction. In the momentum representation, eq. (1.1) takes the form

$$i\frac{\partial}{\partial t}\tilde{u}(\boldsymbol{p})=\sum_{i=1}^{3}\sigma_i\sin p_i\cdot\tilde{u}(\boldsymbol{p})\,.$$

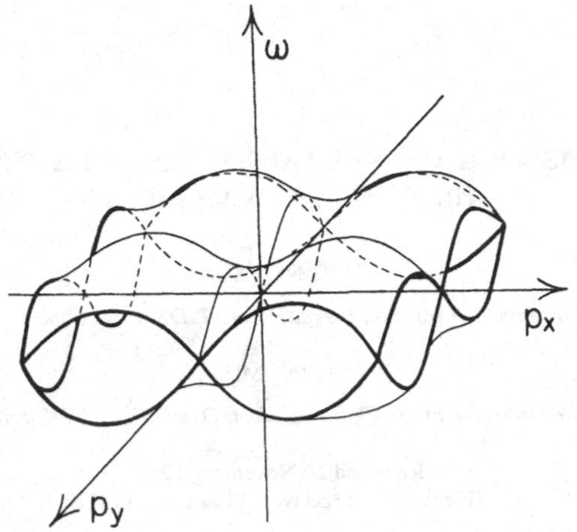

Fig. 1. The dispersion relation for the lattice fermion theory given by (1.1). The P_z dimension has been suppressed, and the drawing corresponds to $P_z = 0$. Just one Brillouin zone is drawn.

Since each sine function has two zeros in its dispersion relation for a period in a Brillouin zone in $\boldsymbol{p}$-space, there appear eight zeros of energy $\omega(\boldsymbol{p})$ at eight momentum values, as is depicted in fig. 1. These zeros represent Weyl particles in the dispersion relation in the long-wavelength limit.

It is the purpose of this article to formulate and prove a *no-go theorem*: the appearance of equally many right- and left-handed species (types) of Weyl particles with given quantum numbers is an unavoidable consequence of a lattice theory under some *mild assumptions*. Here species of neutrino means ν_e, ν_μ or ν_τ, etc. It should be stressed that our no-go theorem concerns *the number of species* of Weyl particles, but says *nothing about the actual number of Weyl particles* in the cosmos. The latter number is, of course, determined from how the states are filled or unfilled. The most important consequence of our no-go theorem is that *the weak interaction cannot be put on the lattice.*

Our "mild assumptions" include such important hypotheses as locality and exact conservation of discrete valued quantum number(s). Also, the charges are assumed to have a density defined from a finite region.

In a somewhat less general (Wilson model) and less rigorous form of this no-go theorem has very recently been put forward by Karsten and Smit [3a]. Firstly they argue for it from the Adler anomaly, but that would only give a theorem still allowing the standard $SU(3) \times SU(2) \times U(1)$ model with quarks and leptons to be put on a lattice. Since this will not be allowed by our theorem, the latter must be stronger. Secondly, they give, mainly for the Wilson model, a topological argument more similar to that applicable in $1+1$ dimensions, which we shall exhibit in our next article [12].

There are in the literature [2, 3, 4, 4a] various ingenious ways of getting rid of some of the extra and at first unwanted fermions in the low energy regime. Nobody seems, however, to have been able to get rid of all of them so that only one charged (with an exactly conserved quantum number) Weyl particle remains. If, however, no exactly conserved charge is required, a model with an odd number of Weyl particles can be built and we shall, in fact, present one in the successive article [12]. If one is interested in *strong interactions* and wants the Dirac particle rather than the Weyl particle, there should be no unsurmountable problem. A Dirac particle can be thought of as a composite of the components of two Weyl particles, a right-handed one and a left-handed one.

According to our no-go theorem it is *not*, however, *possible** even in the strong interaction models, to keep chiral invariance conserved on the scale of the fundamental lattice. The important consequence of our work is to *discourage any attempt to construct chiral invariant lattice models for QCD.*

(b) We will consider the general class of lattice fermion theories for which the bilinear part of the action for the N-component complex fermion field $\psi(x)$ is of the form

$$S=-i\int \mathrm{d}t \sum_{x} \dot{\bar{\psi}}(x)\psi(x)-\int \mathrm{d}t \sum_{x,y} \bar{\psi}(x)H(x-y)\psi(y)\,, \tag{1.2}$$

with H a hamiltonian. Interactions of quartic or higher degree in ψ are neglected. These interactions do not change the dispersion relation that we are interested in. We in fact define the number of species of Weyl fermions so that it depends only on the bilinear part of the action and the dispersion relation and not on the interaction, which just causes scattering processes. The kinetic term in (1.2) is not the most general one, since we could have

$$\sum_{x,y} \dot{\bar{\psi}}(x)T(x-y)\psi(y)\,. \tag{1.3}$$

With such a term we may risk the appearance of the singularities (poles) in the dispersion relation and, thus, particles with unbounded velocity. For our argument we shall need only the assumption that there are no such singularities. With eq. (1.2) we obtain the linear equation of motion

$$i\dot{\psi}(x)=\sum_{y} H(x-y)\psi(y)\,.$$

(If, however, we include a term like (1.3), the hamiltonian is effectively H/T.)

(c) We have assumed the following three conditions on the action (1.2):

(i) Locality of interaction, i.e. the hamiltonian satisfies $H(x-y)\to 0$, when $|x-y|\to$ large, fast enough in the sense that the Fourier transform of $H(x)$ has continuous first derivative.

* See, as examples of non-chirally invariant ways [2, 3, 4a].

(ii) Translational invariance on the lattice (invariance under group translations by an integer number of lattice constants).

(iii) Hermiticity of the $N \times N$ matrix hamiltonian H (reality of S).

In momentum space the field takes the form

$$\tilde{\psi}(\boldsymbol{p}) = \frac{1}{V} \sum_{\boldsymbol{x}} e^{-i\boldsymbol{p}\cdot\boldsymbol{x}} \psi(\boldsymbol{x}) .$$

Since the momentum $\boldsymbol{p}$ is only unique modulo $2\pi\boldsymbol{n}$ ($\boldsymbol{n}$: set of integers) the independent field variables are only those inside the Brillouin zone with an interval, e.g.

$$-\pi \leq p_i \leq \pi , \qquad (i = 1, 2, 3) .$$

We should here mention the work of Drell, Weinstein and Yankielowicz [4]. Their method of discretizing the fermion theory is to replace ∇_μ by p_μ, i.e.

$$\partial_\mu \psi(\boldsymbol{x}) \to p_\mu \tilde{\psi}(\boldsymbol{p}) .$$

This may introduce a non-local interaction and violate our assumption (i).

The assumptions made for the charges Q (lepton number, say) of the theory are the following:

(i) Exact conservation of Q, even at scales where the lattice cutoff is relevant. Charge conservation means that the energy and momentum eigenstates are also charge eigenstates.

(ii) Q is locally defined, i.e.

$$Q = \sum_{\boldsymbol{x}} j^0(\boldsymbol{x}) .$$

The charge density $j^0(\boldsymbol{x})$ is a function of the field variables $\psi(y)$ related to y within a bounded distance from x.

(iii) Q is quantized. This is, for instance, the case if it generates an abelian closed subgroup of a compact group.

(iv) We also assume that Q is bilinear in the fermion field $\psi(x)$.

(d) It appears to be one of the important features of weak interactions and of the standard Weinberg–Salam model in particular, that right- and left-handed particles *do not* have the same hypercharge. In fact, in this way, parity and charge conjugation are broken in weak interaction processes. In the Weinberg–Salam model it is the different quantum number for right- and left-handed particles that prohibits masses for fermions modulo the Higgs mechanism. So if nature were indeed built on a lattice it should be possible to realize Weyl fermions with different quantum numbers for left- and right-handed ones. As we shall show, this is, however, just what cannot be done.

Thus we are faced with the following dilemma: we must give up either the idea that nature is based on a fundamental lattice cutoff, or some of our mild (helping) assumptions, or weak interaction phenomenology, i.e. the usual understanding of parity violation. The last possibility seems unacceptable indeed since the theory of weak interactions is well established on this point. As concerns the second possibility, we may be able to generalize our no-go theorem to an *amorphous lattice*. We hope to eliminate the assumption of locality of charge density and replace it by the assumption that the gauge field is coupled via the charge. Such generalizations will be presented in one of our forthcoming papers.

(e) In discussing the quantum numbers of the individual Weyl particles we mean a kind of chiral charge for the composite Dirac particles. The chiral charge* is, of course, defined as the difference in the number of right- and left-handed Weyl fermions present in the universe and has nothing to do with the number of species of Weyl fermions. Since a particle and its antiparticle have opposite quantum numbers, we should not consider the sector of opposite charge. Doing so would count the antiparticles of the particle once again. So we use the notation that we count only left-handed particles, letting the right-handed ones be represented by their antiparticles. Thus, our no-go theorem says that there are equal numbers of species of left-handed particles with one set of quantum numbers and with the opposite set.

If we have a chiral charge Q_{chiral}, we can consider the dispersion relation for those fermions with, say, $Q_{\text{chiral}} = -1$. This would correspond to left-handed particles and antiparticles of the right-handed particles of $Q_{\text{chiral}} = 1$. Since this dispersion relation contains only left-handed particles, this is against our no-go theorem. We can (somewhat imprecisely but still correctly) restate our no-go theorem: there are no *quantized conserved* chiral charges on a lattice. So we *cannot introduce chirally coupled (weak) gauge fields.*

(f) The ingredients of our proof are mainly topological** in character. In fact we have two topological arguments:

(i) Intuitive topological formulation. This proof will be relegated to the subsequent paper [12].

(ii) Algebraic topology. The argument considers mappings from closed surfaces homeopmorphic to the surface S_2 of a sphere imbedded in the Brillouin zone into the space CP^{N-1}. The space CP^{N-1} consists of rays of states superposed from the N fundamental fermion components. These mappings are classified into homotopy classes making up the group $\pi_2(CP^{N-1})$, which is isomorphic to the additive group of integers Z. We shall show that the classes corresponding to mappings from small S_2 spheres surrounding each degeneracy point in momentum space, which represents one species of Weyl particle, e.g. ν_e, ν_μ or ν_τ etc. are ± 1, depending on the handedness of the neutrinos in question. Then it is shown that the sum of the classes must be zero, i.e. the unit element of $\pi_2(CP^{N-1})$. Our no-go theorem follows from this sum rule.

* In refs. [4–6] chiral charge on a lattice is considered.

** A good textbook on topology is ref. [7] and for an excellent review of homotopy theory see ref. [8].

Both arguments (i) and (ii) are concerned with the topology in momentum space. It is crucial that *the space of momenta i.e. the Brillouin zone, makes up a torus* $S_1 \times S_1 \times S_1$.

This periodicity is the important way in which the lattice comes into the proof.

(g) Sect. 2 is devoted to defining the generic case of the hamiltonian. In sect. 3 we show how the Weyl particle comes out from the degeneracy point of the dispersion law in the continuum limit. Then, in sect. 4, we depict a strategy for the proof of the no-go theorem. Sects. 5 to 10 are devoted to giving a proof according to this strategy. Finally, in sect. 11, we draw the conclusions of this paper.

2. Generic case

Let us consider dispersion relations which look like e.g. fig. 1 [eq. (1.1)] given by the eigenvalue equation

$$H(\boldsymbol{p})\tilde{\psi}(\boldsymbol{p}) = \omega_i(\boldsymbol{p})\tilde{\psi}(\boldsymbol{p}), \qquad (i = 1, \ldots, N). \tag{2.1}$$

We are interested in the types of dispersion relations that may arise from a generic set of hamiltonians. In fact, we do not need all such properties for our no-go theorem to work, but only assume the following:

(i) For almost all values of $\boldsymbol{p}$ there are N non-degenerate $\omega_i(\boldsymbol{p})$ which are ordered:

$$\omega_i(\boldsymbol{p}) > \omega_2(\boldsymbol{p}) > \cdots > \omega_N(\boldsymbol{p}).$$

(ii) But at several separate points $\boldsymbol{p}_{\text{deg}}$, only a couple of levels are degenerate, e.g.

$$\omega_i(\boldsymbol{p}_{\text{deg}}) = \omega_{i+1}(\boldsymbol{p}_{\text{deg}}).$$

Note that the two-level degeneracy is generic but the degeneracy of three or more is not. The reason is as follows. Consider the case of three-level degeneracy. The hamiltonian which is now a 3×3 matrix has the general form

$$H^{(3)}(\boldsymbol{p}) = \sum_{i=1}^{8} c_i^{(3)}(\boldsymbol{p})\lambda_i + d^{(3)}(\boldsymbol{p})1,$$

where λ_i are Gell-Mann's SU(3) matrices. For a three-level degeneracy the eight coefficients $c_i^{(3)}$ should satisfy the eight equations $c_i^{(3)}(\boldsymbol{p}) = 0$ at $\boldsymbol{p} = \boldsymbol{p}_{\text{deg}}$, but, in reality, the parameters to be determined are just three $\boldsymbol{p}$. Thus these conditions are overdeterminant. On the other hand, in the case of two-level degeneracy we may have the 2×2 matrix hamiltonian

$$H^{(2)}(\boldsymbol{p}) = \sum_{i=1}^{3} \sigma^i c_i(\boldsymbol{p}) + d(\boldsymbol{p})1.$$

The three parameters $c_i(\boldsymbol{p})$ should satisfy $c_i(\boldsymbol{p}) = 0$. These are just three equations for three unknowns $\boldsymbol{p}$, fixing the degenerate momentum $\boldsymbol{p}_{\text{deg}}$.

The above two properties (i) and (ii) can find phenomenological support, provided that the lattice theory is to be interpreted with a "renormalization" (or, rather, a redefinition) of the momenta, as will be explained in sect. 3.

Actually we do not fully need assumptions (i) and (ii); but to be able to formulate a theorem about Weyl fermions, we must assume that the low energy physics is described by particles with the same types of degeneracy as the Weyl equation has. That is, we need to assume that (by definition so to speak) each Weyl fermion represents two levels, being degenerate at one point but separate in a neighbourhood (in momentum space).

Then we can complete the proof given below by slightly modifying the hamiltonian into a generic one. It should not be too difficult to see that such a slight modification of the hamiltonian, and thus of the dispersion law, can be made without altering the number and handedness of the species of Weyl particles. So non-generic cases may be brought back to generic ones.

If we want a stable vacuum that is also in agreement with experiment, we should put $\omega_i(\boldsymbol{p}_{\text{deg}})=0$. (However, there is a possibility of having an unstable vacuum and leaving $\omega_i(\boldsymbol{p}_{\text{deg}})\neq 0$.) The following condition is required from phenomenology for getting a neutrino-like relativistically invariant dispersion relation in the continuum limit $p_0^2=\boldsymbol{p}^2$:

(iii) Zero energies in dispersion relations $\omega_i(\boldsymbol{p})=0$ are always achieved at degeneracy points $\boldsymbol{p}_{\text{deg}}$. It should be mentioned that this is not a generic principle.

3. Weyl particle in the continuum limit

We may still have to consider N-component ψ's but for the behaviour near the degeneracy point it is the two linear combinations of field ψ_i and ψ_{i+1} that are of interest. We can therefore restrict ourselves to study a two-component case here. The eigenvalue equation near the degeneracy point $\boldsymbol{p}_{\text{deg}}$ is given by

$$\begin{aligned} H^{(2)}(\boldsymbol{p})u^{(i)}(\boldsymbol{p}) &= \omega_i(\boldsymbol{p})u^{(i)}(\boldsymbol{p})\,, \\ H^{(2)}(\boldsymbol{p})u^{(i+1)}(\boldsymbol{p}) &= \omega_{i+1}(\boldsymbol{p})u^{(i+1)}(\boldsymbol{p})\,, \end{aligned} \tag{3.1}$$

with two component $u^{(i)}$ and $u^{(i+1)}$. The 2×2 matrix $H^{(2)}(\boldsymbol{p})$ can be expanded in a Taylor series around $\boldsymbol{p}_{\text{deg}}$,

$$H^{(2)}(\boldsymbol{p})=\omega_{\text{deg}}(\boldsymbol{p}_{\text{deg}})+(\boldsymbol{p}-\boldsymbol{p}_{\text{deg}})_k\sigma^\alpha V_\alpha^k+(\boldsymbol{p}-\boldsymbol{p}_{\text{deg}})\boldsymbol{a}+\mathrm{O}((\boldsymbol{p}-\boldsymbol{p}_{\text{deg}})^2)\,.$$

Here the constants $\boldsymbol{a}$ and V_α^k depend on the degeneracy point. Thus the eigenvalue equation (3.1) becomes in the lowest order of the expansion

$$(\boldsymbol{p}-\boldsymbol{p}_{\text{deg}})_k\sigma^\alpha V_\alpha^k u(\boldsymbol{p})=\{\omega(\boldsymbol{p})-\omega_{\text{deg}}(\boldsymbol{p}_{\text{deg}})-(\boldsymbol{p}-\boldsymbol{p}_{\text{deg}})\boldsymbol{a}\}u(\boldsymbol{p})\,. \tag{3.2}$$

We may define a new coordinate system the "practical momentum" $(\omega_{\text{pr}}, \boldsymbol{p}_{\text{pr}})$ by

$$\begin{aligned} \omega_{\text{pr}} &= \omega(\boldsymbol{p})-\omega_{\text{deg}}(\boldsymbol{p}_{\text{deg}})\,, \\ \boldsymbol{p}_{\text{pr}} &= \boldsymbol{p}-\boldsymbol{p}_{\text{deg}}\,. \end{aligned} \tag{3.3}$$

We may call the original $\boldsymbol{p}$ the "fundamental momentum". On the introduction of a new coordinate $(P_0, \boldsymbol{P})$ by

$$\begin{aligned} P_0 &= \omega - \omega_{\rm pr} - \boldsymbol{p}_{\rm pr}\boldsymbol{a}\,, \\ P_\alpha &= p_{{\rm pr}k} V^k_\alpha\,, \end{aligned} \tag{3.4}$$

the equation of motion (3.1) becomes

$$\boldsymbol{\sigma P} u(\boldsymbol{p}) = P_0 u(\boldsymbol{p})\,, \tag{3.5}$$

for small P_0 and $\boldsymbol{P}$. Then we get the usual relativistic neutrino-like dispersion relation $P_0^2 = \boldsymbol{P}^2$ in the long-wavelength limit. It should be stressed that each degeneracy point represents one species (type) of neutrino.

Now we have made a "renormalization" of momenta by eq. (3.3) to get eq. (3.5). This means that the momentum concept $\boldsymbol{p}_{\rm pr}$ used by physicists, who may not care for the lattice theory, deviates by an additive constant $\boldsymbol{p}_{\rm deg}$ from the one that is obtained by the Fourier transform of x on the lattice. In fact, we should subtract the momentum $\boldsymbol{p}_{\rm deg}$ in (3.3). If $\omega_i(\boldsymbol{p}_{\rm deg}) \neq 0$ (this is the generic situation) we should also "renormalize" the energy of a Weyl particle by the prescription in eq. (3.3).

Our philosophy corresponds to the assumption that the Dirac particles in nature are built up from Weyl components with different momentum. So the Higgs-like mechanism, to make mass terms, would have to break the conservation of the fundamental momentum p and conserve $\boldsymbol{p} - \boldsymbol{p}_{\rm deg}$.

The spin of the field $u(\boldsymbol{p})$ is to be determined. Eq. (3.5) for $u(\boldsymbol{p})$ is invariant under rotations generated by $\boldsymbol{J} = \boldsymbol{r} \times \boldsymbol{P} + \frac{1}{2}\boldsymbol{\sigma}$ with the definition $\boldsymbol{r} = i\partial/\partial\boldsymbol{P}$ since $[\boldsymbol{J}, \boldsymbol{\sigma P}] = 0$. The spin of u is $\frac{1}{2}\boldsymbol{\sigma}$. The state with $P_0 > 0$ (upper cone) has $\boldsymbol{\sigma p} > 0$ which means $+1$ helicity in the $(P_0, \boldsymbol{P})$ coordinate system.

To end this section we investigate the relations of the coordinate systems between $(P_0, \boldsymbol{P})$ and $(\omega_{\rm pr}, \boldsymbol{p}_{\rm pr})$. Let us take a convention that coordinate $(\omega_{\rm pr}, \boldsymbol{p}_{\rm pr})$ is right-handed. In the p-coordinate system we take a basis vector $(\boldsymbol{e}_1, \boldsymbol{e}_2, \boldsymbol{e}_3)$ with $\boldsymbol{e}_1$ corresponding to $(P_x, P_y, P_z) = (1, 0, 0)$, $\boldsymbol{e}_2$ to $(P_x, P_y, P_z) = (0, 1, 0)$, and so on. Thus

$$\boldsymbol{e}_1 \to \boldsymbol{p}_{\rm pr} = V^{-1}\begin{pmatrix}1\\0\\0\end{pmatrix},$$

$$\boldsymbol{e}_2 \to \boldsymbol{p}_{\rm pr} = V^{-1}\begin{pmatrix}0\\1\\0\end{pmatrix},$$

and so on. These are basis vectors for the $\boldsymbol{p}_{\rm pr}$ coordinate system. Now we compute the handedness of the $\boldsymbol{p}$ coordinate system, that is, given by

$$\begin{aligned} \boldsymbol{e}_3 \cdot (\boldsymbol{e}_1 \times \boldsymbol{e}_2) &= \det\left((\boldsymbol{e}_1)_{p_{\rm pr}}, (\boldsymbol{e}_2)_{p_{\rm pr}}, (\boldsymbol{e}_3)_{p_{\rm pr}}\right) \\ &= \det(V^{-1})\,. \end{aligned} \tag{3.7}$$

Here the external product $\times$ is defined in the usual way in p_{pr} space and the subscript in $(e_1)_{p_{pr}}$ denotes the e_1 vector in p_{pr} space, (3.6). The sign of det V in (3.7) indicates the handedness of the P coordinate system: if det $V>0$ (<0) it is right (left) handed.

4. Strategy of the proof

The eigenvalue equation

$$H(p)|\omega_i(p)\rangle = \omega_i(p)|\omega_i(p)\rangle\,, \qquad (i=1,\ldots,N)\,, \tag{4.1}$$

determines a ray in an N-dimensional complex space for each value of the momentum p except for the cases $\omega_i(p)=\omega_{i+1}(p)$ or $\omega_i(p)=\omega_{i-1}(p)$. Thus p determines a point in the complex projective space CP^{N-1}. In the generic case the ith and $(i+1)$th levels are degenerate at points in the Brillouin zone space. The crucial property of $|\omega_i(p)\rangle$ for our proof is *periodicity*. The idea of the proof is the following:

(i) Draw an infinitesimal S_2 sphere around each degeneracy point (sect. 5) and consider maps determined by $|\omega_i(p)\rangle$ from the S_2 spheres into CP^{N-1} (sect. 6). Calculate $\pi_2(CP^{N-1})$ and show that these maps correspond to the elements ± 1 of Z which depend on the handedness of the particles corresponding to the degeneracy points in question (sect. 7). (We use the notation that S_n stands for a topological space homeomorphic to the sphere in an $n+1$ dimensional euclidean space.)

(ii) Consider the map determined by $|\omega_i(p)\rangle$ from the whole surface of the Brillouin zone box homeomorphic to an S_2 sphere into CP^{N-1} (sect. 8) and show that it belongs to the unit element in $\pi_2(CP^{N-1})$ (sect. 9).

(iii) Prove that the class of map of (ii) is the sum of the classes of eigenray map from the small S_2 spheres mentioned under point (i).

(iv) The consistency of the two expressions for the element $\pi_2(CP^{N-1})$ requires that

$$N_r(i,i+1)-N_r(i-1,i)=N_\ell(i,i+1)-N_\ell(i-1,i) \tag{4.2}$$

(sect. 10), where, e.g. $N_r(i,j)$ denotes the number of right-handed degeneracy points between levels i and j. If i denotes the lowest energy level above the Fermi (or Dirac) sea and $i+1$ the highest level in the sea, the number of right-handed species of Weyl particles thus equals $N_r(i,i+1)$.

For points (i) and (ii) we will compute explicitly the integers in Z isomorphic to $\pi_2(CP^{N-1})$ by making use of the following explicit isomorphism of $\pi_2(CP^{N-1})$ to Z [7]:

$$\pi_2(CP^{N-1}) \xleftarrow{j_*} \pi_2(S_{2N-1},S_1) \xrightarrow{\partial} \pi_1(S_1) \xrightarrow{\Delta} Z\,. \tag{4.3}$$

The homeomorphisms j_*, ∂ and Δ are naturally defined and we shall give the definitions below (sect. 5). In fact, they turn out to be isomorphisms.

5. $\pi_2(\mathrm{CP}^{N-1})$ for the degeneracy point

Consider the infinitesimal S_2 sphere around a degeneracy point in the Brillouin zone space. The mapping

$$f: S_2 \to \mathrm{CP}^{N-1} \tag{5.1}$$

determines the homotopy class $[f]$, an element in $\pi_2(\mathrm{CP}^{N-1})$. The function f is determined by the given H and the ith level. That is, for

$$\begin{gathered} p \in S_2 , \\ S_2 = \{p \,||p| = \varepsilon \text{ (infinitesimal)}\} , \end{gathered} \tag{5.2}$$

f is given by

$$f(p) = \{Z|\omega_i(p)\rangle | Z \in \mathrm{C}\backslash\{0\}\} . \tag{5.3}$$

Now there is an isomorphism between $\pi_2(\mathrm{CP}^{N-1})$ and the relative homotopy group $\pi_2(S_{2N-1}, S_1)$ [7]. The elements of the latter are homotopy classes of functions of the type

$$g: E_2, S_1 \to S_{2N-1}, S_1 , \tag{5.4}$$

where E_2 is the disk

$$E_2 = \{x \in \mathrm{R}^2 | x^2 \leq 1\} ,$$

and the circle S_1 appearing on the left-hand side of the arrow in (5.4) is the circumference of E_2, i.e.

$$S_1 = \{x \in \mathrm{R}^2 | x^2 = 1\} .$$

S_{2N-1} is the space of complex N-tuples

$$S_{2N-1} = \{u \in \mathrm{C}^N \,||u|^2 = 1\} ,$$

and S_1 appearing on the right-hand side of (5.4) is a subset of proportional vectors

$$S_1 = \{e^{i\delta} u_0 \in \mathrm{C}^N \,| \delta \in \mathrm{R}\} ,$$

where u_0 is a fixed unit norm in the complex N-tuple, i.e.

$$u_0 \in S_{2N-1} .$$

The isomorphism between $\pi_2(\mathrm{CP}^{N-1})$ and $\pi_2(S_{2N-1}, S_1)$ is induced by the relation

$$f = \pi \circ g . \tag{5.5}$$

Here π is the projection

$$\pi: S_{2N-1} \to \mathrm{CP}^{N-1} ,$$

i.e. π assigns ray in CP^{N-1} to which that element belongs to an element in S_{2N-1}. The mapping (5.5) from g into f induces a mapping

$$j_*: \pi_2(S_{2N-1}, S_1) \to \pi_2(CP^{N-1}), \tag{5.6}$$

which is defined by

$$j_*([g]) = [f] = [\pi \circ g]. \tag{5.7}$$

In (5.5) it is understood that the infinitesimal S_2 sphere is identified with the manifold which is obtained from E_2 in such a way that all the points on the boundary $S_1 = \partial E_2$ are identified into one point, e.g. the south pole S. Thus we have a one to one correspondence between $S_2/\{S\}$ and the interior $\mathring{E}_2$ of E_2

$$\mathring{E}_2 = \{x \in R \,|\, |x|^2 < 1\}.$$

S is the point with

$$\frac{1}{\varepsilon} P_z = -1, \qquad P_x = P_y = 0. \tag{5.8}$$

The point on the S_2 sphere is expressed in spherical coordinates $(\varepsilon, \theta, \phi)$, with the restriction $0 \leq \theta \leq \pi$, $0 \leq \theta \leq 2\pi$, by

$$\begin{aligned} P_x &= \varepsilon \sin\theta \cos\phi, \\ P_y &= \varepsilon \sin\theta \sin\phi, \\ P_z &= \varepsilon \cos\theta. \end{aligned} \tag{5.9}$$

This is unique except for the north pole N ($\theta = 0$) and the south pole S ($\theta = \pi$). In the coordinate system (θ, ϕ) the cartesian coordinate (x_1, x_2) or E_2 are expressed by

$$x_1 = \frac{\theta}{\pi} \cos\phi, \qquad x_2 = \frac{\theta}{\pi} \sin\phi. \tag{5.10}$$

This representation is unique except for the centre $\theta = 0$.

The above stated identification is

$$\text{N-pole of } S_2 \to \text{centre of } E_2,$$

$$\text{S-pole of } S_2 \to \text{whole boundary of } E_2.$$

Note that the north-pole and the centre of E_2 have the same non-uniqueness in coordinates. Thus, except for the correspondence of the south pole to the whole boundary $S_1 = \partial E_2$, there is a homeomorphism between S_2 and E_2.

6. $\pi_2(S_{2N-1}, S_1)$

We now construct an explicit form of g corresponding to f. Let us choose our basis in the space C^N so that $H(\boldsymbol{p})$ becomes diagonal at $\boldsymbol{p}=\boldsymbol{p}_{\rm deg}$ and expand $H(\boldsymbol{p})$ around $\boldsymbol{p}=\boldsymbol{p}_{\rm deg}$ in a Taylor series:

$$
H(\boldsymbol{p}) = H(\boldsymbol{p}_{\rm deg}) + (\boldsymbol{p}-\boldsymbol{p}_{\rm deg})\mathbf{V}
\begin{array}{c}
\begin{array}{cccccc} & & & i & i+1 & \end{array}\\
\begin{pmatrix}
0 & & & & & \\
 & \ddots & & & & 0 \\
 & & 0 & & & \\
 & & & \boldsymbol{\sigma} & & \\
 & & & & \ddots & \\
 & & & & & 0 \\
 & 0 & & & & \\
 & & & & & 0
\end{pmatrix}
\begin{array}{c} \\ \\ \\ i \\ i+1 \\ \\ \\ \\ \end{array}
\end{array}
$$

$$
+(\boldsymbol{p}-\boldsymbol{p}_{\rm deg})\begin{pmatrix} b_i & & 0 \\ & \ddots & \\ 0 & & b_N \end{pmatrix}
+(\boldsymbol{p}-\boldsymbol{p}_{\rm deg})
\begin{array}{c}
\begin{array}{cccccc} & & & i & i+1 & \end{array}\\
\begin{pmatrix}
0 & & & & & \\
 & \ddots & & & * & \\
 & & 0 & & & \\
 & & & 0 & 0 & \\
 & & & 0 & 0 & \\
 & & & & 0 & \\
 & * & & & & \ddots \\
 & & & & & 0
\end{pmatrix}
\begin{array}{c} \\ \\ \\ i \\ i+1 \\ \\ \\ \\ \end{array}
\end{array}
$$

$$
+\mathrm{O}((\boldsymbol{p}-\boldsymbol{p}_{\rm deg})^2)\,.
$$

Note that the second term is important to get the eigenvector in lowest order in perturbation theory. We thus may have

$$
|\omega_i(\boldsymbol{p})\rangle = \begin{pmatrix} 0 \\ \vdots \\ 0 \\ u \\ 0 \\ \vdots \\ 0 \end{pmatrix}
\begin{array}{c} \\ \\ \\ i \\ i+1 \\ \\ \\ \end{array}, \tag{6.2}
$$

and

$$
u = \begin{pmatrix} u_1 \\ u_2 \end{pmatrix}.
$$

Here

$$
\boldsymbol{\sigma P} u = |\boldsymbol{p}| u\,. \tag{6.3}
$$

That is, we have the positive eigenvalue for the upper level i.

We define the function g by

$$g = |\omega_i(\boldsymbol{p})\rangle \tag{6.4}$$

where $\boldsymbol{p}$ is a point on an infinitesimal sphere S_2 around $\boldsymbol{P}_{deg}$. We choose to put the following restrictions on u:

(i) Normalization $|u| = 1$. This is necessary for $|\omega_i(\boldsymbol{p})\rangle$ to lie on the sphere S_{2N-1}.

(ii) Phase convention $\operatorname{Arg} u_1 = 0$, i.e.

$$\operatorname{Re} u_1 \geq 0\,, \qquad \operatorname{Im} u_1 = 0\,.$$

Under restrictions (i) and (ii), eq. (6.3) has a unique solution. A little calculation leads us to the solution

$$u = \begin{pmatrix} \dfrac{p_x - ip_y}{\sqrt{2|\boldsymbol{p}|(|\boldsymbol{p}| - p_z)}} \\ \sqrt{\tfrac{1}{2}(1 - p_z/|\boldsymbol{p}|)} \end{pmatrix},$$

or in spherical coordinates

$$u = \begin{pmatrix} e^{-i\phi} \sin \tfrac{1}{2}\theta \\ \cos \tfrac{1}{2}\theta \end{pmatrix}. \tag{6.6}$$

It should be noticed that this u is not unique at $\theta = \pi$ (south pole). This phase ambiguity is due to our phase conventions which do not fix the phase for $u_1 = 0$. But since the S-pole corresponds to $S_1 = \partial E_2$ there is no problem. We can make g continuous and thus choose g on S_1 so that its value is the limiting one from an inside point of E_2.

We thus have constructed a representative g for the class $j_*([f])$ of $\pi_2(S_{2N-1}, S_1)$ in the form

$$g(\theta, \phi) = \begin{pmatrix} 0 \\ \vdots \\ 0 \\ e^{-i\phi} \sin \tfrac{1}{2}\theta \\ \cos \tfrac{1}{2}\theta \\ 0 \\ \vdots \\ 0 \end{pmatrix} \begin{matrix} \\ \\ \\ i \\ i+1 \\ \\ \\ \\ \end{matrix}\,. \tag{6.7}$$

7. Helicity for the degeneracy point

We are now equipped to find the image of $[g]$ via the boundary mapping ∂ in the explicit isomorphism (4.3) [7]. Since the map ∂ is induced by the restriction to the boundary $S_1 = \partial E_2$, i.e. $\theta = \pi$, the representative of $\partial([g]) \in \pi_1(S_1)$ is simply given by

$$g(\phi)|_{S_1} = \begin{pmatrix} 0 \\ \vdots \\ 0 \\ e^{-i\phi} \\ 0 \\ \vdots \\ 0 \end{pmatrix} \begin{matrix} \\ \\ \\ i \\ i+1 \\ \\ \\ \end{matrix} \tag{7.1}$$

on putting $\theta = \pi$ in g of eq. (6.7). Here

$$g(\phi)|_{S_1} \in S_1 \subset S_{2N-1}\,.$$

The $S_1 \subset S_{2N-1}$ consists of all phase possibilities for $|\omega_i(\boldsymbol{p})\rangle$ at the south-pole. Obviously the map $g|_{S_1}$ is a homeomorphism from $S_1 \subset E_2$ on to $S_1 \subset S_{2N-1}$. The integer assigned to the class $[g|_{S_1}] \in \pi_1(S_1)$ by the isomorphism ∂ mentioned in eq. (4.3) is by definition nothing but the winding number. The form (7.1) indicates that the winding number is either $+1$ or -1, i.e.

$$\Delta \circ \partial \circ j_*^{-1}([f]) = \pm 1\,.$$

When $\det V_i^\alpha > 0$ at a degeneracy point, the coordinate system $\boldsymbol{P}$ is right-handed. Then this integer is -1. This means that there are left-handed Weyl fermions described by the states in the ith (upper) level. If $\det V_i^\alpha < 0$ the coordinate system $\boldsymbol{P}$ is right-handed. Thus, one should correct the form of $g(\sigma, \phi)$ by changing P_x to $-P_x$ and we obtain

$$g(\phi)|_{S_1} = \begin{pmatrix} 0 \\ \vdots \\ 0 \\ e^{i\phi} \\ 0 \\ \vdots \\ 0 \end{pmatrix} \begin{matrix} \\ \\ \\ i \\ i+1 \\ \\ \\ \end{matrix} \,.$$

The integer assigned to this is $+1$.

So far we have considered the case of the positive eigenvalue of the upper level i given by eq. (6.3). Replacing this by

$$\boldsymbol{\sigma P} u = -|\boldsymbol{p}| u\,,$$

for the lower level $i+1$ has the same effect as switching $\boldsymbol{P}$ to $-\boldsymbol{P}$ in eq. (6.3). This is also equivalent to changing the coordinate system from right- to left-handed, and has the same effect as the sign change of $\det(V_i^\alpha)$. Thus, we conclude the following list of $\pi_2(CP^{N-1})$ integer elements. These correspond to the infinitesimal S_2 spheres

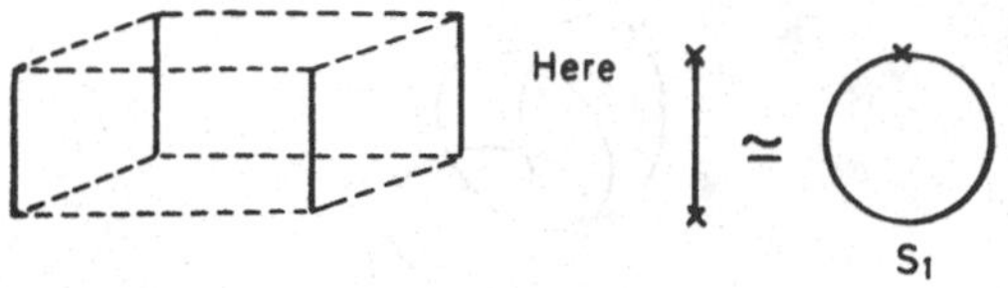

Fig. 2. The box on this figure illustrates the Brillouin zone. Four parallel edges are drawn heavily. They are mapped into a single curve in CP^{N-1} and the curve is indeed a closed one.

around the degeneracy points. These spheres are oriented by a right-handed coordinate system of $\boldsymbol{P}$ for the eigenray of the ith level.

$$\text{degeneracy of level } i \text{ with } i+1 \quad \left.\begin{array}{l}\text{positive helicity of } i\\ \text{and negative one of } i+1\end{array}\right\} -1 \text{ element} \quad \left.\begin{array}{l}\text{negative one of } i\\ \text{and positive one of } i+1\end{array}\right\} +1 . \tag{7.2}$$

8. $\pi_2(CP^{N-1})$ for Brillouin zone surface

We shall show that the map $\hat{f}_{BS}$ from the surface of the Brillouin zone S_2 (box surface) (fig. 2) into CP^{N-1}, which is determined from the eigenrays $|\omega_i(\boldsymbol{p})\rangle$, belongs to the homotopy class corresponding to zero by the isomorphism $\Delta \circ \partial \circ j_*^{-1}$ of $\pi_2(CP^{N-1})$ with Z. The crucial property of this map belonging to a class corresponding to zero is the *periodicity* of $|\omega_i(\boldsymbol{p})\rangle$. The ith eigenrays on each of the six faces in fig. 2 are identical to those on the opposite faces. On the three sets of four parallel edges $|\omega_i(\boldsymbol{p})\rangle$ are the same. Also at the eight corners $|\omega_i(\boldsymbol{p})\rangle$ are the same.

The strategy of the proof is the following:

(i) A continuous deformation into $\hat{f}_d$ of the map $\hat{f}_{BS}$ allows us to replace it by one that maps all the 12 edges and the 8 corners into a single point.

(ii) Using this deformed map $\hat{f}_d$, it is shown to be a sum of six terms (elements) which split up into 3 pairs.

(iii) The sum of the terms in each pair is shown to be zero. Thus the sum of all 6 terms (elements) is zero.

In order to perform the deformation mentioned in (i), let us first consider how the Brillouin zone surface S_2 is imbedded in CP^{N-1} by the map f_{BS} determined from $|\omega_i(\boldsymbol{p})\rangle$. Periodicity implies that each of the six faces of the Brillouin zone surface is mapped by $\hat{f}_{BS}$ into the same piece of surface in CP^{N-1} as the opposite faces. Further, the three sets of four parallel edges of the Brillouin zone cube are mapped by $\hat{f}_{BS}$ into only one closed curve each, depicted in fig. 3. The resulting three curves in CP^{N-1} are closed since periodicity guarantees that all 8 corners of the cube are mapped by $\hat{f}_{BS}$

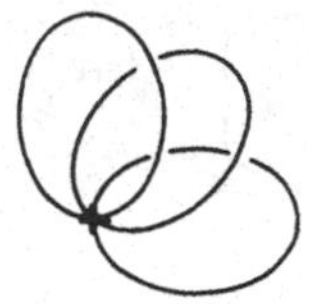

Fig. 3. The images of the three sets of four parallel edges.

into a single point in CP^{N-1}. It would be natural to use this point as a base point, i.e. we restrict our attention to maps mapping the south-pole of S_2 into this point.

Since CP^{N-1} is simply connected, we are guaranteed the existence of a way of deforming each of the three closed curves into the single point which is the image of the corners by the map $\hat{f}_{BS}$. The reader must convince himself that this deformation of $\hat{f}_{BS}$ restricted to the edges can be extended to a deformation of $\hat{f}_{BS}$ itself. By imagining the image of $\hat{f}_{BS}$ as a rubber sheet in CP^{N-1}, this is intuitively possible without spoiling the periodicity inherited by $\hat{f}_{BS}$. Thus, we have seen that $\hat{f}_{BS}$ is homotopic to a deformed mapping $\hat{f}_d$ which maps all the edges into a single point (the base point).

9. The unit element of $\pi_2(CP^{N-1})$

The next step of the proof, (ii), is trivial if we remember the definition of the *group composition of* π_2. Let us split the surface S_2 of the Brillouin zone cube into two parts as is shown in fig. 4a: the first one is just a single face and second one consists of all five other faces. Restriction of the previously constructed deformed map $\hat{f}_d$ to these two parts defines the two maps $\hat{f}_a$ and $\hat{f}_5$. Since $\pi_2(CP^{N-1})$ *is abelian* we denote the group composition law additively and thus we have

$$[\hat{f}_d]=[\hat{f}_a]+[\hat{f}_5]. \tag{9.1}$$

The first map $\hat{f}_a$, corresponding to the first part of the S_2 cube, is a mapping into CP^{N-1} from a bag made out of one face by identifying all pairs of edges as is illustrated in fig. 5, i.e.

$$\hat{f}_a: S_2 \text{ surface of the bag in fig. 5} \rightarrow CP^{N-1}.$$

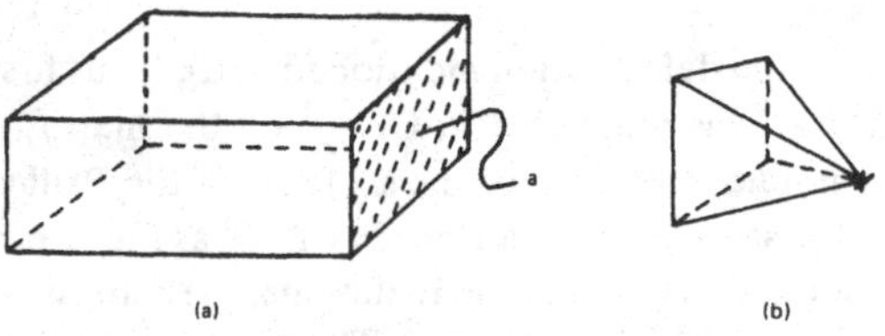

Fig. 4. The split of the surface of the Brillouin zone into a single face (a) and all five other faces illustrated in (b). Here the cross denotes the point into which surface a is contracted.

Fig. 5. The single face is homeomorphic to a bag. The cross in the bag denotes the point with which all pairs of edges are identified.

The second one is

$$\hat{f}_5 : S_2 \text{ surface in fig. 4b} \to CP^{N-1} .$$

Next we split $\hat{f}_5$ into $\hat{f}_b$ and $\hat{f}_4$,

$$[\hat{f}_5] = [\hat{f}_b] + [\hat{f}_4] ,$$

and so on. We here again split the second part of fig. 4b into third and fourth parts depicted in fig. 6. Finally, we have a decomposition

$$[\hat{f}_d] = [\hat{f}_a] + [\hat{f}_{\bar{a}}] + [\hat{f}_b] + [\hat{f}_{\bar{b}}] + [\hat{f}_c] + [\hat{f}_{\bar{c}}] ,$$

where each of the six terms corresponds to a face. We denote by $\hat{f}_a$ and $\hat{f}_{\bar{a}}$ the mappings corresponding to each face a and its antiface ā, as shown in fig. 7.

Step (iii). On the surfaces a and ā the orientations are opposite due to periodicity. Therefore $\hat{f}_a$ and $\hat{f}_{\bar{a}}$ are related to each other by a reflection, i.e.

$$\hat{f}_{\bar{a}} = \hat{f}_a \cdot \xi ,$$

where

$$\xi : S_2 \to S_2 .$$

S_2 is a face and ξ is a reflection in a big circle on S_2:

$$[\hat{f}_{\bar{a}}] = -[\hat{f}_a] . \tag{9.3}$$

Similar relations hold for b and b̄, c and c̄. So from eq. (9.2)

$$[\hat{f}_d] = 0 .$$

Since $[\hat{f}_d]$ is, of course, the same class as $[\hat{f}_{BS}]$, we conclude

$$[\hat{f}_{BS}] = 0 . \tag{9.4}$$

Fig. 6. The split of the surface in fig. 4b into a single face b and all four other faces.

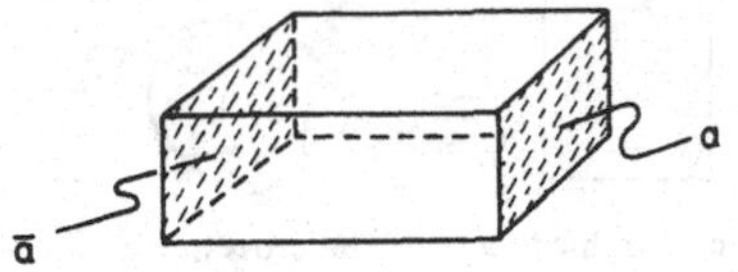

Fig. 7. Illustration of opposite faces.

10. Result

The surface of the Brillouin zone S_2 sphere considered in sect. 4 includes the infinitesimal S_2's around the degeneracy points. We deform all infinitesimal S_2 spheres to have a common base point. By the group composition law of π_2 we decompose

$$[\hat{f}_{BS}] = \sum_j [f_j] \tag{10.1}$$

where f_j denotes the mapping f_j: infinitesimal $S_2 \to CP^{N-1}$ and where j runs over all infinitesimal S_2 spheres. Since we have proved that the left-hand side of (10.1) is zero [eq. (9.4)], eq. (10.1) gives

$$\sum_j [f_j] = 0 .$$

That is to say, from eq. (7.2) the sum of the terms +1 or −1 corresponding to the degeneracy points for the ith and $(i+1)$th or $(i-1)$th levels are zero. Thus we obtain eq. (4.2):

$$N_r(i, i+1) - N_r(i-1, i) = N_\ell(i, i+1) - N_\ell(i-1, i) . \tag{4.2}$$

Here, e.g., $N_r(i, i+1)$ is the number of degeneracy points of the levels i and $i+1$ for which the upper ith has positive helicity.

We want, in fact, to show that

$$N_r(i-1, i) = N_\ell(i-1, i) \tag{10.2}$$

by induction using eq. (4.2). Now for the highest level $i = 1$, we have trivially

$$N_r(0, 1) = N_\ell(0, 1) = 0 ,$$

because there is simply no level number $i = 0$. Assuming that the eq. (10.2) is true for $i-1$, we find from eq. (4.2) that

$$N_r(i, i+1) = N_\ell(i, i+1) . \tag{10.3}$$

Especially, this equation is true for the number of i for which level i is unfilled while level $i+1$ is filled.

Since in this case each degeneracy point between ith and $(i+1)$th levels represents one species of Weyl particle (neutrinos), eq. (10.3) states our no-go theorem: there appear equal numbers of species of left- and right-handed Weyl particles.

11. Discussions

(a) It is easy to generalize our theorem to the case of many conserved charges Q (lepton number) associated with the fermion fields. Here we *assumed* that these Q's are represented by bilinear form in ψ

$$Q = \sum_{x} j_0(x),$$

where

$$j_0(x) = \sum_{y} \bar{\psi}(x)\tilde{Q}(x-y)\psi(y).$$

For each combination of charges we have a separate class of fermion fields. There can be no transition from one such class to another, and to the approximation of a linear equation of motion there is no connection between these classes at all. Each class is therefore to be considered as a separate system of fermions. We can thus apply our no-go theorem to each class. This generalization is very worrisome. In fact, it threatens any hope of putting weak interactions on a lattice as we already discussed in sect. 1(d).

(b) In fact, were it not for the spontaneously broken gauge symmetries of the weak interaction, there would be no mass terms for any of the known fermions in the standard Weinberg–Salam model. It is precisely because of the different quantum numbers for different handed Weyl particles that mass terms are forbidden. That is, unless there are mass terms, no pair of handed Weyl particles can be combined into an ordinary Dirac four-component massive fermion. In discussing our no-go theorem we talked about the number of species of Weyl particles, i.e. neutrino-like particles. However, in nature many fermions—perhaps all, even neutrinos—may have masses. Massive fermions can be described as Weyl particles in the following two ways:

(i) one handed particle pairs up with the same handed one to become a Majorana fermion, or

(ii) one handed particle pairs up with the opposite handed one to become a Dirac particle.

These can precisely be done according to our no-go theorem.

In the case of (ii) there must be one right-handed and one left-handed Weyl particle and a mass term in the hamiltonian in order to cause a transition from one to the other. The alternative statement of our no-go theorem is that there are no quantized chiral charges due to the possible mass terms. For a low energy observer it is natural to redefine the momentum by the prescription (3.3). The subtraction constants ω_{deg} and $\boldsymbol{P}_{\text{deg}}$ in (3.3) may be different for different Weyl particles in one pair. The conservation of the "fundamental" momentum $\boldsymbol{p}$ (i.e. the momentum before subtraction of $\boldsymbol{P}_{\text{deg}}$) may exclude a mass term. We must say here that we do not know in detail how a necessary mass term comes about via the spontaneous breakdown of the "fundamental" momentum and this mass problem is an open question. Also, unless we know the mass term, the chiral charge is ill-defined. However, we do not know if the appropriate spontaneous breakdown of the "fundamental" momentum is unnatural, but it should at least be possible.

(c) It is of interest to note that our theorem *prevents any problem* with the Adler, Bell–Jackiw anomaly [10] in a *trivial manner*: there simply are no anomalies since they are cancelled between oppositely handed particles. But, of course, it should be remembered that in the correct theory of weak interactions anomaly-free conditions are *not* satisfied in such a trivial manner.

(d) There may be a possibility of evading our no-go theorem by abandoning the helping assumptions described in sect. 1(c) in the following ways:

(i) *Taking Q to be unquantized* may not be a proper way, since Q then does not go into a compact group generating a closed subgroup. One would thus have to claim that the guage group of the standard model is a low energy approximation only. It would be a serious complication contrary to the suggestion from phenomenology.

(ii) *Giving up locality of charge.* This will be the subject of our forthcoming paper [12] as was mentioned at the end of sect. 1(d). In that work we still need continuity of the fermion dispersion relation and thus cannot give up locality for the free fermion hamiltonian. If we did that we would allow the Drell, Weinstein and Yankielowicz model [4] as a counter example to our no-go theorem. Tom Banks has pointed out to Foerster how our theorem could be circumvented if the locality of fermion dispersion relation is not required. Then one can let the *light velocity of one or more neutrinos go to infinity* so that the light-cone steepens and effectively disappears.

(e) We have throughout this article discussed the Kogut–Susskind lattice. Can our results be taken over to the Wilson lattice [11]? Yes, *if* we understand it as a theory with a spatial lattice and discrete *imaginary* time. In fact, we can construct an operator e^{-H} which gives the development by one lattice unit along the imaginary time. Here H can be considered as the hamiltonian. In our dicussion of the Kogut–Susskind lattice we only used the existence of such a hamiltonian.

Had we, however, wanted a lattice theory with a discrete real time we would instead have constructed an operator e^{-iH}. Thus H is only unique modulo 2π, and so energy would be defined modulos 2π. The concept that one energy is lower than another then loses its meaning. This would make the question of how to fill a Dirac sea more delicate and may complicate the argument. However, this case of discrete real time seems not to be popular and we shall not go into it further.

(f) In concluding the paper we would like to mention the following point which will be investigated in our succeeding paper [12]:

In the present article we have considered the case where the fermion field $\psi(x)$ is *complex. A real field formulation for* $\psi(x)$ may open the possibility of describing a larger class of physical systems. In fact there exist lattice models with only one two-component fermion in the real field formulation, which cannot be described as a complex field. We shall illustrate this counter example to our no-go theorem. But, if we have a conserved non-zero charge, the charged fermions can be formulated in terms of complex fields.

It is a pleasure to thank very strongly D. Foerster, who participated in our early stage of this work, but was on vacation during a large part of the production time of the article. We are grateful for discussions between D. Foerster and T. Banks on the infinite velocity of Weyl particles. M. Peskin played an important role in stimulating this work and we benefitted from several helpful discussions with him. We thank K. Johnson for encouragement. For discussions on topology we are grateful to B. Duurhus and B. Felsager. We acknowledge helpful discussions with S. Chadha in connection with the generic degeneracy. We also thank P. Scharbach for useful discussions at a late stage. On of us (H.B.N.) acknowledges discussions with H. Hellsten (at an early stage) and J. Greensite (at a late stage). One of us (M.N.) wants to acknowledge the extremely kind hospitality he has received during his stay at the Niels Bohr Institute and discussions with all members of high energy theory group at Rutherford and Appleton Laboratories. We are grateful to P. Scharbach for his extremely hard work on correcting our terrible English.

References

[1] J. Kogut and L. Susskind, Phys. Rev. D11 (1975) 395

[2] L. Susskind, Phys. Rev. D16 (1977) 3031

[3] K.G. Wilson, Erice lecture notes (1975).

[3a] L.H. Karsten and J. Smit, Nucl. Phys. B183 (1981) 103

[4] S.D. Drell, M. Weinstein and S. Yankielowicz, Phys. Rev. D14 (1976) 487

[4a] T. Banks and A. Casher, Nucl. Phys. B169 (1980) 103

[5] V. Baluni and J.F. Willemsen, Phys. Rev. D13 (1976) 3342

[6] M. Peskin, Cornell University preprints CLNS-395, -396 (1976) unpublished

[7] N. Steenrod, The topology of fibre bundles (Princeton University Press)

[8] B. Felsager, Homotopy theory, *in* Lecture on topological invariants in the theory of classical fields. Niels Bohr Institute report (1977) unpublished

[9] H.B. Nielsen, Dual strings, Catastrophe theory programme, *in* Fundamentals of quark models, p. 528, Proc. 17th Scottish Universities Summer School in Physics 1976, ed. I.M. Barbour and A.T. Davies

[10] S. Adler, Phys. Rev. 177 (1969) 2426;
J.S. Bell and R. Jackiw, Nuovo Cim. 60A (1969) 47

[11] K.G. Wilson, Phys. Rev. D10 (1974) 2445

[12] H.B. Nielsen and M. Ninomiya, Absence of neutrinos on a lattice II, Niels Bohr Inst. preprint, NBI-HE-81-1 (1980)

Nuclear Physics B195 (1982) 541–542

ERRATA

H.B. Nielsen and M. Ninomiya, Absence of neutrinos on a lattice (I). Proof by homotopy theory, Nucl. Phys. B185 (1981) 20.

There are some small mistakes, which do not affect the conclusion.

Page 30
The 3rd line below eq. (5.7):
the south pole S → the north pole N.
The next line and 2 lines below:
S → N.
Eq. (5.8):
$-1 \to 1$.
2 lines below:
$0 \leq \theta \leq 2\pi \to 0 \leq \phi \leq 2\pi$.
Eq. (5.10):
$\theta \to \pi - \theta$.
Next line:
$0 \to \pi$.
2 lines below:
N and S should be permuted.
Next 2 lines
north and south should be permuted.

Page 32
6th, 7th lines from the top and 2 lines below eq. (6.6):
$u_1 \to u_2$ and Re $u_1 \geq 0 \to$ Re $u_2 > 0$.
Eqs. (6.6) and (6.7):
$\sin \frac{1}{2}\theta$ and $\cos \frac{1}{2}\theta$ should be permuted.
Next line:
$\theta = \pi$ (south pole) $\to \theta = 0$ (north pole).
2 lines below:
S-pole → N-pole.

Page 33
The 1st line and the 1st line below the eq. (7.1):
$\theta = \pi \to \theta = 0$.

542 *Errata*

2 lines below:
south → north.

7 lines below and the next line:
left and right should be permuted.

Next line:
$\sigma \rightarrow \theta$

Nuclear Physics B192 (1981) 100–124

EFFECTIVE LAGRANGIAN AND DYNAMICAL SYMMETRY BREAKING IN STRONGLY COUPLED LATTICE QCD

N. KAWAMOTO[1] and J. SMIT

Institute of Theoretical Physics, University of Amsterdam, Valckenierstraat 65, 1018 XE Amsterdam, The Netherlands

Received 11 June 1981

A method is proposed for computing effective lagrangians in QCD with N colors using lattice regularization. The meson field lagrangian is worked out in detail in the strong coupling limit with various lattice fermion formulations. For generalized Susskind fermions the spontaneous breakdown $U(n) \otimes U(n) \to U(n)$ (diagonal) is found at large N and a generalized version of the non-linear σ model emerges in a natural way. The Nambu–Goldstone spectrum is investigated and a continuous transition is made to Wilson fermions, for which the effective potential and the $\pi\pi$ scattering amplitude are tested on chiral symmetry. Large d (=dimension) approximations are compared with the large N limit and applied to $N = 3$.

1. Introduction

Effective lagrangians have been very useful in the past in summarizing low energy properties of hadrons, especially with respect to chiral symmetry [1]. Recently, there has been a revival of interest in them in connection with chiral dynamics at large N and the U(1) problem [2]. These lagrangians are constructed in such a way that they satisfy the assumed symmetry constraints which are nowadays attributed to the underlying theory, QCD.

This article describes a method for computation of effective actions in the framework of lattice QCD. It is used at (but not limited to) strong coupling where confinement is explicit, giving a crude description of bound states of fermions taking into account chiral symmetry. It also provides a useful calculational device which incorporates and extends previous lattice work in the euclidean formulation [3–6]. In particular, meson and baryon propagators were obtained in ref. [5], using random walk techniques and the effective lagrangian describes in addition to this the interaction vertex functions.

Strongly coupled lattice gauge theory has recently shed new light on the dynamics of spontaneous chiral symmetry breaking [6–10]. In the hamiltonian formulation of QCD at large N (N = number of colors), spin-wave excitations are found to give an explicit realization of Nambu–Goldstone bosons accompanying the spontaneous breakdown $U(n) \to U(\frac{1}{2}n) \otimes U(\frac{1}{2}n)$ [8, 9]. This $U(n)$ symmetry is reduced

[1] Address after September, Niels Bohr Inst. Blegdamsvej 17, DK-2100, Copenhagen Ø, Denmark.

to chiral $U(N_f) \otimes U(N_f)$ by the addition of special symmetry breaking terms in the SLAC lattice hamiltonian, for which real space renormalization group calculations have shown the existence of a massless excitation in the case $N = N_f = 1$ [7].

In the euclidean formulation Goldstone bosons have also been found explicitly in QCD at large N [4, 6]. Here the nature of the symmetry breakdown, as well as the approximations involved need clarification, which will be given in this article.

We find that the results of Blairon et al. [6] may be justified by a large d approximation, where d is the dimension of euclidean space time. However, for large N we do not need this large d approximation. It will also be shown that for $N \to \infty$ the symmetry aspect is given by the spontaneous breakdown

$$U(n) \otimes U(n) \to U(n) \qquad \text{(diagonal)},$$

with corresponding Goldstone bosons ($n = 4$ in ref. [6]). This symmetry breaking pattern was recently proved necessary by Coleman and Witten [11], given certain assumptions which are satisfied in our strong coupling description. In particular, we find that the effective potential describing the zero-momentum meson dynamics has indeed no accidental degeneracy at large N. The derivation of the effective potential makes use of the large N results of Brézin and Gross on the $U(N)$ group integral with external sources [12]. Another assumption in ref. [11] was the existence of a local order parameter, on which we now comment.

One of the tools in an effective action description for mesonic bound states is the representation of a composite Bose field $\mathcal{M}(x) = \psi(x)\bar{\psi}(x)$ by an elementary Bose field $\mathcal{M}'(x)$. There is a potential clash involved here between the strengths of $\mathcal{M}$ and $\mathcal{M}'$. At given x the composite field $\mathcal{M}(x)$ is bounded, since raising it to a sufficiently high power gives zero, but a customary elementary field $\mathcal{M}'(x)$ is unbounded ($-\infty \leq \mathcal{M}'(x) \leq \infty$ in the Schrödinger representation or in a partition function (path integral)). There are several possibilities for circumventing this problem: the order parameter idea, the large N limit and "bosonization". Order parameters $\mathcal{M}'(x)$ (used in the theory of critical phenomena, Landau theory) may be thought as averages of $\mathcal{M}(y)$ over small regions around x containing a large ($\to\infty$) number of y's. In the large N limit $\mathcal{M}(x) = \psi'(x)\bar{\psi}_1(x) + \cdots + \psi^N(x)\bar{\psi}_N(x)$ becomes itself unbounded. This property is used in the theory of spinwaves (see [8] and references therein). With "bosonization" we mean a procedure which typically replaces $\mathcal{M}(x) \to \mathcal{M}'(x) = v \exp iH(x)$, with $H(x)$ a hermitian Bose field and v some number. We shall use this method and give it a precise meaning. However, large N simplification will also be used in the sense that it gives a justification for using a gaussian approximation around a saddle point in the partition function. A "strength" problem exists also between composite and elementary baryon fields, but we shall reserve the details of this to a separate publication [13].

Fermion fields require special consideration in lattice regularization. There appear to be deep connections between the number of fermion particles (species doubling) and the triangle anomaly [14], and with topology [15, 16]. Several fermion methods have been used in practice in order to reduce the number of particles, namely the non-local method [17] advocated by the SLAC lattice group, Susskind's method [18] and Wilson's method [3]. Of these only Wilson's method seems to avoid all species doubling at both weak and strong coupling [14]. It breaks chiral symmetry explicitly and therefore it has difficulties in application to the electroweak interactions. For QCD, however, there are good indications [14] that in the continuum limit chiral $SU(N_f) \otimes SU(N_f)$ symmetry is restored but that U(1) axial symmetry remains violated, thereby avoiding a U(1) problem. The explicit demonstration at strong coupling of this resolution of the U(1) "problem" requires loop calculations [8] which are awkward in hamiltonian formulation. The Lagrange formulation of the present paper is a first step towards such a calculation.

For Wilson's lattice action a phase structure was proposed in ref. [5] where the existence of a phase boundary was pointed out. The phase boundary develops from the strong coupling limit to the weak coupling limit in the phase plane. The pion mass vanishes on the boundary line. Our standpoint in this paper is that we can best approach the continuum limit on this phase boundary line and expect to have chiral symmetry in the continuum limit in its "spontaneously broken form". One important point which was left to be worked out in refs. [5] and [8] for Wilson's fermions is the question whether there is full chiral symmetry on the phase line at strong coupling. Full chiral symmetry requires not only the vanishing of the pion mass but also of pion scattering at zero momentum, i.e. the effective potential should be chirally symmetric. With the methods in this paper we are able to answer this question unambiguously.

We shall consider both Wilson's and Susskind's fermion method, extending the latter to the euclidean description. This extension brings about a surprise: the $U(n) \otimes U(n)$ chiral symmetry instead of the $U(n)$ symmetry of the hamiltonian formulation. Strictly speaking, $n = 1$ for Susskind fermions. In this paper n is the number of non-color indices of the basic fermion field. It is natural to extend the $n = 1$ case to arbitrary n. It will be shown that when $n = 4N_f$, $N_f = 1, 2, \ldots$, Susskind fermions are equivalent to N_f "naive" Dirac fermions ("naive" = using a nearest neighbor gradient in the usual Dirac lagrangian). For $n = 4N_f$ we can introduce the Wilson mass term used in that method and interpolate smoothly to Wilson's method. A parameter r varying from 0 ("naive") to 1 (Wilson) describes the interpolation. Actually, we start with this latter description and work back to Susskind fermions for ease of presentation.

The remainder of this paper is organized as follows. Sect. 2 defines the effective action. Initially, baryon fields are taken along to indicate the generality of the approach but they are dropped at a later stage. Sect. 3 contains the derivation of the effective lagrangian for $N \to \infty$ using the results of Brézin and Gross. Sect. 4 explains the $U(n) \otimes U(n)$ symmetry. In sect. 5 the effective action is used to determine the type of spontaneous symmetry breaking and the particle spectrum ($r = 0$). In sect. 6 the Wilson mass term is turned on slowly and the change in the spectrum is described. Previous $r=1$ results are reobtained and the effective potential and the $\pi\pi$ scattering amplitude are tested on chiral symmetry. Sect. 7 deals with finite N effects using large d approximations. Comments are in sect. 8. The appendix clarifies some technicalities connected with "bosonization" having to do with group integration.

2. The effective action

We start with the SU(N) gauge theory described by

$$S = S_\psi + S_U + S_{J\eta}\,, \tag{2.1}$$

where

$$S_\psi = \sum_{x,\mu} [\bar\psi(x)P^-_\mu U_\mu(x)\psi(x+a_\mu) + \bar\psi(x+a_\mu)P^+_\mu U^\dagger_\mu(x)\psi(x)] - \sum_x \bar\psi(x)M\psi(x)\,, \tag{2.2}$$

$$P^\pm_\mu = \tfrac{1}{2}(r \pm \gamma_\mu)\,, \qquad 0 \leq r \leq 1\,; \tag{2.3}$$

the action S_U is the standard gauge field action of the schematic form

$$S_U = \frac{1}{g^2} \sum_{\text{plaq}} \mathrm{Tr}\, UUU^\dagger U^\dagger + \text{c.c.} \tag{2.4}$$

and $S_{J\eta}$ contains sources $J(x)$ and $\eta(x)$, $\bar\eta(x)$ for the meson and baryon fields

$$N\mathcal{M}^\alpha_\beta(x) = \psi^{a\alpha}(x)\bar\psi_{a\beta}(x)\,, \tag{2.5}$$

$$N!\mathcal{B}^{\alpha_1\cdots\alpha_N}(x) = \varepsilon_{a_1\cdots a_N}\psi^{a_1\alpha_1}(x)\cdots\psi^{a_N\alpha_N}(x)\,, \tag{2.6}$$

$$N!\bar{\mathcal{B}}_{\alpha_1\cdots\alpha_N}(x) = \varepsilon^{a_1\cdots a_N}\bar\psi_{a_N\alpha_N}(x)\cdots\bar\psi_{a_1\alpha_1}(x)\,, \tag{2.7}$$

$$S_{J\eta} = \sum_x (NJ^\alpha_\beta\mathcal{M}^\beta_\alpha + \bar\eta_{\alpha_1\cdots\alpha_N}\mathcal{B}^{\alpha_1\cdots\alpha_N} + \bar{\mathcal{B}}_{\alpha_1\cdots\alpha_N}\eta^{\alpha_1\cdots\alpha_N})\,. \tag{2.8}$$

We attempt to evaluate the partion function

$$Z = \int \mathrm{D}\bar\psi\mathrm{D}\psi\mathrm{D}U\, \mathrm{e}^S\,, \tag{2.9}$$

by first integrating over U with external sources $A_\mu(x)$ and $\bar A_\mu(x)$:

$$\int \mathrm{D}U\, \mathrm{e}^{S_U + \sum_{x,\mu}\mathrm{Tr}(\bar A_\mu U_\mu + U^\dagger_\mu A_\mu)} = \mathrm{e}^{W(\bar A, A)}\,. \tag{2.10}$$

Then Z can be written as

$$Z = \int \mathrm{D}\bar\psi\mathrm{D}\psi\, \mathrm{e}^{W(\bar A, A) + S_{J\eta}}\,, \tag{2.11}$$

where we have temporarily included the ordinary mass term $\bar{\psi}M\psi$ in the source J and

$$A_\mu(x)^a_b = \bar{\psi}_b(x+a_\mu)P^+_\mu\psi^a(x)\,, \tag{2.12}$$

$$\bar{A}_\mu(x)^a_b = \bar{\psi}_b(x)P^-_\mu\psi^a(x+a_\mu)\,. \tag{2.13}$$

The resulting expression can be written in a form depending explicitly only on the locally color invariant meson and baryon fields (2.5)–(2.7),

$$Z = \int \mathrm{D}\bar{\psi}\mathrm{D}\psi\ \mathrm{e}^{S_1(\mathcal{M},\bar{\mathcal{B}},\mathcal{B})+S_{J\eta}}\,. \tag{2.14}$$

Next the fermion fields ψ are integrated out using

$$\int \mathrm{D}\bar{\psi}\mathrm{D}\psi\ \mathrm{e}^{S_{J\eta}} = Z_0(J,\bar{\eta},\eta)\,, \tag{2.15}$$

$$Z(J,\bar{\eta},\eta) = \exp\left[S_1\left(\frac{1}{N}\frac{\partial}{\partial J}, -\frac{\partial}{\partial\eta}, \frac{\partial}{\partial\bar{\eta}}\right)\right]Z_0(J,\bar{\eta},\eta)\,. \tag{2.16}$$

In order to be able to perform the differentiations explicitly, Z_0 must be written in the form

$$Z_0(J,\bar{\eta},\eta) = \int \mathrm{D}\mathcal{M}'\mathrm{D}\bar{\mathcal{B}}'\mathrm{D}\mathcal{B}'\tilde{Z}_0(\mathcal{M}',\bar{\mathcal{B}}',\mathcal{B}')\ \mathrm{e}^{NJ\mathcal{M}'+\bar{\eta}\mathcal{B}'+\bar{\mathcal{B}}'\eta}\,, \tag{2.17}$$

where now $\mathcal{M}'$ is a dummy Bose field and $\bar{\mathcal{B}}'$ and $\mathcal{B}'$ are dummy fermion fields. We have suppressed the summations in the exponent in (2.17): they are identical to those in $S_{J\eta}$ in (2.8). In the following the primes on $\mathcal{M}'$, $\mathcal{B}'$ and $\bar{\mathcal{B}}'$ will be omitted. The final result can then be written as

$$Z = \int \mathrm{D}\mathcal{M}\mathrm{D}\bar{\mathcal{B}}\mathrm{D}\mathcal{B}\tilde{Z}_0(\mathcal{M},\bar{\mathcal{B}},\mathcal{B})\ \mathrm{e}^{S_1(\mathcal{M},\bar{\mathcal{B}},\mathcal{B})+NJ\mathcal{M}+\bar{\eta}\mathcal{B}+\bar{\mathcal{B}}\eta}\,. \tag{2.18}$$

The effective action we are looking for is $\ln\tilde{Z}_0+S_1$ provided a meaning can be given to $\ln\tilde{Z}_0$ and to the manipulations above.

Consider first the U integral (2.10). In the strong coupling limit $S_U\to 0$ and consequently $W(\bar{A},A)$ reduces to a sum of one-link contributions $w(\bar{A}_\mu(x),A_\mu(x))$ which are known in a number of cases. The function $\exp w$ is known [19] for the groups SU(2), U(2), SU(3), U(3) and w itself is known for the group U(N) in the limit $N\to\infty$ [12, 20]. Series expansions for w have been obtained up to quite high order in $\bar{A}$, A [21]. For the group SU(N) the first few terms of the expansion in $\bar{A}$, A read:

$$w(\bar{A},A) = \frac{1}{N}\mathrm{Tr}\,\bar{A}A + \frac{1}{2N(N^2-1)}\left[-\mathrm{Tr}\,(\bar{A}A)^2 + \frac{1}{N}(\mathrm{Tr}\,\bar{A}A)^2\right] + \cdots$$

$$+\frac{1}{N!}(\det A + \det\bar{A}) + \cdots\,, \tag{2.19}$$

which after the substitution (2.12), (2.13) gives

$$S_1(\mathcal{M}, \bar{\mathcal{B}}, \mathcal{B}) = N \sum_{x,\mu} \Big\{ -\mathrm{tr}\, \mathcal{M} P_\mu^- \mathcal{M}' P_\mu^+$$

$$+ \frac{N^2}{2(N^2-1)} \Big[\mathrm{tr}\, (\mathcal{M} P_\mu^- \mathcal{M}' P_\mu^+)^2 + \frac{1}{N} (\mathrm{tr}\, \mathcal{M} P_\mu^- \mathcal{M}' P_\mu^+)^2 \Big] + \cdots \Big\}$$

$$+ \sum_{x,\mu} [\bar{\mathcal{B}} (\otimes P_\mu^-)^N \mathcal{B}' + \bar{\mathcal{B}}' (\otimes P_\mu^+)^N \mathcal{B} + \cdots] , \tag{2.20}$$

where $\mathcal{M} = \mathcal{M}(x)$, $\mathcal{M}' = \mathcal{M}(x + a_\mu)$ etc., tr means a trace over the flavor–Dirac indices and

$$\bar{\mathcal{B}}' (\otimes P_\mu)^N \mathcal{B} = \bar{\mathcal{B}}'_{\alpha_1 \cdots \alpha_N} P^{\alpha_1}_{\beta_1} \cdots P^{\alpha_N}_{\beta_N} \mathcal{B}^{\beta_1 \cdots \beta_N} . \tag{2.21}$$

In this way the action $S_1(\mathcal{M}, \bar{\mathcal{B}}, \mathcal{B})$ can be found in the strong coupling limit and higher order corrections in $1/g^2$ can in principle be computed.

Consider next the fermion integral (2.15), which is a product of single site integrations,

$$Z_0(J, \bar{\eta}, \eta) = \prod_x z_0(J(x), \bar{\eta}(x), \eta(x)) . \tag{2.22}$$

We now concentrate on the case $\bar{\eta} = \eta = 0$. Then

$$z_0(J) = \int \mathrm{d}\bar{\psi}\, \mathrm{d}\psi\, \mathrm{e}^{-\bar{\psi} J \psi} = (\det J)^N , \tag{2.23}$$

where the power N comes from the color degree of freedom and (2.17) simplifies to

$$(\det J)^N = \int \mathrm{d}\mathcal{M}\, \mathrm{e}^{N\, \mathrm{tr}\, J \mathcal{M}} \tilde{z}_0(\mathcal{M}) . \tag{2.24}$$

The problem is to give a meaning to $\int \mathrm{d}\mathcal{M}$ and to find $\tilde{z}_0(\mathcal{M})$. Note that the left-hand side of (2.24) is a polynominal in the $n \times n$ matrix J, whereas the integrand on the right-hand side is an infinite series in J. In the $n = 1$ case J is just a number and the solution is a Cauchy integral encircling the origin,

$$J^N = \oint \frac{\mathrm{d}z}{2\pi i z} \mathrm{e}^{Jz} z^{-N} N! , \tag{2.25}$$

which provides the identification

$$N\mathcal{M} = z , \qquad \mathrm{d}\mathcal{M} = \frac{\mathrm{d}z}{2\pi i z} , \tag{2.26a}$$

$$\tilde{z}_0(\mathcal{M}) = \exp[-N \ln \mathcal{M} + \ln (N!/N^N)] . \tag{2.26b}$$

The Cauchy integral (2.25) may be evaluated by parametrizing $z = R\mathrm{e}^{i\phi}$, with

$-\pi \leqslant \phi \leqslant \pi$ and fixed R which may be chosen as $R=N$; then $d\mathcal{M} = d\phi/2\pi$. This suggests the following generalization for the $n \times n$ matrix case. Make a polar decomposition

$$N\mathcal{M} = RU, \tag{2.27}$$

where R is hermitian and positive definite and U is unitary. Then (2.25) generalizes to

$$\int dU \, e^{\mathrm{tr}\, JRU} (\det RU)^{-N} = c_N^n (\det J)^N, \tag{2.28}$$

where the integration is over the group U(n) and dU is the Haar measure. Eq. (2.28) is proved in the appendix where the constant c_N^n is determined as

$$c_N^n = \frac{0!1! \cdots (n-1)!}{N!(N+1)! \cdots (N+n-1)!}. \tag{2.29}$$

Note that the left-hand side of (2.28) is independent of R.

The results above can be generalized to the case of non-zero η and $\bar{\eta}$. However, the baryonic variables are somewhat less straightforward to deal with and we prefer to deal with them in a separate publication [13]. For this reason we now switch to the gauge group U(N) which does not admit baryons at strong coupling, as the terms involving det A, det $\bar{A}$ in (2.19) are absent; so we simply put $\bar{\eta} = \eta = \bar{\mathcal{B}} = \mathcal{B} = 0$ everywhere and conclude from (2.27), (2.28) that in the expression (2.18) for Z

$$D\mathcal{M} = \prod_x dU, \qquad \tilde{Z}_0 = e^{S_0(\mathcal{M})}, \tag{2.30}$$

$$S_0(\mathcal{M}) = N \sum_x [-\mathrm{tr} \ln \mathcal{M}(x) + \mathrm{tr}\, M\mathcal{M}(x) + \mathrm{const.}], \tag{2.31}$$

where the fermion mass matrix is again made explicit. The effective meson field action

$$S_{\mathrm{eff}}(\mathcal{M}) = S_0(\mathcal{M}) + S_1(\mathcal{M}) \tag{2.32}$$

separates naturally in the single site term S_0 and the site coupling term S_1. In the next section S_1 will be obtained to all orders in $\mathcal{M}$ for the case $N \to \infty$.

3. U(∞) action

The logarithm of the one link integral

$$w(\bar{A}, A) = \ln \int dU \, e^{\mathrm{Tr}(\bar{A}U + U^{\dagger}A)}, \tag{3.1}$$

which determines the strong coupling form of $S_1(\mathcal{M})$ is known [12, 20] for the group U(N) in the limit $N \to \infty$. At $N = \infty$ there is a phase transition when the strength of A and $\bar{A}$, as measured by a suitable parameter, increases beyond a critical value [12, 22]. In the application here A and $\bar{A}$ are initially composed of fermion fields (cf. (2.12), (2.13)) and the series expansion form of $w(\bar{A}, A)$ must be used. This means that the so-called strong coupling phase expression for $w(\bar{A}, A)$ has to be used. It is given by [12]

$$w(\bar{A}, A) = N^2\left\{-\frac{3}{4} - c + \frac{2}{N}\sum_a (c + x_a)^{1/2}\right.$$

$$\left. - \frac{1}{2N^2}\sum_{a,b} \ln\left[(c + x_a)^{1/2} + (c + x_b)^{1/2}\right]\right\}, \tag{3.2}$$

where x_a are the eigenvalues of $\bar{A}A/N^2$ and c is defined implicitly by

$$1 = \frac{1}{2N}\sum_a (c + x_a)^{-1/2}. \tag{3.3}$$

Expanding in x_a, these expressions can be rewritten as a series in

$$\sum_a (x_a)^k = N, \qquad k = 0, \tag{3.4a}$$

$$= \mathrm{Tr}\left(\frac{\bar{A}A}{N^2}\right)^k, \qquad k > 0. \tag{3.4b}$$

Rearranging the fermion fields in $\bar{A}$ and A (cf. (2.12), (2.13)), gives

$$\mathrm{Tr}\left(\frac{\bar{A}A}{N^2}\right)^k = -\mathrm{tr}\left[\mathcal{M}(x)P_\mu^- \mathcal{M}(x + a_\mu)P_\mu^+\right]^k \tag{3.4c}$$

(we are considering the link $(x, x + a_\mu)$). So for a function f which has a series expansion,

$$\frac{1}{N}\sum_a f(x_a) = f(0) + \frac{1}{N}\sum_{k>0}\frac{1}{k!} f^{(k)}(0)\,\mathrm{Tr}\left(\frac{\bar{A}A}{N^2}\right)^k$$

$$= f(0) - \frac{1}{N}\mathrm{tr}\left[f(-\tfrac{1}{4}\lambda) - f(0)\right], \tag{3.5}$$

where

$$\lambda = \lambda_\mu(x) = -4\mathcal{M}(x)P_\mu^- \mathcal{M}(x + a_\mu)P_\mu^+ \tag{3.6}$$

(the factor -4 is included for later convenience). For example eq. (3.3) reads

$$1 = \tfrac{1}{2}c^{-1/2} - \frac{1}{2N}\mathrm{tr}\left[(c - \tfrac{1}{4}\lambda)^{-1/2} - c^{-1/2}\right], \tag{3.7}$$

which gives

$$2c^{1/2} = 1 - \frac{1}{N}\mathrm{tr}\left[(1 - \lambda)^{-1/2} - 1\right] + O(N^{-2}). \tag{3.8}$$

In this way we obtain a relatively simple form for S_1:

$$S_1(\mathcal{M}) = N \sum_{x,\mu} \operatorname{tr} F(\lambda_\mu(x)) + O(N^0), \tag{3.9}$$

$$F(\lambda) = 1 - (1-\lambda)^{1/2} + \ln\{\tfrac{1}{2}[1 + (1-\lambda)^{1/2}]\}. \tag{3.10}$$

The first few terms of the expansion of $F(\lambda)$ indeed coincide with (2.20) in the limit $N \to \infty$. The function $F(\lambda)$ was guessed before by Bars [21] (except for an overall sign due to the fermionic character of $\bar{A}$ and A).

4. U(*n*)⊗U(*n*) symmetry

When the mass terms are zero ($M = r = 0$) the action (2.2) has $U(n) \otimes U(n)$ symmetry. To see this, it is useful to make the unitary transformation

$$\psi(x) = T(x)\chi(x), \qquad \bar{\psi}(x) = \bar{\chi}(x)T^\dagger(x), \tag{4.1}$$

$$T(x) = (\gamma_1)^{x_1}(\gamma_2)^{x_2}(\gamma_3)^{x_3}(\gamma_4)^{x_4}. \tag{4.2}$$

Expressing S_ψ in terms of χ and $\bar{\chi}$ for $r = 0$ gives

$$S_\psi = S_\chi = \sum_{x,\mu} \tfrac{1}{2}\eta_\mu(x)[-\bar{\chi}(x)U_\mu(x)\chi(x+a_\mu)$$

$$+\bar{\chi}(x+a_\mu)U^\dagger_\mu(x)\chi(x)] - \sum_x \bar{\chi}(x)M\chi(x), \tag{4.3}$$

where

$$\eta_\mu(x) = T^\dagger(x)\gamma_\mu T(x+a_\mu) = T^\dagger(x+a_\mu)\gamma_\mu T(x), \tag{4.4}$$

$$\eta_1(x) = 1, \qquad \eta_2(x) = (-1)^{x_1}, \qquad \eta_3(x) = (-1)^{x_1+x_2}, \qquad \eta_4(x) = (-1)^{x_1+x_2+x_3}. \tag{4.5}$$

We shall refer to the form (4.3) as the χ representation and the original form as the ψ representation. The action (4.3) for $M = 0$ is invariant under the transformations

$$\chi(x) = W_{\varepsilon(x)}\hat{\chi}(x), \qquad \bar{\chi}(x) = \hat{\bar{\chi}}(x)W^\dagger_{-\varepsilon(x)}, \tag{4.6}$$

$$\varepsilon(x) = (-1)^{x_1+x_2+x_3+x_4}, \tag{4.7}$$

with $W_\pm \in U(n)$. So $S_\psi = S_\chi$ is invariant under the group $U(n) \otimes U(n)$. For non-zero flavor symmetric M the symmetry reduces to the diagonal $U(n)$ subgroup for which $W_+ = W_-$. Chiral transformations

$$\psi(x) = e^{i\varphi\gamma_5}\hat{\psi}(x), \qquad \bar{\psi}(x) = \hat{\bar{\psi}}(x)\, e^{i\varphi\gamma_5} \tag{4.8}$$

fit in the off-diagonal type $W_+ = W^\dagger_-$:

$$W_{\varepsilon(x)} = T^\dagger(x)\, e^{i\varphi\gamma_5}T(x) = e^{i\varphi\gamma_5\varepsilon(x)}. \tag{4.9}$$

All reference to the original Dirac matrices is eliminated in the χ-representation and the Dirac-flavor index α which originally took $n=4N_f$ values, may for $r=0$ be considered to take values $\alpha = 1, 2, \ldots, n$ with $n = 1, 2, \ldots$. A generalization to an arbitrary number of dimensions d is also immediate from (4.5):

$$\eta_\mu(x) = (-1)^{x_1+\cdots+x_{\mu-1}}, \quad \mu = 1, \ldots, d. \tag{4.10}$$

Let us note in passing the connection with the fermion species doubling phenomenon: the action (4.3) for $d=4$ is a sum of four identical actions, one for each value of the Dirac index. Thus each of these four actions should describe a number of Dirac particles with mass M in the weak coupling ($U_\mu \to 1$) continuum limit. This number is $4N_f$ as the total action S_ψ describes $16N_f$ Dirac particles for $r=0$ [14]. So even for $n=1$ ("euclidean Susskind fermions") S_χ describes 4 Dirac particles at weak coupling.

The effective action has the same $U(n)\otimes U(n)$ symmetry at $M=r=0$ (more precisely: $SU(n)\otimes SU(n)\otimes U(1)_A$ as the meson field has zero fermion number), which can be made explicit by transforming to the χ representation:

$$\mathcal{M}(x) \to T(x)\mathcal{M}(x)T^\dagger(x) \tag{4.11}$$

(we use the same notation for $\mathcal{M}$ in the χ representation),

$$S_0(\mathcal{M}) \to S_0(\mathcal{M}), \tag{4.12}$$

$$S_1(\mathcal{M}) \to N \sum_{x,\mu} \left\{ \tfrac{1}{4} \operatorname{tr} \mathcal{M}\mathcal{M}' + \frac{N^2}{32(N^2-1)} \left[\operatorname{tr}(\mathcal{M}\mathcal{M}')^2 + \frac{1}{N} (\operatorname{tr} \mathcal{M}\mathcal{M}')^2 \right] + \cdots \right\}, \tag{4.13}$$

(cf. (2.20)). The $U(n)\otimes U(n)$ transformation takes the form

$$\mathcal{M}(x) = W_{\varepsilon(x)} \hat{\mathcal{M}}(x) W^\dagger_{-\varepsilon(x)} \tag{4.14}$$

and $S_0(\mathcal{M})$, $S_1(\mathcal{M})$ as well as the measure $D\mathcal{M}$ defined in (2.27), (2.30) are invariant. To see the invariance at $M=0$ of $S_0(\mathcal{M})$ given in (2.31) clearly, we write it in the form

$$S_0(\mathcal{M}) = -\frac{N}{2d} \sum_{x,\mu} \operatorname{tr} \ln \mathcal{M}\mathcal{M}' + N \sum_x \operatorname{tr} M\mathcal{M}. \tag{4.15}$$

The χ representation brings out the antiferromagnetic features [8] of the meson system: the lattice separates naturally into even sites ($\varepsilon(x)=+1$) and odd sites ($\varepsilon(x)=-1$); at the even sites $\mathcal{M} = W_+\hat{\mathcal{M}}W_-^\dagger$ and at the odd sites $\mathcal{M} = W_-\hat{\mathcal{M}}W_+^\dagger$. The latter reads like the even site transformation if we write $\mathcal{M}^\dagger$ instead of $\mathcal{M}$ at the odd sites. There is close similarity with effective chiral lagrangians and we shall refer to the $U(n)\otimes U(n)$ symmetry as chiral symmetry.

5. Spontaneous symmetry breaking ($r=0$ case)

For large N the partition function Z may be evaluated by the saddle-point method. The successive terms of this asymptotic expansion correspond to the loop expansion in terms of Feynman diagrams. Here we shall investigate the tree graph approximation and use the convenient $N\to\infty$ form (3.9), (3.10) for the action $S_1(\mathcal{M})$. The loop expansion is, however, well defined for any N and only the actual evaluation of loop diagrams can teach us how accurate it is, for example, for $N=3$ using the $N=3$ version of $S_1(\mathcal{M})$.

In the tree graph approximation $Z(J)$ is given by

$$\ln Z(J) = S_{\rm eff}(\mathcal{M}) + N\sum_x \operatorname{tr} J(x)\mathcal{M}(x)\,, \tag{5.1}$$

with $\mathcal{M}(x)$ determined by that stationary point of $\ln Z$ giving maximal Z. In the $r=0$ case the χ representation is most useful, for which $P_\mu^\pm \to \pm\frac{1}{2}\eta_\mu$ and (cf. (3.10), (4.15))

$$S_{\rm eff}(\mathcal{M}) = N\sum_x \operatorname{tr}\mathcal{M}\mathcal{M} + N\sum_{x,\mu}\operatorname{tr}\left[-\frac{1}{2d}\ln\lambda_\mu(x) + F(\lambda_\mu(x))\right], \tag{5.2}$$

$$\lambda_\mu(x) = \mathcal{M}(x)\mathcal{M}(x+a_\mu) \tag{5.3}$$

(compare with (3.6)).

The first thing to determine is the vacuum expectation value of $\mathcal{M}$. We assume that at a stationary point $\lambda_\mu(x)=\lambda=$ independent of x and μ $(J=0)$. Substituting this ansatz into the effective action leads to an effective potential

$$V = -S_{\rm eff}/\text{volume}\,, \tag{5.4}$$

which for $M=0$ has the form

$$\begin{aligned} V &= N\operatorname{tr}[\tfrac{1}{2}\ln\lambda - dF(\lambda)] \\ &= N\sum_{\alpha=1}^{n}[\tfrac{1}{2}\ln\lambda_\alpha - dF(\lambda_\alpha)]\,, \end{aligned} \tag{5.5}$$

where the λ_α are the eigenvalues of λ. As noticed by Coleman and Witten [11], V is for $N\to\infty$ just a sum of n identical potentials. The stationary values λ_α have to satisfy

$$\frac{1}{2\lambda_\alpha} = dF'(\lambda_\alpha) = \tfrac{1}{2}d\frac{1}{1+(1-\lambda_\alpha)^{1/2}}\,, \tag{5.6}$$

with the unique solution

$$\lambda_\alpha = \frac{2}{d} - \frac{1}{d^2}\,. \tag{5.7}$$

Hence λ is a multiple of the identity matrix. This implies that $\mathcal{M}(x)$ is the same for all the even sites (say $\mathcal{M}(x)=\mathcal{M}$), and the same for all the odd sites ($\mathcal{M}(x)=\mathcal{M}^{-1}\lambda$). Note that $\mathcal{M}$ has to be non-singular. We now recall that in the polar decomposition (2.27) the radial part R may be chosen at will, since the identity (2.28) does not depend on it. Choosing $R\propto\mathbb{1}$ and the same for the odd and even sites gives a standard form for the stationary points:

$$\mathcal{M}(x)=vU_0, \qquad \text{even sites}, \\ = vU_0^\dagger, \qquad \text{odd sites}, \tag{5.8}$$

with

$$v=\left(\frac{2}{d}-\frac{1}{d^2}\right)^{1/2} \tag{5.9}$$

and U_0 an arbitrary unitary matrix. Thus $\mathcal{M}(x)$ has the form of a chiral transformation (cf. (4.9)). Any U_0 can be transformed into the identity matrix by the $\mathrm{U}(n)\otimes\mathrm{U}(n)$ transformation (4.14). Then $\mathcal{M}(x)$ is only invariant under the diagonal $\mathrm{U}(n)$ transformations, so there is spontaneous symmetry breaking of the type

$$\mathrm{U}(n)\otimes\mathrm{U}(n)\rightarrow\mathrm{U}(n)\ (\text{diagonal}), \tag{5.10}$$

as predicted by Coleman and Witten [11].

For very small flavor symmetric M the situation is essentially unchanged, except that the stationary point is now uniquely determined by $U_0=\mathbb{1}$ for $M>0$ or $U_0=-\mathbb{1}$ for $M<0$ since these U_0's have the largest S_{eff} in the respective cases ($v>0$). We shall always take $M>0$.

As M increases v will change and its value is determined from the stationary point of the effective potential appropriate to $M^\alpha_\beta=M\delta^\alpha_\beta$, $\mathcal{M}^\alpha_\beta=v\,\delta^\alpha_\beta$,

$$V=Nn[\ln v-Mv-dF(v^2)]. \tag{5.11}$$

The solution is now given by

$$v=[-M(d-1)+d(M^2+2d-1)^{1/2}]/(d^2+M^2); \tag{5.12}$$

it vanishes for $M\rightarrow\infty$ or for $d\rightarrow\infty$.

The potential (5.11) has a maximum at the stationary point instead of a minimum, contrary to what one might expect. See fig. 1. This seems odd, but there is no problem with stability, because of the special properties of the $\mathcal{M}$-integration. Consider for example the case $n=1$, where $\mathcal{M}$ = complex number and the integration is along a closed path encircling the origin (cf. (2.25), (2.26)). The contour may be deformed and chosen to pass through the saddle point along the stable directions (the directions along which the saddle point looks like a minimum). If we put

$$\mathcal{M}(x)=\sigma+i\pi, \qquad \text{even sites}, \\ =\sigma-i\pi, \qquad \text{odd sites}, \tag{5.13}$$

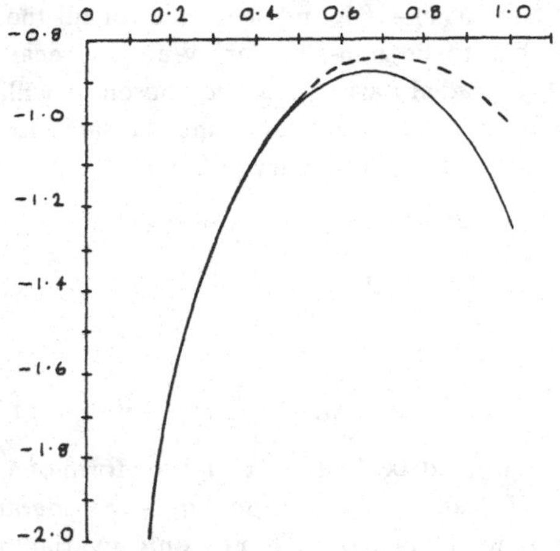

Fig. 1. Plot of the effective potential V/Nn in $d=4$ dimensions as a function of v. The dashed line corresponds to the quadratic approximation to $S_1(\mathcal{M})$.

then the potential becomes

$$V = N[\tfrac{1}{2}\ln(\sigma^2+\pi^2) - M\sigma - dF(\sigma^2+\pi^2)]\,, \qquad (5.14)$$

which is illustrated in fig. 2. Thus V looks like an upside-down version of the usual quartic degree σ-model potential [23], except for the logarithmic singularity at the origin. However, the similarity with the linear σ-model is deceptive, since the integration in that model is over the whole σ-π plane instead of over a closed contour. The situation here is essentially that of the non-linear σ-model [23, 1].

For $M>0$ the stationary point of (5.14) is at $\sigma=v$, $\pi=0$. A circular path passing through this point goes along the π direction, which is stable for $M>0$ (flattening out as $M\to 0$), whereas the σ direction is unstable. For $n>1$ the generalized circular contour defined by (5.8) also passes through the saddle point along the stable directions. This is fortunate, since we do not know how to define integration along arbitrary contours in complex matrix space. The correspondence with the non-linear σ-model is stressed by the parametrization (cf. (4.7), (5.8))

$$\mathcal{M}(x) = v\,\mathrm{e}^{i\varepsilon(x)H(x)}\,, \qquad H(x)=H^{\dagger}(x)\,. \qquad (5.15)$$

Expanding in H we find

$$S_{\text{eff}} = -V\times(\text{volume}) - \frac{v}{n}\frac{\partial V}{\partial n}\sum_x i\varepsilon(x)\,\text{tr}\,H(x)$$

$$+Nv^2\Big[-A\sum_x \text{tr}\,H(x)^2 + B\tfrac{1}{4}\sum_{x,\mu}\text{tr}\,H(x)H(x+a_\mu)\Big] + \mathrm{O}(H^3)\,, \qquad (5.16)$$

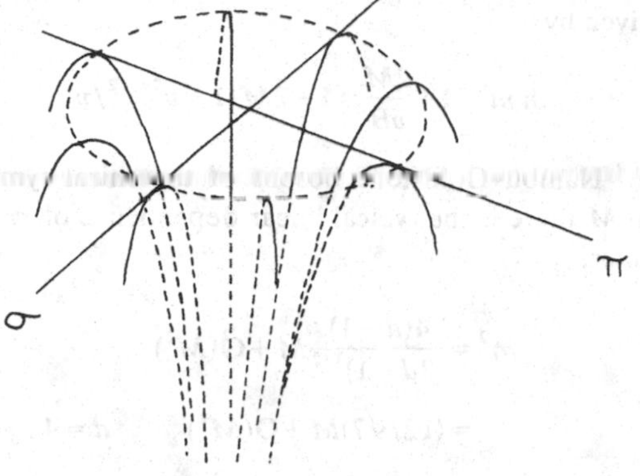

Fig. 2. Illustration of the effective potential for $n = 1$ in the σ-π plane.

where V is given in (5.11) and

$$A = \frac{M}{2v} + \tfrac{1}{4}dB\,, \tag{5.17}$$

$$B = 4[F'(v^2) + v^2F''(v^2)] = (1-v^2)^{-1/2}\,. \tag{5.18}$$

The inverse propagator $D(p)$ for the H-field is given in momentum space by

$$D(p) = 2Nv^2\Big(A - B\tfrac{1}{4}\sum_{\mu} \cos p_\mu\Big)\,. \tag{5.19}$$

Since $A \geq \frac{1}{4}Bd$, all modes are indeed stable.

Let us at this point introduce in $d=4$ dimensions the heuristic definition of particles used in [8]. The large time behavior of the propagator is determined by the positions $b(p)$ of the poles of $D(p)^{-1}$ as a function of $\cos p_4$. This is illustrated by the example

$$\int_{-\pi}^{\pi} \frac{dp_4}{2\pi}\, e^{ip_4x_4} \frac{1}{\cos p_4 - b} = (\pm 1)^{x_4}\, e^{-Ex_4}(b^2-1)^{-1/2}\,, \tag{5.20}$$

where the energy E is determined by

$$\mathrm{ch}\, E(p) = \pm b(p)\,, \qquad \text{for } b \gtrless 0\,. \tag{5.21}$$

We define the energy of a state by eq. (5.21) and a particle by a local minimum of the energy surface; its mass is the minimum value of E at the local minimum. From (5.17)–(5.19) follows the spectrum

$$\mathrm{ch}\, E(p) = 1 + \frac{2M}{vB} + 3 - \sum_{j=1}^{3} \cos p_j \tag{5.22}$$

and we see that there is only one particle per H-mode, situated at the origin $\boldsymbol{p}=0$, with a mass m given by

$$\operatorname{ch} m = 1+\frac{2M}{vB} = 1+2M(1-v^2)^{1/2}/v\,. \tag{5.23}$$

These are the n^2 Nambu–Goldstone bosons of the chiral symmetry breakdown (5.10). For small M there is the typical linear dependence of m^2 on the symmetry breaking parameter,

$$\begin{aligned} m^2 &= \frac{4(d-1)}{(2d-1)^{1/2}}M + \mathrm{O}(M^2) \\ &= (12/\sqrt{7})M + \mathrm{O}(M^2)\,, \qquad d=4\,. \end{aligned} \tag{5.24}$$

6. Turning on the Wilson mass term ($r \neq 0$ case)

For non-zero r we change from the χ representation back to the ordinary ψ representation. Imagine that r increases slowly from $r=0$ to $r=1$ and that M can be adjusted such that the stationary point keeps its form (5.8) with $U_0 = \mathbb{1}$, which looks just the same in the ψ representation:

$$\mathcal{M}^{\alpha}_{\beta}(x) = v\,\delta^{\alpha}_{\beta}\,, \qquad v>0\,. \tag{6.1}$$

A consistency check on this assumption is the stability of the excitation modes: the lowest particle mass squared should be positive. We expect [5, 8] that for a given r the mass gap vanishes at a certain value $M = M_c(r)$. For $g \neq \infty$, the line $M = M_c(r, g)$ provides a phase boundary in the M-g plane [5] (or equivalently, the K-g plane, $K=(2M)^{-1}$).

Chiral symmetry is explicitly broken for $r \neq 0$ but at $M = M_c$ we expect a partial symmetry restoration when the Wilson mass term and the ordinary mass term cancel at zero momentum in certain vertex functions [8, 14]. So the mass gap corresponds to pseudoscalar excitations, as will be verified below.

Substitution of the ansatz (6.1) into the effective action (2.31), (3.10) leads to the effective potential ($d=4$)

$$V = Nn[\ln v - Mv - 4F((1-r^2)v^2)] \tag{6.2}$$

and the equation for the stationary value of v is

$$1 = Mv + \frac{4(1-r^2)v^2}{1+[1-(1-r^2)v^2]^{1/2}}\,. \tag{6.3}$$

The solution is again given by (5.12), provided we make the substitutions $v^2 \to (1-r^2)v^2$, $M \to M/(1-r^2)^{1/2}$, $d=4$. As r increases from 0 to 1 with M fixed, v varies smoothly between (5.12) and $1/M$. However, we must take care that $M \geqslant$

$M_c(r)$, otherwise this v will not describe the vacuum state. To determine $M_c(r)$, the mass gap must be obtained from the meson propagator.

In determining the inverse meson propagator we shall now use the parametrization

$$\mathcal{M}(x) = v + iN^{-1/2}\phi(x), \tag{6.4}$$

$$2\phi = \mathbb{1}S + \gamma_5 P + i\gamma_\alpha\gamma_5 A_\alpha + \gamma_\alpha V_\alpha + \tfrac{1}{2}\sigma_{\alpha\beta}T_{\alpha\beta}, \tag{6.5}$$

which is just as good as (5.15) since the physics is parametrization invariant (recall the arbitrariness of the interpolating field). The i in (6.4) secures the passage along the stable directions through the saddle point, assuming $S, P, \ldots$ to be hermitian. At the stationary point S_{eff} takes the form

$$S_{\text{eff}} = \text{const} - A\sum_x \text{tr}\,\phi(x)^2 + B\sum_{x,\mu} \text{tr}\,\phi(x)P_\mu^-\phi(x+a_\mu)P_\mu^+ + O(\phi^3), \tag{6.6}$$

where, using (6.3),

$$A = \frac{M}{2v} + (1-r^2)B, \tag{6.7}$$

$$B = 4[F' + (1-r^2)v^2F''] = [1-(1-r^2)v^2]^{-1/2}. \tag{6.8}$$

The inverse meson propagator separates into three block matrices corresponding to a scalar channel (S), a pseudoscalar-axial vector channel (PA) and a vector-tensor channel (VT). The procedure is described in detail in [5]. For the S channel

$$D_{SS}(p) = 2A + B\tfrac{1}{2}(1-r^2)\sum_\mu \cos p_\mu. \tag{6.9}$$

For the PA channel

$$D_{PP}(p) = 2A - B\tfrac{1}{2}(1+r^2)\sum_\mu \cos p_\mu, \tag{6.10a}$$

$$D_{PA_\alpha}(p) = -D_{A_\alpha P}(p) = -Br\sin p_\alpha, \tag{6.10b}$$

$$D_{A_\rho A_\alpha}(p) = \delta_{\rho\alpha}\left\{2A + B\left[\tfrac{1}{2}(1-r^2)\sum_\mu \cos p_\mu - \cos p_\alpha\right]\right\} \tag{6.10c}$$

and for the VT channel

$$D_{V_\rho V_\alpha}(p) = \delta_{\rho\alpha}\left\{2A - B\left[\tfrac{1}{2}(1+r^2)\sum_\mu \cos p_\mu - \cos p_\rho\right]\right\}, \tag{6.11a}$$

$$D_{V_\rho T_{\alpha\beta}}(p) = -D_{T_{\alpha\beta}V_\rho}(p) = Br(\delta_{\rho\alpha}\sin p_\beta - \delta_{\rho\beta}\sin p_\alpha), \tag{6.11b}$$

$$D_{T_{\rho\sigma}T_{\alpha\beta}}(p) = (\delta_{\rho\alpha}\delta_{\sigma\beta} - \delta_{\rho\beta}\delta_{\sigma\alpha})\left\{2A + B\left[\tfrac{1}{2}(1-r^2)\sum_\mu \cos p_\mu - \cos p_\rho - \cos p_\sigma\right]\right\}. \tag{6.11c}$$

At $r=0$ the inverse propagator is diagonal and there are again the $n_\chi^2=16N_f^2$ Goldstone bosons with their mass given by (5.23). In the ψ representation only the pseudo-scalar particle is situated at the origin $\boldsymbol{p}=0$, the other particles are at the corners of the Brillouin zone: eq. (6.10a) shows that E_P is still given by (5.22), but for example (6.10c) gives a different form for E_{A_4}:

$$\begin{aligned} \text{ch}\, E_P &= 1+2M/vB+3-\sigma\,, \\ \text{ch}\, E_{A_4} &= 1+2M/vB+3+\sigma\,, \end{aligned} \tag{6.12}$$

where

$$\sigma=\sum_{j=1}^{3}\cos p_j\,. \tag{6.13}$$

So the A_4 particle is at $\boldsymbol{p}=(\pi,\pi,\pi)$ (momenta are modulo 2π). Similarly, the A_3 particle is at $\boldsymbol{p}=(\pi,\pi,0)$, etc.

As r increases from 0 to 1 the excitations get more and more strongly coupled within the channels and an analytic determination of the energy surfaces becomes too complicated. However, at values $p_j=0,\pi$ where the particles are situated, most of the off-diagonal terms vanish and we can still determine the particle masses. To put order into a complicated situation we shall first determine M_c as the value of M, where the mass gap vanishes in the PA channel.

At momenta 0 or π, the PA channel reduces to a 2×2 matrix (PP, PA_4, A_4P, A_4A_4) and a diagonal part (A_1A_1, A_2A_2, A_3A_3). The equation $\det D=0$ determines the position of the poles of the propagator as a function of $\cos p_4=\pm\text{ch}\, E$ (cf. (5.21)). The 2×2 determinant leads to the equation

$$\begin{aligned} 0 &= (1-r^2)^2\,\text{ch}^2 E-2(1+r^2)(4\rho-r^2\sigma)\,\text{ch}\, E \\ &\quad +4[(2\rho-\tfrac{1}{2}(1+r^2)\sigma)(2\rho+\tfrac{1}{2}(1-r^2)\sigma)+r^2]\,, \end{aligned} \tag{6.14}$$

where

$$\rho=A/B \tag{6.15}$$

and σ is supposed to take only the values $\sigma=(-3,-1,1,+3)$. Setting $\text{ch}\, E=1$ we find from (6.14) that the pseudoscalar mass ($\boldsymbol{p}=0, \sigma=3$) vanishes at $\rho=1+r^2$ or $\rho=-\frac{1}{2}+r^2$. Continuity with the $r=0$ case selects the first value, so that M_c follows from

$$1+r^2=\rho=\frac{M_c}{2v[1-(1-r^2)v^2]^{1/2}}+1-r^2\,. \tag{6.16}$$

Eliminating M_c between (6.16) and (6.3) gives v as a function of r at $M=M_c$:

$$v^2=\tfrac{1}{32}[23+gr^2-(81+414r^2+81r^4)^{1/2}]\,, \tag{6.17}$$

showing that v decreases smoothly from $\frac{1}{4}\sqrt{7}$ to $\frac{1}{2}$ as r increases from 0 to 1. From this and (6.16) we see that M_c varies smoothly between the limits

$$M_c \sim \tfrac{3}{4}\sqrt{7}r^2 , \qquad r \to 0 \tag{6.18}$$

and

$$M_c = 2 , \qquad r = 1 . \tag{6.19}$$

From now we shall assume $\rho \geq 1 + r^2 (M \geq M_c)$.

The particle masses follow from the solutions of (6.14):

$$\operatorname{ch} E_{\pm} = (1-r^2)^{-2}\{4\rho - r^2\sigma \pm [64r^2\rho - 32r^4\rho\sigma + 4r^6\sigma^2 + (1-r^2)^2(\sigma^2 - 4r^2)]^{1/2}\} . \tag{6.20}$$

Since $E_- < E_+$, the E_- surface should contain the particles. This interpretation is supported by going back to the $r=0$ case where (6.20) reduces to

$$\operatorname{ch} E_{\pm} = 4\rho \pm |\sigma| = 1 + 2M/vB + 3 \pm |\sigma| . \tag{6.21}$$

At $r=0$ there is no coupling and this formula should be valid for arbitrary momenta ($-3 \leq \sigma \leq 3$ continuous). Comparing with (6.12) we see that $E_+ = E_{A_4}$ and $E_- = E_P$ for $\sigma > 0$, whereas $E_+ = E_P$ and $E_- = E_{A_4}$ for $\sigma < 0$. Since the P particle is at $\sigma = 3$ and the A_4 particle at $\sigma = -3$, E_- indeed contains both particles.

Eq. (6.20) shows again that $m_P = E_-(\sigma = 3)$ vanishes at $\rho = 1 + r^2$, or $M = M_c$. As $r \to 1$, $E_+ \to \infty$ but E_- remains finite and real (no tachyons). The particle at $\boldsymbol{p} = (\pi, \pi, \pi)$ $(\sigma = -3)$ disappears for sufficiently large r (i.e., the local minimum disappears). It is possible to show, taking into account momentum dependence, that $\boldsymbol{p} = (\pi, \pi, \pi)$ is not a local minimum, in fact we believe that it is a maximum.

The masses of the A_1, A_2 and A_3 particles are equal and given by

$$\operatorname{ch} m_{A_3} = 1 + \frac{2}{1-r^2}\left\{\frac{M}{v[1-(1-r^2)v^2]^{1/2}} - r^2\right\} . \tag{6.22}$$

From (6.16) we see that the quantity in curly brackets is positive for $M \geq M_c$, so that $m_{A_3} \to \infty$ as $r \to 1$. The mass of the scalar particle becomes infinite too as $r \to 1$, $M \geq M_c$.

We shall only briefly summarize our analysis of the VT channel. At $r = 0$ the particles are situated at the boundary of the Brillouin zone. Letting $r \to 1$ with $M \geq M_c$, these particles either become infinitely heavy or disappear. Three new particles appear at $\boldsymbol{p} = \boldsymbol{0}$ for sufficiently large r, with equal non-zero mass and these are the only ones remaining at $r = 1$. The energies are real and non-zero so that the mass gap is truly given by the pseudoscalar mass.

The upshot is that at $r = 1$ there are only four particles per flavor combination, one pseudoscalar and three vectors. Their masses are given by [3–5]:

$$\operatorname{ch} m_P = 1 + \frac{(M^2-4)(M^2-1)}{2M^2-3} , \tag{6.23}$$

$$\mathrm{ch}\, m_V = 1 + \frac{(M^2-3)(M^2-2)}{2M^2-3} . \tag{6.24}$$

The physics described above is qualitatively the same as in the hamiltonian formulation [8].

The results at $r=1$ are true for any N and apply equally well to $N=3$. This is because $P_\mu^\pm$ become orthogonal projectors at $r=1$, $P_\mu^+ P_\mu^- = 0$, so that $S_1(\mathcal{M})=0$ for $\mathcal{M} = v\mathbb{1}$. Hence v is determined solely by $S_0(\mathcal{M})$ and the contribution from $S_1(\mathcal{M})$ to the propagator involves only terms quadratic in $\mathcal{M}$. By (2.20) these are simply proportional to N with no further N dependence.

It is useful to know which values to expect for $M - M_c$ on phenomenological grounds, testing this crude strong coupling approximation on the experimental data. The departure of M from M_c is interpreted as a current quark mass m,

$$m = M - M_c . \tag{6.25}$$

With $\mathrm{ch}\, x \to 1 + \frac{1}{2}x^2$ and neglecting $O(m^2)$ we find at $r=1$:

$$m_P^2 = \tfrac{24}{5} m , \tag{6.26a}$$

$$m_V = (\tfrac{4}{5})^{1/2}(1 + \tfrac{11}{5} m) = 0.894 + 1.97\, m . \tag{6.25b}$$

Approximating $1.97 \to 2$ and keeping track of the flavor dependence ($m \to \frac{1}{2}(m_a + m_b)$ in eq. (6.26); $a, b = \mathrm{u, d, s}$) produces a constituent mass formula for the vectors, which seems reasonable on grounds of the crude physical picture. Fitting m_π/m_ρ, m_π/m_K gives $\frac{1}{2}(m_u + m_d) = 0.0055$, $m_s = 0.13$, to be compared with the "dynamical quark mass" $\frac{1}{2} \times 0.894 = 0.45$. The numerical value of $M - M_c$ is quite small. Absolute values are $\frac{1}{2}(m_u + m_d) = 4.7$ MeV, $m_s = 112$ MeV. It seems remarkable that Wilson's fermion method ($r=1$) gives such a good phenomenological description of the spectrum at strong coupling.

We conclude this section with some remarks on the question of chiral symmetry at non-zero r, say at $r=1$. The M and r mass terms cancel in the pseudoscalar mass at $M = M_c$. This tells us something about the quadratic terms in the effective potential: they vanish for the pseudoscalar fields P. This is necessary for the non-linear realization of chiral symmetry [1]. Consider the restricted field configuration

$$\mathcal{M}(x) = v\, e^{iP\gamma_5} , \tag{6.27}$$

with P independent of x. Chiral transformations $\exp i\varphi\gamma_5$ induce non-linear transformations on $P: P \to \hat{P}$ where

$$e^{i\hat{P}\gamma_5} = e^{i\varphi\gamma_5}\, e^{iP\gamma_5}\, e^{i\varphi\gamma_5} . \tag{6.28}$$

So with this non-linear realization a chirally invariant potential is simply dependent on P. For $N \to \infty$ the effective potential for the configuration (6.27) is easily calculated from (3.9), $r=1$,

$$V=4N\,\mathrm{tr}'\{\ln v - Mv\cos P - 2[1-(1-4v^2\sin^2 P)^{1/2} + \ln \tfrac{1}{2}\{1+(1-4v^2\sin^2 P)^{1/2}]\}\,, \tag{6.29}$$

where tr′ is the reduced trace in flavor space. At $M=M_c=2$, $v=\frac{1}{2}$ and the terms quadratic in P indeed cancel. However, the higher order terms in P do not cancel and V is not chirally invariant. A consequence is that the $\pi\pi$ scattering amplitude for massless pions does not vanish at zero four momentum. For $M=M_c=2$ it is given by

$$\langle p_3\alpha_3 p_4\alpha_4|T|p_1\alpha_1 p_2\alpha_2\rangle = A\,\delta_{\alpha_1\alpha_2}\delta_{\alpha_3\alpha_4} + B\,\delta_{\alpha_1\alpha_3}\delta_{\alpha_2\alpha_4} + C\,\delta_{\alpha_2\alpha_4}\delta_{\alpha_2\alpha_3}\,,$$
$$A=[-108+225\,s-106(t+u)]/(25N)+\mathrm{O}(p^4)+\mathrm{O}(N^{-2})\,,$$
$$A(s,t,u)=B(t,s,u)=C(u,t,s)\,, \tag{6.30}$$

where s, t, u are the customary Mandelstam variables and the normalization is such that the current algebra form of A would be $A=s/f_\pi^2$, $t+u=-s$ ($f_\pi \approx 93$ MeV) [1]. The derivation of (6.30) will be given elsewhere [13]. So in case of full chiral symmetry A should vanish at $s=t=u=0$. The magnitude of A is independent of the choice of parametrization of $\mathcal{M}$, contrary to e.g. the coefficient of P^4 in the effective potential. Note that the symmetry breaking effects due to $m_\pi \neq 0$, $M \neq M_c$ are tiny in comparison with $A(0,0,0)=-108/75$ because $M-M_c=0.0055$ is very small.

7. Finite N; large d approximation

The $N\to\infty$ form of the effective action has the advantage of being relatively simple and the results obtained may be considered as an approximation to the finite N case. However, things can be qualitatively different at finite N, except for $r=1$ (it was noticed in sect. 6 that the $r=1$ results for the spectrum are true for any N). We shall concentrate here on the $r=0$ case and consider a different approximation which starts at the quadratic approximation to S_1 and which is improved by taking into account higher order terms in $\mathcal{M}$. The reason this scheme may work is that the effective potential is a power series in v^2 and at the stationary point $v^2<1$, in fact we obtained $v^2=2/d-1/d^2$ for $M=0$, so $v^2\to 0$ as the number of dimensions d becomes large. The approach can be systematized by a large d expansion, by scaling $\mathcal{M}=\hat{\mathcal{M}}d^{-1/2}$, $M=\hat{M}d^{1/2}$ and expanding in $1/d$ (for $1/d$ expansions in another context, see [24]).

We shall not elaborate on this $1/d$ expansion but instead consider the quadratic and quartic approximation, treating these exactly. The numbers thus obtained may be compared with the exact $N\to\infty$ results in order to get an idea of the accuracy. There are two quantities to compare to which are of special interest, namely v at

$M=0$ and the coefficient in the Goldstone boson mass relation obtained from (5.23),

$$m^2 = \frac{4}{vB} M + \mathrm{O}(M^2) . \tag{7.1}$$

Both quantities will be taken at $d=4$. The exact $N \to \infty$ numbers look like

$$\begin{aligned} v &= \left(\frac{2}{d} - \frac{1}{d^2}\right)^{1/2} = 0.66 , \\ m^2 &\sim 4.54 M . \end{aligned} \tag{7.2}$$

In the quadratic approximation $B=1$ and

$$v = \left(\frac{2}{d} + \frac{M^2}{d^2}\right)^{1/2} - \frac{M}{d}$$

$$= 0.71 \quad \text{at} \quad d=4 , \qquad M=0 ; \tag{7.3}$$

$$m^2 \sim 5.66 M . \tag{7.4}$$

This approximation corresponds to the results of Blairon et al. [6]. Inclusion of the quartic terms in $S_1(\mathcal{M})$ gives for $M=0$,

$$\begin{aligned} v^2 &= [(1+4\gamma/d)^{1/2} - 1]/\gamma , \qquad \gamma = \tfrac{1}{2}(1+n/N)N^2/(N^2-1) ; \\ v &= 0.67 , \qquad N=\infty , \qquad d=4 , \\ &= 0.64 , \qquad N=n=3 , \qquad d=4 . \end{aligned} \tag{7.5}$$

A new feature is the dependence on n/N coming from the $(\mathrm{tr}\,\mathcal{M}\mathcal{M}')^2$ term in $S_1(\mathcal{M})$. Such terms also separate the singlet ($\mathrm{tr}\,\mathcal{M} \neq 0$) from the non-singlet ($\mathrm{tr}\,\mathcal{M} = 0$) excitations. In the quartic approximation

$$B = 1 + \frac{1}{2}\frac{N^2}{N^2-1}\left[1 + (1+\delta)\frac{n}{N}\right] v^2 , \tag{7.6}$$

with $\delta = 0$ ($\delta = 1$) for the non-singlet (singlet) excitations. The Goldstone boson mass relation is now

$$m^2 \sim 4.87 M , \qquad N=\infty , \tag{7.7}$$

which differs 7% from the exact result (7.2); for $N=n=3$,

$$\begin{aligned} m^2 &\sim 4.30 M , \qquad \text{non-singlet} , \\ &\sim 3.72 M , \qquad \text{singlet} . \end{aligned} \tag{7.8}$$

Such effects have been conjectured [8] to solve the U(1) problem, which is present at $N=\infty$. Essential to the argument was that we must have $r \neq 0$ (or better $r=1$, which gives a reasonable phenomenological description of masses, including vector mesons), because at $r=0$ both singlet and non-singlet are Goldstone bosons whose masses vanish simultaneously with M (cf. (7.8)). At $r=1$ singlet splitting occurs first at the one loop level and the idea is that a δM_c is required to obtain a zero non-singlet mass, leaving a non-zero singlet mass due to the splitting.

We conclude this section with another possible effect at finite N: a deviation from the symmetry breaking pattern $U(n)\otimes U(n)\to U(n)$. In general the gradient part $S_1(\mathcal{M})=W(\bar{A},A)$ of the effective action depends on the U(N) invariants $\mathrm{Tr}\,(\bar{A}A)^k = \mathrm{tr}\,(\frac{1}{4}\mathcal{M}\mathcal{M}')^k$, $k=1,\ldots,N$. So eq. (5.5) for the effective potential generalizes to

$$V=N(\tfrac{1}{2}\,\mathrm{tr}\ln\lambda-d\mathcal{F})\,, \tag{7.9}$$

where $\mathcal{F}$ is a function of the invariants

$$\sigma_k=\mathrm{tr}\,(\mathcal{M}\mathcal{M}')^k=\mathrm{tr}\,\lambda^k$$

$$=\sum_{\alpha=1}^{n}(\lambda_\alpha)^k\,, \tag{7.10}$$

with λ_α the eigenvalues of λ. Varying λ_α gives the stationarity conditions

$$1=\sum_{k=1}^{N}C_k(\lambda_\alpha)^k\,,\qquad \alpha=1,\ldots,n\,; \tag{7.11}$$

$$C_k=2kd\;\partial\mathcal{F}/\partial\sigma_k\,. \tag{7.12}$$

For given C_k eg. (7.11) has N solutions, say $x_1,\ldots,x_N$, which are functions of the C_k. We may pick various sets $\{\lambda_\alpha\}$ from the x_k (eg. $\lambda_1=\cdots=\lambda_n=x_1$ or $\lambda_1=\cdots=\lambda_{n-1}=x_1$, $\lambda_n=x_2$) and evaluate the corresponding σ_k as functions of the C's. Then (7.12) gives N constraints on $C_1,\ldots,C_N$. For $N\geq n$ the symmetry breaking pattern $U(n)\otimes U(n)\to(\otimes\,U(1))^n$ is possible.

The procedure above stops at (7.11) in the $N\to\infty$ case, since then $\mathcal{F}$ depends linearly on the σ's and the C_k are constants. For $N=\infty$ eq. (7.11) is a transcendental equation which turns out to have only one solution, given in (5.7).

We have tried the procedure above for the case $n=2$ in the quartic approximation and indeed found a solution with $\lambda_1\neq\lambda_2$: $\lambda_{1,2}=2[1\pm(1+2(N+1)/(N-1)d)^{1/2}](N-1)/N$. The corresponding effective potential is higher than the $\lambda_1=\lambda_2$ solution, so the latter should correspond to the true vacuum. However, the λ-values involved in the non-symmetric solution are too large to trust the quartic approximation. The symmetric solution seems to us the most likely candidate for the true ground state in the finite N case.

8. Comments

We have computed in the strong coupling limit of lattice QCD with N colors and n "non-colors" a class of effective lagrangians depending on the symmetry breaking mass parameters M and r. The symmetric lagrangian led to spontaneous symmetry breaking and the corresponding particle spectrum has been determined. Relativistic results for scattering and decay amplitudes will be derived in a separate publication [13].

The $r = 0$ case needs further elucidation of its fermion method before making contact with phenomenology. At strong coupling it looks natural to identify n with the number of flavors. However, as indicated briefly in sect. 4, the weak coupling limit describes $4n$ fermion particles, so one might naively expect $16n^2$ Goldstone bosons. At strong coupling there are only n^2 Goldstone bosons as dictated by the pattern of spontaneous symmetry breaking. In the hamiltonian version there are $2n$ fermion particles at weak coupling [18] and strong coupling expansions with Padé extrapolation to weak coupling give for $n = 1$ indeed 4 pseudoscalar mesons [25]. However, there is no indication yet that these are Nambu–Goldstone bosons [25].

In the $r = 1$ case $n = 4N_f$ and N_f is simply the number of flavors both at weak (N_f fermion particles) and at strong coupling. The light $I \neq 0$ pseudo scalar and vector mesons (π, ρ, K, K*) fit naturally into the description as well as the $I = 0$ mesons ($\eta, \eta', \omega, \phi$). The latter are degenerate with the $I \neq 0$ mesons, because the tree graph approximation has the expected U(1) problem which should be resolved by loop calculations. The methods of this paper makes such calculations feasible.

The N_f^2 ($N_f^2 - 1$ after inclusion of loop effects) pseudoscalar mesons may be interpreted as Nambu–Goldstone bosons in the sense that the explicit symmetry breaking mass terms cancel in the quadratic part of the effective potential. We found, however, no such cancellation in the higher order terms in the meson fields, and that the $\pi\pi$ scattering amplitude does not satisfy the customary chiral symmetry constraints, at strong coupling. Apparently, for $r = 1$ full chiral invariance can only be expected in the true continuum limit at zero bare gauge coupling. Investigations in weak coupling perturbation theory with Wilson's fermion method [14], as well as the phase structure proposed in ref. [5], support this conjecture.

The effective action presented here may be considered as the zeroth order approximation in a strong coupling expansion. Some higher order corrections for the mass spectrum have already been calculated in the random walk language [4, 5]. These computations are combinatorically rather complicated and the effective lagrangian is intended to provide useful guidance. Calculations using a computer have started.

N.K. is thankful to K. Wilson for clarifying conversations and for encouraging to calculate the $\pi\pi$ scattering amplitude in connection with chiral symmetry. J.S. would like to thank J. Hoek and L.H. Karsten for discussions. This work is part of the research programme of the "Stichting voor Fundamenteel Onderzoek der Materie (FOM)", which is financially supported by the "Nederlandse Organisatie voor Zuiver Wetenschappelijk Onderzoek (ZWO)".

Note added

After this work was completed we learned of the article: H. Kluberg-Stern, A. Morel, O. Napoly and B. Petersson, Spontaneous chiral symmetry breaking for a U(N) gauge theory on a lattice [27], which contains material similar to ours. We thank J. Greensite for bringing this paper to our attention.

Euclidean Susskind fermions have recently been investigated at weak coupling: H.S. Sharatchandra, H.J. Thun and P. Weisz, Susskind fermions on a euclidean lattice [28].

Appendix

First some conventions. Our euclidean γ-matrices are hermitian and satisfy $\{\gamma_\mu, \gamma_\nu\} = 2\,\delta_{\mu\nu}$; furthermore $\gamma_5 = \gamma_1\gamma_2\gamma_3\gamma_4$ and $\sigma_{\mu\nu} = \frac{1}{2}i[\gamma_\mu\, \gamma_\nu]$. For fermion fields $\bar\psi_A$ and ψ^A, the integration at a site is defined as $\mathrm{d}\bar\psi\,\mathrm{d}\psi = \prod_A \mathrm{d}\bar\psi_A\,\mathrm{d}\psi^A$ and for the whole lattice $\mathrm{D}\bar\psi\mathrm{D}\psi = \prod_x \mathrm{d}\bar\psi(x)\,\mathrm{d}\psi(x)$.

Next we derive eq. (2.28). Writing $\mathrm{U} = S\,\mathrm{e}^{i\phi}$ with $S \in S\mathrm{U}(n)$ and integrating over ϕ ($\int_0^{2\pi} \mathrm{d}\phi/2\pi$) gives

$$\frac{(\det R)^{-N}}{(Nn)!}\int \mathrm{d}S\,(\mathrm{tr}\, JRS)^{Nn} \tag{A.1}$$

for the left-hand side of (2.28). Comparing with terms of order B^{Nn} ($B = JR$) in

$$\int \mathrm{d}S\,\mathrm{e}^{\mathrm{tr}\,BS} = \sum_{k=0}^{\infty} c_k^n\,(\det B)^k \tag{A.2}$$

gives (2.28). Eq. (A.2) rests on invariance grounds, the problem is the determination of c_k^n. These have been obtained by Creutz [26]; a somewhat different derivation is sketched below.

Denote the expression (A.2) by $F(B)$. Then $F(B)$ satisfies the differential equation

$$DF(B) = F(B)\,, \tag{A.3a}$$

$$D = \frac{1}{n!}\varepsilon_{\alpha_1\cdots\alpha_n}\varepsilon^{\beta_1\cdots\beta_n}(\partial/\partial B^{\beta_1}_{\alpha_1})\cdots(\partial/\partial B^{\beta_n}_{\alpha_n})\,, \tag{A.3b}$$

with the boundary condition $F(0) = 1$. Eq. (A.3) implies a recursion relation for the coefficients c_k^n. To find this, differentiate $\det B^k = \exp k\,\mathrm{tr}\ln B$ with respect to B:

$$(\partial/\partial B^{\beta_1}_{\alpha_1})\det B^k = k(B^{-1})^{\alpha_1}_{\beta_1}\det B^k\,,$$

$$\partial/\partial B^{\beta_2}_{\alpha_2} \to (k^2(B^{-1})^{\alpha_1}_{\beta_1}(B^{-1})^{\alpha_2}_{\beta_1} - k(B^{-1})^{\alpha_1}_{\beta_2}(B^{-1})^{\alpha_2}_{\beta_1}]\det B^k\,, \tag{A.4}$$

and so on. We observe that in (A.3b) only the completely antisymmetric combination of indices contributes. So (A.4) may be replaced by

$$\to (k^2+k)(B^{-1})^{\alpha_1}_{\beta_1}(B^{-1})^{\alpha_2}_{\beta_2}\det B^k\,.$$

Then it is straightforward to obtain

$$D \det B^k = \frac{(n+k-1)!}{(k-1)!} \det B^{k-1} , \tag{A.5}$$

$$c_{k+1}^n = \frac{k!}{(n+k)!} c_k^n , \qquad c_0^n = 1 , \tag{A.6}$$

from which (2.29) follows.

References

[1] S. Gasiorowicz and D.A. Geffen, Rev. Mod. Phys. 41 (1969) 531;
S. Weinberg, Physica 96A (1979) 327
[2] C. Rosenzweig, J. Schechter and C.G. Trahern, Phys. Rev. D21 (1980) 3388;
P. di Vecchia and G. Veneziano, Nucl. Phys. B171 (1980) 253;
K. Kawarabayashi and N. Ohta, Nucl. Phys. B175 (1980) 477;
P. Nath and R. Arnowitt, Phys. Rev. 23 (1981) 473;
E. Witten, Ann. of Phys. 128 (1980) 363
[3] K.G. Wilson, in New phenomena in subnuclear physics, Erice Lectures 1975, ed. A. Zichichi (Plenum, 1977)
[4] B. de Wit and G. 't Hooft, Topological expansion in a lattice color gauge theory, Harvard preprint (1976) unpublished; Phys. Lett. 69B (1977) 61
[5] N. Kawamoto, Nucl. Phys. B190 [FS3] (1981) 617
[6] J.-M. Blairon, R. Brout, F. Englert and J. Greensite, Nucl. Phys. B180 [FS2] (1981) 439
[7] B. Svetitsky, S.D. Drell, H.R. Quinn and M. Weinstein, Phys. Rev. D22 (1980) 490;
M. Weinstein, S.D. Drell, H.R. Quinn and B. Svetitsky, Phys. Rev. D22 (1980) 1190
[8] J. Smit, Nucl. Phys. B175 (1980) 307
[9] J. Greensite and J. Primack, Nucl. Phys. B180 [FS2] (1981) 170
[10] T. Banks and A. Casher, Nucl. Phys. B169 (1980) 103
[11] S. Coleman and E. Witten, Phys. Rev. Lett. 45 (1980) 100
[12] E. Brézin and D.J. Gross, Phys. Lett. 97B (1980) 120
[13] J. Hoek, N. Kawamoto and J. Smit, Baryons in the effective lagrangian of strongly coupled lattice QCD, Amsterdam preprint ITFA-81-4 (8181).
[14] L.H. Karsten and J. Smit, Nucl. Phys. B183 (1981) 103
[15] H.B. Nielsen and M. Ninomiya, Nucl. Phys. B185 (1981) 20 and to be published
[16] L.H. Karsten, Lattice fermions in euclidean space time, Amsterdam preprint ITFA-81-1 (5/81)
[17] S.D. Drell, M. Weinstein and S. Yankielowicz, Phys. Rev. D14 (1976) 487
[18] L. Susskind, Phys. Rev. D16 (1977) 3031; *in* Weak and electromagnetic interactions at high energy, Les Houche Lectures 1976, ed. S. Balian and C.H. Llewellyn Smith (North-Holland, 1976)
[19] K.E. Eriksson, H. Svarthold and B.S. Skagerstam, On invariant group integrals in lattice QCD, CERN preprint Ref. Th.2974 (10/80)
[20] R.C. Brower and M. Nauenberg, Nucl. Phys. B180 [FS2] (1981) 221
[21] I. Bars and F. Green, Phys. Rev. D20 (1979) 3311;
I. Bars, J. Math. Phys. 21 (1980) 2678;
S. Samuel, J. Math. Phys. 21 (1980) 2695;
J. Hoek, SU(N) one link integral by recursion, Amsterdam preprint ITFA (2/81)
[22] D.J. Gross and E. Witten, Phys. Rev. D21 (1980) 446
[23] M. Gell-Mann and M. Levy, Nuovo Cim. 16 (1960) 705
[24] J.M. Drouffe, G. Parisi and N. Sourlas, Nucl. Phys. B161 (1980) 397;
J.M. Drouffe, Nucl. Phys. B170 (1980) 211
[25] T. Banks et al., Phys. Rev. D15 (1977) 1111
[26] M. Creutz, J. Math. Phys. 19 (1978) 2043
[27] H. Kluberg-Stern, A. Morel, O. Napoly and B, Petersson, Nucl. Phys. B190 [FS3] (1981) 504
[28] H.S. Sharatchandra, H.J. Thun and P. Weisz, Nucl. Phys. B192 (1981) 205

Volume 114B, number 2,3 PHYSICS LETTERS 22 July 1982

THE STRONG-COUPLING LIMIT OF GAUGE THEORIES WITH FERMIONS ON A LATTICE

Hannah KLUBERG-STERN [1], André MOREL
DPh-T, CEN SACLAY, 91191 Gif-sur-Yvette, France
and
Bengt PETERSSON
Department of Theoretical Physics, University of Bielefeld, D-4800 Bielefeld, Fed. Rep. Germany

Received 3 March 1982

Low-energy properties of the lattice gauge theories U(*N*) or SU(*N*) with fermions are derived in the infinite-coupling limit from a $1/d$ expansion of Green's functions. Spontaneous global symmetry breaking is shown to occur. The analytically computed meson spectrum nearly coincides with the one obtained by Monte Carlo methods at finite coupling. Estimates of the baryon mass and pion decay constant are also given.

In this note, we present results which were obtained for U(*N*) and SU(*N*) lattice gauge theories with fermions, in the strong-coupling limit. We want to explore specifically the low-energy content of such theories, in contrast with their short-distance behaviour, which is known to be governed by the weak-coupling regime due to asymptotic freedom. Some very important features of hadron physics, such as the small pion mass or the π-decay constant, are supposed to follow from the spontaneous breakdown of chiral symmetry in QCD. This is a non-perturbative phenomenon, and the limit $\beta \equiv 2N/g^2 = 0$, where the hadrons are automatically confined may be a natural starting point for the study of the long-distance, non-perturbative features. In a previous work [1], motivated by similar considerations, we established spontaneous breakdown of chiral symmetry for a U(*N*) gauge theory and exhibited the corresponding Goldstone bosons, in the large-*N* limit. This approach was initiated by Blairon et al. [2] in the spirit of the mean field techniques [‡1]. In ref. [1], we developed at $\beta = 0$ a saddle-point method, valid for any dimension *d* in the large-*N* limit, without reference to mean field approximations. Similar conclusions were reached by Kawamoto and Smit [3].

The present work is primarily concerned with SU(*N*) for fixed *N*, because we wish of course to extend our previous work to theories including baryons [4], such as QCD. The technique used is an extension of that of ref. [1], modified for the fact that the large parameter is now chosen to be *d* rather than *N*. We use the functional integral formalism on a lattice to exhibit a systematic $1/d$ expansion of the Green's functions. Similar methods are known since a long time for spin models in statistical mechanics where they give an expansion around the mean field result [5]. The results of our calculation may be compared with some of the recent Monte Carlo calculations [6–9].

We consider the partition function

$$Z(m) = \int [\mathrm{d}\mu]\,[\mathrm{d}\chi\,\mathrm{d}\bar{\chi}]\,\exp(S_{\mathrm{E}}), \qquad (1)$$

where the plaquette term is absent from the action S_{E} at $\beta = 0$; for S_{E} we choose a lagrangian euclidean version of the Susskind action [10,3,11], namely

[1] Chargée de Recherche au CNRS.
[‡1] For earlier discussions of chiral symmetry on the lattice, see the references in ref. [1].

$$S_E = -\sum_{\alpha=1}^{f}\sum_{i,j=1}^{N}\sum_r\Big(\frac{1}{2}\sum_{l=1}^{d}\eta_{rl}[\bar{\chi}_i^\alpha(r)\,U_{ij}(rl)\,\chi_j^\alpha(r+l)$$
$$+\bar{\chi}_i^\alpha(r+l)\,U_{ij}^\dagger(rl)\,\chi_j^\alpha(r)] - im\delta_{ij}\bar{\chi}_i^\alpha(r)\chi_j^\alpha(r)\Big). \quad (2)$$

$$\eta_{rl} = (-)^{r_1+r_2+...+r_l-1}$$

is a sign attached to the link $(r, r+l)$ between the site r, of coordinates $r_1, r_2, ..., r_d$ and the neighbouring site in the lth direction. Unless specified, the lattice spacing a is fixed to unity. $U(rl)$ is the $N\times N$ matrix of the fundamental representation of the gauge group, U(N) or SU(N), attached to the link (rl). There are $2Nf$ independent fermionic degrees of freedom per site, $\bar{\chi}_i^\alpha(r)$, $\chi_i^\alpha(r)$, where α is a non-colour index. It is known that for $f = n_f \times 2^{[d/2]}$, this action S_E can be derived from the naively discretized euclidean action [12] for n_f flavours of Dirac spinors [3, 11]. However, this action leads to multiplication of fermionic excitations. In the Susskind interpretation, S_E in fact corresponds to $2^{[d/2]}f$ different flavours [11]. In what follows, f and d are treated as independent parameters. With the fermionic action of eq. (2), we maintain on the lattice at least some part of chiral symmetry, namely a global U(f) ⊗ U(f) continuous symmetry which is broken only by the mass term of eq. (2) to its diagonal subgroup $U_\Delta(f)$.

Once the action is fixed, we perform a $1/d$ expansion of the integral (1) where $[d\mu]$ is the product over all links of the group invariant measure, and $[d\chi d\bar{\chi}]$ the product over all fermion degrees of freedom of the corresponding Grassmann measure. All the derivations and details will be given in an expanded version [13] of this letter. Here, we just enumerate the steps followed and present our main results.

Step (i): Integration over the link variables. We integrate over $d\mu(rl)$. The result can be cast into the form of a new action $S(\mathrm{M}, \mathrm{B}, \bar{\mathrm{B}})$, where M and B, $\bar{\mathrm{B}}$ are gauge invariant, local, composite meson and baryon variables:

$$M^{\alpha\beta}(r) = \text{const.}\sum_i \chi_i^\alpha(r)\bar{\chi}_i^\beta(r),$$
$$B^{\alpha_1\alpha_2...\alpha_N}(r) = \text{const.}\ \epsilon_{i_1 i_2 ... i_N}\chi_{i_1}^{\alpha_1}(r)\chi_{i_2}^{\alpha_2}(r)...\chi_{i_N}^{\alpha_N}(r). \quad (3)$$

Of course B and $\bar{\mathrm{B}}$ are absent for U(N).

Step (ii): Laplace transforms. We introduce the Laplace transform $\exp[\tilde{S}]$ of $\exp[S]$ by writing the identity [5]

$$\exp[S(\mathrm{M},\mathrm{B},\bar{\mathrm{B}})] \equiv \int [d\lambda\, d\Lambda\, d\bar{\Lambda}]$$
$$\times \exp[\tilde{S}(\lambda,\bar{\Lambda},\Lambda) + \lambda\mathrm{M} + \bar{\Lambda}\mathrm{B} + \bar{\mathrm{B}}\Lambda]\,. \quad (4)$$

The new fields $\lambda, \Lambda, \bar{\Lambda}$ are elementary in the sense that they appear as independent integration variables in the new form of the functional integral Z.

Step (iii): Integration over χ and $\bar{\chi}$. In eq. (4), the new action depends *linearly* upon the *single site* combinations M, B, $\bar{\mathrm{B}}$ of the χ fields. This makes an exact integration over the χ variables possible ‡2.

Step (iv): The $1/d$ expansion. After step (iii), the partition function $Z(m)$ has the form of a partition function for the variables $\lambda, \Lambda, \bar{\Lambda}$ conjugate to the mesonic and baryonic currents. The corresponding action has a stationary point $[\lambda = \lambda_s, \Lambda = \bar{\Lambda} = 0]$ around which the quantum fluctuations in $\mu = \lambda - \lambda_s$, Λ and $\bar{\Lambda}$ can be studied. It is proven in ref. [13] that the terms in $\tilde{S}$ which come from terms higher than quadratic in S, the various propagators and interaction terms can be systematically expanded in powers of $1/d$. The loop expansion for the field theory in μ, $\Lambda, \bar{\Lambda}$ can be reorganized as a $1/d$ expansion for the Green's functions.

We now turn to the presentation of the results which follow from this expansion. We first study spontaneous symmetry breaking by computing the order parameter $P = \Sigma_{i,\alpha}\langle\bar{\chi}_i^\alpha(r)\chi_i^\alpha(r)\rangle$. We find

$$iP/Nf = (2/d)^{1/2}[\bar{\lambda} - 2\bar{m} - 1/4d\bar{\lambda}^4 + O(1/d^2)] \quad (5)$$

with

$$\bar{\lambda} = 1 + \bar{m}, \quad \bar{m} = m/(2d)^{1/2}, \quad (6)$$

which is valid for arbitrary f and N, and for both gauge groups U(N) and SU(N) ‡3. For $m = 0$, P does not vanish. We emphasize that the symmetric value $P = 0$ corresponding to the disordered phase never

‡2 We have achieved the integration over χ for U(N) and SU(2) at arbitrary f. For SU(N), $N \geqslant 3$, we have succeeded only for $f = 1$ [4].

‡3 The difference between SU(N) and U(N) appears for P at order $(1/d)^{N-1}$ only.

shows up for any value of N. Spontaneous chiral symmetry breaking thus appears as a very general feature in the strong-coupling limit. Not only that, but at the two lowest orders in $1/d$, the order parameter per degree of freedom is independent of f and N, and insensitive to the presence of baryons in the sense that U(N) and SU(N) lead to the same result. Eq. (5) includes the limiting case $N \to \infty$, where the exact dependence in d is already known [1,3]. A calculation of P at order $1/d^2$ for $U(1), f = 1$, and comparison with $U(\infty)$, shows that the f, N independence of P/Nf does not persist beyond the two first orders. We point out that the first-order correction to P is already small at $d = 4$: 7% at small m. Also, our result (5) explains why the value of P obtained with SU(2) by numerical methods [6] agrees so well with the U(∞) result [1,2]. Concerning the β dependence in the strong-coupling region, we notice that an analytic calculation at leading order in $1/N$ and $1/d$ [14] gives no β dependence at all for P, and that a weak dependence is observed in numerical calculations [6,7]. Conversely, $a(\beta)$, when fixed by fitting to the ρ mass in SU(3) [9], is found nearly constant.

We next study the spectrum of the model, at lowest order in $1/d$. The energy levels are then independent of f, and for simplicity we set $f = 1$. For generic f, the $f = 1$ meson (respectively baryon) spectrum must be replicated f^2 (respectively C^N_{N+f-1}) times. Let us consider the correlation function

$$G(k) = \sum_r e^{ikr} \langle \bar{\chi}(r)\chi(r)\bar{\chi}(0)\chi(0)\rangle \qquad (7)$$

of which the poles in k give the meson spectrum. At lowest order in $1/d$, $G(k)$ is proportional to the propagator for the excitations μ introduced at step (iv) above, and thus directly obtained from the quadratic part of the action in μ. We find

$$G^{-1}(k) \propto \frac{1}{d}\sum_{l=1}^{d} \cos k_l + \bar{\lambda}^2 . \qquad (8)$$

According to the Susskind interpretation of the action (2), we identify the meson masses M in the model as follows. We set

$$k = K + \delta\pi ,$$

where the K components are restricted to

$$-\pi a^{-1}/2 \leqslant K_l \leqslant \pi a^{-1}/2, \quad l = 1, \ldots, d, \qquad (9)$$

and where $\delta_l = 0$ or 1.

This defines 2^d subzones of the first Brillouin zone, and K is interpreted as a momentum in a particular channel. Taking the dth direction as the euclidean time direction, we now look for zeroes in $M = iK_d$ of $G^{-1}(k)$ at $K \equiv (K_1, \ldots, K_{d-1}) = 0$ in each of the subzones, i.e. for each value of the d-dimensional vector δ. There are only d distinct energy levels M, given by

$$(aM_0)^2 = 2(2d)^{1/2} ma ,$$

$$aM_p = 2\log[\sqrt{p} + (p+1)^{1/2}] + ma[d/2p(p+1)]^{1/2} ,$$

$$p = 1, 2, \ldots, d-1, \quad \text{with degeneracy } C^p_{d-1} . \qquad (10)$$

The equations have been linearized in the mass parameter ma because terms of higher order in ma appear with powers of $1/d$ which are consistently neglected in the present lowest-order calculation.

The first observation is of course that there exist Goldstone bosons in the limit $m = 0$. For $m \neq 0$, they have the mass M_0. There are $d-1$ other energy levels given by M_p. Specializing to $d = 4$, we obtain

$$aM_0 = (5.66\, ma)^{1/2} , \quad aM_1 = 1.76 + ma ,$$

$$aM_2 = 2.29 + ma/\sqrt{3} , \quad aM_3 = 2.63 + ma/\sqrt{6} . \qquad (11)$$

Recall that the value of f in fact needs not to be specified: the spectrum is independent of f, and f^2-times degenerate because at lowest order the f^2 μ-variables conjugate to the composite mesons M of eq. (3) are decoupled. In particular, we have the expected f^2 Goldstone bosons associated with the spontaneous breakdown $U(f) \otimes U(f) \to U_\Delta(f)$ [‡4].

At $d = 4$, the model is supposed to describe $4f$ different flavours, but, as stated above, the energy levels are f independent at leading order. We thus tentatively identify the two lowest masses M_0 and M_1 with $m_\pi = 140$ MeV and $m_\rho = 780$ MeV, and determine the values of the two lattice parameters m and a^{-1}.

‡4 The SU(2) gauge model is special (see ref. [13]) because its representations 2 and $\bar{2}$ are equivalent. The breaking pattern is $U(2f) \to O(2f)$. There are f^2 Goldstone mesons and $f(f+1)$ Goldstone baryons or antibaryons.

We find

$$m = 8\text{ MeV}, \quad a^{-1} = 440\text{ MeV}. \tag{12}$$

The two higher levels are then predicted to be

$$M_2 = 1010\text{ MeV}, \quad M_3 = 1160\text{ MeV}. \tag{13}$$

We have also evaluated the baryon masses. For N odd, when the baryons are fermions, they are all degenerate and have the mass formula

$$aM_B = (N/2)\log(2d) + [N/(2d)^{1/2}]\,ma. \tag{14}$$

This result was previously obtained in ref. [4]. With the value a^{-1} of eq. (12), we find a baryon mass

$$M_B = 1380\text{ MeV}$$

for $N = 3$ and $d = 4$. Note that this result is less reliable than the meson masses since it slightly exceeds the momentum cut-off $\pi a^{-1} \sim 1300$ MeV. However, it has a reasonable order of magnitude. The actual value is 20% smaller than that obtained with Wilson fermions at $\beta = 0$ [12], showing the regularization dependence. It may indicate that the chiral invariant regularization is a better starting point in the strong-coupling region.

Finally, we may estimate the π decay constant f_π *assuming* the usual PCAC relation to be valid on the lattice, namely

$$f_\pi^2 m_\pi^2 = \frac{m}{2}\sum_i \langle \bar{\chi}_i \chi_i \rangle_{f=1},$$

where, following ref. [6], the factor 1/2 is introduced in order to reduce the effective number of flavours from 4 to 2. We obtain

$$f_\pi^2 = (3/4d)a^{-2} \tag{15}$$

i.e. $f_\pi = 190$ MeV at $d = 4$. We notice that f_π^2, which measures the rate for π-decay, is of order $1/d$. In fact, any quantity related to strong interactions appears at most at order $1/d$ when calculated with the effective action in $\mu, \Lambda, \bar{\Lambda}$. Hence strong interactions are weak at large d.

A comparison of our results with those obtained by various numerical estimates using SU(2) [6] and SU(3) [7] is presented in table 1. The general agreement found is quite striking, recalling that we work at $\beta = 0$, far away from the region where the scaling behaviour is observed to take place ($\beta \sim 2.2$ and 5.4 for SU(2) and SU(3), respectively).

We conclude with a few comments. We feel that it is important to start with a chiral invariant action in order to ensure the appearance of Goldstone bosons, a significant feature of the hadron spectrum. Our analytic approach at $\beta = 0$ with such an action reveals fruitful. On one hand it shows that the $1/d$ expansion is an efficient tool and suggests that a mean field treatment of lattice gauge theories including fermions is of interest. On the other hand, it is a very economical way of investigating physical quantities which have not been calculated by numerical

Table 1
Comparison of our results for meson and baryon masses, and for f_π, with numerical evaluations in SU(2) [6] and SU(3) [7]. Masses and f_π are given in MeV. Only in the case of Wilson fermions is a spin and isospin assignment straightforward.

	Present model Susskind fermions $g^2 = \infty$	SU(2) Susskind fermions $g^2 = 0.55$ a)	Physical values	SU(3) Wilson fermions $r = 1$ $g^2 = 1$ b)
M_0	$(4\sqrt{2}\,ma)^{1/2}a^{-1}$	$[(6.5 \pm 0.1)\,ma]^{1/2}a^{-1}$	$m_\pi = 140$	$m_\pi = (6ma)^{1/2}a^{-1}$
M_1	$1.76\,a^{-1} + m$	800 ± 80	$m_\rho = 780$	$m_\rho = 800 \pm 100$
M_2	$2.29a^{-1} + m/\sqrt{3} = 1010$	950 ± 100	$m_\delta = 980$	$m_\delta = 1000 \pm 100$
M_3	$2.63\,a^{-1} + m/\sqrt{6} = 1160$	–	$m_{A_1} = 1100$	$m_{A_1} = 1200 \pm 100$
M_B	$(3a^{-1}/2)\log 8 + 3m/\sqrt{8} = 1380$	–	$m_N = 940$ $m_\Delta = 1240$	$m_N = 950 \pm 100$ $m_\Delta = 1300 \pm 100$
f_π	$(\sqrt{3}/4)a^{-1} = 190$	150 ± 10	95	$f_\pi = 95 \pm 10$
Input	m_π, m_ρ	m_π, string tension		m_π, string tension

a) Ref. [6]. b) Ref. [7].

methods and might require a lot of computer time. Finally, since the β dependence of physical quantities seems to be weak for β between zero and the value β_0 where the scaling behaviour takes place, our calculation at $\beta = 0$ and the knowledge of β_0 may lead to reliable estimates of the physical quantities at infinite β.

One of us (B.P.) would like to thank the Saclay theory group, where most of this work was done, for hospitality extended to him. We are especially grateful to J.M. Drouffe, J. Zinn-Justin and E. Brézin for very helpful discussions. We also thank J.B. Zuber for his careful reading of the manuscript.

References

[1] H. Kluberg-Stern, A. Morel, O. Napoly and B. Petersson, Nucl. Phys. B190 [FS3] (1981) 504.
[2] J.M. Blairon, R. Brout, F. Englert and J. Greensite, Nucl. Phys. B180 [FS2] (1981) 439.
[3] N. Kawamoto and J. Smit, Nucl. Phys. B192 (1981) 100.
[4] J. Hoek, N. Kawamoto and J. Smit, Baryons in the effective lagrangian of strongly coupled lattice QCD, preprint ITFA-81-4 (August 1981).
[5] E. Brézin, J.C. Le Guillou and J. Zinn-Justin, in: Phase transitions and critical phenomena, eds. C. Domb and M.S. Green, Vol. 6 (Academic Press, London, 1976), and references therein.
[6] E. Marinari, G. Parisi and C. Rebbi, Phys. Rev. Lett. 47 (1981) 1795.
[7] H. Hamber and G. Parisi, Phys. Rev. Lett. 47 (1981) 1792.
[8] D. Weingarten, Phys. Lett. 109B (1982) 57.
[9] A. Hasenfratz, Z. Kunszt, P. Hasenfratz and C.B. Lang, Phys. Lett. 110B (1982) 289.
[10] L. Susskind, Phys. Rev. D16 (1977) 3031.
[11] H.S. Sharatchandra, H.J. Thun and P. Weisz, Nucl. Phys. B192 (1981) 205.
[12] K.G. Wilson, Phys. Rev. D10 (1974) 2145; in: New phenomena in subnuclear physics, Erice Lectures 1975, ed. A. Zichichi (Plenum, New York, 1977).
[13] H. Kluberg-Stern, A. Morel and B. Petersson, Saclay DPh-T preprint, in preparation.
[14] V. Alessandrini, V. Hakim and A. Krzywicki, Chiral symmetry in the mean field approach to large N lattice QCD, preprint LPTHE 81/29 (November 1981).

Note added

Since this letter was written, some progress has been made in the assignment of quantum numbers to the bound states of Susskind fermions [1,2]. Accordingly in Table 1 above, the levels M_2 and M_3 (1st column) together with their values in our model (2nd column) should be interchanged. Also data from Monte-Carlo calculations have changed [3].

1. H.Kluberg-Stern, A.Morel, O.Napoly, B.Petersson ; "Susskind fermions in Configuration space", Saclay preprint DPh-T/82/56 (July 82) and in preparat:
2. H.Kluberg-Stern, A.Morel, B.Petersson ; "Spectrum of lattice gauge theories with fermions from a 1/d expansion at strong coupling" Nucl.Phys.B (in Press
3. H.Hamber and G.Parisi, Phys. Rev. D27 (1983) 208.

Nuclear Physics B180[FS2] (1981) 369–377

A PROPOSAL FOR MONTE CARLO SIMULATIONS OF FERMIONIC SYSTEMS

F. FUCITO

Istituto di Fisica dell'Università di Roma and INFN, Frascati, Italy

E. MARINARI

Istituto di Fisica dell'Università di Roma and INFN, Roma, Italy

G. PARISI

INFN, Frascati, Italy

C. REBBI[1]

CERN, Geneva, Switzerland

Received 27 October 1980

We suggest a possible extension of the Monte Carlo technique to systems with fermionic degrees of freedom. We study in detail the application to an elementary example.

1. Introduction

Monte Carlo simulations have recently emerged as one of the most powerful methods for obtaining information on pure gauge theories [1]. If this technique is to be used for a direct computation of properties of known particles, however, the effect of fermions must be properly included.

In the Feynman path integral formulation, fermions are described by anticommuting variables. N anticommuting variables span an algebra with 2^N generators: even for fairly small values of N, the amount of space needed to store a single element of this algebra exceeds by far the memory capacity of any possible computer.

Anticommuting variables must be avoided in computer simulations. In many physically interesting cases, this can be accomplished by using the Matthews-Salam

[1] On leave from Brookhaven National Laboratory, Upton, NY, USA.

formula. Let the euclidean action be given by

$$S[\bar{\psi},\psi,A]=\sum_{i,j}\bar{\psi}_i\Delta_{ij}[A]\psi_j+S_0[A], \tag{1.1}$$

where $\bar{\psi}_i$ and ψ_i $(i=1,N)$ are the fermionic fields and A stands for the bosonic fields. It is crucial that the action be bilinear in the fermionic variables (quite often the action is of this form or may be reduced to it by introducing auxiliary fields). Then the integration over the fermionic degrees of freedom can be done analytically:

$$\int d[\bar{\psi}]\,d[\psi]\exp\{-S[\bar{\psi},\psi,A]\}=\det\{\Delta[A]\}\exp\{-S_0[A]\},$$

$$\int d[\bar{\psi}]\,d[\psi]\,\bar{\psi}_i\psi_j\exp\{-S[\bar{\psi},\psi,A]\}=\{\Delta[A]\}_{ji}^{-1}\det\{\Delta[A]\}\exp\{-S_0[A]\}. \tag{1.2}$$

Similar results can be obtained for higher-order Green functions.

If $\det\{\Delta[A]\}$ does not change sign as a function of A (e.g., $\Delta[A]$ is a positive definite operator or its eigenvalues always have even multiplicity), it can be absorbed into the action, producing an effective action for the bosonic field A:

$$S_{\text{eff}}[A]=S_0[A]-\operatorname{Tr}\ln\Delta[A]. \tag{1.3}$$

All of this is well known. In principle, one could do Monte Carlo simulations for $S_{\text{eff}}[A]$: however, the exact computation of the determinant is too slow (it requires N^3 steps). In this paper we propose a simple technique to evaluate approximately the determinant (more precisely the ratio of two determinants), which requires an acceptable amount of computer time. In sect. 2, we describe the method; in sect. 3, we apply it to a simple system, and in sect. 4 we present a few short remarks on the connection of this procedure with equations of the Langevin type.

2. The introduction of pseudofermions

For the reader's convenience, we recall the principles of Monte Carlo simulations [1, 2]. The aim is to compute

$$\langle f[A]\rangle\equiv\frac{\int d[A]\,f[A]\exp\{-S[A]\}}{\int d[A]\exp\{-S[A]\}}. \tag{2.1}$$

As a starting point one constructs an algorithm which, given the configuration A, generates a new, trial configuration $\tilde{A}$, according to a definite probability distribution $P(A \to \tilde{A})$. P must satisfy $P(A \to \tilde{A}) = P(\tilde{A} \to A)$. If A is an unconstrained real variable, the simplest algorithm consists in adding to A a random variable with symmetric distribution.

A sequence of configurations $A^{(i)}$ is then generated in the following way. Starting from $A = A^{(i)}$, the new, trial configuration $\tilde{A}$ is determined and a random number x is extracted with uniform probability distribution over the unit interval. If

$$\exp\{-(S[\tilde{A}] - S[A])\} > x, \tag{2.2}$$

$A^{(i+1)}$ is set equal to $\tilde{A}$; otherwise, $A^{(i+1)}$ is set equal to the old configuration A. (In other words, if the new configuration leads to lower action, the change $A \to \tilde{A}$ is always accepted; if not, it is accepted with the conditional probability $\exp\{-(S[\tilde{A}] - S[A])\}$.) This guarantees that the sequence eventually reaches a regime of statistical equilibrium, where the probability of encountering any definite configuration A is proportional to $\exp\{-S[A]\}$. It follows

$$\langle f[A] \rangle = \lim_{k\to\infty} \left(\frac{1}{k} \sum_{i=1}^{k} f[A^{(i)}] \right). \tag{2.3}$$

Eq. (2.3) holds independently of the choice of probability distribution $P(A \to \tilde{A})$; however, if this choice is not "appropriate", the convergence of the right-hand side of eq. (2.3) may be very slow.

To apply this algorithm to

$$S_{\text{eff}}[A] = S_0[A] - \operatorname{Tr}\ln\{\Delta[A]\},$$

we must compute the ratio of two determinants. A substantial simplification occurs if we choose $P(A \to \tilde{A})$ so that $\tilde{A}$ is close to A. Neglecting terms corresponding to higher powers of $(\tilde{A} - A)$, we obtain

$$S_{\text{eff}}[\tilde{A}] - S_{\text{eff}}[A] = S_0[\tilde{A}] - S_0[A] - \sum_{i,j} (\Delta^{-1})_{ji} \Delta'_{ij} (\tilde{A} - A), \tag{2.4}$$

where $\Delta'_{ij} = \delta\Delta_{ij}/\delta A$.

Unfortunately, the computation of $(\Delta^{-1})_{ji}$ is also impractical: our suggestion is to compute $(\Delta^{-1})_{ji}$ approximately using a Monte Carlo technique. Indeed, if Δ is a positive operator, we can write

$$(\Delta^{-1})_{ji} = \langle \bar{\phi}_i \phi_j \rangle$$

$$= \frac{\int d[\bar{\phi}]\, d[\phi]\, \bar{\phi}_i \phi_j \exp\{-\Sigma_{i,j} \bar{\phi}_i \Delta_{ij} \phi_j\}}{\int d[\bar{\phi}]\, d[\phi] \exp\{-\Sigma_{i,j} \bar{\phi}_i \Delta_{ij} \phi_j\}}, \tag{2.5}$$

where $\bar{\phi}_i, \phi_i$ are complex bosonic fields which will be called pseudofermions. $\bar{\phi}_i$ and ϕ_i interact like the fermions but are ordinary numbers.

[If Δ is real symmetric, real bosonic fields ϕ_i are sufficient. Eq. (2.5) is replaced by

$$(\Delta^{-1})_{ji} = \langle \phi_i \phi_j \rangle$$

$$= \frac{\int d[\phi]\, \phi_i \phi_j \exp\{-\frac{1}{2} \Sigma_{i,j} \phi_i \Delta_{ij} \phi_j\}}{\int d[\phi] \exp\{-\frac{1}{2} \Sigma_{i,j} \phi_i \Delta_{ij} \phi_j\}} \tag{2.6}$$

and all subsequent formulae are changed accordingly.]

The final prescription is the following. We construct Monte Carlo simulations for the coupled system of A, $\bar{\phi}$ and ϕ very much as in the standard application of the method, but with two major differences: for each upgrading of the bosonic field A, the pseudofermionic variables $\bar{\phi}$ and ϕ are upgraded n times. When we upgrade the fields, different actions are used: these are, respectively,

$$S_{\bar{\phi},\phi} = \sum_{i,j} \bar{\phi}_i \Delta_{ij}[A] \phi_j, \tag{2.7}$$

$$S_A = S_0[A] - \sum_{i,j} \overline{\bar{\phi}_i \phi_j}\, \Delta_{ij}[A], \tag{2.8}$$

where the long bar denotes the average over the last n upgradings of the pseudofermionic fields.

For large n, the pseudofermionic dynamics is much faster than the bosonic one, which is relatively slow: $\langle \bar{\phi}_i \phi_j \rangle$ is very near to $\overline{\bar{\phi}_i \phi_j}$. Neglecting errors proportional to $(\tilde{A} - A)^2$, the correct results are obtained for n going to infinity.

This method reproduces the functional averages of a system with fermions as a limit of Monte Carlo-like simulations; in practice it can work only if not too high values of n are needed to extrapolate to $n \to \infty$ (the computer time is linear in n). It is hard to estimate theoretically how large n must be; in sect. 3 we show with an explicit example that good results are also obtained for low n. We notice *en passant* that eqs. (2.7) and (2.8) are easily generalized to the case where N_f fermionic species (flavours) are present, interacting with the A field in an SU(N_f) invariant way. The action for the pseudofermions is not modified; the new bosonic action is

$$S_A = S_0[A] - N_f \sum_{i,j} \overline{\phi_i \phi_j} \Delta_{ij}[A]. \tag{2.9}$$

As expected, for $N_f = 0$ we have no feedback from the pseudofermions on the bosons. With $N_f = -1$ we recover the bosonic theory (-2 charge conjugate bosons are a fermion) and the correct results are obtained also for $n = 1$.

3. A simple example

The ultimate goal would be to apply the method to gauge theories. In this case, A stands for the gauge fields and the fermionic term in the action is $\bar{\psi}(\not{D} + m)\psi$, D being the covariant derivative. The operator $\not{D} + m$ is not positive definite, but we can use the following chain of identities [3]:

$$\det[\not{D} + m] = \left[\det\{(\not{D} + m)^2\}\right]^{1/2}$$

$$= \left[\det\{-\not{D} + m\} \det\{\not{D} + m\}\right]^{1/2} = \left[\det\{-\not{D}^2 + m^2\}\right]^{1/2}. \tag{3.1}$$

The operator $-\not{D}^2 + m^2$ is now positive definite and can be used for the pseudofermionic action. Neglecting colour indices, one would find

$$S_A = \tfrac{1}{4} F_{\mu\nu} F^{\mu\nu} - \tfrac{1}{2} N_f \bar{\phi}(m^2 - \not{D}^2)\phi, \tag{3.2}$$

$$S_{\bar{\phi},\phi} = \bar{\phi}(m^2 - \not{D}^2)\phi. \tag{3.3}$$

Having in mind this future application, we have investigated a model with action

$$S = A^2 + \bar{\psi}\psi(1 + gA^2). \tag{3.4}$$

The integrations over bosonic and fermionic degrees of freedom are trivial. One finds

$$\langle A^2 \rangle = \frac{2 + 3g}{4 + 2g},$$

$$\langle A^4 \rangle = \frac{6 + 15g}{8 + 4g},$$

$$\langle \bar{\psi}\psi \rangle = \frac{2}{2 + g},$$

$$\langle \delta(A - x) \rangle \equiv \rho(x) = \frac{2(1 + gx^2)}{\sqrt{\pi}\,(2 + g)} \exp\{-x^2\}. \tag{3.5}$$

We have tried to reproduce these results with our modified Monte Carlo simulation. The effective actions for the A and (real) ϕ evolutions are

$$S_A = A^2 - 2g\phi^2 A^2 ,$$

$$S_\psi = \phi^2(1 + gA^2) . \qquad (3.6)$$

n upgradings of the pseudofermion (ϕ) are done for each upgrading of the bosonic field A. The expectation value $\langle\bar{\psi}\psi\rangle$ is given by twice the mean value of ϕ^2. Using complex ϕ we would have

$$S_A = A^2 - g\bar{\phi}\phi A^2 ,$$

$$S_{\bar{\phi},\phi} = \bar{\phi}\phi(1 + gA^2) ,$$

$$\langle\bar{\psi}\psi\rangle = \langle\bar{\phi}\phi\rangle . \qquad (3.7)$$

In the simulation we have set $\tilde{A} = A + \eta, \tilde{\phi} = \phi + \eta'$, where η, η' are random variables uniformly distributed over the interval $[-\frac{1}{4}, \frac{1}{4}]$. $\langle\eta^2\rangle = \frac{1}{48} \ll 1$. With such a small value of $\langle\eta^2\rangle$ the bulk of the error should come only from the finite value of n. In order to see the n dependence, we have done long runs (250 000 steps for A, 250 000 $\times$ n steps for ϕ) at different values of g and n. In figs. 1 and 2, we show

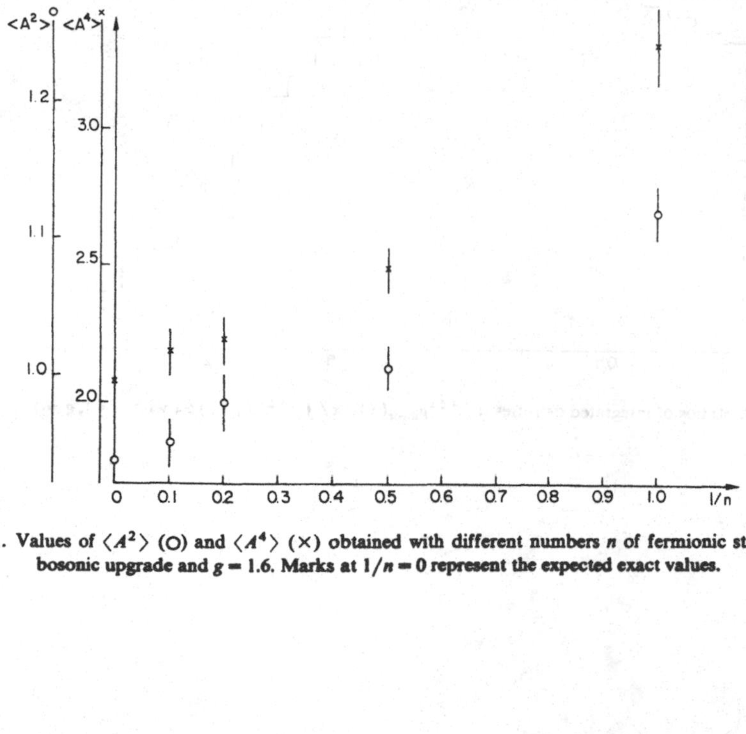

Fig. 1. Values of $\langle A^2\rangle$ (O) and $\langle A^4\rangle$ (×) obtained with different numbers n of fermionic steps per bosonic upgrade and $g = 1.6$. Marks at $1/n = 0$ represent the expected exact values.

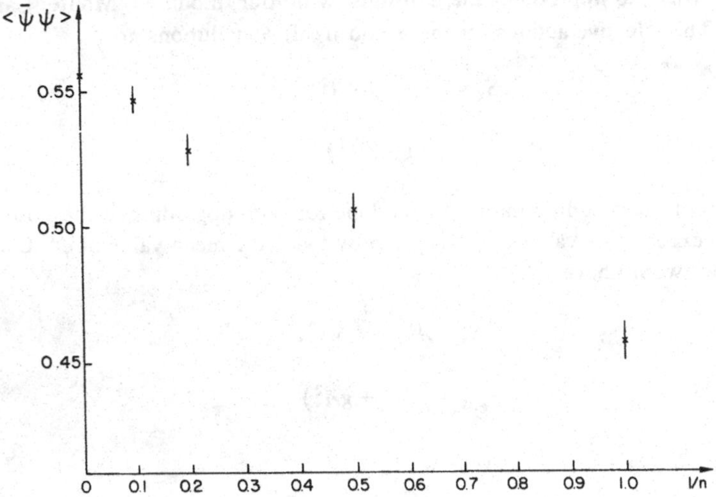

Fig. 2. The same as fig. 1, but values of $\langle\bar{\psi}\psi\rangle$ are reported.

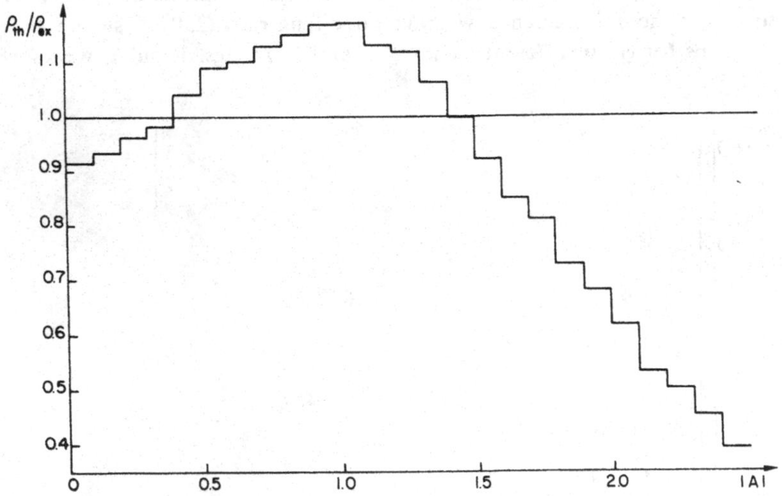

Fig. 3. Ratios of integrated densities $\int_x^{x+\Delta x}\rho_{\text{theor}}(x)\,\mathrm{d}x/\int_x^{x+\Delta x}\rho_{\text{exp}}(x)\,\mathrm{d}x$ with $n=1, g=1.6$.

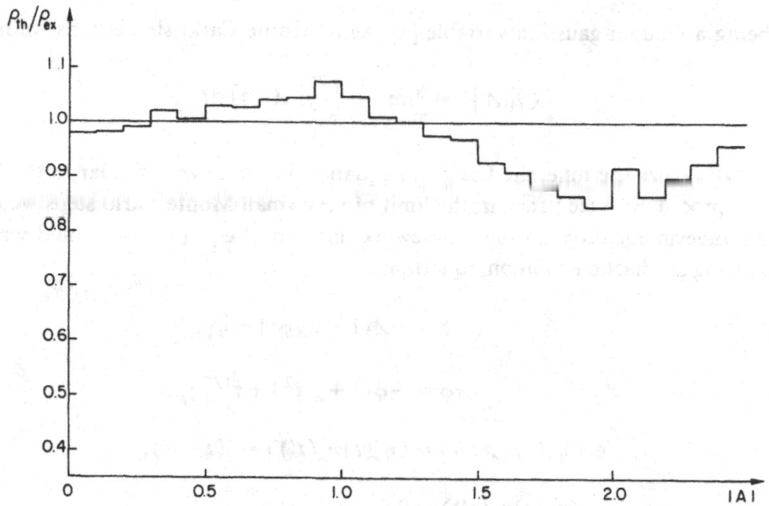

Fig. 4. The same as in fig. 3, but with $n = 5$.

typical results for the moments of A and $\langle\bar{\psi}\psi\rangle$ as functions of n, with $g = 1.6$. In figs. 3 and 4, we display the ratios of the experimental versus theoretical values of the density distributions, integrated over intervals of width 0.1. For large values of $|A|$ (> 2.5), the expected number of events is so small that statistical errors become dominant. In the last interval we reproduce (2.4 – 2.5) only $\sim$400 events are expected, so that even for totally uncorrelated events the above statistical error would be $\sim$5%.

The results are quite satisfactory. Even for $n = 1$ they appear qualitatively correct and the numbers become rather accurate for $n = 2$. We hope that this gratifying feature will survive more interesting applications.

4. The modified Langevin equation

The disadvantage of the Monte Carlo method is that analytic estimates of the rates of convergence and therefore of the errors are difficult, although one can always use the associated master equation. The Langevin equation is not so efficient for numerical simulations (although its use simplifies the computation of correlation functions), but the analytic study of the solution is easy to do.

For conventional systems the Langevin equation is

$$\dot{A} = -\frac{1}{2}\frac{\partial S}{\partial A} + \eta,$$

$$\langle\eta(t)\eta(t')\rangle = \delta(t - t'), \tag{4.1}$$

η being a random gaussian variable [4]. As in Monte Carlo simulations we have

$$\langle f[A]\rangle = \lim_{T\to\infty} \frac{1}{T}\int_0^T f[A(t)]\,dt. \tag{4.2}$$

If we discretize the time, the Langevin equation becomes very similar to the Monte Carlo procedure. Vice versa, in the limit of very small Monte Carlo steps we obtain the Langevin equation. In this framework, instead of eqs. (3.6), we could write the following stochastic evolution equations:

$$\begin{aligned} \dot{A} &= -A(1-2g\phi^2)+\eta_A, \\ \tau\dot{\phi} &= -\phi(1+gA^2)+\tau^{1/2}\eta_\phi, \\ \langle\eta_A(t)\eta_A(t')\rangle &= \langle\eta_\phi(t)\eta_\phi(t')\rangle = \delta(t-t'), \\ \langle\eta_A(t)\eta_\phi(t')\rangle &= 0. \end{aligned} \tag{4.3}$$

It will be shown elsewhere that in the limit $\tau\to 0$, one recovers the results for the fermionic system [5]. Everything in this approach is explicit enough to allow precise estimates of the errors.

It is well known that the correlation functions of a bosonic field theory can be computed using the solution of a stochastic differential equation [6, 7]; it is rather remarkable that the correlation functions of a theory with fermions can also be computed using a stochastic differential equation with commuting variables only.

It is a pleasure to thank F. Guerra for many useful discussions.

References

[1] K. Wilson, Cargese Lecture Notes (1979);
M. Creutz, L. Jacobs and C. Rebbi, Phys. Rev. Lett. 42 (1979) 1390; Phys. Rev. D20 (1979) 1915;
M. Creutz, Phys. Rev. Lett. 43 (1979) 553; Phys. Rev. D21 (1980) 2308;
C. Rebbi, Phys. Rev. D21 (1980) 3350
[2] N. Metropolis, A.W. Rosenbluth, A.H. Teller and E. Teller, J. Chem. Phys. 21 (1953) 1087;
Monte Carlo methods, ed. K. Binder (Springer-Verlag, 1979)
[3] T. Banks and A. Casher, Nucl. Phys. B169 (1979) 103
[4] P. Langevin, Comptes Rendus 146 (1906) 567
[5] E. Marinari, Simulazione numerica di teorie con fermioni, Rome University thesis (1980), unpublished
[6] C. De Dominicis, Nuovo Cim. Lett. 12 (1975) 567
[7] G. Parisi and Wu Yang-shi, ASITP preprint (1980) Sci. Sin., to be published;
G. Parisi and N. Sourlas, Phys. Rev. Lett. 43 (1979) 744

Nuclear Physics B190 [FS3] (1981) 734–750

MONTE CARLO SIMULATION OF THE MASSIVE SCHWINGER MODEL

E. MARINARI

Istituto di Fisica dell' Università di Roma
and
INFN, Roma, Italy

G. PARISI

Istituto di Fisica della Facoltà di Ingegneria, Roma
and
INFN, Frascati, Italy

C. REBBI[1]

CERN, Geneva, Switzerland

Received 15 June 1981

In this article we apply a previously proposed defermionization method to the study of two-dimensional QED (massive Schwinger model). We find good evidence for the spontaneous breaking of axial symmetry, i.e., $\langle\bar{\psi}\psi\rangle \neq 0$ in the massless limit.

1. Introduction

Monte Carlo methods have proved very powerful in the study of quantum gauge theories [1]. In order to apply them to fully realistic situations, fermions must be introduced into the algorithm. The straightforward introduction of anti-commuting variables is impossible. It was, however, suggested in ref. [2] that one could integrate the fermionic degrees of freedom by the Matthews–Salam formula [3] and then perform the bosonic integral using standard Monte Carlo techniques. Unfortunately, if the degrees of freedom of the system are N, the analytic evaluation of the fermionic determinant involves N^3 terms so that the algorithm is slowed down by a factor of N^3: the method cannot be used unless N is very small. As was stressed in ref. [2], the real problem consists of finding a fast algorithm to evaluate the determinant (or, even better, the ratio of two determinants). As far as Monte Carlo methods are concerned, one only needs an approximate estimate of the determinant. It was then suggested that the ratio of the determinants be computed by a Monte Carlo simulation over bosonic variables (which will be called pseudofermions). In this way, irrespective of the number of dynamical variables N, the time spent in

[1] On leave from Brookhaven National Laboratory, Upton, NY, USA.

the algorithm is only increased by a factor of n, n being the number of pseudofermionic iterations needed to achieve, within the desired accuracy, statistical equilibrium of the pseudofermions. Simulations of simple systems have indicated that the error in this method is proportional to $1/n$. The estimate of the magnitude of the error is obtained by considering the associated Langevin equation [4].

Before applying our technique to four-dimensional gauge theories, we decided to test it by studying a simplified model; we have chosen this to be two-dimensional QED [5–10], or massive Schwinger model, for which several analytic results are available. We report here on the outcome of our computations.

In sect. 2 we formulate the lattice action, in the form most convenient for our method, and we discuss how spontaneous breaking of axial symmetry should manifest itself in the massless limit. In sect. 3 we study the system, neglecting fermionic induced vacuum polarization digrams (the quenched case). The case in which $\beta \equiv 1/g^2 \to -\infty$ can be solved analytically and, in this case, axial symmetry is spontaneously broken. In sect. 4 we analyze the quenched case by Monte Carlo computations. In sect. 5 we explain how the Monte Carlo procedure is extended to take into account fermionic vacuum polarization (annealed case, in metallurgic terminology). As a test of the procedure, the results for a 2×2 lattice are obtained and compared with the method based on the exact evaluation of the determinant. Finally, in sect. 6 we present our results for a 16×16 lattice, for various values of the fermionic mass, at $\beta = 3$. Our conclusions are drawn in sect. 7.

2. The lattice action

We need a lattice formulation of the euclidean action

$$S = \int d^2x\{\bar{\psi}(\gamma^\mu D_\mu + m)\psi + \tfrac{1}{4}F_{\mu\nu}F^{\mu\nu}\}, \tag{2.1}$$

where ψ, $\bar{\psi}$ are two-component spinors and D_μ stands for the covariant derivative $\partial_\mu + igA_\mu$. Lattice models with fermions are plagued with a proliferation of modes, the cause of which can be explained in various ways [11], but which are all eventually related to the presence of first order derivatives. On a lattice, a derivative like $\partial_x\psi$ will be approximated, unless one wants to introduce asymmetries and possible non-hermiticities, by a central difference:

$$\partial_x\psi \approx \frac{\psi(x+a)-\psi(x-a)}{2a}, \tag{2.2}$$

a being the lattice spacing. This effectively makes $2a$ the size of the unit cell and the values of the field at the 2^d (d being the dimensionality) points inside the cell turn up as additional degrees of freedom.

An ingenious way of alleviating the problem is the one proposed by Susskind [12], namely assigning single components of the spinors, rather than the full spinors, to the various lattice points inside the cell. In a two-dimensional euclidean system the formulation becomes particularly elegant if one uses a representation

$$\gamma_1 = \sigma_1 , \qquad \gamma_2 = \sigma_2 , \qquad \gamma_5 = -i\gamma_1\gamma_2 = \sigma_3 , \tag{2.3}$$

for the Dirac matrices, so that ψ^+ and ψ^-, the upper and lower components of ψ, are chirality eigenstates. Then, denoting the points of a square lattice by integers i and j, the one-component variables ψ_{ij} with $i+j$ even and odd, respectively, can be conveniently taken to represent the fields ψ^+ and ψ^-.

The free lattice action takes the form

$$S = \sum_{ij} \{\bar{\psi}_{ij}(\psi_{i+1j} - \psi_{i-1j}) - i(-1)^{i+j}\bar{\psi}_{ij}(\psi_{ij+1} - \psi_{ij-1}) + \mu\bar{\psi}_{ij}\psi_{ij}\} . \tag{2.4}$$

(In this formula, as we shall often do in the following, we set $a = 1$, or equivalently the lattice spacing may be thought of as absorbed in a rescaling of ψ.) Notice that the $\gamma^1 \equiv \left(\begin{smallmatrix} 0 & 1 \\ 1 & 0 \end{smallmatrix}\right)$ matrix of the continuum action manifests itself simply in the fact that the first term in eq. (2.4) couples sites of different "chirality", whereas the factor $-i(-1)^{i+j}$ reproduces the alternation of signs of $\gamma^2 \equiv \left(\begin{smallmatrix} 0 & -i \\ i & 0 \end{smallmatrix}\right)$.

To couple in a U(1) gauge field is straightforward: one just follows the standard formulation of lattice gauge theories. Group elements U^x_{ij} and U^y_{ij} (of the form $e^{i\theta}$) are associated with the links between sites ij and $i+1j$, $ij+1$, respectively, and are used for the covariant transport of the complex fields between neighbouring sites (the reversed transports $i+1j \to ij$, $ij+1 \to ij$ being of course performed by $\bar{U}^x_{ij}$ and $\bar{U}^y_{ij}$). The fermionic part of the action then becomes

$$\begin{aligned} S_F &= (\bar{\psi}, G(U)\psi) \\ &\equiv \sum_{ij} \{\bar{\psi}_{ij}\bar{U}^x_{ij}\psi_{i+1j} - \bar{\psi}_{ij}U^x_{i-1j}\psi_{i-1j} \\ &\quad - i(-1)^{i+j}(\bar{\psi}_{ij}\bar{U}^y_{ij}\psi_{ij+1} - \bar{\psi}_{ij}U^y_{ij-1}\psi_{ij-1}) + \mu\bar{\psi}_{ij}\psi_{ij}\} . \end{aligned} \tag{2.5}$$

The pure gauge part of the action is given by

$$S_G = \beta \sum_{\Box ij} (1 - \mathrm{Re}\, U_{\Box ij}) , \tag{2.6}$$

where the sum is extended to all elementary squares of the lattice (plaquettes) and

$$U_{\Box ij} = U^x_{ij}U^y_{i+1j}\bar{U}^x_{ij+1}\bar{U}^y_{ij} . \tag{2.7}$$

The relation between the coupling constant g, the lattice spacing a and the parameter β in eq. (2.6) is given by

$$\beta = 1/g^2a^2 . \tag{2.8}$$

Indeed, writing

$$U^\mu_{ij} = \exp\{igaA_\mu(ia, ja)\} ,$$

the lattice gauge action reduces in the $a \to 0$ limit to the familiar form $\frac{1}{4}\int d^2x F_{\mu\nu}F^{\mu\nu}$.

Integrating out the fermionic degrees of freedom, one obtains an effective action for the U^{μ}_{ij} variables alone:

$$S_{\rm eff}(U) = S_{\rm G}(U) - {\rm Tr}\ln G(U)\,. \tag{2.9}$$

However, as discussed in ref. [2], one needs a hermitian operator, bounded below, to simulate the variation of Tr ln G by pseudofermionic variables. G does not satisfy this requirement, but G and $\bar{G}$ have the same spectrum and determinant. Thus one can use the operator $K = \bar{G}G$ to implement the method of ref. [2] according to the equation

$$S_{\rm eff}(U) = S_{\rm G}(U) - \tfrac{1}{2}{\rm Tr}\ln K(U)\,. \tag{2.10}$$

This greatly corresponds to describing the fermions by a second-order formalism, as done by Feynman and Gell-Mann [13]. The Dirac equation is replaced by the equation

$$\{-D_{\mu}D^{\mu} + g^2\sigma^{\mu\nu}F_{\mu\nu} + m^2\}\psi = 0\,, \tag{2.11}$$

where

$$\sigma^{\mu\nu} = \tfrac{1}{2}[\gamma^{\mu}, \gamma^{\nu}]\,.$$

The form of the lattice operator $K(U)$ is best understood by having in mind the various terms in the right-hand side of eq. (2.11). Simple algebra gives the following structure for the quadratic form

$$S_{\rm PF} = (\bar{\phi}, K(U)\phi)\,. \tag{2.12}$$

(Here and from now on we denote the pseudofermionic variables, which, as we recall from ref. [2], are ordinary c numbers, by ϕ_{ij}.) $S_{\rm PF}$ contains a term

$$-\sum_{ij}\{\bar{\phi}_{i+2j}U^{x}_{i+1j}U^{x}_{ij}\phi_{ij} - \bar{\phi}_{ij}\phi_{ij}\}\,. \tag{2.13}$$

Together with the analogous terms, where ϕ_{ij} is coupled with $\bar{\phi}_{i-2j}$, $\bar{\phi}_{ij+2}$ and $\bar{\phi}_{ij-2}$ (see fig. 1a), this term constitutes the lattice version of the covariant laplacian $-D_{\mu}D^{\mu}$. Notice, however, that the step, in the construction of the discrete laplacian, is by two lattice units rather than by one. This corresponds to the fact that in eq. (2.11) the components ψ^{+} and ψ^{-} appear decoupled.

$S_{\rm PF}$ contains, of course, a term

$$\sum_{ij}\mu^2\bar{\phi}_{ij}\phi_{ij}\,, \tag{2.14}$$

and then terms of the form

$$\sum_{ij} i(-1)^{i+j}\bar{\phi}_{i+1\,j+1}(U^{y}_{i+1j}U^{x}_{ij} - U^{x}_{ij+1}U^{y}_{ij})\phi_{ij}\,. \tag{2.15}$$

In this last expression the field ϕ_{ij} is coupled to its nearest neighbour along the

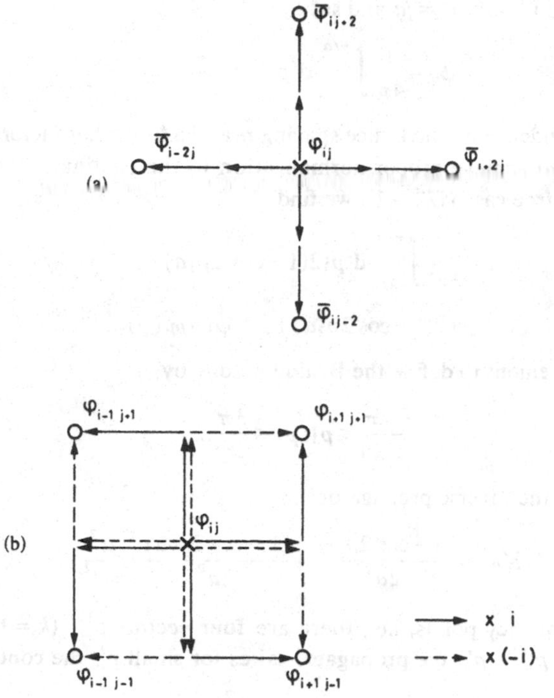

Fig. 1. Diagrammatic representation of the terms in the operator K.

diagonal: $\bar{\phi}_{i+1\,j+1}$. There are two possible transports from site ij to site $i+1\,j+1$ along opposite sides of the plaquette with diagonals between ij and $i+1\,j+1$. The difference between the two corresponding transport factors appears in the coupling. But this difference is related, by a phase factor, to $1-U_{\Box ij}$ and thus measures the electromagnetic field strength on the lattice. There are three additional terms of the form (2.15) in S_{PF}, where ϕ_{ij} is similarly coupled to the other neighbours along the diagonals (see fig. 1b). Altogether these terms reproduce the $\sigma^{\mu\nu}F_{\mu\nu}$ term of eq. (2.11).

Although assigning different chirality components to different sites reduces the severity of the multiplication of modes, the problem is not entirely solved. Heuristically, we may argue that separating the components of ψ reduces the number of degrees of freedom by a factor of 2, whereas there are four points in a cell of site $2a$. Thus we expect the fermions to appear doubled. This is best verified by expressing the action in terms of Fourier conjugate variables. We define vectors

x^μ of components $x^1 = ia$, $x^2 = ja$ and set

$$\phi_{ij} = \frac{1}{4\pi} \int_{-\pi/a}^{\pi/a} d^2p \, e^{ip_\mu x^\mu} \tilde{\phi}(p) \tag{2.16}$$

(the explicit dependence on the lattice spacing a and a few other factors have been introduced here to achieve proper normalization to the continuum). Considering for simplicity the free case ($U_{ij}^\mu = 1$), we find

$$S_{PF} = \frac{1}{4a^2} \int_{-\pi/a}^{\pi/a} d^2p \{2(1-\cos 2p_1 a) + 2(1-\cos 2p_2 a) + \mu^2\} \bar{\tilde{\phi}}(p) \tilde{\phi}(p) . \tag{2.17}$$

It is actually convenient to define the Brillouin zone by

$$-\frac{\pi}{2a} \leqslant p_1, p_2 \leqslant \frac{3\pi}{2a} .$$

We see then that the inverse propagator

$$K_0^{-1} = \frac{(1-\cos 2p_1 a)}{2a^2} + \frac{(1-\cos 2p_2 a)}{2a^2} + \frac{\mu^2}{4a^2} \tag{2.18}$$

has four low-frequency points, i.e., there are four vectors $p_\mu^{(k)}$ ($k = 1 \cdots 4$) such that, setting $p_\mu = p_\mu^{(k)} + p'_\mu$, the propagator takes for small p'_μ the continuum form

$$K_0^{-1} \approx (p')^2 + \frac{\mu^2}{4a^2} . \tag{2.19}$$

(Incidentally, this establishes the relation $\mu = 2ma$ between the mass m of the continuum theory and the dimensionless parameter μ.)

The low-frequency points are the origin of momentum space $p_\mu^{(1)} = 0$, and the other three points obtained by translating any component of p^μ by π/a:

$$p_\mu^{(2)} = \left(\frac{\pi}{a}, 0\right), \quad p_\mu^{(3)} = \left(0, \frac{\pi}{a}\right), \quad p_\mu^{(4)} = \left(\frac{\pi}{a}, \frac{\pi}{a}\right).$$

Let us define

$$\phi^{(k)}(p_\mu) = \tilde{\phi}(p_\mu^{(k)} + p_\mu) \tag{2.20}$$

and consider only neighbourhoods of the low-frequency points (i.e., small p_μ in all $\phi^{(k)}$). Since

$$e^{ip_\mu^{(4)} x^\mu} = e^{i\pi(i+j)} = (-1)^{i+j} ,$$

we see that forming the combinations $\eta_\pm = \phi^{(1)} \pm \phi^{(4)}$ corresponds to selecting even or odd sites, and the same is true of the combinations $\eta'_\pm = \phi^{(2)} \pm \phi^{(3)}$. Thus these linear combinations can be associated, in the continuum limit, to the two chirality components of the Dirac field. However, there are still two independent modes in the theory, η and η', which in the continuum limit appear to be coupled only through their common interaction with the gauge field. Thus in its present formulation, the lattice action really describes a system of two charged massive fermions coupled to the electromagnetic field.

This doubling is not a welcome feature, since some of the properties of the system under investigation crucially depend on the number of fermion species. To eliminate the doubling in the continuum limit we resort to the following stratagem. A theory, where two identical fermions appear with the same quadratic action $(\bar{\psi}, \Delta(U)\psi)$ and are coupled only to the gauge field, produces, upon integration over the fermionic degrees of freedom, an effective action:

$$\begin{aligned} S_{\text{eff}} &= S_{\text{G}} - 2\,\text{Tr}\ln\Delta \\ &= S_{\text{G}} - \ln(\det\Delta)^2 . \end{aligned} \tag{2.21}$$

Assuming that the determinant of the lattice operator G (and correspondingly also $\det K$) undergoes an analogous factorization in the continuum limit, we shall use for effective action

$$\begin{aligned} S_{\text{eff}} &= S_{\text{G}} - \tfrac{1}{2}\,\text{Tr}\ln G \\ &= S_{\text{G}} - \tfrac{1}{4}\,\text{Tr}\ln K . \end{aligned} \tag{2.22}$$

On the lattice, this action will produce a violation of fundamental axioms, but we expect the violation to disappear in the continuum limit and then recover the theory with a simple fermion (although some care may be needed for a proper definition of the axial current). Returning to the method of ref. [2], in practice the upgrading of the pseudofermionic variables will be done (n times for each updating of the U^{μ}_{ij}) with the action $(\bar{\phi}, K(U)\phi)$; the upgrading of the gauge variables will then be done with the action

$$S_{\text{G}} - \tfrac{1}{4} \sum_{ij,i'j'} \overline{\bar{\phi}_{ij}\phi_{i'j'}}\, K(U)_{ij,i'j'} ,$$

where $\overline{\bar{\phi}_{ij}\phi_{i'j'}}$ stand for the averages taken over the n pseudofermionic steps.

It is known that the one-flavour massless Schwinger model (in the continuum) undergoes a breaking of chiral symmetry. This can be seen by noticing that [10]

$$\langle \bar{\psi}(x)\psi(x)\bar{\psi}(0)\psi(0)\rangle \xrightarrow[x\to\infty]{} C , \tag{2.23}$$

where

$$C = \frac{g^2}{8\pi^3} \exp(2\gamma) ,$$

and γ is the Euler–Mascheroni constant. This result breaks the cluster decomposition, i.e., the vacuum is not a pure state.

We can decompose the vacuum as an integral over clustering ω vacua, with

$$\langle\bar{\psi}, \psi\rangle_\omega = A\cos\omega\,, \qquad \langle\bar{\psi}\gamma_5\psi\rangle_\omega = A\sin\omega\,.$$

If the $m=0$ theory is approached as the limit of $m\neq 0$ theories, the $\omega=0$ vacuum is automatically selected:

$$\lim_{m\to 0}\langle\bar{\psi}\psi\rangle = A\,. \tag{2.24}$$

It is also evident that

$$C = \frac{1}{2\pi}\int_{-\pi}^{\pi} \mathrm{d}\omega A^2\cos^2\omega = \tfrac{1}{2}A^2\,. \tag{2.25}$$

We finally get

$$A = \sqrt{2C}\,. \tag{2.26}$$

The relation between the continuum $\langle\bar{\psi}\psi\rangle$ and the lattice expectation value $\langle\bar{\phi}\phi\rangle$ is most easily found by expressing $\langle\bar{\psi}\psi\rangle$ as a logarithmic derivative of the partition function with respect to the continuum mass parameter m. Leaving the further averaging over the gauge field configuration implicit, we have:

$$\langle\bar{\psi}\psi\rangle = \frac{1}{Na^2}\frac{\partial}{\partial m}\ln \mathrm{e}^{-S_{\mathrm{eff}}}$$

$$= \frac{1}{Na^2}\frac{\partial}{\partial m}\ln \mathrm{e}^{(1/4)\mathrm{Tr}\ln K} \tag{2.27}$$

(N denotes the number of points in the lattice). On the other hand,

$$\langle\bar{\phi}\phi\rangle = \frac{1}{N}\frac{\partial}{\partial\mu^2}\ln \mathrm{e}^{\mathrm{Tr}\ln K}\,. \tag{2.28}$$

The relation

$$\langle\bar{\psi}\psi\rangle = 2m\langle\bar{\phi}\phi\rangle = \frac{\mu}{a}\langle\bar{\phi}\phi\rangle \tag{2.29}$$

follows immediately. Thus chiral symmetry breaking will manifest itself with a divergence of $\langle\bar{\phi}\phi\rangle$ as $1/\mu$ when $\mu\to 0$. This pattern of symmetry breaking is rather peculiar to our formulation.

3. Quenched field theory: the staggered case

In the free field case, eq. (2.17) implies the following result for the expectation value $\langle\bar{\phi}\phi\rangle$:

$$\langle\bar{\phi}\phi\rangle = \frac{1}{\pi^2}\int_{-\pi}^{\pi} \mathrm{d}^2p\{2(1-\cos 2p_1) + 2(1-\cos 2p_2) + \mu^2\}^{-1}\,, \tag{3.1}$$

which, in the small μ limit behaves as

$$\langle\bar{\phi}\phi\rangle=\frac{1}{4\pi}\ln\left(\frac{32}{\mu^2}\right)+O(\mu^2)\,. \tag{3.2}$$

Spontaneous breaking of chiral symmetry, i.e., a non-vanishing limit of $\langle\bar{\psi}\psi\rangle\propto\mu\langle\bar{\phi}\phi\rangle$ for $\mu\to 0$, requires a more singular behaviour of $\langle\bar{\phi}\phi\rangle$. This can come from the effect of the gauge field on the propagation of fermions.

If we denote by $\langle\bar{\phi}\phi\rangle_U$ the average value of $\bar{\phi}\phi$ (or, in general, the value of any Green function) in the presence of a *definite* external gauge field U^{μ}_{ij}, the quantum mechanical expectation value $\langle\bar{\phi}\phi\rangle$ is obtained by averaging $\langle\bar{\phi}\phi\rangle_U$ over all gauge field configurations, with a measure given by

$$e^{-S_{\rm eff}}=e^{-S_G+1/4\,{\rm Tr}\ln K} \tag{3.3}$$

[see eq. (2.22)]. In the presence of N species of identical fermions (N flavours), the measure factor would be

$$e^{-S_{\rm eff}}=e^{-S_G+(N/4)\,{\rm Tr}\ln K}\,. \tag{3.4}$$

A first approximation to the effect of the gauge field on the fermion observables may be achieved by setting $N=0$ in eq. (3.4). This amounts to neglecting the contributions from the fermionic vacuum polarization diagrams and, using a terminology developed in the theory of condensed matter, we shall call the expectation values thus obtained "quenched".

For general values of the gauge coupling parameter β, the problem of deriving quenched averages must still be approached numerically although the computation is simplified, and we shall return to it in the next section. In the extreme case $\beta=-\infty$, however, the normal averages reduce to the quenched ones and both can be evaluated analytically, as we now show.

When $\beta=-\infty$, the only gauge field configurations to contribute are those where all plaquettes have maximal internal energy: $E_\square=1-{\rm Re}\,U_\square=2$ (fully "frustrated" case). By choosing a suitable gauge, these field configurations can be brought to the form

$$U^x_{ij}=1\,,\qquad U^y_{ij}=(-1)^i\,. \tag{3.5}$$

The Wilson loop behaves as $\exp i\pi S$, S being its surface in units of the lattice spacing squared. The gauge chosen has the advantage that the field configuration is explicitly invariant under translations of even numbers of lattice sites. In order to use this symmetry to the maximum, we again introduce the fields $\phi^{(k)}(p)$ of eq. (2.20) and restrict the Brillouin zone to $-\frac{1}{2}\pi\le p_1, p_2\le\frac{1}{2}\pi$ (always with $a=1$). In Fourier space the action then takes the form

$$S_{\rm PF}=\int_{-\pi/2}^{\pi/2}{\rm d}^2p\{[2(1-\cos 2p_1)+2(1-\cos 2p_2)+\mu^2]\sum_i\bar{\phi}^{(i)}\phi^{(i)}$$
$$+8i\sin p_1\sin p_2[\bar{\phi}^{(3)}\phi^{(1)}-\bar{\phi}^{(1)}\phi^{(3)}+\bar{\phi}^{(2)}\phi^{(4)}-\bar{\phi}^{(4)}\phi^{(2)}]\}\,. \tag{3.6}$$

From this one easily obtains

$$\sum_i K_{ii}^{-1}(p) = 2\left[\frac{1}{4(\sin p_1 + \sin p_2)^2 + \mu^2} + \frac{1}{4(\sin p_1 - \sin p_2)^2 + \mu^2}\right], \tag{3.7}$$

where K_{ij}^{-1} denotes the momentum space propagator of the fields $\phi^{(i)}(p)$. In the limit $\mu^2 \to 0$ the propagator becomes singular on the two lines $p_1 = p_2$ and $p_1 = -p_2$. $\langle \bar{\phi}\phi \rangle$ is given by

$$\langle \bar{\phi}\phi \rangle = \frac{1}{4\pi^2}\int_{-\pi/2}^{\pi/2} d^2p \sum_i K_{ii}^{-1}(p) . \tag{3.8}$$

If we set $p_1 = p_2 + q$ in the second term, we obtain, for small q, the following integral

$$\int dp_1 \frac{1}{\cos^2 p_1 \sin^2 q + \mu^2}, \tag{3.9}$$

which gives

$$\langle \bar{\phi}\phi \rangle \approx \frac{B}{\mu} \ln\left(\frac{1}{\mu}\right), \qquad B \cong 0.44 . \tag{3.10}$$

Consequently we find a spontaneous symmetry breaking of the chiral symmetry in the sense that $\langle \bar{\psi}\psi \rangle$ goes like $\ln(1/\mu)$ as $\mu \to 0$, i.e., it is divergent and not going to zero. The divergence of $\langle \bar{\psi}\psi \rangle$ when $\mu \to 0$ is a pathology of the $\beta = -\infty$ case and, as we shall see later, disappears for $\beta \neq -\infty$.

The conclusion is that in the limit $\beta \to -\infty$ chiral symmetry is broken. In the next section we will see that this breaking is also present for finite β in the quenched case. Notice that no analytic solution is available for $\beta = 0$ at arbitrary fermionic mass, contrary to what happens in the hamiltonian formalism.

4. Quenched field theory: the general case

For general values of β there is no way of analytically evaluating the quenched average of $\bar{\phi}\phi$ and we must resort to numerical computations. We have considered a 16×16 lattice and, by the standard Monte Carlo procedure, have brought the dynamical variables U_{ij}^{μ} to statistical equilibrium with respect to the measure e^{-S_G}, given by the exponential of the gauge action alone. We have selected $\beta = 3$ as a reasonable intermediate value between the strong coupling domain, where the structure of the lattice would show up too much, and the domain of very weak coupling, where a 16×16 lattice becomes too small to accommodate typical correlation lengths.

The average $\langle \bar{\phi}\phi \rangle$ should first be computed over all points of the lattice and then over several configurations of the external field U. We have assumed that the lattice average already gives a good approximation to the quenched average (indeed, for an infinite lattice, the volume average of $\langle\bar{\phi}\phi\rangle$ *is* the quenched average) and so have computed $\langle\bar{\phi}\phi\rangle$ in the background provided by the U^{μ}_{ij} obtained at the end of a very long Monte Carlo simulation (several thousand iterations) of the pure gauge system.

$$S_{PF} = \sum_{ij,i'j'} \bar{\phi}_{ij} K(U)_{ij,i'j'} \phi_{i'j'} \tag{4.1}$$

[see eqs. (2.12)–(2.15)], or it can be expressed in terms of the Green function for the operator K, according to the equation

$$\langle\bar{\phi}\phi\rangle = \frac{1}{N}\sum_{ij} \{K(U)^{-1}\}_{ij,ij} \tag{4.2}$$

($N = 16^2$ = number of points in the lattice). The inverse operator K^{-1} can in turn be determined recursively, by a relaxation procedure, as the limiting value of the sequence defined by the equations

$$\{K^{-1}\}^{(0)}_{ij,i'j'} = 0\,,$$

$$\{K^{-1}\}^{(l+1)}_{ij,i'j'} = \{K^{-1}\}^{(l)}_{ij,i'j'} + \varepsilon\Big[\delta_{ij,i'j'} - \sum_{i''j''} K_{ij,i''j''}\{K^{-1}\}^{(l)}_{i''j'',i'j'}\Big]\,. \tag{4.3}$$

It is evident that

$$\lim_{l\to\infty} \{K^{-1}\}^{(l)} = K^{-1}\,. \tag{4.4}$$

if the limit exists. We have found that for $0<\varepsilon<\varepsilon_c \simeq 0.2$ (ε_c is a function of β and μ) the sequence $\{K^{-1}\}^{(l)}$ is convergent, while for $\varepsilon>\varepsilon_c$ it diverges, oscillating with period 2. In the actual computation we have used a value of ε slightly smaller than ε_c and iterated the equation until the result appeared to converge with a relative error $<10^{-3}$. The determination of $\langle\bar{\phi}\phi\rangle$ by the relaxation method requires a longer computer time than if one uses the Monte Carlo algorithm (in particular, one must evaluate $\{K^{-1}\}^{(l)}_{ij,i'j'}$ for all pairs ij, $i'j'$, although only the diagonal values $\{K^{-1}\}_{ij,ij}$ are eventually used); the statistical error inherent in the Monte Carlo method is, however, avoided.

In fig. 2 we exhibit the result of the computations for values $\mu = 0.1$, 0.2, 0.4, 0.6, 0.8, 1 and 1.2. From the exact solution of the Schwinger model one expects [see eqs. (2.23)–(2.26)]

$$\langle\bar{\psi}\psi\rangle = \frac{g}{2\pi\sqrt{\pi}}\,e^{\gamma}\,, \tag{4.5}$$

for $m = 0$ in the continuum limit. With eqs. (2.28) and (2.29) this becomes

$$\lim_{\mu\to 0} \mu\langle\bar{\phi}\phi\rangle = \frac{1}{2\pi\sqrt{\pi\beta}}\,e^{\gamma}\,. \tag{4.6}$$

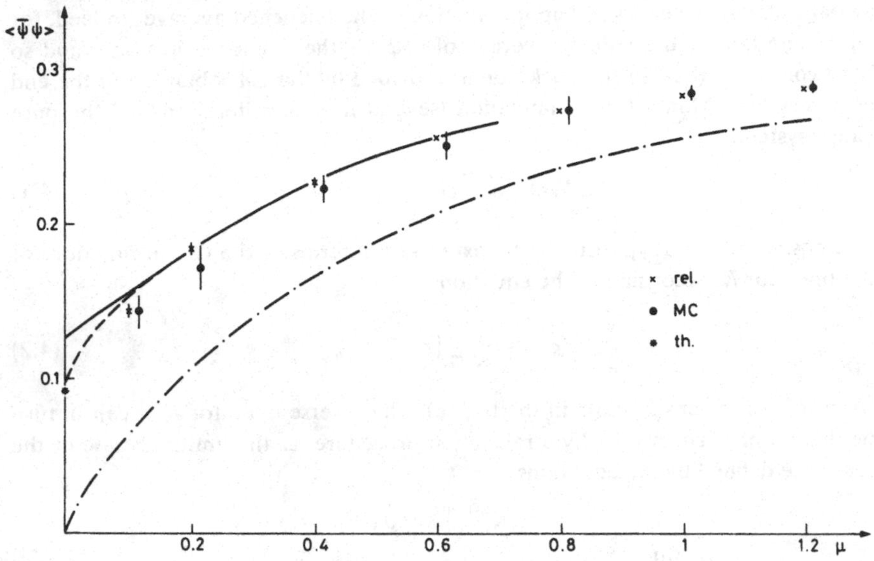

Fig. 2. Results for the quenched expectation value $\langle\bar{\psi}\psi\rangle$: × points obtained with the relaxation method; • points obtained with a Monte Carlo simulation; ∗ theoretical value in the Schwinger model. Full curve: quadratic fit through the points at $m=0.2, 0.4, 0.6$; dashed curve: quadratic fit to $\langle\bar{\psi}\psi\rangle-\langle\bar{\psi}\psi\rangle_{\rm free}$ at the same values of m; dot-dashed curve; expectation value in the free theory.

The value of this last expression for $\beta=3$ is reported on the $\mu=0$ axis in fig. 2 (and fig. 3).

From fig. 2 one notices that the values obtained for $\langle\bar{\phi}\phi\rangle_{\rm quenched}$ with the Monte Carlo method and the relaxation method are consistent, and, moreover, that the numbers clearly tend to a non-vanishing limit for $\mu\to 0$.

The dashed-dotted line in fig. 2 gives the values of $\mu\langle\bar{\phi}\phi\rangle$ in the free case [eq. (3.1)]. The solid line gives the result of a quadratic fit to the values of $\mu\langle\bar{\phi}\phi\rangle$ (relaxation method) at $\mu=0.2$, 0.4 and 0.6. There is, of course, a certain degree of ambiguity in the extrapolation to $\mu=0$, made particularly evident by the rapid variation of the free-field results. To take into account this margin of uncertainty we also present the extrapolation obtained from a quadratic fit not to $\mu\langle\bar{\phi}\phi\rangle$ directly, but to the difference $\mu\langle\bar{\phi}\phi\rangle-\mu\langle\bar{\phi}\phi\rangle_{\rm free}$ (dashed line).

In principle, the ambiguity in the extrapolation could be resolved by going to smaller values of μ, but there is then a conflict with the finite size of the lattice. We have already seen how crucially a non-zero expectation value of $\langle\bar{\psi}\psi\rangle$ for $\mu=0$ depends on the low frequency part of the fermion propagation. A lattice of $L\times L$ sites introduces discretization of the momenta, with a gap of the order $\Delta p\approx 2\pi/L$

(or even $4\pi/L$ if the unit cell indeed contains four sites). Thus one would expect finite size distortions when μ becomes comparable with Δp. In the free case for instance, with periodic boundary conditions, the single zero-mode at the origin produces a divergent result $\mu\langle\bar{\phi}\phi\rangle_{\text{free}}\sim 1/N\mu$ ($N=L^2=$ number of points in the lattice), rather than the expected $\mu \ln \mu$. But using antiperiodic boundary conditions instead, the modes are displaced and $\mu\langle\bar{\phi}\phi\rangle_{\text{free}}$ goes to zero linearly in μ without any $\ln \mu$ factor. We expect the randomness introduced by the gauge field, with its averaging effect, to solve a little the problems associated with the finite size; but in any case we have estimated $\mu=0.2$ to be the lowest reliable value with a 16^2 lattice. We have nevertheless performed computations at $\mu=0.1$, but prefer to attribute them only indicative value.

Returning to the results displayed in fig. 2, the computation at $\mu=0.1$ seems to favour the fit on the unsubtracted values (solid line). Both fits though lead to a non-vanishing value of $\langle\bar{\psi}\psi\rangle$ for $\mu=0$ and to spontaneous breaking of chiral symmetry. If anything, the quenched computation seems to produce too large a value of $\langle\bar{\psi}\psi\rangle_{\mu=0}$: as we shall see in sect. 6, this effect can be attributed to the lack of dynamical feedback from the fermions onto the gauge field.

5. The Monte Carlo method: 2×2 lattice

The goal of the Monte Carlo simulation is to generate U^{μ}_{ij} fields with probability distribution proportional to

$$\exp\{-S_{\text{eff}}\}=\exp\{-S_{\text{G}}+\tfrac{1}{4}\operatorname{Tr}\ln K\}\,.$$

Once this is achieved, the computation of correlation functions of fermions is not a serious problem. The basic idea of ref. [2] is that, although computing the exact variation of $S_{\text{G}}-\frac{1}{4}\operatorname{Tr}\ln K$ in an upgrade $U^{\mu}_{ij}\to U^{\mu}_{ij}+\delta U^{\mu}_{ij}$ is too time consuming, for small δU^{μ}_{ij} the variation of S_{eff} can be linearized:

$$\delta S_{\text{eff}}=\delta S_{\text{G}}-\tfrac{1}{4}\operatorname{Tr}K^{-1}\,\delta K\,, \tag{5.1}$$

and the simulation may proceed rapidly if a fast algorithm for the computation of the Green function is available*. In ref. [2] it was further proposed that sufficiently approximate values for K^{-1} could be evaluated by a Monte Carlo simulation over a parallel system for pseudofermionic, c-number variables ϕ_{ij}. n upgrading steps are done for these variables with action $S_{\text{PF}}=(\bar{\phi},K(U)\phi)$ and the Green functions are approximated by

$$\{K(U)^{-1}\}_{ij,i'j'}\cong\overline{\bar{\phi}_{ij}\phi_{i'j'}}\equiv\frac{1}{n}\sum_{k}\bar{\phi}^{(k)}_{ij}\phi^{(k)}_{i'j'}\,, \tag{5.2}$$

* Although some systems (think for instance of the Ising model) do not allow small steps in the upgrading, δU^{μ}_{ij} can always be taken small if the range of the dynamical variables is continuous; moreover, one expects the small fluctuations to become the relevant ones when one approaches the continuum limit.

where $\phi_{ij}^{(k)}$ denotes the values met in the n pseudofermionic steps. The approximate Green functions are then used to upgrade once all the U_{ij}^{μ}. Exact results are obtained when $n \to \infty$ and the error is expected to become proportional to $1/n$. The value of n necessary to get good results changes from system to system and, as is usually the case with Monte Carlo simulations, the validity of the approximation must be estimated from the stability of the results. Generally speaking, one expects faster convergence if the correlation lengths are not too large.

In our simulation the gauge variables U_{ij}^{μ} have been upgraded by adding to their phase a random angle θ, with a distribution symmetric centered around the origin, and a normal deviation $\Delta\theta \simeq 0.2$. The pseudofermionic variables have been upgraded by adding to their real and imaginary parts independent random variables, $r_{R,I}$, again with a distribution symmetric centered around the origin and a deviation $\Delta r \simeq 0.2$.

On a very small lattice, extending for only two sites in the two directions, the determinant of the 4×4 matrix $K_{ij,i'j'}(U)$ is simple enough to be computed and used in a Monte Carlo simulation. One can then upgrade the δU_{ij}^{μ} variables either by following our method or with a direct Monte Carlo simulation with a measure factor incorporating det K.

To check the validity of the method we have carried out Monte Carlo computations for the 2×2 system either by working with the gauge variables alone, and with measure factors $(\det K)\,e^{-S_G}$ (fermionic case) and $(\det K)^{-1}\,e^{-S_G}$ (bosonic case), or by using the coupled system of gauge and pseudofermionic variables. We have also considered the case where the determinant appears at the denominator to see the relevance of the dynamical feedback from the pseudofermions. In the bosonic case, of course, the computation reduces to the ordinary Monte Carlo simulation of a gauge field coupled to a bosonic matter field and the evolution equations guarantee the exactness of the results with $n=1$. (In these checks we have not used the "demultiplication" trick of taking the fourth root of the determinant; correspondingly $S_{\text{eff}} = S_G \pm \text{Tr} \ln K$.)

The results for $\langle\bar{\phi}\phi\rangle$ are reproduced in table 1 and are rather satisfactory, even for low values of n. (50 000 Monte Carlo iterations have been used to reach good statistical accuracy; on larger lattices the volume factor produces comparable accuracy with a lot less iterations.) The mean value of $\bar{\phi}\phi$ decreases drastically when going from the bosonic system to the fermionic one. Indeed, the fermionic determinant at numerator decreases the probability of encountering those gauge field configurations leading to small eigenvalues of K and to large $\text{Tr}\,K^{-1}$. Conversely these configurations are enhanced in the bosonic case.

6. Monte Carlo simulations: 16×16 lattice

We have performed simulations for the coupled system of gauge variables and pseudofermionic variables on a 16×16 lattice at $\beta=3$. We started from the

TABLE 1

Numbers determined for the expectation value $\langle\bar{\phi}\phi\rangle$ on a small (2×2) lattice either with a direct Monte Carlo simulation of the gauge system with the appropriate effective action or using the algorithm of ref. [2]

β	μ	Fermionic/ bosonic system	Direct	New algorithm	n
3	1	F	0.158 ± 0.002	0.156 ± 0.004	5
3	1	B	0.677 ± 0.007	0.667 ± 0.005	5
0.2	0.4	F	0.267 ± 0.003	0.285 ± 0.006	5
0.2	0.4	B	$3.10\ \pm0.03$	$3.10\ \pm0.4$	1

equilibrium configuration of the pure gauge system, already used in the quenched simulations, and vanishing pseudofermionic fields. 1000 iterations, with $n=5$ (hence altogether 1000 Monte Carlo steps per U^{μ}_{ij}, 5000 Monte Carlo steps per ϕ_{ij}) were made keeping the mass parameter fixed at $\mu=1.2$. The final configuration of the U and ϕ variables was recorded on magnetic tape and used as the initial configuration for a run of 1000 iterations at $\mu=1$. The final configuration was again used

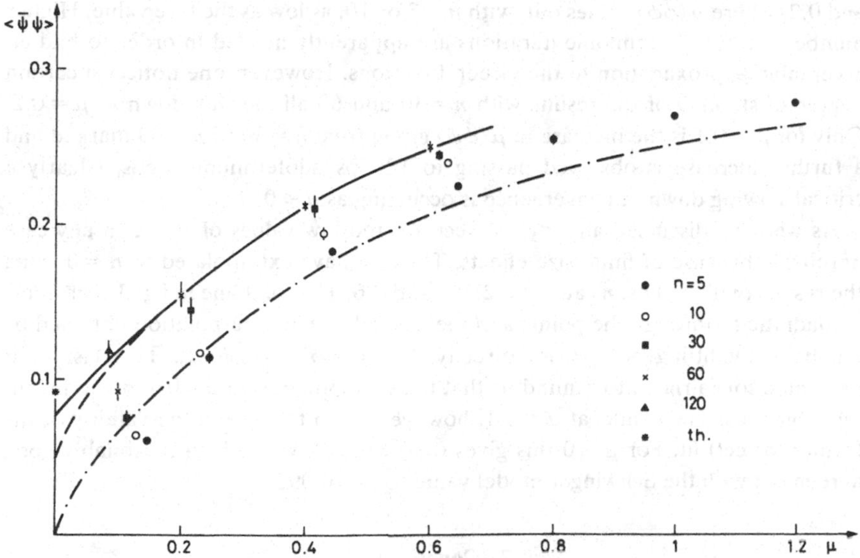

Fig. 3. Results for $\langle\bar{\psi}\psi\rangle$ obtained with the method of ref. [2], and with 5 (•), 10 (○), 30 (■), 60 (×) and 120 (▲) pseudofermionic steps per bosonic iteration (the marks are offset to avoid overlap). The legend for * and the lines is the same as in fig. 2.

as the initial configuration for a run at $\mu = 0.8$ and, repeating the procedure, runs of 1000 iterations, all with $n = 5$, were made at progressively lower values $\mu = 0.6$, 0.4, 0.2 and 0.1. The final configurations from the simulations at $\mu = 0.6$–0.1 were also used as initial configurations for runs at the same values of the mass, but $n = 10$. The final configurations of these simulations were used as starting ones for runs with $n = 30$ and in the same way simulations with $n = 60$ and, for $\mu = 0.1$, $n = 120$ were also made. With the highest values of n smaller numbers of bosonic iterations (from a few hundred down to one hundred for $n = 120$) were used to evaluate the averages.

The details of the upgrading procedure are the same as those described in sect. 5. With a reasonably optimized programme, the CP time requirement on a CDC 7600 was contained within an acceptable $(25+15n)$ milliseconds per iteration.

The results for $\mu\langle\bar{\phi}\phi\rangle$ are presented in fig. 3 (where the free-field values, given by the dashed-dotted line, are also reproduced). Comparing the quenched values (fig. 2) with the values obtained with the full simulation, one immediately notices that the inclusion of the feedback from the fermions tends to lower the expectation value of $\bar{\psi}\psi$. This is in agreement with what one expects from the rôle played by the fermionic determinant (cf. the discussion at the end of the previous section) and is a welcome modification, since in the $\mu \to 0$ limit the quenched values appeared too high. With a small number ($n = 5$, 10) of pseudofermionic steps, however, the algorithm seems to over-correct, in particular for smaller values of the mass ($\mu = 0.1$ and 0.2) where $\mu\langle\bar{\phi}\phi\rangle$ comes out, with $n = 5$ or 10, as low as the free value. Higher numbers of pseudofermionic iterations are apparently needed in order to find an acceptable approximation to the Green functions. However, one notices a certain degree of stability of the results with $n = 30$ and 60 all the way down to $\mu = 0.2$. Only for $\mu = 0.1$ is the increase in $\mu\langle\bar{\phi}\phi\rangle$ going from $n = 30$ to $n = 60$ marked and a further increase is observed passing to 120 pseudofermionic steps. Clearly a critical slowing down of convergence is occurring as $\mu \to 0$.

As we have discussed already (cf. sect. 4), too low values of μ are in any case unreliable because of finite size effects. Thus we have extrapolated to $\mu = 0$ using the results (with highest n) at $\mu = 0.2$, 0.4 and 0.6. The solid line in fig. 3 represents a quadratic fit through the points and the dashed line an interpolation obtained by quadratically fitting not $\mu\langle\bar{\phi}\phi\rangle$ directly, but $\mu\langle\bar{\phi}\phi\rangle - \mu\langle\bar{\phi}\phi\rangle_{\text{free}}$. This last fit is presented for fairness, to remind us that the extrapolation to $\mu = 0$ is by no means unambiguous. The results at $\mu = 0.1$, however, seem to support the validity of the former (direct) fit. For $\mu = 0$ this gives $\langle\bar{\psi}\psi\rangle \cong 0.077$, which is in reasonably good agreement with the Schwinger model value $\langle\bar{\psi}\psi\rangle \cong 0.092$.

7. Conclusions

In this article we have shown that the "defermionization" procedure proposed in ref. [2] can be efficiently used to study fermionic systems with a high number of degrees of freedom. Other, somehow related, methods have recently been presented in the literature [14–16]. The basic ideas of integrating out the fermionic degrees of freedom and of evaluating the ratio of the determinants by means of Green functions are maintained: the alternative proposals differ, however, from ref. [2] in the technique used to compute the Green functions themselves. The relaxation method of ref. [14] seems to require definitely too long CP times. More promising appears the suggestion of refs. [15, 16] of upgrading the Green function, once calculated, by a linearized formula. All these methods, however, imply an increase of CP time by factors at least proportional to the number of degrees of freedom (rather than the fixed number of pseudofermionic steps). Moreover, all the Green functions must be computed and stored and the memory requirements become soon unmanageable as the size of the system increases. For smaller systems, however, the results are satisfactory.

Maybe some combination of our method and these alternative proposals will lead, even for larger systems, to faster and more accurate results, than can be achieved with our technique alone. At present, however, our method has allowed a numerical verification of the spontaneous breaking of chiral symmetry, with a value of $\langle\bar{\psi}\psi\rangle$ consistent with the exact results for the continuum theory.

References

[1] M. Creutz, L. Jacobs and C. Rebbi, Phys. Rev. Lett. 42 (1979) 1390; Phys. Rev. D20 (1979) 1915; M. Creutz, Phys. Rev. Lett. 43 (1979) 553; Phys. Rev. D21 (1980) 2308; Phys. Rev. Lett. 45 (1980) 313;
K. Wilson, Cornell preprint (1980);
C. Rebbi, Phys. Rev. D21 (1980) 3350;
G. Bhanot and C. Rebbi, Nucl. Phys. B180 [FS2] (1981) 469

[2] F. Fucito, E. Marinari, G. Parisi and C. Rebbi, Nucl. Phys. B180 [FS2] (1981) 369

[3] T. Matthews and A. Salam, Nuovo Cim. 12 (1954) 563; 2 (1955) 120

[4] F. Fucito and E. Marinari, Nucl. Phys. B190 [FS3] (1981) 266

[5] J. Schwinger, Phys. Rev. 125 (1962) 397; 128 (1962) 2425

[6] J.H. Lowenstein and J.A. Swieca, Ann. of Phys. 68 (1971) 172

[7] A. Casher, J. Kogut and L. Susskind, Phys. Rev. D10 (1974) 732

[8] S. Coleman, R. Jackiw and L. Susskind, Ann. of Phys. 93 (1975) 267;
S. Coleman, Ann. of Phys. 101 (1976) 239

[9] D.H. Weingarten and J.L. Challifour, Ann. of Phys. 123 (1979) 61;
D.H. Weingarten Ann. of Phys. 126 (1980) 154

[10] B.E. Baaquie, ICTP, Trieste preprint (1980)

[11] H.B. Nielsen and M. Ninomiya, Rutherford preprint (1980); Nordita preprint (1981);
L. Karsten and J. Smit, Stanford preprint (1980)

[12] T. Banks, S. Raby, L. Susskind, J. Kogut, D.R.T. Jones, P.N. Scharbach and D.K. Sinclair, Phys. Rev. D15 (1977) 1111;
L. Susskind, Phys. Rev. D16 (1977) 3031

[13] R.P. Feynman, Phys. Rev. 84 (1951) 108;
R.P. Feynman and M. Gell-Mann, Phys. Rev. 109 (1958) 193

[14] D.N. Petcher and D.H. Weingarten, University of Indiana preprint (1980)

[15] D.J. Scalapino and R.L. Sugar, Phys. Rev. Lett. 46 (1981) 519

[16] A. Duncan and M. Furman, Columbia University preprint (1981)

PHYSICAL REVIEW LETTERS

VOLUME 46 — 23 FEBRUARY 1981 — NUMBER 8

Method for Performing Monte Carlo Calculations for Systems with Fermions

D. J. Scalapino and R. L. Sugar
Institute for Theoretical Physics, University of California, Santa Barbara, California 93106
(Received 1 December 1980)

A method is presented for carrying out Monte Carlo calculations for field theories with fermion degrees of freedom. As an example of this technique, results are given for a simple one-dimensional model.

PACS numbers: 03.70.+k, 11.10.Np

Recently Monte Carlo calculations have been used to study a variety of field-theory problems in condensed matter and high-energy physics. To date this technique has been applied only to boson systems. The difficulty with treating fermions is that in the path-integral formulation of field theory, they are represented by anticommuting c-number fields which do not lend themselves to direct numerical calculations. However, Fucito, Marinari, Parisi, and Rebbi have just made a very interesting proposal for performing Monte Carlo calculations of systems with fermions.[1] In this note we present an alternative, but closely related, method for carrying out such calculations. We illustrate our approach by studying a simple one-dimensional field-theory model.

Let us consider the interaction of a boson field A_i with a fermion field ψ_i. We work on a lattice and the subscripts on the fields refer to the lattice points. For simplicity we suppress spin and internal-symmetry labels. We take the Euclidean action to be

$$S = S_0(A) + \sum_{i,j} \bar{\psi}_i O_{ij}(A)\psi_j. \tag{1}$$

The matrix $\underline{O}$ contains the kinetic-energy terms for the fermion field as well as the coupling terms between the boson and fermion fields. It is crucial that the action be bilinear in the fermion field. Most systems of interest are of this form or can be reduced to it by the introduction of auxiliary fields.

In the usual way, we start by integrating out the fermion field. For example, the fermion correlation function can be written as

$$\begin{aligned}\langle \bar{\psi}_i \psi_j \rangle &= z^{-1} \int \delta A\, \delta\bar{\psi}\, \delta\psi\, \exp(-S)\bar{\psi}_i\psi_j \\ &= z^{-1}\int \delta A \exp[-S_0(A)][\underline{O}^{-1}(A)]_{ij} \det[\underline{O}(A)],\end{aligned} \tag{2}$$

where z is the normalization integral

$$\begin{aligned} z &= \int \delta A\, \delta\bar{\psi}\, \delta\psi\, \exp(-S) \\ &= \int \delta A \exp[-S_0(A)] \det[\underline{O}(A)]. \end{aligned} \tag{3}$$

Clearly, all quantities of interest can be obtained from functional integrals with respect to A with an effective action given by

$$\exp[-S_{\text{eff}}(A)] = \exp[-S_0(A)] \det[\underline{O}(A)]. \tag{4}$$

Of course, Eq. (4) makes sense only if $\det[\underline{O}]$ has a definite sign.

Let us now imagine carrying out the functional integral over A with use of the Metropolis Monte Carlo method.[2] We wish to bring the system into equilibrium with the probability of any field configuration A being proportional to $\exp[-S_{\text{eff}}(A)]$. To this end we repeatedly generate random changes in A, $A \to A + \delta A$, which in turn generate changes in the matrix $\underline{O}$, $\underline{O}(A) \to \underline{O}(A+\delta A) \equiv \underline{O}(A) + \delta\underline{O}$. We accept or reject such a change depending on whether the quantity $\exp[-S_{eff}(A+\delta A) + S_{eff}(A)]$ is greater or less than a random number between 0 and 1. To make this comparison we must evaluate the ratio of determinants

$$\Delta = \det[\underline{O}(A+\delta A)]/\det[\underline{O}(A)] = \det[1 + \underline{G}(A)\delta\underline{O}], \tag{5}$$

where $\underline{G}(A) = \underline{O}^{-1}(A)$. At first glance, the calculation of Δ appears to be prohibitively lengthy because on a lattice with N sites, $\underline{G}$ and $\delta\underline{O}$ are $N \times N$ matrices.

Fucito *et al.*[1] suggest that we restrict ourselves to small variations in A, so that $\delta\underline{O}$ will be small. Then $\Delta \simeq \mathrm{tr}(\underline{G}\,\delta\underline{O})$. They further suggest that the elements of $\underline{G}$ can be obtained from a Monte Carlo calculation by introducing an auxiliary boson field φ governed by an action

$$S_\varphi = \sum_{i,j} \varphi_i^* O_{ij} \varphi_j. \tag{6}$$

It would appear that the major difficulty with the proposal of Fucito *et al.* is the restriction to small variations in A. One can easily imagine systems with more than one local minimum in $S_{eff}(A)$ which it would be difficult to leave by making only a succession of small variations in A. However, we can eliminate this restriction. In practical Monte Carlo calculations, one sweeps through the lattice, making a change in A at one lattice site (or on one lattice link) at a time. In a theory with local couplings a change in A at the site l will induce a change in $O_{ij}(A)$ only for values of i and j in the vicinity of l. By writing Δ in the form $\Delta = \exp[\mathrm{tr}\ln(1+\underline{G}\,\delta\underline{O})]$ and expanding the logarithm as a power series in $\underline{G}\,\delta\underline{O}$, we see that if δO_{ij} is nonzero only for L values of i and of j, then in order to obtain Δ, we need only calculate the determinant of an $L \times L$ matrix. In particular, for nonderivative couplings, a change in A at the site l will induce a change in $\underline{O}$ of the form $\delta O_{ij} = c\delta_{il}\delta_{jl}$, so that

$$\Delta = 1 + cG_{ll}(A). \tag{7}$$

Thus, the restriction to small changes in A is eliminated.

At this point one could simply adopt the proposal of Fucito *et al.* and obtain the needed elements of $G_{ij}(A)$ by a Monte Carlo calculation. However, it may be useful to take further advantage of the locality of $\delta\underline{O}$. Suppose one knows $\underline{G}$ at some particular field configuration A. Then

$$\underline{G}(A+\delta A) = \underline{G}(A) - \underline{G}(A)(\delta\underline{O})\underline{G}(A+\delta A). \tag{8}$$

If $\delta\underline{O}$ has only L^2 nonzero matrix elements, then Eq. (8) can be solved for $\underline{G}(A+\delta A)$ simply by inverting an $L \times L$ matrix. For our special case of nonderivative coupling,

$$G_{ij}(A+\delta A) = G_{ij}(A) - G_{il}(A)G_{lj}(A)c[1 + cG_{ll}(A)]^{-1}. \tag{9}$$

An extreme procedure would be to start the calculation with a trivial field configuration, such as a constant A field, so that $\underline{G}(A)$ could be obtained analytically. Then update G_{ij} each time A is changed with use of the exact formula of Eq. (8) or Eq. (9). We have applied this procedure to the simple one-dimensional model defined by

$$S_0 = \sum_{i=1}^{N} A_i^2, \quad O_{ij}(A) = D_{ij} + (m + gA_i^2)\delta_{ij}, \tag{10}$$

where the derivative matrix, D_{ij}, has $D_{ii} = 2$, $D_{i,i+1} = D_{i+1,i} = D_{1,N} = D_{N,1} = -1$, and all other elements zero. In Table I, we compare typical results of our Monte Carlo calculation of the fermion correlation function with the exact answer, which is easily seen to be

$$c(i-j) = \overline{\psi}_i\psi_j = [\underline{D} + (m + \tfrac{1}{2}g)\underline{I}]_{ij}^{-1}. \tag{11}$$

The results shown in Table I are for $m = g = 1$, with $N = 10$. They were obtained by making 1100 passes through the lattice. The first 100 were to allow the system to reach equilibrium; thereafter, data were collected every tenth pass.

Although the above procedure worked extremely well for our one-dimensional model, we do not expect it to be directly applicable to multidimensional systems. It simply takes too long to update the $N \times N$ matrix $G_{ij}(A)$, even by so simple

TABLE I. The fermion correlation function defined in Eq. (11) for $g = m = 1$. In column 1 we give results of the Monte Carlo calculation described in the text; and in column 2, the exact result.

	Monte Carlo	Exact
$c(0)$	0.3492	0.3482
$c(1)$	0.1099	0.1093
$c(2)$	0.0346	0.0343
$c(3)$	0.0110	0.0109
$c(4)$	0.003 75	0.003 71
$c(5)$	0.002 15	0.002 12

a formula as Eq. (9). We can, however, imagine dividing the lattice into p blocks with N/p sites per block. As we sweep through the lattice, updating A at each site, we can calculate G_{ij} within a block by a Monte Carlo calculation as suggested by Fucito *et al.*, but at each step within the block we can update G_{ij} by Eq. (8) or Eq. (9). We will thus have to make p, rather than N, Monte Carlo calculations during each sweep through the lattice, and in our recent updating of G_{ij} we will only have to work with an $(N/p)\times(N/p)$ matrix. The optimum choice for p will, of course, depend on the system being studied.

Finally, we remark that it may be possible to find alternative methods for calculating the matrix elements of $\underline{G}$. For example, consider the equation

$$\underline{O}\,\underline{g}=\underline{h}, \tag{12}$$

where $\underline{g}$ and $\underline{h}$ are N-component column vectors. If we choose the elements of h to be $h_i=\delta_{ij}$, then $g_i=G_{ij}$. Since $\underline{O}$ will generally be a very sparse matrix, it may be possible to solve for the g_i rapidly. For example, in the model which we have discussed, the elements of $\underline{g}$ satisfy a three-term recursion relation which can be solved in $2N$ steps, a considerable saving over the N^2 steps necessary to update the matrix $\underline{G}$ by Eq. (9). Alternatively, one may be able to solve Eq. (12) rapidly by relaxation methods.

Although it is far from clear that we have arrived at the optimal procedure, it does seem certain that these techniques can be applied to a number of interesting systems, particularly lower-dimensional ones in condensed-matter physics.

We would like to thank J. Richardson, S. Shenker, and D. Toussaint for helpful discussions. This work was supported by the National Science Foundation under Grant No. PHY-77-27084.

[1]F. Fucito, E. Marinari, G. Parisi, and C. Rebbi, CERN Report No. CERN-TH-2960 (to be published).

[2]N. Metropolis, A. W. Rosenbluth, A. H. Teller, and E. Teller, J. Chem. Phys. 21, 1087 (1953).

VOLUME 49, NUMBER 3 PHYSICAL REVIEW LETTERS 19 JULY 1982

Stochastic Method for the Numerical Study of Lattice Fermions

Julius Kuti
Institute for Theoretical Physics, University of California, Santa Barbara, California 93106
(Received 11 January 1982)

A new stochastic method for the numerical study of lattice fermions is presented. Its efficiency is demonstrated on a field-theoretic model in four dimensions with coupled boson and fermion degrees of freedom. The exact fermion propagator is calculated and agrees very accurately with the numerical results of the stochastic procedure on finite lattices of 10^4 and $8^3 \times 16$ sites, respectively. The contribution of fermionic vacuum polarization to mass renormalization is evaluated with precision. The method is directly applicable to quantum chromodynamics.

PACS numbers: 11.15.Ha, 02.70.+d

During the last twelve months we have witnessed considerable effort to develop Monte Carlo methods for the numerical study of quantum systems with fermionic degrees of freedom. This outstanding problem is of great importance for applications in quantum field theories, condensed matter physics, and nuclear physics.

Previous techniques[1-9] were slow when a very large number of fermionic degrees of freedom were involved, since the computational time required for a Monte Carlo sweep through the lattice was always proportional to the square of the crystal volume,[10] or even worse.

Recently, this problem was solved by Hirsch *et al.* in one space and one time dimension.[11] They follow the evolution of fermion world lines along the Euclidean time direction with an update time independent of the lattice volume. The method is fast and efficient in applications. The generalization of this ingenious idea to higher dimensions is desirable.

I will follow here the more standard strategy and work directly with a new effective action of the boson fields when the fermionic degrees of freedom are integrated out. Though the effective action becomes nonlocal in the presence of the fermion determinant, the new procedure maintains the efficiency of the standard Monte Carlo technique where the update time on a site is independent of the lattice volume. The method is applicable in any number of dimensions.

For a general presentation, I will consider now the Euclidean action

$$S=S_0(U)+\sum_{ij}\bar{\psi}_i M_{ij}(U)\psi_j \tag{1}$$

in four dimensions. It describes the interaction of a boson field U_i with a fermion field ψ_i, and the subscripts on the fields refer to the lattice points. Spin and internal symmetry indices are suppressed, for simplicity. The matrix $M_{ij}(U)$ designates both kinetic and mass terms for the fermion field, and couplings to the boson field. $S_0(U)$ describes the pure bosonic part of the Euclidean action. It is important to note that most of the interesting models in quantum field theory, condensed matter physics, and nuclear physics can be brought to a bilinear form in the fermion fields.

The fermion Green's functions can be calculated by inserting sources into the path integral

$$Z(\bar{\eta},\eta)=\int \mathfrak{D}\bar{\psi}\,\mathfrak{D}\psi\,\mathfrak{D}U\exp[-S+\textstyle\sum_i(\bar{\eta}_i\psi_i+\bar{\psi}_i\eta_i)]. \tag{2}$$

By taking the functional derivatives and integrating out the Grassman variables, the fermion correlation function can be written as

$$\langle\bar{\psi}_i\psi_j\rangle=\frac{\delta^2}{\delta\eta_i\,\delta\bar{\eta}_j}\ln Z(\bar{\eta},\eta)|_{\bar{\eta}=\eta=0}$$
$$=Z^{-1}\int\mathfrak{D}U\,M_{ij}^{-1}(U)\exp[-S_{\rm eff}(U)], \tag{3}$$

where Z is the partition function (normalization integral) of the boson-fermion system. The effective action is given by

$$\exp[-S_{\rm eff}(U)]=\det[M(U)]\exp[-S_0(U)], \tag{4}$$

and I assume, for simplicity only, that the fermion determinant has positive sign.

I apply now the Metropolis Monte Carlo method to the evaluation of the functional integral in Eq. (3). Other Euclidean Green's functions can be treated similarly.

It was shown by Scalapino and Sugar,[2] and by Fucito *et al.*,[1] that a local change $U\to U+\delta U$ implies

$$\frac{\exp[-S_{\rm eff}(U+\delta U)]}{\exp[-S_{\rm eff}(U)]}=\det[1+M^{-1}(U)\delta M(U)]\times\frac{\exp[-S_0(U+\delta U)]}{\exp[-S_0(U)]}. \tag{5}$$

With local boson-fermion coupling the nontrivial change δM in the fermion matrix is restricted to the neighborhood of the updated lattice site. Consequently, we need only a few inverse elements of the large matrix M in each Metropolis step.

At that point I depart from standard procedures. Since the results of a Monte Carlo calculation are always subject to some statistical inaccuracy, it is reasonable to evaluate the decision-making step stochastically. The error analysis becomes subtle,[12] but I am not concerned with it here. I will calculate the inverse matrix elements of M by some modification of a stochastic method which was first suggested by J. von Neumann and S. M. Ulam, but never published by them.[13] It is a very efficient method for the approximate summation of the von Neumann series defined by the inverse of the operator M.

Assume that the inverse of a matrix M of order m is desired and let $H=I-M$, where I is the unit matrix. For the method to be applicable, it is necessary and sufficient that the eigenvalues of the matrix $\bar{H}_{ij}=|H_{ij}|$ are less than 1 in absolute value. Note that the above condition can always be arranged by proper normalization. The matrix elements $(M^{-1})_{ij}$ are given by the solutions of the linear system of equations $Mx=b$, with unit driving vectors on the right-hand side. This equation is equivalent to $(2/\mu)M^\dagger Mx=(2/\mu)M^\dagger b$, where μ is the first norm of the matrix $M^\dagger M$. The driving vector $(2/\mu)M^\dagger b$ may be decomposed into a linear combination of unit vectors and, with the replacement $M\to(2/\mu)M^\dagger M$, the method applies even in the worst case.

I decompose the matrix element H_{ik} into $H_{ik}=P_{ik}R_{ik}$ with the restriction that $P_{ik}>0$ and

$\sum_{r=1}^{m} P_{ir} < 1$ for all values of i. Consider a random walk on the domain of integers $1, 2, \ldots, m$. The walk begins at some selected point i and proceeds from point to point with the transition probabilities P_{ik}. The walk stops after k steps at some point s_k with the stop probability $P_{s_k} = 1 - \sum_{r=1}^{m} P_{s_k r}$. When the walk stops, a score S_{ij} is registered for the elements in the ith row of the inverse matrix. It is defined by the product of the residues $R_{s_r s_{r+1}}$ along the trajectory $i \to s_1 \to s_2 \to \cdots \to s_k = j$ divided by the stop probability P_j:

$$S_{ij} = \begin{cases} 0 & \text{if } s_k \neq j \\ R_{is_1} R_{s_1 s_2} \cdots R_{s_{k-1} j} P_j^{-1} & \text{if } s_k = j. \end{cases} \tag{6}$$

I will prove that the expectation value of the random variable S_{ij} is $(M^{-1})_{ij}$. Indeed, the probability of a walk to follow some trajectory $i \to j$ and to stop at j is $P(i \to j) P_j = P_{is_1} P_{s_1 s_2} \cdots P_{s_{k-1} j} \times P_j$. The expected score is given by the sum over all trajectories from i to j:

$$\langle S_{ij} \rangle = \sum_{i \to j} P(i \to j) P_j S_{ij} = \sum_{i \to j} P(i \to j) R(i \to j), \tag{7}$$

where $R(i \to j)$ is the product of the residues along the trajectory. Since $P_{ij} R_{ij} = H_{ij}$, Eq. (7) is recognized as the von Neumann series expansion for $M^{-1} = (I - H)^{-1}$. The term δ_{ij} in the von Neumann series is generated by walks which stop immediately.

It is easy to prove that the variance σ_{ij}^2 of the random variable S_{ij} is given by

$$\sigma_{ij}^2 = (Q^{-1})_{ij} P_j^{-1} - (M^{-1})_{ij}^2,$$

where $Q = (I - K)^{-1}$ with $K_{ij} = H_{ij} R_{ij}$. The variance of S_{ij} is finite, provided the von Neumann series for $Q = (I - K)^{-1}$ exists.

The statistical error on $(M^{-1})_{ij}$ is given by $\sigma_{ij}/\sqrt{N}$ for N walks which all begin at point i. For a given statistical accuracy in the decision-making step of the Metropolis procedure, the required number of walks does not depend on the size of the matrix. Therefore, the update time in this stochastic procedure is independent of the lattice volume. My tests involved matrices of the order of 10^4, or larger.

I will now modify the von Neumann–Ulam algorithm for better efficiency in fermionic Monte Carlo procedures. It is easy to realize that during a walk which started at point i, one can register the product of residues at each pass through the point j. I define a new random variable $\tilde{S}_{ij}$ as the sum of the products of residues, adding a new term to the score at each pass through the point j. The stop probabilities are eliminated from the random variable $\tilde{S}_{ij}$, but they still govern the average length of a walk.

It is straightforward to show that $\langle S_{ij} \rangle = \langle \tilde{S}_{ij} \rangle$ when the stop probability P_j is positive. I have also proved[11] that the expectation value $\langle \tilde{S}_{ij} \rangle$ is equal to $(M^{-1})_{ij}$ when the stop probability P_j vanishes. The original method does not apply in this case.

In order to compare efficiencies, I choose a simple case when all $R_{ij} = 1$ and P_j is positive. A necessary and sufficient condition[12] for the variance of the random variable $\tilde{S}_{ij}$ to be smaller than σ_{ij}^2 is $P_j < e_j/(2 - e_j)$, where e_j designates the escape probability from the point j. In practice, this condition is enforced by the nature of the fermion problem, and the modified method is much more efficient.

In the special case when all stop probabilities vanish, one has to stop by fiat. Some bias is introduced then, since the von Neumann series is truncated after a finite number of terms. The modification described above is probably known to some experts on stochastic methods and the special case when the random walk is stopped by fiat appears in the work of Bakhvalov.[14]

I tested my stochastic fermion method on a four-dimensional boson-fermion model which was first suggested by Scalapino and Sugar.[2,7] The fermion matrix M in Eq. (1) is specified now as

$$M_{ij} = -\Delta_{ij} + (m^2 + gU_i^2)\delta_{ij}, \tag{8}$$

where U_i and ψ_i are a scalar boson field and a spinless fermion field, respectively. Δ_{ij} defines the Laplacian operator on the lattice, m is the bare fermion mass in lattice spacing units, and g designates the dimensional boson-fermion coupling constant. The functional integral is calculable analytically in this model,[2] and one finds

$$D(i-j) = \langle \bar{\psi}_i \psi_j \rangle = (-\Delta + m^2 + \tfrac{1}{2} g)_{ij}^{-1}. \tag{9}$$

The fermion-boson interaction generates a mass term dynamically, and the renormalized fermion mass is given by $m_r = (m^2 + \frac{1}{2} g)^{1/2}$.

Some results of my calculations are presented in Fig. 1. The complete fermion mass was generated dynamically with the choice $m = 0$. The agreement of the numerical points with the exact analytic form is very satisfactory (the statistical errors are practically not visible on the logarithmic plot). The quenched approximation,[15-17] where one neglects the fermionic vacuum polarization effects from the fermion determinant in

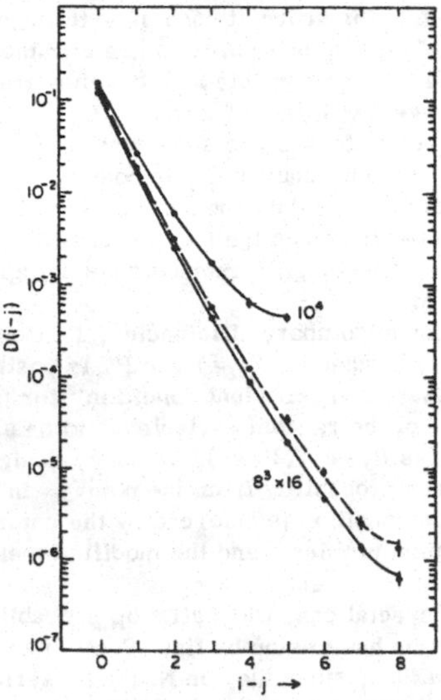

FIG. 1. Some numerical results on the fermion correlation function compared with exact calculations. $D(i-j)$ is depicted for a lattice of $8^3\times 16$ sites with periodic boundary conditions and coupling constant $g=2.6a^{-2}$ (in the actual calculations the lattice spacing a was set to unity). The free fermion propagator, with renormalized mass $m_r^{\,2}=1.3a^{-2}$ on the same lattice size, is represented by the solid line. Results are also presented for a lattice of 10^4 sites with $m_r^{\,2}=0.25a^{-2}$. The dashed line is the fit of a free fermion propagator of mass $m_q^{\,2}=1.02a^{-2}$ to the results of the quenched approximation on the $8^3\times 16$ periodic lattice with $g=2.6a^{-2}$ and $m=0$. The continuous curves for the exact propagators are drawn to guide the eye.

the effective action, is also presented in Fig. 1. The contribution of the fermion loops is clearly seen and accurately calculated: $m_q^{\,2}/m_r^{\,2}=0.78$.

The speed and efficiency of my stochastic method is very promising. It took only about 3 h on a VAX 11/780 to calculate the complete fermion propagator on the $8^3\times 16$ lattice, with relative errors which are only a few percent even for a separation of eight links along the fourth direction. Details of the numerical results will be published elsewhere.[12]

Some work is in progress now on the numerical solution of quantum chromodynamics,[18] and the Hubbard model in two and three spatial dimensions.[19] I am also working on further theoretical improvements in the stochastic method.

I would like to acknowledge very useful discussions with J. E. Hirsch, K. Johnson, J. Richardson, R. L. Sugar, D. Toussaint, and F. Wilczek. I am grateful for the kind hospitality extended to me at the Institute for Theoretical Physics in Santa Barbara. This research was supported in part by the National Science Foundation under Grant No. PHY77-27084.

---

[1]F. Fucito, E. Marinari, G. Parisi, and C. Rebbi, Nucl. Phys. B180 [FS2], 369 (1981).

[2]D. J. Scalapino and R. L. Sugar, Phys. Rev. Lett. 46, 519 (1981).

[3]D. Weingarten and D. Petcher, Phys. Lett. 99B, 333 (1981).

[4]R. Blankenbecler, D. J. Scalapino, and R. L. Sugar, Phys. Rev. D 24, 2278 (1981).

[5]D. J. Scalapino and R. L. Sugar, Phys. Rev. B 24, 4295 (1981).

[6]E. Marinari, G. Parisi, and C. Rebbi, Nucl. Phys. B190 [FS3], 734 (1981).

[7]H. Hamber, Phys. Rev. D 24, 951 (1981).

[8]A. Duncan and M. Furman, Nucl. Phys. B190 [FS3], 767 (1981).

[9]Y. Cohen, S. Elitzur, and E. Rabinovici, to be published.

[10]A possible escape from this difficulty with use of Wilson's hopping parameter expansion was recently advocated by I. O. Stamatescu, Max Planck Institute Report No. MPI-PAE/PTh 60/80, 1980 (to be published); A. Hasenfratz and P. Hasenfratz, Phys. Lett. 104B, 489 (1981).

[11]J. E. Hirsch, D. J. Scalapino, R. L. Sugar, and R. Blankenbecler, Phys. Rev. Lett. 47, 1628 (1981).

[12]J. Kuti, to be published.

[13]G. E. Forsythe and R. A. Leibler, Math. Tables Other Aids Comput. 4, 127 (1950).

[14]N. F. Bakhvalov, *Numerical Methods* (MIR Publishers, Moscow, 1977).

[15]H. Hamber and G. Parisi, Phys. Rev. Lett. 47, 1792 (1981).

[16]E. Marinari, G. Parisi, and C. Rebbi, Phys. Rev. Lett. 47, 1795 (1981).

[17]D. Weingarten, to be published.

[18]J. Kuti, to be published.

[19]J. E. Hirsch and J. Kuti, to be published.

VOLUME 47, NUMBER 25 PHYSICAL REVIEW LETTERS 21 DECEMBER 1981

Numerical Estimates of Hadronic Masses in a Pure SU(3) Gauge Theory

H. Hamber
Department of Physics, Brookhaven National Laboratory, Upton, New York 11973

and

G. Parisi
Istituto Nazionale di Fisica Nucleare, Frascati, Italy, and Istituto di Fisica della Facoltà di Ingegneria, Rome, Italy

(Received 2 October 1981)

In lattice quantum chromodynamics, the hadronic mass spectrum is evaluated by computer simulations in the approximation where closed quark loops are neglected. Chiral symmetry is shown to be spontaneously broken and an estimate of the pion decay constant is given.

PACS numbers: 12.70.+q, 11.10.Np, 11.30.Jw, 12.40.Cc

In this Letter we present results of a computation of the mass spectrum of the lighter hadrons in the SU(3) lattice gauge theory in the approximation of neglecting internal quark loops. Although these effects will have to be taken into account in a full calculation, we shall see that reasonable results for the spectrum can be obtained within this framework. This approximation enforces the Zweig rule for all the flavors, becomes exact in the limit $N\to\infty$, and can easily be justified for the mass spectrum with phenomenological arguments. Some numerical simulations of two-dimensional lattice gauge models also suggest that it might be a reasonable simplification.[1] We have found some evidence for this to be true also in the present case. In this approximation nonet symmetry holds: Closed quark loops are crucial to remove the η-π degeneracy. Similar computations not including the baryons and with only one form of the fermionic action (the Kogut-Susskind action) for the group SU(2) will be published elsewhere.[2]

The mass spectrum of the lighter hadrons can be computed by studying the decay at infinity of the correlation functions of composite operators. The key formulas we use are

$$\langle\bar\psi(x)\psi(x)\bar\psi(0)\psi(0)\rangle = \int d_\mu[A]G(x,0;A)G(0,x;A),$$

$$\langle\bar\psi(x)\bar\psi(x)\bar\psi(x)\psi(0)\psi(0)\psi(0)\rangle \qquad (1)$$

$$= \int d_\mu[A]G(x,0;A)G(x,0;A)G(x,0;A),$$

where we have suppressed flavor, spinor, and color indices, and $G(x,0;A)$ is the inverse of $\not{D}+m$ in a background A_μ gauge-field configuration. D_μ is the covariant derivative and $d_\mu[A]$ is the probability distribution of pure gauge fields. These formulas hold for all operators that do not have the flavor quantum numbers of the vacuum. In the full theory with n_f fermion flavors and vacuum polarization effects included we would have

$$d_\mu[A]=e^{S_G}[\det(\not{D}+m)]^{n_f}\,dU_H,$$

where S_G is the Wilson action for lattice gauge fields given by

$$S_G=g_0^{-2}\sum_{\square}\mathrm{Tr}(U_p+U_p^\dagger-2),$$

and dU_H is the Haar measure for the group SU(3) for each link. The sum is over all elementary squares in the four-dimensional hypercubic lattice of spacing a, and U_p is a product of four SU(3) group elements around each square. In this Letter we will discuss results obtained by setting the determinant equal to 1 ($n_f=0$), which is equivalent to neglecting dynamic fermion loops. The fields A can be extracted by using a standard Monte Carlo simulation technique, while the inverse propagators are computed using iterative matrix inversion methods.[1]

When we implement these methods on the space-time lattice we have to make a choice regarding the fermionic action. In general we can write[3-7]

$$S(\bar\psi,\psi)=\sum_{n,\mu}\bar\psi_n[(\gamma_\mu-r)\psi_{n+\mu}-(\gamma_\mu+r)\psi_{n-\mu}] + M\sum_n\bar\psi_n\psi_n. \qquad (2)$$

If $r=0$ the theory is chirally invariant in the $M\to 0$ limit,[4-6] but unfortunately describes sixteen flavors instead of one. (These can be reduced to four by an appropriate canonical transformation, as discussed in Refs. 4 and 6.) If $r\neq 0$ only one flavor is obtained in the continuum limit, but chiral symmetry is lost on the lattice and can only be recovered in the continuum limit, as dis-

VOLUME 47, NUMBER 25 PHYSICAL REVIEW LETTERS 21 DECEMBER 1981

cussed in Ref. 3.

In this Letter we present results of a computation of the hadronic spectrum in the cases $r=0$ and $r=1$ on lattices $6^3\times 12$ and up to $6^3\times 10$, respectively, and compare the results.

We have used the standard Wilson action for the gauge fields[3] and generated the gauge-field distributions using a modified Monte Carlo method: We did ten trials for each gauge-field variable without changing the others. In this way the Monte Carlo method is very similar to the heat-bath method. We have limited ourselves to a study of the region of β between 5 and 6.2 (we use the notation $\beta=6/g_0^2$). The crossover between the weak- and strong-coupling regimes happens around $\beta\simeq 5$ and in the region we have studied one starts to see the exponential behavior in β of the string tension as predicted by asymptotic freedom. If for definiteness we assume $\Lambda_0=7\times 10^{-3}\sqrt{k}$ (Ref. 7) and $\sqrt{k}=420$ MeV from the ρ-f-g-h trajectory, our inverse lattice spacing ranges from 440 MeV at $\beta=5$ to 1590 MeV at $\beta=6.2$. At $\beta=5.6$ and 6.0 where we have higher statistics we obtain $a^{-1}=660$ MeV and $a^{-1}=1120$ MeV, respectively.

In this interval of β we have computed the expectation value of $\langle\bar\psi\psi\rangle$ at $r=0$ in the mass range $m=0.1$–0.3 on a 6^4 lattice using the Langevin algorithm of Ref. 8 and the iterative method, and then extrapolated to $m=0$ using a quadratic fit. We clearly see evidence for $\langle\bar\psi\psi\rangle\neq 0$ at $m=0$ for all values of β that we have explored. Chiral symmetry is spontaneously broken and the usual Goldstone theorem holds. The results extrapolated to $m=0$ are shown in Fig. 1. The line is a fit of the form

$$\langle\bar\psi\psi\rangle_{m=0}[(3\alpha_B)^{-4/33}R\sqrt{k}]^3 \qquad (3)$$

following the prescription of the renormalization group. Here $\alpha_B=3/2\pi(\beta-2.75)$ and the scale parameter of quantum chromodynamics is defined as

$$\Lambda_{\text{mom}}=\frac{\pi}{a}\left[\frac{8\pi^2}{33}(\beta-2.75)\right]^{51/121}\times\exp\left[-\frac{4\pi^2}{33}(\beta-2.75)\right]. \qquad (4)$$

The data suggest $R=0.90\pm 0.05$ in agreement with phenomenological estimates.[9]

At $r=0$ we have obtained some results for the mass spectrum of the lowest-lying states at $\beta=6.0$ by computing the propagator $G(x,0;A)$ for four different configurations. In the region of m between 0.3 and 0.1 in lattice units we can fit the

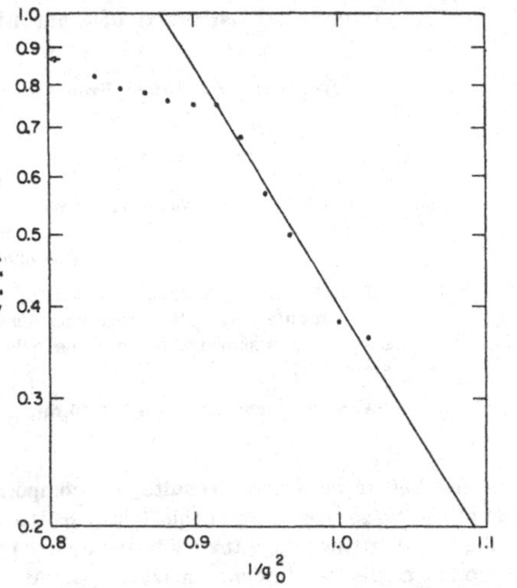

FIG. 1. The quantity $\langle\bar\psi\psi\rangle^{1/3}$ as a function of $1/g_0^2$. The line is a fit that gives $\langle\bar\psi\psi\rangle^{1/3}=1.5(3\alpha_B)^{-4/33}\Lambda_{\text{mom}}$.

data by

$$m_P^2=6m/a,\quad m_V^2=0.5/a^2+6m/a,\quad m_B=1.0/a+12m. \qquad (5)$$

Here m_B^2 is the average mass of the lowest baryonic states [something between the nucleon and the Δ mass; we do not have enough statistics yet to see their mass difference in this formulation ($r=0$)]. This gives the rough estimates

$$m_\rho=800\pm 100\ \text{MeV},\quad m_B=1000\pm 100\ \text{MeV}. \qquad (6)$$

The linear dependence of the pion mass on the bare quark mass is a consequence of the Goldstone theorem. If we use the partially conserved axial-vector current relation

$$m\langle\bar\psi\psi\rangle=f_\pi^2 m_\pi^2, \qquad (7)$$

we get $f_\pi=95\pm 10$ MeV from the value of $\langle\bar\psi\psi\rangle$ at $\beta=6$ which is 0.043 ± 0.01 in lattice units. While the masses of the ρ meson and of the baryon do not appreciably change when varying β, we have noticed that f_π varies from the above value at $\beta=6$ to $f_\pi=140\pm 10$ MeV at $\beta=5.6$. This does not come entirely as a surprise since f_π is more or less the pion wave function at the origin and is therefore expected to be a more sensitive quantity than the masses.

We now come to a discussion of the results for $r=1$ (Wilson's fermions). By doing a fit to the data for m between 0.3 and 0.05 (see Fig. 2), where all the masses are between the low-energy cutoff at 590 MeV and the high-energy cutoff at 3520 MeV, one finds at $\beta = 6$

$$m_P^2 = 6.0m/a, \quad m_V^2 = 0.5/a^2 + 6m/a,$$
$$m_S^2 = 0.8/a^2 + 6m/a, \quad m_A^2 = 1.2/a^2 + 6m/a, \qquad (8)$$
$$m_N \; 0.0[illegible]/a + [illegible]m, \quad m_\Delta = 1.1/a + 5m.$$

Here m is, in Wilson's notation, $(k_c - k)/2k_c^2$ with $k = 1/M$, and $k_c \simeq 0.156$ at $\beta = 6$. These results were obtained with fifty different configurations at several different values for the bare quark mass, and the error in the masses is of order 10%.

Using $a^{-1} = 1120$ MeV we get the following estimates (in MeV) (we use as input $m_\pi = 140$ MeV):

$$m_\rho = 800 \pm 100, \quad m_p = 950 \pm 100,$$
$$m_\delta = 1000 \pm 100, \quad m_\Delta = 1300 \pm 100, \qquad (9)$$
$$m_{A_1} = 1200 \pm 100, \quad f_\pi = 95 \pm 10.$$

Analogous estimates can be obtained for strange mesons and baryons. We could have alternatively chosen to fit both the π and the ρ masses to experiment, and this would have given us roughly the above value for the lattice spacing and $\sqrt{T} = 400 \pm 50$ MeV. The bare quark masses turn out to be

$$(3\alpha_B)^{-4/11}(m_u + m_d) = 8 \text{ MeV},$$
$$(3\alpha_B)^{-4/11}(m_d + m_s) = 100 \text{ MeV}, \qquad (10)$$

which then gives 3, 5, and 100 MeV for the u, d, and s invariant quark masses, in agreement with previous phenomenological estimates.[9] The results at $r = 0$ and $r = 1$ seem therefore to be compatible, within statistical errors. However, the $r = 1$ approach seems to be more promising for the study of the spectrum of hadrons since the separation of operators with different spins and parities can be done on a single site. Using the large number of configurations generated for the fermions, we have also estimated the lowest glueball mass by extracting the plaquette-plaquette correlations from finite-size effects[10] at $\beta = 6$, with the result

$$m_G = (3.6 \pm 0.5)\sqrt{k} = 1500 \pm 200 \text{ MeV}. \qquad (11)$$

We plan to improve further the accuracy of the spectrum calculation by increasing the statistics. Calculations of the spectrum on larger lattices are in progress in order to check the smallness of finite-size effects on the lattice. A comparison between the results on a $5^3 \times 8$ and a $6^3 \times 10$ lattice seems to show that these effects are small when the pion mass is not too close to the high- or low-energy cutoffs. A more detailed account of the spectrum calculation will be published elsewhere. We are also considering the possibility of introducing the effects of closed fermion loops.[1,6,11-13] One way of doing it would be to use chirally invariant fermions at $r = 0$ with one component per site for the fermionic loops and the noninvariant formulation at $r = 1$ for the computation of the propagators.

The authors thank M. Creutz, J. Kogut, and K. Wilson for fruitful discussions, and the organizers of the workshop at the University of California, Santa Barbara, for their hospitality where part of this work was done. One of us (G.P.) is grateful for the hospitality at Brookhaven National Laboratory where this work was completed. The authors would also like to thank A. Omero and G. Martinelli for pointing out an early error in our program. Finally, thanks go also to the Isabelle group and the Rome University Istituto Nazionale di Física Nucleare group for use of their VAX/11780.

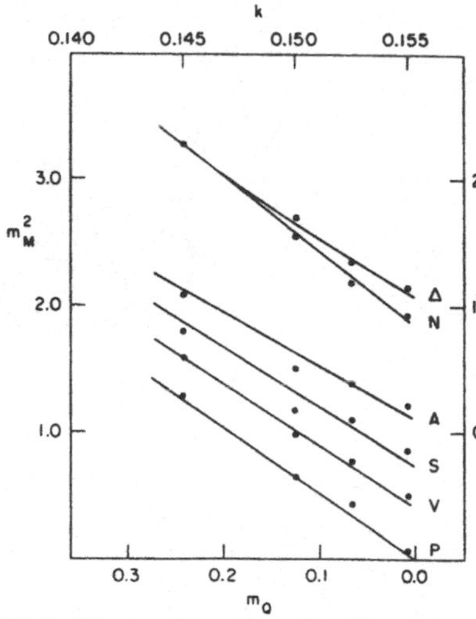

FIG. 2. Meson masses squared and baryon masses as a function of k and the bare quark mass $m_Q = (k_c - k)/2k_c^2$ obtained with use of the Wilson fermion action ($r = 1$) at $\beta = 6$. P, V, S, and A stand for pseudoscalar ($J^{PC} = 0^{-+}$), vector (1^{--}), scalar (0^{++}), and axial vector (1^{++}) masses. N and Δ stand for nucleon ($\frac{1}{2}^+$) and delta ($\frac{3}{2}^+$) masses.

[1]H. Hamber, Phys. Rev. D 24, 951 (1981); E. Marinari, G. Parisi, and C. Rebbi, following Letter [Phys. Phys. Rev. 47, 1795 (1981)]; D. Weingarten, unpublished.

[2]E. Marinari, G. Parisi, and C. Rebbi, to be published.

[3]K. Wilson, Phys. Rev. D 10, 2445 (1974), and in *New Phenomena in Subnuclear Physics*, edited by A. Zichichi (Plenum, New York, 1977).

[4]T. Banks, S. Raby, L. Susskind, J. Kogut, D. Jones, P. Scharbach, and D. Sinclair, Phys. Rev. D 15, 1111 (1977).

[5]M. Weinstein, S. Drell, H. Quinn, and B. Svetitsky, Phys. Rev. D 22, 1190, 490 (1980).

[6]N. Kawamoto and J. Smit, Institute of Theoretical Physics, University of Amsterdam Report No. ITFA-81-2, 1981 (to be published), and references therein.

[7]M. Creutz, Phys. Rev. Lett. 45, 313 (1980); E. Pietarinen, Nucl. Phys. B190, 349 (1981).

[8]F. Fucito, E. Marinari, G. Parisi, and C. Rebbi, Nucl. Phys. B180, 369 (1981).

[9]I. Yoffe, to be published, and references therein.

[10]M. Nauenberg, T. Schalk, and R. Brower, Phys. Rev. D 24, 548 (1981).

[11]D. Weingarten and D. Petcher, Phys. Lett. 99B, 333 (1981).

[12]D. J. Scalapino and R. Sugar, Phys. Rev. Lett. 46, 519 (1981).

[13]A. Duncan and M. Furman, to be published.

VOLUME 47, NUMBER 25 PHYSICAL REVIEW LETTERS 21 DECEMBER 1981

Computer Estimates of Meson Masses in SU(2) Lattice Gauge Theory

E. Marinari
Istituto di Fisica "G. Marconi," Università degli Studi, Roma, Italy, and Istituto Nazionale di Fisica Nucleare, Roma, Italy

and

G. Parisi
Istituto di Fisica, Facoltà di Ingegneria, Università degli Studi, Roma, Italy, and Istituto Nazionale di Fisica Nucleare, Frascati, Italy

and

C. Rebbi
Brookhaven National Laboratory, Upton, New York 11973
(Received 9 October 1981)

It is shown that in an SU(2) lattice gauge theory, in the approximation where internal quark closed loops are neglected, chiral symmetry is broken. With use of partially conserved axial-vector current f_π, the bare masses of the u and d quarks, and the ρ and δ masses are estimated.

PACS numbers: 11.10.Np, 05.50.+q, 12.70.+q, 14.40.-n

Recently some progress has been made in numerical simulations of theories with fermions.[1-3] Although in a complete computation the effects of fermionic closed loops must be taken into account, a reasonable estimate of the hadron spectrum can be obtained by eliminating all internal quark loops (quenched case, see Ref. 2). In this way the Zweig rule is enforced for all flavors. In this note we present a study of chiral-symmetry breaking and of the π, ρ, and δ masses for the SU(2) gauge theory in the quenched approximation. A similar study for the SU(3) gauge theory, including also baryons, can be found in Ref. 4. The results obtained are rather satisfactory.

Let us begin discussing our strategy in the continuum case; later we will adapt it to the lattice version of the model. We consider the fermionic Euclidean action

$$S_f = \int d^D x\, \bar{\psi}(\not{D} + m)\psi\,, \tag{1}$$

where D_μ is the covariant derivative in presence of a gauge field A_μ. If $G(x, 0|A)$ is the fermionic Green function with A_μ as background, and $d\mu[A]$ is the probability distribution of the field A (normalized to 1), the following relations hold:

$$\langle\bar{\psi}(0)\psi(0)\rangle = \int d\mu[A]\,\mathrm{Tr}[G(0,0|A)],$$

$$\langle\bar{\psi}(x)\gamma_5\psi(x)\bar{\psi}(0)\gamma_5\psi(0)\rangle = \int d\mu[A]\,\mathrm{Tr}[G(x,0|A)G^*(x,0|A)]. \tag{2}$$

The fact that $\not{D}$ is an anti-Hermitean operator with spectral density $\rho(i\lambda)$ implies that

$$\lim_{m\to 0}\langle\bar{\psi}(0)\psi(0)\rangle=\frac{\pi}{V}\rho(0),$$

$$\mathrm{Tr}[G(0,0|A)]=m\int d^Dx\,\mathrm{Tr}[G(x,0|A)G^*(x,0|A)]. \quad (3)$$

If $\rho(0)\neq 0$, when $m\to 0$ chiral symmetry is broken, and if the integrand in Eq. (2) is finite in this limit then the Goldstone theorem holds.

A natural procedure to evaluate the expectation values of composite field operators is the following: A_μ field configurations are generated with probability distribution $d\mu[A]$ by a Monte Carlo simulation (suitably generalized if one wants to include the effects of inner fermionic loops[1-3]); the propagator $G(x,0|A)$ is then calculated by Monte Carlo like techniques or by relaxation methods. If the effect of closed loops is included, Eq. (2) (and its obvious generalizations) holds for those operators which do not have the internal quantum numbers of the vacuum.

In the relaxation method one obtains the Green functions as the $t\to\infty$ limit of $G_t(x,0|A)$ satisfying

$$dG_t(x,0|A)/dt=(\not{D}+m)G_t(x,0|A)+\delta(x). \quad (4)$$

On the contrary, direct Monte Carlo simulations cannot be performed with a first-order formalism.[1,2] However, one can adapt the standard Langevin formulation to this case by writing

$$d\varphi_i(x,t)/dt=[(-1)^i\not{D}+m]\varphi_i(x,t)+\eta(x,t)\quad i=1,2, \quad (5)$$

where η is a Gaussian stochastic white noise: $\langle\langle\eta(x,t)\eta(x',t')\rangle\rangle=2\delta(t-t')\delta(x-x')$ (the double angular bracket denotes an average over the noise). It is straightforward to check that

$$\lim_{t\to\infty}\langle\langle\varphi_1(x,t)\varphi_2^\dagger(y,t)\rangle\rangle=\left\langle x\left|\frac{1}{\not{D}+m}\right|y\right\rangle\equiv G(x,y|A). \quad (6)$$

Let us briefly underline the main differences between the Langevin and the relaxation techniques: Using the Langevin equation we can complete $G(x,y|A)$ for all x and y at the same time, while in a comparable computer time the relaxation procedure gives only $G(x,0|A)$. On the other hand the relaxation procedure gives exact results for $G(x,y|A)$, while statistical errors are present with the Langevin method. So we can conclude that to measure $G(x,y|A)$ at $x\sim y$, where G is large, the Langevin approach is the most suitable, whereas for computing G in the large-$|x-y|$ region, where G itself is small, the relaxation method should be used. The second is the situation one encounters in the computation of the mass spectrum of the theory.

On the lattice we used the fermionic action[5-8]

$$S(\psi)=\sum_i\bar{\psi}_i[(D_x\psi)_i+(-1)^x(D_y\psi)_i+(-1)^{x+y}(D_z\psi)_i+(-1)^{x+y+z}(D_t\psi)_i+m\psi_i], \quad (7)$$

where D_i ($i=x,y,z,t$) is the covariant version of the central first derivative ∂_i $\{(\partial_i\psi)_j=\frac{1}{2}[\psi(\vec{j}+\vec{n}_i)-\psi(\vec{j}-\vec{n}_i)]\}$. It is known that this action describes four fermion flavors and is invariant under an SU(4) internal flavor group. As discussed in Ref. 2, the quenched correlation functions for the two-flavor theory can be obtained simply by dividing by a factor of 2 the correlation functions computed with the full action (7).

No multiplicative factor is needed in the computation of the masses. However, if

$$P(\vec{i}-\vec{j})=\int d\mu[A]\,\mathrm{Tr}[G(\vec{i},\vec{j}|A)G^*(\vec{i},\vec{j}|A)], \quad (8)$$

particles with different spin parity will appear as singularities at different corners of the Brillouin zone; this effect is typical of the approach of Refs. 5 and 8.

If π and ρ are the lowest-mass particles it is easy to check that asymptotically

$$\sum_{n_x,n_y,n_z}P(\vec{n})\equiv\Delta_\pi(n_t)\simeq\exp\{-n_tm_\pi\}, \quad (9)$$

$$\sum_{n_x,n_y,n_z}P(\vec{n})\{(-1)^{n_x}+(-1)^{n_y}+(-1)^{n_z}\}\equiv\Delta_\rho(n_t)\simeq\exp\{-n_tm_\rho\}. \quad (10)$$

Similar expressions are also valid for the other particles of the theory.

This completes the description of all the basic machinery we used to perform the computation. As a first step we generated a few equilibrium configurations for the pure gauge theory, defined by the 120-element subgroup of SU(2), $\tilde{Y}$ (the covering group of the symmetry group of the iscosahedron): We worked on an $8\times8\times8\times8$ lattice, with periodic boundary condition and the standard Wilson action.[9]

We concentrated our attention to the range $\beta = 2.1-2.4$, where $\beta = 4/g^2$ is the coupling parameter of the gauge theory. This is the region where the asymptotically free behavior of the string tension appears to set in. To relate lattice spacing a to $\tilde{\Lambda}_{\rm mom}$ we use[10]

$$\tilde{\Lambda}_{\rm mom} = (\pi/a)[(6\pi^2/11)(\beta - 1.08)]^{51/121} \times \exp\{-(3\pi^2/11)(\beta - 1.08)\}. \quad (11)$$

With $\tilde{\Lambda}_{\rm mom} \simeq 250$ MeV, we find that the size of the box goes from 2.7 to 1.3 fm. The momentum cutoff $C_M \equiv \pi/a$ (i.e., the boundary of the Brillouin zone) ranges from 1.8 to 3.9 GeV. In Ref. 9 a parameter

$$\Lambda_{\rm mom} = \tilde{\Lambda}_{\rm mom}(1 - 1.08/\beta)^{-0.42} \quad (12)$$

was used. $\tilde{\Lambda}_{\rm mom}$ and $\Lambda_{\rm mom}$ are asymptotically equal, but in the β range we are considering they differ by about 30%. With the value of the string tension K determined in Refs. 9 and 11 we obtain $\tilde{\Lambda}_{\rm mom} \simeq 250$ MeV if $\sqrt{K} = 500$ MeV.

We treated both Eq. (4) and (6) implementing the time derivative by a second-order Runge-Kutta algorithm[12]; $\langle\bar{\psi}\psi\rangle$ has been computed with both methods, obtaining compatible results (the values found by use of the Langevin equation have small statistical errors). As a check we have computed $\langle\bar{\psi}\psi\rangle$ as function of m at $\beta = 0$. The results are shown in Fig. 1; the continuous line is the prediction from the limit $N \to \infty$.[7,8] The very good agreement implies that the $1/N^2$ corrections are negligible (as expected) for $N = 2$.

In the whole β range we have explored, we find clear evidence of the fact that $\langle\bar{\psi}\psi\rangle \neq 0$ in the limit $m \to 0$ [the value at $m = 0$ is computed by extrapolating the data obtained with m varying in the range $(0.05-0.5)a^{-1}$]. From renormalization-group arguments we expect

$$\langle\bar{\psi}\psi\rangle_{m=0} \sim \tilde{\Lambda}^3 \alpha_B^{-6/11} \quad [\alpha_B = 1/\pi(\beta - 1.08)] \quad (13)$$

for $\beta \to \infty$. In Fig. 2 we plot $(\frac{3}{2}\langle\bar{\psi}\psi\rangle)^{1/3}$ versus β. The continuous line represents the quantity $R\tilde{\Lambda}(2\alpha_B)^{-2/11}$ with $R = 1.75$. The fit is satisfactory, and we can provisionally assume $R = 1.75 \pm 0.1$.

The pion and ρ masses have been estimated by looking at the large-distance decay of the correlation functions; for the computation of these quantities we used the relaxation method. The number of iterations needed for a good convergence ranges from 50 to 500. The accuracy reached with use of the relaxation procedure can be estimated by checking the validity of the sum rule (3); this is also a consistency check for the algorithm. With a lattice of size 8^4, the largest distance at which correlation functions can be computed is 4; in order to remove finite-size effects from the t direction we have constructed lattices of size $8^3 \times 16$ and $8^3 \times 32$, respectively, by duplicating and quadruplicating the same gauge-field configuration. (This procedure is justified by the short range of the gauge field correlations. Indeed an 8^4 lattice would be adequate if one could obtain the exact spectrum for the propagation of fermions. Iterating the gauge-field configuration in time allows a good determination of the lowest masses in this spectrum, through the rate of decay of the Green functions.) Most of our estimates for the values of the masses (in the β range

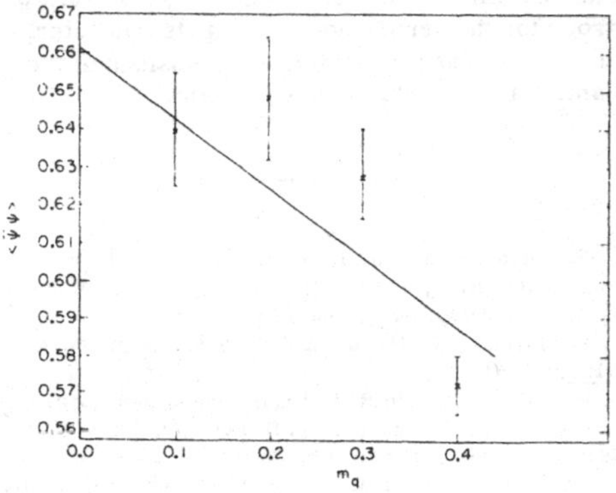

FIG. 1. $\langle\bar{\psi}\psi\rangle$ vs m at $\beta = 0$. The continuous line is the prediction from the limit $N \to \infty$ (Refs. 7 and 8).

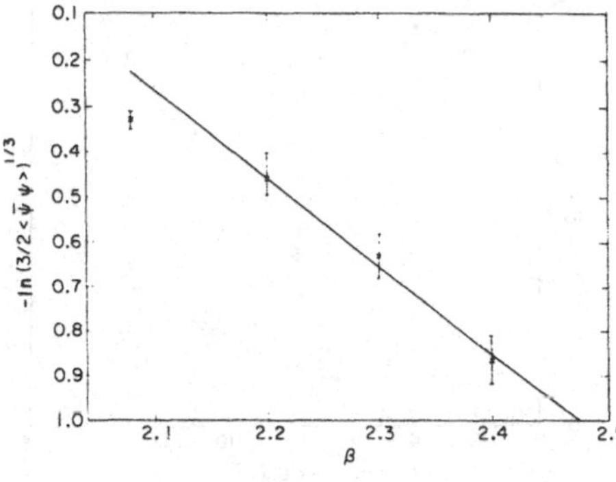

FIG. 2. $(\frac{3}{2}\langle\bar{\psi}\psi\rangle)^{1/3}$ vs β. The continuous line fits $R\Lambda\alpha_B^{-2/11}$ with $R = 1.75$.

we are considering) have been obtained from the $8^3\times 16$ lattice (this could be impossible for higher values of β). So the results from the $8^3\times 32$ lattice were mainly used as a check.

We concentrated our efforts on $\beta=2.2$, which is already in the scaling region for $\langle\bar\psi\psi\rangle$. We estimated the masses by looking at the rate of exponential decay of the correlation functions. We obtained all our results by averaging over four configurations of the gauge fields. We have a good control of the π correlation functions (i.e., $\bar\psi\gamma_5\psi$) at all distances (see Fig. 3), while our statistical accuracy for the ρ ($\bar\psi\gamma_\mu\psi$) and δ ($\bar\psi\psi$) correlation functions is reasonable up to distances 7 and 5, respectively. In our range of quark masses (0.3 – 0.1 in units of a^{-1}) the data can be fitted by

$$m_\pi^2=(6.5\pm 0.1)m_q/a,$$

$$m_\rho^2=(1.0\pm 0.1)/a^2+m_\pi^2,$$

$$m_\delta^2-m_\rho^2=(0.4\pm 0.1)/a^2.$$

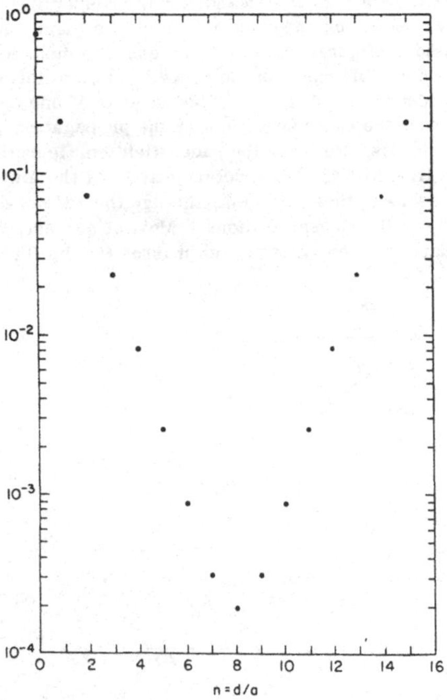

FIG. 3. Correlation function for the π with $m_q=0.2$ and $\beta=2.2$

Using the partial conservation of axial-vector current relation $m_\pi^2 f_\pi^2=\frac{3}{2}\langle\bar\psi\psi\rangle_{SU(2)}m_q$ ($f_\pi^{exp}=0.95$ MeV), we get $f_\pi=(0.19\pm 0.01)a^{-1}$. With $K=(500\text{ MeV})^2$ we finally obtain

$$f_\pi=150\pm 10\text{ MeV},$$

$$m_\rho=800\pm 80\text{ MeV},$$

$$m_\delta=950\pm 100\text{ MeV}.$$

(These errors do not reflect the possible uncertainty in the Monte Carlo determination of the string tension.) $m_I=m_q\alpha^{-6/11}$ is renormalization-group invariant. Its value turns out to be 7 MeV, in agreement with phenomenological estimates.

Data with lower statistics at $\beta=2.4$ seem to indicate

$$m_\rho\sim 710\text{ MeV},\quad f_\pi\sim 120\text{ MeV}.$$

It is clear that one should extend this computation to smaller lattice spacing to check the reliability of our results. Doing this would not present any problem of principle, and the only difficulty would be the larger amount of computer time required. The central processing unit time needed for the computations we described here can be estimated to be about 100 h of VAX 780 (the equivalent of about 10 h of CDC), which is actually not a lot.

We are at present extending our analysis to include the effects of fermionic loops, following the method of Ref. 1.

We are grateful to all the people around the VAX in Frascati and in Rome, and in particular to M. Ferrer, F. Marzano, and E. Valente for their generous help. One of us (G.P.) is happy to thank H. Quinn for the warm hospitality at the Stanford Linear Accelerator Center, the Mark 3 group for the permission of using its computer, and H. Hamber for clarifying discussions at the Santa Barbara 1981 Summer Meeting.

[1]F. Fucito, E. Marinari, G. Parisi, and C. Rebbi, Nucl. Phys. B180, 369 (1981); F. Fucito and E. Marinari, Nucl. Phys. B190, 266 (1981).

[2]E. Marinari, G. Parisi, and C. Rebbi, Nucl. Phys. B190, 734 (1981).

[3]D. J. Scalapino and R. L. Sugar, Phys. Rev. Lett. 46, 519 (1981); D. Weingarten and D. Petcher, Phys. Lett. 99B, 333 (1981); H. Hamber, Phys. Rev. D 24, 951 (1981); A. Duncan and M. Furman, Columbia University Report No. CU-TP-194 (to be published).

[4]H. Hamber and G. Parisi, preceding Letter [Phys.

Rev. Lett. 47, 1792 (1981)].

[5]T. Banks, S. Raby, L. Susskind, J. Kogut, D. R. T. Jones, P. N. Scharbach, and D. K. Sinclair, Phys. Rev. D 15, 1111 (1977); L. Susskind, Phys. Rev. D 16, 3031 (1977).

[6]B. Svetitsky, S. D. Drell, H. R. Quinn, and M. Weinstein, Phys. Rev. D 22, 490 (1980); M. Weinstein, S. D. Drell, H. R. Quinn, and B. Svetitsky, Phys. Rev. D 22, 1190 (1980).

[7]H. Kluberg-Stern, A. Morel, O. Napoly, and B. Petersson, Centre d'Etudes Nucléaires de Saclay Report No. Dph-T/19, 1981 (to be published).

[8]N. Kawamoto and J. Smit, Institute for Theoretical Physics, University of Amsterdam, Report No. ITFA-81-2, 1981 (to be published), and references therein.

[9]G. Bhanot and C. Rebbi, Nucl. Phys. B180, 469 (1981).

[10]A. Hasenfratz and P. Hasenfratz, Phys. Lett. 93B, 103 (1980).

[11]M. Creutz, Phys. Rev. Lett. 45, 313 (1980).

[12]Periodic boundary conditions have been imposed on the fermionic fields.

Volume 108B, number 4,5 PHYSICS LETTERS 28 January 1982

SPECTROSCOPY IN A LATTICE GAUGE THEORY

H. HAMBER
Brookhaven National Laboratory, Upton, NY 11973, USA

E. MARINARI [1]
Brookhaven National Laboratory and INFN-Roma, Italy

G. PARISI
Istituto di Fisica, Facoltá di Ingegneria, Università degli Studi, Roma and INFN-Frascati, Italy

C. REBBI
Brookhaven National Laboratory, Upton, NY 11973, USA

Received 10 November 1981

The masses of the lightest states of the J/ψ family are computed in a SU(2) lattice gauge theory, in an approximation where internal quark loops are neglected.

Very recently it has been possible to obtain reasonably accurate results [1–3] for the mass spectrum of the lightest hadrons, using an approximation to the lattice gauge theory where internal quark loops are neglected (the "quenched approximation" in the terminology of ref. [4]). The success of the computation has prompted us to extend the method to a calculation of the lowest lying states of the J/ψ family. The results, considering the obvious limitations of the formulation on the lattice, have been rather satisfactory and we wish to present them here.

Let us recall the general features of the quenched calculations [1–5]. Quantum averages of products of fermionic operators may be expressed evaluating first the Green's functions for the propagation of the fermions in the background of a fixed gauge field and averaging them over all configurations of the gauge field with suitable measure. This is exemplified by the formula

$$\langle\bar{\psi}(x)\,\Gamma_1\,\psi(x)\,\bar{\psi}(0)\,\Gamma_2\,\psi(0)\rangle = \int \mathrm{d}\mu[A]\ \mathrm{Tr}[\Gamma_1\,G(x,0|A)\,\Gamma_2\,G(0,x|A)]\,, \quad (1)$$

where $\mathrm{d}\mu[A]$ stands for the measure, Γ_1 and Γ_2 are appropriate γ matrices and the Green's function or quark propagator $G(x,0|A)$ solves the equation

$$(\not{D} + m)\,G(x,0|A) = \delta(x)\,, \quad (2)$$

$\not{D}$ being the covariant derivative. In the exact computation $\mathrm{d}\mu[A]$ should be proportional to $\exp\{-S_G(A)\}\ \mathrm{Det}\,(\not{D}+m)$, where S_G is the pure gauge action. The quenched approximation [4,5] consists of leaving out the $\mathrm{Det}\,(\not{D}+m)$ factor in the measure. This is equivalent to neglecting the contributions from inner fermion loops.

For our study of the J/ψ family we have used an SU(2) lattice gauge model, approximating the gauge group with its maximal $\tilde{Y}$-subgroup, which reproduces well the dynamics of the original system throughout the physically relevant range of couplings. The Wilson

[1] Permanent address: Istituto di Fisica "G. Marconi", Università degli Studi, Roma, Italy.

Volume 108B, number 4,5 PHYSICS LETTERS 28 January 1982

form of the action has been assumed and samples of gauge field configurations were obtained by the standard Monte Carlo procedure. Fermions have been introduced on the lattice using the method of refs. [4] and [6], whereby a single component of ψ is associated with each lattice site, and the Green's functions have been evaluated with high accuracy by the relaxation technique.

We define

$$M_\Gamma(x,0) = \langle \Gamma(x)\, \overline{\psi}(x)\, \psi(x)\, \overline{\psi}(0)\, \psi(0)\rangle\,, \tag{3}$$

where the position vector x is restricted to the lattice sites and $\Gamma(x)$ is a suitable linear combination of factors $(-1)^{nx}$, the components of n being zero or one. As is typical in the approach of ref. [6], particles with different spin, parity, charge conjugation etc., manifest themselves as poles of the propagator at different corners of the Brillouin zone. Thus, by choosing Γ suitably, the particles that will be excited from the vacuum by the action of the two-quark operator can be given the quantum numbers of $\eta_C(0^{-+})$, $J/\psi(1^{--})$, $\chi_0(0^{++})$, $P_C(1^{++})$, $P(1^{+-})$. As was done in refs. [1–3], it is actually convenient to define a correlation function depending on time only by summing $M(x,0)$ over space positions:

$$\widetilde{M}_\Gamma(t,0) = \sum_x M_\Gamma(x,t,0)\,. \tag{4}$$

The sum has the effect of projecting over the states with lattice three-momentum equal to zero and thus the mass of the lowest excitation in the spectrum can be read off from the large time behavior

$$\widetilde{M}_\Gamma(t,0) \approx e^{-mt}\,. \tag{5}$$

In principle the masses of the radial recurrences could also be determined from the rate of decay of the subdominant terms; however, due to our limited statistics, we have results only for the η_C'.

We have done our study on an $8\times 8\times 8\times 32$ lattice, using a value of β ($\alpha_B = 1/\pi\beta$) equal to 2.7. From the lattice determination of the string tension [7,8], the corresponding lattice spacing turns out to be

$$a(\beta = 2.7) = 0.083\ \text{fm} = (2.37\ \text{GeV})^{-1}\,. \tag{6}$$

The spatial size of our box is then 0.7 fm, which should be enough to contain the J/ψ without distorting the wave function. The length of the box in time allows us to consider points at a maximum time separation of $(140\ \text{MeV})^{-1}$, which is enough to resolve between the lowest states and their radial excitations. We have fixed the value of the bare quark mass m so as to reproduce the experimentally measured mass of the J/ψ. This gave $m = 0.3a^{-1}$ and the corresponding renormalization invariant mass is $m_R = m(2\alpha_B)^{-6/11} \approx 1.2$ GeV, a reasonable number for the mass of the charmed quark.

We performed averages over eight gauge field configurations to determine the meson propagators. Our results are

$$\begin{aligned}
m_{J/\psi} &= (1.30 \pm 0.02)\,a^{-1}\,,\\
m_{J/\psi} - m_{\eta_C} &= (0.06 \pm 0.01)\,a^{-1}\,,\\
m_{\eta_C'} - m_{\eta_C} &= (0.30 \pm 0.06)\,a^{-1}\,,\\
m_{\chi_0} - m_{\eta_C} &= (0.19 \pm 0.04)\,a^{-1}\,,\\
m_{P_C} - m_{\eta_C} &= (0.27 \pm 0.05)\,a^{-1}\,,\\
m_P - m_{\eta_C} &= (0.33 \pm 0.06)\,a^{-1}\,,
\end{aligned} \tag{7}$$

or, using eq. (6),

$$\begin{aligned}
m_{J/\psi} &= 3.10\ \text{GeV (input)}, && (\text{exp. } 3.097\ \text{GeV}),\\
m_{\eta_C} &= (2.95 \pm 0.03)\ \text{GeV}, && (\text{exp. } 2.979\ \text{GeV}),\\
m_{\eta_C'} &= (3.65 \pm 0.15)\ \text{GeV}, &&\\
m_{\chi_0} &= (3.40 \pm 0.10)\ \text{GeV}, && (\text{exp. } 3.414\ \text{GeV}),\\
m_{P_C} &= (3.60 \pm 0.15)\ \text{GeV}, && (\text{exp. } 3.507\ \text{GeV}),\\
m_P &= (3.75 \pm 0.15)\ \text{GeV}\,.
\end{aligned} \tag{8}$$

These results are reasonably satisfactory, the bare quark mass being the only free parameter.

The errors we quote are purely statistical and it would be possible to decrease them substantially using more computer time. Our calculations took only 20 hours of Vax, i.e. approximately 2 hours of CDC 7600. Using 200 hours of CDC 7600 or 2000 hours of Vax, the statistical errors could be reduced to the order of 10 MeV. Obviously before reaching this level of precision systematic effects, such as the errors introduced by the finite lattice space, the finite site of the box etc., will become important.

References

[1] H. Hamber and G. Parisi, Numerical estimates of hadronic masses in a pure SU(3) gauge theory, Brookhaven preprint BNL 30170 (1981).
[2] E. Marinari, G. Parisi and C. Rebbi, Computer estimates of meson masses in SU(2) lattice gauge theory, Brookhaven preprint BNL 30212 (1981).
[3] D. Weingarten, Monte Carlo evaluation of hadron masses in lattice gauge theories with fermions, Indiana preprint IUHET-69 (1981).
[4] E. Marinari, C. Parisi and C. Rebbi, Nucl. Phys. B190 [FS3] (1981) 734.
[5] H. Hamber, Phys. Rev. D24 (1981) 951.
[6] N. Kawamoto and J. Smit, Amsterdam preprint ITFA-81-2 (1981), and references therein.
[7] M. Creutz, Phys. Rev. Lett. 45 (1980) 313.
[8] G. Bhanot and C. Rebbi, Nucl. Phys. B180 [FS2] (1981) 469.

Volume 109B, number 1,2 PHYSICS LETTERS 11 February 1982

MONTE CARLO EVALUATION OF HADRON MASSES IN LATTICE GAUGE THEORIES WITH FERMIONS

Don WEINGARTEN
Physics Department, Indiana University, Bloomington, IN 47405, USA

Received 13 October 1981

An improved Monte Carlo method is presented for lattice gauge theories with fermions. Taking the pion mass and meson Regge trajectory slope as input, this procedure is used to calculate the rho mass on lattices up to 12^4 for gauge group $\bar{\mathrm{I}}$, the best discrete approximation to SU(2). The final result is $m_\rho = 670 \pm 100$ MeV. Arguments are given to show that this prediction would be changed by replacing $\bar{\mathrm{I}}$ by SU(2) or SU(3).

Path integrals for lattice gauge theories [1] without fermions have been evaluated using Monte Carlo algorithms by Creutz and co-workers [2], Wilson [3], and others [4]. These methods have been extended to theories including fermions by Petcher and the present author [5], Fucito et al. [6] and by other groups [7, 8]. Unfortunately, however, the algorithms in refs. [5–7] require computer time which grows as the eight [5,6] or twelfth [7] power of lattice size [‡1]. As a result, these methods are limited to lattices which are probably too small to give much information concerning the theory's infinite-volume, continuum limit. In particular, it is likely the methods in refs. [5–7] cannot be used to obtain reliable values of low lying hadron masses. The algorithm in ref. [8], on the other hand, can be used on much larger lattices but probably encounters difficulties if the pion mass is not made unrealistically large.

In the present article we will describe a modification of the strategy used in ref. [5] which requires computer time growing as the fourth power of lattice size and therefore can be run on lattices perhaps as large as 16^4. We will then present an evaluation of the rho mass, taking the pi mass and meson Regge trajectory slope as input, for gauge group $\bar{\mathrm{I}}$, the best discrete approximation to SU(2).

Our final prediction for the rho mass is 670 ± 100 MeV for a 12^4 lattice with box size $(88.9\ \mathrm{MeV})^{-1}$ and lattice spacing $(1067\ \mathrm{MeV})^{-1}$. Both cut-offs, we believe, are sufficiently far from the rho and pi masses to leave our results fairly close to the theory's infinite-volume, continuum limit. From comparisons of pure gauge theories for $\bar{\mathrm{I}}$ with those for SU(2) [10], it is clear that our results would be unaffected by replacing $\bar{\mathrm{I}}$ with SU(2). A semi-quantitative argument suggests that if $\bar{\mathrm{I}}$ were replaced by SU(3) our results would not be changed by much (perhaps 10%) [‡2].

We will begin by briefly reviewing Wilson's path integral [1] for vacuum expectation values in a euclidean lattice gauge theory over $\bar{\mathrm{I}}$. The theory lives on a periodic, four-dimensional, hypercubic lattice with spacing a and periodicity N. We choose a system of units with $a = 1$. Residing on each nearest neighbor link (x, y) is a value $U(x, y) \in \bar{\mathrm{I}}$ of the link field U. On each site are a set of anticommuting Grassmann variables $\bar{\psi}^a_{ik}(x)$, $\psi^b_{jl}(x)$, where i and j are spinor indices, k and l are group indices, and b and c are flavor indices. We assume two flavors of fermion, u and d.

An action for the gauge fields is

‡1 The method of ref. [5] probably requires a Metropolis step size which falls as a^2 to retain accuracy as the lattice spacing a becomes small. The result appears to be $O(N^8)$ operations to perform calculations by this method, not $O(N^4)$ as claimed in ref. [9].

‡2 Preliminary results applying our method to SU(3) on a $6^3 \times 10$ lattice are consistent with this expectation [11].

$$S_G = \tfrac{1}{2}\beta \sum_{(w,x,y,z)} \mathrm{tr}[U(w,x)\,U(x,y)\,U(y,z)\,U(z,w)]\,, \tag{1}$$

where the sum is over all closed loops (w, x, y, z) around the boundary of an elementary lattice square and $\beta = 4g_0^{-2}$ for bare gauge coupling g_0. The action for fermion fields is

$$S_F = -\sum_x \bar\psi(x)\psi(x) + K \sum_{x,\mu,\pm} \bar\psi(x)\,U(x, x \pm \hat\mu)(1 \mp \gamma_\mu)\psi(x \pm \hat\mu), \tag{2}$$

where $\hat\mu$ is a unit lattice vector in the $+\mu$ direction, $K = (8 + 2m_0)^{-1}$ for bare mass m_0, and $\gamma_1, ..., \gamma_4$ are hermitean 4×4 euclidean γ-matrices.

The vacuum expectation of a product $G\, \Pi_i \psi(f_i) \times \bar\psi(h_i)$ of smeared fermion fields

$$\psi(f) = \sum_x f^*(x)\,\psi(x), \quad \bar\psi(h) = \sum_x h(x)\,\bar\psi(x) \tag{3}$$

and a term G depending only on gauge fields is

$$\Big\langle G \prod_i \psi(f_i)\,\bar\psi(h_i)\Big\rangle = Z^{-1} \sum_U \int \mathrm{d}\mu_F\, G \prod_i \psi(f_i)\,\bar\psi(h_i) \exp(S_G + S_F), \tag{4}$$

where the sum in (4) is over all possible link fields U, the integral $\int \mathrm{d}\mu_F$ is the usual Grassmann integral (see, for example ref. [1]) and Z is defined by $\langle 1 \rangle = 1$. To evaluate (4) by Monte Carlo integration it is convenient to carry out the integral over fermion fields explicitly. The result is

$$\Big\langle G \prod_i \psi(f_i)\bar\psi(h_i)\Big\rangle = Z^{-1} \sum_U \det_{(i,j)}(f_i, (1 - KA)^{-1} h_j) \times G \det(1 - KA) \exp S_G \tag{5}$$

where the fermion coupling matrix A is defined by (2) combined with

$$S_F = -\sum_x \bar\psi(x)\,\psi(x) + K \sum_{x,y} \bar\psi(x) A(x,y)\,\psi(y). \tag{6}$$

In principle, the sum over gauge variables in (5) could now be evaluated by standard Monte Carlo methods [2–4]: Generate a sequence of random U fields with probability distribution $P(U) = Z^{-1} \times \exp S_G(U) \det(1 - KA)$. Then average $\det_{(i,j)}(f_i, (1 - KA)^{-1} h_j) G$ over this ensemble to obtain an approximation to the vacuum expectation in (5). The obstacle to this procedure is essentially that $\det(1 - KA)$ requires so many arithmetic operations to evaluate that generating a sequence of random U with probability distribution $P(U)$ would be hopelessly slow even on a lattice of size 4^4. In ref. [5] a faster indirect method for Monte Carlo evaluation of (5) was developed. Here we propose a still faster procedure which is simply to replace $\det(1 - KA)$ by 1 yielding

$$\Big\langle G \prod_i \psi(f_i)\,\bar\psi(h_i)\Big\rangle = Z^{-1} \sum_U G \det_{(i,j)}(f_i, (1 - KA)^{-1}, h_j) \exp S_G\,. \tag{7}$$

If β and K for (7) are chosen to give the correct Regge slope for boson trajectories and the correct pion mass, we believe most other masses will have the same values they would have had if the same tuning process (yielding in general distinct β' and K') were applied to the full path integral of (5). We will now present a simple physical picture which supports this hypothesis.

The effect of $\det(1 - KA)$ in strong or weak coupling expansions of (5) is to contribute closed quark loops inside diagrams. With $\det(1 - KA)$ replaced by 1 such loops are removed. With loops removed the gauge field configuration between, for example, the valence quark and antiquark in a meson is expected to be generally string-like, characterized by some string tension T. If we restore $\det(1 - KA)$ to (5) but make no other changes, it is plausible that the field in a meson still remains string-like but now at various points has breaks where the string has been cut open by a quark loop. The string with breaks will have some new effective string tension T' with real part smaller than T, since the breaks tend to save string energy, and perhaps with an imaginary part since, for long strings, some of the quarks and antiquarks occurring along the breaks might condense into real hadrons and contribute physical intermediate states to the string's propagation. The hypothesis that the fields between quarks and antiquarks in mesons remain generally string-like in the presence of quark loops is supported by the fact that meson Regge trajectories in the real world are nearly linear as predicted by classical or quantized forms of the Nambu–Goto string model.

If field configurations remain string-like for (5) with quark loops present, then the theory governed by (7) without loops can be made to reproduce, approximately, the theory including loops, simply by shifting β in (7) to adjust T to be equal to the real part of T'. Assuming the approximate validity of the string model with or without virtual quarks, the required renormalization of β is automatically accomplished if β is chosen to yield the string tension which predicts the correct Regge slope α' according to the string model relation $\alpha' = (2\pi T)^{-1}$. With this choice of β the valence quark interaction predicted by (7) with $\det(1 - KA)$ absent should approximate the valence quark interaction predicted by (5) with $\det(1 - KA)$ present. If K is then adjusted to give the correct pion mass, most of the remainder of the mass spectrum should be the same for (5) and (7).

The only sector of the mass spectrum which may not be reproduced adequately by our approximation is flavor singlets, such as the η. The mass of η is presumed to be largely generated by a process in which the valence quarks annihilate into gluons, followed by gluon annihilation back into quarks.

Most of the preceding discussion of the effect of omitting quark loops can be repeated for the effect of replacing $\bar{\mathrm{I}}$ or SU(2) as gauge group with SU(3). Thus despite the fact that we have used $\bar{\mathrm{I}}$, we believe our mass predictions for mesons are quite close to those which would have been obtained from SU(3) (see footnote 2). This hypothesis is further supported by an explicit comparison of the ratios of loop expectations $W(n, m)$ for SU(2) and SU(3) [2] with β chosen to give the same string tension in both cases. The results for SU(2) and SU(3) agree to within error bars of about 10%.

In any case, we are now left with the problem of evaluating (7) for various choices of G, f_i, and h_i. The feature of (7) which still distinguishes it from Monte Carlo integrals done in pure gauge theories is the presence of the fermion propagator $(1 - KA)^{-1}$. To evaluate (7) in reasonable amounts of computer time, a fast method for generating $(1 - KA)^{-1}$ is required. A satisfactory procedure, proposed in ref. [5], is Gauss–Seidel iteration. Actually once the lattice spacing is made fairly small, we have found, the rate of convergence of the Gauss–Seidel can be improved somewhat by the introduction of an optimally chosen relaxation parameter [13].

To find $(1 - KA)^{-1}h$ for some smearing function h we proceed as follows. Let f be $(1 - KA)^{-1}h$. Then

$$f = KAf + h. \tag{8}$$

For any $\lambda \neq 0$, (8) is equivalent to

$$f = (1 - \lambda)f + \lambda(KAf + h). \tag{9}$$

We solve (9) by iteration. Choose some convenient initial f_1 for f. Possible choices are 0 or, alternatively, the final results of preceding calculations for slightly different values of K or r. Now define the sequence $f_2, ..., f_M$:

$$f_{m+1}(x) = f_m(x) \quad |x| \neq m \bmod N^4$$

$$f_{m+1}(x) = (1 - \lambda)f_m(x)$$
$$+ \lambda[(KAf_m)(x) + h(x)] \quad |x| = m \bmod N^4\,, \tag{10}$$

where $x_1, ..., x_4$ are the components of x and $|x| = \Sigma_i x_i N^{i-1}$. If the sequence of f_m converges, it must approach a solution to (9). For the values of K we have considered, between 100 and 600 full sweeps of the lattice (m between $100N^4$ and $600N^4$) were sufficient to obtain values of the rho and pi masses stable to better than 0.5%. The optimal choice of λ we found to be between 0.7 and 1.0.

Using our Gauss–Seidel to generate $\det_{(i,j)}(f_i, (1 - KA)^{-1}h_j)$ in (7) we can then average over a Monte Carlo ensemble of random U with $P(U) = Z^{-1}\exp S_G(U)$ and obtain an approximation to the vacuum expectation $\langle G\, \Pi_i\, \psi(f_i)\, \bar{\psi}(h_i)\rangle$. The entire calculation on an N^4 lattice takes $O(N^4)$ operations in place of $O(N^8)$ for refs. [5,6] or $O(N^{12})$ for ref. [7]. On the other hand, there is a close relation between the number of Gauss–Seidel sweeps used in our method and the highest power of K included in the expansion in ref. [8]. The large number of Gauss–Seidel sweeps required by our calculation suggests that to obtain accurate results by the methods in ref. [8] will require calculations to higher order than is practically possible.

We will now describe our method for measuring rho and pi masses. Define rho and pi fields to be

$$\rho_\mu^+(x) = \bar{\psi}^u(x)\gamma_\mu \psi^d(x), \tag{11}$$

Volume 109B, number 1,2 PHYSICS LETTERS 11 February 1982

$$\rho_\mu^-(x) = \bar{\psi}^{\rm d}(x)\gamma_\mu\psi^{\rm u}(x),$$

$$\pi^+(x) = \bar{\psi}^{\rm u}(x)\gamma^5\psi^{\rm d}(x),$$

$$\pi^-(x) = \bar{\psi}^{\rm d}(x)\gamma^5\psi^{\rm u}(x), \qquad \text{(11 con'd)}$$

and sum over hypersurfaces with, say, x_4 fixed at t

$$\tilde{\rho}_\mu^\pm(t) = \sum_{x_1,x_2,x_3} \rho_\mu^\pm(x),$$

$$\tilde{\pi}^\pm(t) = \sum_{x_1,x_2,x_3} \pi^\pm(x). \qquad (12)$$

The Monte Carlo algorithm is then used to evaluate the expectations

$$C_\rho(t') = \sum_{i=1,2,3} \langle \tilde{\rho}_i^+(t+t')\tilde{\rho}_i^-(t)\rangle,$$

$$C_\pi(t') = \langle \tilde{\pi}^+(t+t')\tilde{\pi}^-(t)\rangle. \qquad (13)$$

If t' and $N - t'$ are large it is easily shown these averages approach

$$C_\rho(t') = Z_\rho\,\{\exp(-m_\rho t') + \exp[-m_\rho(N-t')]\}, \qquad (14)$$

$$C_\pi(t') = Z_\pi\,\{\exp(-m_\pi t') + \exp[-m_\pi(N-t')]\}, \qquad (15)$$

where Z_ρ and Z_π are positive field strength renormalization constants and m_ρ and m_π are the masses of the lightest states carrying rho and pi quantum numbers, respectively.

The sum over hyperplanes in (12) serves two purposes. First of all, in the processes of generating the values of $(1 - KA)^{-1}$ for some Monte Carlo field U needed to evaluate a correlation of the form $\langle\rho_i^+(x') \times \rho_j^-(x)\rangle$, we automatically generate enough matrix elements of $(1 - KA)^{-1}$ to calculate the summed correlation in (12) for all t'. The summed quantity, however, will have smaller statistical fluctuations. A second advantage of summing over hyperplanes is that it eliminates corrections corresponding to rho or pi states with small but finite momenta which would appear in (14) and (15) for $\langle\rho_i^+(x')\,\rho_i^-(x)\rangle$. These corrections would have the same form as the leading terms in (14) and (15) but in place of m_ρ and m_π, the corrections include slightly larger values giving the energy of the translating states. Thus the corrections fall at large t' and $N - t'$ only a small amount faster than the leading term and their presence would make it more difficult to find the leading term accurately. If confinement holds, on the other hand, the leading corrections to (14) and (15) for the summed correlations will correspond to radial excitations of the rho and pi. These terms will also have the same form as (14) and (15) but now with the masses of the radial excitations in place of m_ρ and m_π. Since the radial excitation masses are significantly above the ground state masses, the effect of these terms dies out rapidly for large t' and $N - t'$. By fitting such higher order corrections, one can obtain additional masses. In particular, confinement can be tested with these higher masses since confinement requires that only a discrete spectrum of masses occur in the full spectral expansions for (14) and (15) with no continuum contributions arising as the physical box size becomes large. If confinement fails at some threshold energy m_t, then above m_t the mass spectrum will become progressively denser as the physical box size approaches infinity.

A number of other tricks can be used to speed up the evaluation of $C_\rho(t')$ and $C_\pi(t)$. By translational invariance we have

$$C_\rho(t') = N^3 \sum_i \langle\tilde{\rho}_i^+(t+t')\rho^-(x_t)\rangle, \qquad (16)$$

$$C_\pi(t') = N^3\langle\tilde{\pi}^+(t+t')\,\pi^-(x_t)\rangle, \qquad (17)$$

for any x_t with $x_4 = t$. To find (16) and (17) for all t' we need values of both $(1 - KA)^{-1}(x, x_t)$ and $(1 - KA)^{-1}(x_t, x)$ for all x. One of these calculations can be eliminated using the relation

$$[(1 - KA)^{-1}]^\dagger = \gamma^5(1 - KA)^{-1}\gamma^5 \qquad (18)$$

which follows from (2) and (6). In addition, it can be shown that (16) and (17) will both be unchanged on each individual Monte Carlo U, except for a factor of 1/2, if we make the replacement

$$\rho_i^-(x_{t'}) \to \bar{\psi}_l^{\rm d}(x_{t'})\gamma_i\psi_l^{\rm u}(x_{t'}) \qquad (19)$$

and a similar replacement for $\pi^-(x_{t'})$, where the subscript l in (19) is a color index. Finally the statistical errors encountered in $C_\rho(t)$ and $C_\pi(t)$ can be minimized by summing $C_\rho(t')$ and $C_\pi(t')$ over all eight possible lattice directions for the t' interval.

We now present our results. Fig. 1 shows values of the rho mass as a function of the lattice size N for lattices with physical size Na nearly (89 MeV)$^{-1}$. The pion mass in each case was tuned to 140 MeV and the string tension tuned to (425 MeV)^2 giving a meson

Volume 109B, number 1,2 PHYSICS LETTERS 11 February 1982

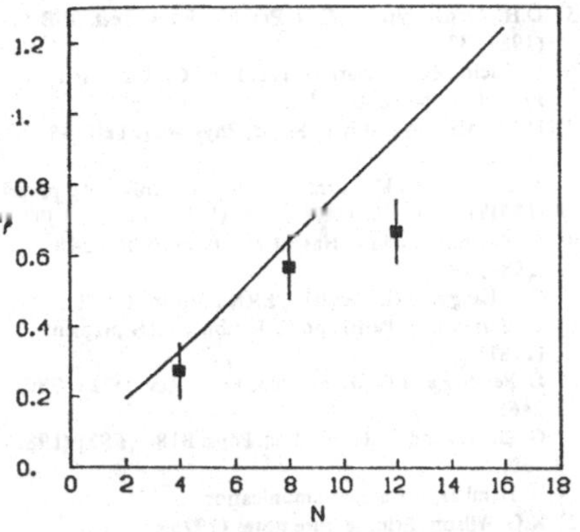

Fig. 1. The rho mass as a function of N with the pion mass and string tension tuned to their physical values. The physical lattice size Na is fixed at $(89\ \mathrm{MeV})^{-1}$.

Regge slope of $(938\ \mathrm{MeV}^{-1})^2$. The values of (K, β) corresponding to these results are (0.2181, 1.0515) for $N = 4$, (0.1687, 2.25) for $N = 8$, and (0.1545, 2.431) for $N = 12$. We have not quoted errors on the pion mass and string tension since statistical fluctuations in our tuning process should be interpreted as giving rise to errors in (K, β) and corresponding errors in the rho mass prediction. The critical values of K at which the pion mass becomes zero are 0.220 ± 0.002 for 4^4, 0.169 ± 0.004 for 8^4, and 0.155 ± 0.003 for 12^4.

These results were obtained from eight gauge configurations for the 4^4 lattice and four configurations for the 8^4 and 12^4 lattices. To minimize correlations among the U we carried out 128 Monte Carlo sweeps between each configuration on the 4^4 lattice and 512 between each on the 8^4 and 12^4 lattices. The calculations for 4^4 used a modified Gauss–Seidel to be described elsewhere, run at the physical hopping constant of 0.2181. For 8^4 and 12^4 the Gauss–Seidel of eq. (10) [5] was used. As we have already mentioned, performing 100 to 600 Gauss–Seidel sweeps of the lattice with a relaxation parameter between 0.7 and 1, we obtained m_ρ and m_π stable to better than 0.5%. Masses were measured at K of 0.157, 0.161 and 0.1625 for 8^4 and 0.149, 0.1505, and 0.152 for 12^4. For $K > 0.163$ on 8^4 and $K > 0.153$ on 12^4, the Gauss–Seidel iteration failed to converge for any value of the relaxation parameter. For both 8^4 and 12^4 lattices m_ρ^2 and m_π^2 over the K range we considered were found to be consistent with linear functions of K to an accuracy of 2% or better. On each lattice a single set of gauge configurations was used for all values of K to minimize point-to-point statistical fluctuations. The results we have quoted were obtained from our linear fits. The errors we have given were determined by repeating our procedures on subensembles of the set of gauge configurations. The total computer time required to obtain the points in fig. 1 is equivalent to approximately 50 hours on a CDC 7600.

The solid line in fig. 1 is the prediction for m_ρ obtained by applying our tuning process to m_ρ and m_π given by Wilson's leading-order strong-coupling expansion in ref. [12]. The value of m_ρ shown in fig. 1 for a 12^4 lattice differs significantly from the leading-order strong-coupling calculation. Fig. 1 shows some tendency for m_ρ to approach a finite limit as the lattice spacing $(N \times 90\ \mathrm{MeV})^{-1}$ goes to zero. The value of m_ρ begins to depart appreciably from the strong-coupling line near $N = 8$ corresponding to β of 2.25, which is the same value of β at which the string tension [2] switches from strong-coupling to weak-coupling behavior in pure gauge theories. The values of the critical hopping constant K_c, on the other hand, are consistent with the predictions of ref. [14]. For the 8^4 and 12^4 lattices, K_c is within 10% of the large-β calculation of ref. [14].

Given that our results for m_ρ are not too far from the predictions of the strong coupling expansion, it seems reasonable to assume the sensitivity of our m_ρ to changes in the value of the lattice periodicity, Na, will be roughly the same as the sensitivity of the strong coupling prediction to Na. If so, our choice of $Na = (89\ \mathrm{MeV})^{-1}$ is more than sufficient for stability. The strong coupling expansion in finite volume yields correlation functions which, if fitted to (14) and (15) give m_ρ and m_π independent of Na for all N large enough to measure a mass ($N \geqslant 2$).

In conclusion, we believe we have found a tractable method for extracting a variety of physical predictions from realistic lattice gauge theories including fermions. A convincing test of QCD for low energy phenomena is perhaps not too far in the future.

I would like to thank the high energy experimental physics group at Indiana University for time on their

VAX 11/780, T. Sulanke for help using the VAX, and Fermi National Accelerator Laboratory for time on Fermilab's system of CDC 175's. I am grateful to the theory groups at Brookhaven and at Fermilab for their hospitality during the summer of 1981 while part of this work was being completed. For helpful conversations I would like to thank W. Bardeen, D.Brydges, W. Celmaster, M. Creutz, M. Feigenbaum, B. Freedman, D. Lichtenberg, F. Paige and D. Petcher. This work was supported in part by the US Department of Energy.

References

[1] K.G. Wilson, Phys. Rev. D10 (1973) 2445.

[2] M. Cretuz, L. Jacobs and C. Rebbi, Phys. Rev. Lett. 42 (1979) 1390; Phys. Rev. D20 (1979) 1915;
M. Creutz, Phys. Rev. Lett. 43 (1979) 553; Phys. Rev. D21 (1980) 2308; Phys. Rev. Lett. 45 (1980) 313.

[3] K.G. Wilson, Cornell preprint (1979).

[4] A partial list includes G.A. Jongeward, G. Stack and G. Jayapraaksh, Phys. Rev. D21 (1980) 3360;
D.N.

[5] D.H. Weingarten and D.N. Petcher, Phys. Lett. 99B (1981) 333.

[6] F. Fucito, E. Marinari, G. Parisi and C. Rebbi, Nucl. Phys. B, to be published.

[7] D.J. Scalapino and R.L. Sugar, Phys. Rev. Lett. 46 (1981) 519;
A. Duncan and M. Furman, Columbia University preprint (1981).

[8] A. Hasenfratz and P. Hasenfratz, Phys. Lett. 104B (1981) 489;
C.B. Lang and H. Nicolai, CERN preprint (1981).

[9] E. Marinari, G. Parisi and C. Rebbi, CERN preprint (1981).

[10] D. Petcher and D. Weingarten, Phys. Rev. D22 (1980) 2465;
G. Bhanot and C. Rebbi, Nucl. Phys. B180 [FS2] (1981) 469.

[11] H. Hamber, private communication.

[12] K.G. Wilson, Erice lecture notes (1975).

[13] R.S. Varga, Matrix iterative analysis (Prentice-Hall, Englewood Cliffs, 1974).

[14] N. Kawamoto, Nucl. Phys. B190, FS3 (1981) 617.

Volume 110B, number 3,4 PHYSICS LETTERS 1 April 1982

HOPPING PARAMETER EXPANSION FOR THE MESON SPECTRUM IN SU(3) LATTICE QCD

A. HASENFRATZ
Central Research Institute for Physics, Budapest, Hungary

Z. KUNSZT
Eötvös University, Budapest, Hungary

and

P. HASENFRATZ and C.B. LANG
CERN, Geneva, Switzerland

Received 13 January 1982

The low-lying pseudoscalar and vector mesons containing u, d, s and c quarks are studied by the method of the hopping parameter expansion combined with a Monte Carlo simulation at $1/g^2 = 0.0, 0.7, 0.9, 0.925, 0.95$ and 1.0. The effect of virtual quark loops is taken into account. Meson masses, quark masses, the lattice scale and K_u, K_s and K_c as functions of g^2 are determined.

In statistical physics the high-temperature expansion in powers of $1/T$ is a valuable tool for investigating critical phenomena [1]. It has a non-vanishing radius of convergence and therefore the result is certainly correct at high enough temperatures. Close to the critical point, however, some information on the singularity structure is necessary. Based on that information (or assumption) some extrapolation method might be applied.

The hopping parameter expansion in lattice gauge theories with fermions in Wilson's formulation [2–10] is analogous to the high-temperature expansion in many respects. The amplitude of moving a quark by one lattice unit is proportional to the hopping parameter K. An expansion in K is equivalent to an expansion in the length of quark paths in configuration space. In their propagation the quarks are constrained by the maximum order of the expansion but they should still gather the essential information on the hadron's structure. This defines the conditions under which the expansion to a given order will be reliable. The size of the hadron measured in lattice units should not be too large, it has to be comparable to the regions covered by possible quark paths. Or, alternatively, the lattice distance a measured in physical units cannot be too small.

The hadron mass is determined by the long-distance behaviour of the propagators. An exponential decay in configuration space implies a pole in momentum space. As we discuss below, this singularity corresponds to a pole in the hopping parameter K. Of course a finite expansion cannot produce a pole. Padé approximants will be used to identify this pole in K. This procedure establishes a relationship between the pole position K and the mass of the hadron for each value of the gauge coupling $1/g^2$.

In Wilson's formulation the lattice action has the following form:

$$S = \sum_n \Big(-\bar{\psi}^{a,i}_\alpha(n)\psi^{a,i}_\alpha(n) + \sum_\mu K_i \bar{\psi}^{a,i}_\alpha(n)(1-\gamma_\mu)_{\alpha\beta} U^{ab}_{n\mu} \psi^{b,i}_\beta(n+\hat{\mu}) + \sum_\mu K_i \bar{\psi}^{a,i}_\alpha(n+\hat{\mu})(1+\gamma_\mu)_{\alpha\beta} U^{+ab}_{n\mu} \psi^{b,i}_\beta(n)\Big) + \frac{1}{g^2}\sum_{\text{plaquettes}} (\operatorname{tr} U_p + \text{c.c.}) , \quad (1)$$

where a, b, α, β and i are colour, Dirac and flavour indices, respectively. In the continuum limit there is a simple relation between the hopping parameter K_i and the more familiar mass parameter m_i:

$$K_i^{-1} = 2m_i a + 8\,, \quad i = \mathrm{u, d, s, c}, \ldots\,, \tag{2}$$

S is not chiral invariant; there is no symmetry which would prevent the occurrence of mass counterterms even for massless quarks. The mass counterterm receives perturbative and non-perturbative contributions and thus K_i should be renormalized. $K_i = K_i(g^2)$ is not known a priori.

The action is quadratic in the fermion fields and one may integrate over these fields. The result is an effective gauge theory. One possibility is to study this effective gauge theory by direct numerical methods [11–14]. Preliminary studies indicated that it is very time-consuming to take the contribution of virtual quark loops in the effective action into account. However, there are qualitative arguments supporting the approximation where the virtual quark loops are neglected. Recently exciting results have been published in this approximation [15–17].

Another possibility for studying the effective gauge theory is the hopping parameter expansion combined with a Monte Carlo simulation [7,8]. In this paper we report on a calculation for the spectrum of low-lying s wave, pseudoscalar and vector mesons containing u, d, s and c quarks [‡1]. It is based on an expansion of tenth order in K and includes the contribution of virtual quark loops. The gauge group is $\mathrm{SU(3)}_{\mathrm{colour}}$. Our resolution at $1/g^2 = 0.90, 0.925, 0.95$ and 1.0 (these are the points we studied) is crude for mesons containing c quarks (especially concerning the perturbative contribution to the hyperfine splitting), therefore our results on these mesons should be considered as a first approximation.

Consider a propagator in momentum space: $D(p; K, g)$ and assume it has a pole. Close to this pole we have in the continuum limit

$$D(p_0 = \mathrm{i}E, \boldsymbol{p} = 0; K, g) \sim [E^2 - M^2(K, g)]^{-1}. \tag{3}$$

Consider now D as a function of K at a fixed E (and g). Let us define K^* by the equation $E^2 \doteq M^2(K^*, g)$.

[‡1] We consider only contributions where the valence quarks do not annihilate which is well-justified for the particles discussed.

Then by using the expansion

$$M^2(K, g) = E^2 + (\partial/\partial K) M^2(K, g)|_{K^*}(K - K^*) + \ldots, \tag{4}$$

one finds that a pole in momentum space implies a pole in K.

Any operator with the correct quantum numbers may serve to obtain the masses of the lowest lying states. For pseudoscalar and vector mesons we use $\bar\psi(x)\Gamma\psi(x)$, where Γ is a matrix in Dirac and flavour space, e.g., $\gamma_5\,\sigma_a$ for the $\pi(0^{-+})$ and $\gamma_\mu \sigma_a$ for the $\rho(1^{--})$. The corresponding local operators for the other meson states [e.g., $\delta(0^{++})$, $\mathrm{B}(1^{+-})$, $\mathrm{A}_1(1^{++})$] lead to vanishing expansion coefficients at $1/g^2 = 0$, and for larger $1/g^2$ the results change significantly as the order of the calculation is increased (from eighth to tenth order for instance). One possibility for studying these states in our framework is to go to higher orders. Another way would be to define spatially extended meson operators which might lead to faster convergence. We discuss only the results for pseudoscalar and vector mesons.

The propagator $D_\Gamma(n) = \langle \bar\psi(0)\Gamma\psi(0)\bar\psi(n)\Gamma\psi(n)\rangle$ can be written in the following way [‡2]

$$D_\Gamma(n; K, g) = \Bigg[\int \mathrm{D}U \sum_{\substack{\text{closed paths}\\ \text{connecting 0 with } n\ 0}}$$

$$\times \exp\left(\frac{1}{g^2}\sum_{\mathrm{p}} (\mathrm{tr}\, U_{\mathrm{p}} + \mathrm{c.c.}) + \sum_{\substack{\text{all closed}\\ \text{paths}}}$$
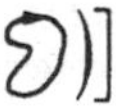
$$\Bigg)\Bigg]$$

$$\times \Bigg[\int \mathrm{D}U \exp\Bigg(\frac{1}{g^2}\sum_{\mathrm{p}} (\mathrm{tr}\, U_{\mathrm{p}} + \mathrm{c.c})$$

$$+ \sum_{\substack{\text{all closed}\\ \text{paths}}}$$
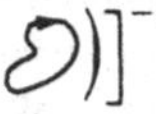
$$\Bigg)\Bigg]^{-1}, \tag{5}$$

where $\underset{m\quad m+\hat\mu}{\bullet\!\!\longrightarrow\!\!\bullet}$ contributes $K(1 - \gamma_\mu)U_{m\mu}$, the matrix Γ is included at the points 0 and n and finally the trace is taken in flavour, Dirac and colour space. The Boltzmann factor $\exp(S_{\mathrm{eff}})$ is to be expanded in terms of K, therefore (5) gives $D(n)$ as a systematic power series in K:

$$D(n; K, g) = \sum_{l=l_{\min}}^{\infty} C_l^{(n)}(g) K^{2l}\,. \tag{6a}$$

[‡2] For more details on the graphical rules see refs. [2,6–8].

Volume 110B, number 3,4 PHYSICS LETTERS 1 April 1982

Consider first the contribution to $C_l^{(n)}(g)$ from those graphs which do not contain vacuum loops. All the closed loops of total length $2l$ running through 0 and n should be constructed. By calculating the Dirac trace one is left with a trace over the product of U matrices along the closed loop, i.e., a Wilson loop. There are many different paths and corresponding Wilson loops. However, the number of topologically different Wilson loops is much less; all the others are the translated, rotated or reflected versions of these independent ones.

The calculation of Dirac traces and the topological classification of loops is done on a computer on a lattice of infinite extension. This part of the calculation is independent of the gauge group (and the coupling constant of course) such that

$$D(n;K,g) = \sum_{\substack{\text{topologically inequivalent} \\ \text{closed loops C}}} d^{(n)}(K,\mathrm{C}) \operatorname{tr} U_{\mathrm{C}} . \qquad \text{(6b)}$$

Here d is independent of the gauge group and exact up to the given order in K; U_{C} is the product of link variables along the path C.

In the next step the expectation value of the topologically independent Wilson loops is calculated by a usual Monte Carlo calculation for pure gauge theory for different g^2 values [‡3]. The number of configurations at a given g^2 was chosen to be between 15 and 20 on an 8^4 lattice. We believe that the systematical errors (mainly due to the truncation of the K series) are larger than the statistical ones, therefore we did not attempt to further reduce the statistical errors.

In tenth order the algebraic part [determination of all the $d^{(n)}(K,\mathrm{C})$] required ~2500 seconds on the CDC 7600, while the Monte Carlo time was ~3 hours at a given g^2 on the CERN IBM.

Including the vacuum loop contributions up to tenth order is rather simple. These contributions can be expressed in terms of the derivatives (with respect to $1/g^2$) of the plaquette and of the three topologically independent Wilson loops of length six.

Tables 1 and 2 give the Fourier transform of the π and ρ propagators at $1/g^2 = 0.925$, taking

‡3 We are indebted to Pietarinen [18] for the computer program we used to create equilibrium configurations.

Table 1
The K expansion of the Fourier coefficients of the π propagator at $p = (\mathrm{i}E,0,0,0)$ is given for $1/g^2 = 0.925$ in the form $D_\pi = C_0 + \Sigma_{t=1}^{5} C_t(\mathrm{e}^{Et} + \mathrm{e}^{-Et})$. The contribution of the origin has been discarded. The errors have been determined by treating all contributions to the coefficients as uncorrelated, which provides a conservative estimate.

	C_0	C_1	C_2	C_3	C_4	C_5
K^2	144	24	–	–	–	–
K^4	2903.81 ± 0.72	869.95 ± 0.31	96	–	–	–
K^6	43558.58 ± 24.17	24465.93 ± 11.27	5866.21 ± 2.82	384	–	–
K^8	506003.7 ± 1771.8	616671.9 ± 769.5	242173.5 ± 81.3	34398.9 ± 19.2	1536	–
K^{10}	4429340 ± 147086	15061550 ± 65351	8390367 ± 10307	1945945 ± 563	184403 ± 116	6144

Table 2
The same as table 1, but for the ρ meson

	C_0	C_1	C_2	C_3	C_4	C_5
K^2	96	24	–	–	–	–
K^4	1355.90 ± 0.72	677.95 ± 0.31	96	–	–	–
K^6	4611.04 ± 23.41	13515.03 ± 10.67	5098.21 ± 2.82	384	–	–
K^8	−382531.1 ± 1746.0	197041.0 ± 762.2	175707.4 ± 75.74	31326.9 ± 19.2	1536	–
K^{10}	−15998000 ± 156294	1529173 ± 64254	4800184 ± 10270	1581902 ± 529	172115 ± 116	6144

Volume 110B, number 3,4 PHYSICS LETTERS 1 April 1982

$p = (iE, 0, 0, 0)$[‡4]. The Fourier transform is defined as

$$D(p) = \sum_n \exp(-ipn)\, D(n)\,. \quad (7)$$

It is straightforward to modify the calculation outlined above for mesons containing quarks with different K's. (We did not distinguish K_u from K_d however.)

Considerable effort has been spent on verifying the correctness of the expansion coefficients. The propagators can be calculated by other methods in the strong and weak coupling limits. The computer result also agrees with a sixth order calculation obtained "by hand".

Analysis; results. The pole of the propagator in Fourier space is built up from the long-distance part of the propagator in configuration space. We decided to leave out the contribution of the origin $D(n = (0,0,0,0))$ in performing the Fourier transform. $D(0)$ has nothing to do with real propagation, in addition it is divergent in the continuum limit. [$D(0)$ is the expectation value of the product of four fermion fields taken at the same point.]

For every n there are special graphs contributing to $D(n)$ containing loops stuck to the end points and thus independent of n. These end point contributions factorize, giving rise to $D(n;K) \sim Z^2(K)\tilde{D}(n;K)$ where Z is independent of n. In order to cancel this factor we considered the ratio $D(p;K)/D(n = (1,0,0,0))$. [The distance = 1 propagator $D(1000)$ contains the Z factor up to the highest order in a K expansion.] $D(p;K)/D(1000)$ is then a series of eighth order.

In order to consider the analysis sensible, it should satisfy a few obvious requirements. First, it should work in the strong-coupling limit, where the exact result is known (exact though without virtual quark loops). Second, the different Padé ratios must give consistent results and finally the results should change smoothly with a signal of convergence as the order of the calculation is increased.

Our analysis satisfies these consistency requirements. The Padé analysis for $D(p;K)/D(1000)$ works reasonably well in the strong-coupling limit. We switched off the vacuum quark contribution in order to compare it

‡4 Tables at other $1/g^2$ values and also the corresponding tables for mesons containing quarks with different K's can be obtained from the authors on request.

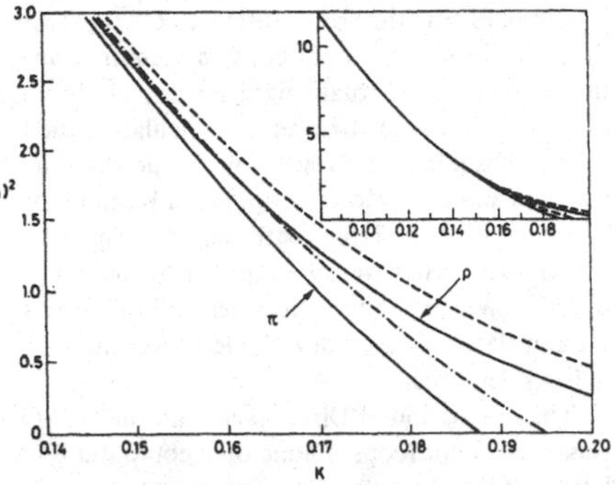

Fig. 1. $(m_\pi a)^2$ and $(m_\rho a)^2$ are plotted as the functions of K_u at $1/g^2 = 0.925$ (solid lines). For comparison, the corresponding curves without vacuum loop corrections are also given (dotted lines). In the insert the same curves are given over an extended K region.

with the exact results obtained in this limit. The different Padé's are almost identical. The predicted critical K^{crit} (where $m_\pi = 0$) is between 0.254 and 0.255 (the exact value is 0.25), while $(m_\rho a)$ is between 0.843 and 0.86 (the exact value is 0.867) in tenth order for that K^{crit}. The Padé table is also stable for non-zero $1/g^2$ values. The following tables and figures refer to the 3/1 Padé. Fig. 1 gives the predicted $(m_\pi a)^2$ and $(m_\rho a)^2$ as the function of K_u at $1/g^2 = 0.925$. For comparison, the corresponding curves without vacuum loops are also given. By taking the experimental value for m_π and m_ρ, $K_u = K_u(g^2)$ and $a = a(g^2)$ are obtained as given in figs. 2 and 3 respectively. $K_s(g^2)$ and $K_c(g^2)$ are also plotted in fig. 2. They are obtained by fixing the mass of K(495) and D(1868).

Let us recapitulate the number of input parameters: one pseudoscalar meson mass for each of the different quark species [i.e., three masses for K_u (= K_d), K_s and K_c] corresponding to the unknown quark mass, and one additional mass (the ρ mass in our case) in order to fix the scale. All other meson masses as well as $K_u(g^2)$, $K_s(g^2)$, $K_c(g^2)$, the critical $K^{crit}(g^2)$ (where $m_\pi = 0$) and the lattice spacing $a(g^2)$ in physical units are results.

The lattice distance a (fig. 3) changes very slowly until the crossover and falls down rapidly beyond this point. Between 0.95 and 1.0 it breaks away to a slow variation again. This behaviour is similar to that of ratios of Wilson loops of a given size, and indicates the

Volume 110B, number 3,4 PHYSICS LETTERS 1 April 1982

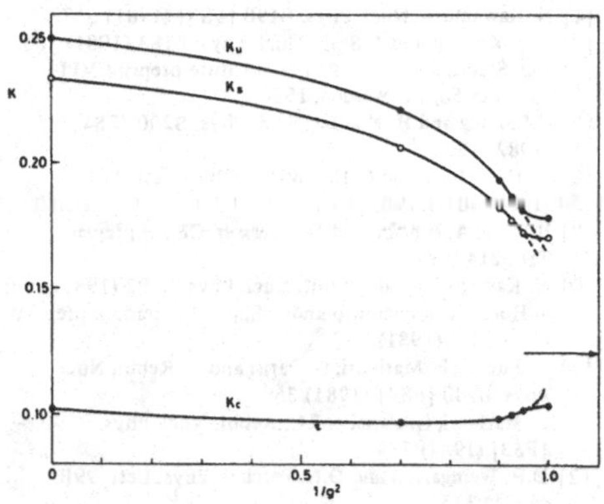

Fig. 2. The hopping parameter $K_i = K_i(g^2)$, i = up, strange and charmed. They are obtained by fixing m_π/m_ρ, m_K/m_ρ and m_D/m_ρ. The curves have been drawn to guide the eye.

region of reliability of the results. Our tenth-order calculation could not reproduce the slope expected from the renormalization group behaviour for $a(g^2)$. Correspondingly, the predicted value for Λ^{latt} changes with g^2, giving $\Lambda^{\mathrm{latt}} = 0.9$ MeV ($\Lambda^{\mathrm{MOM}}_{\alpha=1} = 95$ MeV [19]) at $1/g^2 = 0.925$ and $\Lambda^{\mathrm{latt}} = 0.77$ MeV ($\Lambda^{\mathrm{MOM}}_{\alpha=1} = 81$ MeV) at $1/g^2 = 0.95$. Here Λ^{latt} ($\Lambda^{\mathrm{MOM}}_{\alpha=1}$) is defined by using $n_f = 3$, assuming that the number of relevant flavour channels is three at the cut-off values entering our calculation.

The mass predictions are almost independent on the $1/g^2$ value taken. For the mass of K* and ϕ we obtained:

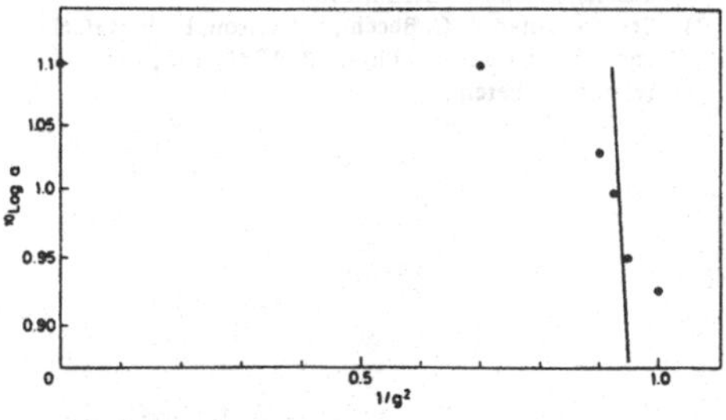

Fig. 3. The lattice distance a as the function of the gauge coupling $1/g^2$. a is determined by fixing the ρ meson mass. The straight line indicates the expected RG behaviour for $n_f = 3$.

$$m_{K^*(892)} = 860 \text{ MeV}\,, \quad m_{\phi(1020)} = 950 \text{ MeV}\,. \tag{8}$$

Our lattice is too coarse grained for the charm sector where there are important perturbative contributions as well. Our mass predictions are:

$$m_{F(2030)} = 1960 \text{ MeV}\,, \quad m_{D^*(2000)} = 1910 \text{ MeV}\,,$$
$$m_{F^*(2140)} = 1980 \text{ MeV}\,, \quad m_{\eta_c(2980)} = 2870 \text{ MeV}\,,$$
$$m_{\psi(3097)} = 2870 \text{ MeV} \tag{9}$$

The effect of the statistical errors of the Wilson loops on these numbers is small (~1%). A conservative estimate of the systematic errors related to the Padé analysis is ~5%. It is difficult to guess the systematic error due to the finite order of the expansion, which depends on the $1/g^2$ value taken and will be more important for the higher mass mesons.

Finally, let us collect the results on the quark masses. In Wilson's formulation the free quark propagator taken at $p = (im, 0)$ has a pole determined by the equation

$$\tfrac{1}{2}(1/K - 1/K^{\mathrm{crit}}) = \exp(ma) - 1\,, \tag{10}$$

where for the free case $K^{\mathrm{crit}} = 1/8$. This equation can also be considered to be valid when, due to the gauge interaction, K^{crit} (the value of K when $m_\pi = 0$) is different from 1/8. The mass parameters we obtained are summarized in table 3.

The renormalization group invariant mass is obtained by multiplying these mass values by $A \equiv (\mathrm{const} \cdot g^2)^{-4/11}$ [20]. Choosing the constant is a matter of convention [21]. For the possibility of comparison we choose const. = $3/4\pi$ and replace g^2_{latt} by g^2_{MOM} as in ref. [17] ‡5. In the equation

$$g^{-2}_{\mathrm{latt}} - 2\beta_0 \ln(\pi^{-1}\Lambda^{\mathrm{MOM}}/\Lambda^{\mathrm{latt}}) = g^{-2}_{\mathrm{MOM}}\,, \tag{11}$$

‡5 This last step is arbitrary. In the continuum limit Am is both RG and scheme invariant.

Table 3
The mass parameters m_u (which is not distinguished from m_d), m_s and m_c in MeV for two different gauge couplings. (These quantities are not the renormalization group invariant masses.)

$1/g^2$	$m_{u,d}$	m_s	m_c
0.925	5.6	144	1213
0.95	5.7	141	1209

the relating constant $c = 2\beta_0 \ln(\pi^{-1}\Lambda^{\rm MOM}/\Lambda^{\rm latt})$ depends weakly on the number of flavours ($c = 0.458$ for $n_f = 0$, $c = 0.400$ for $n_f = 3$ [19]). In our coupling constant range one obtains $A = 1.3$ which gives, for the renormalization group invariant u(d), s and c masses ≈7.5 MeV, ≈190 MeV and ≈1600 MeV respectively.

Some of our numbers can be compared with those obtained in ref. [17] by switching off the virtual quark loop contribution. Unfortunately the only point investigated in ref. [17] is $1/g^2 = 1.0$, where our series become too short. Without quark loops we get

$$K_{\rm u}(g^{-2} = 0.90) = 0.201\,, \quad K_{\rm u}(g^{-2} = 0.925) = 0.195\,,$$
$$K_{\rm u}(g^{-2} = 0.95) = 0.186\,, \qquad (12)$$

and by a smooth extrapolation one estimates $K_{\rm u}(1/g^2 = 1.0) \sim 0.165 \pm 0.01$ which is slightly above the value $K_{\rm u} = 0.156$ of ref. [17]. Our u quark mass also seems to be larger by almost a factor of two compared to the number obtained by Hamber and Parisi.

The authors are indebted to A. Frenkel, I. Montvay, G. Martinelli, H. Nicolai, C. Omero, D. Petcher and R. Petronzio for valuable discussions.

References

[1] H.E. Stanley, Introduction to phase transitions and critical phenomena (Clarendon, Oxford, 1971), and references therein;
M. Wortis, in: Phase transitions and critical phenomena, eds. C. Domb and S. Green, Vol. 3 (Academic Press, New York, 1974), and references therein.
[2] K.G. Wilson, Phys. Rev. D10 (1974) 2445; in: New phenomena in subnuclear physics (Erice, 1975) ed. A. Zichichi (Plenum, New York, 1977).
[3] K. Osterwalder and E. Seiler, Ann. Phys. (NY) 110 (1978) 440.
[4] N. Kawamoto, Nucl. Phys. B190 [FS3] (1981) 617.
[5] L.H. Karsten and J. Smit, Nucl. Phys. B183 (1981) 103.
[6] J.O. Stamatescu, Max Planck Institute preprint MPI-PAE/Pth 60/80 (Munich, 1980).
[7] C.B. Lang and H. Nicolai, Nucl. Phys. B200 [FS4] (1982) 135.
[8] A. Hasenfratz and P. Hasenfratz, Phys. Lett. 104B (1981) 489.
[9] B. Berg, A. Billoire and D. Foerster, CERN preprint TH-3214 (1981).
[10] N. Kawamoto and J. Smit, Nucl. Phys. B192 (1981) 100;
J. Hoek, N. Kawamoto and J. Smit, Amsterdam preprint ITFA-81-4 (1981).
[11] F. Fucito, E. Marinari, G. Parisi and C. Rebbi, Nucl. Phys. B180 [FS2] (1981) 369;
E. Marinari, G. Parisi and C. Rebbi, Nucl. Phys. B190 [FS3] (1981) 734.
[12] D.H. Weingarten and D.N. Petcher, Phys. Lett. 99B (1981) 333.
[13] D.J. Scalapino and R.L. Sugar, Phys. Rev. Lett. 46 (1981) 519.
[14] A. Duncan and M. Furman, Nucl. Phys. B190 [FS3] (1981) 767.
[15] E. Marinari, G. Parisi and C. Rebbi, Brookhaven preprint BNL-30212 (1981).
[16] D.H. Weingarten, Indiana University preprint IUHET-69 (1981).
[17] H. Hamber and G. Parisi, Brookhaven preprint BNL-30170 (1981), revised version.
[18] E. Pietarinen, Nucl. Phys. B190 [FS3] (1981) 349.
[19] A. Hasenfratz and P. Hasenfratz, Phys. Lett. 93B (1980) 165; Nucl. Phys. B193 (1981) 210;
R. Dashen and D. Gross, Phys. Rev. D23 (1981) 2340;
H. Kawai, R. Nakayama and K. Seo, Nucl. Phys. B189 (1980) 40;
P. Weisz, DESY preprint 80/118 (1980);
A. Gonzales-Arroyo and C.P. Korthals Altes, Marseille preprint CPT-81/p.1303 (1981).
[20] For a detailed discussion see: E. de Rafael, in: Lecture Notes in Physics, Vol. 118, eds. J.L. Alonso and R. Tarrach (Springer, Berlin, 1980).
[21] See, for instance: C. Becchi, S. Narison, E. de Rafael and F.J. Yndurain, Z. Phys. C8 (1981) 335, and references therein.

Nuclear Physics B210[FS6] (1982) 407–421

HADRON SPECTROSCOPY IN LATTICE QCD

F. FUCITO

Ist. di Fisica G. Marconi, INFN, Sezione di Roma, Italy

G. MARTINELLI

Lab. Nazionali di Frascati, INFN, Frascati, Italy

C. OMERO

Ist. di Fisica, Trieste
and
INFN, Sezione di Trieste, Italy

G. PARISI

Ist. di Fisica, Univ. di Tor Vergata, Roma
and
Lab. Nazionali di Frascati, Italy

R. PETRONZIO

CERN, Geneva, Switzerland

F. RAPUANO[1]

Ist. di Fisica G. Marconi, INFN, Sezione di Roma, Italy

Received 24 May 1982

A Monte Carlo computation of meson and hadron spectroscopy within lattice QCD is made. We give a detailed discussion of the statistical and systematic errors of the results and analyze the present limitations of our approach. The results are in agreement with the observed spectrum. We also estimate the values of up, down and strange quark masses.

1. Introduction

After the first promising results of computer simulations of hadronic states in the quenched approximation [1] (no fermionic vacuum polarization diagrams) we have started a more detailed analysis in order to have both the systematic and statistical errors under control.

The general pattern of this kind of computation consists of first finding the quark propagator in the presence of an external gauge field and building up from it various species of operator-operator correlations. In this paper we will not deal with all possible correlations which can be formed: we have focused our attention on the

[1] Partially supported by the A. Della Riccia Foundation.

masses of the lightest pseudoscalar and vector mesons and of the lowest lying states of the spin $\frac{1}{2}$ octet and of the spin $\frac{3}{2}$ decuplet of both intrinsic parities.

The purpose of this paper is twofold: on the one hand we want to see if there is agreement, within errors, between our results and the experimental values; on the other hand we aim to clarify the data analysis procedure and the role of systematics and volume effects which limit the accuracy of the results. We believe that the experience gained by this work will be useful for future developments of this kind of computation.

The paper is organized as follows. Sect. 2 contains the basic notations and formulae in sect. 3 we discuss the method of analysis of the data. The results are presented in sect. 4 and concluding remarks in sect. 5.

2. Basic formulae and notations

In this section we recall the basic formalism [1] we used to evaluate correlation functions and hadron masses on the lattice.

In the Wilson formulation [2], the action for interacting quarks and gluons on a euclidean lattice is given by

$$
\begin{aligned}
S &= \sum_{\mathrm{f}} \bar{\psi}^{\mathrm{f}} \Delta(U, K_{\mathrm{f}}) \psi^{\mathrm{f}} + S_{\mathrm{Gluon}}(U) \\
&= \sum_{\mathrm{f}} \sum_{x} \Big\{ \sum_{\mu} [K_{\mathrm{f}} \bar{\psi}^{\mathrm{f}}(x)(1-\gamma^{\mu}) U_{\mu}(x) \psi^{\mathrm{f}}(x+\hat{\mu}) \\
&\quad + K_{\mathrm{f}} \bar{\psi}^{\mathrm{f}}(x+\hat{\mu})(1+\gamma^{\mu}) U^{+}_{\mu}(x) \psi^{\mathrm{f}}(x)] - \bar{\psi}^{\mathrm{f}}(x)\psi^{\mathrm{f}}(x) \Big\} \\
&\quad + \tfrac{1}{6}\beta \sum_{\mathrm{Plaquettes}} \mathrm{Tr}\,[U_{\mathrm{P}} + U^{+}_{\mathrm{P}}] .
\end{aligned}
\tag{2.1}
$$

All colour and spinor indices have been omitted for simplicity. The quantity $\psi^{\mathrm{f},A}_{\alpha}(x)$ represents the quark field with flavour f, colour A and spinor index α; $(\gamma^{\mu}_{\alpha\beta})$ are Dirac matrices in euclidean space-time defined as

$$
\boldsymbol{\gamma} = -i\boldsymbol{\gamma}_{\mathrm{BD}}, \qquad \gamma_0 = -\gamma_{0\mathrm{BD}}, \qquad (\gamma^{+}_{\mu} = \gamma_{\mu}),
$$

where the subscript BD indicates the basis defined in ref. [3], $U^{AB}_{\mu}(x)$ is the SU(3) gauge field defined on the link with direction μ and U_{P} is the product of gauge fields belonging to the elementary plaquette P. The quantity β is related to the inverse of the lattice bare coupling constant g_0: $\beta = 6/g_0^2$. In the limit where the lattice spacing $a \to 0$, the action (2.1) describes the interactions of n_{f} coloured quarks, where f denotes the type of flavour (u, d, s . . .). The action we used avoids the problem of fermion doubling [4] but contains an explicit breaking of the chiral symmetry. In the free case, for $a \to 0$ one gets $K_{\mathrm{f}} \to \frac{1}{8}$ and correspondingly $m_{\mathrm{f}} = (1-8K)/2a \to 0$ (m_{f} is the bare quark mass for flavour f): the chiral symmetry is restored in the low momentum region. We expect that also in the interacting case, in the limit $a \to 0$, there will be a value of K_f for which chiral symmetry is restored, corresponding to a zero physical mass $m_f^{\rm phys}$ for the quark. However, if dynamical spontaneous symmetry breaking in the continuum theory occurs, for $m_f^{\rm phys} = 0$ we should observe a Goldstone boson in the spectrum, which is in fact what happens.

We are interested in computing from the action (2.1) vacuum expectation values of operators containing quark and gluon fields. Since the action is quadratic in the fermionic fields, these can be formally integrated out:

$$\langle \hat{O}(\psi, \bar{\psi}, U)\rangle \equiv \frac{\int d[U]\, d[\psi]\, d[\bar{\psi}] \exp(-S) O(\psi, \bar{\psi}, U)}{\int d[U]\, d[\psi]\, d[\bar{\psi}] \exp(-S)}$$

$$= \frac{\int d[U][\prod_f \det(\Delta(U, K_f))] \exp(-S_G(U)) \tilde{O}(U)}{Z} \tag{2.2}$$

If $\prod_f \det(\Delta(U, K_f))$ has a definite sign (which definitely happens for two equal mass flavours) one can write

$$\prod_f \det[\Delta(U, K_f)] \exp(-S_G(U)) \to \exp(-S_G(U)) + \sum_f \mathrm{Tr}[\ln|\Delta(U, K_f)|]$$

$$\equiv \exp(-S_{\rm eff}(U)) . \tag{2.3}$$

The problem is then to compute, with standard Monte Carlo methods, the expectation values of functions depending on the gauge fields U only, but with an effective action $S_{\rm eff}(U)$. In a Monte Carlo simulation the functional integral over U variables in eq. (2.2) is replaced by a sum over independent link configurations extracted by the measure $\exp(-S_{\rm eff}(U))\, d[U]$ using the standard Metropolis algorithm.

The quantity $\det[\Delta(U, K_f)]$ represents the feedback of closed quark loops on the gluon Green functions as shown in fig. 1. The computation of $\det[\Delta(U, K_f)]$ at each Monte Carlo step is very time (or computer memory) consuming because it requires the evaluation of a non-local quantity. For the purposes of this paper, we will make the approximation $\det[\Delta(U, K_f)] = 1$. If the number of flavours n_f is small compared to N_c, this is a reasonable approximation: it gives an exact Zweig

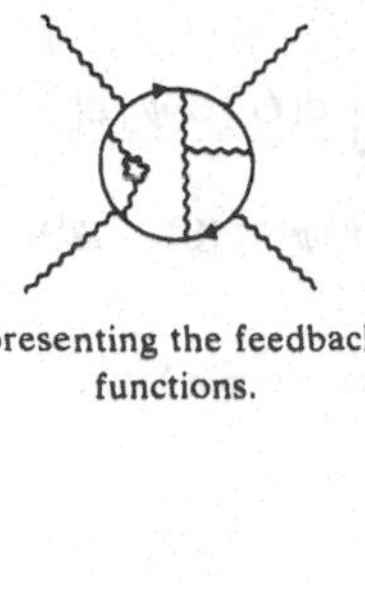

Fig. 1. Example of a typical diagram representing the feedback of closed quark loops on gluon Green functions.

rule and hadrons made of valence quarks and gluons only, which is not so far from the real world. The result becomes exact when the number of colours N_c tends to infinity at fixed n_f ($n_f/N_c \to 0$). Moreover, a recent paper based on the hopping parameter expansion [5] shows that the inclusion of quark loops (with $n_f = 3$) only produces small co·rections to the relevant physical quantities. It has been argued [1] that the effect of including quark loops roughly amounts to a shift of the value of β.

The hadron masses are computed by measuring the expectation values of correlation functions for operators carrying the same quantum numbers of the hadrons. A possible operator corresponding to a meson is, for example,

$$M(x) = \sum_{f_1, f_2} C_{f_1 f_2} \bar{\psi}^{f_1, A}(x) \Gamma \psi^{f_2, A}(x), \tag{2.4}$$

where Γ is one of the 16 Dirac matrices. $C_{f_1 f_2}$ are the appropriate Clebsch–Gordan coefficients and the sum over colour indices A is performed. For a spin $\frac{3}{2}$ baryon one can write [7]*

$$B_{\mu,\delta}(x) = \sum_{f_1 f_2 f_3} C_{f_1 f_2 f_3} [\psi^{f_1, A}(x) C \gamma_\mu \psi^{f_2, B}(x)] \psi_\delta^{f_3, C}(x) \varepsilon_{ABC}, \tag{2.5}$$

where ε_{ABC} is the colour antisymmetric tensor and C is the charge conjugation matrix.

Finally, for a spin $\frac{1}{2}$ baryon

$$B_\delta(x) = \sum_{f_1 f_2 f_3} C_{f_1 f_2 f_3} [\psi^{f_1, A}(x) C \gamma_5 \psi^{f_2, B}(x)] \psi_\delta^{f_3, C}(x) \varepsilon_{ABC}. \tag{2.6}$$

Note that B_δ is antisymmetric for the exchange $f_1 \leftrightarrow f_2$. Using eqs. (2.4)–(2.6), we write

$$\begin{aligned} \pi^+(x) &= \bar{u}^A(x) \gamma_5 d^A(x), \\ \rho_\mu^+(x) &= \bar{u}^A(x) \gamma_\mu d^A(x), \\ P_\delta(x) &= [u^A(x) C \gamma_5 d^B(x)] u_\delta^C(x) \varepsilon_{ABC}, \\ \Delta_{\mu,\delta}^{++}(x) &= [u^A(x) C \gamma_\mu u^B(x)] u_\delta^C(x) \varepsilon_{ABC}. \end{aligned} \tag{2.7}$$

Starting from the meson and baryon operators, it is straightforward to compute their Green functions in the quenched approximation. In the case of a meson

$$\begin{aligned} G(x) &\equiv \langle M(x) M(0)^+ \rangle \\ &= \frac{1}{Z} \sum_{f_1 \cdots f_4} C_{f_1 f_2} C^*_{f_3 f_4} \int d[U]\, d[\psi]\, d[\bar{\psi}] \exp(-S) \\ &\quad \times \bar{\psi}^{f_1, A}(x) \Gamma \psi^{f_2, A}(x) \bar{\psi}^{f_4, B}(0) \Gamma^+ \psi^{f_3, B}(0) \end{aligned}$$

* Note that the field $B_{\mu,\sigma}(x)$ also contains a spin $\frac{1}{2}$ state.

$$=\frac{1}{Z}\sum_{f_1f_2}|C_{f_1f_2}|^2\int d[U]\left\{\prod_f \det[\Delta(U,K_f)]\right\}\exp(-S_G(U))$$
$$\times \mathrm{Tr}\,[G^{f_1}(0,x)\Gamma G^{f_2}(x,0)\Gamma^+]$$
$$\Rightarrow\frac{1}{Z}\sum_{f_1f_2}|C_{f_1f_2}|^2\int d[U]\exp(-S_G(U))\,\mathrm{Tr}\,[G^{f_1}(0,x)\Gamma G^{f_2}(x,0)\Gamma^+], \quad (2.8)$$

where [1]

$$G^{f,AB}_{\alpha\beta}(x,0)=\Delta^{-1}(U,K_f,x,0)^{AB}_{\alpha\beta},$$
$$G^{AB}_{\alpha\beta}(x,0)=[\gamma_5 G^{BA}(0,x)^*\gamma_5]_{\beta\alpha}. \quad (2.9)$$

Notice that in the last term of eq. (2.8) the contribution of quark loops has been dropped. At the lowest order in n_f/N_c and for flavour singlet mesons, we neglected a term proportional to

$$\mathrm{Tr}\,[G^{f_1}(x,x)\Gamma]\,\mathrm{Tr}\,[G^{f_2}(0,0)\Gamma^+] \quad (2.10)$$

corresponding to the diagram shown in fig. 2.

In this approximation the pion and the η' are degenerate. It is interesting to note that even in the quenched case it is possible to compute the mass difference between π and η' because fig. 2 is the first non-trivial diagram responsible for the mass difference [5].

In a perfect analogy with mesons we can compute the baryon Green functions. The knowledge of the propagator gives us the mass of the corresponding particle. In fact, if there is only one pole corresponding to some stable meson, its Green function, integrated over the space hyperplane, behaves as

$$G(t)\equiv\sum_{x} G(x,t)\sim\exp(-mt), \quad (2.11)$$

which is the euclidean propagator for a scalar particle at rest. On a periodic lattice with period T in the time direction, one finds

$$G(t)\sim\cosh[m(t-\tfrac{1}{2}T)]+\cdots, \quad (2.12)$$

where the dots indicate terms which are down by at least a factor $\exp(-mT)$. In the case of a baryon, the most general form of the propagator is

$$G(t)=(1-\gamma_0)\{\theta(t)C_+\exp[-m_+t]+\theta(-t)C_-\exp[m_-t]\}$$
$$+(1+\gamma_0)\{\theta(-t)C_+\exp[m_+t]+\theta(t)C_-\exp[-m_-t]\}, \quad (2.13)$$

Fig. 2. Example of a diagram contributing, even in the quenched approximation, to the $\eta'-\pi$ mass difference.

which, with periodic boundary conditions, becomes

$$G(t)=(1-\gamma_0)[C_+\exp(-m_+t)+C_-\exp(-m_-(T-t))]$$
$$+(1+\gamma_0)[C_+\exp(-m_+(T-t))+C_-\exp(-m_-t)]\,. \tag{2.14}$$

$m_\pm$ corresponds to particles with opposite intrinsic parity [1, 5] [e.g., the proton and the N*(1535)].

By fitting the meson and baryon propagators with the expressions (2.12) and (2.14), one measures the masses in units of the lattice spacing. In general, we expect that more than only one [eq. (2.12)] or two [eq. (2.14)] particles corresponding to the lowest lying states will contribute to the propagator. In fact, we can have other contributions due to higher mass particles or many particle states. In practical cases, the best one can do in order to give an estimate of the systematic error on the mass induced by these effects is to parametrize the propagator by adding new terms of the same form as in eqs. (2.12) and (2.14). For example, in the case of the meson propagator

$$G(t)=C\cosh m(t-\tfrac{1}{2}T)+C'\cosh m'(t-\tfrac{1}{2}T)\,. \tag{2.15}$$

The relevance of these corrections will be discussed in the section on numerical results (sect. 5). By neglecting them, one systematically increases the value of the lowest lying state mass. At large distances ($t\to\infty$) these effects are expected to become completely negligible [$m'>m$ in eq. (2.15)].

In order to give predictions in physical units, after the determination of the hadron masses in units of the lattice spacing, one has to fix a fundamental hadronic scale and a mass parameter (in our case K_f) for each quark flavour. We decided to set the scale and the masses of the quarks by fixing the squared mass difference between the ρ and the π ($m_\rho^2-m_\pi^2$) and the mass of the lightest pseudoscalar states containing one quark of a certain flavour ($m_\pi, m_K, \ldots$)*. In the Wilson formulation, the bare mass of the quark can be defined as [1]

$$\exp(m_f^B a)-1=\frac{1}{2}\left(\frac{1}{K_f}-\frac{1}{K_{crit}}\right). \tag{2.16}$$

K_{crit} is the value of K for which the pseudoscalar meson becomes massless. The value of the lattice coupling constant g_0 is another free parameter as long as the renormalization group equation relating the lattice spacing and Λ_{MOM} is not checked:

$$\Lambda_{MOM}\simeq\frac{\pi}{a}[\tfrac{8}{33}\pi^2(\beta-2.75)]^{51/121}\exp[-\tfrac{4}{33}\pi^2(\beta-2.75)]\,. \tag{2.17}$$

Λ_{MOM} is a fixed physical quantity which can be extracted from experiment. Relation (2.17) ensures that the lattice spacing $a\to 0$ in physical units when $\beta\to\infty$. The

* We work in the approximation of an exact isospin symmetry.

scaling behaviour predicted by eq. (2.17) has been measured in several Monte Carlo experiments [8]; in this paper we will work at a fixed value of $\beta = 6$.

3. The experiment and the analysis of the data

We have carried out a Monte Carlo experiment using a $5^3 \times 10$ lattice obtained by duplicating the gauge fields in the time direction. We computed the quark propagator at $\beta = 6$ and $K = 0.130$, 0.145, 0.1475, 0.150, 0.1525 for 32 gauge field configurations using the Gauss–Seidel algorithm*. The average number of iterations required to obtain a 1% accuracy for the quark propagator at any distance ranges from 20 ($K = 0.130$) to 100 ($K = 0.1525$). In the case of $K = 0.1525$, for some configurations, it was not possible to make the Gauss–Seidel method converge with any number of iterations.

Each link configuration was obtained by the previous one by performing 100 Monte Carlo sweeps (×10 upgradings of the same link) after an initial thermalization of the system of 500 sweeps with a hot start. We found a long-range correlation of the order of 400–500 sweeps, between successive link configurations for the value of hadron masses. This is illustrated in fig. 3 where we report on the mass squared of the ρ extracted by a single configuration at $K = 0.1475$ as a function of the number of Monte Carlo sweeps. The knowledge of the correlation between different link configurations avoids underestimating the size of statistical errors. In

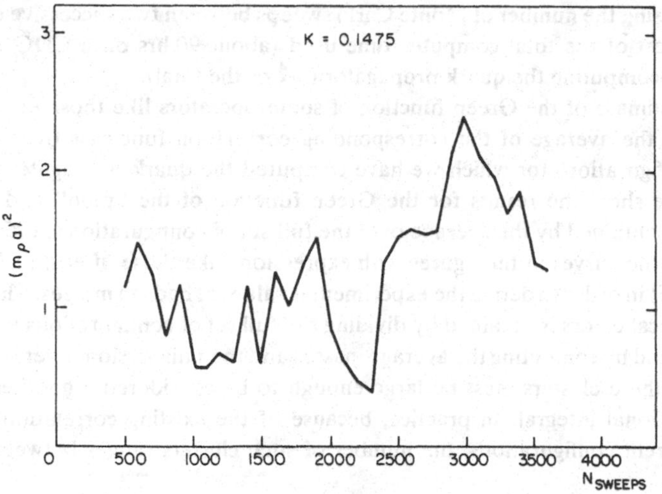

Fig. 3. The value of m_ρ^2 in lattice units is plotted at $K = 0.1475$ as a function of the number of sweeps made on the gauge configurations.

* To generate SU(3) configuration and to compute the quark propagator, we have basically used the programme written by Hamber [1].

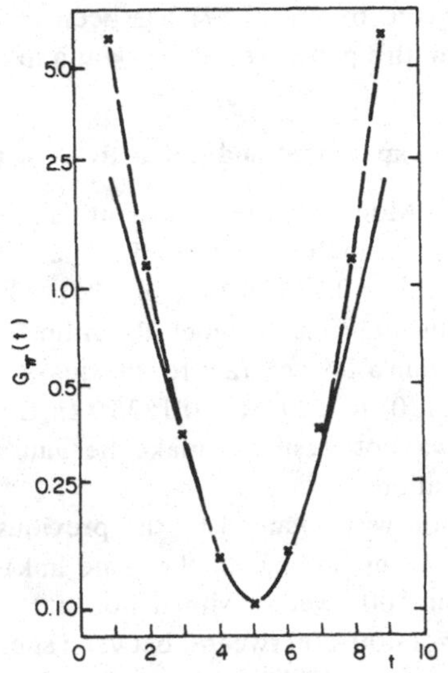

Fig. 4. The crosses are the data points of the pion propagator as a function of time. The full curve represents the fit of eq. (2.12) to the five central points and the dashed curve the fit of eq. (2.15).

general, with a given number of configurations, one can reduce the statistical errors by increasing the number of Monte Carlo sweeps between two successive configurations. Most of the total computer time used (about 90 hrs on a CDC 7600) was spent on computing the quark propagator ($\sim\frac{4}{5}$ of the total).

The estimate of the Green function of some operators like those of eq. (2.7) is given by the average of the corresponding correlation functions over the gauge field configurations for which we have computed the quark propagator. In figs. 4 and 5 we show the results for the Green function of the "pion" and "proton" operator obtained by the average over the full set of configurations at $K = 0.1475$.

We fit the curves in the figures with expressions like those of eqs. (2.12), (2.13) and (2.15) in order to derive the experimental values of hadron masses. The estimate of statistical errors is obtained by dividing the full set of configurations into various clusters and by computing the average masses and their dispersions over the clusters. Each of these clusters must be large enough to be considered a good estimate of the functional integral. In practice, because of the existing correlations between the different configurations, the number of such clusters ranges between one and four.

Given the small number of clusters, a better estimate of the average value of the masses is obtained by averaging over different partitions of the configurations at a fixed number of clusters. Of course, in this case the error is given by the average error. In fig. 6 we plot the results for the proton, Δ^{++}, ρ and pion-like

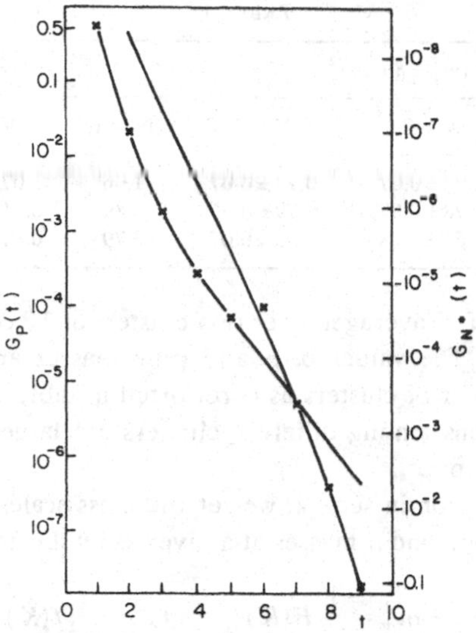

Fig. 5. The crosses for $t \leqslant 5$ refer to the experimental data for the "proton" propagator as a function of time. The corresponding curve is obtained with the fit of eq. (2.14). For $t \geqslant 6$ the scale is on the right-hand side: the corresponding curve is obtained with the fit of eq. (2.14) and represents the opposite intrinsic parity hadron propagating backwards in time.

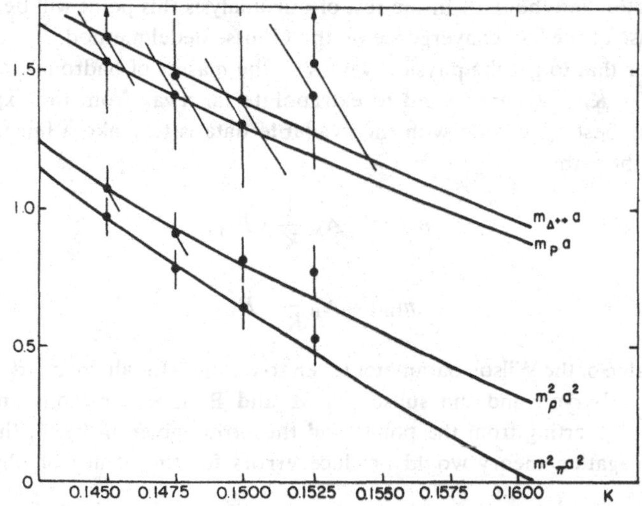

Fig. 6. The experimental points and their linear fit (in $1/K$) for pion, ρ, proton and Δ^{++} states. The point at $K = 0.1525$ has been plotted for completeness but not included in the fit.

TABLE 1

K	$(m_\pi^2 \pm \delta m_\pi^2)a^2$			$(m_\rho^2 \pm \delta_\rho^2)a^2$		
	$N_c=1$	$N_c=2$	$N_c=3$	$N_c=1$	$N_c=2$	$N_c=3$
0.1450	0.96	0.97 ± 0.07	0.98 ± 0.07	1.06	1.07 ± 0.08	1.09 ± 0.07
0.1475	0.77	0.78 ± 0.07	0.79 ± 0.07	0.89	0.91 ± 0.08	0.92 ± 0.07
0.1500	0.62	0.64 ± 0.08	0.65 ± 0.07	0.79	0.81 ± 0.08	0.82 ± 0.08

masses as a function of K averaged over two clusters of 16 configurations with the corresponding errors. The values of ρ and pion masses are very stable against variations of the number of clusters as is reported in table 1. For the proton and the Δ^{++} the fluctuations among different clusters are larger, but still within the estimated statistical errors.

Following the discussion in sect. 2, we set the mass scales for the lightest (u, d) quarks by fixing the pion and ρ masses at a given β by the equations

$$m_\rho^2 - m_\pi^2 = \frac{1}{a^2}F(K)\,, \qquad m_\pi^2 = \frac{1}{a^2}f(K)\,, \tag{3.1}$$

where the functions F and f are experimentally measured. The value one obtains by solving these equations is $K \sim 0.16$ which is rather far from the accessible region: given the size of our lattice we cannot go to values of K larger than ~0.1525 because of volume effects due to nearly massless excitations*. Already at $K = 0.1525$ we note a clear set up of these volume effects, as one can see from fig. 6, especially for the proton and the Δ^{++}. In the rest of our analysis this point will be neglected, also because of the bad convergence on the Gauss–Siedel method.

It is clear that to get the physical values of the masses of hadrons made of light quarks ($K_f = K_{u,d}$) we are forced to extrapolate far away from the experimental points. The best we can do with the available data is to make a linear fit to the masses of the form

$$\begin{aligned} m_M^2 a^2 &= A_M \frac{1}{K} + B_M\,, \\ m_B a &= A_B \frac{1}{K} + B_B\,. \end{aligned} \tag{3.2}$$

K is the value of the Wilson parameter taken to be equal for all the quarks contained in a meson/baryon and the subscripts M and B denote mesons and baryons respectively. Starting from the points and the errors given in fig. 6, the standard error propagation theory would produce errors for the values of physical light

* The points of ref. [1] for $K = 0.1525$ and 0.155 are affected by the same convergence problems of the Gauss–Siedel method; see also ref. [5].

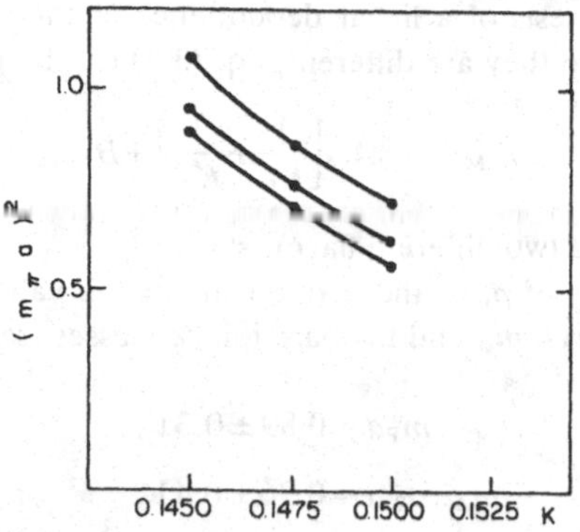

Fig. 7. The experimental values of m_π^2 in lattice units for three different clusters of configurations as a function of K. The full lines are only to guide the eyes.

hadron masses of the order of hundred per cent. This happens if one does not take into account the strong correlation existing, for a given cluster, between the slope and the intercept of the linear fit as can be seen from fig. 7 where, for the squared pion mass, the behaviour of three clusters is separately displayed. The procedure which minimizes the statistical errors entails first making the fit of eqs. (3.2) cluster by cluster, computing the masses for each cluster at $K = K_{u,d}$ and then averaging the masses over the clusters. In order to apply such a procedure, the same configuration set must be used for all values of K. There are quantities, such as the ratio of the baryon $\frac{3}{2}^+$ over the corresponding vector meson, which are found to be almost independent of the mass of the quark; for them we can directly extract the physical values from our data without any extrapolation. We used this method as an independent estimate of the p, Δ^{++} and Ω^- masses.

4. Results

In this section we present our numerical results. All the results refer to a subdivision of the configuration set into two clusters.

In table 2 we give the values of A_M, B_M, A_B, B_B for the pseudoscalar and the vector mesons and for the $\frac{3}{2}^+$ and $\frac{1}{2}^+$ baryons. In the same table we give the value

TABLE 2

K_{crit}	A_{PS}	B_{PS}	A_v	B_v
0.1590 ± 0.0010	1.22 ± 0.09	-7.6 ± 0.6	1.01 ± 0.07	-6.1 ± 0.5
	$A_{3^+/2}$	$B_{3^+/2}$	$A_{1^+/2}$	$B_{1^+/2}$
	1.01 ± 0.36	-5.4 ± 2.7	0.98 ± 0.24	-5.2 ± 1.8

of $K_{\rm crit}$. In the hypothesis of a linear dependence of the meson mass near $K_{\rm crit}$, even in the case where they are different, eq. (3.2) can be generalized as follows:

$$m_{\rm M}^2 a^2 = \tfrac{1}{2} A_{\rm M}\left[\frac{1}{K_{\rm f_1}} + \frac{1}{K_{\rm f_2}}\right] + B_{\rm M}\,, \tag{4.1}$$

where $K_{\rm f_1}$, $K_{\rm f_2}$ refer to two different flavours.

By fixing the values of ρ, π and κ one can get the value of $\Lambda_{\rm lattice}$, $m_{\rm p}$, $m_{\Delta^{++}}$, $m_{\rm N^*}$, $m_{\Delta(3^-/2)}$, m_{Ω^-}, $m_{\rm K^*}$, m_ϕ and the bare quark masses [see eq. (2.16)]. In lattice units we get

$$\begin{aligned} m_{\rm P} a &= 0.89 \pm 0.31\,,\\ m_{\Delta(3^+/2)} a &= 0.96 \pm 0.41\,,\\ m_{\Omega^-} a &= 1.18 \pm 0.38\,,\\ m_{\rm K^*}^2 a^2 &= 0.39 \pm 0.06\,,\\ m_\phi^2 a^2 &= 0.48 \pm 0.05\,, \end{aligned} \tag{4.2}$$

and for the bare quark masses* $m_{\rm f}^{\rm B}$

$$\begin{aligned} m_{\rm u,d}^{\rm B} a &= (3.2 \pm 0.4) \times 10^{-3}\,,\\ m_{\rm s}^{\rm B} a &= (8.17 \pm 0.50) \times 10^{-2}\,, \end{aligned}$$

by using the central values of the results reported above, we get

$$\begin{aligned} \frac{m_\phi^2 - m_\rho^2}{2(m_{\rm K}^2 - m_\pi^2)} &= 0.82\,, \qquad (\exp \sim 0.96)\,,\\ \frac{m_\Omega^2 - m_\Delta^2}{3(m_{\rm K}^2 - m_\pi^2)} &= 2.01\,, \qquad (\exp \sim 1.89)\,. \end{aligned} \tag{4.3}$$

As we have already pointed out in the previous section, it is possible to give an independent estimate of baryon masses by making their ratio with the vector meson mass under the verified hypothesis of K independence. In this case we get

$$\frac{m_{\rm P}}{m_\rho} = 1.46 \pm 0.23\,, \qquad (\exp \sim 1.20)$$

$$\frac{m_{\rm N^*}}{m_\rho} = 2.41 \pm 0.26\,, \qquad (\exp \sim 1.97)\,,$$

from which follows

$$\frac{m^{\rm N^*}}{m_{\rm P}} = 1.65\,, \qquad (\exp \sim 1.65)\,;$$

* Using the relation between the bare mass on the lattice and the renormalization group invariant mass ($\hat{m}$) in the continuum and eq. (4.5), one finds [9] $\hat{m}_{\rm u,d} = 7.3 \pm 1.0$ MeV; $\hat{m}_{\rm s} = 188 \pm 10$ MeV.

TABLE 3

m_P	$m_{\Delta(3^+/2)}$	m_{Ω^-}	m_{K^*}	m_ϕ	$m^B_{u,d}$	m^B_S
1.27±0.44	1.37±0.60	1.68±0.50	0.89±0.07	0.99±0.05	(4.5±0.6)10⁻³	0.116±0.007

$$\frac{m_{\Delta(3^+/2)}}{m_p} = 1.55 \pm 0.24\,, \qquad (\exp \sim 1.58)\,,$$

$$\frac{m_{\Delta(3^-/2)}}{m_p} = 2.44 \pm 0.14\,, \qquad (\exp \sim 2.14)\,, \tag{4.4}$$

from which follows

$$\frac{m_{\Delta(3^-/2)}}{m_{\Delta(3^+/2)}} = 1.57\,, \qquad (\exp \sim 1.36)\,,$$

$$\frac{m_{\Omega^-}}{m_\phi} = 1.60 \pm 0.25\,, \qquad (\exp \sim 1.64)\,,$$

$$a^2[m^2_{\Delta(3^+/2)} - m^2_P] = 0.25 \pm 0.11\,.$$

With this method we get better results for the baryon masses. In particular the ratio m_{N^*}/m_P turns out to be the right one in spite of the fact that the absolute values are rather large. This is also the only way we can give a reasonable error to the mass difference between the proton and the Δ^{++}. The translation of the results into physical units is given in tables 3 and 4. The inverse of the lattice spacing a turns out to be

$$a^{-1} = 1.42 \pm 0.12 \text{ GeV} \tag{4.5}$$

corresponding to a value for $\Lambda_{\overline{\text{MS}}} = 0.075 \pm 0.008$ GeV in reasonable agreement with the results from deep inelastic scattering.

The results which have been presented have been obtained by fitting the meson propagator by eq. (2.12) using only the five central points in the plot in fig. 4. Using eq. (2.15) and all the points (except that at zero distances) one obtains masses shifted downwards by less than 13%. This is an estimate of the systematic error for meson masses and for the value of Λ. For the baryons, a similar analysis has been made: in this case it is more difficult to disentangle the proton contribution to the propagator because, besides the effects of higher excitations, there is a

TABLE 4

m_P	m_{N^*}	$m_{\Delta(3^+/2)}$	$m_{\Delta(3^-/2)}$	m_{Ω^-}	$m_{\Delta(3^+/2)} - m_P$
1.14±0.18	1.88±0.20	1.21±0.19	1.90±0.11	1.58±0.24	0.22±0.09

contamination of the opposite parity hadron. All these effects are reduced by using larger lattices. In fact, as can be seen from the fit to the "proton" propagator in fig. 5, it is only at a distance of about five lattice spacings that the proton pole stems out. This kind of effect can partially account for the systematically large values we obtained for baryons and for the larger statistical errors. We also computed the masses of the mesons $\delta(980)$, A_1 and B(1235). We observe that for small values of K there is a clear splitting between these states and the pion and the ρ, which decreases very quickly for increasing K. For $K = K_{u,d}$ we would extrapolate for these states a value for the mass smaller than the ρ mass. The reason for this behaviour is not clear to us: it could be connected to the fact that the operators we used to estimate their masses are zero in the non-relativistic limit. Another possible reason is that our lattice is too coarse-grained to resolve the structure of more extended objects ($L = 1$ states).

5. Concluding remarks

From the analysis of the results of our Monte Carlo experiment we learned the following lesson. First of all, the statistics required to obtain reasonable errors for the masses is rather large and many configurations are needed for a precision of the order of the percentage. Furthermore, all the difficulties one encounters are connected with volume effects: (i) the strong correlation between different configurations may be due to finite temperature effects (or to the existence of metastable states); (ii) the anomalous behaviour of the baryon masses for large values of K which forbids us to go close to the critical value of K; (iii) the contamination in the determination of the masses of the lowest lying states due to higher excited states (remember that the effect is stronger for baryons). We finally observe that an efficient and fast algorithm for computing the quark propagator is needed: in fact most of the computer time was spent on the propagator and, close to K_{crit}, it was impossible to get a sensible result with the Gauss–Seidel method.

In spite of all the difficulties mentioned, the results are in agreement, within errors, with the observed pattern of hadron spectroscopy and confirm the validity of the lattice technique for the investigation of the hadronic world in the low-energy regime.

We thank C.B. Lang and P. Hasenfratz for fruitful discussions and H. Hamber for providing us with the programme that we basically used in our computation. One of us (G.M.) thanks the CERN Theory Division and in particular the senior staff for the extension of his Fellowship at CERN where most of this work was carried out. F.F., C.O. and F.R. thank the CERN Theory Division for support. Particular help has been given to us by the CERN Computer Division. We also thank Drs. L.M. Barone, F. Ferroni and E. Valente for advice.

References

[1] H. Hamber and G. Parisi, Phys. Rev. Lett. 47 (1981) 1792;
E. Marinari, G. Parisi and C. Rebbi, Phys. Rev. Lett. 47 (1981) 1795;
H. Hamber, E. Marinari, G. Parisi and C. Rebbi, Phys. Lett. 108B (1982) 314;
D. Weingarten, Phys. Lett. 109B (1982) 57

[2] K.G. Wilson, New phenomena in subnuclear physics, ed. A. Zichichi (Erice, 1975) (Plenum Press, New York, 1977)

[3] J.D. Bjorken and S.D. Drell, Relativistic quantum mechanics (McGraw-Hill, 1965)

[4] L.H. Karsten and J. Smit, Nucl. Phys. B183 (1981) 103, and references therein

[5] A. Hasenfratz, P. Hasenfratz, Z. Kunszt and C.B. Lang, Phys. Lett. 110B (1982) 289

[6] H. Hamber and G. Parisi, BNL preprint 31322 (April, 1982) Phys. Rev. D, submitted

[7] I. Ioffe, Nucl. Phys. B188 (1981) 317

[8] M. Creutz, Phys. Rev. Lett. 43 (1979) 553; 45 (1980) 313;
G. Bhanot and C. Rebbi, Nucl. Phys. B188 (1981) 469;
E. Pietarinen, Nucl. Phys. B190 [FS3] (1981) 349

[9] A. Gonzales-Arroyo, G. Martinelli and F.J. Yndurain, in preparation